ENCYCLOPEDIA
OF
VIBRATION

Access for a limited period to an online version of the Encyclopedia of Vibration is included in the purchase price of the print edition.

This online version has been uniquely and persistently identified by the Digital Object Identifier (DOI)

10.1006/rwvb.2001

By following the link

http://dx.doi.org/10.1006/rwvb.2001

from any Web Browser, buyers of the Encyclopedia of Vibration will find instructions on how to register for access.

If you have any problems with accessing the online version, e-mail: **idealreferenceworks@harcourt.com**

ENCYCLOPEDIA OF VIBRATION

Editor-in-Chief

S BRAUN

Editors

D EWINS

S S RAO

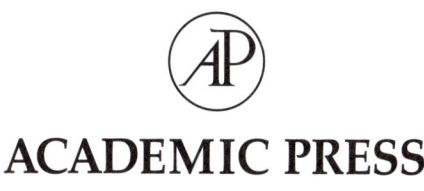

ACADEMIC PRESS
A Division of Harcourt, Inc.

San Diego　San Francisco　New York　Boston
London　Sydney　Tokyo

This book is printed on acid-free paper.

Copyright © 2002 by ACADEMIC PRESS

All Rights Reserved.
No part of this publication may be reproduced or transmitted in any form or by any means, electronic or mechanical, including photocopying, recording, or any information storage and retrieval system, without permission in writing from the publisher.

Academic Press
A division of Harcourt, Inc.
Harcourt Place, 32 Jamestown Road, London NW1 7BY, UK
http://www.academicpress.com

Academic Press
A division of Harcourt, Inc.
525 B Street, Suite 1900, San Diego, California 92101-4495, USA
http://www.academicpress.com

ISBN 0-12-227085-1

Library of Congress Catalog Number: 2001092782

A catalogue record for this book is available from the British Library

---

Access for a limited period to an online version of the Encyclopedia of Vibration is included in the purchase price of the print edition.

This online version has been uniquely and persistently identified by the Digital Object Identifier (DOI)

10.1006/rwvb.2001

By following the link

http://dx.doi.org/10.1006/rwvb.2001

from any Web Browser, buyers of the Encyclopedia of Vibration will find instructions on how to register for access.

If you have any problems with accessing the online version, e-mail:
idealreferenceworks@harcourt.com

Typeset by Bibliocraft, Dundee, Scotland
Printed and bound in Great Britain by MPG Books Ltd, Bodmin, Cornwall, UK

02 03 04 05 06 07 MP 9 8 7 6 5 4 3 2 1

Editors

EDITOR-IN-CHIEF

S Braun
Technion – Israel Institute of Technology
Faculty of Mechanical Engineering
Haifa 32000
Israel

EDITORS

D Ewins
Imperial College of Science, Technology and Medicine
Department of Mechanical Engineering
Exhibition Road
London SW7 2BX, UK

S S Rao
University of Miami
Department of Mechanical Engineering
PO Box 248294, Coral Gables,
FL 33124-0624, USA

Editorial Advisory Board

R Bigret
22 Rue J Varnet
93700 Drancy
France

P Cawley
Imperial College of Science, Technology & Medicine
Department of Mechanical Engineering
Exhibition Road
London SW7 2BX, UK

R Craig
University of Texas
Aeronautical Engineering and Engineering Mechanics Department
Austin TX 78712, USA

B Dubuisson
33 Rue Saint Hubert
60610 La Croix Saint Ouen
France

R Eshleman
Vibration Institute
6262 S. Kingery Highway
Willowbrook
IL 60514, USA

M Geradin
Universite de Liège
LTAS Dynamique des Structures
Institut de Mecanique et Genie Civil
Chemin des Chevreuils 1
4000 Liège, Belgium

J Hammond
University of Southampton
Institute of Sound and Vibration Research
Southampton SO9 5NH, UK

S Hayek
Pennsylvania State University
112 EES Building
University Park
PA 16802-6812, USA

D Inman
Virginia Polytechnic Institute & State University
Department of Engineering Science and Mechanics
310 NEB, Mail Code 0261
Blacksburg
VA 24061-0219, USA

M Lalanne
Laboratorie de Mecaniques des Structures
Institut National des Sciences Appliquees de Lyon
LMSt - Batiment 113 20
avenue Albert Einstein
69621 Villeurbanne Cedex, France

M Link
Universitat Gesamthoschule Kassel
Fachgebiet Leichtbau
Moenchebergstrasse 7,
D34109 Kassel, Germany

K McConnell
Iowa State University
AEEM
Black Engineering Building
Ames, IA 50011, USA

D E Newland
University of Cambridge
Department of Engineering
Trumpington Street
Cambridge CB2 1PZ, UK

N Okubo
Chuo University Faculty of Science & Engineering
Department of Precision Mechanical Engineering
1-37-27 Kasuga
Bunkyo-ku, Tokyo, Japan

M Radeş
Universitatea Politehnica Bucuresti
Department of Engineering Sciences
313 Splaiul Independentei
79590 Bucuresti
Romania

S Rakheja
Concordia University
Department of Mechanical Engineering
1455 de Masonneuve Blvd W
Montreal, Quebec H3G 1M8
Canada

G Rosenhouse
Technion – Israel Institute of Technology
Faculty of Civil Engineering
Technion City
Haifa 32000, Israel

EDITORIAL ADVISORY BOARD

S Shaw
Michigan State University
Department of Mechanical Engineering
A321 Engineering Building
East Lansing, MI 48824-1226, USA

M Sidahmed
Université de Technologie de Compiègne
Heuristique et Diagnostic des Systèmes Complexes
Centre de Recherches de Royallieu
BP 20529
60205 Compiègne, France

H S Tzou
University of Kentucky
Department of Mechanical Engineering
Dynamics and Systems Laboratory
Lexington, KY 40506, USA

Foreword

Vibration is usually defined in dictionaries as rhythmic motion back and forth. It has attracted the curiosity of humans since people have had time to contemplate the natural world. In the sixth century BC, Pythagoras had related harmonic intervals to ratios of lengths of a vibrating string. In 1581, Galileo had observed that the period of a simple pendulum is (nearly) independent of the amplitude of vibration. A century later the basic principles of dynamics were put on a firm basis by I. Newton. Further development of the science of mechanics was led by L. Euler (1707–83). Most of the analytical tools now used in vibration studies were available by the year 1788, when *Mécanique Analytique* was published by J.L. Lagrange. The first book devoted entirely to the theory of vibration was volume I of Lord Rayleigh's *Theory of Sound* (1887). It became the model for the classical vibration textbooks of Timoshenko (1928), and DenHartog (1934). Subsequent texts have followed much the same pattern, with the addition of matrix notation and linear algebra after the engineering-science revolution of the sixties.

Vibration is a fascinating physical phenomenon, well worth studying on its own merits. In many applications, vibration is harnessed for useful purposes, e.g., to make music, to drive vibrational transport systems, or provide frequency standards in clocks and precision instruments. In many other applications, vibration is an undesired intruder that interferes with the normal operation of the system, creating noise, and developing stresses that may cause fatigue failure.

Engineers have been dealing with vibration problems since the beginning of the industrial revolution and the introduction of the steam engine. With each new mode of transportation and each new technology, there often appears an unexpected vibrational challenge. The interaction of steam-driven trains with relatively flexible metal bridges produced unexpected problems of vibration and fatigue in the 19th century. At the beginning of the 20th century, cities began to install central electric power stations and engineers were faced with a variety of vibration problems associated with the rotor dynamics of turbine generator sets. By the end of World War I, the diesel engine had become a popular medium-power prime mover. A rash of fatigue failures in diesel engine shafting spurred work on the torsional vibration of crank-shafts. A major technological effort of the period between World War I and II was the development of airplanes and helicopters. The unexpected vibrational challenge for aircraft was the phenomenon of wing flutter, while the new problem for helicopters was the phenomenon of ground resonance. In 1940 the Tacoma Narrows suspension bridge developed large self-excited torsional vibrations in a moderate wind and failed dramatically. The challenge to explain this completely unexpected result kept vibration experts busy for decades.

The latter half of the 20th century saw a great expansion of high technology. As new technologies proliferated, job opportunities for vibration engineers multiplied. The new technologies also provided vastly improved tools for the vibration engineer. Particularly important has been the development of improved sensors and actuators. It is interesting to compare the vast array of optical, electrical, magnetic, piezoelectric, eddy-current, and laser transducers available to the engineer today, with the situation faced by the pioneers in torsional vibration, who had no way of detecting the presence of large vibrations, superposed on the steady rotation of the shaft, until the shaft failed.

Perhaps the most important technical advance of the last half-century has been the development of the digital computer. The computer is now an essential part of most vibration measurement systems, and the computational power it has unleashed has provided important support for many theoretical advancements.

After World War II, the development of rocket propulsion and space flight introduced vibration engineers to the topic of random vibration and the requirements for large wide-band shakers and sophisticated data-processing instrumentation. Instrumentation based on analog principles was soon replaced by digital processors. The fast Fourier transform (FFT) was conceived in 1965, and a decade later, commercial instruments based on the FFT were on the market. Soon digital analyzers capable of performing experimental modal analysis became available. At the same time, computer software for performing finite element analysis and boundary element analysis was being applied to the dynamic analysis of complicated real structures. The difficulty in reconciling the experimental measurements with the analytical predictions, beyond the first few modes of a structure, remains a challenge for vibration engineers.

During the past half-century, important advances in the theory of nonlinear vibrations have been made, including the clarification of the concept of chaotic response from a deterministic system. Much of the supporting work would not have been possible without high-speed digital computers.

A present trend in vibration technology is the inclusion of vibration considerations in the preliminary design of products. The old tradition of waiting until a prototype was available, and hoping that any vibration problem that arose could be cured with a quick fix (which did not impact the primary design goals) is gradually giving way to early consideration of the vibrational problems of the product. In some industries, machinery is being delivered with vibration sensors already installed, so that the customer can monitor the performance, or diagnose the cause of a malfunction. Another present trend is the miniaturization of sensors and actuators using microelectromechanical elements (MEMS).

The field of vibration is now so broad that no single person can keep up-to-date in all its branches. Furthermore, it is impossible to squeeze all the essential information into a single book. The editors of the present *Encyclopedia of Vibration* have performed a valuable service to the field by dividing the subject matter into 168 chapters and selecting experts to create authoritative chapters on the individual topics. The result is a three-volume encyclopedia, which will be of great value to practicing vibration engineers and theorists. The encyclopedia will also be useful as a general reference, and as a guide to students, and professionals in neighboring fields.

Stephen H. Crandall
Ford Professor of Engineering, Emeritus
Massachusetts Institute of Technology

Preface

We are surrounded by vibrations, some of them generated by nature, others by human-built devices. While some are harmless, many have adverse effects. These can affect human well-being or health, cause malfunctioning of devices, or result in life-threatening situations. Vibrating patterns may also carry information, which when deciphered correctly can convey important knowledge.

This three-volume reference work focuses on aspects of vibration which are of interest mainly to practicing and research engineers. Even within this specific audience, vibration problems span multidisciplinary aspects, and we thus believe there is a need for a reference work, geared to the specific area of vibration, covering as many relevant examples as possible.

The more classical topics of vibration engineering are naturally covered: basic principles, sound radiation by vibrating structures, vibration isolation, damping principles, rotating systems, wave propagation, nonlinear behavior, and stability. Modern aspects developed in the last two decades also have a prominent place. To mention just a few, these include software and computational aspects, testing and specifically modal testing/modeling, active vibration control, instrumentation, signal processing, smart materials, and vibration standards. Specific entries address analysis/design aspects as encountered in aeronautical, civil, and mechanical engineering: vibration abatement, seismic vibrations, civil engineering structures, rotating machines and surveillance, aircraft flutter, and the effect of vibrations on humans.

In view of the large scope of topics addressed, the type and depth of presentation had to be adapted to the style of concise alphabetical entries, and some compromises were necessary in the choice of topics. More 'classical' aspects suffer somewhat at the expense of more modern topics, due to our opinion that more information sources are already available for them. Carefully chosen references appearing under the 'further reading' sections are however given for them. In addition, at the end of each entry there is a cross-referencing 'see also' section where the reader is directed to other entries within the *Encyclopedia* which contain additional and/or related information. In view of the large scope of topics chosen, the type and depth of presentation had to be adapted to specific entries. Unified symbols and notation have been used to some extent, but when deemed preferable, superceded by those commonly accepted in specific areas. The table of contents lists entries in an intuitive alphabetical form, and any topic not found through the contents list can be located by referring to the index which appears in each volume. Useful reference data can be found in the appendices, which also appear in each volume.

An online version of the *Encyclopedia* will be available to all purchasers of the print version and will have extensive hypertext links and advanced search tools. Leading authorities have contributed to this work. Together with our advisory board they should receive the main credit for this important source of knowledge. My personal thanks are naturally extended to them, but even more so to my colleagues and co-editors, Dave Ewins and Singiresu Rao, without whose constant help this task would have been impossible. Special thanks are due to Dr Carey Chapman and Lorraine Parry from the Academic Press Major Reference Work team for their encouragement and help extended to us at every stage.

Simon Braun
Editor-in-Chief

Guide to Use of the Encyclopedia

Structure of the *Encyclopedia*

The material in the *Encyclopedia* is arranged as a series of entries in alphabetical order. Some entries comprise a single article, whilst entries on more diverse subjects consist of several articles that deal with various aspects of the topic. In the latter case the articles are arranged in a logical sequence within an entry.

To help you realize the full potential of the material in the *Encyclopedia* we have provided three features to help you find the topic of your choice: contents list, cross-references, and index.

Contents lists

Your first point of reference will probably be the contents list. The complete contents list appearing in each volume will provide you with both the volume number and the page number of the entry. On the opening page of an entry containing more than one article, a contents list is provided so that the full details of the articles within the entry are immediately available.

Alternatively, you may choose to browse through a volume using the alphabetical order of the entries as your guide. To assist you in identifying your location within the *Encyclopedia*, a running headline indicates the current entry and the current article within that entry.

You will find dummy entries in the following instances:

1. Where obvious synonyms exist for entries.
   For example, a dummy entry appears for Structural Damping which directs you to Hysteretic Damping where the material is located.
2. Where we have grouped together related topics.
   For example, a dummy entry appears for Magnetorheological Fluids which leads you to Electrorheological and Magnetorheological Fluids where the material is located.
3. Where there is debate over the given entry title and whether readers would intuitively find the topic they are trying to locate under that title.
   For example, a dummy entry appears for Vibration Absorbers which directs you to Absorbers, Vibration where the material is located.
4. Where there is debate over whether diverse subjects should comprise several single articles, or one entry consisting of several articles.
   For example, a dummy entry appears for Smart Materials which directs you to the entries Electrorheological and Magnetorheological Fluids, Electrostrictive Materials, Magnetostrictive Materials, Piezoelectric Materials, and Shape Memory Alloys.

Dummy entries appear in both the contents list and the body of the text.

Example
If you were attempting to locate material on diagnostics via the contents list the following information would be provided:

DIAGNOSTICS *See* BEARING DIAGNOSTICS; DIAGNOSTICS AND CONDITION MONITORING, BASIC CONCEPTS; GEAR DIAGNOSTICS; NEURAL NETWORKS, DIAGNOSTIC APPLICATIONS

The page numbers of these entries are given at the appropriate location in the contents list.

If you were trying to locate the material by browsing through the text and you looked up Diagnostics then the following would be provided:

> **DIAGNOSTICS** *See* **BEARING DIAGNOSTICS; DIAGNOSTICS AND CONDITION MONITORING, BASIC CONCEPTS; GEAR DIAGNOSTICS; NEURAL NETWORKS, DIAGNOSTIC APPLICATIONS**

Cross-references

To direct the reader to other entries on related topics, a 'see also' section is provided at the end of each entry.

Example
The entry Wave propogation, Waves in an Unbounded Medium includes the following cross-references:

See also: **Nondestructive testing**, Sonic; **Nondestructive testing**, Ultrasonic; **Ultrasonics**; **Wave propagation**, Guided waves in structures; **Wave propagation**, Interaction of waves with boundaries.

Index

The index will provide you with the volume number and page number of where the material is located, and the index entries differentiate between material that is a whole article, is part of an article, or is data presented in a table. On the opening page of the index detailed notes are provided. Any topic not found through the contents list can be located by referring to the index.

Color Plates

The color figures for each volume have been grouped together in a plate section. The location of this section is cited both in the contents list and before the 'See also' list of the pertinent articles.

Appendices

The appendices appear in each volume.

Contributors

A full list of contributors appears at the beginning of each volume.

Contributors

Agnes, G
1489 Fudge Drive
Beavercreek
OH 45434
USA

Ahmadian, M
Virginia Tech
Advanced Vehicle Dynamics Laboratory
Department of Mechanical Engineering
Blacksburg, VA 24061-0238
USA

Ahmed, A K W
Concordia University
Department of Mechanical Engineering
Montreal
Quebec
Canada

Bajaj, A
Purdue University
Department of Mechanical Engineering
West Lafayette
IN 47907-1288
USA

Banks, H T
North Carolina State University
Center for Research - Science Comput.
324 Harrelson Hall, Box 8205, CRSC
Raleigh, NC 27695-8105
USA

Bauchau, O
Georgia Institute of Technology
Atlanta
GA 30332-0710
USA

Baz, A
University of Maryland
Department of Mechanical Engineering
2137 Engineering Building
College Park, MD 20742
USA

Benson, D J
University of California, San Diego
Division of Mechanical Engineering
Department of Applied Mechanics and
Engineering Sciences
9500 Gilman, La Jolla, CA 92093
USA

Bert, C W
University of Oklahoma
Department of Mechanical Engineering
865 Asp Avenue, Room 202
Norman, OK 73019
USA

Bigret, R
22 Rue J Varnet
93700 Drancy
France

Book, W
Georgia Institute of Technology
School of Mechanical Engineering
Room 472 Manufacturing Research Center
813 Ferst Drive, Atlanta, GA 30332-0405
USA

Braun, S
Technion - Israel Institute of Technology
Faculty of Mechanical Engineering
Haifa 32000
Israel

Cai, G Q
Florida Atlantic University
Center for Applied Stochastics
Boca Raton
FL 33431
USA

Cardona, A
INTEC
Grupo de Technologia Mecanica
Guemes 3450
RA-3000, Santa Fe
Argentina

Castellini, P
Universitá di Ancona
Dipartimento Di Meccanica
Via Brecce Bianche
1-60131 Ancona
Italy

Chou, C S
National Taiwan University
Taipei
Taiwan
Republic of China

Constantinides, T
Imperial College of Science, Technology and Medicine
Engineering Department, Room 812
Exhibition Road
London SW7 2BT
UK

Cooper, J E
School of Engineering
University of Manchester
Oxford Road
Manchester M13 9PL
UK

Craig, Jr R R
University of Texas
Aeronautical Engineering and Engineering Mechanics Department
Austin TX 78712
USA

D'Aubrogio, W
University of L'Aquila
Dipartimento di Energetica
Roio Poggio
67040 L'Aquila
Italy

Dalpiaz, G
University of Bologna
Faculty of Mechanical Engineering
2, 40136 Bologna
Italy

CONTRIBUTORS

David, A
Auburn University
Nonlinear Systems Research Laboratory
Department of Mechanical Engineering
Auburn, AL 36849
USA

Devloo, P
Universidade Estadual de Campinas
Faculdade de Engenharia Civil
Caixa Postal 6021
13083-970 Campinas São Paulo
Brazil

Dimentberg, M F
Worcester Polytechnic Institute
Mechanical Engineering Department
Worcester
MA 01609
USA

Doebling, S
Los Alamos National Laboratory
ESA - EA, M/S P946, PO Box 1663
NM 87545
USA

Drew, S J
The University of Western Australia
Department of Mechanical and Materials Engineering
Nedlands 6709, Perth
Western Australia
Australia

Dubuisson, B
33 rue Saint Hubert
60610 La Croix Saint Ouen
France

Dyne, S
University of Southampton
Institute of Sound and Vibration Research
Southampton
SO17 1BJ
UK

Elishakoff, I
Florida Atlantic University
Department of Mechanical Engineering
Boca Raton
FL 33431-0991
USA

Elliott, S J
The University of Southampton
Institute of Sound and Vibration Research
Southampton SO17 1BJ
UK

Ewins, D J
Imperial College of Science, Technology and Medicine
Department of Mechanical Engineering
Exhibition Road
London SW7 2BX
UK

Farhat, C
University of Colorado
Department of Aerospace Engineering Sciences
Campus Box 429
Boulder, CO 80309
USA

Farrar, C
Los Alamos National Laboratory
ESA - EA, M/S P946, PO Box 1663
NM 87545
USA

Fassois, S D
University of Patras
Stochastic Mechanical Systems (SMS) Group
Faculty of Mechanical and Aeronautical Engineering
GR 265 00 Patras
Greece

Feeny, B F
Michigan State University
Department of Mechanical Engineering
2555 Engineering Building
East Lansing, MI 48824
USA

Feldman, M
Technion - Israel Institute of Technology
Faculty of Mechanical Engineering
Haifa 32000
Israel

Flatau, A
National Science Foundation
Dynamic Systems and Control Program
4201 Wilson Blvd. Suite 545
Arlington VA 22230
USA

Fuller, C R
Virginia Tech
Department of Mechanical Engineering
Blacksburg
VA 24061-0238
USA

Gandhi, F
The Pennsylvania State University
Department of Aerospace Engineering
233 Hammond Building
University Park, PA 16802
USA

Gern, F H
Center for Intelligent Material Systems and Structures (CIMSS)
Virginia Polytechnic Institute and State University
New Engineering Building, Room 303
Blacksburg, VA 24061-0261
USA

Giurgiutiu, V
University of South Carolina
Department of Mechanical Engineering
300 S. Main Street, Room A222
Columbia, SC 29208
USA

Griffin, M J
The University of Southampton
Human Factors Research Unit
Institute of Sound and Vibration Research
Southampton SO17 1BJ
UK

Griffin, S
AFRL/VSSV
3550 Aberdeen Ave SE
Kirtland AFB
NM 87117-5748
USA

Haddow, A
Michigan State University
Department of Mechanical Engineering
East Lansing
MI 48824
USA

CONTRIBUTORS

Hallquist, J
Livermore Software Technology Corporation (LSTC)
97 Rickenbacker Circle
Livermore
CA 94550
USA

Hann, F
University of Notre Dame
Department of Civil Engineering
Notre Dame
IN 46556-0767
USA

Hartmann, F
University of Kassel
Dept. Baustatik, Fb 14
Kurt-Wolters-Str 3
D-34109 Kassel
Germany

Hayek, S I
Pennsylvania State University
112 EES Building
University Park
PA 16802-6812
USA

Holmes, P J
Princeton University
Department of Mechanical and Aerospace Engineering
Princeton
New Jersey 08544
USA

Ibrahim, R
Wayne State University
Mechanical Engineering
Detroit, MI 48202
USA

Inman, D
Virginia Polytechnic Institute and State University
Department of Engineering Science and Mechanics
310 NEB, Mail Code 0261
Blacksburg, VA 24061-0219
USA

Kapania, R
Virginia Polytechnic Institute and State University
Department of Aerospace and Ocean Engineering
Blacksburg
VA 24061-0203
USA

Kareem, A
University of Notre Dame
Department of Civil Engineering
Notre Dame
IN 46556-0767
USA

Kijewski, T
University of Notre Dame
Department of Civil Engineering
Notre Dame
IN 46556-0767
USA

Klapka, I
Université de Liège
Laboratoire des Techniques Aéronautiques et Spatiales
Dynamique des Structures
rue E Solvay 21, B-4000, Liège
Belgium

Kobayashi, A S
University of Washington
Department of Mechanical Engineering
Seattle, WA 98195
USA

Krishnan, R
University of Maryland at College Park
Department of Aerospace Engineering
3180 Engineering Classroom Building
College Park
MD 20742-3015
USA

Krousgrill, C M
Purdue University
School of Mechanical Engineering
West Lafayette
IN 47907-1288
USA

Leissa, A W
Ohio State University
Department of Mechanical Engineering
206 W. 18th Ave
Columbus, OH 43210-1154
USA

Lesieutre, G A
Pennsylvania State University
Department of Aerospace Engineering
153G Hammond Building
University Park, PA 16802
USA

Li, C J
Rensselaer Polytechnic Institute
Department of Mechanical Engineering,
Aeronautical Engineering, and Mechanics
110 8th Street
Troy, NY 12180
USA

Lieven, N A J
Bristol University
Department of Aerospace Engineering
Queens Building, University Walk
Bristol, BS8 1TR
UK

Lin, Y K
Florida Atlantic University
Center for Applied Stochastics
Boca Raton
FL 33431
USA

Link, M
Universität Gesamthoschule Kassel
Fachgebiet Leichtbau
Mönchebergstrasse 7
D34109 Kassel
Germany

Lowe, M J S
Imperial College of Science, Technology and Medicine
Department of Mechanical Engineering
Exhibition Road
London, SW7 2BX
UK

Ma, F
University of California
Department of Mechanical Engineering
Berkeley
California, CA 94720
USA

CONTRIBUTORS

Maddux, G E
PO Box 203
Tipp City
OH 45371 - 9435
USA

Maia, N M M
Instituto Superior Technico
Department de Engenharia Mecanica
Av Rovisco Pais
1049-001 Lisboa
Portugal

Marcondes, J
San Jose University
Packaging Department
One Washington Square
San Jose, CA 95192-0005
USA

McConnell, K
Iowa State University Ames
AEEM
Black Engineering Building
Iowa 50011
USA

McKee, K
Rensselaer Polytechnic Institute
Department of Mechanical Engineering, Aeronautical Engineering, and Mechanics
110 8th Street
Troy, NY12180
USA

Mucino, V H
West Virginia University
Mechanical and Aerospace Engineering
Room 351
PO Box 6106, Morgantown WV 26506-6106
USA

Naeim, F
John A Martin & Associates, Inc.
1212 S. Flower Street
Los Angeles, CA 90015
USA

Natori, M C
Institute of Space and Astronautical Science
Kanagawa
Japan

Niemkiewicz, J
Techonology Transfer Manager
Maintenance and Diagnostic (M&D) LLC
440 Baldwin Tower
Eddystone, PA 19022
USA

Norton, M P
The University of Western Australia
Department of Mechanical and Materials Engineering
Nedlands 6709, Perth
Western Australia
Australia

Pan, J
University of Western Australia
Department of Mechanical and Materials Engineering
Nedlands 6709, Perth
Western Australia
Australia

Perkins, N C
University of Michigan
Mechanical Engineering and
Applied Mechanics Department
Ann Arbor
MI 48109
USA

Peterka, F
Academy of Sciences of the Czech Republic
Institute of Thermomechanics
182 00 Prague 8, Doplejskova 5
The Czech Republic

Pierre, C
University of Michigan
Mechanical Engineering and Applied Mechanics
College of Engineering, 2202 GG Brown Building
2350 Hayward Street, Ann Arbor, MI 48109-2125
USA

Pinter, G A
North Carolina State University
Center for Research - Science Comput.
324 Harrelson Hall, Box 8205, CRSC
Raleigh, NC 27695-8105
USA

Prasad, M G
Stevens Institute of Technology
Department of Mechanical Engineering
Hoboken
NJ 07030
USA

Rade, D
Federal University of Uberlandia
Mechanical Engineering Department
Campus Santa Monica
PO Box 593, 38400-902 Uberlandia, MG
Brazil

Radeş, M
Universitatea Politehnica Bucuresti
Department of Engineering Sciences
313 Splaiul Independentei
79590 Bucuresti
Romania

Ram, Y M
Louisiana State University
Mechanical Engineering Department
Baton Rouge
LA 70803
USA

Ramulu, M
University of Washington
Department of Mechanical Engineering
Seattle, WA 98195
USA

Randall, R B
University of New South Wales
School of Mechanical and Manufacturing Engineering
Sydney 2052, New South Wales
Australia

Rao, S S
University of Miami
Department of Mechanical Engineering
PO Box 248294, Coral Gables
FL 33124-0624
USA

Revel, G M
Università di Ancona
Dipartimento di Meccanica
Via Brecce Bianche
1-60131 Ancona
Italy

Rivin, E
Wayne State University
Mechanical Engineering
2100W Engineering Building
Detroit, Michigan 48202
USA

Rixen, D
Delft University of Technology
PO Box 5
2600 AA Delft
The Netherlands

Robert, G
Samtech SA
Rue des Chasseurs-Ardennais, 8
B-4031 Liège
Belgium

Rosenhouse, G
Technion - Israel Institute of Technology
Faculty of Civil Engineering
Technion City
Haifa 32000
Israel

Schmerr, Jr, L W
Iowa State University
Center for NDE, Aerospace Engineering and Engineering Mechanics
211A Applied Science Complex II, 1915 Scholl Road
1915 Scholl Road, Ames, IA 50011
USA

Sciulli, D
5725 Cedar Way #301
Centerville
VA 20121
USA

Scott, R A
University of Michigan
Department of Mechanical Engineering and Applied Mechanics
2206 G. G. Brown Building
Ann Arbor, MI 48109
USA

Sestieri, A
Università Degli Studi di Roma
Dipartimento Meccanica e Aeronautica
Via Eudossinia 18
I - 00184 Roma
Italy

Shaw, S
Michigan State University
Department of Mechanical Engineering
A321 Engineering Building
East Lansing, MI 48824-1226
USA

Shteinhauz, G D
The Goodyear Tire Rubber Company
Tire-Vehicle Engineering Technology
Technical Center D/460G
PO Box 353, Akron, Ohio 44309-3531
USA

Sidahmed, M
Université de Technologie de Compiègne
Heuristique et Diagnostic des Systèmes Complexes
Centre de Recherches de Royallieu
BP 20529
60205 Compiègne
France

Sieg, T
Paulstra Industries Inc.
Carlsbad, CA
USA

Silva, J M M
Instituto Superior Tecnico
Departmento de Engenharia Mecanica
Av Rovisco Pais
1049-001 Lisboa
Portugal

Sinha, S
Auburn University
Nonlinear Systems Research Laboratory
Department of Mechanical Engineering
Auburn, AL 36849
USA

Smallwood, D
Sandia National Laboratories
PO Box 5800
Albuquerque
NM 87185-0865
USA

Snyder, R
University of Maryland at College Park
Department of Aerospace Engineering
3180 Engineering Classroom Building
College Park, MD 20742-3015
USA

Soedel, W
Purdue University
School of Mechanical Engineering
West Lafayette
IN 47907
USA

Soong, T T
State University of New York at Buffalo
MCEER
107 Red Jacket Quadrangle
Buffalo, NY 14261-0025
USA

Spencer, Jr, B F
University of Notre Dame
Department of Civil Engineering
Notre Dame
IN 46556-0767
USA

Stanway, R
The University of Sheffield
Department of Mechanical Engineering
Mappin Street
Sheffield S1 3JD
UK

Steffen, Jr, V
Federal University of Uberlandia
Mechanical Engineering Department
Campus Santa Monica
PO Box 593, 38400-902 Uberlandia, MG
Brazil

CONTRIBUTORS

Steindl, A
Vienna University of Technology
Wiedner Hauptstrasse 8-10
A-1040 Vienna
Austria

Stiharu, I
Concordia University
Department of Mechanical Engineering
1455 de Masonneuve Blv W
Montreal, Quebec H3G 1M8
Canada

Sun, J-Q
University of Delaware
Department of Mechanical Engineering
Newark, DE 19716
USA

Sunar, M
King Fahd University of Petroleum and Minerals
Department of Mechanical Engineering
PO Box 1205
Dhahran 31261
Saudi Arabia

Tang, J
The Pennsylvania State University
Department of Mechanical Engineering
157 E Hammond Building, University Park
PA 16802
USA

Tomasini, E P
Università di Ancona
Dipartimento Di Meccanica
Via Brecce Bianche
1-60131 Ancona
Italy

Tordon, M J
University of New South Wales
School of Mechanical and
Manufacturing Engineering
Sydney 2052, New South Wales
Australia

Troger, H
Vienna University of Technology
Wiedner Hauptstrasse 8-10
A-1040 Vienna
Austria

Tzou, H S
University of Kentucky
Department of Mechanical Engineering
Dynamics and Systems Laboratory
Lexington, KY 40506
USA

Uchino, K
The Pennsylvania State University
134 Materials Research Laboratory
University Park
PA, 16802-4800
USA

Ungar, E E
Acentech Incorporated
33 Moulton Street
Cambridge
MA 02138-1118
USA

Vakakis, A
University of Illinois
Department of Mechanical and Industrial Engineering
140 Mechanical Engineering Building,
1206 West Green Street
Urbana, IL 61801
USA

Varoto, P S
Escola de Engenharia de São Carlos, USP
Dept. Engenharia Mecanica
Av.Dr. Carlos Botelho, 1465, CP 359
São Carlos - SP - 13560-250
Brasil

Vorus, W S
University of New Orleans
School of Naval Architecture and Marine Engineering
College of Engineering
New Orleans, LA 70148
USA

Wang, K W
The Pennsylvania State University
Department of Mechanical Engineering
157 E Hammond Building, University Park
PA 16802
USA

Wereley, N M
University of Maryland at College Park
Department of Aerospace Engineering
3180 Engineering Classroom Building
College Park, MD 20742-3015
USA

White, P
The University of Southampton
Institute of Sound and Vibration Research (ISVR)
Southampton, SO9 5NH
UK

Wickert, J
Carnegie Mellon University
Department of Mechanical Engineering
Pittsburg
PA 15213-3890
USA

Wright, J
University of Manchester
School of Engineering
Oxford Road
Manchester M13 9PL
UK

Yang, B
University of Southern California
Department of Mechanical Engineering
Los Angeles
CA 90089-1453
USA

Zacksenhouse, M
Technion - Israel Institute of Technology
Faculty of Mechanical Engineering
Haifa 32000
Israel

Zu, J W
University of Toronto
Department of Mechnical and Industrial Engineering
5 King's College Road
Toronto, Ontario
Canada M5S 3G8

CONTENTS

VOLUME 1

A

ABSORBERS, ACTIVE *G Agnes*	1
ABSORBERS, VIBRATION *V Steffen, Jr D Rade*	9
ACTIVE ABSORBERS See ABSORBERS, ACTIVE	26
ACTIVE CONTROL OF CIVIL STRUCTURES *T T Soong B F Spencer, Jr.*	26
ACTIVE CONTROL OF VEHICLE VIBRATION *M Ahmadian*	37
ACTIVE ISOLATION *S Griffin D Sciulli*	46
ACTIVE VIBRATION CONTROL See ACTIVE CONTROL OF VEHICLE VIBRATION; ACTIVE ISOLATION; ACTIVE VIBRATION SUPPRESSION; ACTUATORS AND SMART STRUCTURES; DAMPING, ACTIVE; FEED FORWARD CONTROL OF VIBRATION; FLUTTER, ACTIVE CONTROL; HYBRID CONTROL.	48
ACTIVE VIBRATION SUPPRESSION *D Inman*	48
ACTUATORS AND SMART STRUCTURES *V Giurgiutiu*	58
ADAPTIVE FILTERS *S J Elliott*	81
AEROELASTIC RESPONSE *J E Cooper*	87
ANTIRESONANCE See RESONANCE AND ANTIRESONANCE	98
AVERAGING *S Braun*	98

B

BALANCING *R Bigret*	111
BASIC PRINCIPLES *G Rosenhouse*	124
BEAMS *R A Scott*	137
BEARING DIAGNOSTICS *C J Li K McKee*	143
BEARING VIBRATIONS *R Bigret*	152
BELTS *J W Zu*	165
BIFURCATION See DYNAMIC STABILITY	174
BLADES AND BLADED DISKS *R Bigret*	174
BOUNDARY CONDITIONS *G Rosenhouse*	180
BOUNDARY ELEMENT METHODS *F Hartmann*	192
BRIDGES *S S Rao*	202

C

CABLES *N C Perkins*	209
CEPSTRUM ANALYSIS *R B Randall*	216
CHAOS *P J Holmes*	227
CHATTER See MACHINE TOOLS, DIAGNOSTICS	236
COLUMNS *I Elishakoff C W Bert*	236

COMMERCIAL SOFTWARE G Robert	243
COMPARISON OF VIBRATION PROPERTIES	256
Comparison of Spatial Properties M Radeş	256
Comparison of Modal Properties M Radeş	265
Comparison of Response Properties M Radeş	272
COMPONENT MODE SYNTHESIS (CMS) See THEORY OF VIBRATION, SUBSTRUCTURING	278
COMPUTATION FOR TRANSIENT AND IMPACT DYNAMICS D J Benson J Hallquist	278
COMPUTATIONAL METHODS See BOUNDARY ELEMENT METHODS; COMMERCIAL SOFTWARE; COMPUTATION FOR TRANSIENT AND IMPACT DYNAMICS; CONTINUOUS METHODS; EIGENVALUE ANALYSIS; FINITE DIFFERENCE METHODS; FINITE ELEMENT METHODS; KRYLOV-LANCZOS METHODS; LINEAR ALGEBRA; OBJECT ORIENTED PROGRAMMING IN FE ANALYSIS; PARALLEL PROCESSING; TIME INTEGRATION METHODS.	286
CONDITION MONITORING See DIAGNOSTICS AND CONDITION MONITORING, BASIC CONCEPTS; ROTATING MACHINERY, MONITORING.	286
CONTINUOUS METHODS C W Bert	286
CORRELATION FUNCTIONS S Braun	294
CRASH V H Mucino	302
CRITICAL DAMPING D Inman	314

D

DAMPING IN FE MODELS G A Lesieutre	321
DAMPING MATERIALS E E Ungar	327
DAMPING MEASUREMENT D J Ewins	332
DAMPING MODELS D Inman	335
DAMPING MOUNTS J-Q Sun	342
DAMPING, ACTIVE A Baz	351
DATA ACQUISITION R B Randall M J Tordon	364
DIAGNOSTICS AND CONDITION MONITORING, BASIC CONCEPTS M Sidahmed	376
DIAGNOSTICS See BEARING DIAGNOSTICS; DIAGNOSTICS AND CONDITION MONITORING, BASIC CONCEPTS; GEAR DIAGNOSTICS; NEURAL NETWORKS, DIAGNOSTIC APPLICATIONS	380
DIGITAL FILTERS A G Constantinides	380
DISCRETE ELEMENTS S S Rao	395
DISKS D J Ewins	404
DISPLAYS OF VIBRATION PROPERTIES M Radeş	413
DISTRIBUTED SENSORS AND ACTUATORS See SENSORS AND ACTUATORS	431
DUHAMEL METHOD See THEORY OF VIBRATION: DUHAMEL'S PRINCIPLE AND CONVOLUTION	431
DYNAMIC STABILITY A Steindl H Troger	431

E

EARTHQUAKE EXCITATION AND RESPONSE OF BUILDINGS F Naeim	439
EIGENVALUE ANALYSIS O Bauchau	461

ELECTRORHEOLOGICAL AND MAGNETORHEOLOGICAL FLUIDS *R Stanway*	467
ELECTROSTRICTIVE MATERIALS *K Uchino*	475
ENVIRONMENTAL TESTING, OVERVIEW *D Smallwood*	490
ENVIRONMENTAL TESTING, IMPLEMENTATION *P S Varoto*	496
EQUATIONS OF MOTION *See* THEORY OF VIBRATION: EQUATIONS OF MOTION	504

VOLUME 2

F

FATIGUE *A S Kobayashi M Ramulu*	505
FE MODELS *See* DAMPING IN FE MODELS; FINITE ELEMENT METHODS	513
FEEDFORWARD CONTROL OF VIBRATION *C R Fuller*	513
FFT METHODS *See* TRANSFORM METHODS	520
FILTERS *See* ADAPTIVE FILTERS; DIGITAL FILTERS; OPTIMAL FILTERS	520
FINITE DIFFERENCE METHODS *S S Rao*	520
FINITE ELEMENT METHODS *S S Rao*	530
FLUID/STRUCTURE INTERACTION *S I Hayek*	544
FLUTTER *J Wright*	553
FLUTTER, ACTIVE CONTROL *F H Gern*	565
FORCED RESPONSE *N A J Lieven*	578
FOURIER METHODS *See* TRANSFORM METHODS	582
FREE VIBRATION *See* THEORY OF VIBRATION: FUNDAMENTALS	582
FRICTION DAMPING *R Ibrahim*	582
FRICTION INDUCED VIBRATIONS *R Ibrahim*	589

G

GEAR DIAGNOSTICS *C J Li*	597
GROUND TRANSPORTATION SYSTEMS *A K Waizuddin Ahmed*	603

H

HAND-TRANSMITTED VIBRATION *M J Griffin*	621
HELICOPTER DAMPING *N M Wereley R Snyder R Krishnan T Sieg*	629
HILBERT TRANSFORMS *M Feldman*	642
HUMAN RESPONSE TO VIBRATION *See* GROUND TRANSPORTATION SYSTEMS; HAND-TRANSMITTED VIBRATION; MOTION SICKNESS; WHOLE-BODY VIBRATION	649
HYBRID CONTROL *J Tang K W Wang*	649
HYSTERETIC DAMPING *H T Banks G A Pinter*	658

I

IDENTIFICATION, FOURIER-BASED METHODS *S Braun*	665
IDENTIFICATION, MODEL-BASED METHODS *S D Fassois*	673
IDENTIFICATION, NON-LINEAR SYSTEMS *See* NON-LINEAR SYSTEM IDENTIFICATION	685
IMPACTS, NON-LINEAR SYSTEMS *See* VIBRO-IMPACT SYSTEMS	686

IMPULSE RESPONSE FUNCTION See THEORY OF VIBRATION: IMPULSE RESPONSE FUNCTION	686
INTENSITY See VIBRATION INTENSITY	686
INVERSE PROBLEMS Y M Ram	686
ISOLATION, ACTIVE See ABSORBERS, ACTIVE; ACTIVE CONTROL OF VEHICLE VIBRATION; ACTIVE ISOLATION	690
ISOLATION VIBRATION – APPLICATIONS AND CRITERIA See VIBRATION ISOLATION, APPLICATIONS AND CRITERIA	690
ISOLATION VIBRATION – THEORY See VIBRATION ISOLATION THEORY	690

K

KRYLOV-LANCZOS METHODS R R Craig Jr	691

L

LAGRANGE METHOD See BASIC PRINCIPLES; THEORY OF VIBRATION, ENERGY METHODS	699
LAPLACE TRANSFORMS See TRANSFORM METHODS	699
LASER BASED MEASUREMENTS E P Tomasini G M Revel P Castellini	699
LINEAR ALGEBRA C Farhat D Rixen	710
LINEAR DAMPING MATRIX METHODS F Ma	721
LIQUID SLOSHING R A Ibrahim	726
LOCALIZATION C Pierre	741

M

MACHINERY, ISOLATION See VIBRATION ISOLATION, APPLICATIONS AND CRITERIA.	753
MAGNETORHEOLOGICAL FLUIDS See ELECTRORHEOLOGICAL AND MAGNETORHEOLOGICAL FLUIDS	753
MAGNETOSTRICTIVE MATERIALS A Flatau	753
MATERIALS, DAMPING See DAMPING MATERIALS	762
MEASUREMENT See LASER BASED MEASUREMENT; SEISMIC INSTRUMENTS, ENVIRONMENTAL FACTORS; STANDARDS FOR VIBRATIONS OF MACHINES AND MEASUREMENT PROCEDURES; TRANSDUCERS FOR ABSOLUTE MOTION; TRANSDUCERS FOR RELATIVE MOTION	762
MEMBRANES A W Leissa	762
MEMS, APPLICATIONS I Stiharu	771
MEMS, DYNAMIC RESPONSE I Stiharu	779
MEMS, GENERAL PROPERTIES I Stiharu	794
MODAL ANALYSIS, EXPERIMENTAL	805
Basic Principles D J Ewins	805
Measurement Techniques J M M Silva	813
Parameter Extraction Methods N M M Maia	820
Construction of Models from Tests N M M Maia	824
Applications D J Ewins	829
MODE OF VIBRATION D J Ewins	838
MODEL UPDATING AND VALIDATING M Link	844
MODELS, DAMPING See DAMPING MODELS	856

MODES, NON-LINEAR SYSTEMS See NON-LINEAR SYSTEMS MODES	856
MODES, ROTATING MACHINERY See ROTATING MACHINERY, MODAL CHARACTERISTICS	856
MONITORING See DIAGNOSTICS AND CONDITION MONITORING, BASIC CONCEPTS, ROTATING MACHINERY, MONITORING	856
MOTION SICKNESS *M J Griffin*	856

N

NEURAL NETWORKS, DIAGNOSTIC APPLICATIONS *M Zacksenhouse*	863
NEURAL NETWORKS, GENERAL PRINCIPLES *B Dubuisson*	869
NOISE	877
Noise Radiated from Elementary Sources *M P Norton J Pan*	877
Noise Radiated by Baffled Plates *M P Norton J Pan*	887
NONDESTRUCTIVE TESTING	898
Sonic *S Doebling C Farrar*	898
Ultrasonic *L W Schmerr Jr*	906
NONLINEAR NORMAL MODES *A F Vakakis*	918
NONLINEAR SYSTEM IDENTIFICATION *B F Feeny*	924
NONLINEAR SYSTEM RESONANCE PHENOMENA *A Bajaj C M Krousgrill*	928
NONLINEAR SYSTEMS, OVERVIEW *N C Perkins*	944
NONLINEAR SYSTEMS ANALYSIS *A Bajaj*	952

O

OBJECT ORIENTED PROGRAMMING IN FE ANALYSIS *I Klapka A Cardona P Devloo*	967
OPTIMAL FILTERS *S J Elliott*	977

P

PACKAGING *J Marcondes*	983
PACKAGING, ELECTRONIC See ELECTRONIC PACKAGING	990
PARALLEL PROCESSING *D Rixen*	990
PARAMETRIC EXCITATION *S C Sinha A David*	1001
PERTURBATION TECHNIQUES FOR NONLINEAR SYSTEMS *S Shaw*	1009
PIEZOELECTRIC MATERIALS AND CONTINUA *H S Tzou M C Natori*	1011
PIPES *S S Rao*	1019
PLATES *A W Leissa*	1024

VOLUME 3

R

RANDOM PROCESSES *M F Dimentberg*	1033
RANDOM VIBRATION See RANDOM VIBRATION, BASIC THEORY; RANDOM PROCESSES, STOCHASTIC SYSTEMS	1040
RANDOM VIBRATION, BASIC THEORY *M F Dimentberg*	1040
RESONANCE AND ANTIRESONANCE *M Radeş*	1046

RESONANCE, NON LINEAR SYSTEMS *See* NON LINEAR SYSTEM RESONANCE PHENOMENA; STOCHASTIC SYSTEMS	1055
ROBOT VIBRATIONS *W Book*	1055
ROTATING MACHINERY *See* ROTATING MACHINERY, ESSENTIAL FEATURES; ROTATING MACHINERY, MODAL CHARACTERISTICS; ROTATING MACHINERY, MONITORING; ROTOR DYNAMICS; ROTOR STATOR INTERACTIONS; BALANCING; BLADES AND BLADED DISKS	1064
ROTATING MACHINERY, ESSENTIAL FEATURES *R Bigret*	1064
ROTATING MACHINERY, MODAL CHARACTERISTICS *R Bigret*	1069
ROTATING MACHINERY, MONITORING *R Bigret*	1078
ROTOR DYNAMICS *R Bigret*	1085
ROTOR–STATOR INTERACTIONS *R Bigret*	1107

S

SEISMIC INSTRUMENTS, ENVIRONMENTAL FACTORS *K McConnell*	1121
SENSORS AND ACTUATORS *H S Tzou* *C S Chou*	1134
SHAPE MEMORY ALLOYS *H S Tzou* *A Baz*	1144
SHELLS *W Soedel*	1155
SHIP VIBRATIONS *W S Vorus*	1167
SHOCK *J Marcondes* *P. Singh*	1173
SHOCK ABSORBERS *See* SHOCK ISOLATION SYSTEMS	1180
SHOCK ISOLATION SYSTEMS *M Radeş*	1180
SIGNAL GENERATION MODELS FOR DIAGNOSTICS *M Sidahmed* *G Dalpiaz*	1184
SIGNAL INTEGRATION AND DIFFERENTIATION *S Dyne*	1193
SIGNAL PROCESSING, CEPSTRUM *See* CEPSTRUM ANALYSIS	1199
SIGNAL PROCESSING, MODEL BASED METHODS *S Braun*	1199
SMART MATERIALS *See* ELECTRORHEOLOGICAL AND MAGNETORHELEOLOGICAL FLUIDS; ELECTROSTRICTIVE MATERIALS; MAGNETOSTRICTIVE MATERIALS; PIEZOELECTRIC MATERIALS; SHAPE MEMORY ALLOYS	1208
SOUND *See* VIBRATION GENERATED SOUND, FUNDAMENTALS; VIBRATION GENERATED SOUND, RADIATION BY FLEXURAL ELEMENTS	1208
SPECTRAL ANALYSIS, CLASSICAL METHODS *S Braun*	1208
SPECTRAL ANALYSIS, MODEL BASED METHODS *See* SIGNAL PROCESSING, MODEL BASED METHODS	1223
SPECTRAL ANALYSIS, WINDOWS *See* WINDOWS	1223
STABILITY *See* DYNAMIC STABILITY	1224
STANDARDS FOR VIBRATIONS OF MACHINES AND MEASUREMENT PROCEDURES *J Niemkiewicz*	1224
STOCHASTIC ANALYSIS OF NON LINEAR SYSTEMS *Y K Lin* *G Q Cai*	1238
STOCHASTIC SYSTEMS *M F Dimentberg*	1246
STRUCTURAL DAMPING *See* HYSTERETIC DAMPING	1252
STRUCTURAL DYNAMIC MODIFICATIONS *A Sestieri* *W D'Ambrogio*	1253
STRUCTURE-ACOUSTIC INTERACTION, HIGH FREQUENCIES *A Sestieri*	1265

STRUCTURE-ACOUSTIC INTERACTION, LOW FREQUENCIES *A Sestieri*	1274
SUBSTRUCTURING See THEORY OF VIBRATION, SUBSTRUCTURING	1283
SUPERPOSITION See THEORY OF VIBRATION, SUBSTRUCTURING	1283
SVD See LINEAR ALGEBRA	1283

T

TESTING, MODAL See MODAL ANALYSIS, EXPERIMENTAL: APPLICATIONS; MODAL ANALYSIS, EXPERIMENTAL: BASIC PRINCIPLES; MODAL ANALYSIS, EXPERIMENTAL: CONSTRUCTION OF MODELS FROM TESTS; MODAL ANALYSIS, EXPERIMENTAL: MEASUREMENT TECHNIQUES; MODAL ANALYSIS, EXPERIMENTAL: PARAMETER EXTRACTION METHODS	1285
TESTING, NONLINEAR SYSTEMS *A Haddow*	1285
THEORY OF VIBRATION	1290
Fundamentals *B Yang*	1290
Superposition *M G Prasad*	1299
Duhamel's Principle and Convolution *G Rosenhouse*	1304
Energy Methods *S S Rao*	1308
Equations of Motion *J Wickert*	1324
Substructuring *M Sunar*	1332
Impulse Response Function *R K Kapania*	1335
Variational Methods *S S Rao*	1344
TIME–FREQUENCY METHODS *P White*	1360
TIRE VIBRATIONS *G D Shteinhauz*	1369
TOOL WEAR MONITORING *M Sidahmed*	1379
TRANSDUCERS FOR ABSOLUTE MOTION *K G McConnell*	1381
TRANSDUCERS FOR RELATIVE MOTION *G E Maddux K G McConnell*	1398
TRANSFORM METHODS *S Braun*	1406
TRANSFORMS, HILBERT See HILBERT TRANSFORMS	1419
TRANSFORMS, WAVELETS *P White*	1419
TRANSMISSION See VIBRATION TRANSMISSION	1435
TRANSPORTATION SYSTEMS See GROUND TRANSPORTATION SYSTEMS	1435

U

ULTRASONICS *M J S Lowe*	1437
ULTRASONICS, NONDESTRUCTIVE TESTING See NONDESTRUCTIVE TESTING: ULTRASONIC	1441

V

VARIATIONAL METHODS See THEORY OF VIBRATION: VARIATIONAL METHODS	1443
VEHICLES, ACTIVE VIBRATION CONTROL See ACTIVE CONTROL OF VEHICLE VIBRATION	1443
VIBRATION ABSORBERS See ABSORBERS, VIBRATION	1443
VIBRATION GENERATED SOUND	1443
Fundamentals *M P Norton S J Drew*	1443
Radiation by Flexural Elements *M P Norton S J Drew*	1456
VIBRATION INTENSITY *S I Hayek*	1480

VIBRATION ISOLATION THEORY *E Rivin*	1487
VIBRATION ISOLATION, APPLICATIONS AND CRITERIA *E Rivin*	1507
VIBRATION PROPERTIES, COMPARISON *See* COMPARISON OF VIBRATION PROPERTIES: COMPARISON OF MODAL PROPERTIES; COMPARISON OF VIBRATION PROPERTIES: COMPARISON OF RESPONSE PROPERTIES; COMPARISON OF VIBRATION PROPERTIES: COMPARISON OF SPATIAL PROPERTIES	1521
VIBRATION TRANSMISSION *S I Hayek*	1522
VIBRO-IMPACT SYSTEMS *I F Peterka*	1531
VISCOUS DAMPING *F Gandhi*	1548

W

WAVE PROPAGATION	1551
Guided Waves in Structures *M J S Lowe*	1551
Interaction of Waves with Boundaries *M J S Lowe*	1559
Waves in an Unbounded Medium *M J S Lowe*	1565
WAVELETS *See* TRANSFORMS, WAVELETS	1570
WHOLE-BODY VIBRATION *M J Griffin*	1570
WIND-INDUCED VIBRATIONS *T Kijewski F Hann A Kareem*	1578
WINDOWS *S Braun*	1587

Z

Z TRANSFORMS *See* TRANSFORM METHODS	1597

GLOSSARY	Gi–Gvi
APPENDICES	Ai–Axviii
INDEX	Ii–Ixxxii
COLOUR PLATE SECTIONS	
Volume 1	292–293
Volume 2	812–813
Volume 3	1308–1309

ABSORBERS, ACTIVE

G Agnes, Beavercreek, OH, USA

Copyright © 2001 Academic Press

doi:10.1006/rwvb.2001.0190

Active vibration absorbers combine the benefits of mechanical vibration absorbers with the flexibility of active control systems. Mechanical vibration absorbers use a small mass coupled to the structure via a flexure (in some form) to add a resonant mode to the structure. By tuning this resonance, the vibration of the structure is reduced. Two limitations on the performance of mechanical vibration absorbers are the mechanism's strokelength and the added mass.

The active vibration absorber replaces the flexure and mass with electronic analogs. Active vibration absorbers can achieve larger effective strokelengths with less mass added to the system. In addition, they can be implemented via strain actuators located in regions of high strain energy instead of large displacement as required by mechanical vibration absorbers – a benefit for some applications.

The drawback to active vibration absorbers is the need for power and electronics. Higher cost and complexity results from the custom analog circuits which must be built or the digital controllers which must be implemented. Finally, unlike mechanical vibration absorbers, active vibration absorbers can lead to spillover, destabilizing the system.

In the following sections, the equations of motion for three common implementations are discussed: the piezo-electric vibration absorber, positive position feedback control and the active vibration absorber. Other combinations of position, velocity, and acceleration feedback are possible (depending on sensor availability) but will not be discussed herein.

Piezo-electric Vibration Absorber

Piezo-electric materials act as a transformer between mechanical and electrical energy. Common forms include a ceramic (PZT) and a polymer (PVDF). When a piezo-electric material undergoes strain, electrical charge is produced on its electrode. By creating a resonant electrical shunt via a resistor and inductor, an electrical resonator is formed. (The piezo-electric material acts as a capacitor in an L–R–C circuit.) By tuning the resonant frequency, the vibration of the structural system is reduced. In practice, inductances on the order of kilohenries are required for low-frequency modes. The inductor is therefore often implemented as an active circuit, requiring power to operate. Hence the piezo-electric vibration absorber is considered an active vibration absorber.

Equations of Motion

A modal model of a structure containing piezo-electric materials can be idealized as shown in **Figure 1**. The base system consists of a mass constrained by a structural spring, K_S, and a piezo-electric spring, K_P, arranged in parallel. The displacement of the structure, X, is to be minimized by a tuned resonant circuit, with charge Q, on the piezo-electric electrodes. The system thus has two-degrees-of-freedom or four states.

The linear constitutive equations for piezo-electric materials, simplified for one-dimensional transverse actuation are:

$$\left\{ \begin{array}{c} E_3 \\ T_{11} \end{array} \right\} = \left[\begin{array}{cc} 1/\varepsilon^S & -h_{31} \\ -h_{31} & c^D \end{array} \right] \left\{ \begin{array}{c} D_3 \\ S_{11} \end{array} \right\} \quad [1]$$

Figure 1 The piezo-electric vibration absorber.

The views expressed in this article are those of the author and do not reflect the official policy or position of the United States Air Force, Department of Defense, or the US Government.

Here, the standard IEEE notation is used (i.e., E is electric field, T is stress, D is electrical displacement, S is strain, ε^S is electrical permittivity, h is the piezo-electric coupling constant, and c^D is the elastic modulus). Assuming a standard patch-like application, these equations may be rewritten in terms of variables more convenient for this study. The equations for piezo-electric spring are thus:

$$\begin{Bmatrix} V \\ F \end{Bmatrix} = \begin{bmatrix} 1/C_P^S & -H \\ -H & K_P^D \end{bmatrix} \begin{Bmatrix} Q \\ X \end{Bmatrix} \quad [2]$$

where V is the voltage or the piezo-electric electrode, F is the force of the spring, Q is the charge flowing into the patch electrodes, X is the displacement of the spring, C_P^S is the capacitance of the patch under constant strain, K_P^D is the stiffness of the piezo-electric spring under constant charge, and H is the electromechanical coupling parameter. Note that coefficients in these equations can be modified for more complicated geometries, but would assume a similar form.

Placing an inductive–resistive (LR) shunt across the electrodes of the piezo-electric spring, the equations of motion for the mass in **Figure 1** are:

$$M\ddot{X} + C\dot{X} + K_P^D X + K_S X - HQ+ = F(t) \quad [3a]$$

$$L\ddot{Q} + R\dot{Q} + \frac{1}{C_P^S}Q - HX = 0 \quad [3b]$$

Here, L is the shunt inductance; R, the shunt resistance; M, the structural mass; and $F(t)$ is an external disturbance.

These equations are next nondimensionalized:

$$\ddot{x} + 2\zeta\dot{x} + x - \frac{\omega_e}{\omega^D}\alpha q = f(t) \quad [4a]$$

$$\ddot{q} + r\left(\frac{\omega_e}{\omega^D}\right)^2 \dot{q} + \left(\frac{\omega_e}{\omega^D}\right)^2 q - \frac{\omega_e}{\omega^D}\alpha x = 0 \quad [4b]$$

with the nondimensional quantities used in eqns [4a] and [4b] defined as:

$$\omega^D = \sqrt{\left(\frac{K_S + K_P^D}{M}\right)} \quad \omega_e = \frac{1}{\sqrt{(LC_P^S)}}$$

$$\alpha^2 = \tilde{K}_{31}^2 = \frac{K^E}{K + K^E}\frac{k_{31}^2}{1 - k_{31}^2} \quad r = RC_P^S \omega^D \quad [5]$$

$$q = \sqrt{(L)}Q \quad x = \sqrt{(M)}X$$

Time has been nondimensionalized such that $t = \omega^D T$. Note that eqns [4a], [4b] and [5] differ from those in the literature since the constant charge (or shorted) stiffness, K^D, of the piezo-electric spring is used in place of the usual constant voltage stiffness, K^E. This simplifies the tuning. Also note, that the coupling term is the generalized electromechanical coupling coefficient, $\tilde{K}_{31}$, which can be determined experimentally – as a modal quantity.

These equations are of the same form as those for the mechanical vibration absorber. The coordinates x and q are proportional to the system displacement and shunt charge, respectively, thus maintaining physical significance. In the next section, the equations of motion for a single-degree-of-freedom structure under positive position feedback control will be derived of this same form. First, however, the tuning and response of the PVA will be considered.

Controller Design

The design of a piezo-electric vibration absorber involves three factors: $\tilde{K}_{ij}$, L and R. The value of $\tilde{K}_{ij}$ is maximized by locating the piezo-electric material in areas of high strain energy. This constant may be determined as a modal constant by considering either the open- and short-circuit resonant mode:

$$\tilde{K}_{ij} = \sqrt{\left[\frac{(\omega^D)^2 - (\omega^E)^2}{(\omega^E)^2}\right]} \quad [6]$$

or analytically by:

$$\tilde{K}_{ij} = \sqrt{\left[\frac{(\omega^E)^2 - (\omega^*)^2}{(\omega^E)^2}\frac{k_{ij}^2}{1 - k_{ij}^2}\right]} \quad [7]$$

where ω^* is the natural frequency with the mass of the piezo-electric device included, but its stiffness neglected.

Given the piezo-electric coupling coefficient, a broadband vibration absorber analogous to Den-Hartog's equal peak implementation can be formed by setting:

$$\omega_e = \omega^E \sqrt{\left(1 + K_{ij}^2\right)} = \omega^D \quad [8]$$

$$r = \sqrt{(2)}\frac{K_{ij}}{\left(1 + K_{ij}\right)^{3/2}} \quad [9]$$

For multimodal applications, numerical optimization must be used to determine the proper electrical network to suppress the vibration of the structure.

System Response

The response of a single-degree-of-freedom system to harmonic excitation is shown in **Figure 2**. The tuning

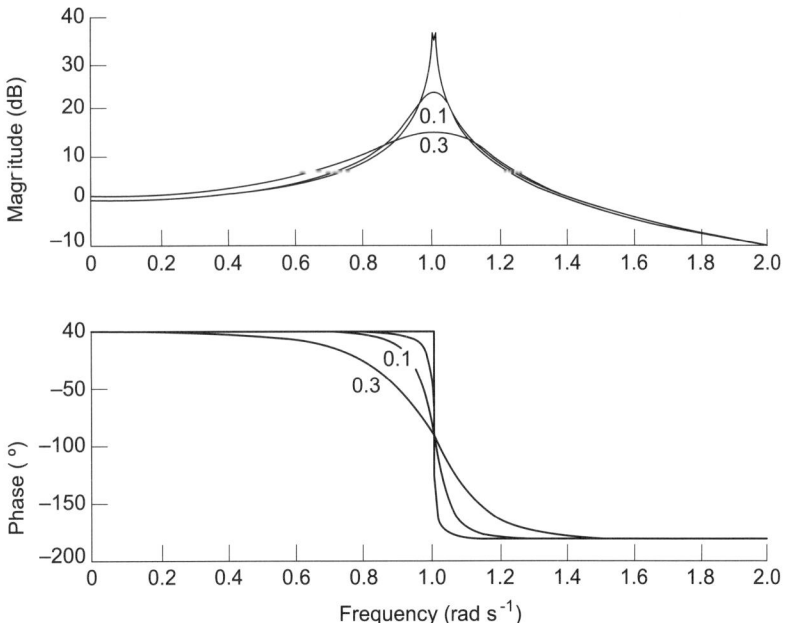

Figure 2 PVA frequency response as the piezo-electric coupling varies.

in eqns [8] and [9] was used. As the piezo-electric coupling is increased, the response of the system decreases. In practice, the piezo-electric coupling is generally limited to at most 0.3.

The impulse response of the system with the $\alpha = 0.2$ and the shunt tuned as presented above is shown in **Figure 3**. Note the system response is damped. The low coupling factors and the fragile nature of piezo-ceramics limit the application of piezo-electric vibration absorbers when higher damping is desired. In the next section, another implementation of an active vibration absorber is discussed: positive position feedback control.

Positive Position Feedback

Modern control design is traditionally performed using first-order dynamical equations. The positive position feedback (PPF) algorithm developed by Goh and Caughey and implemented by Fanson and Caughey uses second-order compensation, allowing physical insight to vibration control by active modal addition. In this algorithm, a position signal is compensated by a second order filter for feedback control. For linear systems, the PPF controller is stable even in the presence of unmodeled actuator dynamics. In addition it is possible to transform the dynamical equations to modal space and design independent second-order feedback compensators for individual modes. Many numerical and experimental implementations of the PPF control scheme may be found in the vibrations literature.

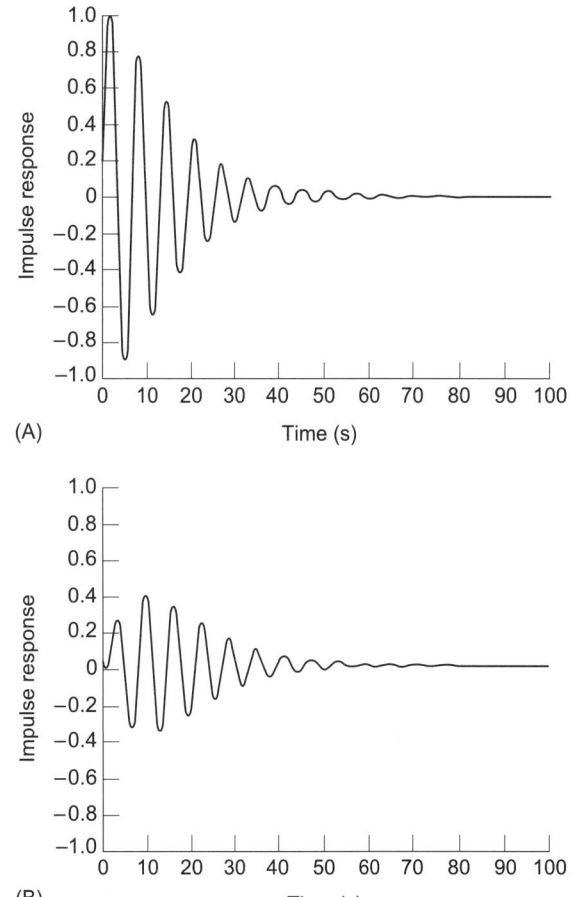

Figure 3 PVA results for a single-degree-of-freedom system with $\alpha = 0.2$. (A) System; (B) controller.

Equations of Motion

A modal model of a structure containing an actuator can be idealized as shown in **Figure 4**.

A single-degree-of-freedom with mass, M, viscous damping, C, and stiffness, K, is driven by an external force, F. The displacement, X, of M is controlled by an actuation force, U. The equations of motion are:

$$M\ddot{X} + C\dot{X} + KX = F + U \quad [10]$$

Introducing the usual nondimensional parameters:

$$\frac{C}{M} = 2\zeta\omega, \quad \frac{K}{M} = \omega^2, \quad \frac{f}{M} = f, \quad \frac{U}{M} = u \quad [11]$$

Equation [10] can be nondimensionalized. For positive position feedback, U is defined:

$$u = g\omega_c^2 x_c \quad [12a]$$

$$\ddot{x}_c + 2\zeta_c\omega_c\dot{x}_c + \omega_c^2 x_c = g\omega_c^2 x \quad [12b]$$

The sensor and controller gains, H and G, have been set equal (as is conventional) and defined as $G = H = g\omega_c^2$. The equations of motion for the combined system are therefore:

$$\ddot{x} + 2\omega\zeta\dot{x} + \omega^2 x = g\omega_c^2 x_c + f \quad [13a]$$

$$\ddot{x}_c + 2\zeta_c\omega_c\dot{x}_c + \omega_c^2 x_c = g\omega_c^2 x \quad [13b]$$

Controller Design

Using eqns [13a] and [13b], a controller with prescribed closed loop damping ratio ζ_p can be found. The design requires three factors: g, ω_c and ζ_c. The value of g is determined such that actuator saturation is avoided. In practice, this may require experimentally setting g for the worst case disturbance.

Again a broadband vibration absorber analogous to the equal peak tuning can be determined. Given the feedback gain, the controller is tuned by setting:

$$\omega_c = \omega\sqrt{(1+g^2)} \quad [14]$$

$$\zeta_c = \sqrt{\left[\frac{(\omega_c g)^2}{*\left(1+(\omega_c g)^2\right)^3}\right]} \quad [15]$$

For low values of g this leads to 'equal peak' results similar to the piezo-electric vibration absorber. Using this tuning law, the variance in system natural frequency, damping ratio, and pole locations are plotted in **Figure 5** as the gain is varied. Note that this law is not an equal peak law at higher values of g.

For multimodal application, numerical optimization must be used to determine the proper parameter

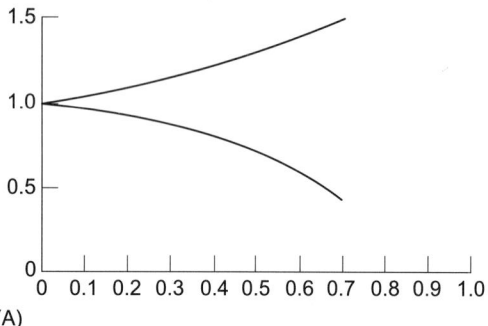

(A)

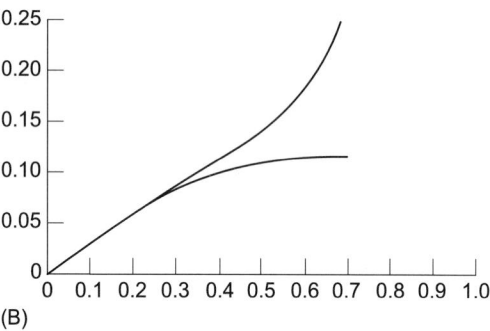

(B)

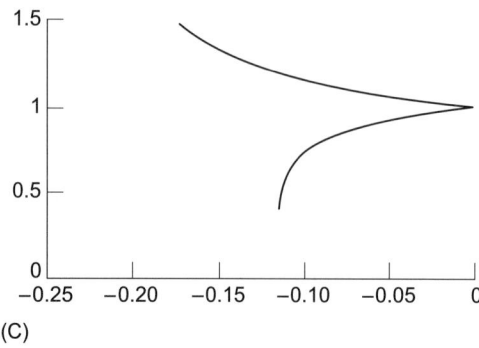

(C)

Figure 5 Positive position feedback results for a single-degree-of-freedom system as feedback gain is increased from 0 to 0.7. (A) Natural frequencies; (B) damping ratio; (C) pole locations.

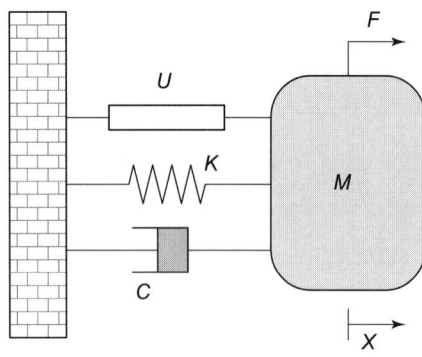

Figure 4 An actuated structure.

set which suppresses the vibration of the structure. Often traditional output control optimization algorithms are used to design the controller.

System Response

The response of a single-degree-of-freedom system to harmonic excitation is shown in **Figure 6**. The tuning discussed in the previous section was used. As the gain of the PPF system increased, the response of the system decreases. However, with higher gains, the damping of the two modes does not remain equal.

The impulse response of the system with $g = 0.2$ and the controller tuned as presented above is shown in **Figure 7**. Note the system response is damped; however, also note that the actual control force required is $g\omega_c^2$ times the controller state which is plotted. For the system this amounts to dividing the plot in **Figure 7B** by 5.

Active Vibration Absorber

The active vibration absorber (AVA) is another implementation of a second-order compensator. It can be generalized to use a combination of position, velocity and acceleration feedback, but, in this section, only acceleration feedback is considered. Positive position feedback is a version of the more general implementation with the direct feedthrough term neglected. The advantage of AVA is that unlike PPF, the gain can be increased without fear of instability of the controlled mode. However, in contrast to the PPF controller, the AVA control law does not roll off at higher frequencies. This can lead to spillover (instability) when implemented in multimodal systems, limiting the achievable gain.

Equations of Motion

Again, a modal model of a structure containing an actuator can be idealized as shown in **Figure 4**. The displacement, X, of M is controlled by an actuation force, U. The equations of motion are:

$$M\ddot{X} + C\dot{X} + KX = F + U \quad [16]$$

Introducing the usual nondimensional parameters:

$$\frac{C}{M} = 2\zeta\omega \quad \frac{K}{M} = \omega^2 \quad \frac{f}{M} = f \quad \frac{U}{M} = u \quad [17]$$

Equation [16] can be nondimensionalized. For the active vibration absorber, u is defined:

$$u = \ddot{x}_c - g^2\ddot{x} \quad [18a]$$

$$\ddot{x}_c + 2\zeta_c\omega_c\dot{x}_c + \omega_c^2 x_c = gx \quad [18b]$$

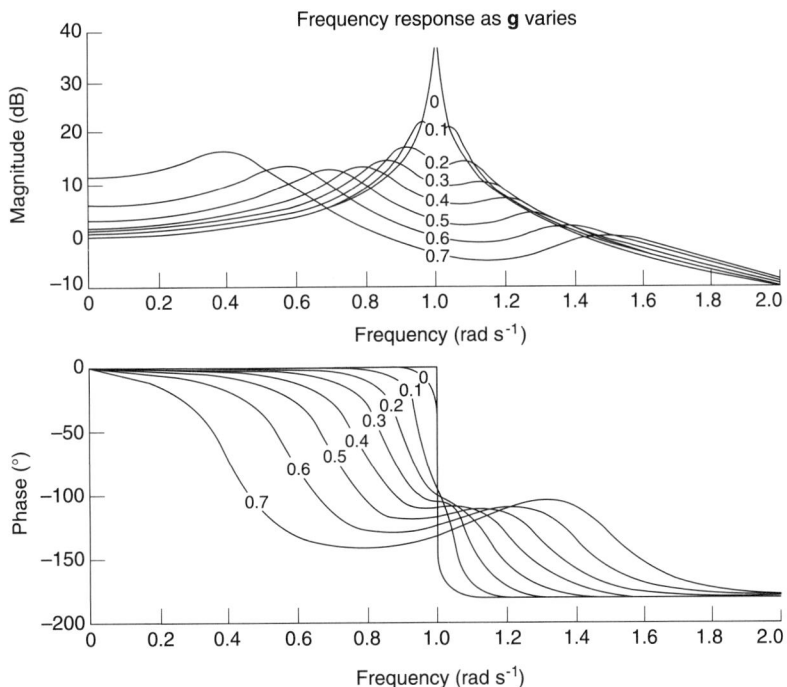

Figure 6 PPF frequency response as the gain varies.

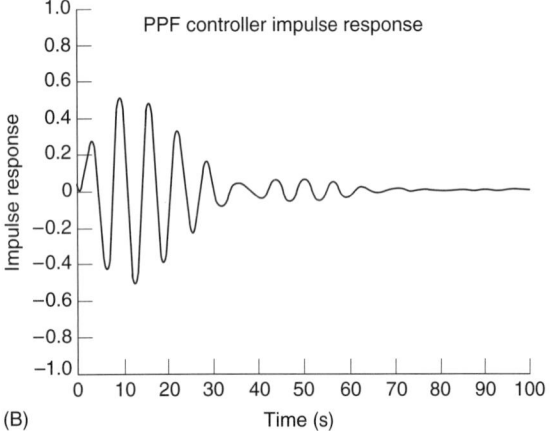

Figure 7 PPF results for a single-degree-of-freedom system with $g = 0.2$. (A) System; (B) controller.

The equations of motion for the combined system are therefore:

$$(1 + g^2)\ddot{x} - g\ddot{x}_c + \omega^2 x = f \quad [19a]$$

$$-g\ddot{x} + \ddot{x}_c + 2\zeta_c\omega_c\dot{x}_c + \omega_c^2 x_c = 0 \quad [19b]$$

This equation can be transformed into a mechanical analogy by performing a similarity transformation:

$$\begin{Bmatrix} x \\ x_c \end{Bmatrix} = \begin{bmatrix} 0 & 1 \\ 1 & g \end{bmatrix} \begin{Bmatrix} q_c \\ q \end{Bmatrix} \quad [20]$$

The nondimensional equations written for comparison are:

$$\ddot{q}_c + 2\zeta_c\omega_c\dot{q}_c + g2\zeta_c\omega_c\dot{q} + \omega_c^2 q_c + g\omega_c^2 q = 0 \quad [21a]$$

$$\ddot{q} + g2\zeta_c\omega_c\dot{q}_c + g^2 2\zeta_c\omega_c\dot{q} + g\omega_c^2 q_c \\ + (\omega^2 + g^2\omega_c^2)q = 0 \quad [21b]$$

Note that while q still represents the motion of the mass, x, $q_c = x_c - gx$. Thus the motion of the mass is combined with the electrical degree-of-freedom. The control force required is $g\ddot{q}_c$.

Controller Design

Using eqns [19a] and [19b], a controller which minimizes the resonant amplitude of the system (acting as an optimal damped mechanical vibration absorber) can be obtained.

$$\omega_c = \frac{\omega}{1 + g^2} \quad [22a]$$

$$\zeta_c = \frac{1}{g^2}\sqrt{\left[\frac{g^2}{4(1 + g^2)^3}\right]} \quad [22b]$$

The larger g^2 is, the lower the amplitude of the response in a manner analogous to the mass ratio of a mechanical vibration absorber. These values lead to 'equal peak' results similar to the piezo-electric vibration absorber. Using these values, the variance in system natural frequency, damping ratio, and pole locations are plotted in **Figure 8**.

System Response

The response of a single-degree-of-freedom system to harmonic excitation is shown in **Figure 9**. The tuning discussed in the previous section was utilized. As the gain of the AVA system increased, the response of the system decreases. The damping of the two peaks is uniform unlike the damping of the PPF controller.

The impulse response of the system is shown in **Figure 10**. This response is similar to that PPF response. Recall however, that the control force required is a function of acceleration for the AVA controller.

Summary

The piezoelectric vibration absorber, positive position feedback, and active vibration absorbers all allow electrical analogies of mechanical vibration absorbers. The choice amongst these implementations will depend on the knowledge of the system dynamics, availability of suitable actuators or sensors, and other application specific criteria. Each provides an effective method of vibration suppression without the strokelength and mass limitation of the mechanical vibration absorber albeit with greater cost and complexity.

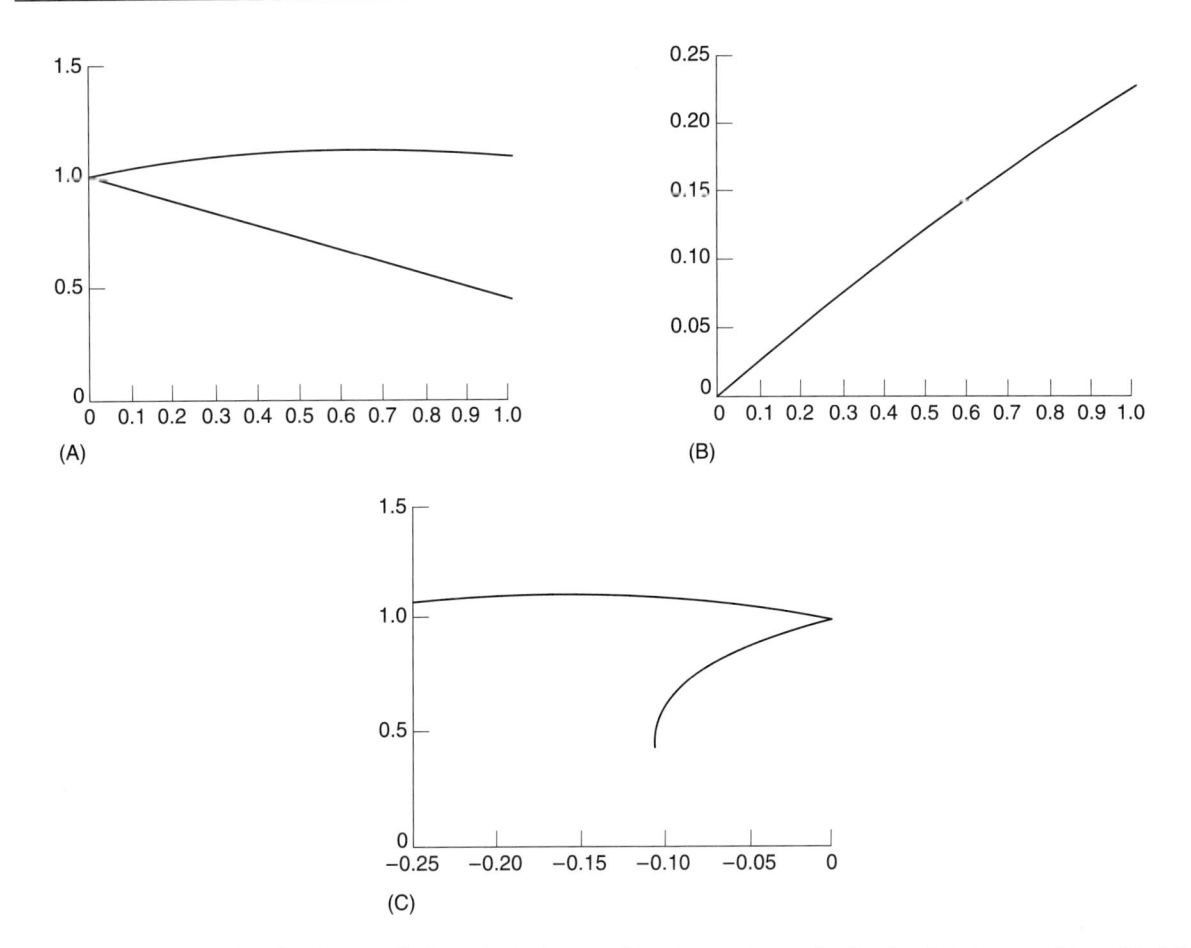

Figure 8 Active vibration absorber results for a single-degree-of-freedom system as feedback gain is increased from 0 to 1.0. (A) Natural frequencies; (B) damping ratio; (C) pole locations.

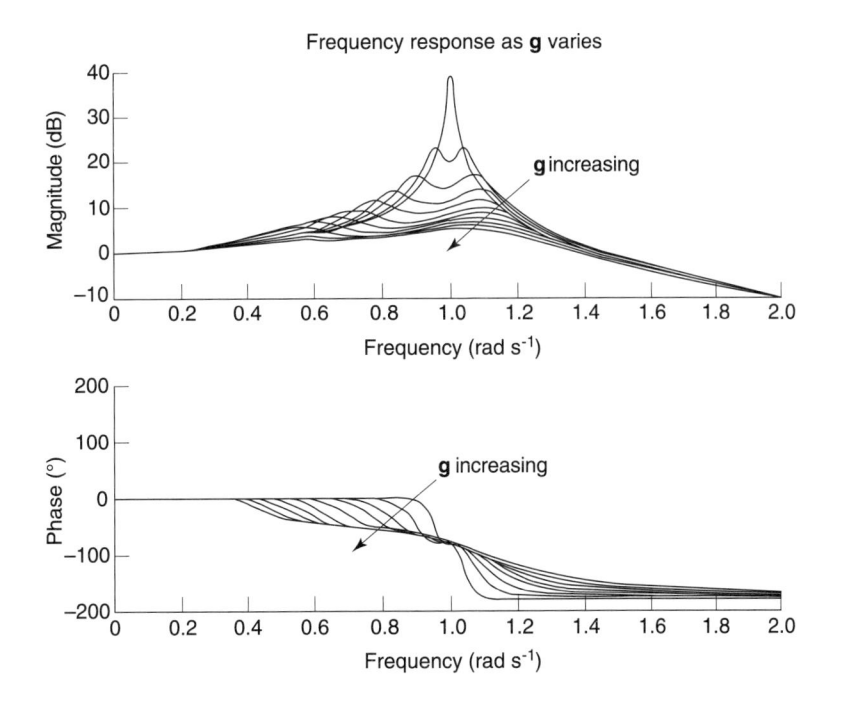

Figure 9 AVA frequency response as the gain varies.

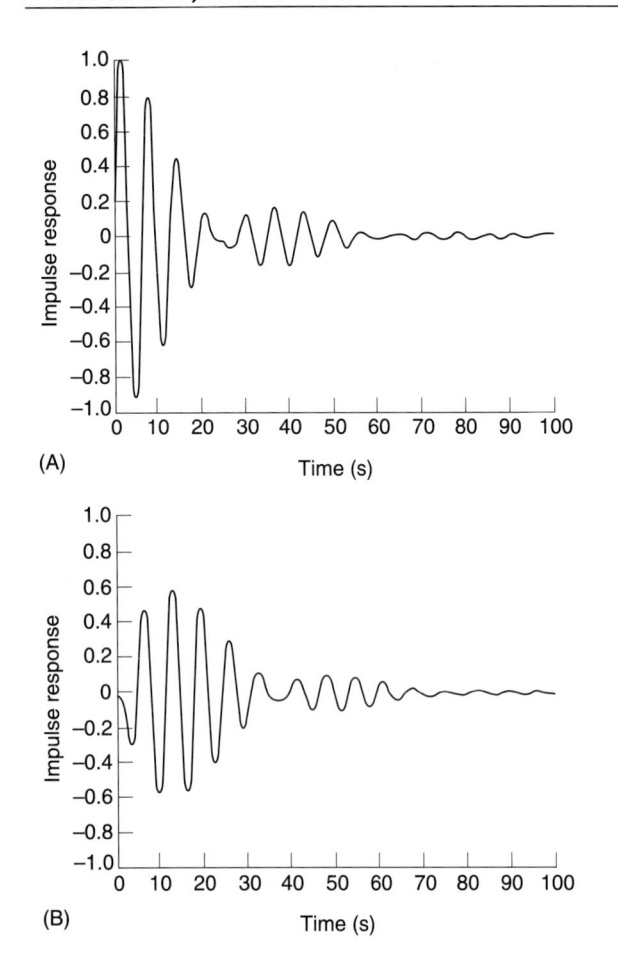

Figure 10 AVA results for a single-degree-of-freedom system with $g = 0.2$. (A) System; (B) controller.

Nomenclature

c^D	elastic modulus
C	capacitance
D	electrical displacement
E	electric field
F	force
$F(t)$	external disturbance
G	controller gain
h	piezo-electric coupling constant
H	sensor gain
K	stiffness
L	shunt inductance
M	structural mass
Q	charge
R	shunt resistance
S	strain
T	stress
U	actuation force
V	voltage
X	displacement
ε	permittivity
ξ	damping ratio

See also: **Absorbers, vibration; Active control of civil structures; Active control of vehicle vibration; Vibration isolation, applications and criteria; Vibration isolation theory**

Further Reading

Agnes GS (1995) Active/passive piezoelectric vibration suppression. *Journal of Intelligent Materials, Systems, and Structures* 6:482–7.

American Institute of Aeronautics and Astronautics (1987) *Proceedings of the 28th AIAA/ASME/ASCE/AHS/ACS Structures, Structural Dynamics and Materials Conference*, April, Monterey, CA.

American Institute of Aeronautics and Astronautics (1994) *Proceedings of the 35th AIAA/ASME/ASCE/AHS/ACS Structures, Structural Dynamics and Materials Conference*, April, Hilton Head, SC.

Baz A, Poh S, Fedor J (1992) Independent modal space control with positive position feedback. *Transactions of the ASME* 114:96–103.

Caughey TK (1995) Dynamic response of structures constructed from smart materials. *Smart Materials and Structures* 4:A101–A106.

DenHartog JP (1985) *Mechanical Vibrations*. New York: Dover Books.

Dusch JJ (1995) *Active Vibration Suppression: Stability and Design in Second Order Form*. PhD thesis, SUNY at Buffalo.

Fanson JL, Caughey TK (1987) Positive position feedback control for large space structures. In *Proceedings of the 28th AIAA/ASME/ASCE/AHS/ACS Structures, Structural Dynamics and Materials Conference* 87–0902. April, Monterey, CA.

Fanson JL, Caughey TK (1990) Positive position feedback control for large space structures. *AIAA Journal* 28(4).

Flotow AH Von, Beard A, Bailey D (1994) Adaptive tuned vibration absorbers: tuning laws, tracking agility, sizing, and physical implementations. *Proceedings Noise-Conference 94*.

Forward RL (1979) Electronic damping of vibrations in optical structures. *Journal of Applied Optics*, 18:690–7.

Goh CJ, Caughey TK (1985) On the stability problem caused by finite actuator dynamics in the collocated control of large space structures. *International Journal of Control* 41(3):787–802.

Hagood NW, Flotow A Von (1991) Damping of structural vibrations with piezoelectric materials and passive electrical networks. *Journal of Sound and Vibration* 146:243–68.

Hollkamp JJ (1994) Multimodal passive vibration suppression with piezoelectric materials and resonant shunts. *Journal of Intelligent Materials Systems and Structures* 5:49–57.

Hollkamp JJ, Starchville TF (1994) A self-tuning piezoelectric vibration absorber. In *Proceedings of the 35th AIAA/ASME/ASCE/AHS/ACS Structures, Structural Dynamics and Materials Conference* 94–1790. April, Hilton Head, SC.

Jaffe B, Cook R, Jaffe H (1971) *Piezoelectric Ceramics*. New York: Academic.

Lee-Glauser G, Juang J-N, Sulla JL (1995) Optimal active vibration absorber: Design and experimental results. *Journal of Vibration and Acoustics* 117:165–171.

Sun JQ, Jolly MR, Norris MA (1995) Passive, adaptive, and active tuned vibration absorbers. *Trans. ASME Combined Anniversary Issue Journal of Mechanical Design and Journal of Vibration and Acoustics*, 117(B):234–42.

ABSORBERS, VIBRATION

V Steffen, Jr and D Rade, Federal University of Uberlandia, Uberlandia, Brazil

Copyright © 2001 Academic Press

doi:10.1006/rwvb.2001.0176

Introduction

Dynamic vibration absorbers (DVAs), also called Vibration Neutralizers or Tuned Mass Dampers, are mechanical appendages comprising inertia, stiffness, and damping elements which, once connected to a given structure or machine, named herein the primary system, are capable of absorbing the vibratory energy at the connection point. As a result, the primary system can be protected from excessively high vibration levels. In practice, DVAs can be included in the original system design or can be added to an existing system, often as part of a remedial course of action.

Since their invention by Frahm at the beginning of the twentieth century, dynamic vibration absorbers have been extensively used to mitigate vibrations in various types of mechanical systems. A very well-known application is the so-called Stockbridge damper, widely used to reduce wind-induced vibrations in overhead power transmission lines. In a remarkable engineering application, a 400-ton absorber has been designed for Citicorp Center, a 274-m high office building in New York City, for suppressing primarily the contribution of the first vibration mode in wind-induced oscillations. In a similar application, two 300-ton DVAs have been installed in the John Hancock Tower, in Boston, Massachussets. The dynamics of television towers are particularly favorable for the use of pendulum-like DVAs, which have been applied, for example, to the towers of Alma-Ata and Riga, in the former Soviet Union.

Due to their technological relevance both in the academic and industrial domains, DVAs are still a subject of permanent interest. New applications include devices used to stabilize ship roll motion, to improve the comfort of users when walking on pedestrian bridges, to attenuate vibrations transmitted from the main rotor to the cockpit of helicopters, and to improve machine tool operation conditions, to mention just a few examples. Military applications have also been developed. The use of DVAs to reduce the dynamic forces transmitted to an aircraft due to high rates of fire imposed on the canon motion can be mentioned as another example.

In practical applications, DVAs can be found in various configurations, intended for the attenuation of either rectilinear or angular motion. The simplest setup is that formed by a single mass attached to the primary system through a linear spring. This configuration is named the 'undamped dynamic vibration absorber'. As will be shown later, in designing an undamped DVA to attenuate harmonic vibrations, the values of its physical parameters (stiffness and inertia) must be chosen according to the value of the excitation frequency and it is then said that the DVA is tuned. The undamped DVA may become ineffective when the excitation frequency deviates, even slightly, from the nominal tuning frequency. In order to provide a mechanism for energy dissipation and to enlarge the effective bandwidth of the absorber, damping can be introduced into the DVA. In most applications, a viscous damping model is used, although viscoelastic and Coulomb-type dampers can be found in certain cases. In general, a DVA is designed to attenuate vibrations generated by a purely harmonic excitation. However, in several situations, vibrations are produced by periodic forces containing various harmonic components. In this case, multiple DVAs can be used, each one tuned to a specific frequency component. It is also possible to use distributed-parameter structural elements, such as beams or plates, as dynamic absorbers. Besides the ease of physical realization, the main interest in using these configurations is related to the fact that the DVA can be tuned to various frequency values simultaneously.

All the configurations mentioned above form the class of 'passive' DVAs, defined as those containing exclusively passive, time-independent, components. For this type of absorber, tuning can be achieved only by physically constructing inertia, stiffness and damping elements with adequate values. When the excitation frequency changes, which is likely to occur in many cases, the absorber becomes mistuned and less effective. To overcome this limitation, active DVAs have been developed. Besides the passive elements, they contain an actuator which applies a control force calculated according to an adequate control law. This strategy provides self-tuning capability to the DVA, over a finite frequency band.

In the following sections the basic theory of passive and active dynamic vibration absorbers are presented, as well as some special configurations.

Undamped Dynamic Vibration Absorbers

Figure 1 illustrates an undamped two-degree-of-freedom system, where the subsystem (m_p, k_p) represents the primary system, whose vibrations are to be attenuated, and the subsystem (m_a, k_a) represents the dynamic vibration absorber. The primary system is assumed to be excited by an external harmonic force with constant amplitude and constant circular frequency, given by:

$$f(t) = F_0 e^{i\omega t} \quad [1]$$

Using Newton's laws, the following equations of motion are obtained for the two-degree-of-freedom system, in terms of the coordinates defined in **Figure 1**:

$$m_p \ddot{x}_p + (k_p + k_a) x_p - k_a x_a = f \quad [2a]$$

$$m_a \ddot{x}_a + k_a (x_a - x_p) = 0 \quad [2b]$$

The steady-state harmonic responses are written:

$$x_p(t) = X_p e^{i\omega t} \quad [3a]$$

$$x_a(t) = X_a e^{i\omega t} \quad [3b]$$

Upon substitution of eqns [3] in eqns [2], the following set of frequency-dependent algebraic equations involving the amplitudes of the harmonic responses is obtained:

$$X_p(-m_p\omega^2 + k_p + k_a) - k_a X_a = F_0 \quad [4a]$$

$$-k_a X_p + X_a(-m_a\omega^2 + k_a) = 0 \quad [4b]$$

Solving eqns [4], the following expression is obtained for the amplitudes:

$$\frac{X_p}{(X_p)_{st}} = \frac{\left[1 - (\omega/\omega_a)^2\right]}{\left[1 + (k_a/k_p) - (\omega/\omega_p)^2\right]\left[1 - (\omega/\omega_a)^2\right] - (k_a/k_p)} \quad [5a]$$

$$\frac{X_a}{(X_p)_{st}} = \frac{1}{\left[1 + (k_a/k_p) - (\omega/\omega_p)^2\right]\left[1 - (\omega/\omega_a)^2\right] - (k_a/k_p)} \quad [5b]$$

where:

$$\omega_p = \sqrt{\left(\frac{k_p}{m_p}\right)} \quad \omega_a = \sqrt{\left(\frac{k_a}{m_a}\right)} \quad (X_p)_{st} = \frac{F_0}{k_p} \quad [6]$$

are, respectively, the natural frequency of the primary system, the natural frequency of the absorber, when both are considered as separate single-degree-of-freedom systems, and the static displacement of the primary mass.

In eqn [5a], it can be seen that the amplitude X_p vanishes when the excitation frequency ω coincides with the natural frequency of the DVA, ω_a. In this situation, the amplitude of the response of the DVA mass is obtained by introducing $X_p = 0$ in eqn [4a]:

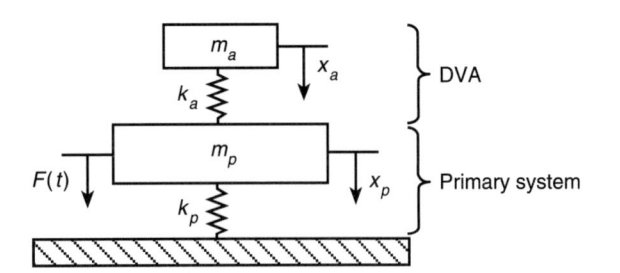

Figure 1 Schematic representation of an undamped DVA connected to a primary system.

$$X_a|_{\omega=\omega_a} = -\frac{F_0}{k_a} \quad [7]$$

In eqn [7] the minus sign indicates that there is a phase shift of 180° between the excitation force and the response of the DVA mass. **Figure 2** illustrates the variation of X_p with ω, according to eqn [5a]. It can be seen that an antiresonance is generated at $\omega = \omega_a$. Thus, to achieve complete attenuation of harmonic vibrations with a given frequency ω, the values of the inertia and stiffness parameters of the undamped DVA must be selected so as to satisfy $\omega = \sqrt{(k_a/m_a)}$. **Figure 2** also shows the amplitudes of the response of the primary system without the DVA. It can be seen that with the addition of the DVA, two resonance peaks are generated in the frequency response and vibration reduction is achieved only within the frequency band limited by points A and B. Since this bandwidth is generally small, even slight modifications in the forcing frequency and/or in the DVA parameters can lead to a significant decrease in the attenuation capability of the absorber. This is the major drawback of the undamped DVAs.

At this point, the following comments are made regarding the practical design of undamped DVAs. In any physical realization, the interest is to have a DVA with a small mass. This means that the mass ratio, defined as $\mu = m_a/m_p$ should be kept as small as possible (values of μ up to 5 percent are generally acceptable). However, eqns [6b] and [7] state that small values of the secondary mass correspond to small values of the DVA stiffness and high amplitude of vibration of the secondary mass. This last fact has direct implication in the fatigue life of the resilient element of the DVA.

In many circumstances the interest is to reduce the amplitude of vibration of the primary system in the vicinity of its resonance frequency. In these cases, the DVA must be tuned so that its natural frequency coincides with the natural frequency of the primary system, that is:

$$\frac{k_a}{m_a} = \frac{k_p}{m_p} \quad [8]$$

In this case, eqns [5] can be rewritten in terms of dimensionless parameters, as follows:

$$\frac{X_p}{(X_p)_{st}} = \frac{1-g^2}{(1-g^2)(1-g^2+\mu)-\mu} \quad [9a]$$

$$\frac{X_a}{(X_p)_{st}} = \frac{1}{(1-g^2)(1-g^2+\mu)-\mu} \quad [9b]$$

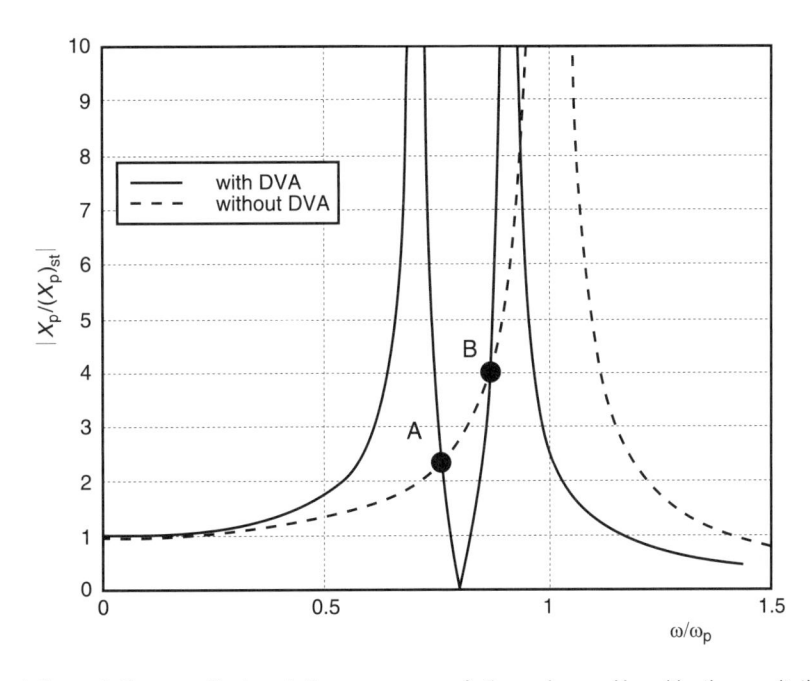

Figure 2 Typical variation of the amplitude of the response of the primary X_p with the excitation frequency ω for $\omega_a/\omega_p = 0.8$, $m_a/m_p = 0.1$.

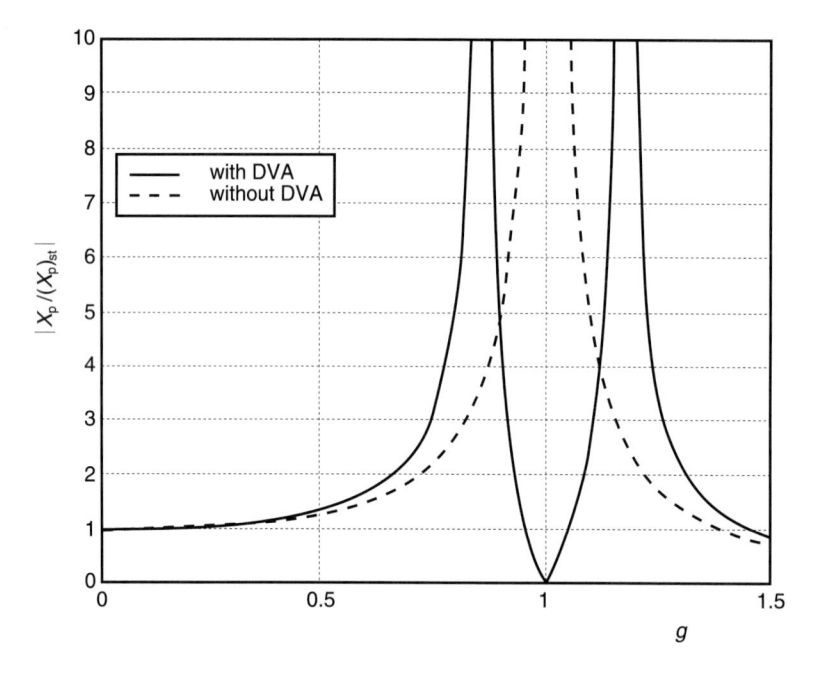

Figure 3 Typical variation of the amplitude of the response of the primary X_p with the excitation frequency ω for $\omega_a/\omega_p = 1.0$, $\mu = 0.1$.

where the dimensionless parameters are defined as $g = \omega/\omega_p$ (forcing frequency ratio) and $\mu = m_a/m_p$ (mass ratio).

The roots of the denominator of eqns [9] define the natural frequencies of the two-degree-of-freedom system (primary system + DVA), given by:

$$g^2 = 1 + \frac{\mu}{2} \pm \sqrt{\left[\mu + \left(\frac{\mu^2}{4}\right)\right]} \quad [10]$$

Eqns [9] show that for $g = 1$ the primary mass will not vibrate while the absorbing mass will vibrate with amplitude $X_a = -(X_p)_{st}/\mu$. A typical plot of eqn [9a] is shown in **Figure 3**.

Viscously Damped Dynamic Vibration Absorbers

The effective bandwidth of a DVA can be enlarged by introducing a damping element responsible for energy dissipation. Moreover, damping facilitates a reduction of the amplitude of relative motion between the primary and the secondary masses, thus rendering less critical the fatigue of the resilient element of the DVA. **Figure 4** illustrates a viscously damped DVA (m_a, c_a, k_a) attached to an undamped primary system

(m_p, k_p). Assuming that a harmonic excitation force given by $f(t) = F_0 e^{i\omega t}$ acts on the primary mass, the equations of motion for this two-degree-of-freedom system are written:

$$m_p \ddot{x}_p + k_p x_p + k_a(x_p - x_a) + c_a(\dot{x}_p - \dot{x}_a) = F_0 e^{i\omega t} \quad [11a]$$

$$m_a \ddot{x}_a + k_a(x_a - x_p) + c_a(\dot{x}_a - \dot{x}_p) = 0 \quad [11b]$$

The resulting steady-state amplitude for the primary and DVA masses are found to be given by:

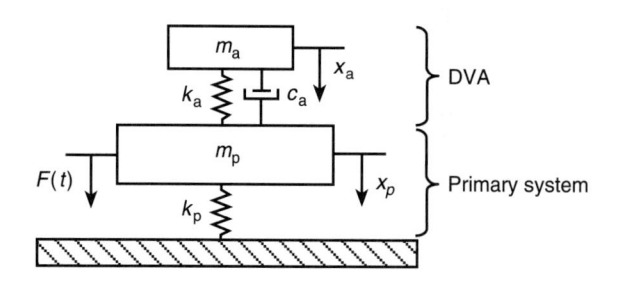

Figure 4 Scheme of a viscously damped DVA connected to a primary system.

$$X_p = F_0 \frac{k_a - m_a\omega^2 + i\omega c_a}{(k_p - m_p\omega^2)(k_a - m_a\omega^2) - m_a k_a \omega^2 + i\omega c_a(k_p - m_p\omega^2 - m_a\omega^2)} \quad [12a]$$

$$X_a = F_0 \frac{k_a + i\omega c_a}{(k_p - m_p\omega^2)(k_a - m_a\omega^2) - m_a k_a \omega^2 + i\omega c_a(k_p - m_p\omega^2 - m_a\omega^2)} \quad [12b]$$

Similarly to what has been done when developing the formulation for the undamped DVA, the following parameters are introduced: $\mu = m_a/m_p$, mass ratio; $\omega_a = \sqrt{(k_a/m_a)}$, undamped natural frequency of the DVA considered separately; $\omega_p = \sqrt{(k_p/m_p)}$, undamped natural frequency of the primary system considered separately; $f = \omega_a/\omega_p$, turning factor; $g = \omega/\omega_p$, forcing frequency ratio; $c_c = 2m_a\omega_p$, critical damping; $\zeta = c_a/c_c$, damping ratio; $(X_p)_{st} = F_0/k_p$, static displacement of the primary mass. Thus, eqns [12] can be rewritten in terms of the dimensionless parameters as follows:

$$\frac{|X_p|}{(X_p)_{st}} = \sqrt{\left\{\frac{(2\zeta g)^2 + (g^2 - f^2)^2}{(2\zeta g)^2(g^2 - 1 + \mu g^2)^2 + [\mu f^2 g^2 - (g^2 - 1)(g^2 - f^2)]^2}\right\}} \quad [13a]$$

$$\frac{|X_a|}{(X_p)_{st}} = \sqrt{\left\{\frac{(2\zeta g)^2 + f^4}{(2\zeta g)^2(g^2 - 1 + \mu g^2)^2 + [\mu f^2 g^2 - (g^2 - 1)(g^2 - f^2)]^2}\right\}} \quad [13b]$$

Figure 5 illustrates a typical variation of the amplitude X_p with the forcing frequency ratio, for different values of the damping factor ζ. As can be seen, for $\zeta = 0$ the system behaves like an undamped two-degree-of-freedom system, with response amplitudes tending to increase indefinitely at each of the two resonance frequencies. As the amount of damping is progressively increased, the system behaves like a typical damped two-degree-of-freedom system and eventually exhibits the apparent behavior of a single-degree-of-freedom system with mass $(m_1 + m_2)$ when the two masses become virtually connected through the dashpot. It can also be seen that all curves intercept at points P and Q, named 'invariant points'. Eqn [13b] can be expressed as:

$$\frac{|X_a|}{(X_p)_{st}} = \sqrt{\frac{A\zeta^2 + B}{C\zeta^2 + D}} \quad [13c]$$

where A, B, C, and D are functions of f and g, only. The characterization of the invariant points is based on the fact that the identity $A/C = B/D$ holds, regardless of the value of the damping factor. **Figures 6** and **7** show the influence of the damping ratio ζ and the tuning factor f on the frequency response of the primary mass. It can be seen that the response amplitudes at the invariant points vary when the tuning factor is changed. For the purpose of optimal design of a damped DVA, it is desired to find a set of values (ζ_{opt}, f_{opt}) to ensure a response curve which is as flat as possible. Based on the behavior illustrated in **Figures 5–7**, this optimal configuration is achieved when both the invariant points are adjusted to equal heights and the response curve presents null slope at one of them. It makes marginal difference which invariant point is taken. According to the development originally presented by Den Hartog, imposing these conditions to the frequency response given by eqn [13a], the following expression for the optimal tuning ratio is obtained:

$$f_{opt} = \frac{1}{1 + \mu} \quad [14]$$

The expressions for the optimal damping ratios that ensure zero slope at each invariant point are:

$$\zeta_P^2 = \frac{\mu[3 - \sqrt{(\mu/\mu + 2)}]}{8(1 + \mu)^3} \qquad \zeta_Q^2 = \frac{\mu[3 + \sqrt{(\mu/\mu + 2)}]}{8(1 + \mu)^3}$$

Den Hartog suggests to take the average between the two values provided by the equations immediately above as the optimal value of the damping ratio, as given by:

$$\zeta_{opt} = \sqrt{\left[\frac{3\mu}{8(1 + \mu)^3}\right]} \quad [15]$$

An optimally shaped frequency response curve is indicated in **Figure 7**.

DVAs for Viscously Damped Primary Systems

Dynamic vibration absorbers are likely to be added only to lightly damped systems, since highly damped

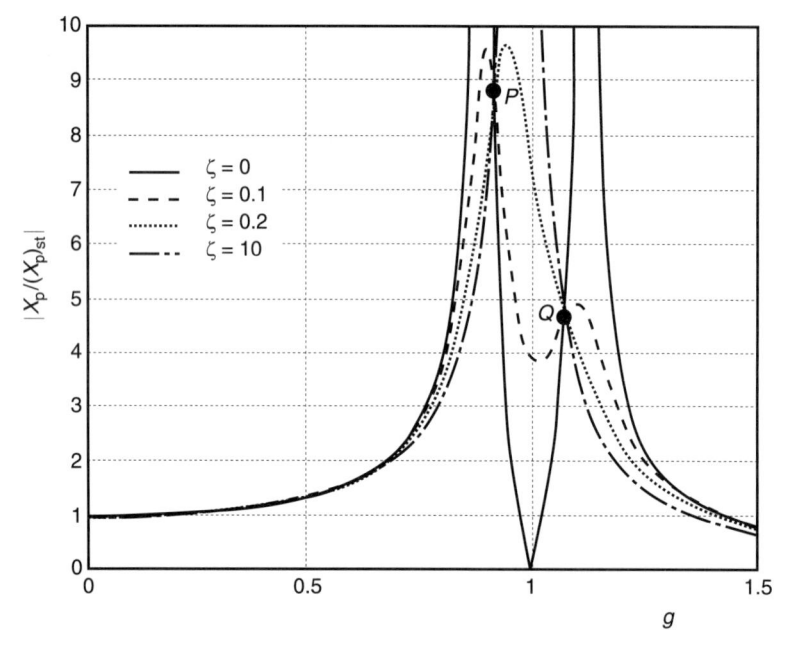

Figure 5 Influence of damping on the frequency response of the primary mass $f = 1$, $\mu = 1/20$.

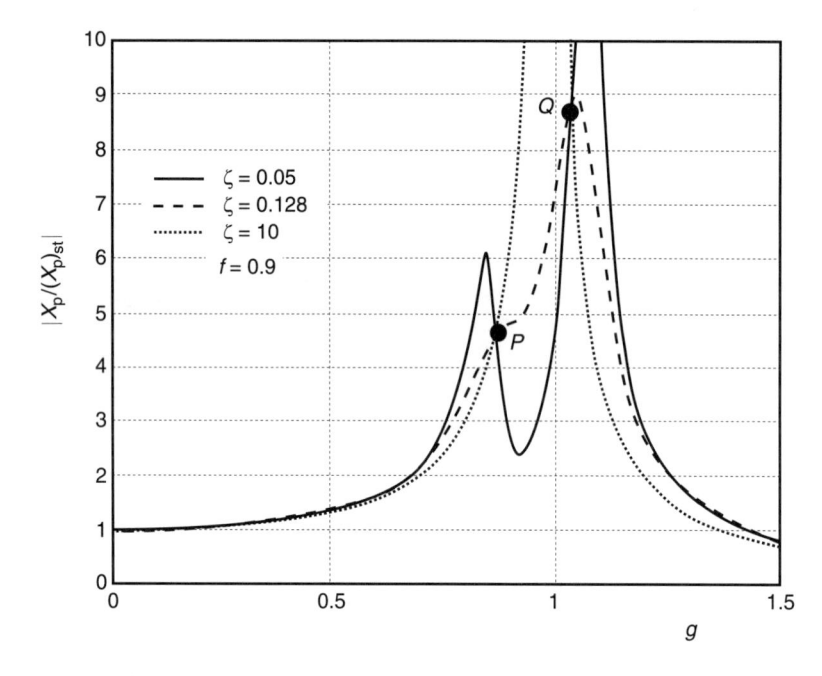

Figure 6 Frequency response of the primary mass for different values of the damping ratio and $f = 0.90$.

systems usually present moderate vibration levels which do not require any additional attenuation. However, special situations may be found where it is intended to design DVAs taking into account the damping of the primary system. Such a situation is illustrated in **Figure 8**. For this two-degree-of-freedom system, the following equations of motion are written:

$$m_p \ddot{x}_p + (c_p + c_a)\dot{x}_p - c_a \dot{x}_a + (k_p + k_a)x_p - k_a x_a = F \quad [16a]$$

$$m_a \ddot{x}_a + c_a \dot{x}_a - c_a \dot{x}_p + k_a x_a - k_a x_p = 0 \quad [16b]$$

From eqns [16], and using the same procedure as the one adopted for the undamped primary system, the

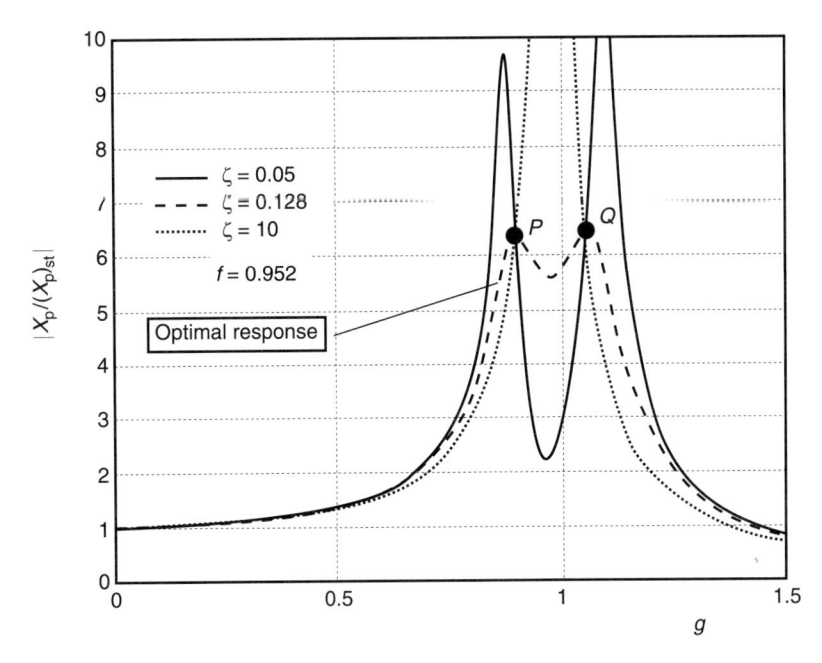

Figure 7 Frequency response of the primary mass for different values of the damping ratio and $f = 0.952$.

following expression is obtained for the frequency response of the primary system in terms of dimensionless parameters.

$$\frac{|X_p|}{(X_p)_{st}} = \sqrt{\left\{\frac{(2\zeta_a g)^2 + (g^2 - f^2)^2}{[2\zeta_a g(1 - g^2 - \mu g^2) + 2\zeta_p \mu g(f^2 - g^2)]^2 + [\mu f^2 g^2 - (g^2 - 1)(g^2 - f^2)]^2}\right\}} \quad [17]$$

where:

$$\zeta_a = \frac{c_a}{2m_a \omega_p} \quad \text{and} \quad \zeta_p = \frac{c_p}{2m_a \omega_p}$$

and the other dimensionless parameters remain the same as those already defined for the case of the undamped primary system.

As opposed to eqn [13a], eqn [17] does not admit the existence of invariant points in the frequency response curves, since Eqn [17] cannot be put in the form of Eqn [13c] with $A/C = B/D$. As a result, Den Hartog's optimization procedure does not apply. In this case, to obtain the optimal values of the tuning ratio and absorber damping ratio, numerical optimization has to be carried out. For this purpose an objective function related to the maximum response amplitude must be defined and minimized with respect to the DVA parameters. Such an optimization procedure was implemented by Warburton and Yorinde, resulting in the values presented in **Table 1**. It can be seen that, for small values of the primary system damping, little influence of this damping on the values of the optimal parameters f_{opt} and ζ_a^{opt} is noticed.

Optimal Design of Damped DVAs Applied to Multi-degree-of-freedom Primary Systems

Single-degree-of-freedom primary systems are rarely encountered in practical applications. Instead, real-world mechanical systems for which vibration protection is pursued are more conveniently modeled either as multi-degree-of-freedom discrete systems or continuous distributed parameter systems. The procedure for the optimum design of damped DVAs applied to

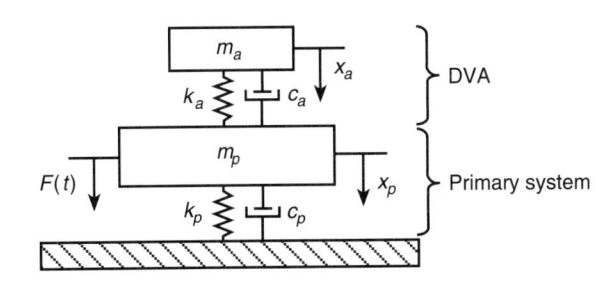

Figure 8 Scheme of a damped DVA connected to a damped primary system.

Table 1 Optimal values of DVA parameters for viscously damped primary systems

Mass ratio	Primary system damping $\zeta_p = c_p/2m_p\omega_p$	Optimal values f_{opt}	$\zeta_{opt} = c_a/2m_a\omega_a$	Values of g for equal peaks g_1	g_2
0.01	0	0.9901	0.061	0.960	1.030
	0.01	0.9886	0.062	0.956	1.032
	0.02	0.9869	0.064	0.953	1.033
	0.05	0.9807	0.068	0.942	1.034
	0.1	0.9663	0.073	0.923	1.030
0.1	0	0.9091	0.185	0.848	1.059
	0.01	0.9051	0.187	0.843	1.058
	0.02	0.9009	0.188	0.838	1.058
	0.05	0.8875	0.193	0.823	1.054
	0.1	0.8619	0.199	0.795	1.043
1.0	0	0.499	0.448	0.487	0.928
	0.01	0.494	0.448	0.481	0.924
	0.02	0.489	0.449	0.476	0.921
	0.05	0.473	0.454	0.462	0.904
	0.1	0.446	0.455	0.434	0.882

Adapted with permission from Warburton GB and Yorinde EO (1980) Optimum absorber parameters for simple systems. *Earthquake Engineering and Structural Dynamics*, 8: 197–217. John Wiley.

single-degree-of-freedom undamped primary systems can be extended to these types of systems, using a modal approach, developed in the following.

The main idea is to apply an optimization criterion in the vicinity of a particular natural frequency of the primary system. Several DVAs can be designed independently for each individual vibration mode. For this purpose it is assumed that the natural frequencies of the primary system are sufficiently well separated and that the masses of the DVAs are small enough not to modify significantly the natural frequencies of the primary system. **Figure 9** shows schematically a damped DVA attached to an undamped multi-degree-of-freedom primary system, modeled by inertia matrix **M** and stiffness matrix **K**. The indicated coordinates x_c and x_f correspond to the coordinates to which the DVA is attached and the excitation force is applied, respectively. In the general case of multi-dimensional systems, these coordinates may correspond to either displacements or rotations. To attenuate the vibrations in the vicinity of the nth natural frequency of the primary system, it is assumed that the responses are dominated by this particular mode. Thus, the vector of time responses of the primary system can be written:

$$\mathbf{x}(t) = \psi_n q_n(t) \quad [18a]$$

and, in particular, for the forcing and coupling coordinates:

$$x_f(t) = \psi_{fn} q_n(t) \quad [18b]$$

$$x_c(t) = \psi_{cn} q_n(t) \quad [18c]$$

In the equations above, ψ_{fn} and ψ_{cn} designate the components of the nth eigenvector corresponding to the excitation and coupling coordinates, respectively,

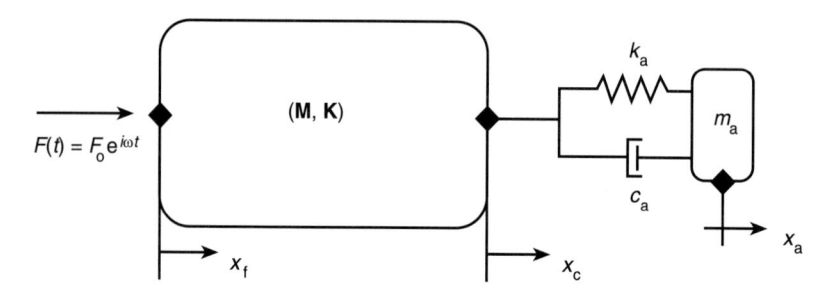

Figure 9 Schematic representation of a damped DVA attached to a multi-degree-of-freedom primary system.

and $q_n(t)$ is the generalized coordinate associated with the nth vibration mode. It is also assumed that the nth eigenvector is normalized so as to satisfy:

$$\psi_n^T \mathbf{M} \psi_n = M_n \quad [19\text{a}]$$

$$\psi_n^T \mathbf{K} \psi_n = M_n \omega_n^2 \quad [19\text{b}]$$

where M_n and ω_n are, respectively, the generalized mass and the natural frequency associated with the nth vibration mode of the primary system. In order to formulate a generalized substructuring theory, the following quantities are defined for the coupled system (primary system + DVA):

- Kinetic energy: $T = \frac{1}{2}\dot{\mathbf{x}}^T \mathbf{M} \dot{\mathbf{x}} + \frac{1}{2} m_a \dot{x}_a^2$ [20a]
- Strain energy: $V = \frac{1}{2}\mathbf{x}^T \mathbf{K} \mathbf{x} + \frac{1}{2} k_a (x_a - x_c)^2$ [20b]
- Rayleigh dissipation function: $F = \frac{1}{2} c_a (\dot{x}_a - \dot{x}_c)^2$ [20c]
- Virtual work of the excitation force: $\delta W^{nc} = F e^{i\omega t} \delta x_f$ [20d]

Introducing eqns [18] into eqns [20], taking into account relations [19], and employing Lagrange's equations, the following equations of motion are obtained:

$$M_n \ddot{q}_n + c_a \psi_{cn}^2 \dot{q}_n - c_a \psi_{cn} \dot{x}_a \\ + (M_p \omega_n^2 + k_a \psi_{cn}^2) q_n - k_a \psi_{cn} x_a = \psi_{fn} F e^{i\omega t} \quad [21\text{a}]$$

$$M_a \ddot{x}_a - c_a \psi_{cn} \dot{q}_n + c_a \dot{x}_a - k_a \psi_{cn} q_n + k_a x_a = 0 \quad [21\text{b}]$$

At this point the concepts of effective mass $(M_{\text{eff}})_n$ and effective stiffness $(K_{\text{eff}})_n$ are introduced, according to:

$$\tfrac{1}{2}(M_{\text{eff}})_n \dot{x}_c^2 = \tfrac{1}{2} \dot{\mathbf{x}}^T \mathbf{M} \dot{\mathbf{x}} \quad [22\text{a}]$$

$$\tfrac{1}{2}(k_{\text{eff}})_n x_c^2 = \tfrac{1}{2} \mathbf{x}^T \mathbf{K} \mathbf{x} \quad [22\text{b}]$$

As can be seen from the eqns [22], the effective mass can be interpreted as the mass that, once placed at the connection point, yields the same value of kinetic energy as that of the primary system. Similar interpretation is reserved to the effective stiffness, in terms of strain energy. Introducing eqns [18] into eqns [22]

and taking into account relations [19], the effective parameters are expressed as:

$$(M_{\text{eff}})_n = \frac{M_n}{\psi_{cn}^2} \quad [23\text{a}]$$

$$(K_{\text{eff}})_n = M_n \omega_n^2 \quad [23\text{b}]$$

Assuming steady-state harmonic motion with frequency ω, by manipulating eqn [21] and using eqns [18], the following expression for the harmonic amplitude at the coupling coordinate is obtained:

$$\frac{|X_c|}{(X_{\text{est}})_n} = \psi_{cn} \psi_{fn} \left\{ \frac{(2\zeta_n g_n)^2 + (g_n^2 - f_n^2)^2}{(2\zeta_n g_n)^2 [g_n^2 - 1 + (\mu_{\text{eff}})_n g_n^2]^2 + [(\mu_{\text{eff}})_n f_n^2 g_n^2 - (g_n^2 - 1)(g_n^2 - f_n^2)]^2} \right\} \quad [24]$$

where the parameters are defined as follows:

$$(\mu_{\text{eff}})_n = \frac{m_a}{(M_{\text{eff}})_n} \quad g_n = \frac{\omega}{\omega_n} \quad f_n = \frac{\omega_a}{\omega_n}$$

$$(X_{\text{est}})_n = \frac{F_0}{(K_{\text{eff}})_n} \quad \omega_a = \sqrt{\frac{k_a}{m_a}} \quad \zeta_n = \frac{c_a}{2 m_a \omega_n}$$

[25]

Comparing eqns [24] and [13a], it can be seen that the same dependence on the excitation frequency and dimensionless parameters is present in both. Thus, using the correspondences $\zeta \Longleftrightarrow \zeta_n$ and $\mu \Longleftrightarrow (\mu_{\text{eff}})_n$ it is possible to extend the optimization procedure developed for single-degree-of-freedom primary systems to discrete multi-degree-of-freedom or continuous distributed parameter systems, by treating each vibration mode separately.

In the example illustrated in **Figure 10**, the primary system is a simply-supported beam, simulated by finite elements, the properties of which are given in **Table 2**. The values of the first three natural frequencies and the components of the mode shapes corresponding to the vertical displacement of point B are given in **Table 3**. The mode shapes are normalized so that the generalized masses M_n are numerically equal to the mass of the beam. A damped DVA is optimally designed to attenuate the vibrations in the vertical direction at point B in the frequency band neighboring the second natural frequency. Choosing the effective mass ratio $\mu_{\text{eff}} = 0.02$, according to the procedure previously described, the following computations lead to the optimal values of the DVA parameters:

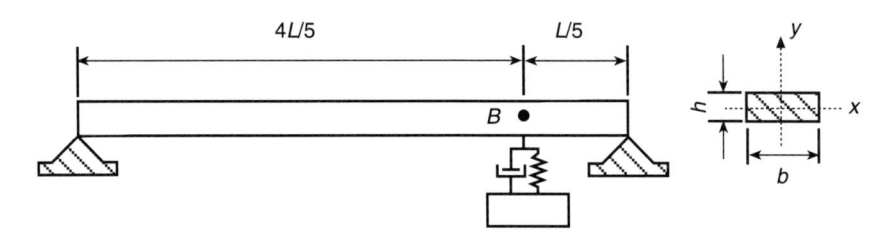

Figure 10 Beam-like primary system.

$$M_{\text{eff}} = \frac{M_p}{\psi_{\sigma c}^2} = \frac{2.54}{1.36^2} = 1.37$$

$$\zeta_{\text{opt}} = \sqrt{\frac{3\mu_{\text{eff}}}{8(1+\mu_{\text{eff}})}} = \sqrt{\frac{3 \times 0.02}{8(1+0.02)}} = 0.09$$

$$f_{\text{opt}} = \frac{\omega_a}{\omega_2} \Rightarrow k_a = m_a(f_{\text{opt}}\omega_2)^2 = 3.85 \times 10^4 \text{ N m}^{-1}$$

$$f_{\text{opt}} = \frac{1}{1+\mu_{\text{eff}}} = \frac{1}{1+0.02} = 0.98$$

$$m_a = \mu_{\text{eff}} M_{\text{eff}} = 0.02 \times 1.37 = 0.03 \text{ kg}$$

$$\zeta_{\text{opt}} = \frac{c_a}{2m_a\omega_2} \Rightarrow c_a = 6.25 \text{ N m}^{-1}\text{s}$$

Figure 11 shows the frequency response for the co-ordinate to which the optimal DVA is attached, superimposed on the same frequency response of the primary system without the DVA. It can be seen that the resonance peak corresponding to the second natural frequency has been significantly damped. **Figure 12** shows the plots of the frequency response in the vicinity of the second natural frequency for different values of the damping ratio and a unique value of the tuning ratio, demonstrating the existence of the invariant points.

Table 2 Physical and geometrical characteristics of the beam-like primary system

Property	Value
Young modulus	$E = 2.1 \times 10^{11}$ N m^{-2}
Mass density	$\rho = 7800$ kg m^{-3}
Beam length	$L = 763.0$ mm
Cross-section width	$b = 36.6$ mm
Cross-section height	$h = 11.4$ mm
Second moment of area about x axis	$I = 4518.7$ mm^4
Total mass	2.54 kg

Special Configurations of Dynamic Vibration Absorbers

Although the theory presented above has been developed for the attenuation of translational motion of vibrating systems, it can be readily extended to DVAs intended for reducing rotational vibrations. Moreover, besides their classical representation as shown by **Figure 1**, DVAs can assume various constructive forms according to the specific application desired. In the following, some of these special configurations are reviewed.

Torsional Dynamic Vibration Absorbers

Torsional vibrations of internal combustion engines and other rotating systems can be controlled by using torsional vibration absorbers. Such an arrangement is shown in **Figure 13A**. The primary system is represented by inertia J_p and torsional stiffness k_{T_p}, and the absorber is represented by inertia J_a and torsional stiffness k_{T_a}. Viscous damping is provided by oil inside a housing rigidly connected to the primary system, in such a way that a dissipative torque given by $c_T(\dot{\theta}_p - \dot{\theta}_a)$ is generated. An equivalent translational two-degree-of-freedom system is shown in **Figure 13B**. Since the dynamic equations of motion are similar for both systems, the formulae obtained for the translational DVA and the procedure to obtain its optimal design remain applicable for the torsional system. The equivalence between the parameters of the translational and torsional systems is indicated in **Table 4**.

Table 3 Modal characteristics of the beam-like primary system

Mode	Natural frequency (Hz)	Mode-shape component
1	46.10	-0.84
2	184.2	1.36
3	414.3	-1.33

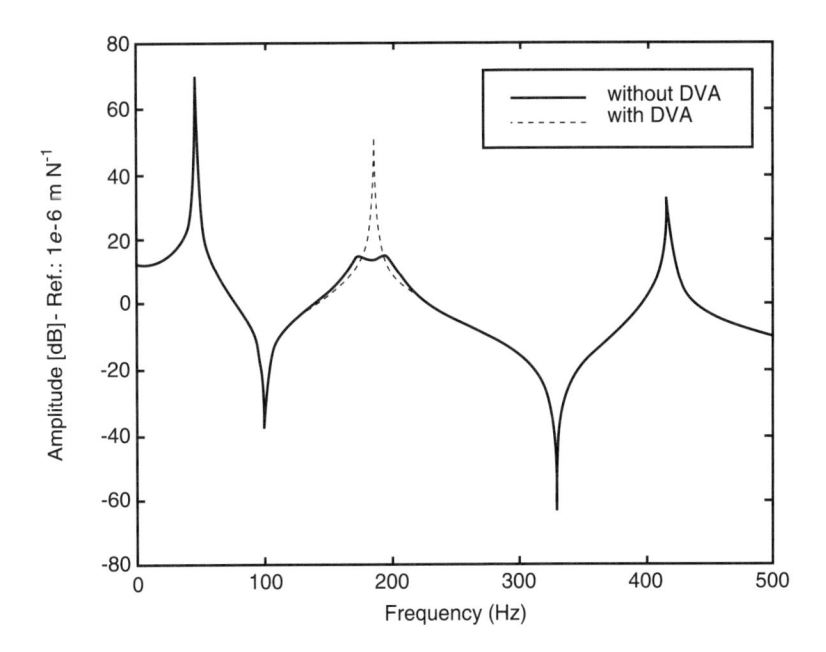

Figure 11 Frequency response corresponding to the coupling coordinate.

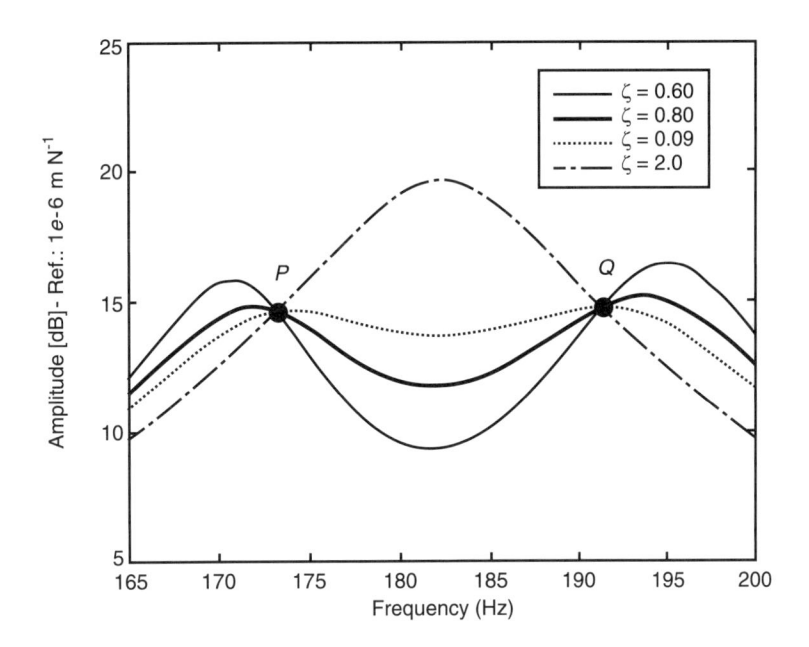

Figure 12 Illustration of the invariant points in the vicinity of the second natural frequency.

The Gyroscopic Dynamic Vibration Absorber

The roll motion of ships can be reduced by installing a large gyroscopic fixed to the hull, as shown in **Figure 14**. This arrangement is called gyroscope of Schlick and consists of a heavy gyroscope rotating at a high speed about a vertical axis. The roll motion of the ship induces the gyroscope to precess in the plane of symmetry along the length of the ship. Optionally, the precession motion can be damped by introducing an energy dissipation device. Neglecting damping, the equations of motion for the coupled system (ship + gyro) are written:

$$J_s \ddot{\phi} + J\Omega \dot{\theta} + k_r \phi = \tau_s \quad [26a]$$

$$J_g \ddot{\theta} - J\Omega \dot{\phi} + Wa\theta = 0 \quad [26b]$$

where $a =$ distance between precession axis and gyro's center of gravity; $J =$ polar moment of inertia

20 ABSORBERS, VIBRATION

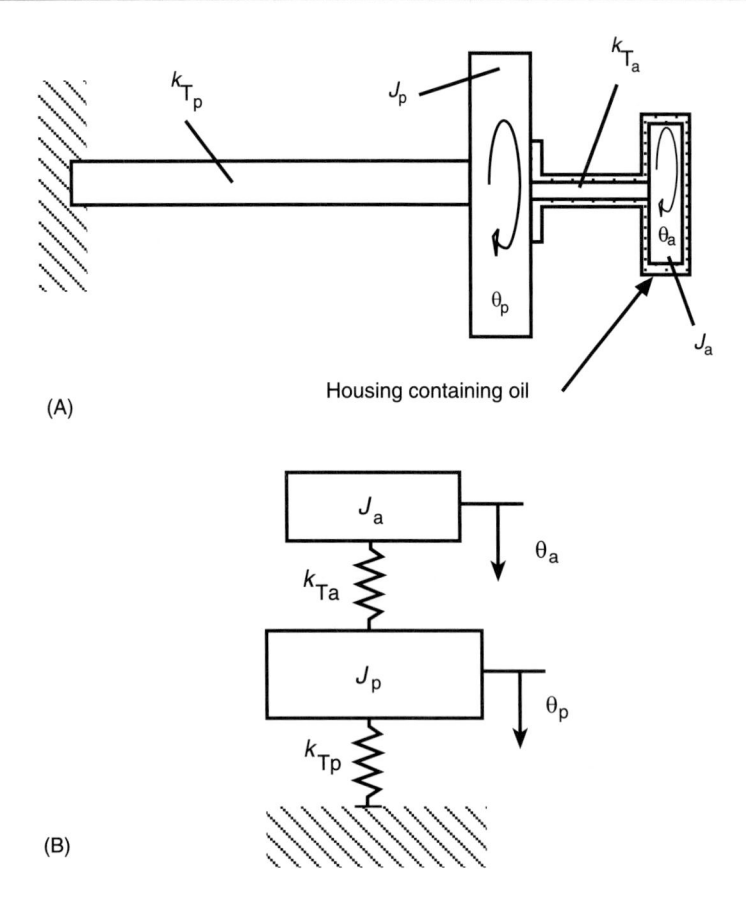

Figure 13 (A) Scheme of a torsional system with DVA and (B) equivalent rectilinear system.

Table 4 Equivalence between translational and torsional parameters

	Translational system	Torsional system
Inertia of the primary system	m_p (kg)	J_p (kg m^2)
Stiffness of the primary system	k_p (N m^{-1})	k_{Tp} (N.m rad^{-1})
Inertia of the DVA	m_a (kg)	J_a (kg m^2)
Stiffness of the DVA	k_a (N m^{-1})	k_{Ta} (N.m rad^{-1})
Damping of the DVA	c_a (N.s m^{-1})	c_{Ta} (N.m.s rad^{-1})

of the gyro; J_s = mass moment of inertia of the ship about its longitudinal axis; J_g = mass moment of inertia of the gyro about the precession axis; k_r = stiffness associated with roll motion; W = gyro's weight; ϕ = roll angle of the ship; θ = precession angle of the gyro; Ω = rotational speed of the gyro; τ_s = external excitation torque applied to the ship. As indicated in **Figure 14**, the gyro's center of gravity is situated below the precession axis. This ensures that the gyro is submitted to a restoring gravitational torque about the precession axis.

Assuming harmonic excitation:

$$\tau_s = T_s\, e^{i\omega t} \quad [27]$$

the steady-state solution to eqns [26] is written:

$$\phi = \Phi\, e^{i\omega t} \quad [28a]$$

$$\theta = \Theta\, e^{i\omega t} \quad [28b]$$

After some algebraic manipulation, the following expression is obtained for the frequency response corresponding to the roll motion of the ship:

$$\Phi = \frac{Wa - \omega^2 J_g}{(k_r - \omega^2 J_s)(Wa - \omega^2 J_g) - (J\Omega\omega)^2} T_s \quad [29]$$

It can be readily seen from eqn [29] that the roll motion is completely eliminated when the numerator vanishes, i.e.:

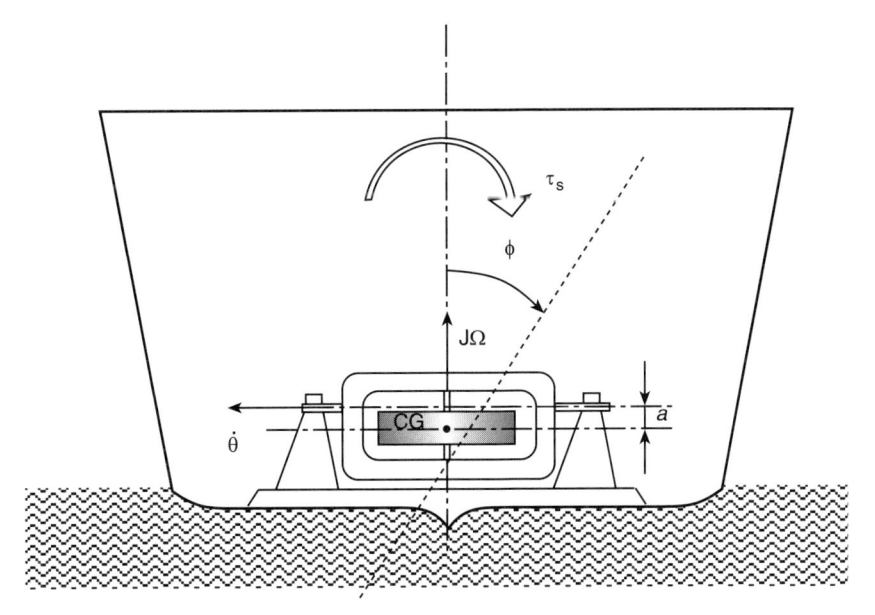

Figure 14 Sketch of a gyroscopic DVA used to attenuate ship roll motion.

$$\omega = \sqrt{\left(\frac{Wa}{J_g}\right)} \quad [30]$$

Thus, the gyroscope parameters can be chosen so as to achieve tuning to the excitation frequency according to eqn [30].

It can be demonstrated that, when the precession motion is viscously damped, the expression for the frequency-response function associated to the roll motion is similar to eqn [13a], and allows for invariant points. Thus, the gyroscopic system can be optimally designed by using Den Hartog's procedure in the same way as it is applied to rectilinear damped DVAs.

The Centrifugal Pendulum Vibration Absorber

In the torsional vibration of rotating systems, it is generally the case that the excitation occurs at the same frequency as the rotational speed or at a multiple n of this frequency. For example, a shaft that drives a propeller can be subjected to torsional vibrations whose frequency is given by the number of blades of the propeller times the rotation speed. In such systems, a configuration of vibration absorber which has been frequently used is the centrifugal pendulum, depicted in **Figure 15**. The equations of motion for the two-degree-of-freedom system (flywheel + pendulum) can be written as:

$$\left(\frac{J}{mRl} + \frac{R}{l} + \frac{l}{R} + 2\right)\ddot{\theta} + \left(\frac{l}{R} + 1\right)\ddot{\psi} + \frac{k_T}{mRl}\theta = \frac{\tau}{mRl} \quad [31a]$$

$$\left(1 + \frac{R}{l}\right)\ddot{\theta} + \ddot{\psi} + \frac{\Omega^2 R}{l}\psi = 0 \quad [31b]$$

where J = mass moment of inertia of the flywheel; k_T = torsional stiffness of the shaft; l = length of the pendulum; m = mass of the pendulum; n = ratio between the excitation frequency and the angular velocity of the flywheel (order of vibration); R = distance between the center of the flywheel and the pivoting point of the pendulum; θ = angular coordinate describing the torsional vibration of the flywheel; ψ = angular coordinate describing the oscillation of the pendulum; Ω = angular velocity of the flywheel. Assuming steady-state harmonic vibration, one writes:

$$\tau = T e^{in\Omega t} \quad \theta = \Theta e^{in\Omega t} \quad \psi = \Psi e^{in\Omega t} \quad [32]$$

and manipulation of eqns [31] leads to the following relation between the amplitudes of torsional vibration of the shaft and oscillation of the pendulum:

$$\frac{\Theta}{\Psi} = \frac{R - ln^2}{n^2(l + R)} \quad [33]$$

It can be readily seen that if the geometry of the system is designed such that $R/l = n^2$, the torsional vibration of the flywheel is completely cancelled. In this case, the pendulum behaves as a DVA tuned to a given multiple of the excitation frequency.

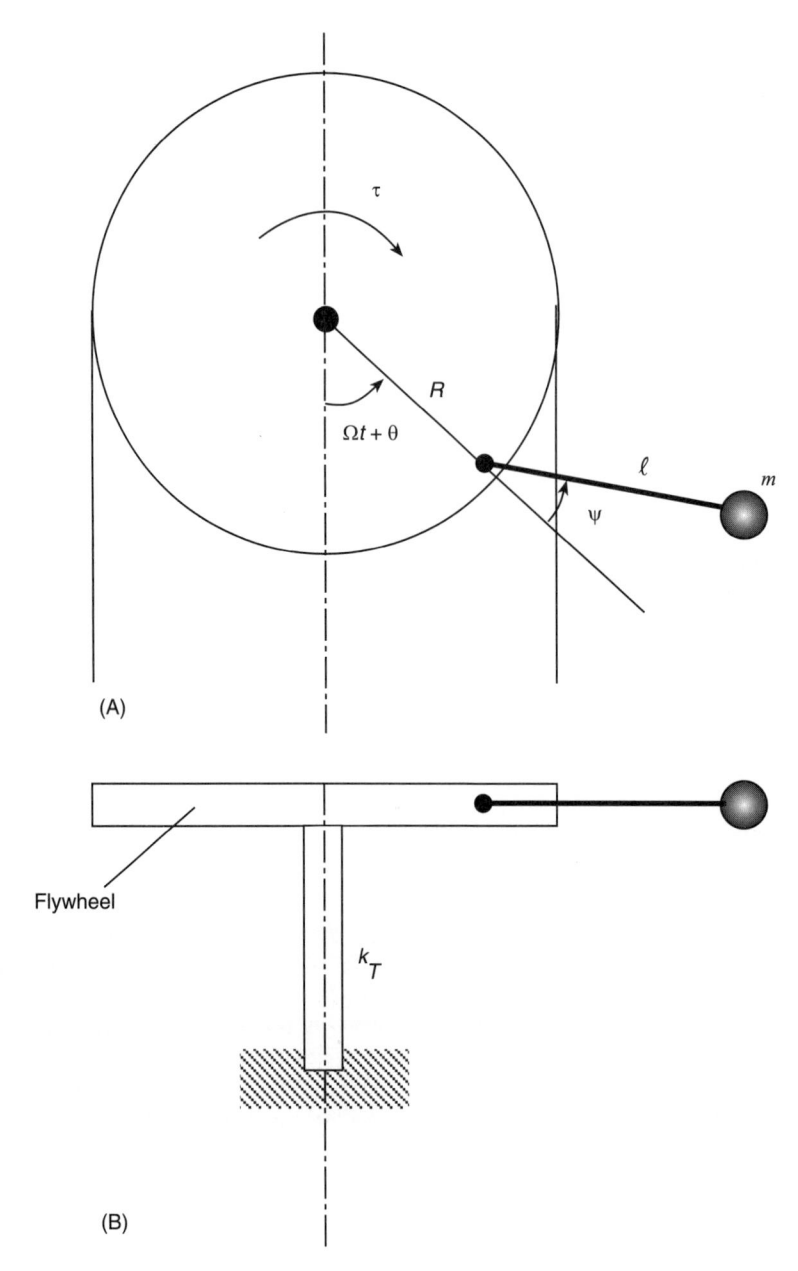

Figure 15 Scheme of a centrifugal pendulum DVA applied to a rotating system. (A) Top view; (B) side view.

Active Dynamic Vibration Absorbers

In applications requiring attenuation capability over a broad frequency band, active DVAs can be a very interesting solution. Moreover, active DVAs offer the possibility of automatic real-time tuning to the excitation frequency varying within a frequency band. Active DVAs are understood as those having an active element (actuator) installed in parallel with the passive elements supporting the reactive mass, as shown in **Figure 16**. The force impressed by the actuator is calculated through a previously defined control law. It is important to point out, however, that active DVAs have some drawbacks, such as energy consumption and instability.

Various control laws can be used, involving either absolute or relative dynamic responses. In the following, for illustration, the theory of an active DVA based on a control law expressed as a linear combination of relative displacement, velocity and acceleration responses is developed. In the system represented by **Figure 16** the control force applied by the actuator is assumed to be expressed as:

$$u(t) = -[\alpha(\ddot{x}_a - \ddot{x}_p) + \beta(\dot{x}_a - \dot{x}_p) + \gamma(x_a - x_p)] \quad [34]$$

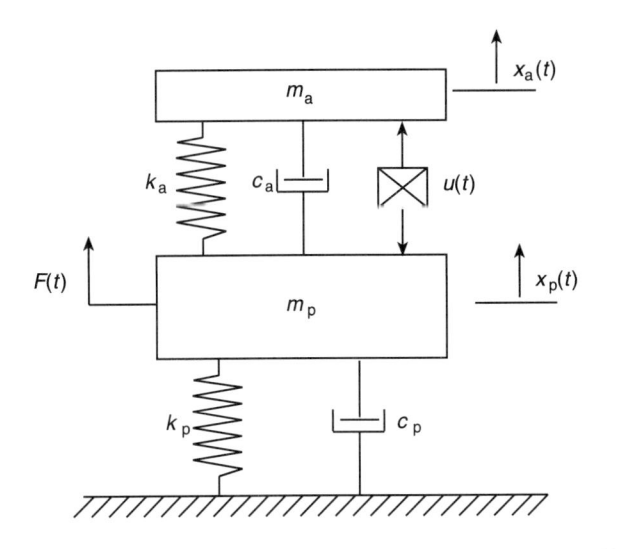

Figure 16 Schematic representation of a primary system with an active DVA.

where:

$$\begin{cases} a_0 = m_a\alpha + m_p(m_a + \alpha) \\ a_1 = (m_p + m_a)(c_a + \beta) + c_p(m_a + \alpha) \\ a_2 = k_p(m_a + \alpha) + c_p(c_a + \beta) + (k_a + \gamma)(m_p + m_a) \\ a_3 = c_p(k_a + \gamma) + k_p(c_a + \beta) \\ a_4 = k_p(k_a + \gamma) \end{cases} \quad [37]$$

As in every application of active control, it is important to carry out a study of the stability of the system. Using the Routh–Hurwitz stability criterion, based on the expressions of the coefficients of the characteristic equation, as given by eqn [37], it can be shown that stability is ensured, provided that the following inequalities are satisfied by the feedback gains:

$$\alpha > \frac{-m_p m_a}{(m_p + m_a)} \quad [38a]$$

$$\beta > -c_a \quad [38b]$$

$$\gamma > -k_a \quad [38c]$$

The feedback gains can be adjusted so as to tune the absorber to an arbitrarily chosen value of the excitation frequency. Using the equations of motion [35] and assuming harmonic excitation $F = F_0 e^{i\omega t}$, the frequency response of the primary mass is found to be expressed as:

$$\frac{X_a}{F_0} = \frac{-(m_a + \alpha)\omega^2 + i\omega(c_a + \beta) + k_a + \gamma}{a_0\omega^4 - ia_1\omega^3 - a_2\omega^2 + ia_3\omega + a_4} \quad [39]$$

For the case of an undamped DVA without velocity feedback ($c_2 = \beta = 0$), from [39] it can be seen that the response vanishes when:

$$\omega^2 = \frac{k_a + \gamma}{m_a + \alpha} \quad [40]$$

Thus, it is possible to cancel the harmonic vibrations at a given frequency ω by adjusting the values of the feedback gains γ and/or α so as to satisfy eqn [40] and the stability conditions given by [38a] and [38c].

Considering the particular case of an undamped primary system and a control law without acceleration feedback ($c_p = 0, \alpha = 0$), eqn [39] can be expressed as follows:

where α, β, and γ are, respectively, the feedback gains for acceleration, velocity, and displacement signals.

The matrix equation of motion for the two-degree-of-freedom system is written as:

$$\mathbf{M}\ddot{\mathbf{x}} + \mathbf{C}\dot{\mathbf{x}} + \mathbf{K}\mathbf{x} = \mathbf{F} \quad [35]$$

where:

$$\mathbf{M} = \begin{bmatrix} m_a + \alpha & -\alpha \\ -\alpha & m_p + \alpha \end{bmatrix}$$

$$\mathbf{C} = \begin{bmatrix} c_p + c_a + \beta & -(c_a + \beta) \\ -(c_a + \beta) & c_a + \beta \end{bmatrix}$$

$$\mathbf{K} = \begin{bmatrix} k_p + k_a + \gamma & -(k_a + \gamma) \\ -(k_a + \gamma) & k_a + \gamma \end{bmatrix}$$

$$\mathbf{x} = \begin{Bmatrix} x_p \\ x_a \end{Bmatrix} \quad \mathbf{F} = \begin{Bmatrix} f \\ 0 \end{Bmatrix}$$

As can be seen in the previous equations, the parameters β and γ can be interpreted as damping and stiffness parameters, which are added to the corresponding passive elements of the absorber. The same interpretation cannot be given the parameter α, since it also appears in the off-diagonal positions of the mass matrix.

The characteristic equation of the system is written in the Laplace domain as:

$$a_0 s^4 + a_1 s^3 + a_2 s^2 + a_3 s + a_4 = 0 \quad [36]$$

$$\left(\frac{X_p k_p}{F_0}\right)^2 =$$

$$\frac{(2\zeta g)^2 + (g^2 - f^2)^2}{(2\zeta g)^2 (g^2 - 1 + \mu g^2)^2 + [\mu g^2 f^2 - (g^2 - 1)(g^2 - f^2)]^2}$$

[41]

where:

$$\mu = \frac{m_a}{m_p}; \quad \omega_a = \sqrt{\left(\frac{k_a + \gamma}{m_a + \alpha}\right)}; \quad \omega_p = \sqrt{\left(\frac{k_p}{m_p}\right)}$$

$$g = \omega/\omega_p; \quad f = \omega_a/\omega_p; \quad \zeta = \frac{c_a + \beta}{2 m_a \omega_p}$$

[42]

Eqn [41] has the same form as eqn [13a] developed for purely passive DVAs. As a result, the optimization procedure based on the invariant points also holds for the active DVA. Thus, assuming that the values of the passive elements are not varied, the optimization of the DVA can be achieved by finding an optimal set of feedback gains. Introducing eqns [42] into eqns [14] and [15], the following expressions for the optimal gains are obtained:

$$\gamma_{opt} = \left(\frac{\omega_p}{1 + \mu}\right)^2 m_a - k_a \qquad [43a]$$

$$\beta_{opt} = 2\sqrt{\left[\frac{3\mu}{8(1 + \mu)^3}\right]} \omega_p m_a - c_a \qquad [43b]$$

To illustrate the procedure for optimal design of an active DVA, consider an undamped primary system, for which the parameters take the following values:

$$m_p = 1.0 \text{ kg}, \quad k_p = 10\,000 \text{ N m}^{-1}, \quad c_p = 0$$

The passive parameters of the DVA are chosen as:

$$m_a = 0.1 \text{ kg}, \quad k_a = 1000 \text{ N m}^{-1}, \quad c_a = 0$$

Eqns [43] are used to determine the following optimal values of the feedback gains γ and β:

$$\gamma_{opt} = -173.55 \text{ N m}^{-1} \quad \beta_{opt} = 3.36 \text{ Ns m}^{-1}$$

Figure 17 shows the influence of the active DVA on the frequency system response of the primary system as compared to the response of this system without DVA and with the purely passive undamped DVA.

Final Remarks and Future Perspectives

In the previous sections, only DVAs comprising stiffness and damping elements exhibiting linear behavior have been considered. However, studies have demonstrated that nonlinear DVAs generally provide a suppression bandwidth much larger than linear absorbers. As a result, in spite of a more involved theory and design procedure, nonlinear vibration absorbers have received much attention lately.

Although only harmonic excitations were considered here, the reader should be aware of the fact that dynamic vibration absorbers have been extensively used to attenuate other types of vibrations, such as transient and random. In such cases, the optimal design is generally carried on by using time domain-based procedures.

The study and development of techniques related to smart materials represent new possibilities of vibration reduction in mechanics and mechatronics. The physical properties of such materials can be modified by controlled modifications of some environmental parameters. To mention a few examples, the viscosity (damping capacity) of electrorheological and magnetorheological fluids can be varied by applying external electric and magnetic fields, respectively. The geometry of components made of shape memory alloys can be changed by applying temperature variations. Some researchers have considered the possibility of using such smart materials to conceive self-tunable adaptive vibration absorbers. Furthermore, the possibility of dissipating mechanical energy with piezoelectric material, such as piezoelectric ceramics, shunted with passive electrical components has been investigated by various authors in this decade. The four basic kinds of shunt circuits are: inductive, resistive, capacitive, and switched. If a piezoelectric element is attached to a structure, it is strained as the structure deforms and part of the vibration energy is converted into electrical energy. The piezoelectric element behaves electrically as a capacitor and can be combined with a so-called shunt network in order to perform vibration control. Shunting with a resistor and inductor forms a RLC circuit introducing an electrical resonance which, in the optimal case, is tuned to structural resonances. The scheme of such an arrangement is depicted in **Figure 18**. The inductor is used to tune the shunt circuit to a given resonance of the structure and the resistor is responsible for peak amplitude reduction of a particular mode. The inductive shunt or resonant circuit shunt presents a vibration suppression effect that is very similar to the classical dynamic vibration absorber. The classical DVA stores part of the kinetic energy of the primary

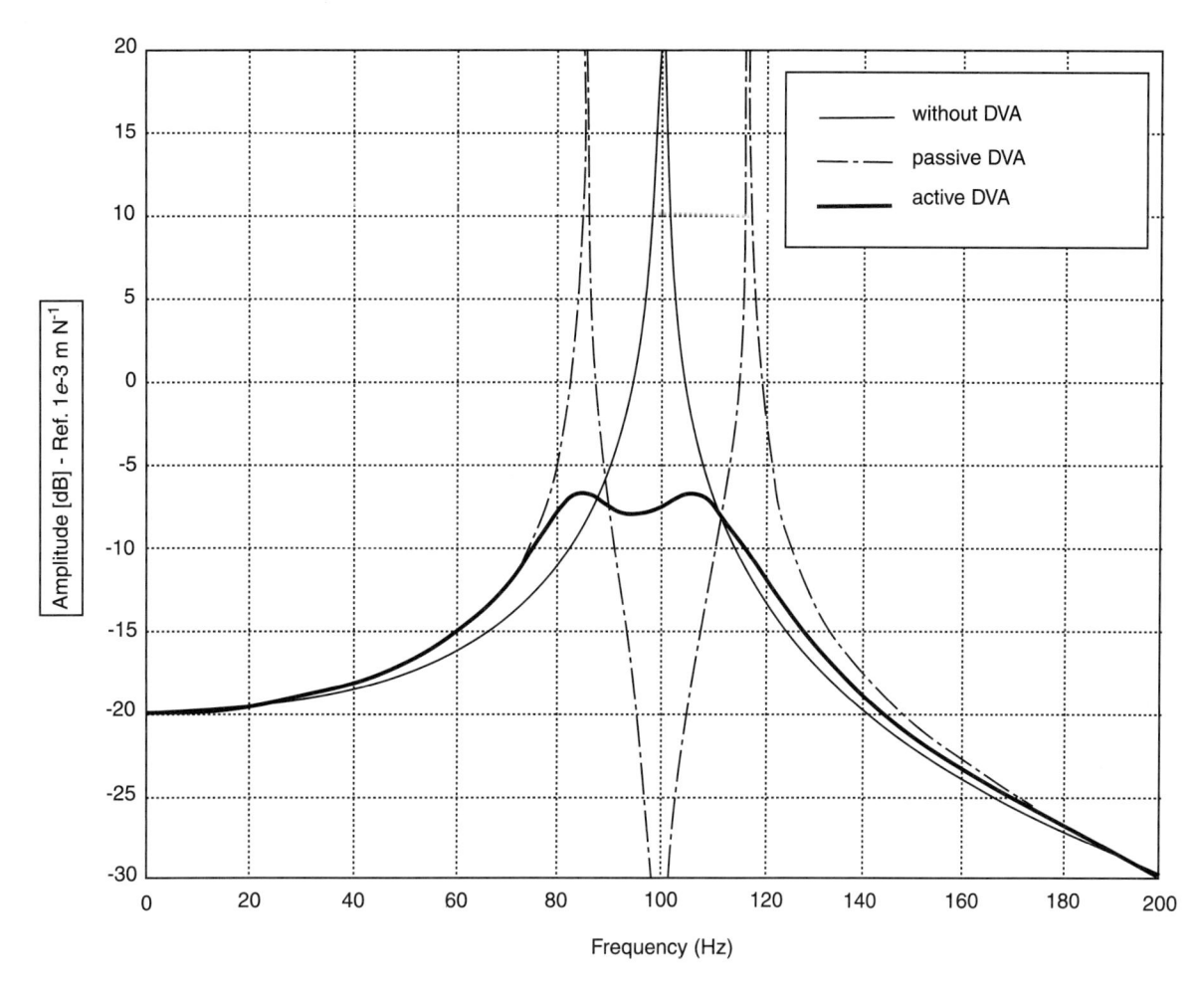

Figure 17 Frequency response of the primary system.

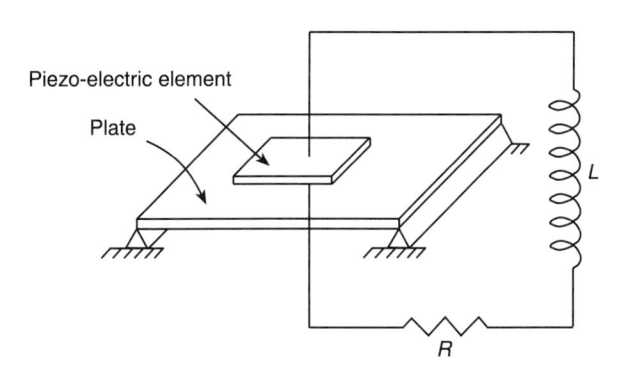

Figure 18 Scheme of a resonant circuit shunt used for vibration attenuation.

Nomenclature

f	tuning factor
g	forcing frequency ratio
J	inertia
l	length
W	weight
X	amplitude
μ	mass ratio
ψ	angular coordinate
θ	precession angle
τ	external excitation torque
Φ	roll angle
Ω	rotational speed/angular velocity

system, while the resonant circuit shunt is designed to dissipate the electrical energy that has been converted from mechanical energy by the piezoelectric. A multimode damper can be obtained by adding a different shunt for each suppressed mode in such a way that attenuation can be obtained for a given number of frequencies.

See also: **Absorbers, active**; **Active control of civil structures**; **Active control of vehicle vibration**; **Active isolation**; **Damping, active**; **Flutter, active control**; **Ship vibrations**; **Shock isolation systems**; **Theory of vibration**, Fundamentals; **Vibration isolation theory**; **Viscous damping**.

Further Reading

Den Hartog JP (1956) *Mechanical Vibrations*, 4th edn. McGraw-Hill Book Company.

Hagood NW and von Flotow A (1991) Damping of structural vibrations with piezoelectric materials and passive electrical networks. *Journal of Sound and Vibration* 146: 243–268.

Harris CM (1988) *Shock and Vibration Handbook*, 3rd edn, McGraw-Hill Book Company.

Inman DJ (1989) *Vibration with Control, Measurement and Stability*. Prentice-Hall.

Korenev BG and Reznikov LM (1993) *Dynamic Vibration Absorbers. Theory and Practical Applications*. John Wiley.

Nashif AD, Jones DIG and Henderson JP (1985) *Vibration Damping*. John Wiley.

Newland DE (1989) *Mechanical Vibration Analysis and Computation*. Longman Scientific and Technical.

Snowdon JC (1968) *Vibration and Shock in Damped Mechanical Systems*. John Wiley.

Sun JQ, Jolly MR and Norris MA (1995) Passive, adaptive and active tuned vibration absorbers – a survey. *Trans. ASME Combined Anniversary Issue Journal of Mechanical Design and Journal of Vibration and Acoustics* 117: 234–242.

ACTIVE ABSORBERS

See **ABSORBERS, ACTIVE**

ACTIVE CONTROL OF CIVIL STRUCTURES

T T Soong, State University of New York at Buffalo, Buffalo, NY, USA

B F Spencer, Jr., University of Notre Dame, Notre Dame, IN, USA

Copyright © 2001 Academic Press

doi:10.1006/rwvb.2001.0189

Introduction

In civil engineering structural applications, active, semiactive, and hybrid structural control systems are a natural evolution of passive control technologies such as base isolation and passive energy dissipation. The possible use of active control systems and some combinations of passive and active systems, so-called hybrid systems, as a means of structural protection against wind and seismic loads has received considerable attention in recent years. Active/hybrid control systems are force delivery devices integrated with real-time processing evaluators/controllers and sensors within the structure. They act simultaneously with the hazardous excitation to provide enhanced structural behavior for improved service and safety. Remarkable progress has been made over the last 20 years. As will be discussed in the following sections, research to date has reached the stage where active systems have been installed in full-scale structures. Active systems have also been used temporarily in construction of bridges or large-span structures (e.g., lifelines, roofs) where no other means can provide adequate protection.

The purpose of this article is to provide an assessment of the state-of-the-art and state-of-the-practice of this exciting, and still evolving, technology. Also included in the discussion are some basic concepts, the types of active control systems being used and deployed, and their advantages and limitations in the context of seismic design and retrofit of civil engineering structures.

Active, Hybrid, and Semiactive Control Systems

An active structural control system has the basic configuration shown schematically in **Figure 1A**. It consists of: (1) sensors located about the structure to measure either external excitations, or structural response variables, or both; (2) devices to process the measured information and to compute necessary control force needed based on a given control algorithm; and (3) actuators, usually powered by external sources, to produce the required forces.

When only the structural response variables are measured, the control configuration is referred to as feedback control since the structural response is continually monitored and this information is used to make continual corrections to the applied control forces. A feedforward control results when the control forces are regulated only by the measured excitation, which can be achieved, for earthquake inputs, by measuring accelerations at the structural base. In the case where the information on both the response quantities and excitation is utilized for control design, the term feedback–feedforward control is used.

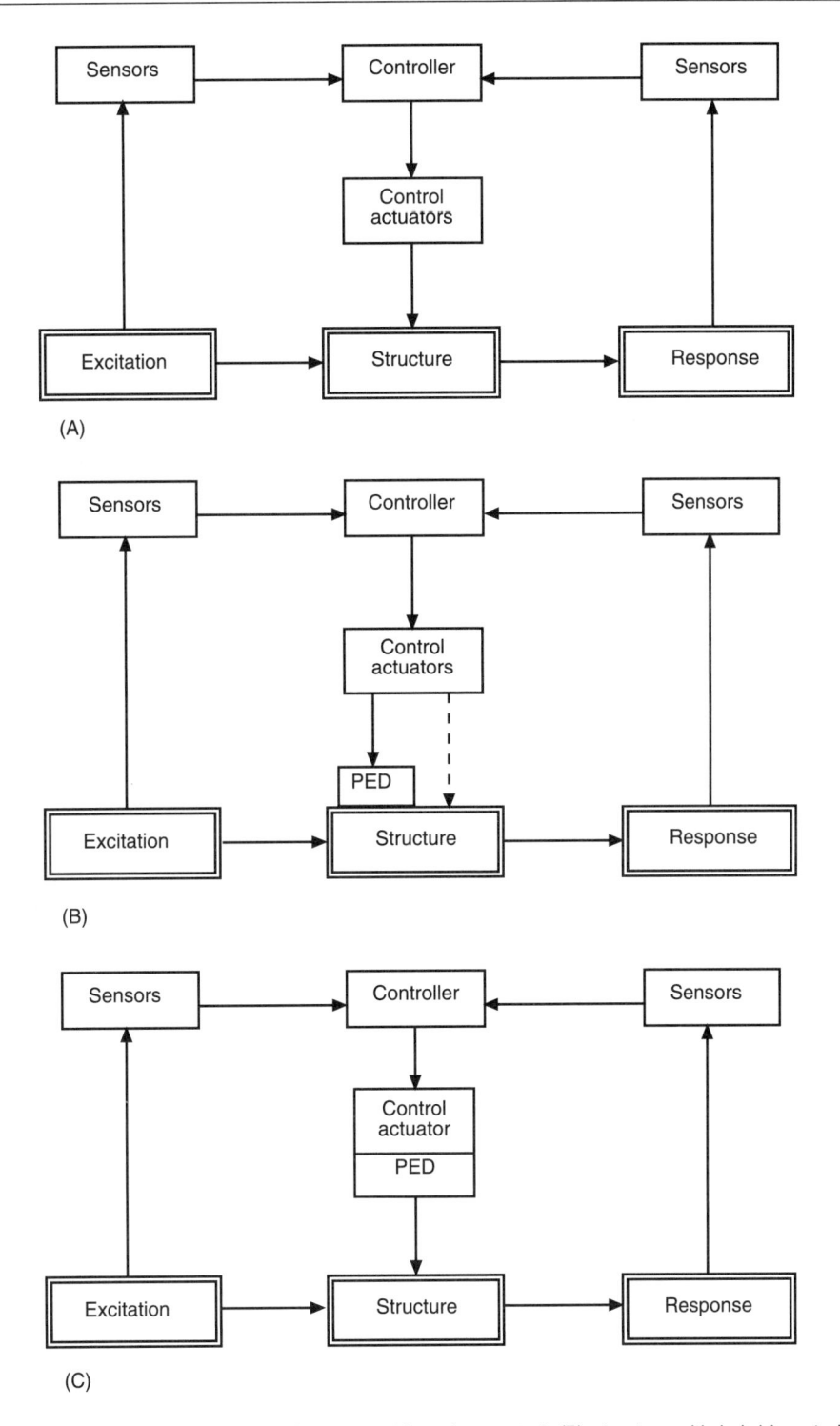

Figure 1 Structure with various schemes. (A) Structure with active control; (B) structure with hybrid control; (C) structure with semiactive control. PED, passive energy dissipation.

To see the effect of applying such control forces to a linear structure under ideal conditions, consider a building structure modeled by an n-degree-of-freedom lumped mass-spring-dashpot system. The matrix equation of motion of the structural system can be written as:

$$\mathbf{M}\ddot{x}(t) + \mathbf{C}\dot{x}(t) + \mathbf{K}x(t) = \mathbf{D}u(t) + \mathbf{E}f(t) \quad [1]$$

where $\mathbf{M}$, $\mathbf{C}$, and $\mathbf{K}$ are the $n \times n$ mass, damping and stiffness matrices, respectively, $x(t)$ is the n-dimensional displacement vector, the m-vector $f(t)$ represents the applied load or external excitation, and

r-vector $u(t)$ is the applied control force vector. The $n \times r$ matrix **D** and the $n \times m$ matrix **E** define the locations of the action of the control force vector and the excitation, respectively, on the structure.

Suppose that the feedback–feedforward configuration is used in which the control force $u(t)$ is designed to be a linear function of measured displacement vector $x(t)$, velocity vector $\dot{x}(t)$ and excitation $f(t)$. The control force vector takes the form:

$$u(t) = \mathbf{G}_x x(t) + \mathbf{G}_{\dot{x}} \dot{x}(t) + \mathbf{G}_f f(t) \qquad [2]$$

in which $\mathbf{G}_x$, $\mathbf{G}_{\dot{x}}$, and $\mathbf{G}_f$ are respective control gains which can be time-dependent.

The substitution of eqn [2] into eqn [1] yields:

$$\mathbf{M}\ddot{x}(t) + (\mathbf{C} - \mathbf{DG}_{\dot{x}})\dot{x}(t) + (\mathbf{K} - \mathbf{DG}_x)x(t) \\ = (\mathbf{E} + \mathbf{DG}_f)f(t) \qquad [3]$$

Comparing eqn [3] with eqn [1] in the absence of control, it is seen that the effect of feedback control is to modify the structural parameters (stiffness and damping) so that it can respond more favorably to the external excitation. The effect of the feedforward component is a modification of the excitation. The choice of the control gain matrices $\mathbf{G}_x$, $\mathbf{G}_{\dot{x}}$, and $\mathbf{G}_f$ depends on the control algorithm selected.

In comparison with passive control systems, a number of advantages associated with active control systems can be cited; among them are: (1) enhanced effectiveness in response control; the degree of effectiveness is, by and large, only limited by the capacity of the control systems; (2) relative insensitivity to site conditions and ground motion; (3) applicability to multihazard mitigation situations; an active system can be used, for example, for motion control against both strong wind and earthquakes; and (4) selectivity of control objectives; one may emphasize, for example, human comfort over other aspects of structural motion during noncritical times, whereas increased structural safety may be the objective during severe dynamic loading.

While this description is conceptually in the domain of familiar optimal control theory used in electrical engineering, mechanical engineering, and aerospace engineering, structural control for civil engineering applications has a number of distinctive features, largely due to implementation issues, that set it apart from the general field of feedback control. In particular, when addressing civil engineering structures, there is considerable uncertainty, including nonlinearity, associated with both physical properties and disturbances such as earthquakes and wind, the scale of the forces involved can be quite large, there are only a limited number of sensors and actuators, the dynamics of the actuators can be quite complex, the actuators are typically very large, and the systems must be failsafe.

It is useful to distinguish between several types of active control systems currently being used in practice. The term hybrid control generally refers to a combined passive and active control system, as depicted in **Figure 1B**. Since a portion of the control objective is accomplished by the passive system, less active control effort, implying less power resource, is required.

Similar control resource savings can be achieved using the semiactive control scheme sketched in **Figure 1C**, where the control actuators do not add mechanical energy directly to the structure, hence bounded-input bounded-output stability is guaranteed. Semiactive control devices are often viewed as controllable passive devices.

A side benefit of hybrid and semiactive control systems is that, in the case of a power failure, the passive components of the control still offer some degree of protection, unlike a fully active control system.

Full-scale Applications

As alluded to earlier, the development of active, hybrid, and semiactive control systems has reached the stage of full-scale applications to actual structures. **Figure 2** shows that, up to 1999, there have been 43 installations in building structures and towers, most of which are in Japan (**Table 1**). In addition, 15 bridge towers have employed active systems during erection. Most of these full-scale systems have been subjected to actual wind forces and ground motions and their observed performances provide invaluable information in terms of: (1) validating analytical and simulation procedures used to predict actual system performance; (2) verifying complex electronic–digital–servohydraulic systems under actual loading conditions; and (3) verifying the capability of these systems to operate or shut down under prescribed conditions.

Described below are several of these systems together, in some cases, with their observed performances. Also addressed are several practical issues in connection with actual structural applications of these systems.

Hybrid Mass Damper Systems

As seen from **Table 2**, the hybrid mass damper (HMD) is the most common control device employed in full-scale civil engineering applications. An HMD is a combination of a passive tuned mass damper (TMD) and an active control actuator. The ability of

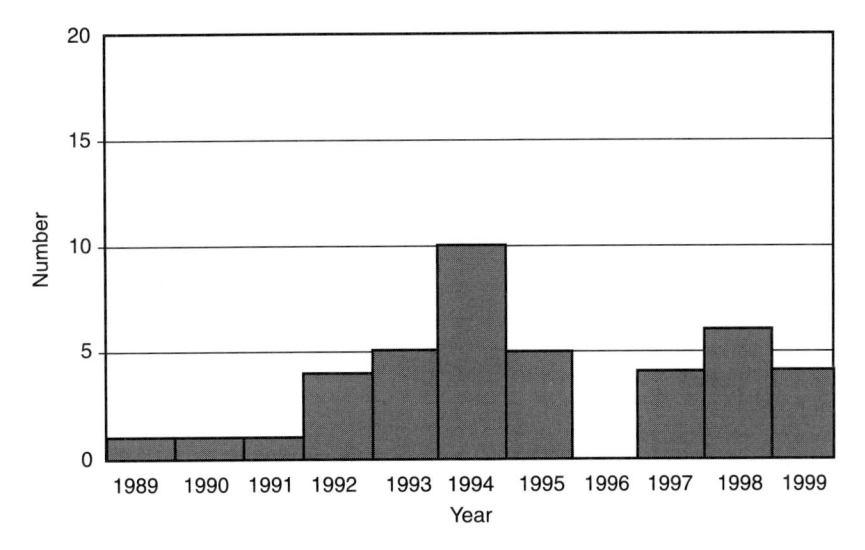

Figure 2 Active control – number of installations.

Table 1 System applications

Country	Number[a]
Japan	39
Others	4

[a] Up to 1999.

Table 2 System configurations

Type	Number[a]
Hybrid mass damper	33
Others	10

[a] Up to 1999.

this device to reduce structural responses relies mainly on the natural motion of the TMD. The forces from the control actuator are employed to increase the efficiency of the HMD and to increase its robustness to changes in the dynamic characteristics of the structure. The energy and forces required to operate a typical HMD are far less than those associated with a fully active mass damper system of comparable performance.

An example of such an application is the HMD system installed in the Sendagaya INTES building in Tokyo in 1991. As shown in **Figure 3**, the HMD was installed atop the 11th floor and consists of two masses to control transverse and torsional motions of the structure, while hydraulic actuators provide the active control capabilities. The top view of the control system is shown in **Figure 4**, where ice thermal storage tanks are used as mass blocks so that no extra mass needs to be introduced. The masses are supported by multistage rubber bearings intended for reducing the control energy consumed in the HMD and for insuring smooth mass movements.

Sufficient data were obtained for evaluation of the HMD performance when the building was subjected to strong wind on 29 March 1993, with peak instantaneous wind speed of $30.6\,\mathrm{m\,s^{-1}}$.

An example of the recorded time histories is shown in **Figure 5**, giving both the uncontrolled and controlled states. Their Fourier spectra using samples of 30-s durations are shown in **Figure 6**, again showing good performance in the low-frequency range. The response at the fundamental mode was reduced by 18% and 28% for translation and torsion, respectively.

Variations of such an HMD configuration include multistep pendulum HMDs (as seen in **Figure 7**), which have been installed in, for example, the Yokohama Landmark Tower in Yokohama, the tallest building in Japan, and in the TC Tower in Kaohsiung, Taiwan. Additionally, the DUOX HMD system which, as shown schematically in **Figure 8**, consists of a TMD actively controlled by an auxiliary mass, has been installed in, for example, the Ando Nishikicho Building in Tokyo.

Active Mass Damper Systems

Design constraints, such as severe space limitations, can preclude the use of an HMD system. Such is the case in the active mass damper or active mass driver (AMD) system designed and installed in the Kyobashi Seiwa Building in Tokyo and the Nanjing Communication Tower in Nanjing, China.

The Kyobashi Seiwa Building, the first full-scale implementation of active control technology, is an ary AMD has a weight of 1 ton and is employed to reduce torsional motion. The role of the active system is to reduce building vibration under strong winds and moderate earthquake excitations and consequently to increase comfort of occupants in the building.

Semiactive Damper Systems

Control strategies based on semiactive devices appear to combine the best features of both passive and active control systems. The close attention received in this area in recent years can be attributed to the fact that semiactive control devices offer the adaptability of active control devices without requiring the associated large power sources. In fact, many can operate on battery power, which is critical during seismic events when the main power source to the structure may fail. In addition, as stated earlier, semiactive control devices do not have the potential to destabilize (in the bounded input/bounded output sense) the structural system. Extensive studies have indicated that appropriately implemented semiactive systems perform significantly better than passive devices and have the potential to achieve the majority of the performance of fully active systems, thus allowing for the possibility of effective response reduction during a wide array of dynamic loading conditions.

One means of achieving a semiactive damping device is to use a controllable, electromechanical, variable-orifice valve to alter the resistance to flow of a conventional hydraulic fluid damper. A schematic of such a device is given in **Figure 10**. Such a system was implemented, for example, in a bridge to dissipate the energy induced by vehicle traffic.

More recently, a semiactive damper system was installed in the Kajima Shizuoka Building in Shizuoka, Japan. As seen in **Figure 11**, semiactive hydraulic dampers were installed inside the walls on both sides of the building to enable it to be used as a disaster relief base in postearthquake situations. Each

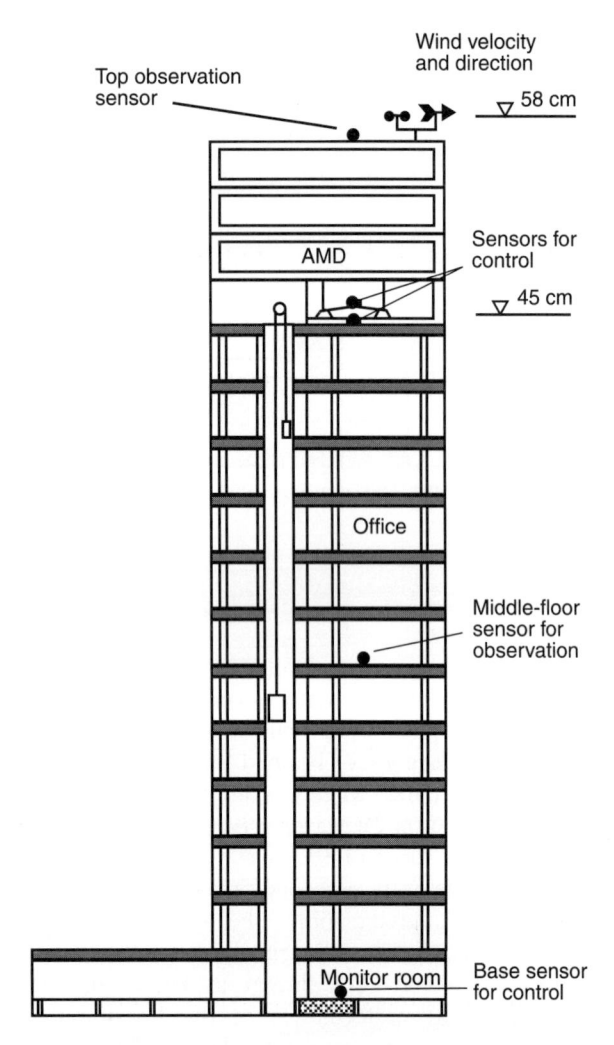

Figure 3 Sendagaya INTES building with hybrid mass damper. AMD, active mass damper.

11-story building with a total floor area of 423 m². As seen in **Figure 9**, the control system consists of two AMDs where the primary AMD is used for transverse motion and has a weight of 4 tons, while the second-

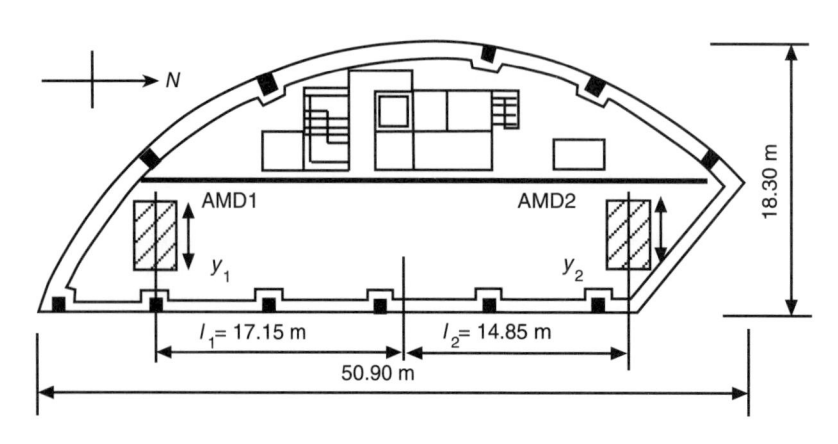

Figure 4 Top view of hybrid mass damper. AMD, active mass damper.

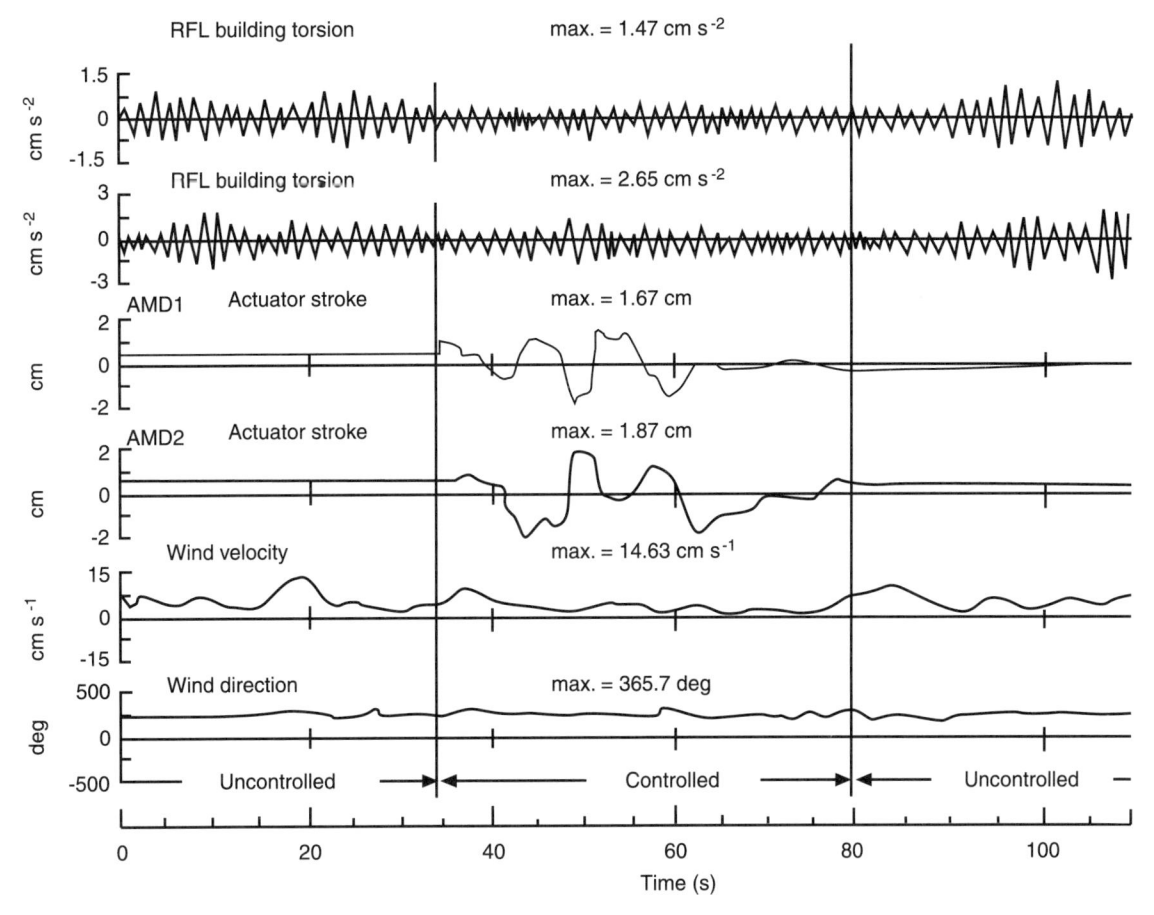

Figure 5 Response time histories (23 March 1993). AMD, active mass damper.

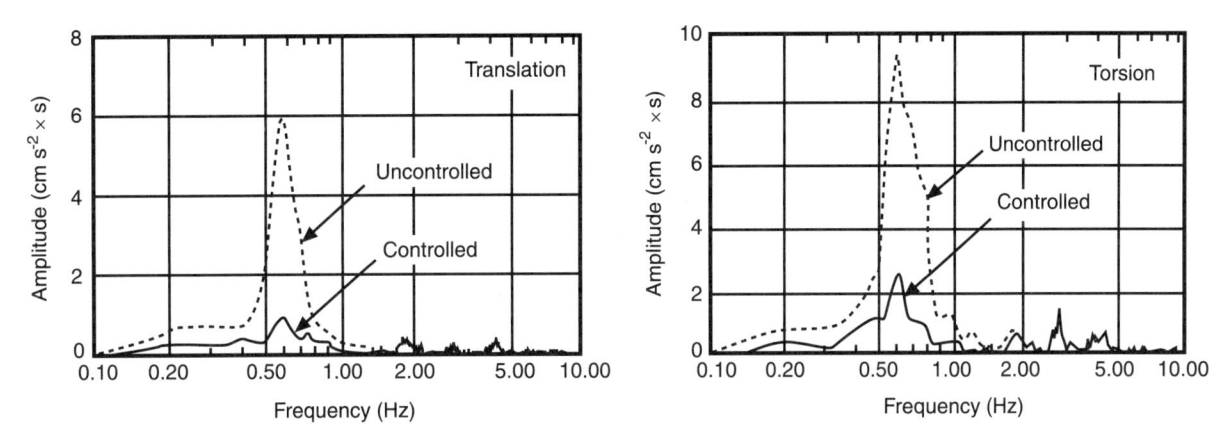

Figure 6 Response Fourier spectra (23 March 1993).

damper contains a flow control valve, a check valve, and an accumulator, and can develop a maximum damping force of 1000 kN. **Figure 12** shows a sample of the response analysis results based on one of the selected control schemes and several earthquake input motions with the scaled maximum velocity of 50 cm s^{-1}, together with a simulated Tokai wave. It is seen that both story shear forces and story drifts are greatly reduced with the control system activated. In the case of the shear forces, they are confined within their elastic-limit values (indicated by E-limit) while, without control, they would enter the plastic range.

Semiactive Controllable Fluid Dampers

Another class of semiactive devices uses controllable fluids, schematically shown in **Figure 13**. In comparison with semiactive damper systems described

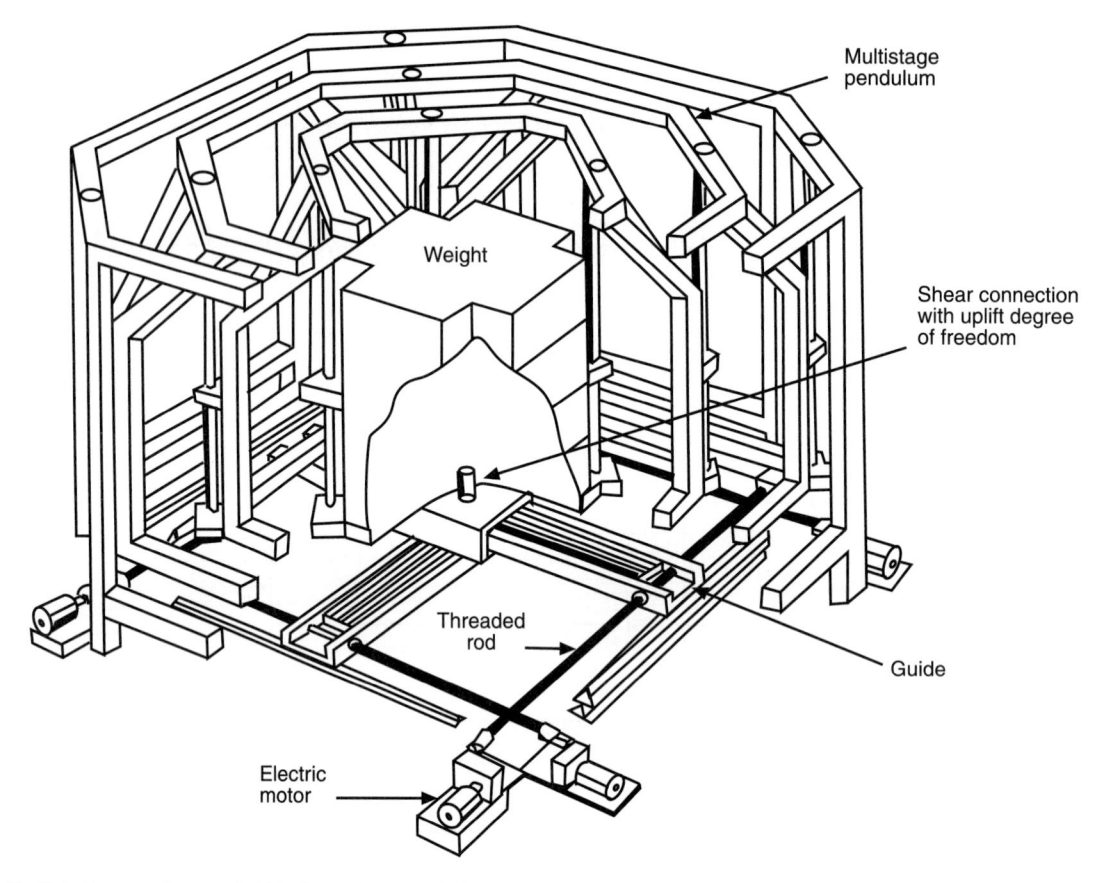

Figure 7 Hybrid mass damper in Yokohama Landmark Tower.

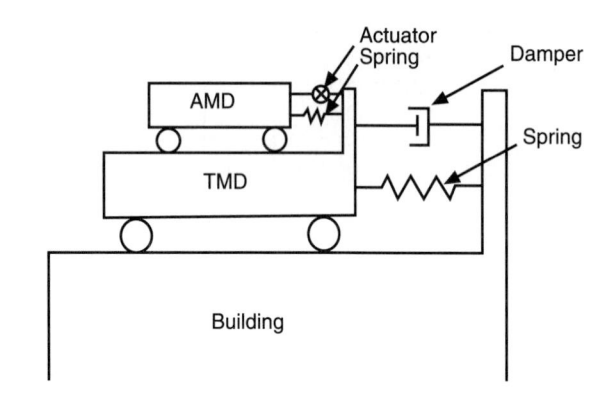

Figure 8 Principle of DUOX system. AMD, active mass damper; TMD, tuned mass damper.

above, an advantage of controllable fluid devices is that they contain no moving parts other than the piston, which makes them simple and potentially very reliable.

The essential characteristics of controllable fluids is their ability to change reversibly from a free-flowing, linear, viscous fluid to a semisolid with a controllable yield strength in milliseconds when exposed to an electric (for electrorheological (ER) fluids) or magnetic (for magnetorheological (MR) fluids) field.

In the case of MR fluids, they typically consist of micron-sized, magnetically polarizable particles dispersed in a carrier medium such as mineral or silicone oil. When a magnetic field is applied to the fluid, particle chains form, and the fluid becomes a semisolid and exhibits viscoplastic behavior. Transition to rheological equilibrium can be achieved in a few milliseconds, allowing construction of devices with high bandwidth. Additionally, it has been indicated that high yield stress of an MR fluid can be achieved and that MR fluids can operate at temperatures from $-40°C$ to $150°C$ with only slight variations in the yield stress. Moreover, MR fluids are not sensitive to impurities such as are commonly encountered during manufacturing and usage, and little particle/carrier fluid separation takes place in MR fluids under common flow conditions. Further, a wider choice of additives (surfactants, dispersants, friction modifiers, antiwear agents, etc.) can generally be used with MR fluids to enhance stability, seal life, bearing life, and so on, since electrochemistry does not affect the magnetopolarization mechanism. The MR fluid can be readily controlled with a low-voltage (e.g., 12–24 V), current-driven power supply outputting only 1–2 A.

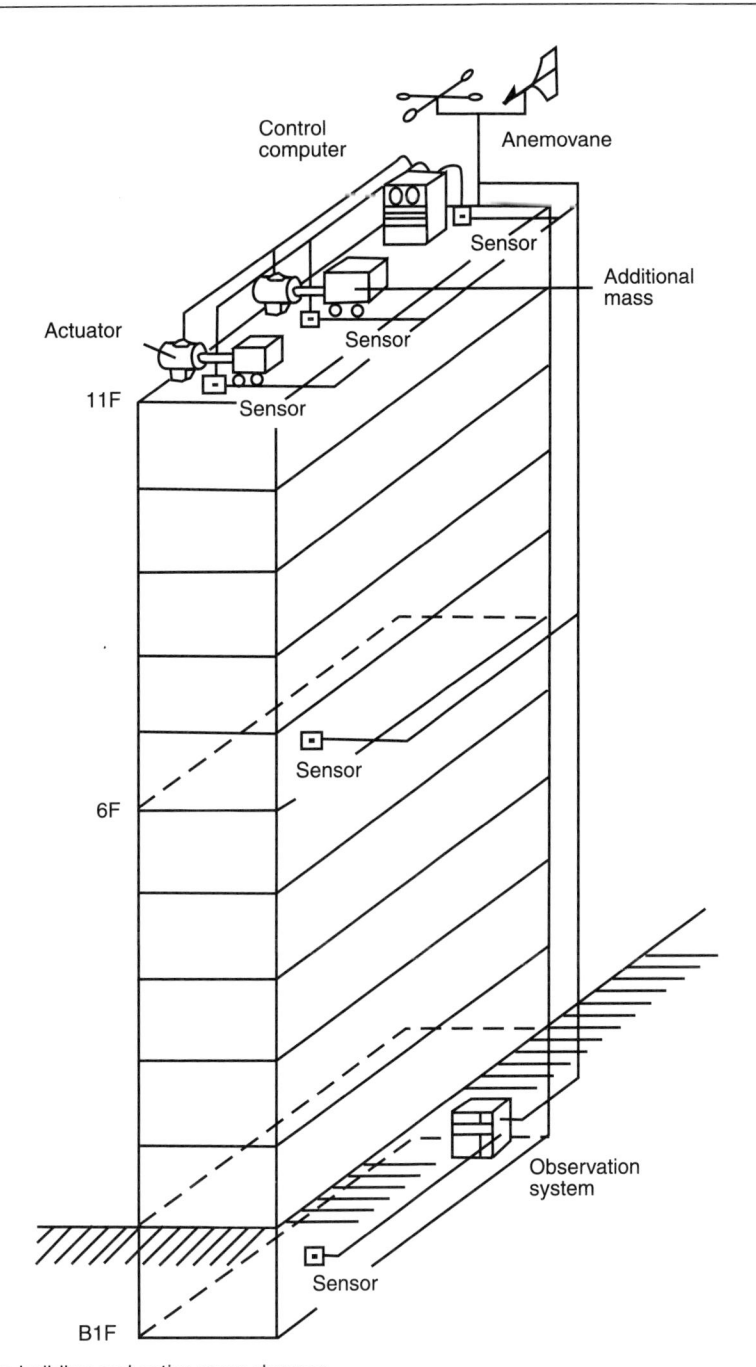

Figure 9 Kyobashi Seiwa building and active mass damper.

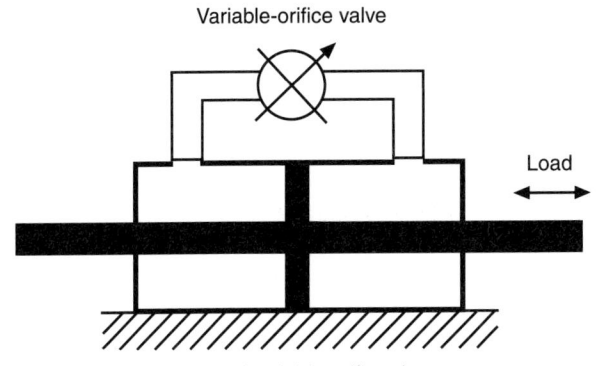

Figure 10 Schematic of variable-orifice damper.

While no full-scale structural applications of MR devices have taken place to date, their future for civil engineering applications appears to be bright. There have been published reports on the design of a full-scale, 20-ton MR damper, showing that this technology is scalable to devices appropriate for civil engineering applications. At design velocities, the dynamic range of forces produced by this device is over 10 (**Figure 14**), and the total power required by the device is only 20–50 W.

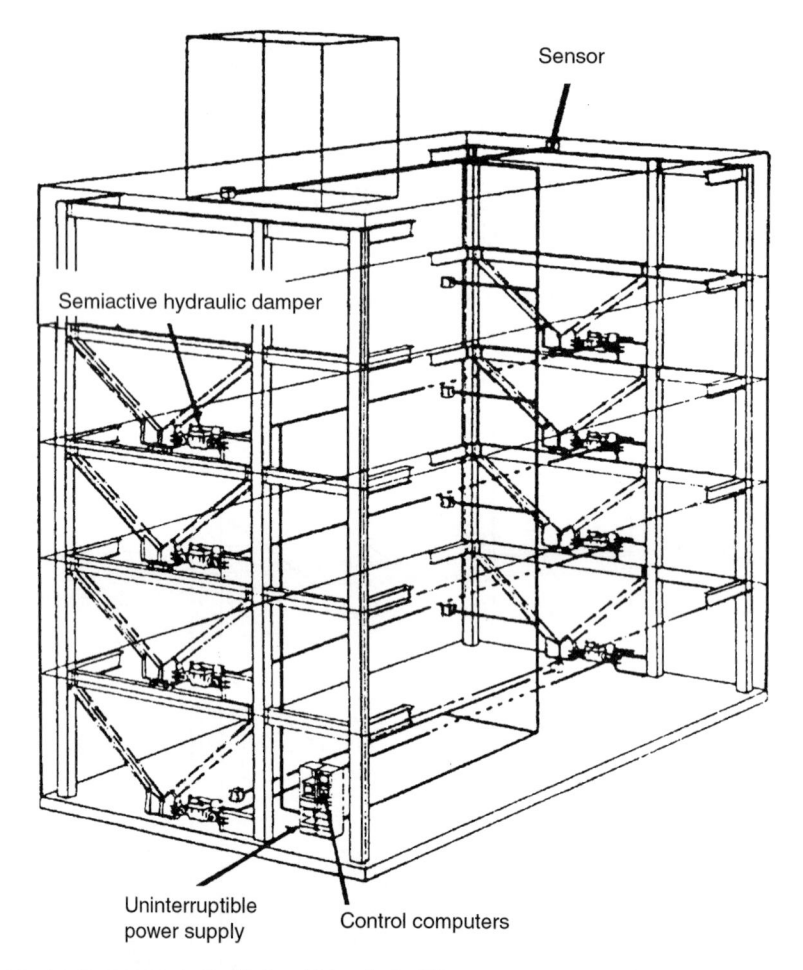

Figure 11 Semiactive hydraulic damper in the Kajima Shizuoka building.

Concluding Remarks

An important observation to be made in the performance observation of control systems such as those described above is that efficient active control systems can be implemented with existing technology under practical constraints such as power requirements and stringent demand of reliability. Thus, significant strides have been made, considering that serious implementational efforts began less than 15 years ago. On the other hand, there remains a significant distance between the state-of-the-art of active control technology and some originally intended purposes for developing such a technology. Two of these areas are particularly noteworthy and they are highlighted below.

1. Mitigating higher-level hazards. In the context of earthquake engineering, one of the original goals for active control research was the desire that, through active control, conventional structures can be protected against infrequent, but highly damaging earthquakes. The active control devices currently deployed in structures and towers were designed primarily for performance enhancement against wind and moderate earthquakes and, in many cases, only for occupant comfort. However, active control systems remain to be one of only a few alternatives for structural protection against near-field and high-consequence earthquakes.

   An upgrade of current active systems to this higher level of structural protection is necessary, since only then can the unique capability of active control systems be realized.

2. Economy and flexibility in construction. Another area in which great benefit can be potentially realized by the deployment of active control systems is added economy and flexibility to structural design and construction. The concept of active structures has been advanced. An active structure is defined here as one consisting of two types of load-resisting members: the traditional static (or passive) members that are designed to support basic design loads, and dynamic (or active) members whose function is to augment the structure's capability in resisting extraordinary dynamic

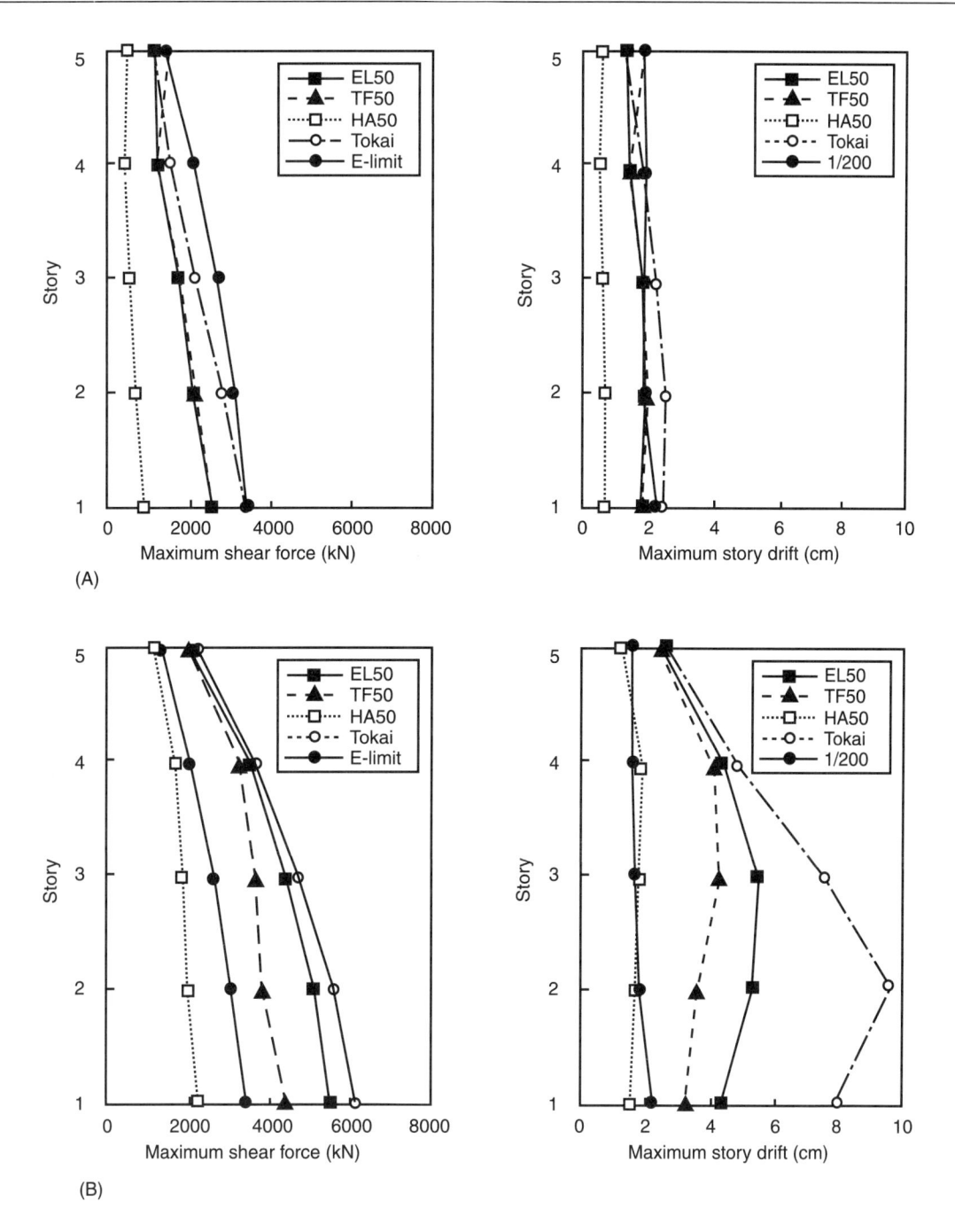

Figure 12 Maximum responses (El Centro, Taft, and Hachinohe waves with 50 cm s^{-1} and assumed Tokai waves). (A) With semi-active hydraulic damper control; (B) without control.

loads. Their integration is done in an optimal fashion and produces a structure that is adaptive to changing environmental loads and usage.

Note that an active structure is conceptually and physically different from a structure that is actively controlled, as in the cases described above. In the case of a structure with active control, a conventionally designed structure is supplemented by an active

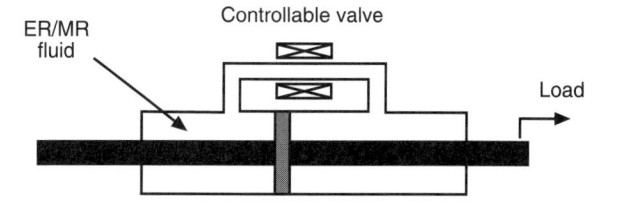

Figure 13 Schematic of controllable fluid damper. ER, electrorheological; MR, magnetorheological.

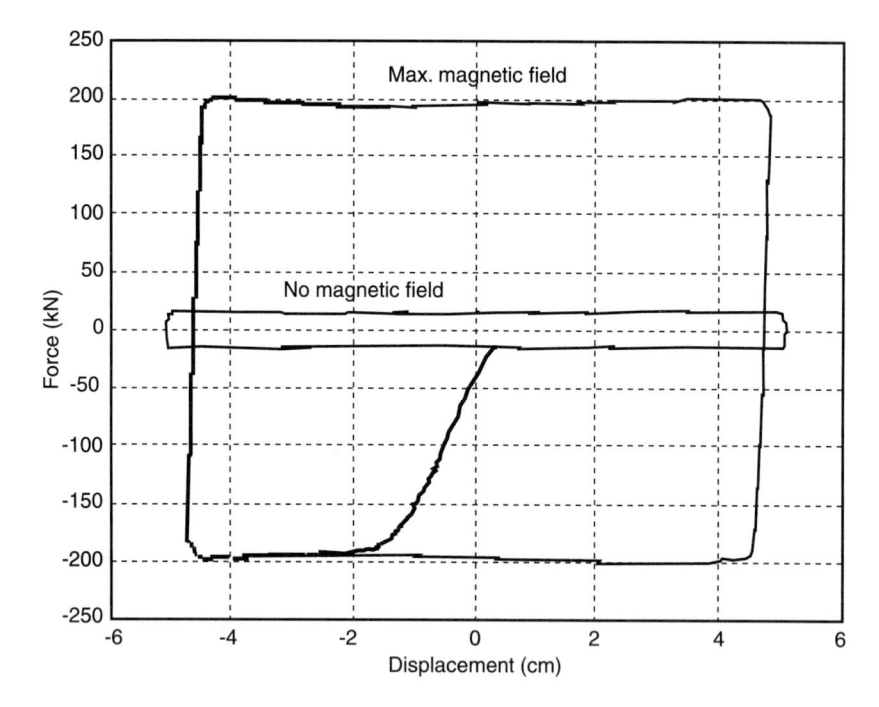

Figure 14 Force–displacement loops at maximum and zero magnetic fields.

control device that is activated whenever necessary in order to enhance structural performance under extraordinary loads. Thus, the structure and the active control system are individually designed and optimized. An active structure, on the other hand, is one whose active and passive components are integrated and simultaneously optimized to produce a new breed of structural systems. This important difference makes the concept of active structures exciting and potentially revolutionary. Among many possible consequences, one can envision greater flexibilities which may lead to longer, taller, slender, or more open structures and structural forms.

See Plates 1,2,3.

See also: **Damping, active; Hybrid control.**

Further Reading

Carlson JD and Spencer BF Jr. (1996) Magneto-rheological fluid dampers for semi-active seismic control. In: *Proceedings of the 3rd International Conference on Motion and Vibration Control*, Chiba, Japan. Japan Society of Mechanical Engineers, Tokyo, Japan, Vol. III, pp. 35–40.

Carlson JD and Weiss KD (1994) A growing attraction to magnetic fluids. *Machine Design*, pp. 61–64.

Housner GW, Bergman LA, Caughey TK *et al.* (1997) Structural control: past, present, and future. *Journal of Engineering Mechanics* 123: 897–971.

Kobori T (1994) Future direction on research and development of seismic-response-controlled structure. *Proceedings 1st World Conference on Structural Control* (GW Houser, SF Masri and AG Chassiakos, eds), International Association for Structural Control, Los Angeles, CA, pp. 19–31.

Kobori T (1999) Mission and perspective towards future structural control research. *Proceedings of the 2nd World Conference on Structural Control*, Kyoto, Japan. (T Kobori, Y Inoue, K Seto, H Iemura and A Nishitani, eds), John Wiley, Chichester, UK, pp. 25–34.

Patten WN (1999) The I-35 Walnut Creek bridge: an intelligent highway bridge via semi-active structural control. In: *Proceedings of 2nd World Conference on Structure Control*, Kyoto, Japan (T Kobori, Y Inoue, K Seto, H Iemura, A Nishitani, (eds), John Wiley, Chichester, UK, pp. 427–436.

Soong TT (1990) *Active Structural Control: Theory and Practice.* New York, NY: Wiley.

Soong TT and Manolis GD (1987) Active structures. *Journal of Structural Engineering* 113: 2290–2301.

Soong TT, Reinhorn AM, Aizawa S and Higashino M (1994) Recent structural applications of active control technology. *Journal of Structural Control* 1: 5–21.

Spencer BF Jr, Carlson JD, Sain MK and Yang G (1997) On the current status of magnetorheological dampers: seismic protection of full-scale structures. In: *Proceedings of American Control Conference*, American Automatic Control Council, pp. 458–462, Albuquerque, NM.

Spencer BF Jr, Yang G, Carlson JD and Sain MK (1999) Smart dampers for seismic protection of structures: a full-scale study. In: *Proceedings of 2nd World Conference on Structure Control*, Kyoto, Japan (T Kobori, Y Inoue, K Seto, H Iemura, A Nishitani, eds), John Wiley, Chichester, UK, pp. 417–426.

ACTIVE CONTROL OF VEHICLE VIBRATION

M Ahmadian, Virginia Tech, Blacksburg, VA, USA

Copyright © 2001 Academic Press

doi:10.1006/rwvb.2001.0193

Introduction

Perceived comfort level and ride stability are two of the most important factors in a vehicle's subjective evaluation. There are many aspects of a vehicle that influence these two properties, most importantly the primary suspension components, which isolate the frame of the vehicle from the axle and wheel assemblies. In the design of a conventional primary suspension system, there is a tradeoff between the two quantities of ride comfort and vehicle stability, as shown in **Figure 1**. If a primary suspension is designed to optimize the handling and stability of the vehicle, the operator is often subjected to a large amount of vibration and perceives the ride to be rough and uncomfortable. On the other hand, if the primary suspension is designed to be soft and 'cushy', the vibrations in the vehicle are reduced, but the vehicle may not be too stable during maneuvers such as cornering and lane change. As such, the performance of primary suspensions is always limited by the compromise between ride and handling. Good design of a passive suspension cannot eliminate this compromise but can, to some extent, optimize the opposing goals of comfort and handling.

Passive Suspensions

A passive suspension system is one in which the characteristics of the components (springs and dampers) are fixed. These characteristics are determined by the suspension designer, according to the design goals and the intended application. A passive suspension, such as shown in **Figure 2**, has the ability to store energy via a spring and to dissipate it via a damper. **Figure 2** represents one-quarter of a vehicle, and therefore is commonly referred to as 'quarter-car model'. The mass of the vehicle body (sprung mass) and tire–axle assembly (unsprung mass) are defined respectively by m_b and m_a, with their corresponding displacements defined by x_b and x_a. The suspension spring, k_s, and damper, c_s, are attached between the vehicle body and axle, and the stiffness of the tire is represented by k_t.

The parameters of a passive suspension are generally fixed to achieve a certain level of compromise between reducing vibrations and increasing road holding. Once the spring has been selected, based on the load-carrying capability of the suspension, the damper is the only variable remaining to specify. Low damping yields poor resonance control at the

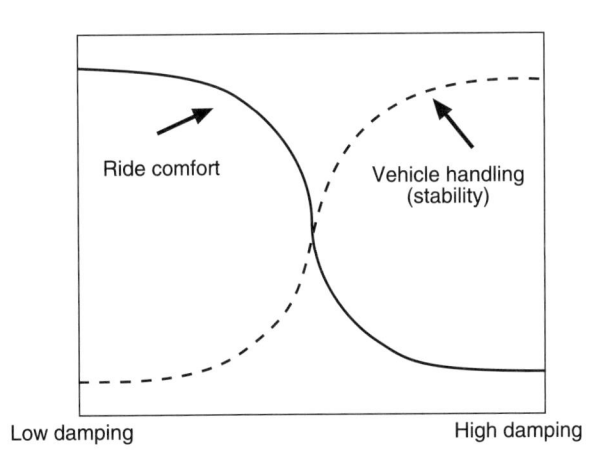

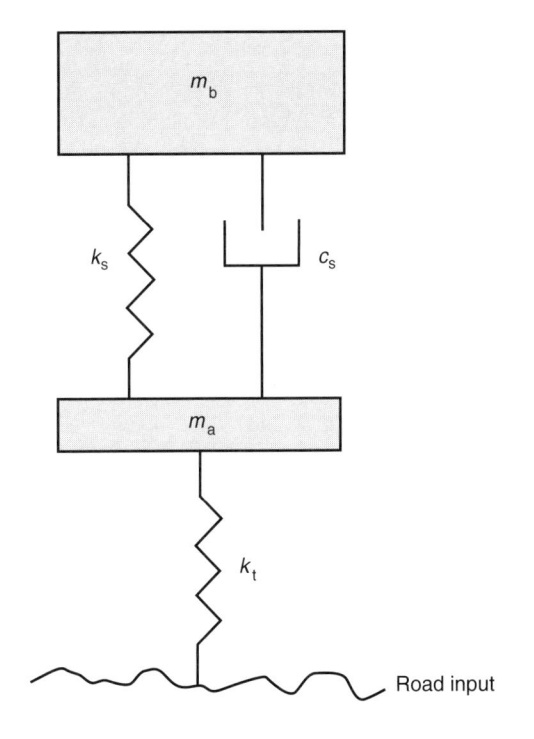

Figure 1 Vehicle vibration and handling compromise due to suspension damping.

Figure 2 Passive suspension.

natural frequencies of the body (sprung mass) and axle (unsprung mass), but provides the necessary high-frequency isolation required for lower vibrations and a more comfortable ride. Conversely, large damping results in good resonance control at the expense of lower isolation from the road input and more vibrations in the vehicle.

Adjustable Suspensions

An adjustable suspension system combines the passive spring element found in a passive suspension with a damper element whose characteristics can be adjusted by the operator. As shown in **Figure 3**, the vehicle operator can use a selector device to set the desired level of damping based on a preference for the subjective feel of the vehicle. This system has the advantage of allowing the operator occasionally to adjust the dampers according to the road characteristics. It is, however, unrealistic to expect the operator to adjust the suspension system to respond to time inputs such as potholes, turns, or other common road inputs.

Active Suspensions

In an active suspension, the passive damper, or both the passive damper and spring, are replaced with a force actuator, as illustrated in **Figure 4**.

The force actuator is able both to add and dissipate energy to and from the system, unlike a passive damper, which can only dissipate energy. With an active suspension, the force actuator can apply force independent of the relative displacement or velocity across the suspension. Given the correct control strategy, this results in a better compromise between ride comfort and vehicle stability as compared to a passive system, as shown in **Figure 5** for a quarter-car model.

Semiactive Suspensions

Semiactive suspensions were first proposed in the early 1970s. In this type of system, the conventional spring element is retained, but the damper is replaced with a controllable damper, as shown in **Figure 6**.

Whereas an active suspension system requires an external energy source to power an actuator that controls the vehicle, a semiactive system uses external power only to adjust the damping levels and operate an embedded controller and a set of sensors. The controller determines the required damping force based on a control strategy, and automatically commands the damper to achieve that damping force. The force achieved by the damper can simply be in two levels: a minimum and a maximum damping force, as shown in **Figure 7A**. This type of system is typically referred to as an on–off (bang–bang) semiactive suspension. Alternatively, the damping force can be adjusted in a range of damping bound by the minimum and maximum damping, as shown in **Figure 7B**. This is commonly called damping. Several studies have shown that one can get nearly all of the benefits of an active suspension with a continuously variable semiactive system, without the complications and costs inherent to active suspensions.

Semiactive Control Methods

In semiactive suspensions, the damping force is adjusted by a controller that may be programmed with any number of control schemes. The control schemes that are commonly used for vehicle suspensions include:

- on–off skyhook control
- continuous skyhook control
- on–off groundhook control
- hybrid control
- fuzzy logic damping control

The succeeding paragraphs describe these control methods in more detail.

On–Off Skyhook Control

In on–off skyhook control, the damper is controlled by two damping values. Illustrated earlier in **Figure 7A**, these are referred to as maximum and minimum damping. The determination of whether the damper is to be adjusted in real time to either its maximum or minimum damping state depends on the product of the relative velocity across the damper and

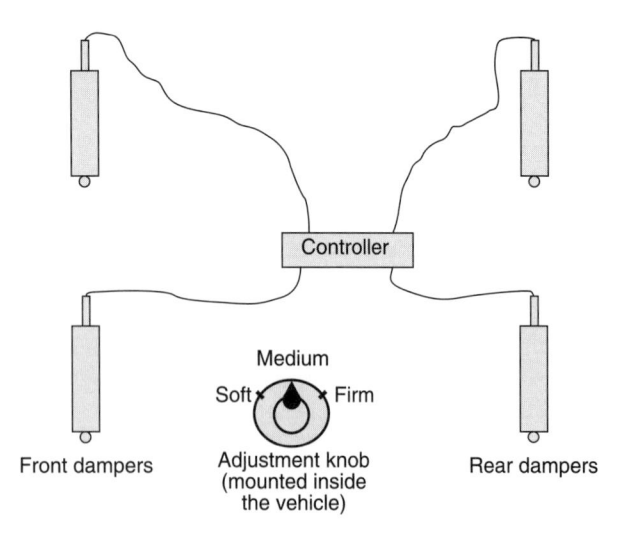

Figure 3 Schematics of a driver-adjustable suspension.

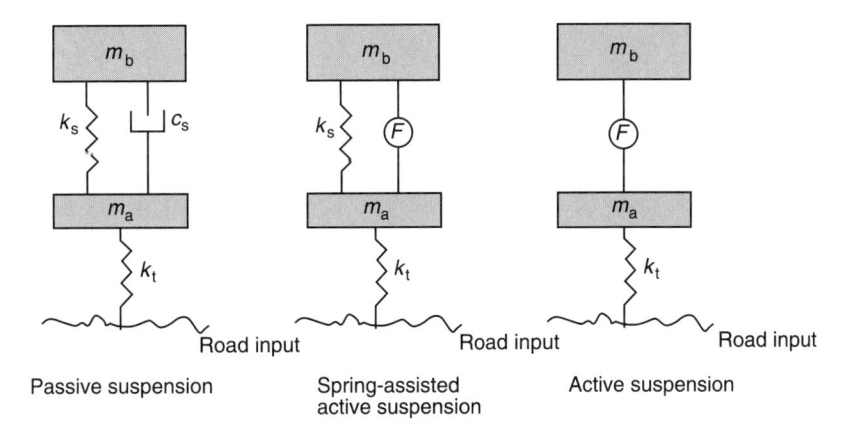

Figure 4 A schematic comparison of passive and active suspensions.

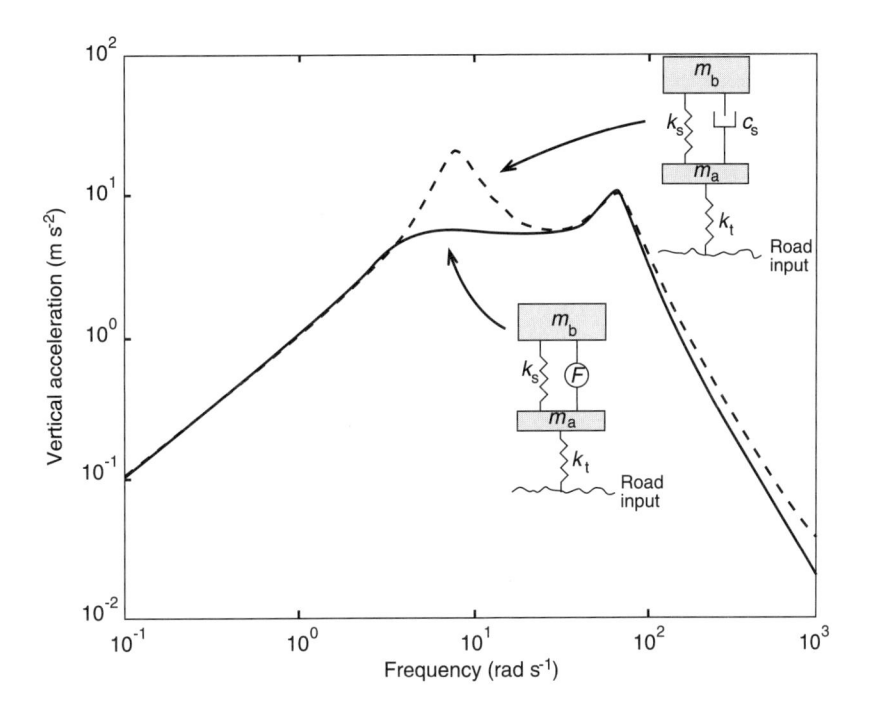

Figure 5 Passive and active suspension comparison.

the absolute velocity of the vehicle body, as illustrated in **Figure 8**. If the product is positive or zero, the damper is adjusted to its maximum state, otherwise the damper is set to the minimum damping. For the quarter-car model of **Figure 6**, this concept is summarized by:

$$v_b v_{rel} > 0, \quad c_s = \text{maximum damping} \quad [1a]$$

$$v_b v_{rel} < 0, \quad c_s = \text{minimum damping} \quad [1b]$$

The variables v_b and v_{rel} represent the body velocity and the relative velocity across the suspension (i.e., between the body and the axle), respectively. The logic of the on–off skyhook control policy is as follows. When the relative velocity of the damper is positive, the force of the damper acts to pull down on the vehicle body; when the relative velocity is negative the force of the damper pushes up on the body. Thus, when the absolute velocity of the vehicle body is negative, it is traveling downwards and the maximum value of damping is desired to push up on the body, while the minimum value of damping is desired to continue pulling down on the body. If, however, the absolute velocity of the body is positive and it is traveling upward, the maximum value of damping is desired to pull down the body, while the minimum

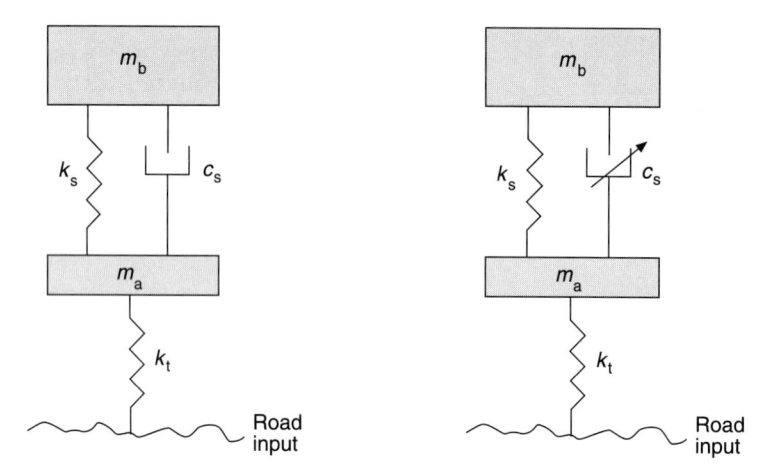

Figure 6 A schematic comparison of passive and semiactive suspensions.

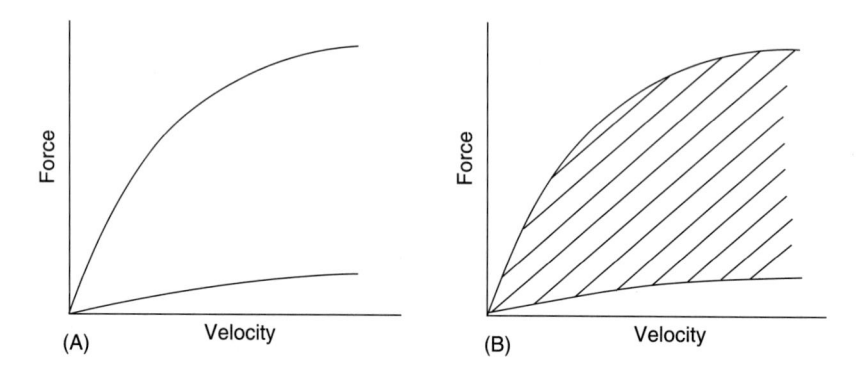

Figure 7 Range of damping values. (A) On–off semiactive; (B) continuously variable semiactive.

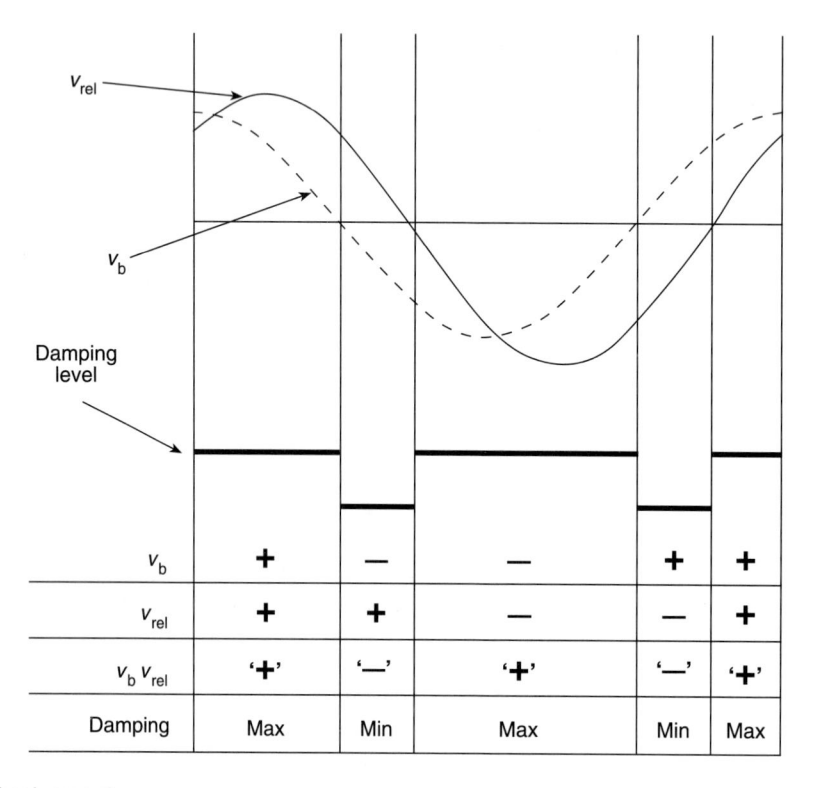

Figure 8 On–off skyhook control.

value of damping is desired to further push the body upward. The on–off skyhook semiactive policy emulates the ideal body displacement control configuration of a passive damper 'hooked' between the body mass and the 'sky' as shown in **Figure 9**; hence, the name 'skyhook damper'.

Continuous Skyhook Control

In continuous control, the damping force is not limited to the minimum and maximum states alone, as was the case for the on–off skyhook control. As illustrated in **Figure 7B** the damper can provide any damping force in the range between the minimum and maximum limits. This will enable the semiactive suspension to achieve a performance that is closer to the ideal skyhook configuration shown in **Figure 9**.

In continuous skyhook control, the low state remains defined by the minimum damping value, while the high state is set equal to a constant gain value multiplied by the absolute velocity of the vehicle body, bounded by the minimum and maximum damping force of the damper:

$$v_b v_{rel} \geq 0, \quad c_s = \max\{\text{Minimum damping}, \min[G \times v_b, \text{maximum damping}]\} \quad [2a]$$

$$v_b v_{rel} < 0, \quad c_s = \text{minimum damping} \quad [2b]$$

The constant gain G in eqn [2a] is selected, empirically, such that the allowable damping range shown in **Figure 7B** is fully utilized.

On–Off Groundhook Control

In on–off groundhook control, the damper is also controlled by two damping values referred to as minimum and maximum damping. The determination of which damping value the damper needs to adjust to is made based on the product of the relative velocity across the suspension and the absolute velocity of the axle. As shown in **Figure 10,** if the product of the relative velocity across the suspension and absolute velocity of the axle is negative or zero, the damper is adjusted to its maximum damping. Otherwise, the damper is adjusted to its minimum value. For the quarter-car model of **Figure 2**, this concept is summarized by:

$$v_a v_{rel} < 0, \quad c_s = \text{maximum damping} \quad [3a]$$

$$v_a v_{rel} > 0, \quad c_s = \text{minimum damping} \quad [3b]$$

The variable v_a represents the axle velocity. The logic for the on–off groundhook control policy is similar to the on–off skyhook control policy, except that control is based on the unsprung mass. When the relative velocity of the damper is positive, the force of the damper acts to pull up on the axle; when the relative velocity is negative, the force of the damper pushes down on the tire mass. When the absolute velocity of the axle is negative, however, it is traveling downward and the maximum value of damping is desired to pull up on the axle, while the minimum value of damping is desired to continue pushing down on the axle. But, if the absolute velocity of the axle is positive and it is traveling upward, the maximum value of damping is desired to push down on the axle, while the minimum value of damping is desired to pull the axle upward. The on–off groundhook semiactive policy emulates the ideal axle displacement control configuration of a passive damper 'hooked' between the axle and the 'ground' as shown in **Figure 11**, hence, the name 'groundhook damper'.

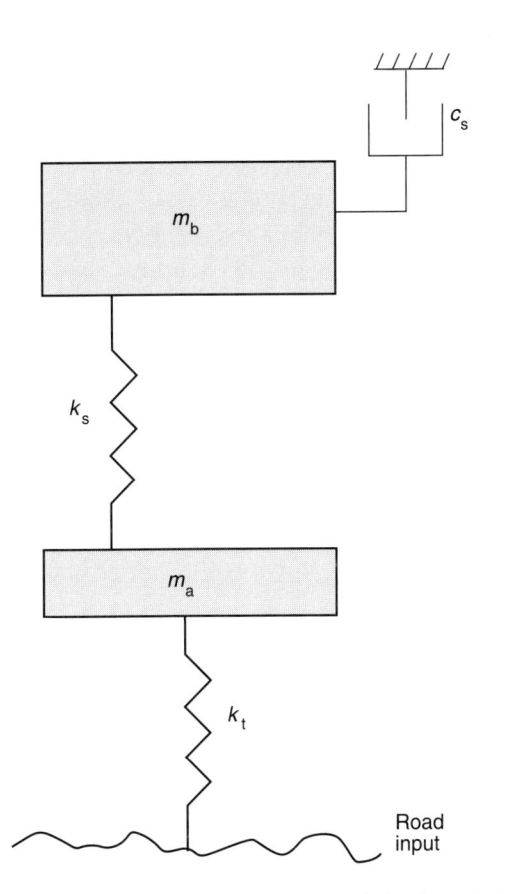

Figure 9 Idealized illustration of semiactive skyhook control.

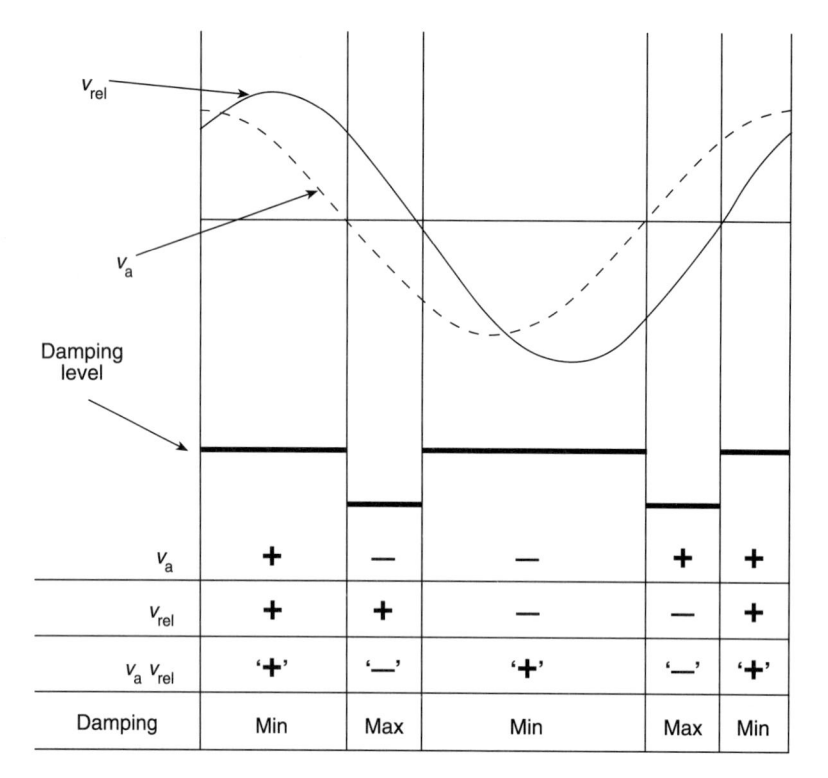

Figure 10 On–off groundhook control.

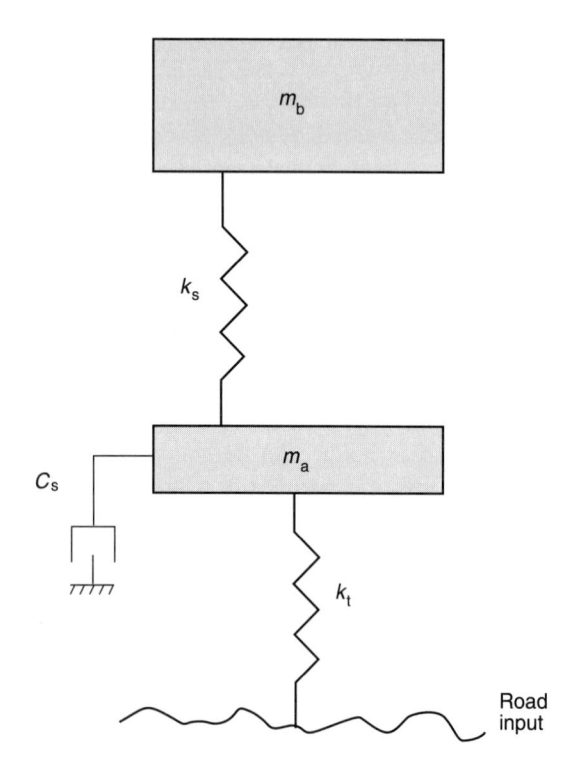

Figure 11 Idealized illustration of semiactive groundhook control.

Hybrid Control

An alternative semiactive control policy, known as hybrid control, combines the concept of skyhook and groundhook control to take advantage of the benefits of both. With hybrid control, the system can be set up to function as a skyhook or groundhook controlled system, or a combination of both, as shown in **Figure 12**. Mathematically, hybrid control policy is a linear combination of the formulation of the skyhook and groundhook control, and can be expressed as:

$$\begin{cases} v_b v_{rel} > 0 & \sigma_{SKY} = v_b \\ v_b v_{rel} < 0 & \sigma_{SKY} = 0 \end{cases}$$

$$F_{SA} = G[\alpha \sigma_{SKY} + (1-\alpha)\sigma_{GND}]$$

$$\begin{cases} -v_a v_{rel} > 0 & \sigma_{GND} = v_a \\ -v_a v_{rel} < 0 & \sigma_{GND} = 0 \end{cases}$$

[4]

The variables σ_{SKY} and σ_{GND} are the skyhook and groundhook components of the damping force, α is the relative ratio between the skyhook and groundhook control, and G is a constant gain that is chosen in the same manner as described earlier for eqn [2a]. When $\alpha = 1$, hybrid control reduces to pure skyhook control, and when $\alpha = 0$, it becomes groundhook control.

ACTIVE CONTROL OF VEHICLE VIBRATION 43

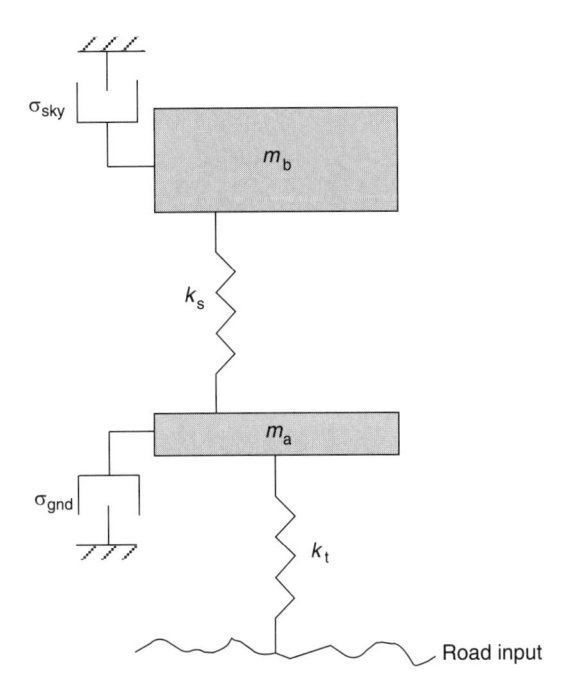

Figure 12 Idealized illustration of semiactive hybrid control concept.

an example of a triangular-shaped membership function used to describe the controller input, x, is shown in **Figure 13**. The number of memberships assigned may be any number of linguistic variables.

The three linguistic variables assigned to the example of **Figure 10** are negative (N), zero (Z), and positive (P), and are correspondingly assigned values of -1, 0, and $+1$. Once the membership function is assigned for each input, the actual fuzzification of the inputs is performed. First, the input is read as a crisp value. Say, for example, that the input, x, is entered as -0.6. A line is drawn from the x-axis at -0.6 to indicate its point of intersection with each component of the membership function, as detailed in **Figure 14**.

In **Figure 14**, the intersection of the x value of -0.6 crosses Z at a weighting function of 0.4 and crosses N at a weighting function of 0.6. In linguistic terms, an input of -0.6 is considered to be 40% zero and 60% negative. These are the fuzzified values of the crisp input, x. Once this process has been performed for all the inputs into the controller, the fuzzification step of the fuzzy logic controller is complete and the rules of the controller are executed.

Fuzzy Logic Control

Fuzzy logic control of semiactive dampers is another example of continuous control illustrated in **Figure 7B**. The output of the controller as determined by the fuzzy logic that may exist anywhere between the minimum and maximum damping states. Fuzzy logic is used in a number of controllers because it does not require an accurate model of the system to be controlled. Fuzzy logic works by executing rules that correlate the controller inputs with the desired outputs. These rules are typically created through the intuition or knowledge of the designer regarding the operation of the controlled system. No matter what the system, there are three basic steps that are characteristic to all fuzzy logic controllers. These steps include the fuzzification of the controller inputs, the execution of the rules of the controller, and the defuzzification of the output to a crisp value to be implemented by the controller. These steps will be explained in succeeding paragraphs.

Step One: Fuzzification

The first step of the fuzzy logic controller is the fuzzification of the controller inputs. This is accomplished through the construction of a membership function for each of the inputs. The possible shapes of these functions are infinite, though very often a triangular or trapezoidal shape is used. For simplicity,

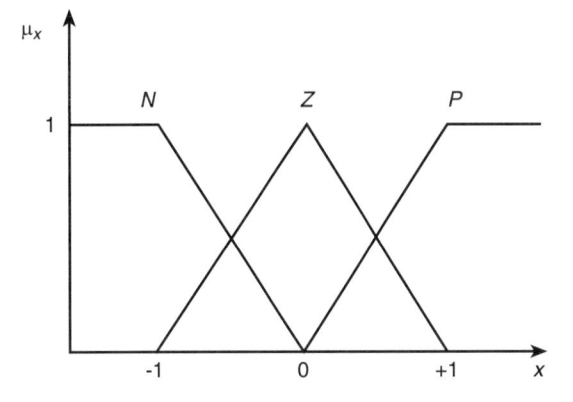

Figure 13 Triangular-shaped input membership function example.

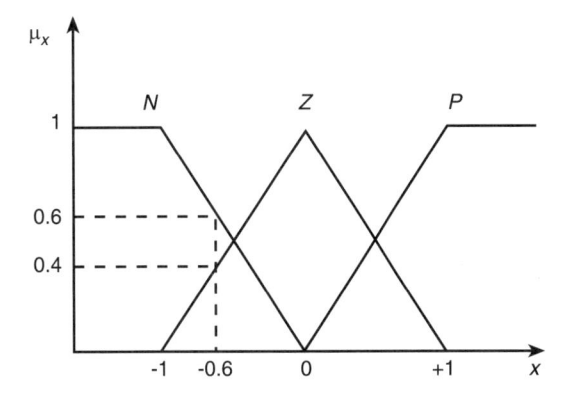

Figure 14 Input intersection with memberships.

Step Two: Execution of Rules

In order to create the rule-base of the controller, the membership function of the output must first be defined. Take, for example, the triangular-shaped membership function for the output, y, shown in **Figure 15**. The linguistic variables assigned to the output are defined as: small (S), medium small (MS), medium (M), medium large (ML), and large (L), and are given the values of 2, 4, 6, 8, and 10, respectively.

Now that the derivation of the controller output membership function is complete, the rule-base of the controller may be created. Suppose two inputs exist, x_1 and x_2, that are each defined by the membership function of **Figure 13**; and one output exisits, y, defined by **Figure 15**. For each possible input combination of x_1 and x_2, a value for the output, y, is linguistically defined. An example of a possible rule-base is shown in **Figure 16**.

The rules of **Figure 16** may also be described as a series of IF–THEN statements. For instance:
IF
$$x_1 = N$$
and
$$x_2 = N$$
THEN
$$y = S$$
and:
IF
$$x_1 = N$$
and
$$x_2 = Z$$
THEN
$$y = MS$$

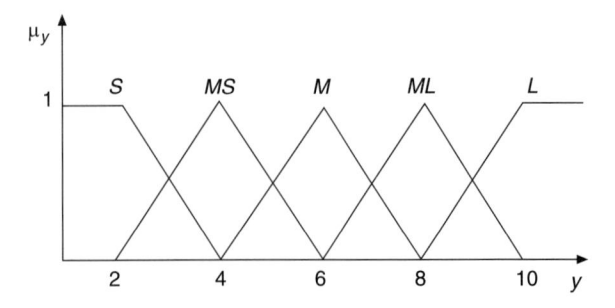

Figure 15 Triangular-shaped output membership function example.

Figure 16 Rule table example.

and so forth, until all the rules of the table are described. The table in **Figure 16** has a total of nine rules.

Now suppose that, after fuzzification, x_1 is found to be 40% zero and 60% positive, and x_2 is found to be 50% zero and 50% positive. In other words: $\mu_{x_1, N} = 0$, $\mu_{x_1, Z} = 0.4$, $\mu_{x_1, P} = 0.6$, $\mu_{x_2, N} = 0.5$, $\mu_{x_2, Z} = 0.5$, and $\mu_{x_2, P} = 0.5$. The rules are applied by assigning the output variable with the minimum (or maximum, depending on the defuzzification method to be applied) weighting function described by the rules. Applying the rule:
IF
$$x_1 = P$$
and
$$x_2 = P$$
THEN
$$y = L$$

with the fuzzified values of x_1 and x_2, the rule becomes:
IF
$$x_1 = 0.6P$$
and
$$x_2 = 0.5P$$
THEN
$$y = \min(0.6, 0.5)L = 0.5L$$

In other words, $y(0.6P, 0.5P)$ is $0.5L$, where the output weighting function $\mu_y(x_1 = P, x_2 = P)$ is 0.5. Applying this procedure to the inputs defined by this example, the fuzzy outputs become:

$$y(0.0N, 0.0N) = 0.0S \quad y(0.0N, 0.5Z) = 0.0MS \quad y(0.0N, 0.5P) = 0.0S$$
$$y(0.4Z, 0.0N) = 0.0M \quad y(0.4Z, 0.5Z) = 0.4M \quad y(0.4Z, 0.5P) = 0.4M$$
$$y(0.6P, 0.0N) = 0.0L \quad y(0.6P, 0.5Z) = 0.5ML \quad y(0.6P, 0.5P) = 0.5L$$

These fuzzy outputs now go through the defuzzification process to determine a single, or crisp, controller output value.

Step Three: Defuzzification

Defuzzification is the process of taking the fuzzy outputs and converting them to a single or crisp output value. This process may be performed by any one of several defuzzification methods. Some common methods of defuzzification include the max or mean–max membership principles, the centroid method, and the weighted average method. The weighted average method was used for this research. It should be noted that the weighted average method is only valid for symmetrically shaped output membership functions.

The weighted average method for finding the crisp output value, y^*, is accomplished by taking the sum of the multiplication of each weighting function, μ_y, with the maximum value of its respective membership value, $\bar{y}$, and dividing it by the sum of the weighting functions. This concept is presented in eqn [5]:

$$y^* = \frac{\sum \left[\mu_y(\bar{y}) \times \bar{y} \right]}{\sum \mu_y(\bar{y})} \quad [5]$$

Applying the weighted average defuzzification method to the example discussed above, the crisp output, $y*$, of the controller is calculated by:

$$y^* = \frac{(0.4M + 0.4M + 0.5ML + 0.5L)}{(0.4 + 0.4 + 0.5 + 0.5)} = 7.67 \quad [6]$$

Recall from **Figure 16** that the maximum values of the linguistic variables M, ML, and L were respectively defined as 6, 8, and 10. A crisp output of 7.67 is thus calculated and applied to the system being controlled. At this point, the current controller inputs are read, and the steps of the fuzzy logic controller are repeated.

Nomenclature

G	constant gain
L	large
M	medium
ML	medium large
MS	medium small
N	negative
P	positive
S	small
y^*	crisp output value
Z	zero

See also: **Damping, active**; **Damping measurement**.

Further Reading

Ahmadian M (1999) On the isolation properties of semiactive dampers. *Journal of Vibration and Control*, 5: 217–232.

Inman DJ (1994) *Engineering Vibration*. Prentice Hall.

Ioannou PA, Sun J (1996) *Robust Adaptive Control*. Prentice Hall.

Jamshidi M, Nader V, Ross TJ (1993) *Fuzzy Logic and Control, Software and Hardware Applications*. New Jersey: Prentice-Hall.

Karnopp D, Crosby MJ, Hardwood RA (1974) Vibration control using semi-active force generators. *Journal of Engineering for Industry*, 96: 619–626.

Milliken WF, Milliken DL (1995) *Race Car Vehicle Dynamics*. Warrendale, PA: Society of Automotive Engineers.

Ross TJ (1995) *Fuzzy Logic with Engineering Applications*. New York: McGraw-Hill.

Widrow B, Stearns SD (1985) *Adaptive Signal Processing*. Prentice Hall.

Widrow B, Walach E (1996) *Adaptive Inverse Control*. Prentice Hall.

Zadeh LA (1965) Fuzzy sets. *Information and Control* 8: 338–353.

ACTIVE ISOLATION

S Griffin, Kirtland AFB, NM, USA

D Sciulli, Centerville, VA, USA

Copyright © 2001 Academic Press

doi:10.1006/rwvb.2001.0191

Control Approaches – Feedforward versus Feedback

Active isolation control approaches can be broadly grouped into one of two categories: feedback and feedforward control. There are also control approaches which include some combination of both. For instance, it is often necessary to consider secondary or feedback paths in feedforward control and, under special circumstances, adaptive feedforward control has an equivalent feedback interpretation.

Feedforward Control

In the case where active isolation of periodic or otherwise predictable excitations is desired, feedforward control is usually the simplest approach to attaining performance goals. In order for feedforward control to be effective, a disturbance sensor signal must be available that is well correlated with the motion of the system to be controlled. In addition, application of feedforward control requires an error sensor and an actuator. The necessary components of feedforward control when the goal is single-degree-of-freedom active isolation of sensitive equipment on a moving base are shown in **Figure 1**.

In the case of **Figure 1**, the actuator provides the secondary force, f_s, and the feedforward controller commands the secondary force in a way that minimizes the error signal, e. The measure of how well correlated the disturbance sensor signal is with the motion of the system to be controlled is the coherence, γ_{dx}, between the error signal, e, and the disturbance signal, x. The explicit relationship between controlled-output power spectral density, O_c, and uncontrolled-output power spectral density, O_u, and coherence is given by:

$$\frac{O_c}{O_u} = 1 - \gamma_{dx}^2 \quad [1]$$

For a linear time-invariant system with no external noise on either the error or disturbance signal, the coherence is 1 at all frequencies and it is theoretically possible to obtain a zero-controlled power spectral density at all frequencies. In realistic implementations, sensor noise and system nonlinearities constrain the actual value of coherence to a value of less than 1. Eqn [1] provides a convenient measure of achievable feedforward control performance before control implementation.

Feedback Control

If the excitation is random or unpredictable and a disturbance sensor cannot make a coherent measurement, feedback control is the favored approach. In order to implement a feedback control approach, it is necessary to have a mathematical model that characterizes the structural dynamics of the system. Unlike feedforward control, there is no predictive measurement of the maximum achievable performance attainable with feedback control. Control performance is adversely affected by model errors and limited by stability margins. A control design and simulation must be accomplished to estimate maximum achievable control performance within stability margins. In general, feedback control requires a control sensor, an error sensor, and an actuator. The necessary components of feedback control when the goal is active isolation of sensitive equipment on a moving base are shown in **Figure 2**. In cases when only local isolation is required or when the components to be isolated can be considered as rigid, the feedback sensor can also serve as the error sensor. In the case of **Figure 2**, the actuator provides the secondary force, f_s and the feedback controller commands the secondary force in a way that minimizes the control signal, y. The error signal, e, provides a secondary measure of control performance.

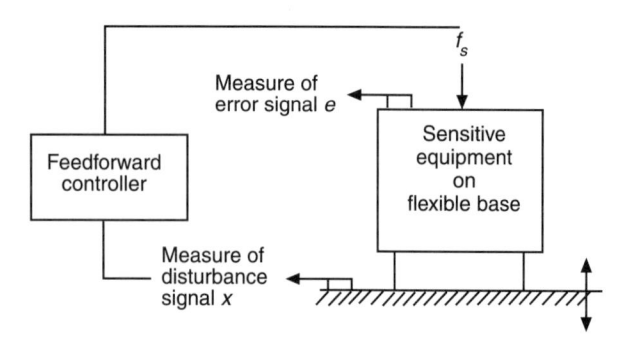

Figure 1 Feedforward control schematic.

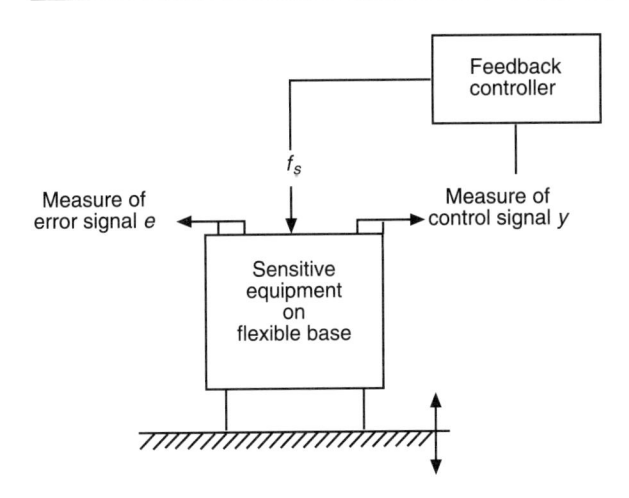

Figure 2 Feedback control schematic.

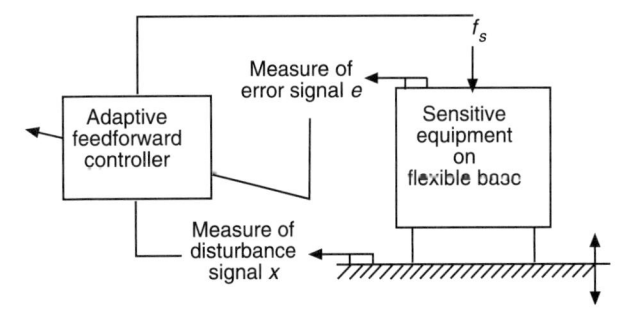

Figure 3 Feedforward control with adaptive path.

Equivalence of Feedback and Feedforward Control

Since the form of the feedforward control filter is dependent on both the path between the disturbance and the system to be controlled and the nature of the disturbance, it is often necessary to make the feedforward control filter-adaptive. Adaptation of the feedforward controller is accomplished by feeding back the error sensor signal to an adaptive feedforward filter, as shown in **Figure 3**, applied to the same system considered in **Figure 1**.

Since the adaptation of the feedforward controller is dependent on the error signal, a feedback path is introduced into the system. For the case of sinusoidal disturbance, there is an equivalent feedback controller which exhibits exactly the same performance characteristics as the feedforward controller.

Actuation Approaches

Many different actuators are available and have found application in both practical and experimental active isolation systems. Traditional active isolation actuators include hydraulic and electromagnetic drives. Force and displacement limits of actuators based on hydraulic drives depend on the energy-producing mechanism and the fluidic circuit which makes up the actuator. As such, hydraulic actuators have been employed in systems as large as earthquake simulators and as small as miniature valves. Linear force and displacement limits of actuators based on electromagnetic voice coils of 100 lb (45 kg) and 0.5 in (12.5 mm) are readily available commercially. Greater performance is possible but may require a custom design. An example of a hydraulic active isolation system is the fully active suspension on high-performance Lotus race cars. In these systems, hydraulic actuators are located at each wheel and force is provided by a reservoir and pump system. The active suspension car is programmed to keep the car parallel to the road at all times, thus minimizing roll and pitch. An example of an electromagnetic active isolation system is the vibration isolation and suppression system VISS experiment. In this system, voice coil actuators form a hexapod mount which actively isolates an infrared telescope from the spacecraft bus.

Active isolation actuators which incorporate active materials include piezoceramic magnetostrictive and magnetorheological materials. **Table 1** compares the actuation properties of these commercially available active materials. Piezoceramic properties are for PZT 5H and magnetostrictive properties are for Terfenol-D. Blocked stress is the product of maximum strain and modulus.

Since piezoceramic materials tend to exhibit relatively high force and low stroke, they are often combined with a hydraulic or mechanical load-coupling mechanism to multiply motion at the expense of applied force.

An example of an active isolation system that uses piezoceramic actuators is the satellite ultraquiet isolation technology experiment (SUITE). In this

Table 1 Comparison of actuation properties of piezoceramic, magnetostrictive, and magnetorheological materials

	Material	Activation	Maximum strain	Blocked stress
Piezoceramic	Ceramic	Electric field	0.13%	79 MPa
Magnetostrictive	Metal alloy	Magnetic field	0.2%	59 MPa
Magnetorheological	Fluid	Magnetic field	Not applicable	Not applicable

experiment, viscoelastic damped piezoceramic actuators form a hexapod mount to isolate a sensitive instrument from a spacecraft bus. An example of an actuator that uses a magnetostrictive material is the Terfenol-D-based reaction mass actuator (RMA) made by SatCon Technologies. This actuator is being investigated for use in helicopter noise and vibration control. Finally, an example of a linear actuator that uses a magnetorheological fluid is the linear pneumatic motion control system made by LORD Corporation. This technology is also available in dampers and brakes.

In applications where maximum power draw is constrained, stored electrical energy is limited, or weight is constrained, it is also necessary to compare the relative efficiency of each of the actuation approaches. To make a fair comparison of efficiency between actuation materials, it is necessary to include control and power electronics and energy storage. This becomes a complicated question, since the choice of power electronics will drastically affect the power or energy efficiency metric that is selected. If the actuator that is incorporated into the active isolation system is considered in terms of its closed-loop electrical impedance, the real part of the impedance is directly related to the mechanical work and mechanical losses that are associated with the actuator. This represents the minimum amount of energy necessary to accomplish active isolation, and the ideal amplifier would supply this energy with 100% efficiency. Taking the example of a piezoceramic actuator, the closed-loop electrical impedance is highly capacitive and contains a large complex component. A linear amplifier is very inefficient at driving a capacitive load, so the total system weight using this approach would have to include sufficient batteries and passive or active cooling to allow for inefficiency.

Conversely, a properly designed switching amplifier is much more efficient at driving a capacitive load and would result in a lower system weight.

See also: **Actuators and smart structures**; **Feedforward control of vibration**.

Further Reading

Anderson E, Evert M, Glaese R *et al.* (1999) Satellite Ultraquiet Isolation Technology Experiment (SUITE): electromechanical subsystems. *Proceedings of the 1999 SPIE Smart Structures and Materials Conference*, 3674: 308–328.

BEI Kimco Magnetics Division. *Voice Coil Actuators: An Applications Guide*. BEI Sensors and Systems Company.

Cobb R, Sullivan J, Das A *et al.* (1999), Vibration isolation and suppression system for precision payloads in space. *Smart Materials and Structures* 8: 798–812.

Fenn R, Downer J, Bushko D *et al.* (1996) Terfenol-D driven flaps for helicopter vibration reduction. *Smart Materials and Structures* 5: 49–57.

Fuller C, Elliot S and Nelson P (1996) *Active Control of Vibration*. San Diego, CA: Academic Press.

Maciejowski J (1989) *Multivariable Feedback Design*. Addison-Wesley.

Ross CF (1980) Active control of sound. PhD thesis, University of Cambridge, England.

Sievers L and von Flotow A. Comparison and extensions of control methods for narrowband disturbance rejection. *IEEE Transactions Signal Processing* 40: 2377–2391.

Warkentin, D (1995) Power amplification for piezoelectric actuators in controlled structures. PhD thesis, Massachusetts Institute of Technology.

Widrow B and Stearns SD (1985) *Adaptive Signal Processing*. Englewood Cliffs, NJ: Prentice-Hall.

Williams D and Haddad W (1997) Active suspension control to improve vehicle ride and handling. *Vehicle System Dynamics* 28: 1–24.

ACTIVE VIBRATION CONTROL

See **ACTIVE CONTROL OF VEHICLE VIBRATION; ACTIVE ISOLATION; ACTIVE VIBRATION SUPPRESSION; ACTUATORS AND SMART STRUCTURES; DAMPING, ACTIVE; FEED FORWARD CONTROL OF VIBRATION; FLUTTER, ACTIVE CONTROL; HYBRID CONTROL.**

ACTIVE VIBRATION SUPPRESSION

D Inman, Virginia Polytechnic Institute and State University, Blacksburg, VA, USA

Copyright © 2001 Academic Press

doi:10.1006/rwvb.2001.0192

Vibration suppression is a constant problem in the design of most machines and structures. Typically vibration reduction is performed by redesign. Redesign consists of adjusting mass and stiffness values or adding passive damping in an attempt to reduce vibration levels to acceptable values. Vibration isolation, vibration absorbers and constrained layer damping treatments are all traditional methods of controlling vibration levels by passive means. Indeed, if passive redesign or add-on techniques allow desired vibration levels to be met, then a passive approach to vibration suppression should be used. However, if passive techniques cannot achieve desired vibration levels within design and operational constraints then an active control method should be attempted as addressed in this article.

Active control consists of adding an applied force to the machine, part or structure under consideration in a known way to improve the response of a system: in this case to suppress vibrations. Control concepts for mechanical systems originated with radar work during World War II and the subject has developed its own special jargon. The object to be controlled (machine, part or structure) is often called the 'plant'. Control methods for linear plants (the only type considered here) can be divided into two main categories: frequency domain methods (also called classical control) and state space methods (also called modern control). Control methods are further classified as open-loop or closed-loop. In an open-loop control, the control force is independent of the response of the system, while in a closed-loop system the control force applied to the system (called the input) depends directly on the response of the system. Closed-loop control can be further divided into feedforward control or feedback control. Feedforward control is most often used in acoustic and wave applications while feedback is most often used in vibration suppression. Here we focus on closed-loop feedback control.

Closed-loop feedback control consists of measuring the output or response of the system and using this measurement to add to the input (control force) to the system. In this way the control input is a function of the output called closed-loop control. There are a variety of methods for choosing the control force and several common methods are presented here at an introductory level.

Single-Degree-of-Freedom Systems

Single-degree-of-freedom systems are used here to introduce the basic concepts for control. Consider first a single-degree-of-freedom system with an applied force (mass normalized), denoted by $u(t)$ rather than the usual notation of $f(t)$ to remind us that the force is a control force derived from an actuator rather then a disturbance. The equation of motion is written as:

$$\ddot{x}(t) + 2\zeta\omega_n \dot{x}(t) + \omega_n^2 x(t) = Ku(t) \quad [1]$$

where ζ and ω_n are the usual damping ratio and undamped natural frequency, respectively. The scalar constant K is called the gain and is a parameter that can be adjusted to change the magnitude of the applied force. Taking the Laplace transform of eqn [1] with zero initial conditions yields:

$$\frac{X(s)}{U(s)} = \frac{K}{s^2 + 2\zeta\omega_n s + \omega_n^2} = KG(s) \quad [2]$$

Here the capital letters indicate the Laplace transform of the variable and s is the complex valued Laplace variable.

Eqn [2] also defines the transfer function $G(s)$ that characterizes the dynamics of the plant. Eqn [2] is represented symbolically by the block diagram of **Figure 1**. Block diagrams have formal rules of manipulation that reflect the equations they represent. Open-loop control design consists of choosing the form of $u(t)$ (or $U(s)$ in the Laplace domain) and adjusting the gain K until a desired response characteristic is obtained. If the control force, $u(t)$, is chosen by looking at the response in $x(t)$ in the time domain, called time domain design, the value of K can be seen to directly adjust the maximum value of the steady-state response as well as the amplitude of the transient terms. Picking the form of $U(s)$ and K in the Laplace domain is referred to as frequency domain design. In frequency domain design, the magnitude and phase plots are examined as $U(s)$ and K are adjusted. Choosing the form of $U(s)$ (or $u(t)$) and the value of K are referred to as designing the control law. Note that compared to passive control there are more parameters to choose from in shaping the response.

As an example of open-loop control design suppose that $x(t)$ corresponds to the angular motion of a shaft and the $G(s)$ is the dynamics of a motor-load system. Suppose also that it is desired to move the shaft from its current position to 20°. Then the control law design consists of choosing K and $u(t)$ to force $x(t) = 20°$. The solution would be to choose $u(t)$ to be a step function and K to have a value determined

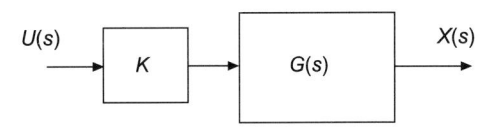

Figure 1 The block diagram of the open-loop system represented by eqn [2].

by setting the magnitude of the steady-state response equal to 20°.

The concept of a closed-loop system is illustrated in **Figure 2**. In the closed-loop case the form of the control law $u(t)$ depends on the response $x(t)$. A common choice for this is called position and derivative (PD) control. This assumes that the velocity and position can be measured and then used to form the control force. This is illustrated in **Figure 2** where the control law $H(s)$ has the following form:

$$H(s) = g_1 s + g_2 \qquad [4]$$

The constant g_1 is called the velocity gain and g_2 is called the position gain. These gains reflect measurement factors and other constants involved in the electronics yet are adjustable to produce a desired response. The equation of motion corresponding to the closed-loop system of **Figure 2** is:

$$\ddot{x}(t) + (Kg_2 + 2\zeta\omega_n)\dot{x}(t) + (Kg_1 + \omega_n^2)x(t) = Ku(t) \qquad [5]$$

In the frequency domain the input–output transfer function becomes:

$$\frac{X(s)}{U(s)} = \frac{K}{s^2 + (Kg_2 + 2\zeta\omega_n)s + (Kg_1 + \omega_n^2)} \qquad [6]$$

Note from eqns [4] and [5] that the effective damping ratio and natural frequency are adjusted by tuning the gains g_1 and g_2. Thus the response of the system to the input $u(t)$ can be adjusted.

To see the impact that PD control has on the design process, define an equivalent damping ratio, ζ^e and natural frequency ω^e. Then from eqn [5]:

$$\omega_n^e = \sqrt{Kg_1 + \omega_n^2} \text{ and } \zeta^e = \frac{Kg_2 + 2\zeta\omega_n}{2\omega_n^e} \qquad [7]$$

Thus the effective frequency and damping ratio can be tuned by adjusting the gains g_1 and g_2, providing two additional design parameters.

The passive design process will chose the best possible values of m, c, and k so that the response has the desired properties. However, these values and hence ω_n and ζ, will be subject to physical constraints such as static deflection, mass limitations, and material properties. Thus the desired frequency and damping ratio may not be obtained because of these limitations. In this case active control is implemented to provide two additional parameters, g_1 and g_2, to adjust in order to produce the desired response. Active control, or PD control as introduced here, allows additional freedom in shaping the response to meet desired specifications. The control gains are not subject to the same sorts of physical constraints that the mass, damping and stiffness are, so that the desired response can often be achieved when passive design fails. The control gains are usually adjusted electronically or hydraulically. However, g_1 and g_2 are expensive, as they require sensor and actuator hardware as well as electronics. The gains are also subject to other physical constraints such as available current and voltage, actuator stroke length and force limitations. Active control provides additional help in cases where passive design fails, but at a cost.

An additional problem with implementing active control can be seen by realizing that an active control device is 'adding' energy to the system to cancel vibration. However, if an error is made in sign or calculation, then the closed-loop system could become unstable. For example if the design requirements call for a reduced frequency, then the coefficient of g_1 would be negative. If in addition, the exact value of m or k were not known, the coefficient of $x(t)$ could be driven negative causing the response to grow without bound and become unstable.

Velocity and position feedback control, also called PD control, is an example of full-state feedback. For a single-degree-of-freedom system there are two state variables: the position and the velocity. The idea of state variables comes from the thought in mathematics that a single nth order differential equation may be represented as n-first-order differential equations. The state variables come from the transformation from nth order to first order. For example eqn [1] can be written as the two, coupled first-order equations:

$$\begin{aligned}\dot{x}_1(t) &= x_2(t) \\ \dot{x}_2(t) &= -2\zeta\omega_n x_2(t) - \omega_n^2 x_1(t) + Ku(t)\end{aligned} \qquad [8]$$

Here the new variables $x_1(t)$ and $x_2(t)$ are called the state variables and are the position and velocity, respectively:

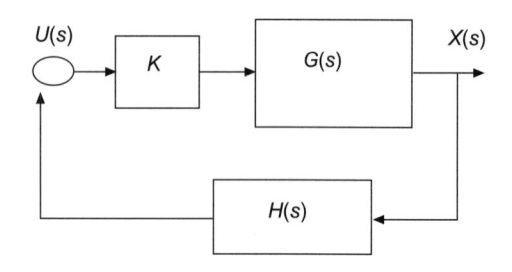

Figure 2 Block diagram for closed-loop feedback control representing eqn [5].

$x_1(t) = x(t)$ and $x_2(t) = \dot{x}(t)$

In this case the use of both state variables in the feedback loop is called full-state feedback because all of the states are used to form the control law. Thus for a single-degree-of-freedom system, full-state feedback and PD control are the same. In this way, the example of PD control may be easily extended to multiple-degree-of-freedom systems. The equations of motion represented in the form of eqn [8] is called the state space formulation and allows a productive connection between dynamics, vibration and control with the mathematics of linear algebra.

Multiple-Degree-of-Freedom Systems

Multiple-degree-of-freedom systems arise in both machines and structures. Many of these have dynamics governed by linear, time invariant, ordinary differential equations that may be written in the following form:

$$\mathbf{M\ddot{q} + D\dot{q} + Kq = f} \qquad [9]$$

Here $\mathbf{M}$, $\mathbf{D}$ and $\mathbf{K}$ are the usual mass, damping and stiffness matrices defined in such a way as to be positive definite and symmetric and $\mathbf{f}$ is the vector of applied forces. Note that $\mathbf{D}$ is used instead of the traditional $\mathbf{C}$, because of the need to interface with the notation of control theory. The vector $\mathbf{q}$ is used to denote the generalized displacement vector (with velocity and acceleration denoted by overdots) commonly used in the Lagrangian formulation of the equations of motion. These coefficient matrices can be derived from Newton's laws in the case of lumped parameter systems, or from some approximation of a set of partial differential equations in the case of a distributed mass system. Eqn [9] is a system of coupled second-order, ordinary differential equations with constant coefficients. Because of the symmetry and definiteness the coefficient matrices for structures, the solution to eqn [9] is stable, perhaps asymptotically stable.

The majority of control theory has developed around systems of first-order differential equations cast in what is called state space form introduced in eqn [8]. For more than one degree-of-freedom, the state equations are also given in matrix form by:

$$\mathbf{\dot{x} = Ax + Bu} \qquad [10]$$

$$\mathbf{y = Cx} \qquad [11]$$

$$\mathbf{u = Gy} \qquad [12]$$

Here $\mathbf{y}$ is the vector representing the measured combinations of states $\mathbf{x}$, called a state vector. The matrix $\mathbf{B}$ denotes positions of the applied control forces contained in the vector $\mathbf{u}$, and is often called the control input matrix. The matrix $\mathbf{C}$ contains the locations where sensors are placed that measure the various states. Control law refers to how the vector $\mathbf{u}$ is chosen. A common choice, called output feedback, is to choose $\mathbf{u}$ according to the law eqn [12] and measurement eqn [11], where $\mathbf{G}$ is a gain matrix chosen to satisfy some performance or stability condition. If the matrix $\mathbf{C}$ reflects measurements of every state (i.e., if $\mathbf{C}$ is an identity matrix), then the control law becomes $\mathbf{u = GCx = Gx}$ and is called full-state feedback.

The various theories and numerical algorithms that are centered on the state space formulation expressed by eqns [10]–[12] overshadow results developed in the physical coordinates of eqn [9]. In fact, numerous high level programs exist for solving control problems associated with the state space formulation. None the less, in order to put equations derived from Newton's law or Euler–Lagrange formulations of eqn [9] into a first-order or state space form, the natural properties of the coefficient matrices (bandedness, definiteness and symmetry) are lost. Furthermore, the order of the vector $\mathbf{x}$, is twice that of the vector $\mathbf{q}$ and the physical importance of the entries may be lost. The relationship between the physical coordinate description given by eqn [9] and the state space formulation given by eqn [10] is simply:

$$\mathbf{x} = \begin{bmatrix} \mathbf{q} \\ \mathbf{\dot{q}} \end{bmatrix} \mathbf{A} = \begin{bmatrix} \mathbf{I} & \mathbf{0} \\ -\mathbf{M}^{-1}\mathbf{K} & -\mathbf{M}^{-1}\mathbf{D} \end{bmatrix} \qquad [13]$$

Note that if there are n degrees-of-freedom, then the vector $\mathbf{x}$ is $2n \times 1$ and the state matrix $\mathbf{A}$ is $2n \times 2n$. Also note that the identity matrix $\mathbf{I}$ is $n \times n$ and the $\mathbf{0}$ here denotes and $n \times n$ matrix of zeros. Thus given the physical model in eqn [9], the state space model can be easily constructed.

Modal Control

Modal control is a popular method of providing control laws for structures because it appeals to the simplistic and intuitive notions of single-degree-of-freedom systems. In addition, much of structural dynamics analysis and testing centers around modal methods making modal approaches a natural place to start understanding control. Mathematically, modal control derives its power from consideration of matrix decoupling transformations. Modal control methods developed in two distinct disciplines: control theory and structural dynamics. Basically the use of modes in control theory is different from that in structural dynamics and hence from that in structural control. In control, the theory of modal control refers to collapsing the problem into a series of first-order, ordinary differential equations that are decoupled and hence easily solved. Modal methods in control theory center around the Jordan canonical form of the state space matrix given by eqn [13]. In structural dynamics, modal control refers to a series of decoupled second-order ordinary differential equations, motivated by an understanding of single-degree-of-freedom/oscillators which provides both a set of decoupled equations to analyze and the physical intuition and experience of (structural) modal analysis.

The goal of both the state space (first-order) and the physical space (second-order) modal control approaches is to essentially find a nonsingular, coordinate transformation that takes the equations of motion into a canonical, or decoupled form. In the state space this amounts to looking for the Jordan form of a matrix and the resulting system will be completely decoupled as long as there are no repeated (nonsimple) eigenvectors. That is the system of eqn [10] can be written as a set of first-order decoupled equations by simply applying the Jordan transformation. In the physical space, the attempt to look for a transformation that decouples the equations of motion amounts to looking at whether or not the mass weighted eigenvectors of the stiffness matrix can be used as system eigenvectors. The existence of modally decoupled control laws depends on the conditions under which two matrices share common eigenvectors. This condition is that two symmetric matrices share the same eigenvectors if and only if they commute.

First, let $\mathbf{f}(t)$ become the control-input vector $\mathbf{B}\mathbf{u}(t)$. Then assume that the mass matrix $\mathbf{M}$ is nonsingular. Then use the transformation:

$$\mathbf{q} = \mathbf{M}^{-1/2}\mathbf{r} \qquad [14]$$

where the exponent refers to the positive definite matrix square root. (Numerically it is better to use the Cholesky factors of the mass matrix rather than the matrix square root.) Substitution of eqn [14] into eqn [9] and multiplying by $\mathbf{M}^{-1/2}$ from the left yields:

$$\ddot{\mathbf{r}} + \tilde{\mathbf{D}}\dot{\mathbf{r}} + \tilde{\mathbf{K}}\mathbf{r} = \mathbf{M}^{-1/2}\mathbf{B}\mathbf{u}(t) \qquad [15]$$

Here, the coefficients are:

$$\tilde{\mathbf{D}} = \mathbf{M}^{-1/2}\mathbf{D}\mathbf{M}^{-1/2} \text{ and } \tilde{\mathbf{K}} = \mathbf{M}^{-1/2}\mathbf{K}\mathbf{M}^{-1/2} \qquad [16]$$

which we call the mass normalized damping and stiffness matrices, respectively. Two symmetric matrices commute if and only if they have the same eigenvectors. These eigenvectors form a complete set because the matrices in eqn [16] are carefully constructed to be symmetric. Thus the matrix $\mathbf{P}$ formed by combining the normalized mode shapes as its columns becomes a modal matrix which will decouple both of the coefficient matrices in eqn [15]. To see this, substitute $\mathbf{r} = \mathbf{P}\mathbf{z}$ into eqn [15] and premultiply by $\mathbf{P}^T$ to obtain the n modal equations

$$\ddot{z}_i + 2\zeta_i\omega_i\dot{z}_i + \omega_i^2 z_i = (\mathbf{P}^T\mathbf{M}^{-1/2}\mathbf{B}\mathbf{u})_i \qquad [17]$$

Here, ζ_i and ω_i are the modal damping ratios and natural frequencies, respectively, and the subscript refers to the ith element of the vector $\mathbf{z}$. This decoupled form results if and only if the matrices in eqn [9] commute, or if and only if

$$\mathbf{D}\mathbf{M}^{-1}\mathbf{K} = \mathbf{K}\mathbf{M}^{-1}\mathbf{D} \qquad [18]$$

This last condition is a necessary and sufficient condition for the system of eqn [9] to possess normal modes (which are real valued) and to decouple as indicated in eqn [15]. Thus with condition [18] satisfied, the system may be analyzed using modal analysis. Furthermore, the undamped mode shapes are, in fact, also the physical mode shapes of the damped system.

The idea behind modal control is to assume that the damping in the system is such that eqn [18] is satisfied, use the transformation $\mathbf{P}$ to decouple the equations and then choose the individual control inputs $(\mathbf{P}^T\mathbf{M}^{-1/2}\mathbf{u}(t))_i$ to shape each individual mode to have a desired behavior. In particular, suppose that state feedback is chosen as the form of the control law. Then the closed loop system will have the form:

$$\mathbf{M}\ddot{\mathbf{q}} + \mathbf{D}\dot{\mathbf{q}} + \mathbf{K}\mathbf{q} = -\mathbf{G}_v\dot{\mathbf{q}} - \mathbf{G}_p\mathbf{q} \qquad [19]$$

where G_v and G_p are gain matrices to be chosen in the control design. The gain matrices are now chosen to force the closed-loop system to have the desired modal behavior. Under the modal transformation, and using the idea of full-state feedback, the modal equations all reduce to the form given in eqn [7] for the singe-degree-of-freedom case. This procedure can also be carried out in the state space coordinate system of eqns [10]–[12].

For simplicity, consider the single-input single-output case (siso) which collapses the input matrix $\mathbf{B}$ to the vector $\mathbf{b}$, and the measurement matrix $\mathbf{C}$ to the vector $\mathbf{c}$. Furthermore, assume that gain matrix is the identity matrix ($\mathbf{G} = \mathbf{I}$). Let $\mathbf{P}$ denote the matrix of eigenvectors that transforms the state matrix $\mathbf{A}$ into a diagonal matrix of eigenvalues (denoted Λ) and perform the transformation $\mathbf{x} = \mathbf{P}\mathbf{z}$ on eqns [10] and [11]. Substitution of eqn [12] into eqn [11] and substituting the result into eqn [10], yields:

$$\dot{\mathbf{z}}(t) = \Lambda \mathbf{z}(t) + \mathbf{b}\mathbf{c}^T \mathbf{z}(t) \qquad [20]$$

Here the vector $\mathbf{z}$ is the modal state vector and the matrix Λ is a diagonal (or Jordan form) matrix of the (complex) eigenvalues of the system. If the structure is underdamped, the eigenvalues appear in complex conjugate pairs of the form:

$$\begin{aligned}\lambda_i &= -\zeta_i \omega_i - \omega_i \sqrt{(1-\zeta_i^2)}\mathrm{j} \\ \lambda_{i+1} &= -\zeta_i \omega_i + \omega_i \sqrt{(1-\zeta_i^2)}\mathrm{j}\end{aligned} \qquad [21]$$

where j is the imaginary unit. The elements of the matrix $\mathbf{b}\mathbf{c}^T$ are now chosen to modify the eigenvalues of the closed loop system to provide the desired response. To see this possibility note that eqn [20] can be written as:

$$\dot{\mathbf{z}}(t) = \underbrace{\left(\Lambda + \mathbf{b}\mathbf{c}^T\right)}_{\hat{\mathbf{A}}} \mathbf{z}(t) \qquad [22]$$

The dynamics of the closed-loop system are characterized by the closed-loop state matrix:

$$\hat{\mathbf{A}} = \Lambda + \mathbf{b}\mathbf{c}^T \qquad [23]$$

Various methods of choosing the values of $\mathbf{b}\mathbf{c}^T$ exist in the control literature.

Controllability and Observability

An important aspect of controlling structures is the placement of sensors and actuators. This can be seen by examining eqn [17] and focusing on the control-input term:

$$(\mathbf{P}^T \mathbf{M}^{-1/2} \mathbf{B} \mathbf{u})_i \qquad [24]$$

If this term is small, not much energy is being put into the ith mode. This means that the control effort $\mathbf{u}(t)$ will have a very difficult time changing the ith mode. On the other hand, if the value of the term given in eqn [24] is large, the choice of $\mathbf{u}(t)$ will have a great effect on the ith mode. This leads to the concept of controllability. Basically, if the value of eqn [24] is zero for some value of i, then the ith mode is said to be uncontrollable because the applied control has no effect on this mode. On the other hand, if it is non-zero, then the ith mode is said to be controllable. If every mode is controllable then the system is said to be controllable. Most structures are controllable by the nature of their connectivity. However, the real issue is how controllable is a given mode. The larger the value of eqn [24] the more likely the control effort will be able to change the desired mode to its desired state. The concept of controllability is defined without reference to modes, but modes are used here to connect the idea to vibrations. Note that if $\mathbf{B}$ happens to be orthogonal to a column of the modal transformation, then that particular mode will not be controllable. This corresponds to the concept known to most structural test engineers that exciting a mode by placing a shaker at a node of a mode will not excite the mode at all. Thus one significant control design task is the placement of actuators so that the modes of interest have high values of modal controllability.

A similar concept called observability applies to sensor placement. The locations of the sensors, as captured by the matrix $\mathbf{C}$, determine if a given mode can be measured by the sensors. If the mode cannot be measured it is also difficult to control.

Spillover

Another significant consideration in controlling the vibration of structures is that of spillover. Spillover addresses the issue of designing a control law to suppress the response of a group of modes and neglecting other modes. Such is the case if the model used in the control design is a reduced-order model as is common practice in structural dynamics. Control and observation spillover can be explained by a simple partitioning of eqn [20] into two groups of modes at index k. This yields:

$$\begin{bmatrix}\dot{\mathbf{z}}_k \\ \dot{\mathbf{z}}_{2n-k}\end{bmatrix} = \begin{bmatrix}\Lambda_k & 0 \\ 0 & \Lambda_{2n-k}\end{bmatrix}\begin{bmatrix}\mathbf{z}_k \\ \mathbf{z}_{2n-k}\end{bmatrix} \\ + \begin{bmatrix}\mathbf{b}_k \mathbf{c}_k^T & \mathbf{b}_k \mathbf{c}_{2n-k}^T \\ \mathbf{b}_{2n-k}\mathbf{c}_k^T & \mathbf{b}_{2n-k}\mathbf{c}_{2n-k}^T\end{bmatrix}\begin{bmatrix}\mathbf{z}_k \\ \mathbf{z}_{2n-k}\end{bmatrix} \qquad [25]$$

Here, the vectors $\mathbf{P}^{-1}\mathbf{b}$ and $\mathbf{c}^T\mathbf{P}$ are partitioned according to:

$$\mathbf{P}^{-1}\mathbf{b} = \begin{bmatrix}\mathbf{b}_k \\ \mathbf{b}_{2n-k}\end{bmatrix} \quad \text{and} \quad \mathbf{c}^T\mathbf{P} = \begin{bmatrix}\mathbf{c}_k^T & \mathbf{c}_{2n-k}^T\end{bmatrix} \qquad [26]$$

Note that while the state matrix transforms into a diagonal form, the second matrix containing the control hardware parameters has clear off-diagonal terms. If $\mathbf{b}_{2n-k}$ and $\mathbf{c}_{2n-k}$ both happen to be zero, then the closed-loop system also decouples into modal form. Unfortunately, if these are not zero, the off-diagonal terms couple the first k modes with the last $2n - k$ modes, spoiling the modal form. The term $\mathbf{b}_{2n-k}$ gives rise to modal coupling due to control action (called control spillover) and the term $\mathbf{c}_{2n-k}$ gives rise to modal coupling due to observation or measurement (called observation spillover). The coupled term $\mathbf{b}_{2n-k}\mathbf{c}_{2n-k}^T$ represents the control action that 'spills over' into the uncontrolled modes $(2n - k)$ in the situation where the control law is designed considering only the first k modes. If this term is large, then the performance, or even stability, of the closed-loop system will be lost. While this explanation is made in state variable form, the argument also holds true in the physical coordinates.

Eqn [26] also provides an opportunity to comment on model reduction. One simple model reduction method is to transform the system into modal form and then drop the higher modes $(2n - k)$. The question of model reduction then becomes one of asking how many modes should we keep in the model (i.e., how large should k be)? For open-loop analysis this question may be answered satisfactorily by looking at the modal participation factors and picking k accordingly. The difficulty in closed-loop control is that a satisfactory choice of k in the open-loop case may not be satisfactory in the closed-loop case because of the influence of the control and observation spillover terms. Thus, it is desired to find means of control that drive these coupling and spillover terms to zero. Using a large number of sensors and actuators can eliminate or minimize the spillover terms.

Pole Placement

State feedback refers to requiring the feedback control law to be a linear combination of only the state variables (velocity and position). Once state feedback is chosen as the form for the control force, there are several other possible control laws. One possibility is to use pole placement. Pole placement in structural control consists of determining desired modal damping ratios and natural frequencies. Once these are specified, eqns [21] are used to change these into eigenvalues, called the desired eigenvalues and denoted by λ_{di}. These are then formed into a diagonal matrix denoted by $\mathbf{\Lambda}_D$. Let $\mathbf{P}$ denote the matrix of eigenvectors that transforms the state matrix $\mathbf{A}$ into a diagonal matrix of eigenvalues (denoted $\mathbf{\Lambda}$) and perform the transformation $\mathbf{x} = \mathbf{Pz}$ on eqns [10] and [11]. Substitution of eqns [12] into [11] and substituting the result into eqn [10], yields:

$$\dot{\mathbf{z}}(t) = (\mathbf{\Lambda} + \mathbf{BGC}^T)\mathbf{z}(t) \qquad [27]$$

which is in modal coordinates. The goal of the pole placement technique is to force the eigenvalues, or poles, of eqn [27] to have the desired eigenvalues by choosing, $\mathbf{B}$, $\mathbf{G}$, and $\mathbf{C}$ so that:

$$\begin{aligned}(\mathbf{\Lambda} + \mathbf{P}^T\mathbf{BGC}^T\mathbf{P}) &= \mathbf{\Lambda}_D \quad \text{or} \\ \mathbf{BGC}^T &= \mathbf{P}(\mathbf{\Lambda}_D - \mathbf{\Lambda})\mathbf{P}^T\end{aligned} \qquad [28]$$

The last expression guides the designer for the choice of the gain matrix ($\mathbf{G}$) and the placement of the sensors ($\mathbf{C}$) and the actuators ($\mathbf{B}$). Here $\mathbf{P}$ and $\mathbf{\Lambda}$ are known from the open-loop state matrix, and $\mathbf{\Lambda}_D$ is given by the desired modal information. For full-state feedback, the matrices $\mathbf{C}$ and $\mathbf{B}$ may be taken as the identity matrix and eqn [28] yields a simple solution for the gain matrix that will provide a closed-loop system with exactly the desired damping ratios and natural frequencies.

The calculation for the gain matrix looks straightforward, but solving eqn [28] for $\mathbf{G}$ can be difficult unless $\mathbf{B}$ and $\mathbf{G}$ are nonsingular and well conditioned (requiring full-state feedback). There are a variety of pole placement algorithms each solving specialized problems and dealing with the difficulty of using a small number of sensors and actuators ($\mathbf{B}$ and $\mathbf{C}$ singular, or output feedback). Another difficulty is partial pole placement. This refers to the case where it is desired to assign only a few of the poles and to leave the other poles alone, as is often the case in design. It should also be pointed out that the larger the difference between the open-loop poles, $\mathbf{\Lambda}$, and the desired poles, $\mathbf{\Lambda}_D$, the more gain and hence the more power and force required by the controller.

A more radical version of pole placement is called eigenstructure assignment. Eigenstructure assignment refers to a control law that aims to produce a closed-loop system with both desired eigenvalues (poles or natural frequencies and damping ratios) and desired eigenvectors (mode shapes). The calculation for the gain matrix $\mathbf{G}$ to perform eigenstructure assignment is not presented here. Rather it is important to note that one cannot exactly specify any desired mode shape for the closed-loop system, but rather the assigned mode shapes are somewhat determined by the nature of the original state matrix, $\mathbf{A}$. Thus, while one can completely specify desired natural frequencies and damping ratios and compute a closed-loop control providing these specified poles, one cannot do the same for desired mode shapes. Rather, the desired mode shapes must be modified according to the original system dynamics. This is not surprising to structural engineers who have worked in model updating as it is well known there that the analytical model and the test data must be fairly close before updating is successful.

Optimal Control

The question of whether or not there is a best control system is somewhat answered by methods of optimal control. Optimal control is a closed-loop method devised using variational methods to find a control law $\mathbf{u}(t)$ that minimizes a quadratic 'cost function' containing the response of the system. One simple approach to optimal control is to compute the control, $\mathbf{u}(t)$ that minimizes the square of the response, $\mathbf{x}(t)$, subject to the constraint that the response satisfies the equation of motion. This constrained optimization problem produces controls that are independent of the modal model of the structure. The optimization procedure also allows one to include the control force as part of the cost function. In this case the cost function becomes:

$$J = \int_0^t \left(\mathbf{x}^T(t)\mathbf{Q}\mathbf{x}(t) + \mathbf{u}^T(t)\mathbf{R}\mathbf{u}(t) \right) dt \quad [29]$$

Here the matrices $\mathbf{Q}$ and $\mathbf{R}$ are weighting matrices that are chosen by the designer to balance the effect of minimizing the response with that of minimizing the control effort. Optimal control with this sort of cost function is referred to as the linear quadratic regulator (LQR) problem. Specifically the LQR problem is to minimize eqn [29] subject to eqn [10] and initial conditions. If full-state feedback is used (that is if $\mathbf{u} = -\mathbf{G}\mathbf{x}$), then the solution to the LQR problem leads to:

$$\mathbf{u}(t) = -\mathbf{R}\mathbf{B}^T S(t) \mathbf{x}(t) \quad [30]$$

Here $S(t)$ is the solution to the Riccati equation:

$$\mathbf{Q} - S(t)\mathbf{B}\mathbf{R}^{-1}S(t) + \mathbf{A}^T S(t) + S(t)\mathbf{A} + \frac{dS(t)}{dt} = 0 \quad [31]$$

which must be solved numerically. Steady-state solutions are found by assuming the derivative of $S(t)$ to be zero and that the system is controllable (more precisely: that pair A, B are controllable). The restrictions are that the matrix $\mathbf{R}$ must be symmetric and positive definite and that the matrix $\mathbf{Q}$ must be symmetric and positive semidefinite. Fortunately there are several numerical algorithms for solving the steady-state-matrix Riccati equation.

Optimal control is the only control methodology discussed here that allows the designer some clear method of restricting the amount of control energy expended while designing the controller. The other methods, such as pole placement, can potentially lead to solutions that call for more control effort than is practically available. The $\mathbf{R}$ term in the LQR cost function allows the excessive use of control energy to be penalized during the optimization. The designer adjusts $\mathbf{Q}$ to reduce the vibration to acceptable levels and adjusts $\mathbf{R}$ to reduce the control effort to meet available control forces. Commercial codes are available for computing the optimal control given the equations of motion, sensor locations and actuator locations.

Compensators

Compensation is a technique of essentially adding dynamics to a system to compensate for its open-loop performance. The idea is very similar to that of a passive vibration absorber. In the case of a vibration absorber an additional spring-mass system is added to the plant. This additional degree-of-freedom is used to absorb a harmonic input, causing the base system or plant to remain free of vibration. In control theory, the compensator, as it is called, adds dynamics to a system to change its performance to obtain a more desirable response.

A good example of the difference between state feedback and compensation is to consider the vibration isolation design. It is well known that isolators are designed by choosing the stiffness to be low enough to avoid the resonance peak yet high enough to provide low static deflection. Increased damping, which lowers the peak, unfortunately raises the magnitude in the region of isolation. Thus the designer is often left with the dilemma of adding damping to improve shock isolation and at the same time reducing the performance of the vibration (steady-state) isolation. If state feedback is used in an attempt to provide an active isolator that provides good performance across the entire spectrum, the same difficulty arises because state feedback for a single-degree-of-freedom system can only change damping and stiffness. To solve this problem, new dynamics must be added to the system. These new dynamics are effectively a compensator.

One popular example of compensation in structural control is called positive position feedback (PPF). To illustrate the formulation consider the single-degree-of-freedom system (or alternately, a single mode of the system):

$$\ddot{x} + 2\zeta\omega_n \dot{x} + \omega_n^2 x = bu \quad [32]$$

where ζ and ω_n are the damping ratio and natural frequency of the structure, and b determines the level of force into the mode of interest. The PPF control is implemented using an auxiliary dynamic system (compensator) defined by:

$$\ddot{\eta} + 2\zeta_f \omega_{nf} \dot{\eta} + \omega_{nf}^2 \eta = \omega_{nf}^2 x$$
$$u = \frac{g}{b} \omega_{nf}^2 \eta \quad [33]$$

Here ζ_f and ω_{nf} are the damping ratio and natural frequency of the controller and g is a constant. The particular form of eqn [33] is that of a second-order compensator. The idea is to choose the PPF frequency and damping ratio so that the response of this mode has the desired behavior. Combining eqns [32] and [33] gives the equations of motion in their usual form, which are, assuming no external force:

$$\begin{bmatrix} \ddot{x} \\ \ddot{\eta} \end{bmatrix} + \begin{bmatrix} 2\zeta\omega_n & 0 \\ 0 & 2\zeta_f\omega_{nf} \end{bmatrix} \begin{bmatrix} \dot{x} \\ \dot{\eta} \end{bmatrix} + \begin{bmatrix} \omega_n^2 & -g\omega_{nf}^2 \\ -\omega_{nf}^2 & \omega_{nf}^2 \end{bmatrix} \begin{bmatrix} x \\ \eta \end{bmatrix} = \begin{bmatrix} 0 \\ 0 \end{bmatrix} \quad [34]$$

This is a stable closed-loop system if the 'stiffness' matrix is positive definite. That is if the eigenvalues of the displacement coefficient matrix are all positive. This condition is satisfied if:

$$g\omega_{nf}^2 < \omega_n^2 \quad [35]$$

Notice that the stability condition only depends on the natural frequency of the structure, and not on the damping or mode shapes. This is significant in practice because when building an experiment, the frequencies of the structure are usually available with a reasonable accuracy while mode shapes and damping ratios are much less reliable.

This stability property is also important because it can be applied to an entire structure eliminating spillover by rolling off at higher frequencies. That is, the frequency response of the PPF controller has the characteristics of a low-pass filter. The transfer function of the controller is:

$$\frac{U(s)}{X(s)} = \frac{g\omega_{nf}^4}{b\left(s^2 + 2\zeta_f\omega_{nf} s + \omega_{nf}^2\right)} \quad [36]$$

illustrating that it rolls off quickly at high frequencies. Thus the approach is well suited to controlling a mode of a structure with frequencies that are well separated, as the controller is insensitive to the unmodeled high frequency dynamics. Thus if cast in state space, the term $\mathbf{b}_{2n-k}$ in eqn [25] is zero and no spillover results. In many ways, the PPF active feedback control scheme is much like adding a vibration absorber that targets each mode of interest.

Stability and Robustness

Two important considerations in designing a control system are those of stability and robustness. Most of the structures and machines that mechanical engineers deal with are open-loop stable. However, adding control to such systems has the potential to make the closed-loop system unstable. This usually occurs because the physical parameters in the system and/or the disturbance forces are not known exactly. This uncertainty leads in a natural way to the concept of stability and performance robustness. As an example of robustness, consider the simple suspension design that produces a damping ratio resulting in a smooth ride and not too much static deflection. Now if the vehicle is overloaded, then the damping ratio decreases and the static deflection increases, both serving to ruin the performance (ride). The amount the vehicle can be loaded (mass change) before acceptable performance is lost is a measure of the performance robustness of the design.

Stability robustness refers to how much the parameters of the system may be changed before stability is lost. For instance, suppose a control system is designed to reduce damping in the system in order to provide a really fast responding machine. The effective control reduces damping based on the amount of natural damping in the system. However, if that amount of natural damping is greatly overestimated, it is possible to design a controller that will subtract too much damping and render the closed-loop system unstable. The amount of damping in the original system then becomes a measure of stability robustness. Stability robustness indicates the amount of error that can be tolerated in the system's parameters before the system is in danger of losing stability.

As an example of one robustness condition, consider a control law that regulates the difference between the closed-loop response and the desired response of a system. A good measure of performance of this control system is then whether or not the steady-state error is zero. The system is said to be robust if there exists a control that regulates the system with zero steady-state error when subjected to variations in the state matrix $\mathbf{A}$, the control input matrix $\mathbf{B}$, or the sensor matrix $\mathbf{C}$.

An example of a stability robustness condition consider the closed-loop system with full-state feedback defined by:

$$\dot{\mathbf{x}} = \underbrace{(\mathbf{A} + \mathbf{BG})}_{\mathbf{A}'}\mathbf{x} \qquad [37]$$

Next suppose that the closed-loop matrix $\mathbf{A}'$ is subject to some error (either in $\mathbf{A}$, $\mathbf{B}$ or $\mathbf{G}$), defined by the matrix of errors $\mathbf{E}$. The exact value of $\mathbf{E}$ is not known, but it is reasonable to assume that bounds on the elements of $\mathbf{E}$ are known. Let each element of the uncertainty matrix be bounded in absolute value by the same number ε, so that:

$$|\mathbf{E}_{ij}| < \varepsilon \qquad [38]$$

Then it can be shown that the closed-loop system of eqn [37] will be asymptotically stable if:

$$\varepsilon < \frac{1}{2n}\left(\frac{1}{\sigma_{\max}(\mathbf{F})}\right) \qquad [39]$$

Here n denotes the number of degrees of freedom in the system and $\sigma_{\max}(\mathbf{F})$ is the largest singular value of the matrix $\mathbf{F}$. The matrix $\mathbf{F}$ is a solution of the Liapunov matrix equation

$$\mathbf{A}'\mathbf{F} + \mathbf{F}\mathbf{A}' = -2\mathbf{I} \qquad [40]$$

Here the matrix $\mathbf{I}$ is the $2n \times 2n$ identity matrix. Numerical algorithms must be used to solve the Liapunov equation. In this way the maximum allowable variation in either $\mathbf{A}$ or $\mathbf{B}$ can be determined. Variations falling within less then ε will not spoil the stability of the closed-loop response. However, this particular result does not give any information about performance.

The concept of robust control designs has been the focus of a great deal of research. Unfortunately a majority of the literature focuses on stability robustness, while structural control is usually more concerned with performance robustness. None the less, useful methods exist, which are beyond the scope of the introduction presented here.

Limitations

Numerous other control methods exist that may be adapted to the structural control problem. However, one item often ignored in the literature and in design is the need to match the actuation and sensing requirements of the control law to physically available hardware. In particular, the force and stroke and time constant of each actuator must be matched with the force, stroke and time constant required of the structure to be controlled. For instance if the actuator takes of the order of a second to respond, then it is doubtful that the actuator will be capable of controlling transient vibrations that are over in less then a second. The key elements in control, beyond the theoretical methods discussed here, are force, stroke, time constant, and location of actuators and sensors. These represent the hardware issues that cannot be ignored in any successful application of feedback control for vibration suppression of structures and machines.

Another limitation of control design is that of model order. Most control designs are straightforward to compute for plants that are of low order (say less than five or six degrees-of-freedom). However, in order to obtain good models of the plant, larger-order finite element models are often used. Hence many control designs focus on ways to control a reduced-order model without losing performance or stability through spillover. Several methods for dealing with control in the presence of model reduction exist and others are still being researched.

The question of control when the plant or actuators have significant nonlinear effects has also been addressed in the literature but is not summarized here. Nonlinear behavior forms a constant concern and several methods exist for controlling structures with nonlinear deformations, rigid body motions and nonlinear damping elements. Just as nonlinear structural analysis is very case dependent, so is nonlinear control.

Nomenclature

A	state matrix
B	control input matrix
C	sensor matrix
D	damping matrix
E	matrix of errors
g_1	velocity gain
g_2	position gain
G	gain matrix
I	identity matrix
Q, R	weighting matrices
Λ	diagonal matrix

See also: **Active control of civil structures; Active control of vehicle vibration; Absorbers, active; Damping models; Vibration isolation, applications and criteria; Vibration isolation theory.**

Further Reading

Clark RL, Saunders WR, Gibbs GP (1998) *Adaptive Structures: Dynamics and Control.* New York: Wiley.

Fuller CR, Elliot SJ, Nelson PA (1996) *Active Control of Vibration.* London: Academic Press.

Inman DJ (1989) *Vibration with Control Measurement and Stability.* Englewood Cliffs: Prentice Hall.

Meirovitch L (1990) *Dynamics and Control of Structures.* New York: Wiley.

Skelton RE (1988) *Dynamic Systems and Control.* New York: Wiley.

Soong TT (1990) *Active Structural Control: Theory and Practice.* Harlow: Longman and New York: Wiley.

ACTUATORS AND SMART STRUCTURES

V Giurgiutiu, University of South Carolina, Columbia, SC, USA

Copyright © 2001 Academic Press

doi:10.1006/rwvb.2001.0197

Introduction

The subject of smart structures and induced-strain actuators has attracted considerable attention in recent years. Smart structures offer the opportunity to create engineered material systems that are empowered with sensing, actuation, and artificial intelligence features. The induced-strain active materials actuators are the enabling technology that makes adaptive vibration control of smart structures realizable in an optimal way. Some generic concepts about smart structures and induced-strain active materials actuators will be given first. Then, details will be presented about piezo-electric, electrostrictive, magnetostrictive, and shape memory alloy materials that are used in the construction of induced-strain actuators. Guidelines for the effective design and construction of induced-strain actuation solutions will be provided. Details about sensory, actuatory, and adaptive smart structures will be provided, together with some examples. Conclusions and directions for further work are given last.

Smart Structures: Concepts

The discipline of adaptive materials and smart structures, recently coined as adaptronics, is an emerging engineering field with multiple defining paradigms. However, two definitions are prevalent. The first definition is based upon a technology paradigm: 'the integration of actuators, sensors, and controls with a material or structural component'. Multifunctional elements form a complete regulator circuit resulting in a novel structure displaying reduced complexity, low weight, high functional density, as well as economic efficiency. This definition describes the components of an adaptive material system, but does not state a goal or objective of the system. The other definition is based upon a science paradigm, and attempts to capture the essence of biologically inspired materials by addressing the goal as creating material systems with intelligence and life features integrated in the microstructure of the material system to reduce mass and energy and produce adaptive functionality.

It is important to note that the science paradigm does not define the type of materials to be utilized. It does not even state definitively that there are sensors, actuators, and controls, but instead describes a philosophy of design. Biological systems are the result of a continuous process of optimization taking place over millennia. Their basic characteristics of efficiency, functionality, precision, self-repair, and durability continue to fascinate scientists and engineers alike. Smart structures have evolved by biomimesis; they aim at creating man-made structures with embedded sensing, actuation, and control capabilities. **Figure 1** shows how the modern engineer might try to duplicate nature's functionalities with man-made material systems: composite materials to replicate the biological skeleton; piezo and optical sensors to duplicate the five senses; piezo and shape memory alloy actuators to replicate the fast twitch and slow twitch muscles; artificial intelligence networks to mimic the motor control system, etc. Such innovative developments have been spurred by the revolutionary emergence of commercially available smart active materials and their sensing and actuation derivatives.

Active Materials Actuators

The term 'smart materials' incorporates a large variety of revolutionary material systems that exhibit sensing and actuation properties similar to that of the living world. Of these, some smart materials may have only sensing properties, others may exhibit both sensing and actuation. An example of the former are the optical fiber sensors, and composite materials incorporating such fibers in their fibrous structure. Of the latter, a most obvious example is that of piezo-electric ceramics that can both sense and create mechanical strain. We will not cover the smart sensing materials here. Concentrating our attention on actuating smart materials, we notice that their list is

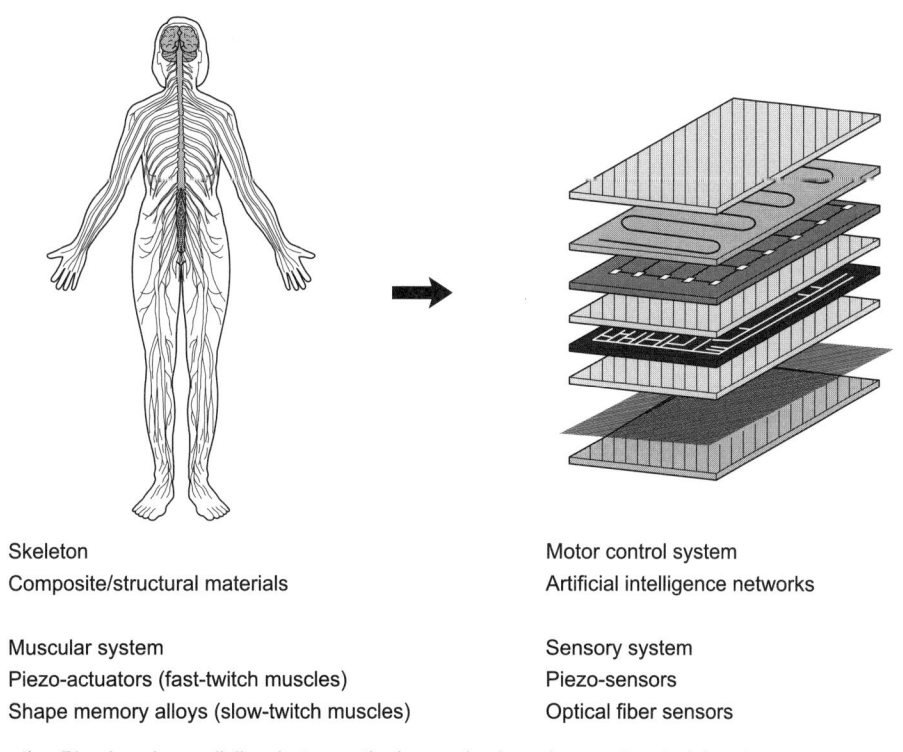

Figure 1 (see Plate 4). Biomimesis parallelism between the human body and a smart material system.

Skeleton
Composite/structural materials

Muscular system
Piezo-actuators (fast-twitch muscles)
Shape memory alloys (slow-twitch muscles)

Motor control system
Artificial intelligence networks

Sensory system
Piezo-sensors
Optical fiber sensors

long and varied. A raw categorization may distinguish several large classes:

1. Piezo-electric (PZT), electrostrictive (PMN), and magnetostrictive (Terfenol-D) materials that directly convert an externally applied electric or magnetic field into induced strain through a direct physical effect at microstructural level (similar to the thermal expansion effect). Such materials come as piezo and electrostrictive ceramics, piezo-electric polymers, and magnetostrictive alloys.
2. Shape memory alloys (SMA) that produce induced-strain actuation through a metallurgical phase transformation triggered by the crossing of certain temperature thresholds (Martensitic to Austenitic transformation). The induced-strain effect of SMA materials is the result of the strain states not being carried over from one metallurgical phase into the other phase. Thus, a material deformed (say, stretched) in the Martensitic phase will return to the undeformed state when transformed into the Austenitic phase, with the overall result being a net change in shape (i.e., length).
3. Electrochemical actuators, which are base on the change of volume that takes place during electrically controlled chemical reactions.
4. Chemomechanical actuators, which start with biological muscle and continue with several 'artificial muscle' materials based on net-linked collages, polyelectrolyte gels, conducting polymers, etc.

One example is provided by the electroactive polymers Nafion® and Flemion® that display remarkable large strains when used as wet membranes.

Also mentioned by some authors are certain smart materials that change their effective damping properties in response to electric and magnetic stimuli:

1. Electrorheological fluids (ERF), which are suspensions of micron-sized high-dielectric-strength particles in an insulating base oil, able to modify their effective viscosity and shear strength in response to electric fields.
2. Magnetorheological fluids (MRF), which are suspension of micron-sized ferromagnetic particles able to modify their effective viscosity and shear strength in response to magnetic fields.

In the present discussion, we will focus our attention on the piezo-electric, electrostrictive, magnetostrictive, and shape memory alloy materials, and explain how they can be used in the construction of active-materials actuators.

Actuators are devices that produce mechanical action. They convert input energy (electric, hydraulic, pneumatic, thermic, etc.) to mechanical output (force and displacement). Actuators rely on a 'prime mover' (e.g., the displacement of high-pressure fluid in a hydraulic cylinder, or the electromagnetic force in an electric motor) and on mechanisms to convert the prime mover effect into the desired action. If the prime mover can impart a large stroke, direct actuation can be effected (e.g., a hydraulic ram). When the prime mover has small stroke, but high frequency bandwidth, a switching principle is employed to produce continuous motion through the addition of switched incremental steps (e.g., some electric motors). Like everywhere in engineering, the quest for simpler, more reliable, more powerful, easier to maintain, and cheaper actuators is continuously on. In this respect, the use of active-materials solid-state induced-strain actuators has recently seen a significant increase. Initially developed for high-frequency, low-displacement acoustic applications, these revolutionary actuators are currently expanding their field of application into many other areas of mechanical and aerospace design. Compact and reliable, induced-strain actuators directly transform input electrical energy into output mechanical energy. One application area in which solid-state induced-strain devices have a very promising future is that of translational actuation for vibration and aeroelastic control. At present, the translational actuation market is dominated by hydraulic and pneumatic pressure cylinders, and by electromagnetic solenoids and shakers.

Hydraulic and pneumatic cylinders offer reliable performance, with high force and large displacement capabilities. When equipped with servovalves, hydraulic cylinders can deliver variable stroke output. Servovalve-controlled hydraulic devices are the actuator of choice for most aerospace (**Figure 2**), automotive, and robotic applications. However, a major drawback in the use of conventional hydraulic actuators is the need for a separate hydraulic power unit equipped with large electric motors and hydraulic pumps that send the high-pressure hydraulic fluid to the actuators through hydraulic lines. These features can be a major drawback in certain applications. For example, a 300-passenger airplane has over a kilometer of hydraulic lines spanning its body, from the engines to the most remote wing tip. Such a network of vulnerable hydraulic piping can present a major safety liability, under both civilian and military operation. In ground transportation, similar considerations have spurred automobile designers to promote the 'brake-by-wire' concept that is scheduled to enter the commercial market in the next few years. In some other applications, the use of conventional actuation is simply not an option. For example, the actuation of an aerodynamic servo-tab at the tip of a rotating blade, such as in helicopter applications, cannot be achieved through conventional hydraulic or electric methods due to the prohibitive high-g centrifugal force field environment generated during the blade rotation.

At present, electro-mechanical actuation that directly converts electrical energy into mechanical energy is increasingly preferred in several industrial applications. The most widely used high-power electro-mechanical actuators are the electric motors. However, they can deliver only rotary motion and need to utilize gearboxes and rotary-to-translational conversion mechanisms to achieve translational motion. This route is cumbersome, leads to additional weight, and has low-frequency bandwidth. Direct conversion of electrical energy into translational force and motion is possible, but its practical implementation in the form of solenoids and electrodynamic shakers is marred by typically low-force performance. The use of solenoids or electrodynamic shakers to perform the actuator duty-cycle of hydraulic cylinders does not seem conceivable.

Solid-state induced-strain actuators offer a viable alternative (**Figure 3**). Though their output displacement is relatively small, they can produce remarkably high force. With well-architectured displacement amplification, induced-strain actuators can achieve

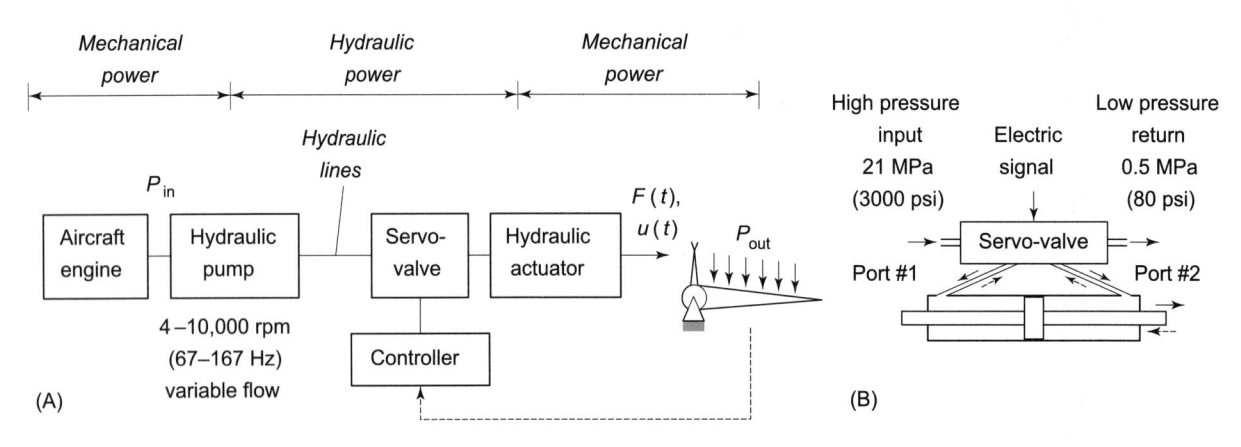

Figure 2 Conventional hydraulic actuation system: (A) flight controls of modern aircraft: (B) details of the servo-valve controlled hydraulic cylinder.

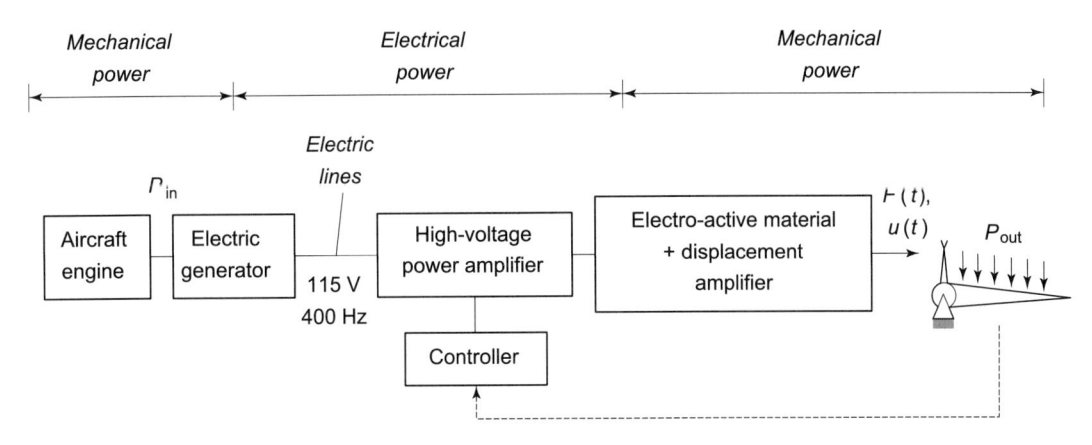

Figure 3 Schematic representation of a solid-state induced-strain actuated flight control system using electro-active materials.

output strokes similar to those of conventional hydraulic actuators, but over much wider bandwidth. Additionally, unlike conventional hydraulic actuators, solid-state induced-strain actuators do not require separate hydraulic power units and long hydraulic lines, and use the much more efficient route of direct electric supply to the actuator site.

The development of solid-state induced-strain actuators has entered the production stage, and actuation devices based on these concepts are likely to reach the applications market in the next few years. An increasing number of vendors are producing and marketing solid-state actuation devices based on induced-strain principles.

Piezo-electric and Electrostrictive Materials

Piezo-electricity

Piezo-electricity (discovered in 1880 by Jacques and Pierre Curie) describes the phenomenon of generating an electric field when the material is subjected to a mechanical stress (direct effect), or, conversely, generating a mechanical strain in response to an applied electric field. Piezo-electric properties occur naturally in some crystalline materials, e.g., quartz crystals (SiO_2), and Rochelle salt. The latter is a natural ferroelectric material, possessing an orientable domain structure that aligns under an external electric field and thus enhances its piezo-electric response. Piezo-electric response can also be induced by electrical poling of certain polycrystalline materials, such as piezo-ceramics (**Figure 4**).

The application of a high poling field at elevated temperatures results in the alignment of the crystalline domains; this alignment is locked in place when the high temperatures are removed. Subsequently, the poled ceramics response to the application of an applied electric field or mechanical stress with typical piezo-electric behavior. The distortion of the crystal domains produces the piezo-electric effect. Lead zirconate titanate, $PbZrO_3$, is commercially known as PZT. To date, many PZT formulations exist, the main differentiation being between 'soft' (e.g., PZT 5-H) and 'hard' (e.g., PZT 8). Within the linear range, piezo-electric materials produce strains that are proportional to the applied electric field or voltage. Induced strains in excess of 1000 μstrain (0.1%) have become common. These features make piezo-electric materials very attractive for a variety of sensor and actuator applications.

Modeling of Piezo-electric Behavior

For linear piezo-electric materials, the interaction between the electrical and mechanical variables can be described by linear relations (ANSI/IEEE Standard

Table 1 Mechanic and dielectric properties for piezo-electric (PZT-5, PZT-8) and electrostrictive (PMN EC-98) materials

Property	Soft PZT-5 Navy Type VI	Hard PZT-8 Navy Type III	PMN EC-98
ρ (kg m^{-3})	7600	7600	7850
k_{31}	0.36	0.31	0.35
k_{33}	0.71	0.61	0.72
k_{15}	0.67	0.54	0.67
$d_{31}(\times 10^{-12}$ m V^{-1})	−270	−100	−312
$d_{33}(\times 10^{-12}$ m V^{-1})	550	220	730
$d_{15}(\times 10^{-12}$ m V^{-1})	720	320	825
$g_{31}(\times 10^{-3}$ Vm N^{-1})	−9.0	−11.3	−6.4
$g_{33}(\times 10^{-3}$ Vm N^{-1})	18.3	24.9	15.6
$g_{15}(\times 10^{-3}$ Vm N^{-1})	23.9	36.2	17
$s_{11}^E(\times 10^{-12}$ m^2 N^{-1})	15.9	10.6	16.3
Poisson ratio	0.31	0.31	0.34
$s_{33}^E(\times 10^{-12}$ m^2 N^{-1})	20.2	13.2	21.1
Curie temp. (°C)	200	350	170
Mechanical Q_m	75	900	70

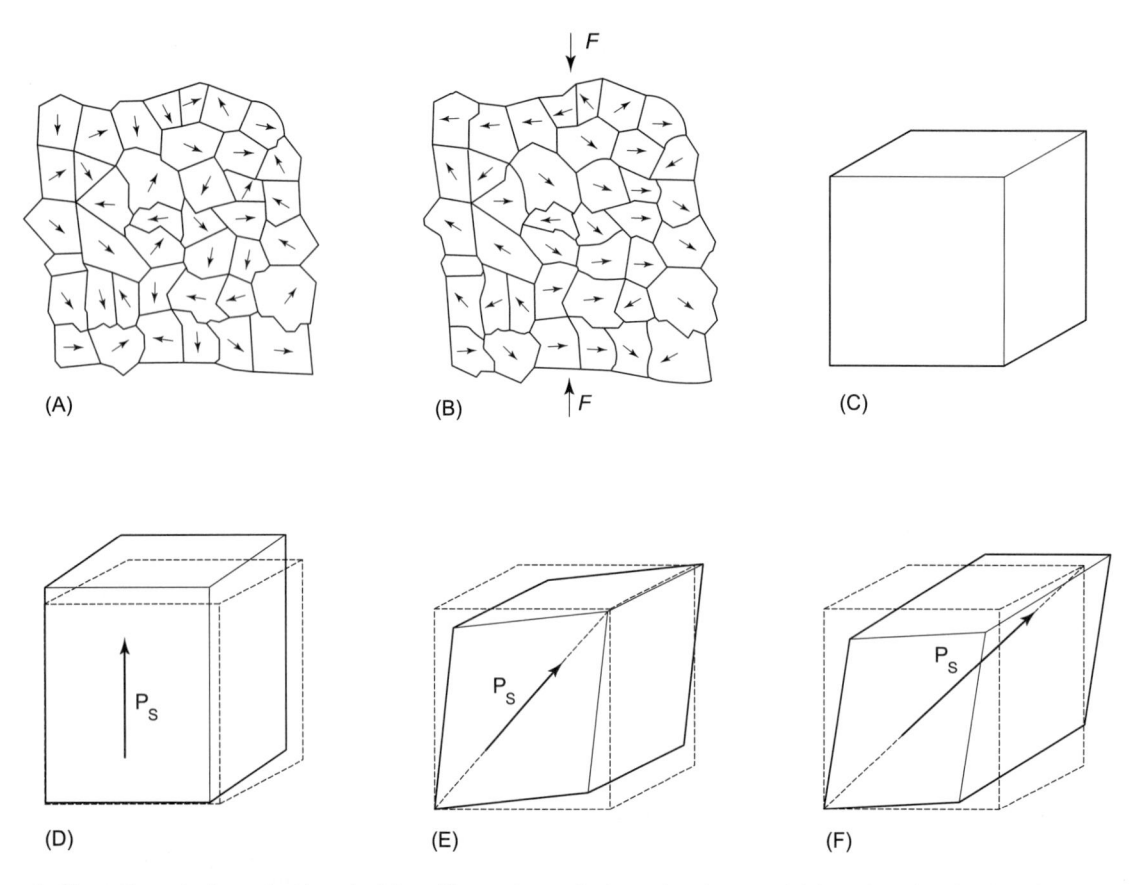

Figure 4 Illustration of piezo-electric principles: (A) un-stressed piezo-electric material has the electric domains randomly oriented; (B) stressing produces orientation of the electric domains perpendicular to the loading direction; (C) cubic phase, stable above the Curie temperature, no piezo-electricity; (D) tetragonal phase; (E) orthorhombic phase; (F) rhombohedral phase.

176–1987). A tensorial relation between mechanical and electrical variables is established in the form:

$$S_{ij} = s_{ijkl}^E T_{kl} + d_{kij} E_k + \alpha_i^E \theta$$
$$D_j = d_{jkl} T_{kl} + \varepsilon_{jk}^T E_k + \tilde{D}_m \theta \quad [1]$$

where S_{ij} is the mechanical strain, T_{ij} is the mechanical stress, E_i is the electrical field, D_i is the electrical displacement (charge per unit area), s_{ijkl}^E is the mechanical compliance of the material measured at zero electric field $(E = 0)$, ε_{jk}^T is the dielectric constant measured at zero mechanical stress $(T = 0)$; d_{kij} is the piezo-electric displacement coefficient that couples the electrical and mechanical variables, θ is the absolute temperature, α_i^E is the coefficient of thermal expansion under constant electric field; $\tilde{D}_j$ is the electric displacement temperature coefficient. The stress and strain are second-order tensors, while the electric field and the electric displacement are

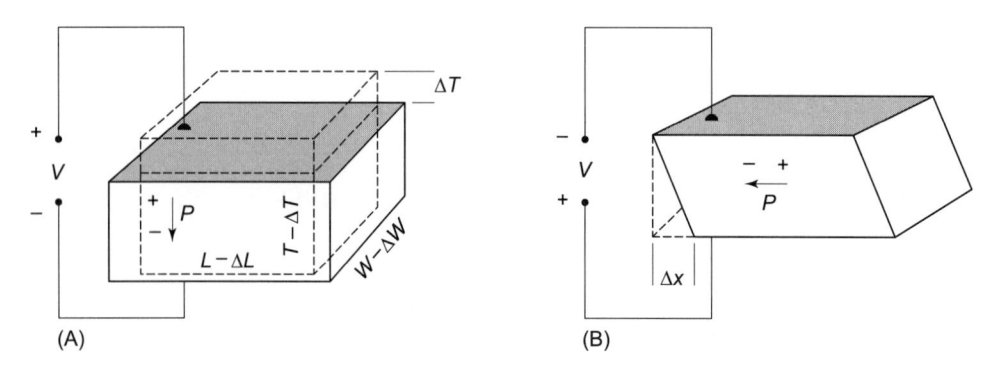

Figure 5 Basic induced-strain responses of piezo-electric materials: (A) axial and transverse strain; (B) shear strain.

first-order tensors. The superscripts, T, D, E, signify that the quantities are measured at zero stress, or zero current, or constant electric field, respectively. The second equation reflects the direct piezo-electric effect, while the first equation refers to the converse piezo-electric effect.

In engineering practice, the tensorial eqn [1] is rearranged in matrix form using the six-component stress and strain vectors, $(T_1, \ldots, T_6)$ and $(S_1, \ldots, S_6)$, where the first three components represent direct stress and strain, while the last three represent shear. **Figure 5** illustrates the physical meaning of these equations in the case of simple polarization. **Figure 5A** shows that an electric field $E_3 = V/t$ applied parallel to the direction of polarization $(E_3 \| P)$ induces thickness expansion $(\varepsilon_3 = d_{33}E_3)$ and transverse retraction $(\varepsilon_1 = d_{31}E_3)$. **Figure 5B** shows that, if the field is perpendicular to the direction of polarization $(E_3 \perp P)$, shear deformation is induced $(\varepsilon_5 = d_{15}E_3)$.

Electromechanical coupling coefficient is defined as the square root of the ratio between the mechanical energy stored and the electrical energy applied to a piezo-electric material. For direct actuation, we have $\kappa_{33} = d_{33}/\sqrt{(s_{11}\varepsilon_{33})}$, while for transverse actuation $\kappa_{31} = d_{31}/\sqrt{(s_{11}\varepsilon_{33})}$.

Electrostrictive Materials

Electrostrictive ceramics are ferroelectric ceramics that contain both linear and quadratic terms:

$$S_{ij} = s^E_{ijkl}T_{kl} + d_{kij}E_k + M_{klij}E_kE_l \quad [2]$$
$$D_j = d_{jkl}T_{kl} + \varepsilon^T_{jk}E_k$$

Note that the first two terms in the first equation are the same as for piezo-electric materials. The third term is due to electrostriction, with M_{klij} being the electrostrictive coefficient. In general, the piezo-electric effect is possible only in noncentrosymmetric materials, whereas the electrostrictive effect is not limited by symmetry and is present in all materials. A common electrostrictive compound is lead–magnesium–niobate, or PMN. Commercially available PMN formulations are internally biased and optimized to give quasilinear behavior, but do not accept field reversal. **Figure 6A** compares induced-strain response of some commercially available piezo-electric, electrostrictive, and magnetostrictive actuation materials. It can be seen that the electrostrictive materials have much less hysteresis, but more nonlinearity.

Piezo-polymers

Polyvinylidene fluoride (PVDF or PVF_2), is a polymer with strong piezo-electric and pyroelectric properties. In the α phase, PVDF is not polarized and is used as a common electrical insulator among many other applications. To impart the piezo-electric properties, the α phase is converted to the β phase and polarized. Stretching α-phase material produces the β phase. After surface metallization, a strong electric field is applied to provide permanent polarization. The flexibility of the PVDF overcomes some of the drawbacks associated with the piezo-electric ceramics brittleness. As a sensor, PVDF provides higher voltage/electric field in response to mechanical stress. Their piezo-electric g-constants (i.e., the voltage generated per unit mechanical stress) are typically 10–20 times

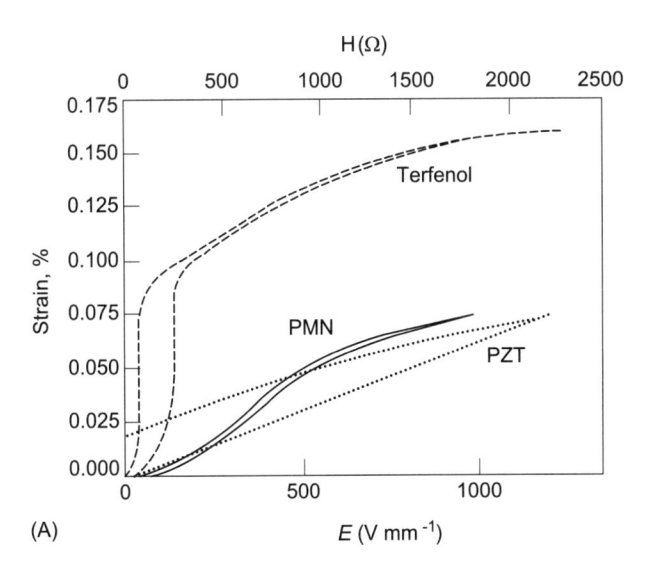

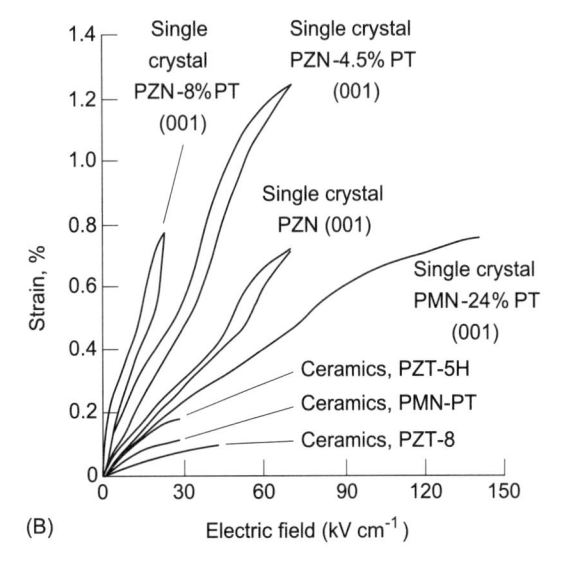

Figure 6 Strain vs electric field behavior of induced-strain materials: (A) currently available materials; (B) new <001> oriented rhombohedral crystals of PZN–PT and PMN–PT compared to current piezo-electric ceramics.

larger than for piezo-ceramics. PVDF film also produces an electric voltage in response to infrared light due to its strong pyroelectric coefficient. However, the use of PVDF materials as actuators is inappropriate for most structural applications due to their low elastic modulus which cannot generate large actuation forces.

New Piezo-electric Materials

New piezo-electric materials with much larger induced-strain capabilities are currently being developed in research laboratories. Very promising are the single crystal materials: single crystals of PZN–PT, a relaxor pervoskite $Pb(Zn_{1/3}Nb_{2/3})O_3$–$PbTiO_3$, have been studied. Strain levels of up to 1.5 percent, and reduced hysteresis have been reported. The response of these new materials can be an order of magnitude larger than that of conventional PZT materials (see curves in **Figure 6B**. When building actuators from PZN–PT material, one must take into account the strong dependence of the piezo-electric properties on the crystal orientation. This imposes certain design restrictions in comparison with conventional piezo-ceramics. Commercial production of induced strain actuators based on these new materials has been undertaken by TRS Ceramics, Inc., and a prototype PZN–PT actuator with a maximum strain of around 0.3% has been reported.

Advantages and Limitations of Piezo-electric and Electrostrictive Actuation Materials

Piezo-electric ceramics, e.g., PZT, are essentially small-stroke large-force solid-state actuators with very good high-frequency performance. However, they also display certain limitations. The most obvious limitation is that, in many engineering applications, some form of mechanical amplification is required. Other limitations are associated with electrical breakdown, depoling, Curie temperature, nonlinearity, and hysteresis.

1. Electrical breakdown may happen when an electric field applied in the poling direction exceeds the dielectric strength of the material, resulting in electrical arcing through the material and short-circuit. Electrical breakdown also destroys the piezo-electric properties of the material.
2. Depoling may happen when an electric field is applied opposite to the poling direction, resulting in degradation of the piezo-electric properties or even polarization in the opposite direction. The depoling field (or coercive field) may be as low as half of the electrical breakdown field.
3. Curie temperature. At temperatures close to the Curie temperature, depoling is facilitated, aging and creep are accelerated, and the maximum safe mechanical stress is decreased. For typical PZT materials, the Curie temperature is about 350°C. The operating temperature should generally be at least 50°C lower than the Curie temperature.
4. Nonlinearity and hysteresis. Actual piezo-ceramics are nonlinear and hysteretic (**Figure 6**). Hysteresis is due to internal sliding events in the polycrystalline piezo-electric material. Upon removal of the electric field, remnant mechanical strain is observed. Hysteresis of common piezo-electric may range from 1 to 10%. Under high frequency operation, hysteresis may generate excessive heat, and loss of performance may occur if the Curie temperature is exceeded.

The main advantage of electrostrictive materials over piezo-electric materials is their very low hysteresis. This could be especially beneficial in high-frequency dynamic applications, which could involve considerable hysteresis-associated heat dissipation. The main disadvantage of electrostrictive materials is the temperature dependence of their properties.

Magnetostrictive Materials

Magnetostrictive materials expand in the presence of a magnetic field, as their magnetic domains align with the field lines. Magnetostriction was initially observed in nickel cobalt, iron, and their alloys but the values were small (<50 μstrain). Large strains ($\sim 10\,000$ μstrain) were observed in the rare-earth elements terbium (Tb) and dysprosium (Dy) at cryogenic temperatures (below 180 K). The compound Terfenol-D ($Tb_{.3}Dy_{.7}Fe_{1.9}$), developed at Ames Laboratory and the Naval Ordnance Laboratory (now Naval Surface Weapons Center), displays magnetostriction of up to 2000 μstrain at room temperature and up to 80°C and higher.

Modeling of the Magnetostrictive Materials

The magnetostrictive constitutive equations contain both linear and quadratic terms:

$$S_{ij} = s^E_{ijkl}T_{kl} + d_{kij}H_k + M_{klij}H_kH_l$$
$$B_j = d_{jkl}T_{kl} + \mu^T_{jk}H_k \qquad [3]$$

where, in addition to the already defined variables, H_k is the magnetic field intensity, B_j is the magnetic flux density, and μ^T_{jk} is the magnetic permeability under constant stress. The coefficients d_{kij} and M_{klij} are defined in terms of magnetic units. The magnetic field intensity, H, in a rod surrounded by a coil with n turns per unit length is related to the current, I, through the relation:

$$H = nI \qquad [4]$$

Advantages and Limitations of Magnetostrictive Actuators

Magnetostrictive materials, like Terfenol-D, are essentially small-stroke large-force solid-state actuators that have wide frequency bandwidth. However, they also display certain limitations. The most obvious one is that, in actuation applications, some form of mechanical amplification is required. The main advantage of magnetoactive actuation materials over electroactive materials may be found in the fact that it is sometimes easier to create a high intensity magnetic field than a high intensity electric field. High electric fields require high voltages, which raise important insulation and electric safety issues. According to eqn [4], high magnetic fields could be realized with lower voltages using coils with a large number of turns per unit length, through which a high-amperage current flows.

An important limitation of the magnetoactive materials is that they cannot be easily energized in the two-dimensional topology, as when applied to a structural surface. This limitation stems from the difficulty of creating high-density magnetic fields without a closed magnetic circuit armature. For the same reason, the magnetoactive induced-strain actuators will always require additional construction elements besides the magnetoactive materials. While a bare-bones electroactive actuator need not contain anything more than just the active material, the bare-bones magnetoactive actuator always needs the energizing coil and the magnetic circuit armature. For this fundamental reason, the power density (either per unit volume or per unit mass) of magnetoactive induced-strain actuators will always remain below that of their electroactive counterparts.

Shape Memory Alloys

Another class of induced-strain actuating materials, with much larger strain response but low frequency bandwidth, is represented by shape memory alloys. Shape memory alloys (SMA) materials are thermally activated ISA materials that undergo phase transformation when the temperature passes certain values. The metallurgical phases involved in this process are the low-temperature martensite and the high-temperature austenite. When phase transformation takes place, the SMA material modifies its shape, i.e., it has 'memory'. The SMA process starts with the material being annealed at high-temperature in the austenitic phase (**Figure 7**). In this way, a certain shape is 'locked' into the material. Upon cooling the material transforms into the martensitic phase, and adopts a twinned crystallographic structure. When mechanical deformation is applied, the twinned crystallographic structure switches to a skew crystallographic structure. Strains as high as 8% can be achieved through this de-twinning process. This process gives the appearance of permanent plastic deformation, though no actual plastic flow took place. In typical actuator applications, this process is used to store mechanical energy by stretching SMA wires.

Upon heating, the martensitic phase changes into austenite and the shape initially imposed by annealing is recovered. In this way, the permanent deformation created through de-twinning of the martensitic phase is removed, and the material returns to its initial state

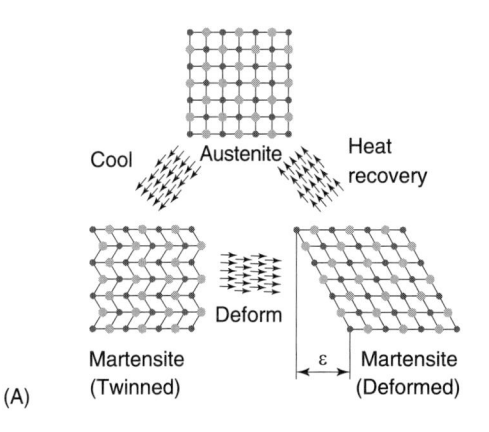

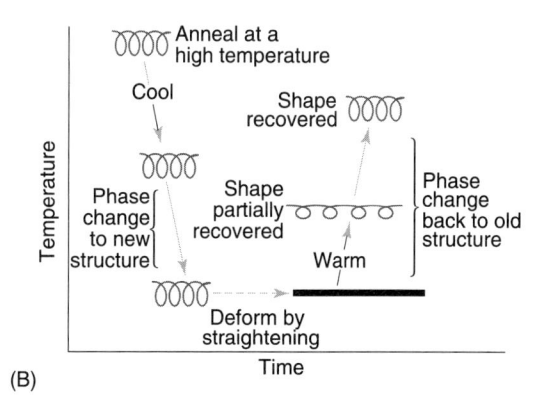

Figure 7 (see Plate 5). Principles of SMA materials: (A) change in crystallographic structure during cooling and heating; (B) associated component-shape changes, using a coil spring as an example.

memorized during annealing at austenite temperatures. If recovery is mechanically prevented, restraining stresses of up to 700 MPa can be developed. This is the one-way shape memory effect, which is a one-time only deployment at the end of which full recovery of the initial state is obtained. For dynamic operations, a two-way shape memory effect is preferred, in order that cyclic loading and unloading is attained. Two-way shape memory effect can be achieved through special thermomechanical 'training' of the material. In such cases, repeated activation of SMA strains is possible in synch with heating–cooling cycles (**Figure 8**). Many materials are known to exhibit the shape–memory effect. They include the copper alloy systems of Cu–Zn, Cu–Zn–Al, Cu–Zn–Ga, Cu–Zn–Sn, Cu–Zn–Si, Cu–Al–Ni, Cu–Au–Zn; Cu–Sn, the alloys of Au–Cd, Ni–Al; Fe–Pt; and others. The most common shape–memory alloy is the nickel–titanium compound, Nitinol™, discovered in 1962 by Buehler *et al.* at the Naval Ordnance Laboratory (NOL).

During heating, the change from martensite to austenite starts at temperature A_s and finishes at temperature A_f. During cooling, the reverse change from austenite to martensite starts at temperature M_s and finishes at temperature M_f. Note that $A_s < A_f$, while $M_s > M_f$. In the interval (A_s, A_f) on heating, and (M_s, M_f) on cooling, two phases coexist in temperature-dependent volume fractions. The values of the phase-transition temperatures can be varied with alloy composition and stress level. In fact, in SMA materials, stress and temperature act in opposition such that an SMA material brought to austenite by heating can be isothermally forced back to martensite by stressing. Shape–memory alloys composition can be tuned to start the austenite transformation at almost any predetermined temperature in the range $-50°C$ to $62°C$.

In addition to 'memory', SMA materials have other remarkable properties: phase-dependent elastic modulus, superelasticity, and high internal damping. In most metals, the elastic modulus decreases with temperature: in SMA materials it actually increases, since the modulus of the high-temperature austenitic phase can be up to three times larger than that of the low-temperature martensitic phase. Superelasticity is associated with the fact that strains of up to 12% can be accommodated before actual plastic deformation starts to take place. Superelasticity is displayed by low-A_f SMA materials, which are austenitic at room temperature. As stress is applied, the superelastic SMA material undergoes phase transformation and goes into martensite. Upon unloading, the austenite phase is recovered, and the material returns to the zero-stress zero-strain state, i.e., displays superelastic properties (**Figure 9**). Though full strain recovery has been achieved, the recovery is nonlinear and follows a hysteretic path. The internal damping of the SMA materials is associated with the hysteresis curve enclosed by the loading–unloading cycle. The area within this cycle is orders of magnitude larger than for conventional elastic materials.

Modeling of SMA Materials

Since the mechanical behavior of SMA materials is closely related to the microscopic martensitic phase transformation, the constitutive relations developed for conventional materials such as Hooke's law and plastic flow theory are not directly applicable. Hence, specific constitutive relations, which take into consideration the phase transformation behavior of SMA, have been developed. Two approaches are generally used: (i) the phenomenological (macroscopic) approach, based on extensive experimental work; and (ii) the physical (microscopic) approach using fundamental physical concepts. Hybrid approaches that combine both approaches to obtain a more accurate description and prediction of the SMA material behavior have also been used. Tanaka's model, based on the concept of the free-energy driving force, considers a one-dimensional metallic material undergoing phase transformation. The state variables for the material are strain, ε, temperature, T, and martensitic fraction, ξ. Then, a general state variable, Λ, is defined as:

$$\Lambda = (\bar{\varepsilon}, T, \xi) \qquad [5]$$

The Helmholtz free energy is a function of the state variable Λ. The general constitutive relations are then derived from the first and second laws of thermodynamics as:

$$\bar{\sigma} = \rho_0 \frac{\partial \Phi}{\partial \varepsilon} = \sigma(\bar{\varepsilon}, T, \xi) \qquad [6]$$

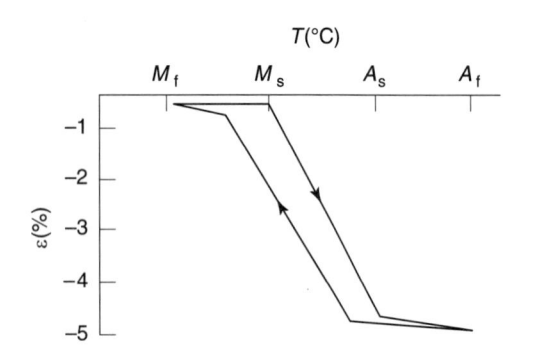

Figure 8 Schematic representation of strain recovery through the SMA effect in a two-way shape memory alloy.

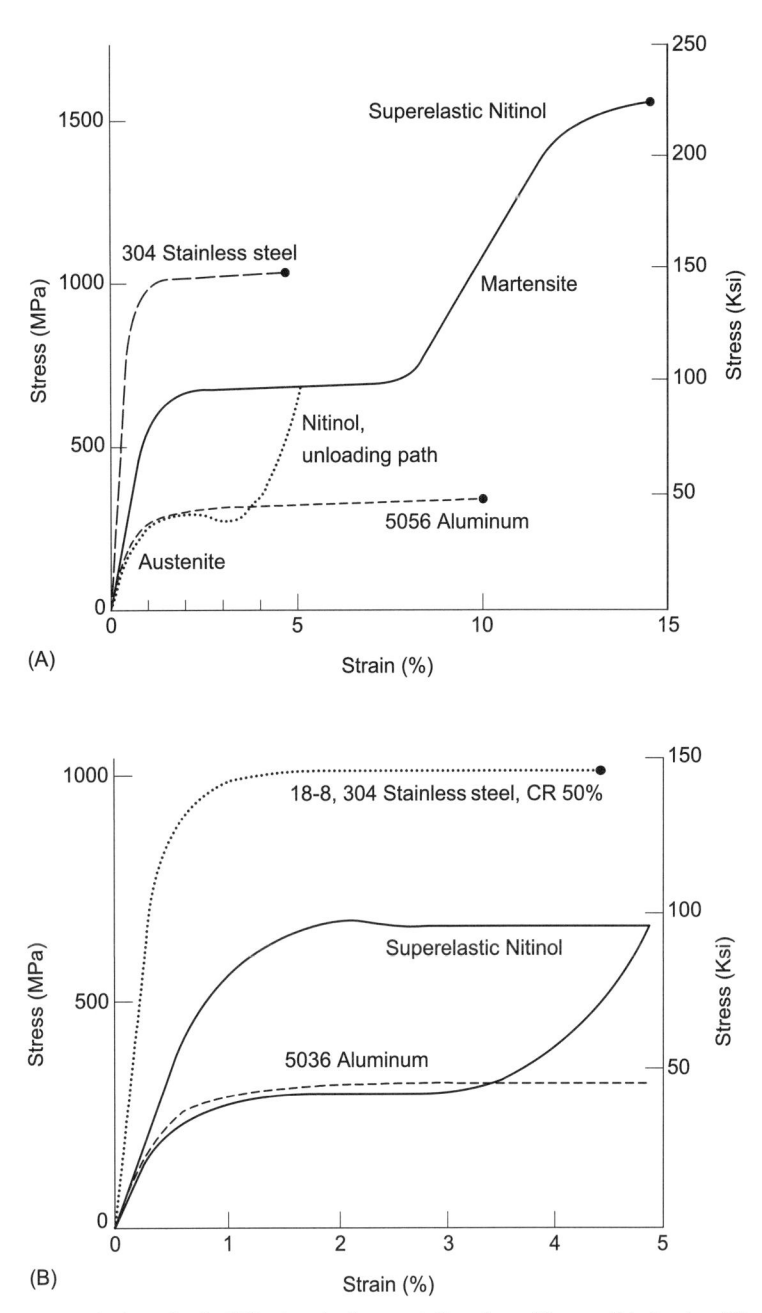

Figure 9 Stress–strain curves of superelastic Nitinol and other metallic wires: (A) overall behavior; (B) zoom-in on the low-strain range.

The stress is a function of the martensite fraction, an internal variable. Differentiation of eqn [6] yields the stress rate equation:

$$\bar{\sigma} = \frac{\partial \sigma}{\partial \bar{\varepsilon}} \dot{\bar{\varepsilon}} + \frac{\partial \sigma}{\partial T} \dot{T} + \frac{\partial \sigma}{\partial \xi} \dot{\xi} = D \dot{\bar{\varepsilon}} + \Theta \dot{T} + \Omega \dot{\xi} \quad [7]$$

where D is Young's modulus, Θ is the thermoelastic tensor, and Ω is the transformation tensor, a metallurgical quantity that represents the change of strain during phase transformation. The Martensite fraction, ξ, can be assumed as an exponential function of stress and temperature:

$$\begin{aligned}\xi_{M \to A} &= \exp\left[A_a(T - A_s) + B_a \sigma\right], \\ \xi_{A \to M} &= 1 - \exp\left[A_m(T - M_s) + B_m \sigma\right]\end{aligned} \quad [8]$$

where A_a, A_m, B_a, B_m are material constants in terms of the transition temperatures, A_s, A_f, M_s, M_f. Time integration of eqn [7] yields:

$$\sigma - \sigma_0 = D(\varepsilon - \varepsilon_0) + \theta(T - T_0) + \Omega(\xi - \xi_0) \quad [9]$$

It is convenient to adopt a cosine model for the phase transformation, i.e.:

$$\xi_{A \to M} = \frac{\xi_M}{2} \{ \cos[a_A(T - A_s) + b_A \sigma] + 1 \},$$

$$\xi_{M \leftarrow A} = \frac{1 - \xi_A}{2} \{ \cos[a_M(T - M_s) + b_M \sigma] + 1 \}$$
$$+ \frac{1 + \xi_A}{2} \quad [10]$$

where

$$a_A = \frac{\pi}{A_f - A_s}, \quad a_M = \frac{\pi}{M_s - M_f},$$
$$b_A = -\frac{a_A}{C_A}, \quad b_M = -\frac{a_M}{C_M}$$

while C_M and C_A are slopes

Advantages and Limitations of SMA Actuation

The main advantage of SMA materials is their capability to produce sizable (up to 8%) actuation strains. In addition, they have inherent simplicity since only heating (readily available through the electric Joule effect) is needed for actuation. Main limitations of the SMA actuators are the poor energy conversion efficiency, and the low bandwidth of the heating/cooling process which can only achieve a few Hz, at the very best.

Effective Implementation of Induced-strain Actuation

Design and Construction of Piezo-electric and Electrostrictive Actuators

An electroactive solid-state actuator consists of a stack of many layers of electroactive material (PZT or PMN) alternatively connected to the positive and negative terminals of a high-voltage source (**Figure 10A**). Such a PZT or PMN stack behaves like an electrical capacitor. When activated, the electroactive material expands and produces output displacement. The PZT or PMN stacks are constructed by two methods. In the first method, the layers of active material and the electrodes are mechanically assembled and glued together using a structural adhesive. The adhesive modulus (typically, 4–5 GPa) is at least an order of magnitude lower than the modulus of the ceramic (typically, 70–90 GPa). This aspect may lead to the stack stiffness being significantly

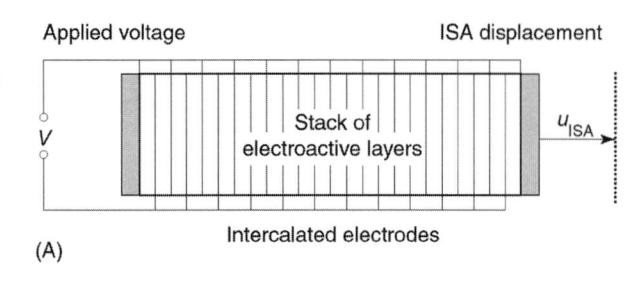

Figure 10 Induced strain actuator using a PZT or PMN electroactive stack: (A) schematic; (B) typical commercially available cofired stack from EDO Corporation; (C) a range of Polytec PI actuators heavy-duty actuators.

lower than the stiffness of the basic ceramic material. In the second method, the ceramic layers and the electrodes are assembled in the 'green' state. Then, they are fired together (co-fired) under a high isostatic pressure (HIP process) in the processing oven. This process ensures a much stiffer final product and, hence, a better actuator performance. However, the processing limitations, such as oven and press size, etc., do not allow the application of this process to anything else but small-size stacks.

The PZT and PMN stacks may be surrounded by a protective polymeric or elastomeric wrapping. Lead wires protrude from the wrapping for electrical connection. Steel washers, one at each end, are also provided for distributing the load into the brittle ceramic material. When mounted in the application structure, these stacks must be handled with specialized knowledge. Protection from accidental impact damage must be provided. Adequate structural support and alignment are needed. Mechanical connection to the application structure must be such that neither tension stresses nor bending are induced in the stack since the active ceramic material has low tension strength. Hence, the load applied to the stack must always be compressive and perfectly centered. If tension loading is also expected, adequate prestressing must be provided through springs or other means. For applications, the stack can be purchased as such (**Figure 10B**), or encapsulated into a steel casing which provides a prestress mechanism and the electrical and mechanical connections (**Figure 10C**).

Figure 11 compares the response of two commercially available actuators, one piezo-electric, the other electrostrictive. It can be seen that both have about the same maximum induced-strain value, as it can be readily verified by dividing the maximum displacement by the actuator length. Both material types display quasilinear behavior. The piezo-electric actuator allows some field reversal (up to 25%, according to manufacturer), which make the total stroke larger. On the other hand, the electrostrictive actuator has much less hysteresis, i.e., less losses and less heating in high-frequency regime.

Design and Construction of Magnetostrictive Actuators

Magnetostrictive materials can be also used to produce an effective actuator. **Figure 12A** shows the typical layout of a magnetostrictive actuator. It consists of a Terfenol-D bar surrounded by an electric coil and enclosed into an annular magnetic armature. The magnetic circuit is closed through end caps. In this arrangement, the magnetic field is strongest in the cylindrical inner region filled by the Terfenol-D bar. When the coil is activated, the Terfenol-D expands and produces output displacement. The Terfenol-D bar, the coil, and the magnetic armature are assembled with prestress between two steel-washers and put inside a protective wrapping to form the basic magnetoactive induced-strain actuator. Though the Terfenol-D material has been shown to be capable of up to 2000 μstrain, its behavior is highly nonlinear in both magnetic field response and the effect of compressive prestress. Manufacturers of magnetostrictive actuators optimize the internal prestress and magnetic bias to get a quasilinear behavior in the range of 750–1000 μm m^{-1}. **Figure 12B** shows the displacement–magnetic field response for a typical large-power magnetostrictive actuator (ETREMA AA–140J025, 200 mm long, ≈1 kg weight, 0.140 mm peak-to-peak output displacement).

Principles of Induced-Strain Structural Actuation

In structural applications, the induced ISAs must work in direct relation with the actuated structure,

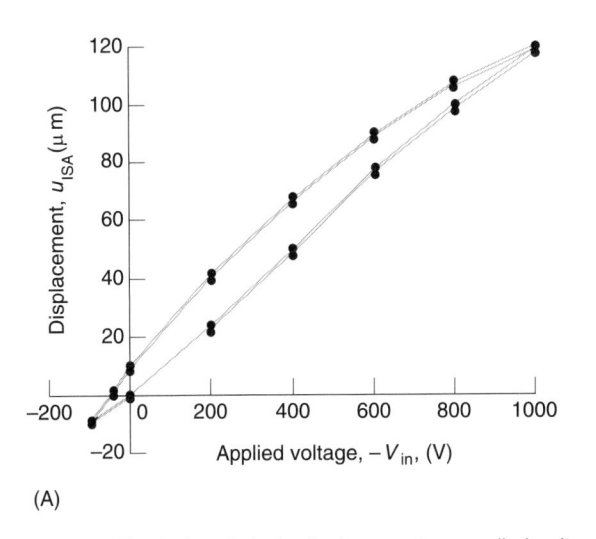

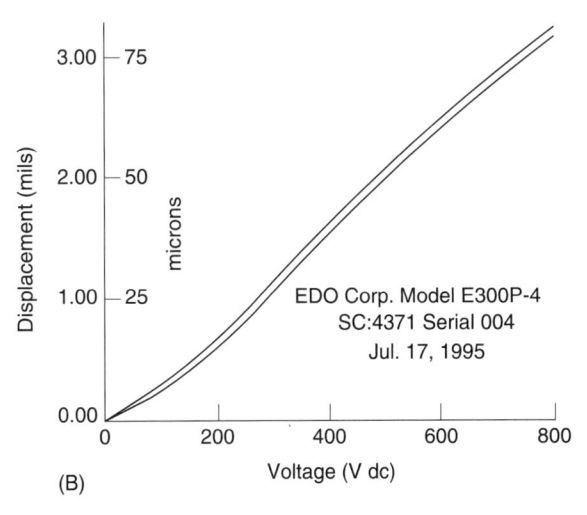

Figure 11 The induced-strain displacement vs applied voltage for typical electroactive actuators: (A) Polytec PI P-245.70 PZT stack, 99 mm long (Giurgiutiu *et al.*, 1997); (B) EDO Corporation model E300P PMN stack, 57 mm long.

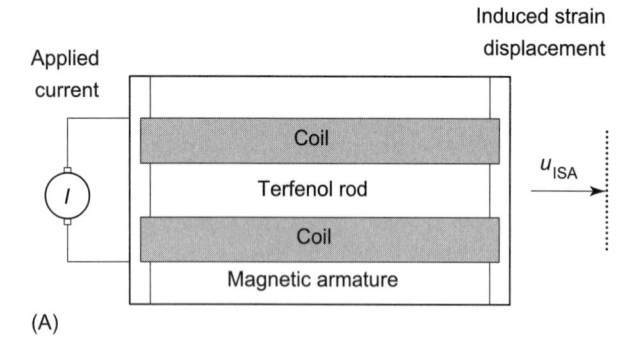

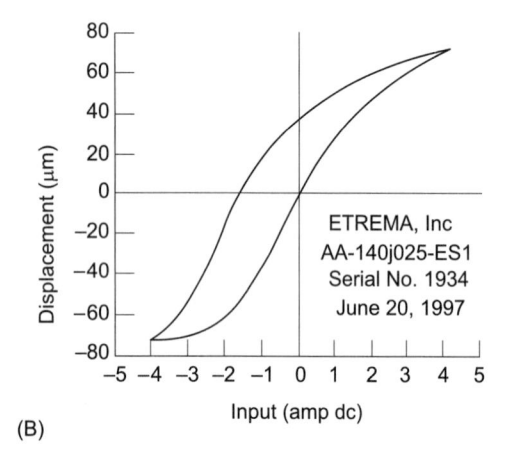

Figure 12 (A) Schematic construction of a magnetostrictive (Terfenol-D) solid-state actuator; (B) typical response curve (actuator model AA140J025-ES1, ETREMA, Inc.).

and special attention must be given to their interaction. The main differentiating feature between active materials actuators and conventional actuators lies in the amount of available displacement. The induced-strain effect present in active materials results in output displacements that seldom exceed 100 µm (0.1 mm). In a conventional actuator, e.g., a hydraulic cylinder, displacement of the order of several millimeters can be easily achieved, and if more displacement is needed, additional hydraulic fluid could be pumped in. In contrast, an ISA has at its disposal only the very limited amount of displacement generated by the induced-strain effect. This limited displacement needs to be carefully managed, if the desired effect is to be achieved. Under reactive service loads, the internal compressibility of the active materials actuator 'eats-up' part of the induced-strain displacement, and leads to reduced output displacement (**Figure 13A**). If the external stiffness, k_e, is reduced, the force in the actuator is also reduced, and more displacement is seen at the actuator output end. For a free actuator, i.e., under no external reaction, the output displacement is maximum. However, no active work is being done in this case since the force is zero. At the other extreme, when the actuator is fully constrained, the force is maximum, but no work is again performed since the displacement is zero. An optimum is attained between these two extremes, and this optimum can be best described in terms of stiffness match principle. Under static conditions, the stiffness match principle implies that the external stiffness (i.e., the stiffness of the application) and the internal stiffness of the actuator are equal, which gives a stiffness ratio with value $r = 1$. As shown in **Figure 13B**, $r = 1$ corresponds to maximum energy situation.

Under dynamic conditions, this principle can be expressed as dynamic stiffness match or impedance match. **Figure 13C** shows a solid-state ISA operating against a dynamic load with parameters $k_e(\omega)$, $m_e(\omega)$, and $c_e(\omega)$. The actuator is energized by a variable voltage, $v(t)$, which sends a time-varying current, $i(t)$. The actuator output displacement and force are, $u(t)$ and $F(t)$, respectively. The complex stiffness ratio is given by

$$\bar{r}(\omega) = \bar{k}_e(\omega)/\bar{k}_i \quad [11]$$

where $\bar{k}_e(\omega) = (k_e - \omega^2 m_e) + i\omega c$ is the complex external stiffness of the dynamic load, and $\bar{k}_i$ is the complex internal stiffness of the ISA.

Power and Energy Extraction from ISAs

Using the definition, mechanical power = force × velocity, and assuming harmonic motion, the complex power is expressed as $\bar{P} = \frac{1}{2}\bar{F}\bar{v}^*$, where, $\bar{v}^*$ is the complex conjugate of $\bar{v}$, and $\bar{v} = i\omega\bar{u}$. Output mechanical power and energy is written as:

$$\bar{P}_{mech} = \tfrac{1}{2}\bar{k}_e \bar{u}(i\omega\bar{u})^*$$
$$= -i\omega \frac{\bar{r}(\omega)}{[1+\bar{r}(\omega)][1+\bar{r}(\omega)]^*}\left(\tfrac{1}{2}\bar{k}_i \hat{u}_{ISA}^2\right) \quad [12]$$

$$\bar{E}_{mech} = \tfrac{1}{2}\bar{k}_e \bar{u}(\bar{u})^* = \frac{\bar{r}(\omega)}{[1+\bar{r}(\omega)][1+\bar{r}(\omega)]^*}\left(\tfrac{1}{2}\bar{k}_i \hat{u}_{ISA}^2\right) \quad [13]$$

The maximum output power and energy are attained at the dynamic stiffness match condition:

$$P_{out}^{max} = \omega \tfrac{1}{4} E_{mech}^*, \quad E_{out}^{max} = \tfrac{1}{4} E_{mech}^* \quad [14]$$

where:

$$E_{mech}^* = \tfrac{1}{2} k_i \hat{u}_{ISA}^2$$

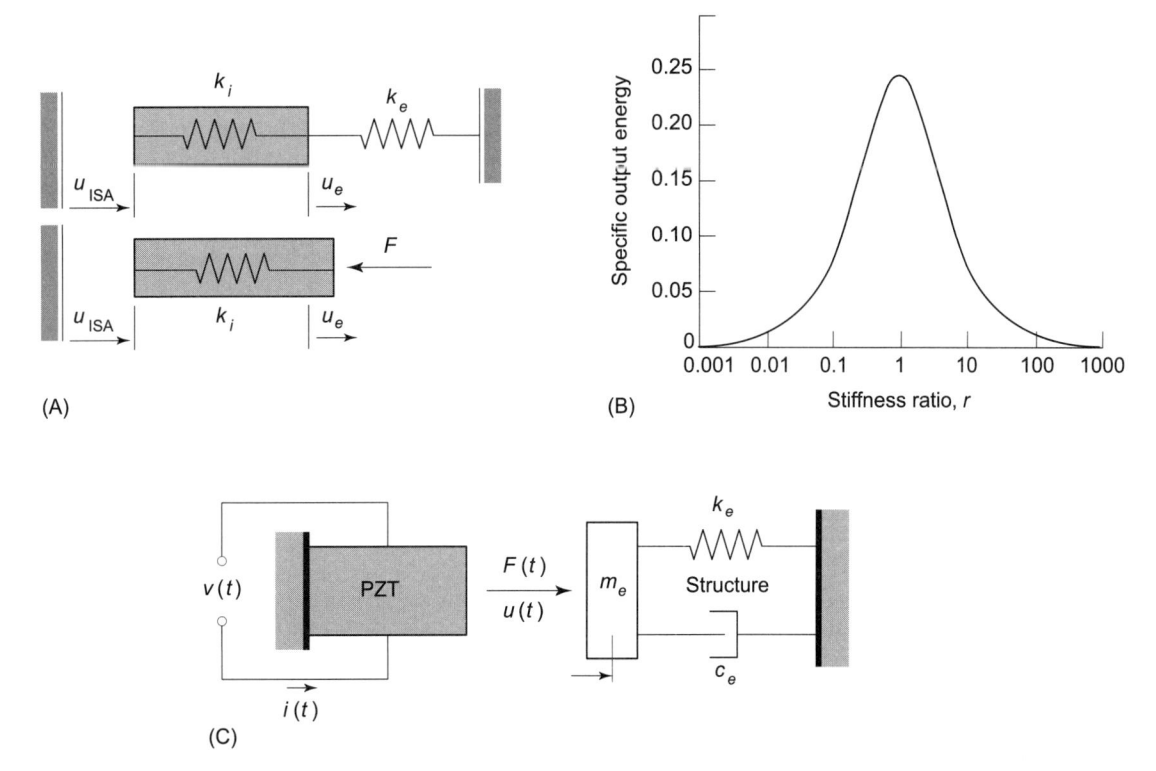

Figure 13 Schematic of the interaction between an active material actuator and an elastic structure: (A) the active displacement, u_{ISA}, is partially lost due to the actuator internal compressibility, such that the resulting external displacement, u_e, is always less; (B) maximum energy extraction is attained when the internal and external stiffness match; (C) under dynamic conditions, the frequency dependent dynamic stiffness must be used.

Electrical Input Power and Energy

From the electrical side, note that, for frequencies well below the actuator free resonance (typically, 1–10 kHz, depending on length), the wave propagation effects can be ignored, and the equivalent input admittance as seen at the actuator terminals is:

$$Y_C(\omega) = i\omega C \left[1 - \kappa_C^2 \frac{\bar{r}(\omega)}{1 + \bar{r}(\omega)} \right] \qquad [15]$$

where:
$\kappa^2 = d^2/(\bar{s}\bar{\varepsilon})$ and $\bar{s} = (1 - i\delta_s)s$, $\bar{\varepsilon} = (1 - i\delta_\varepsilon)\varepsilon$, δ_s is the hysteresis internal damping coefficient of the actuator, δ_ε is the dielectric loss coefficient of the actuator, $d = (u_{ISA}/V)(t/l)$, $\varepsilon = (C/A)(t^2/l)$, $s = A/lk_i$, l is the stack length, t is the layer thickness, A is the stack cross-sectional area, and u_{ISA} is the dynamic free stroke at voltage V, while $\kappa_C^2 = d^2/s\varepsilon = k_i u_{ISA}^2/CV^2$ is the effective full-stroke electromechanical coupling coefficient of the actuator. For magnetoactive actuators, $\kappa_L^2 = k_i u_{ISA}^2/LI^2$ and

$$Z_L(\omega) = i\omega L \left[1 - \kappa_L^2 \frac{\bar{r}(\omega)}{1 + \bar{r}(\omega)} \right] \qquad [16]$$

The input electrical power can be evaluate as $(1/2)Y_C V^2$ for capacitive loads, and $(1/2)Z_L I^2$ for inductive loads, where the frequency dependent Y_C and Z_L are given by eqns [15] and [16], respectively. When the DC bias effects are included, expressions that are slightly more complicated need to be used.

Design of Effective Induced-strain Actuators

The design of effective piezo actuators is particularly difficult due to the small displacement generated by these materials. However, the forces that can be generated by a piezo actuator can be very large; are only limited by the inherent stiffness and compressive strength of the piezo material; by its effective area. For most practical applications, displacement amplification of the induced strain displacement is employed. The effective design of the displacement amplifier can 'make or break' the practical effectiveness of a piezo actuator. The most important parameter that needs to be optimized during such a design is the energy extraction coefficient defined as the ratio between the effective mechanical energy delivered by the actuator and the maximum possible energy that can be delivered by the actuator in the stiffness–match condition.

Of extreme significance in the design of an effective ISA is the amount of energy available for performing external mechanical work. **Figure 14** presents comparison of the energy density per unit volume for some commercially available ISAs under static and dynamic regimes. Typical energy density values are in the range 2.71–$15.05\,\text{J}\,\text{dm}^{-3}$ (0.292–$1.929\,\text{J}\,\text{kg}^{-1}$) under static conditions, and 0.68–$3.76\,\text{J}\,\text{dm}^{-3}$ (0.073–$0.482\,\text{J}\,\text{kg}^{-1}$) under dynamic conditions. Power densities of up to $23.6\,\text{kW}\,\text{dm}^{-3}$ ($3.0\,\text{kW}\,\text{kg}^{-1}$) at 1 kHz are predicted. The overall efficiency of active-material actuation depends, to a great extent, on the efficiency of the entire system that includes the active-material transducer, the displacement amplification mechanisms, and the power supply.

Power Supply Issues in Induced Strain Actuation

Since ISAs can operate in the kHz range, eqn [14], in conjunction with **Figure 14**, indicates the opportunity for the generation of large output power densities. However, a number of practical barriers need to be overcome before this can be achieved: (i) the capability of power supply to deliver kVA reactive power; (ii) the dissipation of the heat generated in the active material due to internal losses (hysteresis) that can be in the range 5–10% of the nominal reactive input power; and (iii) the electromechanical system resonance that sets upper frequency limits.

The power supply aspects of induced-strain actuation are currently being addressed by using specialized power supplies (switching amplifiers) that are able to handle reactive loads much better than conventional linear amplifiers. The switching amplifiers (**Figure 15**) utilize high frequency pulse width modulation (PWM) principles and solid-state switching technology to drive current in and out of the capacitive load presented by the induced-strain piezo-electric actuator. In this process, a high-value inductance ballast is utilized (**Figure 15A**). To simultaneously achieve optimal actuator performance and minimum weight, the switching amplifier design has to be performed using a complete model that includes adequate representation of the actuator and external load dynamics (**Figure 15B**).

The difficulty connected with the electric-supply may seem less severe when using magnetoactive devices, such as Terfenol-D actuators, because they are current driven and require lower voltages at low frequencies. At present, the low-voltage power supply technology is predominant and cheaper than the high-voltage technology, which somehow facilitates the use of magnetostrictive actuators.

The effect of adaptive excitation on the current and power requirements, and on the displacement output from a heavy-duty piezo-electric actuator driving a smart structure near the electromechanical resonance is shown in **Figure 16**. The actuator was assumed to have $k_i = 370\,\text{kN}\,\text{mm}^{-1}$ internal stiffness, $C = 5.6\,\mu\text{F}$ internal capacitance, 3.5 kHz resonance, and output displacement $u_m \pm u_a = 22.5 \pm 37.5\,\mu\text{m}$ when driven by a voltage $V_m \pm V_a = -375 \mp 625\,\text{V}$. The external mechanical load was assumed to have a matched static stiffness, a mechanical resonance of 30 Hz,

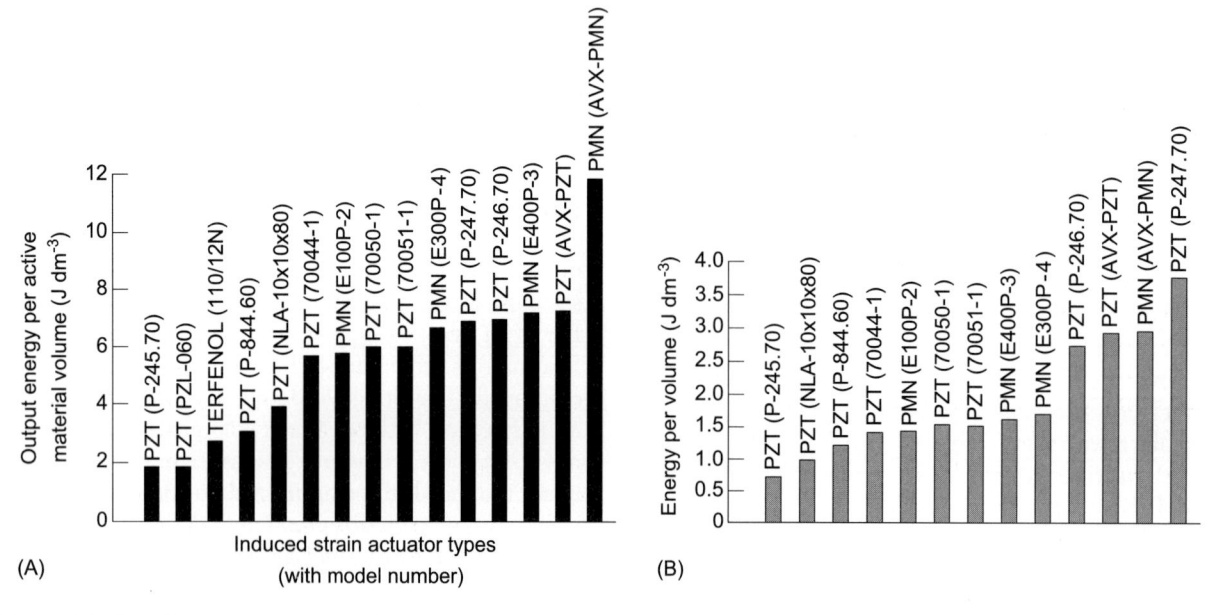

Figure 14 Specific energy output capabilities of a selection of commercially available induced-strain actuators: (A) static operation; (B) dynamic operation.

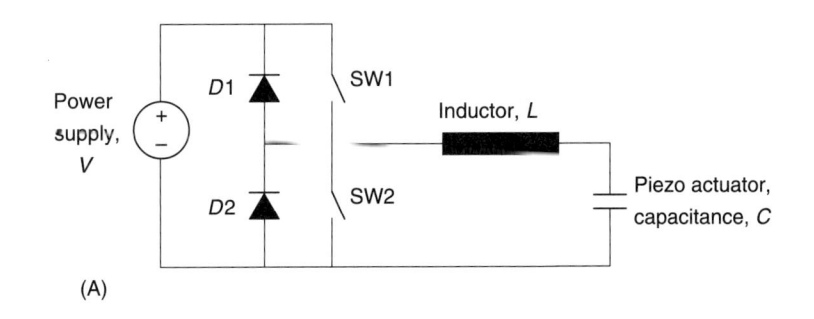

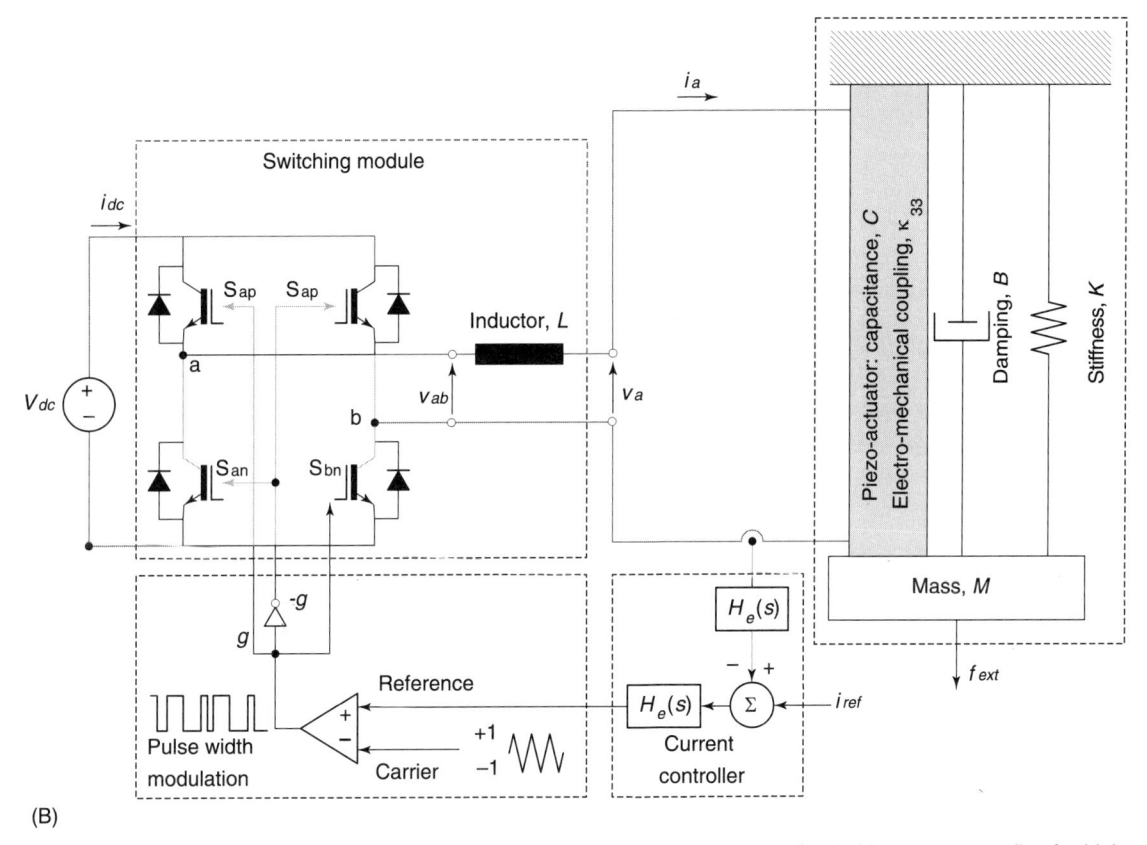

Figure 15 (see Plate 6). Power supplies for active material actuators: (A) principle of switching power supplies for high reactive load; (B) schematic of the supply system incorporating the switching module, current controller, pulse width modulator, and the piezo-actuator-external load assembly.

and a 1% internal damping. The system was assumed driven by a 1 kVA amplifier, with up to 1000 V voltage and 1 A current. As show in **Figure 16**, the electromechanical system incorporating the structure and the embedded actuator displays an electromechanical resonance at 42.42 Hz. Without adaptive excitation, i.e., under constant voltage supply, the current demands are very large, the actuator experiences a displacement peak that may lead to its destruction, and the required power is excessive. When adaptive excitation was simulated, both the displacement and the current could be kept within bounds, while the power requirements became more manageable and kept within the 1 kVA capability of the power amplifier.

Smart Structures

The concept of smart structures has largely evolved by biomimesis and under the influence of Asimov's three laws of robotics. In the biological world, plants and animals alike react to the environment in order to protect their existence, or to acquire the much needed nourishment. For example, the sharp heat from an open flame would instantly make someone retract his hand. It would also, most probably, make the person shout 'ouch!'. In the engineering world, smart structures are viewed as adaptive systems fitted with sensors, actuators, and command-control processors that could take automatic actions without specific human interventions.

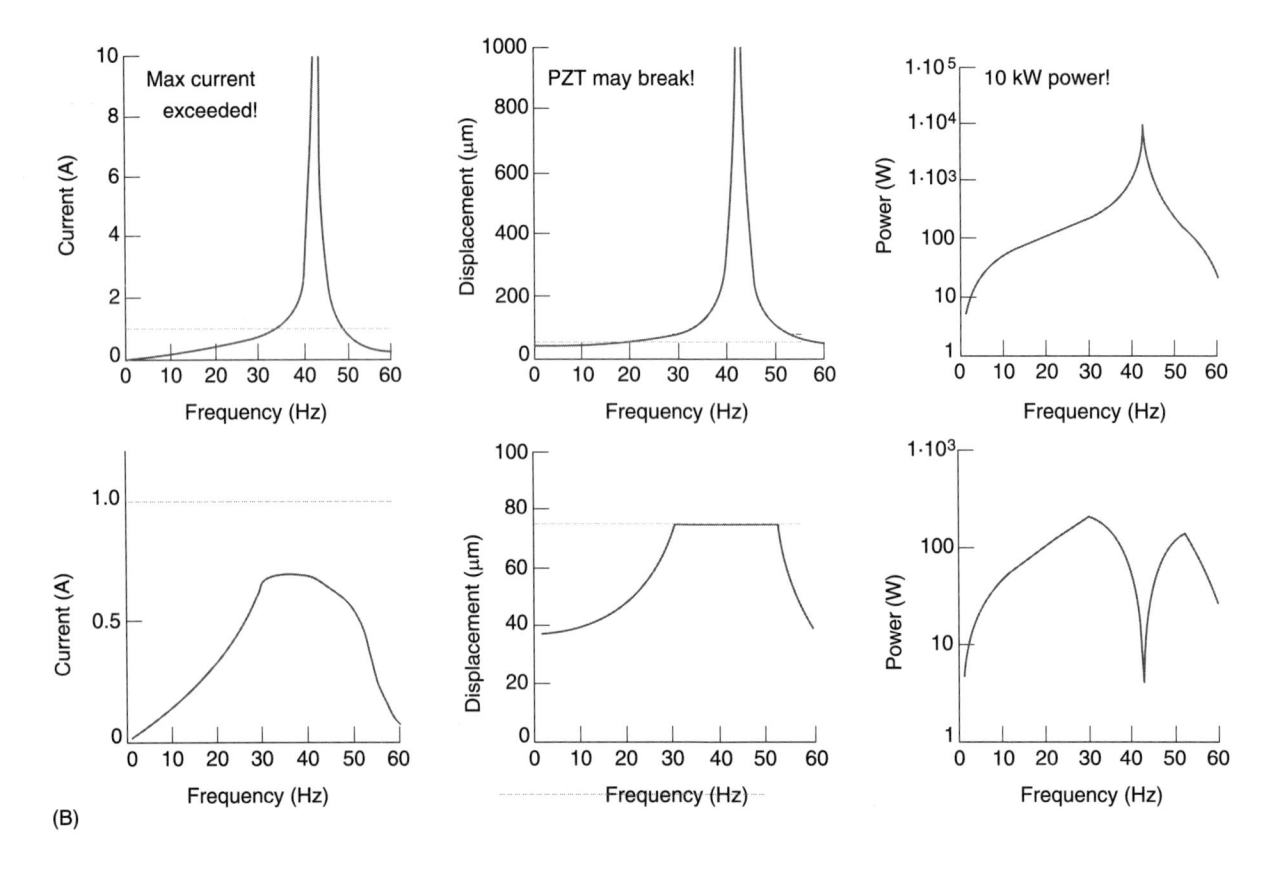

Figure 16 Adaptive excitation of piezo-electric actuators near embedded electromechanical resonance: (A) under constant voltage excitation, the current demands are very large, the actuator may break, and the required power is excessive; (B) with adaptive excitation, both the displacement and the current are kept within bounds, and the power becomes manageable.

Sensory Smart Structures

Sensory smart structure attributes are found in even the simplest single-cell micro-organism (**Figure 17A**). A bridge fitted with smart structure characteristics has been conceptualized. Such a 'smart' bridge (**Figure 17B**) would be expected to 'sense' the environment, react accordingly, and 'tell' what is happening by sending alarm signals that announce that strength and safety are diminishing, and appropriate action is needed. In this example, a smart structure is seen that is capable of automatic health monitoring, damage detection, and failure prevention.

Adaptive Actuation Smart Structures

Another class of smart structure is that fitted with adaptive actuation. Nature offers the ideal example of adaptive actuation. A pair of antagonistic muscles (*musculus biceps brachii* and *musculus triceps brahii*) ensures exact and precise position control during the most difficult maneuvers of our arms. The skeletal muscles of the human arm are attached at a small-displacement high-force position, well-suited for the induced-strain muscle actuation (**Figure 18A**). As a catapulting twitch is performed, the nonlinear skeletal kinematics effects a favorable response during which the distance between the thrown weight and the elbow joint decreases while the perpendicular distance between the muscle tendon and the joint increases, such that the overall force required in the muscular actuator decreases as the catapulting motion develops.

In fact, once the weight has been put in motion, the biceps muscle can reduce the contracting signal and even stop before the end of the catapulting motion. In the same time, the antagonistic triceps muscle can start its braking action before the end position has been achieved, such that a smooth transition is achieved. This remarkable feat is accomplished through complex adaptive control architecture, as depicted in **Figure 18B**. An engineering building incorporating adaptive structural response is presented in **Figure 18C**. This smart structure not only senses and processes the external stimuli, but also takes mechanical action. When the vibration excitation from a gust of wind or earthquake is endangering the structural integrity through excessive resonance response, action is taken in the variable stiffness

ACTUATORS AND SMART STRUCTURES

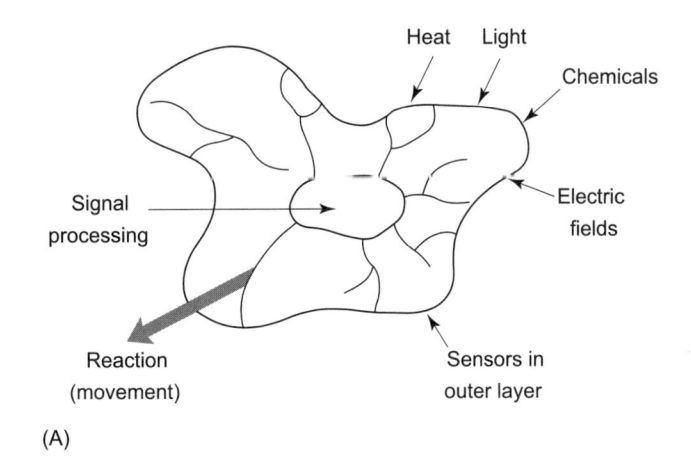

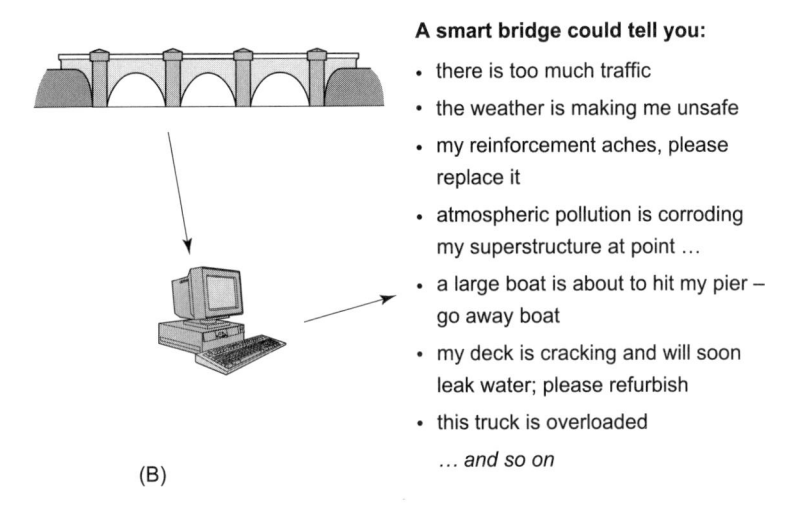

Figure 17 Sensory smart structures: (A) single cell micro-organism viewed as a sensory smart structure; (B) smart structure concept applied to a bridge.

members such that the structure's natural frequency is shifted and the resonance is avoided. This natural frequency shift achieved through stiffening or relaxing of active structural members resembles the bracing of one's muscles when trying to attain steadiness in a challenging situation. Besides resonance avoidance, smart structures can also attempt to dissipate energy through active and passive mechanisms, or to prevent a nonlinear vibrations or aeroelastic effect from building up through vibration cancellation.

Applications of Shape Memory Alloys to Vibration Control

Because of its biocompatibility and superior resistance to corrosion, shape memory alloys such as Nitinol have gained wide usage in the medical field as bone plates, artificial joints, orthodontic devices, coronary angioplasty probes, arthroscopic instrumentation, etc. In engineering, these materials have been used as force actuators and robot controls. They also offer vibration control potentials based on three important principles: (i) the three to four times increase in elastic modulus in the transition from martensitic to austenitic phase; (ii) the creation of internal stresses; and (iii) the dissipation of energy through inelastic hysteretic damping. These effects can be practically realized either as additional components to be retrofitted on existing structures, or as hybrid composite materials containing embedded SMA fibers.

The increase in elastic modulus is used in the active properties tuning (APT) vibration control method. As SMA wires are activated by heating with electric current or other methods, their modulus increases threefold from 27 GPa to 82 GPa. Depending on the structural architecture and on how much SMA material is used, this may result in a sizable change in the effective structural stiffness, and a considerable frequency shift away from an unwanted resonance.

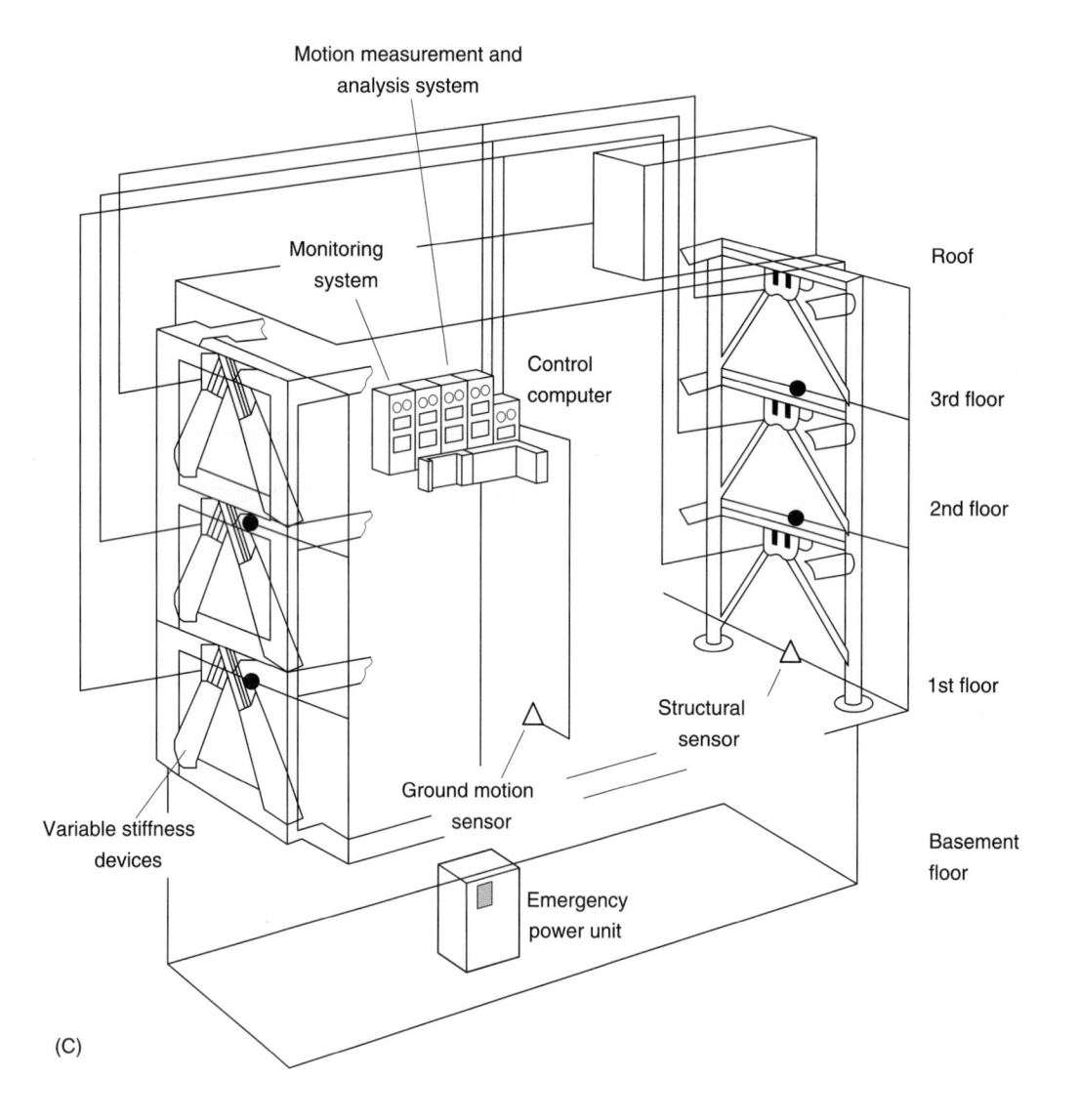

Figure 18 Concepts in adaptive smart structures actuation: (A) kinematic of an arm muscle; (B) muscle control system with sensory adaptation; (C) schematic of an earthquake-resistant variable-stiffness building (KaTRI No.21 in service in Tokyo, Japan).

The activation speed, which depends on the heating rate, is usually sufficient to achieve acceptable structural control. The recovery, however, depends on the rate of cooling, and hence takes place much more slowly.

The creation of internal stresses is used in the active strain energy tuning (ASET) method. The activation of stretched SMA fibers can make them shrink by 4–8% and thus create considerable contractile stress in the support structure. If the SMA fibers are placed inside beams or plates, active frequency control can be readily achieved, since the presence of inplane compressive stresses can considerably change the beams' and plates' natural frequencies. One ready application of this effect is the avoidance of critical speeds during the run-up and run-down of high-speed shafts.

Applications of Electro- and Magnetoactive Materials to Vibration Control

Active materials are well suited for vibration and aeroelastic control. Electroactive (PZT and PMN) and magnetoactive (Terfenol-D) actuators can be used as translational actuators to replace conventional devices, or as surface-bonded actuators to induce axial and bending strains in the host structure. In the former case, the strain induced parallel to the field direction is utilized, as, for example, in piezoelectric stacks. In the latter case, the strain induced transverse to the direction of the applied field is used. Translational ISAs are ideal in retrofit situations when the replacement of a conventional actuator with a 'smart' actuator is sought. Surface-bonded actuation is a completely new engineering concept, which is specific to the smart structures world. Bonded electroactive actuator wafers have been used successfully to control the shape of deformable mirrors. In the stack configuration, they have been used for impact dot-matrix printing. Advantages over conventional electromagnetic actuators included order-of-magnitude higher printing speeds, order-of-magnitude lower energy consumption; and reduced noise emissions. A tunable ultrasonic medical probe composed of electroactive elements embedded in a polymer matrix has also been developed. Electro- and magneto-active materials can be used to enhance structural damping and reduce vibrations. Two mechanisms are available: (i) direct approach, in which the vibration energy is dissipated directly through the electromechanical interaction between the active material and the host structure; (ii) indirect approach, in which the active material is used to enhance the damping properties of a conventional damping treatment. The dissipation of vibration energy through the electromechanical interaction of the active materials with the host structure can be achieved either passively or actively. Since the active material is connected to the structure undergoing vibration, the deformation of the active material follows the deformation of the structure. As the structure deforms during vibrations, the active material takes up the strain and transforms it into an oscillatory electrical field.

For passive active-material vibration suppression, the induced electric field is used to drive currents into an external resistance thus dissipating the energy through Ohmic heating (**Figure 19**). In order to dissipate selected frequencies, RCL tuning principles are applied. In active vibration suppression, the active material is used to produce vibration input in antiphase with the external disturbance, thus resulting in noise and vibrations cancellation. In this case, the energy is dissipated in the heat sink of the driver–amplifier circuit. The active material can be also used to enhance the damping properties of a conventional damping material through the 'constraint layer damping effect'. Conventional vibration damping treatments utilize the dissipation properties of viscoelastic materials that mainly operate in shear. The shear can be enhanced if, on top of the damping layer, an extra layer of active material is added that deforms in antiphase with the base structure. Thus, the damping layer between the active material layer and the base structure is subjected to a much larger differential shear strain than in the absence of the active layer.

For illustration, two current aerospace smart structure projects are discussed. One project is aimed at the reduction of noise and vibrations in helicopter rotors (**Figure 20**). The other project is addressing the buffet vibrations alleviation in a fixed wing aircraft (**Figure 21**). The smart materials actuation rotor technology (SMART) rotor blade program, under way at Boeing (Mesa), is tasked to test the feasibility of using active materials actuators for rotor blade control to reduce noise and vibrations, improve ride qualities, and extend the service life. The conceptual design of the SMART rotor blade program calls for the simultaneous satisfaction of two important operational requirements: (i) reduction of blade vibration through in-flight rotor track and balance adjustments; and (ii) reduction and counteraction of aerodynamically induced noise and vibration through an actively controlled aerodynamic surface. The first objective is achieved with a slow-moving trim tab controlled through a bidirectional SMA actuator (**Figure 20**). The second objective is met with a fast-moving control flap actuated by piezo-ceramic stacks through a stroke-amplifier.

Figure 19 Tuned shunt circuit for vibration control. L, inductor; R, resistor.

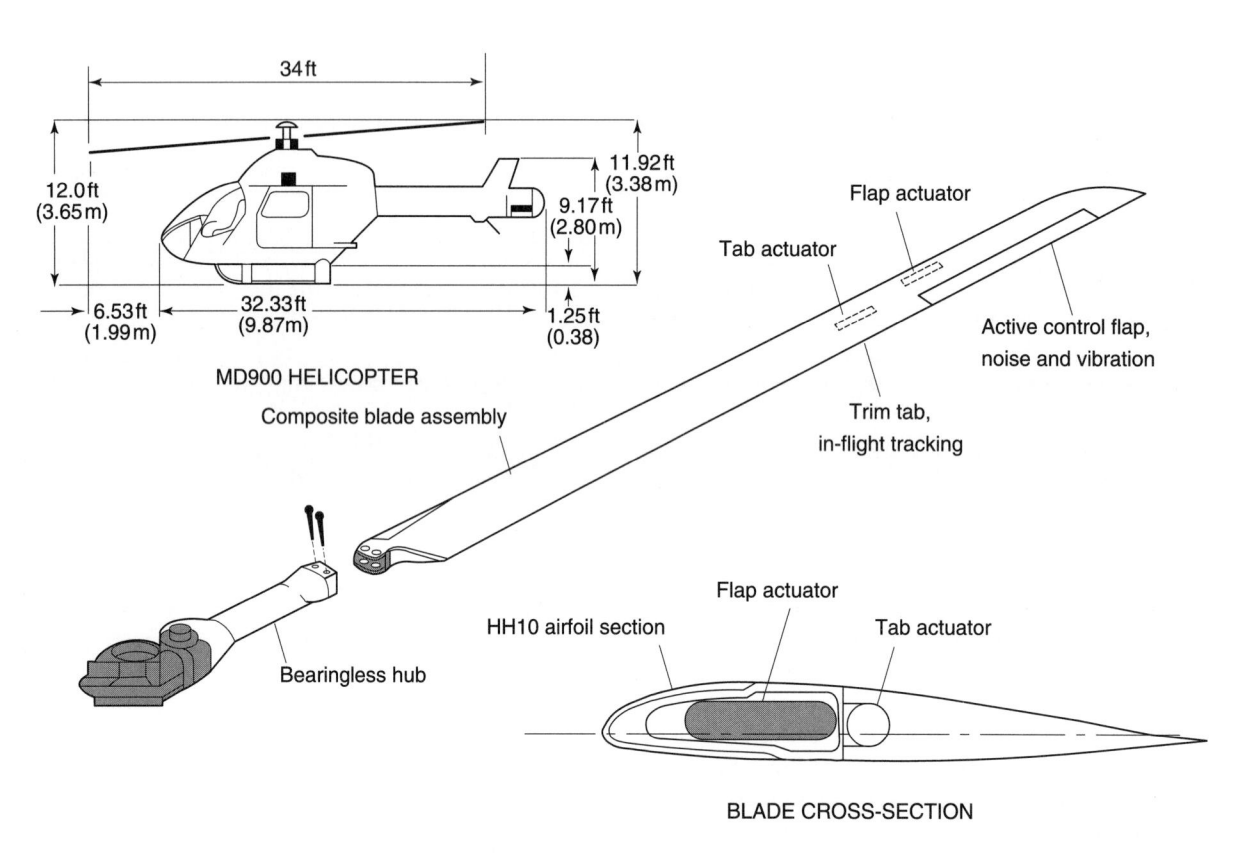

Figure 20 MD 900 helicopter and hingeless blade displaying the planned trim tab for in-flight tracking and active control flap for noise and vibration reduction.

The second smart structures project to be discussed here relates to aircraft tail buffeting. Tail buffeting is a significant concern for aircraft fatigue and maintenance. The actively controlled response of buffet affected tails (ACROBAT) program studied active materials solutions to resolve the buffet problems of the F/A–18 twin-tail aircraft. A 1/6-scale full-span model was tested in the NASA Langley transonic dynamics tunnel. The portside vertical tail was equipped with surface-bonded piezo-electric wafer actuators, while the starboard vertical tail had an active rudder and other aerodynamic devices. During

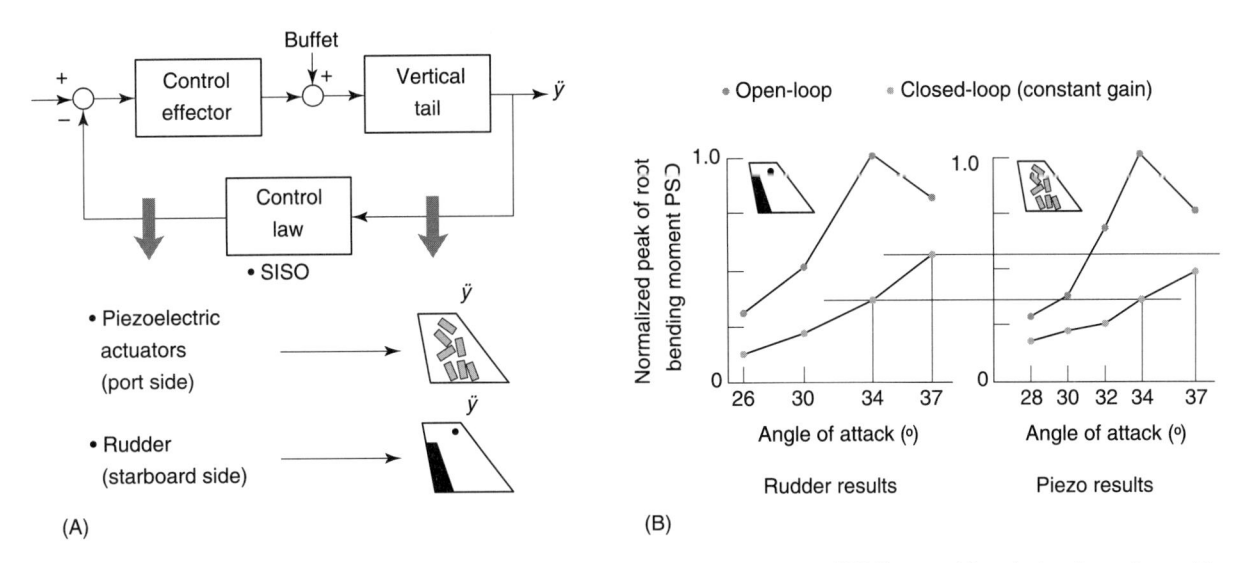

Figure 21 ACROBAT tail buffet alleviation experiments: (A) single-input single-output (SISO) control law design for active rudder and piezo-electric wafers excitation; (B) power spectrum density (PSD) peak values for the root bending moment at the first bending resonance.

the wind tunnel tests, constant-gain active control (**Figure 21A**) was able to reduce power spectral density of the first bending resonance by as much as 60% (**Figure 21B**).

Adaptive Algorithms for Smart Structures Control

Implementation of control algorithms in smart structures architecture is subject to attentive scrutiny. Conventional application of classical control algorithms is only the first step in this process. Much better results are obtained if modern adaptive control is used, such that the resulting smart structure can react to changes in the problem-definition parameters. Actual structural designs are very complex, nonlinear in behavior, and subject to load spectra that may be substantially modified during the structures service life. Under such adverse situations, the resulting uncertainty in the controlled plant dynamics is sufficient to make 'high-performance goals unreachable and closed-loop instability a likely result'. To address this problem, at least three adaptive control approaches are advocated: (i) adaptive signal processing methods; (ii) model reference adaptive control (MRAC); and (iii) self tuning regulators (STR). Though different in detail, all three aim at the same goal, i.e., to eliminate the effect of variations in disturbance signature and plant dynamics on the smart structure's performance. Many adaptive signal processing algorithms operate in the feedforward path. For example, an acoustic cost function was used in conjunction with structural inputs in controlling far-field acoustic pressures to achieve aircraft cabin noise cancellation. MRAC and STR are mainly associated with adaptive feedback structural control. **Figure 22** illustrates the adaptive feedback and feedforward concepts in an adaptively controlled smart structure. The disturbance input, $u_d(z)$, is fed through the feedforward compensator, $K_{ff}(z)$, and summated with the error signal, $e(z)$, modulated through the feedback compensator, $K_{fb}(z)$, to generate the control signal, $u_c(z)$. The error signal, $e(z)$, is obtained in the classical control fashion by subtracting the resulting output, $y_e(z)$, from the required reference signal, $r(z)$. The blocks $P_{de}(z)$ and $P_{ce}(z)$ represent the plant's response to disturbance and control, respectively. The feedforward compensator, $K_{ff}(z)$, is adaptively modified in response to trends noticed in the output signal, $y_e(z)$, by the adaptive algorithm block that could typically be a least-mean-squares (LMS) method applied to the coefficients of a finite impulse response (FIR) filter. A cost functional incorporating a measure of the output, $y_e(z)$, or a time-averaged gradient (TAG) descent algorithm may be used. Use of hybrid control architecture combines the feedback and feedforward principles to achieve best performance when complex disturbances incorporating both persistent and impulsive components are present.

Conclusion

The concepts of induced-strain actuation and smart structures have been reviewed and briefly discussed. ISAs are based on active materials that display dimensional changes when energized by electrical, magnetic, or thermal fields. Piezo-electric, electrostrictive, magnetostrictive, and shape memory alloy

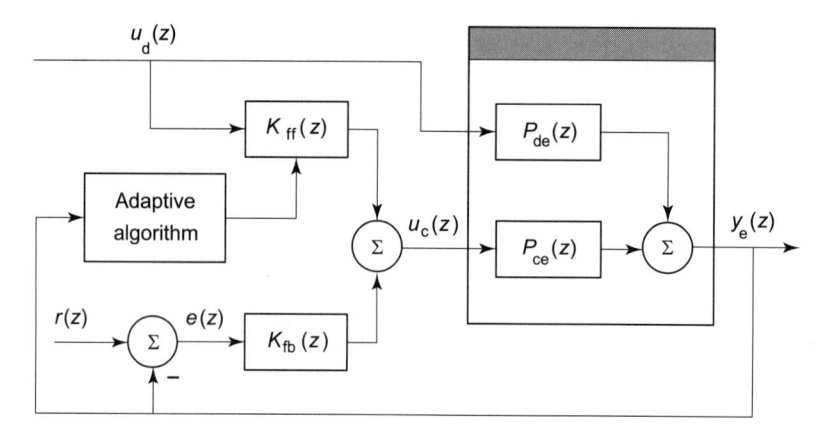

Figure 22 Schematic representation of smart structure with adaptive hybrid control capabilities.

materials have been presented and analyzed. Of these, piezo-electric (PZT), electrostrictive (PMN), and magnetostrictive (Terfenol-D) materials have been shown to have excellent frequency response (1–10 kHz, depending on actuator length), high force (up to 50 kN on current models), but small induced-strain stroke capabilities (typically, 0.1 mm for 0.1% strain on a 100 mm actuator). With this class of ISAs, displacement amplification devices need always to be incorporated into the application design. In contrast, the shape memory alloy (Nitinol) materials have large induced-strain stroke capabilities (typically, 4 mm for 4% strain on a 100 mm actuator), but low-frequency response (< 1 Hz).

The effective implementation of induced strain actuation was discussed and guidelines for achieving optimal energy extraction were presented. As different from conventional actuation techniques, induced-strain actuation can only rely on a limited amount of active stroke, and this has to be carefully managed. It has been shown here that the stiffness and impedance matching principles can produce the maximum energy extraction from the ISA and ensure its transmission into the external application. Details of these principles, together with typical energy density values for various ISAs have been given. It was found that, for dynamic applications, as much as 3.76 J dm^{-3} (0.482 J kg^{-1}) could be extracted under dynamic conditions. Power densities of up to 23.6 kW dm^{-3} (3.0 kW kg^{-1}) were predicted at 1 kHz.

In summary, one can conclude that the potential of smart materials in structural applications has been clearly demonstrated through laboratory research in many institutions around the world. However, the field is still in its infancy and further research and development is required to establish smart materials as reliable, durable, and cost-effective materials for large-scale engineering applications.

Nomenclature

A	cross-sectional area
B	magnetic flux density
D	electrical displacement; Young's modulus
E	electrical field
F	force
H	magnetic field intensity
I	current
l	length
M	electrostrictive coefficient
P	power
S	strain
t	thickness
T	stress; temperature
V	voltage
ε	strain
μ	magnetic permeability
ξ	martensitic fraction
Λ	state variable
Θ	thermoelastic tensor
Ω	transformation tensor

See Plates 4, 5, 6.

See also: **Electrostrictive materials; Magnetostrictive materials; Piezoelectric materials and continua; Shape memory alloys.**

Further Reading

Anon (1988) IEEE standard on piezoelectricity, ANSI/IEEE Std 176–1987, Institute of Electrical and Electronics Engineers, Inc., New York.

Bank R (1975) *Shape Memory Effects in Alloys*, p. 537. New York: Plenum.

Clark AE (1992) High power rare earth magnetostrictive materials. In: Rogers CA and Rogers RC (eds). *Proc. Recent Advances in Adaptive and Sensory Materials*, pp. 387–397. Lancaster, PA: Technomic Publishing.

Clark RL, Saunders, WR, Gibbs G (1998) *Adaptive Structures – Dynamics and Control*. Wiley.

Culshaw B (1996) *Smart Structures and Materials*. Artech House Publishers.

Duerig TW, Melton KN, Stockel D, Wayman CM (1990) *Engineering Aspects of Shape Memory Alloys*. Butterworth-Heinemann.

Fuller CR, Elliot SJ, Nelson PA (1996) *Active Control of Vibration*. London: Academic Press.

Funakubo H (1987) *Shape Memory Alloys*. New York: Gordon and Breach.

Giurgiutiu V (2000) Active-materials induced-strain actuation for aeroelastic vibration control. *Shock and Vibration Digest* 32(5): 335–368.

Giurgiutiu V, Craig A, Rogers CA (1997) Power and energy characteristics of solid-state induced-strain actuators for static and dynamic applications. *Journal of Intelligent Material Systems and Structures* 8: 738–750.

Janocha H (1999) *Adaptronics and Smart Structures*. Springer-Verlag.

Lachisserie, E du Tremolet de (1993) *Magnetostriction – Theory and Applications*. CRC Press.

Moses RW (1997) Active vertical tail buffeting alleviation on a twin-tail fighter configuration in a wind tunnel. *CEAS International Forum on Aeroelasticity and Structural Dynamics 1997*, Rome, Italy, June 17–20.

Nachtigall W (1999) Adaptronic systems in biology and medicine. In: Janocha H (ed.), *Adaptronics and Smart Structures*. Springer.

Park S-E, Shrout TR (1997) Ultrahigh strain and piezoelectric behavior in relaxor ferroelectric single crystals. *Journal of Applied Physics* 4(82): 1804–1811.

Rao SS, Sunar M (1999) Recent advances in sensing and control of flexible structures via piezoelectric materials technology. *Applied Mechanics Reviews* 52(1): 1–16.

Straub FK, King RJ (1996) Application of smart materials to control of a helicopter rotor. *SPIE Symposium on Smart Structures and Materials*, February 26–29, San Diego, CA.

Tanaka K, Iwasaki R (1985) A phenomenological theory of transformation superplasticity. *Engineering Fracture Mechanics* 21(4): 709–720.

Udd E (1995) *Fiber Optic Smart Structures*. Wiley.

ADAPTIVE FILTERS

S J Elliott, The University of Southampton, Institute of Sound and Vibration (ISVR), Southampton, UK

Copyright © 2001 Academic Press

doi:10.1006/rwvb.2001.0059

In the article on optimal filters (see **Optimal filters**) we saw that the optimum finite impulse response (FIR) filter, which minimizes the mean-square error for the model problem shown in **Figure 1** of that article, can be directly calculated from a knowledge of the autocorrelation properties of the reference signal and the cross-correlation between the reference and desired signal. In a practical problem, these auto- and cross-correlation functions would have to be estimated from the time histories of these signals, which would require a considerable amount of data in order to calculate accurately. It was also assumed that the reference and desired signals are stationary, since otherwise their correlation properties will change with time. The calculation of the optimal filter with I coefficients involves the inversion of the $I \times I$ autocorrelation matrix. Although this matrix has a special form (it is symmetric and Toeplitz), and efficient algorithms can be used for its inversion, the computational burden is still proportional to I^2 and so can be significant, particularly for long filters. The matrix inversion may also be numerically unstable if the matrix is ill-conditioned.

Another approach to determining the coefficients of such a filter would be to make them adaptive. Instead of using a set of data to estimate correlation functions, and then using these to calculate a single set of optimal filter coefficients, the data are used sequentially to adjust the filter coefficients gradually so that they evolve in a direction which minimizes the mean-square error. Generally, all the filter coefficients are adjusted in response to each new set of data, and the algorithms used for this adaptation use a considerably smaller number of calculations per sample than the total number of calculations required to compute the true optimal coefficients. As well as converging towards the optimal filter for stationary signals, an adaptive filter will also automatically readjust its coefficients if the correlation properties of these signals change. The adaptive filter is thus capable of tracking the statistics of nonstationary signals, provided the changes in the statistics occur slowly compared with the convergence time of the adaptive filter.

Steepest Descent Algorithm

The most widely used algorithm for adapting FIR digital filters is based on the fact that the error surface for such filters has a quadratic shape, as shown in **Figure 2** of the article on optimal filters (see **Optimal filters**). This suggests that, if a filter coefficient is adjusted by a small amount, which is proportional to the negative of the local gradient of the cost function with respect to that filter coefficient, then the coefficient is bound to move towards the global minimum of the error surface. If all the filter coefficients are simultaneously adjusted using this gradient descent method, the adaptation algorithm for the vector of filter coefficients may be written as:

$$\mathbf{w}(\text{new}) = \mathbf{w}(\text{old}) - \mu \frac{\partial J}{\partial \mathbf{w}}(\text{old}) \quad [1]$$

where μ is a convergence factor and $\partial J/\partial \mathbf{w}$ was defined in eqn [14] of the article on optimal filters (see **Optimal filters**).

For the model problem shown in **Figure 1** of that article, the vector of derivatives is given by eqn [15], and using the definitions given in eqns [7] and [8] of that article, this can be written as:

$$\frac{\partial J}{\partial \mathbf{w}} = -2E[\mathbf{x}(n)\mathbf{x}^T(n)\mathbf{w} - \mathbf{x}(n)d(n)] \quad [2]$$

The measured error signal is given by:

$$e(n) = d(n) - \mathbf{x}^T(n)\mathbf{w} \quad [3]$$

and so the vector of derivations may also be written as:

$$\frac{\partial J}{\partial \mathbf{w}} = -2E[\mathbf{x}(n)e(n)] \quad [4]$$

To implement the true gradient descent algorithm, the expectation value of the product of the error signal with the delayed reference signals would need to be estimated to obtain eqn [4], probably by time-averaging over a large segment of data, and so the filter coefficients could only be updated rather infrequently.

The suggestion made in the seminal paper by Widrow and Hoff was that, instead of infrequently updating the filter coefficients with an averaged estimate of the gradient, the coefficients be updated at every sample time using an instantaneous estimate of the gradient, which is sometimes called the stochastic gradient. This update quantity is equal to the derivative of the instantaneous error with respect to the filter coefficients:

$$\frac{\partial e^2(n)}{\partial \mathbf{w}} = -2\mathbf{x}(n)e(n) \quad [5]$$

The adaptation algorithm thus becomes:

$$\mathbf{w}(n+1) = \mathbf{w}(n) + \alpha\mathbf{x}(n)e(n) \quad [6]$$

where $\alpha = 2\mu$ is the convergence coefficient, which is known as the LMS (least mean-square) algorithm, and this has been very widely used in a variety of practical applications, for example in adaptive electrical noise cancellation, adaptive modeling and inversion and adaptive beam forming. A block diagram indicating the operation of the LMS algorithm is shown in **Figure 1**.

Properties of the LMS Algorithm

In order to guarantee convergence it is not only necessary to ensure that the mean value of the filter coefficients converge but also that their mean square value converges. In theory, this can be ensured provided the convergence coefficient α is positive, but below a certain value which is proportional to the mean-square value of the reference signal, i.e.:

$$0 < \alpha < \frac{2}{I\bar{x}^2} \quad [7]$$

where I is the number of coefficients and $\bar{x}^2 = E[x^2(n)]$ is the mean-square value of $x(n)$. This has been found to give a reasonable estimate of the maximum stable value of the convergence coefficient in many practical simulations.

The convergence rate of the LMS algorithm is not uniform, however, but depends on the shape of the multidimensional error surface formed by plotting the mean-square error against each of the FIR filter coefficients. If a section is taken through this surface, the resulting graph is guaranteed to be quadratic, but its shape depends on exactly which combination of filter coefficients is included in the section, and the autocorrelation properties of the reference signal. If we consider the combination of filter coefficients corresponding to the eigenvectors of the autocorrelation matrix (eqn [10] of the article on optimal filters; see **Optimal filters**) then the steepness of the quadratic sections along these directions, which are called the principal axis of the error surface, are defined by the

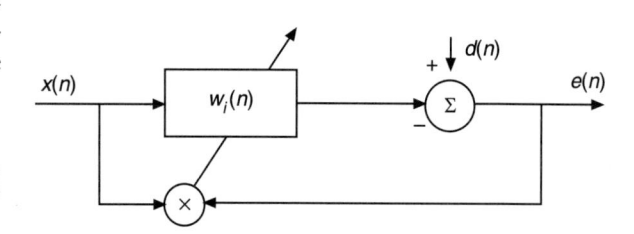

Figure 1 Diagrammatic representation of an FIR filter, with coefficients w_i at the nth sample time, being adapted using the LMS algorithm.

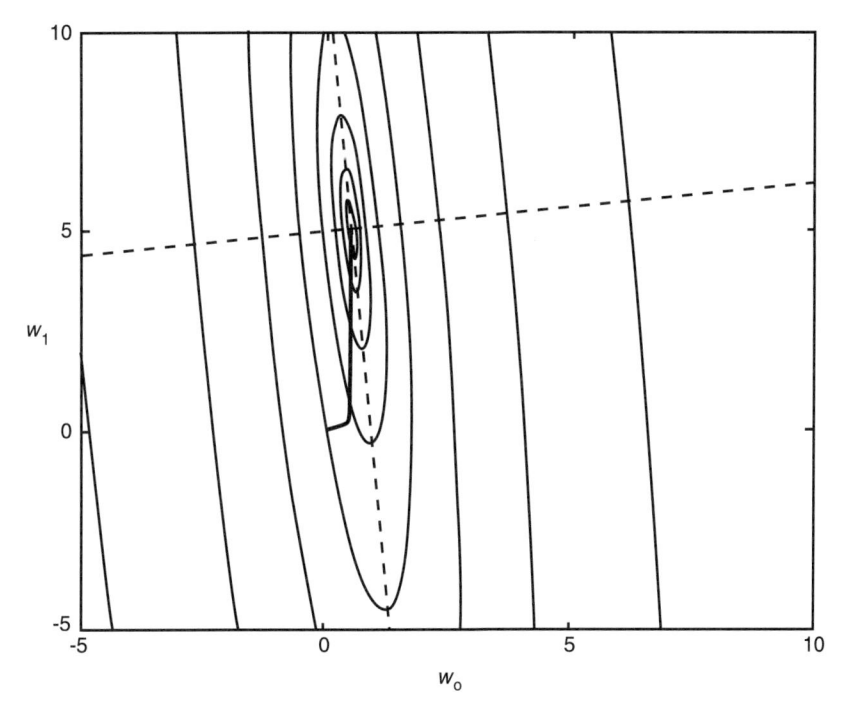

Figure 2 Contour map of an example error surface showing the trajectory of the average behavior of the LMS algorithm. Also shown are the principal axes of the error surface.

eigenvalues of the autocorrelation matrix, as illustrated in **Figure 2**. The LMS algorithm will, on average, follow the steepest descent path down this error surface, as also illustrated in **Figure 2**. The average convergence of the LMS algorithm along one of the principle axes would be exponential, with a time constant (in samples) given by:

$$\tau_i \approx \frac{1}{2\alpha\lambda_i} \qquad [8]$$

where λ_i is the eigenvalue of the autocorrelation matrix corresponding to this principal axis. In practice, the convergence of the algorithm is associated with all principal axes at once, although clearly that associated with the largest eigenvalue, λ_{max}, decays fastest and that associated with the smallest eigenvalue, λ_{min}, decays most slowly. **Figure 3** shows the average reduction in the level of the average mean-square error as the LMS descends the error surface, shown in **Figure 2**. This has two exponential decays, corresponding to the two eigenvalues of the autocorrelation matrix in this case, which are different by a factor of 100. The ratio of the smallest time constant to the largest for such a decay can be expressed using eqn [8] as:

$$\frac{\tau_{min}}{\tau_{max}} = \frac{\lambda_{min}}{\lambda_{max}} \qquad [9]$$

The convergence properties of the LMS algorithm are thus determined by the eigenvalue spread of the autocorrelation matrix of the reference signal. A physical interpretation of the eigenvalue spread can be obtained from an interesting relationship between the eigenvalues of the autocorrelation matrix and the power spectral density, which can be used to show that:

$$\frac{\lambda_{max}}{\lambda_{min}} < \frac{S_{xx,max}}{S_{xx,min}} \qquad [10]$$

where $S_{xx,\,max}$ and $S_{xx,\,min}$ are the maximum and

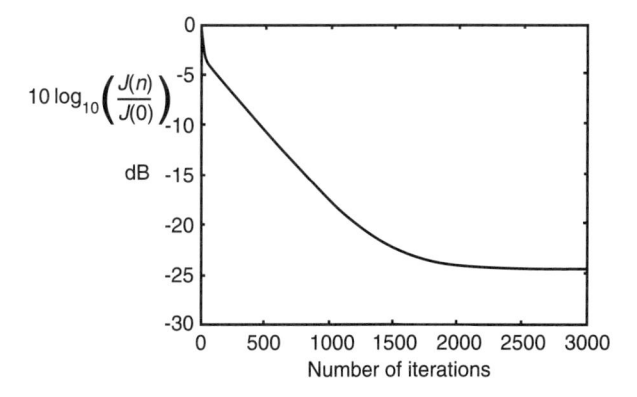

Figure 3 The average convergence behavior of the LMS algorithm in the case in which two modes converge at different rates.

minimum values of the power spectral density of the reference signal, $x(n)$.

The Recursive Least-squares (RLS) Algorithm

In order to overcome the slow convergence of the steepest descent-based LMS algorithm when faced with a large eigenvalue spread, algorithms based on faster optimization techniques, such as Newton's method, have been developed. Writing the mean-square error (J) in the quadratic form used in the article on optimal filters (see **Optimal filters**), $J = \mathbf{w}^T\mathbf{A}\mathbf{w} + 2\mathbf{w}^T\mathbf{b} + c$ where $\mathbf{A}$ is the autocorrelation matrix of the reference signal, Newton's algorithm may be written as an extension of eqn [1] as:

$$\mathbf{w}(\text{new}) = \mathbf{w}(\text{old}) - \mu \mathbf{A}^{-1} \frac{\partial J}{\partial \mathbf{w}}(\text{old}) \quad [11]$$

The RLS algorithm is a practical form of Newton's algorithm which at each new sample time, n, minimizes the exponentially weighted mean-square error given by:

$$J(n) = \sum_{l=0}^{n} \lambda^{n-l} \varepsilon^2(l) \quad [12]$$

where λ is the forgetting factor and $\varepsilon(l)$ is the error which would be produced if the current values of the filter coefficients had been used for all previous samples, i.e.:

$$\varepsilon(l) = d(l) + \mathbf{w}^T(n)\mathbf{x}(l) \quad [13]$$

By calculating the values of the $\mathbf{A}$ matrix and $\partial J/\partial \mathbf{w}$ associated with the cost function given by eqn [12] and using these in eqn [11], the RLS algorithm may be derived which can be written as:

$$\mathbf{w}(n+1) = \mathbf{w}(n) - \alpha(n)\mathbf{A}^{-1}(n-1)\mathbf{x}(n)e(n) \quad [14]$$

where the time-varying convergence coefficient is equal to:

$$\alpha(n) = \frac{1}{\lambda + \mathbf{x}^T(n)\mathbf{A}^{-1}(n-1)\mathbf{x}(n)} \quad [15]$$

and in order to avoid the inverse of a matrix having to be calculated at every sample, the inverse of the $\mathbf{A}$ matrix is calculated intervals using the matrix inversion lemma, to give:

$$\mathbf{A}^{-1}(n) = \lambda^{-1}\mathbf{A}^{-1}(n-1) \\ - \alpha(n)\lambda^{-1}\mathbf{A}^{-1}(n-1)\mathbf{x}(n)\mathbf{x}^T(n)\mathbf{A}^{-1}(n-1) \quad [16]$$

There are also various approaches to fast RLS algorithms, in which the number of operations required to implement the adaptive filter is of the order of I where I is the number of filter coefficients, instead of the order of I^2, as required to implement eqns [14], [15], and [16] above.

Another potential method of improving the convergence characteristics of the time domain LMS algorithm is to use the normalized frequency domain LMS algorithm. This algorithm can also significantly reduce the computation requirements for the adaptive filter if the filter has a large number of coefficients. There are also several other benefits to frequency–domain adaptation when considering feedforward and feedback controllers.

Block LMS Algorithm

Let us return to eqn [1], which describes the general philosophy of adaptation using the method of steepest descent:

$$\mathbf{w}(\text{new}) = \mathbf{w}(\text{old}) - \mu \frac{\partial J}{\partial \mathbf{w}}(\text{old}) \quad [17]$$

and the expression for the vector of derivatives, eqn [4]:

$$\frac{\partial J}{\partial \mathbf{w}} = -2E[\mathbf{x}(n)e(n)] \quad [18]$$

In deriving the normal LMS algorithm, we used the instantaneous version of these derivatives to update the filter coefficients. Another approach would be to calculate the average value of $\mathbf{x}(n)e(n)$ over N samples and use this to update the filter coefficients only every N samples, so that:

$$\mathbf{w}(n+N) = \mathbf{w}(n) + \frac{\mu}{N}\sum_{l=n}^{n+N-1} \mathbf{x}(l)e(l) \quad [19]$$

This algorithm is called the block LMS algorithm. This can have similar convergence properties to the LMS algorithm, since the filter coefficients change less frequently but by larger amounts, provided the filter does not converge too quickly compared with the block size, N. The main difference between the LMS algorithm and the block LMS algorithm is that the former uses a recursive averaging of the gradient estimate whereas the latter uses a finite moving average. The quantity used to update the filter coefficients in eqn [19] can be regarded as an estimate of the cross-correlation function between the reference signal, $x(n)$, and the error signal, $e(n)$, which can be written as:

$$\hat{R}_{xe}(m) = \frac{1}{N} \sum_{l=n}^{n+N-1} x(l-m)e(l) \qquad [20]$$

This estimate of the cross-correlation function needs to be calculated for $m = 0$ to $I - 1$, where I is the number of filter coefficients being adapted. The most efficient implementation of the block LMS algorithm occurs when $N = I$, and under these conditions the block LMS algorithm requires N^2 multiplications every N samples and thus has about the same computational requirements as the normal LMS algorithm. If N is large, then it would be more computationally efficient to calculate the estimated cross-correlation function from an estimate of the power spectral density, using the fast Fourier transform (FFT) to calculate the discrete Fourier transform (DFT) of blocks of reference and error signals. In order for this estimate of the cross-correlation function to be unbiased, however, some care needs to be taken to prevent circular correlation effects, such as the use of the overlap–save method. This involves taking $2N$-point FFTs and adding N zeros to the error data block before taking the FFT. Only the causal part of the cross-correlation function is used to update the filter, and so half the length of the point cross-correlation function is discarded. This requires an operation similar to that denoted $\{\}_+$ in eqn [29] of the article on optimal filters (see **Optimal filters**) and involves an inverse FFT of the spectral density, the setting to zero or windowing of the noncausal part of the resulting cross-correlation function, and taking an FFT of the result. The adaptation algorithm for the filter coefficient in the kth frequency bin can thus be written as:

$$W_{\text{new}}(k) = W_{\text{old}}(k) + \alpha\{X^*(k)E(k)\}_+ \qquad [21]$$

where $X(k)$ is the DFT of $2N$ points of $x(n)$, * denotes complex conjugation, and $E(k)$ is the DFT of N points of $e(n)$ padded with N zeros, assuming the overlap save method is used to prevent circular correlation. This algorithm is called the fast LMS (FLMS) algorithm, and is exactly equivalent to the block LMS algorithm.

Once the reference signal and filter response are available in the frequency domain, further computational savings can be made by also performing the convolution required to obtain the filter output in the frequency domain, so that:

$$Y(k) = W(k)X(k) \qquad [22]$$

from which the time history must be obtained by taking the inverse FFT (IFFT) and only using the last N points to avoid circular convolution effects. The block diagram of the complete frequency domain LMS algorithm is shown in **Figure 4**.

The implementation of the frequency domain adaptation of the filter, and the convolution of the reference signal with the reference signal thus

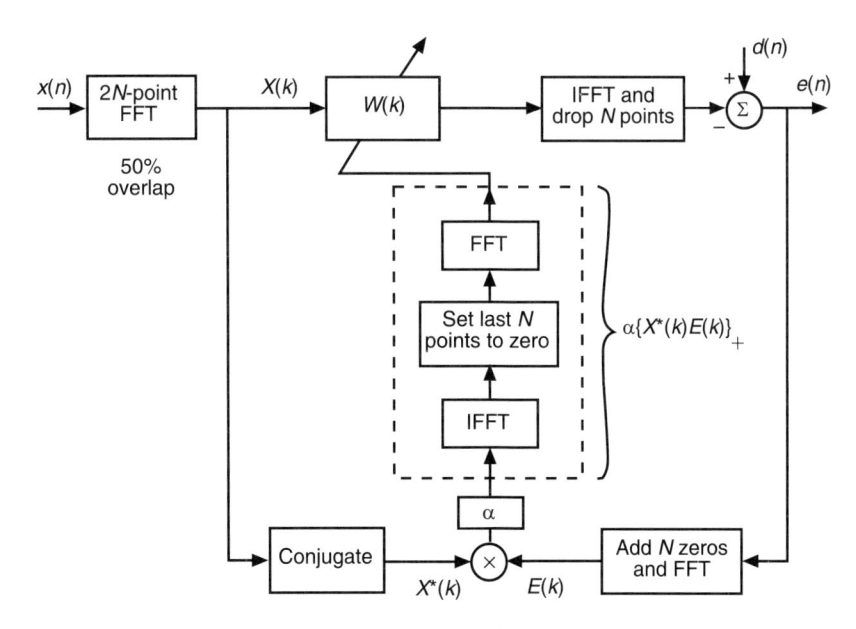

Figure 4 Block diagram of a frequency domain implementation of the LMS algorithm.

uses five 2N-point FFTs, where $N = I$ and I is the number of adaptive filter coefficients. If each 2N point FFT requires $2N \log_2 2N$ operations, then frequency domain adaptive filtering requires about $10N \log_2 2N$ operations every N samples, or $10 \log_2 2N$ operations per sample. This can be compared with N operations per sample to adapt the coefficients using the normal LMS, and N operations per sample, for the normal time domain convolution of the reference signal and filter coefficients. The frequency domain implementation of the LMS algorithm thus requires a factor of about $N/5 \log_2 2N$ fewer computational operations compared with its direct implementation. For a 512-point filter this corresponds to a computational saving of about a factor of 10.

Frequency-dependent Convergence Coefficients

Apart from the computational advantages, it may also be possible to use the frequency domain implementation to improve the convergence properties of the LMS algorithm. This was originally suggested by Ferrara, who argued that in the frequency domain the error signal in a given frequency bin, $E(k)$, is only a function of the filter coefficient in the same bin, $W(k)$, and so each of the frequency domain filter coefficients converge independently. If this were the case the convergence coefficient could be selected independently for each bin, so that the adaptation algorithm becomes:

$$W_{\text{new}}(k) = W_{\text{old}}(k) + \{\alpha(k)X^*(k)E(k)\}_+ \quad [23]$$

The convergence coefficients used in the adaptation of the individual frequency bins could be normalized by the average power in each frequency bin, for example, so that:

$$\alpha(k) = \frac{\alpha'}{\mathrm{E}\left[|X(k)|^2\right]} \quad [24]$$

where α' is a single normalized convergence coefficient. For some applications, this normalization of the frequency-dependent convergence coefficient has been found to improve considerably the convergence rate of the adaptive filter.

Unfortunately, if a frequency-dependent convergence coefficient is used in eqn [23], the adaptive filter is not guaranteed to converge towards the optimum causal Wiener filter. This problem is particularly severe when the adaptive filter is used as a linear predictor, and is caused by spurious compo-nents in the causal part of the Fourier transform of $\alpha(k)X^*(k)E(k)$, in eqn [23], which arises because of the interaction between the causal parts of the Fourier transform of $\alpha(k)$ with the noncausal parts of the Fourier transform of $X^*(k)E(k)$.

A solution to this problem can be obtained by deriving a frequency domain adaptive algorithm directly based on Newton's method, which can be written as:

$$W_{\text{new}}(k) = (1 - \alpha)W_{\text{old}}(k) + \alpha \hat{W}_{\text{opt}}(k) \quad [25]$$

where α is the convergence coefficient and $\hat{W}_{\text{opt}}(k)$ is an estimate of the optimum causal filter. We have seen in the article on optimal filters (see **Optimal filters**) that the optimum causal filter can be written in the frequency domain as:

$$W_{\text{opt}}(e^{j\omega T}) = \frac{1}{F(e^{j\omega T})} \left\{ \frac{S_{xd}(e^{j\omega T})}{F^*(e^{j\omega T})} \right\}_+ \quad [26]$$

where $F(e^{j\omega T})$ and $F^*(e^{j\omega T})$ are the spectral factors of $S_{xx}(e^{j\omega T})$. In the discrete frequency domain we can obtain an estimate of the optimum causal filter from the current block of data as:

$$\hat{W}_{\text{opt}}(k) = \frac{1}{\hat{F}(k)} \left\{ \frac{X^*(k)D(k)}{\hat{F}^*(k)} \right\}_+ \quad [27]$$

where $\hat{F}(k)$ is the estimated spectral factor of $S_{xx}(k)$, which could in practice be obtained by taking the Hilbert transform of $\sqrt{S_{xx}(k)}$.

We can now write $D(k)$ as $E(k) + W(k)X(k)$, so that eqn [27] can be expressed as:

$$\hat{W}_{\text{opt}}(k) = \frac{1}{\hat{F}(k)} \left\{ \frac{X^*(k)E(k)}{\hat{F}^*(k)} \right\}_+ \\ + \frac{1}{\hat{F}(k)} \left\{ \frac{W(k)X^*(k)X(k)}{\hat{F}^*(k)} \right\}_+ \quad [28]$$

The average value of the term $X^*(k)X(k)/\hat{F}^*(k)$ in this equation is equal to $\hat{F}(k)$, and since $W(k)\hat{F}(k)$ is entirely causal, the final term in eqn [28] will, on average, be equal to $W(k)$. Using eqn [28] in eqn [25], a form of the frequency domain Newton's algorithm is obtained, which may be written as:

$$W_{\text{new}}(k) = W_{\text{old}}(k) + \alpha^+(k)\{\alpha^-(k)X^*(k)E(k)\}_+ \quad [29]$$

where $\alpha^+(k) = \sqrt{2}/\hat{F}(k)$, which corresponds to an entirely causal time sequence, and $\alpha^-(k) = (\alpha^+(k))^*$

which corresponds to an entirely noncausal time sequence. In this algorithm the bin-normalized convergence coefficient given by eqn [24] has essentially been split into two parts and since only the noncausal part is used inside the causality constraint in eqn [29], this does not cause noncausal components of the Fourier transform of $X^*(k)E(k)$ to affect the adaptation of the filter, which thus converges to the optimal causal filter.

Adaptive filters are used to cancel electrical interference in measurement systems and also as the basis for practical feedforward and feedback controllers in active vibration control.

Nomenclature

A	autocorrelation matrix
$e(n)$	error signal
I	number of filter coefficients
J	mean-square error
S_{xx}	power spectral density
$x(n)$	reference signal
α	convergence coefficient
λ	eigenvalue
μ	convergence factor
τ	time constant

See also: **Adaptive filters**; **Digital filters**; **Signal processing, model based methods**.

Further Reading

Clarkson PM (1993) *Optimal and Adaptive Signal Processing*. Boca Raton, FL, CRC Press.
Elliott SJ (2001) *Signal Processing for Active Control*. Academic Press.
Elliott SJ, Rafaely B (2000) Frequency-domain adaptation of causal digital filters. *IEEE Transactions on Signal Processing*, 48(5).
Ferrara ER (1985) *Frequency-domain Adaptive Filtering in Adaptive Filters*. Cowan CFN and Grant PM (eds) 145–179. Englewood Cliffs, NJ: Prentice Hall.
Haykin S (1996) *Adaptive Filter Theory*, 3rd edn. Englewood Cliffs, NJ: Prentice Hall.
Rabiner LR, Gold B (1975) *Theory and Application of Digital Signal Processing*. Englewood Cliffs, NJ: Prentice Hall.
Shynk JJ (1992) Frequency-domain and multirate adaptive filtering. *IEEE Signal Processing Magazine*, January, 14–37.
Widrow B, Hoff ME Jr (1960) Adaptive switching circuits. *IRE WESCON Conv. Rec.* Pt.4. 96–104.
Widrow B, Stearns SD (1985) *Adaptive Signal Processing*. Englewood Cliffs, NJ: Prentice Hall.

AEROELASTIC RESPONSE

J E Cooper, University of Manchester, Manchester, UK

Copyright © 2001 Academic Press

doi:10.1006/rwvb.2001.0125

Introduction

Aircraft are subjected to a wide range of static (i.e., steady) and dynamic loads in flight and also on the ground. Combined with the inherent flexibility of the structure, these dynamic loads result in vibration. Gusts and dynamic maneuvers can cause the limit loads to be exceeded; however, in most cases structural failure would occur due to fatigue. Consideration should also be made regarding the discomfort caused by the responses to the pilot or passengers, and also the consequent malfunctioning of equipment and systems.

Care must be taken at the design stage to ensure that the static and dynamic responses stay within acceptable limits of deformation and load on production aircraft, otherwise costly redesigns will have to be made. Both the civil and military airworthiness regulations, that dictate how aircraft must be certified to fly, have several sections devoted to loads. If new technologies or ideas are being applied that are not accounted for in the regulations, then the engineers have to discuss the way forward for certification with the authorities.

This item provides a brief overview of an immense subject area, with a wealth of literature and work in the aerospace community being devoted to each of the topics covered. The problems associated with a number of critical phenomena are described. The classical flutter phenomenon is covered in a different item (see **Flutter**).

Flight Loads

Flight loads include all loads that can occur in any part of the flight envelope. Typically, the most critical design cases occur in flight.

Steady Maneuvers

The loads on an aircraft increase during maneuvers and these are often the critical design cases, particularly for military aircraft. The loads must be calculated, first by determining the lift and inertia forces (obtained from application of D'Alembert's principle), and then determining the internal loads (e.g., bending moments, torques) from which the resulting stresses can be found.

Balance Calculations

If an aircraft is considered in accelerated flight (**Figure 1**), the overall lift due to the wings and tail $L = L_W + L_T$ is balanced by the inertia force nW, where W is the weight and n is known as the load factor, defined as

$$\text{Load factor}$$
$$n = \text{(total lift developed)/(weight of aircraft)}$$

The load factor n is produced by the aircraft accelerating upwards at $(n-1)g$, where g is the acceleration due to gravity. In straight and level flight, $n = 1$ and thrust $T = \text{drag } D$. In a particular maneuver, it is possible to calculate the value of the load factor from which the total lift can be found. The distribution of lift between the wing and tailplane and the thrust required in the maneuver may be determined from moment equilibrium.

Internal Loads

Although the external loads discussed above are in overall equilibrium, they do not have the same distribution over the structure. Consider the aircraft in **Figure 2** with a fuselage mass of typically two times that of the wings (dependent on the position of the power plant and the fuel storage) whereas nearly all of the lift is distributed over the wings. The difference in load distribution gives rise to internal loads, namely shear force, torque and bending moment distributions along the wings, fuselage, etc. It is traditional to consider the aircraft in sections (**Figure 3**). The inertia and lift forces can be estimated from the results of the balanced calculations. The shear force, torque and bending moment can then be determined at the junction of each section by equilibrium considerations. Nowadays, once the lift distribution is known, the loads and resultant stresses can be found using finite element (FE) models. The designer needs to consider the critical parts (most highly stressed) of the structure and to ensure that they do not exceed ultimate stresses during any desired maneuvers. These highly stressed areas are also going to be the areas where failure due to fatigue may occur. During the design process, trade-off studies may be conducted to change the design if these stresses are considered too high.

Maneuver Envelope

Rather than specifying a whole range of maneuvers of varying severity and speed, the airworthiness requirements define a maneuver envelope that show the combinations of load factor n and speed v that an aircraft has to be able to withstand. **Figure 4** shows a typical envelope for a commercial aircraft.

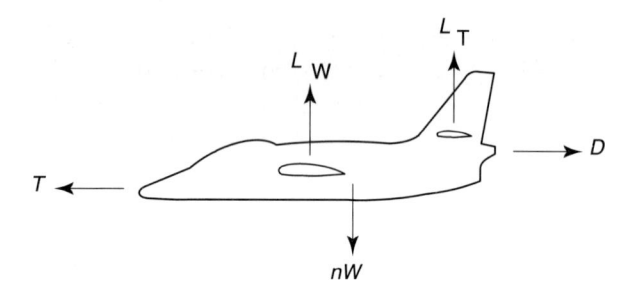

Figure 1 Balance forces on aircraft in accelerated flight.

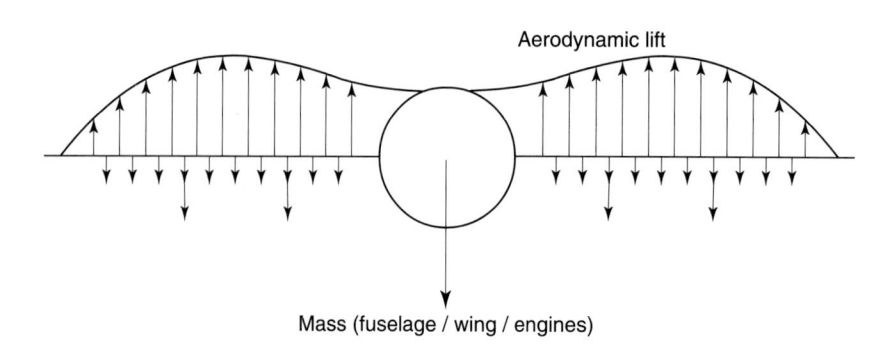

Figure 2 Aerodynamic and mass load distributions.

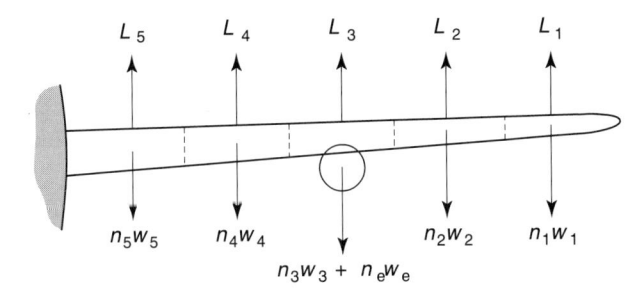

Figure 3 Sectional distribution of loads.

The envelope is constructed knowing the design cruise and dive speeds v_C, v_D, as well as the maximum maneuvering load factor (usually 2.5 for commercial aircraft).

Dynamic Maneuvers

Pilot Induced Maneuvers

The airworthiness regulations define a number of maneuvers that must be considered, e.g., (i) step input; (ii) sinusoidal motion of the controls at various frequencies.

An aeroelastic model that includes both rigid-body and flexible modes must be used in the form of

$$\begin{bmatrix} A_{rr} & A_{rf} \\ A_{rf} & A_{ff} \end{bmatrix} \begin{bmatrix} \ddot{q}_r \\ \ddot{q}_f \end{bmatrix} + \begin{bmatrix} B_{rr} & B_{rf} \\ B_{rf} & B_{ff} \end{bmatrix} \begin{bmatrix} \dot{q}_r \\ \dot{q}_f \end{bmatrix} + \begin{bmatrix} C_{rr} & C_{rf} \\ C_{rf} & C_{ff} \end{bmatrix} \begin{bmatrix} q_r \\ q_f \end{bmatrix} = \begin{bmatrix} U_r \\ U_f \end{bmatrix}$$

where subscripts r and f refer to rigid and flexible terms, respectively. These equations can be solved in time for given inputs U, and consequently nonlinear terms can be catered for.

On some aircraft, maneuver load alleviation systems are implemented to reduce the loads on critical parts of the aircraft structure. By sensing the maneuvers that the aircraft is being asked to perform, the control surfaces may be moved in such a way as to reduce the loads while still undertaking the maneuver.

Gusts

Aircraft are often subjected in flight to motion of the air in the form of gusts. These gusts can impose considerable loads on aircraft and may come from all directions:

1. vertical gusts load the wing, fuselage and horizontal tail (**Figure 5**);
2. lateral or 'side' gusts, loading the fuselage, vertical tail and pylons;
3. longitudinal or 'head-on' gusts which may cause important loads on flap structures.

Since the recognition during the First World War that these effects produced significant loads, gust design criteria have been formulated, which have evolved over the years and are still under development.

For civil aircraft, gust load cases are often the most critical for load design, and are also one of the main fatigue loading sources for the structure. A further important consideration is the aircraft dynamic response due to gusts in terms of passenger comfort. Although military type aircraft structures are generally maneuver load critical, for specific parts of the structure, such as flexible outer wing sections and store pylons, gusts may determine the critical design load cases, particularly at high speed/low level conditions.

The analysis of the response to, and resulting loads from, gusts takes one of two possible approaches; (i) each gust is considered as a separate discrete event; (ii) the gusts are considered as a random turbulent

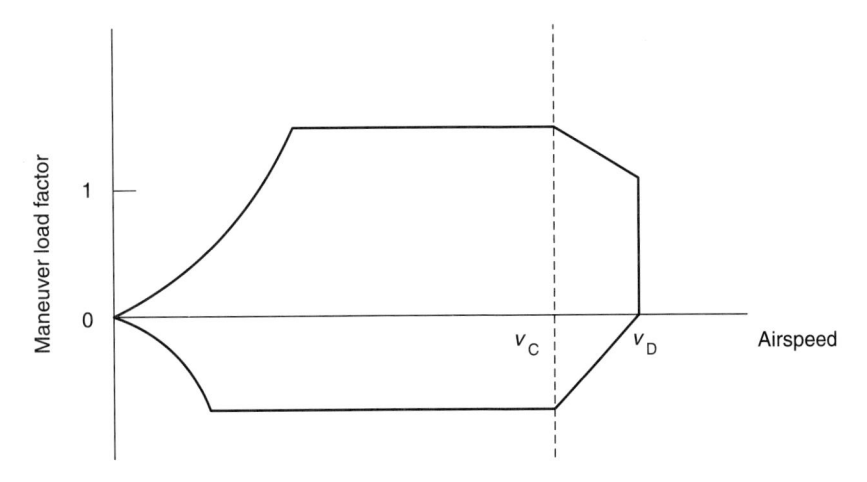

Figure 4 Maneuver envelope.

Figure 5 Aircraft encountering vertical gusts.

sequence. All major current airworthiness codes include two sets of gust criteria based on concepts. The discrete gust criterion is typically 'worst case': the highest load resulting from a particular time sequence must be taken. This discrete case is included to safeguard against sudden 'stand-alone' gusts that can occur in practice. The continuous turbulence criterion is based upon a consistent model of the atmospheric turbulence defining design loads based on an averaged response and considering all possible gust lengths that prevail in random turbulence.

Discrete gusts The basic loading mechanism of gusts is schematically illustrated in **Figure 6** for motion in the vertical plane. A rigid aircraft flying with speed v, entering an upward gust with velocity w, experiences a sudden change in angle of incidence $\Delta \alpha = w/v$. The gust gives rise to an additional lift proportional to v^2 and $\Delta \alpha$.

Such an analysis was first considered in the 1920s. In reality, the above model is a gross oversimplification of a number of effects that must be considered in order to give accurate response and loads predictions:

1. The abrupt or 'sharp-edged' gust indicated in **Figure 6** is physically impossible. A real gust must have some distance over which its velocity builds up. A number of different shapes have been considered over the years, with the '1 − cos' gust shape (shown in **Figure 7**) being included in almost all current airworthiness regulations.
2. Due to unsteady aerodynamic effects, a sudden change in angle of incidence does not immediately

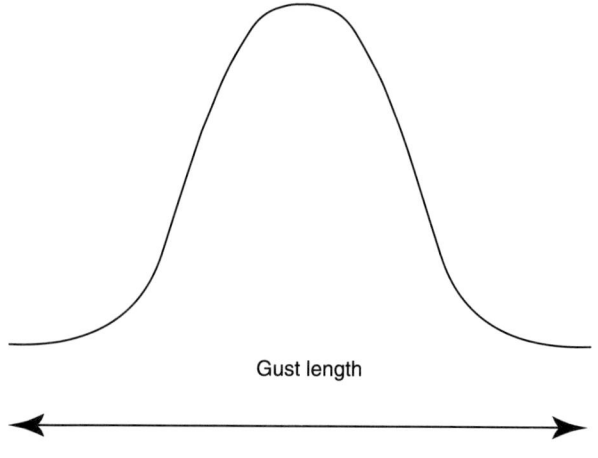

Figure 7 '1 − Cosine' gust.

result in a proportional change in lift. These effects are modeled by the Kussner function shown in **Figure 8**. Consequently, the aircraft has time to respond to the gust, thus reducing the gust-imposed load in comparison to the 'sharp-edged' gust case.

3. The gust is not only going to impart a lift force on the wings, but also onto the tail surfaces at some fixed time delay (depending on the speed and the geometry). This so-called 'penetration effect' will result in a pitch motion being imparted to the aircraft.
4. The aircraft is flexible, thus the resulting motion and load will depend on the size and frequency characteristics of the gust as well as the vibration characteristics of the aircraft.
5. Modern aircraft are becoming increasingly nonlinear in their behavior. The nonlinearities can be structural (e.g., free-play), aerodynamic (e.g., transonic shock effects), or control system based (e.g., nonlinear control laws) and these effects must be considered in the gust design calculations. The prediction of maximum gust loads is much more complicated for nonlinear aircraft as the law of superposition does not hold in any analysis.

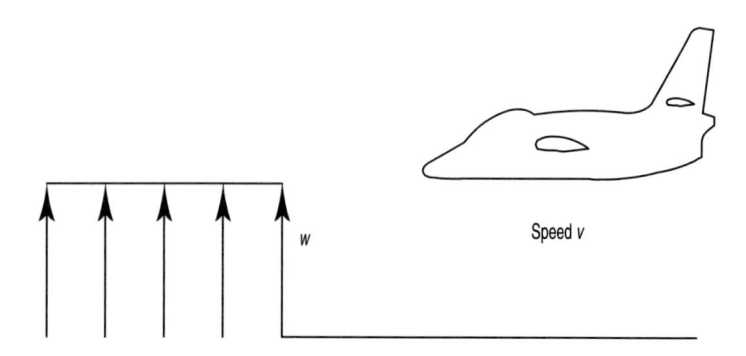

Figure 6 Aircraft encountering a sharp-edged gust.

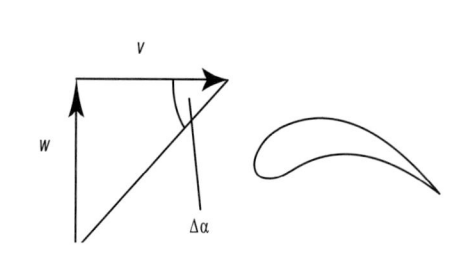

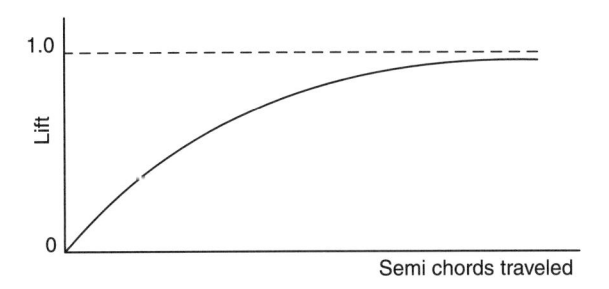

Figure 8 Kussner function: lift due to flight in a sharp-edged gust.

For many years, the main airworthiness codes included simplifying assumptions with regard to the length of the gust (e.g., a '1 − cos' gust of 25 wing chords). The response of a rigid aircraft in heave due to a sharp-edged gust resulted in very simple gust-response expressions. The well-known 'Pratt formula' introduced a gust load alleviation factor (multiply the predicted loads by 0.7 typically) to allow crudely for gust velocity build-up and unsteady aerodynamic effects. The gust envelope shown in **Figure 9** shows the flight points that needed to be considered.

Discrete tuned gusts With the growing size and increasing flexibility of aircraft, the above assumptions became more and more unacceptable. Hence, the major airworthiness codes currently demand a full dynamic response calculation. The mathematical model used to calculate the gust response is similar to that used for flutter calculations (see **Flutter**) but includes additional unsteady aerodynamic terms to account for the gust loading. The response is determined by integration of the equations in the time domain.

As the physical length of the gust affects the frequency content of the load time history, and hence the structural response, a range of gust lengths has to be considered. The gusts giving the highest loads (the discrete tuned gust) is taken for design calculations. The magnitude of the peaks reduces at shorter gust wavelengths and higher altitudes because the energy in the turbulence decreases in these cases.

Recent work has investigated methods of finding the gust sequence (e.g., a combination of several '1 − cos' gusts) that gives rise to the greatest response and loads. The techniques have been extended to consider nonlinear aircraft.

Continuous turbulence methods Measurements in flight reveal a pattern of gusts resembling a random process of continuous turbulence. This notion led in the early sixties to the development of additional design guidelines. Over short periods of time, such a random process may be considered as stationary with gaussian properties. Over longer periods of time, the standard deviation or gust intensity is not a constant, but varies randomly with a given probability function.

The turbulence is often considered to have a 'von Karman' power spectral density function derived from matching experimental data, shown in **Figure 10**, describing how the energy in the process is distributed over frequency.

The response power spectral density can be found by multiplying the known aircraft transfer function (derived from the full aeroelastic model; see **Flutter**) by the 'von Karman' power spectral density. This operation gives the RMS of the response parameter. It is then possible to determine load RMS values that can be used for design purposes. Recent advances have extended the continuous turbulence approach to deal with nonlinear aircraft.

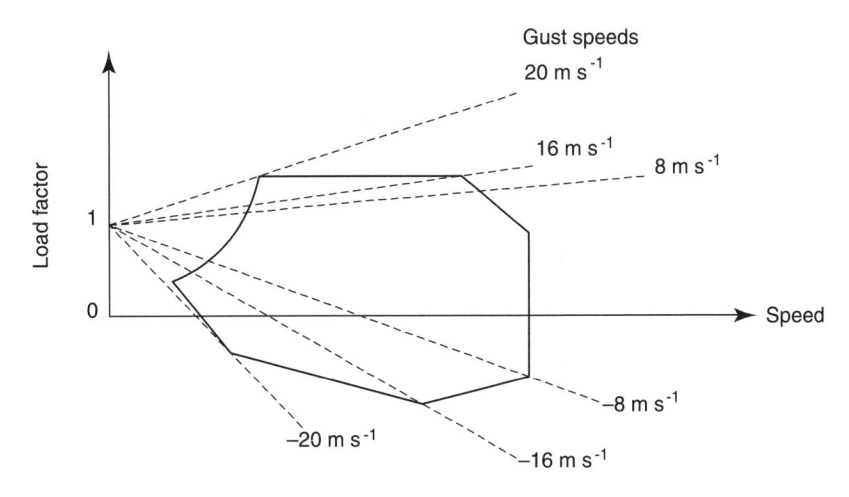

Figure 9 Gust envelope.

92 AEROELASTIC RESPONSE

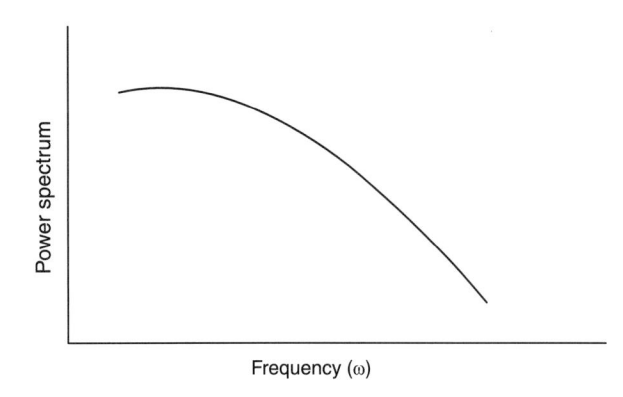

Figure 10 Von Karman gust spectrum.

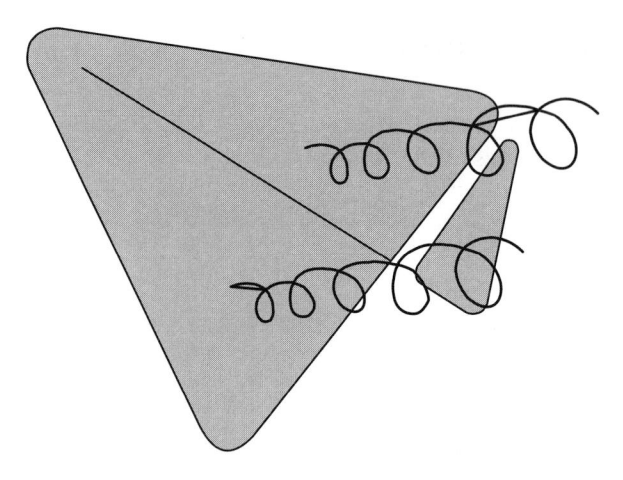

Figure 11 Tail buffet on delta wing fighter aircraft.

Active Gust Alleviation Systems

A common approach to reduce gust loads, particularly in terms of increasing passenger comfort, is to make use of an active gust alleviation system. By sensing the response of the aircraft as it encounters a gust, the control surfaces may be moved to counteract the effect and reduce the loads. Any flight control system (FCS) may well perform this function anyway. It should be noted that by alleviating the wing root bending moment, the fatigue life of the control surfaces and attachments are often drastically reduced.

Buffet and Buffeting

Buffet is defined as the aerodynamic excitation due to a separated flow. It is usually random and covers a wide frequency range, but is dependent upon the geometry and the flight condition. In some cases the buffet can be periodic and exist at individual frequencies.

Buffeting is the response resulting from the excitation of the buffet. The term originated in 1930 following a fatal accident that was attributed to the aircraft being subjected to gust loading. These gusts led to a sharp increase in the effective angle of incidence (see section on gusts) with the consequent formation of flow separation over the wing. The tail was subjected to intense vibrations (given the name buffeting) due to the turbulent wake and this resulted in structural failure. Buffeting will occur on aircraft if the vibration modes of the tail or wing structure are excited by the buffet. It rarely produces an instant catastrophic failure; however, the loads can be severe and consequently fatigue problems can result.

Fin buffet/buffeting of delta wing fighter aircraft When high angles of attack maneuvers are undertaken, turbulent flow detaches itself from the wings and meets the fin (**Figure 11**). Although this effect is desirable to enable lateral control of the aircraft, the fin can experience severe loads (up to 400g has been measured) due to the buffeting. The fatigue life of the fin can be used up in a few hours. Buffet can also be caused due to airbrakes and cavities in the structure. Twin-fin fighter aircraft are particularly susceptible to buffet and there is a significant effort underway worldwide to reduce/eliminate the difficulty. Approaches under investigation include:

1. The use of MEMS on the wing surfaces to eliminate the turbulent flow (see **MEMS, general properties**)
2. SMART devices on the fin to reduce the resulting vibration
3. Aerodynamic devices on the fin to counteract the buffeting
4. Aerodynamic devices upstream to influence the vortex characteristics

Buffet/buffeting of commercial aircraft Although commercial aircraft do not reach such high angles of attack as fighter aircraft, the buffeting problem can still occur. In this case, passenger comfort also needs to be taken into consideration. Separated flow can occur on the main wing surfaces, particularly on application of wing control surfaces, and if this impinges on the tail, then not only the tail modes may be excited but also those of the fuselage.

One major difficulty with design against buffeting is that it is still extremely difficult and impractical to produce a predictive model of buffet. Consequently, wind tunnel model testing must be undertaken in order to predict the buffet behavior. Once an appreciation of the frequency characteristics of the buffet and the geometry of the flow has been gained, it is possible to make response predictions using the aircraft mathematical model (assumed mode or finite element) and to construct any design changes that may be required.

Acoustic Excitation

High levels of acoustic excitation due to jet efflux and turbulent fluid flow have been experienced on the skin panels of modern jet aircraft, particularly on short take-off and vertical landing (STOVL) aircraft such as the Harrier. This high intensity noise environment is often combined with very high temperatures, e.g., at some points on the structure the sound pressure could be in the region of 180 dB with panel temperatures over 1500 °C. There is a high likelihood of fatigue damage occurring in such an environment.

Similar problems are likely to occur with the new generation of stealth aircraft as power-plant and stores will be internal. When the cavities have to be opened, for instance when a store is to be dropped, extremely high noise levels will occur. Civil aircraft are also susceptible to such difficulties on the flaps and rear fuselage, etc. Much work is being undertaken to investigate approaches (e.g., passive damping technology, smart materials and devices, and MEMS) in order to reduce the severe problems that will have to be overcome.

Gunfire Loads

Repetitive gunfire blasts impart pulses that can cause vibration levels typically twice that occurring in normal flight. These vibrations occur at distinct frequencies as shown in **Figure 12**. The aircraft structure, and the equipment within, must be cleared to withstand the vibration levels due to gunfire.

Store Release

When stores are released from military aircraft (or even dropping aid parcels from transport aircraft) an impulse is imparted onto the structure and the aircraft will respond dynamically. The designer needs to ensure that this response is not significant, bearing in mind that the mass distribution and aerodynamic characteristics have changed. Care has also to be taken in the design to guarantee that the store jettisons safely rather than rebounding back onto the aircraft.

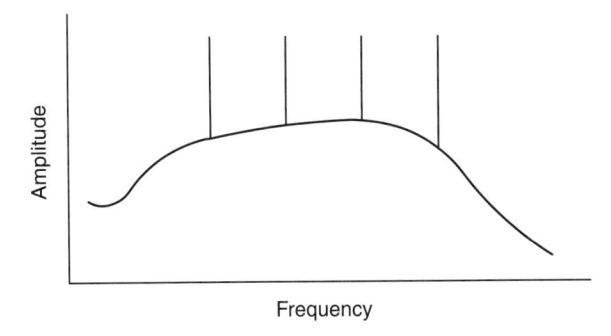

Figure 12 Typical flight vibration spectra including gunfire.

Birdstrike

All aircraft fly with the likelihood of hitting one or more birds in flight. The airworthiness regulations lay down strength criteria that every aircraft must meet. Certification is usually met through ground testing. The leading edges of the wings and tailplane, as well as the canopy and radome, must be constructed so that they can survive such an incident. Ingestion of birds into jet engines is also a problem that must be considered particularly for large diameter civil aircraft engines.

Hammershock

Jet engines cannot operate in supersonic flow conditions, therefore the intakes and ducts of supersonic aircraft need to be designed so that subsonic conditions occur at the engine (**Figure 13**). At the limits of engine performance, it is possible for the airflow into the engine to distort, resulting in an engine surge. This surge in turn causes a pressure wave called hammershock to occur in the duct. This wave advances at high speed (typically 400 m s^{-1}) down the duct in the opposite direction to the airflow. The wave causes a pressure in the duct of up to three times the usual steady-state pressure that in turn leads to a dynamic response.

The engine intake and duct structure are designed to withstand the stresses due to maneuver loads, the steady-state operating pressure, and critically the hammershock loads. It is not possible at present to predict the hammershock pressures, so experimental measurements on the ground and in flight are taken. These measurements can then be used with a structural FE model to predict the resultant stresses and to make any design modifications that may be necessary.

Ground Loads

The landing gear of an aircraft undertakes two functions: to absorb the energy due to the vertical descent on landing, and to enable the aircraft to maneuver on the ground. A complete analysis of the behavior of the aircraft during both phases of operation is essential so that the undercarriage is designed to be strong enough and also to ensure that no other component fails. The airworthiness regulations dictate how the undercarriage should be cleared for each aircraft.

Landing Loads

Aircraft are subjected to significant forces during landing, especially in the special case of landing on an aircraft carrier. The characteristics of these loads are dependent upon the landing gear. There are a number of variables that must be considered:

Figure 13 Typical intake of a jet fighter.

- limit descent velocity at the design landing weight (typically $3 \, \text{m s}^{-1}$ for transports and land based fighters (not trainers), $6 \, \text{m s}^{-1}$ for carrier based aircraft),
- undercarriage and tyre energy absorption characteristics (nonlinear),
- aircraft attitude relative to the ground,
- distribution of the aircraft mass,
- lift acting on the aircraft at the impact point,
- friction coefficient between the tyre and the ground,
- rotation of the aircraft on take-off.

The analysis of the aircraft response should consider the following behavior:

- landing gear dynamic characteristics,
- spin up loading due to forces that accelerate the wheel up to the ground speed,
- spring back once the wheel has reached the ground speed and the sliding friction has gone to zero, the strain energy stored by the undercarriage's rearward bending motion is released and the undercarriage returns to its original orientation with a sizeable inertia load,
- rigid-body response of the aircraft,
- dynamic response of the aircraft.

A typical commercial aircraft would make a conventional two-point landing (nose gear does not contact the ground until the main undercarriage has absorbed all the energy from the descent) as shown in **Figure 14**. However, the cases of three-wheel (nose gear critical case) and one-wheel landings must also be considered. In each case the loads and resulting motions must be analyzed to ensure that structural integrity is maintained.

Runway Loads

These are usually defined as the loads resulting from taxiing including turning, braking applied during take-off (e.g., an aborted take-off) or landing, as well as towing, jacking and tethering (tying down an aircraft in a very high wind). Of particular interest are the loads on the nose undercarriage when the brakes have been applied suddenly, causing an increase in the vertical load on the nose gear due to the pitching moment of the aircraft. Similarly, the forces due to the application of the spoilers, brakes, thrust reversers, brake parachute deployment and arrestor wire engagement need to be considered.

Shimmy

In the 1930s, the air pressure in car tyres was reduced with the object of producing a smoother ride. This change produced an effect known as 'shimmy' that consisted of a self-excited unstable rotational oscillation of the front wheels about the vertical axis. The problem was eventually cured through independent spring suspension of the front wheels. Shimmy is not restricted solely to pairs of wheels on a common axis, but has been observed on single castored wheels such as on supermarket trolleys and, more importantly, on aircraft nose and main undercarriages where its

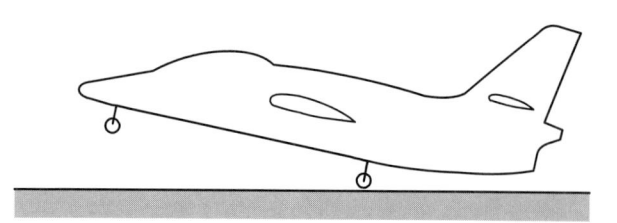

Figure 14 Conventional two-point (main undercarriage) landing.

occurrence can be disastrous. The coupling of the torsion and bending modes, analogous to flutter, leads to an unstable motion.

The analysis of the shimmy problem is made very complex due to the dynamics of the tyre (viscoelastic material, nonlinear friction and deformation characteristics). Experimental tests are performed to verify the design calculations.

Other Aircraft Aeroelastic Phenomena

Classical linear flutter is considered in **Flutter**. In this section, a number of related phenomena that have nonlinear response characteristics are described.

Limit Cycle Oscillations

If an aircraft that behaves linearly is flown fast enough, flutter will occur. In practice though, aircraft are nonlinear in their vibration behavior and these nonlinearities can produce phenomena known as limit cycle oscillations (LCOs). LCOs are characterized as bounded instabilities, as shown in **Figure 15**, where it can be seen that the amplitude initially increases but then stops growing. This effect can arise due to nonlinearities occurring in the:

- structure – freeplay, backlash, cubic stiffening,

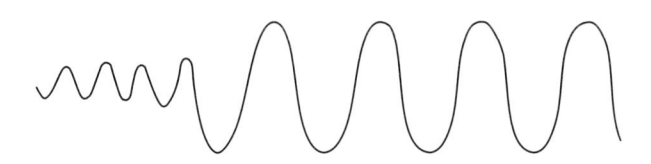

Figure 15 Typical limit cycle oscillations.

- aerodynamics – moving shocks, other transonic aerodynamic effects,
- control system – time delays, nonlinear control laws.

As LCOs are not immediately destructive, they can be considered as a fatigue problem. However, they are extremely difficult to predict as an accurate nonlinear model is required. At present, a large amount of computer simulation is needed to do this. However, as modern aircraft are becoming more flexible and nonlinear, consequently, LCOs are becoming more of a problem. Research is currently being directed towards methods that can predict LCOs without the need for vast amounts of testing or computer simulation. Such methods will speed up the flight test procedure whilst improving safety by giving an improved predictive capability.

Aeroservoelasticity

Modern aircraft contain a range of active control technology including flight control systems (FCSs), gust alleviation systems, flutter suppression systems, etc., that can combine with the usual aeroelastic phenomena. The FCS detects a motion (which it assumes is the uncommanded aircraft rigid-body response) and tries to correct it. In the worst case, the frequency of activation commanded coincides with the structural mode frequency being corrected, leading to a potentially unstable condition. This interaction is illustrated by the 'aeroservoelastic pyramid' in **Figure 16**. The entire system including feedback must be included in all stability calculations and the siting of motion sensor units is critical. It is

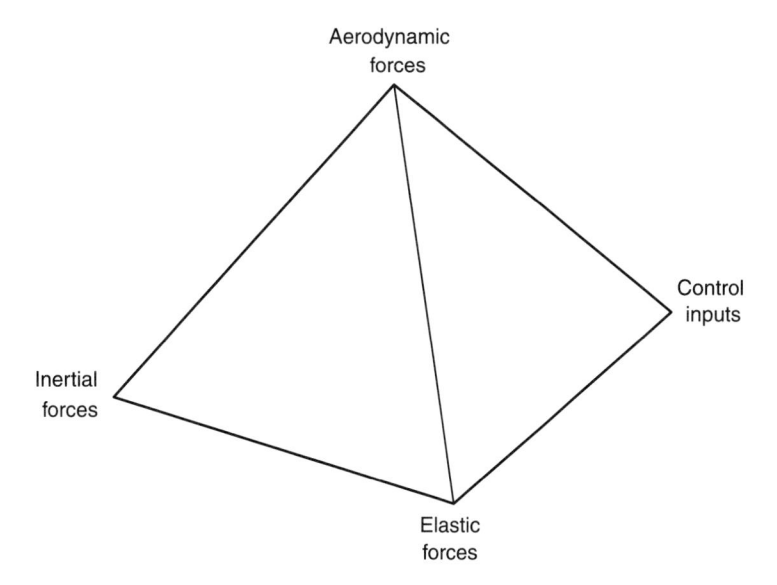

Figure 16 Aeroservoelastic pyramid.

possible for the control system to interact with the structure in such a way that unstable oscillations can occur. The control system is likely to contain nonlinearities and these can give rise to phenomenon such as LCOs.

Panel Flutter

When there is supersonic flow past a thin panel, there is the possibility of self-excited oscillations of the panel normal to its plane. Consider the panel in **Figure 17**. The pressure acting on the plate is proportional to the dynamic pressure and the local slope of the surface. Therefore, a symmetric slope distribution leads to an unsymmetrical pressure loading which tends to deform the surface of the plate into a more complicated shape. As the dynamic pressure increases, the plate loses its ability to maintain equilibrium and a response limited LCO occurs due to nonlinear stiffening effects at higher amplitudes. The solution to this problem is to increase the stiffness of the panel through the use of larger or more stringers/ribs. Other solutions include increasing the tension in the panel. Recent work has investigated the use of shape memory alloys to do this, using the temperature increase associated with supersonic flight to take the alloys beyond the activation temperature.

Control Surface Buzz

This is a LCO-type phenomenon that occurs in the transonic flight regime. At transonic speeds there will be a shock wave acting on the airfoil. Any movement of the shock will result in a difference in aerodynamic forces with a consequent change in deflection of the aerofoil, but in particular of the control surface itself.

Consider the aerofoil with flap in **Figure 18**. In this case, the flap is entirely in subsonic flow as the shock is on the aerofoil surface. A similar type of response occurs at slightly higher speeds when the shock is on the control surface. Oscillation of the flap causes a fore-and-aft motion of the shock waves that changes the aerodynamic moments acting on the control surface. The resulting change in deflection of the control surface results in further movement of the shock waves. The resulting motion is amplitude limited, but can result in deflections of up to ±10 degrees causing severe problems for the pilot. The phenomenon also causes undesirable vibration and fatigue damage.

Stall Flutter

If the angle of incidence of an airfoil exceeds its stall angle, then stall will occur. Stall flutter is a LCO-type phenomenon that can occur in propellers, helicopter rotors, compressor/turbine blades as well as aircraft wings. It is characteristically a torsion type motion shown in **Figure 19** and consists of the following stages:

1. the aerofoil angle of incidence θ increases dynamically, increasing the lift;
2. the angle of attack exceeds the stalling angle, separation occurs, and the lift reduces;
3. with the lift reduced, the angle of attack reduces, until is passes the stalling angle;
4. the flow reattaches and the cycle is repeated.

Stall flutter is not immediately catastrophic, but may lead to fatigue problems. For structures such as aircraft wings, or turbine blades, the phenomenon will occur at the tip, where the twist will be largest.

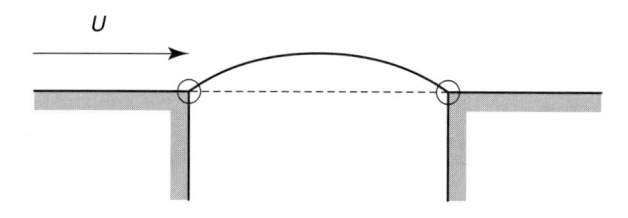

Figure 17 Panel flutter.

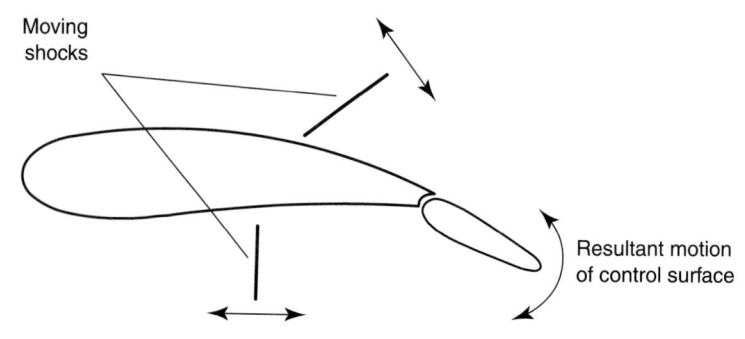

Figure 18 Control surface buzz.

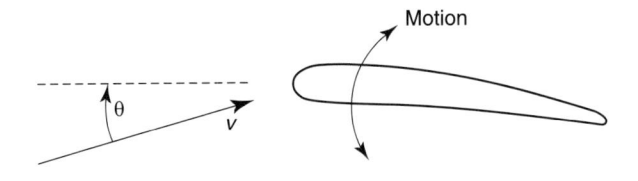

Figure 19 Typical stall flutter.

Another Aeroelastic Phenomenon

For completeness, a number of important aeroelastic phenomenon are included, however, these are not directly relevant to aircraft.

Aerodynamically Induced Negative Damping

It can be shown that most simple cross-section shapes, except a smooth circle, produce an aerodynamic force in the direction of motion. This is equivalent to having a negative angle of lift curve slope and will produce an unstable response. Such shapes can occur through the build-up of ice on the structure. The phenomenon can occur on electric transmission lines giving rise to the phenomenon known as 'galloping'.

Vortex Shedding

Under the correct flow conditions, the flow around a bluff body produces a regular pattern of alternating vortices known as a von Karman vortex street (**Figure 20**). These vortices produce a sinusoidal force acting on the body perpendicular to the flow. For a cylinder, the frequency ω (in Hz) of the shedding is related to the diameter D(m) and wind speed v (m s^{-1}) via the Strouhal number S such that

$$S = \frac{\omega D}{v} = 0.21$$

If the frequency of the shed vortices is the same as one of the natural frequencies of the structure, then large vibrations can result. Although it is possible for this loading to cause the structure to fail through the ultimate load being exceeded, failure is more likely to occur through fatigue.

The effect is exacerbated by an effect known as 'lock-in' where the vortex shedding frequency is forced to remain at the critical structural frequency at a range of speeds below and above the critical speed. The design of bridges, chimneys, submarine periscopes, etc. must take this phenomenon into account. Aerodynamic devices are used commonly nowadays to either ensure that the flow remains attached, or to break up the regular formation of the vortexes.

See also: **Flutter**; **Structural dynamic modifications**.

Further Reading

AGARD–SMP Report 815 (1997) Loads and requirements for Military Aircraft.

Bisplinghoff RL, Ashley H, Halfman RL (1996) *Aeroelasticity*. Dover.

Dowell EH, Curtis HC, Scanlan RH, Sisto F (1989) *A Modern Course in Aeroelasticity*. Kluwer.

ESDU Item 84020 *An Introduction to time-dependent aerodynamics of aircraft response, gusts and active controls*.

Fung YC (1993) *An Introduction to the Theory of Aeroelasticity*. Dover.

Lomax TL (1995) *Structural Loads Analysis for Commercial Transport Aircraft: Theory and Practice*. AIAA Education Series.

Niu MCY (1988) *Airframe Structural Design*. Conmilit Press Ltd.

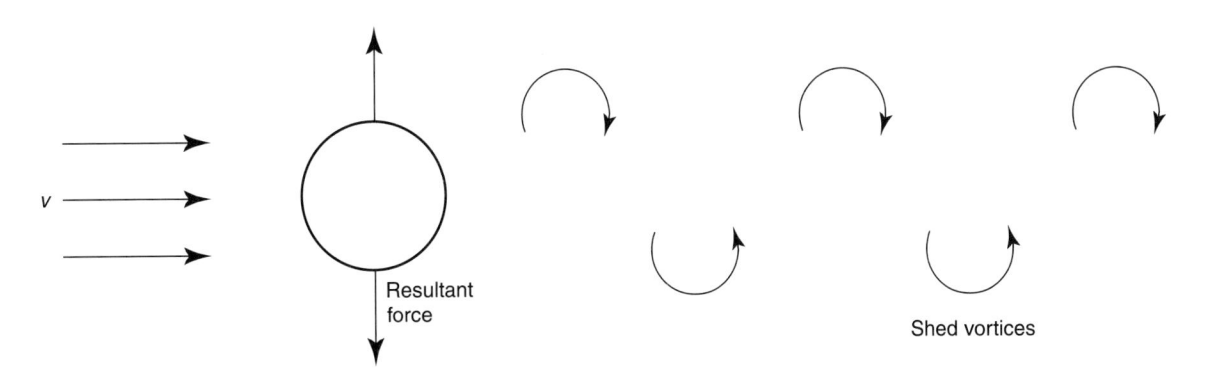

Figure 20 Vortex shedding.

ANTIRESONANCE

See **RESONANCE AND ANTIRESONANCE**

AVERAGING

S Braun, Technion – Israel Institute of Technology, Haifa, Israel

Copyright © 2001 Academic Press

doi:10.1006/rwvb.2001.0049

General

Computations of averages are necessary in many signal-processing tasks, whenever uncertainties result from random components in the analyzed signal. Averaging operations are thus performed in simple computations like those of root mean square (RMS) values, up to operations on vectors like time domain (synchronous) averaging (TDA) and power spectral density (PSD) averaging. Batch and recursive realizations are possible. The recursive procedures also enable the obtaining of properties of decaying memory, more suitable for situations involving some non-stationarity.

All these averaging procedures are extensively used in the processing of vibration signals, and available on most dedicated analyzing systems.

Basic operations

Given a sequence x_n, the most basic operation of averaging is given by:

$$y = \frac{1}{N}\sum_{n=1}^{N} x_n \qquad [1a]$$

where N is the number of elements in the sequence. Often a different weight is given to the samples, as per:

$$y = \sum_{n=1}^{N} w_n x_n \qquad [1b]$$

Eqn [1a] corresponds to the more general case of eqn [1b] with $w_n = 1/N$. The weights w_n are often chosen according to the importance or confidence of the data.

For dynamic signals, we often use moving averages (MAs). This is given by:

$$y_n = \sum_{k=0}^{N-1} w_n x_{n-k} \qquad [2]$$

and shown in **Figure 1**. The sequence w is called a window (see **Windows**) and often is intended to concentrate the (moving) averaging operation around the windows duration.

Specific cases are shown in **Figure 2**. The first case (**Figure 2A**) shows a rectangular window with $w_n = 1/N$; an average is computed each time for the data within the sliding window. A typical example is the continuous monitoring of the RMS value of a signal. For random signals, the random error of the computed RMS is approximately inversely proportional to $1/\sqrt{N}$, and can be controlled by averaging N elements of:

$$X_{\text{RMS}} = \left(\frac{1}{N}\sum_{k=0}^{N-1} x_k^2\right)^{0.5}$$

or by the equivalent MA expression for time-varying RMS values:

$$X_{\text{RMS}_n} = \left(\frac{1}{N}\sum_{k=0}^{N-1} x_{n-k}^2\right)^{0.5}$$

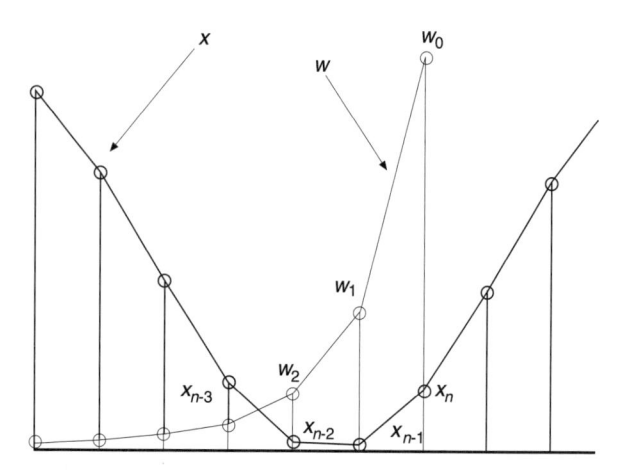

Figure 1 Moving average with window.

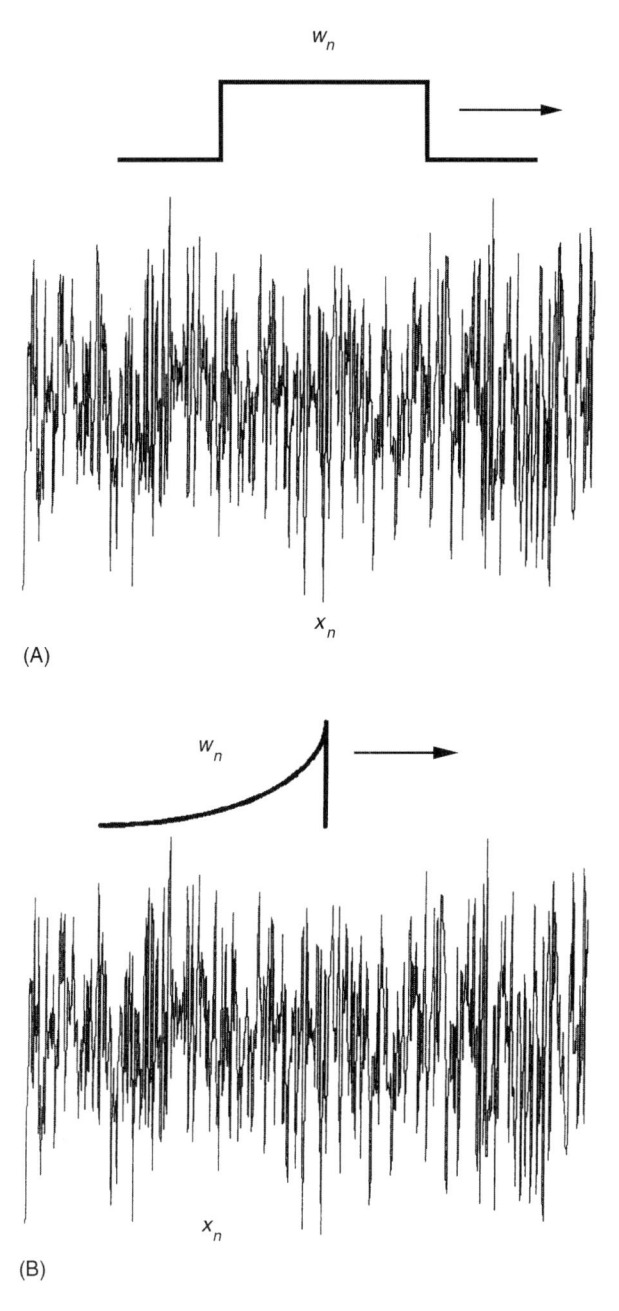

Figure 2 Moving average: (A) rectangular window; (B) exponential window.

An exponential window is shown in **Figure 2B**, where a forgetting effect is achieved for past data, less and less weight being given to data more distant in the past.

The Frequency Response of the Moving Average

It is convenient to consider the MA as a linear filter, where a new y_n is produced for each new input x_n. For the windows shown in **Figure 2** we obviously get a smoothing effect, hence the operation is equivalent to some low pass filter. The filter's bandwidth must be inversely proportional to the windows effective duration (which must be defined according to a specific criterion).

The MA filter can be analyzed in the frequency domain, describable by $H(\omega)$, its frequency response function (FRF). This can be conveniently undertaken via the Z transform (see **Transform methods**).

For the rectangular window, N increases as each new x_n occurs:

$$Y(z) = \frac{1}{N} \sum_{k=0}^{N-1} X(z) z^{-k} \qquad [3]$$

and the FRF is computed by substituting $z = \exp(j\omega\Delta t)$, Δt the time interval between samples and computing:

$$Y(\omega) = \frac{1}{N} \sum_{k}^{N-1} X(\omega) \exp(-j\omega k \Delta t) \quad [4a]$$

$$H(\omega) = \frac{1}{N} \sum_{k}^{N-1} \exp(-j\omega k \Delta t) \quad [4b]$$

The filter's response is shown in **Figure 3**. The magnitude of H consists of a main lobe and secondary lobes attenuated with frequency. Zeros occur at locations $f = 1/N$.

In addition to the finite-length MA filter, an infinitely operating MA can be used, as shown in **Figure 4**. The average is still computed via:

$$y_n = \frac{1}{n} \sum_{k=1}^{n} x_k \quad [5a]$$

but with n increasing with each incoming data sample. This is equivalent to a filter with time-dependent parameters. Comparing this to **Figure 3**, the width of the main lobe will decrease as N increases (see **Figure 5**).

Recursive Averaging Filters

Recursive computations are often preferred to batch ones, as close to real-time processing is possible. For n increasing as each new x_n occurs, eqn [5b] can be cast is the recursive form:

$$y_n = y_{n-1} + \frac{x_n - y_{n-1}}{n} \quad [5b]$$

the average y_n is updated at each step, and a diminishing weight (with n) is applied to the difference between the incoming data and the prior average. The frequency response of **Figure 5** obviously applies.

A modified averaging process can be used by replacing the variable n in eqn [5b] by a constant parameter N_1. The corresponding recursive equation can be rewritten as:

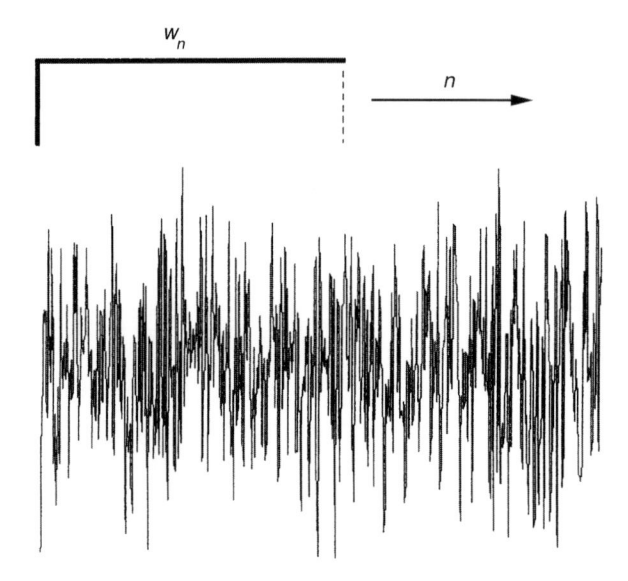

Figure 4 Running moving average with increasing number of samples.

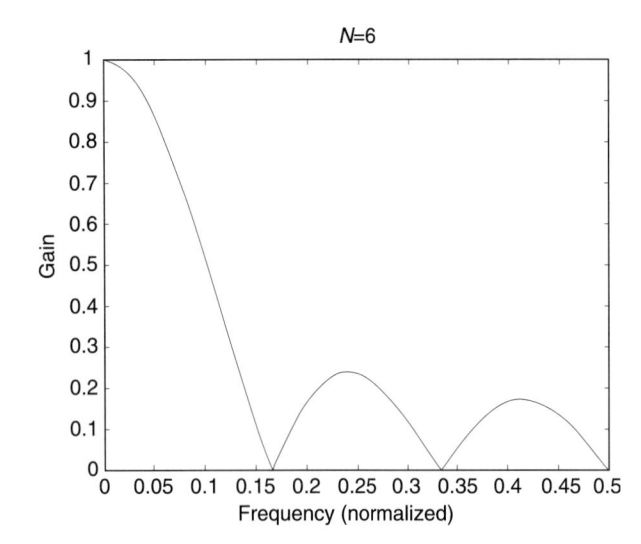

Figure 3 Frequency response function of moving average with rectangular window.

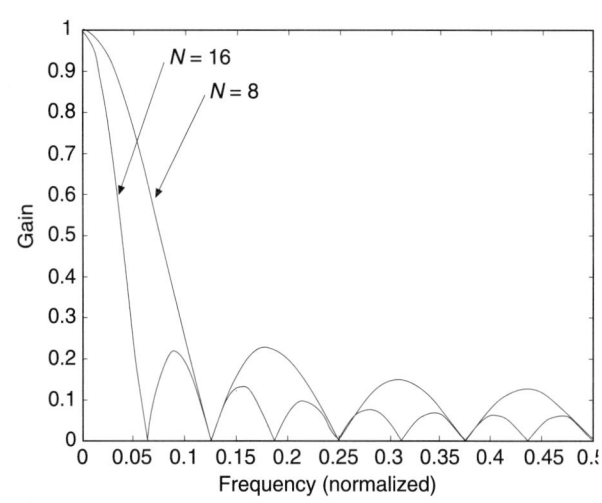

Figure 5 Frequency response function of running moving average for $N = 8$ and $N = 16$.

$$y_n = (1-\alpha)y_{n-1} + \alpha x_n \qquad [6]$$

where $\alpha = 1/N_1$. This is sometimes called exponential recursive averaging. The FRF of the equivalent filter:

$$H(z) = \frac{\alpha}{1-(1-\alpha)z^{-1}} \qquad [7a]$$

is:

$$H(\omega) = \frac{\alpha}{1-(1-\alpha)\exp(-j\omega\Delta t)} \qquad [7b]$$

is shown in **Figure 6**, the impulse response in **Figure 7**. The equivalent time constant for $\alpha \gg 1$ is approximately equal to N_1. In this averaging process, gradually decreasing weight is given to past samples. The bandwidth of the FRF is inversely proportional to α, and no zeros or secondary lobes result.

Time Domain Averaging (TDA)

One of the most powerful and practical methods of extracting periodic signals from a composite signal is based on averaging signal sections of the period (frequency) sought. An *a priori* knowledge of the frequency sought is obviously necessary. This method is also known as synchronous averaging.

Regular TDA

The principle is shown in **Figure 8**, whereby signal elements separated by one (known) period are averaged. Signal elements which are of the frequency sought will be unaffected by the averaging process. All other components will be attenuated, and will converge to zero as the process is repeated indefinitely.

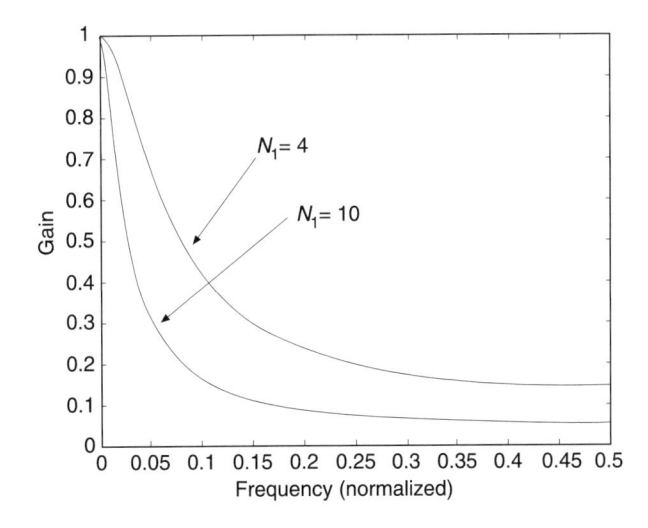

Figure 6 Frequency response function of exponential moving average for $N_1 = 4$ and $N_1 = 10$.

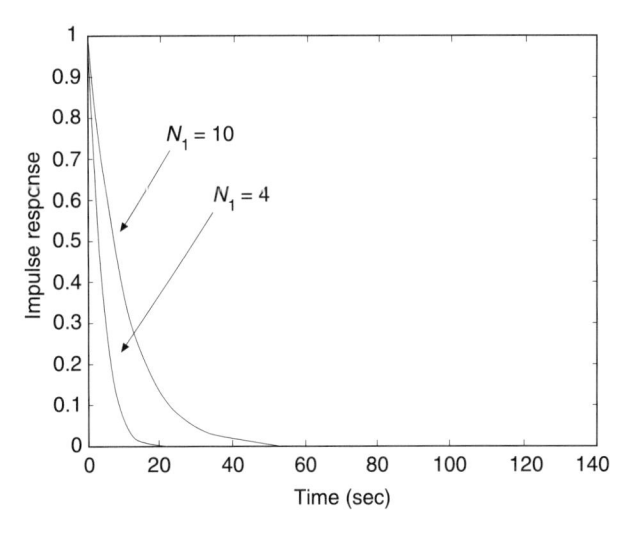

Figure 7 Impulse response of exponential moving average for $N_1 = 4$ and $N_1 = 10$.

Knowledge of the period can also be based on appropriate acquisition of a synchronizing signal. The method is thus extensively used in analyzing vibration (and other) signals acquired from rotating machinery, where synchronizing trigger signals are easily available via suitable hardware solutions (see **Data acquisition**).

The mathematical description of the averaging process is given by:

$$y(n\Delta t) = \frac{1}{N}\sum_{r=0}^{N-1} x(n\Delta t - rM\Delta t) \qquad [8]$$

where M is the number of elements per period, and N the number of sections averaged.

This has similarities to eqn [3], and the process can be analyzed as a filtering process. The frequency response can be computed via the Z transform as:

$$H(z) = \frac{1}{N}\frac{1-z^{-MN}}{1-z^{-N}} \qquad [9]$$

$$H(f/f_p) = \frac{1}{N}\frac{\sin(\pi Nf/f_p)}{\sin(\pi f/f_p)} \qquad [10a]$$

$$\phi_H(f/f_p) = -\pi(N-1)f/f_p \qquad [10b]$$

where:

$$f_p = \frac{1}{MN\Delta t} \qquad [10c]$$

is the frequency of the periodic component to be extracted.

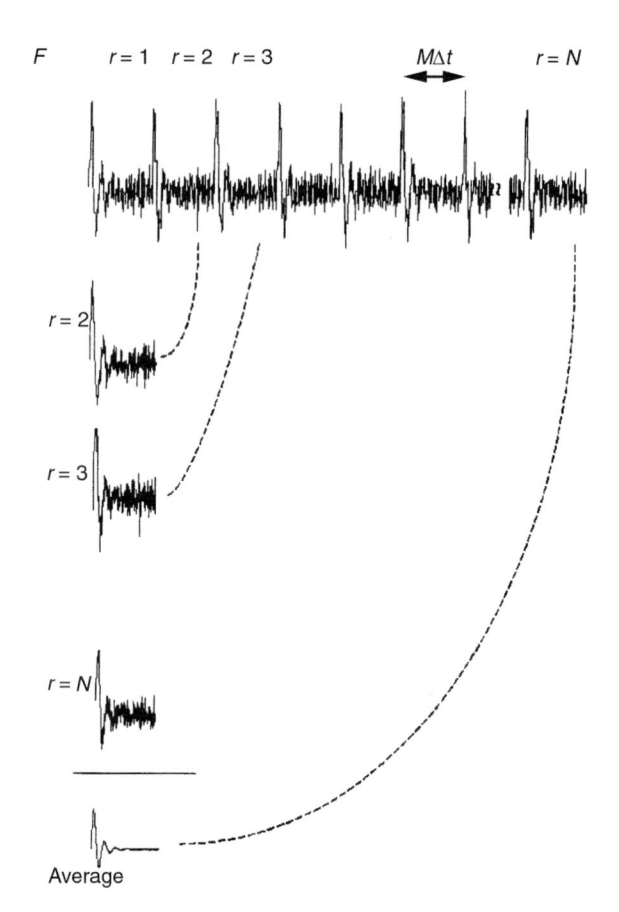

Figure 8 Principle of time domain averaging.

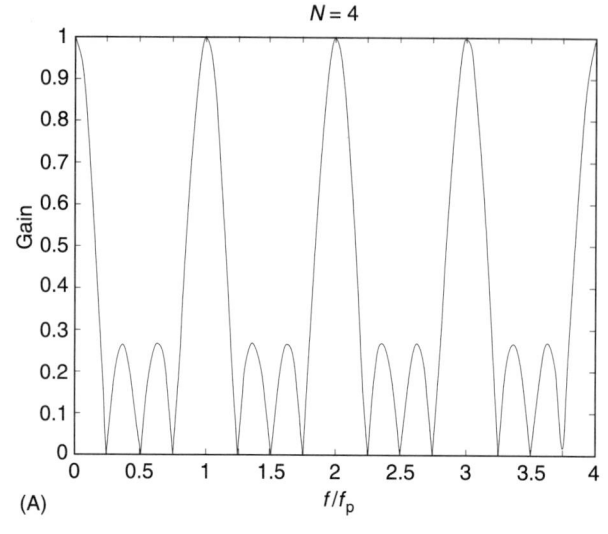

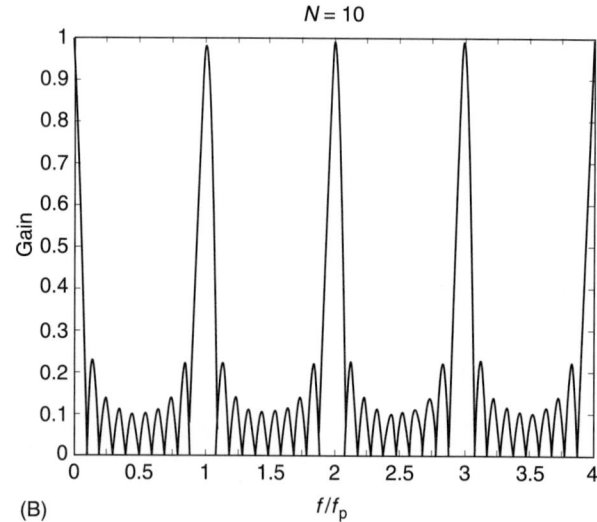

Figure 9 Frequency response function of exponential TDA, $H(f/f_p)$: (A) $N_1 = 4$; (B) $N_1 = 10$.

The FRF is shown in **Figure 9**. It has the form of a 'comb' filter, with main lobes centered around integer multiples of the synchronizing frequency f_p. The gain at these frequencies is 1, and the phase shift zero, hence the fundamental and harmonic components of the periodic signal are extracted exactly. All other components however, are attenuated. The zeros of the FRF are separated by:

$$f/f_p = k/N \quad k = 1, 2, \ldots$$

The bandwidth of the main lobe is thus inversely proportional to the number of sections averaged.

Recursive Schemes for TDA

The recursive algorithms described above can be applied as well. The algorithm:

$$y_r(n\Delta t) = y_{r-1}(n\Delta t) + \frac{x_r(n\Delta t) - y_{r-1}(n\Delta t)}{r} \quad [11a]$$

where y_r is the running average at the rth period. The algorithm is very efficient, only two vector additions and one division are necessary per averaging, and real-time operating systems (for frequency ranges typical of mechanical vibration tasks) exist. The results are obviously equivalent to those described above. Additional versatility can be achieved by using the exponential averaging of eqn [6], resulting in:

$$y_r(n\Delta t) = y_{r-1}(n\Delta t) + \frac{x_r(n\Delta t) - y_{r-1}(n\Delta t)}{N_1} \quad [11b]$$

The resulting FRF is shown in **Figure 10**. It has again the form of a comb filter, but only one lobe exists around each center frequency. Their lobes function is given by eqn [4b], the bandwidth dictated by the parameter N_1.

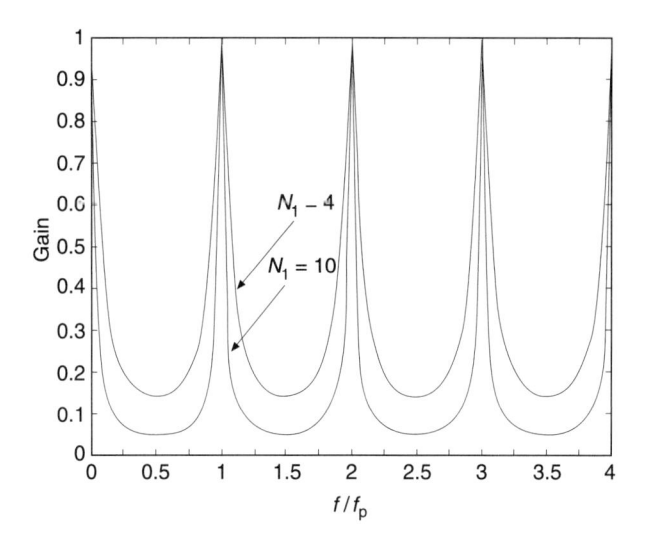

Figure 10 Frequency response function of exponential TDA, $H(f/f_p)$ for $N_1 = 4$ and $N_1 = 10$.

Performance of TDA

Rejection of unwanted components The rejection of unwanted components is a function of their properties. Denoting the signal $x(n\Delta t)$ as a sum of a periodic component $s(n\Delta t)$ and noise $e(n\Delta t)$, where the noise includes all the signal elements not periodic within $1/f_p$:

$$x(t) = s(t) + e(t) \quad [12]$$

we may estimate the attenuation of $e(t)$ for the cases when it is broadband random or harmonic.

For additive broadband random noise, the averaging process will attenuate the RMS of the noise according to eqn [14a]. An example is given in **Figure 11**, where a periodic signal in the form of decaying oscillations is corrupted by random noise. This shows the result of the averaging, considered as a filtering process, where a new output is generated for each new input. In practice, only M points of the last averaging operation, i.e., one period only, would be shown, as in **Figure 8**. The improvement in the signal-to-noise ratio as the number of averages increases is evident.

For additive harmonic noise, the attenuation of $e(t)$ is a function of its frequency, according to the frequency response of the averaging filter (**Figure 9**). An upper bound of this attenuation can be based on the maxima of the secondary lobes, resulting in:

$$H_{\max}(k) = \left[N \sin\left(\pi f/f_p\right)\right]^{-1} \quad [13a]$$

$$\frac{f}{f_p} = (k - 0.5)/N \quad [13b]$$

An example is shown in **Figure 12**. It can be noted that the increase in the attenuation of the noise with N has an oscillatory character. The number of averages N, necessary to achieve a given attenuation, is given by:

$$\left.\begin{array}{ll} \text{Broadband noise} & N > (1/NR)^2 \\ \text{Harmonic noise} & N > \left[NR \sin\left(\pi f/f_p\right)\right]^{-1} \end{array}\right\} \quad [14]$$

where NR is the desired noise reduction of the interfering signal components.

Comparisons between regular and exponential averaging The exponential averaging achieves a fading memory property. For regular averaging see eqn [11a], as the number of incoming data samples (and hence N, the number of available periods) increases, the bandwidth of the comb filter main lobes becomes narrower and narrower, with increased rejection of unwanted components. With exponential averaging (see eqn [11b]) the bandwidth is fixed, and after a transient stabilizing time, the rejection of the interfering signals stays constant. This is shown in **Figure 13**.

One advantage of exponential averaging is due to the decaying memory of the process. The effect of a single artifact or time-localized interference will slowly disappear as future data become available. An example is shown in **Figure 14**, extracting a periodic signature from broadband noise, while a single disturbing transient occurred once. Comparing exponential averaging to the regular one, the fading effect of the transient can easily be controlled by N_1 (see eqn [11b]), the approximate decaying memory length.

Variable periods and the effect of jitter In many practical cases, the period of the signal of interest may not be constant. An example could be the vibratory response of an impacted structure, where many transient responses could be averaged. A triggered mode of data acquisition (see **Data acquisition**) could then be used, where a limited and fixed number of data points would be acquired after each trigger. In such a case, the interval between the transient responses would not inhibit the use of the averaging method. An example is shown in **Figure 15**. Comparison with **Figure 8** shows that the averaged result would extract the exactly repeating component following the trigger. Obviously both regular and exponential averaging could be used.

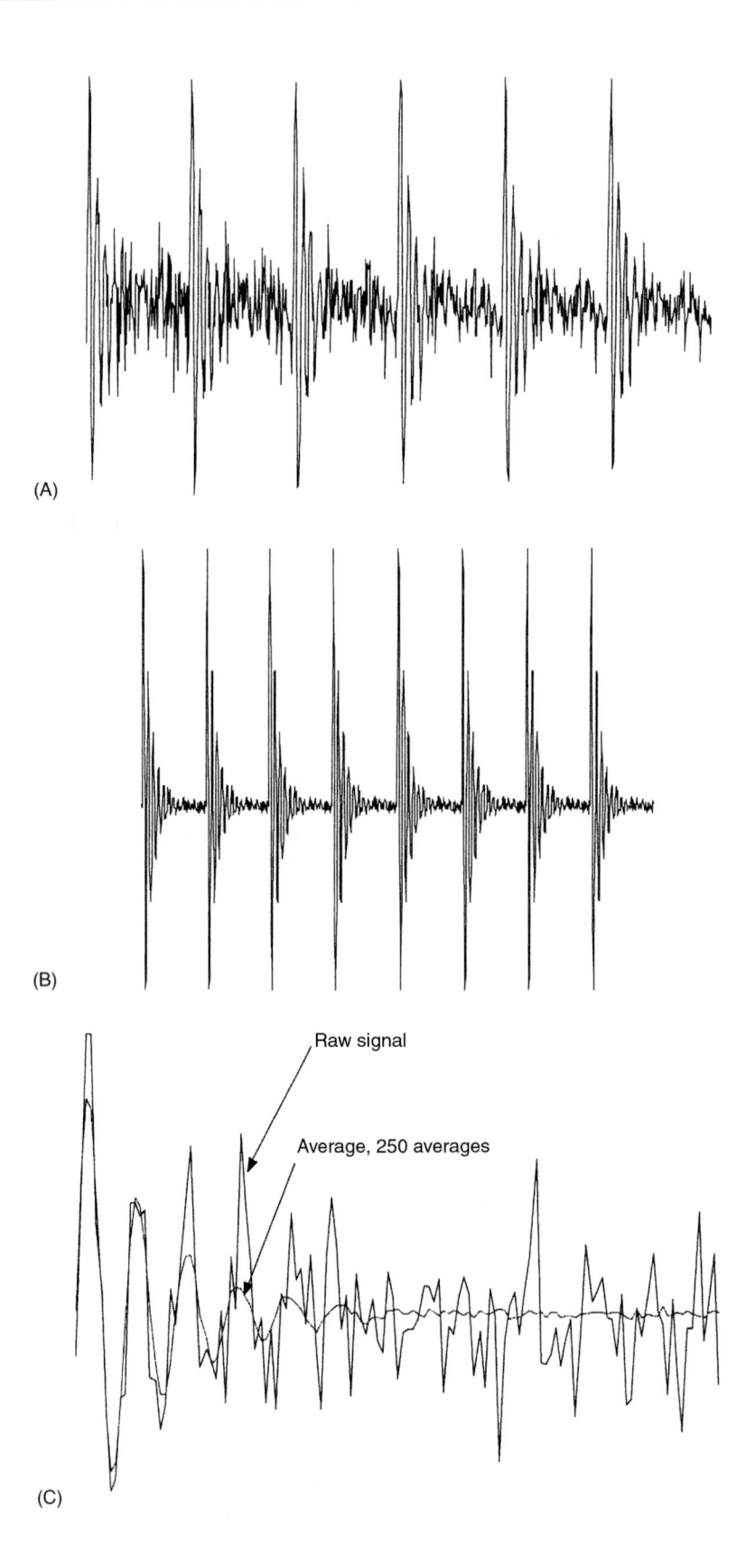

Figure 11 Rejection of broadband random interfering signal by TDA: (A) signal for periods 2 to 8; (B) signal for periods 200 to 208; (C) single raw signal period and result of 250 averages.

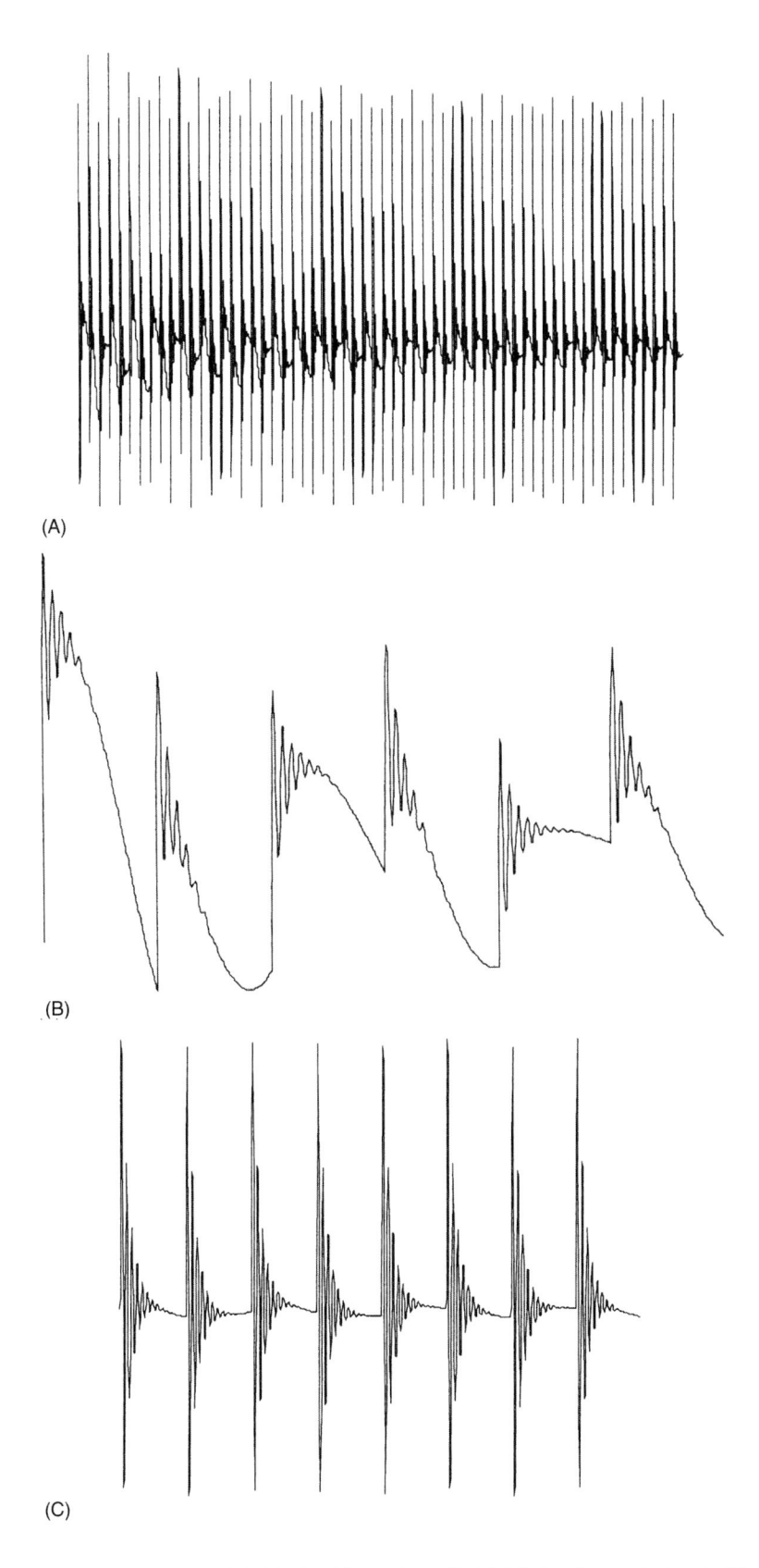

Figure 12 Rejection of harmonic interference by TDA: (A) signal for periods 20–80; (B) signal for periods 2–8; (C) signal for periods 200–208.

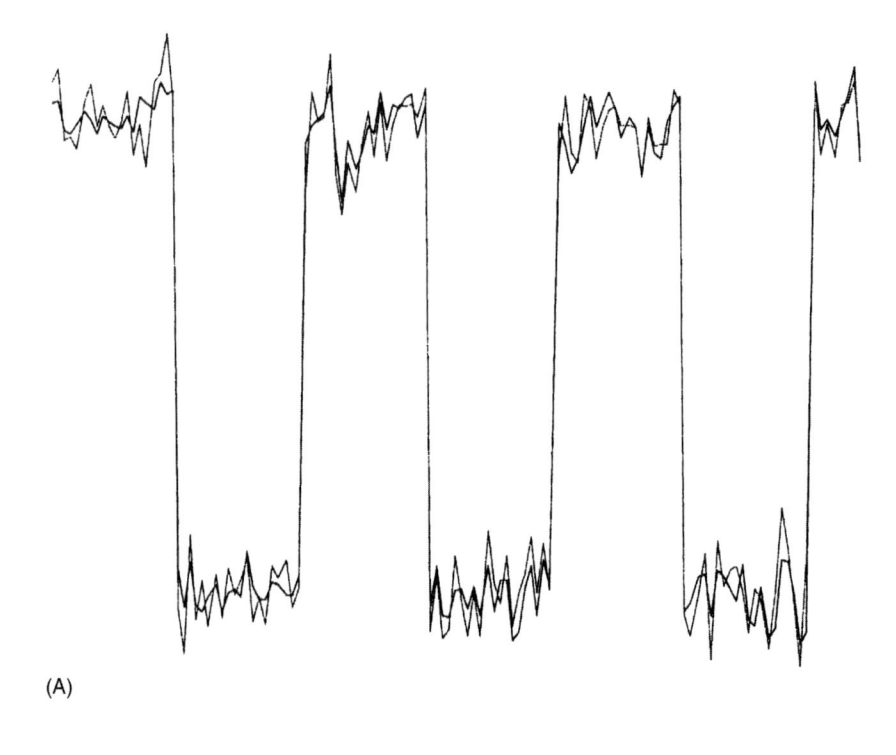

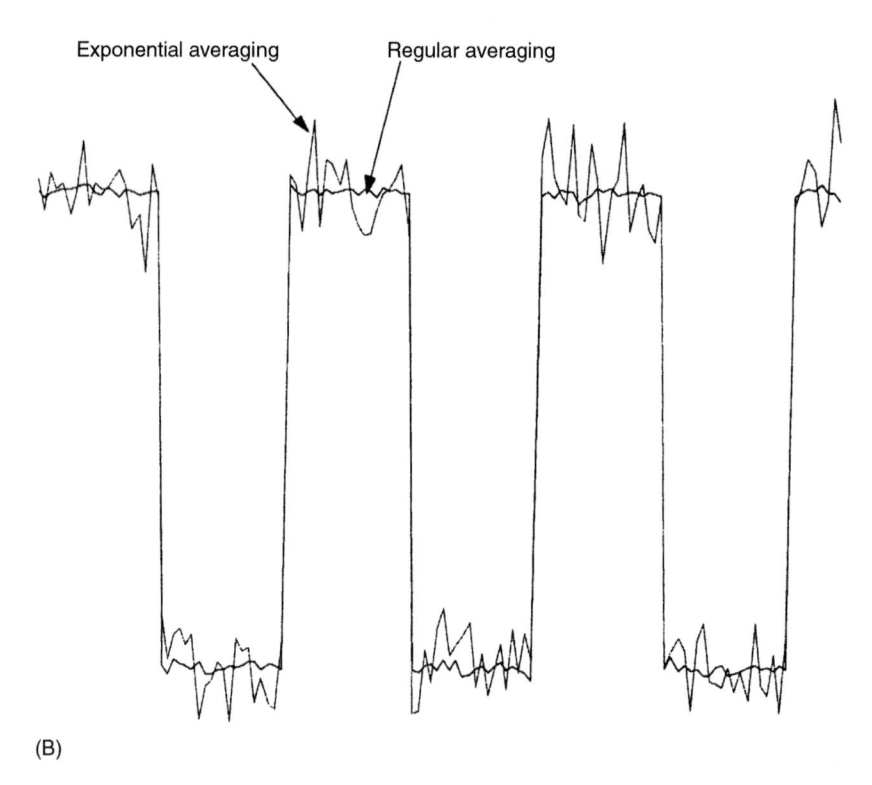

Figure 13 Regular and exponential TDA for square signal and broadband random interference: (A) N_1 and N same order of magnitude; (B) $N \gg N_1$.

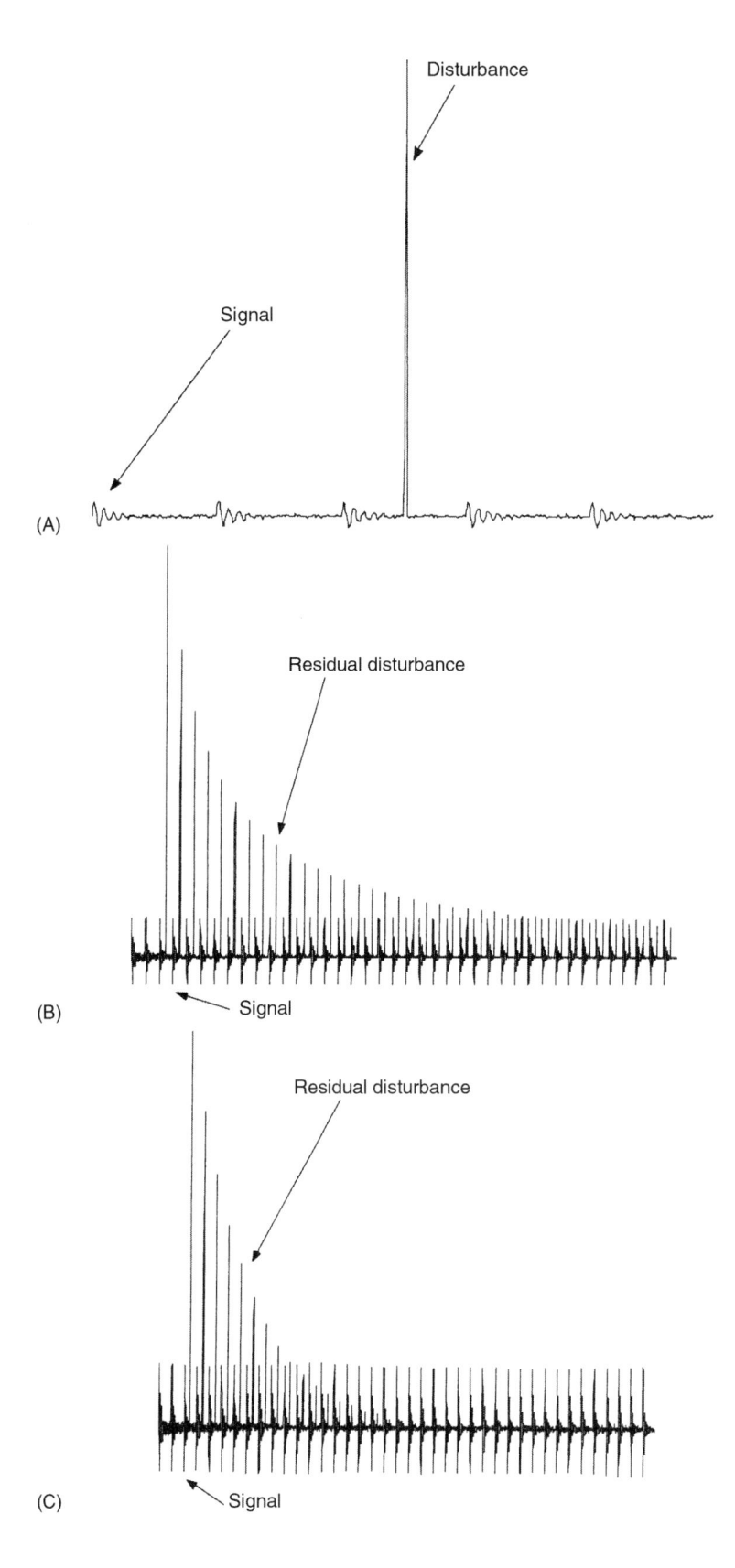

Figure 14 Rejection of single artifact disturbance by TDA: (A) occurrence of artifact disturbance; (B) rejection with regular TDA; (C) rejection with exponential TDA.

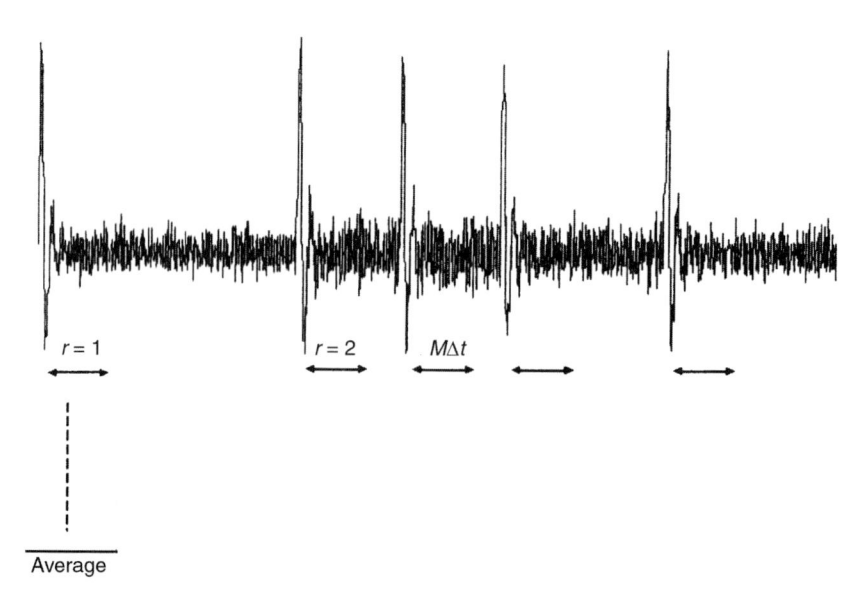

Figure 15 TDA with triggered data acquisition.

Another example occurring in practice is that of small variations between consecutive periods. As an example, when dealing with vibrations of rotating machinery, speed fluctuations may be induced by changes of loads and other affecting variables. Large fluctuations can decrease and even nullify the magnitude of extracted periodic components.

An example is shown in **Figure 16A**. A random fluctuation of 2% (RMS) in the period is assumed. Also shown is the average of 100 periods, compared to the case of zero fluctuation, and the resulting attenuation is evident.

The effect of jitter can be recognized by averaging signal sections which correspond to a multitude of the desired period. At least two periods should be extracted to notice the effect. **Figure 16B** shows the result of averaging 100 sections. It can be noticed that the magnitude of the first half of the averaged signal is larger than that of the second one. This effect is traced to the fact that the effect of the jitter is larger for data points more distanced from the start of the averaged section. The larger the number of periods in the averaged section, the more pronounced the effect towards the end of the section. Obviously the price paid for increasing this section length is that more data points are needed for the same number of averaged sections.

Averaging in the Frequency Domain

Computation of PSD

Often there is a need to average descriptors of data in the frequency domain. In the computations of the PSD, segmentation of the data (with possible overlapping processing) is used in order to decrease random errors whenever random components exist in the signal. The PSDs of the segmented sections are averaged (see **Windows; Spectral analysis, classical methods**).

Recursive schemes are usually implemented, and the averaging effect can often be followed in real time. Both regular and exponential averaging can be used, obviously in the frequency domain. The exponential averaging is better suited to situations involving nonstationarities. **Figure 17** shows the PSD of a stationary random process with an additive component – a harmonic signal with time-dependent increasing frequency. Instantaneous PSD, based on one segment at a time, can follow the frequency changes, but the random error of the PSD would not be controlled. In regular averagings, the 'smearing' due to changing frequency is evident. This is much less pronounced in the exponential averaging scheme, the decaying memory 'forgets' some of the initial frequencies, the corresponding peak would be much stronger. Commercial analyzers sometimes have a 'peak' averaging mode. This is not an averaging operation; the global maximum values of segment's PSD is retained.

Extraction of Periodic Signals via Averaging in Time and Frequency

The time domain averaging is aimed at extracting periodic signals from a composite signal. Averaging the PSD is aimed at reducing the random error of the PSD estimate. However, when the frequency domain description of the periodic component only is sought,

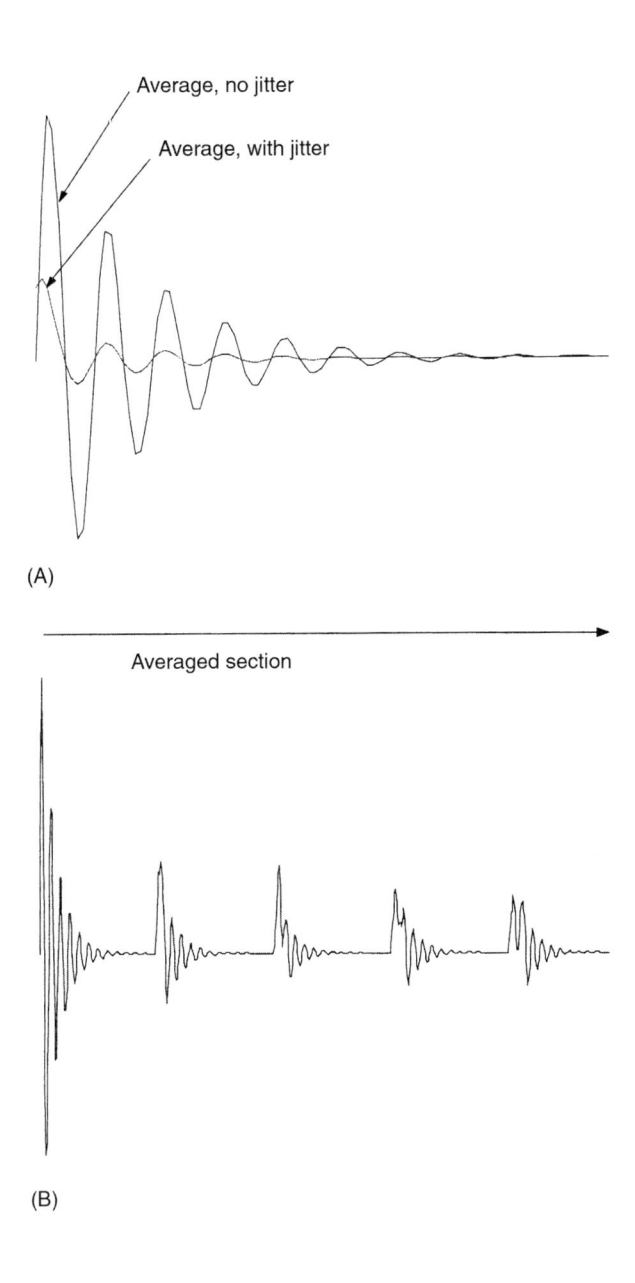

Figure 16 TDA in the presence of jitter: (A) TDA with and without jitter – sections of single period averaged; (B) TDA with jitter – sections of multiple periods averaged.

then two equivalent procedures are possible. Both assume knowledge of the frequency of the signal sought, whether via hardware triggering or external knowledge. Both procedures are shown in **Figure 18**. In the first, an FFT computation follows extraction via TDA. In the second, FFT operations (without squaring for PSD) are performed on each segment, and then averaged. Both procedures reject components not synchronous with the frequency sought (whether deterministic or random). The first procedure is slightly more efficient (one averaging vs one FFT operation).

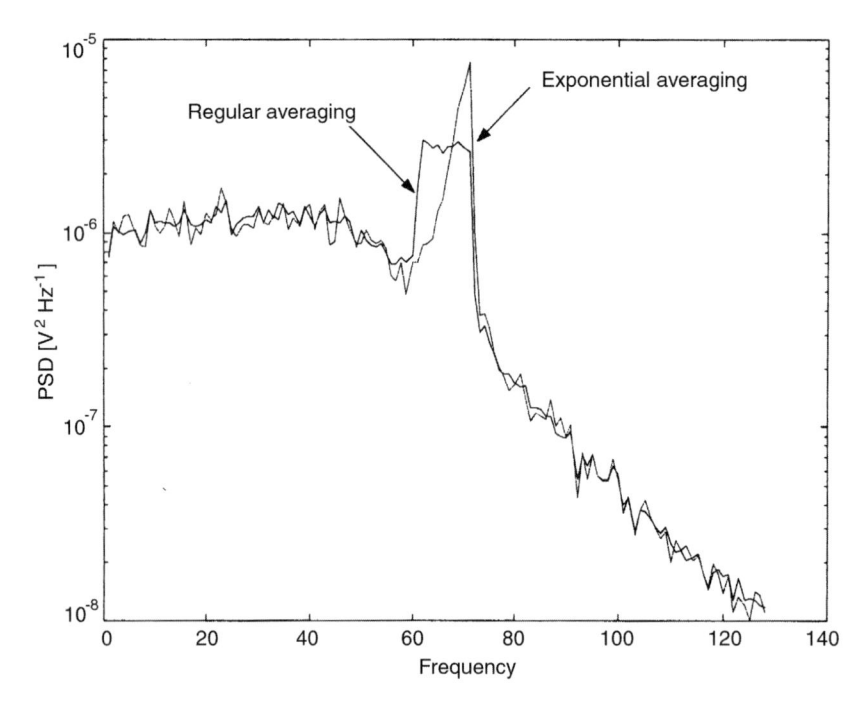

Figure 17 Averaging for computation of PSD – composite signal including harmonic component of varying frequency: regular and exponential averaging.

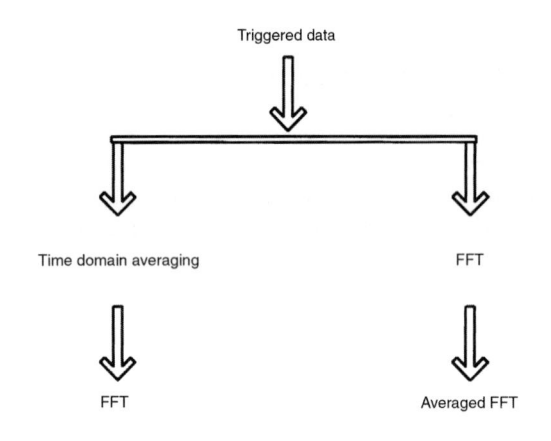

Figure 18 Equivalent procedure for the DFT of the periodic component.

Nomenclature

f frequency (Hz)
f_p frequency of periodic component
ω radial frequency (rad s^{-1})

See also: **Data acquisition**; **Spectral analysis, classical methods**; **Transform methods**; **Windows**.

Further Reading

Braun S (1986) *Mechanical Signature Analysis*, London: Academic Press.

B

BALANCING

R Bigret, Drancy, France

Copyright © 2001 Academic Press

doi:10.1006/rwvb.2001.0093

Introduction

The center of gravity of a rotor does not generally coincide with the geometrical axis of rotation. Neither are the axial moments on inertia of these elements parallel to the rotation axis. These deviations from the ideal situation result from inaccuracies during the machining and assembly process, and result in vibrations whose frequency equal the frequency of rotation.

The vibrations can be attenuated by a balancing procedure by i masses attached to the rotor. These m_i masses of small dimension, at distances d_i from the axis of rotation and angular position θ_i, generate couples $m_i d_i$. The centrifugal forces generated oppose the forces and couples, and will reduce the magnitude of vibrations (unless the malfunction causing the vibrations are too severe).

When the maximum rotation frequency Ω_{max} is less than that of the first natural flexural mode, ω_1, say $\Omega_{max} \leq 0.8\omega_1$, the rotor is considered as being rigid. The balancing can be achieved by $i = 2$ balancing masses, which can be determined (by an appropriate procedure) at any speed inferior to Ω_{max}. When $\Omega_{max} \geq 0.8\omega_i$, the rotor is considered as flexible. Balancing may necessitate $i > 2$ masses, determined at more than one rotational speed.

Balancing is undertaken by dedicated balancing machines by the manufacturer. It can also be performed on site, sometimes with an improved attenuation of vibrations. The balancing is aimed at attenuated vibration levels, as determined by contract (or standard), but need to consider technological constraints.

Set of Forces and Rotating Couples: Balancing

In the article on definitions and characteristics of rotating machines (see **Rotating machinery, essential features**) elementary unbalance and unbalance torsion around the line of rotation are defined. In the chapter on dynamics of rotating machinery (see **Rotor dynamics**) the performance of the De Laval rotor which is subject to centrifugal force caused by elementary unbalance is studied. Moment due to run out on the disk is defined.

Definitions: Eccentricity and Elementary Run Out

Figure 1 shows elements which induce rotating forces. The axis Cx_R is part of the rotor axis, the geometric centers of section profiles. In rotation under the effect of centrifugal forces, as well as other forces, this axis becomes a left line around which the sections rotate. The terms associated with the element are: C, geometric center of the profile of the median section or center of gravity; $Cx_Ry_Rz_R$, frame of reference for the element; G_g, element mass center of gravity; $G_g x_n y_n z_n$, principal axes of central inertia ellipsoid; $\overrightarrow{CG_g}$, eccentricity (e, θ); $\overrightarrow{CG_g}m$, unbalance (em, θ); $\overrightarrow{CG_g}m\Omega^2$, rotating force unbalance $(em\Omega^2, \theta)$. The moments of mass intertia produced in $Cx_Ry_Rz_R$ (close to $G_g x_Ry_Rz_R$) Ix_Ry_R, Ix_Rz_R, induce rotating couples which are in proportion to

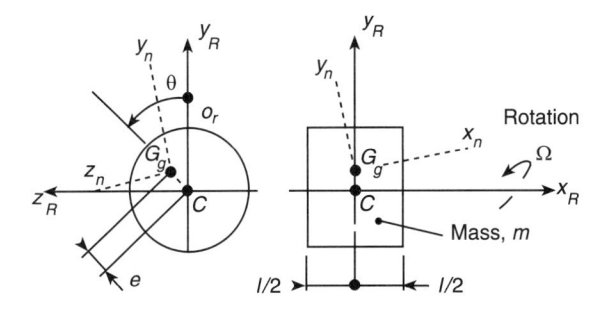

Figure 1 Rotor element. C is the geometric center of the median section profile (barycenter); $Cx_Ry_Rz_R$ is the referential linked to element; G_g is the mass element barycenter; $G_g x_n y_n z_n$ is the principal axes of central inertia ellipsoid.

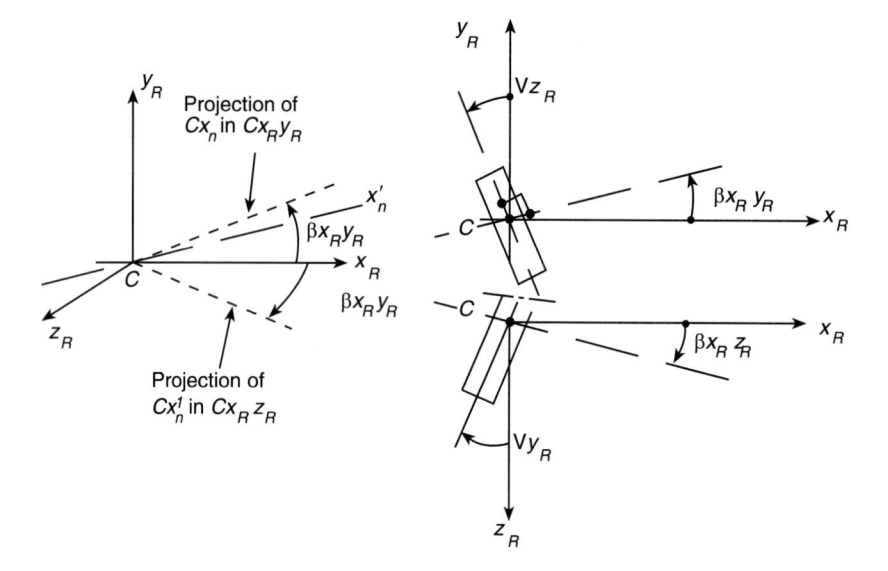

Figure 2 Rotor element. Angles linked to inertia products.

Ω^2. Inertia products arise from the angles $\beta x_R y_R$ and $\beta x_R z_R$ (**Figure 2**). For an element which is thin compared to its radial dimensions

$$\beta_{x_R y_R} = \frac{I x_R y_R}{I x_R x_R - I y_R y_R} \quad \beta_{x_R z_R} = \frac{I x_R z_R}{I x_R x_R - I z_R z_R} \quad [1]$$

where $I x_R x_R$, $I y_R y_R$, $I z_R z_R$ are moments of mass inertia in relation to $C x_R$, $C y_R$, $C z_R$. The small angles β characterize run out which are generally designated $V z_R$ and $V y_R$. $\overrightarrow{CG_g}$, $I x_R y_R$, $I x_R z_R$ are caused by manufacturing or assembly faults, transformations of materials, or deteriorations. For a thin disk of mass m and radius R:

$$I x_R y_R = -\frac{mR^2}{4} V z_R, \quad I x_R z_R = -\frac{mR^2}{4} V y_R$$

Definitions: Set of Unbalance and Run Out

Let there be a rotor of axis $0x_R$ in a rigid state, without any deformations. The geometric centers of sections perpendicular to right-angle $0x_R$ are at $0x_R$ (**Figure 3**). Eccentricity $e(x)$ of an element of size $\Delta x (\to dx)$ defines the position of the mass center of gravity $0x_g$. The arrangement $e(x)$ defines the set which can be characterized by a resultant, and a resultant moment which on rotation induces excitations at frequency Ω. In the frame of reference $G_r x_R y_R z_R$, unbalance which results from set of eccentricity $e(x)$ has the following components:

$$\text{For } G_r y_R: \quad m y_G = \int_{x_0}^{x_l} e_y(x) \mu(x) \, dx$$

$$\text{For } G_r z_R: \quad m z_G = \int_{x_0}^{x_l} e_z(x) \mu(x) \, dx \quad [2]$$

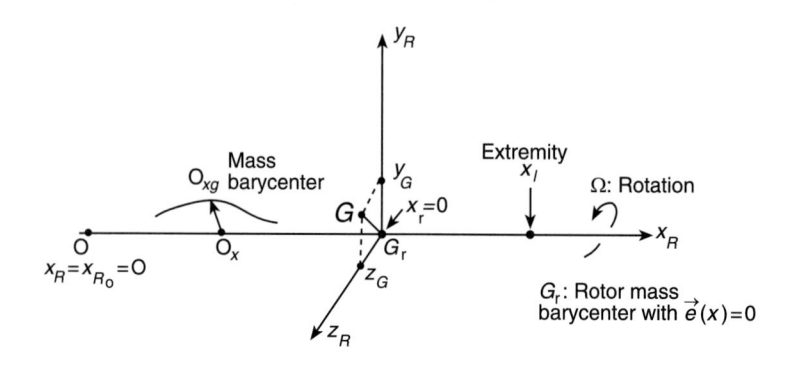

Figure 3 Rotor eccentricities. $G_r x_R y_R z_R$ is linked to the rotor.

where y_G, z_G are coordinates of the center of gravity of mass G; m is the rotor mass; $\mu(x)$ is the mass by unit of length; my_G and mz_G are components of initial unbalance resultants. They are independent of the position of G_r, but dependent on the direction of $G_r y_R$ and $G_r z_R$. The resultant is expressed by:

$$m\,\overrightarrow{G_r G} \to \begin{cases} \text{module } m\sqrt{y_G^2 + z_G^2} \\ \text{angular position } \theta \to \operatorname{tg}\theta = z_G/y_G \end{cases}$$

In the frame of reference $G_r x_R y_R z_R$ the moment resultant from the set of eccentricities has the components:

$$\text{For } G_r y_R: \quad M_{r y_R} = \int_{x_0}^{x_l} e_z(x)\mu(x)(x - x_r)\,dx$$

$$\text{For } G_r z_R: \quad M_{r z_R} = \int_{x_0}^{x_l} e_y(x)\mu(x)(x - x_r)\,dx \quad [3]$$

$M_{r y_R}$ and $M_{r z_R}$ are components of the initial moment of unbalance. They depend on the position of G_r and the direction of $G_r y_R$ and $G_r z_R$. The moment can be represented by a vector $\overrightarrow{M_r}$ in the same way as the resultant. In general, let $\overrightarrow{V}_1 \ldots \overrightarrow{V}_i$ (where $i = 1 \pm n$) a set of vectors applied to $0_1 \ldots 0_i \ldots$. At 0 this set can be reduced to a resultant:

$$\overrightarrow{R}_0 = \sum_{i=1}^{n} \overrightarrow{V}_i$$

and at a moment:

$$\overrightarrow{M}_0 = \sum_{i=1}^{n} \overrightarrow{00_i} \wedge \overrightarrow{V}_i$$

At:

$$0';\quad \overrightarrow{R}'_0 = \overrightarrow{R}_0;\quad \overrightarrow{M}'_0 = \overrightarrow{M}_0 + \overrightarrow{00'} \wedge \overrightarrow{R}_0$$

The result is independent of the point of reduction. The resultant moment depends on it.

Forces arise from the vectors of eccentricity. Couples are a result of run out vectors. They are connected to the line of rotation (**Figure 4**).

In G_r, the central inertia ellipsoid is defined by: the moments of inertia $I_{x_R x_R}$, $I_{y_R y_R}$, $I_{z_R z_R}$, moments of inertia produced $I_{x_R y_R}$, $I_{x_R z_R}$. During rotation the excitations at frequency Ω are caused by $I_{x_R y_R}$ and $I_{x_R z_R}$.

$$\begin{aligned} I_{x_R y_R} &= \int\limits_{\text{volume}} xy\rho(x,y)\,dV \\ I_{x_R z_R} &= \int\limits_{\text{volume}} xz\rho(x,z)\,dV \end{aligned} \quad [4]$$

where x, y, z are coordinates of the center of an element of volume dV and specific mass $\rho(x, y)$; $\rho(x, z)$.

The moments of inertia depend on the direction of axes $G_r y$ and $G_r z$. They can be defined at any point.

Balancing

Balancing is a procedure which arranges on the rotor elements of slight dimensions (of small mass) which, creating unbalance and so rotating forces, reduce forces and initial rotating couples. Balancing can be carried out at a rotating speed which is much less than the first critical flexion speed, in a flexible condition. The quality of balancing is appreciated by rotor vibration amplitude, nonrotating situations, forces perpendicular to the links, levels of sound, or other features. Their modules must be less than the limits which may be agreed under contract. The vocabulary of balancing is the subject of standards ISO 1925 and AFNOR E 90–002.

Rigid States

Introduction to Balancing

Figure 5 shows an isotrope arrangement of rotor and links which is affected by a line of eccentricity, which is obscured in the fixed plane $0xy$. The line of eccentricity is also projected in the plane $0xz$ which is perpendicular to $0xy$. Taking into account only the

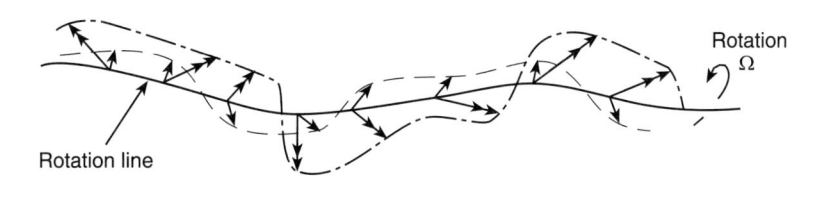

Figure 4 Eccentricities leading to displacement.

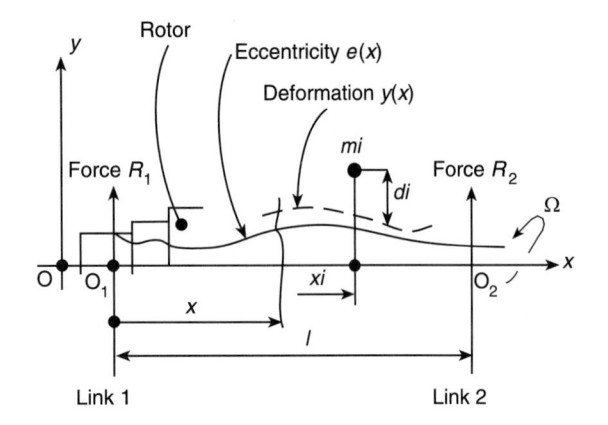

Figure 5 Rotor links: eccentricity.

plane Oxy does not restrict the conclusions that can be drawn.

Balance results in the solution to the two equations:

$$\int_{O_1}^{O_2} [e(x) + y(x)]\Omega^2 \mu(x)\,dx + \sum_i m_i(d_i + y(x))\Omega^2 + R_1 + R_2 = 0 \quad [5]$$

$$\int_{O_1}^{O_2} [e(x) + y(x)]x\Omega^2 \mu(x)\,dx + \sum_i m_i(d_i + y(x))x_i\Omega^2 + R_2 l = 0 \quad [6]$$

where $\mu(x)$ is the linear mass; m_i is the mass of corrective element i; and x_i is the position of the chosen balancing plane. Balancing is satisfactory when:

$$R_1 \leq R_{1L}; \quad R_2 \leq R_{2L} \quad [7]$$

The condition, which is theoretically strictly rigid, is expressed by:

$$y(x) = y_1 + ax$$
$$a = \frac{y_2 - y_1}{l} \quad [8]$$

where y_1 and y_2 are amplitudes at right-angles to the links; $y_1 = R_1/Z_1$; $y_2 = R_2/Z_2$ where Z_1 and Z_2 are impedances perpendicular to the links; $R_1 = R_{1L}$; $R_2 = R_{2L}$; $y_1 = y_{2L} = R_{2L}/Z_2$; $\Omega = \Omega_{max}$.

Eqns [5] and [6] make it possible to calculate corrective balancing masses:

$$m_1 d_1, m_2 d_2$$

where d_1 and d_2 are chosen to be large values in order to have m_1 and m_2 small.

Where $R_{1L} = R_{2L} = 0$, the two corrective balancing masses are independent of Ω. Practically, taking into account rotor deformation, it is necessary to verify that quality criteria are satisfied whatever the value of Ω in the rigid state.

Balancing of Eccentricities and Shadows: Vectoral Expressions

With run out and eccentricities, two corrective balancing masses make it possible to balance in a strictly rigid condition following the relations (**Figure 6**):

$$\vec{B}_{c1} + \vec{B}_{c2} + \vec{B}_0 = 0$$
$$\vec{B}_{c1} \wedge \overrightarrow{C_0 O_1} + \vec{B}_{c2} \wedge \overrightarrow{C_0 O_2} + \vec{M}_0 = 0$$

The positions of the balancing planes are fixed: they must allow the corrective elements, or masses, to be attached.

The set can be carried to a point $G_0'(C_0')$:

$$\vec{B}_0' = \vec{B}_0$$
$$\vec{M}_0' = \overrightarrow{C_0 C_0'} \wedge \vec{B}_0 + \vec{M}_0$$

At O_1, for balancing with $\vec{B}_{c1}$ and $\vec{B}_{c2}$.

$$\vec{B}_{01} = \vec{B}_0 = -\left(\vec{B}_{c1} + \vec{B}_{c2}\right)$$
$$\vec{M}_0 + \overrightarrow{C_0 O_1} \wedge \vec{B}_0 = -\vec{M}_{01} = -\left(\overrightarrow{O_1 O_2} \wedge \vec{B}_{c2}\right)$$

The module of $\vec{B}_{c2}$ is large if $|\overrightarrow{O_1 O_2}|$ is small. Taking into account technological limitations, it may not be possible to achieve balancing, in particular when the run out of a thin disk must be compensated by two mass elements arranged on each of its faces.

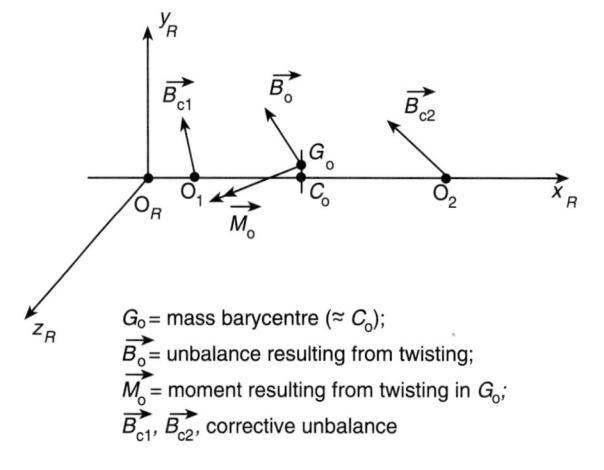

G_0 = mass barycentre ($\approx C_0$);
$\vec{B}_0$ = unbalance resulting from twisting;
$\vec{M}_0$ = moment resulting from twisting in G_0;
$\vec{B}_{c1}, \vec{B}_{c2}$, corrective unbalance

Figure 6 Unbalance and moments.

At the first balancing time, the direction of resulting unbalance was determined by setting out the two journal bearings on narrow horizontal tempered steel tracks to reduce friction moment caused by roller bearings. The mass center of gravity was placed under the tracks in a vertical direction. After several attempts, suitable corrective unbalance placed the rotor in the position fixed by the operator. Static unbalance was also corrected. When the size of the elements or disks on the shaft was significant, a moment could exist. Its correction required rotation in order to remedy dynamic unbalance. Today, for thin disks and for mediocre qualities, static balance is determined in this manner.

Balancing

In the planes P_1 and P_2, which are perpendicular to $0x_R$, let us arrange the elements E_1 and E_2 of mass M_1 and M_2 at $\vec{d}_1$ and $\vec{d}_2$ of $0x_R$ (**Figure 7**); this *a priori* arrangement of two elements is justified *a posteriori* by the solution which can be obtained thus.

The components of unbalance are:

For P_1: $m_1 d_{1y} = b_{1y}$; $m_1 d_{1z} = b_{1z}$
For P_2: $m_2 d_{2y} = b_{2y}$; $m_2 d_{2z} = b_{2z}$

The set of unbalance, which is corrective for perfect balancing, must:

- Cancel the resultant of the initial set of eccentricities, defined by eqn [2]
- Cancel the moment of initial set of eccentricities, defined by eqn [3] and the moment caused by products of inertia defined by eqn [4]

With eqn [2] canceling will necessitate:

$$my_G + m_1 d_{1y} + m_2 d_{2y} = 0$$
$$mz_G + m_1 d_{1z} + m_2 d_{2z} = 0 \quad [9]$$

Using eqns [3] and [4] canceling the moment is achieved when:

$$-I_{x_R z_R} + M_{rz} + m_1 d_{1y} x_1 + m_2 d_{2y} x_2 = 0$$
$$I_{x_R y_R} + M_{ry} + m_2 d_{1z} x_1 + m_2 d_{2z} x_2 = 0 \quad [10]$$

x_1 and x_2 are fixed in the position of the balancing planes. In the referential $G_r yz$, connected to the rotor, it is demonstrated that excitation along $G_r y_R$ and $G_r z_R$ is defined by $I_{x_R y_R} \Omega^2$ and $I_{x_R z_R} \Omega^2$. Balancing masses $\vec{b}_1$ and $\vec{b}_2$ modify $I_{x_R y_R}$ and $I_{x_R z_R}$, and cancel them to achieve perfect balance. In eqn [10], unbalance appears in explicit form.

Eqns [9] and [10] make it possible to calculate corrective unbalances:

$$m_1 d_{1y} = b_{1y} \qquad m_1 d_{1z} = b_{1z}$$
$$m_2 d_{2y} = b_{2y} \qquad m_2 d_{2z} = b_{2z}$$
$$d_{1y}, d_{1z} \;\rightarrow\; \sqrt{d_{1y}^2 + d_{1z}^2} = |\vec{d}_1|$$
$$d_{2y}, d_{2z} \;\rightarrow\; \sqrt{d_{2y}^2 + d_{2z}^2} = |\vec{d}_2|$$

must be chosen: large values in order that m_1 and m_2 are small, according to technological possibilities. y_G, z_G, $I_{x_R y_R}$, $I_{x_R z_R}$, M_{rz}, M_{ry} are not known, but are derived from the balancing procedure.

Specific examples

- $y_G = z_G = 0$

Using eqn [9]:

$$m_1 d_{1y} = -m_2 d_{2y} \rightarrow b_{1y} = -b_{2y}$$
$$m_1 d_{1z} = -m_2 d_{2z} \rightarrow b_{2z} = -b_{2z}$$

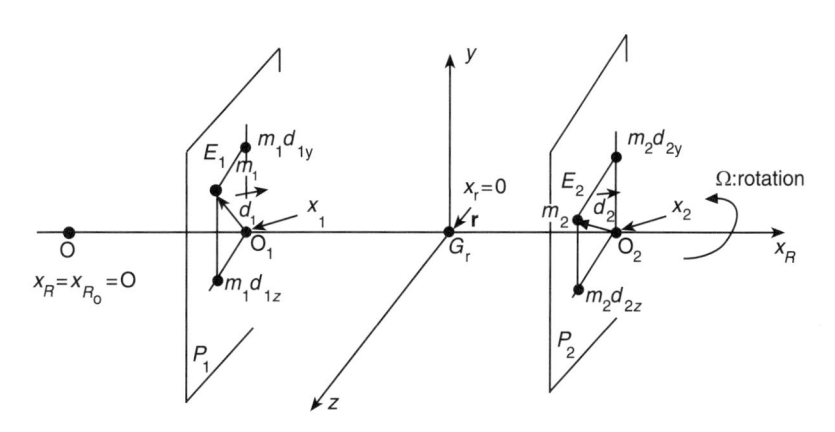

Figure 7 Unbalance $m_1 \vec{d}_1 = \vec{b}_1$, $m_2 \vec{d}_2 = \vec{b}_2$.

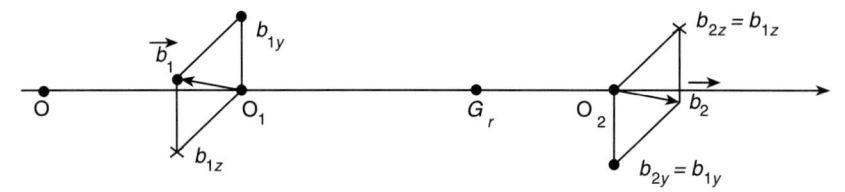

Figure 8 Corrective balancing masses: $\vec{b}_1$ and $\vec{b}_2$ for $y_G = z_G = 0$.

Using eqn [10]:

$$I_{x_R z_R} + M_{rz} + b_{1y}(x_1 - x_2) = 0 \\ I_{x_R y_R} + M_{ry} + b_{1z}(x_1 - x_2) = 0 \quad [11]$$

Figure 8 shows corrective balancing masses $\vec{b}_1$ and $\vec{b}_2$ which form a couple. Their direction is identical; their sense is opposite. They depend on x_1 and x_2.

- $y_G = z_G = 0$, $M_{rz} = M_{ry} = 0$

Using eqn [11]:

$$\frac{I_{x_R z_R}}{x_2 - x_1} = b_{1y} \qquad \frac{I_{x_R y_R}}{x_2 - x_1} = b_{1z}$$

Let us consider a thin disk ($2l$) placed on a rotor, with a run out defined by the angle V_z (**Figure 9**), corresponding to a rotation vector $\vec{V}_z$, following $0z_R$.

It is easy to show:

$$I_{x_R y_R} \approx -\frac{mR^2}{4} V_z \quad [13]$$

Similarly:

$$I_{x_R z_R} \approx \frac{mR^2}{4} V_y$$

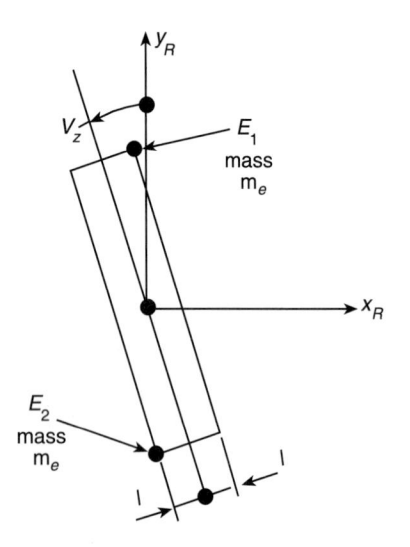

Figure 9 Disk: radius R, mass m, run out V_z.

The balancing which corrects $I_{x_R y_R}$ can be done using elements E_1 and E_2 to produce a product of inertia:

$$2m_e R \left(-V_z + \frac{l}{R}\right) R \quad [14]$$

Considering the products of inertia according to x_R and y_R, balancing is achieved taking into account eqns [13] and [14] using elements E_1 and E_2 of mass m_e, so that:

$$m_e = \frac{m}{8} \frac{1}{\dfrac{1}{V_z}\dfrac{l}{R} - 1}$$

If $l/R \approx V_z$, m_e is large, and so balancing cannot be achieved. A similar situation may arise with regard to $I_{x_R z_R}$.

- $I_{x_R z_R} = I_{x_R y_R} = 0$

The components of corrective balancing mass:

$$m_1 d_{1y} = b_{1y}; \quad m_1 d_{1z} = b_{1z} \\ m_2 d_{2y} = b_{2y}; \quad m_2 d_{2z} = b_{2z}$$

can be determined using eqns [9] and [10]. The two corrective balancing masses $\vec{b}_1$ and $\vec{b}_2$ depend on x_1 and x_2.

In vector form, balancing is expressed in $G_r, 0_1, 0_2$ by:

$$m\overrightarrow{G_r G} + \vec{b}_1 + \vec{b}_2 = 0 \\ \vec{M}_r + \overrightarrow{G_r 0_1} \wedge \vec{b}_1 + \overrightarrow{G_r 0_2} \wedge \vec{b}_2 = 0$$

In another form:

$$m\overrightarrow{G_r G} + \vec{b}_c = 0 \quad \vec{M}_r + \vec{M}_c = 0$$

At a point G_r':

$$m\overrightarrow{G_r G} + \vec{b}_c = 0 \\ \vec{M}_r + \overrightarrow{G_r G_r'} \wedge m\overrightarrow{G_r G} + \vec{M}_c + \overrightarrow{G_r G_r'} \wedge \vec{b}_c = 0$$

Notes In industrial applications, perfect balancing is not obtained. Eqns [9] and [10] are not satisfied.

The quality of balancing is characterized by residual unbalance, after balancing. They are defined by standards which propose modules of residual unbalance for a crusher which are greater than those of a gyroscope.

Standards

Standards ISO 1940/1, 1940/2 and AFNOR NF E 90600 propose criteria and methods to estimate the quality by residual amplitudes of force and vibration. A factor of quality G is also used. Modules of residual unbalance at right-angles to links 1 and 2 must be less than:

$$|b_{Li}|(\text{kg mm}^{-1}) = G\frac{M_i}{\Omega} \quad i = 1, 2 \quad [15]$$

where $M_i \times 9.81\,\text{m s}^{-2}$ is the weight on link i(N); Ω is the maximum rotation speed (rad s^{-1}); G is defined as a function of the machine type (mm s^{-1}); $G = 630$ for crankshaft arrangements; and $G = 0.4$ for precision mills and gyroscopes.

When set results from eccentricity e of mass center of gravity G_g (**Figure 10**), unbalance is equal to Me. Initial unbalance at right-angles to the links is:

At 1: $M_1 e$; at 2: $M_2 e$ where $M_1 = M(l_2/l); M_2 = M(l_1/l)$.
If $e \leq G/\Omega$ balancing is not necessary.

Using eqn [15], at the limit residual eccentricities at G_g at 1 and 2 are equal to G/Ω. Its relation to initial eccentricity is equal to $G/\Omega e$. For a compressor, pump, or electric motor, $G = 6.3$. Where $M_1 = M_2 = 500\,\text{kg}$, $\Omega = 314\,\text{rad s}^{-1}$ (3000 rpm), $|b_{L_1}| = |b_{L_2}| \approx 10\,\text{kg mm}^{-1}$. Let there be 20 g at 500 mm neighboring eccentricity of 20 µm. Quality can be defined perpendicular to the links by the relationship R_f of amplitude of residual rotating force, to static force caused by gravity. Where $l_1 = l_2$ and Me at G_g (**Figure 10**):

$$R_f = \frac{e\Omega^2}{9.81\,\text{m s}^{-2}} \quad [16]$$

R_f is independent of rotor mass. For $R_f = 0.1$, $\Omega = 314\,\text{rad s}^{-1}$ ($\to$3000 rpm); $G = 3.13$:

$$e \approx 10\,\mu\text{m}$$

When torsion is reduced to Me and $M_1 = M_2$:

$$G = \frac{9810\,R_f}{\Omega}$$

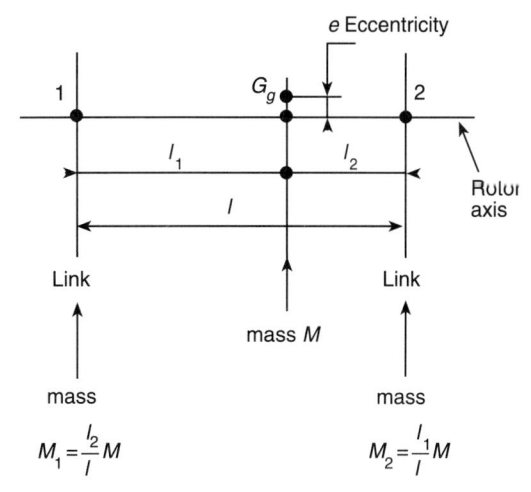

Figure 10 Unbalance M_l at the mass barycenter G_g.

So that the quality fixed by R_f is identical to that fixed by G, G has to vary as $1/\Omega$. Where R_f is 0.1; $\Omega = 314$ rad s^{-1}; $G \approx 3.13$.

When the set is a result of a central axis of inertia at angle γ of the axis (**Figure 11**), where γ is small and a thin disk, l_d, in relation to radius R:

$$R_f = \frac{2(I_{x_R x_R} - I_{y_R y_R})\gamma\,\Omega^2}{9.81\,\text{m s}^{-2} Ml} \quad [17]$$

For couple unbalance which, at right-angles to the links, balance couple due to γ are:

$$|b_{L1}| = |b_{L2}| = \frac{(I_{x_R x_R} - I_{y_R y_R})\gamma}{l} = \frac{9.81}{2}\frac{M}{\Omega^2}R_f$$

where:

$$I_{y_R y_R} = \frac{I_{x_R x_R}}{2} = \frac{MR^2}{4}; \quad M = \pi R^2 l_d l$$

$$R_f = \frac{\Omega^2 \gamma}{2 \times 9.81}\frac{R^2}{l}$$

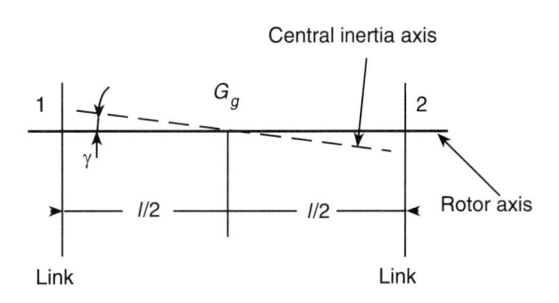

Figure 11 Couple due to γ.

where $R_f = 1$, $\Omega = 314\,\text{rad s}^{-1}$; $l = 2\,\text{m}$; $R = 0.5\,\text{m}$; $\gamma = 1.59 \times 10^{-4}\,\text{rad} \to \approx 0.092°$. Where $l_d = 0.1\,\text{m}$; $\rho = 7800\,\text{kg m}^{-3}$; $M = 612\,\text{kg}$. $|b_{L_1}| = |b_{L_2}| = 305 \times 10^{-5}\,\text{kg m} \to 61\,\text{g}$ at $0.5\,\text{m}$ in two planes $2\,\text{m}$ apart.

To obtain $R_f = 0$, after balancing, two opposing unbalances equal to $122\,\text{g} \times 0.5\,\text{m}$ must be set out in two planes at $l/4 = 0.5\,\text{m}$ on both sides of G_g, the mass of corrective elements must be equal to $1.22\,\text{kg}$. It is probable that technological limitations do not permit their implantation, despite the low value of γ (**Figure 12**).

Balancing Machines

Rotor balancing in a rigid state is carried out by dedicated machines which generate by themselves a minimum amount of vibration. They permit to determine the residual of unbalance and uncertainties associated with them. Vibration sensors are mounted on nonrotating parts of the machine. The corrective unbalance is determined by a computer. Standard ISO 2953 deals with balancing machines.

Rotors can be arranged in simple machines or in the stator of the affected machine.

Flexible State

Modal Description: General Case

The article on modal characteristics (see **Rotating machinery, modal characteristics**) defines the components of vibrations due to excitation:

$$U_k = \sum_l \frac{U^d_{kl}}{\alpha_{ll}(i\Omega - p_l)} \left[\ldots + U^g_{ql}(me)_q \exp i\psi_q \right. \quad [18]$$
$$\left. + U^g_{(q+1)l}\left(-i(me)_q \exp i\psi_q\right) + \ldots \right] \Omega^2$$

where Ω is the rotation speed; U_k is the component related to displacement $u_k = |U_k|\cos(\Omega t + \varphi_k)$; where φ_k is the argument for U_k. k marks out the direction point (in the plane $0xy$ or $0xz$) of observation of relative vibration between rotor and bearing, absolute of the nonrotating structure. Other components may be considered, for example $\dot{u}_k = -\Omega|U_k|\sin(\Omega t + \varphi_k)$ expressed by $-i\Omega U_k$, and forces perpendicular to the links; p_l is the complex natural frequency: $p_l = \delta_l + i\omega_l$ and $p_l = \delta_l - i\omega_l$, $i^2 = -1$; $\delta_l < 0$: attenuation factor (stability); ω_l is the natural damped frequency; U^d_{kl} is the component of natural mode at right-angles associated with p_l in the point-direction k, associated with U_k; α_{ll} is the component determined by biorthogonality relations: $U^d_{kl}U^g_{.l}/\alpha_{ll}$ is independent of arbitrary standards defining natural modes. $U^g_{.l}$ is the component of natural left-hand mode associated with p_l in the point-direction $\cdot$, $U^g_{.l}$ can be composed of different components from that of U^{dl}_k, an angle, for example, when it expresses excitation by a couple due to run outs of a thin disk characterized by angles V_y, V_z.

Components of unbalance in rotor radial sections $q, s, t, \ldots$ are in general counterbalanced by displacement components of left-hand modes of geometric centers of these sections $(q, s, t, \ldots)$:

$$U^g_{ql}(me)_q \exp i\psi_q$$
$$U^g_{(q+1)l}\left[-i(me)_q \exp i\psi_q\right]$$

where U^g_{ql} and $U^g_{(q+1)l}$ are displacement components of the geometric center of section q; $(me)_q \exp i\psi_q$; $(me)_q \exp i\psi_q$: components of unbalance $(me)_q.\psi_q$ in section q; terms of eqn [18] between parentheses, modal excitation, is designed by modal unbalance. For each natural frequency p_l, it intervenes in displacement U_k in the same way: it is independent of k. $B_l = [\ldots U^g_{ql}(me)_q \exp i\psi_q + \ldots]$.

Eqn [18] assumes linear components: proportionality between displacement and exciters.

Slow Rotation Speed: Rigid State

For the first natural frequency ($\omega_1 < \omega_2 < \omega_3 \ldots \delta_1$ small), modules of $i\Omega - p_1$ and $i\Omega - p_1^*$ are: $[(\Omega - \omega_1)^2 + \delta_1^2]^{1/2}$. Where $\Omega \ll \omega_1$, if these terms are small, in relation to terms corresponding to $\omega_2 \ldots$ eqn [18] leads to expression of U_k in a rigid state:

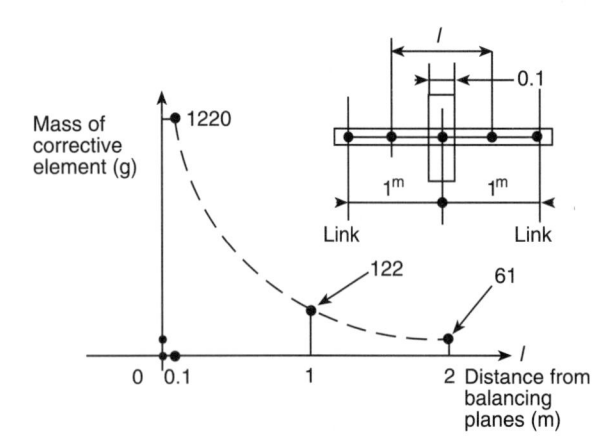

Figure 12 Mass of corrective element at $0.5\,\text{m}$ from rotation axis.

$$U_k \approx \frac{U_{k1}^d B_1 \Omega^2}{\alpha_{ll}[i(\Omega - \omega_1) - \delta_1]} + \frac{U_{k1}^{d*} B_1^{\oplus} \Omega^2}{\alpha_{ll}^*[i(\Omega + \omega_1) - \delta_1]} \quad [19]$$

$\tilde{B}_l^{\oplus}$ is a result of B_l by substituting U^{g*} for U^g. In a rigid state the first natural left-hand mode is written:

$$U_{.l}^g = a_1^g x. + b_1^g$$
$$U_{.l}^{g*} = \bar{a}_1^g x. + b_1^{g*}$$

where a_1^g, b_1^g are complex constants and x is the abscissa of sections $q, q+2, q+4,\ldots$.

Two unbalances at abscissa x_{c1} and x_{c2} (balancing planes) make it possible to reduce, and even cancel at speed $\Omega|B_1|$ and $|B_1|$: this is the balancing in a rigid state.

Reduction to Two Terms

When $\omega_{l-1} \ll \Omega \ll \omega_{l+1}$; $\delta_l \ll \omega_l$, $\delta_{l-1} \ll \omega_{l-1}$; $\delta_{l+1} \ll \omega_{l+1}$, the modal decomposition eqn [18] is reduced to two terms only defined in the article on modal characteristics (see **Rotating machinery, modal characteristics**). The model is of a De Laval rotor.

Speed Close to Critical Speed

When rotation speed is close to natural frequency ($\Omega \approx \omega_l$) and thus close to critical speed and:

$$\delta_l \ll \omega_l; \quad \omega_{l-1} \ll \omega_l \ll \omega_{l+1}$$

modal decomposition is reduced to two terms (see **Rotating machinery, modal characteristics**) distinguishing between unbalance $(me)_r \exp i\psi_r$ and $(me)_s \exp i\psi_s$. In the radial planes r and s:

$$U_k \approx \frac{U_{kl}^d}{\alpha_{ll}(i\Omega - p_l)} \Big[E_l + \big(U_{rl}^g - iU_{(r+1)l}^g\big)(me)_r$$
$$\times \exp i\psi_r + \big(U_{sl}^g - iU_{(s+1)l}^g\big)(me)_s \exp i\psi_s\Big]\Omega^2$$
$$+ \frac{U_{kl}^{d*}}{\alpha_{ll}(i\Omega - p_l^*)} \Big[E_l^{\oplus} + \big(U_{rl}^{g*} - iU_{(r+1)l}^{g*}\big)(me)_r$$
$$\times \exp i\psi_r \big(U_{sl}^{g*} - iU_{(s+1)l}^{g*}\big)(me)_s \exp i\psi_s\Big]\Omega^2$$

where E_l = modal excitation which performs on unbalance and run out factored with Ω^2: it is modal unbalance; E_l can perform from initial peaks which are not factored with Ω^2. $E_l^{\oplus}$ is a result of E_l substituting $U_{..}^{g*}$ and $U_{..}^g$.

When $\Omega = \Omega_0 \approx \omega_l$ $|U_k|$ can be canceled by $(me)_r \exp i\psi_r$ and $(me)_s \exp i\psi_s$; that is to say, by two unbalances in an identical way balancing in a rigid state. When initial peaks are zero, the domain of variation Ω, around Ω_0 in which amplitude is acceptable is in general more drawn out than when there exist initial peaks.

Balancing Procedure by Influence Coefficients

Direct Definition of Influence Coefficients

Vibration V_n – relative, displacement absolute of displacement speed (see **Rotating machinery, monitoring**) – of a force in point-direction n is due to unbalance test B_{mt} in a section plan m of rotor at speed Ω by an influence coefficient:

$$C_{nm}(\Omega) = \frac{|V_{nt}| \exp i\varphi_{nt}}{|B_{mt}| \exp i\varphi_{mt}} \quad [20]$$

According to **Figure 13**:

$$|V_n| = A_{nt}; \quad |B_{mt}| = m_t R$$

Taking into account the influence of initial vibration:

$$C_{nm}(\Omega) = \frac{|V_{nt}| \exp i\varphi_{nt} - |V_{no}| \exp i\varphi_{no}}{|B_{mt}| \exp i\psi_{mt}} \quad [21]$$

Two tests are necessary to measure V_{no} and V_{nt} with B_{mt}. $C_{nm}(\Omega)$ is independent of B_{mt} if the performance of the arrangement of machine and measuring system and analysis is linear and repeatable.

Definition of Influential Coefficients by Modal Decomposition

Modal decomposition (eqn [18]) permits expression of $C_{nm}(\Omega)$ to be defined by eqn [20]:

$$C_{nm}(\Omega) = \sum_l \frac{U_{nl}^d}{\alpha_{ll}(i\Omega - p_l)} \big(U_{ml}^g - U_{(m+1)l}^g\big)\Omega^2 \quad [22]$$

In general, $|C_{nm}(\Omega)|$ is maximal in the neighborhood of critical speeds: $\Omega \approx \omega_l$.

Corrective Unbalance and Residual Vibration Forces

Vibration is connected to unbalance by matrix **C** of influence coefficients:

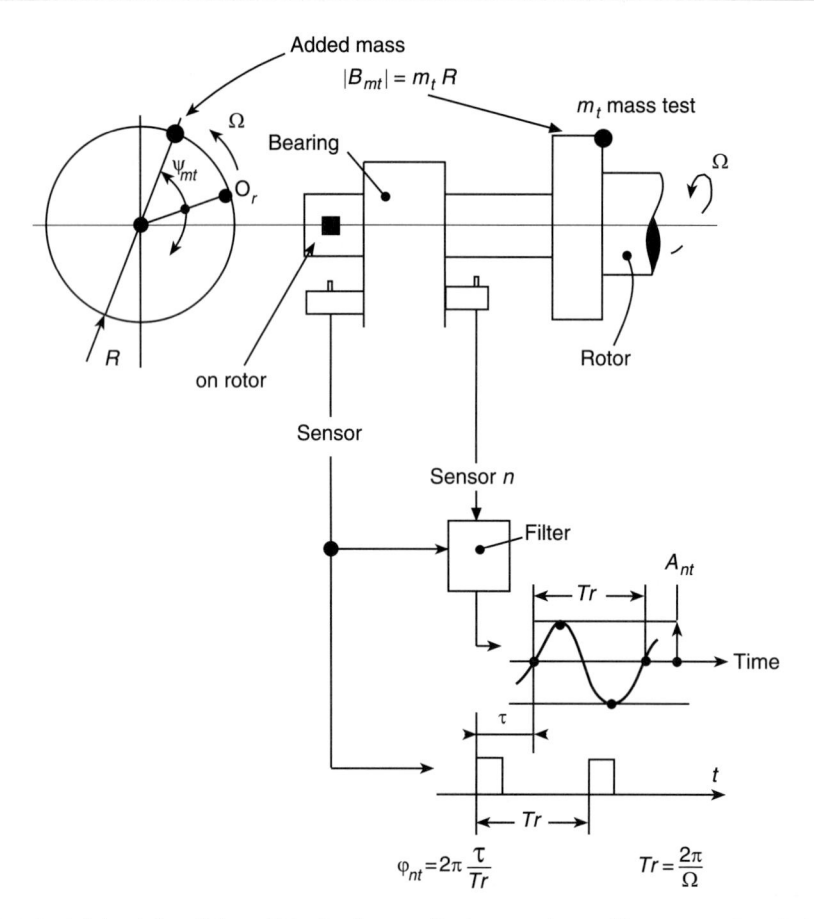

Figure 13 Elements defining influential coefficients. A_{nt}, amplitude; φ_{nt}, phase difference; $m_t R$, unbalance module; φ_{mt}, angular position of unbalance.

$$\begin{array}{cc} \text{Speed} & \text{Sensor} \\ \Omega_1 & C_1 \\ & C_2 \\ \cdots & \vdots \\ \Omega_q & C_k \\ \cdots & \vdots \\ & C_l \\ \Omega_p & \vdots \end{array} \begin{bmatrix} V_1 \\ V_2 \\ \vdots \\ V_3 \\ \vdots \\ \vdots \\ V_N \end{bmatrix} = \begin{bmatrix} C_{11} & \cdots & C_{1m} \\ & \vdots & \\ & C_{nm} & \\ & \vdots & \\ C_{1M} & \cdots & C_{NM} \end{bmatrix} \begin{bmatrix} B_1 \\ \vdots \\ B_m \\ \vdots \\ B_M \end{bmatrix} \rightarrow \mathbf{V} = \mathbf{CB}$$

$$\underbrace{\phantom{\begin{bmatrix} C_{11} & \cdots & C_{1m} \\ C_{1M} & \cdots & C_{NM} \end{bmatrix}}}_{\text{Coefficients of influence}} \underbrace{\phantom{\begin{bmatrix} B_1 \\ B_M \end{bmatrix}}}_{\text{Unbalance}}$$

Certain speeds can be critical speeds. Nominal speed is generally taken into account. The number of vibrations is equal to the number of speeds when these speeds are all different.

Forces perpendicular to the links can be introduced. In the matrix $\mathbf{V}$ given in eqn [23], relative vibrations $V^1_{Li} V^2_{Li}$ measured in the same radial plane at right-angles to link Li in two perpendicular directions, noted 1 and 2, are marked out to form:

$$\mathbf{V}_L = \begin{bmatrix} V^1_{L1} V^2_{L1} & \cdots & V^1_{Li} V^2_{Li} & \cdots \end{bmatrix}^T$$

Forces are expressed by impedance matrices $\mathbf{Z}_i$ (see **Bearing vibrations**).

$$\begin{bmatrix} I & | & & O \\ - & - & - & - \\ & | & Z_1 & \\ & | & & \ddots \\ O & | & & Z_l \end{bmatrix} \begin{bmatrix} V \\ V_L \end{bmatrix}$$

$$= \begin{bmatrix} I & | & & O \\ - & - & - & - \\ & | & Z_1 & \\ & | & & \ddots \\ O & | & & Z_l \end{bmatrix} \begin{bmatrix} C \\ C_L \end{bmatrix} \begin{bmatrix} B_1 \\ \vdots \\ B_M \end{bmatrix} = \begin{bmatrix} V \\ F_L \end{bmatrix}$$

where I is the unit matrix. Lines of the matrix C_L are the lines of the matrix C associated to the vibrations: $V_{Li}^1 V_{Li}^2 \ldots V_{Li}^1 V_{Li}^2$.

$$F_L = \begin{bmatrix} Z_1 & & \\ & \ddots & \\ & & Z_i \end{bmatrix} V_L$$

forces perpendicular to the links.

The matrices Z_i can be calculated. They allow the determination of the forces which are necessary to the constructor to define the machine.

Corrective unbalances B_c in the corrective planes $1 \ldots m \ldots M$ which cancel, or reduce, initial vibratory amplitude V_0 resulting from eqn [23]:

$$B_C = C^{-1} V_0 \exp i\pi \qquad [24]$$

Residual vibration is zero:

$$V_r = V_0 + CB_C = 0 \qquad [25]$$

C^{-1} exists if $N = M$ and if the determinant of $C \neq 0$. If $M > N$ it is necessary to augment N where $M \leq N$. If $N > M$, the number of equations is greater than the unknown number B.

Optimization permits control of vibrations. With matrix A at M lines and N columns, eqn [23] leads to:

$$B_C = [AC]^{-1} A V_0 \exp i\pi$$
$$V_r = \left[I + [AC]^{-1} A \exp i\pi\right] V_0$$

where I is unit matrix. Matrix A is acceptable if the modules of residual vibration and corrective unbalance are less than limits characterizing the quality:

$$\frac{|V_{nr}|}{|V_{nr}|_L} < 1 \qquad n = 1, 2 \ldots N$$
$$\frac{|B_{mc}|}{|B_{mc}|_L} < 1 \qquad m = 1, 2 \ldots M \qquad [26]$$

A first solution, which is often satisfactory, is obtained with $A = C^T$ which expresses optimization by least-squares, where T denotes transposition.

When the inequalities [26] are not satisfied, improvements can be achieved by diagonal weighting matrices P_V on vibration, P_b on unbalance:

$$B_C = [C^T C]^{-1} C^T P_V V_0 \exp i\pi$$
$$V_r = \left[I + P_b C [C^T C]^{-1} C^T P_V \exp i\pi\right] V_0$$

When C is square ($N = M$):

$$V_r = [I + P_b P_V \exp i\pi] V_0$$

residual amplitudes are generally not zero, unless $P_b P_V = I$. The use of weightings does not always appear to be justified.

As a complement to the weightings, vibration and unbalance can be suppressed. This degeneration brings about suppression of lines or columns in matrix C.

The diagonal terms of P_V and P_b can be adjusted by servo-control subjected to eqn [26]. An automatic iterative process can lead to a solution or instability characterized by variations of eqn [26] around 1. In this case a modification of the law of servo-control can be carried out, or the limits $|V_{nr}|_L$, $|B_{mc}|_L$ can be increased or modified.

Instability can lead to the assumption that the problem posed does not have a solution: $N - M$ can seem too large, the limits too weak, the positions of the balancing planes may be bad, initial amplitudes are too raised.

Maintenance

When the performance is linear and repeatable, matrix C of influencing coefficients, which are intrinsic to a machine, on memory on a computer disk, thanks to quality maintenance can be used so that the rotor is in a rigid or flexible state. The minimum number of tests is equal to 2 (it is equal to $M + 2$ when the influencing coefficient matrix is to be determined).

An abnormal situation can be seen when the matrix C used does not permit reduction of vibration. It is then possible that the procedure modifies vibration amplitude which arise from other excitations than unbalance. A balancing procedure must be engaged after a diagnosis which makes it possible to find out the causes of variations in vibratory amplitude.

Measuring and Analysis

Vibration is measured and analyzed to find out components whose frequency is equal to rotation frequency. Discrete Fourier transforms are edited by Bode graphs or polar (Nyquist) graphs: amplitudes and phases as a function of rotation speed. Measurement is carried out in steady state, at constant speed, or transient state where there is increase or reduction in rotation speed Ω. The transitory mode proves useful for extracting amplitude and phases at critical speed with reasonable accuracy, so long as the gradient $|d\Omega/dt|$ is less than a limit which enables the machine to be in a state which is close, or tangent, to a permanent state. The measuring system must acquire signals at speeds $\Omega_0 + n\Delta\Omega (n = 0, 1, 2 \ldots)$ with an increment $\Delta\Omega$ which is even smaller for small damping ($\xi = 1/2Q_{3dB}$). The damping controls the amplitude around critical speed (**Figure 14**). $\Delta\Omega < \Delta\Omega_{cr}/6$ is generally suitable.

Discretization of signals can be done in synchronous or free mode. Using discrete Fourier transform, the synchronous mode makes it possible to determine components whose frequencies are multiples ($n = 1 \ldots K$) of rotation speed (harmonic spectrum). K is independent of speed Ω. Using discrete Fourier transform, free mode permits determination of components whose frequencies are between f_{mini} and f_{maxi} (spectra). f_{mini} and f_{maxi} are independent of speed Ω.

Measuring the signal by a magnetic sensor of displacement, fixed on the nonrotating structure, a look at a rotor measuring track (**Figure 15**) results from movement (vibration) of the geometric center C profiled and:

- Gaps ($\varepsilon_e, \varepsilon_i \ldots$) between this profile and the circumference of the average profile
- Variations in the magnetic permeability and resistivity between 0 and 360° of the measuring track
- Permanent magnetic spot on the measuring track

These three components constitute a noise (runout) which it is necessary to deduct from the signal measured to obtain vibration, used for preliminary diagnosis before balancing and in its realization.

The noise is measured at slow rotation speed when the displacement from the geometric center C is negligible. Balancing carried out with noisy signals can lead to increases or decreases in vibration. For example, the maximum acceptable peak-to-peak level, resulting from all spectral components, is, according to standard ISO 7919–3, for industrial machines, equal to $L(\mu m)$ peak-to-peak $= 4800/\sqrt{N(rpm)}$ where N is the nominal speed on a turbine, a compressor, a pump at 10 000 rpm; $L = 48 \mu m$ peak-to-peak. Balancing carried out without subtraction of a noise equal to $10 \mu m$ peak-to-peak can lead – when the spectrum is made up by the only component of rotation frequency – to an amplitude of vibration between 19 and 29 μm ($24 \pm 5 \mu m$) peak-to-peak 38 and 58 μm. The combination of response to rotating forces and speed of rotation Ω with the fundamental component of noise can produce antiresonances. Measuring and acquisition of signals must be made simultaneously on the arrangement of Q sensors and the sensor which indicates

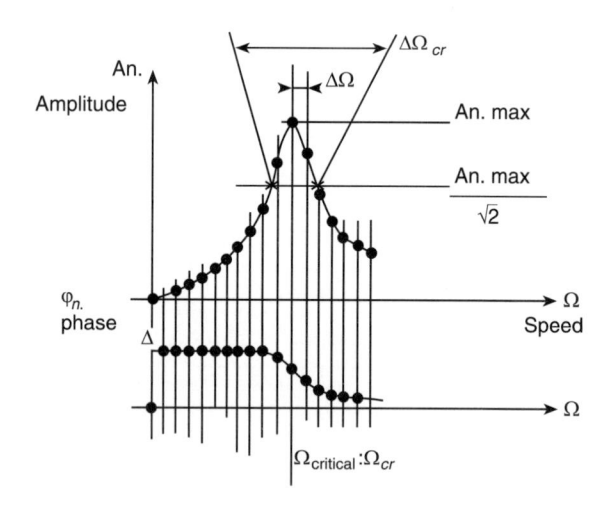

Figure 14 Increment $\Delta\Omega$ of signal acquisition. $Q_{3dB} = \Omega_{cr}/\Delta\Omega_{cr}$ = amplifiction factor at 3 dB.

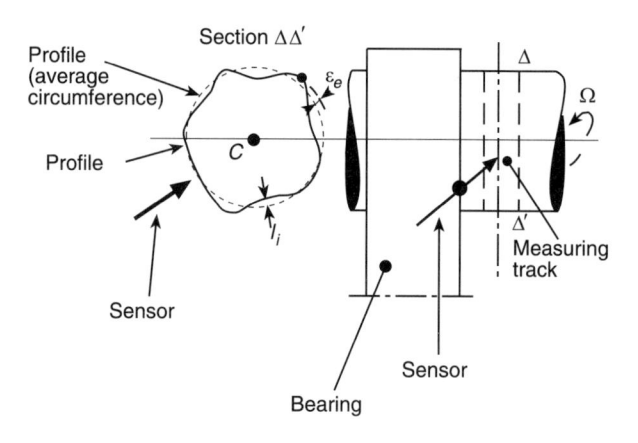

Figure 15 Fault of form: ($\varepsilon_e, \varepsilon_i$).

rotation (measurement of speed and determination of phases) in order to reduce uncertainty caused by faults in repeatability: a system capable of measuring simultaneously $(Q+1)/R$ signals lead to k tests on conditions which, in general, may not be identical, and so affected by repeatability faults. In addition, the difference in price of a system of $Q+1$ tracks and to $(Q+1)/k$ tracks must be compared to the difference in price between k tests and a test.

Software

The balancing procedure necessitates dedicated software. This makes it possible to take into account a large number of vibrations (50 ...) and balancing planes (10 ...); the signal components, chosen as a function of rotation speed on Bode graphs, automatically pass from the measuring, acquisition, and analysis system to the computer which carries out the subtraction of noise and simultaneously treats absolute vibration (nonrotating parts) and relative vibration (rotor and nonrotating parts) following the bases defined above, proposed corrective unbalance and associated residual vibration.

As for the rotors in a rigid state, the corrective unbalances for normal situations are small in general, at critical speeds, particularly since damping is weak. Complementary software enables these procedures: composition and decomposition of unbalance, average signal, unbalance imposed to calculate vibration; composition of signals to determine elliptical pathways, evaluation of the quality of balancing by the length of the ellipse axis; coefficients of nonrepeatability and nonlinearity.

It is still useful to manage and monitor a large number N of vibrations, even when the number M of corrective unbalance is low (1, 2). $N \gg M$ is useful when the rotor is in a flexible state in a special casing under vacuum, or in the structure of the machine on site.

Diagnostics: Limit Values

The procedure of amplitude reduction (balancing) results from diagnostics capable of pinpointing its need and/or that of other operations.

Reduction in amplitude may make it possible to obtain linear performance (following reduction in amplitude of harmonic components of rotation speed), passing critical speed without risk of contact between the rotor and the stator, rotating forces which are weaker than static forces at right-angles to the bearings, and an increase in the lifespan of ball bearings or roller bearings (see **Bearing vibrations**).

The limit values $|V_{nr}|_L$ of vibration amplitude permits quality apprecation in respect of limit values $|B_{mc}|_L$ of unbalance modules, imposed by technology. The textbook on charges and standards (see **Rotating machinery, monitoring**) must be consulted for the definition of limit values. The fuzziness of certain standards makes it possible to negotiate and make agreements between users and constructors.

Example

The machine comprises an electric motor, including two rotors. Two elastic couplings are arranged between the rotors. Links are established by ball bearings. Vertical vibration is measured by accelerometers on the motor stator (numbers 1 and 2) and on each of four rotor bearings (numbers 3–6). Corrective unbalances are placed in four planes: one close to the motor coupling (P_1) and one and two on the rotors (P_2, P_3, and P_4). Measurements are carried out when there is an increase in speed (temporary rate of flow) after maintaining speed at a constant level to obtain a permanent thermal rate of flow. The matrix of influential coefficients comprises 72 terms. Corrective unbalance and residual unbalance are calculated by the least-squares matrix (matrix $\mathbf{C}^T$). There is little uncertainty caused by repeatablity faults – several percent on amplitude. **Table 1** shows conditions and results. Ten critical speeds are used. Reduction is important at critical speed; theoretical and measured amplitudes are similar.

See also: **Bearing vibrations; Rotating machinery, essential features; Rotating machinery, modal characteristics; Rotating machinery, monitoring; Rotor dynamics; Spectral analysis, classical methods; Standards for vibrations of machines and measurement procedures.**

Further Reading

Adams ML Jr (1999) *Rotating Machinery Vibration. From Analysis to Troubleshooting.* New York: Marcel Dekker.

Bigret R (1980) *Vibrations des Machines Tournantes et des Structures.* [*Vibrations of Rotating Machines and Structures.*] Paris: Technique et Documentation Lavoisier.

Bigret R (1997) *Stabilité des Machines Tournantes et des Systèmes.* [*Stability of Rotating Machines and Systems.*] Simlis, France: CETIM.

Bigret R, Feron J-L (1995) *Diagnostic, Maintenance, Disponibilité des Machines Tournantes.* [*Diagnostics, Maintenance, Availability of Rotating Machines.*] Paris: Masson.

Boulenger A, Pachaud S (1998) *Diagnostic Vibratoire en Maintenance Préventive.* [*Vibratory Diagnostics and Preventive Maintenance.*] Paris: Dunod.

Childs D (1993) *Turbomachinery Rotordynamics.* Chichester: John Wiley.

Table 1 Balancing

Sensor accelerometer no.	Speed (rpm)	Critical speed	Amplitude ($mm\,s^{-1}$ peak) Initial (to be reduced)	Residual Theoretical	Residual Measured
1	1496	Yes	3.32	0.77	1
1	1696	Yes	5.40	1.78	1
1	1960	Yes	10.36	1.10	1.1
2	1408	No	0.80	1.08	0.6
2	1680	Yes	2.16	1.78	0.4
2	1960	Yes	4.90	2.30	0.8
3	1660	Yes	3.20	1.10	0.1
3	1856	Yes	2.90	0.33	0.2
3	2048	No	1.60	1.39	1.1
4	1612	Yes	7.36	0.13	1.7
4	1856	No	6.92	0.65	0.8
4	2048	No	9.80	1.11	1.2
5	1328	No	2.76	1.02	0.5
5	1564	Yes	11.20	0.37	1.2
5	1984	No	11.62	1.64	1.2
6	1328	No	5.40	1.84	0.2
6	1548	Yes	18.60	1.55	1.5
6	1984	No	13.60	2.32	1.8

Corrective unbalance (g mm): P_1=161; P_2=195; P_3=450; P_4 = 203.

Dimentberg FM (1961) *Flexural Vibrations of Rotating Shafts*. London: Butterworths.

Frêne J, Nicolas D, Dequeurce, Berthe D, Godet M (1990) *Lubrification Hydrodynamique – Paliers et Butées. [Hydrodynamic Lubrication – Bearings and Thrust.]* Paris: Eyrolles.

Gasch R, Pfützner (1975) *Rotordynamik, Eine Einführung*. Berlin: Springer Verlag.

Genta G (1999) *Vibration of Structures and Machines*. Berlin: Springer Verlag.

Lalanne M, Ferraris G (1990) *Rotordynamics Prediction in Engineering*. Chichester: John Wiley.

Morel J (1992) *Vibrations des Machines et Diagnostic de leur Etat Mécanique. [Machine Vibrations and Diagnostics of their Mechanical Condition]* Paris: Eyrolles.

Tondt A (1965) *Some Problems of Rotor Dynamics*. London: Chapman & Hall.

Vance JM (1988) *Rotordynamics of Turbomachinery*. Chichester: John Wiley.

BASIC PRINCIPLES

G Rosenhouse, Technion - Israel Institute of Technology, Haifa, Israel

Copyright © 2001 Academic Press

doi:10.1006/rwvb.2001.0088

Vibrations of bodies including motions and deformation waves under the action of forces can be approximated by the application of physical models and mathematical tools. Here the two main branches of the theory, which are the vectorial approach and the calculus of variations, are presented. Next, the basic rules of the vectorial approach are reviewed. Then the development and variants of the calculus of variations is linked to the theory of vibrations. Some examples help to understand the use of the formulae.

Background

Mechanical vibrations involve phenomena that are periodic in time, and can also be generalized to nonperiodic phenomena. In order to predict them it is necessary to be familiar with their basic parameters, including:

1. The independent variables, the time, t, and the location, $\mathbf{r}$, in a spatial reference frame. In a Cartesian orthogonal system of coordinates $Oxyz$ (**Figure 1A**) the radius vector is given by its components as:

$$\mathbf{r} = x\mathbf{i} + y\mathbf{j} + z\mathbf{k} \qquad [1]$$

2. The kinematics parameters that depend on t and $\mathbf{r}$: displacements, $\mathbf{u}$ velocities, $\mathbf{v}$, and accelerations, $\mathbf{a}$. In a Cartesian orthogonal system of coordinates $Oxyz$ (**Figure 1B**) the velocity is given by its components as:

$$\mathbf{v} = \lim_{\Delta t \to 0} \frac{\Delta \mathbf{r}}{\Delta t} = \lim_{\Delta t \to 0} \frac{\mathbf{r}(t+\Delta t) - \mathbf{r}(t)}{\Delta t} = \frac{d\mathbf{r}}{dt} = D\mathbf{r}$$
$$\Rightarrow \mathbf{v} = \dot{\mathbf{r}} = \dot{x}\mathbf{i} + \dot{y}\mathbf{j} + \dot{z}\mathbf{k} \qquad [2]$$

3. Similarly, the acceleration is:

$$\mathbf{a} = \dot{\mathbf{v}} = \lim_{\Delta t \to 0} \frac{\Delta \mathbf{r}}{\Delta t} = \lim_{\Delta t \to 0} \frac{\dot{\mathbf{r}}(t+\Delta t) - \dot{\mathbf{r}}(t)}{\Delta t} = \frac{d\dot{\mathbf{r}}}{dt}$$
$$= D^2 \mathbf{r} \Rightarrow \mathbf{a} = \ddot{\mathbf{r}} = \ddot{x}\mathbf{i} + \ddot{y}\mathbf{j} + \ddot{z}\mathbf{k} \qquad [3]$$

4. The physical measures: mass, m, force, $\mathbf{F}$, inertia, spring constant, k, damping factor.
5. Parameters that are specific to vibrations: period, T, frequency, $f = 1/T$, amplitude, A, generalized coordinates, q, and number of degrees-of-freedom.
6. Specific terms: particle, rigid body, matter, harmonic motion, small displacements, finite displacements.

The basic principles on which the theory of vibration relies are from two main branches: the vectorial approach, based on the laws of Newton (1643–1727), and the principle of virtual work, mentioned by Leonardo da Vinci (1452–1519). The fundamentals of dynamics were proposed in the 17th century by Galileo Galilei (1564–1642). Descartes (1596–1650) recognized the basic idea of infinitely small motions, and John Bernoulli (1667–1748) made the final formulation of the principle of virtual displacements. D'Alembert (1718–1783) developed a basic principle in dynamics, Lagrange (1736–1813) developed a differential approach, and Hamilton (1805–1865) used an integral form, describing the energy principles of mechanics concerning vibrations, and extended the principle of virtual work to kinetics. Simplified solutions of dynamics problems where derived by energy considerations and approximating functions have been developed by Rayleigh (1842–1919) and Ritz (1878–1909). The above-mentioned outstanding research workers are only some of the many who have contributed to the area of mechanical vibrations.

The Vectorial Approach

Vibration modeling is based on the laws of dynamics. These laws are mostly axiomatic, being generalized results of observations over many centuries. The laws of dynamics were summarized in the monumental book *Principia Mathematica*, written by Isaac Newton in 1687. These laws, together with the principle of conservation of mass and the relevant constitutive law, combine the equations of motion and the boundary conditions in the examined material space. The three basic laws of dynamics are:

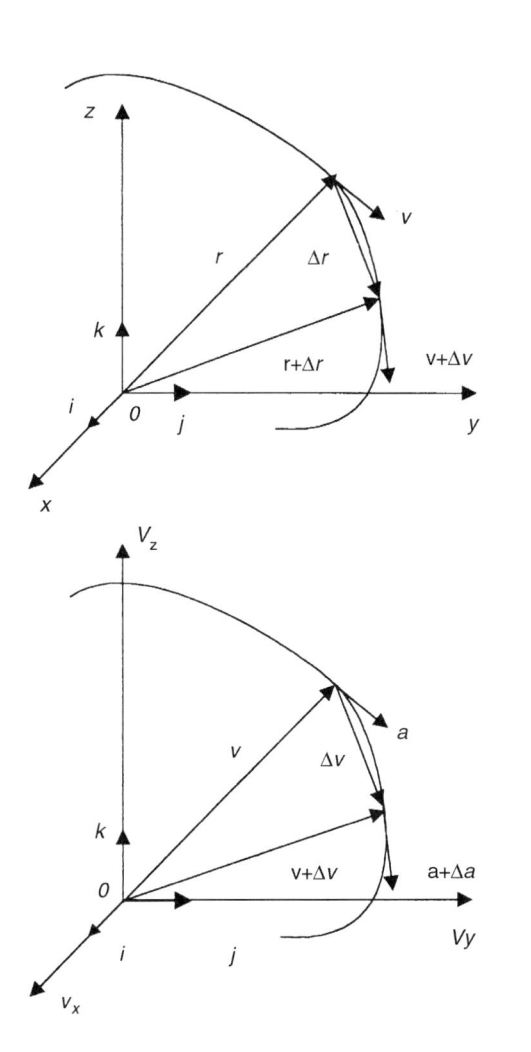

Figure 1 Independent variables of mechanics in a spatial Cartesian orthogonal reference frame.

Law 1: Every body preserves at its state of rest, or of uniform motion in a right line, unless it is compelled to change that state by forces impressed thereon.

Galileo first discovered this law in 1638.

Law 2: The alteration of motion is ever proportional to the motive force impressed, and is made in the direction of the right line in which that force is impressed.

In this fundamental law of dynamics, Newton refers to the quantitative value of motion, or momentum $m\mathbf{v}$ change under the action of the force $\mathbf{F}$ on a body. m is the instantaneous proportionality coefficient of the material body that is defined as mass. $\mathbf{v}$ is the instantaneous velocity of the body. If the direction of motion is $\mathbf{r}$ then the second law implies that:

$$\mathbf{F} = m\frac{d(m\mathbf{v})}{dt} = \frac{d\left(m\frac{d\mathbf{r}}{dt}\right)}{dt} \quad [4]$$

An example of varying mass occurs, for example, to a rocket, at the stage where it loses fuel, during flight. Where the mass remains constant, the second law becomes:

$$\mathbf{F} = m\frac{d^2\mathbf{r}}{dt^2} = m\mathbf{a}; \quad \mathbf{a} = \frac{d^2\mathbf{r}}{dt^2} \quad [5]$$

where $\mathbf{a}$ is the acceleration of the body.

Newton's second law can be extended to rotation, where the force $\mathbf{F}$ along $\mathbf{x}$ is replaced by a moment M_T about $\mathbf{r}$, $\mathbf{r}$ is replaced by the rotation θ about $\mathbf{r}$, and mass m is replaced by the polar moment of inertia I_p about the x axis.

The second law can also be extended to the action of resultant forces on the body.

Law 3: To every action there is always an equal opposed reaction, or the mutual actions of two bodies upon each other are always equal, and directed to contrary parts.

An important principle linked to Newton's second law is the dynamic equilibrium concept suggested by D'Alembert (1743). D'Alembert's principle says that when a body in a fixed system is forced by a resultant $\mathbf{R} = \sum_{i=1}^{n} \mathbf{F}_i$ of n forces, then Newton's second law defines its absolute acceleration, $\mathbf{a}$. However, if the reference system moves at the same velocity as the body, the observer will note that the body is in rest, or equilibrium. This can occur since $\mathbf{R}$ is balanced by the inertia force $-m\mathbf{a}$ (**Figure 2**):

$$\mathbf{P} = \mathbf{R} - m\mathbf{a} = 0 \quad [6]$$

For a system of n bodies (**Figure 2**):

$$\sum_{i=1}^{n} \mathbf{P}_1 = \sum_{i=1}^{n} (\mathbf{R}_i - m\mathbf{a}) = 0 \quad [7]$$

The resultant force is obtained by the vectorial rule of the parallelogram, as presented by Newton (1687). The rule follows Aristotle's assumption that a force is proportional to speed: 'A body by two forces conjoined will describe the diagonal of a parallelogram, in the same time that it would describe the sides, by these forces apart'. At that time the parallelogram of velocities was known. In fact, Leonardo da Vinci had already defined in his way the terms 'force' and 'moment', and suggested combinations of forces into a resultant force (**Figure 3**).

Varignon's (1654–1722) contribution was the superposition of moments. He introduced the rule that the moment of the result of two forces, at a chosen point, equals the algebraic sum of moment of the two forces.

An important rule in vibrations of deformable bodies is that when a deformable free body is subjected to action of forces in equilibrium, this equilibrium is kept when the body becomes rigid.

Calculus of Variations

The theory of variations has many applications, and it allows for the derivation of basic principles. It deals with maximal and minimal problems determining the values of independent variables $(x_1, x_2, x_3 \ldots x_n)$ for which the function $y(x_1, x_2, x_3 \ldots x_n)$ has either a maximum or minimum value. In its fundamental description, a definite integral that has an integrand composed of unknown functions (one or more), and their derivatives, is given. These unknown functions are solved in such a way that the integral, usually of the form, if F contains derivatives of y to the first order:

$$I = \int_{x_1}^{x_2} [F(x, y, y')] \, dx; \quad y' = \frac{dy}{dx} \quad [8]$$

will be of either minimum or maximum value.

Euler's Equations

Given two points, A_1 and A_2 in the (x, y) plane, it is necessary to find for eqn [8] the admissible arc $A_1 A_2$ that minimizes I (**Figure 4A**). Such admissible arcs are usually continuous and consist of a finite number of arcs. The tangent of each arc varies continuously, but the whole curve may have corners. The admissible curves joining A_1 and A_2 can be given as:

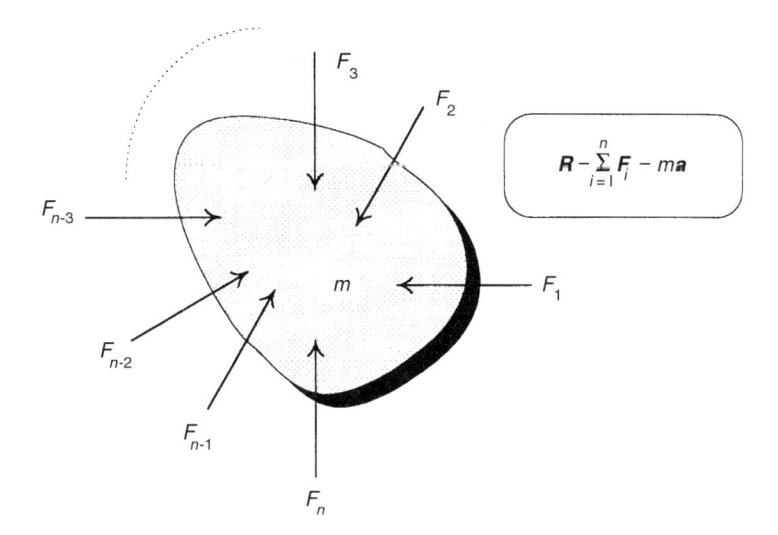

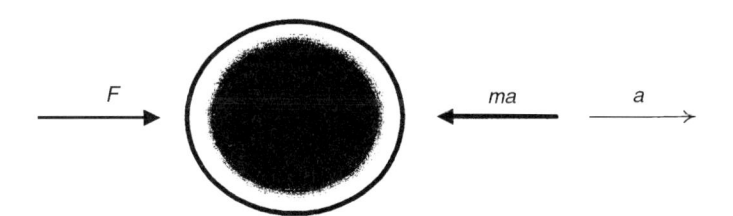

Figure 2 D'Alembert's principle.

$$y = y(x) + \alpha\eta(x) \quad [9]$$

where $\eta(x)$ is an arbitrary function that vanishes at A_1 and A_2. $y = y(x)$ is the minimizing function. Substituting eqn [8] in eqn [9] yields:

$$I(\alpha) = \int_{x_1}^{x_2} F[x; y + \alpha\eta(x); y' + \alpha\eta'(x)]\,dx \quad [10]$$

To minimize $I(\alpha)$ for $\alpha = 0$, it is necessary that:

$$I'(0) = \int_{x_1}^{x_2} [F_y \eta(x) + F_{y'}\eta'(x)]\,dx = 0;$$

$$F_y = \frac{\partial F(x,y,y')}{\partial y}; \quad F_{y'} = \frac{\partial F(x,y,y')}{\partial y'} \quad [11]$$

By integrating the parts and satisfying the end conditions, the following version of Euler's equation results in:

$$F_y - \frac{dF_{y'}}{dx} = 0 \quad [12]$$

The last equation must be satisfied conditionally by every function $y = y(x)$ that minimizes or maximizes eqn [8]. There are additional conditions, which are more complicated, that are needed to secure the maximization or minimization of eqn [8], which in most cases are suspended. A more explicit version of Euler's equation is:

$$F_{y'y'}y'' + F_{y'y}y' + F_{y'x} - F_y = 0; \quad F_{y'y'} = \frac{\partial^2 F}{\partial y'^2};$$

$$F_{y'x} = \frac{\partial^2 F}{\partial y' \partial x} \quad [13]$$

The solutions of eqn [12] are called extremals and the resulting curves are called extremal arcs. In either case, y must able to be differentiated twice and F three times in order to discover whether the result is extremal.

(A) A superposition of forces
Leonardo da Vinci
Codex Arundel.
The Trustees of the British Museum,
London

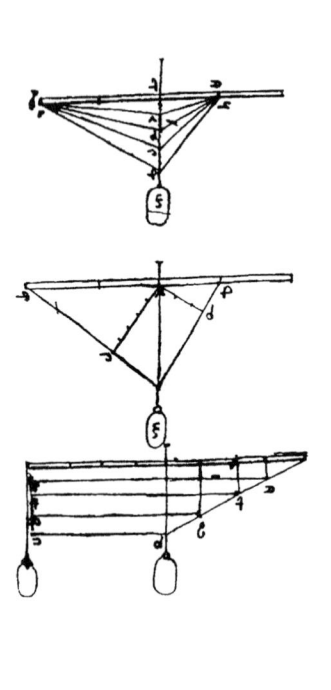

(B) A study equilibrium
Leonardo da Vinci
Codex Madrid I.
Biblioteca National, Madrid
and McGraw-Hill

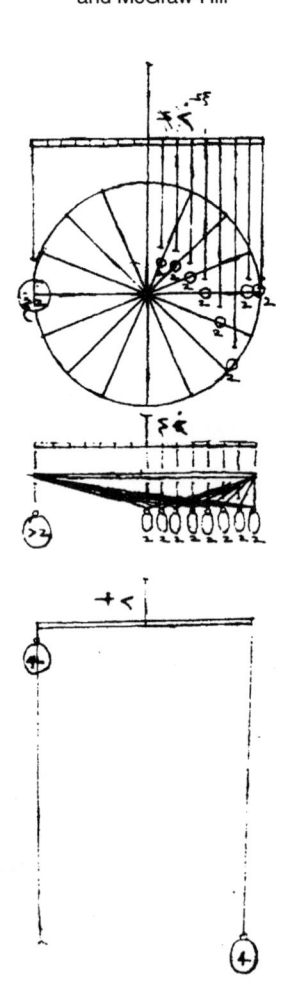

Figure 3 (A) Superposition and (B) equilibrium by Leonardo da Vinci.

The functional may depend on more then one function:

$$I = \int_{A_1}^{A_2} F(x, y_1, y_2 \ldots y_n, y'_1, y'_2 \ldots y'_n) \, dx \quad [14]$$

and the end conditions at A_1 and A_2. As a result, and since it is possible to vary only one function at a time, a set of n Euler equations results in:

$$F_{y_i} - \frac{d}{dx} F_{y'_i} = 0; \quad i = 1, 2, 3 \ldots n \quad [15]$$

Euler's equation can be further generalized to include derivatives of higher order:

$$I = \int_{A_1}^{A_2} F\left(x, y(x), y'(x) \ldots y^{(n)}(x)\right) dx \quad [16]$$

and the end conditions at A_1 and A_2. For the extremum of the functional I, the following Euler–Poisson equation must be satisfied:

$$\frac{\partial F}{\partial y} - \frac{d}{dx} \frac{\partial F}{\partial y'} + \frac{d^2}{dx^2} \frac{\partial F}{\partial y''} - \frac{d^3}{dx^3} \frac{\partial F}{\partial y'''} \\ + \cdots + (-1)^n \frac{d^n}{dx^n} \frac{\partial F}{\partial y^{(n)}} = 0 \quad [17]$$

This is an equation of the order $2n$, and y must be able to be differentiated $2n$ times, while F must be able to be differentiated $n + 2$ times: once in order to examine if the result yields an externum.

BASIC PRINCIPLES

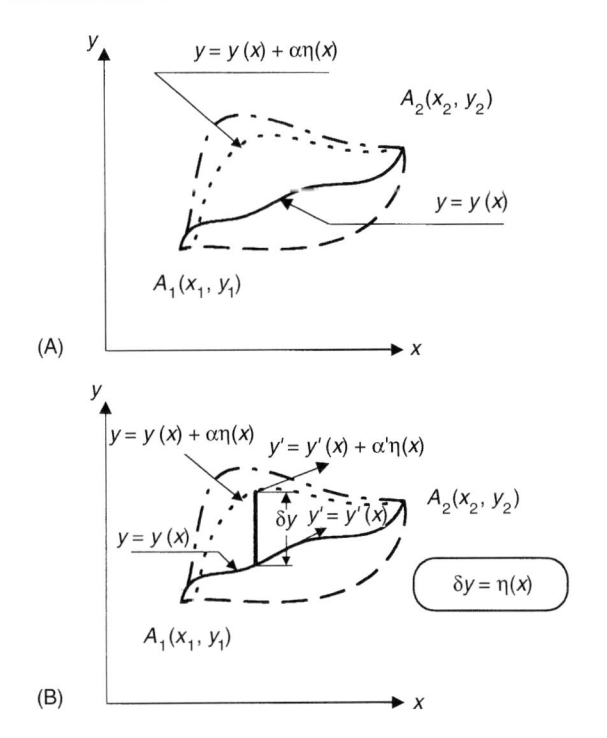

Figure 4 Variations in a curve between two stationary points.

If the functional depends on several functions and their higher derivatives:

$$I = \int_{A_1}^{A_2} F\left(x_1, y_1, y_1' \ldots y_1^{(n1)}, x_2, y_2, y_2' \ldots y_2^{(n2)} \right.$$
$$\left. \ldots x_m, y_m, y_m' \ldots y_m^{(nm)}\right) dx \quad [18]$$

and the end conditions at A_1 and A_2 then, for extremum, the following set of equations has to be satisfied:

$$\frac{\partial F}{\partial y_i} - \frac{d}{dx}\frac{\partial F}{\partial y_i'} + \frac{d^2}{dx^2}\frac{\partial F}{\partial y_i''} - \frac{d^3}{dx^3}\frac{\partial F}{\partial y_i'''} + \cdots$$
$$+ (-1)^{ni}\frac{d^{ni}}{dx^{ni}}\frac{\partial F}{\partial y_i^{(ni)}} = 0; \quad i = 1, 2 \ldots m \quad [19]$$

The functional can also be dependent on several independent variables:

$$I[z(x_1, x_2 \ldots x_n)] = \iint \cdots \int_D F(x_1, x_2 \ldots x_n, z, p_1,$$
$$p_2 \ldots p_n)\, dx_1 dx_2 \ldots dx_n;$$
$$p_i = \frac{\partial z}{\partial x_i} \quad [20]$$

and the end conditions at A_1 and A_2. Then for extremum, variation of I is used:

$$\delta I = \delta \iint \cdots \int_D F(x_1, x_2 \ldots x_n, z, p_1, p_2 \ldots p_n)$$
$$dx_1\, dx_2 \ldots dx_n$$
$$= \iint \cdots \int_D (F_z \delta z + F_{p1}\delta p_1 + F_{p2}\delta p_2 + \cdots$$
$$+ F_{pn}\delta p_n) dx_1\, dx_2 \ldots dx_n = 0 \quad [21]$$

and the following set of equations must be satisfied:

$$F_z - \sum_{i=1}^{n} \frac{\partial}{\partial x_i} F_{p_i} = 0 \quad [22]$$

In the specific case of two independent variables, $z = z(x_1, x_2)$, Euler–Lagrange or Ostrogradsky (1801–1861) equation results. Likewise, other cases may be developed.

Notes

1. As mentioned by Bolza (1931), Weierstrass has proven that a continuous function $y(x)$ that has isolated discontinuities in its first or second derivatives has a minimum value for I if it satisfies Euler's equation in any interval between successive discontinuities. This is in addition to conditions that must be satisfied at the points of discontinuity of the derivatives.
2. If F has several unknown functions ($y, z \ldots$), it is necessary to satisfy the separate Euler's equation for $y, z \ldots$ for a stationary value of I.

Another way of proving Euler's equation is by the method of variations. Now $\eta(x)$ is defined as an incremental change, δy, in the neighborhood of the function $y(x)$ that minimizes or maximizes the integral [8] (**Figure 4B**). δy is defined as variation, and the analysis involved with δy is called the calculus of variations. Important relations in the calculus of variations are at each x between $A_1 A_2$:

$$\eta'(x) = \frac{d(\delta y)}{dx} = \delta \frac{dy}{dx};$$
$$\delta F(y) = F(y + \delta y) - F(y) = \frac{\partial F}{\partial y}\delta y = F_y \delta y;$$
$$\delta F(x, y, y') = \delta F(y, y') = \frac{\partial F}{\partial y}\delta y + \frac{\partial F}{\partial y'}\delta y'$$
$$= F_y \delta y + F_{y'} \delta y' \quad [23]$$

Since $y(x)$ obeys the minimax condition it is stationary, and:

$$\delta \int_{A_1}^{A_2} F(x, y, y') \, dx = \int_{A_1}^{A_2} \delta F(x, y, y') \, dx$$

$$= \int_{A_1}^{A_2} \left[F_y \delta y + F_{y'} \delta y' \right] dx = 0 \quad [24]$$

then, using integration by parts and the fact that the variation is an arbitrary function of x, this leads to Euler's equation [12].

The calculus of variations can be used in dynamics in general and in the theory of vibrations specifically if the time t is introduced as an independent variable. In a three-dimensional space (x_1, x_2, x_3) we have:

$$I = \int_{t_0}^{t_1} F(t, x_1, x_2, x_3, \dot{x}_1, \dot{x}_2, \dot{x}_3) \, dt \quad [25]$$

An extremum is sought for this functional by finding functions that satisfy the end conditions: $x_i(t_0) = x_{i0}$; $x_i(t_1) = x_{i1}$; $i = 1, 2, 3$ at each specific problem.

In many cases the spatial coordinates are constrained (e.g., the length in a spherical pendulum is: $L = \sqrt{x_1^2 + x_2^2 + x_3^2}$). This constraint can be written as: $C(t, x_1, x_2, x_3) = 0$, and for $\partial C/\partial x_3 \neq 0$, x_3 can be extracted: $x_3 = f(t, x_1, x_2)$. This relation leads to the unconstrained extremum problem:

$$\hat{I} = \int_{t_0}^{t_1} \hat{F}(t, x_1, x_2, \dot{x}_1, \dot{x}_2) \, dt \quad [26]$$

with the relation:

$$\hat{F}(t, x_1, x_2, \dot{x}_1, \dot{x}_2)$$
$$= \Psi\left(t, x_1, x_2, f(t, x_1, x_2), \dot{x}_1, \dot{x}_2, \frac{d}{dt} f(t, x_1, x_2)\right) \quad [27]$$

For an extremum the functions x_i must satisfy the Euler's equations:

$$\hat{F}_{x_i} - \frac{d}{dt} \hat{F}_{\dot{x}_i} = 0; \quad i = 1, 2 \quad [28]$$

Also, the extremalizing functions and the involved Lagrange multiplier function must satisfy the Euler's equations:

$$\hat{F}_{x_i} - \frac{d}{dt} \hat{F}_{\dot{x}_i} = \lambda C_{x_i}; \quad i = 1, 2, 3 \quad [29]$$

The Equation of Flexural Motion Obtained by the Calculus of Variations

To illustrate the use of the calculus of variations in the theory of vibrations, we find the equation of motion for a beam (**Figure 5**). The total virtual work obtained by the variation δ along the beam is:

$$\delta I = \int_0^L (-M\delta v'' - q\delta v + m\ddot{v}\delta v) \, dx = 0, \quad [30]$$

$$\zeta' = \frac{\partial \zeta}{\partial x}; \quad \dot{\zeta} = \frac{\partial \zeta}{\partial t}$$

We modify the first term in eqn [30]:

$$\int_0^L (-M\delta v'') \, dx = -M\delta v'\big|_0^L + M_{,x}\delta v\big|_0^L$$

$$-\int_0^L M_{,xx}\delta v \, dx = -\int_0^L M_{,xx}\delta v \, dx$$

and get:

$$\delta I = \int_0^L (-M_{,xx} - q + m\ddot{v})\delta v \, dx = 0$$

Since $M = -EIv' \Rightarrow M_{,xx} = -(EIv'')''$, and the virtual work equals zero whatever δy is, the resulting equation becomes:

$$-M_{,xx} - q + m\ddot{v} = 0 \quad \text{or} \quad (EI_{,xx})_{,xx} - q + m\ddot{v} = 0$$
$$\quad [31]$$

Generalized Coordinates of a System

If the displacements **u** of a body are defined in terms of m coordinates, and if there are r equations of constraints among these displacements, then the rest of the coordinates, $n = m - r$, are independent. These n independent coordinates, that specify completely the configuration (displacements and forces) of the system, are defined as generalized coordinates, **q**. The number of generalized coordinates of a system is also its number of degrees-of-freedom. If it is possible to extract a set of independent coordinates in a mechanical system, then the system is holonomic. However, if the resulting set of equations of a system does not permit extraction of the constraints, then the system is nonholonomic. Usually, the systems are holonomic. In a holonomic system it is possible to transform **u** into **q** by a transform matrix **C**: **u** = **Cq**. For

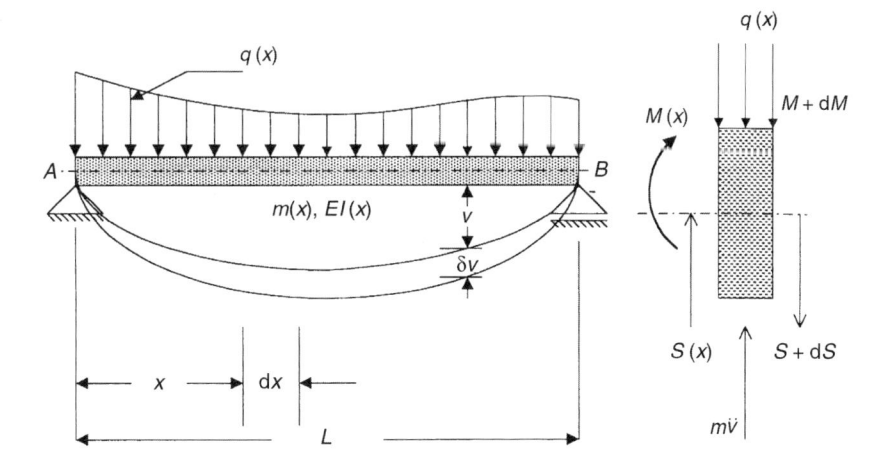

Figure 5 Variation in a deflection line of a beam.

example, if the kinetic energy of a holonomic set of linked rigid bodies m_j is $T = \frac{1}{2}\sum_j m_j \dot{u}_j^2$, then the displacements u_j can be given in term of generalized coordinates $u_j = u_j(q_1, q_2, q_3 \ldots)$:

$$\mathbf{du} = \sum_k \frac{\partial u_j}{\partial q_k} dq_k; \quad \mathbf{du} = \mathbf{C}\, \mathbf{dq}; \quad C_{jk} = \frac{\partial u_j}{\partial q_k} \quad [32]$$

A virtual displacement is an arbitrary small displacement. A system of n degrees-of-freedom has n possible independent virtual displacements. A virtual work is the work done by forces that act during the virtual displacement: $\delta W = \mathbf{Q}^T \delta \mathbf{q} = \mathbf{F} \delta \mathbf{u}$. $\mathbf{Q}$ are generalized forces in accordance with the generalized coordinates, and $\mathbf{F}$ are the forces that correspond to $\mathbf{u}$.

As an example, D'Alembert's principle can be given in terms of virtual displacements:

$$\delta W = \sum_j m_j \ddot{u}_j \delta u_j = \sum_j F_j \delta u_j \quad [33]$$

Hamilton's Principle

Hamilton's principle is one of the most fundamental principles in vibration analysis. It leads to the basic equations of dynamics and elasticity. It is based on the assumption that when a system moves from a state at a time t_1 to a new state at the time t_2, in a Newtonian route, then the actual route out of all the possible ones obeys stationarity. This condition leads to Hamilton's principle:

$$\delta \int_{t_1}^{t_2} (T - V)\, dt = \delta \int_{t_1}^{t_2} L\, dt = 0; \quad [34]$$

$$L = T - V = \text{Lagrangian}$$

where T is the kinetic energy of the system and V is the potential energy of the system. The Lagrangian L can be written in that case of elasticity in the form $L = U - K + A$, where A is the potential energy of the system, U is the strain energy of the system, and K is the overall kinetic energy of the system.

In terms of generalized coordinates, Hamilton's principle obtains the form:

$$\int_{t_1}^{t_2} \sum_j \left\{ -\frac{d}{dt}\left(\frac{\partial T}{\partial \dot{q}_j}\right) + \frac{\partial T}{\partial q_j} + Q_j \right\} \partial q_j\, dt = 0 \quad [35]$$

It is noted here that, for calculation of the kinetic energy of a system, Koenig's theorem can be used. This theorem states that the total kinetic energy of a rigid body of mass M is the kinetic energy of a particle of mass M that moves with the center of gravity of the body, plus the kinetic energy of the motion relative to the center of gravity of the body (as if it were fixed).

Lagrange Equations

While Hamilton's principle has an integral form, Lagrange equations are differential and describe an instantaneous situation. Lagrange equations carry the form:

$$\frac{\mathrm{d}}{\mathrm{d}t}\left(\frac{\partial T}{\partial \dot{q}_j}\right) - \frac{\partial T}{\partial q_j} + Q_j = 0; \quad j = 1, 2 \ldots n \quad [36]$$

The generalized forces can include external forces, Q_{Aj}, internal elastic forces, $Q_{Ej} = -\partial U/\partial q_j$, and damping forces, Q_{Dj}. Hence:

$$\frac{\mathrm{d}}{\mathrm{d}t}\left(\frac{\partial T}{\partial \dot{q}_j}\right) - \frac{\partial T}{\partial q_j} + \frac{\partial U}{\partial q_j} - Q_{Dj} = Q_{Aj}; \quad j = 1, 2 \ldots n \quad [37]$$

In a conservative system, the potential energy due to potential sources is: $V(q_1, q_2, q_3 \ldots)$, and Lagrange's equation becomes:

$$\frac{\mathrm{d}}{\mathrm{d}t}\left(\frac{\partial T}{\partial \dot{q}_j}\right) - \frac{\partial T}{\partial q_j} + \frac{\partial V}{\partial q_j} = 0 \quad \text{or} \quad \frac{\mathrm{d}}{\mathrm{d}t}\left(\frac{\partial L}{\partial \dot{q}_j}\right) - \frac{\partial L}{\partial q_j} = 0;$$
$$L = T - V; \quad j = 1, 2 \ldots n \quad [38]$$

Lagrange's equation is also useful in the case of dependent coordinates. Assume a set of m coordinates x_i; r of which are dependent and have the relations:

$$C_j(x_1, x_2, x_3 \ldots x_m) = 0; \quad j = 1, 2 \ldots r \quad [39]$$

then, by using Lagrange multipliers, the Lagrangian can be formulated by coordinates, not all of which are independent:

$$L = L_0 + \sum_j \lambda_j C_j(x_1, x_2 \ldots x_m) \quad [40]$$

The equations of motion become:

$$\frac{\partial L}{\partial x_i} - \frac{\mathrm{d}}{\mathrm{d}t}\frac{\partial L_0}{\partial \dot{x}_i} + \sum_j \lambda_j \frac{\partial C_j}{\partial x_i} = 0; \quad i = 1, 2 \ldots n \quad [41]$$

Hence, an equation of motion that represents the D'Alembertian sum of forces exists for each coordinate. Each multiplication of force by virtual displacement generates a contribution to the total virtual work that should be zero. Also, in the case of dependent coordinates, it is necessary that the virtual work will be zero. However, two separate conditions are necessary to insure:

$$\left(\frac{\partial L}{\partial x_i} - \frac{\mathrm{d}}{\mathrm{d}t}\frac{\partial L_0}{\partial \dot{x}_i} + \sum_j \lambda_j \frac{\partial C_j}{\partial x_i}\right)\delta x_i = 0; \quad i = 1, 2 \ldots n$$

$$[42]$$

One is the set of $n = m - r$ equations of motion and the second is involved with the r constraint equations.

Rayleigh's Principle

Rayleigh's principle and the fundamental equations of dynamics enable derivation of the theory of vibrations. When a conservative system vibrates freely, the total mechanical energy is constant. If the initial state is 0 and another state is defined by 1, then the sum of the kinetic and potential energies in the two states is the sum:

$$T_0 + V_0 = T_1 + V_1 \Rightarrow T_0 = V_1 \quad [43]$$

In simple harmonic motion the average energies are $\bar{T} = 0.5 T_0$; $\bar{V} = 0.5 V_0 \Rightarrow \bar{T} = \bar{V}$. Since, in linear vibrations:

$$T = 0.5 \sum_{i=1}^{n} (\dot{u}_i^2);$$
$$V = 0.5 \sum_{i=1}^{n} (\omega_i^2 u_i^2) \Rightarrow \quad [44]$$
$$\bar{T} = \bar{V} = 0.25 \sum_{i=1}^{n} (A_i^2 \omega_i^2)$$

then, using $\bar{T} = \bar{V}$ or $T_0 = V_1$ enables one to determine the natural frequency for a conservative 1-degree-of-freedom system. When the system has several degrees of freedom, the equation $\bar{T} = \bar{V}$ or $T_0 = V_1$ can be used for each mode, provided its modal shape can be approximated. This usually leads to frequencies that are higher by some percent than the natural frequencies.

A scheme of a one-dimensional straight element of length l, loaded by a distributed mass $m(x)$ and discrete masses M_i, is given in **Figure 6** to illustrate Rayleigh's principle.

Assuming a deflection shape $\phi(x)$ that approximates one modal shape, such that the deflection is given by $y = y_0 \phi(x)$, then the system is approximated into a single-degree-of-freedom one, with an equivalent mass, $M_{\mathrm{eq}}(a)$ at $x = a$. The equivalent mass is calculated as:

$$\frac{1}{2} M_{\mathrm{eq}}(a) \dot{y}_0^2 \phi^2(a)$$
$$= \frac{1}{2}\left\{\sum_{i=1}^{n} (M_i \dot{y}_0^2 \phi^2(i)) + \int_{(l)} (m(x) \dot{y}_0^2 \phi^2(x))\,\mathrm{d}x\right\}$$

or

$$M_{\mathrm{eq}}(a) = \frac{1}{\phi^2(a)}\left\{\sum_{i=1}^{n}(M_i \phi^2(i)) + \int_{(l)} (m(x)\phi^2(x))\,\mathrm{d}x\right\}$$

$$[45]$$

Obviously:

$$\frac{M_{eq}(a)}{M_{eq}(b)} = \frac{\phi^2(b)}{\phi^2(a)}; \quad \frac{K_{eq}(b)}{K_{eq}(a)} = \frac{\phi^2(b)}{\phi^2(a)}$$

and, the approximate frequency is calculated as:

$$\omega^2(a) = \omega^2(b) = \frac{K_{eq}(a)}{M_{eq}(a)} = \frac{K_{eq}(b)}{M_{eq}(b)} \quad [46]$$

Example: given a fixed steel beam of a profile INP 30 (a standard I profile made of steel, common in Europe, with moment of inertia $I = 9800 \, cm^4$, and Young's modulus $E = 2.1 \times 10^4 \, kN\,m^{-1}$). See **Figure 7**. It is assumed that the deflection line carries the form:

$$y(x) = \frac{1}{2}y_0 \left[1 - \cos\left(\frac{2\pi x}{L}\right)\right] = y_0 \phi(x)$$

The maximum deflection of the beam under concentrated load P at its center is $f_{max} = PL^3/192EI$, and the equivalent spring constant becomes:

$$K_{eq} = \frac{192EI}{L^3} = 3.165 \times 10^5 \, N\,m^{-1}$$

The equivalent mass is derived from the equation: $\frac{1}{2}M_{eq}\dot{y}_0^2 = \frac{1}{2}m\dot{y}_0^2 \int_0^L \phi^2 \, dx$. The result is $M_{eq} = \frac{3}{8}mL = 3750 \, Ns^2\,m$. Hence, following eqn [43], the radian frequency becomes:

$$\omega^2 = 84 \, rad^2\,s^{-2} \rightarrow \omega = 9.165 \, rad\,s^{-1}$$

The equivalent spring constant relates the force to displacement and to the static deflection of the vibrating element at the generalized coordinate. Hence it is important in this case to calculate the static deflection at that coordinate. In many cases the examined structure is indeterminate and its static analysis can be supported by the theorem of Castigliano (1847–1884) and of Menabrea (1809–1896). This theorem

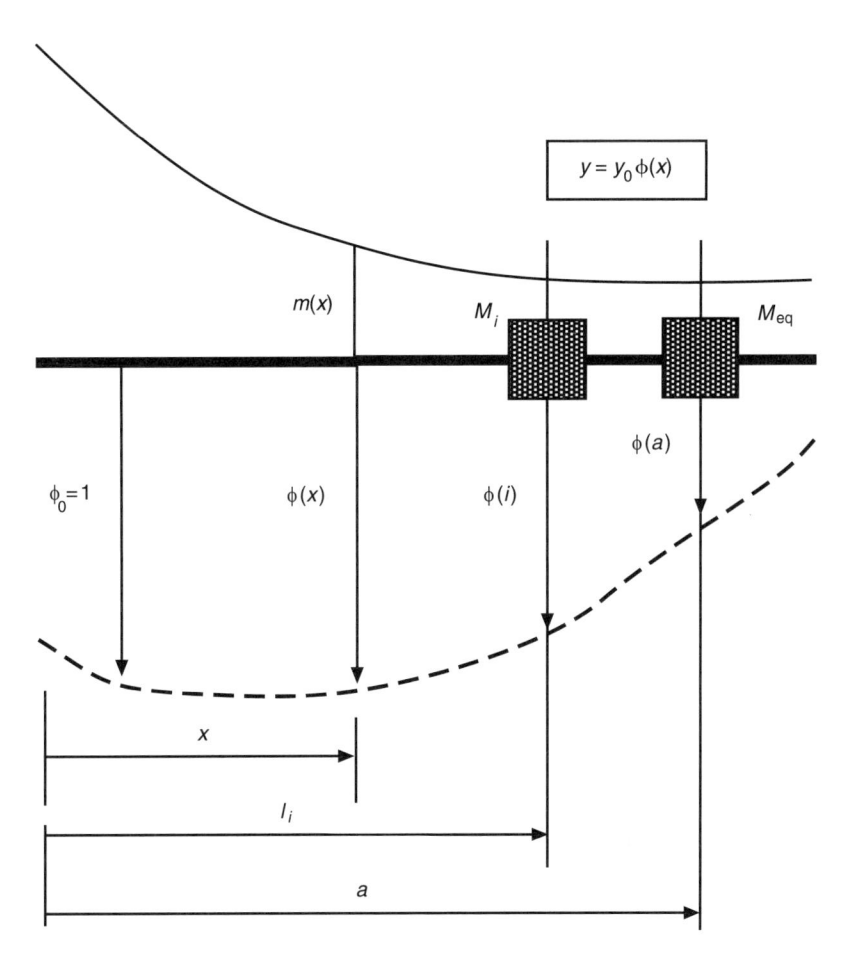

Figure 6 A scheme to illustrate Rayleigh's method.

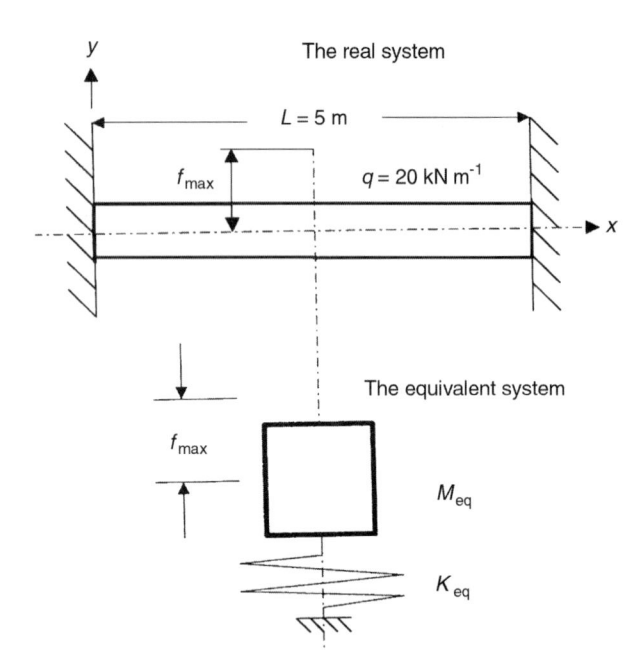

Figure 7 Equivalent mass and spring constant.

of the minimum strain energy says that each system brings its internal (potential) energy to the minimum.

Rayleigh's principle can also be used for the first and the higher modes $\phi_j(x)$ of a beam. The maximum kinetic energy of the beam in the jth mode (**Figures 5–7**) is:

$$T_{\max j} = \frac{1}{2}\int_0^L m\dot{v}_j^2 \, dx = \frac{1}{2}f_{\max j}^2 \int_0^L m\omega_j^2\phi_j^2(x)\,dx$$

$$= \frac{1}{2}f_{\max j}^2 \omega_j^2 M_{\text{eq}j} \qquad [47]$$

$$M_{\text{eq}j} = \int_0^L m\phi_j^2(x)\,dx$$

The maximum potential energy of the beam is:

$$U_{\max} = \frac{1}{2}\int_0^L \frac{M_j^2}{EI}\,dx = \frac{1}{2}\int_0^L EI\left(\frac{d^2 v_j}{dx^2}\right)^2 dx$$

$$= \frac{1}{2}f_{\max j}^2 \int_0^L EI\left(\frac{d^2\phi_j}{dx^2}\right)^2 dx = \frac{1}{2}K_{\text{eq}j}f_{\max j}^2 \qquad [48]$$

$$K_{\text{eq}} = \int_0^L EI\left(\frac{d^2\phi_j}{dx^2}\right)^2 dx$$

Hence, the natural frequency of the jth mode is:

$$\omega_j^2 = \frac{K_{\text{eq}j}}{M_{\text{eq}j}} = \frac{\int_0^L EI(x)\left[\phi_j''(x)\right]^2 dx}{\int_0^L m_j(x)\left[\phi_j(x)\right]^2 dx} \qquad [49]$$

Rayleigh–Ritz or Ritz Method

The Rayleigh–Ritz method enables one to reduce an infinite number of degrees-of-freedom of a system into a finite number, which makes analysis possible and easier. The method relies on the approximation of the possible deformation shapes of the system, following the basic idea beyond Rayleigh's principle. A proper combination of these deformation shapes is supposed to reproduce roughly the expected overall vibration pattern and natural frequencies. Following the Ritz method the family of curves over which the values of the functional are calculated is obtained by a linear combination of functions:

$$\phi(x) = \sum_{i=1}^n \alpha_i \varsigma_i(x); \quad i = 1, 2, 3 \ldots n \qquad [50]$$

Here, α_i are constants and ς_i are functions that satisfy certain end conditions. Hence, the shape of the function ϕ is determined by ς_i while its values are determined by the coefficients α_i. The functional I turns to be a function of the coefficients α_i:

$$I = \Psi(\alpha_1, \alpha_2 \ldots \alpha_n) \qquad [51]$$

where the condition of obtaining an extremum dictates the following set of n equations:

$$\frac{\partial \Psi}{\partial \alpha_i} = 0; \quad i = 1, 2 \ldots n \qquad [52]$$

The solution of the coefficients α_i yields the values of the function $y(x)$ for which the functional obtains the approximate extremal value. The more suitable the choice of the family of function, the more exact the solution will be.

The method also applies to the solution of problems with functionals that depend on several functions of one or more independent variables. For example, the Rayleigh–Ritz method can be used for the analysis of flexural vibrations of a beam. The jth natural frequency of the beam satisfies the ratio given in eqn [49]. Using this equation and eqns [51] and [52] results in:

$$\frac{\partial}{\partial \alpha_1} \frac{\int_0^L EI(x)\left[\phi_j''(x)\right]^2 dx}{\int_0^L m_j(x)\left[\phi_j(x)\right]^2 dx} = 0 \qquad [53]$$

Substituting the expression [50] in eqn [53] leads to a set of linear equations that solve the α_is. However, in order to avoid a trivial solution, the determinant of coefficients of this set of equations should be zero, which yields the frequency equation for the various modes of vibrations. Now eqn [50] is again used to find the mode shape, having solved the jth natural frequency and the ratios of the α_is.

Galerkin's Method

In a similar way to that of Ritz's method, Galerkin's method combines a family of curves for which the values of the functional I are obtained follow eqn [50]. The difference is in that each function has to satisfy all the end conditions of the problem. This means that, for all the combinations $\delta v = 0$, the solution is that of a problem with fixed ends. The condition for extremum obtains in this case the form:

$$\delta I = \int \Psi(\alpha_1, \alpha_2 \ldots \alpha_n) \delta y \, dx = 0 \qquad [54]$$

This equation has to be satisfied for any arbitrary δy. A set of n equations for the n unknowns α_i is obtained by a sequential choice of z_i for δy. In many cases Galerkin's method coincides with that of Ritz. However, whenever the solution is not the same, Galerkin's method is more exact for the same number of terms. The reason is that the end conditions are satisfied *a priori* by this method.

The method also applies to the solution of problems with functionals that depend on several functions of one or more independent variables.

We demonstrate now this method for the natural vibrations of a beam to illustrate the point. The virtual work of the beam of constant EI is:

$$\int_0^L \left[(EI_{,xx})_{,xx} + m\ddot{v}\right] \delta v \, dx = 0 \qquad [55]$$

A deflection curve of the beam is assumed to have the form and variation:

$$v(x, t) = \zeta(x) \exp(i\omega t)$$
$$\delta v(x, t) = \delta\zeta(x) \exp(i\omega t) \qquad [56]$$

Substitution in eqn [55] yields:

$$\exp(i\omega t) \int_0^L \left[(EIv_{,xx})_{,xx} - m\omega^2 v\right] \delta v \, dx = 0 \qquad [57]$$

Assume a solution of the form:

$$\bar{v}(x) = \sum_{j=1}^n \alpha_j \zeta_j(x);$$

and:

$$\delta \bar{v}(x) = \sum_{i=1}^n \delta\alpha_j \zeta_j(x) \qquad [58]$$

Since $\delta\alpha_j$ is arbitrary, it is possible to make all the virtual displacements zero. Then for each k the following equation is obtained:

$$\int_0^L \left[(EI\bar{v}_{,xx})_{,xx} - m\omega_k^2 \bar{v}\right] \bar{v}_k \, dx = 0 \qquad [59]$$

After calculating the integral and using the orthoganality rules, the result for the natural frequency becomes:

$$\omega_k^2 = \frac{4k^4 \pi EI}{mL^4} \qquad [60]$$

Dunkerley's Method

While Rayleigh's method yields the upper bound of the natural frequencies, Dunkerley's method (1894) suits the lower bound. The method originally estimates the fundamental frequency in a multidegree-of-freedom system, with the eigenvalue problem:

$$L(u) = \omega^2 M(u) = 0 \qquad [61]$$

For:

$$M(u) = \sum_{i=1}^m M_i(u) \qquad [62]$$

the partial problem is self-adjoint and fully defined if its fundamental frequency is known. Hence:

$$L(u) = \omega_i^2 M_i(u) = 0 \qquad [63]$$

and, using Galerkin's method, Dunkerley's frequency, ω_D, becomes:

$$\frac{1}{\omega_{1D}^2} = \sum_{i=1}^m \frac{1}{\omega_{1i}^2}; \quad \omega_{1D}^2 \leq \omega_1^2 \qquad [64]$$

Here, ω_1 is the exact fundamental mode, and ω_{li} are the fundamental frequencies of the partial problems.

The natural frequencies of the second and higher modes are often much higher than the fundamental one. A scheme of an example is given in **Figure 8**. A cantilever of length L has evenly distributed mass m, with $mL = M$, and a concentrated mass at its end. The first natural frequency for the distributed mass is $\omega_{11}^2 = 3.515^2(EI/ML^3)$, and the first natural frequency for the concentrated mass is $\omega_{12}^2 = 3.00(EI/ML^3)$. The first frequency of the whole system is calculated following eqn [64]:

$$\frac{1}{\omega_{1D}^2} = \frac{1}{\omega_{11}^2} + \frac{1}{\omega_{12}^2} \rightarrow \omega_{1D}^2 = \frac{\omega_{11}^2 \omega_{22}^2}{\omega_{11}^2 + \omega_{12}^2}$$
$$= \frac{3.515^2 \times 3.00}{3.515^2 + 3} \frac{EI}{ML^3} = 2.41 \frac{EI}{ML^3}$$

This value is the lower bound. The upper bound, ω_{1R}, is found by Rayleigh's method:

$$\omega_{1R}^2 = \frac{3EI}{\left(1 + \frac{33}{140}\right)ML^3} = 2.43 \frac{EI}{ML^3}$$

The exact first frequency is somewhere between ω_{1D} and ω_{1R}.

Maxwell's Theorem of Reciprocity

Maxwell's theorem of reciprocity (1831–1879) states that deflection at a point A in the direction **a**, due to a unit force acting along **b**, is equal to deflection at the point B, in the direction **b**, caused by a unit force acting along **a**. This theorem is useful in solving vibration problems since it does not deny D'Alembertian forces.

As an example, we seek the two natural frequencies of a weightless simply supported beam that supports two masses at A and B (**Figure 9**).

If f_{ij} is the deflection at j due to the force at i, then the deflection at A is:

$$\begin{aligned} v_A &= -m_A \ddot{v}_A f_{AA} - m_B \ddot{v}_B f_{BA}; \\ v_B &= -m_A \ddot{v}_A f_{AB} - m_B \ddot{v}_B f_{BB} \end{aligned} \quad [65]$$

and following Maxwell's theorem, $f_{AB} = f_{BA}$. The assumed solutions are:

$$v_i = \bar{v}_i \sin(\omega t), \quad i = A, B \quad [66]$$

and by substitution into eqn [65] we obtain:

$$\begin{aligned} (1 - m_A \omega^2 f_{AA})\bar{v}_A - m_B \omega^2 f_{AB} \bar{v}_B &= 0; \\ -m_A \omega^2 f_{AB} \bar{v}_A + (1 - m_B \omega^2 f_{BB})\bar{v}_B &= 0 \end{aligned} \quad [67]$$

In order to avoid a trivial solution, the determinant of coefficients has to be zero, and this yields two natural frequencies:

$$\omega_{1,2} = \sqrt{\frac{1}{m_A \chi_{AB}}}$$

$$\chi_{AB} = \frac{f_{AA} + \vartheta f_{BB}}{2} \pm \sqrt{\frac{(f_{AA} - \vartheta f_{BB})^2}{4} + \vartheta f_{AB}^2} \quad [68]$$

$$\vartheta = \frac{m_B}{m_A}$$

Nomenclature

a	accelerations
C	transform matrix
F	force
l	length
L	Lagrangian; length
M	mass
Q	generalized forces
r	radius vector
u	displacements
v	velocities
ω_D	Dunkerleys frequency

See Plate 7.

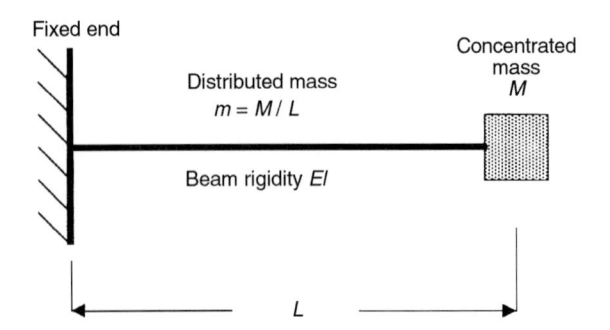

Figure 8 A cantilever beam supporting a mass at its end.

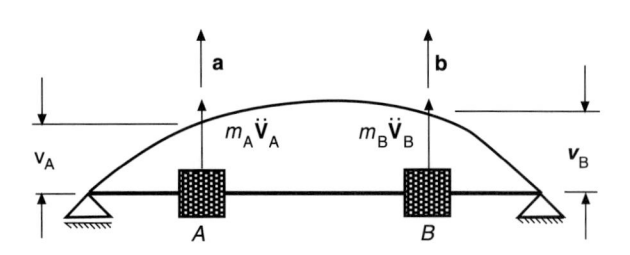

Figure 9 A simply supported beam with two concentrated masses.

See also: **Theory of vibration**, Equations of motion; **Theory of vibration**, Variational methods; **Wave propagation**, Waves in an unbounded medium.

Further Reading

Argyris JH and Kelsey S (1960) *Energy Theorems and Structural Analysis*. London: Butterworth.

Bolza O (1931) *Lectures on the Calculus of Variations*. New York: Stechert-Hafner.

Collatz L (1948) *Eigenvertprobleme und ihre numerische Behandlung*. New York: Chelsea.

D'Alembert J (1743) *Traite de Dynamique* (see also Dugas, 1955)

Dugas RA (1955) *History of Mechanics*. New York: Central Book.

Duhem PMM (1905-1906) Les Origines de la Statique (two volumes). Paris. See also: *The Evolution of Mechanics*, translated by M Cole. The Netherlands: Sijhoff of Noordhoff, Alphen an den Rijn. (Translation of *L'Evolution de la Mechanique*).

Dunkerley S (1894) On the whirling and vibration of shafts. *Philosophical Transactions of the Royal Society of London (Series A)* 185: 279-360.

Galerkin BG (1915) Bars and plates. Series in certain problems of elastic equilibrium of bars and plates. *Vestnik inzhenerov i tekhnikov* 19: 17-26.

Hart IB (1925) *The Mechanical Investigations of Leonardo da Vinci*. Chicago: Open Court

Keller EG (1947) *Mathematics of Modern Engineering*, vol. II. New York: Dover.

Lagrange L (1788) *Mechanique Analytique*. Paris. (The complete works of Lagrange were edited by Serret and Darboux, *Oeuvres de Lagrange*, 14 vol. (1867-1892)).

Lanczos C (1970) *The Variational Principles of Mechanics*, 4th edn. Toronto: University of Toronto Press.

Langhaar HL (1962) *Energy Methods in Applied Mechanics*. New York: John Wiley.

Newton I (1995: originally 1687) *The Principia*. New York: Prometheus Books.

Ritz W (1908) Ueber eine neue Methode zur Luesung gewisser Variationsprobleme der Mathematichen. *Physik J. für reine und angewandte Mathematik* 135: 1-61.

Temple G and Bickley WG (1956) *Rayleigh's Principle*. New York: Dover (by special arrangement with Oxford University Press).

Weinstock R (1952) *Calculus of Variations*. New York: McGraw-Hill.

Wittaker ET (1944) *A Treatise on the Analytical Dynamics of Particles and Rigid Bodies*, 4th edn. Dover (by special arrangement with the Cambridge University Press and the Macmillan Co. first edition 1904).

BEAMS

R A Scott, University of Michigan, Ann Arbor, MI, USA

Copyright © 2001 Academic Press

doi:10.1006/rwvb.2001.0128

Introduction

The topic treated is the transverse vibrations of single-span uniform beams with constant cross-sections. Detailed results are given for Euler–Bernoulli beams subjected to various boundary conditions. Results for a Timoshenko beam are presented for the pin–pin case.

Transverse Beam Vibrations

Euler–Bernoulli Beam Theory

Treated are external loads, cross-sections, and materials such that the motions are planar. The beam transmits a shear force, V, and a bending moment M (the x-axis is along the undeformed neutral axis of the beam and $f(x, t)$ is the external lateral force per unit length). Neglecting shear deformations and rotary inertia, the equation of motion for the homogeneous beam of cross-sectional area A is:

$$\frac{\partial^4 v}{\partial x^4} + \frac{\rho A}{EI}\frac{\partial^2 v}{\partial t^2} = \frac{f(x,t)}{EI} \quad [1]$$

Here, v is the transverse deflection of the neutral axis, E and ρ are Young's modulus and density, respectively, and I is the area moment of inertia about the z-axis. V and M are related to v by:

$$M = EI\frac{\partial^2 v}{\partial x^2} \quad [2]$$

$$V = -EI\frac{\partial^3 v}{\partial x^3} \quad [3]$$

Free motions The solution to eqn [1] for free vibrations ($f = 0$) is:

$$v(x,t) = h(x) \sin(\omega t) \quad [4]$$

where:

$$h(x) = c_1 \sin \beta x + c_2 \cos \beta x + c_3 \sinh \beta x + c_4 \cosh \beta x$$

$$\beta = (\rho A \omega^2 / EI)^{1/4}$$

and c_1, c_2, c_3, c_4 are arbitrary constants. Application of boundary conditions gives an equation – the frequency equation – for the determination of β (and hence ω) and expressions for the ratios of the cs, which then yield the associated mode shapes. Some commonly occurring boundary conditions are found using eqns [2], [3], and [4]: (i) pinned edge: $h = d^2h/dx^2 = 0$; (ii) fixed edge: $h = dh/dx = 0$; (iii) free edge: $d^2h/dx^2 = d^3h/dx^3 = 0$. Results for some standard configurations follow. Tabulation of the first five natural frequencies are given, together with a sample of selected mode shapes.

1. *Pin–pin*:
   Frequency equation: $\sin \beta_n L = 0 \Rightarrow \beta_n L = n\pi$, $n = 1, 2, \ldots$.
   Mode shapes: $h_n(x) = c_n \sin(n\pi x/L)$
   The c_n are arbitrary scale factors and here and in the sequel are set equal to 1. L is the length of the beam.

2. *Fixed ($x = 0$) – free($x = L$)*:
   Frequency equation: $\cos \beta_n L \cosh \beta_n L = -1$
   Mode shapes:

   $$h_n(x) = \sin \beta_n x - \sinh \beta_n x - \left(\frac{\sin \beta_n L + \sinh \beta_n L}{\cos \beta_n L + \cosh \beta_n L} \right)(\cos \beta_n x - \cosh \beta_n x).$$

See **Table 1**.

3. *Fixed–fixed*:
   Frequency equation: $\cos \beta_n L \cosh \beta_n L = 1$
   Mode shapes:

   $$h_n(x) = \sin \beta_n x - \sinh \beta_n x + \left(\frac{\sin \beta_n L - \sinh \beta_n L}{\cos \beta_n L - \cosh \beta_n L} \right)(\cos \beta_n x - \cosh \beta_n x).$$

See **Table 2**.

4. *Free–Free*:
   Frequency equation: $\cos \beta_n L \cosh \beta_n L = 1$
   This is the same as in the fixed–fixed case. However the mode shapes are different.
   Mode shapes:

   $$h_n(x) = \sin \beta_n x + \sinh \beta_n x + \left(\frac{\sin \beta_n L - \sinh \beta_n L}{\cos \beta_n L - \cosh \beta_n L} \right)(\cos \beta_n x + \cosh \beta_n x)$$

5. *Fixed ($x = 0$) – pin($x = L$)*:
   Frequency equation: $\sin \beta_n L \cosh \beta_n L = \sinh \beta_n L \cos \beta_n L$
   Mode shapes:

   $$h_n(x) = \sin \beta_n x + \left(\frac{\sinh \beta_n L - \sin \beta_n L - \sinh \beta_n x}{\cos \beta_n L - \cosh \beta_n L} \right)(\cos \beta_n x - \cosh \beta_n x).$$

See **Table 3**.

Table 1 Roots for fixed–free case

n	$\beta_n L$	Mode shape
1	1.8751	First mode
2	4.6941	Second mode
3	7.8548	Third mode
4	10.9955	Fourth mode
5	14.1372	Fifth mode

6. *Pin (at $x=0$)–linear spring (at $x=L$; rate k):*
   The boundary conditions at $x=L$ are $d^2h/dx^2 = 0$, $d^3h/dx^3 = (k/EI)h$
   Frequency equation: $(\beta_n L)^3 (\sin \beta_n L \cosh \beta_n L - \cos \beta_n L \sinh \beta_n L) - 2q_k \sin \beta_n L \sinh \beta_n L$
   where q_k, which is essentially the ratio of the spring stiffness and beam stiffness, is given by $q_k = k/(EI/L^3)$. See **Table 4**.
   Mode shapes:

   $$h_n(x) = \sin \beta_n x + \frac{\sin \beta_n L}{\sinh \beta_n L} \sinh \beta_n x$$

   Shown in **Figure 1** is a plot of the lowest root $\beta_1 L$ vs q_k. It is seen that, after $q_k > 50$, $\beta_1 L$ rapidly approaches the pin–pin value: $\beta_1 L = \pi$. For example, for $q_k = 350$, the difference is of the order of 1.4 percent.

7. *Fixed (at $x=0$)–linear spring (at $x=L$):*
   Frequency equation: $(1 + \cos \beta_n L \cosh \beta_n L)(\beta_n L)^3 = (\cos \beta_n L \sinh \beta_n L - \cosh \beta_n L \sin \beta_n L) q_k$

   Mode shapes:

   $$h_n(x) = \sin \beta_n x - \sinh \beta_n x - \left(\frac{\sin \beta_n L + \sinh \beta_n L}{\cos \beta_n L + \cosh \beta_n L}\right)(\cos \beta_n x - \cosh \beta_n x).$$

   See **Table 5**.
   Note that for $q_k = 0.1$ the difference between $\beta_1 L$ and the fixed–free root is only about 0.8 percent.

8. *Pin (at $x=0$) – rotational spring (at $x=L$; rate G):*
   The boundary conditions at $x=L$ are $d^3h/dx^3 = 0$, $d^2h/dx^2 = -(G/EI)\,dh/dx$
   Frequency equation: $(-\cos \beta_n L \sinh \beta_n L + \cosh \beta_n L \sin \beta_n L)(\beta_n L) = 2q_r \cos \beta_n L \cosh \beta_n L$
   where $q_r = G/(EI/L)$
   Mode shapes:

   $$h_n(x) = \sin \beta_n x + \frac{\cos \beta_n L}{\cosh \beta_n L} \sinh \beta_n x.$$

   See **Table 6**.

Table 2 Roots for fixed–fixed case

n	$\beta_n L$	Mode shape
1	4.7300	First mode
2	7.8532	Second mode (zero at 0.5)
3	10.9956	Third mode (zeros at 0.36, 0.64)
4	14.1372	Fourth mode (zeros at 0.28, 0.50, 0.72)
5	17.2788	Fifth mode (zeros at 0.23, 0.41, 0.59, 0.77)

Table 3 Roots for fixed–pin case

n	$\beta_n L$	Mode shape
1	3.9266	First mode
2	7.0686	Second mode (zero at 0.57)
3	10.2102	Third mode (zeros at 0.39, 0.69)
4	13.3518	Fourth mode (zeros at 0.30, 0.53, 0.76)
5	16.4934	Fifth mode (zeros at 0.24, 0.42, 0.61, 0.81)

9. *Fixed (at $x = 0$)–rotational spring (at $x = L$):*
   Frequency equation: $(1 + \cos \beta_n L \cosh \beta_n L)(\beta_n L) = -q_r(\cos \beta_n L \sinh \beta_n L + \cosh \beta_n L)$
   Mode shapes:
   $$h_n(x) = \sin \beta_n x - \sinh \beta_n x + \left(\frac{\cos \beta_n L + \cosh \beta_n L}{\sin \beta_n L - \sinh \beta_n L}\right)(\cos \beta_n x - \cosh \beta_n x)$$

   See **Table 7**.
   Note that for $q_r = 0.1$, the difference between $\beta_1 L$ and the lowest fixed–free root is only about 1.4 percent.

10. *Fixed (at $x = 0$)–mass loaded (at $x = L$; rigid mass m):*
    The boundary conditions at $x = L$ are $d^2 h/dx^2 = 0$, $d^3 h/dx^3 = -(m\omega^2/EI)h$
    Frequency equation: $1 + \cos \beta_n L \cosh \beta_n L = q_m \beta_n L(\cos \beta_n L \sinh \beta_n L - \sin \beta_n L \cosh \beta_n L$
    where q_m, which is the ratio of the external mass to the mass of the beam, is given by $q_m = m/\rho AL$.
    Mode shapes:
    $$h_n(x) = \sin \beta_n x - \sinh \beta_n x - \left(\frac{\sin \beta_n L + \sinh \beta_n L}{\cos \beta_n L + \cosh \beta_n L}\right)(\cos \beta_n x - \cosh \beta_n x).$$

    See **Table 8**.
    If the inertia of the beam is ignored and it is simply treated as a spring with a spring rate $k_e = 3EI/L^3$, then $\omega_1 = (k_e/m)^{1/2}$. This gives $\beta_1 L = (3/q_m)^{1/4}$. A correction to this expression is $\beta_1 L = [3/(q_m + 0.23)]^{1/4}$. The exact values of $\beta_1 L$ and the two approximate values are shown in **Figure 2** as a function of q_m. Note that for $q_m > 2$, the approximate value is very accurate.

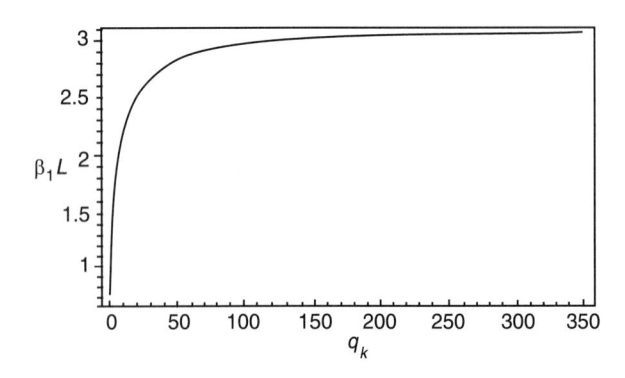

Figure 1 Lowest root vs stiffness ratio.

Table 4 Roots for pin–linear spring case

$q_k = 0.1$		$q_k = 1.0$		$q_k = 10.0$	
n	$\beta_n L$	n	$\beta_n L$	n	$\beta_n L$
1	0.7397	1	1.3098	1	2.2313
2	3.9283	2	3.9432	2	4.0954
3	7.0689	3	7.0714	3	7.0974
4	10.2103	4	10.2111	4	10.2120
5	13.3518	5	13.3522	5	13.3560

Table 5 Roots for fixed–linear spring case

$q_k = 0.1$		$q_k = 1.0$		$q_k = 10.0$	
n	$\beta_n L$	n	$\beta_n L$	n	$\beta_n L$
1	1.8901	1	2.0100	1	2.6389
2	4.6951	2	4.7038	2	4.8757
3	7.8550	3	7.8568	3	7.0974
4	10.9956	4	10.9963	4	11.0031
5	14.1372	5	14.1375	5	14.1407

Table 6 Roots for pin–rotational spring case

$q_r = 0.1$		$q_r = 1.0$		$q_r = 10.0$	
n	$\beta_n L$	n	$\beta_n L$	n	$\beta_n L$
1	0.7314	1	1.1916	1	1.5007
2	3.95132	2	4.1197	2	4.5298
3	7.08250	3	7.1901	3	7.5856
4	10.2199	4	10.2985	4	10.6609
5	13.3592	5	13.4210	5	13.7503

Table 7 Roots for fixed–rotational spring case

$q_r = 0.1$		$q_r = 1.0$		$q_r = 10.0$	
n	$\beta_n L$	n	$\beta_n L$	n	$\beta_n L$
1	1.9022	1	2.0540	1	2.2912
2	4.7156	2	4.8686	2	5.2887
3	7.8673	3	7.9657	3	8.3531
4	11.0045	4	11.0782	4	11.4321
5	14.1442	5	14.2029	5	14.5243

11. *Fixed (at $x = 0$)–rigid rotor (at $x = L$; mass moment of inertia J):*
    The boundary conditions at $x = L$ are $d^2 h/dx^2 = 0$, $d^2 h/dx^2 = (J\omega^2/EI) \, dh/dx$
    Frequency equation: $1 + \cos \beta_n L \cosh \beta_n L = q_J(\beta_n L)^3(\sinh \beta_n L \cos \beta_n L + \sin \beta_n L \cosh \beta_n L)$
    where $q_J + J/\rho AL^3$

Mode shapes:

$$b_n(x) = \sin \beta_n x - \sinh \beta_n x + \left(\frac{\cos \beta_n L + \cosh \beta_n L}{\sin \beta_n L + \sinh \beta_n L}\right)(\cos \beta_n x - \cosh \beta_n x).$$

See **Table 9**.

Forced motions For $f \neq 0$ in eqn [1], the steady-state response can be obtained using modal analysis. With the exception of boundary conditions involving mass and inertia loads, the mode shapes are orthogonal in the sense:

$$\int_0^L b_n(x)b_m(x)\,\mathrm{d}x = \begin{cases} 0, & n \neq m \\ I_n \equiv \int_0^L b_n^2(x)\,\mathrm{d}x, & n = m \end{cases}$$

Using this property, the steady-state response is given by:

$$v(x,t) = \sum_{j=1}^{\infty} \eta_j(t)b_j(x)$$

where the $\eta_j(t)$ are determined from:

$$\ddot{\eta}_j + \omega_j^2 \eta_j = \frac{1}{\rho A I_j} \int_0^L f(x,t)b_j(x)\,\mathrm{d}x$$

the ω_j being the natural frequencies of the system. For example if $f(x,t) = f_0 \sin \omega t$, f_0 constant:

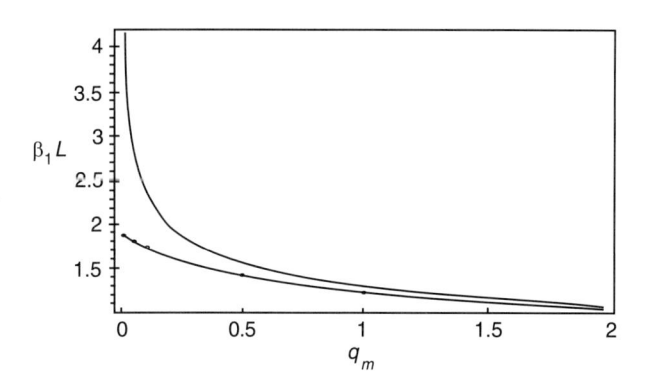

Figure 2 Lowest root vs mass ratio.

$$\ddot{\eta}_j + \omega_j^2 \eta_j = \frac{f_0}{\rho A I_j} \sin \omega t \int_0^L b_j(x)\,\mathrm{d}x \equiv \Theta_j f_0 \sin \omega t$$

Thus:

$$\eta_j = \frac{f_0 \Theta_j}{\omega_j^2 - \omega^2} \sin \omega t$$

and:

$$v(x,t) = f_0 \left\{ \sum_{j=1}^{\infty} \frac{\Theta_j}{\omega_j^2 - \omega^2} b_j(x) \right\} \sin \omega t$$

clearly showing the occurrence of resonances at $\omega = \omega_j$.

Timoshenko Beam Theory

In Timoshenko beam theory, rotary inertia and deflections due to shear are taken into account. The coupled equations of motion for v and the bending slope ψ are (lateral force = 0):

Table 8 Roots for fixed–mass loaded case

$q_m = 0.01$		$q_m = 0.05$		$q_m = 0.1$		$q_m = 0.5$		$q_m = 1$		$q_m = 2$	
n	$\beta_n L$	n	$\beta_n L$	n	$\beta_n L$	n	$\beta_n L$	n	$\beta_n L$	n	$\beta_n L$
1	1.8568	1	1.7912	1	1.7227	1	1.4200	1	1.2479	1	1.0762
2	4.6497	2	4.5127	2	4.3995	2	4.1111	2	4.0311	2	3.9826
3	7.7827	3	7.5863	3	7.4511	3	7.1903	3	7.1341	3	7.1027
4	10.8976	4	10.6609	4	10.5218	4	10.2984	4	10.2566	4	10.2340
5	14.0149	5	13.7503	5	13.6142	5	13.4210	5	13.3878	5	13.3701

$$A\rho \frac{\partial^2 v}{\partial t^2} - A\kappa G\left(\frac{\partial^2 v}{\partial x^2} - \frac{\partial \psi}{\partial x}\right) = 0 \quad [5]$$

$$EI\frac{\partial^2 \psi}{\partial x^2} + A\kappa G\left(\frac{\partial v}{\partial x} - \psi\right) - I\rho\frac{\partial^2 \psi}{\partial t^2} = 0 \quad [6]$$

where G is the shear modulus and κ is the Timoshenko shear coefficient. Vibration solutions to eqns [5] and [6] are of the form: $v = h(x) \sin \omega t$, $\psi = g(x) \sin \omega t$, where:

$$h(x) = c_1 \cos \delta s^2 x/L + c_2 \sin \delta s^2 x/L \\ + c_3 \cosh \gamma s^2 x/L + c_4 \sinh \gamma s^2 x/L \quad [7]$$

$$Lg(x) = c_1 \Gamma_t \sin \delta s^2 x/L - c_2 \Gamma_t \cos \delta s^2 x/L \\ + c_3 \Gamma_h \sinh \gamma s^2 x/L + c_4 \Gamma_h \cosh \gamma s^2 x/L \quad [8]$$

$$\delta, \gamma = \left\{ \pm\frac{1}{2}(s_h^2 + R^2) + \frac{1}{2}\left[(S_h^2 - R^2)^2 + \frac{4}{s^4}\right]^{1/2} \right\}^{1/2},$$

$$s = \beta L, \quad R^2 = \frac{I}{AL^2}, \quad s_h^2 = \frac{EI}{\kappa G A L^2} = \frac{ER^2}{\kappa G}$$

$$\Gamma_t = s^2(s_h^2 - \delta^2)/\delta, \quad \Gamma_h = s^2(s_h^2 + \gamma^2)/\gamma$$

In arriving at the above solutions it has been assumed that:

$$\left[(s_h^2 - R^2)^2 + \frac{4}{s^4}\right]^{1/2} > (s_h^2 + R^2)$$

If this condition (which should be monitored in numerical work) is not true, then the terms involving c_3 and c_4 in eqns [7] and [8] have to be replaced by:

$$c_3 \cos \varepsilon s^2 x/L + c_4 \sin \varepsilon s^2 x/L,$$
$$c_3 \Gamma_n \sin \varepsilon s^2 x/L - c_4 \Gamma_n \cos \varepsilon s^2 x/L$$

where:

$$\varepsilon = \left\{\frac{1}{2}(s_h^2 + R^2) - \frac{1}{2}\left[(s_h^2 - R^2)^2 + \frac{4}{s^4}\right]^{1/2}\right\}^{1/2}$$

$$\Gamma_n = s^2(s_h^2 - \varepsilon^2)/\varepsilon$$

Some common boundary conditions are: (i) pinned edge: $h = \mathrm{d}g/\mathrm{d}x = 0$; (ii) fixed edge: $h = 0, g = 0$; (iii) free edge: $\mathrm{d}g/\mathrm{d}x = 0$, $\mathrm{d}h/\mathrm{d}x - Lg = 0$.

Pin–Pin

- Frequency equation: $\sin \delta s^2 = 0$.
- Mode shapes:

$$h_n(x) = c_n \sin \delta s^2 x/L - \left(\frac{\sin \delta s^2}{\sinh \gamma s^2}\right) \sinh s^2 \gamma x/L$$

Results for the Timoshenko beam are not universal; definite properties and geometries have to be specified. For a solid circular beam of radius a, $R = a/L$ and κ is taken to be 0.75. For steel $E/G = 2(1 + \nu)$, ν being Poisson's ratio, taken to be 0.33. **Figure 3** shows a plot of the ratio of the lowest root for the Timoshenko beam divided by that for the Euler beam as a function of the aspect ratio R. For large L, there is excellent agreement between the two roots. Note that for $R = 0.2$, there is about a 20% drop-off, the Euler root being the smaller.

Table 9 Roots for fixed–inertia loaded case

$q_J = 0.01$		$q_J = 0.05$		$q_J = 0.1$		$q_J = 0.5$		$q_J = 1$		$q_J = 2$	
n	$\beta_n L$	n	$\beta_n L$	n	$\beta_n L$	n	$\beta_n L$	n	$\beta_n L$	n	$\beta_n L$
1	1.8396	1	1.7075	1	1.5771	1	1.1597	1	0.9873	1	0.8357
2	3.7818	2	2.8882	2	2.6482	2	2.4219	2	2.3932	2	2.3791
3	5.8267	3	5.5602	3	5.5286	3	5.5039	3	5.5009	3	5.4993
4	8.7207	4	8.6550	4	8.6472	4	8.6409	4	8.6402	4	8.6398
5	11.8122	5	11.7871	5	11.7840	5	11.7816	5	11.7813	5	11.7811

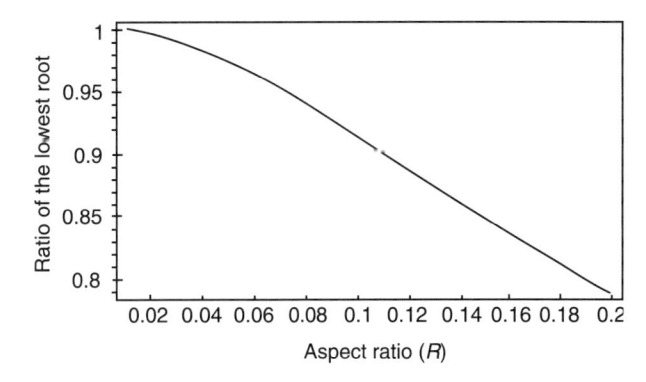

Figure 3 Timoshenko/Euler lowest roots vs aspect ratio.

Nomenclature

a	radius
A	cross-sectional area
E	Young's modulus
G	shear modulus
I	inertia
M	bending moment
R	aspect ratio
v	transfer deflection
V	shear force
ψ	bending slope
ρ	density
κ	Timoshenko shear coefficient

See Plate 8.

See also: Finite element methods; Shock.

Further Reading

Blevins RD (1984) *Formulas for Natural Frequency and Mode Shape*. Malabar, FL: Krieger.

Cook RD, Malkus DS, Plesha ME (1989) *Concepts and Applications of Finite Element Analysis*, 3rd edn. New York: John Wiley.

Harris CM (ed.) (1996) *Shock and Vibration Handbook*, 4th edn. New York: McGraw-Hill.

Huang, TC (1961) The effect of rotatory inertia and of shear deformation on the frequency and normal mode equations of uniform beams with simple end conditions. *Journal of Applied Mechanics, Transactions of the ASME* 579–584.

Zu JW, Han RPS (1992) Natural frequencies and normal modes of a spinning Timoshenko beam with general boundary conditions. *Journal of Applied Mechanics, Transactions of the ASME* 59: 197–204.

BEARING DIAGNOSTICS

C J Li and K McKee, Rensselaer Polytechnic Institute, Troy, NY, USA

Copyright © 2001 Academic Press

doi:10.1006/rwvb.2001.0200

Roller Bearing

Figure 1A illustrates the cross-section of an angular contact ball bearing. It shows the angle of contact, and the diameters of outer race, rollers, inner race and pitch circle. **Figure 1B** illustrates the cross-section of a taper roller bearing. It shows the angle of contact, and the diameters of outer race (cup), inner race (cone), and rollers. The primary excitations of vibration are the contact forces between the rollers and the raceways, and the rollers and the cage. The frequencies of these contacts are directly related to the roller passing frequency on the raceways, the roller spinning frequency, and the cage spinning frequency which can be theoretically estimated with the equations in **Table 1**.

Failure Modes

Bearing failure modes include corrosion, wear, plastic deformation, fatigue, lubrication failures, electrical damage, fracture, and incorrect design. Out of these, classical localized defects result from fatigue, and certain distributed defects have fairly distinct vibration patterns which will be detailed below. However, the vibration patterns of other failure modes are, in general, not easy to predict, let alone detect and diagnose at onset. While some of them do eventually lead to localized defects, others give almost no vibration warning and therefore call for means other than vibration analysis.

Localized Defects

The most common failure mode of a properly installed and operated rolling element bearing is localized defects such as fatigue spall, which occur when a sizable piece of material on the contact surface is dislodged during operation, mostly by fatigue

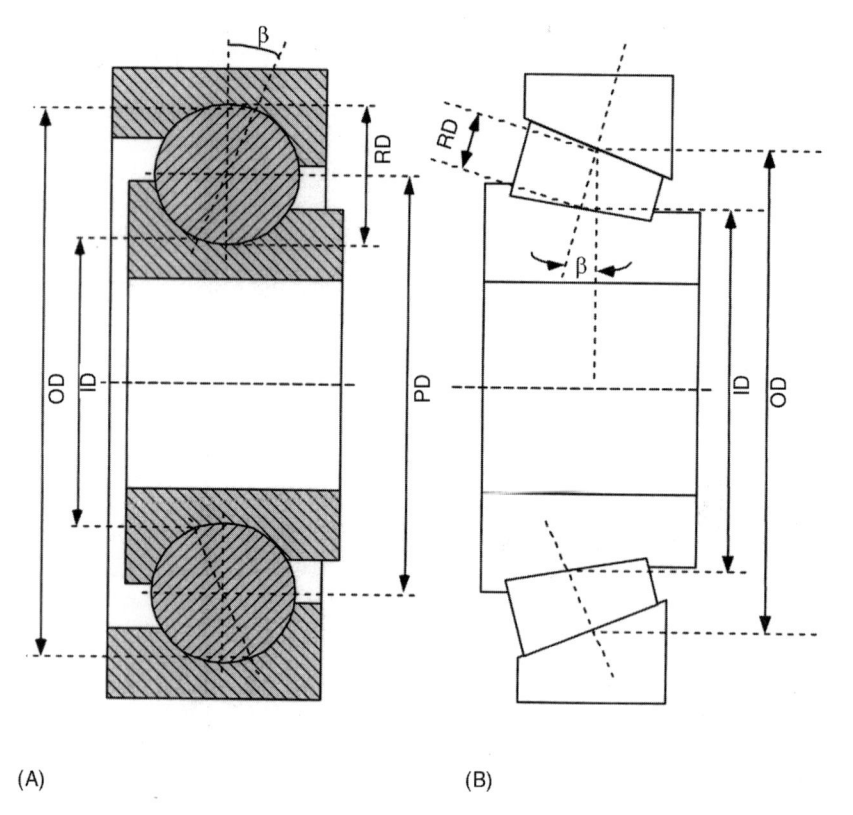

Figure 1 Cross-section. (A) Ball bearing; (B) taper roller bearing.

cracking under cyclic contact stressing. In addition, the bearing cage can fail due to overload or poor lubrication. Thus, a failure alarm for a rolling element bearing is often based on the detection of the onset of localized defects.

Signature-generating Mechanism

During bearing operation, wide-band impulses are generated when rollers pass over a defect at a frequency determined by shaft speed, bearing geometry, and defect location. The frequency, which is often termed characteristic defect frequency, can be estimated as f_{rpi}, f_{rpo} and $2xf_{rs}$ (**Table 1**) for an inner race, outer race and roller defect, respectively. Some of the vibration modes of the bearing and its surrounding structure, or even the sensor, will be excited by the periodic impulses, and a distinct signature will be generated by such ringings. The leading edge of each ringing usually comprises a sharp rise that corresponds to the impact between a roller and the defect. The ringing then decays due to damping. (It is like driving over a pothole.) Therefore, the signature of a damaged bearing consists of exponentially decaying vibration ringings that occur quasiperiodically at the

Table 1 Important bearing frequencies

	Ball bearing	Taper roller bearing
Inner race roller passing frequency (f_{rpi})	$\dfrac{n(f_o - f_i)(1 + (RD/PD)\cos\alpha)}{2}$	$n(f_o - f_i)\dfrac{OD}{OD + ID}$
Outer race roller passing frequency (f_{rpo})	$\dfrac{n(f_o - f_i)(1 - (RD/PD)\cos\alpha)}{2}$	$n(f_o - f_i)\dfrac{ID}{OD + ID}$
Cage rotating frequency (f_c)	$\dfrac{f_i(1 - (RD/PD)\cos\alpha)}{2} + \dfrac{f_o(1 + (RD/PD)\cos\alpha)}{2}$	$\dfrac{f_i \times ID + f_o \times OD}{OD + ID}$
Roller spinning frequency (f_{rs})	$\dfrac{(f_o - f_i)}{2}\dfrac{PD}{RD}(1 - ((RD/PD)\cos\alpha)^2)$	$(f_o - f_i)\dfrac{ID}{RD}\dfrac{OD}{OD + ID}$

n = number of rollers, f_o = outer race rotating speed (rps), f_i = inner race rotating speed (rps), OD = outer race diameter, ID = inner race diameter, RD = roller diameter, PD = pitch circle diameter, α = contact angle.
These equations were derived under the assumption that it is pure rolling contact (no slipping, skidding, etc.) and perfect geometry even under load. One would expect some deviations under actual conditions.

characteristic defect frequency. **Figure 2** shows an example of bearing vibration which has the distinct pattern of periodic ringings. It is sometimes considered as the resonance being amplitude-modulated at the characteristic defect frequency. However, this pattern will begin to fade and become more random as localized defects develop around the races and the rollers. (Then, it is like driving a car on a gravel road.)

The above model does not include many factors such as: radial loading, variation due to transmission path between the signal source and the transducer, misalignment, unbalance, pre-load, and manufacturing defects.

The dominant effect of radial loading is modulation. Assuming a stationary outer race, an inner-race defect rotates around at shaft frequency. Under a radial load, the bearing vibration will be modulated periodically at shaft frequency as the defect goes in and out of the loading zone. For a defect on a roller revolving at the cage frequency, a radial load will result in a modulation at cage frequency. For a defect on the stationary outer race, no modulation will be produced.

Cage fault is different from other localized defects in its signal generating mechanism. It does not usually produce sharp impacts that excite narrow-band bearing resonance. When a cage develops a weak point such as deformation or breakage, a modulation of the wide-band bearing vibration at the cage's rotating speed is common.

Diagnostic Algorithms

Statistical parameters When periodic ringings due to a localized defect become dominant, bearing vibration becomes impulsive. The following statistical parameters have been used as diagnostic parameters for detecting bearing localized defects. For a discrete vibration signal, $x(n)$, let N and $\bar{x}$ denote the length and the mean, respectively. Then:

$$\text{Peak-to-valley} = \max(x(n)) - \min(x(n)) \quad [1]$$

$$\text{RMS} = \sqrt{\left(\frac{1}{N}\sum_{i=1}^{N}(x(i) - \bar{x})^2\right)} \quad [2]$$

$$\text{Crest Factor} = \frac{\text{Peak to valley}}{\text{RMS}} \quad [3]$$

$$\text{Kurtosis} = \frac{\frac{1}{N}\sum_{i=1}^{N}(x(i) - \bar{x})^4}{\text{RMS}^4} \quad [4]$$

probability density function
$$p(x \leq x(n) \leq x + \Delta x) \\ = \frac{\text{No. of } x(n) \text{ between } x \text{ and } (x + \Delta x)}{N} \quad [5]$$

A common drawback of these time domain techniques is that they cannot provide information about the location of a defect because they do not take advantage of the characteristic defect frequencies. While the peak-to-valley value is expected to increase due to the periodic ringings, a single noise spike can throw it off and it is also sensitive to changes in operating conditions such as loading and speed. The root mean square (RMS) is not sensitive at the initial stage of damage because a few short-lived, not very large ringings would not change the overall RMS much and operating conditions also affect it. The crest factor provides a normalized peak-to-valley which should be somewhat less sensitive to changes in operation conditions. Kurtosis' quadruple power emphasizes larger amplitudes in defect-induced ringings and therefore is more sensitive to bearing defects than the RMS. The probability density function is expected to have larger tails for a damaged bearing. However, kurtosis, crest factor, and probability density function will fall back to undamaged levels as localized defects spread around races and rollers.

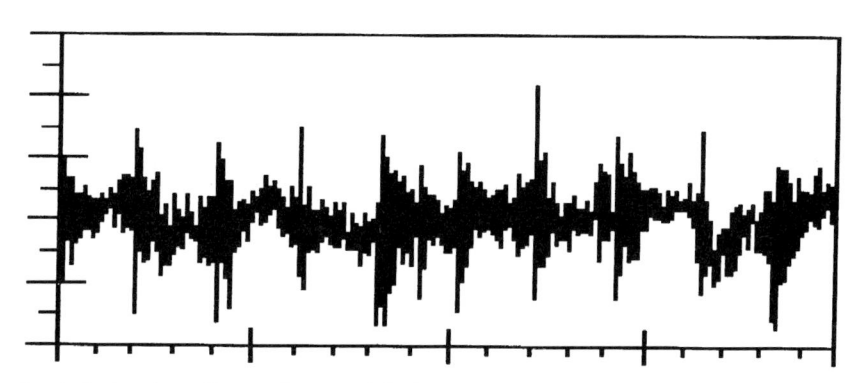

Figure 2 Vibration signal of a bearing with an outer-race defect.

In general, the success of these parameters depends on finding a frequency band that is dominated by bearing ringings. Otherwise, they indicate a change in something other than the condition of the bearing. It was reported that good results can be obtained from higher frequency bands such as 10–20 kHz and 20–40 kHz, where the structure resonance is usual. While it is easy to find some band that works to some extent, the optimal result is difficult to obtain because there is no simple way to predict a bearing structure's natural frequencies, and which one will be excited.

Although trending can be used with these statistical parameters, some of them have been toted as single-shot diagnostic variables that have an absolute threshold and therefore do not need historical data. For example, kurtosis is said to have a value of about 3.0 for a relatively random vibration from an undamaged bearing. Any value significantly higher than 3, say 4.5, is therefore considered as a sign of a bearing fault. Crest factor and probability density function are sometimes considered as single-shot variables. Realistically, these variables are just too simple to offer single-shot diagnosis under all circumstances.

Shock pulse counting This method takes advantage of transducer resonance. When a roller runs over a localized defect, the impact produces a pressure wave. When this pressure wave reaches the transducer, it frequently rings the resonance of the transducer. The number of vibration peaks above a certain threshold during a fixed length of time is then used as a damage indicator. Typically, the transducer is selected to have a resonance frequency between 25 and 40 kHz to avoid the usual machine noise. Due to its simplicity, it is implemented in a couple of off-the-shelf bearing monitoring systems. However, there are few theoretical guidelines about setting the threshold and it has limited use when there are sources of shock pulses other than a roller bearing. Furthermore, as a time domain method, it cannot determine the location of the defect.

High-frequency resonance technique (HFRT) This is probably one of the best known bearing diagnostic algorithms: it is also known as amplitude demodulation, demodulated resonance analysis, and envelope analysis. The technique band-pass filters a bearing signal to remove low-frequency mechanical noise and then estimates the envelope of the filtered signal. (Enveloping can be accomplished with a rectifier followed by a smoothing operation, or a squaring operation followed by an analog or digital low-pass filter. In addition, the Hilbert transform can also be used if a computer is available.)

The periodicity of the envelope signal, like the one shown in **Figure 3**, is then estimated by spectral analysis or autocorrelation, and compared with the characteristic defect frequencies. If a match is found, the bearing is declared to be damaged.

Obviously, this technique will give the best result if the passing band is chosen to include one or more excited resonances. For simple machines, good results can often be obtained with a fixed passing band with low cut-off placed between 5 and 10 kHz and high cut-off between 25 and 40 kHz. For complex machines such as helicopter transmission, selecting a passing band to include resonances and avoid gear meshing harmonics may not be easy. Varying speed, and therefore meshing harmonics, further compounds the problem.

Synchronized averaging This algorithm can be used as a preprocessing technique to enhance the signal-to-noise ratio. The technique consists of first

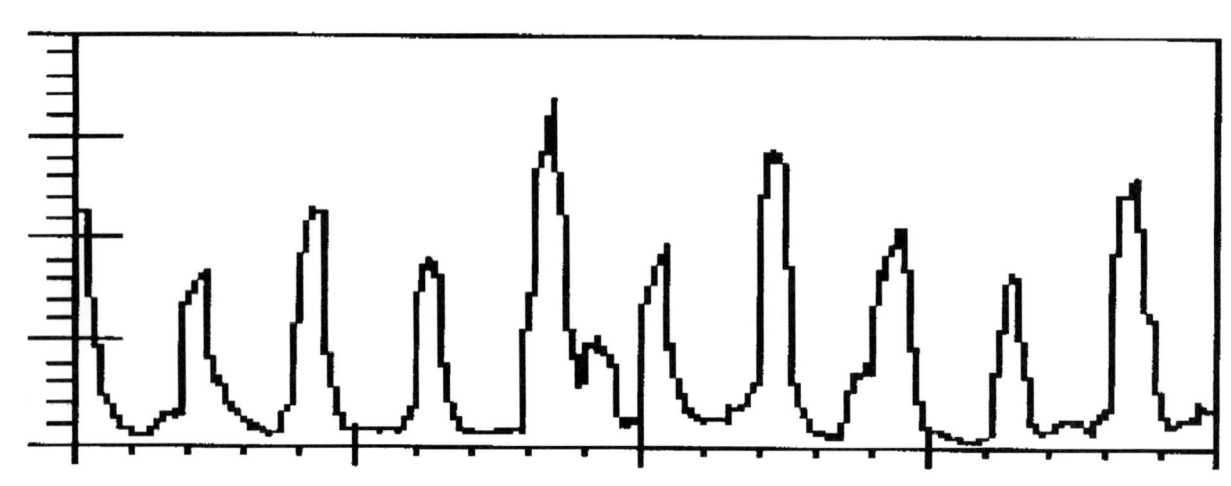

Figure 3 Envelope of the band-pass-filtered signal in **Figure 2**.

calculating the period of repeated ringings by inverting the characteristic defect frequency and then ensemble averaging consecutive segments of a bearing vibration, each one period long. Signals that cannot fit full waves in the period will be attenuated while signals that do will remain. In the original work, only the RMS of the average was calculated. Other diagnostic algorithms can certainly be applied as appropriate.

Cepstral analysis As shown in **Figure 2**, localized defects produce periodic ringings. The spectrum of such a vibration contains a characteristic defect frequency and its harmonics with the larger magnitude around the excited resonance(s). Since the energy of the bearing vibration is spread across a wide frequency band, it can easily be buried by noise. Therefore, spectral analysis will not be effective. The cepstrum, which was initially defined as the 'power spectrum of the logarithmic power spectrum', is useful in detecting spectrum periodicities, such as families of harmonics of bearing defect frequencies, by reducing a whole family of harmonics into a single cepstral line:

$$x_c(\tau) = F\left[\ln|X(\omega)|^2\right] \quad \text{(original definition)}$$

or:

$$= F^{-1}[\ln[X(\omega)]] \text{ where } X(\omega) = F(x(t))$$

[6]

Cepstral analysis has been shown to be very effective in bearing diagnosis. However, it has not received much acceptance, perhaps because it is more difficult to compute and interpret than the output of, say, the HFRT.

Wavelet transform For a continuous signal $x(t)$, the wavelet transform (WT) is defined as:

$$W_x(a,b) = \int g^{*(a,b)}(t)x(t)\,dt$$

[7]

where * denotes the complex conjugate, and $g(t)$ represents the mother wavelet, e.g.:

$$g(t) = \exp(-\sigma t)\sin(\omega_0 t) \quad \text{for } t = 0$$
and: $\quad g(t) = -g(-t) \quad \text{for } t < 0$

$$g^{(a,b)}(t) = \frac{1}{\sqrt{a}}g\left(\frac{t-b}{a}\right)$$

[8]

where a is the dilation parameter which defines a baby wavelet for a given value, and b is the shifting parameter.

For a given a, carrying out WT over a range of b is like passing the signal through a filter whose impulse response is defined by the baby wavelet. Therefore, one may consider WT as a bank of band-pass filters defined by a number of a's. The salient characteristic of the WT is that the width of the passing band of the filters is frequency-dependent. Therefore, the WT can provide a high-frequency resolution at the low-frequency end while maintaining good time localization at the high-frequency end. This is advantageous for processing transient bearing ringings.

When applied to a bearing signal as a preprocessing tool, the passing band of one or more of the filters could overlap with some of the resonances that are being excited by the roller defect impacts. This results in an enhanced signal-to-noise ratio. For example, **Figure 4** shows a bearing vibration measured from a roller-damaged bearing. (Note that it is already high-pass-filtered.) The periodic ringings are not obvious. **Figure 5** is the result of WT with one of the baby wavelets. The periodic ringings can be seen more readily and therefore easily identified by, say, an envelope analysis.

By breaking up a broad-band bearing signal into a number of narrow-band subsignals and then scanning them for evidence of bearing defect, WT avoids the risk of selecting a wrong band that does not include any resonance and then missing the defect. The price is that one has to repeat the same bearing diagnostic algorithm on more than one subsignal.

Time–frequency distribution for bearing monitoring The periodic ringings can be revealed or accentuated with time–frequency distributions which give an account of how energy distribution over frequencies changes from one instant to the next. Examples of such distributions include the spectrogram (short-time Fourier transform), Wigner–Ville distribution, and Choi–Williams distribution (CWD).

Figure 6 is a CWD of a bearing signal. Horizontal stripes correspond to the periodic ringing due to a ball passing on an outer-race defect. It is clear that the resonance occurs at 10, 18, and 22 kHz. When compared to WT, which has a better time localization of the ringing, CWD gives a better frequency resolution at high frequencies at a higher computational cost.

The drawback of time–frequency distributions is the medium-to-high computational cost and two-dimensional image-like output, which is usually troublesome for a computer to interpret.

Bicoherence As shown in **Figure 2**, localized defects produce repetitive ringings. The spectrum of such a vibration contains a characteristic defect frequency and its harmonics, with the largest magnitude around

148 BEARING DIAGNOSTICS

Figure 4 Bearing vibration with inner-race defect.

Figure 5 Wavelet transform of the vibration in **Figure 4**.

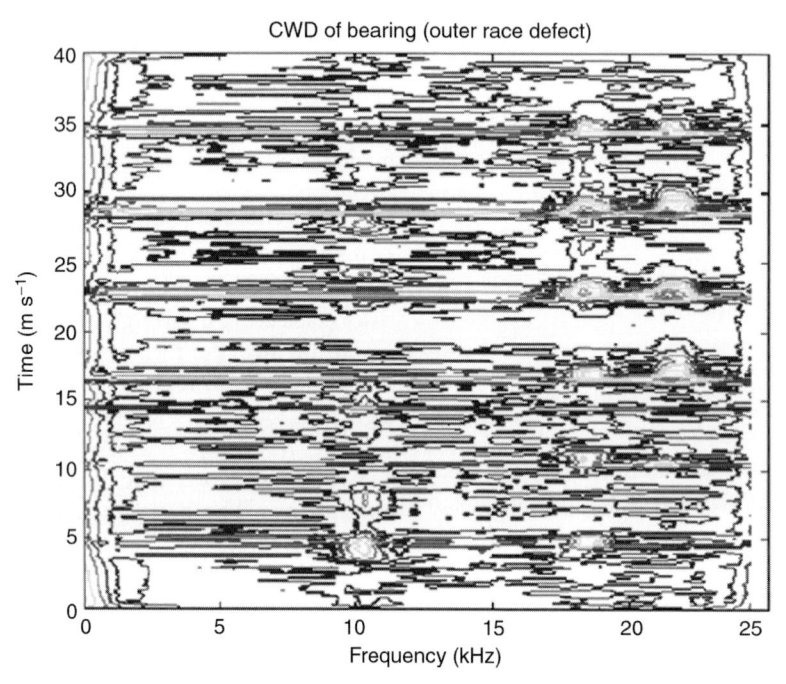

Figure 6 Choi–Williams distribution of bearing vibration with outer-race defect.

the excited natural frequencies. These harmonics are the Fourier components of a periodic signal, and therefore have a fixed phase relationship with one another.

Bicoherence, defined below, quantifies the phase dependency among the harmonics of the characteristic defect frequency, which increases when a bearing is damaged:

$$b^2(f_1, f_2) = \frac{|E[X(f_1)X(f_2)X^*(f_1+f_2)]|^2}{E\left[|X(f_1)X(f_2)|^2\right] E\left[|X(f_1+f_2)|^2\right]} \quad [9]$$

where $E[\ ]$ denotes the expected value, and $X(f_i)$ denotes the complex frequency component from the Fourier transform. In this case, f_i would be one of the harmonics of the characteristic defect frequency. For example, f_1 and f_2 can be chosen as f_{rpi} and $2f_{rpi}$.

Because bicoherence is based on phase, it works even when the ringings are less than dominant. For example, **Figure 7** shows bearing vibration signals from a good bearing and another with a damaged roller. Although one cannot see the repeated ringing patterns in the bad bearing signal, bicoherence plots can tell them apart (**Figure 8**). While bicoherence outperfoms traditional algorithms such as HFRT, it

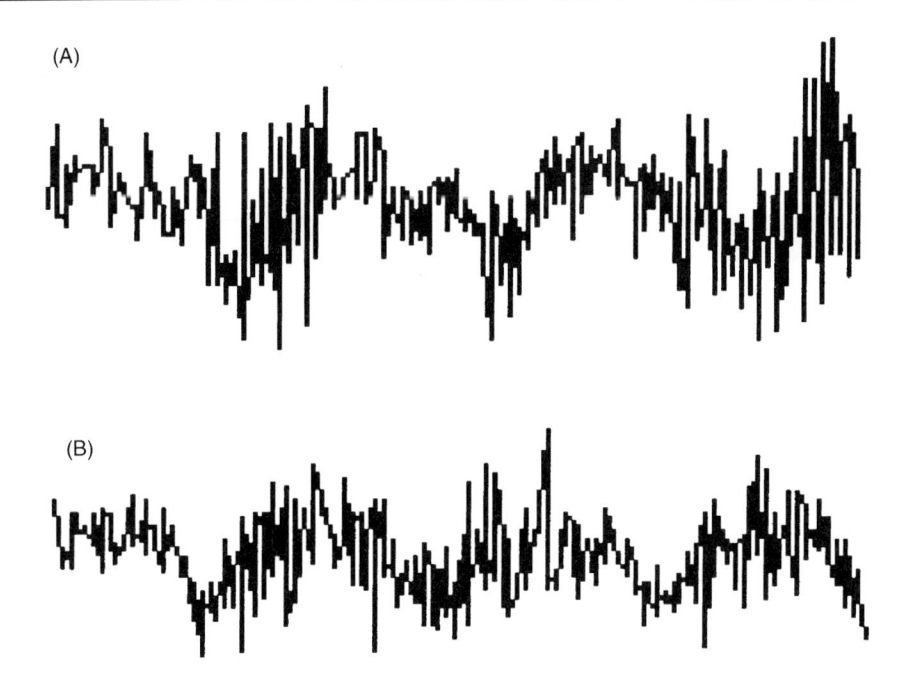

Figure 7 Bearing vibrations. (A) Good bearing signal; (B) bad bearing signal.

is one of the most expensive algorithms in terms of computation cost and data length.

Cage fault When a cage develops a weak point such as deformation or breakage, a modulation of the wide-band bearing vibration at the cage's rotating frequency is common. Traditional bearing diagnostic algorithms such as narrow-band envelope analysis are not effective in detecting this kind of wide-band modulation. One way to determine if a modulation is happening at the cage rotating frequency is to perform a synchronized average, at the cage rotating period, on the envelope of the bearing vibration. The average is then examined for once-per-cage-rotation phenomena. Alternatively, bandwidth-weighted demodulation can be used.

Summary of localized defect diagnostic algorithms

A significant number of the afore-mentioned algorithms look for the distinct pattern of repeated ringings (**Figure 2**) associated with the onset and the early stage of damage. Consequently, they suffer from the limitation that they will fall back to the undamaged levels as localized defects spread and the distinct pattern fades away as rollers strike different local defects almost simultaneously all the time. In other words, they do not trend well while damage evolves and this is a problem for the prognosis of bearing life. In addition, since a resonance excited at the onset of a defect may no longer be excited at a later stage, using a band-pass filter to zoom on to a resonance can have its downside of missing another defect-excited resonance later.

Distributed Defects

In addition to localized defects, distributed defects such as off-size rollers and waviness of components are another class of bearing problems. While distributed defects are not generally failures *per se*, they often lead to excessive contact forces and vibrations which in turn lead to premature failures. Fortunately, analysis of vibration spectra is often adequate to diagnose this kind of faults.

Figure 8 Bicoherence values of good bearing and roller-damaged bearings at four frequency pairs.

A bearing without any distributed defect is considered first. Under a thrust load, roller forces act on the races to produce a deflection. **Figure 9** shows how roller forces act on the inner race. As the rollers and races rotate, the forces and the deflection become periodic. Thus, a vibration sensor located at the inner race would sense vibrations at f_{rpi} and its harmonics. Similarly, a sensor at the outer race would sense vibration at f_{rpo} and its harmonics. **Table 2** gives the frequency associated with a number of conditions while the inner race is stationary.

Table 2 Frequencies associated with various distributed defects

Conditions	Frequencies (sensor at inner race)
One roller off size (k: 1, 2, ... etc.)	$k(f_{rpi}/n)$
Misaligned outer race	Side band $k(f_{rpi}/n) \pm f_o$
Waviness in outer race (m = number of full waves around the circumference)	Side band $k(f_{rpi}/n) \pm mf_o$

From Meyer et al. (1980).

Journal Bearing

Figure 10 shows the three basic parts to a journal bearing: the outer housing, the journal, and the lubricant. The outer housing is a cylindrical shaft, which contains a hollow core large enough to create a close fit between itself and the journal. In addition to confining the path of the journal's orbit, it provides radial support to the journal through direct contact or by aiding in the creation of the oil wedge. A small clearance space between the outer housing and the journal is necessary: in order to assist in the assembly of the journal and the bearing, to provide space for the addition of the lubricant, to accommodate thermal expansion of the journal, and to anticipate any journal misalignment. Within this small clearance space resides the lubricant which provides the basic function of lubricating the journal and outer housing contact, as well as producing the load-carrying capability of the journal bearing and possibly attenuating the vibrations of the rotors. This load-carrying capability is a result of the pressure that is developed by the viscous effects within the thin film lubricant. A more complex journal bearing that is commonly used is the tilt-pad journal bearing. This type of journal bearing contains the same components as the simple bearing, with the added feature of tilt pads. Tilt pad journal bearings are used for their innate ability better to handle rotor-dynamic instability problems; however, they provide less damping than the simple bearing. Consequently, if vibrations occur due to other sources aside from the bearing, then the amount of damping provided by the bearing may arise as an issue.

Diagnostic Algorithms

Unlike roller bearings, whose vibration has simple and distinct patterns that can mostly be predicted from the geometry, journal bearings, while simple-looking, are rather complicated in their dynamics. Since journal bearings are only one part of a rotating machine, the analysis of the bearings must take into consideration vibrations caused by the flexible journal, fluid and rotor dynamics, and any outside sources of vibration acting on the system. Consequently, this type of analysis usually becomes complicated, and results in no clear vibration patterns associated with most of the failure modes. This is why the tools for the diagnosis of a journal bearing stagger are lacking.

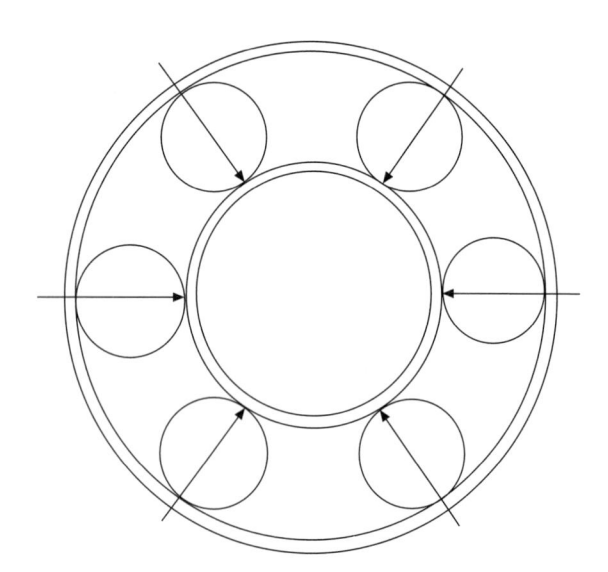

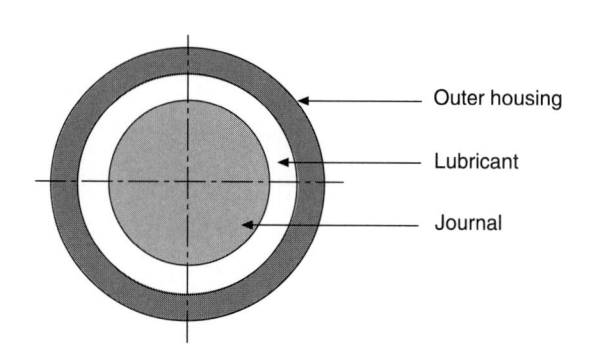

Figure 9 Roller forces acting on inner race.

Figure 10 Simple journal bearing.

'Whirl' refers to a circulating path the journal takes within the bearing resulting from large vibrations of the journal. Proximity sensors can be placed at perpendicular positions to the orbit of the shaft. These sensors would allow the monitoring of the position of the orbit, and notify the user of the misalignment, the amount of clearance existing, and the range of motion of the journal. The user would be warned if the motion of the journal exceeds a certain allowable envelope of motion.

An accelerometer can be placed on the outer housing to measure the frequency and amplitude of the housing vibrations which do not have a straightforward relationship with journal whirl. Based on the vibrations taken from the bearing's normal conditions, abnormal conditions can be detected. For example, in cases where the bearing's housing is supported on springs rather than secured directly to an immobile object, chaotic motion is found at intermediate speed ranges, and disappears at low and high speeds. At low and high speeds, different distinct subharmonic frequency components in the X- and Y-direction of the bearing are excited. However, during intermediate speeds, there is a rich spectrum of excited frequencies in both directions, which results in vibrations with comparatively large amplitudes that may induce fatigue failure.

Misalignment occurs when the journal's centerline does not coincide with the bearing's centerline. This is caused by combinations of rotational movements about a pivot point in the longitudinal cross-section and translational movements of the journal in the vertical and horizontal axis in the radial cross-section. Misalignment results from assembling or manufacturing errors, off-centric loads, shaft deflection such as elastic and thermal distortions, and externally imposed misaligned moments. A side-effect of the misalignment is the creation of a converging wedge geometry, known as the oil wedge, between the journal and the outer housing. In addition, misalignment contributes to the whirl by changing the threshold speed at which instability occurs. It is able to change the journal bearing's load-carrying capability, increase its frictional power loss, alter the fluid film thickness, change the dynamic characteristics such as system damping and critical speeds, and modify the vibrations as well as the overall stability of the system. One of the most significant effects of misalignment is its ability to produce a substantial amount of vibration when the frequency of the rotor's vibrations is a harmonic of the rotational speed of the journal.

Hot spots found on a journal, which ultimately develop into thermal bends, are results of the Newkirk effect. These hot spots are quite common, resulting from the contact of the rotor with the bearing or due to a temperature difference across the diameter of the journal. In the latter case, the temperature difference is a result of the differential shearing in the oil film. This phenomenon, in conjunction with the system running near a critical speed of the journal, can generate unstable vibrations in the bearing.

Under normal, low journal velocity conditions, the journal resides at an equilibrium position that is determined by its velocity. However, as the speed increases and approaches a threshold speed of instability, the journal's stability becomes compromised. Speeds above the threshold speed cause a self-excited oscillation to occur, during which the whirling motion of the journal is increased by its own rotational energy. This is dangerous if the oscillations are of large magnitude. If the journal suddenly becomes unstable, this is called subcritical bifurcation. On the other hand, if the journal gradually becomes unstable, it is called supercritical bifurcation. Subcritical bifurcation is also possible under the threshold speed when the rotor is given small perturbations by an outside force. Consequently, factors such as constant and imbalance loads on the journal are considered important when preventing bifurcation.

Vibrations can also occur from the lack of an oil wedge. Oil wedges, which are responsible for the load-carrying capability of the journal bearing, can be prevented from forming if the load is too heavy, the journal speed is too slow, or there is a lack of lubrication. In all three cases, metal-to-metal contact occurs, thus causing vibrations in the bearing.

Seizure, which results in a complete halt of the journal's movement, is a serious common problem. This mode of failure can result from the lack of an oil wedge (dry rubbing) which leads to highly localized heating, inadequate heat release from the system, and thermal expansion of the journal. In the third case, the journal may thermally expand faster than the bearing housing, causing the clearance between the journal and the bearing to disappear and metal-to-metal contact to occur. In the case of tilt pad bearings, thermal expansion of the tilt pads can produce the same phenomenon.

See also: **Balancing; Diagnostics and condition monitoring, basic concepts**.

Further Reading

Arumugam P, Swarnamani S and Prabhu BS (1997) Effects of journal misalignment on the performance characteristics of three-lobe bearings. *Wear* 206: 122–129.

Braun S and Datner B (1979) Analysis of roller/ball bearing vibration. *Journal of Mechanical Design* 101: 118–125.

Chen CL and Yau HT (1998) Chaos in the imbalance response of a flexible rotor supported by oil film bearings with non-linear suspension. *Nonlinear Dynamics* 16: 71–90.

de Jongh FM and Morton PG (1996) The synchronous instability of a compressor rotor due to bearing journal differential heating. *Journal of Engineering for Gas Turbines and Power–Transactions of the ASME* 118: 816–823.

Deepak JC and Noah ST (1998) Experimental verification of subcritical whirl bifurcation of a rotor supported on a fluid film bearing. *Journal of Tribology – Transactions of the ASME* 120: 605–609.

Elkholy AH and Elshakweer A (1995) Experimental analysis of journal bearings. *Journal of Engineering for Gas Turbines and Power–Transactions of the ASME* 117: 589–592.

Howard I (1994) *A Review of Rolling Element Bearing Vibration 'Detection, Diagnosis, and Prognosis'*. DSTO-RR-0013. Melbourne, Australia: DSTO Aeronautical and Maritime Research Laboratory.

Jang JY, Khonsari MM and Pascovici, MD (1998) Thermohydrodynamic seizure: Experimental and theoretical analysis. *Journal of Tribology–Transactions of the ASME* 120: 8–15.

Li CJ and Ma J (1992) Bearing localized defect detection through wavelet decomposition of vibrations. In: Pusey H and Pusey S (eds), *Proceedings of 46th General Meeting of Mechanical Failures Prevention Group*, 7–9 April, Virginia Beach, Virginia, pp. 53–62. Willowbrook, IL: Vibration Institute.

Li CJ, Ma J, Hwang B and Nickerson GW (1991) Pattern recognition based bicoherence analysis of vibrations for bearing condition monitoring. In: Liu TI, Meng CH and Chao NH (eds), *Proceedings of Symposium on Sensors, Controls and Quality Issues in Manufacturing*, ASME Winter Annual Meeting, 1–6 December, Atlanta, GA, PED – vol. 55, pp. 1–11. New York: ASME.

Limmer JD and Li CJ (1997) A case study in cage fault detection using bandwidth-weighted demodulation. In Burroughs CB (ed.) *Proceedings of NOISE-CON 97*, 15–17 June, State-College, PA, pp. 189–194. Poughkeepsie, New York: Noise Control Foundation.

Meyer LD, Ahlgren FF and Weichgrodt B (1980) An analytic model for ball bearing vibrations to predict vibrations response to distributed defects. *Journal of Mechanical Design* 102; 205–210.

Monmousseau P, Fillon M and Frene J (1998) Transient thermoelastohydrodynamic study of tilting-pad journal bearings – application to bearing seizure. *Journal of Tribology – Transactions of the ASME* 120: 319–324.

Su YT and Lin SJ (1992) On initial fault detection of a tapered roller bearing: frequency domain analysis. *Journal of Sound and Vibration* 155(1): 75–84.

White MF and Chan SH (1992) The subsynchronous dynamic behavior of tilting-pad journal bearings. *ASME Journal of Tribology* 114: 167–173.

BEARING VIBRATIONS

R Bigret, Drancy, France

Copyright © 2001 Academic Press

doi:10.1006/rwvb.2001.0092

Introduction

Rotors in rotating machinery are guided by links. They are classified into three categories:

1. Fluid
2. Roller bearings
3. Magnetic field

Forces are transmitted to nonrotating parts. They are a result of associated impedance:

- To links
- To sealing systems
- To elements connected to the rotor, such as wheels.

Sealing systems and elements connected to the rotor play a part in guiding the rotor. Coupling between rotors falls into the link category. This article describes the characteristics of links.

Functions

Links must allow:

- Fixed-precision guidance to avoid rotor–stator contact
- Minimum wear and tear to reduce machine unavailability
- Removal of heat, in particular from the rotor
- Minimization of leaks
- Neutralization or removal of impurities
- Minimum loss through friction, for economy and low-level heating

In 1987, in the USA, the American Society of Mechanical Engineering (ASME) estimated that 11% of consumption was lost through friction; in 1996 France it was valued at 1.6% of gross national product.

Through their impedance, links contribute to the dynamic behavior of rotating machines, their stability and vibratory levels, in particular at critical speed. Their optimization complements that of rotors and stators, their function and performance.

Fluid Links

Fluid links offer contacts which are dry or hydrodynamically lubricated. In 1902 Richard Stribech (1861–1950) proposed a function which makes it possible to distinguish four zones within a bearing (**Figure 1**). Zone 1 falls within the province of tribology. The term 'tribology' was invented in 1968 and encompasses the study of friction, wear, lubrication, and the design of bearings.

Contacts may be:

- Surface: friction is oily; pressure is 1–1000 bar. This is the sphere of bearings, thrust, and sealing systems.
- Hertzian: surfaces are separated by films formed in a chemical reaction between materials and additives contained in the fluid (the oil). This process is elastohydrodynamic. The pressure is 1000–25 000 bar. This is the sphere of gears and roller bearings.
- Dry, nonlubricated: there may be microsoldering and particles may become loose, which reduces friction.

In zone 2, a hydrodynamic effect separates the surfaces which remain in contact at their rough patches. In zone 3, a hydrodynamic laminary film, in which the fluid trickles are parallel, separates the surfaces when:

$$\Re = \frac{\rho V}{\mu} J < 1000$$

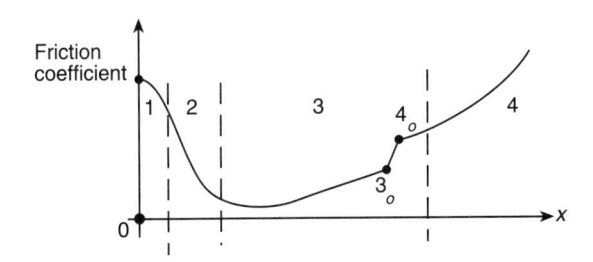

Figure 1 Stribech curve. x, film thickness; $\mu\Omega/F$ where $\mu =$ viscosity; $\Omega =$ rotation speed; $F =$ force.

where $\Re =$ average Reynolds number, $\rho =$ fluid volume mass; $V =$ speed, ΩR, where $\Omega =$ rotation speed and $R =$ radius; $\mu =$ fluid viscosity; and $J =$ radius clearance. (Reynolds (1842–1912) was an English engineer who defined the principles of lubrification in 1884–86 and coefficients of similarity in 1883.)

In zone 4, the flow is turbulent hydrodynamic with $\Re > 2000$, even 1000. Between 3_0 and 4_0, so-called Taylor rotors may be seen in the direction of the bearing axis. This instability is due to radial centrifugal forces of inertia which are related to tangential forces of viscosity.

Hydrodynamic laminary lubrication is currently used in rotating machines. Lubrication limits in zones 1 and 2 occur when the machine is started up and stopped.

At the time of construction of the pyramid of Kheops in Egypt (around 2700 years BC), the Egyptians did not know about the wheel. They used to pull blocks of stone on ramps tilted at most at 9°; these ramps were made of bricks lubricated by silt from the Nile. Some 2400 years BC, this was how the statue of Ti was moved. In Mesopotamia, 2500 years BC, the palace doors of the sultans of the Ottoman empire would turn on *crapaudines*, using steam pressure. In China around the year 900, iron bearings were used, and in the Middle Ages iron and bronze axles and bearings were developed. Leonardo da Vinci (1452–1519) used an alloy comprising 30% copper and 70% tin. Today antifriction bearings contains 80% tin. Research continued with the Frenchman Amontons, who in 1699 published *On Resistance Caused in Machines*; Euler (1707–83), who was Swiss, distinguished between static and dynamic friction, while Coulomb (in 1782) discussed simple machines and proposed the friction coefficients which are used today.

Bearings

Fluid bearings – which are often oil bearings – are useful and necessary in the basic rotating machinery used in industry. They allow:

- Optimization of impedance to the rotor for stability; passage of critical speed and normal function with vibrations and force transmitted to low-amplitude bearings
- Adaptation to moderate misalignment
- The possibility to remove calories by fluid, of which a small part is used for lubrication

Precautions must be taken to:

- Avoid impurity and debris by filtering
- Insure constant temperature of the fluid

154 BEARING VIBRATIONS

- Avoid wear and tear at low speed when the machine starts up and stops when the film is not formed
- Insure emergency supply

The cylindrical bearing shown in **Figure 2** enables to define the elements which permit characterization of normal performance. The rotor section turns around C. The surface of the rotor section is cylindrical and parallel to the surface of the bearing. A thin fluid film is formed between the rotor section and the bearing. In general, the surface is affected by geometrical faults, which require precautions for machine monitoring and diagnostics. Colinearity $\overrightarrow{C_o C_m}$ results from the static form: $\overrightarrow{W}$ possibly results from rotating force $\overrightarrow{W_t}$. If $|\overrightarrow{W_t}|$ varies over one revolution, the path of C is influenced by the variation in C_m.

If $\overrightarrow{C_o C_m}$ is constant and if $h(\theta, t)$ is included in the limits which permit linear performance, the path of the center C of the whirl is an ellipse. C_m is fixed. The

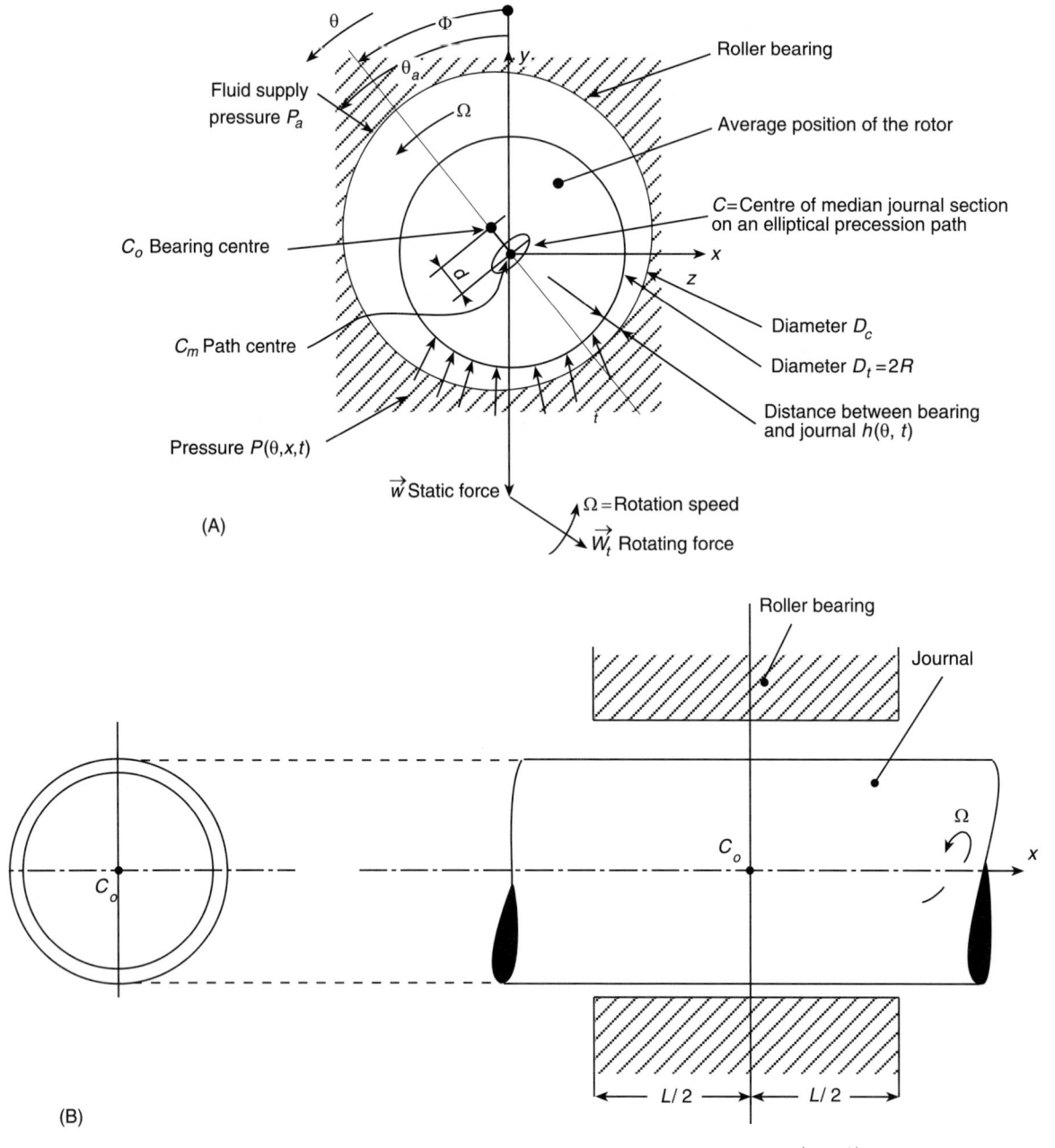

Figure 2 Rotor bearing–journal, where t = time, L = bearing length; unbalance of the axis: $|\overrightarrow{C_o C_m}| = d$; Radius clearance: $D_c/2 - R = J$; $\overrightarrow{W}$ and $\overrightarrow{W_t}$ are applied to C.

minimum value of $h(\theta, t)$ depends on the nature of the materials of the rotor and the bearing. It may be close to 30 μm, or may even be less. The field of pressure (p, θ, x, t) balances the forces applied on C.

Supply is generally between 1.5, and 6 bars absolute, in a weak pressure zone opposed to its maximum [average $p(\theta, x)$]. In this zone, when d is significant $(d/J > 0.5)$, theoretical pressure is negative; practically, there exists cavitation and pressure is equal to the pressure of saturating steam, which may be ≈ 0.4 bar at 30°C; when $d = 0$, there is no cavitation but the arrangement of rotor and bearing may become unstable. An increase in supply pressure reduces the cavitation zone.

When $\Re < 1000$, the angle ϕ and $|\overrightarrow{C_oC_m}|$ which characterize the average position, the stiffness k_{ij} and damping coefficient f_{ij} (either positive or negative) expressed by $K_{ij} = k_{ij}(J/W), F_{ij} = f_{ij}(J/W)\Omega$ $(i = x, z; j = y, z)$ and characterize dynamics, depending on $L/2R$ and Summerfeld number:

$$S = \frac{\mu \Omega R L}{\pi W}\left(\frac{R}{J}\right)^2 \qquad [1]$$

where μ is the viscosity and the Reynolds number $\Re = \rho \Omega R J/\mu$ must be less than 1000. When $\Re > 1000$ laminary, the flow becomes turbulent. When $\Re > 2000$, it is extremely turbulent.

Stiffness and damping coefficients which form the impedance in a radial plane are obtained by the Reynolds equation:

$$\frac{\partial}{\partial \theta}\left[(1 + \varepsilon \cos \theta)^3 \frac{\partial p}{\partial \theta}\right] + R^2 \frac{\delta}{\delta x}\left[(1 + \varepsilon \cos \theta)^3 \frac{\partial p}{\partial x}\right]$$
$$= -6\mu \left(\frac{R}{J}\right)^2 \left[\left(\Omega - 2\frac{d\phi}{dt}\right)\varepsilon \sin \theta - 2\frac{d\varepsilon}{dt}\cos \theta\right] \qquad [2]$$

where $h = J(1 + \varepsilon \cos \theta); \varepsilon = d/J$.

This equation is developed via the Navier equations with the following hypotheses:

- Constant viscosity
- Negligible mass of the fluid
- Incompressible fluid
- Constant pressure over the thickness of the film
- Speed of the fluid in contact with the journal equal to ΩR, zero in contact with the bearing

Many types of boundary conditions may be assumed:

- That of Gümbel
$p(\theta, x) = 0$

where $\pi < \theta - \theta_a < 2\pi$
$p(\theta = \theta_a, x) = p_a = 1$ bar absolute
- That of Reynolds, which is more often used:
$\delta p/\delta x = \delta p/\delta z = 0$ and $p = p_a$ when $z = z_s \rightarrow$ function of x
$p(\theta = \theta a, x) = p_a$; $p = p_a$ for supply, $p = 0$ at $x = \pm L/2$

The solution to eqn [2] may be obtained analytically for large L (= length of bearing) is $(\partial p/\partial x = 0)$ and small L is $(\partial p/\partial \theta = 0)$. This solution was obtained in 1952 by Ocvirk for the so-called Ocvirk bearing where $L/2R \leq 0.5$.

This is more easily expressed by the method of finite elements which, balancing energy, allows viscosity corresponding to temperature to be used for each element. A more complete formulation makes it possible to take into account the moment and rotation around the axes C_{my} and C_{mz} (**Figure 2**); the link is represented by eight stiffnesses and eight damping coefficients.

Bearing and Corner

A bearing made up of a thin fluid film tolerates significant charges. The same applies to a fluid corner. **Figure 3** shows a fluid corner and a bearing composed of a partial arc. For the corner, the Reynolds equation leads to:

$$W = \frac{6\mu V l L_1^2}{J_2^2(r-1)^2}\left(\ln r - 2\frac{r-1}{r+1}\right)$$

For the partial bearing, d, θ, stiffness k_{ij}, and the damping coefficients f_{ij} are defined by the Sommerfeld number S. Where $\mu = 0.1$ Pa s (for water at 20°C, $\mu = 0.001$); $V = 31.4$ m s^{-1} ($R = 0.1$ m; $\Omega = 314$ rad s$^{-1} \rightarrow 3000$ rpm); $J_2 = 0.1833$ mm; $r = 2.2$ (for W_{maxi}; $J_1 = 0.40326$ mm); $L_1 = 209$ mm; $l = 153$ mm ($l/L_1 = 0.732$; $S = 320$ cm^2); $\beta = 10^{-3}$ rad (0.06 deg); $\alpha = 100$ deg; $J = 0.2$ mm; $L = 2R = 0.2$ m ($L/2R = 1$) Length $2\pi R 100/360 = 174$ mm. Force W is close to 10^5 N. For the corner, $\Re \approx 92$. Average pressure $\approx 3.13 \times 10^6$ Pa. Power lost through viscosity is defined by: force × speed $= \mu \Omega R \omega$ surface $\times \Omega R \approx 17\,200$ W. Average pressure $\approx 3.13 \times 10^6$ Pa.

For the partial arc, $\Re \approx 63$; $S = 0.5$. Average pressure on the projected section (between 1 and 2) $\approx 3.5 \times 10^6$ Pa; $d \approx 0.02$ mm ($d/J = 0.1$), $\theta \approx 55$ deg; $d_f = 0.18$ mm ($d_f/J = 0.9$) where $W = 10^5$ N. The force W is significant, if the film is thin.

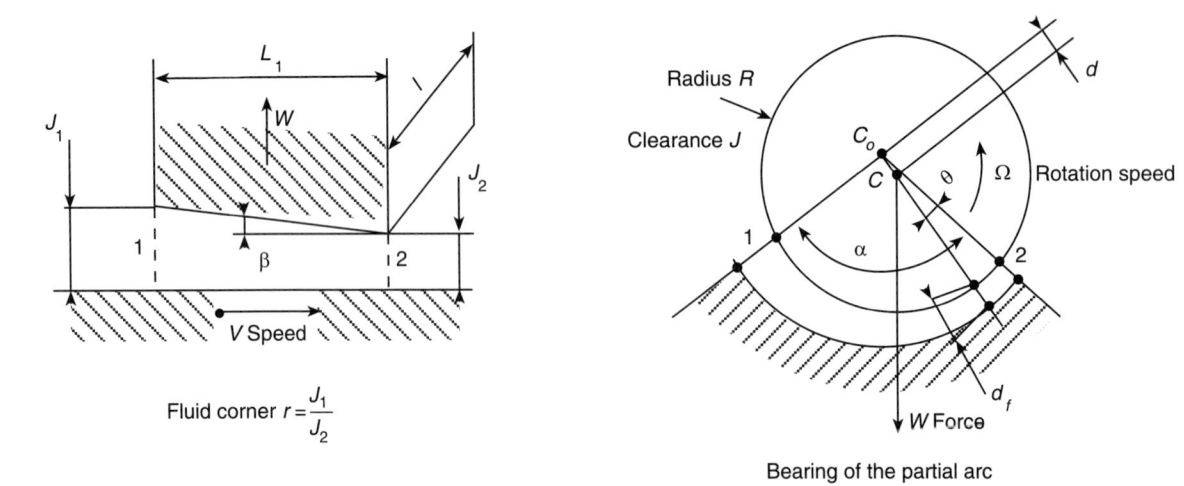

Figure 3 Bearing corner. Pressure in 1 = pressure in 2 (atmospheric pressure). Fluid viscosity, μ. Disregarded marginal leak.

Diversity of Bearings

Figure 4 shows forms used in industry: at fixed geometry (with two, three, or four lobes) with oscillating runners (one, two, three, four, five runners). When the mass of the runners is negligible, there is no static force:

$$k_{xy} = k_{yx} = f_{xy} = f_{yx} = 0$$

This property facilitates the achievement of stability which is discussed in **Rotor dynamics**. The oscillating runner was introduced in the USA by Kinsbury in 1905.

The minimum recommended radial clearance J_m is a function of speed and diameter D: for 60 rpm, $D = 710$ mm, $J_m = 0.16$; $2J_m/D \to 0.04\%$. For 40 000 rpm, $D = 25$; $J_m = 0.04$; $2J_m/D \to 0.16\%$. Temperature can be measured by thermocouples placed in the neighborhood of contact surfaces of the bearings and runners.

In linear performance, forces and displacements are linked by a transfer matrix, displacement impedance, and isochrome for frequency ω:

$$\begin{bmatrix} F_y \\ F_z \end{bmatrix} = \begin{bmatrix} Z_{yy} & Z_{yz} \\ Z_{zy} & Z_{zz} \end{bmatrix} \begin{bmatrix} y \\ z \end{bmatrix} \to \mathbf{F} = \mathbf{ZD} \qquad [3]$$

The terms used in eqn [3] are complex members: Z depends on ω and on rotation speed Ω. When the velocities are substituted for the displacements, the square matrix is one of impedance: its inverse is an admittance, or mobility. The isotropy is expressed by:

$$Z_{yy} = Z_{zy} \quad Z_{yz} = -Z_{zy}$$

The term Z_{00} depends on the viscosity ($\to$ temperature) and on the vector of colinearity $\overrightarrow{C_o C_m}$ (**Figure 2**) affected by misalignment (see **Rotator stator interactions**). When variations in $\overrightarrow{C_o C_m}$ are significant, the

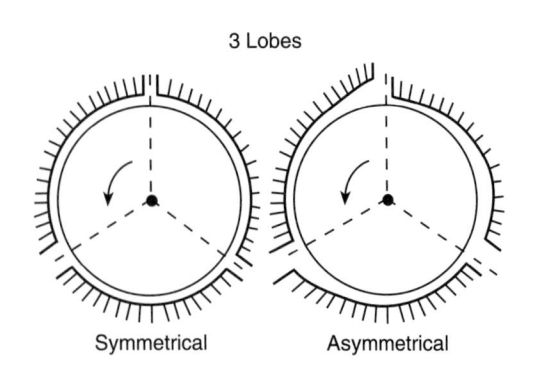

Figure 4 Bearings at fixed geometry, with oscillating runners.

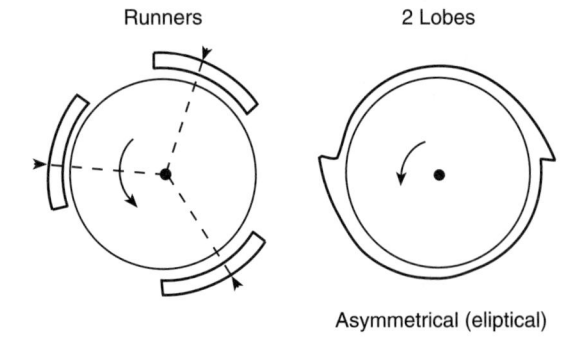

term Z_{00} causes a nonlinear behavior which increases the number of spectral components.

Specificities

Links play a large part in achieving a significant margin of stability ($\Omega_{max} \ll \Omega_{limit}$) at low-level vibration amplitudes and critical speed.

If the bearing material is correctly chosen, even with a lack of oil, then the amount of damage may be limited.

Guidance of the rotors and their vertical support may be obtained hydrostatically by bearings and thrust, in which fluid and oil are introduced under pressure; these devices enable the machine to operate at very slow rotational speed. In particular, they are used to insure, without wear and tear, slow rotor rotation before and after their use in turbomachines submitted to high temperatures.

The experimental determination of stiffness and damping coefficients of bearings and thrust requires special installations which permit to control excitation forces for constant rotation speed.

Thrust with fluid enables axial guidance of the rotors, classically for hydraulic turbines on a vertical axis and for turbomachines. Thrust utilizing fluid or oil corners formed by 'tilt pads' rotation on a pellet fixed between them are handled by the blocks at fixed geometry or by oscillating runners, which were first introduced into Australia by Michell in 1905 (**Figure 5**). Thrust may bring two fixed pellets, partly rotating pellets. Axial force may reach 5×10^6 N where $D \approx 1800$ mm and axial clearance ≈ 0.7 mm.

Roller Bearings

In Malta, which was inhabited from the end of the fifth millennium BC, stone rollers enabled stone blocks to be moved 5 km when megaliths were being constructed. The Egyptians were not aware of the wheel, but were able to use wooden spheres ($\phi \approx 75$ mm) to move stones for the pyramids, as well as statues. In Central America from the fourth to the 15th century the Mayans used logs to move heavy objects. A roller bearing comprising eight marbles was drawn by Leonardo da Vinci in 1501–2. Suriray gave it its current configuration in 1870: this is an example of art preceeding manufacturing. This then became technical in 1940 and then in 1960 was associated with lubrication and heat removal.

Roller bearings using ball bearings, rollers, and needles are shown in **Figure 6**. Manufacturing tolerance may be close to 1 μm. The impedance to the rotor is large. Special arrangements permit an angular distances between the rotor axis and the axis of the outer cage. Displacements and mislalignments between bearings induce radial forces which may lead to damage.

Roller Bearings

Between rotors guided by roller bearings, the use of flexible couplings avoids damage. Damping arises because of elastomer or rubber rings between the outer cage and the bearing, or through a fluid film of oil supplied by a pump. This arrangement requires specific and precise mechanical adaptations which must not affect machine reliability. The coefficients damping and stiffness are combined to form an impedance matrix.

Certain roller bearings require lubrication by oil or the flow of a fluid, which may also remove heat generated by the rotor.

The lifespan of a rolling bearing depends on:

- Rotation speed
- Static force
- Dynamic or rotating force
- The environment: temperature and displacement

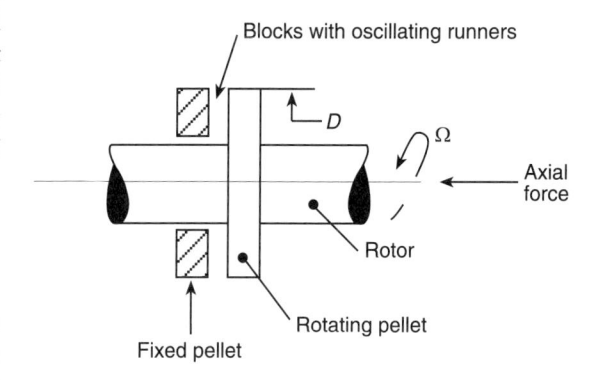

Figure 5 Thrust.

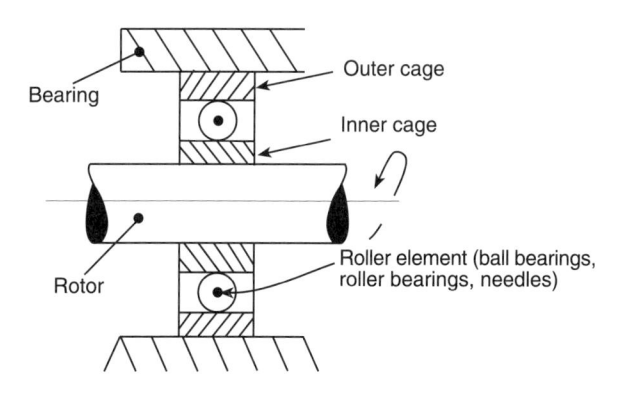

Figure 6 Roller bearings.

Uncertainties concerning dynamic force may be important. The reduction of dynamic force by balancing increases the lifespan. For example, the lifespan as determined by a statistic, drawn from experimental studies, may be defined for ball bearings by:

$$D = \frac{a}{60} \frac{10^6}{N} \left(\frac{C}{P}\right)^3$$

where a is the constant related to reliability; to material, and to the environment; N = speed (rpm); C = dynamic charge; P = equivalent dynamic charge determined according to operating conditions. Balancing which, for a bearing at 3600 rpm (ϕ_{ext} = 190 mm; ϕ_{int} = 75) reduces C to 1/5C, leads to an increase in lifespan from 230 to 430 days. The stiffness of a ball bearing is defined by:

$$K_r \left(\frac{\text{da N}}{\text{mm}}\right) = \frac{W}{\delta_r} = \frac{\cos \alpha}{2.5 \times 10^{-3}} (WD_b)^{1/3}$$

where W = radial force (da N); D_b = diameter of the ball bearings; α = contact angle. Stiffness results from Hertzian theory where there are fewer deformations at the elastic limits; K_r is not independent of W, where $W = 5 \times 10^4$ N; D_b = 10 mm (outer diameter $\approx$ 110 mm; inner diameter $\approx$ 60 mm); $\alpha = 0$; $K_r \approx 1.5 \times 10^8$ (N m^{-1}). Speed must be less than 10 000 rpm, depending on the desired lifespan. The stiffness of a rolling bearing is greater than that of a ball bearing.

The performance of a roller bearing is defined by its N dm (rpm × average diameter in mm), which is close to 1.5×10^6; at a maximum, it is 3×10^6, which may be 46 mm at 65 000 rpm. The outer ring diameter may be several meters: it was 4 m for the drill head for the Channel Tunnel. Roller bearings make it possible to handle thrust.

Sensors may be installed in roller bearings to monitor their state of vibration, to measure useful rotation speed for the Antilocking System (ABS), which avoids wheel locking. A car uses 50 roller bearings; a high-speed train uses 500, a plane 800, and a paper machine 2000.

Aeronautical motors use numerous roller bearings which permit low-speed rotation without causing damage, and which leave a considerable time between their wear, tear (when they are being monitored) and their destruction and which time they are quickly and easily replaced.

Magnetic Links

Figure 7 shows the relations between mechanical, magnetic, and electric phenomena. Magnetic flow and battery voltage are coupled to the mechanical. Magnetoelectric phenomena falls within this category.

Figure 8 shows the principle of the radial lift of a rotor submitted to force F (gravity or others). When I is constant and e is small:

$$F_m = \frac{B^2 S}{2\mu_0} \quad [4]$$

where B is the magnetic induction in the air gap of length e (in Tesla); S = active surface (penetration of flow in the rotor); μ_0 = permeability of space = 1.25×10^{-6} (system: meter, kg, second, Tesla).

When the permeability of rotor and stator, within a magnetic field, is very great in relation to μ_0:

$$B \approx \frac{NI\mu_0}{2e} \quad [5]$$

where N = number of electromagnetic spikes and I = current in amps.

From eqns [4] and [5], it follows that:

$$F_m = \frac{(NI)^2 S}{8e^2} \mu_0; \quad \frac{dF_m}{de} = \frac{-(NI)^2 S \mu_0}{4e^3} = \frac{-2F_m}{e} de$$

When e decreases ($de \to \Delta e < 0$), F_m increases ($dF_m \to \Delta F_m > 0$): the system is intrinsically unstable. This instability is a result of a negative stiffness of the order of -5×10^7 N m^{-1}. Stability is obtained by a loop which takes its information from a sensitve sensor at $\Delta e_m = \Delta e$; a secondary loop uses magnetic flow. With sheet metal (iron silicium 3%), B_{max} = 1.5 T; S = 1 cm^2; F_m = 9 da N; where $e = 0.5$ mm, $(dF_m/de) = 360$ N^{-1}.

The pressure is equal to 90 da N cm^{-2} $\to 9 \times 10^5$ Pa. The pressure is small relative to the pressure which can be achieved with fluids. NI = 600 A rev^{-1} where I = 20 A (2 < I < 40 at 300 V); N = 30 rev. When F_m = 100 000 N, using eqn [4] S_i = 1100 cm^2. With the arrangement shown in **Figure 9**, $S = \frac{4}{6}(\pi D/4) l_m k$ where $k = 0.9$; ($\to$ curvature) and l_m: breadth, where D = 0.2 m (partial arc bearing, l_m = 1.17 m).

The classical arrangement of a magnetic circuit (containing rotor and stator) is shown in **Figure 10**. Magnetic forces generated as per the principle in **Figure 8** oppose static and rotating forces such that

BEARING VIBRATIONS 159

Magnetism
Magnesia stone

Compass (Gilbert, 1600)

Attraction–repulsion ⟶ **Flux**

Medicine: Wave of life (≈1800)

Electricity
Lightning, thunder

Friction of transparent bodies

Emanation of bituminous and resinous bodies

Battery (Volta, 1800)

Voltage

Current

Flux ϕ

Mechanical
Motor (Davenport, 1832–50)

Force (air, water, steam)

Flux variation $\frac{d\phi}{dt}$
(Faraday, 1831)

Voltage (induction)

Current

Magnetic force
(Oersted 1819–20:
Biot–Savart–Laplace)

Generator
(magnetic)
(Piwi, 1832)
Mechanical

Voltage

Current

Mechanical
Iron (ferromagnetic)

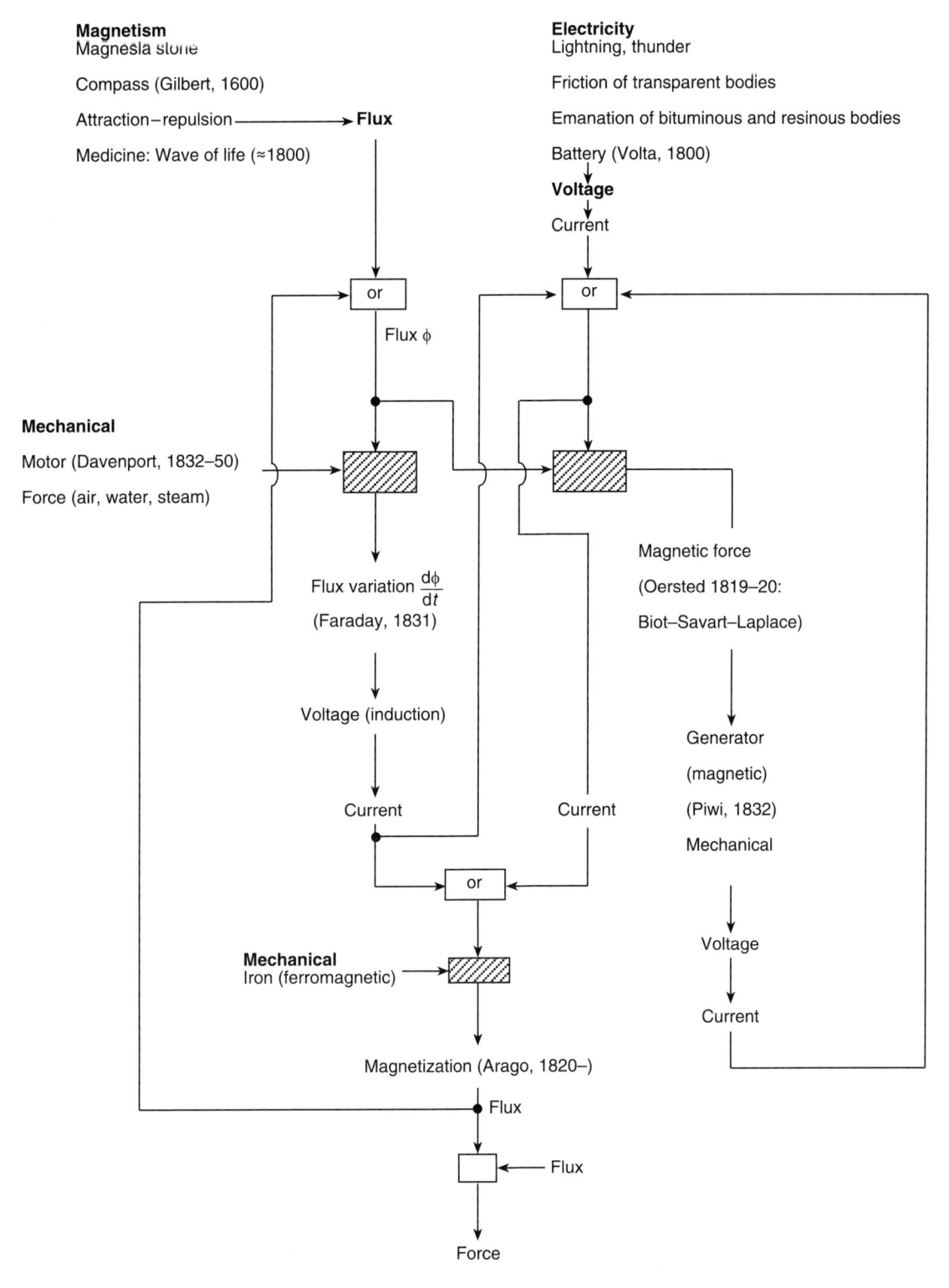

Magnetization (Arago, 1820–)

Flux

Flux

Force

Mechanical: electromagnetic (1820–)

Figure 7 Magnetic–electric–mechanical synergy.

the displacement relative to center C (**Figure 8**) is slight; the maximum force (static and dynamic) is close to $F_m(\text{N}) \approx \alpha D\,(\text{cm})\,l_m\,(\text{cm})$; $\alpha = 40$ for silicium sheet metal, 80 for cobalt sheet metal; $0.3 \leq e \leq 0.6$ mm where $D \leq 100$ mm; $0.6 < e \leq 2$ when $100 < D \leq 1000$; $N_{\text{maxi}} = 180\,000$ rpm.

Speed is limited to $200\,\text{m s}^{-1}$ (if diameter is 127 mm at 30 000 rpm then it is 1270 mm at 3000 rpm) by constraints in the rotor ferromagnetic sheet metal, which are less at the elastic limits. $F_{m\text{maxi}} \approx 100\,000$ N. The peripheral speed V of a disk of constant thickness is related to the constraint σ at the center (tangential = radial) by the following relation: $V^2 = [\gamma/(3+v)](\sigma/\rho)$ where $v = 0.3$: Poisson coefficient and ρ: volume mass. When $V = 200\,\text{m s}^{-1}$, $\rho = 7800\,\text{kg m}^{-3}$, constraint σ is equal to $1.29 \times 10^8\,\text{N m}^{-2}$. This value is less than the elastic limit.

Stability and performance which permit operation beyond critical speed at low-level vibrations are obtained by the corrective circuit and secondary loop (**Figure 8**). Interactions between the circuits associated with each of the electromagnets are possible, even between the circuits associated with n machine bearings, where $n = 2, 3, \ldots$. For horizontal rotors, when the rotating forces are less than the force of gravity, the number of magnetic forces may be less than four.

When rotors are supported and guided by links, fluid, or roller bearings, static magnetic forces may be zero. Losses result from joule losses in coils. Losses by Foucault currents depend on the circumferential speed of permeability and thickness e_t of sheet metal of the foliated circuits (loss $= \beta(l_t)^2$ where $e_t = 0.1 - 1$ mm). Losses by hysteresis are slight in comparison to loss by Foucault current.

The power, P, which is lost by Foucault current is in proportion to the force F_m:

$$\frac{P}{F_m}\left(\frac{\text{W}}{\text{N}}\right) \approx 5 \times \frac{e_t^2(\text{m})}{h(\text{m})}[\pi D(\text{m})f_r(\text{Hz})]^2 \lambda$$

where $f_r =$ rotation frequency; $3 \leq \lambda \leq 4$ obtained through tests; where $e_t = 5 \times 10^{-4}(\text{m})$; $h = 5 \times 10^{-2}(\text{m})$; $fr = 50\,(\text{Hz})$; $\lambda = 3.5$; $D = 0.2$ m; $P/F_m = 8.63 \times 10^{-2}$ (W/N) when $F_m = 10^5$ N; $P \approx 8630$ (W).

Magnetic sensors are arranged in radial planes, close to the electromagnetic planes. They are supplied by high-frequency tension, modulated by variations in l_m (**Figure 8**). Their sensitivity is in the range of 20 and 50 mV μm^{-1}. Signals developed by captors at $180°$ (a couple) are treated by a Wheatstone bridge assembly. Couples at $60°$ and others, may be used. These arrangements enable to eliminate certain imperfections of the measuring track.

The canceling out of vibration at frequency f_a of the magnetic stator circuit – and thus of the bearing on which it is fixed – can be obtained by canceling, at f_a, the corrective circuit module and function of transfer (**Figure 8**). Consequently, rotor vibration at f_a is a result of excitatory forces at f_a.

Landing devices avoid destruction caused by failure of the magnetic link; in general, they are comprised of ball bearings mounted on the rotor or the stator. The radial clearance is close to half of magnetic clearance; this is useful when assembling, disassembling.

Magnetic links permit axial guidance by thrust and magnetic axial bearing. It uses an identical principle to that of radial guidance. The landing arrangement comprises ball bearings or roller bearings. The max-

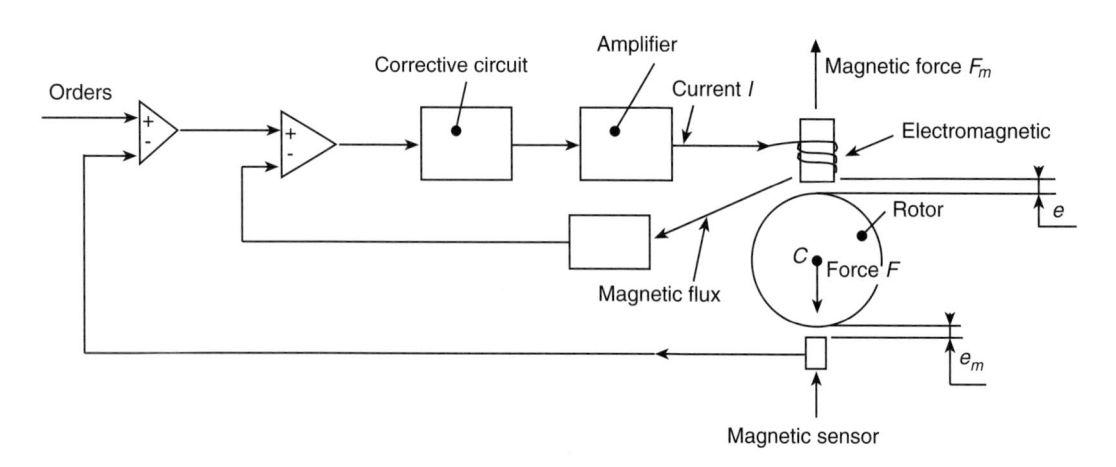

Figure 8 Principles of rotor magnetic lift.

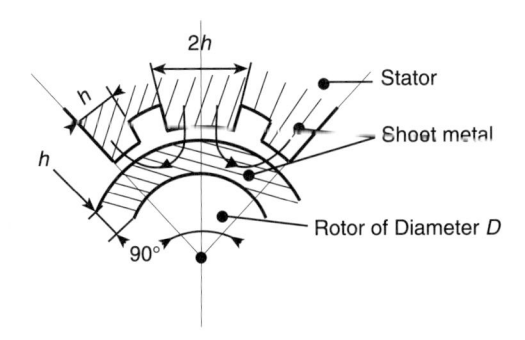

Figure 9 Magnetic circuits.

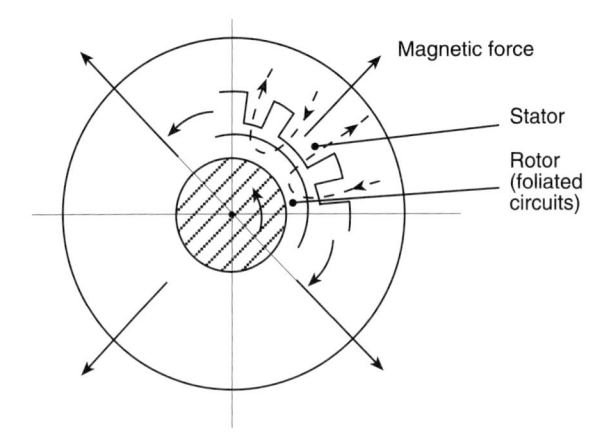

Figure 10 Magnetic circuits.

imum speed is close to 350 m s^{-1}; the maximum force is close to 350 000 N.

To monitor vibratory components, signals delivered by the sensors are processed and compared to the limits; the temperature may also be measured (see **Rotating machinery monitoring**).

Magnetic links are used in gyroscopes, machine tools, cryogenic turbopumps, compressors, vacuum pumps, and centrifuges. They may be associated with fluid links or bearings to optimize dynamic characteristics. This results in an increase in the margin of stability, and a reduction of amplitude at resonance.

As an example, for aluminum manufacturing an electric pin at 40 000 rpm permits 1200 cm^3 of turnings per minute and very slight faults on the manufactured surface. This result is obtained by optimizing the transfer function of the servo-control in order to eliminate spectral components caused by tool impact.

Comparison

Table 1 enables to compare different properties of two fluid links and a magnetic link. Variations are possible to obtain optima which would take into account other constraints.

For the corner shown in **Figure 3**, the force W expressed as a function of $r = J_1/J_2$ has a maximum for $r = 2.2$. This calculation is adopted here. As a result, $\Delta W/\Delta J_1 \approx 0$ where $r > 2.2$: $\Delta W/\Delta J_1 < 0$ and $r < 2.2$: $\Delta W/\Delta J_1 > 0$.

Sealing Systems

Leaks reduce the output and the machine performance. They may pollute the environment. Harmful leaks are essentially seen between rotor and stator at the rotor edges and perpendicular to the disk, the bladed motor or receptive wheels. **Figure 11** shows four arrangements.

The clearance J must be minimum, but sufficient to avoid contact between rotor and stator undergoing displacements and relative vibration. In the arrangement 1, contact is tolerable, to the detriment of an increase in clearance. The contact may lead to local rotor overheating, as a result of flexural deformation which may increase vibration.

Table 1 Comparison of radial links at charge force: 10^5 N, speed: $V = 31.4$ m s^{-1}

	Corner (**Figure 3**)	*Fluid* *Partial arc* (**Figure 3**)	*Magnetic* (**Figures 8** and **9**)
Rotation speed (rpm)	0	3000	Whatever ($V < 200$ m s^{-1})
Viscosity (Pa s)	0.1	0.1	
Clearance (mm)	0.29 (average)	0.2 (to radius)	0.5 (to radius)
Supporting surface (cm^2)	320	288 (projected 307)	1100
Length (mm)	209	174 ($\phi = 200$)	105 ($\phi = 200$)
Width (mm)	153	200	1170
Stiffness (N m^{-2})	≈ 0	1.5×10^9	3.6×10^5
Loss (W)	$\approx 10\,760$ viscosity	$\approx 17\,200$ viscosity	$\approx 8630 - 3000$ rpm Foucault currents

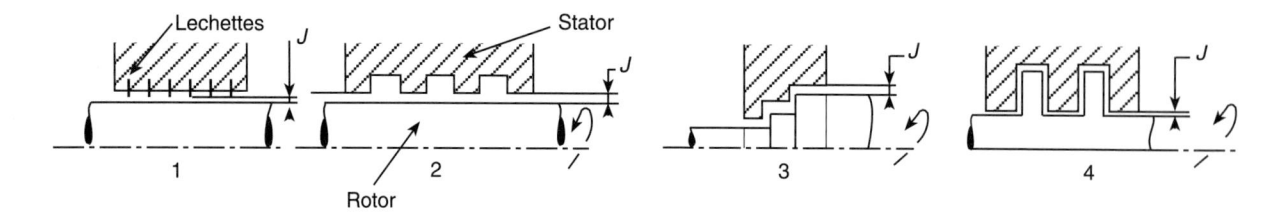

Figure 11 Sealing systems.

Figure 12 shows a mobile element device, with seals. When the slight clearance J is cancelled, the seal is displaced radially without harm. Oil films between seals and rotor insure sealing. **Figure 13** shows *lechettes* (so-called in France) used on disks.

Sealing systems are characterized by impedance matrices which may strongly influence the static and dynamic performance, and may cancel the stabilizing effect of the bearings. Such situations can be seen on boiler supply pumps which force back the flow of water at several hundred bars, and on the turbopumps of satellite launchers.

Since the study of Lomakin in 1957, research was undertaken in how to realize impedance matrices. For a blindfold steam turbine disk, the following matrix relation is assumed:

$$\begin{bmatrix} f_y \\ f_z \end{bmatrix} = \begin{bmatrix} 0 & \dfrac{0.6T}{DH} \\ \dfrac{-0.6T}{DH} & 0 \end{bmatrix} \begin{bmatrix} y \\ z \end{bmatrix}$$

where f_y, f_z are components of force in two perpendicular directions; y, z are displacement components; T = disk couple; D = average blade diameter; and H = blade height.

For a pump with radial diffusion:

$$\begin{bmatrix} f_y \\ f_z \end{bmatrix} = \frac{R}{\rho A V^2} \begin{bmatrix} -2 & 0.7 \\ -0.7 & 2 \end{bmatrix} \begin{bmatrix} y \\ z \end{bmatrix}$$

where R is the outer wheel radius; l is the fluid volume mass; A is the section of fluid exit; $V = \Omega R$, where Ω is the rotation speed. The terms of the second diagonal which are of equal and opposing sign express a tangential force which reduces the margin of stability.

Figure 12 Sealing system with lagging.

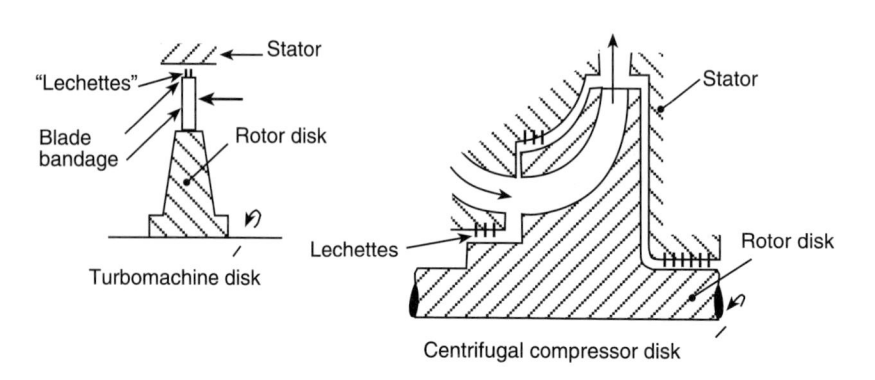

Figure 13 Sealing systems on disks.

Coupling

Couplings create a link between two rotors. The coupling comprises two elements (disks and coupling element) fixed on rotors and by one of the link elements: membrane, disk, blades, cable, cogs, elastomers, universal joints, magnetic field, fluid. Couplings:

- facilitate rotor construction as well as rotor assembly and transport
- enable to fix limits of furniture from different constructors: electric motor, pump
- can enable relative displacement between bearings, without significant variation of forces perpendicular to bearings; in this regard roller bearings are generally not very tolerant (see **Rotator stator interactions**)
- enable decoupling of rotors without displacing them
- enable organization of devices for damping torsion vibrations
- can reduce the transmission of axial forces

Couplings use various technological arrangements which makes them flexible (supple) or rigid. Flexible couplings are used for low-level power: for example, 3 MW at 40 000 rpm with cogs. An acceptable transmitted couple depends on the rotation speed, for example, 5 MW at 600 rpm (80 000 da N m^{-1}), 2.5 MW at 950 rpm. Rigid couplings are used for high power: 10 MW at 3000 rpm, 1500 MW at 1500 rpm.

Coupling may be permanent – the more classic type – or temporary (such as a clutch).

Coupling may be affected by faults which modify the vibratory performance, the transient and spectral signal forms, which may then detect them.

Coaxiality faults between rigid coupling plateaux are a result of wear and tear or slippage between disks caused by significant transient torque. They induce rotating forces at the rotation speed.

Deterioration and significant linear misalignments (coaxiality fault) and angular misalignments result in spectral components of frequency $(p/q)\Omega$, where p and q are integers, and Ω is rotation speed.

Example: Disc Contact of a Flexible Coupling

A flexible coupling is shown schematically in **Figure 14**. The amplitude of the fundamental component of horizontal vibration of a bearing containing a ball bearing is shown in **Figure 15**.

Around critical speed (≈ 26.2 Hz) of the initial state, the slope, $dA/d\Omega$, is great when $d\Omega/dt > 0$; $dA/d\Omega$ is normal when $d\Omega/dt < 0$. After balancing, the slopes are normal, hysteresis subsists.

The amplitude of the fundamental component of relative vertical vibration close to a bearing next to the coupling is shown in **Figure 16**. Significant slopes $dA/d\Omega$ and hysteresis appear. This amplitude is shown in **Figure 17** where $d = 3$ mm: the maximum amplitude is reduced (from $\approx 700\ \mu$m to $\approx 300\ \mu$m), critical speed is reduced (from ≈ 26 Hz to ≈ 2520 Hz), and the slope is normal. The anomaly, which balancing blurs, disappears on structuring d from 1 to 3 mm to suppress nonlinearity.

The solutions to the Duffing equation with damping, presented in the article on **Rotor–stator interactions** compared to the measures shown in **Figures 15** and **16** support the hypothesis of nonlinearity in elastic forces.

See also: **Rotating machinery, essential features; Rotating machinery, monitoring; Rotor dynamics; Rotor stator interactions**.

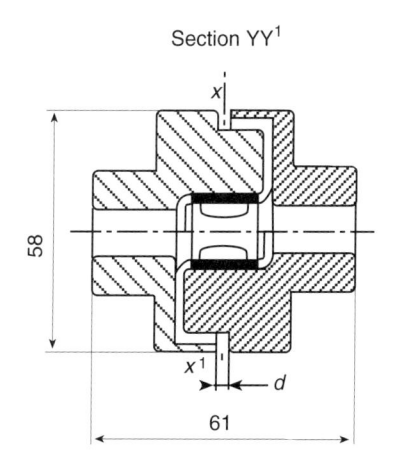

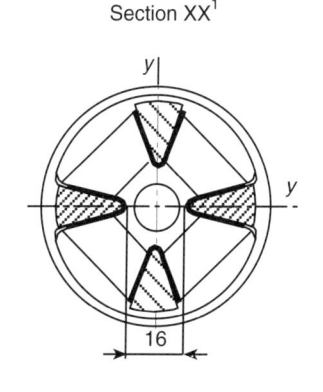

Figure 14 Coupling.

164 BEARING VIBRATIONS

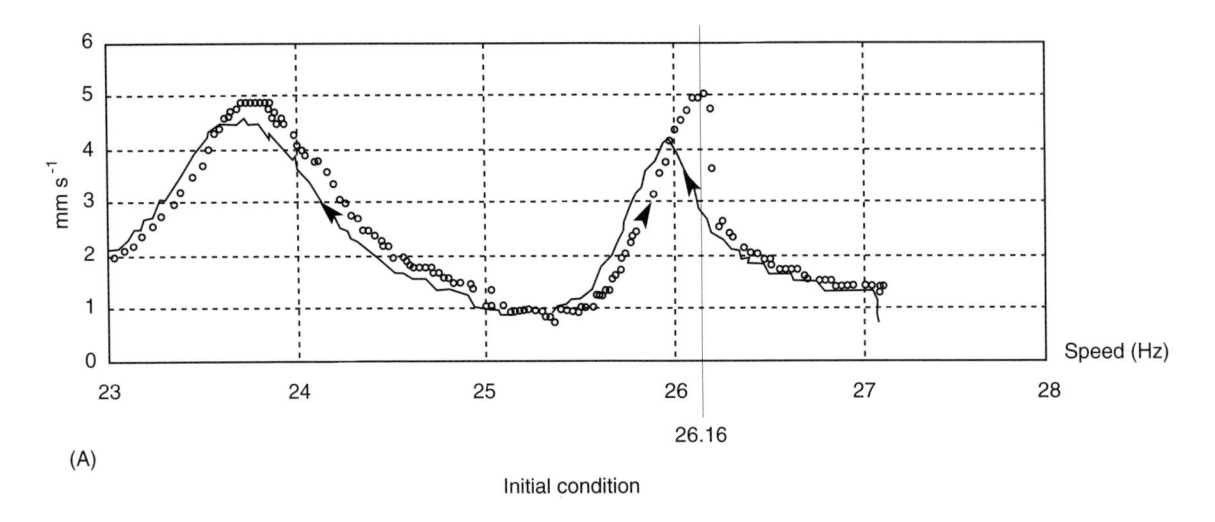

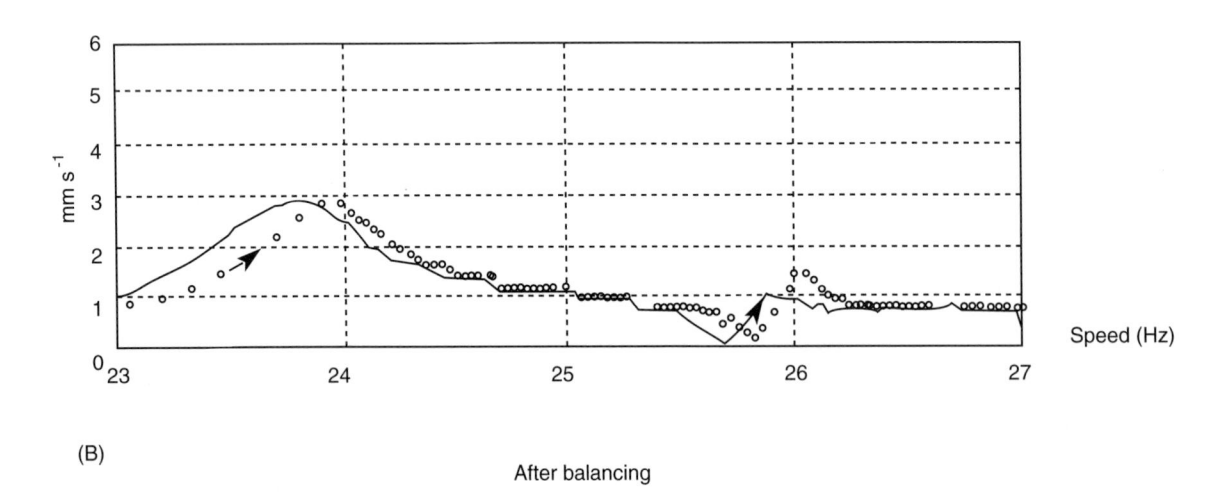

(A) Initial condition

(B) After balancing

Figure 15 (A, B) Bearing vibration (fundamental component).

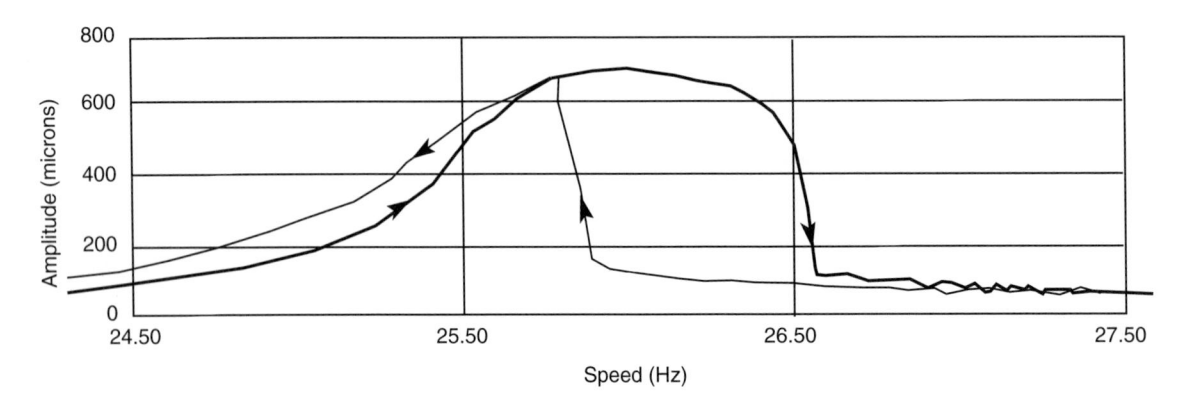

Figure 16 Relative speed (fundamental component). $d = 1$ mm (see **Figure 14**).

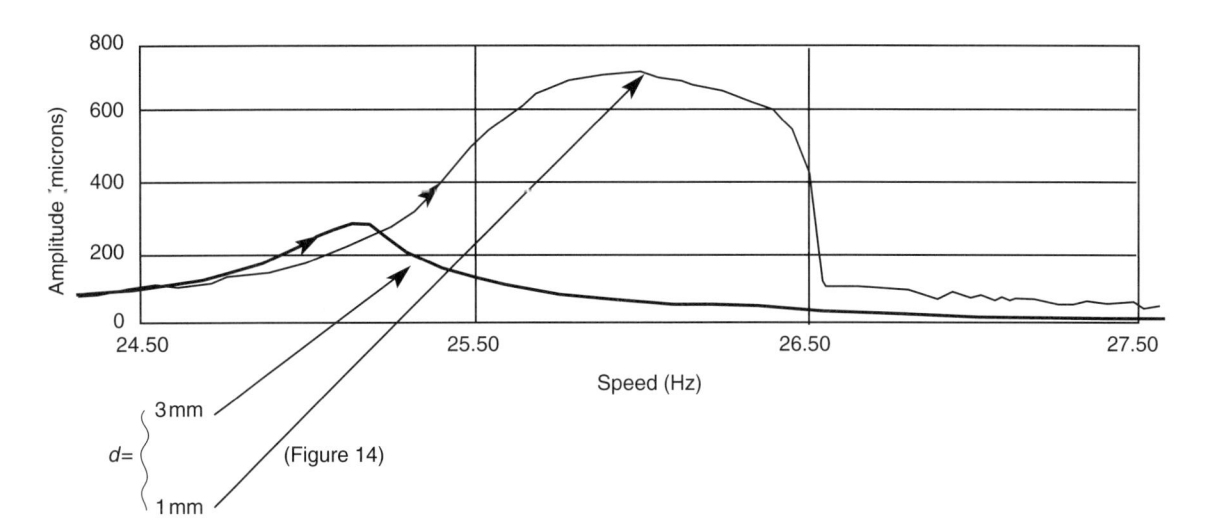

Figure 17 Relative vibration (fundamental component).

Further Reading

Adams ML Jr (1999) *Rotating Machinery Vibration. From Analysis to Troubleshooting.* New York: Marcel Dekker.

Bigret R (1980) *Vibrations des Machines Tournantes et des Structures. [Vibrations of Rotating Machines and Structures.]* Paris: Technique et Documentation Lavoisier.

Bigret R (1997) *Stabilité des Machines Tournantes et des Systèmes. [Stability of Rotating Machines and Systems.]* Simlis, France: CETIM.

Bigret R, Feron J-L (1995) *Diagnostic, Maintenance, Disponibilité des Machines Tournantes. [Diagnostics, Maintenance, Availability of Rotating Machines.]* Paris: Masson.

Boulenger A, Pachaud S (1998) *Diagnostic Vibratoire en Maintenance Préventive. [Vibratory Diagnostics and Preventive Maintenance.]* Paris: Dunod.

Childs D (1993) *Turbomachinery Rotordynamics.* Chichester: John Wiley.

Dimentberg FM (1961) *Flexural Vibrations of Rotating Shafts.* London: Butterworths.

Frêne J, Nicolas D, Dequeurce, Berthe D, Godet M (1990) *Lubrification Hydrodynamique – Paliers et Butées. [Hydrodynamic Lubrication – Bearings and Thrust.]* Paris: Eyrolles.

Gash R, Pfützner (1975) *Rotordynamik. Eine Einführung.* Berlin: Springer Verlag.

Genta G (1999) *Vibration of Structures and Machines.* Berlin: Springer Verlag.

Lalanne M, Ferraris G (1990) *Rotordynamics Prediction in Engineering.* Chichester: John Wiley.

Morel J (1992) *Vibrations des Machines et Diagnostic de leur Etat Mécanique. [Machine Vibrations and Diagnostics of their Mechanical Condition.]* Paris: Eyrolles.

Tondt A (1965) *Some Problems of Rotor Dynamics.* London: Chapman & Hall.

Vance JM (1988) *Rotordynamics of Turbomachinery.* Chichester: John Wiley.

BELTS

J W Zu, University of Toronto, Toronto, Ontario, Canada

Copyright © 2001 Academic Press

doi:10.1006/rwvb.2001.0130

Belts are commonly used in power transmission through belt drives. A power transmission belt is a flexible element subject to initial tension moving in the axial direction. Belts and belt drive systems may exhibit complex dynamic behavior, severe vibrations, and excessive slippage of belts under certain operating conditions.

The vibration analysis of a single belt span assumes that the belt span is uncoupled from the rest of the system. The vibration analysis of belt drives considers the whole drive system, including each belt span and all the pulleys in the system.

Stationary Belts

The simplest model to describe the transverse vibration of a belt is to assume that the belt is stationary by ignoring the transport speed of the belt. The string theory and the Euler–Bernoullie beam theory are combined to obtain the equation of motion for free vibration of a stationary belt with length L as:

$$\rho \frac{\partial^2 w}{\partial t^2} - T \frac{\partial^2 w}{\partial x^2} + EI \frac{\partial^4 w}{\partial x^4} = 0 \quad [1]$$

where w is the transverse displacement of the belt, ρ is the mass per unit length, T is the initial tension of the belt and EI is the bending rigidity. If the belt span is assumed to be constrained under simply supported boundary conditions, the nth natural frequency ω_n is obtained as:

$$\omega_n^2 = \omega_{sn}^2 + \omega_{bn}^2 \quad [2]$$

where:

$$\omega_{sn} = \left(\frac{n\pi}{L}\right)\sqrt{\frac{T}{\rho}} \quad [3]$$

is the nth natural frequency of a string and:

$$\omega_{bn} = \left(\frac{n\pi}{L}\right)^2 \sqrt{\frac{EI}{\rho}} \quad [4]$$

is the nth natural frequency of a simply supported beam. It is shown from experiments that the fundamental natural frequency of a stationary belt calculated by the string theory alone can accurately predict the relationship between natural frequencies and tension, except for short belt spans. **Figure 1** shows the influence of the bending rigidity on the natural frequencies where the correction factor ψ_n is defined as:

$$\omega_n = \omega_{sn}\psi_n \quad [5]$$

where:

$$\psi_n^2 = 1 + n^2\pi^2\varepsilon^2 \quad \text{and} \quad \varepsilon^2 = EI/(TL^2)$$

Moving Belts

A moving belt belongs to a class of axially moving continua. There are three types of vibrations for a moving belt, transverse vibration, torsional vibration, and axial vibration, as shown in **Figure 2**. Among the three vibration modes, transverse vibration is the dominant one and is of main interest to industry. Torsional and axial vibrations in a moving belt also exist, but are much less important in most cases.

A Moving String Model for Transverse Vibration

The bending rigidity of a belt may often be neglected with small errors. In the absence of bending stiffness,

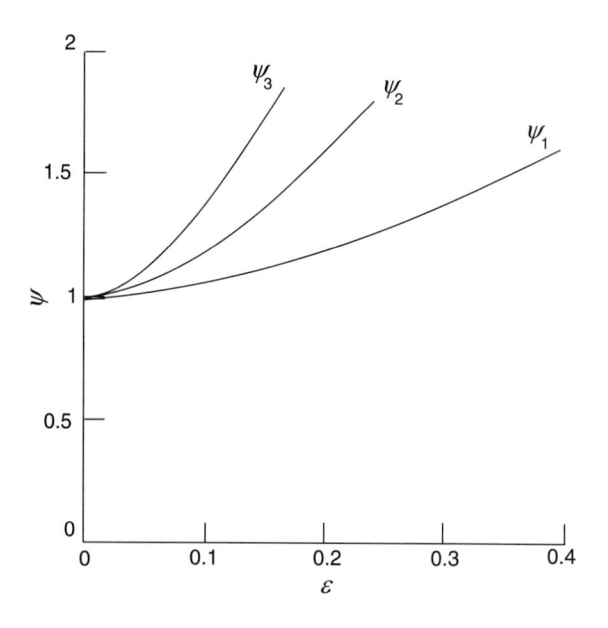

Figure 1 The influence of bending rigidities.

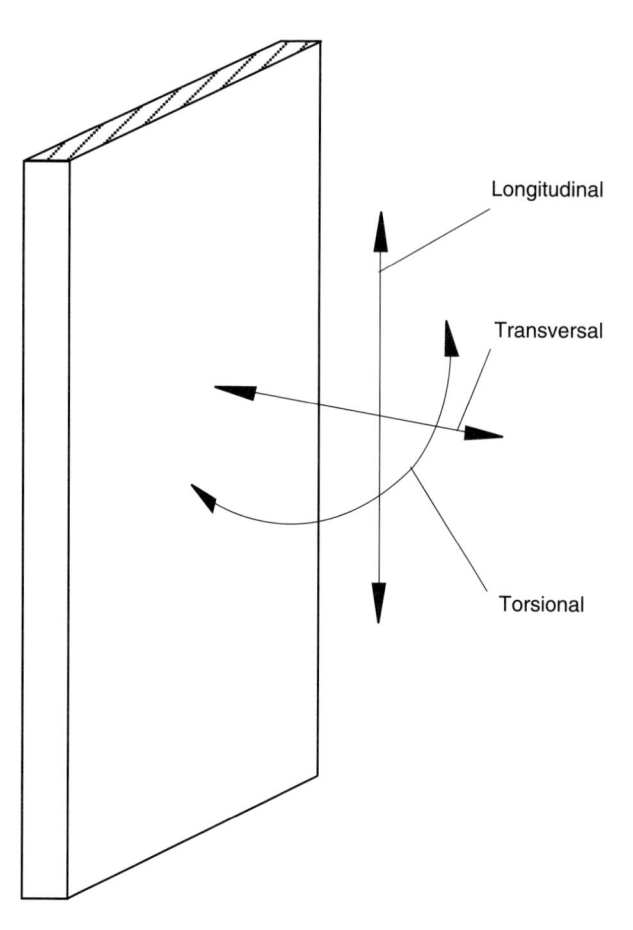

Figure 2 Three vibration modes of a belt segment.

$$\omega_n = n\pi(1 - v^2) \quad [8]$$

where $v = c\sqrt{\rho A/T}$ is the nondimensional transport speed. As the critical transport speed parameter $v_c = 1$ at which the transport speed c equals the transverse wave speed in the corresponding stationary string, the natural frequency of each mode vanishes. In **Figure 4**, the first three natural frequencies of a moving belt as a function of the transport speed parameter in the range of a stationary string and a critical string are presented. The eigenfunctions governing free vibration of a moving belt are complex and speed-dependent resulting from the Coriolis acceleration term. The real and imaginary components of the orthonormal eigenfunctions are given by:

$$\psi_n^R(x) = \frac{1}{n\pi}\sqrt{\frac{2}{1-v^2}}\,\sin(n\pi x)\cos(n\pi v x) \quad [9]$$

$$\psi_n^I(x) = \frac{1}{n\pi}\sqrt{\frac{2}{1-v^2}}\,\sin(n\pi x)\sin(n\pi v x) \quad [10]$$

Figure 5 shows the first three complex, orthonormal eigenfunctions of a moving belt at $v = \frac{1}{2}$. The real component is represented by solid lines and the imaginary component is represented by dotted lines.

Nonlinear Effect on Transverse Vibration

There exist primarily two nonlinear sources in a moving belt: geometric nonlinearity and material nonlinearity.

For geometric nonlinearity, it is observed from experiments that at very small amplitudes the belt behaves linearly, but nonlinearities occur at amplitudes well within the range of those observed in practical operations. The equation of motion to consider large transverse vibration of a moving string may be given by:

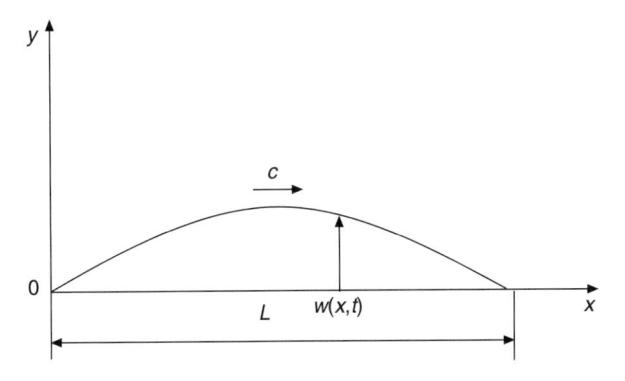

Figure 3 A prototypical model a moving belt.

the belt can be modeled as an axially moving string. A moving string model is commonly used to describe the transverse vibration of a moving belt as long as the thickness of the belt is relatively small and the belt material is flexible enough such as automotive serpentine belts.

One important characteristic of a moving belt is that the axial velocity of the moving belt introduces two convective Coriolis acceleration terms which are not present in the equivalent stationary belt. As a result, a moving belt is a gyroscopic system described by one symmetric and one skew-symmetric operator. The linear equation of motion for free vibration of a moving belt shown in **Figure 3** is given by:

$$\rho\frac{\partial^2 w}{\partial t^2} + 2\rho c\frac{\partial^2 w}{\partial x \partial t} + (\rho c^2 - T)\frac{\partial^2 w}{\partial x^2} = 0 \quad [6]$$

where c is the transport speed of the belt and A is the area of cross-section of the belt. The boundary conditions are:

$$w = 0 \quad \text{at } x = 0 \text{ and } x = L \quad [7]$$

The nth natural frequency is obtained as:

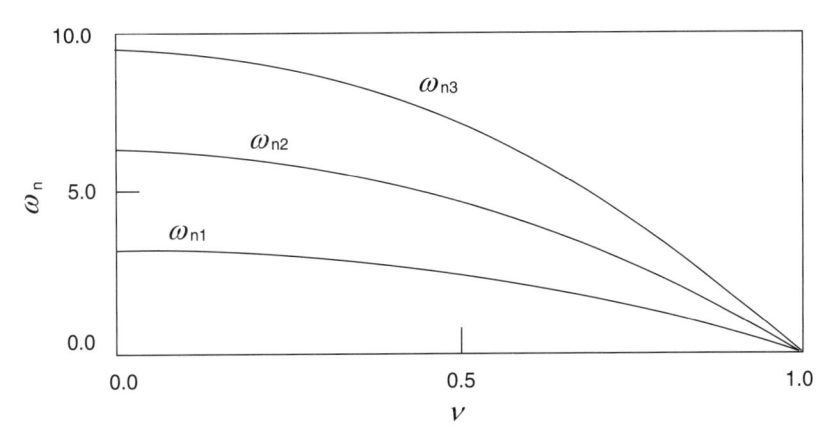

Figure 4 The first three natural frequencies of a moving belt with the change of the transport speed.

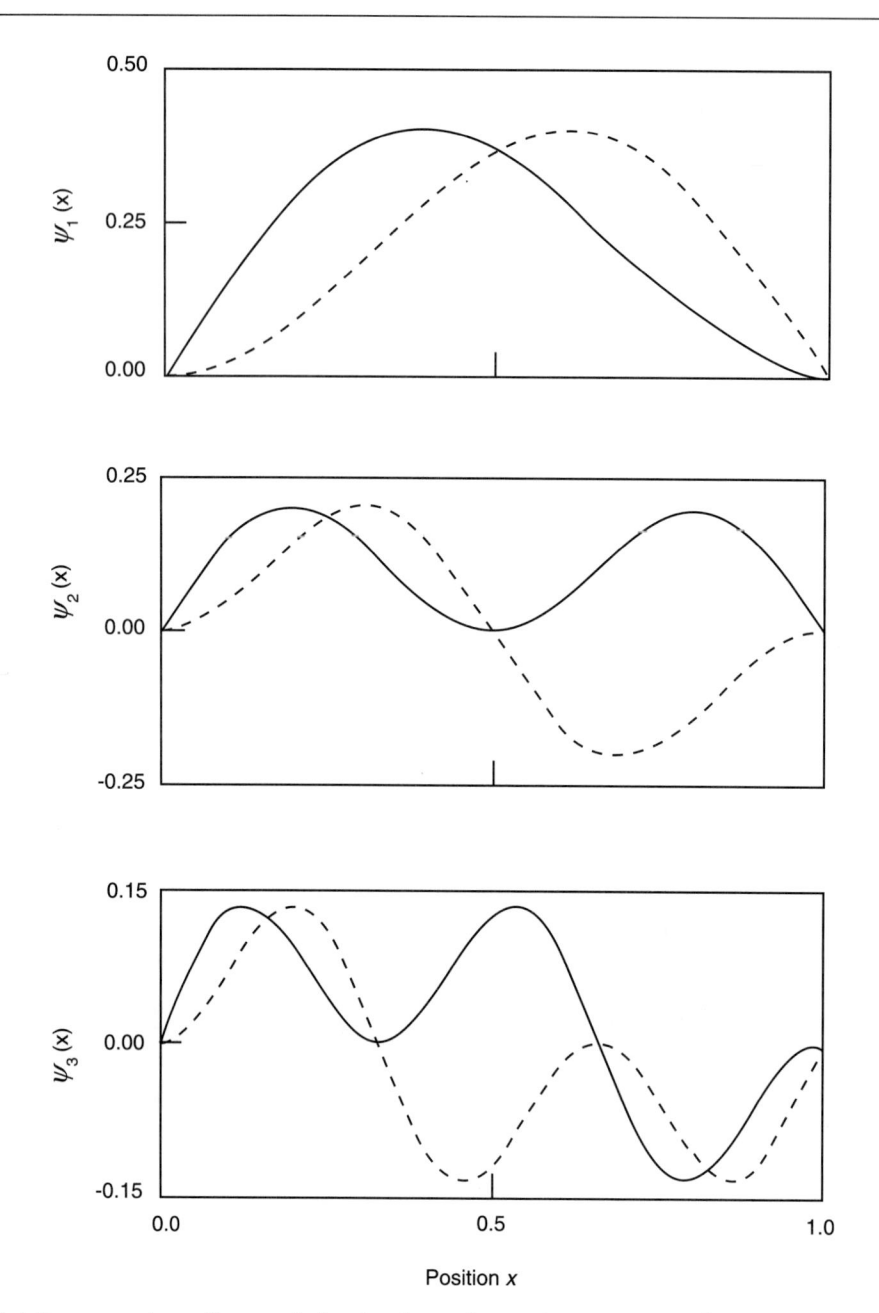

Figure 5 The first three complex, orthonormal eigenfunctions of a moving belt at $v = \frac{1}{2}$: real component (solid), imaginary component (dashed).

$$\rho\frac{\partial^2 w}{\partial t^2} + 2\rho c\frac{\partial^2 w}{\partial x \partial t} + (\rho c^2 - T)\frac{\partial^2 w}{\partial x^2}$$
$$+ \frac{3}{2}\frac{\partial^2 w}{\partial x^2}\left(\frac{\partial w}{\partial x}\right)^2 (T - AE) = 0 \quad [11]$$

The natural frequencies of a nonlinear moving string depend on the initial tension, the transport speed and the amplitude of the vibration. An approximate expression to predict the nth nonlinear frequency is as follows:

$$\omega_n^2 = n^2\pi^2(1 - v^2) + \frac{9}{32}n^4\pi^4\frac{EA}{T}(w^*)^2 \quad [12]$$

where $w*$ is the nondimensional amplitude defined as the amplitude of transverse vibration divided by the span length. The second term in eqn [12] represents the effect of nonlinearities on the natural frequencies. The influence of this term becomes larger as the transport speed increases. The second term is also directly proportional to the belt tensile modulus EA. Furthermore, eqn [12] indicates that nonlinear effects are much more pronounced for higher modes and that nonlinear effects can only be neglected for small amplitudes and high tensions.

Another source of nonlinearity comes from the belt material. The belt material is usually assumed to be linear elastic. However, experiments show that the stiffness of the belt is a function of the belt peak-to-peak strain. An empirical formula for belt stiffness as a function of peak-to-peak strain is in the form of a linear term plus an exponential decaying term. The nonlinear stiffness makes the resonant frequency shifted to smaller value and the amplitude of the resonant vibration smaller.

Parametrically Excited Transverse Vibration

Tension fluctuations cause parametrically excited vibrations and they are the main source of excitation in automotive belts. For parametrically excited moving belts, the constant tension T in eqn [6] becomes a function of time and its fluctuation is characterized as a small periodic perturbation $T_1 \cos \Omega t$ on the steady state tension T_0, i.e.,

$$T = T_0 + T_1 \cos \Omega t \qquad [13]$$

where Ω is the frequency of excitation. For automotive belt drives, these belt tension variations arise from the loading of the pulley by the belt-drive accessories (e.g., air conditioning compressor). They may also arise from pulley eccentricities. These tension fluctuations may parametrically excite large amplitude transverse belt vibrations and adversely impact belt life.

For nonlinear moving belts, there are one trivial solution and two nontrivial solutions (limit cycles) for steady-state dynamic response of parametrically excited vibrations. There exists a lower boundary for the nontrivial limit cycles.

There are two possible parametric resonances: the summation resonance, where the excitation frequency is the sum of any two natural frequencies; and the difference resonance, where the excitation is the difference between any two natural frequencies. Primary resonance, where the excitation frequency is twice the natural frequency, is a special case of summation resonance. For summation resonance, it is found that the first nontrivial limit cycle is always stable while the second nontrivial limit cycle is always unstable.

Transverse Vibration of Viscoelastic Moving Belts

Belts are usually composed of some metallic or ceramic reinforced materials like steel-cord or glass-cord and polymeric materials such as rubber. Most of these materials exhibit inherently viscoelastic behavior. The viscoelastic characteristic can effectively reduce the vibration of moving belts without suffering from greatly reduced natural frequencies. It is found that for nonlinear moving belts, viscoelasticity leads to the upper boundary of nontrivial limit cycles in addition to the lower boundary observed for elastic moving belts. In other words, viscoelasticity narrows the stable region of the first limit cycle and the unstable region of the second limit cycle.

A Moving Beam Model for Transverse Vibration

A moving beam model can be adopted to predict the transverse vibration of a moving belt if the belt is stiff and thick, such as V belts. The linear equation of motion of a moving Euler–Bernoullie beam model is given by:

$$\rho \frac{\partial^2 w}{\partial t^2} + 2\rho c \frac{\partial^2 w}{\partial x \partial t} + (\rho c^2 - T) \frac{\partial^2 w}{\partial x^2} + EI \frac{\partial^4 w}{\partial x^4} = 0 \qquad [14]$$

The natural frequency for a simply supported moving beam can be calculated by:

$$\omega = p^2 \left(\frac{EI}{\rho L^4} \right)^{1/2} \qquad [15]$$

where p^2 is a parameter determined by solving the equation:

$$c^2 \sigma \delta (\cos 2c - \cosh \delta \cos \sigma) + (c^2 - \delta^2) \sinh \delta \sin \sigma = 0 \qquad [16]$$

in which:

$$C = \frac{c}{2} \left(\frac{\rho L^2}{EI} \right)^{1/2}, \quad \sigma^2 = p^2 + c^2 \quad \text{and} \quad \delta^2 = p^2 - C^2$$

The variation of the first natural-frequency parameter p^2 with nondimensional velocity C is shown in **Figure 6**.

Torsional Vibration

Torsional vibration is usually not important compared with the transverse vibration of a moving belt, particularly for a single belt. However, if the width of a belt is much bigger compared with a string (this might be the case for toothed belt used in sychronous drives), or if a whole belt drive system is considered, the belt may exhibit twisting or torsional vibrational modes. In this case, the belt is observed to twist along its longitudinal center-line. By ignoring

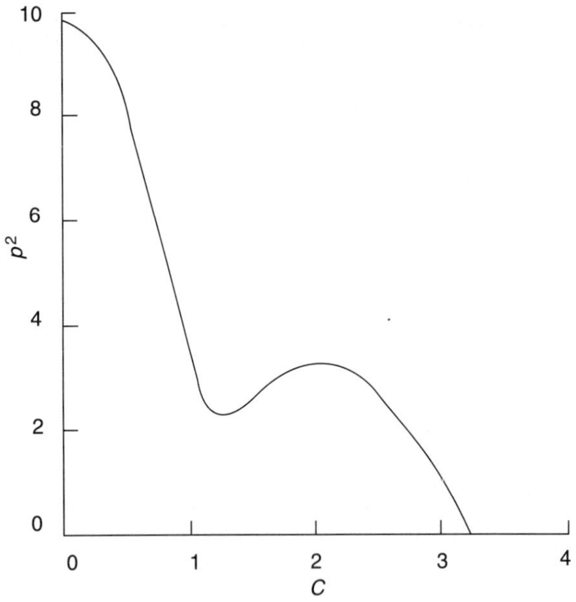

Figure 6 The variation of the first natural-frequency parameter p^2 of a moving beam with the transport speed parameter C.

the coupling between the transverse vibration and the torsional vibration, the equation of motion for torsional vibration θ of a moving belt is given by:

$$\frac{\partial^2 \theta}{\partial t^2} + 2c\frac{\partial^2 \theta}{\partial x \partial t} + (c^2 - c_0^2)\frac{\partial^2 \theta}{\partial x^2} = 0 \quad [17]$$

in which the torsional wave velocity c_0 is:

$$c_0^2 = 4\left(\frac{b}{h}\right)^2 \frac{G}{\rho} + \frac{T}{\rho b h} \quad [18]$$

where b and h are the width and height of the cross-section of a belt. Note that eqn [18] is identical in form to eqn [6] for transverse vibration of a moving string. Therefore, analysis and results from the transverse vibration apply to the torsional vibration.

Axial Vibration

Axial vibration of a moving belt is very small compared with transverse vibration. In most cases, axial motion is not important, since axial vibration has much higher natural frequencies. Therefore, a quasi-static assumption can be employed to describe axial motion.

Belt Drives

Belt drives are used in many mechanical systems for power transmission. **Figure 7** shows a typical automotive belt drive system where a single polyvee belt drive powers all the accessories, including alternator, power steering, water pump, tensioner, and so on.

Two distinct types of vibration modes occur for a belt drive: (i) rotational modes and (ii) transverse modes. In rotational modes, the pulleys rotate about their spin axes and the belt spans act as axial springs. In transverse modes, the belt spans vibrate transversely, as discussed in the previous section on moving belts. Both types of vibration are harmful to the system performance. Rotational motions induce dynamic belt tension and the attendant problem of belt fatigue, dynamic bearing reaction, structure-borne noise, and bearing fatigue. Transverse vibrations also induce dynamic tensions, and may directly radiate noise.

For belt drive systems with fixed-center pulleys, the rotational and transverse modes are uncoupled and thus can be modeled by a multi-degree-of-freedom rotational system and by a moving belt model for each belt span, respectively. However, for belt drives with a nonfixed-center pulley, such as the tensioner pulley in the automotive serpentine belt drive system, the rotational and transverse modes are linearly coupled through the rotation of the tensioner arm.

Two commonly adopted models for vibration analysis of automotive serpentine belt drive systems are presented here.

Discrete Spring–Mass Model

The discrete spring–mass model ignores the transverse vibration of the belt and treats each belt spring as an axial spring. Thus, the belt drive is approximated as a multi-degree-of-freedom system where each pulley's moment of inertia is associated with one rotational degree-of-freedom, as shown in **Figure 8**. Such a discrete spring–mass model is used to describe the rotational vibration mode of the belt drive system. The prescribed harmonic crankshaft speed fluctuation due to the cylinder engine operation serves as the excitation of the system. Based on this model, the equation of motion for each fixed center pulley is given by:

$$I_j \ddot{\theta}_j = R_j(T_{j-1} - T_j) - c_j \dot{\theta}_j + Q_j, \quad j = 2, 3, \ldots, n \quad [19]$$

where R_j is the jth pulley's radius ($j = 1, 2, \ldots, n$), I_j is the jth pulley's moment of inertia including the accessory driven by this pulley and any shaft extension and couplings, c_j is the jth pulley's rotational

Figure 7 A typical automotive belt drive system.

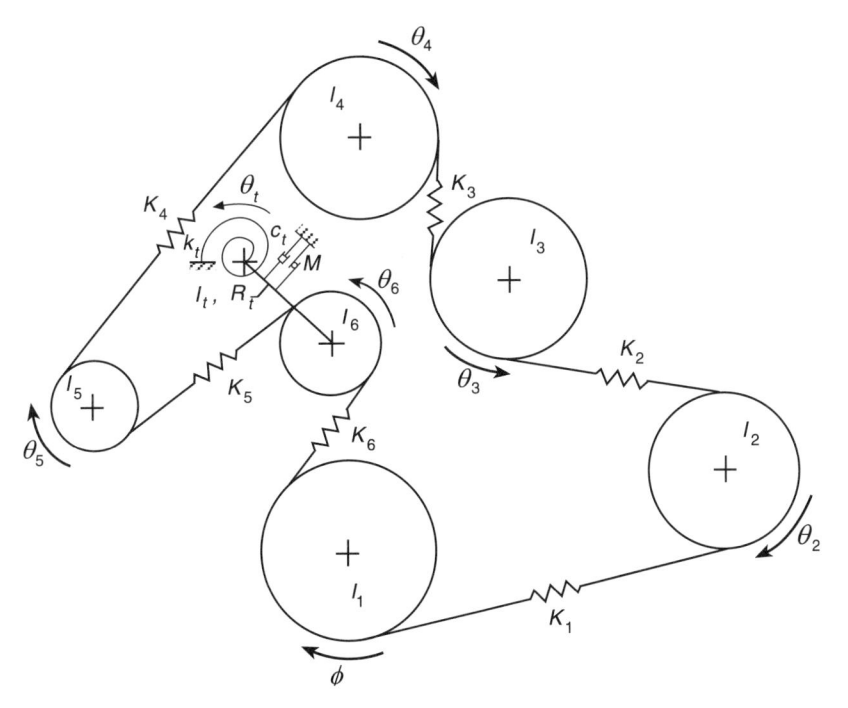

Figure 8 A discrete automotive belt drive system.

damping, Q_j is the dynamic torque requirement of the jth pulley and T_j is the jth belt span tension. The tensioner pulley can be any number from 2 to n and the first pulley is always defined as the crankshaft pulley.

The equation of motion for the tensioner arm is given by:

$$I_t \ddot{\theta}_t = -T_{j-1} L_t \sin\left(\theta_t - \beta_j\right) - T_j L_t \sin\left(\theta_t - \beta_j\right) - c_t \dot{\theta}_t - k_t(\theta_t - \theta_{t0}) + Q_{t0}$$

[20]

where I_t is the moment of inertia of the tensioner arm and T_{j-1} and T_j are the two belt tensions acting on the tensioner pulley. In addition, β_j is the angle between the tensioner arm and the vertical line at equilibrium state, R_t is the length of the tensioner arm, θ_t is the dynamic angle of the tensioner arm, c_t is the external viscous damping acting on the tensioner arm, k_t is the rotational stiffness of the tensioner arm, and M denotes the dry friction torque (Coulomb damping torque) acting on the tensioner arm.

For a practical accessory belt drive system, the magnification ratio of the dynamic response of each component with the change of the excitation frequency is shown in **Figure 9**. It is shown that the alternator has the largest magnification ratio.

Hybrid Discrete–Continuous Model

The hybrid model for an automotive belt drive system captures the coupling between discrete rotational vibration of pulleys and the continuous transverse vibration of belt spans. **Figure 10** shows a prototypical belt drive system where θ_i ($i = 1, 4$) is the rotation from equilibrium of the ith discrete element (pulleys or tensioner arm), w_i ($i = 1, 3$) is the transverse deflection of span i from equilibrium, J_i and r_i are the mass moment of inertia and radius of the ith discrete element, and l_i is the length of belt span i. The linear equations of motion of the coupled system are composed of three equations of motion for the three continuous belt spans, three equations of motion for the discrete pulleys 1, 2, 4 and the equation of motion for the tensioner arm rotation θ_4. The linear coupling between the transverse vibration of the continuous belt and the rotational vibration of the pulleys are reflected from the equation of motion of the tensioner arm.

Figure 11 shows a prototypical three-pulley serpentine belt drive system called baseline system. The physical properties of the system are given in **Table 1**. The first transverse and rotational vibration mode shapes of the system are shown in **Figures 12** and **13**. It is seen from the mode shapes that there is a significant coupling between the span's transverse vibration and the tensioner arm's rotational vibration.

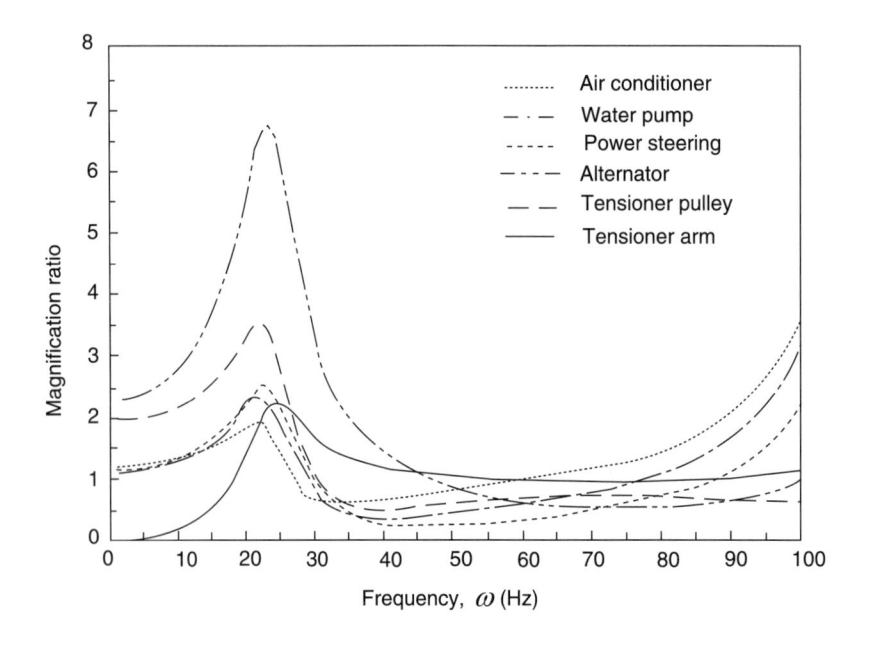

Figure 9 Dynamic response of an automotive belt drive system.

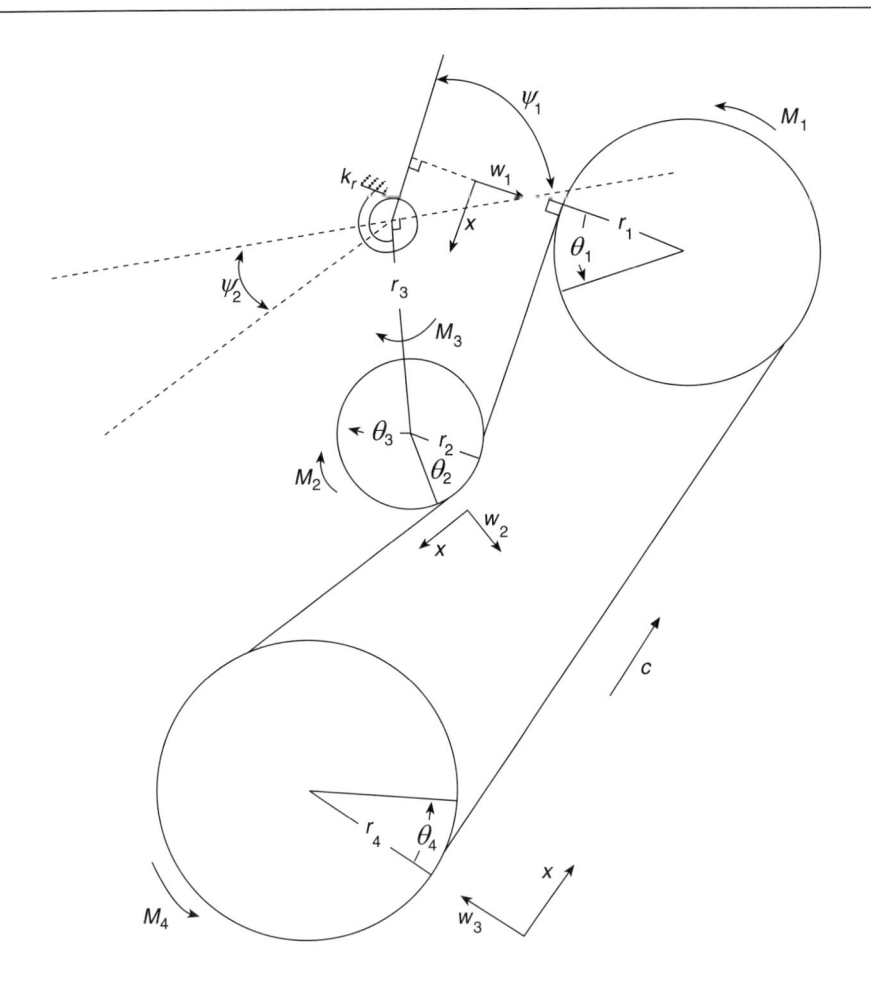

Figure 10 A prototypical belt drive system.

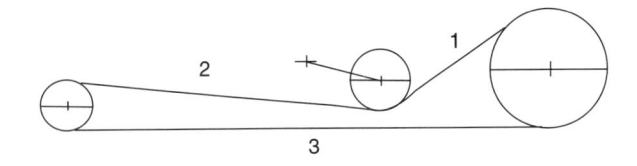

Figure 11 A three-pulley baseline system.

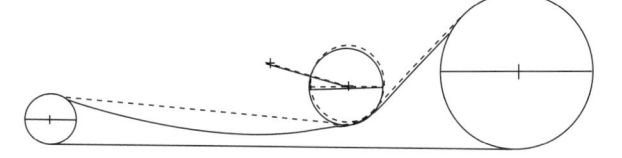

Figure 12 Transverse vibration mode of span 2 of the baseline system (50.53 Hz). Continuous line, mode shape; dashed line, baseline system.

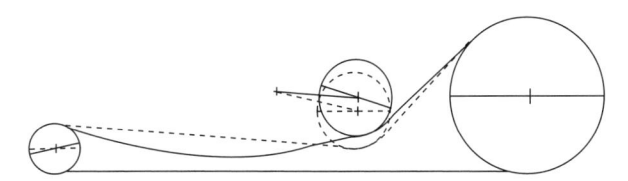

Figure 13 Rotational vibration mode of the baseline system (62.18 Hz). Continuous line, mode shape; dashed line, baseline system.

Nomenclature

A	area of cross section
c	torsional wave velocity
C	nondimensional velocity
EA	belt tensile modulus
EI	bending rigidity
I	inertia
L	length
M	dry friction torque
T	tension
ψ	correlation factor
ρ	correction factor
Ω	fequency of excitation

Table 1 The physical properties of a three-pulley baseline system.

	Pulley 1	Pulley 2	Tensioner arm	Pulley 4
Spin axis coordinates	(0.5525, 0.0556)	(0.3477, 0.05715)	(0.2508, 0.0635)	(0.0, 0.0)
Radii (m)	0.0889	0.0452	0.097	0.02697
Rotational inertia	0.07248	0.00293	0.001165	0.000293
Other physical properties		Belt modulus: $EA = 170\,000\,\text{N}$, $m = 0.1029\,\text{kg m}^{-1}$, Tensioner spring constant: $k_r = 54.37\,\text{N m rad}^{-1}$, Tensioner pulley mass: 0.302 kg (baseline) 0.378 kg (modified)		

See also: **Nonlinear normal modes**; **Nonlinear system identification**; **Nonlinear systems analysis**; **Nonlinear systems, overview**.

Further Reading

Abrate S (1992) Vibrations of belts and belt drives. *Mechanism and Machine Theory* 27: 645–659.

Mockensturm EM, Perkins NC, Ulsoy AG (1996) Stability and limit cycles of parametrically excited, axially moving strings. *ASME Journal of Vibration and Acoustics* 118: 346–351.

Wickert JA, Mote CD Jr (1990) Classical vibration analysis of axially moving continua. *ASME Journal of Applied Mechanics* 57: 738–744.

Zu JW *Static and Dynamic Analysis and Software Development of Automotive Belt Drive Systems*. Internal Final Report to Tesma International Inc.

BIFURCATION

See **DYNAMIC STABILITY**

BLADES AND BLADED DISKS

R Bigret, Drancy, France

Copyright © 2001 Academic Press

doi:10.1006/rwvb.2001.0094

Introduction and Definition

In a turbomachine, a fluid flows across blades mounted on a rotor which is made up of one or more disks with moving blades. When fluid is released, it provides energy to the blades and the rotor works as an engine. On compression, the fluid receives energy from the blades and the rotor acts as a receptor.

In hydraulic machines the function of the blade is as a paddle, whereas when it is larger it works as a blade. Blades are either acted upon by the fluid (motor) or act on the fluid (receptor). Based on the laws of aerodynamics and thermodynamics, blades serve a particular function with great efficiency. Centrifugal, aerodynamic, and hydrodynamic forces all set blades in motion.

A study carried out by VGB (Essen, Germany) from 1973 to 1977 on 76 steam turbines operating at 64–365 MW noted that there were 50 breakdowns in 28 of the 76 turbines. Of these, 80% were caused by moving blades. Were the blades the cause of the breakdown? Blade reliability is an important factor in machine reliability.

In general, stationary blades which are linked to the stator are connected to moving blades which direct the fluid. These fixed, stationary blades are also acted on by aerodynamic and hydrodynamic forces.

Instability

Bladed disks are acted on by three types of aerodynamic force:

1. Anisotropy
2. Particular phenomena
3. Erratic fluctuations in pressure

Anisotropy arises as a result of flow from the radial field of pressure and wakes, which are as wide as the thickness of the blades and stator obstacles. At speed Ω the bladed disks are excited by frequencies:

$$F_{cx} = h\Omega/2\pi$$
$$h = 1, 2, 3 \ldots$$

The amplitude of the components can be high $(vn_d \to h)$:

$$vn_d\Omega/2\pi \quad (v = 1, 2, 3 \ldots)$$

where n_d is the number of blades, distributors and stator elements above the disk.

Particular phenomena may be seen at low flow rates, when the fluid does not follow the path designated by the blades. This stalling produces rotating fields at speed $\pm\Omega_t$ in relation to the stator, and at speed $\Omega_{tr} = \Omega \pm \Omega_t$ in relation to the disks which are excited at the following frequencies:

$$f_{ex,t} = m\frac{\Omega \pm \Omega_t}{2\pi}$$

where $m = 1, 2, 3 \ldots$. Observation of high-power turbines shows that $0.3 \leq |\Omega_t|/\Omega \leq 0.7$.

These three forms of excitation may create resonances when:

1. The frequency of excitation is close to a natural disk frequency $\omega_q/2\pi$ (where natural frequency $p_q = \delta_q + i\omega_q$ where $i^2 = -1$; $q = 1, 2, 3 \ldots$) at speed Ω: $h\Omega \approx \omega_q$ or $m(\Omega \pm \Omega_t) \approx \omega_q$
2. The number of nodal diameters n of the natural mode q of the rotational disk, connected to ω_q is equal to h or m.

During resonance, vibration waves, which are fixed in relation to the rotating stator, are rotating in relation to the bladed disk at speed ω_q. The component at ω_1, without a nodal diameter, cannot cause resonance. There may be instability of the limit cycle resulting from a coupling between disk vibration and fluid, in particular in axial compressors. It may be difficult to distinguish between this instability and a forced rate of flow from aerodynamic excitation caused by a rotating field.

Strains

Dynamic strains on the body of the blade, and on its base which is in contact with the disk, are due to a forced rate of flow, resonances, or eventual instability. For a fixed lifespan, which is often important, the maximum acceptable level of dynamic strain is even less when the static strain increases as a result of centrifugal and aerodynamic force (driving or resisting torque).

Table 1 shows the static strains on two turboalternator blades.

The lifespan of high-power steam turbines used by Electricité de France is close to 40 years. A machine working at 600 MW at 3000 rpm, using fossil fuel, contains 4468 moving blades. Machines operating at 1000 MW and 1500 rpm and using nuclear fuel contain 9749 blades, while machines working at 1500 MW, which were first brought into use in 1996, contain 4350 blades. This is a notable reduction.

For low-pressure steam expansion towards the condenser (whose temperature is dependent on that of the cold source, for example 35°C gives rise to 55 mbar), the necessary section for the steam flow imposes a compromise between the number of low-pressure rotors, the diameter of the last disk, and the length of its blades. This section is longer where there are two low-pressure flows in parallel, than when there are three flows. The natural frequencies of the blades and their strains play an essential role in this compromise.

Table 1 Static strains on two turboalternator blades

Steam turbine		Bladed disk: last stage, low pressure						
Power (MW)	Speed (rpm)	Power (MW)	No. of blades	Blade length (mm)	Blade mass (kg)	Base centrifugal force (N)	Maximum static base strain (N m^{-2})	
							Centrifugal force	Torque
860 (fossil)	3600	17	78	885	23	2×10^6	1.5×10^8	0.12×10^8
1000 (nuclear)	1500	19.6	138	1220	39	1.22×10^6	0.278×10^8	0.123×10^8

Causes and Consequences of Breakdowns

The driving forces and thus the strains are weak when the distances between the trailing and leading edges of the stator and rotor blades are sufficient for a trickle of fluid to reorganize, beyond the journal zone, as also happens when there are obstacles in the flow. Large gaps make it possible to avoid siren noises, which are particularly annoying in ventilation circuits. However, large gaps may reduce efficiency. In general, stator blades are subject to less dangerous vibration than rotor blades.

The blades of compressors and fans which function at low flow and under unstable conditions (pumping) at frequencies of several hertz are subject to significant strain which may lead to breakdown.

A blade breaking may lead to the machine being destroyed and this may even spark off a greater catastrophe. Breakdown is caused by fatigue as a result of large dynamic strains which are manifested as cracks, and transiently raised thermal gradients. Where there is instability (flutter), the strain may be greater than that seen with the forced rate of flow caused by the harmonics of the rotation speed, and thus the lifespan may be noticeably shorter. If part of a broken blade falls free in the stator, it may destroy other blades, and may well end in catastrophe.

The unbalance created by a broken blade generates a rotating force. For example, for a blade of the last disk of a low-pressure rotor, with 8.5 bar absolute downstream of a 15 MW turbine operating at 7600 rpm: in this case, the rotating force is equal to $mr\Omega^2 = 0.3 \times 0.28(\pi.7600/30)^2 = 53150\,N$.

The ratio of the rotating force to the static charge at right-angles to the bearings can be high: for a turbine operating at 15 MW, it is $53150/3826 \approx 14$. Deterioration in the bearings, in the sealants, and of the contacts between the rotor and the stator may lead to chaos.

Misalignments and rotating forces from other causes – such as the receiving structure being deformed, coaxiality problems, rotor shaft, cracked rotor, instability, short axial distance between the rotor and stator as a result of differential expansion – can lead to contact between the blades and the stator, as a consequence of deterioration and chaos.

The presence of water within steam, or cavitation within the pump and turbine, may cause erosion of the blades. These erosions affect the natural frequencies; they may cause harmful resonance and greater unbalance, leading to a breakdown in the links. Flowing fluids can leave deposits on the blades, such as soot on fan blades; when these deposits come loose they create unbalance. The blades need to be cleaned: introducing grains of rice to the machine helps loosen any deposits.

Technology

Blades are fixed by an integrated base radiating out from the edge of the disk, in various ways. Axial rivets may be used. Where gas turbines work with gas at high temperature (800–1000°C), the play between the base and the disk permits transient differences in expansion, and low levels of strain.

Blades may be connected together at the top with shrouds which are in contact with each other. Wires or shrouds attached to the body of the disk may also be used. Such connectivities between blades may prevent dangerous vibration resonance. This connectivity may not be acceptable for particularly long blades, because of the strain involved.

In some sets of blades, in particular in turboreactors, masses – so-called 'candies' – are placed on the platform at the base of the disks. These masses, which come into contact through centrifugal force, create friction which reduces vibration.

Natural frequencies and their associated natural modes characterize the disk and blade set-up and even the rotor. When the natural frequencies of the disks and rotor are much greater than those frequencies which are liable to provoke unacceptable resonance, only those blades, whether connected or standing alone, fixed on the disks are taken into account when evaluating the vibratory rate of flow. This situation is common in fans where only a few blades (≤ 10) are soldered on to a massive rotor element.

On certain axial compressors the influence of the disks and rotors may have a minimal effect on the vibration of thin blades.

Hydraulic turbine wheels, pumps, and radial compressors may be extracted from the same metallic blocks from which the blades are cut. The frequencies and natural modes are related to the total wheel assembly.

Technological considerations are taken into account when defining procedures to determine frequencies and natural modes by theoretical models and experiments. As a manufacturer's control or to define the limits of a study, approaches are generally made on isolated elements, such as a free fitted blade or a nonbladed disk.

Pulsations and Natural Modes of a Disk

Figure 1 shows a naturally axially vibrating disk. This is characterized by two nodal diameters and one nodal circle, on which the level of vibration is very low.

The number of diameters and nodal circles increases with natural frequency. The natural frequencies, ω_{nc}, of a thin cylindrical disk without damping, are defined by:

$$\omega_{nc} = \frac{\alpha_{nc}}{r^2}\sqrt{\frac{D}{\rho h}}$$

where r = radius, $D = Eh^3/12(1-v^2)$; h = thickness; E = elasticity modulus, v = Poisson coefficient; ρ = volume mass; n = number of nodal diameters; and c = number of nodal circumferences.

Where $n = 1$, whether the center of the disk is fixed or free, $\alpha_{11} = 20.52$ when $c = 1$; $\alpha_{12} = 59.86$ when $c = 2$.

Neglecting the increase in natural frequencies with rotation speed Ω, resonance can be shown by:

$$\Omega = \begin{cases} \omega_{11} \\ \omega_{12} \end{cases}$$

when the disk rotates in a field of fixed force which has a component of frequency Ω ($h = 1$; first harmonic for $n = 1$).

Technological requirements lead to bladed disks whose form is more complex than that of the disk shown in **Figure 1**. Their modal shape is defined by components in axial and tangential directions. On the whole, the radial components are weak. Nodal lines may be moving: they constitute nodal zones.

When the natural frequencies of rotating disks are high in comparison to frequencies from driving forces, isolated blades are resonant when the frequency of the exciting force is the same as the natural eigenvalues of the blades.

Example

The following example is an evaluation of the resonant frequencies of a low-pressure bladed disk from an 860-MW turbine. The characteristics of this disk are shown in **Table 1**.

In a test rotor ($l = 2990$ mm; two bearings) 78 blades are mounted. They are linked together by arches of wire (stanchion: 200 mm^2 section). The outside diameter is 3140 mm; the rotor is contained in a casing under vacuum (pressure $\approx$ 3 mbar absolute). The blades pass through two fixed magnetic fields at 180° which excite them. The deformations (strains) are measured by extensometers.

Figure 2 is a schematic of the measuring system. **Figure 3** shows a Campbell diagram, dating from 1924. The resonances at the intersection $f = hN$,

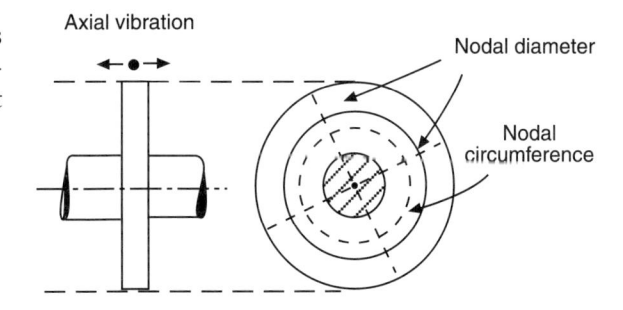

Figure 1 Vibrating disk.

and the frequency of the bladed disk. These are close to the natural frequencies, ω_q, which are associated to the mode waves at order h (nodal diameters). Nodal circles may exist in the four groups depicted.

Figure 4 shows a resonance at the nominal speed of 3600 rpm.

Analyses during Rotation and at Rest ($f_r = 0$)

Measuring systems incorporating rotating contacts to transmit signals are used for on-site analysis and in workshops. **Figure 2** shows the structure of one such system. Natural frequencies may be adjusted to shorten the blades mounted on the disks, preventing resonance during machine use. On-site, dynamic strains are measured with reference to power, flow, and pressure at the condenser. As an example, a reduction in power of the nominal value W_0 at $0.13 W_0 = W_r$ can lead to an increase in strain: $\sigma_r/\sigma_0 \approx 6.3$. For flow from Q_0 to $Q_r = 0.2 Q_0$, $\sigma_r/\sigma_0 \approx 2$. For pressure from P_0 to $P_r = 0.2 P_0$, $\sigma_r/\sigma_0 \approx 5$.

The increase in strain arises from rotor and flow. These phenomena may cause instability occurring on natural frequencies related to the order of a wave.

The frequencies and natural waves of the bladed disks are established at zero speed; they are excited by harmonic forces and the responses measured with a seismic or magnetic sensor. Modal analysis procedures using specific software can also be used; the frequencies and natural modes are obtained faster than with simple methods.

Great uncertainty occurs for the value of natural frequency computed from natural frequencies which are determined at zero speed. This is the influence of the centrifugal forces which directly act as stiffening and modify the contacts between the blades, between the blades and the disk, and even between the disk and the rotor shaft.

The influence may be approximated for the first natural frequencies by:

$$f_{dr} = \sqrt{f_{d0}^2 + \alpha f_r^2}$$

where f_{dr} = natural frequency of the rotating disk; f_{d0} = natural frequency of the disk at zero speed; and f_r = rotation speed. The value of α can be estimated by comparing the natural frequency f_{d0} calculated at zero rotation speed, to the frequency of resonance f_{dr}, which is related to the rotation speed f_{rr}:

$$\alpha = \frac{f_{dr}^2 - f_{d0}^2}{f_{rr}^2}$$

Measured natural frequencies can be used to tune the mathematical model. An algorithm can then calculate the natural frequencies from rotation speed.

For the bladed disk shown in the Campbell diagram (**Figure 3**), at resonance on the sixth harmonic, $\alpha \approx 10$.

The modification of the contacts, often more important if there are significant imperfections (tightenings or loosenings, which give rise to anisotropy), can cause important variations in natural frequencies as a function of rotation speed. In this regard, the connectivity between the blades (beyond their stiffening role) causes homogenization, which minimizes the influence of local imperfections.

The results of on-site rotation measurement, or measurements made in the casing, at zero speed can be used to tune theoretical models.

Studies are currently underway to monitor the vibratory conditions of blades *in situ*. These studies incorporate measurement and analysis of the movement of the blade edges by the radial captors, as well as vibrations of the fixed parts (bearings) and the rotors to detect the vibrating components of both fixed and moving blades.

When breaking blades cause violent variations in vibration intensity, whether increasing or decreasing, the machine must be shut down to avert catastrophe.

Models

Mathematical models are obtained by three-dimensional finite-element methods. The disks are generally taken into account in this, as is the rotor. Uncertainty about conditions at the limits, at right-angles to the contacts, and variations in rotation speed, temperature, and time (creep) may be important. Models make it possible to estimate their influence, for example, variation in link stiffness of some of the disk blades.

Experimental studies on elementary physical models make it possible to extract useful sizes for resetting models.

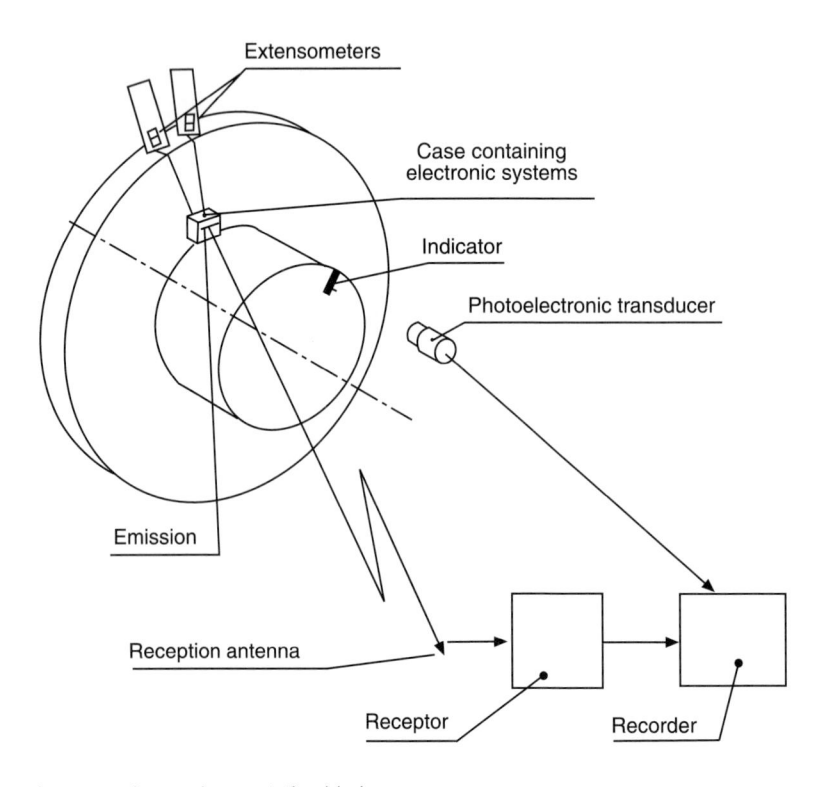

Figure 2 System for measuring strain on rotating blades.

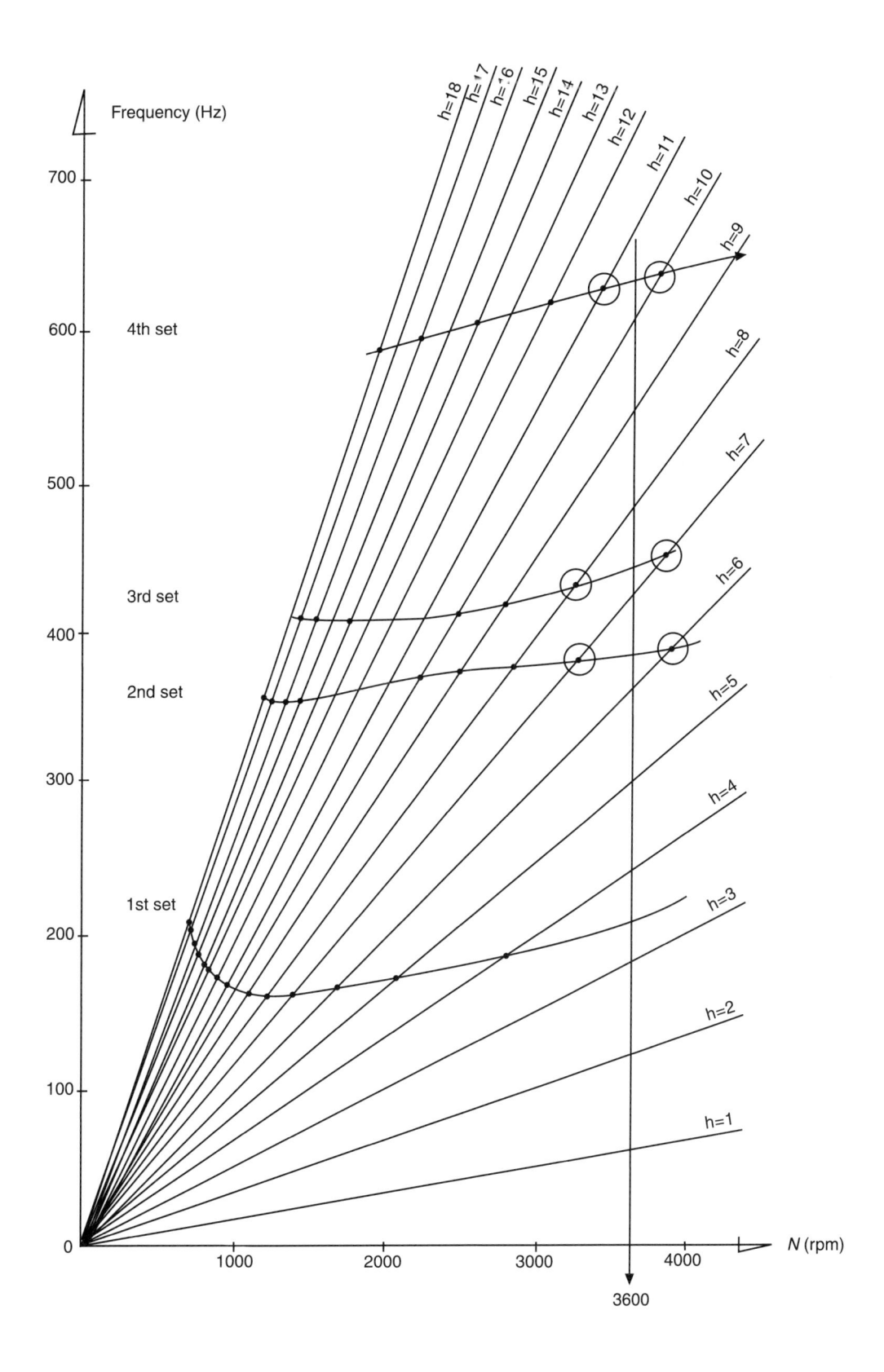

Figure 3 Resonances (●) of a bladed disk. *h*: Harmonic order of speed. Modified from Campbell.

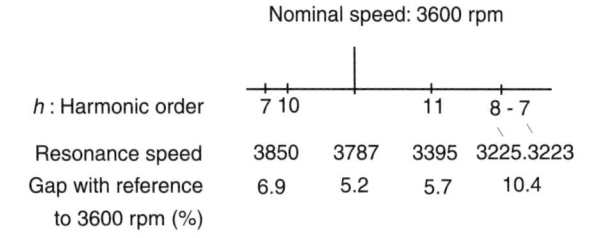

Nominal speed: 3600 rpm				
h : Harmonic order	7 10	11	8 - 7	
Resonance speed	3850 3787	3395	3225.3223	
Gap with reference to 3600 rpm (%)	6.9 5.2	5.7	10.4	

Figure 4 Resonances.

Aeroelasticity can be introduced to study behavior, which is particularly relevant for conditions of turbomachine stability, as well as aviation compressors and pumps, which are widespread because of their hydroelasticity.

See also: **Discs; Rotating machinery, essential features; Rotating machinery, monitoring; Rotor dynamics.**

Further Reading

Adams ML Jr (1999) *Rotating Machinery Vibration. From Analysis to Troubleshooting.* New York: Marcel Dekker.

Bigret R (1980) *Vibrations des Machines Tournantes et des Structures.* [Vibrations of Rotating Machines and Structures.] Paris: Technique et Documentation Lavoisier.

Bigret R (1997) *Stabilité des Machines Tournantes et des Systèmes.* [Stability of Rotating Machines and Systems.] Simlis, France: CETIM.

Bigret R, Feron J-L (1995) *Diagnostic, Maintenance, Disponibilité des Machines Tournantes.* [Diagnostics, Maintenance, Availability of Rotating Machines.] Paris: Masson.

Boulenger A, Pachaud S (1998) *Diagnostic Vibratoire en Maintenance Préventive.* [Vibratory Diagnostics and Preventive Maintenance.] Paris: Dunod.

Childs D (1993) *Turbomachinery Rotordynamics.* Chichester: John Wiley.

Dimentberg FM (1961) *Flexural Vibrations of Rotating Shafts.* London: Butterworths.

Frênc J, Nicolas D, Dequeurce, Berthe D, Godet M (1990) *Lubrification Hydrodynamique – Paliers et Butées.* [Hydrodynamic Lubrication – Bearings and Thrust.] Paris: Eyrolles.

Gasch R, Pfützner (1975) *Rotordynamik. Eine Einführung.* Berlin: Springer Verlag.

Genta G (1999) *Vibration of Structures and Machines.* Berlin: Springer Verlag.

Lalanne M, Ferraris G (1990) *Rotordynamics Prediction in Engineering.* Chichester: John Wiley.

Morel J (1992) *Vibrations des Machines et Diagnostic de leur Etat Mécanique.* [Machine Vibrations and Diagnostics of their Mechanical Condition.] Paris: Eyrolles.

Tondt A (1965) *Some Problems of Rotor Dynamics.* London: Chapman & Hall.

Vance JM (1988) *Rotordynamics of Turbomachinery.* Chichester: John Wiley.

BOUNDARY CONDITIONS

G Rosenhouse, Technion - Israel Institute of Technology, Haifa, Israel

Copyright © 2001 Academic Press

doi:10.1006/rwvb.2001.0089

The present article reviews the effect of boundaries and obstacles on generation of vibrations in solid bodies. We distinguish between a point motion and vibrations in one-, two-, and three-dimensional spaces. The article also deals with special effects, such as change from wave propagation into the state of vibrations via boundary conditions, flanking through joints, coupling between longitudinal and flexural waves, elastic mountings, nonreflecting boundary conditions, moving boundaries and semi-definite systems.

Background

A homogeneous isotropic elastic infinite medium does not reflect waves. Hence, when a source radiates mechanical (acoustic) energy into an infinite medium it does not turn into vibration in the sense of standing waves. On the other hand, existence of ends or inhomogeneities of the medium influence the displacement field within the medium by causing reflections that can lead to vibration patterns. Another effect that occurs generally at the boundaries is the transmission of mechanical energy into or out of the medium. Even along a rigid boundary, the mechanical energy is not totally reflected and some of it disappears through a thermodynamic process of absorption. Vibrations can occur at a point and in any spatial dimension.

A typical point vibration is the motion of a mass of a single-degree-of-freedom 'mass-spring' unit. A one-dimensional vibration occurs longitudinally in a rod by an axial excitation force, axially by torque and laterally in a beam by a dynamic transverse load. Two-dimensional vibrations occur in-plane and out-of-plane in plates. Vibrations of shells can be like that of vibration of surfaces that are described in three dimensions. Vibrations also occur in three-dimensional bodies, as in a piezoelectric transducer built of a layered block. Determination of vibrations in any of these dimensions involves satisfying both the governing field equations and the boundary conditions.

Boundary Conditions for a Longitudinal Wave

If a longitudinal mechanical wave propagates in x-direction, and a boundary exists at x_0, it causes reflection. The simplest boundary conditions are:

1. A rigid end that does not allow for displacements, u:

$$u(x_0, t) = 0 \qquad [1]$$

2. A free end where no stress, σ_x, exists on the surface:

$$\left.\frac{\partial u(x,t)}{\partial x}\right|_{x_0} = 0 \qquad [2]$$

See **Figure 1**. If E is Young's modulus, and x_0 is an interface between two media, 1 and 2, then continuity of displacements and equality of stresses on both sides of the interface yield interface conditions:

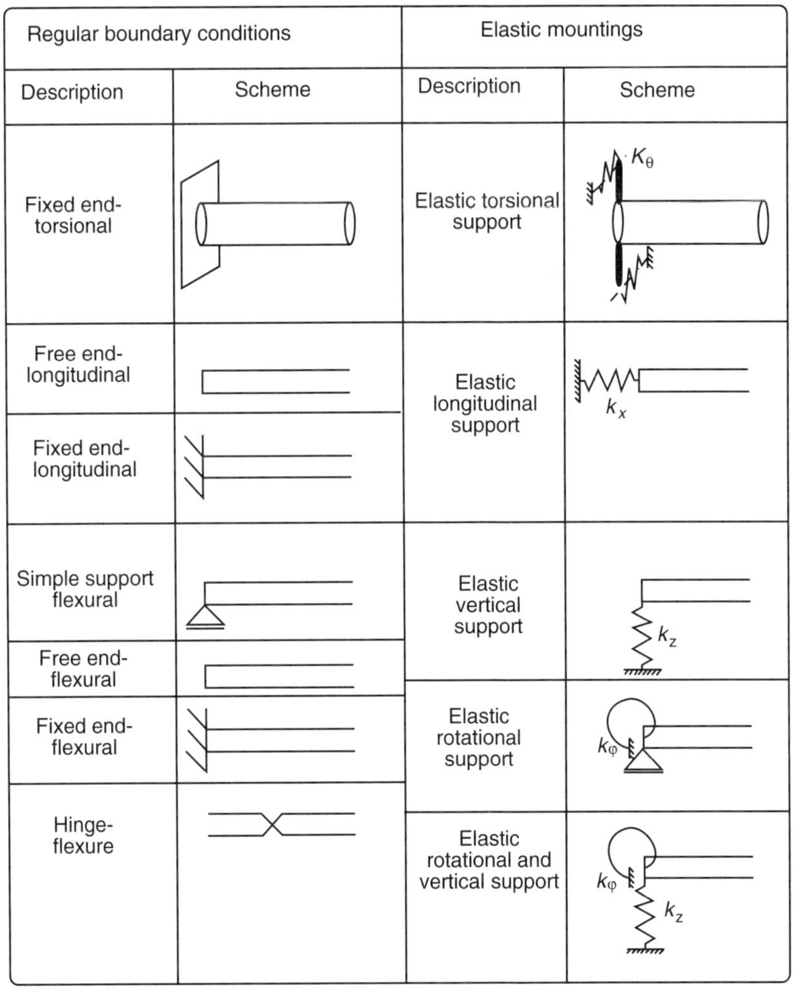

Figure 1 Ends and elastic mountings of rods and beams.

$$(u_1 - u_2)_0 = 0, \quad E_1 \frac{\partial u_1}{\partial x} - E_2 \frac{\partial u_2}{\partial x}\bigg|_{x_0=0} = 0 \quad [3]$$

Eqn [3] also expresses the interface conditions for longitudinal waves.

A specific important case is the boundary interaction of a harmonic wave of radian frequency ω. Such a wave in a complex form is:

$$\begin{aligned} u(\omega, x, t) &= \hat{C}_1(\omega) \exp[i(kx - \omega t + \psi_1)] + \hat{C}_2(\omega) \\ &\quad \times \exp[-i(kx + \omega t + \psi_2)] \\ &= C_1(\omega) \exp[i(kx - \omega t)] + C_2(\omega) \\ &\quad \times \exp[-i(kx + \omega t)] \end{aligned} \quad [4]$$

where $k = \omega/c$ is the wave number, and c is the velocity of propagation of the wave. Only the real part is to be taken into account in the last equation for physical interpretation. The first term in the right-hand side of eqn [4] is a wave propagating in x-direction, while the second one is propagating in the opposite direction. ψ_1 and ψ_2 are the phases of the two terms respectively. A whole signal can be built of harmonic components, and when this dependence is continuous, the Fourier integral is used:

$$u(x,t) = \int_{-\infty}^{\infty} \left\{ \begin{array}{l} C_1(\omega) \exp[i(kx - \omega t)] \\ + C_2(\omega) \exp[-i(kx + \omega t)] \end{array} \right\} d\omega \quad [5]$$

One harmonic term can suffice to illustrate the effect on the boundary conditions. We consider one term that propagates towards $x = 0$ within the domain $-\infty < x \leq 0$:

$$u(\omega, x, t) = C_1(\omega) \exp[i(kx - \omega t)] \quad [6]$$

Assume a rigid boundary at $x = 0$, then by satisfying eqn [1] at $x_0 = 0$, a reflected wave that propagates opposite x in the domain $-\infty < x \leq 0$ occurs. This wave component has the same frequency and magnitude of amplitude as the incident wave, but the reflected amplitude has a negative sign. Hence, due to the specific boundary condition, the displacement function of the rod becomes:

$$u(\omega, x, t) = 2iC_1 \sin(kx) \exp(-i\omega t); \quad -\infty < x \leq 0 \quad [7]$$

For a free end at $x = 0$, and the same incident wave, eqn [2] has to be satisfied. Again, a reflection occurs with the same frequency and magnitude of amplitude, but with a positive sign of the amplitude, and the result becomes:

$$u(\omega, x, t) = 2C_1 \cos(kx) \exp(-i\omega t); \quad -\infty < x \leq 0 \quad [8]$$

Given two semiinfinite rods that are firmly connected at $x = 0$, the properties of the whole rod are Young's modulus E_1 and density ρ_1 at $x < 0$ and E_2 and ρ_2 at $x > 0$. $u(\omega, x, t) = C_1(\omega) \exp[i(kx - \omega t)]$ of eqn [6] propagates within the domain $-\infty < x \leq 0$ towards $x = 0$. In this case a reflection wave with a reflection coefficient R and a transmitted wave with a transmission coefficient Tr are needed in order to satisfy the boundary conditions (eqn [3]), while the frequency of these added waves is the same as that of the incident wave. R and Tr are defined by Fresnel formulae:

$$R = \frac{\sqrt{\rho_1 E_1} - \sqrt{\rho_2 E_2}}{\sqrt{\rho_1 E_1} + \sqrt{\rho_2 E_2}} = \frac{n - m}{n + m};$$

$$Tr = \frac{2n}{n + m}; \quad n = \frac{c_1}{c_2}; \quad m = \frac{\rho_2}{\rho_1} \quad [9]$$

From Wave Propagation to Vibrations via Boundary Conditions

Disturbing a rod from its rest conditions by $u(x, 0) = 0; 0 < x < L$, without external excitation when $t > 0$, then the rod should obey the longitudinal homogeneous wave equation:

$$\frac{\partial^2 u}{\partial t^2} - c^2 \frac{\partial^2 u}{\partial x^2} = 0; \quad t > 0 \quad [10]$$

This homogeneous second-order differential equation with constant coefficients has a solution of the form:

$$u(x, t) = \hat{u}(x) \exp(-i\omega t) \quad [11]$$

Substituting eqn [11] into eqn [10] yields:

$$\frac{\partial^2 \hat{u}}{\partial x^2} + k^2 \hat{u} = 0; \quad k = \frac{\omega}{c} \quad [12]$$

The solution to the last equation is:

$$\hat{u} = C \sin(kx) \quad [13]$$

Eqn [13] has to satisfy the boundary conditions of the rod. Say that the rod is fixed at $x = 0$ and $x = L$, which means $\hat{u}(0) = u(L) = 0$, then the boundary condition at $x = 0$ is satisfied identically. However, the boundary condition at $x = L$ dictates $\sin(kx) = 0$, which leads to a series of discrete possible motions satisfying:

$$k_n = \frac{n\pi}{L}; \quad n = 1, 2, 3 \ldots$$

or:

$$\omega_n = \frac{n\pi c}{L}; \quad n = 1, 2, 3 \ldots \quad [14]$$

Hence, due to the boundary conditions also under excitation, the vibrations of the rod can carry in general only a combination of specific mode shapes $\phi_n(x)$:

$$\hat{u}_n(x) = \hat{C}_n \phi_n(x) = \hat{C}_n \sin\left(\frac{n\pi x}{L}\right) \quad [15]$$

For other boundary conditions other mode shapes are suitable.

The total solution for free vibrations carries the form:

$$u(x,t) = \sum_{n=1}^{M} [C_n \phi_n(x) \exp(i\omega_n t)]; \quad [16]$$
$$C_n = \hat{C}_n \exp(i\psi_n)$$

$\hat{C}_n$ and ψ_n are arbitrary and depend on the initial conditions.

Longitudinal Elastic Mounting

Between the states of free and fixed-end conditions, a state of elastic support with a reactive force that limits the motion can be constructed. The larger the displacement at the end, then the larger is the reactive force. Such a situation can occur if, for example, a spring is attached to the end of the rod and constrained at its other end (**Figure 1**). If k_x is the spring constant in x-direction, then the boundary condition is:

$$\left.\frac{\partial u(x,t)}{\partial x}\right|_{x_0} = \frac{k_x u(x_0, t)}{ES} \quad [17]$$

where S is the cross-section area of the rod.

Boundary Conditions for a Torsion Wave

Torsion vibration of a rod is one-dimensional, like longitudinal vibration. In fact, the angle of rotation $\theta(x,t)$ replaces the longitudinal displacement $u(x,t)$, the torque $M_T(x,t)$ replaces the axial force $F(x,t)$, the polar cross-sectional moment of inertia I_p replaces the cross-section area S, the shear modulus of elasticity G replaces Young's modulus E, and the velocity of torsion wave propagation c_T replaces the velocity of propagation of the longitudinal wave c.

The rest is analogous. Hence, the regular torsion boundary conditions are:

1. A rigid end that does not allow for rotation:

$$\theta(x_0, t) = 0 \quad [18]$$

2. A free end where no torque, $M_T = GI_p(\partial\theta/\partial x)$, exists:

$$\left.\frac{\partial \theta(x,t)}{\partial x}\right|_{x_0} = 0 \quad [19]$$

The homogeneous equation of torsion motion is also analogous to the longitudinal one:

$$\frac{\partial^2 \theta}{\partial x^2} + \frac{1}{c_T^2}\frac{\partial^2 \theta}{\partial t^2} = 0 \quad [20]$$

Boundary Conditions for a Beam

Lateral (Flexural) Vibrations

The effect of boundary conditions on both natural frequencies and normal modes is shown in this section for a straight beam with different end conditions. Generally, in a three-dimensional problem there are six resultant generalized force components that can act on the cross-section of the beam (three forces and three moments); **Figure 2** (bottom). However, most two-dimensional models of beams are satisfactory, as shown in **Figure 2** (top). The beam in this case has a length L, cross-section area S, rigidity EI, density ρ, and mass per unit length m. I is the moment of inertia of the cross-section, around the z-axis, through the center of gravity.

The free (without external excitation) deflection $v(x,t)$ of the beam is given by the following fourth-order differential equation:

$$EI\frac{\partial^4 v}{\partial x^4} + \rho S\frac{\partial^2 v}{\partial t^2} - \rho I\frac{\partial^4 v}{\partial t^2 \partial x^2} = 0 \quad [21]$$

In the case of negligible rotary inertia, which is a common situation, the equation of motion becomes:

$$EI\frac{\partial^4 v}{\partial x^4} + \rho S\frac{\partial^2 v}{\partial t^2} = 0; \quad \text{or} \quad r_0^2 \frac{E}{\rho}\frac{\partial^4 v}{\partial x^4} + \frac{\partial^2 v}{\partial t^2} = 0;$$
$$r_0 = \sqrt{\frac{I}{S}} \quad [22]$$

The most common boundary conditions, ignoring axial forces, are defined **Table 1** and **Figure 1**.

The general solution to eqn [22] for a wave propagating in x-direction is $v(x,t) = C_1 \phi(x) \exp(-i\omega t)$

184 BOUNDARY CONDITIONS

Table 1 Various flexural boundary conditions

Type of boundary or joint	Conditions
Free boundary	$d^2\phi/dx^2 = 0$; $d^3\phi/dx^3 = 0$
Clamped boundary	$\phi = 0$; $d\phi/dx = 0$
Pinned or simply supported boundary	$\phi = 0$; $d^2\phi/dx^2 = 0$
Sliding boundary	$d\phi/dx = 0$; $d^3\phi/dx^3 = 0$
An internal hinge	$d^2\phi/dx^2 = 0$

and, when substituted in the equation of motion, leads to the ordinary differential equation for $\phi(x)$:

$$\frac{d^4\phi(x)}{dx^4} - k^4\phi = 0; \quad k = \sqrt[4]{\frac{\rho}{Er_0^2}}\sqrt{\omega} \quad [23]$$

The general solution for the ω is:

$$v(x,t) = C_1 \exp[i(kx - \omega t)] + C_2 \exp[-i(kx - \omega t)] + C_3 \exp[kx - i\omega t]C +_4 \exp[-kx - i\omega t];$$

$$k = \sqrt[4]{\frac{\rho}{Er_0^2}}\sqrt{\omega}$$

[24]

The two last terms on the right-hand side of eqn [24] are nonpropagating harmonic oscillations, that appear mostly near the beam ends or obstructions along the beam. The first two are propagating in x-direction and in the opposite, at the phase velocity:

$$c_{\text{ph}} = \pm\frac{\omega}{k} = \sqrt[4]{\frac{Er_0^2}{\rho}}\sqrt{\omega} \quad [25]$$

As in the case of longitudinal waves, boundary conditions cause reflection of bending waves. $v(\omega, x, t) = C_1(\omega) \exp[i(kx - \omega t)]$ eqn [24] propagates within the domain $-\infty < x \leq 0$ towards $x = 0$,

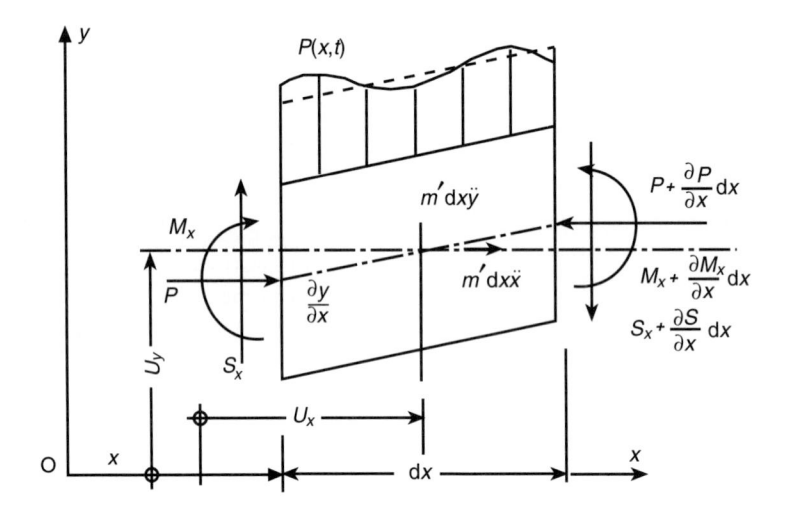

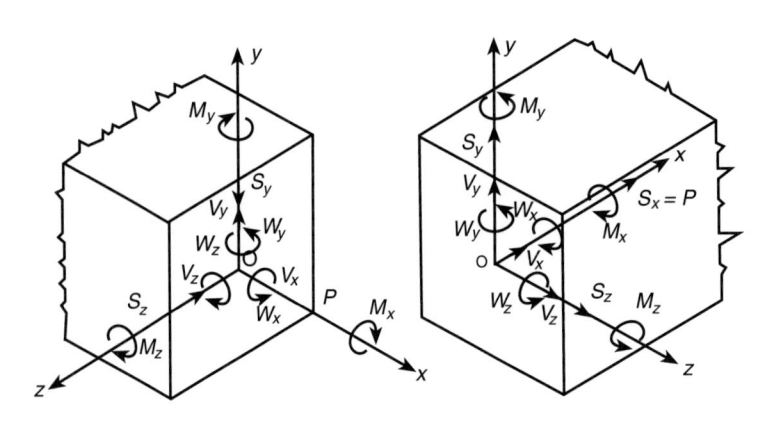

Figure 2 Notation of displacements and forces.

where the beam is clamped ($\phi = 0$; $d\phi/dx = 0$). In this case a reflected wave with a reflection coefficient R and the evanescent wave of a coefficient V are needed in order to satisfy the specified boundary conditions, while the frequency of these added waves is the same as that of the incident wave. Introducing into eqn [24]:

$$C_1 = C; \quad C_2 = RC; \quad C_3 = VC; \quad C_4 = 0 \quad [26]$$

and using the boundary conditions of a fixed end, the reflection and evanescence coefficients are:

$$R = \frac{1-i}{1+i}; \quad V = \frac{-2i}{1+i} \quad [27]$$

and finally, the solution of the sound field for the fixed end becomes:

$$v(x,t) = C\left\{\exp[i(kx - \omega t)] + \frac{1-i}{1+i}\right.$$
$$\left.\times \exp[-i(kx - \omega t)] - \frac{2i}{1+i}\exp[kx - i\omega t]\right\}$$
$$[28]$$

Excluding the cases where the boundary or support is elastically held against deflection and rotation and ignoring longitudinal effects, the lateral mode shapes can be defined as $\phi_n(x/L)$; $n = 1, 2, 3\ldots$ In the cases presented in the example the mode shapes carry the form:

$$\phi_n\left(\frac{x}{L}\right) = \cosh\left(\vartheta_n \frac{x}{L}\right) + \cos\left(\vartheta_n \frac{x}{L}\right)$$
$$- \xi_n\left[\sinh\left(\vartheta_n \frac{x}{L}\right) + \sin\left(\vartheta_n \frac{x}{L}\right)\right]; \quad [29]$$
$$n = 1, 2, 3\ldots$$

and the nth natural frequency:

$$f_n = \frac{\vartheta_n^2}{2\pi L^2}\sqrt{\frac{EI}{m}}; \quad n = 1, 2, 3\ldots \quad [30]$$

The values of ϑ_n and ξ_n depend on the boundary conditions and on the nth mode shape.

A relatively simple example is of a simply supported beam of length L, where $\phi = 0$; $d^2\phi/dx^2 = 0$ at $x = 0$ and $x = L$. These boundary conditions lead to the following two characteristic equations:

$$\begin{bmatrix} \sin(kL) & \sinh(kL) \\ -\sin(kL) & \sinh(kL) \end{bmatrix}\begin{bmatrix} C_1 \\ C_2 \end{bmatrix} = \begin{bmatrix} 0 \\ 0 \end{bmatrix} \quad [31]$$

For a nonzero solution we have:

$$\sin(kL) = 0 \rightarrow k_n = \frac{n\pi}{L}; \quad n = 1, 2, 3\ldots$$
$$\rightarrow \omega_n = \sqrt{\frac{Er_0^2}{\rho}}\left(\frac{n\pi}{L}\right)^2 \quad [32]$$

and:

$$v_n = C\sin(k_n x)\exp(-i\omega t) \quad [33]$$

Illustration of the Effect of Boundary Condition of a Beam

The coefficients ϑ_n and ξ_n for various boundary conditions and modes are given in **Table 2**. A comparison of the effect of the various boundary conditions considered in **Table 2** shows how different boundary conditions lead to significantly different modes and natural frequencies.

Elastically Mounted Beams

Between the states of free and fixed supports there are states of elastic mountings. The elastic mounting allows for deflection and rotation under the action of forces and moments respectively (**Figure 1**). The boundary condition in this case equalizes the action and reaction at the mounting. The elastic reaction can be against the linear motion, rotation or both, as can be seen from **Figure 1**. The use of such boundary conditions is demonstrated by the following example.

Given a beam as shown in **Figure 3**; m is the mass per unit length of the beam, and L is the span of the beam. The beam is simply supported at A and elastically mounted at B. k_z is the spring constant of the beam. The natural frequency of the beam is sought. If the reaction of the beam at B is $R_B = mgL/2$, then $v_B = R_B k_z$ is the deflection at B. The natural radian frequency can be estimated by Rayleigh's method. If the midspan of the beam is considered as the location of the equivalent mass, then the equivalent mass and spring constant are respectively:

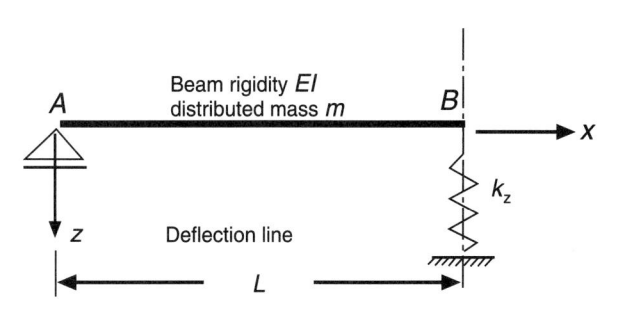

Figure 3 Elastically mounted beam.

Table 2 The coefficients ϑ_n and ξ_n for various boundary conditions and modes

Boundary conditions →	Free–free	Clamped–free	Clamped–pinned	Clamped–clamped	Clamped–sliding
ϑ_1	4.73004074	1.87510407	3.92660231	4.73004074	2.36502037
ϑ_2	7.85320462	4.69409113	7.06858275	7.85320462	5.49780392
ϑ_3	10.9956078	7.85475744	10.21017612	10.9956079	8.63937983
ϑ_4	14.1371655	10.99554073	13.35176878	14.1371655	11.78097245
ϑ_5	17.2787597	14.13716839	16.49336143	17.2787597	14.92256510
$\vartheta_n, n > 5$	$0.5\pi(2n+1)$	$0.5\pi(2n-1)$	$0.25\pi(4n+1)$	$0.5\pi(2n+1)$	$0.25\pi(4n-1)$
ξ_1	0.982502215	0.734095514	1.000777304	0.982502215	0.982502207
ξ_2	1.000777312	1.018467319	1.000001445	1.000777312	0.999966450
ξ_3	0.999966450	0.999224497	1.000000000	0.999966450	0.999999933
ξ_4	1.000001450	1.000033553	1.000000000	1.000001450	0.999999993
ξ_5	1.999999937	0.999998550	1.000000000	0.999999937	0.999999993
$\xi_n, n > 5$	~ 1.0	~ 1.0	1.0	1.0	1.0

$$M_{eq} = \frac{m^2 gL\left\{\dfrac{25L^6}{2 \times 384^2 EI} + \dfrac{1}{12k_z^2} + \dfrac{5}{384}\dfrac{L^2}{\pi EI k_z}\right\}}{\left(\dfrac{5}{384}\dfrac{L^2}{EI} + \dfrac{1}{4k_z}\right)^2};$$

$$K_{eq} = \frac{48EIk_z}{L^3 k_z + 12EI} \quad [34]$$

and:

$$\omega^2 = \frac{K_{eq}}{M_{ek}} \quad [35]$$

It should be noted that, if the spring's constant at the support is complex, then it absorbs vibration energy. In this case we talk about damped supports.

Joints that Couple Longitudinal and Flexural Waves

Structures encounter a variety of joints, and examples are presented in **Figure 4**. a-1 and a-2 describe obstructions, a-3 to a-5 illustrate additional elements that differ from the original elements, a-6 is a corner, a-7 depicts columns within a wall, a-8 and a-9 are schemes of T joints. All the other examples stand for a variety of branching elements, including hinged ones. There have been results where coupling during flanking is ignored (**Figure 5**; see Further reading).

An extended solution has been developed, taking into account simultaneously axial forces shear and moments at a joint in a two-dimensional problem (**Figure 6**). This extension was verified experimentally, and used to analyze a variety joints and structures (**Figure 7**).

Three-dimensional Boundary Value Problems in Mechanics

The fundamental boundary value problems of elasticity involve prescribed distribution of data over the surface of the body. These can be of two types:

1. Displacement: u_i, $i = 1, 2, 3$
2. External forces: $t_{(n)i}$, $i = 1, 2, 3$

Hence, it is reasonable to present the problem to be solved in terms of either displacements or stresses. The state of stress and deformation in an anisotropic material involves a stress (τ_{kl}, $k, l = 1, 2, 3$)–strain (e_{ij}, $i, j = 1, 2, 3$) relation using repeated indices for the summation convention, as follows:

$$e_{ij} = d_{ijkl}\tau_{kl} \quad [36]$$

where d_{ijkl} is the tensor of elastic compliance. In the case of an isotropic material we have:

$$e_{ij} = \frac{1}{2\mu}\left(\tau_{ij} - \frac{\lambda}{2\mu + 3\lambda}\tau_{kl}\delta_{ij}\right) \quad [37]$$

where λ and μ are the Lamé coefficients and δ_{ij} is the delta of Kronecker.

As mentioned before, the prescribed boundary conditions are to be expressed either by displacements or stresses. In the first case the strains are given in terms of displacements:

$$e_{ij} = 0.5\left(u_{i,j} - u_{j,i}\right) \quad [38]$$

In the second case, the following constitutive relations give the strains:

$$e_{ij} = d_{ijkl}\tau_{kl} \text{ or } e_{ij} = \frac{1}{2\mu}\left(\tau_{ij} - \frac{\lambda}{2\mu + 3\lambda}\tau_{kl}\delta_{ij}\right) \quad [39]$$

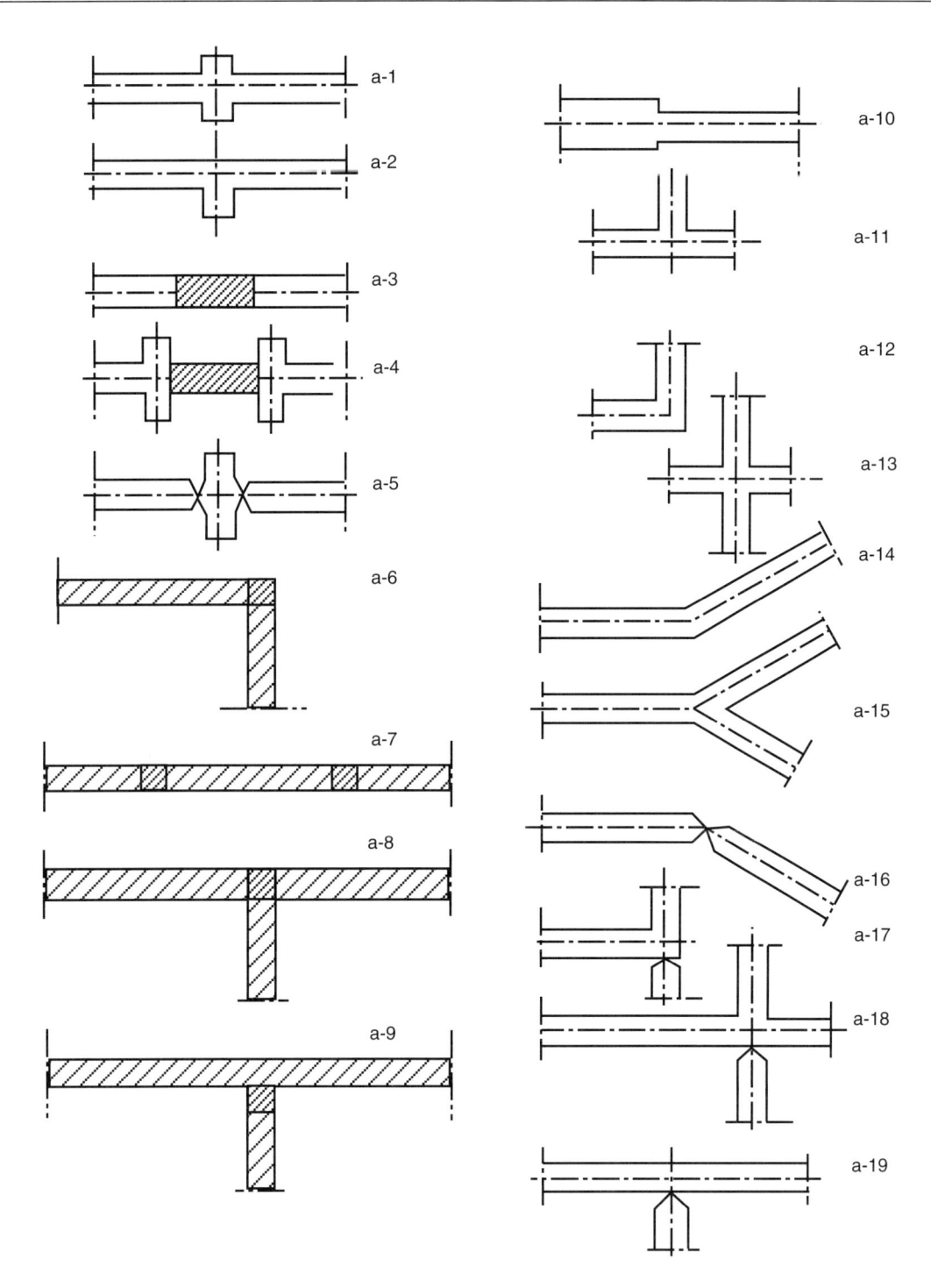

Figure 4 Various kinds of single joints.

The compatibility equations are:

$$e_{ij,kl} + e_{kl,ij} - e_{ik,jl} - e_{jl,ik} = 0 \qquad [40]$$

They are satisfied if the displacements are sufficiently regular functions of position. The dynamic equilibrium equations are:

$$\tau_{ij,i} + F_i = \rho u_{i,tt} \qquad [41]$$

F_i designates body force, ρ is density and t denotes time. For unknown displacements only, the dynamic equilibrium equations become:

$$\mu u_{i,kk} + (\lambda + \mu) u_{kk,i} + F_i = \rho u_{i,tt} \qquad [42]$$

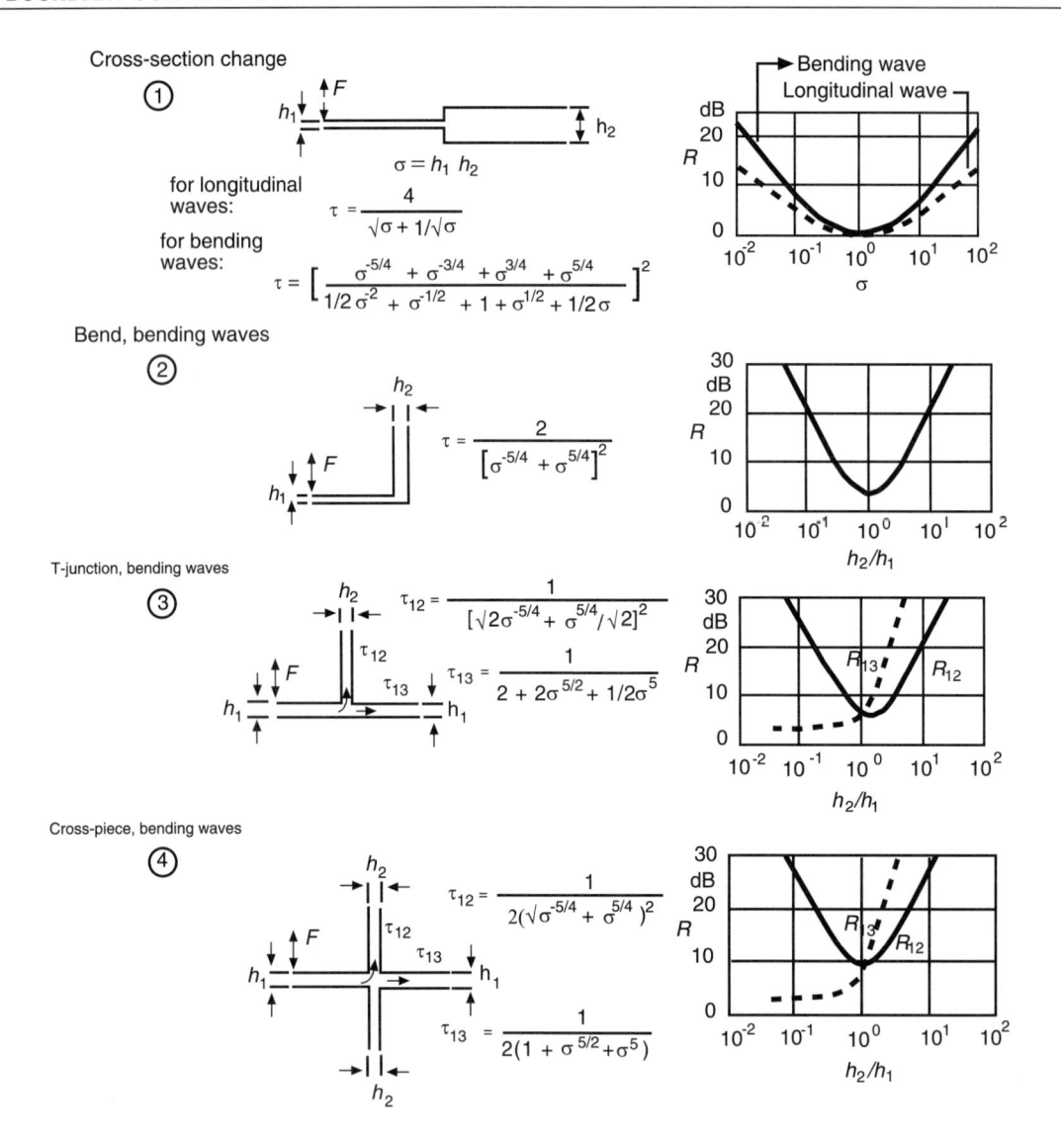

Figure 5 Flanking through joints. From Heckl M (1981) The tenth Richard Fairey memorial lecture: sound transmission in buildings. *Journal of Sound Vibration* 77: 165–180 with permission

Consequently, the boundary value problem includes the last field equation and the boundary conditions of displacements:

$$u_i = u_{i0} \text{ over } \Gamma_u = \Gamma \qquad [43]$$

Γ is the boundary of the investigated domain Ω. Γ coincides in this case with the boundary Γ_u of the prescribed displacements.

If prescribed tractions are imposed on the boundary, the equilibrium equation [41] and the compatibility condition [40] are to be obeyed, using Hooke's law for compatibility. Yet, satisfaction of the dynamic equilibrium equations does not automatically satisfy the deformation equations. This problem of mismatch between the equilibrium equation and compatibility has already been addressed (Further reading). At the boundary surface, the following equation has to be satisfied:

$$t_{(n)i} = \tau_{ij} \text{ over } \Gamma_\tau = \Gamma \qquad [44]$$

where n_i is the external normal. Γ is the boundary of the investigated domain Ω, which coincides in this case with boundary, Γ_τ, of prescribed tractions.

There are also cases where over a part, Γ_u, of the boundary Γ, the displacements are prescribed, and on the other part, Γ_τ, the tractions are prescribed. In the last case, we have $\Gamma_u + \Gamma_\tau = \Gamma$.

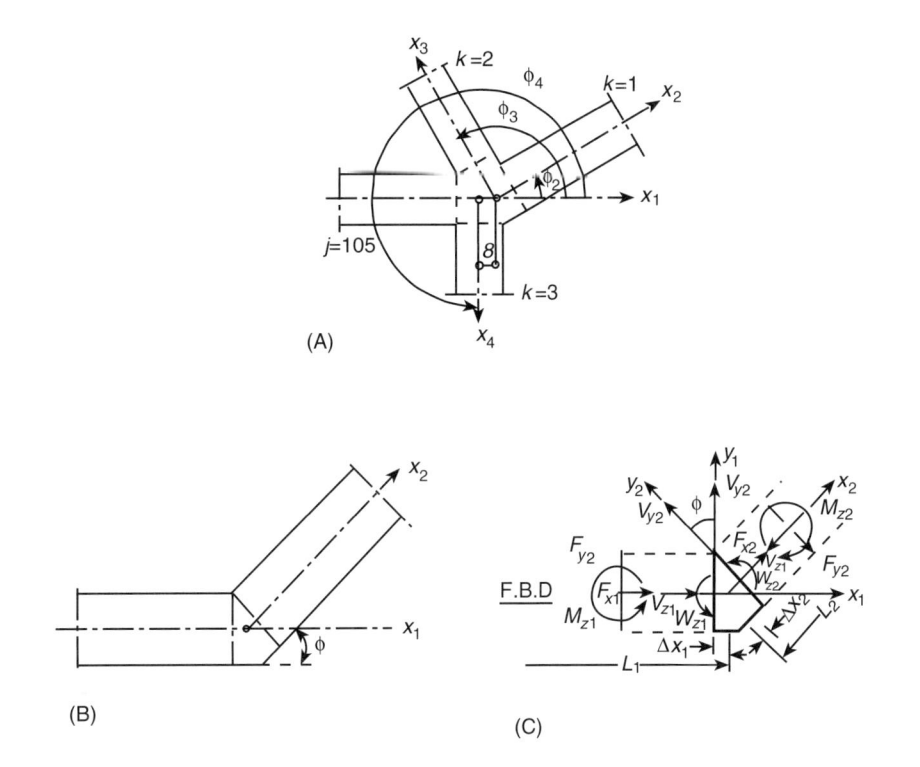

Figure 6 Flanking through a joint: (A) Branched joint; (B) bent one-dimensional wave guide; (C) free body diagram of a bend. From Rosenhouse G (1979) Acoustic wave propagation in bent thin-walled wave guides. *Journal of Sound Vibration* 67: 469–486 with permission.

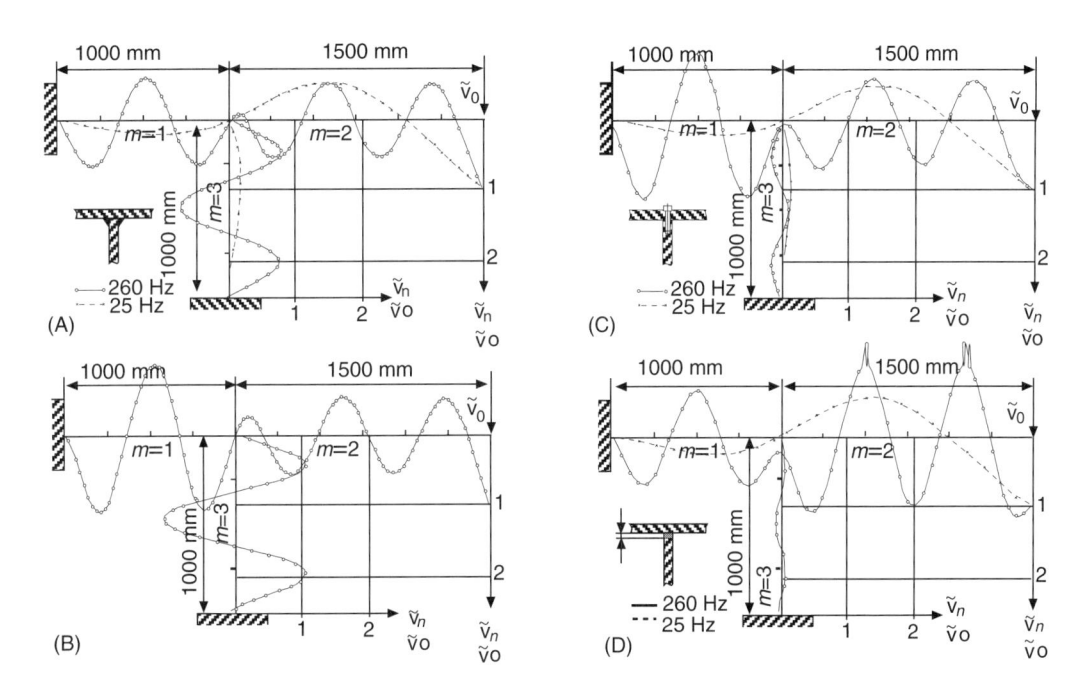

Figure 7 Sound transmission through a joint – theory and experiment. From Rosenhouse G, Ertel H and Mechel FP (1981) Theoretical and experimental investigations of structure-borne sound transmission through a 'T' joint in a finite system. *Journal of the Acoustic Society of America* 70: 492–499.

Moving Boundaries

As has been shown up to now, the boundary conditions are a major influence on the resulting vibrating field. If they are forced to move in certain ways, such as in the case of longitudinally moving support, they can form sources of vibration and sound. Such special boundary conditions have specific effects. Typical moving boundaries of technological interest are rotating blades. This is a rather complicated motion that sometimes involves a longitudinal motion as well, as in airplanes and helicopters. The forces that act on the blades enforce vibrations. This subject relates to the study of aerodynamics (see Further reading).

Moving boundaries also appear in contact problems of elastic bodies. When two elastic bodies collide, their contact area varies as a function of the contact force. Compatibility, equilibrium, and constitutive relations are to be satisfied at the contact area, and the definition of these begins with the simple models of Hertz and are supported today by numerical techniques.

Moving boundaries may be a consequence of nonmechanical processes. An example is the definition of domains in a melting/solidification of a certain material. The definition of the moving melting front is Stefan's problem. This varying boundary (interface) problem can combine with a vibration problem.

Nonreflecting boundaries

Earthquakes are geophysical mechanical sources of vibrations. Such sources radiate energy from distant locations in a domain that can be considered for the purpose of structural analysis to be theoretically infinite. However, estimates of vibrations generated in such large domains are difficult, especially if they are done numerically. Ways of confining the domain of analysis were sought as a result. A finite domain, where the vibration field is calculated, is cut out from the infinite one, yet defining boundary conditions that replace the removed infinite space at the cut or artificial boundary, without changing the vibration field within the investigated domain (**Figure 8**).

Nonreflecting boundaries mean local impedance matching. However, this necessity can lead to a more complicated formulation in order to secure nonreflecting transmission of all the components of stress and displacement along the whole artificial boundary. This leads to an integral (nonlocal) formulation of the boundary conditions.

Semidefinite systems

In many problems the whole system is moving under certain excitation forces. It might be a moving elastic elongated body. Ignoring air resistance, the boundary

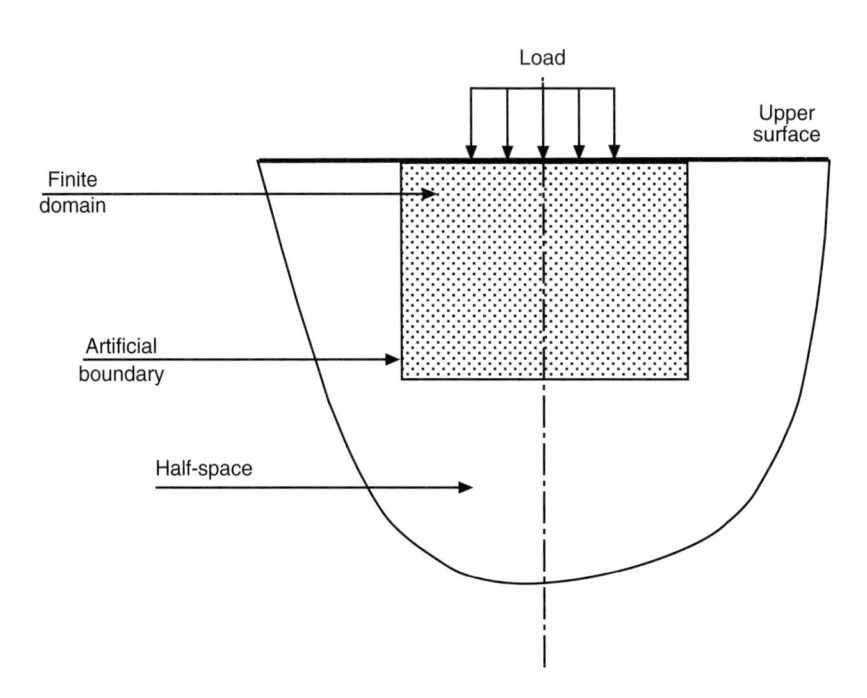

Figure 8 Nonreflecting boundaries in a half-space.

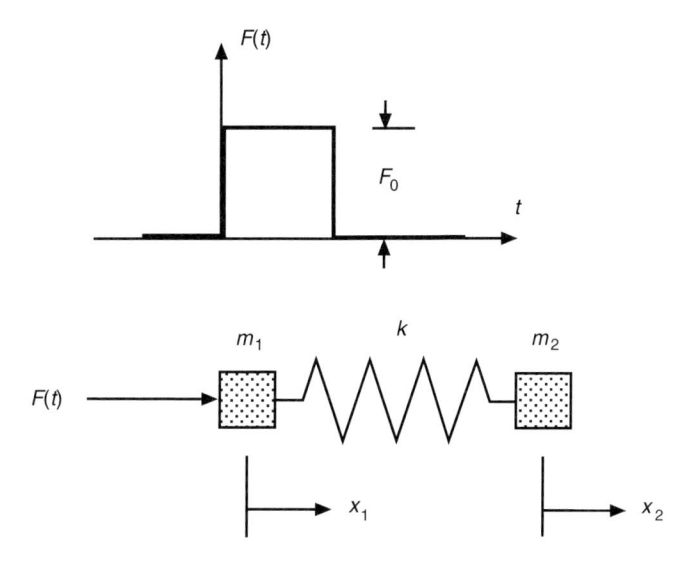

Figure 9 A semidefinite system of two masses linked by a spring.

conditions at the ends of the body can be defined as free–free. A simple example is shown in **Figure 9**. Two masses, m_1 and m_2, are linked together by a spring with a constant k. The behavior of this system depends on the excitation function. If the excitation is a rectangular pulse of constant amplitude in the range $0 < t < t_0$, then, within the range of excitation, the average velocity of the whole system will rise linearly, and at the same time the masses will oscillate about the center of gravity of the system. Later, when no additional energy is supplied to the system, the average velocity will not change, but the masses will continue to oscillate.

Nomenclature

E	Young's modulus
EI	rigidity
$F(x, t)$	axial force
G	shear modulus of elasticity
I	moment of inertia
L	length
R	reflection coefficient
S	cross-section area of rod
Tr	transmission coefficient
ρ	density
$\theta(x, t)$	angle of rotation

See also: **Basic principles**; **Beams**; **Structural dynamic modifications**; **Theory of vibration**, Variational methods; **Vibration isolation theory**; **Wave propagation**, Interaction of waves with boundaries.

Further Reading

Blevins RD (1977) *Flow-Induced Vibration*. New York: Van Nostrand Reinhold.

Chang TC and Craig RR (1969) Normal modes of uniform beams. *Journal of Engineering Mechanics Division, American Civil Engineers* 95: 1027–1031.

Crank J (1984) *Free and Moving Boundary Problems*. Oxford: Clarendon Press.

Goldsmith W (1960) *Impact*. London: Edward Arnold.

Heckl M (1981) The tenth Richard Fairey memorial lecture: sound transmission in buildings. *Journal of Sound Vibration* 77: 165–189.

Johnson KL (1987) *Contact Mechanics*. New York: Cambridge University Press.

Keller JB and Givoli D (1989) Exact non-reflecting boundary conditions. *Journal of Computational Physics* 82: 172–192.

Prager W and Synge JL (1947) Approximations in elasticity based on the concept of function space. *Quarterly Journal of Applied Mathematics* 5: 241–269.

Rosenhouse G (1979) Acoustic wave propagation in bent thin-walled wave guides. *Journal of Sound Vibration* 67: 469–486.

Rosenhouse G, Ertel H and Mechel FP (1981) Theoretical and experimental investigations of structureborne sound transmission through a 'T' joint in a finite system. *Journal of the Acoustic Society of America* 70: 492–499.

BOUNDARY ELEMENT METHODS

F Hartmann, University of Kassel, Kassel, Germany

Copyright © 2001 Academic Press

doi:10.1006/rwvb.2001.0111

The boundary element method is an integral equation method. Approximations are made only on the boundary or surface of the problem domain. No interior mesh of elements need be specified and at each interior point the governing differential equation is satisfied exactly.

This reduction of the dimension of the problem is the main advantage of the method. In cases where the domain extends to infinity, as in potential flow problems or in seismic wave problems, this advantage is even more striking because the equation governing the infinite domain is reduced to an integral equation over the boundary of the domain.

The boundary element method is applicable to most of the classical problems of mechanics. The method requires that the problem is linear – although the method nowadays is also applied successfully to nonlinear problems – and that a fundamental solution of the governing equation is known.

In the following sections we shall concentrate on the fundamental principles behind the boundary element method and show how it is applied to time harmonic and transient problems in vibration analysis.

Fundamental Solutions

From a physical point of view, the boundary element method is a method of sources or potentials. The smallest source is a point load whose magnitude oscillates in time (harmonic problems) or a short impulse which hits the structure at a time, T, for an infinitesimal small time step dt as in transient problems.

Mathematically speaking the point sources are Dirac δ functions and the solution of the corresponding field equations, for example, the Poisson equation:

$$-\Delta u(y) = \delta(y - x)$$

are called fundamental solutions:

$$g_0(y, x) = \frac{1}{2\pi} \ln r \quad r = |y - x|$$

They are homogeneous solutions of the differential equation at all points y which do not coincide with the source point, x. Typically these functions are symmetric with respect to the source point x and field point y, so that $g_0(y, x) = g_0(x, y)$. At the source point, x, they exhibit the $1/r$ (or higher) behavior in the stresses – here it is actually the slope:

$$\frac{\partial}{\partial v} g_0(y, x) = \frac{1}{2\pi} \frac{1}{r} (r_{,y1} v_1 + r_{,y2} v_2)$$

which guarantees that in the limit the integral of the stresses over any circle centered at the source point x tends to the unit point load (**Figure 1**).

$$\lim_{r \to 0} \int_{\Gamma_r} \frac{1}{2\pi r} ds = \int_0^{2\pi} \frac{1}{2\pi r} r \, d\varphi = 1$$

Fundamental solutions are also called free space Green's functions because they formulate the response of the infinite medium to the action of a point load. If the same load is applied to, say, a circular plate, then boundary conditions must be met and we speak of a Green's function, $G_0(y, x)$. If we know what happens to a structure under the action of a point load, we can instantly formulate the answer to a load p spread over any patch Ω_p:

$$u(x) = \int_{\Omega_p} G_0(y, x) p(y) \, d\Omega y$$

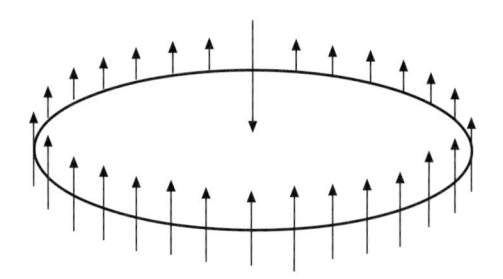

Figure 1 Point source.

This explains why the boundary element method is also called a method of influence functions.

Influence functions are linear by nature: they are L_2-scalar products between kernel functions and domain or boundary layers. For the boundary element method to be applicable the physical nature of the problem must therefore be linear and we must know the fundamental solution of the governing equation – the response of the infinite medium to a point load or point excitation.

The Direct Method

Most of the linear differential operators in physics, for example, the Laplacian, are of even order, say $2m$, they are self-adjoint and we have an identity such as:

$$G(u, \hat{u}) = \int_\Omega -\Delta u \hat{u} \, d\Omega + \int_\Gamma \frac{\partial u}{\partial n} \hat{u} \, ds - \int_\Omega \nabla u \cdot \nabla \hat{u} \, d\Omega = 0$$

which allows us to formulate Green's second identity as well:

$$B(u, \hat{u}) = G(u, \hat{u}) - G(\hat{u}, u)$$
$$= \int_\Omega -\Delta u \hat{u} \, d\Omega + \int_\Gamma \frac{\partial u}{\partial n} \hat{u} \, ds - \int_\Gamma u \frac{\partial \hat{u}}{\partial n} \, ds - \int_\Omega u(-\Delta \hat{u}) \, d\Omega = 0$$

on any domain Ω with boundary Γ where u and $\hat{u}$ are smooth enough, $u, \hat{u} \in C^2(\Omega)$. The B stands for Betti. In elastodynamics the equivalent identity goes under the name of Graffi's theorem.

If we substitute for $\hat{u}$ the fundamental solution, consider the identity on the punctured domain $\Omega_\varepsilon = \Omega - N_\varepsilon(x)$ (**Figure 2**) and let the radius ε of the cut-out N_ε tend to zero:

$$\lim_{\varepsilon \to 0} B(u, g_0[x])_{\Omega_\varepsilon} = 0 \quad [1]$$

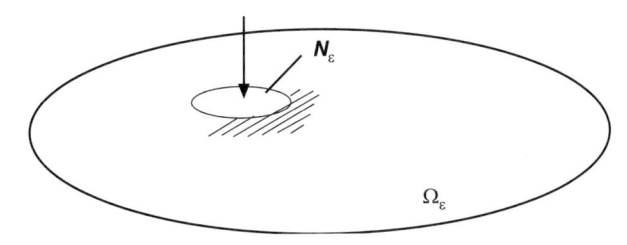

Figure 2 Punctured domain.

then we obtain an integral representation for the field u:

$$c(x)u(x) = \int_\Gamma \left[g_0(y, x) \frac{\partial u(y)}{\partial v} - \frac{\partial g_0(y, x)}{\partial v} u(y) \right] ds_y$$
$$+ \int_\Omega g_0(y, x) p(y) \, d\Omega_y \quad [2]$$

where $c(x)$ is the characteristic function:

$$= c(x)$$
$$\begin{cases} 1 & x \in \Omega \\ \Delta\varphi/2\pi & x \in \Gamma, \quad \Delta\varphi = \text{angle at the boundary point} \\ 0 & x \in \Omega^c \quad \text{(outside)} \end{cases}$$
$$[3]$$

Hence we can compute the solution u of the field equation at any interior point x by a series of L_2-scalar products of known kernel functions and boundary layers and (one or more) domain functions p.

In the case of the Laplacian where $2m = 2$, the two kernel functions

$$\left(g_0(y, x), \; \frac{\partial}{\partial v} g_0(y, x) \right)$$

are problem-independent, while the $2m$ boundary layers $(u(y), \partial u(y)/\partial v)$ are not. They are the boundary values of the solution u. For a boundary value problem to be consistent with the physical nature of the problem, only m of these $2m$ boundary values can be prescribed. If a boundary value, say the slope $\partial u/\partial n$ is given, then the conjugated boundary value – the function u in this case – is unknown.

To find these unknown boundary layers we have to take the point x in the integral representation (2) to the boundary. By this maneuver the integral representation becomes an integral equation. It formulates a coupling condition between the trace of u on the boundary and the slope $t = \partial u/\partial n$ of u, which means that on each portion of the boundary either one of these boundary functions determines the other.

To solve this integral equation numerically we subdivide, as in a finite element method, the boundary curve (2-D) or the surface (3-D) into a mesh of small boundary elements (**Figure 3**) and approximate the boundary functions by piecewise linear or quadratic nodal functions $\phi_i(x), \psi_i(x)$:

$$u(x) = u_i \varphi_i(x) \quad t(x) = t_i \psi_i(x)$$

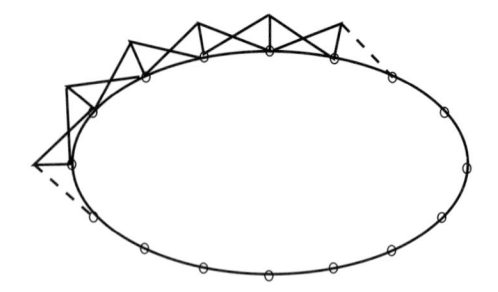

Figure 3 Boundary elements.

and we determine the unknown nodal values of u or t respectively either by a collocation method or by Galerkin's method and so obtain a linear system of equations:

$$\mathbf{H}\mathbf{u} = \mathbf{G}\mathbf{t} + \mathbf{p} \quad [4]$$

where (in a collocation method) the coefficient matrices:

$$H_{kj} = c(x^k)\delta_{kj} + \int_\Gamma \frac{\partial}{\partial \nu} g_0(y, x^k) \varphi_j(y)\,ds_y$$

$$G_{kj} = \int_\Gamma g_0(y, x^k) \psi_j(y)\,ds_y$$

contain the influence of the boundary layers $\phi_j(x)$ and $\psi_j(x)$ on the collocation point x^k on the boundary. The vector $\mathbf{p}$ lists the influence of the distributed load p on the same points:

$$p_k = \int_\Omega g_0(y, x^k) p(y)\,d\Omega_y$$

Fortunately, in most applications we can transform this domain or volume integral into an equivalent boundary integral or we have vanishing body forces or zero initial conditions so that the computation of this volume integral is not needed.

After we have solved this system (4) for the unknown nodal values we can calculate with [2] the field $u(x)$. The technique outlined here is basically the direct boundary element method.

If we multiply eqn [4] from the left with the inverse of the matrix $\mathbf{G}$ and the matrix $\mathbf{F}$, where:

$$F_{ij} = \int_\Gamma \varphi_i \varphi_j\,ds$$

then we obtain a nonsymmetric finite element stiffness matrix $\mathbf{K} = \mathbf{F}\mathbf{G}^{-1}\mathbf{H}$.

Symmetric Formulations

Because the kernel $\partial g_0 / \partial \nu$ depends on the normal vector ν at the integration point, the matrix $\mathbf{H}$ is not symmetric and in addition both matrices $\mathbf{G}$ and $\mathbf{H}$ are fully populated.

If the lack of symmetry is a problem, then we can mix the influence function for u with the influence function for the slope:

$$c_1(x)u_{,1}(x) + c_2(x)u_{,2}(x) = \int_\Gamma \left[g_1(y,x) \frac{\partial u}{\partial \nu}(y) \right.$$
$$\left. - \frac{\partial}{\partial \nu} g_1(y,x)(u(y) - u(x)) \right] ds_y + \int_\Omega g_1(y,x) p(y)\,d\Omega_y$$
$$\quad [5]$$

where the c_i are the two characteristic functions:

$$c_i(x) = \begin{cases} n_i, & x \in \Omega \\ \dot{c}_i(x), & x \in \Gamma \\ 0, & x \in \Omega^c \end{cases}$$

with the following boundary values:

$$\dot{c}_1(x) = \frac{1}{2\pi}\left[(\varphi + \tfrac{1}{2}\sin 2\varphi)n_1 + \sin^2 \varphi n_2\right]_{\varphi_2}^{\varphi_1} \quad [6]$$

$$\dot{c}_2(x) = \frac{1}{2\pi}\left[\sin^2 \varphi n_1 + (\varphi - \tfrac{1}{2}\sin 2\varphi)n_2\right]_{\varphi_2}^{\varphi_1} \quad [7]$$

The kernel g_1 is the normal derivative of the kernel g_0:

$$g_1(y,x) = \frac{\partial}{\partial n_x} g_0(y,x) = -\frac{1}{2\pi r}(r_{,x1} n_1 + r_{,x2} n_2)$$

If we put the point x in eqn [5] on the boundary we obtain an additional integral equation or coupling condition between the two boundary values u and $\partial u / \partial n$. In more abstract terms the two coupling conditions in eqns [2] and [4] can be stated as:

$$\frac{1}{2}\begin{bmatrix} \partial_y^0 u \\ \partial_y^1 u \end{bmatrix} = \int_\Gamma \begin{bmatrix} \partial_y^0 \partial_x^0 g_0 & \partial_y^1 \partial_x^0 g_0 \\ \partial_y^0 \partial_x^1 g_0 & \partial_y^1 \partial_x^1 g_0 \end{bmatrix} \begin{bmatrix} +\partial^1 u \\ -\partial^0 u \end{bmatrix} ds_y$$
$$+ \int \begin{bmatrix} \partial_x^0 g_0 \\ \partial_x^1 g_0 \end{bmatrix} p\,d\Omega_y$$

The kernels of the single integral operators are products of the boundary operators ∂_y^i and ∂_x^j and the zero-order fundamental solution g_0. By a proper combination of these two integral equations – on Γ_t where the slope t is unknown we formulate the first integral equation and on Γ_u where u is unknown the second integral equation – and by using Galerkin's method instead of simple collocation, we obtain a symmetric system matrix for the unknown nodal values.

This system of integral equations is typical for the boundary element method. Given an equation of order $2m$ there are $2m$ boundary values $\partial^j u$ which satisfy a system of $2m$ integral equations on the boundary. The higher we go in this hierarchy of integral equations, the more singular the kernels become. Note that the calculation of the coefficients G_{kj} and H_{kj} near the diagonal, if the collocation point x^k lies near or even on the patch of elements where the shape function ϕ_j lives, is not a real problem. On straight elements we can usually integrate analytically and on curved elements we can still use some semianalytical procedures which render well-behaved expressions for even hypersingular integrals. In this regard the limit [1] holds all the answers.

The Indirect Method

In mathematical terms the influence function [2] is the sum of a single-layer potential, a double-layer potential, and a volume potential. The first two potentials are homogeneous solutions of the governing equation while the volume potential has a so-called reproducing kernel. It solves the equation $-\Delta u = p$.

To have a homogeneous solution, either the single-layer or double-layer potential would therefore suffice:

$$u(x) = \int_\Gamma g_0(y,x) a(y) \, ds_y \quad v(x) = \int_\Gamma \frac{\partial g_0(y,x)}{\partial \nu} b(y) \, ds_y$$

Techniques based on this approach are called indirect methods because the boundary layers $a(y)$ and $b(y)$ are not the boundary values of the solution:

$$\lim_{x \to \Gamma} u(x) = \int_\Gamma g_0(y,x) a(y) \, ds_y$$

$$\lim_{x \to \Gamma} v(x) = \frac{\Delta \varphi}{2\pi} b(x) + \int_\Gamma \frac{\partial g_0(y,x)}{\partial \nu} b(y) \, ds_y$$

We would like to have $u(x) \to a(x)$ and $v(x) \to b(x)$. This makes it difficult to control and predict the behavior of the boundary layers. But the indirect method has a long tradition: it is identical with the approach of classical potential theory. The sudden popularity of boundary element methods must certainly be attributed to the switch from the indirect method to the direct method. The direct method is much more transparent than the indirect method because it allows engineers to state problems in mechanical terms rather than using artificial boundary layers. Many of the features we have outlined here are common to all boundary element formulations. In the following we shall look at specific differential equations and see how the boundary element approach evolves.

Harmonic Oscillations

If the excitation is harmonic, $p(x,t) = p(x) \cos(\omega t + \varphi)$, then the response of the structure, the medium, is also harmonic. This is an important case, not only because transient problems are easier to solve but also because transient problems can be reduced to harmonic-type problems by applying a Fourier or a Laplace transform.

Helmholtz Equation

An important part in dynamics is played by the wave equation which governs, for example, the vibration of a stretched membrane. The response of a membrane to harmonic excitations:

$$-\Delta v + \rho v = b(x,t) = p_1(x) \cos \omega t + p_2(x) \sin \omega t$$

can be stated in the form:

$$v(x,t) = v_1(x) \cos \omega t + v_2(x) \sin \omega t$$

If we write:

$$b(x,t) = \Re \left\{ p(x) e^{-i\omega t} \right\} \quad p(x) = p_1(x) + i p_2(x)$$

and:

$$v(x,t) = \Re \left\{ u(x) e^{-i\omega t} \right\} \quad u(x) = u_1(x) + i u_2(x)$$

then this leads to the Helmholtz equation:

$$-\Delta u(x) - \lambda u(x) = p(x) \quad \lambda = \rho \omega^2$$

for the complex valued amplitude $u(x)$. Associated with these differential equations is the identity:

$$G(\hat{u}, u) = \int_\Omega (-\Delta \hat{u} - \lambda \hat{u}) u \, d\Omega + \int_\Gamma \frac{\partial \hat{u}}{\partial n} u \, ds$$

$$- \int_\Omega \nabla \hat{u} \cdot \nabla u \, d\Omega + \int_\Omega \lambda \hat{u} u \, d\Omega = 0$$

If we formulate the second identity $B(\hat{u}, u) = G(\hat{u}, u) - G(u, \hat{u}) = 0$ with the fundamental solutions:

$$g_0(y,x) = \frac{-i}{2} H_0^{(1)}(\lambda r) \quad \text{(2-D) (Hankel function)}$$

$$g_0(y,x) = -\frac{1}{4\pi r} e^{i\lambda r} \quad \text{(3-D)}$$

$$g_1(y,x) = \frac{\partial}{\partial n_x} g_0(y,x)$$

then we obtain the influence function for $u(x)$:

$$c(x)u(x) = \int_\Gamma \left[g_0(y,x) \frac{\partial u}{\partial \nu}(y) - \frac{\partial}{\partial \nu} g_0(y,x) u(y) \right] ds_y$$
$$+ \int_\Omega g_0(y,x)(-\Delta u(y) - \lambda u(y))\, d\Omega_y$$

and the normal derivative:

$$c_j(x) u_{,j}(x)$$
$$= \int_\Gamma \left[g_1(y,x) \frac{\partial u}{\partial \nu}(y) - \frac{\partial}{\partial \nu} g_1(y,x)(u(y) - u(x)) \right] ds_y$$
$$+ \int_\Omega g_1(y,x)(-\Delta u(y) - \lambda u(y))\, d\Omega_y$$

The characteristic functions are the same functions as in eqns [3] and [6]. Note that in exterior problems Sommerfeld's radiation condition:

$$\lim_{r\to\infty} \left(\frac{\partial u}{\partial n} - i\lambda u \right) = 0 \quad r = |y - x|$$

is satisfied automatically. The Helmholtz equation appears in many wave propagation problems in acoustics as well as in fluid dynamics.

Elastic Solids (2-D and 3-D)

The vibrations v of an isotropic, homogeneous linear elastic solid satisfy the differential equations:

$$-c_2^2 \Delta v - (c_1^2 - c_2^2) \nabla \operatorname{div} v + \ddot{v} = \frac{1}{\rho} b(x,t) \quad [8]$$

where the constants:

$$c_2 = \left(\frac{\mu}{\rho} \right)^{1/2} \quad c_1 = \left(\frac{2\mu}{\rho} \left(\frac{1-\nu}{1-2\nu} \right) \right)^{1/2}$$

are the isochoric velocity $c_2 (= c_s)$, or s-wave velocity, and the irrotational velocity $c_1 (= c_p)$, or p-wave velocity (**Figure 4**). Harmonic excitations:

$$\frac{1}{\rho} b(x,t) = p_1(x) \cos \omega t + p_2(x) \sin \omega t$$

and harmonic displacement fields:

$$v(x,t) = v_1(x) \cos \omega t + v_2(x) \sin \omega t$$

can be considered as the real parts of complex-valued functions:

$$\frac{1}{\rho} b(x,t) = \Re\left\{ p(x) e^{-i\omega t} \right\} \quad v(x,t) = \Re\left\{ u(x) e^{-i\omega t} \right\}$$

where:

$$p(x) = p_1(x) + i p_2(x) \quad u(x) = v_1(x) + i v_2(x)$$

If we substitute these expressions into eqn [8], then we obtain the following system of equations for the complex-valued amplitude:

$$-c_2^2 \Delta u - (c_1^2 - c_2^2) \nabla \operatorname{div} u - \omega^2 u = p(x)$$

To this system belong the identities:

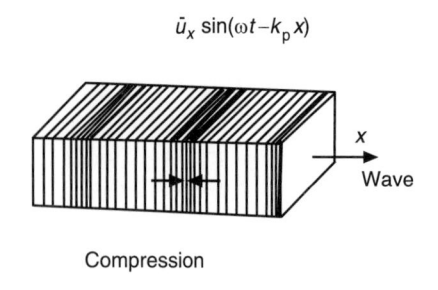

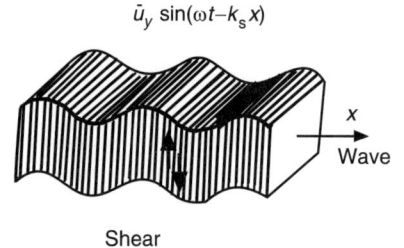

Figure 4 Compression and shear waves in a solid.

$$G(\hat{u}, u) = \int_\Omega (-L\hat{u} - \omega^2 \hat{u}) \cdot u \, d\Omega + \int_\Gamma \tau(\hat{u}) \cdot u \, ds$$
$$- E(\hat{u}, u) + \int_\Omega \omega^2 \hat{u} \cdot u \, d\Omega = 0$$
$$B(\hat{u}, u) = G(\hat{u}, u) - G(u, \hat{u}) = 0$$

The operator $-L$ denotes the operator in eqn [8] without the inertial forces and $E(\hat{u}, u)$ is the associated energy product. The fundamental solutions of eqn [8] are:

$$U_{ij} = \frac{1}{4\pi\rho c_2^2}[\psi \delta_{ij} - \chi r_{,i} r_{,j}]$$

where:

$$\psi = \frac{1}{r}\left[e^{-i\omega r/c_2}\left(1 - \frac{c_2^2}{\omega^2 r^2} + \frac{c_2}{i\omega r}\right)\right.$$
$$\left.- \frac{c_2^2}{c_1^2}\left(-\frac{c_1^2}{\omega^2 r^2} + \frac{c_1}{i\omega r}\right)e^{-i\omega r/c_1}\right]$$

$$\chi = \left(-\frac{3c_2^2}{\omega^2 r^2} + \frac{3c_2}{i\omega r} + 1\right)\frac{e^{-i\omega r/c_2}}{r}$$
$$- \frac{c_2^2}{c_1^2}\left(-\frac{3c_1^2}{\omega^2 r^2} + \frac{3c_1}{i\omega r} + 1\right)\frac{e^{-i\omega r/c_1}}{r}$$

and the components of the associated traction vectors are:

$$T_{ij} = \frac{1}{4\pi}\left[\left(\frac{d\psi}{dr} - \frac{1}{r}\chi\right)(\delta_{ij}r_\nu + r_{,j}\nu_i)\right.$$
$$- \frac{2}{r}\chi(\nu_j r_{,i} - 2r_{,i}r_{,j}r_\nu)$$
$$\left. - 2\frac{d\chi}{dr}r_{,i}r_{,j}r_\nu + \left(\frac{c_1^2}{c_2^2} - 2\right)\left(\frac{d\psi}{dr} - \frac{d\chi}{dr} - \frac{2}{r}\chi\right)r_{,i}\nu_j\right]$$

If we formulate with the three fundamental solutions $g_0^i = \{U_{ij}\}$ the limit:

$$\lim_{\varepsilon \to 0} B(g_0^i, u)_{\Omega_\varepsilon} = 0$$

then we obtain three influence functions for the complex-valued amplitude:

$$C_{ij}(x)u_j(x) = \int_\Gamma [U_{ij}(y,x)t_j(y) - T_{ij}(y,x)u_j(y)] \, ds_y$$
$$+ \int_\Omega U_{ij}(y,x)p_j(y) \, d\Omega_y$$

The terms $C_{ij}(x)$ are the same characteristic functions as in steady-state problems. The method can easily be extended to include material damping, viscous elastic effects, anisotropy, and thermoelasticity. It has found widespread use in the analysis of soil–structure interaction, beginning with the calculation of dynamic stiffness matrices of rigid foundations with as little as 6 degrees-of-freedom up to the complete analysis in the frequency domain of the interaction between the soil and a complete building.

Kirchhoff Plates

We mention Kirchhoff plates because they are governed by a fourth-order equation. The application of boundary element methods differs little from the standard approach based on Betti's principle (or Rayleigh–Green identity in plate bending). The response of a plate to a harmonic excitation:

$$K\Delta\Delta v + \rho \ddot{v} = b(x,t) = p_1(x) \cos \omega t + p_2(x) \sin \omega t$$

where ρ = specific mass × plate thickness is a harmonic oscillation:

$$v(x,t) = v_1(x) \cos \omega t + v_2(x) \sin \omega t$$

If we formulate as before:

$$v(x,t) = \Re\left\{w(x)e^{-i\omega t}\right\} \quad b(x,t) = \Re\left\{p(x)e^{-i\omega t}\right\}$$

where:

$$w(x) = w_1(x) + iw_2(x) \quad p(x) = p_1(x) + ip_2(x)$$

then the differential equation for the complex-valued amplitude $w(x)$ becomes:

$$K\Delta\Delta w(x) - \rho\omega^2 w(x) = p(x) \qquad [9]$$

Associated with this equation are the identities:

$$G(\hat{w}, w) = \int_\Omega (K\Delta\Delta\hat{w} - \rho\omega^2\hat{w})w \, d\Omega$$
$$+ \int_\Gamma \left(\hat{V}_n w - \hat{M}_n \frac{\partial w}{\partial n}\right) ds$$
$$+ \sum_e \left[\hat{F}(x^e)w(x^e) - \hat{w}(x^e)F(x^e)\right] - E(\hat{w}, w)$$
$$+ \int_\Omega \rho\omega^2 \hat{w} w \, d\Omega = 0$$
$$B(\hat{w}, w) = G(\hat{w}, w) - G(w, \hat{w}) = 0$$

The integral $E(\hat{w}, w)$ is the energy product of the static problem. The fundamental solution of the differential eqn [9] is:

$$g_0(\lambda r) = icJ_0(\lambda r) + cY_0(\lambda r) + dK_0(\lambda r)$$

where:

$$\lambda = \frac{\omega^2}{\rho K} \quad c = \frac{1}{8\lambda^2} \quad d = \frac{1}{4\pi\lambda^2}$$

With these tools one can easily solve time-harmonic problems in plate bending for arbitrarily shaped plate geometries and arbitrary boundary conditions. Note that the boundary element method is also applicable to Mindlin–Reissner plates.

Eigenvalues

Formulations of harmonic problems (formulations in the frequency domain) with homogeneous boundary conditions and zero body forces result after a proper rearrangement of the system $\mathbf{Hu} = \mathbf{Gt} + \mathbf{p}$ in a homogeneous system of equations:

$$\mathbf{A}(\omega)x = 0$$

Because the coefficients of $\mathbf{A}$ are complex-valued functions of ω, free vibration analysis by the boundary element method must be done by a determinant search method. This is an indication of the fact that in the boundary element method continuous mass distribution is preserved, resulting in an infinite number of natural frequencies.

As a result, the boundary element method is not particularly efficient at calculating, say, the eigenmodes of a plate or more generally of a bounded domain Ω (infinite domains do not have eigenmodes). The method still retains its edge over domain-type methods but the search for the eigenvalues is certainly cumbersome and standard numerical routines for such tasks are not applicable.

To reformulate the problem as a finite element-like eigenvalue problem, one could either do a Taylor expansion of the matrix $\mathbf{A}$:

$$[\mathbf{A}(0) + \omega \mathbf{A}'(0)]u = 0$$

or one could use the mass matrix method which leads to the system:

$$\mathbf{Fu} + \mathbf{M\ddot{u}} = \mathbf{Gt}$$

The mass matrix approach is a general method for both steady-state and transient problems in which use is made of the static fundamental solutions and the resulting volume integrals are converted into surface integrals which eventually lead to the mass matrix.

Transient Problems

In transient problems the boundary element method can fully utilize its potential. In infinite domains the radiation condition is automatically satisfied and there are no artificial boundaries with reflecting waves.

In steady-state problems and time harmonic problems, the derivation of Betti's principle starts with the integral:

$$\int_0^l -EAu''\hat{u}\,dx$$

In transient dynamics, the L_2-scalar product is replaced by a convolution:

$$\int_0^l (-EAu'' + \mu\ddot{u}) * \hat{u}\,dx$$

If we apply integration by parts to this integral then we obtain the first identity for the axial displacement of a beam under dynamic loads:

$$G(u,\hat{u}) = \int_0^l (-EAu'' + \mu\ddot{u}) * \hat{u}\,dx + [N * \hat{u}]_0^l$$
$$- \int_0^l \mu[\dot{u}(x,t)\hat{u}(x,0) - \dot{u}(x,0)\hat{u}(x,t)]\,dx$$
$$- \int_0^l \left(\frac{N * \hat{N}}{EA} - \mu\dot{u} * \dot{\hat{u}}\right)dx = 0$$

and therefore also the second identity, Betti's principle, $B(\hat{u}, u) = G(\hat{u}, u) - G(u, \hat{u}) = 0$. The fundamental solution, which now has a singularity in space and time:

$$-EAg_0''(y,x;t,\tau) + \mu\ddot{g}_0(y,x;t,\tau) = \delta(y-x)\delta(t-\tau)$$

is the response of an infinite bar to a concentrated force $\hat{P} = 1$ which acts at the time mark t and at the point x. If we formulate the second identity with this function and a regular function u, then we obtain an influence function for u and thus automatically the coupling condition between the boundary data of u. Formally this condition differs from the coupling condition of the static case only by the two stars:

$$\mathbf{H} * \mathbf{u} = \mathbf{G} * \mathbf{f} + \mathbf{d}(t)$$

which indicate the convolution. Explicitly, these conditions read:

$$\int_0^t H_{ij}(t-\tau)u_j(\tau)\,d\tau = \int_0^t G_{ij}(t-\tau)f_j(\tau)\,d\tau + d_i(t)$$
$$i = 1, 2 \qquad [10]$$

The vector $\mathbf{d}(t) = \{d_i(t)\}$ represents the influence of the distributed forces and the initial conditions.

The system (10) is a system of two Volterra integral equations for the end displacements $u_i(t)$ and end forces $f_i(t)$. This is the typical situation in transient boundary element analysis. (In 2-D and 3-D problems, the number of Volterra integral equations is equal to the number of nodal values.) In time-dependent problems, influence means convolution. We have to verify that at each time step t_i the wave is compatible with its own history and this means that the boundary displacements and boundary stresses have to satisfy Volterra integral equations at each point of the time axis.

The analysis therefore consists of two steps: first, a discretization of the time axis into a sequence of equally spaced time intervals with a constant or linear variation of displacements and tractions over each time interval, and second, a discretization of the boundary Γ into boundary elements over each of which a constant or linear distribution of displacements and tractions is assumed.

The Wave Equation

To the wave equation:
$$-c^2 \Delta u(x,t) + \ddot{u}(x,t) = p(x,t)$$

belong the identities:

$$G(u, \hat{u}) = \int_\Omega (-c^2\Delta u + \ddot{u}) \times \hat{u}\,d\Omega + \int_\Gamma c^2 \frac{\partial u}{\partial n} \times \hat{u}\,ds$$
$$- c^2 \int_\Omega \nabla u \times \nabla \hat{u}\,d\Omega - \int_\Omega (u \times \hat{u})\,d\Omega$$
$$+ \int_\Omega \left[\dot{u}_0 \hat{u} + u_0 \hat{\dot{u}}\right] d\Omega = 0$$

and $B(\hat{u}, u) = G(\hat{u}, u) - G(u, \hat{u}) = 0$. Next, we consider a load case where a concentrated force $\hat{P}$ acts at point x. We assume that the magnitude of the force changes with time according to a given function $f(\tau)$, of which we only require that it has a 'quiet past':

$$f(\tau) = 0 \quad \tau \le 0 \quad \dot{f}(0) = 0$$

The corresponding solution of the wave equation is:

$$\hat{u}(y, x, \tau) = \frac{1}{c^2 4\pi r} f\left(\tau - \frac{r}{c}\right)$$

Let us assume that $u = u(y, \tau)$ is a smooth solution of the differential equation:

$$-c^2 \Delta u + \ddot{u} = p(y, \tau)$$

If we formulate Betti's principle with these two solutions and let the function $f(t)$ converge uniformly to a Dirac function $\delta(t-\tau)$, then we obtain the result:

$$c(x)u(x,t) = \int_\Gamma \frac{r_v}{4\pi r^2}\left[u\left(y, t - \frac{r}{c}\right) + \frac{r}{c}\dot{u}\left(y, t - \frac{r}{c}\right)\right]ds_y$$
$$+ \int_\Gamma \frac{1}{4\pi r}\frac{\partial u}{\partial v}\left(y, t - \frac{r}{c}\right)ds_y$$
$$+ \int_\Omega \frac{1}{4\pi c^2 r}\delta\left(t - \frac{r}{c}\right)\dot{u}(y, 0)\,d\Omega_y$$
$$+ \int_\Omega \frac{1}{4\pi c^2 r}\dot{\delta}\left(t - \frac{r}{c}\right)u(y, 0)\,d\Omega_y$$
$$+ \int_\Omega \hat{u}(y, x, \tau) \times p(y, \tau)\,d\Omega_y$$
$$[11]$$

The integrals of the initial data are equivalent to:

$$\int_\Omega \frac{1}{4\pi c^2 r}\delta\left(t - \frac{r}{c}\right)\dot{u}(y, 0)\,d\Omega_y = t M_{x;ct}[\dot{u}(y,0)]$$

and:

$$\int_\Omega \frac{1}{4\pi c^2 r}\dot{\delta}\left(t - \frac{r}{c}\right)u(y, 0)\,d\Omega_y = \frac{\partial}{\partial t}(t M_{x;ct}[u(y,0)])$$

where:

$$M_{x;ct}[u] = \frac{1}{4\pi}\int_0^{2\pi}\int_0^\pi u(x + ct\nabla_y r, 0)\sin\vartheta\,d\vartheta\,d\varphi$$
$$r = |y - x|$$

is the average value of $u(y, 0)$ on a sphere with radius

Dynamic Displacement Fields

Associated with the system of differential equations:

$$-c_2^2 \Delta u - (c_1^2 - c_2^2)\nabla \operatorname{div} u + \ddot{u} = p(y, \tau) \quad [12]$$

are the identities:

$$G(u, \hat{u}) = \int_\Omega \left(-c_2^2 \Delta u - (c_1^2 - c_2^2)\nabla \operatorname{div} u + \ddot{u}\right) \times \hat{u} \, d\Omega$$
$$+ \int_\Gamma \tau(u) \times \hat{u} \, ds - \int_\Omega (u \times \hat{u}) \, d\Omega$$
$$+ \int_\Omega \left[\dot{u}_0 \cdot \hat{u}(y, t) + u_0 \cdot \dot{\hat{u}}(y, t)\right] d\Omega$$
$$- E^*(u, \hat{u}) = 0$$

and $B(\hat{u}, u) = G(\hat{u}, u) - G(u, \hat{u}) = 0$, where:

$$E^*(u, \hat{u}) = \int_\Omega \sigma_{ij} \times \hat{\varepsilon}_{ij} \, d\Omega = \int_\Omega S(u) \times E(\hat{u}) \, d\Omega$$
$$= \int_\Omega E(u) \times S(\hat{u}) \, d\Omega$$

denotes the convolution of the energy product between the stress and strain tensors $S(u)$ and $E(\hat{u})$. By proceeding as in the case of the wave equation, we obtain the influence function for the transient displacement field of a solid:

$$C(x)u(x, t) + \int_\Gamma \left[T(y, x, \tau) \times u(y, \tau)\right.$$
$$\left. - U(y, x, \tau) \times t(y, \tau)\right] ds_y$$
$$- \int_\Omega U(y, x, \tau) \times p(y, \tau) \, d\Omega_y$$
$$+ \int_\Omega [U(y, x, t) \dot{u}_0(y, t)$$
$$- \dot{U}(y, x, t)u_0(y, t)] \, d\Omega_y = 0$$

where the convolutions can be expressed as:

$$U_{ij} \times t_j = u_{ij}(y, x, t; t_j(y, t))$$
$$T_{ij} \times u_j = t_{ij}(y, x, t; u_j(y, t))$$
$$U_{ij} \times p_j = u_{ij}(y, x, t; p_j(y, t))$$

This results in

$$U_{ij} \times t_j = u_{ij}(y, x, t; t_j)$$
$$= \frac{1}{4\pi \rho r}\left\{(3r_{,i}r_{,j} - \delta_{ij})\int_{1/c_1}^{1/c_2} \lambda t_j(x, t - \lambda r) \, d\lambda \right.$$
$$+ r_{,i}r_{,j}\left[\frac{1}{c_1^2}t_j\left(y, t - \frac{r}{c_1}\right) - \frac{1}{c_2^2}t_j\left(y - \frac{r}{c_2}\right)\right]$$
$$\left.+ \frac{\delta_{ij}}{c_2^2}t_j\left(x, t - \frac{r}{c_2}\right)\right\}$$
$$= \frac{1}{4\pi \rho r}\left\{(3r_{,i}r_{,j} - \delta_{ij})\left[H\left(t - \frac{r}{c_1}\right)\right.\right.$$
$$\times \int_{r/c_1}^{t} \tau t_j(y, t - \tau) \, d\tau$$
$$\left. - H\left(t - \frac{r}{c_2}\right)\int_{r/c_2}^{t} \tau t_j(y, t - \tau) \, d\tau\right]\frac{1}{r^2}$$
$$+ r_{,i}r_{,j}\left[\frac{1}{c_1^2}H\left(t - \frac{r}{c_1}\right)t_j\left(y, t - \frac{r}{c_1}\right)\right.$$
$$\left. - \frac{1}{c_2^2}H\left(t - \frac{r}{c_2}\right)t_j\left(y, t - \frac{r}{c_2}\right)\right]$$
$$\left.+ \frac{\delta_{ij}}{c_2^2}H\left(t - \frac{r}{c_2}\right)t_j\left(y, t - \frac{r}{c_2}\right)\right\}$$

where:

$$H(x) = \begin{cases} 1, & 0 < x \\ 0, & x < 0 \end{cases}$$

is the Heaviside function which acts like a cut-off function and expresses the causality condition. To reach points at a distance r the wave needs the time $t_r = r/c_i$ and as long as t is smaller, $t - t_r = t - r/c_i < 0$, all is quiet at that point, so that only those points y influence the point x, whose distance r satisfies the inequality $t - r/c_i > 0$. Note that, to insure the causality condition numerically for nonconvex domains, the discretization must not be too coarse.

Fourier and Laplace transforms

We can eliminate from eqn [12] the time variable either by a Fourier transformation or by a Laplace transformation:

$$\tilde{u}(x,\omega) = \frac{1}{2\pi} \int_{-\infty}^{+\infty} u(x,t)\, e^{i\omega t}\, dt$$

$$\tilde{u}(x,s) = \int_0^{\infty} u(x,t)\, e^{st}\, dt$$

In this respect we speak of the transition from the time domain into the frequency domain. The Fourier transformation leads to the following system of differential equations:

$$-c_2^2 \Delta \tilde{u} - (c_1^2 - c_2^2)\nabla \operatorname{div} \tilde{u} - \omega^2 \tilde{u} = \tilde{p}(x,\omega)$$

and the Laplace transformation to the system:

$$-c_2^2 \Delta \tilde{u} - (c_1^2 - c_2^2)\nabla \operatorname{div} \tilde{u} + s^2 \tilde{u} = \tilde{p} + s u_0 + \dot{u}_0$$

which is identical to the first system if we substitute for $s = i\omega$. The fundamental solutions are the fundamental solutions of the harmonic problem and therefore we can solve the boundary value problem of the Laplace transform by the boundary element method as well.

The only problem is that we must solve the boundary value problem for a whole range of complex-valued parameters s and then apply a (numerical) inverse transformation from the frequency domain back into the time domain:

$$u(x,t) = \frac{1}{2\pi i} \int_{\beta - i\infty}^{\beta + i\infty} \tilde{u}(x,s)\, e^{st}\, ds$$

to obtain the original solution.

One of the computational problems in the Fourier-transformed domain is that of fictitious eigenfrequencies, which naturally holds also true for time harmonic problems. These are frequencies corresponding to the eigenvalues of the associated interior problem that render the system matrix singular.

Summary

The boundary element method is a numerical method for the approximate solution of partial differential equations but it is deeply rooted in the mechanical and mathematical properties of the governing equations.

The method can be applied successfully to a wide range of dynamic problems (**Figure 5**), of which we have only mentioned a few: soil–structure interaction problems, fluid–structure interaction problems, the dynamic analysis of underground structures, of pile groups, of vibration isolation devices, of the scattering of waves, of wave diffraction. It can also be applied to study the dynamics of structural elements such as beams and plates. The main advantages of the boundary element method are the reduction of the dimensionality of the problem and the high accuracy of results.

See also: **Continuous methods**; **Eigenvalue analysis**; **Fluid/structure interaction**; **Transform methods**.

Further Reading

Antes H and Panagiotopoulos PD (1992) *Boundary Integral Approach to Static and Dynamic Contact Problems*. Basel: Birkhäuser Verlag.

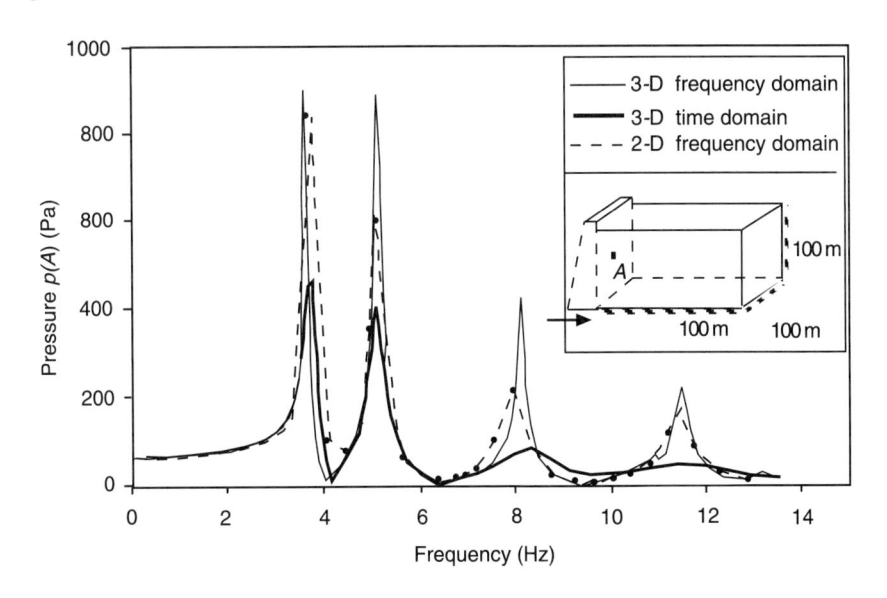

Figure 5 Fluid–structure interaction.

Balaš J, Sládek J and Sládek V (1989) *Stress Analysis by Boundary Element Methods.* Amsterdam: Elsevier

Banerjee PK (1994) *Boundary Element Method in Engineering.* London: McGraw Hill.

Banerjee PK and Kobayashi S (1992) (eds) *Advanced Dynamic Analysis by Boundary Element Methods. Developments in Boundary Element Methods*, vol. 7: London: Elsevier Applied Science.

Beer G and Watson JO (1992) *Introduction to Finite and Boundary Element Methods for Engineers.* Chichester: John Wiley.

Beskos DE (1987) Boundary element methods in dynamic analysis. *Applied Mechanical Review* 40: 1–23.

Beskos DE (1997) Boundary element methods in dynamic analysis. part II (1986–1996). *Applied Mechanical Review* 50: 149–197.

Bonnet M (1999) *Boundary Integral Equation Methods for Solids and Fluids.* Chichester: John Wiley.

Brebbia CA, Telles JCF and Wrobel LG (1984) *Boundary Element Techniques.* Berlin: Springer Verlag.

Do Rego Silva (1994) *Acoustic and Elastic Wave Scattering Using Boundary Elements.* Southampton: Computer and Mechanical Publishing.

Dominguez J (1993) *Boundary Elements in Dynamics.* Southampton: Computer and Mechanical Publishing/ London: Elsevier Applied Science.

Hartmann F (1989) *Introduction to Boundary Elements.* Berlin: Springer-Verlag.

Kitahara M (1985) *Boundary Integral Equation Methods in Eigenvalue Problems of Elastodynamics and Thin Plates.* New York: Elsevier.

Manolis GD and Beskos DE (1988) *Boundary Element Methods in Elastodynamics.* London: Unwin Hyman.

Manolis GD and Davies TG (eds) (1993) *Boundary Element Techniques in Geomechanics.* London: Elsevier Applied Science.

BRIDGES

S S Rao, University of Miami, Coral Gables, FL, USA

Copyright © 2001 Academic Press

doi:10.1006/rwvb.2001.0127

Introduction

The natural frequencies of vibration play a major role in determining the impact factors of highway bridges. The dynamic response of bridges under traveling loads is important in the design of highway and railway bridges. The study of vibration of bridges under traveling loads was started in the middle of the nineteenth century when railroad bridge construction began. The problem was studied by several investigators including Krylov, Timoshenko, and Inglis. The problem of vibration of a uniform beam subject to a constant transverse force moving with constant velocity was studied by Krylov in 1905. The solution of a beam subject to a traveling pulsating load was given by Timoshenko in 1908. Inglis (1938) presented a systematic analysis of the vibration of bridges by considering the influence of factors such as moving loads, damping and spring stiffness of the suspension of the locomotive. When a steam locomotive crosses a bridge, the balance weights attached to the driving wheels, for the purpose of minimizing the inertia effects of the reciprocating parts, will cause hammer-blows on the rails that result in the vibration of the bridge. Although hammer-blows will not be present when an electric locomotive crosses a bridge, there will be some vibration, with a smaller amplitude, due to the moving load. Vibration is also caused by the rail joints and track irregularities.

Natural Frequencies of Bridges

The first flexural frequency of vibration of a bridge is required for the computation of the dynamic load allowance, also called the impact factor, for its major components. A bridge can be modeled as a uniform beam when the width is uniform, the span is much larger than the width and the angle of skew is less than 20°. Typical bridges that can be approximated as uniform beams are shown in **Figure 1**. The first flexural natural frequency of a simply supported single-span uniform beam, in Hertz, is given by:

$$f_s = \frac{\pi}{2L^2}\sqrt{\left(\frac{EI}{m}\right)} \qquad [1]$$

where L is the span, E is Young's modulus, I is the moment of inertia and m is the mass per unit length. The first flexural natural frequency of a multispan bridge, with symmetric span about its center line and uniform flexural rigidity, can be determined as:

$$f_m = cf_s \qquad [2]$$

where f_s is the first flexural natural frequency of the largest span single bridge in isolation, given by eqn static load is determined by multiplying the actual load by (1 + IF) where IF is the impact factor or the dynamic load allowance. A typical variation of the impact factor with the first natural frequency of the bridge is shown in **Figure 3**.

Dynamic Response using Harmonic Analysis

The bridge is modeled as a uniform simply supported beam for simplicity. When the beam is subject to the time-varying distributed load:

$$f(x,t) = f(x) \sin \omega t \quad [3]$$

where the function, $f(x)$, can be expanded into a sum of harmonic components using Fourier series as:

$$f(x) = \sum_{n=1}^{\infty} f_n \sin \frac{n\pi x}{L} \quad [4]$$

where:

$$f_n = \frac{2}{L} \int_0^L f(x) \sin \frac{n\pi x}{L} \, dx \quad [5]$$

When the load on the beam is uniform with $f(x) = f_0$, the Fourier coefficients are given by:

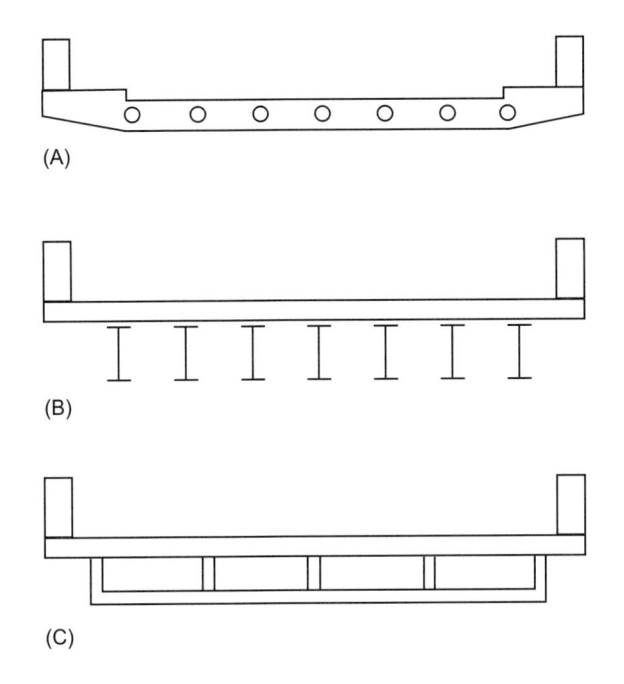

Figure 1 Cross-sections of typical bridges that can be approximated as uniform bridges. (A) Voided slab bridges; (B) slab-on-girder bridges; (C) cellular bridge.

[1], and c is a constant that depends on the number of spans and the ratio of spans. The constant c is shown in **Figure 2** for two-, three-, and four-span bridges.

Many highway bridge codes permit the use of static analysis for moving loads provided that the loads are converted into equivalent static loads. The equivalent

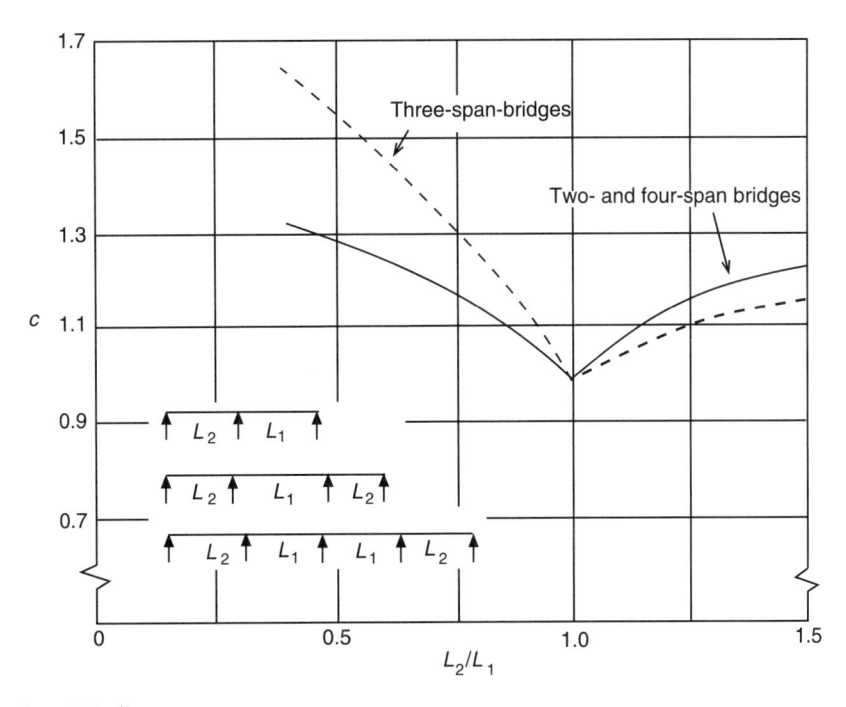

Figure 2 Variation of c with L_2/L_1.

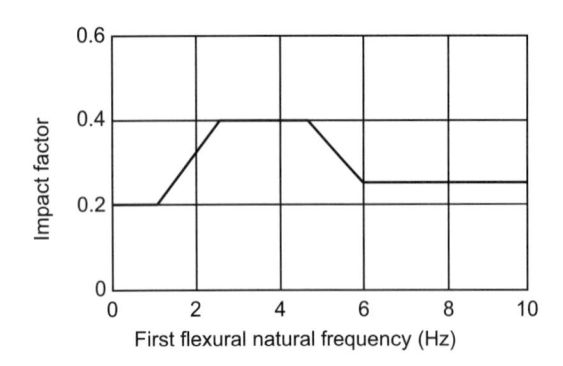

Figure 3 Variation of impact factor for nonwood components.

$$f_n = \begin{cases} 4f_0/n\pi, & n = 1, 3, 5, \ldots \\ 0, & n = 2, 4, 6, \ldots \end{cases} \quad [6]$$

If the load acts at one point of the beam (at $x = a$) with a sinusoidal variation as:

$$f(x) = \begin{cases} F_0, & x = a \\ 0, & x \neq a \end{cases} \quad [7]$$

then the Fourier coefficients are given by:

$$f_n = \frac{2F_0}{L} \sin \frac{n\pi a}{L} \quad [8]$$

By finding the response of the beam to individual harmonic components of the load, the total response of the beam can be determined using superposition. This method is known as harmonic analysis. The equation of motion of the beam subject to a harmonically varying load is given by:

$$EI \frac{\partial^4 y}{\partial x^4} + \rho A \frac{\partial^2 y}{\partial t^2} = f_n \sin \frac{n\pi x}{L} \sin \omega t \quad [9]$$

where y is the transverse deflection, E is Young's modulus, I is the moment of inertia, ρ is the density, A is the cross-sectional area and L is the length of the beam. Using the initial conditions:

$$y(x, 0) = \frac{\partial y}{\partial t}(x, 0) = 0 \quad [10]$$

the solution of eqn [9] can be expressed as:

$$y(x, t) = \frac{L^4 f_n}{EI\pi^4 \left[n^4 - \left\{(\omega/\omega_1)^2\right\}\right]} \sin \frac{n\pi x}{L} \\ \times \left(\sin \omega t - \frac{\omega}{n^2 \omega_1} \sin n^2 \omega_1 t \right) \quad [11]$$

where ω_n is the nth natural frequency of the beam:

$$\omega_n = \frac{n^2 \pi^2}{L^2} \sqrt{\left(\frac{EI}{\rho A}\right)} \equiv n^2 \omega_1; \\ n = 1, 2, 3, \ldots \quad [12]$$

The total response of the beam, considering all harmonic components of the load, is given by:

$$y(x, t) = \frac{L^4}{EI\pi^4} \sum_{n=1}^{\infty} \frac{f_n \sin (n\pi x/L)}{\left[n^4 - (\omega/\omega_1)^2\right]} \\ \times \left(\sin \omega t - \frac{\omega}{n^2 \omega_1} \sin n^2 \omega_1 t \right) \quad [13]$$

Vibration due to Concentrated Traveling Load

When a concentrated load, F_0, moves along a uniform beam with constant velocity, v_0, as shown in **Figure 4**, the time-varying load can be expressed as, using $a = v_0 t$ and $\omega_0 = (\pi v_0/sL)$:

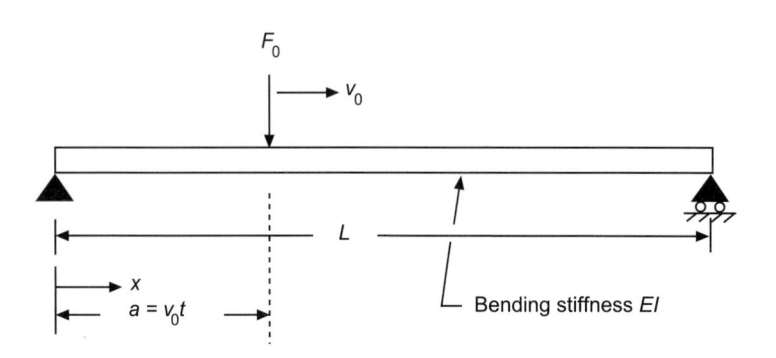

Figure 4 Constant load moving along a beam.

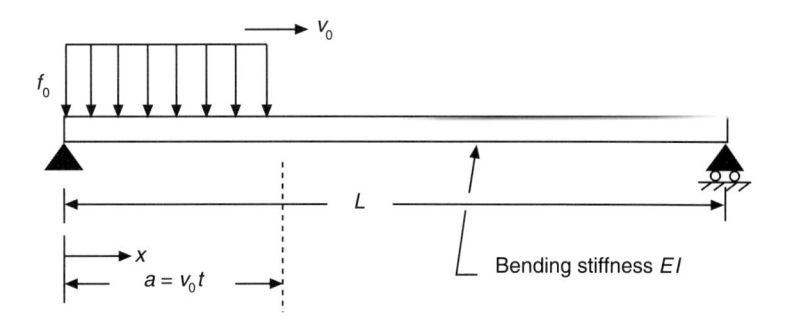

Figure 5 Distributed moving load.

$$f(x,t) = \frac{2F_0}{L} \sum_{n=1}^{\infty} \sin \frac{n\pi x}{L} \sin n\omega_0 t \qquad [14]$$

and the response of the beam is given by:

$$y(x,t) = \frac{2F_0 L^3}{EI\pi^4} \sum_{n=1}^{\infty} \frac{\sin(n\pi x/L)}{\{n^4 - (n\omega_0/\omega_1)^2\}}$$
$$\times \left(\sin n\omega_0 t - \frac{\omega_0}{n\omega_1} \sin n^2 \omega_1 t \right) \qquad [15]$$

Vibration due to Distributed Traveling Load

When a uniformly distributed load of intensity f_0 per unit length moves along a uniform beam with constant velocity, v_0, as shown in **Figure 5**, the time-varying load can be represented as, assuming that the leading edge of the load reaches a distance $a = v_0 t$ from $x = 0$ in time t:

$$f(x,t) = \frac{4f_0}{\pi} \sum_{n=1}^{\infty} \frac{1}{n} \sin \frac{n\pi x}{L} \sin^2 n\omega_0 t$$
$$= \frac{2f_0}{\pi} \sum_{n=1}^{\infty} \frac{1}{n} \sin \frac{n\pi x}{L} (1 - \cos 2n\omega_0 t) \qquad [16]$$

with:

$$\omega_0 = \frac{\pi v_0}{L} \qquad [17]$$

The response of the beam can be expressed as:

$$y(x,t) = \frac{2f_0 L^4}{EI\pi^5} \sum_{n=1}^{\infty} \left\{ \frac{1}{n^5} \sin \frac{n\pi x}{L} \right.$$
$$\left. + (-1)^n \frac{1}{n} \frac{1}{[n^4 - (n\omega_0/\omega_1)^2]} \sin \frac{n\pi x}{L} \cos 2n\omega_0 t \right\} \qquad [18]$$

If ω_0/ω_1 is very small, eqn [18] can be approximated as:

$$y(x,t) \approx \frac{4f_0 L^4}{EI\pi^5} \sum_{n=1}^{\infty} \frac{1}{n^5} \sin \frac{n\pi x}{L} \sin^2 n\omega_0 t$$
$$\approx \frac{2f_0 L^4}{EI\pi^5} \sum_{n=1}^{\infty} \frac{1}{n^5} \sin \frac{n\pi x}{L} (1 - \cos 2n\omega_0 t) \qquad [19]$$

Vibration due to Pulsating Traveling Load

When a harmonically varying concentrated load moves along a uniform beam with constant velocity as shown in **Figure 6**, the time-varying load can be expressed as, using $\omega_0 = (\pi v_0/L)$:

$$f(x,t) = \frac{2F_0}{L} \left(\sum_{n=1}^{\infty} \sin \frac{n\pi x}{L} \sin n\omega_0 t \right) \sin \omega t \qquad [20]$$

The response of the beam can be approximated by the first harmonic term of the infinite series as:

$$y(x,t) \approx \frac{\delta}{2} \sin \frac{\pi x}{L} \left\{ \frac{\cos(\omega - \omega_0)t - \cos \omega_1 t}{1 - [(\omega - \omega_0)/\omega_1]^2} \right.$$
$$\left. - \frac{\cos(\omega + \omega_0)t \cos \omega_1 t}{1 - [(\omega + \omega_0)/\omega_1]^2} \right\} \qquad [21]$$

where δ represents the deflection of the beam at the middle due to the load F_0 given by:

$$\delta = \frac{2F_0 L^3}{EI\pi^4} \qquad [22]$$

Vibration of Railroad Track

The railroad track can be modeled as an infinitely long uniform beam on a Winkler foundation

Figure 6 Harmonically varying moving load.

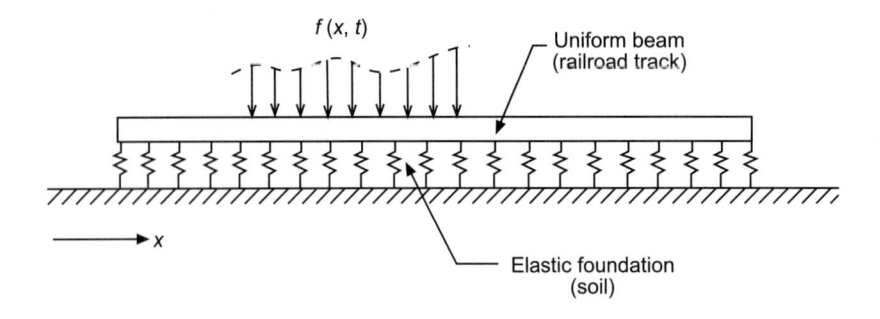

Figure 7 Modeling of a railroad track.

(**Figure 7**). When a distributed load $f(x,t)$ acts on the railroad track, the transverse deflection of the track is governed by the equation:

$$EI\frac{\partial^4 y}{\partial x^4} + m\frac{\partial^2 y}{\partial t^2} + ky = f(x,t) \quad [23]$$

where EI is the bending stiffness and m is the mass per unit length of the track, k is the foundation modulus and $f(x,t)$ is the external load:

$$f(x,t) = f(x - v_0 t) \quad [24]$$

Using $z = x - v_0 t$, eqn [23] can be rewritten as:

$$EI\frac{d^4 y(z)}{dz^4} + mv_0^2\frac{d^2 y(z)}{dz^2} + ky(z) = f(z) \quad [25]$$

If $f(x,t)$ is a concentrated load F_0 moving along the track with a constant velocity v_0, the equation of motion is taken as:

$$EI\frac{d^4 y}{dz^4} + mv_0^2\frac{d^2 y}{dz^2} + ky = 0 \quad [26]$$

and the load F_0 is incorporated into the solution as a known shear force at $z = 0$. The solution of eqn [26] can be expressed as:

$$y(z) = a_1 e^{(\alpha+i\beta)z} + a_2 e^{(\alpha-i\beta)z} + a_3 e^{-(\alpha+i\beta)z} + a_4 e^{-(\alpha-i\beta)z} \quad [27]$$

where $i = \sqrt{(-1)}$:

$$\alpha = d\sqrt{(1-c)}, \quad \beta = d\sqrt{(1+c)},$$

$$c = \left(\frac{v_0}{v_{cri}}\right)^2, \quad v_{cri} = \left(\frac{4EIk}{m^2}\right)^{1/4},$$

$$d = \left(\frac{k}{4EI}\right)^{1/4}$$

and the constants a_i, $i = 1, 2, 3, 4$, are determined using the conditions:

$$y = \frac{d^2 y}{dz^2} = 0 \quad \text{at } z = \infty \quad [28]$$

$$\frac{dy}{dz} = 0, \quad EI\frac{d^3 y}{dz^3} = \frac{F_0}{2} \quad \text{at } z = 0 \quad [29]$$

Thus the response of the railroad track can be expressed as:

$$y(z) = \frac{F_0 e^{-\alpha(x-v_0 t)}}{4EI(\alpha^2 + \beta^2)}$$
$$\times \left[\frac{1}{\alpha} \cos \beta(x - v_0 t) + \frac{1}{\beta} \sin \beta(x - v_0 t)\right] \quad [30]$$

From this solution, the maximum dynamic deflection of the track can be found to occur at $x = v_0 t$ with a magnitude $y_{\max}$:

$$y_{\max} = \frac{dF_0}{2k\left[\left(1 - v_0^2/v_{\text{cri}}^2\right)\right]^{1/2}} \quad [31]$$

Nomenclature

A	cross-sectional area
c	constant
E	Young' modulus
EI	bending stiffness
f	frequency
F_0	concentrated load
I	moment of inertia
k	foundation modulus
L	span length of beam
m	mass
v	velocity
y	transverse deflection
δ	deflection of beam
ρ	density

See also: **Damping, active; Damping materials; Damping measurement.**

Further Reading

American Association of State Highway and Transportation Officials (AASHTO) (1977). *Standard Specifications for Highway Bridges*. Washington DC.

Bakht B, Jaeger LG (1985). *Bridge Analysis Simplified*. New York: McGraw-Hill.

Cain BS (1940). *Vibration of Rail and Road Vehicles*. New York: Pitman.

Cusens AR, Loo YC (1974). Application of the finite strip method in the analysis of concrete box bridges. *Proceedings of the Institute of Civil Engineers* 57(II): 251–273.

Inglis CE (1934). *A Mathematical Treatise on Vibrations in Railway Bridges*. Cambridge: Cambridge University Press.

Inglis CE (1938–1939). The vertical path of a wheel moving along a railway track. *Journal of the London Institution of Civil Engineers* 11: 262–288.

Ministry of Transportation and Communications (1983). *Ontario Highway Bridge Design Code (OHBDC)*, 2nd edn. Ontario, Canada: Downsview.

Rao SS (1995). *Mechanical Vibrations*, 3rd edn. Reading, Mass: Addison-Wesley.

Timoshenko S (1927–1928). Vibration of bridges. *Transactions of the American Society of Mechanical Engineers* 49–50 (II): 53–61.

Volterra E, Zachmanoglou EC (1965) *Dynamics of Vibrations*, Columbus, OH: C. E. Merrill Books.

C

CABLES

N C Perkins, University of Michigan, Ann Arbor, MI, USA

Copyright © 2001 Academic Press

doi:10.1006/rwvb.2001.0135

Introduction

Cables can be viewed as the one-dimensional structural analogs to membranes. As with membranes, the structural stiffness of cables derives from the equilibrium tension created by applied preloads (see **Membranes**). In contrast to other one-dimensional structural elements such as beams and rods, cables are generally considered to lack torsional, flexural, and shear rigidities (see **Columns**). Under static loading, cables are also curved and the equilibrium curvature induces behaviors in cables that are also found in other curved structural elements including arches (one-dimensional) and shells (two-dimensional) (see **Shells**).

The overall flexibility of cables renders them useful in a wide range of applications, particularly those requiring long, flexible, or readily deployable structural elements. In addition to carrying loads as structural elements, cables are also commonly used to transmit power, electrical signals, and optical signals across long distances. Consider, for instance, the diverse uses of cables in ocean engineering applications, some of which are depicted in **Figure 1**. These include mooring lines for vessels, platforms and buoys, towing lines, umbilicals, tethers, instrumentation arrays, etc. Cables play a dominant role in power distribution systems where they serve as conducting elements. They also serve as major load-transmitting elements when used in guyed towers, as cable trusses in suspended roofs, and in cable-stayed bridges (see **Bridges**).

The objective of this article is to review the fundamental characteristics of cable vibration and to summarize the associated analytical models that may be used to predict dynamic cable response. To this end, a general model for a suspended cable is reviewed that is subsequently used to describe linear vibration characteristics. Extensions to this linear theory of cable vibration are then mentioned that generalize results to include a broader range of applications and also nonlinear effects.

Nonlinear Model of a Suspended Cable

Consider an elastic cable of length L that is suspended between two supports, as shown in **Figure 2**. Under the action of gravity (g) the cable will sag in the vertical plane and achieve an equilibrium shape in the form of a catenary (see dashed curve in **Figure 2**). Following a disturbance, the cable may then vibrate about this equilibrium and achieve the three-dimensional shape shown by the solid curve in **Figure 2**.

To describe this dynamic response from equilibrium, we will introduce three unit vectors that are defined by the equilibrium shape. These include the unit tangent vector l_1, the unit normal vector l_2, and the unit binormal vector l_3. The dynamic response about equilibrium is then decomposed as:

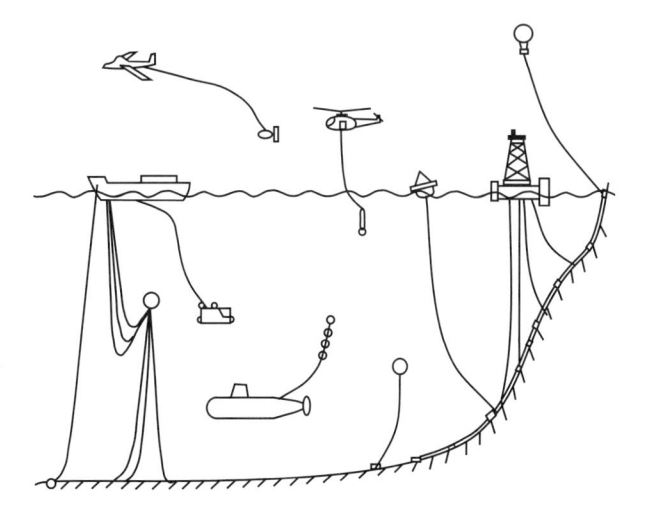

Figure 1 Applications of cables in ocean engineering. Adapted with permission from Choo Y, and Casarella MJ (1973) A survey of analytical methods for dynamic simulation of cable-body systems. *Journal of Hydronautics* 7: 137–144.

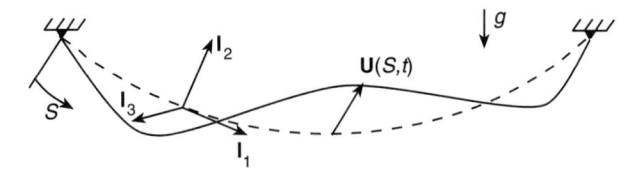

Figure 2 Definition diagram for suspended cable. Dashed curve represents the (planar) equilibrium shape of the cable and the solid curve represents a three-dimensional motion about equilibrium. The orthonormal triad (l_1, l_2, l_3) consists of the tangential, normal, and binormal unit vectors defined by the equilibrium shape.

$$\mathbf{U}(S,T) = U_1(S,T)\mathbf{l}_1 + U_2(S,T)\mathbf{l}_2 + U_3(S,T)\mathbf{l}_3 \quad [1]$$

in which S denotes the arc length coordinate measured along the equilibrium shape of the cable starting from the left support, T denotes time, and U_1, U_2, and U_3 denote the dynamic displacement of the cable in the tangential, normal, and binormal directions, respectively. The components (U_1, U_2) define the in-plane motion of the cable, while the component (U_3) defines the out-of-plane motion.

The cable is assumed to be a one-dimensional continuum obeying a linear elastic law for cable extension and possessing no flexural, shear or torsional rigidity. With these assumptions, the strain – displacement relation for cable extension is given by:

$$\varepsilon = P(S)/EA + \varepsilon_d(S,T) \quad [2a]$$

where $P(S)$ denotes the equilibrium cable tension, EA is the cross-sectional stiffness, and:

$$\varepsilon_d(S,T) = U_{1,S} - KU_2 \\ + \frac{1}{2}\left[(U_{1,S} - KU_2)^2 + (U_{2,S} + KU_1)^2 + (U_{3,S})^2\right] \quad [2b]$$

is the (nonlinear) dynamic strain. In addition, K denotes the equilibrium curvature and the notation $(\ )_{,S}$ denotes partial differentiation with respect to S. The nonlinear equations of motion describing three-dimensional response about equilibrium are:

(tangential direction)

$$\left[P(U_{1,S} - KU_2)\right]_{,S} \\ + \left[EA\varepsilon_d(1 + U_{1,S} - KU_2)\right] \\ - \left[(P + EA\varepsilon_d)K(U_{2,S} + KU_1)\right] = \rho U_{1,TT} \quad [3a]$$

(normal direction)

$$\left[(P + EA\varepsilon_d)(U_{2,S} + KU_1)\right]_{,S} \\ + PK(U_{1,S} - KU_2) \\ + EA\varepsilon_d K(1 + U_{1,S} - KU_2) = \rho U_{2,TT} \quad [3b]$$

(binormal direction)

$$\left[(P + EA\varepsilon_d)U_{3,S}\right]_{,S} = \rho U_{3,TT} \quad [3c]$$

in which ρ denotes the cable mass/length. For the suspension illustrated in **Figure 2**, the boundary conditions:

$$U_i(0,T) = U_i(L,T) = 0, \quad i = 1, 2, 3 \quad [3d]$$

describe the illustrated fixed supports.

The nonlinear equations of cable motion above contain the nonconstant coefficients $P(S)$ and $K(S)$ that describe the equilibrium state of the cable. For the catenary of **Figure 2** with level supports, these are given by:

$$P(S) = \sqrt{P_o^2 + [\rho g(S - L/2)]^2} \quad [4a]$$

$$K(S) = \frac{\rho g P_o}{P_o^2 + [\rho g(S - L/2)]^2} \quad [4b]$$

in which P_0 is the (characteristic) tension at the midspan of the equilibrium cable.

Linear Model of Shallow Sag Cable

We shall now focus on the technically important case of a suspended cable under relatively high tension such that the amount of sag at the midspan does not exceed (approximately) one-eighth of the distance between the supports. Such a cable is frequently referred to as a shallow sag cable for which there is a well established linear theory of vibration.

To obtain this theory, we begin by approximating the equilibrium tension eqn [4a] and curvature eqn [4b] as constants (retain first term in Taylor series expansion about the cable midspan):

$$P(S) \approx P_O \quad [5a]$$

$$K(S) \approx K_O = \rho g/P_O \quad [5b]$$

Next, we introduce two wave speeds:

$$c_t = \sqrt{(p_O/\rho)} \quad [6a]$$

$$c_l = \sqrt{(EA/\rho)} \quad [6b]$$

which have the interpretation of the speed of propagation of transverse waves along a taut string and longitudinal waves along an elastic rod, respectively (see **Wave propagation**, Guided waves in structures). Clearly the longitudinal wave speed is orders of magnitude greater than the transverse wave speed as $EA \gg P_O$. Recognizing this, it then follows that the equation of motion in the longitudinal direction (eqn [3a]) can be approximated by an equation of statics (eliminate the inertia term $\rho U_{1,TT}$) since longitudinal waves will propagate orders of magnitude faster than transverse waves. This assumption is referred to as the quasistatic stretching assumption and is valid for low-order modes of cable vibration. Finally, we eliminate all the nonlinear terms in the equations of motion to obtain the linear theory. The linear equations are then simplified further by ordering each term and retaining the largest of these. The resulting equations of linear free vibration are:

(Normal direction)

$$P_O U_{2,SS} - \frac{K_O^2 EA}{L} \int_0^L U_2 \, dS = \rho U_{2,TT} \quad [7a]$$

(Binormal direction)

$$P_O U_{3,SS} = \rho U_{3,TT} \quad [7b]$$

with the boundary conditions:

$$U_i(0, T) = U_i(L, T) = 0, \quad i = 2, 3 \quad [7c]$$

describing fixed supports. Note that the equation of motion in the binormal (out-of-plane) direction (eqn [7b]) reduces to the classical wave equation for a taut string. Similarly, the equation of motion in the normal direction (eqn [7a]) is also that for a taut string with the added integral term that accounts for the first-order stretching of the cable centerline. The equation of motion in the tangential direction can be integrated to yield:

$$U_1(S, T) = K_O \left[\int_0^S U_2(\eta, T) \, d\eta - \frac{S}{L} \int_0^L U_2(\eta, T) \, d\eta \right] \quad [7d]$$

Thus, the tangential displacement U_1 can now be computed from knowledge of the normal displacement U_2. This reduction from three variables to two follows from the use of the quasistatic stretching approximation.

Natural Frequencies and Vibration Modes

Eqns [7a] and [7b] are uncoupled in the normal and binormal displacement components U_2 and U_3. Thus, there exists two families of vibration modes, one for in-plane modes associated with eqn [7a] and one for out-of-plane modes associated with eqn [7b] (see **Figure 3**). Three-dimensional linear motions about equilibrium can therefore be decomposed into these two families of vibration modes as suggested in **Figure 3**.

Out-of-plane modes As noted above, the equation describing the out-of-plane vibration of a shallow sag cable is identical to that of a taut string with static tension P_O (the static tension at the midspan of the cable). The natural frequencies and mode shapes can be found by solving the eigenvalue problem associated with the equation of motion (eqn [7b]) and the boundary conditions (eqn [7c]). Doing so yields the natural frequencies:

$$\omega_n = (n\pi c_t/L) \, \text{rad s}^{-1}, \quad n = 1, 2, \ldots \quad [8a]$$

and mode shapes:

$$v_n(S) = C_n \sin(n\pi S/L), \quad n = 1, 2, \ldots \quad [8b]$$

in which C_n is an arbitrary constant (frequently selected to satisfy a normalization condition).

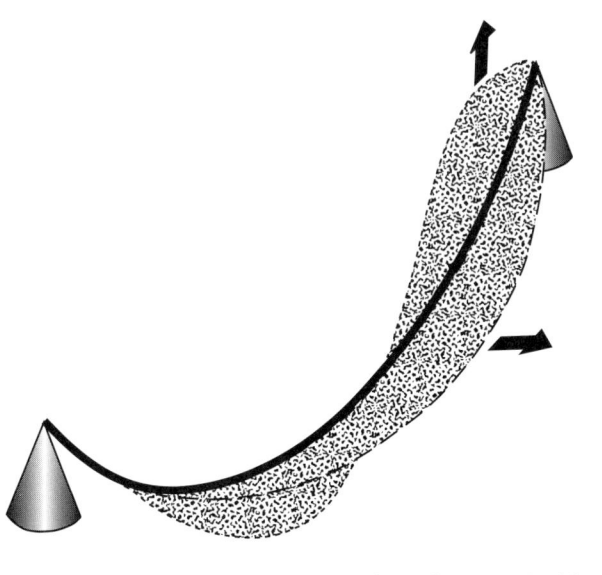

Figure 3 Three-dimensional, linear motions of a sagged cable about equilibrium can be described using two families of vibration modes, namely in-plane modes and out-of-plane modes.

In-plane modes The in-plane modes of a shallow sag cable can be found by formulating the eigenvalue problem associated with eqn [7a] and the boundary conditions in eqn [7b]. Doing so reveals that there are two distinct classes of in-plane modes. One class constitutes all mode shapes that are antisymmetric with respect to the midspan of the cable; that is, all mode shapes that are odd functions about the point $S = L/2$. For these modes, the second term in eqn [7a] vanishes (since an odd function integrated over symmetric limits is zero). Physically, these mode shapes do not induce first-order stretching of the cable centerline. As a result, the equation of in-plane motion is now identical to that of the out-of-plane modes and therefore the antisymmetric in-plane modes and natural frequencies are equivalent to those of the antisymmetric out-of-plane vibration modes.

Antisymmetric in-plane vibration modes

$$\omega_n = (n\pi c_t/L) \text{ rad s}^{-1}, \quad n = 2, 4, 6, \ldots \quad [9a]$$

with mode shapes:

$$v_n(S) = C_n \sin(n\pi S/L), \quad n = 2, 4, 6, \ldots \quad [9b]$$

in which C_n is again an arbitrary constant.

The second class of in-plane modes constitutes all mode shapes that are symmetric with respect to the midspan of the cable; that is, all mode shapes that are even functions about the point $S = L/2$. For these modes, the second term in eqn [7a] does not vanish in general and therefore these modes induce first-order stretching of the cable centerline.

Symmetric in-plane vibration modes The mode shapes for the symmetric in-plane modes are:

$$v_n = C_n \left[\sin\left(\frac{\omega_n}{2c_t} S\right) \sin\left(\frac{\omega_n}{2c_t}(S-L)\right) \right], \\ n = 1, 3, 5, \ldots \quad [10a]$$

in which C_n is an arbitrary constant and ω_n is the corresponding natural frequency that satisfies the characteristic equation:

$$\tan\left(\frac{\omega_n L}{2c_t}\right) = \frac{1}{2}\left[\frac{\omega_n L}{c_t} - \left(\frac{\omega_n L}{c_t}\right)^3 / \lambda^2\right] \quad [10b]$$

Here, the quantity:

$$\lambda^2 = \frac{c_l^2}{c_t^6}(gl)^2 \quad [10c]$$

is a nondimensional quantity commonly referred to as the cable parameter. The cable parameter represents a ratio of two sources of stiffness, namely, stiffness due to elasticity (through the longitudinal wave speed c_l) and stiffness due to tension (through the transverse wave speed c_t). This single parameter can be used to classify the vibration properties of cable suspensions as described below.

Figure 4 illustrates how the natural frequencies of the in-plane cable modes depend upon the cable parameter. For convenience, the natural frequencies plotted in this figure are nondimensionalized by dividing by the natural frequency of the fundamental out-of-plane cable mode ($n = 1$ in eqn [8a]). The frequencies are plotted as functions of the parameter λ/π. Note that small values of this parameter correspond to a highly tensioned cable and/or small axial stiffness. For example, large P_0 implies large c_t, which implies small λ/π. Conversely, large values of this parameter correspond to less tension (greater sag) and larger axial stiffness. For example, large EA implies large c_l, which implies large λ/π. Note that the natural frequencies of the out-of-plane modes are independent of the parameter λ/π, as are the natural frequencies of the antisymmetric in-plane modes. In both of these cases, the natural frequencies are simply those of a taut string. By contrast, the natural frequencies of the symmetric in-plane modes depend on λ/π. To understand this dependence, we discuss first the vibration characteristics for the two limiting cases $\lambda/\pi \to 0$ and $\lambda/\pi \to \infty$.

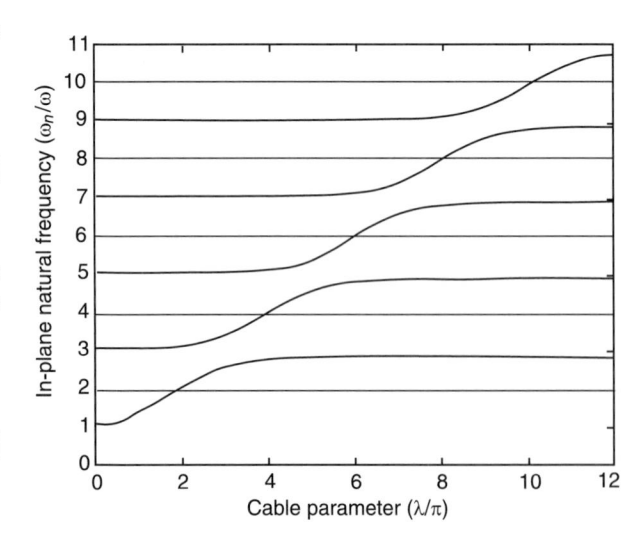

Figure 4 Cable in-plane natural frequency spectrum as a function of the parameter λ/π. The natural frequencies are normalized with respect to the fundamental out-of-plane natural frequency, $\omega = \pi c_t/L$.

Limiting case $\lambda/\pi \to 0$: **the taut string** For this limiting case, the characteristic equation for the symmetric in-plane modes, eqn [10b], reduces to:

$$\tan\left(\frac{\omega_n L}{2c_t}\right) \to \infty \qquad [11a]$$

and admits the roots:

$$\omega_n = (n\pi c_t/L) \text{ rad s}^{-1}, \quad n = 1, 3, 5, \ldots \qquad [11b]$$

Thus, the symmetric in-plane modes of an elastic and sagged cable reduce to those of a taut string in the limit of large tension (vanishing sag). Consequently the classical taut string model is fully recovered in this limit. In **Figure 4**, the natural frequencies for all of the in-plane modes correspond to the simple integer values (1, 2, 3 ...) in the limit $\lambda/\pi \to 0$. The natural frequencies for all of the out-of-plane modes are, of course, independent of λ/π and correspond to those of a taut string as mentioned above.

Limiting case $\lambda/\pi r \to \infty$: **the inextensible cable (chain)** For this limiting case, the characteristic equation for the symmetric in-plane modes (eqn [10b]) reduces to:

$$\tan\left(\frac{\omega_n L}{2c_t}\right) = \frac{1}{2}\left[\frac{\omega_n L}{c_t}\right] \qquad [12]$$

This is precisely the characteristic equation of an inextensible cable or chain with small sag. The roots of this equation are the limiting values of the natural frequencies for the symmetric in-plane modes as $\lambda/\pi \to \infty$. In this limit, the cable no longer stretches and all mode shapes induce deformations that satisfy an inextensibility constraint (derivable from the condition $\varepsilon_d = 0$ in eqn [2b]).

General case: the elastic sagged cable The model above provides a continuous transition between the two limiting cases. For small values of λ/π, the model describes a simple taut string. For large values of λ/π, the model describes a sagged but inextensible cable. It is important to note that the (simpler) theory of a sagged but inextensible cable cannot reduce to that of a taut string in the limit of vanishing sag. In this limit, the cable must stretch in response to first-order (linear) deformations. Thus, cable elasticity must be included in any cable theory that might be used near this limit.

The transition of the elastic cable model from a taut string to an inextensible cable is clearly seen in **Figure 4**. For example, follow the curve that defines the first (symmetric) mode of a taut string for $\lambda/\pi \to 0$ starting at the left of **Figure 4** and then proceeding to the right. There is a rapid increase in this quantity near $\lambda/\pi = 2$ where the first natural frequency is also equivalent to the second natural frequency. This point is often referred to as the first crossover point. Second- and higher-order crossover points at $\lambda/\pi = n$, $n = 4, 6, 8 \ldots$ define the transition of the second- and higher-order symmetric in-plane modes. These crossover points represent the values of the cable parameter where a particular symmetric in-plane mode for an elastic cable is evolving from that of a taut string to that of an inextensible cable (chain). During this transition, the mode shape induces cable stretching and undergoes qualitative changes, as discussed below.

Figure 5 illustrates the transition of the fundamental symmetric in-plane mode as one varies the parameter λ/π through the first crossover point. Starting at the bottom of this figure where $\lambda/\pi = 1$, this mode resembles the fundamental mode of a taut string and appears as one-half of a sine wave. As λ/π increases, the tangent at the boundaries rotate to the degree that they become horizontal when the suspension is tuned to the crossover point $\lambda/\pi = 2$. With any further increase in λ/π, the tangents at the boundaries rotate more and create two interior nodes near the boundaries; see mode shape corresponding to $\lambda/\pi = 2.5$. These nodes migrate inwards with further increases in λ/π, until they reach their limiting positions as observed for the case $\lambda/\pi = 4$. At this point, this mode shape now resembles the third mode of a taut string, yet it corresponds to the second mode of a

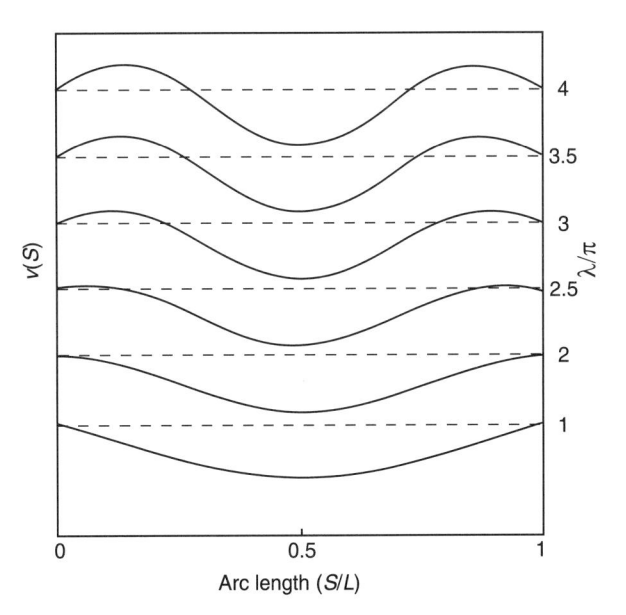

Figure 5 Evolution of fundamental symmetric in-plane mode as the parameter λ/π is varied through the crossover point $\lambda/\pi = 2$.

sagged cable. This is the limiting shape of the first symmetric mode and it induces no stretching of the cable centerline in keeping with the fact the cable model for this mode is equivalent to that of a inextensible cable (chain). Similar transitions occur for all higher-order symmetric modes; namely, each of these gain two interior nodes as the parameter λ/π is increased through the corresponding crossover point.

Extensions of Linear Theory

The linear theory for a shallow sag cable reviewed above provides the basis for many extensions. For instance, specialized theories exist for shallow sag cables that have inclined supports, multispan cables (e.g., transmission lines), translating cables, and cables supporting attached masses, to name a few. In addition, numerous studies have considered possible nonlinear responses of cables, and in doing so have revealed response characteristics that are qualitatively different from predictions based on the linear theory above. A brief review of some nonlinear characteristics is provided here as they significantly differ from what is described above.

To begin, consider a simple experiment as depicted in the schematic of **Figure 6**. In this experiment, a small length of cable is suspended between a fixed support at the right and a movable support at the left. The motion of the left support is controlled by an electromechanical shaker that provides harmonic motion along the cable tangent. This excitation causes the cable to oscillate about its equilibrium configuration. An optical probe positioned somewhere along the cable records these oscillations in the normal and binormal directions (see **Figure 2**).

In this experiment, the tension (sag) of the cable is adjusted so that $\lambda/\pi \approx 2$, i.e., the suspension is tuned to the first crossover. As a consequence, the natural frequency for the fundamental symmetric in-plane mode is approximately twice that of the fundamental out-of-plane mode. The experiment proceeds by adjusting the excitation frequency to be equal to the frequency of the fundamental symmetric in-plane mode. Thus, this mode is resonantly excited and one would anticipate that the cable oscillates in this mode within the equilibrium plane. This expectation, however, is only partially met as can be seen in the experimental results of **Figure 7**.

Figure 7 illustrates the orbit traced by a representative cross-section of the cable as viewed in the normal–binormal plane. Thus, planar motion (motion restricted to the equilibrium plane) will appear as a vertical line in this figure. As the excitation amplitude is slowly increased, the motion begins as planar and then becomes decidedly nonplanar. **Figure 7A** corresponds to the lowest level of excitation and the response is planar as predicted by the linear theory. A modest increase in this excitation leads to a proportional increase in the planar response as seen in **Figure 7B**. A further increase in the excitation,

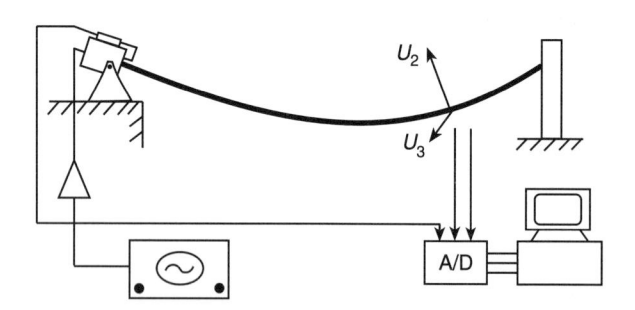

Figure 6 Schematic of a laboratory experiment illustrating nonlinear cable response. The cable is suspended between a fixed support at the right and an electromechanical shaker at the left. Reproduced with permission from Perkins NC (1992) Modal interactions in the non-linear response of elastic cables under parametric/external excitation. *International Journal of Non-linear Mechanics* 27(2): 233–250.

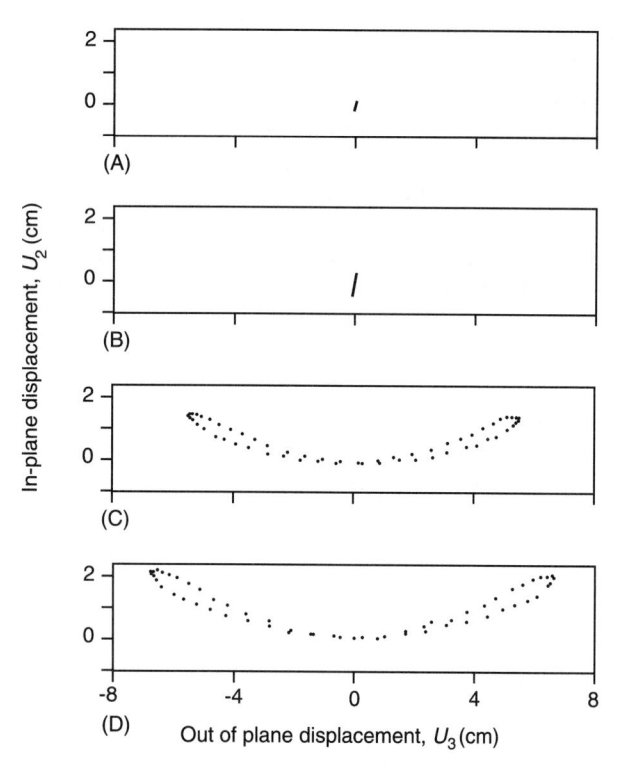

Figure 7 Experimental measurements showing motion of cable cross-section in the normal–binormal plane. Excitation amplitude increases from (A) to (B) to (C) to (D). Reproduced with permission from Perkins NC (1992) Modal interactions in the non-linear response of elastic cables under parametric/external excitation. *International Journal of Non-linear Mechanics* 27(2): 233–250.

however, generates a sizable out-of-plane motion component and the closed orbit shown in **Figure 7C**. Increasing the excitation further yet magnifies this nonplanar motion which cannot be predicted by the linear theory.

Note from **Figures 7C** and **7D** that the nonplanar response forms a closed loop (periodic motion) in the normal–binormal plane. For this particular loop, the cable completes two cycles of motion in the normal direction for every one cycle of motion in the binormal direction. This two-to-one relation in the response frequencies suggests the source of this interesting motion. Further experimental evidence reveals that this motion is produced by two cable modes; namely the fundamental symmetric in-plane mode and the fundamental out-of-plane mode. As mentioned above, these two modes have natural frequencies in a two-to-one ratio when the cable is at the first crossover point as in the experiment. The nonplanar motion observed here results from a nonlinear coupling of these two cable modes. The support excitation resonantly drives the in-plane mode and the in-plane mode is strongly coupled to the out-of-plane mode through the nonlinearities associated with nonlinear (finite) stretching of the cable centerline, (see **Nonlinear system identification**). In particular, there exists a two-to-one internal resonance of these two cable modes that leads to the resulting nonplanar motion (see **Nonlinear system resonance phenomena**). This internal resonance destabilizes the (linear) planar motion through a pitchfork bifurcation. This fact is illustrated in the experimental results of **Figure 8** which shows how the amplitudes of the in-plane displacement (a_2) and the out-of-plane displacement (a_3) vary with the excitation amplitude. Notice that the planar (linear) motion corresponds to the straight line in this figure that begins at the origin. This planar motion ultimately loses stability and is replaced by a periodic nonplanar response that is actually dominated by the out-of-plane motion component.

Numerous studies similar to this have revealed a rich variety of nonlinear responses. These include other classes of internal resonances including one-to-one internal resonances and internal resonances involving multiple (more than two) cable modes. It is important to recognize that these motions develop precisely because of the influence of nonlinear stretching. Thus, they cannot be predicted using a linear theory for cable dynamics.

Nomenclature

EA	cross-sectional stiffness
g	gravity
K	equilibrium curvature
$P(S)$	equilibrium cable tension
T	time
ρ	cable mass/length

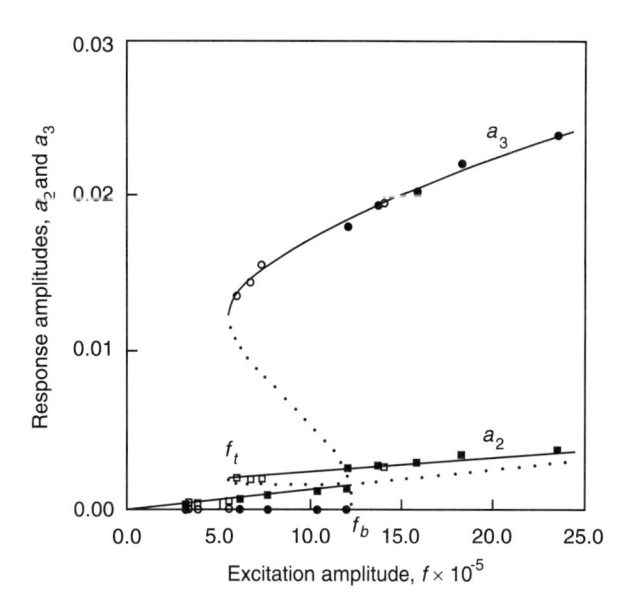

Figure 8 Summary of experimental results showing the amplitudes of the in-plane (a_2) and out-of-plane motion (a_3) components as the excitation amplitude is varied. Solid (open) symbols denote data collected while increasing (decreasing) the excitation amplitude. Reproduced with permission from Perkins NC (1992) Modal interactions in the non-linear response of elastic cables under parametric/external excitation. *International Journal of Non-linear Mechanics* 27(2): 233–250.

See Plates 11, 12.

See also: **Bridges**; **Columns**; **Membranes**; **Nonlinear system identification**; **Nonlinear system resonance phenomena**; **Shells**; **Wave propagation**, Guided waves in structures.

Further Reading

Cheng SP, Perkins NC (1992) Closed-form vibration analysis of sagged cable/mass suspensions. *ASME Journal of Applied Mechanics* 59: 923–928.

Choo Y, Casarella MJ (1973) A survey of analytical methods for dynamic simulation of cable-body systems. *Journal of Hydronautics* 7: 137–144.

Irvine HM (1981) *Cable Structures*. Cambridge, MA: MIT Press.

Irvine HM, Caughey TK (1974) The linear theory of free vibrations of a suspended cable. *Proceedings of the Royal Society of London* A341: 299–315.

Perkins NC (1992) Modal interactions in the non-linear response of elastic cables under parametric/external excitation. *International Journal of Non-linear Mechanics* 27(2): 233–250.

Perkins NC, Mote CD Jr. (1987) Three-dimensional

vibration of travelling elastic cables. *Journal of Sound and Vibration* 114(2): 325–340.

Simpson A (1966) Determination of the in-plane natural frequencies of multispan transmission lines by a transfer matrix method. *IEE Proceedings* 113(5): 870–878.

Triantafyllou MS (1984) Linear dynamics of cables and chains. *Shock and Vibration Digest* 16: 9–17.

Triantafyllou MS (1984) The dynamics of taut inclined cables. *Quarterly Journal of Mechanics and Applied Mathematics* 37: 421–440.

Triantafyllou MS (1987) Dynamics of cables and chains. *Shock and Vibration Digest* 19: 3–5.

Triantafyllou MS (1987) Dynamics of cables, towing cables and mooring lines. *Shock and Vibration Digest* 23: 3–8.

CEPSTRUM ANALYSIS

R B Randall, University of New South Wales, Sydney, Australia

Copyright © 2001 Academic Press

doi:10.1006/rwvb.2001.0055

Introduction

The cepstrum has a number of variants, definitions, and realizations, but all involve a (Fourier) transform of a logarithmic spectrum, and are thus a 'spectrum of a spectrum'. This is in fact the reason for the name 'cepstrum' and a number of related terms coined, by reversing the first syllable, in the original paper by Bogert, Healy, and Tukey, and discussed here in the section on terminology. However, the autocorrelation function is the inverse Fourier transform of the corresponding autospectrum and so is equally a spectrum of a spectrum. What really distinguishes the cepstrum is the logarithmic conversion of the spectrum before the second transform. In response spectra, this converts the multiplicative relationship between the forcing function and transfer function (from force to response) into an additive one which remains in the cepstrum. This gives rise to one of the major applications of the cepstrum. Another property of the logarithmic conversion is that it often makes families of uniformly spaced components in the spectrum, such as families of harmonics and sidebands, much more evident, so that the final transform is able to reveal and quantify them and their spacing. This gives rise to a further range of applications of the cepstrum in vibration analysis. Note, however, that the cepstrum, being based on logarithmic conversions of dimensionless ratios, in general gives no information on the absolute scaling of signals. All such information is contained in the first (or zero 'quefrency') component in the cepstrum, which is often modified, or a combination of several factors.

Terminology

In the same way as 'cepstrum' is formed from 'spectrum' by reversing the phoneme of the first syllable, the original authors proposed a number of terms which are still found in the cepstrum literature, and which are useful to distinguish properties or operations associated with or carried out in the cepstrum domain. The most useful, which are used in this section, are 'quefrency' the x-axis of the cepstrum, which has the units and dimensions of time, 'rahmonics' a series of uniformly spaced components in the cepstrum, and often coming from a family of harmonics in the log spectrum, and a 'lifter', the equivalent of a filter, but realized by windowing in the cepstrum domain. Thus, a 'shortpass lifter' is analogous to a lowpass filter. Other terms such as 'gamnitude' and 'saphe' are of dubious usefulness.

Definitions and Formulae

The original definition of the cepstrum by Bogert, Healy, and Tukey was the 'power spectrum of the logarithm of the power spectrum', but this has been largely superseded by the definition as the 'inverse Fourier transform of the logarithm of a spectrum'. If the spectrum is a power spectrum, there are two differences with respect to the original definition:

1. The second transform is inverse rather than forward, but since the power spectrum is a real, even function, this only gives a difference in scaling. It is more logical to carry out an inverse transform on a function of frequency.
2. Forming the power (amplitude squared) spectrum of the result makes it irreversible and puts more weight on the largest peaks. It precludes applications involving liftering in the cepstrum, followed by transformation back to the log spectrum.

Moreover, the new definition can be extended to the case where the spectrum (and thus the logarithmic spectrum) is complex, and the whole process is reversible back to the time domain.

Thus, the cepstrum is defined as:

$$C(\tau) = F^{-1}\{\log[X(f)]\} \quad [1]$$

where:

$$X(f) = F[x(t)] = A(f)\exp[j\phi(f)] \quad [2]$$

in terms of its amplitude and phase, so that:

$$\log[X(f)] = \ln[A(f)] + j\phi(f) \quad [3]$$

When $X(f)$ is complex, as in this case, the cepstrum of eqn [1] is known as the 'complex cepstrum' although since $\ln[A(f)]$ is even and $\phi(f)$ is odd, the complex cepstrum is real-valued. Note that, by comparison, the autocorrelation function can be derived as the inverse transform of the power spectrum, or:

$$R_{xx}(\tau) = F^{-1}\left[|X(f)|^2\right] = F^{-1}\left[A^2(f)\right] \quad [4]$$

When the power spectrum is used to replace the spectrum $X(f)$ in eqn [1], the resulting cepstrum, known as the 'power cepstrum' or 'real cepstrum', is given by:

$$C_{xx}(\tau) = F^{-1}\{2\ln[A(f)]\} \quad [5]$$

and is thus a scaled version of the complex cepstrum where the phase of the spectrum has been set to zero.

Another type of cepstrum which is useful in some cases is the 'differential cepstrum', which is defined as the inverse transform of the derivative of the logarithm of the spectrum. It is most easily defined in terms of the Z-transform (which can replace the Fourier transform for sampled functions) as:

$$C_d(n) = Z^{-1}\left\{z\frac{(d/dz)[H(z)]}{H(z)}\right\} \quad [6]$$

where n is the quefrency index, and can be directly calculated from a time signal as:

$$C_d(n) = F^{-1}\left\{\frac{F[nx(n)]}{F[x(n)]}\right\} \quad [7]$$

Among other things this has the advantage that the phase of the (log) spectrum does not have to be 'unwrapped' to a continuous function of frequency, as is the case with the complex cepstrum of eqn [1].

Where the frequency spectrum $X(f)$ in eqn [1] is a frequency response function (FRF) which can be represented in the Z-plane by a gain factor K and the zeros and poles inside the unit circle, a_i and c_i, respectively, and the zeros and poles outside the unit circle, $1/b_i$ and $1/d_i$, respectively (where $|a_i|, |b_i|, |c_i|, |d_i| < 1$), then it has been shown by Oppenheim and Schafer that the complex cepstrum is given by the analytical formulae:

$$\begin{aligned} C(n) &= \ln(K), & n &= 0 \\ C(n) &= -\sum_i \frac{a_i^n}{n} + \sum_i \frac{c_i^n}{n}, & n &> 0 \\ C(n) &= \sum_i \frac{b_i^{-n}}{n} - \sum_i \frac{d_i^{-n}}{n}, & n &< 0 \end{aligned} \quad [8]$$

in terms of quefrency index n.

Since the cepstrum is real, the complex exponential terms in eqn [8] can be grouped in complex conjugate pairs so that a typical pair of c_i terms, for example, can be replaced by $(2/n)C_i^n\cos(n\alpha_i)$ where $C_i = |c_i|$ and $\alpha_i = \angle c_i$. This represents an exponentially damped sinusoid, further damped by the hyperbolic function $1/n$. **Figure 1** compares the cepstrum with the impulse response function (IRF) for a single-degree-of-freedom (SDOF) system which has one pair of poles and no zeros. On a logarithmic amplitude scale, zeros of the FRF (antiresonances) are like inverted poles (resonances) so it is no surprise that the corresponding terms in the cepstrum have inverted sign.

Taking the derivative of the log spectrum in the Z-domain to obtain the differential cepstrum results in multiplication by n in the cepstrum, so that a typical term becomes $2C_i^n\cos(n\alpha_i)$, an exponentially damped

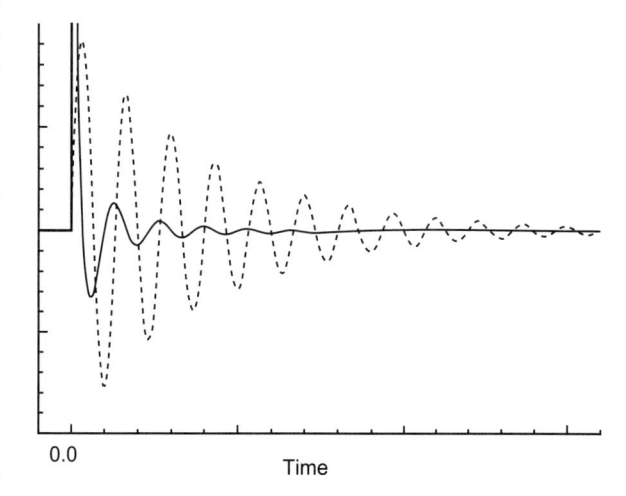

Figure 1 The impulse response (dashed line) of a single-degree-of-freedom system and the corresponding cepstrum (continuous line).

sinusoid without the hyperbolic weighting, so that its form is similar to that of the IRF. This is useful in that techniques which have been developed to curve fit parameters to the IRF can be directly applied to the differential cepstrum. This is the second advantage of the differential cepstrum.

Note that for functions with minimum phase properties, which applies to FRFs for many physical structures, there are no poles or zeros outside the unit circle (the b_i and d_i vanish) and thus there are no negative quefrency terms in eqn [8], so that the cepstrum (and differential cepstrum) are causal. By normal Hilbert transform relationships (see **Hilbert transforms**) this means that the real and imaginary parts of the corresponding Fourier transform, the log amplitude and phase of the spectrum, are related by a Hilbert transform, and only one has to be measured. It also means that the complex cepstrum can be obtained from the corresponding power cepstrum (which is real and even) by doubling positive quefrency terms and setting negative quefrency terms to zero. In this case also, the phase of the spectrum does not have to be measured or unwrapped.

Phase Unwrapping

As mentioned above, the spectrum phase in eqn [1] must be unwrapped to a continuous function of frequency before the inverse transform is carried out, whereas often the phase is obtained as a principal value between $\pm\pi$. **Figure 2** shows a typical example. Simple phase-unwrapping algorithms make a decision based on whether the phase jump between adjacent samples is $>\pi$ or $<\pi$. This can be in error in regions of rapid phase change. Tribolet has devised a more reliable phase-unwrapping algorithm, but perhaps a simpler way of avoiding problems is to use a finer interpolation in the spectrum, by padding time records with zeros up to a sufficient multiple of their original length that the phase jumps between adjacent frequency samples are no longer ambiguous. After unwrapping, the phase function can be decimated back to the original sample rate.

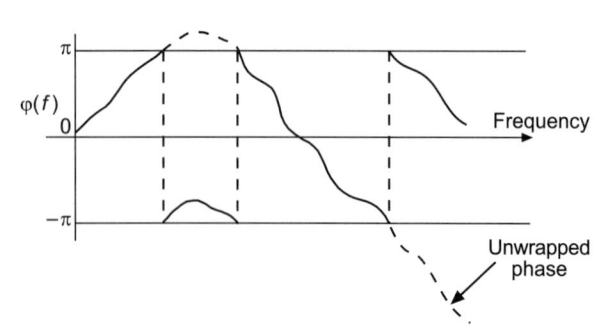

Figure 2 An example of unwrapped phase as determined from the principal values between $\pm\pi$.

Applications of the Power Cepstrum

The major application of the power cepstrum in machine vibrations is to detect and quantify families of uniformly spaced harmonics, such as arise from periodic added impulses (bearing faults, missing turbine blades, faulty valve plate in a reciprocating compressor) and sidebands, such as arise from amplitude and phase modulation of discrete carrier frequencies (faults in gears which modulate the common gearmesh frequency at lower frequencies corresponding to the individual gear rotational speeds).

Faults in Bearings

Figure 3 gives an example of the development of an outer race fault in a ball bearing in a high-speed gearbox driven by a gas turbine, and shows the (log) spectra on the left and the cepstra on the right. Even at the early stages of the fault (24 August 1981) there is a dramatic change in the cepstrum, with a new series of rahmonics appearing in addition to the component corresponding to the shaft speed (RPM). In addition to this detection sensitivity, the other advantages given by the cepstrum are:

1. Since the position of the first rahmonic represents (the reciprocal of) the average harmonic spacing throughout the whole spectrum, the value is much more accurate than can be obtained by measuring the spacing between individual harmonics. Of course, the same accuracy can be obtained by adjusting a finely tunable harmonic cursor on to the spectrum pattern, but even then the cepstrum may be useful in suggesting spacings to try.
2. The fact that the shaft speed quefrency is 4.1 times the quefrency of the unknown component means that its corresponding frequency is 4.1 times the shaft speed. In this case it immediately identified the source as corresponding to the outer race frequency for a particular bearing in the gearbox (which had 10 balls and an effective ball-diameter-to-pitch-diameter ratio of 0.18).
3. Once again, because of the averaging effect across the whole spectrum, the first rahmonic exhibits much less variation with time than the individual harmonics in the spectrum, and thus makes a more valid trend parameter when tracing the course of the fault development. **Figure 4** illustrates this. The higher rahmonics are affected by a number of artifacts and do not add much more information except to confirm the periodicity.

By way of contrast, **Figure 5** compares the (log) power spectrum and its corresponding cepstrum for one of the cases in **Figure 3**, with the (linear) power spectrum for the same case, and its corresponding

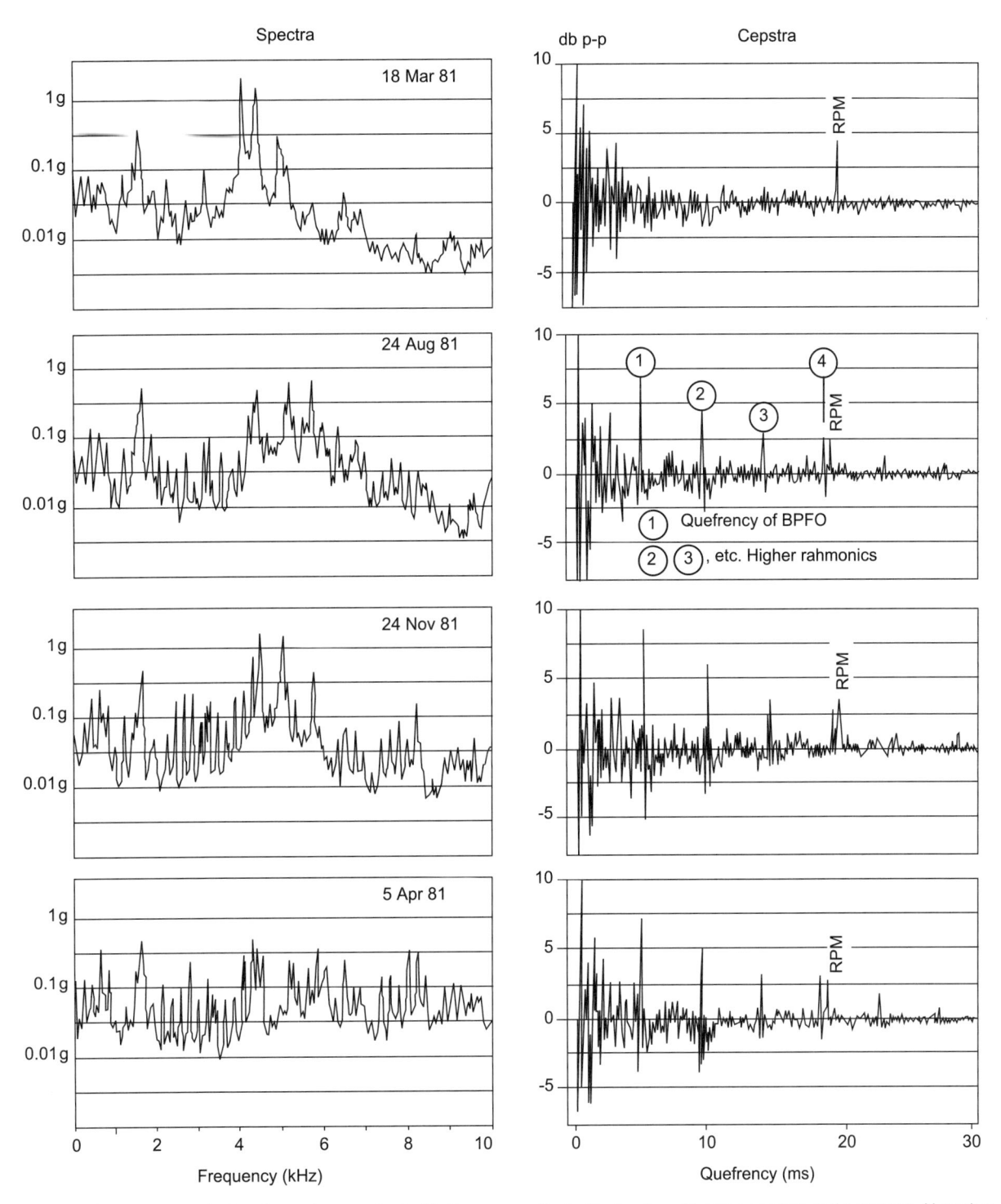

Figure 3 Development of a bearing outer race fault with time as manifested in the (logarithmic) spectrum and cepstrum. Note that variation in load affects some unrelated discrete frequency components, particularly in the range, 4 − 5 kHz.

autocorrelation function. The latter contains no information about the bearing fault; only a beat corresponding to the frequency difference between the two highest spectral peaks.

Thus the cepstrum can be useful in all three phases of condition monitoring: fault detection, diagnosis, and prognosis.

Note that the cepstrum can only be used for bearing fault diagnosis when the fault generates discrete harmonics in the spectrum. This is usually the case for high-speed machines, where resonances excited by the fault represent a relatively low harmonic order of the ballpass frequencies involved, but is often not the case for slow-speed machines, where this order

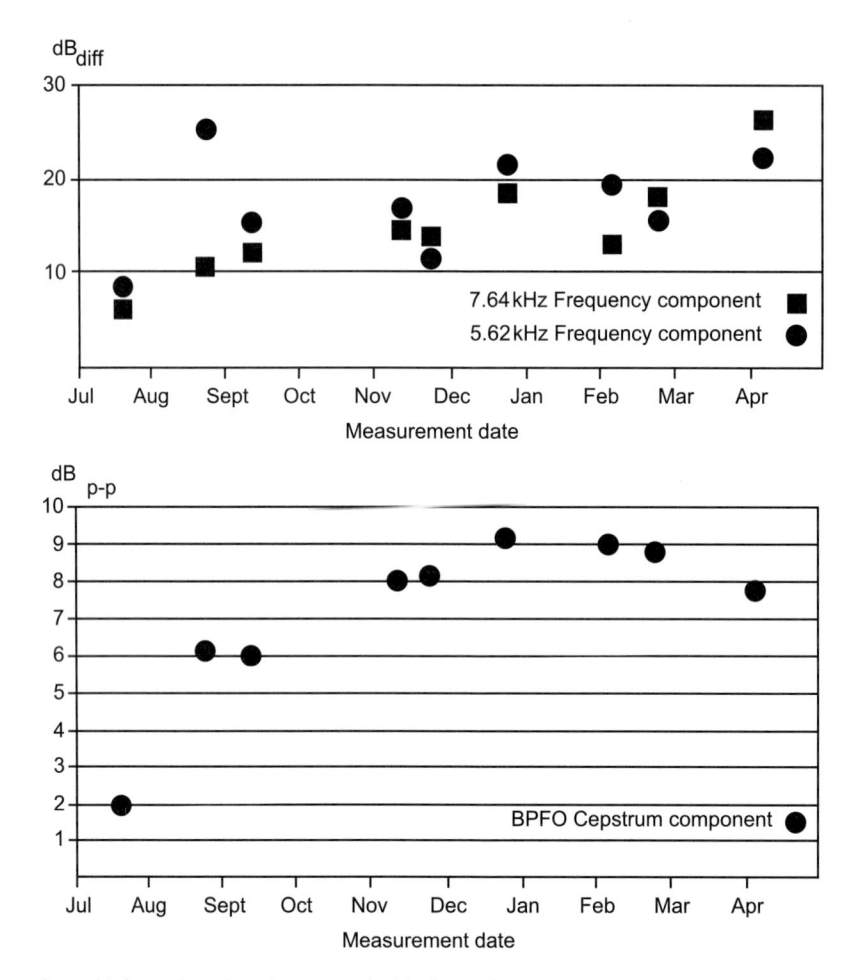

Figure 4 Comparison of trend information given by two typical ballpass frequency harmonics and the corresponding first rahmonic in the cepstrum. BPFO, ballpass frequency, outer race.

may be in the hundreds or even thousands, and these high harmonics are often smeared together. It should be noted that 'envelope analysis', where the envelope obtained by amplitude demodulation of the band-pass-filtered signal is frequency-analyzed, can be used in either case.

Note also that the cepstra in this case have been scaled in terms of 'dB peak-to-peak'. Such practical points are discussed below.

Faults in Turbomachines

The French electrical authority (Electricité de France, EDF) has demonstrated the application of the power cepstrum to the detection of missing blades in a steam turbine. Each missing blade gives rise to an impulse once per revolution as the misdirected steam flow interacts with the stator at the measurement point. This results in the growth of a large number of harmonics of the shaft speed (50 Hz) in the mid-frequency range, and a corresponding growth in the rahmonics of 20 ms in the cepstrum. Only the first rahmonic of the cepstrum needs to be monitored to detect this pattern in the spectrum.

Faults in Gears

Figure 6 illustrates a number of ways in which the power cepstrum can be useful for gear analysis. The degree of modulation of the gearmesh signal by each of the meshing gears is indicated by the corresponding families of rahmonics in the cepstrum, although to separate the sidebands from low harmonics of the shaft speeds (which may have another cause), it is advisable to edit the log spectrum before calculating the cepstrum, for example by only retaining that part of the spectrum above half the toothmesh frequency (but perhaps extending to several harmonics of it). The comparison of **Figure 6B** with **6A** shows that such editing considerably reduces the influence of one gear (121 Hz speed) so that the other (50 Hz speed) is dominant. However, some time later, when the

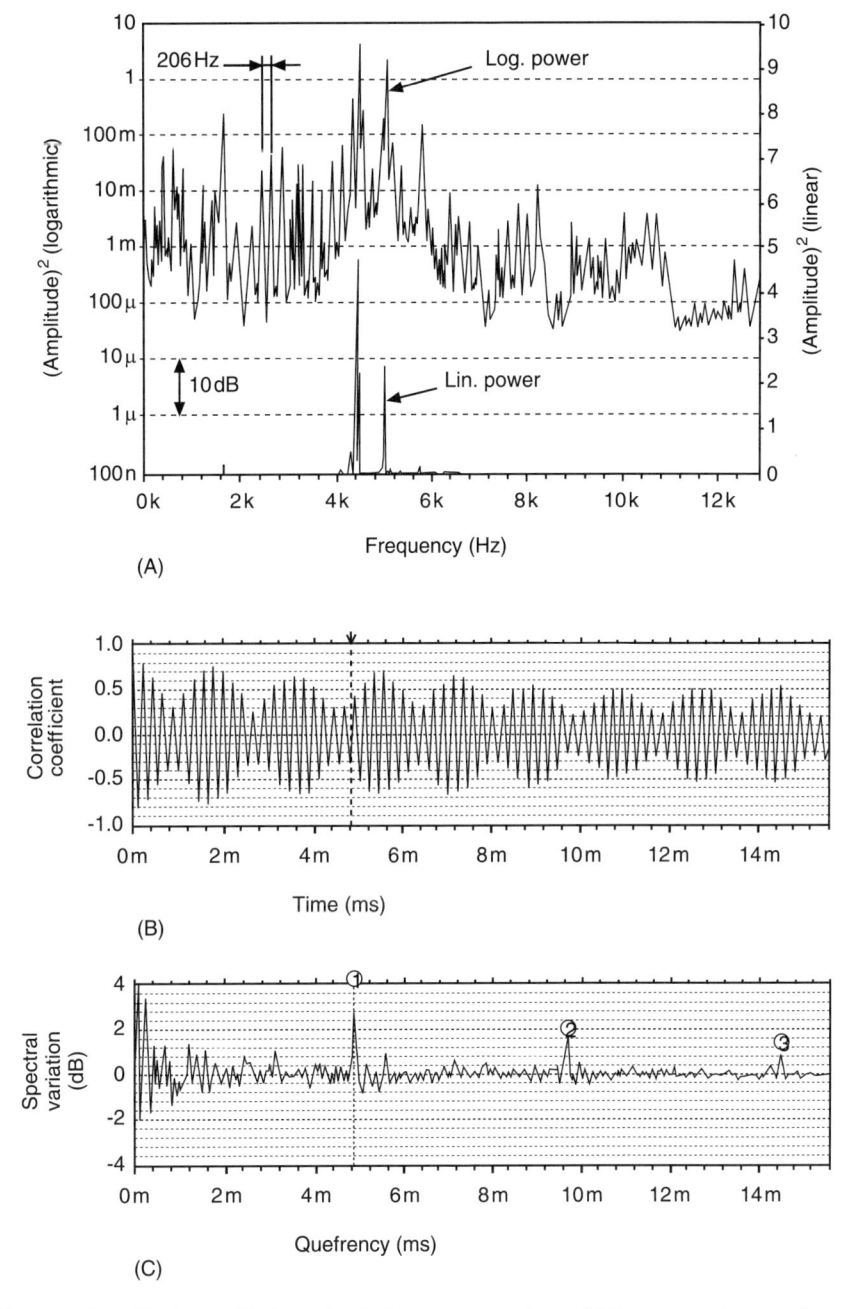

Figure 5 Effect of linear vs logarithmic amplitude scales in the power spectrum. (A) Power spectrum on linear scale (lower curve) and on logarithmic scale (upper curve). (B) Autocorrelation function (obtained from linear representation). (C) Cepstrum (obtained from logarithmic representation). The circled numbers are rahmonics of 4.85 ms, which corresponds to the 206 Hz spacing of the BPFO ballpass frequency harmonics that can be seen in the logarithmic spectrum (but not the linear spectrum). This frequency is 4.1 times the shaft speed.

121 Hz shaft developed some misalignment, it is seen to give increased components in the cepstrum, even with the same editing (**Figure 6C**). **Figure 6D** shows how liftering in the cepstrum can be used to remove one family of sidebands, allowing the other to be more easily visualized. The same can be achieved by synchronous averaging, but requires a tacho signal to synchronize the averaging.

Practical Points in Calculating the Power Cepstrum

Log amplitude spectra are normally represented on a dB scale, and the dB units can be retained for the cepstrum (as there is no interaction with the units of phase). As in **Figure 3**, the amplitude of the cepstrum can be scaled in 'dB peak-to-peak' on the tacit assumption that the harmonic pattern is continuous and stationary, but in any case that the values obtained are the average values for the section of spectrum analyzed. The value represents in some sense the average protrusion of the harmonic/sideband pattern above the base noise level, and is thus very signal-dependent as well as depending on such artifacts as the analysis bandwidth (relative to the harmonic/sideband spacing) and the type of window function used for the original analysis (e.g. Hanning, flattop). Even so, it is often meaningful to make comparisons between cepstra measured under the same operating conditions on the same machine (where the base noise level could be taken to be constant) and analyzed in the same way. Note that the reference level for the dBs in principle only affects the zero quefrency value of the cepstrum, so in practice it is often convenient to place it in the middle of the range of dB values so that the zero quefrency component in the cepstrum does not dominate the dynamic range and reduce the accuracy of higher quefrency values. This can be achieved in practice by taking the dBs with respect to any reference, and then subtracting the mean dB value.

Increase in a family of harmonics/sidebands will only be detected if the lower amplitude limit is a constant noise level rather than a 'bridging' between adjacent components a fixed number of dB below the peaks. This can occur if the spacing between the latter is not sufficiently greater than the analysis bandwidth. As a rule of thumb, the spacing between adjacent components should be at least 6–8 spectral lines if Hanning weighting is used.

This latter requirement will often mean that it is necessary to use a zoom rather than baseband spectrum in order to obtain sufficient resolution, and then there is another practical point to be aware of. **Figure 7** shows the cepstra obtained from two zoom spectra, slightly displaced from each other in center frequency. Because the harmonic pattern no longer passes through the effective 'zero' frequency, it is seen that there is no longer necessarily a positive peak in the cepstrum corresponding to the harmonic spacing, but there can even be positive and negative peaks on either side of a zero crossing. The same effect can occur with baseband spectra, in cases where a sideband spacing is not an exact subharmonic of the carrier frequency, such as in signals from rolling element bearings and planetary gearboxes. This problem can be very easily solved by making use of Hilbert transform theory (see **Hilbert transforms**). If the cepstrum calculation is carried out on the one-sided spectrum (positive frequencies only), then the resulting cepstrum will be analytic and complex (not to be confused with the 'complex cepstrum', which is real) and a peak will always be found at the correct quefrency in the amplitude of this complex function. **Figure 7** illustrates this for the two zoom spectra. Note that any editing or 'liftering' would have to be done on the complex function. The dB values of the one-sided spectrum should be doubled to maintain unchanged scaling, and it should be zero-padded to the same size as the two-sided spectrum.

Applications of the Complex Cepstrum

For a linear single-input multiple-output (SIMO) system, the relationship between the input (force) and the transfer function (or FRF) in measured signals for each response point is given by:

$$x(t) = f(t) \otimes h(t) \quad [9]$$

in the time domain, where $\otimes$ represents convolution;

$$X(f) = F(f) \cdot H(f) \quad [10]$$

in the frequency domain, where $\cdot$ represents multiplication;

$$\log[X(f)] = \log[F(f)] + \log[H(f)] \quad [11]$$

after taking logs, and:

$$C_X(\tau) = C_F(\tau) + C_H(\tau) \quad [12]$$

in the cepstrum (or differential cepstrum).

Thus, subtraction of one component in the cepstrum corresponds to deconvolution or inverse filtering in the time domain.

One of the applications of this is the removal of echoes and reflections in signals, as these can be modeled as a convolution with a delayed delta function. Thus, if the primary signal is $x(t)$, and it has an echo scaled by factor $a(<1)$ and with delay time t_o, it can be represented as $x(t) \otimes [\delta(t) + a\delta(t - t_o)]$ and its spectrum as the product of the two Fourier transforms, or $X(f) \cdot [1 + a\exp(-2\pi f t_o)]$, whose log amplitude and phase are those of $\log(X(f))$ with an additive periodic component in both, which varies with period $1/t_o$ Hz. The cepstrum is thus the sum of the cepstrum of the original function and a series of rahmonics corresponding to the added periodic function, with a spacing of t_o. Provided the original cepstrum is shorter than t_o, these rahmonics can easily be removed by liftering, and the echo thus removed. **Figure 8** gives an example where two equispaced echoes have been removed even though they overlap the original function.

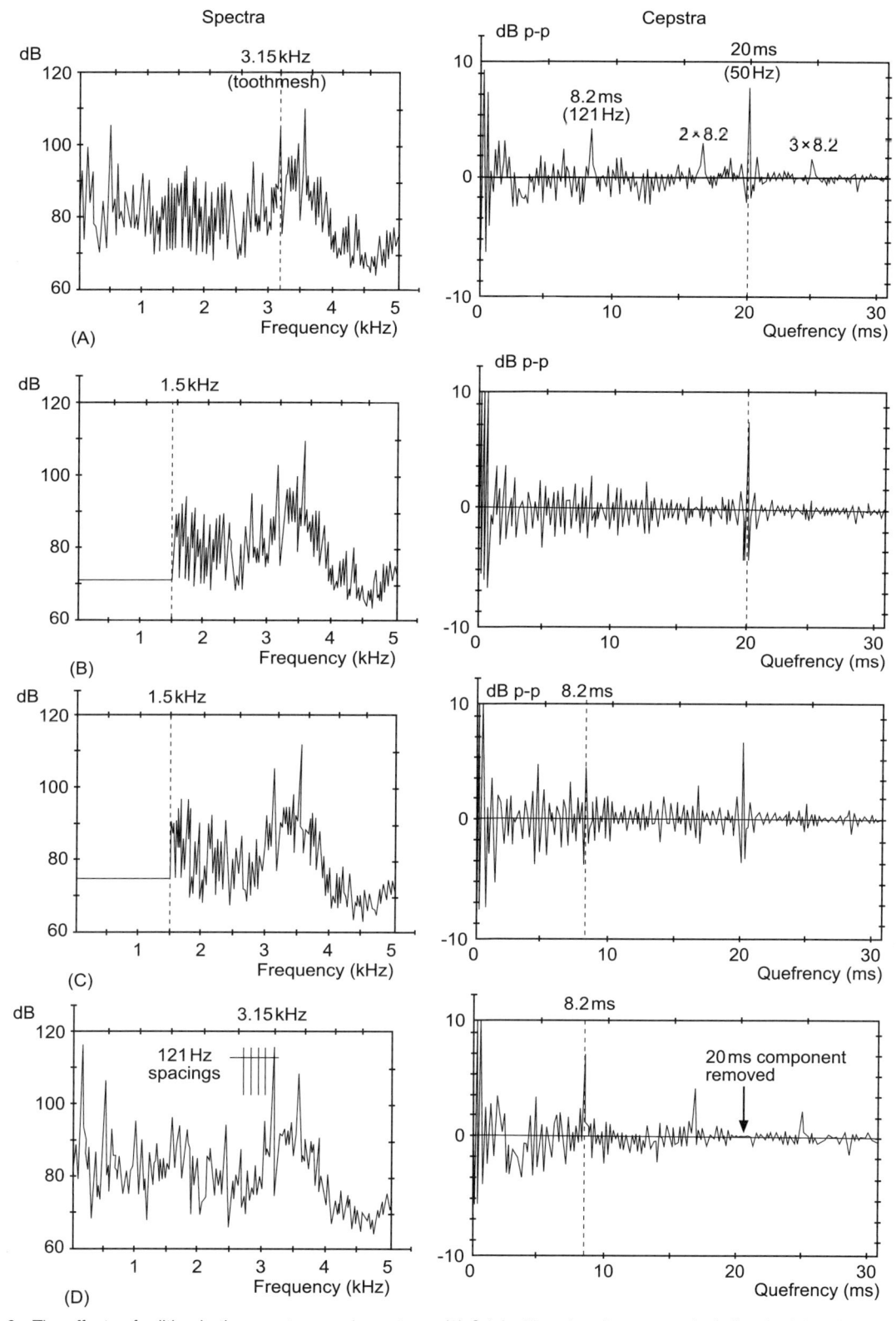

Figure 6 The effects of editing in the spectrum and cepstrum. (A) Original baseband spectrum including both low harmonics and sidebands around a gearbox toothmesh frequency. Cepstrum shows effects from both gear speeds (50 Hz and 121 Hz). (B) Effect of removing low harmonics up to approximately half the toothmesh frequency. The 50 Hz gear now dominates in the cepstrum. (C) Effect of deterioration in alignment of 121 Hz shaft. The effects of this shaft are now evident in the cepstrum, even after editing the spectrum. (D) Effect of removing the cepstral components of one gear (50 Hz) in the cepstrum (from the unedited baseband spectrum) and transforming back to the log spectrum, where the harmonics and sidebands from the other gear (121 Hz) are made more evident.

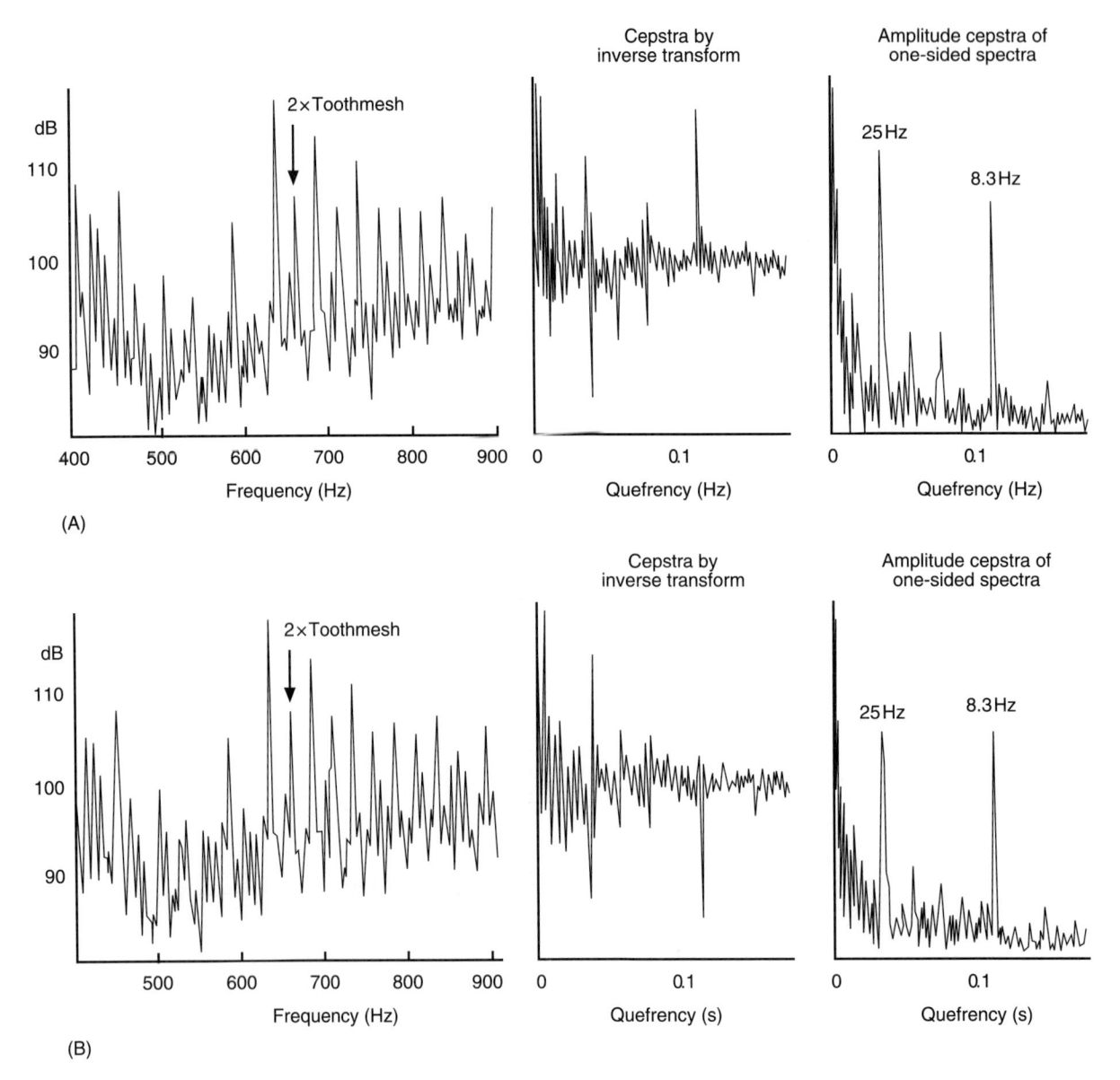

Figure 7 Obtaining cepstra from zoom spectra, by the formal definition (inverse transform of log spectrum) and as the amplitude of the analytic signal obtained from the one-sided spectrum. (A) and (B) represent two slightly displaced zoom spectra from the same signal. Note that the amplitude cepstra indicate the sideband spacings more clearly.

In vibration signals from gears it can be shown that the force at the mesh and the transfer function from the mesh to the measurement point, largely separate in the cepstrum, in that the forcing function is periodic and most of it concentrates at rahmonics corresponding to the toothmesh frequency and individual shaft speeds. Removing these with a suitable 'comb lifter' allows the remaining part of the log spectrum, dominated by the transfer function, to be reproduced by a forward transform. This can reveal whether resonance peaks have changed, and thus whether measured changes are due to changes at the source or in the signal transmission path. **Figure 9** shows the results of such a manipulation in a case where a small number of teeth on the drive pinion of a ball mill were cracked. The resonances in the transfer function are virtually unchanged, demonstrating that the entire change is due to the cracked teeth affecting the mesh force.

Another case where the forcing function and transfer function are well separated in the cepstrum is when a structure is excited by a forcing function with a relatively flat and smooth log spectrum such as the impulse from a hammer blow. In this case the force cepstrum is very short, and the higher-quefrency part of the response cepstrum dominated by the transfer function. The poles and zeros of the FRF can be extracted from this region of the cepstrum (or

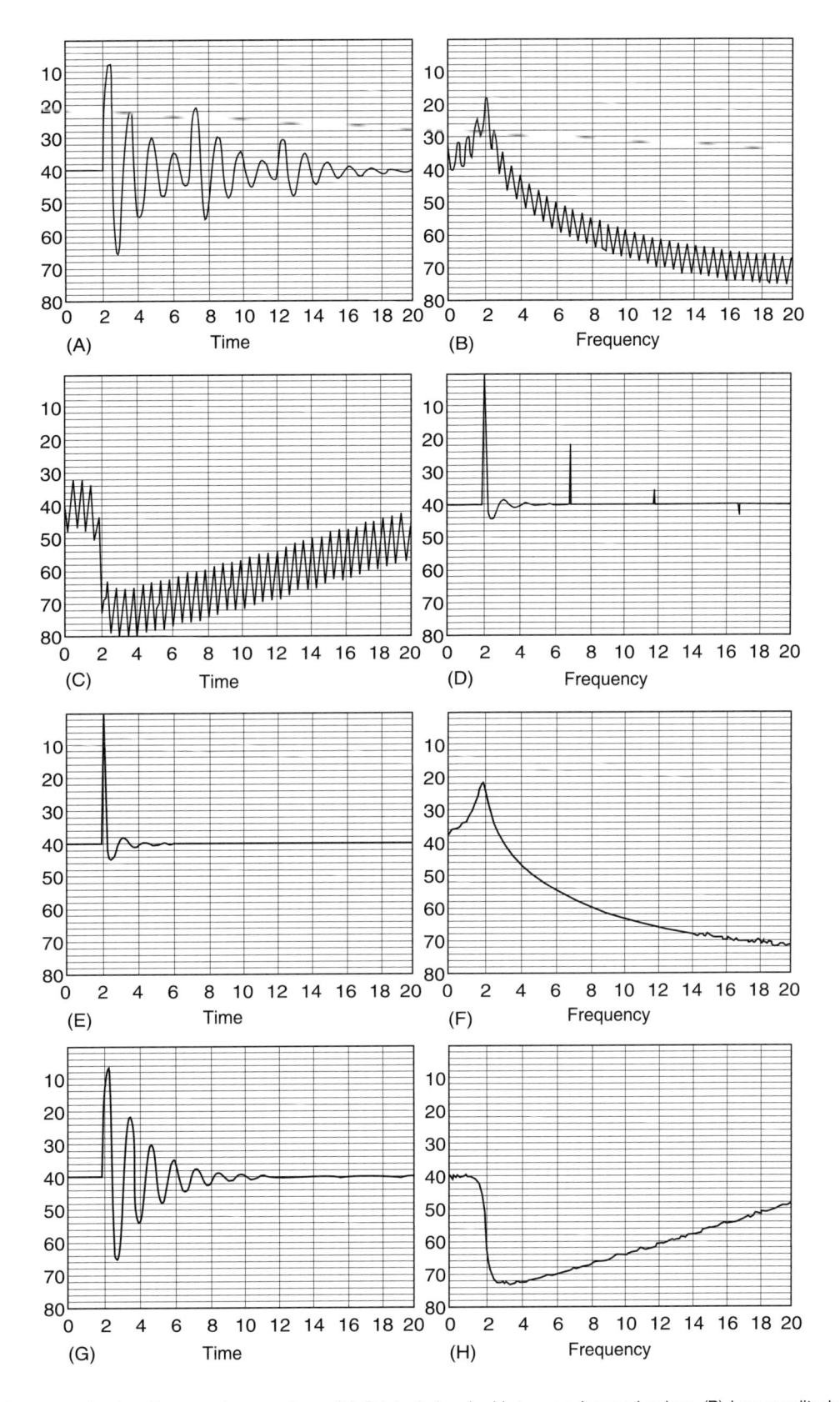

Figure 8 Echo removal using the complex cepstrum. (A) Original signal with two equispaced echos. (B) Log amplitude and (C) phase spectra from (A). (D) Complex cepstrum from (B) and (C). (E) Edited cepstrum after removing rahmonics due to delay. (F) Log amplitude and (G) phase spectra after forward transformation of (E). (H) Time signal by inverse transformation of the complex spectrum obtained by exponentiation of (F) and (G).

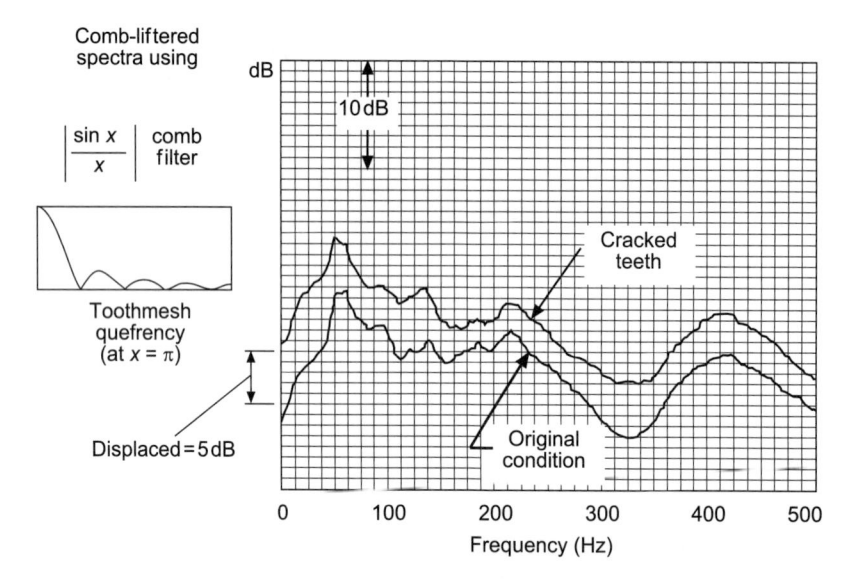

Figure 9 Use of liftering in the cepstrum to separate the effects of the forcing function (gearmesh signal with and without cracked teeth) and the transfer function in the response spectra. Toothmesh rahmonics have been removed by a tailored $|\sin x/x|$ comb lifter (of which the $1/x$ part is a shortpass lifter). The resulting comb-liftered spectra (displaced 5 dB for ease of comparison) indicate that resonance frequencies are unchanged.

differential cepstrum) by curve-fitting expressions of the form of eqn [8], using a nonlinear least-squares optimization algorithm, or from the differential cepstrum by treating it in the same way as a free decay impulse response using the Ibrahim time domain (ITD) method. Note that the latter cannot distinguish between poles and zeros, as there is no absolute time zero, but if measurements are made at several points, use can be made of the fact that the poles are global parameters while the zeros are unique to the different FRFs. It has been found that the poles and zeros within the measurement range are not sufficient in themselves to regenerate the FRF, as the shape is also affected by unmeasured out-of-band modes. However, these effects can be compensated for by in-band 'phantom zeros' (as they usually are in normal modal analysis) and are relatively insensitive to small changes in the pole and zero positions. This means that, once an initial measurement has been made (or perhaps an estimate by finite element modeling), changes in the modal properties of the object can be tracked using response measurements only. **Figure 10** gives an example where phantom zeros determined from FRF measurements on a free–free beam were used in conjunction with updated poles and zeros, extracted by curve-fitting response cepstra, to make estimates of the new FRFs in a case where a milled slot in the middle of the beam had changed some natural frequencies by as much as 10%. An initially determined scaling constant was also used in this

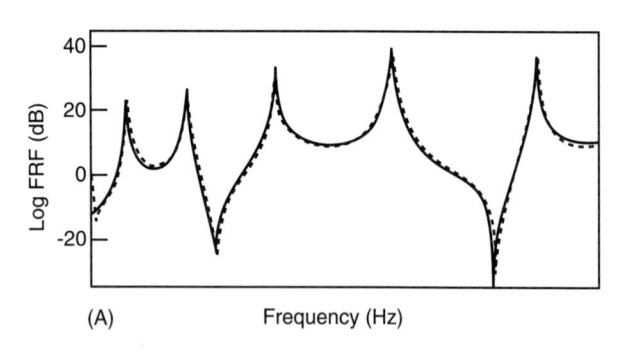

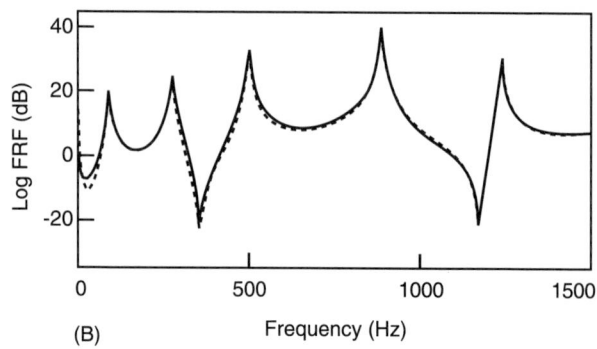

Figure 10 Updating frequency response functions (FRFs) from response measurements obtained by impulsive excitation of a free–free beam. (A) Original measurement. (B) Measurement with a half-depth slot at midspan. Dotted lines – measured FRFs. Solid lines – FRFs reconstructed from poles and zeros extracted by curve-fitting response cepstra. The reconstructed FRF in (B) uses phantom zeros and a scaling factor obtained from the original measured FRF in (A). Note the reduction in frequency of the symmetric (i.e., odd-numbered) modes in (B).

case, as the scaling factor of the FRF cannot be separated in the zero quefrency value of the response cepstrum.

The complex cepstrum has been used by Lyon and others to aid in the inverse filtering process of reconstituting diesel engine cylinder pressure signals from external measurements, typically acceleration of the engine block or head. Small changes in the pole/zero positions mean that they often do not cancel each other in the inverse filtering process, and the resulting pole/zero pair disrupts the estimated pressure signal. Short-pass liftering in the cepstrum smooths the result, giving improved estimates.

Scaling the Complex Cepstrum

Note from eqn [3] that the complex cepstrum has components from both the log amplitude and phase of the spectrum, so the log amplitude should be scaled in nepers (natural log of the amplitude ratio) to agree with the radians of the phase function. The complex cepstrum can then also be scaled in nepers. There are 8.7 dB per neper.

Nomenclature

K	gain factor
n	quefrency index
$\otimes$	convolution

See also: **Gear diagnostics**; **Hilbert transforms**; **Signal processing, model based methods**.

Further Reading

Berther T, Davies P (1991) Condition monitoring of check valves in reciprocating pumps. *Tribology Trans.* 34: 321–326.

Bogert BP, Healy MJR, Tukey JW (1963) The quefrency alanysis of time series for echoes: cepstrum, pseudo-autocovariance, cross-cepstrum, and saphe cracking. In *Proceedings of the Symposium on Time Series Analysis*, by Rosenblatt M (ed), pp. 209–243, New York: John Wiley.

Childers DG, Skinner DP, Kemerait RC (1977) The cepstrum: a guide to processing. *Proceedings of the IEEE* 65: 1428–1443.

Gao Y, Randall RB (1996) Determination of frequency response functions from response measurements. Part I: Extraction of poles and zeros from response cepstra. Part II: Regeneration of frequency response functions from poles and zeros. *Mechanical Systems and Signal Processing* 10: 293–317, 319–340.

Lyon RH, Ordubadi A (1982) Use of cepstra in acoustical signal analysis. *ASME J. Mech. Des.* 104: 303–306.

Oppenheim AV, Schafer RW (1989) *Discrete Time Signal Processing*. New Jersey: Prentice-Hall.

Randall RB (1987) *Frequency Analysis*, 3rd edn, Chapt. 8, Cepstrum Analysis. Copenhagen: Bruel and Kjaer.

Sapy G (1975) Une application du traitement numérique des signaux au diagnostic vibratoire de panne: la détection des ruptures d'aubes mobiles de turbines. *Automatisme* XX: 392–399.

Tribolet JM (1977) A new phase-unwrapping algorithm. *IEEE Trans. Acoust. Speech Signal Proc.* ASSP-25: 170–177.

CHAOS

P J Holmes, Princeton University, Princeton, NJ, USA

Copyright © 2001 Academic Press

doi:10.1006/rwvb.2001.0039

Introduction

Basic ideas and techniques from the theory of dynamical systems are reviewed and applied to analyze and understand chaotic vibrations. Single-degree-of-freedom, periodically forced, nonlinear oscillators are treated; the canonical examples being the pendulum, the Duffing and the van der Pol equations. After sketching some history, the key ideas of Poincaré maps and invariant manifolds are introduced, followed by a simple mathematical example (the doubling map), which illustrates deterministic chaos and a major tool for its analysis: symbolic dynamics. Then, using regular perturbation methods, it is shown that chaotic solutions occur in a broad class of nonlinear oscillators, including Duffing's equation, and the difficulty of proving the existence of 'strange attractors' – motions displaying sensitive dependence on initial conditions that attract almost all initial conditions – is discussed. The article ends with a brief note on sources and types of nonlinearity likely to lead to chaos, and some pointers to the (enormous) literature.

A Brief History

Henri Poincaré's studies of celestial mechanics, in particular of the three-body problem, led him to discover complex motions in deterministic Hamiltonian classical mechanics; he also provided the groundwork for much of the modern qualitative theory of dynamical systems (chaos theory). Poincaré's work was followed by that of George Birkhoff in the US, Andronov and Pontryagin in the USSR, and a remarkable paper on the van der Pol equation by Cartwright and Littlewood, arising from British radar development work in World War II. This latter was the first explicit example of a periodically forced nonlinear oscillator having chaotic solutions; subsequently studied by Levinson, it led Smale in 1960 to create the horseshoe map, and thus complete part of the story begun by Poincaré. Shortly after Smale's work (but before its publication), Ueda, working with an analog computer, independently discovered chaotic motions in a variant of the forced van der Pol system, and Lorenz published his now celebrated example of an autonomous three-dimensional system. See Further Reading.

Poincaré Maps and Invariant Manifolds

Dynamical systems theory addresses nonlinear differential equations and iterated mappings, bringing a geometrical and topological approach to complement perturbative and other analytical methods (see **Perturbation techniques for nonlinear systems**). The study of qualitative behavior is emphasized; solutions of the differential equation:

$$\dot{x} = f(x); \ x \in R^n \quad [1]$$

are viewed as flowlines evolving in the state or phase space, R^n. A key idea is that the behavior of a nonlinear system near a nondegenerate equilibrium or periodic orbit can be deduced by linearization and successive Taylor series approximations; geometrically, local stable and unstable manifolds exist. These manifolds are smooth (hyper-) surfaces, tangent at the equilibrium or periodic orbit to the eigenspaces belonging to exponentially decaying and growing linearized solutions, and invariant under the flow defined by eqn [1]. This is the main consequence of the stable manifold theorem. The local manifolds, which are related to nonlinear normal modes, are defined in a neighborhood of the orbit in question, but they can be extended globally by following solutions backwards and forwards in time, and their structure determines the asymptotic behavior of solutions starting nearby. See **Figure 1A**, which shows the stable and unstable manifolds (= separatrices, here) of the saddle point $(\theta, v) = (\pm\pi, 0)$ corresponding to the inverted (unstable) equilibrium of the damped pendulum, whose governing equation can be written in nondimensional form as:

$$\dot{\theta} = v, \ \dot{v} = -\sin\theta - \delta v \quad [2]$$

Note that the local stable manifold of the 'downward' equilibrium $(\theta, v) = (0, 0)$ includes a full neighborhood of that point: it has no unstable manifold; indeed, almost all solutions eventually approach $(0, 0)$; those that do, belong to its domain of attraction. In the above, 'nondegenerate' means that all eigenvalues of the system linearized at the fixed point have nonzero real parts; such points are also called hyperbolic. Both equilibria are hyperbolic in **Figure 1A**.

Periodically forced oscillators, such as the Mathieu, Duffing and van der Pol equations, or the pendulum itself, require a three-dimensional phase space for their definition as dynamical systems. We think of time, t, as a third 'dependent' variable, ϕ, as here, for the negative stiffness Duffing equation:

$$\dot{x} = y, \quad \dot{y} = x - \delta y - x^3 + \gamma\cos(\phi),$$
$$\dot{\phi} = \omega; \quad (x, y, \phi) \in R^2 \times S^1 \quad [3]$$

Since the excitation $\gamma\cos(\omega t)$ is periodic, we can identify time or ϕ-slices equally spaced by 2π and roll up the ϕ-axis into a circle S^1. The phase space therefore resembles a solid torus: see **Figure 1B**.

In this phase space we fix a cross-section $\Sigma_0 = \{\phi = 0\}$, and consider the Poincaré map, P, obtained by integrating eqn [3] with initial conditions $(x(0), y(0)) = (x_0, y_0) \in \Sigma$ until the solution first returns to Σ_0. This implicitly defines a difference equation or discrete dynamical system:

$$P(x_0, y_0) = (x_1, y_1)$$

or, in general

$$(x_{n+1}, y_{n+1}) = P(x_n, y_n) \quad [4]$$

Orbits $\{P^k(x)\}_{k=0}^{\infty}$ of eqn [4] are discrete sequences of points, not smooth curves, as in the usual phase portraits like that of **Figure 1A**; see **Figure 1B**. Harmonic responses of period $(2\pi/\omega)$ of eqn [3] correspond to fixed points of P, and $(2m\pi/\omega)$-periodic subharmonics to m-periodic cycles of P. Periodic orbits can be attractors (asymptotically stable), neutrally stable (Liapunov stable but not asymptotically stable), of saddle type, or repellors (both unstable). A saddle-type orbit, being itself one-dimensional (topologically a circle), has a two-dimensional sheet of solutions approaching it and a two-dimensional sheet of solutions leaving its stable and unstable manifolds, which intersect Σ_0 in curves (**Figure 1B**). Hence the invariant manifolds of a saddle point p in a two-dimensional

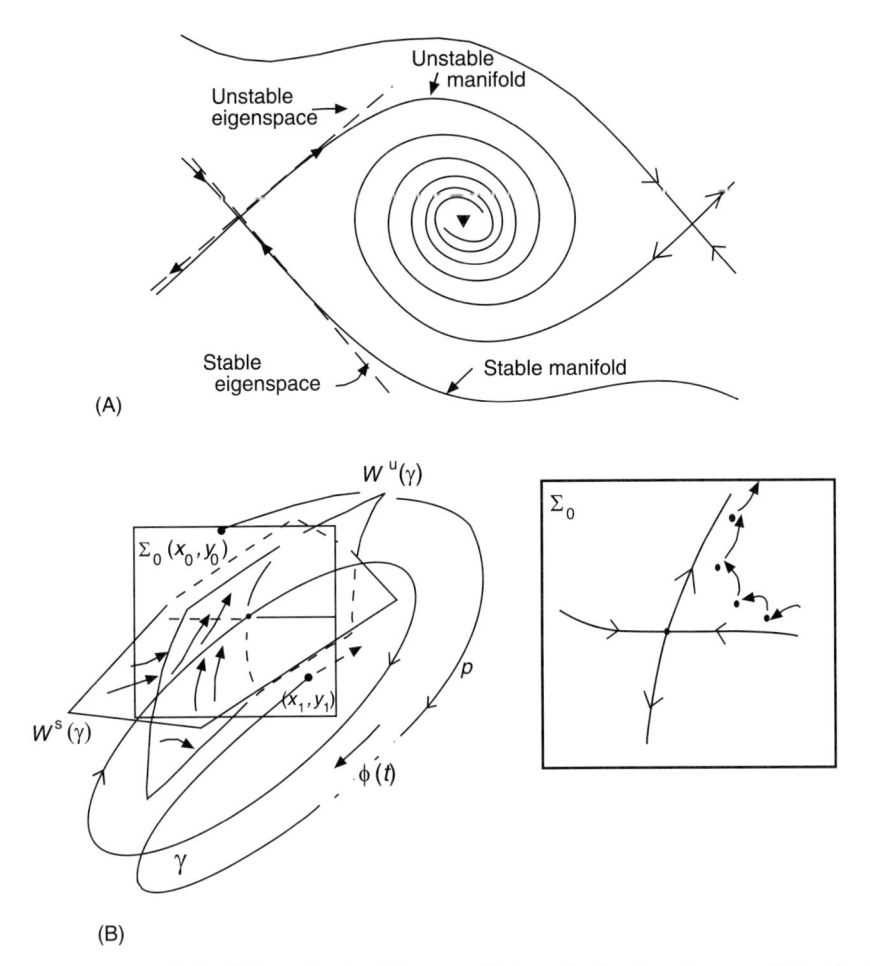

Figure 1 Stable and unstable manifolds (A) for a fixed point: an equilibrium of a 'free' nonlinear oscillator, the damped pendulum; (B) for a periodic orbit, illustrating a cross-section Σ_0 and the Poincaré map P.

map P resemble saddle separatrices for a two-dimensional flow (**Figure 1A**), but there is a crucial difference: the stable and unstable manifolds of a saddle point p for a map can cross at homoclinic points. These are points q which approach (= incline towards) p under both forward and backward iteration: $P^n(q) \to p$ as $|n| \to \infty$. A heteroclinic point is one which approaches distinct fixed points p_1 and p_2 under forward and backward iteration. If the tangents to the manifolds at the intersection point are distinct, the homoclinic (heteroclinic) point is called transverse. As Poincaré realized, such points are associated with extremely sensitive dependence on initial conditions, and what is now commonly called chaos. Before illustrating homoclinic points and their consequences, a simple motivating example is described.

Symbolic Dynamics and Chaos

Consider a mathematical toy: the one-dimensional piecewise linear map defined on the interval $[0, 1] \subset R$ by:

$$h(x) = \begin{cases} 2x & \text{if } 0 \leq x < \frac{1}{2} \\ 2x - 1 & \text{if } \frac{1}{2} < x \leq 1 \end{cases} \quad [5]$$

and illustrated in **Figure 2**. An orbit of h is the sequence $\{x_n\}_{n=0}^{\infty}$ obtained by successively doubling the initial value x_0 and subtracting the integer part at each step; for example: $0.2753 \mapsto 0.5506 \mapsto 1.1012 = 0.1012 \mapsto 0.2024 \mapsto 0.4048 \mapsto \ldots$

To understand the sensitive dependence on initial conditions and its consequences, we represent the numbers in $[0, 1]$ (the phase space of this dynamical system) in binary form. This is the idea of symbolic dynamics. Let:

$$x_0 = \frac{a_1}{2} + \frac{a_2}{2^2} + \frac{a_3}{2^3} + \cdots + \frac{a_k}{2^k} + \cdots \quad [6]$$

where each coefficient a_j takes the value 0 or 1. It follows that:

$$x_1 = h(x_0) = 2\left(\sum_{j=1}^{\infty} \frac{a_j}{2^j}\right) = a_1 + \frac{a_2}{2} + \frac{a_3}{2^2} + \frac{a_4}{2^3} + \cdots$$

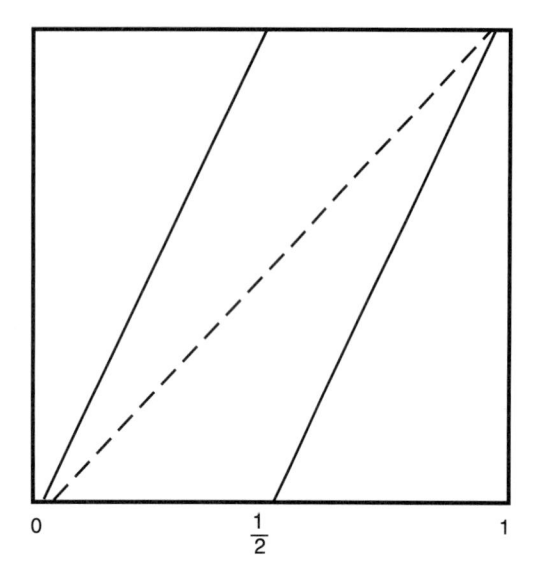

Figure 2 The doubling map, h.

Since each time the integer part is removed (a_1, here), the kth iterate gives:

$$x_k = h^k(x_0) = \sum_{j=k+1}^{\infty} \frac{a_j}{2^{j-k}} \quad [7]$$

Thus, applying h is equivalent to shifting the binary point and dropping the leading coefficient in the binary representation: $(a_1 a_2 a_3 a_4 \ldots) \mapsto (a_2 a_3 a_4 \ldots)$, just as multiplication by 10 shifts the decimal point. In this operation the leading symbol is removed and information is lost. If x_0 were known to infinite accuracy, with all a_ks specified for $1 \leq k < \infty$, at each step the current state x_k would still be known exactly. But, given knowledge of only the first N binary places $(a_1, a_2, \ldots a_N)$, after N iterations, it cannot even be determined whether the state x_{N+1} lies above or below $\frac{1}{2}$. Moreover, even if two initial conditions differ only at the Nth binary place, so that the points lie within $\left(\frac{1}{2}\right)^{N-1}$, after N iterations they lie on opposite sides of $\frac{1}{2}$ and thereafter behave essentially independently. Here is the sensitive dependence: the dynamics amplifies small errors.

The binary representation shows that to every infinite sequence of 0s and 1s there corresponds a number between zero and one, and vice versa. Thus, for any random sequence (e.g., obtained by tossing a coin: heads = 0, tails = 1), there is an initial state $x_0 \in [0, 1]$ such that the orbit $h^k(x_0)$ realizes the chosen sequence. Hence the map has infinitely many random orbits. There are also infinitely many periodic orbits, corresponding to repeating sequences such as 001001001.... However, since a number picked at random is almost always irrational (the irrational numbers form a set of full measure), typical behavior is chaotic rather than periodic.

The binary representation also shows that a dense orbit exists. Consider the sequence a^* formed by concatenating all possible sequences of lengths 1, 2, 3, ... end to end: $a^* = 0\ 1\ 00\ 01\ 10\ 11\ 000\ 001\ldots$. As one iterates and drops leading symbols, every possible subsequence of any given length appears at the head. This implies that the orbit of the point x^* corresponding to a^* contains points which approximate, to any desired accuracy, every point in the interval $[0, 1]$ and so $\{h^k(x^*)\}_{k=0}^{\infty}$ is a dense orbit. The symbol sequences also allow one to enumerate the periodic orbits simply by listing all distinct (i.e., non-shift-equivalent) sequences of lengths 1, 2, 3, ... which do not contain lower period subsequences (Table 1). Asymptotically, there are $\approx 2^N/N$ orbits of period N.

This example embodies the three key characteristics of a chaotic invariant set $V \subset R^n$ for a map $P: R^n \to R^n$:

1. Sensitive dependence on initial conditions: There is a $\beta > 0$ such that, for any $x \in V$ and any neighborhood $U \ni x$, no matter how small, there exists a point $y \in U$ and an integer $k > 0$ such that $|P^k(y) - P^k(x)| > \beta$; i.e., almost all orbits eventually separate.
2. The periodic points are dense in V.
3. There is a dense orbit in V.

The final condition implies that V cannot be decomposed into simpler elements. The dense orbit contains segments passing arbitrarily close to any given point in V and so almost all orbits must 'fill out' V without getting 'stuck' in simple periodic motions.

Below it is shown that the presence of transverse homoclinic orbits in a Poincaré map implies that chaotic dynamics occurs of precisely the same type as that just described.

Table 1 Enumerating periodic orbits for h

Length	Sequences	Number of orbits
1	0, 1	2
2	01 (=10)	1
3	001, 011	2
4	0001, 0011, 0111	3
5	00001, 00011, 00101, 00111, etc.	6
6	......	9
.	......	.
.	......	.
25	......	1 342 176
.	......	.

Perturbing Separatrices: Melnikov's Method

Melnikov developed a perturbative method for detecting homoclinic orbits in periodically perturbed nonlinear systems that have smooth separatrices connecting saddle points prior to perturbation, such as eqns [2] and [3] with $\delta = 0$ and $\gamma = 0$. Letting x denote the two-dimensional vector (x, y) or (θ, v), one may write the perturbed equation as:

$$\dot{x} = f(x) + \varepsilon g(x, t) ; \quad 0 \leq \varepsilon \ll 1 \qquad [8]$$

noting that $g(x, t)$ is T-periodic in t. Assume that for $\varepsilon = 0$ the unperturbed system is Hamiltonian, with $f(x) = (\partial H/\partial y, -\partial H/\partial x)^T$ and (energy) function $H(x, y)$ constant on solutions. (This assumption is not essential, but leads to a simpler expression in eqn [11] below.) Assume further that $x = p_0$ is a hyperbolic saddle point with eigenvalues $\pm \lambda \neq 0$, and that there is a homoclinic (separatrix) loop $q_0(t)$ to p_0: see **Figure 3A**. The fact that p_0 is nondegenerate (hyperbolic), along with perturbation and invariant manifold theory, implies that it perturbs to a small T-periodic orbit $\gamma_\varepsilon = p_0 + O(\varepsilon)$, and that solutions $x_\varepsilon^s, x_\varepsilon^u$ in the stable and unstable manifolds respectively of γ_ε can be written as power series in ε: $x_0 + \varepsilon x_1^{s,u} + O(\varepsilon^2)$, uniformly convergent on semi-infinite time intervals. Substituting these expansions into eqn [5] and equating zeroth and first-order terms yields:

$$\dot{x}_0 = f(x_0), \quad \dot{x}_1^{s,u} = Df(x_0) x_1^{s,u} + g(x_0, t) \qquad [9]$$

where Df denotes the matrix of first partial derivatives.

Now consider the quantity:

$$d(t_0) = f(x_0(0))^\perp \cdot (x_\varepsilon^s - x_\varepsilon^u) \qquad [10]$$

which measures the distance between the perturbed stable and unstable manifolds at a selected point $x_0(0)$, projected on to the normal $f(x_0(0))^\perp$ to the unperturbed solution (= level sets of H), on the cross-section Σ_{t_0} (**Figure 3B**). $d(t_0)$ may be approximated to first-order via the power series for x^s and x^u, and a short computation using eqn [9] leads to the expression:

$$d(t_0) = \varepsilon M(t_0) + O(\varepsilon^2)$$
$$M(t_0) = \int_{-\infty}^{\infty} f(q_0(t))^\perp \cdot g(q_0(t), t + t_0) \, dt \qquad [11]$$

It follows that, if the Melnikov function $M(t_0)$, has

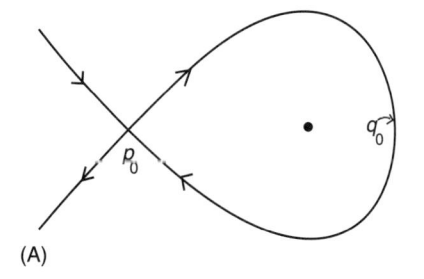

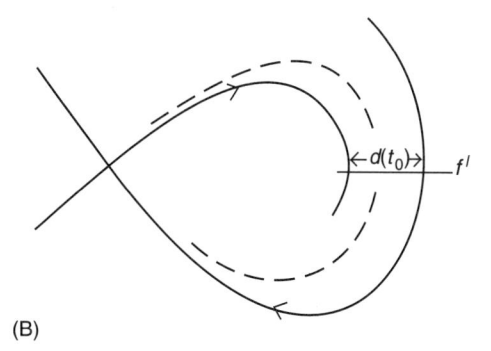

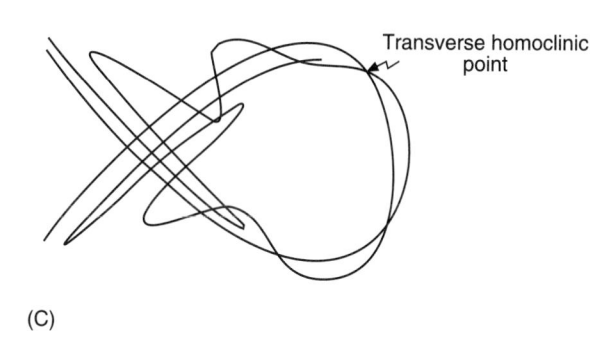

Figure 3 (A) The unperturbed homoclinic loop; (B) the distance estimate for splitting of stable and unstable manifolds in the perturbed Poincaré map; (C) the homoclinic tangle.

simple zeroes, then the stable and unstable manifolds of the fixed point, p_ε, of the Poincaré map corresponding to γ_ε, intersect transversely. As shown below, $M(t_0)$ may be explicitly calculated in specific examples, even though the perturbed Poincaré map, P_ε, is effectively uncomputable (except numerically).

If one (transverse) homoclinic point, q, exists, then every image, $P^k(q)$, is also homoclinic: there is a homoclinic orbit. In fact, considering images of small arcs contained in $W^s(p_\varepsilon)$ and $W^u(p_\varepsilon)$ near q under the map P, we see that the global manifolds must intersect repeatedly in a homoclinic tangle (**Figure 3C**). Poincaré described this 'tissue ... with infinitely fine mesh' and noted that nothing could better represent the complexity of three-body problem. But he went no further in analyzing the structure of orbits associated with the tangle: that awaited the work of Birkhoff and Smale.

Smale's Horseshoe and Chaos

Before presenting an example from nonlinear oscillations, the connection between homoclinic points and chaos, promised above, is made. **Figure 4A** shows a rectified version of **Figure 3B**. Consider the effect of the map P_ε on a rectangular strip, S, containing p_ε, a homoclinic point q_ε, and partially bounded by pieces of $W^s(p_\varepsilon)$ and $W^u(p_\varepsilon)$. Successive applications of P_ε compress S along $W^s(p_\varepsilon)$ and stretch it along $W^u(p_\varepsilon)$, so that for sufficiently large n, the nth image $P_\varepsilon^n(S)$ assumes the horseshoe shape, also shown in **Figure 4A**.

This suggests the following piecewise-linear map F, which approximates P_ε^n. F is defined on the unit square, and takes the lower rectangle H_0 of height γ^{-1} into the left-hand strip V_0 of width λ, and the upper rectangle H_1 of height γ^{-1} into the right-hand strip V_1 of width λ, rotating the latter by π. Here $\lambda < 1/2$, $\gamma > 2$. The image of the central strip is the arch connecting V_0 and V_1 (**Figure 4B**). The key to understanding F (and hence P_ε) is in the description of the set, Λ, of points which never leave S under forward or backward iteration of F. These points must lie in either H_0 or H_1 and V_0 or V_1 (e.g., if they lie in the middle strip, they leave S on the next application of F). Moreover, for their second forward and backward iterates to lie in S, they must inhabit the intersections of four horizontal and four vertical strips H_{ij}, V_{ij} of widths γ^{-2} and λ^2 respectively, also shown in **Figure 4B**, or else they will land in a middle strip after one iterate. Continuing inductively, we find that Λ is a Cantor set: an uncountably infinite 'cloud' of points, containing no open sets, and every point of which is a limit point of other points in the set.

As shown above, symbol sequences may be uniquely assigned to the points of Λ, but now they are doubly infinite, since F is an invertible map.

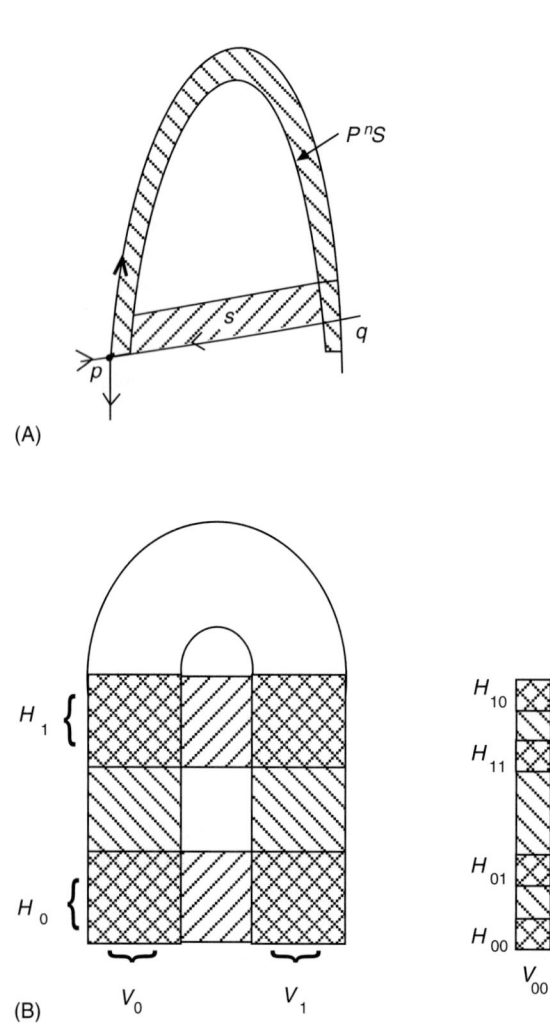

Figure 4 Smale's horseshoe: (A) S and $P^n(S)$; (B) the strips H_j and V_j.

Specifically, associate a sequence $a(x) = \{a_k(x)\}_{-\infty}^{+\infty}$ with each $x \in \Lambda$, via the rule:

$$a_k(x) = \begin{cases} 0 & \text{if } F^k(x) \in H_0 \\ 1 & \text{if } F^k(x) \in H_1 \end{cases} \quad [12]$$

The shift map σ defined on the space S of doubly infinite sequences of 0s and 1s captures the behavior of F restricted to Λ much as in the semiinfinite binary arithmetic shown above. In particular, to every sequence there corresponds a unique point $x \in \Lambda$ and vice versa, and the action of F on Λ is equivalent or topologically conjugate to that of shifting sequences under σ on S. We therefore have the same accounting of periodic points presented in **Table 1**, along with an uncountable infinity of nonperiodic orbits, and infinitely many homoclinic and heteroclinic orbits corresponding to sequences having periodic heads (tails) in the infinite past (future). All these orbits are of unstable saddle type, due to the linear horizontal contraction and vertical expansion of F on H_0 and H_1, and Λ displays sensitive dependence on initial conditions.

The fact that F is linear on H_0 and H_1 is not crucial to this analysis; all one needs is uniform bounds on contraction and expansion, and so it can be shown that all these conclusions follow for the nonlinear map, P_ε^n, resticted to a neighborhood of the homoclinic point q_ε. This leads to the:

Smale–Birkhoff homoclinic theorem: Let $P:\Sigma \mapsto \Sigma$ be a smooth invertible map with a transverse homoclinic point q to a hyperbolic saddle point p. Then, for some $n < \infty$, P has a hyperbolic invariant set Λ on which P^n is topologically conjugate to a shift on two symbols.

Corollary: Λ contains a countable infinity of periodic, homoclinic and heteroclinic orbits, an uncountable infinity of 'chaotic' orbits, and a dense orbit.

Birkhoff's name properly belongs here, since he proved that near every homoclinic point there is an infinite set of periodic orbits: essentially those that circulate near the homoclinic orbit $\{P^k(q)\}_{k=-\infty}^{\infty}$, marking time for arbitrarily long near the saddle point. In symbolic terms, these have sequences with periodically repeating blocks of the form 000 ... 0001 and 000 ... 0011.

We cannot however conclude that almost all, or even a set of positive measure, of orbits of the map behave chaotically, for in constructing Λ we eliminated all orbits which leave S. These orbits may approach stable sinks or attracting periodic cycles.

Chaos in Duffing's Equation: Strange Attractors?

Holmes and Moon provided early analyses of Duffing-type equations modeling the forced vibrations of a buckled beam. Experiments were carried out with a cantilever buckled by magnetic forces and subject to transverse sinusoidal oscillation at the base (**Figure 5**). This led to a considerable amount of research on chaotic vibrations. Including only the fundamental beam mode, and modeling the magnetic forces on the tip by a simple cubic term with negative linear stiffness, one arrives at the nondimensionalized equation:

$$\dot{x} = y, \quad \dot{y} = x - x^3 + \varepsilon(\gamma \cos(\omega t) - \delta y); \quad 0 \leq \varepsilon \ll 1 \quad [13]$$

This is eqn [3], with an explicit small parameter included to denote weak damping and forcing. Note that the nonlinearity is 'large' and present in the unperturbed ($\varepsilon = 0$) limit: this is crucial so that there is a homoclinic loop to perturb from. The unperturbed phase portrait is shown in **Figure 6A**; note the two potential wells surrounding the 'buckled' equilibria and the saddle separatrices bounding these wells. Solutions in these separatrices or homoclinic loops are given by:

$$(x_0(t), y_0(t)) = \left(\pm\sqrt{2} \operatorname{sech} t, \mp\sqrt{2} \operatorname{sech} t \tanh t\right) \quad [14]$$

Inserting these expressions in the Melnikov function (eqn [11]) and evaluating in this case we obtain:

$$M(t_0) = \int_{-\infty}^{\infty} y_0(t)[\gamma \cos(\omega(t + t_0)) - \delta y_0(t)] \, dt \quad [15]$$
$$= -\frac{4\delta}{3} + \sqrt{2}\gamma\pi\omega \operatorname{sech}\left(\frac{\pi\omega}{2}\right) \sin \omega t$$

This calculation yields a critical force/damping ratio:

$$\left.\frac{\gamma}{\delta}\right|_{\text{crit}} = \frac{4 \cosh(\pi\omega/2)}{3\sqrt{2}\pi\omega} \quad [16]$$

which must be exceeded for the stable and unstable manifolds to intersect and transverse homoclinic orbits to exist. It represents a necessary condition for homoclinic chaos involving orbits which pass back and forth from one (unperturbed) potential well to the other, and, via the Smale–Birkhoff theorem, a sufficient condition for the existence of a horseshoe. Note that it is obtained by an explicit calculation.

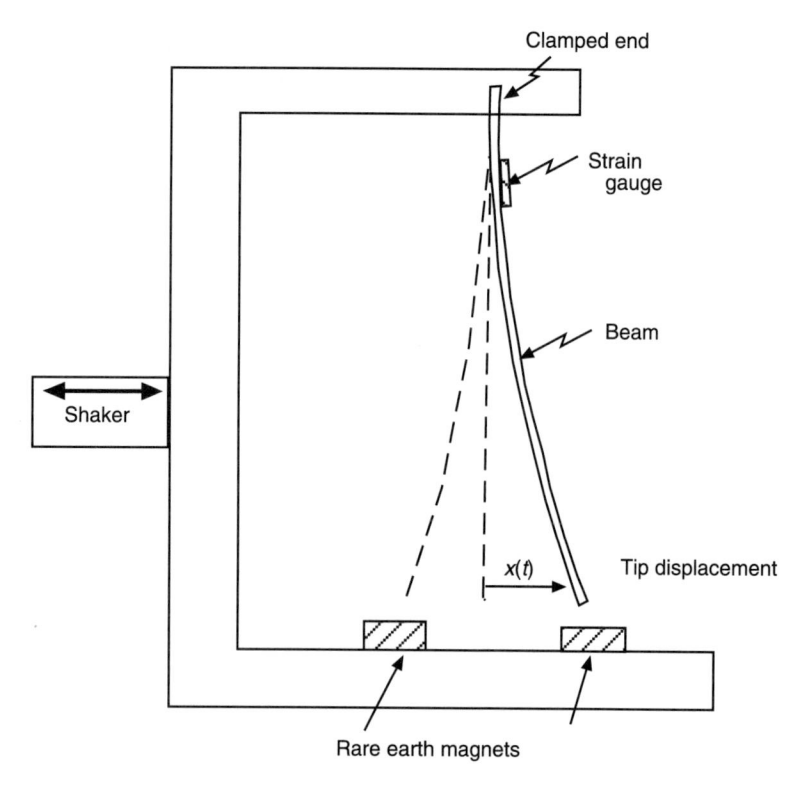

Figure 5 The magnetically buckled cantilever beam.

Figure 6B shows numerical computations of segments of the stable and unstable manifolds; the transverse intersections are clearly visible. For $\omega = 1$ the critical ratio is $0.7530\ldots$ and for $\varepsilon\delta = 0.25$, the manifolds are observed to touch at $\varepsilon\gamma \approx 0.19$, within 1% of the predicted value $0.18825\ldots$.

Along with the chaotic orbits associated with the horseshoe, the damping term $-\varepsilon\delta y$ of eqn [13] makes possible the construction of a large trapping region: a bounded region D (a topological disk) into which the Poincaré map eventually draws all orbits (Liapunov functions are helpful in this construction). We may then define the attracting set $A \subset D$ as:

$$A = \bigcap_{k=0}^{\infty} P^k(D)$$

More generally, an attracting set is a compact invariant set which attracts a full neighborhood of solutions (its stable manifold has dimension equal to that of the phase space).

Here A clearly contains the horseshoe Λ, but it may also contain (many) 'simple' attractors, such as stable periodic orbits. Indeed, numerical studies of the Duffing equation for parameter values above but near the critical ratio (eqn [16]) reveal that almost all orbits eventually approach one or other of the two stable periodic orbits which develop from the fixed points $(x, y) = (\pm 1, 0)$ for $\varepsilon\gamma \neq 0$; this is sometimes called transient chaos. As $\varepsilon\gamma$ increases for fixed $\varepsilon\delta$ and ω, numerical simulations indicate that these orbits lose stability and eventually almost all solutions behave chaotically forever. **Figure 7A** shows an example, for the same parameter values as **Figure 6B**; the corresponding time series is shown in **Figure 7B**; such signals display broad band power spectra. Comparing **Figures 7A** and **6B**, note how the orbit appears to lie 'along' the unstable manifold $W^u(p_\varepsilon)$: it appears that A is the closure of $W^u(p_\varepsilon)$ (the manifold and its limit points).

To qualify as a genuine strange attractor, one has to show that almost all orbits explore the whole of A: this is accomplished if the existence of an orbit dense in A can be proven, not merely one dense in Λ. This has not been proved for the Duffing equation; it is currently impossible to rule out the possibility of stable periodic cycles, and the persistently chaotic behavior of **Figure 7** may be a numerical artefact. It has been proved for closely related Poincaré maps, including a specific example – the quadratic Hénon map – but there is currently no generally applicable method available to achieve it, nor a computable condition that provides both necessary and sufficient conditions for chaos in specific differential equations such as eqn [13].

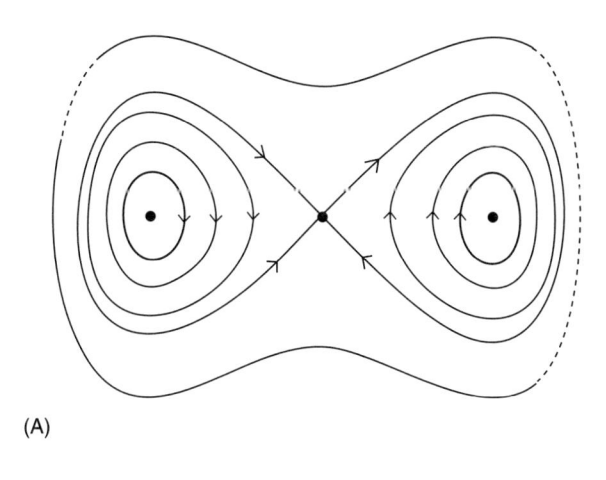

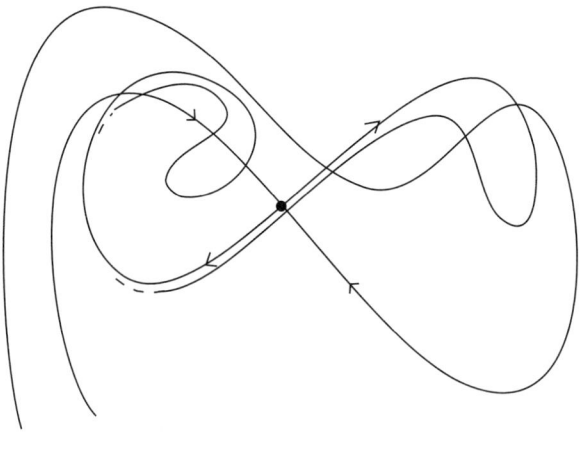

Figure 6 The unperturbed phase portrait ($\varepsilon = 0$) (A) and the perturbed Poincaré map (B) of the Duffing equation (eqn [13]): $\varepsilon\gamma = 0.30, \varepsilon\delta = 0.25, \omega = 1$, showing intersections of stable and unstable manifolds.

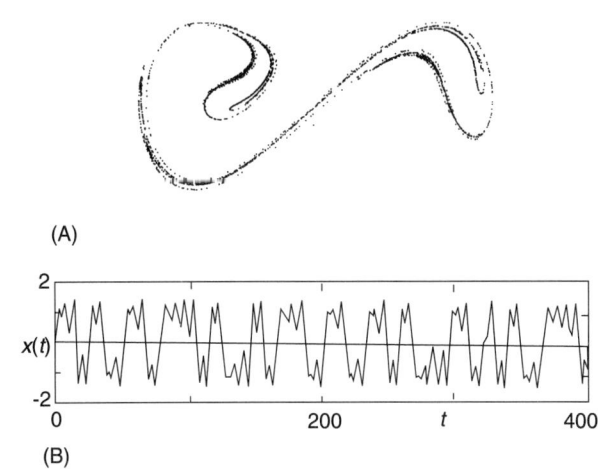

Figure 7 (A) An orbit of the Poincaré map of the Duffing equation (eqn [13]): $\varepsilon\gamma = 0.30, \varepsilon\delta = 0.25, \omega = 1$; (B) the corresponding time series.

Sources of Nonlinearity and Chaos

We have already met one source of nonlinearity in the example above: restoring forces in postbuckling behavior. Large amplitudes in general lead to nonlinear effects, through geometry alone (as in the pendulum), via nonlinear constitutive effects due to large strains, or both. Contact problems often lead to restoring forces of piecewise linear type, for example the wheel flange/rail contact in rail vehicle dynamics, or gears and mechanisms with play. Dissipation can also be strongly nonlinear, a common example being stick–slip friction. Inherent (convective) nonlinearities in fluid dynamics lead to nonlinear effects in fluid–structure interactions. Bang-bang control, or smoothed versions thereof, also leads to strongly nonlinear effects (see article **Nonlinear systems, overview**).

From a phase space perspective, the key properties necessary for chaos are present in the simple doubling map example. One must have (exponential) expansion in at least one dimension, coupled with a cutting or folding mechanism which prevents orbits from simply escaping to infinity. Strong nonlinearity or large amplitudes are generally required for this. Chaos can occur in undamped (conservative, Hamiltonian) systems, where it is associated with behavior in the 'gaps' between quasiperiodic motions associated with the Kolmogorov–Arnold–Moser theorem, but the presence of damping is generally necessary for trapping regions and strange attractors.

Nomenclature

Df	matrix of first partial derivatives
$M(t_0)$	Melnikov function
P	Poincaré map
R^n	state or phase space
Λ	Cantor set

See Plates 13, 14, 15.

See also: **Nonlinear normal modes; Nonlinear system identification; Nonlinear systems, overview; Perturbation techniques for nonlinear systems; Testing, nonlinear systems.**

Further Reading

Cartwright ML and Littlewood JE (1945) On nonlinear differential equations of the second order, i: The equation: $\ddot{y} + k(1 - y^2)\dot{y} + y = b\lambda k \cos(\lambda t + a)$, k large. *Journal of the London Mathematical Society* 20: 180–189.

Glendinning P (1995) *Stability, Instability and Chaos.* Cambridge, UK: Cambridge University Press.

Guckenheimer J and Holmes PJ (1983) *Nonlinear Oscillations, Dynamical Systems and Bifurcations of Vector Fields.* New York: Springer-Verlag.

Holmes PJ (1979) A nonlinear oscillator with a strange attractor. *Philosophical Transactions of the Royal Society of London A* 292: 419–448.

Holmes PJ and Moon FC (1980) A magneto-elastic strange attractor. *Journal of Sound and Vibration* 65: 275–296.

Moon FC (1987) *Chaotic Vibrations.* New York: John Wiley.

Palis J and Takens F (1993) *Hyperbolicity and Sensitive Chaotic Dynamics at Homoclinic Bifurcations.* Cambridge, UK: Cambridge University Press.

Poincaré HJ (1892–1899) *Les méthodes nouvelles de la mécanique celeste*, vols 1–3. Paris: Gauthiers-Villars. Translation (1993) edited by Goroff D: New York: American Institute of Physics.

Smale S (1967) Differentiable dynamical systems. *Bulletin of the American Mathematical Society* 73: 747–817.

Strogatz S (1994) *Nonlinear Dynamics and Chaos.* Reading, MA: Addison Wesley.

Ueda Y (1993) *The Road to Chaos.* Santa Cruz, CA: Ariel Press.

CHATTER

See **MACHINE TOOLS, DIAGNOSTICS**

COLUMNS

I Elishakoff, Florida Atlantic University, Boca Raton, FL, USA

C W Bert, University of Oklahoma, Norman, OK, USA

Copyright © 2001 Academic Press

doi:10.1006/rwvb.2001.0131

A column is a slender structural member subjected to axial compressive force. If the compressive load is excessive, a column may fail by structural instability, called buckling. Buckling may be by general column instability or, in the case of a thin-walled member (angle, box, channel, or I cross-sections), by local buckling.

The purpose of this section is not to discuss column buckling, but rather to discuss vibration of columns. Free vibration analysis of prismatic columns, tapered columns, and thin-walled sections is covered in separate subsections.

It should be mentioned at the outset that compressive loading decreases the natural frequencies of lateral vibration relative to their unloaded values. In fact, vibration is one of the ways to determine experimentally the buckling load by a nondestructive means. In contrast, application of loading in tension increases the natural frequencies of lateral vibration. Such a member is usually called a tie bar but is not covered here.

Free Lateral Vibration of Uniform Columns

Problem Formulation

A slender column of uniform cross-section (area A) and length L is made of a linearly elastic material (elastic modulus E) and subjected to a constant axial compressive force P. The governing partial differential equation of motion for lateral vibration is:

$$EI\frac{\partial^4 w}{\partial x^4} + P\frac{\partial^2 w}{\partial x^2} + \rho A \frac{\partial^2 w}{\partial t^2} = 0 \qquad [1]$$

where I is the centroidal area moment of inertia, t is time, w is the lateral deflection, x is the axial position coordinate, and ρ is the density of the material. Since the problem is linear, the vibration can be taken as harmonic in time, i.e.:

$$w(x,t) = W(x)\cos\omega t \qquad [2]$$

where $W(x)$ is known as the mode shape and ω is the circular natural frequency. Now eqn [1] becomes an ordinary differential equation:

$$EI\frac{d^4 W}{dX^4} + PL^2 \frac{d^2 W}{dX^2} - \rho A L^4 \omega^2 W = 0 \qquad [3]$$

where $X \equiv x/L$.

A combination of two boundary conditions must be specified for each end, making a total of four for the fourth-order differential eqn [3]. The sets of classical boundary conditions for each end, taken here as $X = 0$ just as an example, are:

1. clamped: $W(0) = 0$, $W'(0) = 0$
2. pinned (simply supported): $W(0) = 0$, $W''(0) = 0$ [4]
3. free: $W''(0) = 0$, $W'''(0) + (P/EI)W'(0) = 0$
4. guided (sliding): $W'(0) = 0$, $W'''(0) = 0$

where a prime denotes a derivative with respect to X.

Direct Analytical Solution

The general solution of eqn [3] can be expressed as:

$$W(X) = C_1 \sinh \alpha X + C_2 \cosh \alpha X + C_3 \sin \beta X + C_4 \cos \beta X \quad [5]$$

where $C_1 \ldots C_4$ are constants of integration to be determined from the boundary conditions and:

$$\begin{Bmatrix} \alpha \\ \beta \end{Bmatrix} z = \left[\mp \bar{P} + \sqrt{\bar{P}^2 + \bar{\omega}^2} \right]^{1/2} \quad [6]$$

Here $\bar{P} = PL^2/2EI$ is the dimensionless axial force and $\bar{\omega} = \omega L^2 \sqrt{\rho A/EI}$ is the dimensionless natural frequency. Satisfaction of the four boundary conditions, two at each end, lead to four homogeneous algebraic equations in the coefficients $C_1 \ldots C_4$. To guarantee a nontrivial solution, the determinant of the coefficients must be zero. This leads to a standard eigenvalue problem with a characteristic frequency equation, in which the dimensionless frequencies are the eigenvalues, of which there are an infinite number.

As an example, consider the case of a column that is clamped at $X = 0$ and pinned at $X = 1$. Then the characteristic frequency equation which comes from vanishing of the frequency determinant is:

$$\left(-\bar{P} + \sqrt{\bar{P}^2 + \bar{\omega}^2} \right)^{1/2}$$
$$\times \cosh \left(-\bar{P} + \sqrt{\bar{P}^2 + \bar{\omega}^2} \right)^{1/2} \sin \left(\bar{P} + \sqrt{\bar{P}^2 + \bar{\omega}^2} \right)^{1/2}$$
$$- \left(\bar{P} + \sqrt{\bar{P}^2 + \bar{\omega}^2} \right)^{1/2} \sinh \left(-\bar{P} + \sqrt{\bar{P}^2 + \bar{\omega}^2} \right)^{1/2}$$
$$\times \cos \left(\bar{P} + \sqrt{\bar{P}^2 + \bar{\omega}^2} \right)^{1/2} = 0 \quad [6]$$

The fundamental dimensionless frequency $\bar{\omega}$ as a function of $\bar{P}$ is listed in column 2 of **Table 1**. These are exact values.

Table 1 Dimensionless frequency $\bar{\omega}$; as a function of dimensionless compressive load $\bar{P}$ for a clamped-free homogeneous column

$\bar{P}$	Exact	Lower bound	Upper bound
0.1	14.644	14.627	14.645
0.2	13.823	13.791	13.829
0.3	12.946	12.900	12.961
0.4	12.001	11.943	12.031
0.5	10.971	10.903	11.022
0.6	9.8263	9.7513	9.9124
0.7	8.5229	8.4448	8.6607
0.8	6.9707	6.8954	7.1947
0.9	4.9377	4.8757	5.3403

Bokaian suggested the following expression for $\bar{\omega}$:

$$\bar{\omega} = \bar{\omega}_0 \left(1 - \gamma \bar{P}/\bar{P}_{cr} \right)^{1/2} \quad [7]$$

where $\bar{\omega}_0$ is the dimensionless natural frequency for the same mode but in the absence of axial load, $\bar{P}_{cr}$ is the dimensionless critical axial load for buckling at the same mode, and γ is a dimensionless factor depending only on the set of boundary conditions, as tabulated in **Table 2**.

Hohenemser and Prager's Approximate Lower Bound

Hohenemser and Prager derived the following expression for ω for pinned–pinned boundary conditions:

$$\bar{\omega} = \bar{\omega}_0 \left(1 - \bar{P}/\bar{P}_{cr} \right)^{1/2} \quad [8]$$

It has been shown that the Hohenemser–Prager equation always gives a lower bound, although it is exact in three cases (**Table 2**): pinned–pinned, pinned–guided, and guided–guided. The dimensionless numerical results from the Hohenemser–Prager formula for the clamped-free case are listed in column 3 of **Table 1**.

Table 2 Factor γ for various end conditions

End conditions	γ	End conditions	γ
Free–guided	0.925	Clamped–pinned	0.978
Clamped–free	0.926	Pinned–pinned	1.000
Clamped–clamped	0.970	Pinned–guided	1.000
Clamped–guided	0.970	Guided–guided	1.000
Free–free	0.975	Free–pinned	1.130

Approximate Upper Bound Based on the Energy Method

The Rayleigh quotient (see Continuous methods) for lateral vibration of a homogeneous column is:

$$\omega^2 = \left[\int_0^L EI\left(\frac{d^2w}{dx^2}\right)^2 dx - P\int_0^L \left(\frac{dw}{dx}\right)^2 dx\right] \bigg/ \left[\int_0^L \rho A w^2 \, dx\right] \quad [9]$$

or, in dimensionless form:

$$\bar{\omega}^2 = \left[\int_0^1 (W'')^2 dX - 2\bar{P}\int_0^1 (W')^2 dX\right] \bigg/ \left[\int_0^L W^2 dX\right] \quad [10]$$

In principle, one can use any function $W(X)$ for the mode shape provided that it at least satisfies the geometric boundary conditions. However, if it satisfies the generalized force type as well, greater accuracy can be obtained. A very convenient mode shape function is the eigenfunction for the free vibration of a homogeneous beam without axial force. Such functions and the integrals of their squares and of the squares of their first and second derivatives were first tabulated by Young and Felgar, and Felgar; see also many vibration textbooks.

For the example of a clamped-pinned beam, the Rayleigh results are given in column 4 of **Table 1**. It is seen that the bounds are quite close for low values of $\bar{P}$, but the relative differences increase as $\bar{P}$ is increased.

Extensions and Application

The effect of an elastic foundation may be included by adding a term, $-kw$, to the left side of eqn [1] or by adding a term, $+\int_0^L kw^2 dx$, to the numerator of the right side of eqn [9]. Here k is known as the foundation modulus and has units of force per length squared. The effect, of course, is to increase the natural frequencies.

A distributed axial loading occurs, for instance, when a vertical column is subjected to its own dead weight. This effect can be included by adding a term

$$+\frac{\partial}{\partial x}\left(qx\frac{\partial w}{\partial x}\right)$$

to the left side of eqn [1] or by adding a term

$$-\int_0^L qx\left(\frac{\partial w}{\partial x}\right)^2 dx$$

to the numerator of the right side of eqn [9]. Here q is the distributed axial loading and has units of force per unit length. The result is that the reduction in the natural frequencies is not as severe as it is for a uniform loading equal to the integral $\int_0^L q dx$.

Sometimes columns are loaded by a concentrated axial load applied at an intermediate point within the column. Kunukkasseril and Arumugam studied this effect for both pinned–pinned and clamped–clamped end conditions. In certain applications, it is desirable to use prestressing to resist applied loadings. Kerr showed both analytically and experimentally that, provided the prestressing force passes through the centroids of the beam cross-sections, prestressing has no effect whatsoever on the natural frequencies.

Since the work of Massonet in 1940, the idea of using the lateral vibration of a column to determine the buckling load experimentally in a nondestructive manner has been known. Most of this work has been based on the Hohenemser–Prager formula, eqn [8], rather than the more general Bokaian relation, eqn [7], so caution is advised.

Free Vibration of Stepped and Tapered Columns

To model the vibration of a column with discrete steps, one can use any of the following approaches:

1. Direct analytical solution using mode shapes of the form of eqn [5] for each uniform-cross-section segment
2. The energy method using eqn [9], with A and I being step function of x
3. Galerkin method, using as the governing equation eqn [3], with A and I being step functions of x
4. Finite element method

To model the vibration of a column with a smoothly varying cross-section, direct analytical solution is usually not feasible and may be impossible. For the energy and Galerkin methods, the area A and moment of inertia I must be appropriate smoothly varying functions of x.

Effect of Transverse Shear Flexibility

Slender columns are those having large values of the effective slenderness ratio, L_e/R, where L_e is the effective length (length equivalent to that of a pinned–pinned column) and R is the radius of gyration. For such columns, the ordinary Bernoulli–Euler bending theory is adequate. However, for short columns or slender columns constructed of highly shear-flexible material (such as most composite materials), this theory is inadequate. Instead, Bresse–Timoshenko theory, which includes both transverse shear deformation and rotatory inertia, must be used. For this theory, eqn [9] must be modified as follows:

$$\omega^2 = \left\{ \int_0^L \left[EI\left(\frac{d^2w}{dx^2} + \frac{d\beta}{dx}\right)^2 + KGA\gamma^2 - P\left(\frac{dw}{dx}\right)^2 \right] dx \right\} \bigg/ \left\{ \int_0^L \left[\rho A w^2 - \rho I \left(\frac{dw}{dx} + \beta\right)^2 \right] dx \right\} \quad [11]$$

Here G is the transverse shear modulus, I is the centroidal area moment of inertia, K is known as the shear correction coefficient (usually taken to be $\pi^2/12$ (≈ 0.822) or $5/6$ (≈ 0.833) for homogeneous material), and β is the shear rotation angle. It was shown that the effect of transverse shear deformation is much larger, for practical cross-sections, than the effect of rotatory inertia, but both effects reduce the natural frequency. Thus, it is inconsistent to use Rayleigh bending theory (which considers rotatory inertia but not transverse shear deformation) rather than Bresse–Timoshenko theory.

Free Vibration of Thin-walled Columns

In the preceding subsections, it was tacitly assumed that all cross-sections of a column were doubly symmetric, i.e., symmetric about both the x and y centroidal axes. However, in the cross-sections of many practical column structures, there is often a finite distance between the geometric center (centroid) and the elastic center (shear center). This eccentricity produces a coupling between bending action and twisting action. Also, thin-walled structures are susceptible to warping action.

The following additional symbols are introduced:

- C, C_W, torsional and warping constants
- I_x, I_y, moments of inertia about the x and y axes
- r_0, polar radius of gyration of cross-section with respect to the shear center
- u, v, w, displacements in the x, y, z directions
- x_0, y_0, coordinates of the centroid of the cross-section with respect to the shear center
- ϕ, angle of rotation of cross-section
- $(\ )_{,z}$, $\partial(\ )/\partial z$, etc.

The strain energy can be written as:

$$U = \frac{1}{2} \times \int_0^L \left[EI_y(u_{,zz})^2 + EI_x(v_{,zz})^2 + GC(\phi_{,z})^2 + EC_W(\phi_{,zz})^2 \right] dz \quad [12]$$

The potential energy due to the external compressive load P is:

$$V = -\frac{P}{2} \times \int_0^L \left[(u_{,z})^2 + (v_{,z})^2 + r_0^2(\phi_{,z})^2 - 2y_0 u_{,z}\phi_{,z} + 2x_0 v_{,z}\phi_{,z} \right] dz \quad [13]$$

The kinetic energy is:

$$T = (\rho A/2) \times \int_0^L \left[(u_{,t})^2 + (v_{,t})^2 - 2y_0 u_{,t}\phi_{,t} + 2x_0 v_{,t}\phi_{,t} + r_0^2(\phi_{,t})^2 \right] dz \quad [14]$$

The variational principle is:

$$\delta(U + V - T) = 0 \quad [15]$$

As an illustrative example, the case of pinned (universal-joint) ends is considered. Then:

$$u(z,t)/A_x = v(z,t)/A_y = \phi(z,t)/A_\phi = \sin(\pi z/L)\cos\omega t \quad [16]$$

and the modified Lagrangian functional $\bar{L}$ is:

$$\bar{L} \equiv (4L/\pi^2)(U + V - T) = (P_y - P - \tilde{\omega}^2)A_x^2 \\ + (P_x - P - \tilde{\omega}^2)A_y^2 + (P_\phi - P - \tilde{\omega}^2)r_0^2 A_\phi^2 \\ + 2(P + \tilde{\omega}^2)y_0 A_x A_\phi - 2(P + \tilde{\omega}^2)A_y A_\phi \quad [17]$$

where:

$$P_i \equiv EI_i \pi^2/L^2 \quad \text{for } i = x, y$$
$$P_\phi \equiv GC + EC_W(\pi^2/L^2)$$
$$\tilde{\omega}^2 \equiv \rho A L^2 \omega^2/\pi^2$$

Application of eqn [15] yields three simultaneous homogeneous equations in the amplitudes A_i. To guarantee a nontrivial solution, the following determinant of the coefficients must vanish:

$$\begin{vmatrix} P_y - P - \tilde{\omega}^2 & 0 & (P + \tilde{\omega}^2)y_0 \\ 0 & P_x - P - \tilde{\omega}^2 & -(P + \tilde{\omega}^2)x_0 \\ (P + \tilde{\omega}^2)y_0 & -(P + \tilde{\omega}^2)x_0 & (P_\phi - P - \tilde{\omega}^2)r_0^2 \end{vmatrix} = 0 \quad [18]$$

In the general case ($x_0 \neq 0$, $y_0 \neq 0$, $r_0 \neq 0$), eqn [18] becomes a cubic in $\tilde{\omega}^2$, i.e., there is triple coupling. For the case of a singly symmetric cross-section (say, $y_0 = 0$), eqn [18] is quadratic in $\tilde{\omega}^2$.

Nonlinear Vibration of Columns

In the usual vibration theory of columns, the assumption is made that one end of the column is unrestrained in the axial direction. Hence, deflections are inextensional. In technical practice, one often encounters immovable hinges; then the effect of the axial force on the vibration mode shape must be investigated. Nonlinear free vibration of elastic columns under axial compression was first analyzed by Woinowsky-Krieger. The effect of initial imperfections – deviations from the ideal straight state – was studied by Elishakoff, Birman, and Singer.

The governing differential equation is obtained by replacing the first term in eqn [1] by $EI\partial^4(w - w_0)/\partial x^4$, where $w_0 = w_0(x)$ is the initial imperfection. In new circumstances, in which $w(x, t)$ is the total deflection, the compressive force P in eqn [1] is now given by:

$$P = P_0 - (AE/2L) \int_0^L \left[\left(\frac{\partial w}{\partial x}\right)^2 - \left(\frac{\partial w_0}{\partial x}\right)^2 \right] dx \quad [19]$$

where P_0 is the external compressive force applied to the column. The initial conditions are:

$$w(x, 0) = w_0(x) = a_0 \sin \pi X; \quad \dot{w}(X, 0) = 0 \quad [20]$$

where the dot denotes partial differentiation with respect to time. Assuming that both the initial imperfection and total deflection have the shape of a half-sine wave, one has:

$$w(x, t) = aF(t) \sin \pi X \quad [21]$$

where a is the amplitude of the total deflection. Substitution into eqn [1] suitably modified as described above and using the Bubnov–Galerkin method (nonlinear version of the Galerkin method discussed in the article on **Continuous methods**), one obtains the following ordinary differential equation:

$$\ddot{F} + \left[\omega^2 - 2k^4(a_0/a)^2\right]F + 2k^4F = \frac{P_{cr}}{P_{cr} - P_0}\frac{a_0}{a}\omega^2 \quad [22]$$

For an imperfection-free column, $a equiv 0$ and eqn [22] reduces to the equation for that case. Note that the presence of the initial imperfection makes the equation nonhomogeneous. Eqn [22] was solved exactly in terms of the normal elliptic integral of the first kind by Elishakoff et al.

Following various previous investigators, the term 'throw' is used here for nonlinear vibrations instead of 'amplitude' in linear vibrations. Elishakoff et al. showed that, in some intervals of variation of P_0/P_{cr}, the imperfect column can have frequencies in excess of its perfect-column counterparts. Also, the initial imperfection appears to decrease the frequency of nonlinear free vibration of elastic columns under moderate compression. Nevertheless, there are columns in which the vibration frequency increases due to initial imperfections, when P_0 approaches P_{cr}. Such an increase is characteristic for columns with small values of vibration throw to cross-sectional radius of gyration. Nonlinear vibration of a buckled column under harmonic excitation was studied by Tseng and Dugundji, who found that the buckled column is characterized by soft-spring behavior at small vibration throws before snap-through, and hard-spring-type behavior after snap-through.

A model structure, namely a three-hinge, rigid-element column system constrained laterally by a nonlinear spring, was studied in detail by Elishakoff, Birman, and Singer. They studied the static behavior and then superimposed small vibrations $w(t)$ on it. The spring force was taken as:

$$F = k_1\hat{w} + k_2\hat{w}^2 + k_3\hat{w}^3 \quad [23]$$

where $k_1/2$ constitutes a nondimensional buckling load and $\hat{w}$ is a nondimensional static deflection. The coefficients $a \equiv k_2/k_1$ and $b \equiv (k_2/k_1) - 0.5$ were introduced. If $b = 0$ and $a \neq 0$, the structure is said to be asymmetric; if $a = 0$ and $b \neq 0$, the structure is called symmetric; whereas if $a \neq 0$ and $b \neq 0$, the structure is nonsymmetric.

The following formulas were derived. For the asymmetric case, the nondimensional frequency squared is given by:

$$\bar{\omega}^2 = (1 - P/P_{cr})\left[1 + 4aw_0(P/P_{cr})(1 - P/P_{cr})^{-2}\right]^{1/2} \quad [24]$$

A structure that is imperfection-sensitive for buckling turns out to be imperfection-sensitive for vibration as well, for $aw_0 < 0$, $\bar{\omega}^2 < 1 - P/P_{cr}$. In other words, initial imperfections reduce the natural frequency found by the Hohenemser–Prager formula. For an imperfection-sensitive structure, the natural frequency increases with the magnitude of w_0 compared with that of a perfect structure.

For the symmetric structure, eqn [24] is replaced by:

$$\left(1 - \frac{P}{P_{cr}} + \frac{\bar{\omega}^2}{2}\right)\left[\frac{\bar{\omega}^2 - (1 - P/P_{cr})}{3b}\right]^{1/2} = \frac{3}{2}\frac{P}{P_{cr}}w_0 \quad [25]$$

Eqn [25] implies that for the structure to be vibrationally imperfection-sensitive (i.e., with decreased frequency as a result of imperfection), the coefficient b must be negative, as in the case of static buckling. For $b > 0$, the natural frequency increases with w_0.

Random Vibration of Columns

There are numerous works devoted to random vibration of elastic structures. Apparently the first study devoted to the case of beams was by Eringen in 1957, while random vibration of columns was first studied by Elishakoff in 1987. The governing eqn [1] is modified by adding a damping term, $c\partial w/\partial t$, on the left side and by adding the transverse loading, $q(x, t)$, on the right side.

The autocorrelation function of the loading is taken as the 'rain-on-the-roof' excitation, represented by a space- and time-wise ideal white noise:

$$R_q(x_1, x_2, t_1, t_2) = (R/L)\delta(x_1 - x_2)\delta(\tau) \quad [26]$$

where R is a positive constant, x_1 and x_2 are spatial coordinates, $\delta(x)$ is the Dirac delta function, and $\tau = t_2 - t_1$ is the time lag.

The following result was obtained for the deflection autocorrelation function at zero time lag:

$$R_w(X_1, X_2, 0) = \frac{RL^2}{8\pi^2 EIc\beta}\bigg(2|X_1 - X_2| \\ - (X_1 - X_2)^2 - 2(X_1 + X_2) + (X_1 + X_2)^2 \\ + \frac{2}{\pi\sqrt{\beta}}\frac{1}{\sin\pi\sqrt{\beta}}\bigg\{\left(\cos\pi\sqrt{\beta}\right)[1 - (X_1 + X_2)] \\ - \left(\cos\pi\sqrt{\beta}\right)[1 - |X_1 - X_2|]\bigg\}\bigg) \quad [27]$$

where $X_1 \equiv x_1/L$, $X_2 \equiv x_2/L$, and $\beta \equiv P/P_{cr}$. As the axial compression reaches its critical value, i.e., as β tends to unity, $\sin \pi\sqrt{\beta}$ tends to zero and the autocorrelation function of deflection increases without bound. When β goes to zero, eqn [27] reduces to Eringen's result for a beam (no axial loading):

$$R_W(X, X, 0) = RL^2 X^2 (1 - X)^2 / 6EI \quad [28]$$

Elishakoff showed that the presence of axial compression increases the mean-square response of the column.

The effect of randomness in elastic modulus and/or other material properties on stressed structures was studied by Hart. He developed a method for calculating statistics of natural frequencies and vibrational mode shapes of a structure acted upon by external static loading. In order to formulate a geometric stiffness matrix for the structure, it is necessary, in general, to solve a static response problem to obtain the axial forces in each structural member. Then this geometric stiffness is added to the elastic stiffness matrix and the eigenvalue problem is solved to obtain natural frequencies and mode shapes. The eigenvalue problem of the discretized structure can be written as:

$$\mathbf{M\ddot{W}} + (\mathbf{K}_e - P\mathbf{K}_G)\mathbf{W} = \mathbf{0} \quad [29]$$

where $\mathbf{K}_e \equiv$ elastic stiffness matrix, $\mathbf{K}_G \equiv$ geometric stiffness matrix, $\mathbf{M}$ is the mass matrix, and $\mathbf{W}$ and $\mathbf{\ddot{W}}$ are the vectors of the system's deflections and accelerations, respectively. When $P = P_{cr}$, the system's fundamental natural frequency is zero. Let $r_1, r_2 \ldots r_m$ be random structural parameters. A linear probabilistic structural model uses multiple Taylor series expansions for an arbitrary natural frequency:

$$\omega_j(r_1, r_2, \ldots r_m) = \omega_j[E(r_1) \ldots E(r_m)] \\ + \sum_{\alpha=1}^{m} \left.\frac{\partial \omega_j}{\partial r_\alpha}\right|_{r_\alpha = E(r_\alpha)} [r_\alpha - E(r_\alpha)] \quad [30]$$

where $E(r_\alpha) \equiv$ mean value of the αth random structural parameter, and $\partial\omega_j/\partial r_\alpha \equiv$ sensitivity derivative.

The mean value of the natural frequency can be expressed as:

$$E(\omega_j) = E[\omega_j(r_1, r_2, \ldots r_m)] \quad [31]$$

and the mean-square value is obtained as:

$$\operatorname{Var}(\omega_j) = \sum_{\alpha=1}^{m} \sum_{\beta=1}^{m} \left. \frac{\partial \omega_j}{\partial r_\alpha} \right|_{r_\alpha = E(r_\alpha)} \left. \frac{\partial \omega_j}{\partial r_\beta} \right|_{r_\beta = E(r_\beta)} \operatorname{Cov}(r_\alpha, r_\beta)$$

[32]

where $\operatorname{Cov}(r_\alpha, r_\beta)$ is the covariance given by:

$$\operatorname{Cov}(r_\alpha, r_\beta) = E\{[r_\alpha - E(r_\alpha)][r_\beta - E(r_\beta)]\} \quad [33]$$

and $E(\ )$ denotes mathematical expectation. In a particular example of a two-column structure, Hart showed that the mean value of the natural frequency is a nonlinear function of the ratio $\beta = P/P_{cr}$ and vanishes when $\beta = 1$. The mean-square value of the natural frequency increases with an increase of β.

More recently, use of finite element analysis in the stochastic setting was put forward for eigenvalue problems with uncertainties by Nakagiri.

Nomenclature

A	cross-sectional area
a	amplitude of total deflection
a	k_2/k_1
a_0	amplitude of initial imperfection
A_x, A_y, A_ϕ	coefficients of u, v, and ϕ expressions
b	$(k_2/k_1) - 0.5$
C	torsional constant
b_i	constants of integration
Cov	covariance
C_W	warping constant
c	damping coefficient
E	elastic modulus
$E(r_\alpha)$	expected mean value of $(\ )$
F	dimensionless deflection in eqn [21]
F	spring force
G	shear modulus
I	centroidal area moment of inertia
I_x, I_y	moments of inertia about x and y axes
K	shear correction coefficient
$\mathbf{K}_e$	elastic stiffness matrix
$\mathbf{K}_G$	geometric stiffness matrix
k	foundation modulus
k	$(\pi/L)(Ea^2/\rho)^{1/4}$
k_i	stiffness coefficients in eqn [23]
L	column length
$\overline{L}$	modified Lagrangian functional
$\mathbf{M}$	mass matrix
P	axial compressive force
P_0	external compressive force
P_x, P_y, P_ϕ	coefficient in eqn [17]
$\overline{P}$	dimensionless axial compressive force
$\overline{P}_{cr}$	dimensionless critical axial load
q	distributed axial loading
R	positive constant in equation
R_q	cross-sectional
R_W	deflection autocorrelation function
r_i	random structural parameters
r_0	cross-sectional polar radius of gyration relative to the shear center
T	kinetic energy
t	time
t_1, t_2	time values at start and end
U	strain energy
u, v, w	displacements in the x, y, z directions
V	potential energy due to P
Var	mean-square value of natural frequency
W	mode shape
$\mathbf{W}$	vector of deflections
w	column deflection
$\hat{w}$	nondimensional static deflection
W_0	initial imperfection
X	x/L
X_1, X_2	$X_1/L, X_2/L$
x	axial position coordinate
x, y	centroidal axes in the cross-section
x_1, x_2	spatial coordinates
x_0, y_0	coordinates of cross-sectional centroid relative to the shear center
z	axial position coordinate
α, β	parameters defined in eqn [6]
β	shear rotation angle
β	P/L_{cr}
γ	factor appearing in eqn [7]
δ	Dirac delta-function
ρ	material density
ϕ	angle of rotation of cross-section
τ	$t_2 - t_1$
ω	natural frequency
ω_j	arbitrary natural frequency
$\overline{\omega}$	dimensionless natural frequency
$\overline{\omega}_0$	dimensionless natural frequency in absence of axial force
$(\ )'$	derivative with respect to X
$(\ )_{,z}$	derivative with respect to z

See also: **Continuous methods; Eigenvalue analysis.**

Further Reading

Banerjee JR and Williams FW (1985) Further flexural vibration curves for axially loaded beams with linear or parabolic taper. *Journal of Sound and Vibration* 102: 315–327.

Bokaian A (1988, 1989) Natural frequencies of beams under compressive axial loads. *Journal of Sound and Vibration* 126: 49–65; author's reply 131: 351.

Elishakoff I (1987) Generalized Eringen problem: influence of axial force in random vibration response of simply supported beam. *Structural Safety* 4: 255–265.

Elishakoff I (1999) *Probabilistic Theory of Structures*. New York: Dover.

Elishakoff I, Birman V and Singer J (1984) Effect of imperfections on the vibrations of loaded structures. *Journal of Applied Mechanics* 51: 191–193.

Elishakoff I, Birman V and Singer J (1985) Influence of initial imperfections on nonlinear free vibrations of elastic bars. *Acta Mechanica* 55: 191–202.

Eringen AC (1957) Response of beams and plates to random loads. *Journal of Applied Mechanics* 24: 46–52.

Hart GC (1973) Eigenvalue uncertainty in stressed structures. *Journal of Engineering Mechanics* 99: 481–494.

Hohenemser K and Prager W (1933) *Dynamik der Stabwerke*. Berlin: Springer.

Kerr AD (1976) On the dynamic response of a prestressed beam. *Journal of Sound and Vibration* 49: 569–573.

Kunukkasseril VX and Arumugam M (1975) Transverse vibration of constrained rods with axial force fields. *Journal of the Acoustical Society of America* 57: 89–94.

Mujumdar PM and Suryanarayan S (1989) Nondestructive techniques for prediction of buckling loads – a review. *Journal of the Aeronautical Society of India* 41: 205–223.

Nakagiri S (1984) Structural system with uncertainties and stochastic FEM. *JSME Transactions Series A* 30: 1319–1329.

Tseng W-Y and Dugundji J (1971) Nonlinear vibrations of a buckled beam under harmonic excitation. *Journal of Applied Mechanics* 37: 467–476.

Woinowsky-Krieger S (1950) The effect of an axial force on the vibration of hinged bars. *Journal of Applied Mechanics* 17: 35–36.

COMMERCIAL SOFTWARE

G Robert, Samtech SA, Liege, Belgium

Copyright © 2001 Academic Press

doi:10.1006/rwvb.2001.0010

Introduction

A lot of engineer work is done nowadays using commercial computer software. To deal with large projects, engineers construct numerical models to simulate the real world, thanks to the performances of fast digital processors. Such models allow the engineer to test choices during the design process. In the field of mechanical engineering, finite element analysis (FEA) is the most popular way to investigate structural behavior without experiment. In principle, finite element method (FEM) allows one to build a mathematical model of a given structure and to predict its behavior by numerical computation, given boundary, and initial conditions.

The scope of this article is to present some FEA commercial software. We have limited the choice of software because the goal is not to compare software, but rather to give an overview of FEA software capabilities. The question is not Which software exists? but What does the existing software offer the user? It must be noted that the software is presented at a given moment in its history. It is clear that software is constantly being improved: most have a new version released about once a year.

History

Software dedicated to structural analysis first appeared in the mid 1960s. It was initially used in advanced technology such as aeronautical and space industries. At that time, software focused on efficient computation rather than user-friendliness: it did not matter that the process was tortuous provided that the result achieved were correct. Nevertheless, at this first stage, FEM proved its ability to predict structural behavior. In some situations it appeared as a cost-effective alternative to testing procedure and an efficient way of reducing design cost and time. As a result, these computational techniques spread to more and more engineering sectors.

Factors contributing to software evolution include the following:

Analysis of More and More Complex Structures

The aim is not only to analyze structures with complex geometries, but also to analyze complex assemblies, made up of a large number of parts. The number of degrees of freedom may range from a few to several millions. To solve large systems, new algorithms have been implemented, which use parallel and vector processing. Also Graphical User Interface and database systems have evolved in order to process a large amount of data. This evolution has followed the improvement of graphical and digital processors.

Analysis of More and More Complex Situations

To make an accurate model of a structure, it is necessary to take into account many factors influencing its behavior. For example, to analyze the effects of an explosion on a given structure, a model is made of the structure itself and the surrounding fluid as well as interaction between both. Analysis of such situations led software designers to develop multiphysical simulation. In addition to coupled equations between different physical variables, FEM addresses the problem of interactions between several parts of the same structure, including contact detection and simulation.

Reduction of the Time Devoted to Design

The target of high profitability has reduced the length of the design stage. Adaptation of FEA software to this economic reality requires improvement in usability and of interoperability. That is, software must be easy to use and its integration to the design environment must be as complete as possible.

Change of Software User Itself

During the first years of FEA, the interaction between programmers and users was so strong that often the same people both wrote and ran the software. This time is now passed and it must be borne in mind that the user's background does not include a good command of computer science and a theoretical knowledge of numerical methods. Programmers, and programs, must provide sufficient information both to check input and validate output.

Application Fields in Dynamics

FEM applies to continuous media. It allows one to compute static or dynamic equilibrium of the media subject to initial and boundary conditions. FEM is broadly used to resolve equations related to solid and fluid mechanics, heat transfer, and electromagnetism. In the specific domain of vibration analysis, FEM is used to compute natural frequencies and corresponding vibration modes as well as dynamic response in time, possibly including nonlinearities. Different types of analyses are listed below.

Modal Analysis

The prediction of natural frequencies of a structure is useful so that one can understand fundamental dynamic behavior and avoid the risk of resonance during service. Direct applications of modal analysis are, for example, computation of critical speed of rotating machines and noise reduction by vibration isolation of structures.

Transient Response

This is the most general dynamic response. At a given time, arbitrarily set to zero, initial displacements and velocities are known at any point of the structure and motion or forces are prescribed at some points. Transient response results from the time integration of equations of motion. Solution techniques vary according to the range of natural modes that are excited. At low frequency, modal methods and implicit time integration are generally used, whereas explicit time integration is better suited to high frequency. A typical example of transient loading is increasing pressure in an enclosure. At low speed, implicit integration will give good results. At high speed, if a shock wave ensues, it will become necessary to use explicit software.

Periodic Response

For structures subject to periodic loading, such as an unbalanced mass in rotating machines, it is of interest to predict the response during one of the repetitive load cycles. Excitation may include one or more harmonics. In the case of single harmonics, the response is called harmonic forced response. Modal solution methods are used with linear systems. Techniques such as equivalent linearization or finite elements in time are used in the presence of nonlinearities.

Random Response

In some cases, loading is so erratic that it is better described in a probabilistic way. A typical example is the force profile transmitted by the suspension of a car running on an uneven road. We characterize the excitation – force or motion – by a power spectral density (PSD), which gives the distribution of power in a frequency range. Solution techniques are close to those of the single harmonic case. Response is also expressed in terms of PSD. From the PSD, it is possible to compute statistical moments of the random process and to derive quantities of physical meaning such as root mean square (RMS) value, standard deviation or peak value.

Response Spectrum

This solution technique is widely used in earthquake engineering. It is based on the maximum response of a Single Degree of Freedom (SDOF) oscillator. A response spectrum is obtained by plotting frequency versus the maximum response of several oscillators with different natural frequencies but the same damping to a given input transient signal to a given input transient signal of short duration. Responses may be expressed in terms of displacement, velocity, or acceleration. For a real structure, the response spectrum provides the maximum response for each mode, according to its frequency and damping. Modal responses are then accumulated using combination rules such as Square Root of Sum of Squares (SRSS). The result is the probable maximum response of the structure. Response spectra are also known as shock spectra.

Software Quality Assurance

After initially being research tools, FEA software has now become an industrial product. Its use is conditioned not only by technical aspects, but also by economical aspects. To play its part efficiently, software must fulfill quality requirements. There are several organizations that have established rules and criteria for software quality assurance (QA), including:

- American National Standards Institute (ANSI).
- Federal Aviation Administration (FAA).
- Institute of Electrical and Electronic Engineers (IEEE).
- International Standards Organization (ISO).
- National Agency for Finite Elements and Standards (NAFEMS).
- Nuclear Regulatory Commission (NRC).

Most of these organizations have acquired international status. NAFEMS, based in the UK, is specially dedicated to FEA. It aims 'to promote the safe and reliable use of finite element technology'. Since 1983, this organization has promoted the development of many benchmarks which are used worldwide to qualify and compare FEA software.

The Developer's Side

This section summarizes important aspects contributing to quality that are directly under the developer's responsibility.

Validation Software must provide a correct response to a correct input. It must be reliable. The software vendor should maintain a library of tests related to all its options. Most tests must have accepted references, such as analytical references.

Integration The developer must ensure good integration of the software with the hardware. Products must run efficiently on various available platforms. Efficiency implies that algorithms should take the best advantage of computers, considering the number of operations to perform.

Maintenance FEA software has a long life cycle. Most such software has been on the market for more than 10 years. To continue to be competitive over a long time, software must prove its maintainability and flexibility. Maintainability refers to the ease of bug fixing, while flexibility is the ability to incorporate new functionalities.

Training and support FEA vendors should provide information to support the customer. This includes an installation guide, user's manuals, theoretical references, tutorial and examples, as well as release notes. Documentation must be such that information is easy to find. In addition, it is important to organize training sessions and to propose a hot-line service ready to help users.

The User's Side

There are quality factors that must be checked by the customer, even if the customer has no direct influence on them. Using the same criteria:

Validation The user's responsibility starts with the choice of software. The first rule is that software must meet the user's needs. For instance, if a given problem requires nonlinear analysis, the user must ensure that the software includes adequate formulations. The user's validation is not reduced to a technical choice. It also includes more subjective aspects such as the ease of use.

Integration The customer must make sure that software will be correctly integrated in its working environment. For example, the user should check that it interfaces completely with other software, including office tools and CAD or CAM, including possible testing programs.

Maintenance FEA software may be used throughout a long-term project. The user is then concerned with the availability of new releases and compatibility. For that purpose, it is important that each software output can be clearly identified, with release number, date, and type of result. Possible incompatibilities must be traced and documented. Error reports must be regularly provided. They must include possible bypass when available.

Training and support The fact that a software computes correct results when given the correct input is not a guarantee of success when using it. FEA software is a tool and, as such, one has to learn to use it. Moreover, it is not sufficient to learn about all the possible options: the analyst has to know why and when to use them. Appropriate training must comprehensively establish the correspondence between physics and models. For advanced analysis, FEA vendors generally complete their offer by direct support and by engineering services.

Presentation of FEA Software

As stated in the introduction, this part presents a limited number of commercial FEA software packages. This information has been kindly supplied by software companies which have agreed to contribute to this review. The NAFEMS maintains a more complete list of FEA software and its internet home page contains various links to vendors' sites and other sites related to FEA and numerical computation, including Computational Fluid Dynamics (CFD).

ABAQUS

Developer Hibbitt, Karlsson & Sorensen, Inc., 1080 Main Street, Pawtucket, RI 02860, USA.

General description Advanced, general-purpose, finite element system for the solution of large complex problems.

Main features ABAQUS programs are used throughout the world to simulate the physical response of structures and solid bodies to load, temperature, contact, impact, and other environmental conditions. They are state-of-the-art software for linear and nonlinear analysis of large models. Major market segments include automotive, aerospace, defense, offshore, civil, power generation, consumer products, and medical devices.

First and latest releases

First version Version 1.0, 1978.

Current version Version 5.8, October 1998.

Quality assurance Complies with requirements of ISO 9001 and ASME/ANSI NQA-1.

Documentation Documentation includes user's manual, example problems manual, theory manual, verification manual, and tutorial manual.

Service and support Unlimited technical support is provided to customers, as well as a wide range of training courses, engineering services, and code customization.

Availability Network and nodelock licenses are available. There are offices and representatives in most industrialized countries worldwide.

Computer environment Cray C90/T90 (Unicos 10); Cray J90/YMP (Unicos 10); Cray T90 IEEE (Unicos 10); Digital Alpha (Unix 4.0); Fujitsu VX/VPP (UXP/V); HP-PA7000 (HP-UX 10.20); HP-PA8000 (HP-UX 10.20); HP-PA8000 (HP-UX 11.0); IBM RS6000 (AIX 4.1.5); Intel (Windows-NT 4.0); NEC SX-4 (Super UX R7.2); SGIR4000/R5000 (Irix 6.2); SGI R8000/R10000 (Irix 6.2); Sun UltraSparc (Solaris 2.6).

Interfaces with other software

Geometry IGES, SAT (ACIS), DXF.

CAD and preprocessors Catia, CADFix, Femap, FemGV, HyperMesh, I-DEAS, MSC/Aries, MSC/Patran, Pro/Mesh, TrueGrid.

Postprocessor Catia, CADFix, Ensight, Femap, FemGV, HyperMesh, I-DEAS, MSC/Aries, MSC/Patran, TrueGrid.

Analysis Adams, C-Mold, DADS, Zencrack.

ANSYS

Developer ANSYS, Inc., Southpointe, 275 Technology Drive, Canonsburg, PA 15317, USA.

General description The ANSYS program is a flexible, robust package for design analysis and optimization.

Main features ANSYS is capable of making various types of analyses, including linear and nonlinear structural, steady-state and transient thermal, and coupled-field analysis. In its Multiphysics version, ANSYS allows design engineers and analysts to look at the interaction of structural, thermal, fluid flow, and electromagnetic effects on the same model.

First and latest releases

First release ANSYS, 1970.

Latest release ANSYS 5.4, 1998.

Quality assurance Complies with requirements of ISO 9001 and ASME/ANSI NQA-1.

Documentation Documentation set includes analysis guides (one for each discipline), command and element references, theory manual, workbook, and verification manual. User's guides on advanced topics are also available.

Service and support The customer services department provides service and support through multiple programs that ensure the customer's success. Services range from hot-line support to training courses and include consulting and customization services to tailor the ANSYS program to meet individual engineering requirements.

Availability A network of distributors provides licensing and customer support in the USA, Canada, Brazil, Western Europe, Israel, India, East Asia, and Australia.

Computer environment Digital Alpha UNIX, Digital Alpha NT, HP 7000/8000 Series, HP Exemplar, IBM RS/6000, Silicon Graphics, Sun Solaris, CRAY, Intel PC (Windows NT and 95).

Other products ANSYS/LS-DYNA, explicit dynamic solver.

Interfaces with other software

CAD and preprocessors AutoCAD, CADDS, CATIA, I-DEAS, MicroStation Modeler, Pro/Engineer, Solid Designer, Solid Edge, Unigraphics.

Postprocessor I-DEAS, PATRAN, Pro/Engineer.

Analysis ABAQUS, ALGOR, COSMOS/M, NASTRAN, STARDYNE, WECAN.

COSMOS/M

Developer Structural Research & Analysis Corp., 12121 Wilshire Boulevard, 7th Floor, Los Angeles, CA 90025, USA.

General description A general-purpose finite element analysis package for structural, thermal, fluid flow, and electromagnetics analysis.

Main features

Structural Linear stress, strain, displacement; frequency; buckling; nonlinear static and dynamic; dynamic response; fatigue; design optimization analyses.

Thermal Linear and nonlinear steady-state and transient thermal analysis.

Fluids Laminar/turbulent, incompressible, subsonic/transonic/supersonic compressible, transient and much more.

First and latest releases

First release COSMOS/M 1.0–1985

Latest release COSMOS/M 2.0–1998

Quality assurance A strict quality assurance procedure has been in place since 1990 (over 1800 problems of all types, sizes, and options have been tested and new ones are added continuously).

Documentation A complete set of documents is available, including user's guide, command reference, basic analysis modules, advanced modules manuals, verification and examples, and manuals of theory.

Service and support The training and consulting department offers regular training classes for the basic and advanced modules of the product. The technical support department provides support by phone, e-mail, and fax on all aspects of the program to all customers with maintenance contracts.

Availability Available through distributors, retailers, and directly from SRAC.

Computer environment PCs running Windows NT/95.
UNIX workstations: IBM, SGI, HP, Sun, DEC.

Other products COSMOS/Works, COSMOS/Edge, COSMOS/M for Helix, COSMOS/M DESIGNER II, COSMOS/M for Eureka, COSMOS/M ENGINEER, COSMOS/Wave.

Interfaces with other software

CAD and preprocessors SolidWorks, Solid Edge, Helix Design System, MicroStation Modeler, IronCAD, CADKEY, AutoCAD, Eureka Gold 97, Pro/ENGINEER, DesignWave, Hypermesh, FEMAP.

Postprocessor Pro/FEM-POST (Pro/ENGINEER), FEMAP.

Analysis ANSYS, NASTRAN (MSC and UAI), Patran, I-Deas.

MSC/NASTRAN

Developer MacNeal-Schwendler Corp., 815 Colorado Boulevard, Los Angeles, CA 90041–1777, USA.

General description MSC's principal product, MSC/NASTRAN, is the industry's leading FEA program and has proven its accuracy and effectiveness over many years. It is constantly evolving to take advantage of the newest analytical capabilities and algorithms for structural analysis.

Main features MSC/NASTRAN is a general-purpose analysis program offering a wide variety of analysis types. These include linear statics, normal modes, buckling, heat transfer, dynamics, frequency response, transient response, random response, response spectrum analysis, and aeroelasticity. Most material types can be modeled, including composites and hyperelastic materials. Advanced features include superelements (substructuring), component mode synthesis, and DMAP (a tool kit for creating custom solutions in support of proprietary applications).

In addition to analyzing structures, you can use MSC/NASTRAN to optimize designs automatically. You can optimize statics, normal modes, buckling, transient response, frequency response, acoustics, and aeroelasticity. This can be done simultaneously, with both shape and sizing design variables. Weight, stress, displacement, natural frequency, and many other responses can be used as either the design objective (which can be minimized or maximized) or as design constraints. In addition, you can synthesize the design objective and constraints via user-written equations, making possible capabilities such as updating a model to match test data. MSC/NASTRAN is the only FEA program that does this automatically.

First and latest releases

First release In 1971.

Latest release MSC/NASTRAN V70.5 was released in June 1998.

Quality assurance MSC/NASTRAN's internal QA program is fully documented in a number of publications.

Documentation A full range of documentation is available in both electronic and paper versions. This includes user manuals, theory manual, numerical methods manual, demonstration and verification problem manuals, and many more, including compact user guides for most analysis topics.

Service and support Primary customer support and training is provided by local MSC offices spread worldwide and is backed up by a comprehensive internal system. MSC is also keen to work with customers as part of a 'quickstart' program to make them as effective as possible, as quickly as possible. As well as analysis applications, MSC's engineering service projects range from software integration and customization to a complete company-wide engineering process review.

Availability MSC/NASTRAN is available from over 50 MSC offices in 17 countries plus many more full service distributors and agents spread worldwide.

Computer environment MSC/NASTRAN is available for a wide range of hardware and software platforms, including PC (Windows 95, 98 and NT), SGI, HP, DEC, SUN, IBM unix workstations, and many Supercomputer platforms.

Related products MSC/Acumen, MSC/AFEA, MSC/AKUSMOD, MSC/AMS, MSC/CONSTRUCT, MSC/DropTest, MSC/DYTRAN, MSC/FATIGUE, MSC/FlightLoads, MSC/InCheck, MSC/MVISION, MSC/NVH Manager, MSC/PATRAN, MSC/SuperModel, MSC/SuperForge, MSC/Ultima.

Interfaces with other software

CAD and preprocessors MSC/NASTRAN data formats have become a standard in many industries. This has led to the availability of interfaces to most CAD and preprocessors. MSC also supplies its own interfaces to most major systems via MSC/PATRAN. Direct interfaces are available to Catia, ProEngineer, Euclid, CADDS 5, and Unigraphics packages as well as all Parasolid-based and ACIS-based systems.

Postprocessor MSC/NASTRAN is supported by most commercial postprocessing systems.

Analysis MSC/NASTRAN may be combined with many other analysis packages, either via data transfer using a standard data format, or via a number of programming options that enable direct process-to-process communication.

SAMCEF

Developer SAMTECH S.A., 25 Boulevard Frère Orban, 4000 Liège, Belgium.

General description General-purpose finite element package for linear and nonlinear, structural, and thermal analyses.

Main features SAMCEF is composed of several analysis modules that are interconnected. For the nonlinear part, module MECANO performs static, kinematic, or dynamic analyses on flexible mechanisms modeled with nonlinear structural finite elements and kinematic rigid or flexible joints. The module ROTOR offers a powerful design tool dedicated to rotating machines. All modules are piloted either by menus or by a fully parameterized command language.

First and latest releases

First release   SAMCEF in 1965.

Latest release   SAMCEF 8.0 was released in 1998.

Quality assurance   Since 1990, a quality control and validation procedure has been integrated to the life cycle of the software.

Documentation   The complete documentation is available on HTML format. This includes primer manual, reference manual for the commands, user's manuals (one for each analysis module), and example problems manual.

Service and support   Training sessions are regularly organized at company headquarters and at subsidiaries. Advanced courses are organized either on-site or at SAMTECH. A hot-line is accessible in each subsidiary. In addition, experienced engineers provide consultancy and customization services.

Availability   SAMCEF is available directly from SAMTECH and its subsidiaries. Networks of regional and international distributors are reselling SAMTECH products in many countries.

Computer environment   All products run on UNIX platforms (CRAY, DEC, HP, IBM, SGI, SUN) and on Windows NT (PCs).

Other products

BOSS Quattro   An application manager for optimization, parametric, statistic, and correlation studies.

SAMCEF Design   An innovative technology pre- and postprocessor for several FEA softwares.

Interfaces with other software

Geometry   IGES.

CAD and preprocessors   CATIA, I-DEAS/Master Series, MSC/Patran.

Postprocessor   I-DEAS/Master Series, MSC/Patran.

Analysis   MSC/NASTRAN.

Technical Description of FEA Software

Finite Element Models

Meshing   The mesher divides the geometry into several cells, each cell being a finite element (**Table 1**). The adaptive mesher permits automatic modification of the mesh according to error computation from a previous analysis. The aim is to improve the precision of results (mainly stresses). Rezoning techniques give details of the stress distribution in some regions of an existing mesh.

Standard elements   It is considered that standard elements are those which are required to modelize most mechanical parts. They are trusses, beams, membranes, shells, and solids. Many variants exist (**Table 2**). According to geometry considerations, the formalism used to describe structures can be particularized. We generally distinguish between two- and three-dimensional problems. Axisymmetric structures expanded in Fourier series and semiinfinite media are other forms of idealization. Depending on the software, all standard elements do not necessarily exist for all formalisms.

Nonstandard elements   Nonstandard elements include specialized elements that are useful to express rigid or flexible links between degrees of freedom. The list given in **Table 3** is not exhaustive.

Table 1   Meshing tools

	ABAQUS	ANSYS	COSMOS	NASTRAN	SAMCEF
ID mesh generator	√	√	√	√	√
Plane mesh generator	√	√	√	√	√
Surface mesh generator	√	√	√	√	√
Volume mesh generator	√	√	√	√	√
Mesh editor	√	√	√	√	√
Junction of different meshes	√	√	√	√	√
Error computation	√	√	√	√	√
Adaptive meshing	√	√			
Rezoning, submodeling	√	√	√	√	

Table 2 Standard finite elements

	ABAQUS	ANSYS	COSMOS	NASTRAN	SAMCEF
Axisymmetry	MSV	SV	SV	MSV	MSV
Fourier expansion	SV	SV	S	MSV	BSV
Plane stress	TBM	TBM	M	TM	TBM
Plane strain	V	V	V	MV	MV
Generalized plane strain	V		V	MV	V
3D	TBMSV	TBMSV	TBV	TBMSV	TBMSV
3D p-element		SV		BMSV	
Semiinfinite	V	V			

Table 3 Specialized elements

Element	ABAQUS	ANSYS	COSMOS	NASTRAN	SAMCEF
Pipe	√	√	√	√	√
Cable	√	√	√		√
Gasket	√				
Bushing	√	√		√	√
Bearing					√
Spring	√	√	√	√	√
Dashpot	√	√	√	√	√
Point inertia	√	√	√	√	√
Rigid body element	√	√	√	√	√
Kinematic joints	√	√	√	√	√
Superelement	√	√	√	√	√
Contact	√	√	√	√	√
Airbag	√				
Accelerometer	√				
Monitored volume	√				√
Hydrostatic fluid element	√				
User-defined element	√	√		√	√

Geometric nonlinearity Geometric nonlinearity conditions the amplitude of motion or deformation that a structure can reach during simulation. Formulations such as Eulerian and arbitrary Eulerian Lagrangian (ALE) are adapted to flow analysis (**Table 4**).

Materials The finite element discretization process considers the material at a macroscopic level and supposes that it is continuous. But microscopic structure of material may reveal macroscopic anisotropy (**Table 5**). The FEM allows one to introduce anisotropic properties in constitutive laws. A typical anisotropy is the one present in composite materials.

A material is said to be nonlinear when stresses are not proportional to strains. Material behaviors are divided into three categories: material elasticity supposes that the stresses derive from a strain potential; plasticity expresses the material yielding; and viscoplasticity concerns deformations that depend on strain rate (**Tables 6–10**).

Boundary conditions Once the structure geometry and materials have been described, it is necessary to specify how the structure interacts with the external world. Prescribed conditions refer to known values that are imposed to a set of degrees of freedom (**Table 11**). Loads refer to any forces, whether distributed or not, that are applied to the structure (**Table 12**). Such conditions are given as functions of time or frequency, according to the type of analysis.

Analyses

Linear Analyses

The linearity assumption imposes that the coefficients appearing on both sides of the equation of motion do not depend on the motion itself. This assumption is reasonable if the stresses are proportional to strains and if the displacements remain small during the analysis.

Table 4 Geometric nonlinearities

Formulation	ABAQUS	ANSYS	COSMOS	NASTRAN	SAMCEF
Large displacement	S	S	SM	S	S
Large rotation	S	S	SM	S	S
Large strain	S	S	SM	S	S
Updated Lagrange	S		SM	S	
Eulerian	SM				
Arbitrary Eulerian Lagrangian	SM				

S, Solid structure; M, material flow.

Table 5 Material anisotropy

Orientation	ABAQUS	ANSYS	COSMOS	NASTRAN	SAMCEF
Isotropy	BMSV	BMSV	BMSV	BMSV	BMSV
Orthotropy	BMSV	MSV	MSV	MSV	MSV
Anisotropy	BMSV	MSV	MSV	MSV	MSV
Multilayer	MSV	SV	SV	BMS	MSV

B, Beam; M, membrane; S, shell; V, solid volume.

Table 6 Elastic materials

Elasticity	ABAQUS	ANSYS	COSMOS	NASTRAN	SAMCEF
Linear elasticity	√	√	√	√	√
Time-dependent viscoelasticity	√	√	√	√	√
Frequency-dependent viscoelasticity	√				√
Hypoelasticity	√	√			
Hyperelasticity (Mooney)	√	√	√	√	√
Hyperelasticity (Ogden)	√	√	√		√
Hyperelasticity (Arruda–Boyce)	√				
Hyperelasticity (Bergstrom–Boyce)	√				
Hyperelasticity (polynomial)	√				
Porous elasticity	√				

Table 7 Plastic materials

Plasticity	ABAQUS	ANSYS	COSMOS	NASTRAN	SAMCEF
Von Mises yield criterion	√	√	√	√	√
Drucker–Prager criterion	√	√	√	√	√
Mohr–Coulomb criterion	√			√	
Rankine criterion	√				
Gurson criterion	√				√
Associated flow rule	√	√	√		
Nonassociated flow rule	√	√			
Isotropic hardening	√	√	√	√	√
Kinematic hardening	√	√	√	√	√
Combined hardening	√	√	√	√	√
Soil CAP model	√				
Foam model	√	√	√		√
Concrete cracking/crushing	√				
Jointed rock	√				
Cam-clay plasticity	√				
Ramberg–Osgood plasticity	√	√	√	√	√
Johnson and Cook plasticity	√				
Krieg anisotropic plasticity					

Table 8 Viscoplastic materials

Viscoplasticity	ABAQUS	ANSYS	COSMOS	NASTRAN	SAMCEF
Steady-state hardening	√	√			√
Time hardening	√	√			
Strain hardening	√	√			√
Isotropic hardening	√	√		√	√
Kinematic hardening	√	√		√	√
Damage law					√
Lemaître law	√				√
Chaboche law	√				√

Table 9 Failure models

Failure models	ABAQUS	ANSYS	COSMOS	NASTRAN	SAMCEF
Maximum stress	√	√	√	√	√
Maximum plastic strain	√			√	
Tsai–Hill (composite)	√	√	√	√	√
Tsai–Wu (composite)	√	√	√	√	√
Azzi–Tsai–Hill (composite)	√				
Hashin (composite)	√				√
Breakable connections	√				
User-defined criterion	√			√	√

Table 10 Definition of materials

Facilities	ABAQUS	ANSYS	COSMOS	NASTRAN	SAMCEF
Thermal expansion	√	√	√	√	√
Temperature-dependent coefficients	√	√	√	√	√
Material damping	√	√	√	√	√
Material removal and addition	√				
User-defined material	√	√	√		√
Material database			√		

Table 11 Prescribed conditions

Prescribed conditions	ABAQUS	ANSYS	COSMOS	NASTRAN	SAMCEF
Functions (of time or frequency)	√	√	√	√	√
Local frames	√	√	√	√	√
Fixation	√	√	√	√	√
Prescribed displacement	√	√	√	√	√
Prescribed velocity	√	√	√	√	√
Prescribed acceleration	√	√	√	√	√
Initial displacement	√	√	√	√	√
Initial velocity	√	√	√	√	√
Pretension (bolt loading)	√	√	√	√	√

In linear dynamics, an economical way of solving the equations of motion is to expand the solution (displacements) as a series of lower-frequency modes. The technique applies for relatively low-frequency excitations. The mode acceleration method improves the solution by adding to the response the static contribution of unknown highest-frequency modes. Reactions and stresses benefit dramatically from that correction (**Tables 13** and **14**)

Nonlinear Analyses

The solution of a nonlinear problem usually requires an iterative procedure on all the degrees of freedom of

Table 12 Loads

Loading	ABAQUS	ANSYS	COSMOS	NASTRAN	SAMCEF
Point forces	√	√	√	√	√
Follower point forces	√	√	√	√	√
Pressure	√	√	√	√	√
Hydrostatic pressure	√	√	√	√	√
Global surface tractions			√	√	√
Centrifugal forces	√	√	√	√	√
Coriolis forces	√	√	√		√
Global acceleration (translation)	√	√	√	√	√
Global acceleration (rotation)	√	√	√	√	√

Table 13 Linear structural analyses

Type of analysis	ABAQUS	ANSYS	COSMOS	NASTRAN	SAMCEF
Static	√	√	√	√	√
Vibration mode extraction	√	√	√	√	√
Mode extraction in a prescribed range	√	√	√	√	√
Transient response (modal)	√	√	√	√	√
Transient response (direct solver)	√	√	√	√	√
Harmonic response (modal)	√	√	√	√	√
Harmonic response (direct solver)	√	√		√	√
Response spectrum	√	√	√	√	√
Random response	√	√	√	√	√
Campbel diagrams (rotor dynamics)					√
Buckling	√	√	√	√	√

Table 14 Advanced linear analyses

Options	ABAQUS	ANSYS	COSMOS	NASTRAN	SAMCEF
Mode acceleration method (modal response)			√	√	√
Initial stress stiffening	√	√	√	√	√
Centrifugal stiffening	√	√	√	√	√
Modal damping	√	√	√	√	√
Proportional damping (K, M)	√	√	√	√	√
Viscous damping	√	√	√	√	√
Hysteretic damping (harmonic response)		√		√	√
Gyroscopic matrix		√		√	√
Node-to-node contact	√	√	√	√	√

the structure. In transient static analysis, excitation depends on time- but rate-dependent phenomena (friction, damping, etc.) and inertial effects are neglected. Kinematics studies the motion of bodies, and does not take into account inertial effects and other forces (**Table 15**).

Linear Perturbation Analyses

Linear perturbation analysis means a linear analysis at the vicinity of a nonlinear equilibrium configuration (**Table 16**). An example is the computation of vibration modes of a cable submitted to large deflection in a gravity field.

Interactions

These concern interactions between parts of the model (**Table 17**).

Coupled Analyses

Direct coupling analysis introduces simultaneous coupling between two or more different physical unknowns. It is separate from indirect coupling involving two sequential separate analyses. The commonest indirect coupling case is computational thermomechanics with precomputed temperatures. Here, we consider only coupling with mechanics (**Table 18**).

Table 15 Nonlinear structural analyses

Type of analysis	ABAQUS	ANSYS	COSMOS	NASTRAN	SAMCEF
Static	✓	✓	✓	✓	✓
Kinematic	✓	✓	✓		✓
Transient static	✓	✓	✓	✓	✓
Transient dynamic	✓	✓	✓	✓	✓

Table 16 Linear perturbation analysis

Type of analysis	ABAQUS	ANSYS	COSMOS	NASTRAN	SAMCEF
Static response	✓	✓	✓	✓	✓
Dynamic response	✓	✓	✓	✓	✓
Vibration mode extraction	✓	✓	✓	✓	✓
Eigenvalue buckling prediction	✓	✓	✓	✓	✓

Table 17 Interactions within structures

Interaction type	ABAQUS	ANSYS	COSMOS	NASTRAN	SAMCEF
Node-based contact	✓	✓	✓	✓	✓
Element-based contact			✓	✓	✓
Surface-based contact	✓	✓	✓		✓
Rigid–flexible contact	✓	✓	✓	✓	✓
Flexible–flexible contact	✓		✓	✓	✓
Eroding contact					
Friction	✓	✓	✓	✓	✓
Multipoint constraint	✓	✓	✓	✓	✓
Cyclic symmetry	✓		✓	✓	✓
Periodic symmetry	✓		✓	✓	
Feedback and control systems	✓	✓		✓	✓

Table 18 Coupled analyses

Physics	ABAQUS	ANSYS	COSMOS	NASTRAN	SAMCEF
Compressible fluid flow		I	I		
Incompressible fluid flow		I	I		
Pore fluid flow	D		I		
Acoustic radiation	DI	D	D	D	D
Heat transfer	DI	I	I	I	I
Piezoelectricity	D	D	I		
Magnetic		I	I		
Electromagnetic		I	I		

D, direct coupling; I, indirect coupling.

Solution Methods

Solver

Basically, the solver is that part of the program dedicated to solving a linear system of equations. It allows one to obtain directly the linear static response. To solve nonlinear problems or dynamic ones, it is usually coupled to iterative procedures. Solver efficiency often determines the overall performance of the program. It is why solvers are generally adapted to specific computer architectures (**Table 19**).

Eigenvalue Extraction

The Lanczos method is widely used to compute natural modes and frequencies of large systems. The block Lanczos method is a variant which iterates with a set of vectors instead of a single one. When extracting many modes, the efficiency of the technique is

Table 19 Equation solvers

Solver	ABAQUS	ANSYS	COSMOS	NASTRAN	SAMCEF
Bandwidth reduction	√	√	√	√	√
Iterative (PCG)		√	√	√	√
Frontal (wavefront solver)	√	√		√	√
Multifrontal	√	√	√	√	
Augmented Lagrangian					√
Sparse storage	√	√	√	√	
Nonsymmetric	√	√		√	√
Complex		√	√	√	√
Parallel (shared memory)	√	√	√	√	√
Parallel (distributed memory)	√			√	
Vector processing	√			√	

improved by operating several spectral shifts in order to localize the frequencies in successive subintervals (**Table 20**).

Nonlinear Equilibrium

To converge to an equilibrium at each load increment (statics) or at each time step (dynamics), iterative procedures are used that are based on the well-known Newton or tangent method (**Table 21**). In statics, the loads are usually applied step by step; when there is a risk of buckling or snap-through, increments must be applied according to a continuation method (arc length, Riks, etc.). In dynamics, the size of the time step conditions the size of load increments. In the general case, the loads depend on the configuration reached by the structure and must be computed at each iteration.

Time Integration

In transient dynamics, the displacements are obtained by integrating the equations of motions (**Table 22**). Schemes of the Newmark family are the most widely used. Since the range of excited frequencies may change during the response according to changes in the loading or in the structure itself, automatic time stepping may dramatically reduce the CPU time for a given accuracy.

Substructuring

Substructuring is a technique used to condense the degrees of freedom of a linear structure (**Table 23**). Stiffness, mass, and other matrices and loads are reduced from their initial size to a smaller number of degrees of freedom. Analyses are performed using condensed matrices and results are recovered *a pos-*

Table 20 Eigenvalue algorithms

Algorithm	ABAQUS	ANSYS	COSMOS	NASTRAN	SAMCEF
Lanczos	√	√	√	√	√
Block Lanczos	√	√	√	√	√
Subspace iteration	√	√	√		
Inverse power			√	√	
Nonsymmetric				√	√
Complex		√	√	√	

Table 21 Nonlinear solution techniques

Nonlinear iterations	ABAQUS	ANSYS	COSMOS	NASTRAN	SAMCEF
Newton–Raphson	√	√	√	√	√
Modified Newton–Raphson	√	√	√	√	√
Quasi-Newton	√		√	√	
Line search	√	√	√	√	√
Continuation method	√	√		√	√
Automatic convergence control	√	√	√	√	√
Automatic static stepping	√	√	√	√	√

COMPARISON OF VIBRATION PROPERTIES/Comparison of Spatial Properties

Table 22 Time integration

Time integration	ABAQUS	ANSYS	COSMOS	NASTRAN	SAMCEF
Central difference	√	√	√		
Hilber–Hughes–Taylor	√				√
Newmark	√	√	√	√	√
Automatic time stepping	√	√	√	√	√
Restarts	√	√	√	√	√

Table 23 Substructuring

Substructuring	ABAQUS	ANSYS	COSMOS	NASTRAN	SAMCEF
Guyan reduction	√	√	√	√	√
Craig–Bampton reduction		√		√	√
MacNeal reduction			√	√	
Superelement scaling	√		√		√
Superelement rotation	√	√	√	√	√
Superelement reflection	√	√	√	√	
Prestressed superelement created after a linear or nonlinear analysis	√				√
Use of superelement in nonlinear analysis (small rotations)	√	√		√	√
Use of superelement in nonlinear analysis (large rotations)		√			√

teriori for the whole set of degrees of freedom. The method is statically exact but gives approximate results in dynamics. The condensed matrices form a superelement which is useful to export a model from one software to another. It is possible to impose large rotations to a superelement in a nonlinear analysis.

See also: **Computation for transient and impact dynamics; Eigenvalue analysis; Finite element methods; Krylov-Lanczos methods; Nonlinear systems analysis; Structural dynamic modifications.**

Further Reading

Adams ML Jr (1999) *Rotating Machinery Vibration. From Analysis to Troubleshooting.* New York: Marcel Dekker.

Craveur J-C (1996) *Modélisation des Structures.* Paris: Masson.

Hinton E (ed.) (1992) *Introduction to Nonlinear Finite Element Analysis.* Glasgow: Nafems.

Martin J-P (1987) *La qualité des Logiciels. Du Bricolage à l'Industrialisation.* Paris: Afnor Gestion.

Robert G (1993) Finite element quality control. In: *Proceedings of the Fourth International Conference on Quality Assurance*, pp. 279–290. Glasgow: NAFEMS.

SFM (1990) *Guide de Validation des Progiciels de Calcul des Structures.* Paris: Afnor Technique.

COMPARISON OF VIBRATION PROPERTIES

Comparison of Spatial Properties

M Radeş, Universitatea Politehnica Bucuresti, Bucuresti, Romania

Copyright © 2001 Academic Press

doi:10.1006/rwvb.2001.0162

Spatial properties of structural dynamic models are presented in the form of mass, stiffness, and damping matrices. For gyroscopic systems, such as rotor-bearing systems, one can add gyroscopic and circulatory matrices. The static flexibility matrix, i.e., the inverse of the stiffness matrix, is also of interest because it is measurable.

The mass and stiffness matrices can have the full dimension of the finite element model (FEM) or can be reduced to the size of the test analysis model (TAM), for comparisons based on experimental data. The TAM is a reduced representation of the structure, whose degrees-of-freedom (DOFs) correspond to the instrumented DOFs in modal testing.

A comparison of spatial properties is, in principle, desirable, because it reveals not only the extent of discrepancy between analytical and test-derived models, but also the areas on the structure where these errors might originate.

However, experimentally-derived models are not only truncated or incomplete, but are also nonunique, being computed either from an intermediate truncated modal model, or directly from the small-size response model. The result is that any direct comparison of analytical and test-derived spatial properties at the full size of the FEM is meaningless.

One can consider the comparison of reduced spatial properties, i.e., of the mass and stiffness matrices reduced to the size of the TAM, or of their inverses. Again, a direct comparison is inconclusive. Instead, the experimental mode shapes can be used either to correct the reduced analytical mass and stiffness matrices, or to assess their goodness based on orthogonality constraints.

Comparison of Reduced Mass Matrices

It is useful to compare TAM mass matrices, M_{TAM}, obtained by different model reduction methods, either on the same test modal vectors, or on reduced analytical modal vectors. The comparison is based on the orthogonality of the reduced mass matrices with respect to either the test or the analytical modal vectors.

The test orthogonality matrix (TOR), defined as:

$$\mathbf{TOR} = (\mathbf{\Phi}_{TEST})^T \mathbf{M}_{TAM} \mathbf{\Phi}_{TEST}$$

is used as a measure of the robustness of the TAM reduction method, i.e., the ability of the TAM to provide TOR matrices that resemble the identity matrix, when the FEM has inaccuracies. The matrix $\mathbf{\Phi}_{TEST}$ contains the measured modal vectors as columns.

The generalized mass matrix is:

$$\mathbf{GM} = (\mathbf{\Phi}_{FEM}^R)^T \mathbf{M}_{TAM} \mathbf{\Phi}_{FEM}^R$$

where $\mathbf{\Phi}_{FEM}^R$ is the matrix of target analysis modes reduced to the measured DOFs. It is used as an indicator of the goodness of the TAM mass matrix, revealing errors in the TAM due to the reduction process. It should approximate to the identity matrix. The cross-orthogonality matrix (XOR):

$$\mathbf{XOR} = (\mathbf{\Phi}_{FEM}^R)^T \mathbf{M}_{TAM} \mathbf{\Phi}_{TEST}$$

is a less stringent check of robustness, since the test modal vectors are only used once in the calculation. Note that the TAM accuracy, i.e., its ability to predict the dynamic response of the structure to operating environments, is assessed by comparison of modal frequencies and mode shapes, i.e., of test and TAM modal properties.

Reduced Mass Matrices

In the reduction process, the TAM degrees-of-freedom (DOFs) are referred to as the active (a) DOFs. The other DOFs used in the FEM are called omitted (o) DOFs.

If the FEM coordinate vector, $\mathbf{u}$, is partitioned as:

$$\mathbf{u} = \begin{Bmatrix} \mathbf{u}_a \\ \mathbf{u}_o \end{Bmatrix}$$

where $\mathbf{u}_a$ contains the active DOFs and $\mathbf{u}_o$ contains the omitted DOFs, the transformation from the full set of FEM DOFs to the TAM DOFs can be defined as:

$$\mathbf{u} = \mathbf{T}\mathbf{u}_a$$

The transformation matrix, $\mathbf{T}$, is used to calculate the reduced mass and stiffness matrices:

$$\mathbf{M}_A^R = \mathbf{T}^T \mathbf{M}_A \mathbf{T} \qquad \mathbf{K}_A^R = \mathbf{T}^T \mathbf{K}_A \mathbf{T}$$

where $\mathbf{M}_A$ and $\mathbf{K}_A$ are the full-size analytical FEM matrices, usually partitioned according to the a and o sets as:

$$\mathbf{M}_A = \begin{bmatrix} M_{aa} & M_{ao} \\ M_{oa} & M_{oo} \end{bmatrix} \qquad \mathbf{K}_A = \begin{bmatrix} K_{aa} & K_{ao} \\ K_{oa} & K_{oo} \end{bmatrix}$$

The form of the transformation matrix, $\mathbf{T}$, depends on the reduction method. Four of the most popular reduction methods are presented below.

Guyan–Irons Reduction (Static Condensation)

In the Guyan–Irons reduction, the displacements in the o set are described by a set of static shapes called constraint modes. They are calculated so as to minimize the total strain energy. The transformation matrix has the form:

$$\mathbf{T}_S = \begin{bmatrix} \mathbf{I} \\ -\mathbf{K}_{oo}^{-1}\mathbf{K}_{oa} \end{bmatrix} = \begin{bmatrix} \mathbf{I}_a \\ \mathbf{G}_{oa} \end{bmatrix}$$

where $\mathbf{G}_{oa}$ enters the constraint equation between the a and o DOFs $\mathbf{u}_o = \mathbf{G}_{oa}\mathbf{u}_a$ and $\mathbf{I}_a$ is the identity matrix of size equal to the number of a DOFs.

Improved Reduced System

The improved reduced system (IRS) method is an improved static condensation, involving minimization of both strain energy and the potential energy of applied forces. It includes a static approximation for the inertia terms discarded in the static condensation.

The IRS transformation matrix is:

$$\mathbf{T}_{IRS} = \begin{bmatrix} \mathbf{I}_a \\ \mathbf{G}_{oa} + \mathbf{K}_{oo}^{-1}(\mathbf{M}_{oa} + \mathbf{M}_{oo}\mathbf{G}_{oa})\mathbf{M}_a^{-1}\mathbf{K}_a \end{bmatrix}$$

where:

$$\mathbf{M}_a = \mathbf{T}_S^T \mathbf{M}_A \mathbf{T}_S \quad \mathbf{K}_a = \mathbf{T}_S^T \mathbf{K}_A \mathbf{T}_S$$

are the condensed mass and stiffness matrices obtained using the Guyan–Irons reduction.

Modal TAM

Consider the full-square FEM mode shape matrix partitioned in the form:

$$\mathbf{\Phi} = \begin{bmatrix} \mathbf{\Phi}_a \\ \mathbf{\Phi}_o \end{bmatrix} = [\mathbf{\Phi}_t \quad \mathbf{\Phi}_r] = \begin{bmatrix} \mathbf{\Phi}_{at} & \mathbf{\Phi}_{ar} \\ \mathbf{\Phi}_{ot} & \mathbf{\Phi}_{or} \end{bmatrix}$$

where the subscript t denotes the target modes, i.e., the FEM mode shapes used in the reduction procedure, and the subscript r denotes the residual modes, i.e., all the other analytical modes. Note that $\mathbf{\Phi}_t = \mathbf{\Phi}_A$ and $\mathbf{\Phi}_{at} = \mathbf{\Phi}_A^R$ (see **Comparison of vibration properties: Comparison of modal properties**).

The modal TAM transformation matrix, $\mathbf{T}_M$, is given by:

$$\mathbf{T}_M = \begin{bmatrix} \mathbf{I}_a \\ \mathbf{\Phi}_{ot}\mathbf{\Phi}_{at}^+ \end{bmatrix}$$

where the superscript $+$ indicates the pseudoinverse. The matrix $\mathbf{\Phi}_{at}$ has more rows than columns and is full-column rank. The modal TAM provides an exact reduction provided that there are no modeling errors in the FEM. The modal TAM is a variant of the more general system-equivalent reduction/expansion process (SEREP).

Hybrid TAM

An improved TAM can be derived from the modal TAM by incorporating static modes with the target modes. The hybrid TAM transformation matrix, $\mathbf{T}_H$, is given by:

$$\mathbf{T}_H = \mathbf{T}_S + (\mathbf{T}_M - \mathbf{T}_S)\mathbf{\Phi}_{at}\mathbf{\Phi}_{at}^T\mathbf{M}_{at}$$

where $\mathbf{M}_{at}$ is the modal TAM mass matrix for the target modes satisfying the orthogonality relationship:

$$\mathbf{\Phi}_{at}^T \mathbf{M}_{at} \mathbf{\Phi}_{at} = \mathbf{I}_t$$

where $\mathbf{I}_t$ is the identity matrix of order equal to the number of target modes. The hybrid TAM combines the accuracy of the modal TAM with the robustness of the static TAM.

Test Orthogonality Check

An indirect comparison of the TAM mass matrices, obtained by the four reduction methods presented above, can be made based on the test mode shapes and the mixed orthogonality check TOR. A typical result is illustrated in **Figures 1–4**, presenting the TOR matrices calculated from four different reduced mass matrices. The reference FEM of the structure, having about 45 000 DOFs, has been reduced to a 120-DOFs TAM, having 15 flexible natural modes between 5.2 and 34.3 Hz.

The largest off-diagonal terms occur in the TOR matrix of the modal TAM (**Figure 3**), especially for the higher residual modes. This can be an indication that the spatial resolution given by the selected response measurement points (the a set) is insufficient to make the target modes linearly independent and observable – an important outcome of such a comparison process.

The hybrid TAM shows a slight improvement on the modal TAM, due to the inclusion of static modes with the target modes (**Figure 4**). Surprisingly, the static TAM performs better than the modal TAM, showing smaller off-diagonal terms (**Figure 1**). The IRS TAM yields the best reduced-mass matrix, producing the lowest off-diagonal elements in the TOR matrix (**Figure 2**).

While the modal TAM gives the best match in frequencies and target modes, its prediction capability is low outside the frequency range spanned by the selected target modes. The static TAM, implemented as the Guyan reduction in many computer programs, performs better in orthogonality checks, but is dependent on the selection of a DOFs and generally requires more a DOFs to give comparable accuracy. However, these types of comparisons are usually problem-dependent.

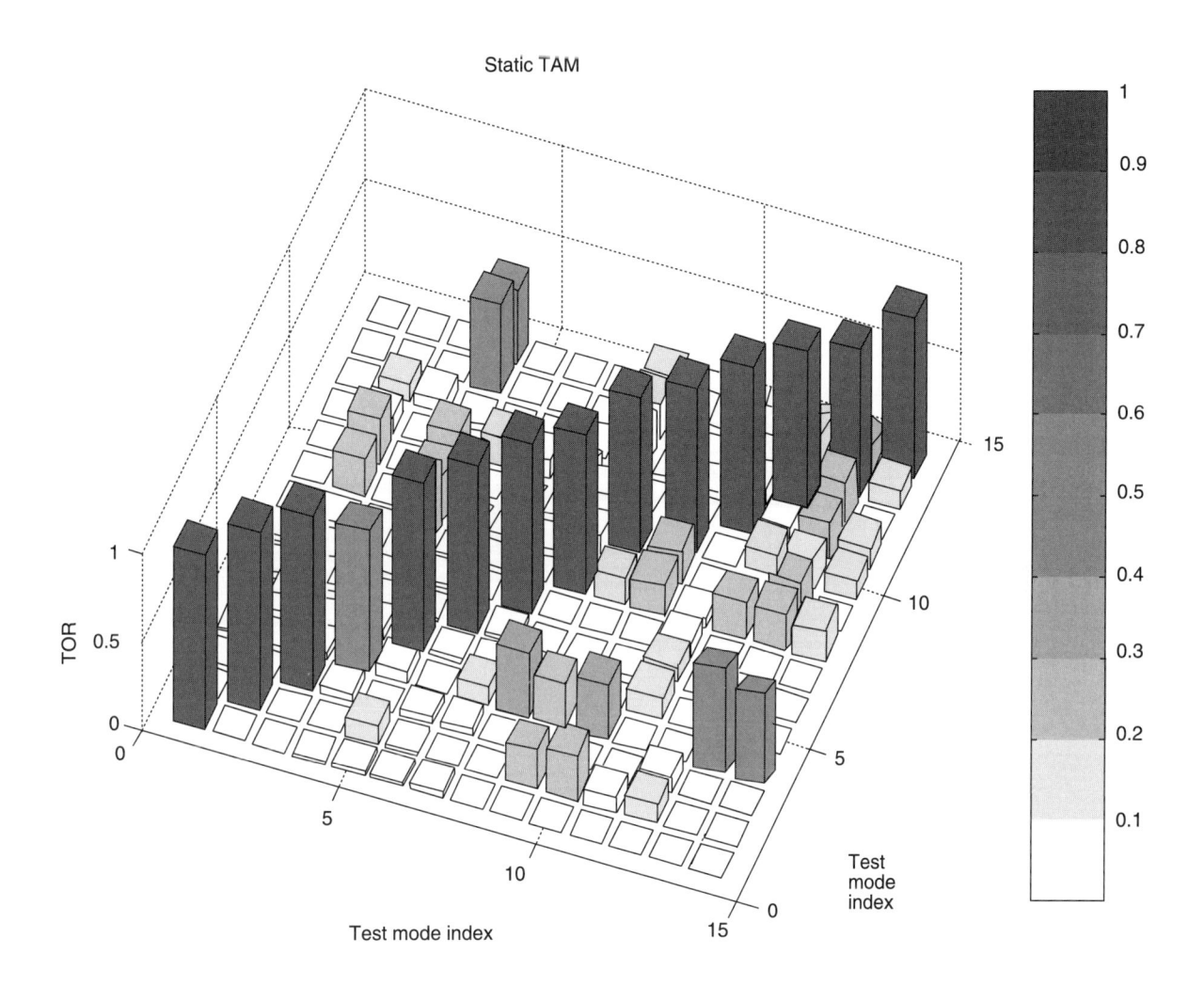

Figure 1

Mass and Stiffness Error Matrices

Different techniques can be used to update the mass and stiffness matrices, based on an incomplete set of modal frequencies and mode shapes derived from measurements. Assuming that the measured modal vectors are correct, the mass matrix can be updated by minimization of the cost function:

$$J_M = \left\| M_A^{-1/2}(M_U - M_A) M_A^{-1/2} \right\|$$

subject to the orthogonality constraint:

$$(\Phi_X^E)^T M_U \Phi_X^E = I$$

where M_A is the original estimate of the full-size FEM mass matrix, M_U is the updated mass matrix, and Φ_X^E is the matrix of expanded modal vectors. The weighting by $M_A^{-1/2}$ allows for differences in the magnitudes of the elements of the two mass matrices.

A 3D plot of the error matrix:

$$\Delta M = M_U - M_A$$

helps to track the updating process and to localize parts of the structure where the inertia properties are poorly defined.

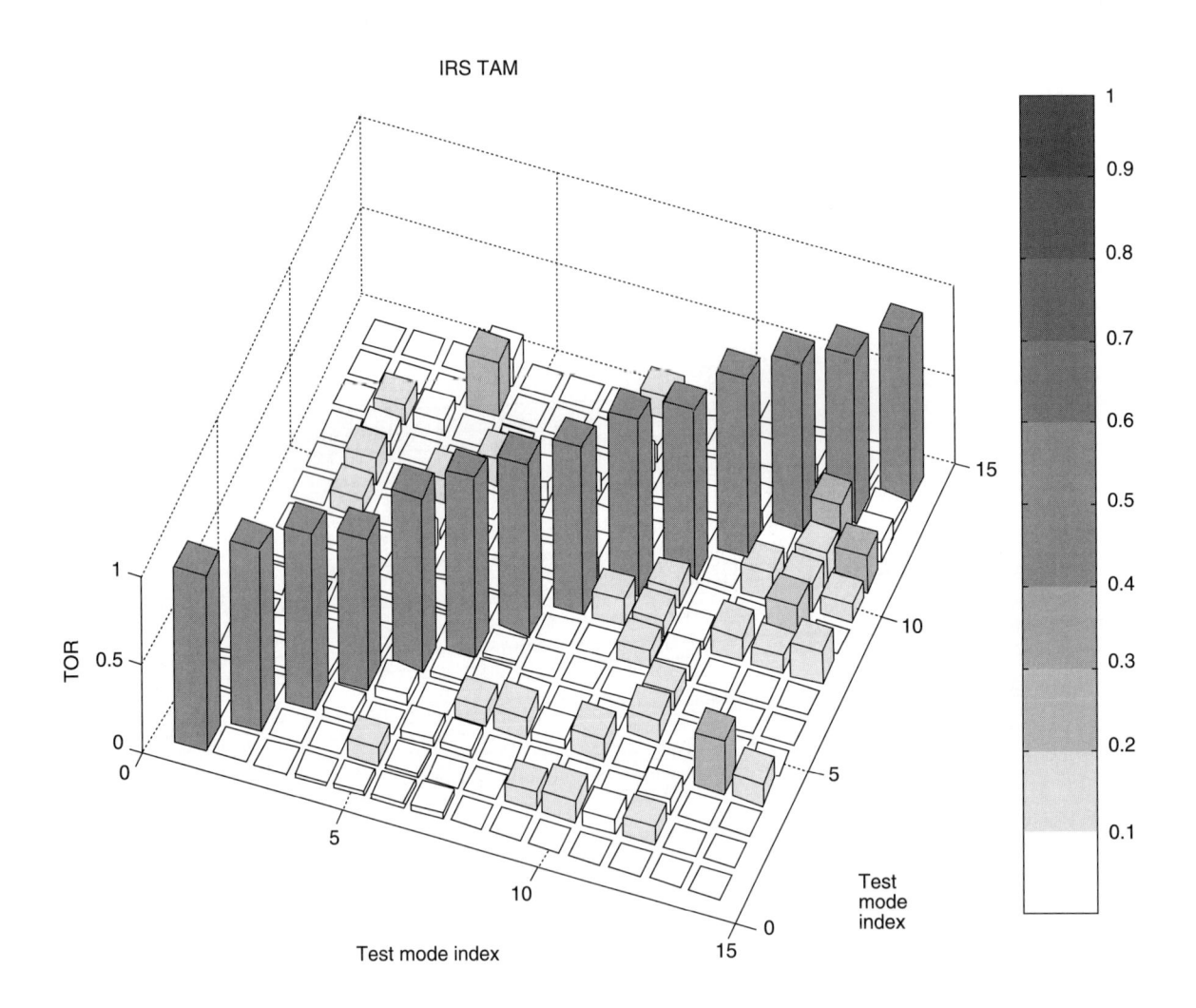

Figure 2

After the calculation of M_U, the stiffness matrix can be obtained by minimization of:

$$J_K = \left\| M_A^{-1/2} (K_U - K_A) M_A^{-1/2} \right\|$$

subject to constraints that enforce the stiffness symmetry, the stiffness orthogonality condition, and the equation of motion.

The stiffness error matrix:

$$\Delta K = K_U - K_A$$

can be presented in a graphical form like the mass error matrix (**Figure 5**).

It must be borne in mind that mode shape expansion can render the direct comparison of spatial properties meaningless. Direct updating methods are usually unable to preserve the connectivity of the structure. The updated matrices are usually fully populated, while analytical matrices are banded or sparse, so that local discrepancies have no physical meaning.

Early updating methods, referred to as error matrix methods, were conceived to estimate directly the

Modal TAM

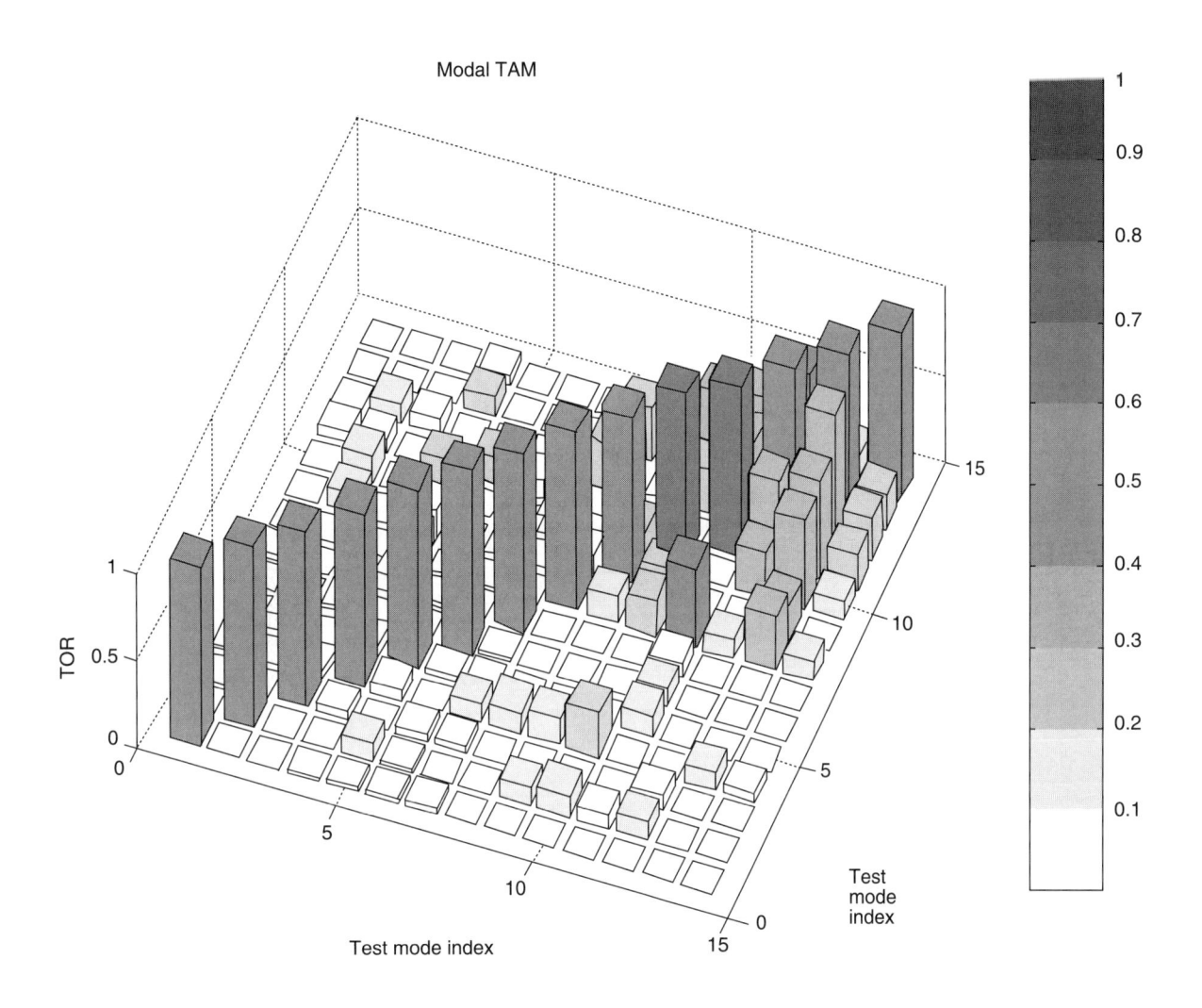

Figure 3

error in the system matrices, expressed as the difference between the measured and the predicted matrix.

Due to the incompleteness of experimental data, the stiffness error matrix:

$$\Delta \mathbf{K} = \mathbf{K}_X - \mathbf{K}_A$$

is calculated either as:

$$\Delta \mathbf{K} \cong \mathbf{K}_A (\mathbf{K}_A^{-1} - \mathbf{K}_X^{-1}) \mathbf{K}_A$$

or, using modal data to express the pseudoflexibility matrices, as:

$$\Delta \mathbf{K} = \mathbf{K}_A \left(\mathbf{\Phi}_A \left[\frac{1}{\omega_A^2} \right] \mathbf{\Phi}_A^T - \mathbf{\Phi}_X \left[\frac{1}{\omega_X^2} \right] \mathbf{\Phi}_X^T \right) \mathbf{K}_A$$

where $\mathbf{\Phi}_X$ is the reduced mass-normalized mode shape matrix from tests and $\mathbf{\Phi}_A$ is calculated with the corresponding modes from the FEM. A similar approach, applied to the mass matrix, yields the mass error matrix:

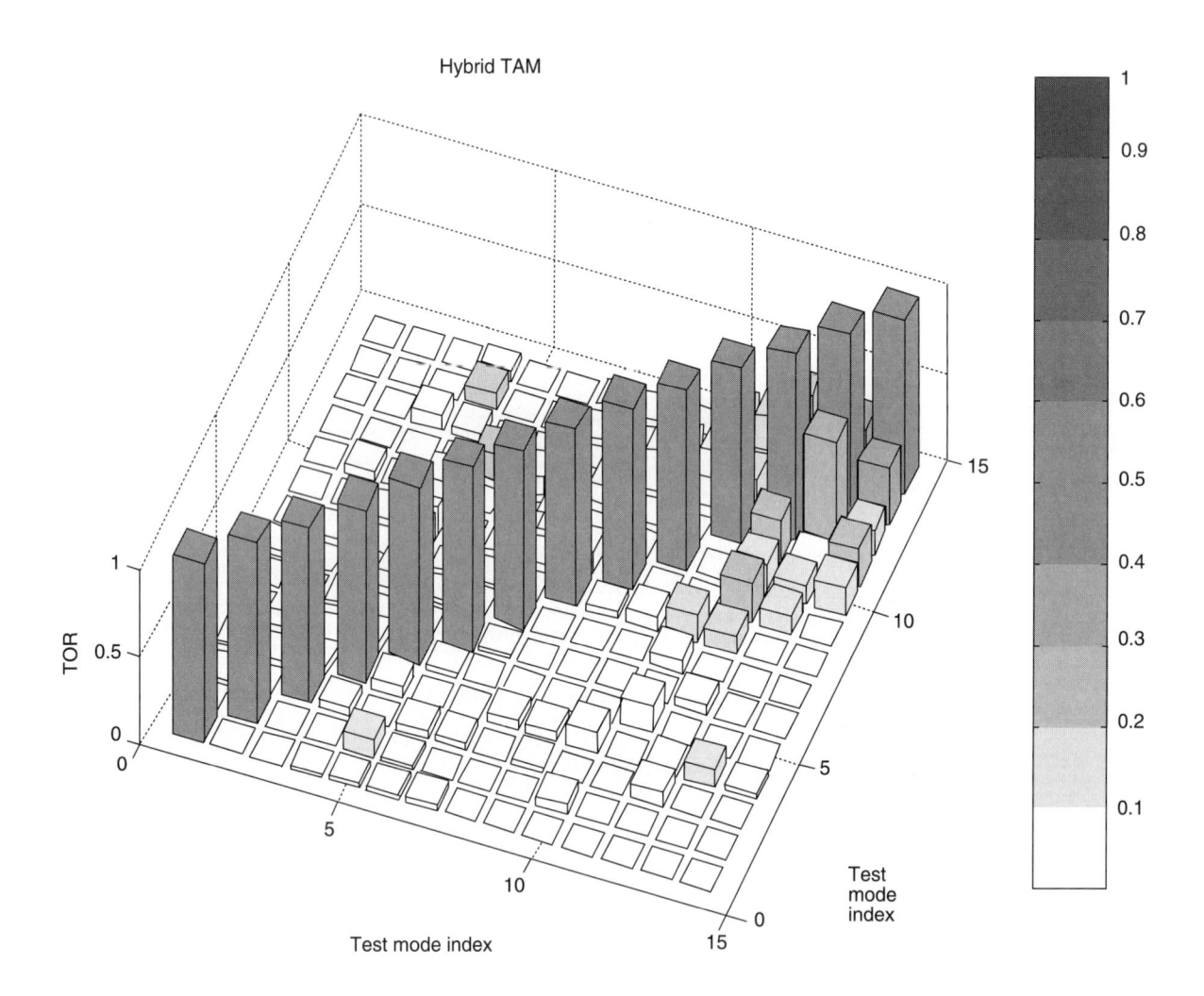

Figure 4

$$\Delta \mathbf{M} = \mathbf{M}_A \left(\mathbf{\Phi}_A \mathbf{\Phi}_A^T - \mathbf{\Phi}_X \mathbf{\Phi}_X^T \right) \mathbf{M}_A$$

The stiffness error matrix has been presented in graphical form using the alternative mesh format from **Figure 6**.

For large FEM, plots like those shown in **Figures 5** and **6** become impractical. Comparisons of spatial properties can be carried out at the reduced size of the TAM. They are only instruments to assess the accuracy of a given procedure and should not be used to locate inconsistencies in the compared models because the reduction process destroys the connectivity of the structure.

Finally, it should be emphasized that in any test/analysis correlation there are three main sources of errors: (1) FEM inaccuracies, (2) test mode shape measurement and identification errors, and (3) reduction/expansion errors. Additionally, errors due to the real normalization of measured complex modes must be taken into account.

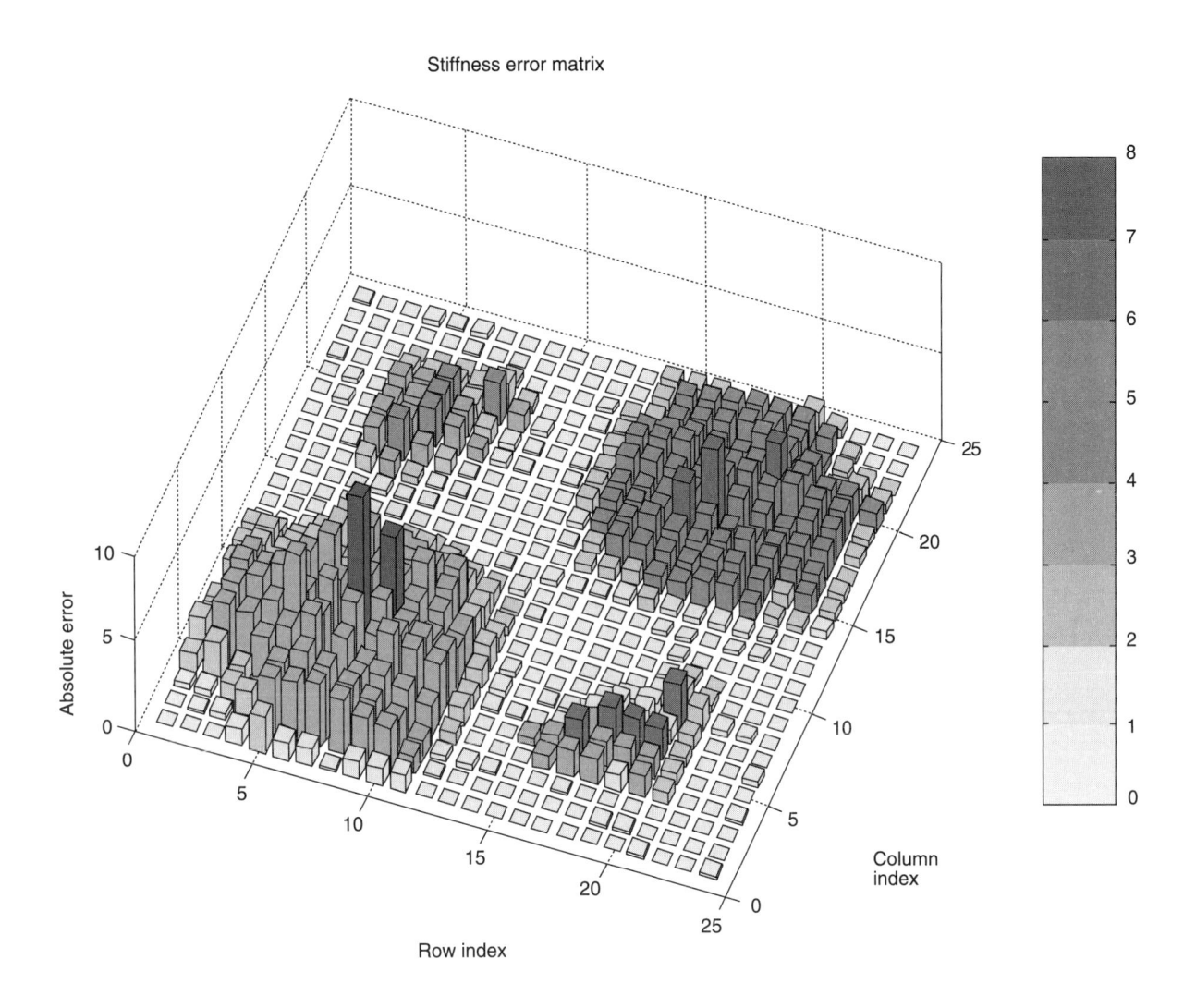

Figure 5

Nomenclature

a	active
GM	generalized mass matrix
I	identity matrix
M	mass matrix
o	omitted
r	residual
t	target
T	transformation matrix
TOR	test orthogonality matrix
XOR	cross orthogonality matrix
ΔK	stiffness error matrix
ΔM	mass error matrix

See also: **Comparison of vibration properties**: Comparison of modal properties; **Comparison of vibration properties**: Comparison of response properties; **Model updating and validating**.

Stiffness error matrix

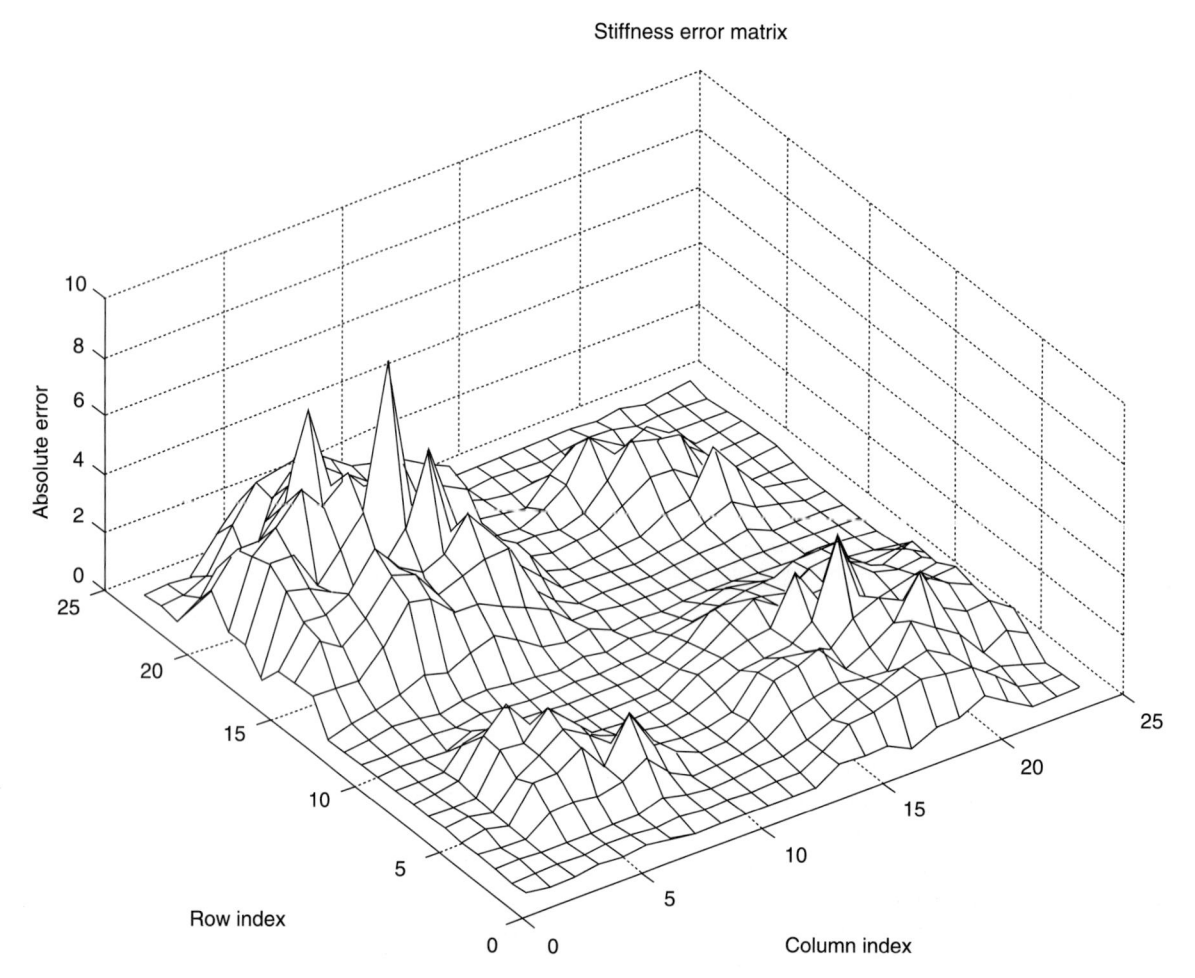

Figure 6

Further Reading

Ewins DJ (2000) *Modal Testing: Theory, Practice and Application*, 2nd edn. Baldock, UK: Research Studies Press.

Freed AM and Flanigan ChC (1991) A comparison of test-analysis model reduction methods. *Sound and Vibration* 25: 30–35.

Friswell MI and Mottershead JE (1995) *Finite Element Model Updating in Structural Dynamics*. Dordrecht: Kluwer.

Guyan RJ (1965) Reduction of stiffness and mass matrices. *AIAA Journal* 3: 380.

Irons BM (1963) Eigenvalue economisers in vibration problems. *Journal of the Royal Aeronautical Society* 67: 526–528.

Kammer DC (1987) Test-analysis model development using an exact model reduction. *International Journal of Analytical and Experimental Modal Analysis* xx: 174–179.

Kammer DC (1991) A hybrid approach to test-analysis modal development for large space structures. *Journal of Vibrations and Acoustics* 113: 325–332.

O'Callahan JC (1989) A procedure for an improved reduced system (IRS) model. *Proceedings of the 7th International Modal Analysis Conference*, Las Vegas, pp. 17–21.

O'Callahan JC (1990) Comparison of reduced model concepts. *Proceedings of the 8th International Modal Analysis Conference*, Kissimmee, Florida, pp. 422–430.

O'Callahan JC, Avitable P, Madden R and Lieu IW (1986) An efficient method of determining rotational degrees of freedom from analytical and experimental modal data. *Proceedings of the 4th International Modal Analysis Conference*, Los Angeles, California, USA, pp. 50–58.

O'Callahan JC, Avitable P and Riemer R (1989) System equivalent reduction expansion process (SEREP). *Proceedings of the 7th International Modal Analysis Conference*, Las Vegas, Nevada, USA, pp. 29–37.

Comparison of Modal Properties

M Radeş, Universitatea Politehnica Bucuresti, Bucuresti, Romania

Copyright © 2001 Academic Press

doi:10.1006/rwvb.2001.0174

Modal properties that are compared usually include: natural frequencies, real mode shape vectors, modal masses, modal kinetic and strain energies. For systems with complex modes of vibration one can add modal damping ratios and complex mode shapes. Left-hand modal vectors, modal participation factors, and reciprocal modal vectors are also considered in some applications. A test-analysis comparison is meaningful only for matched modes, i.e., for correlated mode pairs (CMPs). These are estimates of the same physical mode shape and their entries correspond one-to-one with their counterparts. Mode matching (pairing) is an essential step before any comparison can be undertaken.

In order to make it possible to compare experimental and finite element method (FEM) results, a reduced test-analysis model (TAM) is often used. This is represented by the mass and stiffness matrices computed for the test degrees-of-freedom (DOFs) only. Comparison of modal vectors can be done at the reduced order of the TAM or at the full order of the FEM. Reduction of the physical mass matrix or expansion of test modal vectors bring inherent approximations in the comparison criteria.

There are three main kinds of comparison: (1) analytical-to-analytical (FEM-to-FEM, TAM-to-TAM, and TAM-to-FEM); (2) experimental-to-experimental; and (3) analytical-to-experimental. The third type will be considered in more detail below.

It is useful to compare: (1) measured mode shapes against modal vectors determined by an analytical model; (2) estimates of the same test modal vector obtained from different excitation locations; (3) estimates of the same modal vector obtained from different modal parameter identification processes using the same test data; (4) one test mode shape before and after a change in the physical structure caused by a wanted modification, by damage or by operation over time.

Direct Numerical Comparison

Scalar quantities, such as natural frequencies, modal damping ratios, norms of modal vectors, modal masses, and modal energies are usually compared by simple tabulation of the relative error. Simple comparison of two columns of natural frequencies is meaningless without prior mode matching based on mode shape data. Mode pairing is done using the modal assurance criterion (MAC) described in a later paragraph. If there is one-for-one correspondence between the rth and the qth modes, then the following discrepancy indicators are used:

the relative modal frequency discrepancy:

$$\varepsilon_\omega = \frac{|\omega_r - \omega_q|}{\omega_r} \cdot 100$$

the relative modal damping discrepancy:

$$\varepsilon_\zeta = \frac{|\zeta_r - \zeta_q|}{\zeta_r} \cdot 100$$

the relative mode shape norm difference:

$$\varepsilon_\psi = \text{abs} \frac{|\boldsymbol{\psi}_r^H \boldsymbol{\psi}_r| - |\boldsymbol{\psi}_q^H \boldsymbol{\psi}_q|}{|\boldsymbol{\psi}_r^H \boldsymbol{\psi}_r|} \cdot 100$$

For some aerospace structures, having about 20 flexible modes up to 50 Hz, modeling accuracy criteria typically specify values $\varepsilon_\omega \leq 5\%$. Comparative values for mode shape and damping ratio discrepancies are $\varepsilon_\psi \leq 10\%$ and $\varepsilon_\zeta \leq 25\%$, respectively.

Direct Graphical Comparison

A straightforward way to compare two compatible sets of data is by making an X–Y plot of one data set against the other. The method can be used to compare the natural frequencies from two different models. For well correlated data, the points of the resulting diagram should lie close to a straight line of slope equal to 1. If the approximating straight line has a slope different from 1, this indicates a bias error due to either calibration or erroneous material property data. Large random scatter about a 45° line indicates poor correlation or bad modeling.

The procedure can be applied to the mode shapes of correlated mode pairs. Each element of a test mode shape is plotted against the corresponding element of the analytical modal vector. For consistent correspondence, the points should lie close to a straight line passing through the origin. If both modal vectors are mass-normalized, then the approximating line has a slope of ± 1.

Modal Scale Factor

If the two mode shape vectors have different scaling factors, it is useful to determine the slope of the best line through the data points. This is calculated as the least squares error estimate of the proportionality constant between the corresponding elements of each modal vector.

For real vectors, it is a real scalar referred to as the modal scale factor (MSF), defined as:

$$\text{MSF}(r, q) = \frac{\psi_r^T \psi_q}{\psi_r^T \psi_r}$$

where ψ_r be the test vector and ψ_q the compatible analytical vector. For complex vectors, the superscript T is replaced by H (hermitian) and the MSF is a complex scalar.

The MSF gives no indication on the quality of the fit of data points to the straight line. Its function is to provide a consistent scaling factor for all entries of a modal vector. It is a normalized estimate of the modal participation factor between two excitation locations for a given modal vector.

Orthogonality Criteria

The most relevant way to assess the validity of a set of modal vectors is the orthogonality check. In order to use orthogonality checks, it is necessary to compute: (1) an FEM of the tested structure; (2) an analytical mass matrix reduced to the test DOFs; and/or (3) a set of test mode shapes expanded to the FEM DOFs.

The mass-orthogonality properties of FEM real mode shapes can be written as:

$$\Phi_A^T M_A \Phi_A = I$$

where Φ_A is the modal matrix at the full FEM order, M_A is the FEM full-order mass matrix and I is the identity matrix.

For the TAM, the mass-orthogonality condition becomes

$$\Phi_A^{RT} M_A^R \Phi_A^R = I$$

where M_A^R is the reduced TAM mass matrix (see **Comparison of vibration properties: Comparison of spatial properties**) and Φ_A^R contains the modal vectors reduced to the measured DOFs and mass-normalized with respect to M_A^R.

Modal Auto-orthogonality

A mixed orthogonality test of the set of measured modal vectors Φ_X is often done to check the quality of the measurement data. The test orthogonality (TOR) matrix is defined as:

$$\text{TOR} = \Phi_X^T M_A^R \Phi_X$$

If the measured modal vectors are orthogonal and mass-normalized with respect to the reduced mass matrix M_A^R, then TOR will be the identity matrix. Test guidelines specify $\text{TOR}_{rr} = 1.0$ and $|\text{TOR}_{rq}| < 0.1$.

Modal Cross-Orthogonality

A cross-orthogonality test is performed to compare the paired modal vectors, the measured ones with the analytical vectors.

A cross-generalized mass (CGM) matrix, defining a cross orthogonality (XOR) criterion, can be constructed with mass-normalized modal vectors either at the TAM size:

$$\text{XOR}_{\text{TAM}} = \Phi_X^T M_A^R \Phi_A^R$$

or at the full FEM size:

$$\text{XOR}_{\text{FEM}} = \Phi_X^{ET} M_A \Phi_A$$

where Φ_X^E contains the test modal vectors expanded to the full FEM DOFs.

For perfect correlation, the leading diagonal elements XOR_{rr} should be larger than 0.9, while the off-diagonal entries $|\text{XOR}_{rq}|$ should be less than 0.1. Use of the reduced mass matrix M_A^R raises problems. One must differentiate reduction errors from discrepancies between the FEM and the test model.

When the reduced TAM mass matrix is obtained by the system equivalent reduction expansion process (SEREP) method, XOR is referred to as a pseudo-orthogonality criterion (POC). It is demonstrated that $\text{POC}_{\text{TAM}} = \text{POC}_{\text{FEM}}$. In this case, the full FEM mass matrix is not needed to compute either M_A^R or POC_{TAM} because:

$$M_A^R = \left(\Phi_A^{R+}\right)^T \Phi_A^{R+}$$

where the superscript + denotes the pseudoinverse and Φ_A^R is the rectangular matrix of analytical target modes reduced at the test DOFs.

An average measure of the correspondence between the compared modal vectors is given by the RMS value of either the off-diagonal elements in the **XOR** matrix or of the entries in the matrix calculated as the difference between the **XOR** matrix and the identity matrix. If only the diagonal elements are considered, then the mean deviation from one of these elements:

$$\text{mean}|XOR - 1| = \frac{1}{L}\sum_{l=1}^{L}\left|(\boldsymbol{\phi}_X^E)_l^T \mathbf{M}_A (\boldsymbol{\phi}_A)_l - 1\right|$$

where L is the number of CMPs, gives an indication that the modal expansion fails to provide physically sound modal vectors.

Cross-orthogonality criteria cannot locate the source of discrepancy in the two sets of compared mode shapes. Large off-diagonal elements in the cross-orthogonality matrices may occur simply because they are basically small differences of large numbers. Also, modes having nearly equal frequencies may result in (linear combinations of) analysis modes rotated with respect to the test modes, case in which the off-diagonal elements of **XOR** are skew-symmetric.

Modal Vector Correlation Coefficients

The Modal Assurance Criterion

One of the most popular tools for the quantitative comparison of modal vectors is the modal assurance criterion (MAC). It was originally introduced in modal testing in connection with the MSF, as an additional confidence factor in the evaluation of a modal vector from different excitation locations.

When an FRF matrix is expressed in the partial fraction expansion form, the numerator of each term represents the matrix of residues or modal constants. Each residue matrix is proportional to the outer product of one modal vector and the corresponding vector of the modal participation factors. Each column of the residue matrix is proportional to the respective modal vector. One can obtain estimates of the same modal vector from different columns of the residue matrix. MAC has been introduced as a measure of consistency and similarity between these estimates.

If the elements of the two vectors are used as coordinates of points in an $X - Y$ plot, the MAC represents the normalized least squares deviation of corresponding vector entries from the best straight line fitted to the data, using the MSF. The concept can be applied to the comparison of any pair of compatible vectors. The MAC is calculated as the scalar quantity:

$$\text{MAC}(r,q) = \frac{\left|\boldsymbol{\psi}_r^H \boldsymbol{\psi}_q\right|^2}{\left(\boldsymbol{\psi}_r^H \boldsymbol{\psi}_r\right)\left(\boldsymbol{\psi}_q^H \boldsymbol{\psi}_q\right)}$$

where the form of a coherence function can be recognized. An equivalent formulation is:

$$\text{MAC}(r,q) = \frac{\left|\boldsymbol{\psi}_r^T \boldsymbol{\psi}_q^*\right|^2}{\left(\boldsymbol{\psi}_r^T \boldsymbol{\psi}_r^*\right)\left(\boldsymbol{\psi}_q^T \boldsymbol{\psi}_q^*\right)}$$

where T denotes the transpose and * the complex conjugate. It has been introduced as a mode shape correlation constant, to quantify the accuracy of identified mode shapes. Note that the modulus in the numerator is taken after the vector multiplication, so that the absolute value of the sum of product elements is squared.

The MAC takes values between 0 and 1. Values larger than 0.9 indicate consistent correspondence whereas small values indicate poor resemblance of the two shapes. The MAC does not require a mass matrix and the two sets of vectors can be normalized differently. The division cancels out any scaling of the vectors.

If $\boldsymbol{\psi}_r$ and $\boldsymbol{\psi}_q$ are the rth and qth columns of the real modal matrix $\boldsymbol{\Psi}$, then, using the cross-product (Gram) matrix $\mathbf{G} = \boldsymbol{\Psi}^T\boldsymbol{\Psi}$, the MAC can be written as:

$$\text{MAC}(r,q) = \frac{G_{rq}^2}{G_{rr}G_{qq}} = \cos^2\theta_{rq}$$

where $G_{rq} = \boldsymbol{\psi}_r^T \boldsymbol{\psi}_q$ is the inner product and θ_{rq} is the angle between the two vectors. This equation has the form of a coherence function, indicating the causal relationship between $\boldsymbol{\psi}_r$ and $\boldsymbol{\psi}_q$. The MAC is also a measure of the squared cosine of the angle between the two vectors. It shows the extent to which the two vectors point in the same direction.

Nonconsistency, i.e. near zero MAC values, can be the result of system nonlinearity, nonstationarity, an invalid parameter identification algorithm, and orthogonal vectors. Consistency, i.e., near unity MAC values, can result from different scaling of the same vector, incompleteness of the measured vectors, testing with other sources of excitation than the desired one, and coherent noise.

MAC Matrix

Given two sets of compatible modal vectors, a MAC matrix can be constructed, each entry defining a certain combination of the indices of the vectors belonging to the two sets. The ideal MAC matrix cannot be a unit matrix because the modal vectors are not directly orthogonal, but mass-orthogonal (**Figure 1A**). However, the MAC matrix indicates which individual modes from the two sets relate to each other. If two vectors are switched in one set, then the largest entries of the MAC matrix are no more on the leading diagonal and it resembles a permutation matrix. The two large off-diagonal elements show the indices of the switched vectors, as illustrated in **Figure 1B**. **Figure 2** is the more often used form of **Figure 1B**.

The MAC can only indicate consistency, not validity, so it is mainly used in pretest mode pairing. The MAC is incapable of distinguishing between systematic errors and local discrepancies. It cannot identify whether the vectors are orthogonal or incomplete.

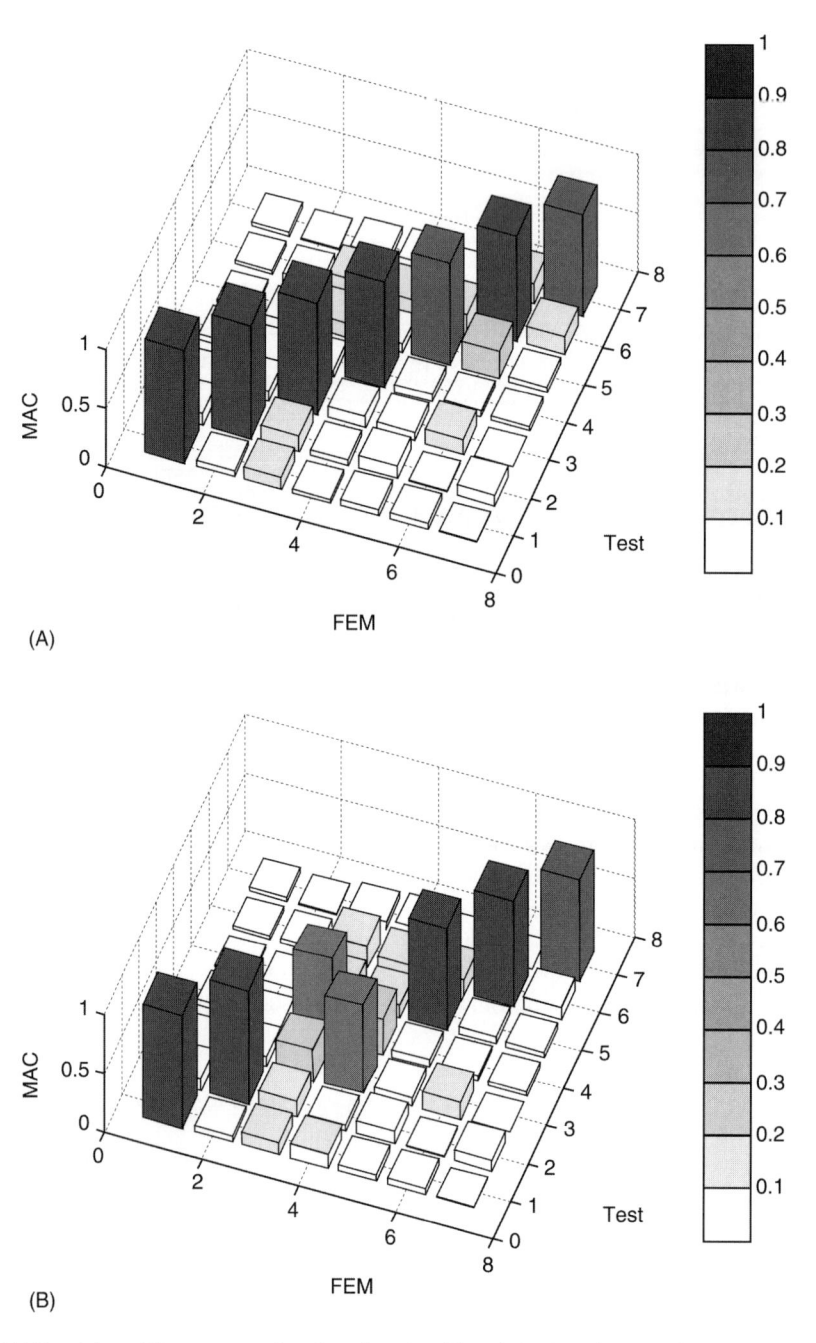

Figure 1 3D plots of MAC matrices: (A) seven correlated mode pairs, (B) switched modes 3 and 4.

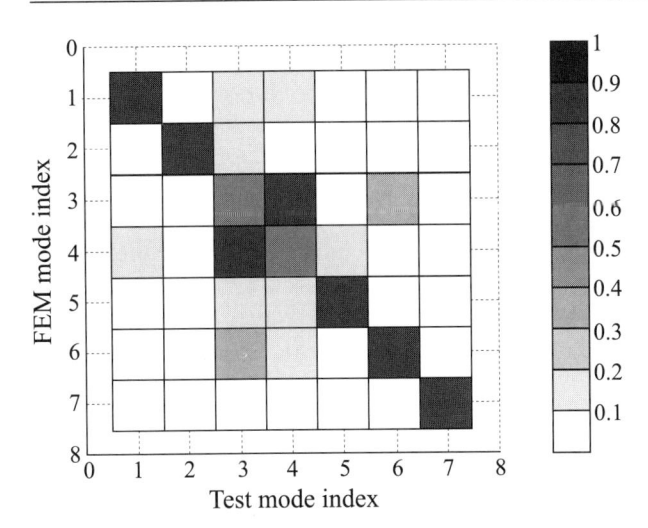

Figure 2 2D plot of MAC matrix.

An overall mode shape error indicator may be calculated from

$$\varepsilon_\phi = \left[1 - \frac{1}{L}\sqrt{\left(\sum_{l=1}^{L}(\mathrm{MAC})_l^2\right)}\right] \cdot 100$$

where L is the number of CMPs.

For axisymmetric structures that exhibit spatial phase shifts between test and analysis mode shapes, improved MAC values can be obtained by the rotation of mode shapes prior to correlation. For test mode shapes that contain multiple diametral orders, a special Fourier MAC criterion has been developed, using the first two primary Fourier indices.

The MAC is often used to assess the experimental vectors obtained by modal testing, especially when an analytical mass matrix is not available, and to compare test modal vectors with those calculated from FEM or TAM. The success of this apparent misuse is explained by two factors. First, test modal vectors usually contain only translational DOFs because rotational DOFs are not easily measured. If rotational DOFs were included in the modal vectors, the MAC would yield incorrect results. In this case it will be based on summations over vector elements of incoherent units, having different orders of magnitude. Second, for uniform structures, the modal mass matrix is predominantly diagonal and with not too different diagonal entries. In these cases, the MAC matrix is a good approximation for a genuine orthogonality matrix.

It should be underlined that a modal vector containing both translational and rotational DOFs cannot be simply scaled, and it has to be normalized with respect to the mass matrix. For example, if all linear displacements are multiplied by a factor of two, then the rotations are not doubled. It is the mass matrix that does the correct job, containing both masses and mass moments of inertia, which multiply linear and angular accelerations, respectively.

Normalized Cross-Orthogonality

A modified MAC, weighted by the mass or the stiffness matrix, referred to as the normalized cross-orthogonality (NCO) is defined as:

$$\mathrm{NCO}(r,q) = \frac{\left|\psi_r^H \mathbf{W} \psi_q\right|^2}{\left(\psi_r^H \mathbf{W} \psi_r\right)\left(\psi_q^H \mathbf{W} \psi_q\right)}$$

where the weighting matrix $\mathbf{W}$ can be either the mass or the stiffness matrix. In the first case, it is sensitive to local modes with high kinetic energy, in the second case it is sensitive to regions of high strain energy. Applying the NCO separately, using the analytical mass and stiffness matrices, it is possible to locate sources of inadequate modeling. However, one must be careful to differentiate inherent reduction errors from discrepancies between the FEM and test data.

The NCO is able to use two arbitrarily scaled modal vectors. It defines the CMPs more clearly than the MAC. The square root version of NCO is being also used as a cross-orthogonality check based on mass-normalized modal vectors.

AutoMAC

The AutoMAC addresses the spatial (or DOF) incompleteness problem. The MAC can show correlation of actually independent vectors. If the number of DOFs is insufficient to discriminate between different mode shapes, it is possible that one analytical modal vector can appear to be well correlated with several experimental vectors.

It is necessary to check if the number of DOFs included in the model is sufficient to define all linearly independent mode shapes of interest. This check can be done using the AutoMAC, which correlates a vector with itself based on different reduced DOF sets. Spatial aliasing is shown by larger-than-usual off-diagonal elements of the AutoMAC matrix.

Degree-of-Freedom Correlation

In the comparison of two sets of modal vectors, one of the issues of interest is the influence of individual DOFs on the vector resemblance. The spatial dependence of the previously presented correlation criteria can be misleading. On one side, unacceptable large off-diagonal terms in cross-orthogonality matrices can correspond to large errors in points of very small shape amplitude. On the other hand, very small off-diagonal elements of the XOR matrix do not necessarily indicate unrelated vectors.

A series of criteria have been developed to reveal the DOF dependence of the discrepancy between modal vectors. Their interpretation is not obvious and caution must be taken in their use as indicators of modeling accuracy.

Coordinate Modal Assurance Criterion

The coordinate MAC (COMAC) is used to detect differences at the DOF level between two modal vectors. The COMAC is basically a row-wise correlation of two sets of compatible vectors, which in MAC is done column-wise. The COMAC for the jth DOF is formulated as:

$$\text{COMAC}(j) = \frac{\left(\sum_{l=1}^{L} \left| (\phi_A)_{jl} (\phi_X)_{jl}^* \right| \right)^2}{\sum_{l=1}^{L} (\phi_A)_{jl}^2 \cdot \sum_{l=1}^{L} (\phi_X)_{jl}^{*2}}$$

where l is the index of the CMP, $(\phi_A)_{jl}$ is the jth element of the lth paired analytical modal vector, and $(\phi_X)_{jl}$ is the jth element of the lth paired experimental modal vector. Both (sets of) modal vectors must have the same normalization.

The COMAC is applied only to CMPs after a mode pairing using the MAC. It is a calculation of correlation values at each DOF, j, over all CMPs, L, suitably normalized to present a value between 0 and 1. The summation is performed on rows of the matrix of modal vectors, in a manner similar to the column-wise summation in the MAC. However, at the numerator, the modulus sign is inside the summation, because it is the relative magnitude at each DOF over all CMPs that matters.

The only thing the COMAC does is to detect local differences between two sets of modal vectors. It does not identify modeling errors, because their location can be different from the areas where their consequences are felt. Another limitation is the fact that COMAC weights all DOFs equally, irrespective of their magnitude in the modal vector.

The simplest output of the computation is a list of COMAC values between 0 and 1, which help to locate the DOFs for which the correlation is low. These DOFs are also responsible for a low value of MAC. The COMAC can be displayed as a bar graph of its magnitude against the DOF index (**Figure 3**).

Enhanced Coordinate Modal Assurance Criterion

A different formulation, also loosely called COMAC, is:

$$\text{COMAC}(j) = 1 - \text{ECOMAC}(j)$$

where the enhanced coordinate modal assurance criterion (ECOMAC) is defined as:

$$\text{ECOMAC}(j) = \frac{\sum_{l=1}^{L} \left| (\phi_A)_{jl} - (\phi_X)_{jl} \right|}{2L}$$

The ECOMAC is the average difference between the elements of the modal vectors. It has low values for correlated vectors. It is sensitive to calibration and phase shifting errors in test data. The ECOMAC is dominated by differences at DOFs with relatively large amplitudes.

Normalized Cross-orthogonality Location

A different criterion that avoids the phase sensitivity is the normalized cross-orthogonality location (NCOL), defined as:

$$\text{NCOL}(j) = \frac{\sum_{l=1}^{L} \left((\phi_X)_{jl} (\phi_A)_{jl} - (\phi_A)_{jl}^2 \right)}{\sum_{l=1}^{L} (\phi_A)_{jl}^2}$$

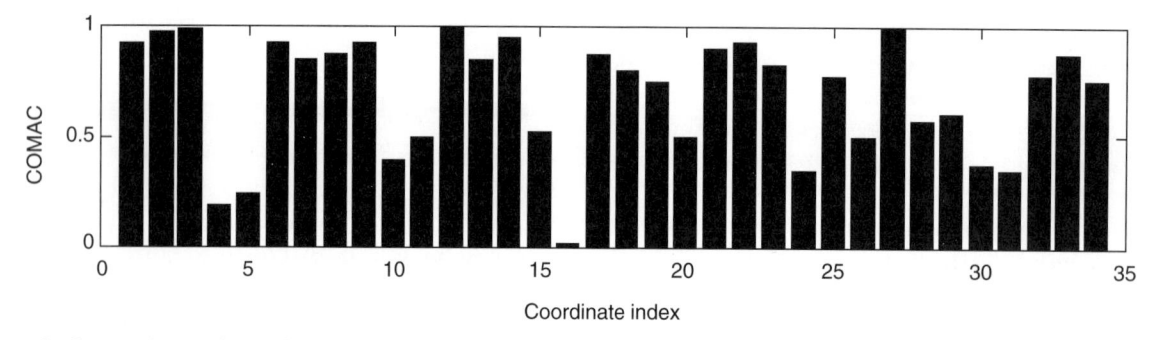

Figure 3 Bar graph plot of the COMAC.

NCOL is a DOF-based normalized cross-orthogonality check which contains neither mass nor stiffness terms. It allows the inclusion of phase inversions that are important near nodal lines.

Modulus Difference

The modulus difference (MD) is defined as the column vector formed by the differences between the absolute values of the corresponding elements of two paired modal vectors:

$$\text{MD}(l) = |(\phi_A)_l| - |(\phi_X)_l|$$

The modulus difference matrix:

$$\mathbf{MDM} = [\text{MD}(1) \quad \text{MD}(2) \quad \ldots \quad \text{MD}(L)]$$

can be displayed as a 3D graph showing the locations of low correlation between two sets of modal vectors.

Coordinate Orthogonality Check

The coordinate orthogonality check (CORTHOG) determines the individual contribution of each DOF to the magnitude of the elements of the cross-orthogonality matrix. If the XOR$_{\text{TAM}}$ for the $r-q$ mode pair and the $k-l$ DOF pair is written in the double sum form:

$$\text{XOR}^{jk}_{rq} = \sum_{j=1}^{L} \sum_{k=1}^{L} \left((\phi_X)_{jr} m_{jk} (\phi_A)_{kq} \right)$$

where m_{jk} are elements of the analytical mass matrix, it can be seen that each off-diagonal element results from a summation of contributions from all DOFs. If (ϕ_X) is replaced by (ϕ_A), then the double sum represents elements of the analytical orthogonality matrix. The CORTHOG is the simple difference of the corresponding triple product terms in the two matrices, summed for the column index of DOFs:

$$\text{CORTHOG}(j)_{rq} = \sum_{l=1}^{L} \left((\phi_X)_{jl} m_{jl} (\phi_A)_{lq} - (\phi_A)_{jr} m_{jl} (\phi_A)_{lq} \right)$$

The CORTHOG can also be displayed as a bar graph of its magnitude against the DOF index.

Other Comparisons

Other comparisons have also been used for FEM/TAM correlation. Modal kinetic energy and modal strain energy comparisons are being used to assess the TAM validity or to locate dynamically important DOFs. Modal effective mass distributions are also used to compare important modes. Orthogonality relations between left and right modal vectors of gyroscopic systems, as well as between inverse modal vectors and their related mode shape vectors are also used in updating procedures.

Nomenclature

I	identity matrix
L	number of CMPs
M	mass matrix
MDM	modulus difference matrix
TOR	test orthogonality matrix
XOR	cross orthogonality matrix
ψ_q	compatible analytical vector
ψ_r	test vector
Φ	modal matrix
Ψ	real modal matrix

See also: **Comparison of Vibration Properties**, Comparison of Spatial Properties; **Modal analysis, experimental**, Applications; **Modal analysis, experimental**, Construction of models from tests; **Modal analysis, experimental**, Parameter extraction methods; **Model updating and validating**.

Further Reading

Allemang RJ (1980) *Investigation of Some Multiple Input/Output Frequency Response Function Experimental Modal Analysis Techniques*. PhD thesis, University of Cincinnati.

Blaschke PG, Ewins DJ (1997) The MAC revisited and updated. *Proceedings of the 15th International Modal Analysis Conference*, Orlando, Florida, USA, pp 147–154.

Chen G, Fotsch D, Imamovic N, Ewins DJ (2000) Correlation methods for axisymmetric structures. *Proceedings of the 18th International Modal Analysis Conference*, San Antonio, Texas, USA, pp 1006–1012.

Ewins DJ (2000) *Modal Testing: Theory, Practice and Application*, 2nd edn. Taunton, UK: Research Studies Press.

Fotsch D, Ewins DJ (2000) Application of MAC in the frequency domain. *Proceedings of the 18th International Modal Analysis Conference*, San Antonio, Texas, USA, pp 1225–1231.

Heylen W, Lammens S, Sas P (1997) *Modal Analysis Theory and Testing*. Leuven, Belgium: K. U. Leuven.

Imamovic N, Ewins DJ (1995) An automatic correlation procedure for structural dynamics. *International Forum on Aeroelasticity and Structural Dynamics*, Manchester, pp 56.1–56.12.

O'Callahan JC, Avitable P, Riemer R (1989) System equivalent reduction expansion process (SEREP). *Proceedings of the 7th International Modal Analysis Conference*, Las Vegas, Nevada, USA, pp 29–37.

Pappa RS, Ibrahim SR (1981) A parametric study of the Ibrahim time domain modal identification algorithm. *Shock and Vibration Bulletin* 51(3): 43–71.

Comparison of Response Properties

M Radeş, Universitatea Politehnica Bucuresti, Bucuresti, Romania

Copyright © 2001 Academic Press

doi:10.1006/rwvb.2001.0175

Compared response functions usually include Frequency Response Functions (FRFs), Operating Deflection Shapes (ODSs), and Principal Response Functions (PRFs). In the following, the presentation will be focused on the comparison of FRFs. ODSs can be compared in the same way as mode shape vectors (see **Comparison of vibration properties: Comparison of modal properties**). There are three main kinds of comparison: (1) analytical-to-analytical; (2) experimental-to-experimental; and (3) analytical-to-experimental. The last is of interest in structural modification (see **Structural dynamic modifications**) and model updating (see **Model updating and validating**) procedures and will be considered as the default case.

Comparison of Individual Response Functions

A typical FRF contains hundreds of values so that a graphical format is the most appropriate for comparisons. Diagrams of the FRF magnitude as a function of frequency are satisfactory for most applications. Bode diagrams, showing both the magnitude and the phase variation with frequency, are often used. Nyquist plots for selected parts of the frequency response are preferred when detailed information around a resonance is required. A visual inspection is usually sufficient to determine the similarities or lack of agreement between two FRFs.

The simplest comparisons may include: (1) FRFs measured using different excitation levels, as a linearity check; (2) FRFs measured or calculated switching the input and output points, as a reciprocity check; (3) FRFs measured or calculated before and after a structural modification, to show its effect on the system response; (4) FRFs calculated for different models and levels of damping; and (5) FRFs calculated before and after a data reduction that is intended to eliminate the noise and the linearly-related redundant information.

It is customary to use an overlay of all the FRFs, measured from all combinations of input and output coordinates, and to count the resonance peaks as a preliminary estimation of the model order.

Comparisons of measured and predicted FRFs may include: (1) FRFs calculated with different numbers of terms included in the summmation, to check the effect of residual terms and whether a sufficient number of modes have been included; (2) a measured FRF and the corresponding regenerated analytical curve, calculated from an identified modal model; (3) an initially-unmeasured FRF curve, synthesized from a set of test data, and the corresponding FRF curve obtained from a later measured set of data, to check the prediction capability of the analytical model.

Three factors must be borne in mind when analytically generated FRF curves are used in the comparison. First, the way the damping has been accounted for in the theoretical model; second, the fact that the analytical FRFs are usually synthesized from the modal vectors of the structure and depend on the degree of modal truncation; and third, when the compared FRFs originate from two models, one model being obtained by a structural modification of the other, the comparison must take into account the frequency shift and the change of the scale factor in the FRF magnitude. For instance, if the reference stiffness matrix is modified by a factor of α, the frequencies in the modified model increase by a factor of $\sqrt{\alpha}$, while the FRF magnitudes of an undamped model decrease by a factor of α.

Generally, in the correlation of measured and synthesized FRFs in an updating process, the pairing of FRFs at the same frequency has no physical meaning. As several physical parameters at the element level are modified, an average frequency shifting exists at each frequency line, so that an experimental frequency, ω_X, will correspond to a different analytical frequency, ω_A.

A global error indicator, calculated as the ratio of the Euclidean norm of the difference of two FRF vectors measured at discrete frequencies and the norm of a reference FRF vector, is of limited practical use. Visual inspection of two overlaid FRF curves can be more effective in localizing discrepancies (**Figure 1**).

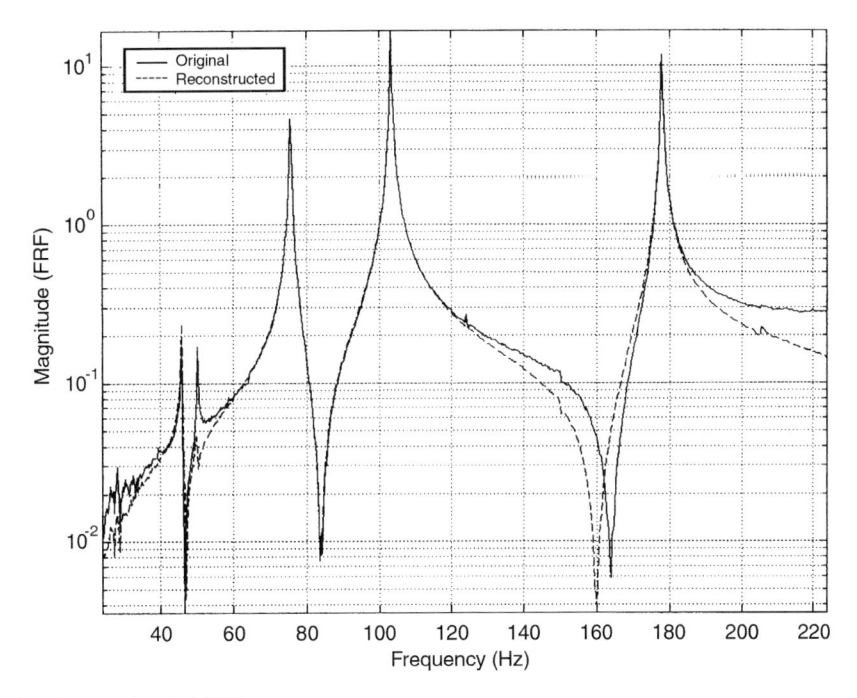

Figure 1 Measured and reconstructed FRFs.

Comparison of Sets of Response Functions

An FRF data set is usually measured at a larger number of response measurement points, N_o, than the number of input (reference) points, N_i. The measured FRF matrix $\mathbf{H}_{N_o \times N_i}(\omega)$ contains values measured at a single frequency, ω. If measurements are taken at N_f frequencies, then a complete set of FRF data is made up of N_f matrices, of row dimension N_o, and column dimension N_i. In a typical modal test, at least one column of the FRF matrix is measured. For structures with close or coincident natural frequencies, FRF elements from several such columns are measured. In order to compare several FRFs simultaneously it is necessary to use some frequency response correlation coefficients.

Frequency Response Assurance Criterion

Consider a complete set of $N_o N_i$ FRFs, measured at N_o response locations and N_i excitation locations, each containing values measured at N_f frequencies. A Compound FRF (CFRF) matrix, of size $N_f \times N_o N_i$ can be constructed such that each row corresponds to different individual FRF values at a specific frequency, and each column corresponds to a different input/output location combination for all frequencies:

$$\mathbf{A}_{N_f \times N_o N_i} = \begin{bmatrix} \mathbf{H}_{11} & \mathbf{H}_{21} & \ldots & \mathbf{H}_{jl} & \ldots \end{bmatrix}$$

where $\mathbf{H}_{jl}$ is an N_f-dimensional FRF column vector, with response at location j due to excitation at l.

Each column of the CFRF matrix is an FRF. If the magnitudes of its elements are plotted as a function of frequency, then an FRF curve is obtained. The columns of the CFRF matrix are (temporal) vectors that can be compared using the Modal Assurance Criterion (MAC) approach (see **Comparison of vibration properties: Comparison of modal properties**), i.e., calculating a correlation coefficient equal to the squared cosine of the angle between the two vectors.

The Frequency Response Assurance Criterion (FRAC), defined as:

$$\mathrm{FRAC}(j,l) = \frac{\left| \mathbf{H}_{Xjl}^{H} \mathbf{H}_{Ajl} \right|^2}{\left(\mathbf{H}_{Xjl}^{H} \mathbf{H}_{Xjl} \right) \left(\mathbf{H}_{Ajl}^{H} \mathbf{H}_{Ajl} \right)}$$

is used to assess the degree of similarity between measured $\mathbf{H}_X$ and synthesized $\mathbf{H}_A$ FRFs, or any compatible pair of FRFs, summed across the frequency range of interest.

The FRAC is a spatial correlation coefficient, similar to the Coordinate Modal Assurance Criterion (COMAC) (see **Comparison of vibration properties:**

Comparison of modal properties), but calculated like the MAC. It is a measure of the shape correlation of two FRFs at each j, l input/output location combination. The FRAC can take values between 0 (no correlation) and 1 (perfect correlation). The FRAC coefficients can be displayed in a FRAC matrix, of size $N_o \times N_i$, which looks different from the usual MAC matrix, the diagram of each column resembling a COMAC plot.

Response Vector Assurance Criterion

The transposed CFRF matrix can be written as:

$$(\mathbf{A}^T)_{N_o N_i \times N_f} = [\mathbf{H}(\omega_1) \quad \mathbf{H}(\omega_2) \quad \ldots \quad \mathbf{H}(\omega_f) \quad \ldots]$$

where each column contains all $N_o N_i$ FRFs, measured at a certain frequency, $\omega_f (f = 1, 2, \ldots, N_f)$, for N_o output locations and N_i input locations.

A temporal vector correlation coefficient can be defined using the columns of the $\mathbf{A}^T$ matrix. If the column vector $\mathbf{H}(\omega_f)$ contains only the N_o entries from the lth input point, then the Response Vector Assurance Criterion (RVAC) is defined as:

$$\text{RVAC}(\omega_f, l) = \frac{\left|\mathbf{H}_l^H(\omega_f)\mathbf{H}_l(\omega_f)\right|^2}{\left(\mathbf{H}_l^H(\omega_f)\mathbf{H}_l(\omega_f)\right)\left(\mathbf{H}_l^H(\omega_f)\mathbf{H}_l(\omega_f)\right)}$$

This contains information from all response degrees-of-freedom simultaneously and for one reference point, at a specific frequency. The RVAC is analogous to the MAC and takes values between 0 (no correlation) and 1 (perfect correlation). Each N_o dimensional column is a response vector, i.e., the vector of displacements at all N_o response measurement points, calculated or measured at a given frequency, so that the RVAC can also be applied to the correlation of ODSs.

When the analytical model is undamped, the complex values of the measured FRFs should be converted into real ones, using an approximation of the type:

$$\mathbf{H}_{\text{real}} = \text{abs}(\mathbf{H}_{\text{complex}}) \otimes \text{sign}(\text{Re}\,\mathbf{H}_{\text{complex}})$$

The RVAC coefficients can be displayed in a plot (**Figure 2**) of the type used for the MAC. However, the RVAC matrix, of size $N_f \times N_f$, yields a much denser diagram, plotted at several hundred frequency values, and hence it is more difficult to interpret. It helps in the selection of frequencies for correlation, within the intervals with high values of RVAC, where the Finite Element Model (FEM) data are close to the test data.

If the analytical FRF is calculated at N_f analytical frequencies, ω_A, and the test FRF is measured at N_f experimental frequencies, ω_X, then a Frequency Domain Assurance Criterion (FDAC) can be defined, whose real version is:

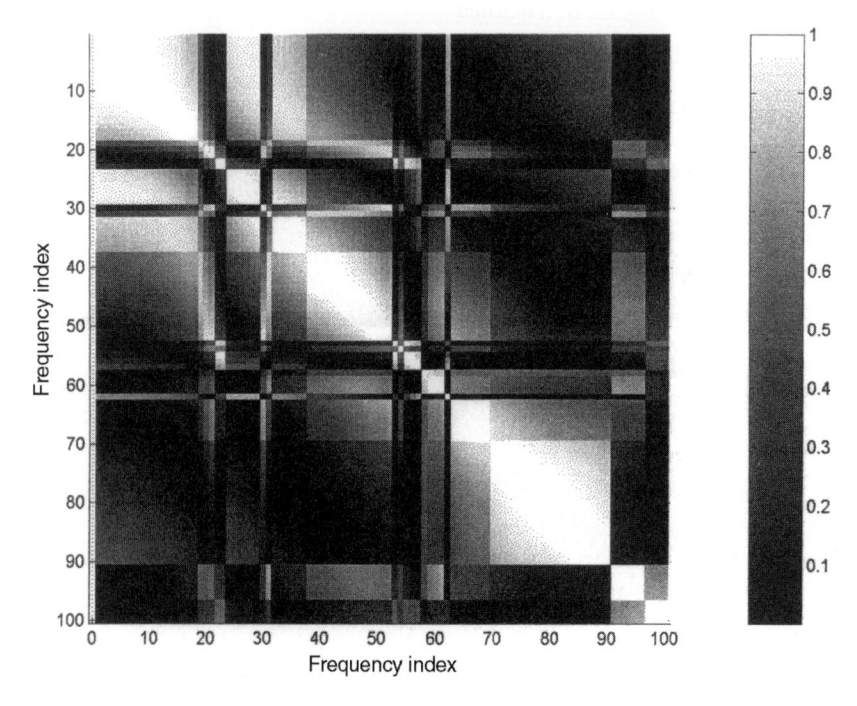

Figure 2 RVAC matrix.

$$\text{FDAC}(\omega_A, \omega_X, l)$$
$$= \frac{\mathbf{H}_X(\omega_X)_l^T \mathbf{H}_A(\omega_A)_l}{\sqrt{\left|\mathbf{H}_X(\omega_X)_l^T \mathbf{H}_X(\omega_X)_l\right| \left|\mathbf{H}_A(\omega_A)_l^T \mathbf{H}_A(\omega_A)_l\right|}}$$

where $\mathbf{H}_A(\omega_A)$ is the analytical FRF at any analytical frequency, ω_A, and $\mathbf{H}_X(\omega_X)$ is the experimental FRF at any experimental frequency, ω_X.

Using experimental FRFs converted to real values, the FDAC is calculated as the cosine of the angle between the FRF column vectors, with values between −1 and 1, to take into account the phase relation between the FRF vectors. Note that the original version of FDAC, still used in many publications, had the numerator squared, like the MAC, and was thus insensitive to the phase lag between the FRFs. A typical FDAC matrix, calculated for the same data as **Figure 2**, free of frequency shifts, is illustrated in **Figure 3**. Lightly shaded zones indicate good frequency correlation.

While the FRAC is a coordinate correlation measure across all frequencies, the RVAC and FDAC coefficients represent the correlation between two sets of FRFs at specific frequencies across the full spatial domain. The RVAC cross-correlates each frequency line with every other measured frequency line, across the spatial domain. In a way, the MAC can be considered as the RVAC evaluated only at the natural frequencies.

The reader must be warned that in some publications the RVAC is loosely referred to as the FRAC and the first variant of FDAC is similar to the RVAC. The FRAC is sometimes compared to the COMAC, but the calculation is different. The modulus in the numerator is taken after the vector multiplication, like in the MAC, and not inside the summation, for each term of the scalar product, as is taken in the COMAC.

Correlation of response properties is a relatively new technique. Frequency response correlation coefficients must be applied with great care, using stiffness factors to adjust for frequency shifts and being aware of the approximations introduced by the inclusion of an arbitrary damping model in the analysis. A global frequency shift between the experimental and predicted FRFs leads to a biased correlation coefficient even if the FRFs are otherwise identical. Selection of frequency points is a key factor in any FRF-based correlation.

Using magnitudes or logarithm values instead of complex values can give better results, especially for lightly-damped structures whose FRFs exhibit large differences in the order of magnitude and the phase angles. When the damping updating is not of interest, it is useful to choose the frequency points away from resonances and antiresonances, though the largest discrepancies noticed visually occur in these regions. The FRAC coefficients are more sensitive to resonances and less sensitive to antiresonances heavily affected by modal truncation.

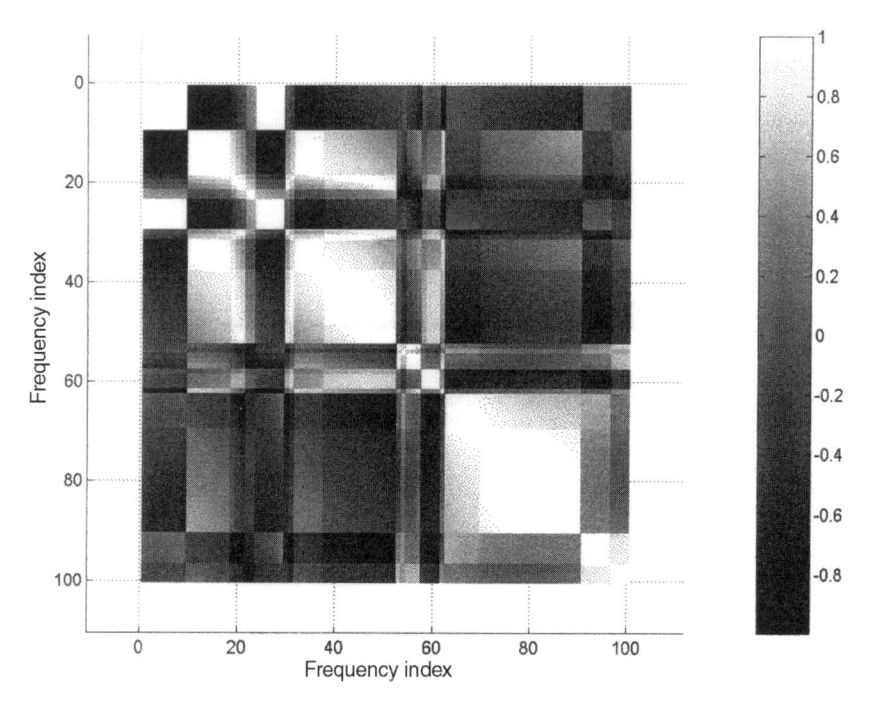

Figure 3 FDAC matrix.

The FRAC and RVAC are useful tools for examining the level of correlation of FRF data used in frequency-based model-updating procedures.

Comparison of Principal Response Functions

The singular value decomposition of the CFRF matrix is of the form:

$$\mathbf{A}_{N_f \times N_o N_i} = \mathbf{U}_{N_f \times N_o N_i} \mathbf{\Sigma}_{N_o N_i \times N_o N_i} \mathbf{V}^{H}_{N_o N_i \times N_o N_i}$$

The columns of the matrix $\mathbf{U}$ are the left singular vectors. They contain the frequency information, being linear combinations of the original FRFs that form the columns of $\mathbf{A}$. The diagonal matrix of singular values, $\mathbf{\Sigma}$, incorporates the amplitude information. The columns of the matrix $\mathbf{V}$ are the right singular vectors. They represent the spatial distribution of the amplitudes from $\mathbf{\Sigma}$.

If $N_r = \text{rank}(A)$, then the PRFs are defined as the first N_r left singular vectors of $\mathbf{A}$ scaled by their associated singular values. The matrix of the PRFs is thus:

$$(\mathbf{P}_r)_{N_f \times N_r} = (\mathbf{U}_r)_{N_f \times N_r} (\mathbf{\Sigma}_r)_{N_r \times N_r}$$

The plot of the left singular vectors of the CFRF matrix is used as the left singular vectors mode indicator function, or the U-Mode Indicator Function (UMIF), to locate frequencies of the dominant modes and to reveal multiple modes.

PRFs are left singular vectors, scale shifted in magnitude by multiplication with the corresponding singular value. They can be used to eliminate redundant, linearly dependent information and noise, and to estimate the rank and condition of the FRF test data.

The first six, twelve and twenty PRFs of a typical CFRF matrix are plotted in **Figures 4–6**. Inspection of such overlays with an increased number of PRFs reveals an upper group of six noise-free curves, more or less clearly separated from a lower group of 'noisy' curves. The number of distinct curves in the upper group is a good estimate of the rank of the CFRF matrix. Retaining only these PRFs, a rank-limited FRF matrix can be reconstructed by multiplying the truncated PRF matrix with the Hermitian of the matrix of corresponding right singular vectors.

Correlation coefficients similar to the FRAC and RVAC can be computed for the PRFs to characterize the average behavior of a structure in a given frequency band, especially in the medium frequency range.

See also: **Comparison of Vibration Properties**, Comparison of Spatial Properties; **Comparison of Vibration Properties**, Comparison of Modal Properties

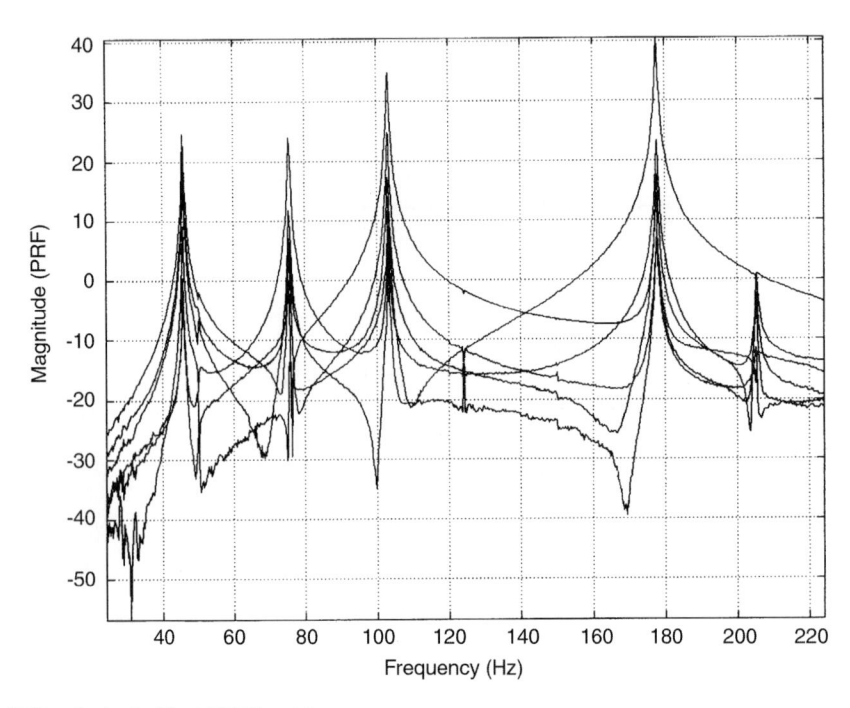

Figure 4 First six PRFs of a typical test CFRF matrix.

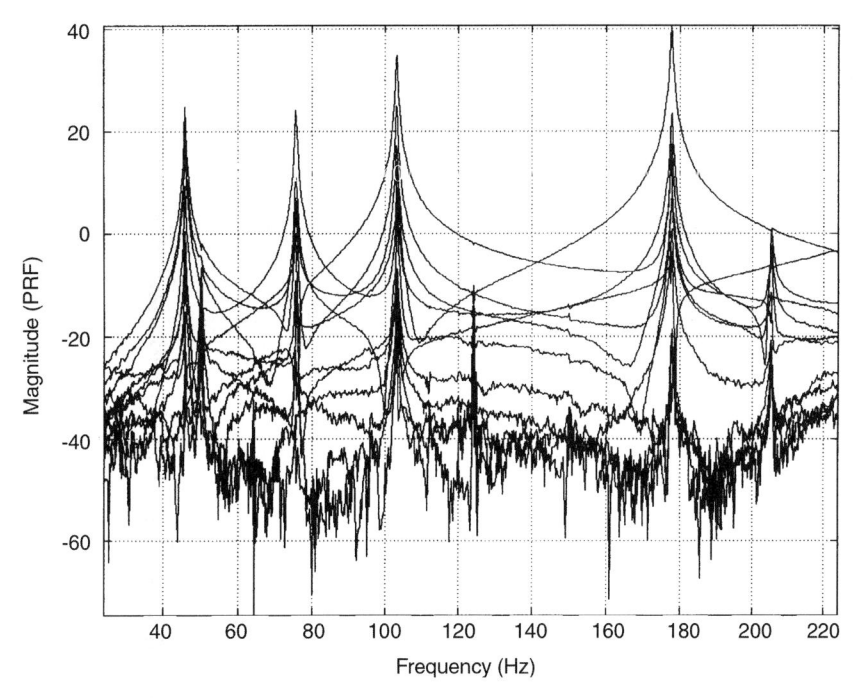

Figure 5 First twelve PRFs of the same matrix as in **Figure 4**.

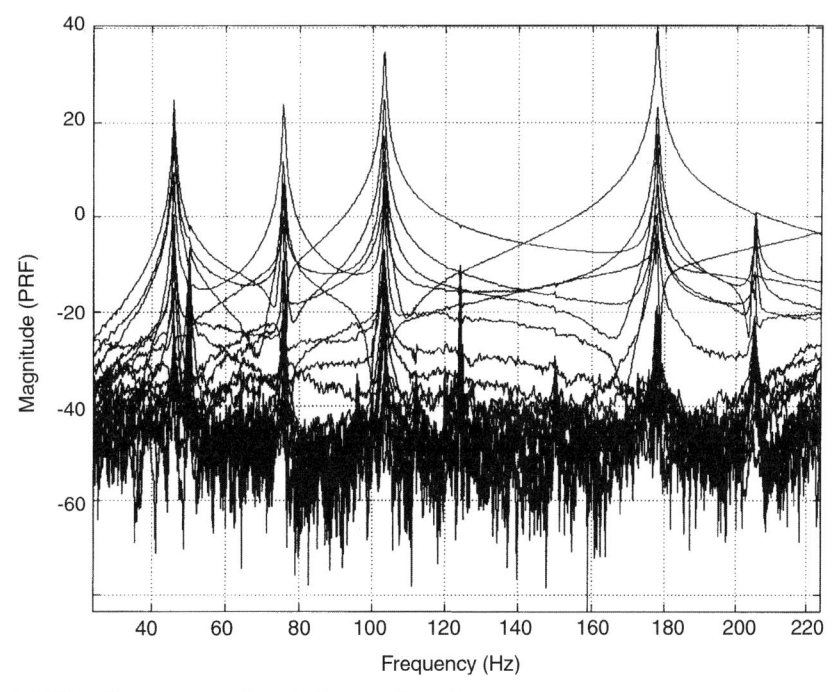

Figure 6 First twenty PRFs of the same matrix as in **Figures 4** and **5**.

Further Reading

Allemang RJ, Brown DL (1996) Experimental modal analysis. In: *Shock and Vibration Handbook*, 4th edn., pp. 21.1–21.74. McGraw Hill.

Avitable P (1999) *Modal Handbook*. Merrimack, NH: Dynamic Decisions Inc.

Ewins DJ (2000) *Modal Testing: Theory, Practice and Application*, 2nd edn. Taunton, UK: Research Studies Press.

Heylen W, Lammens S, Sas P (1997) *Modal Analysis Theory and Testing*. Leuven: K. U. Leuven.

Pascual R, Golinval JC, Razeto M (1997) A frequency domain correlation technique for model correlation and updating. In: *Proceedings of the 15th International Modal Analysis Conference*, Orlando, Florida, pp 587–592.

COMPONENT MODE SYNTHESIS (CMS)

See **THEORY OF VIBRATION, SUBSTRUCTURING**

COMPUTATION FOR TRANSIENT AND IMPACT DYNAMICS

D J Benson, University of California, San Diego, La Jolla, CA, USA

J Hallquist, Livermore Software Technology Corporation (LSTC), Livermore, CA, USA

Copyright © 2001 Academic Press

doi:10.1006/rwvb.2001.0006

Introduction

Transient and impact dynamics problems typically occur over short periods of time, ranging from nanoseconds to milliseconds, and have large deformations and rotations, high strain rates, and nonlinear boundary conditions. Specialized computational methods are used as an efficient solution to problems with this particular combination of characteristics. These methods have evolved from the finite element and finite difference methods used for solving quasistatic and dynamic structural problems. This article focuses on the differences between the computational methods used for transient and impact dynamics and those used to solve more traditional types of problems in structures and solid mechanics.

Typical applications of explicit codes include automotive crashworthiness and occupant protection, bird strike on jet engine fan blades and aircraft structures, industrial processes such as sheetmetal stamping, and defense applications involving ordnance design. To handle such a wide range of problems modern-day explicit codes have many capabilities, including a variety of contact algorithms, a large library of constitutive models for an extensive range of material behavior, equations of state for modeling the response of materials under high pressure, and various forms of adaptive remeshing.

Computational Methods for Transient and Impact Dynamics

Formulations that are intended for transient and impact dynamics share several characteristics that set them apart from methods intended to solve quasistatic and traditional dynamic problems for structures and solid mechanics. Most of the formulations for transient and impact dynamics can be derived from either the finite element or finite difference perspective. In fact, many of the algorithms used in finite element impact calculations by engineers were originally developed by physicists at the national laboratories for the finite difference formulations used to analyze nuclear weapons.

The global solution strategy typically uses:

1. Explicit time integration.
2. Lumped mass matrices.
3. Contact algorithms.
4. Algorithms for mapping solutions from distorted to undistorted meshes.

The elements have:

1. Linear interpolation functions.
2. Uniformly reduced (one-point) spatial integration.
3. Hourglass control to eliminate zero-energy modes.
4. A shock viscosity to resolve stress waves.

Global Solution Strategies

Time Integration

For typical structural dynamics problems, the step size is governed by the accuracy required by the engineer and the truncation error of the time integration method. In contrast, the mechanics of transient and impact calculations governs the time step size. For example, the time step size required to resolve the propagation of a stress wave is the amount of time the wave takes to cross an element. Therefore, computational formulations for impact problems minimize the cost of each time step by using explicit time integration methods.

The second-order accurate central difference method is the time integration method that is most commonly used in codes for impact calculations. Given the accelerations and displacements at time step n and the velocity at $n + 1/2$, the solution is advanced in time to step $n + 1$ with second-order accuracy using the explicit formulas:

$$\mathbf{v}^{n+1/2} = \mathbf{v}^{n-1/2} + \Delta t^n \mathbf{a}^n \quad [1]$$

$$\mathbf{x}^{n+1} = \mathbf{x}^{n-1} + \Delta t^{n+1/2} \mathbf{v}^{n+1/2} \quad [2]$$

subject to the time step size limitation:

$$\Delta t \leq S \cdot \frac{2}{\omega_{max}} \left(\sqrt{\zeta_{max}^2 + 1} - \zeta_{max} \right) \quad [3]$$

where S is a time step safety scale factor, ω_{max} is the maximum natural frequency, and ζ_{max} is the corresponding damping ratio.

The time step size is recalculated every time step because the maximum eigenvalue and the damping change with time. A simple bound:

$$\omega_{max} \leq 2s \max_{e=1}^{\text{Elements}} \left(\frac{c_e}{l_e} \right) \quad [4]$$

on the maximum natural frequency is usually calculated instead of solving the eigenvalue problem, where s is a factor determined by the element formulation, l is a characteristic length for the element and c is the sound speed. This quantity is evaluated for every element, and the largest calculated value is mathematically guaranteed to be an upper bound on the actual value of ω_{max} for the system.

Lumped Mass Matrix

A lumped mass matrix is used instead of the consistent mass matrix because:

1. Inverting a lumped mass matrix is trivial.
2. It requires less storage.
3. It eliminates spurious oscillations in the acceleration.

The accelerations at the nodes are calculated by solving Newton's equation:

$$\mathbf{a} = \mathbf{M}^{-1} \{ \mathbf{F}_{\text{external}} - \mathbf{F}_{\text{internal}} \} \quad [5]$$

where $\mathbf{F}_{\text{external}}$ is the vector of external applied loads due to boundary conditions and gravity, and $\mathbf{F}_{\text{internal}}$ is the vector of internal forces due to the stress in the structure.

The lumped mass matrix M is trivial to invert because it is diagonal. Each acceleration component is simply calculated as $a_i = F_i / M_{ii}$, which avoids the expense of solving a system of simultaneous linear equations, the single largest computational cost in traditional structural calculations.

Since minimizing the cost of each time step is a major concern in developing a formulation for transient and impact problems, the minimization of the storage requirements is also important because the cost of reading and writing information from a hard disk for a traditional finite element formulation is larger than the computational cost of an explicit time step. A lumped mass matrix therefore permits the solution of much larger problems than a consistent mass matrix for a fixed amount of memory.

When an impact applies a sudden load on the boundary of a finite element mesh, an acceleration field that is in the direction of the applied force is expected. A consistent mass matrix, which inertially couples the nodes, often results in an oscillatory acceleration field, with nodes accelerating in the direction opposite of the applied force. This error, which has been analyzed theoretically, can only be eliminated from impact calculations by using a lumped mass matrix.

Contact Algorithms

Impact calculations require contact algorithms to impose the contact forces required to keep exterior surfaces from passing through each other. Ideally, the penetration of surfaces through each other is held to zero by an exact enforcement of the contact constraints, but in practice this is difficult to achieve. Most of the current methods permit the surfaces to penetrate slightly, and the small violation of the contact constraint has no adverse effect for most problems. During the early years of computational mechanics, 'gap elements' required nodes on opposite surfaces to come in direct contact to prevent penetration. Large deformation problems have surfaces that undergo large relative slip, and therefore the nodes that were opposite each other at the beginning of the calculation are remote from each other by the end of it. Furthermore, during large slip, node-on-node contact is impossible to maintain unless the mesh moves relative to the material, which introduces its own complications.

Contact algorithms for impact calculations have two aspects that may be considered independently: calculating the contact locations, and calculating the forces to prevent penetration.

Calculating contact locations Contact is described in terms of a node and its location relative to the exposed edge or face of an element. In the remaining discussion, the exposed element boundary will be referred to as a 'surface segment'. When there are two distinct surfaces, they are called 'master' and 'slave' surfaces in the literature. Buckling calculations have a single surface which contacts itself as the buckles form, and special contact search algorithms have been developed for handling this situation. Some contact algorithms are sensitive to which surface is designated the master surface, while symmetric methods, where the distinction is immaterial, are more robust.

The contact locations are calculated by a combination of global and incremental searches. Global searches are more expensive than the incremental searches, and are performed as infrequently as possible. A brute force search strategy for a global search checks for contact between each node on the slave surface and the surface segments on the master surface. The cost of this strategy is proportional to the number of slave nodes, N_n^s, times the number of master surface segments, N_s^m. Incremental searches check for contact between a slave node and the last surface segment it contacted. If they are no longer in contact, the search continues over the surface segments that are contained in the neighborhood of the last contacting segment. The cost of the incremental search is therefore proportional to N_n^s times the average number of surface segments in the incremental search (which is typically on the order of 4–10 for three-dimensional calculations). A typical computational strategy is to perform a brute force search at the start of a calculation to determine the initial contacts, and to update the contact points with the incremental searches for the duration of the calculation.

Global searches are performed many times during buckling calculations because the incremental searches frequently fail when a surface contacts itself. The brute force approach for the global search is not acceptable because of its cost, and a more sophisticated method, called a bucket sort, is used instead. The central idea is relatively simple, but the actual programming is complicated for an efficient, robust implementation. A bucket sort starts by dividing the space surrounding the mesh into the 'buckets' which are Δx by Δy by Δz. The size of the bucket is problem-dependent, and it is determined automatically by the sorting algorithm. Bucket ijk contains the region $[x_i, x_{i+1}] \times [y_j, y_{j+1}] \times [z_k, z_{k+1}]$ and

$$x_{i+1} = x_i + \Delta x \quad y_{j+1} = y_j + \Delta y$$
$$z_{k+1} = z_k + \Delta z \quad [6]$$

The bucket ijk associated with a point (x, y, z) is calculated by:

$$i = \text{int}(x/\Delta x) + 1 \quad j = \text{int}(y/\Delta y) + 1$$
$$k = \text{int}(z/\Delta z) + 1 \quad [7]$$

and a list of the nodes inside the bucket and contact segments that intersect it is stored. To determine which contact segments are close to a node in bucket ijk, only the contact segments listed in bucket ijk and in the 26 buckets surrounding ijk are considered. The cost of the global search is therefore proportional to the number of nodes and the average number of contact segments stored in a block of 27 buckets. Most of the contact segments within a bucket can be eliminated with simple checks and only a few require a detailed check. The cost of the global search is therefore reduced to a small multiple of the cost of the incremental search.

During the incremental search, the point of contact between a node and the surface segment is calculated in terms of the surface segments' isoparametric coordinates (ξ_1, ξ_2) by solving the closest point minimization problem:

$$J = \frac{1}{2}\left(\left(x^s - \sum x_k^m N_k(\xi_1, \xi_2)\right)^2 + \left(y^s - \sum y_k^m N_k(\xi_1, \xi_2)\right)^2 + \left(z^s - \sum z_k^m N_k(\xi_1, \xi_2)\right)^2 \right) \quad [8]$$

which generates two nonlinear equations. The node is potentially in contact if the isoparametric coordinates lie between -1 and $+1$, otherwise the search proceeds to the next segment. The depth of the penetration, δ, is:

$$\delta = -(\mathbf{x}^s - \mathbf{x}^m(\xi_1, \xi_2)) \cdot \mathbf{n}(\xi_1, \xi_2) \quad [9]$$

where $\mathbf{n}$ is the exterior normal to the contact segment:

$$\mathbf{n}(\xi_1, \xi_2) = \frac{\frac{\partial \mathbf{x}^m}{\partial \xi_1} \times \frac{\partial \mathbf{x}^m}{\partial \xi_2}}{\left\|\frac{\partial \mathbf{x}^m}{\partial \xi_1} \times \frac{\partial \mathbf{x}^m}{\partial \xi_2}\right\|} \quad [10]$$

Calculating the contact force The contact force acts along the surface normal to resist the interpenetration of the two surfaces and its magnitude may be calculated using a penalty method, Lagrange multipliers, or an augmented Lagrangian formulation.

The penalty method applies a force that is a function of the depth of penetration. In most codes, the force is linear, $F = k\delta\mathbf{n}$, which has the desirable property of conserving energy. The value of the penalty stiffness is a function of the material properties of the two contacting surfaces and the sizes of the elements at the contact point. While the idea of using a simple linear spring to resist penetration may seem overly simplistic, it works well in calculations ranging from automobile crashworthiness simulations to the design of munitions.

Until recently, Lagrange multiplier methods were never used in impact calculations because they generate systems of coupled equations over the contacting surfaces in their standard form. Their primary advantage is the contact constraints are enforced exactly. Explicit Lagrange multiplier methods have recently been proposed which avoid this difficulty by introducing assumptions that decouple the equations. In their explicit form, Lagrange multiplier methods resemble a penalty method with a stiffness that is a function of the time step size, and the contact constraint is no longer enforced exactly. The time dependence of the surface stiffness with this method results in small errors in the conservation of energy.

Augmented Lagrangian methods try to combine the Lagrange multiplier method with the penalty method to gain the advantages of both. Although they have enjoyed some success in implicit formulations, they have yet to be used in codes for impact calculations.

Mapping Solutions from Distorted to Undistorted Meshes

The finite element or finite difference mesh distorts as the calculation progresses, which reduces the accuracy of the solution and the time step size. Eventually, the mesh may become too distorted to continue the calculation. A strategy for mapping the solution from the distorted mesh to an undistorted mesh is therefore required. At this time, the most popular strategies are periodic rezoning, arbitrary Lagrangian Eulerian (ALE) formulations, and Eulerian formulations.

While the three approaches differ considerably in their implementation, they possess many similarities. First, the qualities that are desired in the remapping schemes are the same: conservation of solution variables (e.g., momentum); second-order accuracy; and to avoid introducing oscillations into the solution. Second, the simulation time is fixed during the mapping process, i.e., it does not proceed simultaneously with the evolution of the solution variables, an approach that is referred to as 'operator splitting' in the literature on ALE and Eulerian formulations. Third, the functional representation of the solution variables on the old mesh is usually different (and of higher order) than the one used during the evolution of the solution. Fourth, the sequence of the mapping process is first, determine if a new mesh and a mapping are required; second, generate the new mesh; third, project the solution from the old mesh on to the new mesh, and fourth, restart the calculation with the new mesh.

Periodic rezoning The calculation is stopped periodically either by manual intervention or by the program itself (e.g., based on some measure of mesh distortion), and the solution is projected from the old mesh to a completely new mesh. The new mesh has nothing in common with the old mesh other than the shape of the material boundaries. Originally, the new mesh was generated by the analyst, but automatic mesh generation algorithms are advanced enough today that most of the new meshes are generated automatically. If the new mesh is generated by the analyst, the number of rezones in a calculation is typically on the order of 10, while automatic mesh generation schemes may redefine the mesh up to 100 times.

The rezoning scheme has to identify which elements of the old mesh overlap an element in the new mesh. Search algorithms that are similar to the global contact search algorithms are used, and they frequently account for a major part of the mapping cost.

Once the overlapping elements are found, the mapping of the solution usually proceeds in one of two ways. The generality of the periodic rezoning projection makes achieving both conservation and second-order accuracy very difficult. Either the solution is interpolated from the old mesh, which results in a loss of conservation, or a 'completely conservative' mapping scheme calculates the exact integral average for the new solution values. The completely conservative scheme is significantly more expensive and difficult to program than interpolating the new values, but for problems where strong solution gradients are present, it gives a superior answer.

ALE and Eulerian formulations ALE formulations permit the mesh to move relative to the material continuously as the solution evolves. The most common form of the ALE formulation is the simplified ALE or SALE formulation, which permits only one material in an element. This simplification means that the nodes on a material boundary can only move tangentially to the boundary, which limits its usefulness since elements near the boundaries are frequently the most distorted. Eulerian formulations use a spatially fixed mesh, and materials flow through it. An element may therefore contain several materials, as may a general ALE formulation.

Since the material moves relative to the mesh each time step (or sometimes every few time steps), the mapping procedure is performed thousands of times during a calculation. Speed and accuracy are therefore at a premium. While a first-order accurate mapping method may be adequate for 10 periodic rezones, it is too diffusive to be used thousands of times during a calculation, and a minimum of second-order accuracy is a practical necessity. The mapping algorithms, also called transport or advection algorithms, are based on computational fluid dynamics algorithms for the Euler equations. Speed is obtained by keeping the same topology for the new and old meshes and limiting their relative displacement to some fraction of the element width.

Element Technology

Linear Interpolation Functions

The displacements, velocities, and accelerations are interpolated linearly and the stresses are piecewise constants in the elements used in transient and impact calculations. Their advantages over higher-order approximations are:

1. They possess symmetries and antisymmetries that reduce their computational cost relative to higher-order elements far more than a casual inspection would suggest.
2. Linear elements are very robust and are not prone to the singularities that occur in higher-order elements when the nodes are not uniformly spaced.
3. Linear interpolation simplifies the geometric calculations in the contact and mapping algorithms.
4. The zero-energy modes due to reduced integration are more readily suppressed in comparison to higher-order elements.
5. For a given nodal spacing, linear elements permit larger time steps than higher-order elements.

The primary disadvantage of linear elements relative to higher-order elements is that they are too stiff with full integration and too soft with reduced integration. Triangular and tetrahedral elements, in particular, are especially prone to locking with incompressible (plastic) flow for some meshes.

Uniformly reduced integration Only a single integration point is used for linear elements, which reduces the required central processing unit time and memory for the stress by factors of four and eight in two and three dimensions respectively. For quadratic elements, the speed gain would only be a factor of two to three.

Zero-energy mode control One byproduct of uniformly reduced integration is the occurrence of zero-energy or hourglass modes. The shape of the zero-energy modes is a function of the element geometry. Since a zero-energy mode does not produce a strain, no stress is generated to resist it, and the modes may grow without bound unless additional stiffness or damping terms are introduced to resist it. Conversely, since a mode produces no strain, it does not affect the accuracy of the stresses in the calculation. Problems only occur when the modes become large enough to turn the elements inside out or distort the contact geometry.

Zero-energy modes are suppressed by calculating their magnitude, and then adding a force that opposes them. Assuming that the magnitude of the mode, h, is calculated by:

$$h = \mathbf{H} \cdot \mathbf{v} \quad [11]$$

the viscous force, $\mathbf{F}_h$, opposing the zero energy mode is:

$$\mathbf{F}_h = -ch\mathbf{H} = -c\mathbf{H} \otimes \mathbf{H}^\mathrm{T} \mathbf{v} \quad [12]$$

or, if a stiffness form is used, the force is:

$$\mathbf{F}_h = -k \int_0^t h\mathbf{H}\, \mathrm{d}t \quad [13]$$

and it is stored as a solution variable with incremental update performed each time step.

Shock viscosity Shocks propagate via a thermodynamically irreversible process and appear as jump discontinuities in the solution variables. Their accurate resolution and propagation are critical for solving high-speed impact problems. Oscillations will occur behind the jump in the stress unless some form of damping, which is called the shock viscosity, is included in the calculation. The physical thickness of a shock is typically on the order of microns, but in computational practice, they are smeared over three to six elements regardless of the element size. Although this introduces a large error in the shock width itself, the critical aspects of the shock, namely its speed and the stress states on either side of it, are accurately calculated.

The standard form of shock viscosity, q, a function of the volume strain rate, $\dot{\varepsilon}_v$, is:

$$q = -\rho\ell\dot{\varepsilon}_v(c_1 C + c_2|\ell\dot{\varepsilon}_v|) \quad [14]$$

and it differs little from the one originally introduced by von Neumann and Richtmeyer to solve shock problems in the design of the atom bomb (they did not include the linear term). The shock viscosity is treated like a contribution to the pressure for the calculation of the nodal forces.

Example Transient and Impact Calculations

The calculations shown in this section were performed with LS-DYNA3D, developed and marketed by Livermore Software Technology Corporation. It incorporates many advanced capabilities that are not available in the public domain version of DYNA3D, which was originally developed by John Hallquist. While the code runs on everything from PCs to massively parallel computers, the calculations shown here were performed on workstations.

Airbag Deployment

Transportation engineering is currently one of the largest applications areas for explicit finite element methods. Since the safety of the occupants is a major concern, accurate detailed modeling of the impacts between the occupants and the vehicle is a necessity. One of the more challenging modeling aspects is the deployment of the airbags in automobiles. In this example, shown in **Figure 1**, the airbag is initially folded into the center of the steering wheel. A control volume model of the combustion of the propellant located in the steering wheel hub determines the gas pressure in the bag during its inflation. Special element technology and material models were developed to model the dynamic response of the airbag material accurately.

Crashworthiness Simulations

The first crash simulation of a full vehicle model, including the suspension, tires, and other running gear, was performed with DYNA3D in 1986. It had a little over 4000 elements and required over 20 h of computing on the Cray-XMP supercomputer at Lawrence Livermore National Laboratory. Today, calculations with 10–100 times as many elements are routinely performed by automobile manufacturers to enhance the safety of modern vehicles and reduce the number of prototypes required for crashworthiness testing.

Many of the stronger components, such as the engine, are modeled as rigid bodies. In regions removed from the impact, analysts use a coarse computational mesh or rigid bodies to minimize the cost of the calculation. The impact area requires detailed modeling. For example, automobiles typically have thousands of spot welds which may fail during an impact, and each spot weld in the impact area is individually modeled.

As in the previous example, the largest challenge is modeling the contact during the crash. Most of the contact involves interior structural components which are not visible in **Figure 2**. For example, some of the

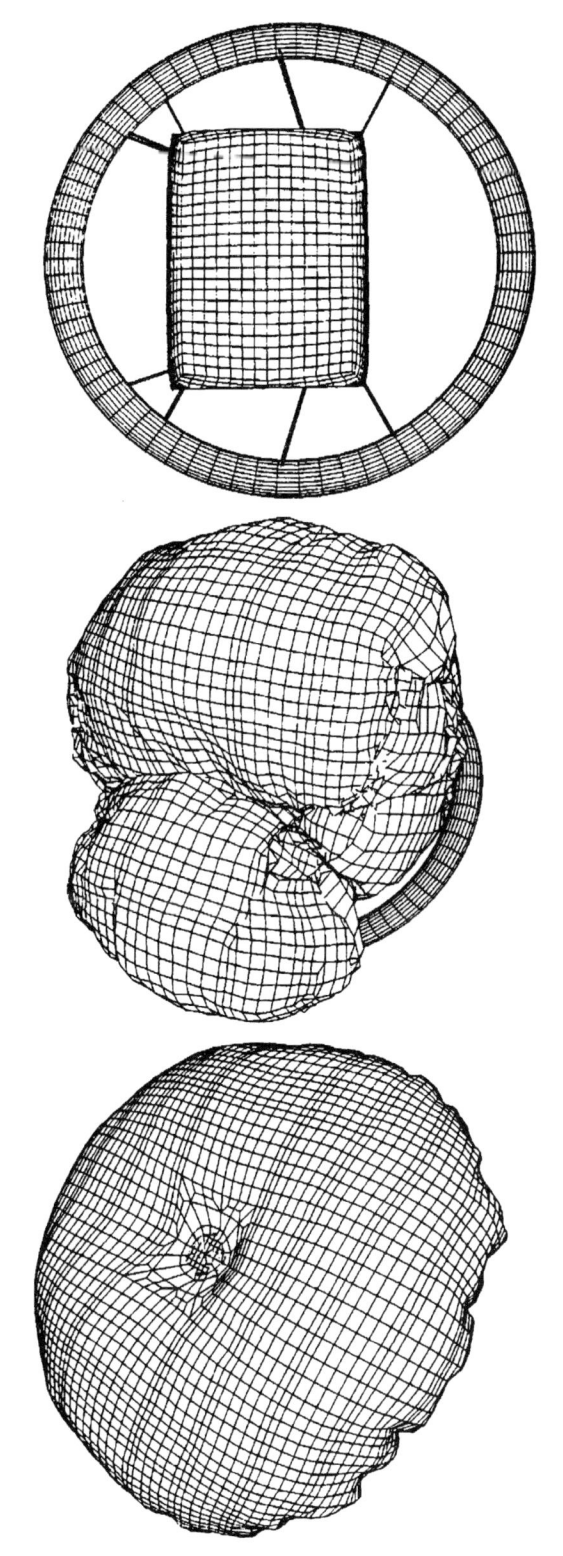

Figure 1 The simulation of the deployment of an airbag.

interior sheetmetal structure is designed to buckle in an accordion mode to absorb the impact energy. The contact interactions are so extensive that all the surfaces on all the front-end components are treated as potential contact surfaces in the calculation.

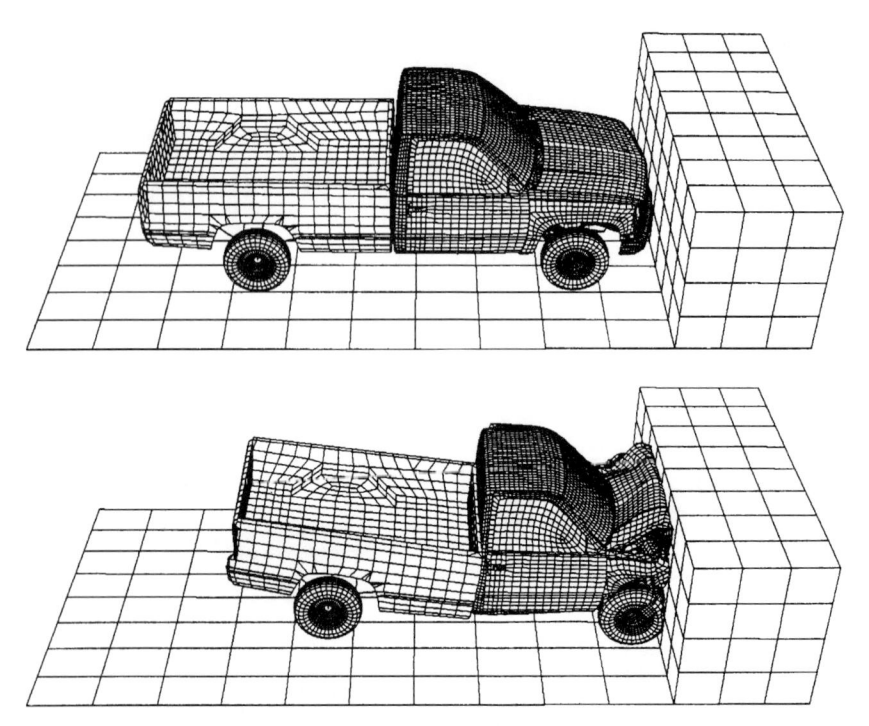

Figure 2 Crashworthiness analysis of a truck. Truck graphics courtesy of National Highway Traffic Safety Administration and National Crash Analysis Center.

Sheetmetal Forming

The buckling patterns, and therefore the energy absorbed by the structure, are very sensitive to the manufacturing process. Current crashworthiness simulations do not account for the plastic work and thinning caused by the sheetmetal forming process; however, they will do so in the near future.

Sheetmetal-forming simulations are used to aid in the design of the dies, which greatly reduces the time and cost to bring a product to market. Before simulations were possible, engineers designed the dies based on their experience and the dies were altered by tool and die makers on the production floor until they produced acceptable parts.

Explicit finite element methods are capable of simulating sheetmetal-forming processes with greater speed and accuracy than current implicit methods. Very fine meshes are required to resolve the sharp bends and the wrinkles caused by inadequate die designs. Adaptive mesh refinement automatically adds elements to the mesh (**Figures 3**

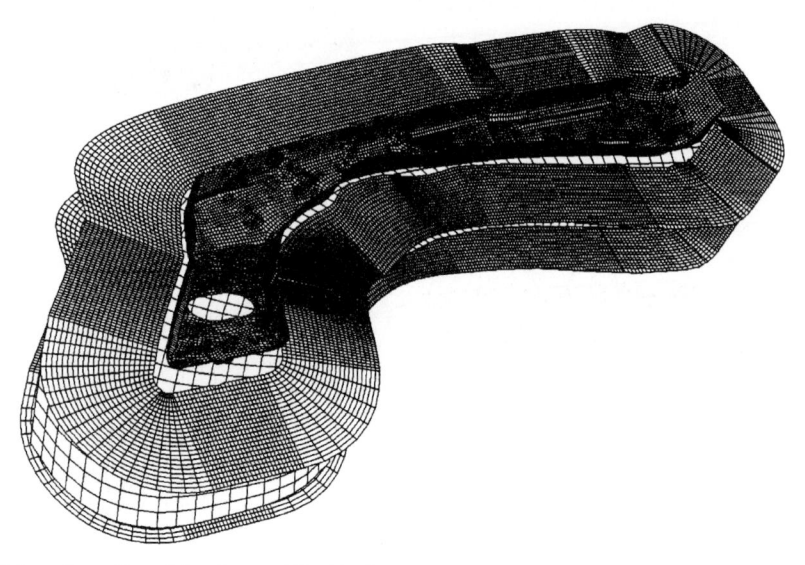

Figure 3 The initial configuration of the dies and the blank.

and 4) during the calculation. Various criteria, e.g., element distortion, determine when a single quadrilateral element should be split into four elements, adding one level of local mesh refinement. The maximum number of levels (typically two to four) is specified in the input file to prevent the adaptive algorithm from generating a mesh that is too large for the available computer resources.

In comparison to the 60 ms duration of a vehicle crash, the timescale of a metal-forming operation is extremely long. Since inertial forces are not important in metal forming, the density of the metal blank can be scaled up to permit the explicit finite element code to take larger time steps. Care must be taken with this strategy, since scaling the density to too large a value will create spurious inertial effects.

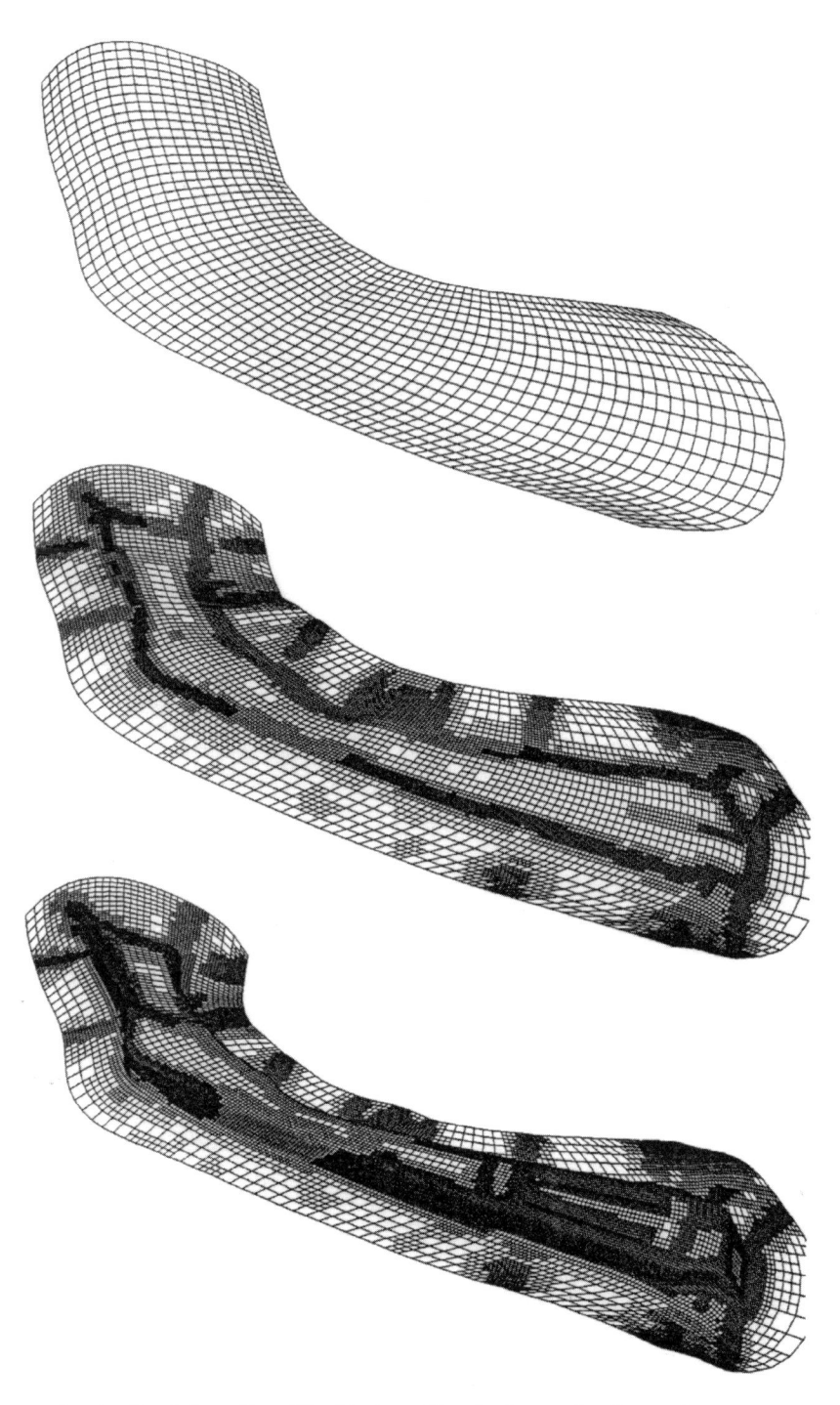

Figure 4 A sequence showing the deformation of the blank and its adaptive mesh refinement.

Nomenclature

c	speed of sound
h	magnitude of the mode
l	length
M	lumped mass matrix
q	shock viscosity
S	time step safety scale factor
δ	depth of penetration
ζ	damping ratio
ω	frequency

See also: **Computation for transient and impact dynamics; Commercial software; Crash; Discrete elements; Dynamic stability; Eigenvalue analysis; Finite difference methods; Finite element methods; Helicopter damping; Krylov-Lanczos methods; Shock; Ship vibrations; Structural dynamic modifications; Tire vibrations; Wave propagation,** Waves in an unbounded medium.

Further Reading

Bathe KJ (1995) *Finite Element Procedures*. Prentice Hall.

Belytschko TB and Hughes TJR (eds) (1983) *Computational Methods for Transient Analysis*, vol. 1. North-Holland.

Hughes TJR (1987) *The Finite Element Method, Linear Static and Dynamic Finite Element Analysis*. Prentice Hall.

Irons B and Ahmad S (1980) *Techniques of Finite Elements*. John Wiley.

Kikuchi N and Oden JT (1988) *Contact Problems in Elasticity: A Study of Variational Inequalities and Finite Element Methods*. SIAM.

Oden JT (1972) *Finite Elements of Nonlinear Continua*. McGraw-Hill.

Simo JC and Hughes TJR (1998) *Computational Inelasticity. Interdisciplinary Applied Mathematics*, vol. 7. Springer Verlag.

Zienkiewicz OC and Taylor RL (1991) *Finite Element Method*, vols 1 and 2, 4th edn. McGraw-Hill.

COMPUTATIONAL METHODS

See **BOUNDARY ELEMENT METHODS; COMMERCIAL SOFTWARE; COMPUTATION FOR TRANSIENT AND IMPACT DYNAMICS; CONTINUOUS METHODS; EIGENVALUE ANALYSIS; FINITE DIFFERENCE METHODS; FINITE ELEMENT METHODS; KRYLOV-LANCZOS METHODS; LINEAR ALGEBRA; OBJECT ORIENTED PROGRAMMING IN FE ANALYSIS; PARALLEL PROCESSING; TIME INTEGRATION METHODS.**

CONDITION MONITORING

See **DIAGNOSTICS AND CONDITION MONITORING, BASIC CONCEPTS; ROTATING MACHINERY, MONITORING.**

CONTINUOUS METHODS

C W Bert, The University of Oklahoma, Norman, OK, USA

Copyright © 2001 Academic Press

doi:10.1006/rwvb.2001.0009

Continuous methods of vibration analysis are applicable to continuous structural elements such as one-dimensional (1-D) members (strings, bars, beams, and columns), 2-D members (membranes, thin plates, and thin shells), and 3-D members (blocks, thick plates, and thick shells). These methods are called continuous to distinguish them from discrete methods, such as finite difference, collocation, finite element, transfer matrix, boundary element, differential quadrature, quadrature element, etc.

The present discussion is limited to linear vibration analysis; see **Nonlinear systems analysis**. Furthermore, since forced vibration of linear systems can always be expanded in terms of the free vibration modes (see **Modal analysis, experimental**, Parameter extraction methods), the present section is limited to free vibration. Finally, although continuous methods can be applied to damped as well as undamped systems, damping introduces considerable complexity and thus is omitted here. In summary, the present section treats free vibration of undamped, linear, continuous systems.

All problems of the present class lead to self-adjoint eigenvalue problems (see **Eigenvalue analysis**). It is rather dry and not very informative to discuss the general free vibration problem, so here a model problem has been selected. It is the axial vibration of a slender bar of rectangular cross-section gently tapered in planform but uniform in thickness so that 1-D theory is adequate.

In order to provide a yardstick with which to evaluate the various approximate methods, first the model problem is solved exactly in closed form. Then it is solved using the ordinary Rayleigh method, the noninteger-power Rayleigh method, the Ritz (or Rayleigh–Ritz) method, several versions of the complementary energy method, and the Galerkin method. All of the approximate methods mentioned yield upper bounds for the fundamental frequency. Thus, several methods of estimating the lower bound are also presented. Finally, the problem is solved by an alternative exact method, the differential transformation method.

Formulation and Classical Solution

Free vibration problems may be formulated in the form of integral equations or differential equations. The latter formulation is more convenient and is used here. The governing differential equation of motion may be obtained using a differential element or by Hamilton's principle. For the present model problem, the result is:

$$\frac{\partial}{\partial x}\left(A(x)E(x)\frac{\partial u}{\partial x}\right) - m(x)\frac{\partial^2 u}{\partial t^2} = 0 \quad [1]$$

where $A(x)=$ cross-sectional area, $E(x)=$ elastic modulus, $m(x) =$ mass per unit length, $t=$ time, $u=$ axial displacement, and $x=$ axial position coordinate.

The classical boundary conditions associated with the problem are:
Fixed end:

$$u(x_B, t) = 0 \quad [2a]$$

Free end:

$$A(x_B)E(x_B)\frac{\partial u}{\partial x}(x_B, t) = 0 \quad [2b]$$

where x_B may be either 0 or L.

In the present case:

$$A(x) = 2A_0 x/L \quad [3]$$

$$m(x) = 2m_0 x/L \quad [4]$$

and E is constant. Here A_0 and m_0 are constants (reference area and reference mass, respectively), and L is the length of the bar.

To be specific, the free end, boundary condition [2b], is taken to be at $x = 0$ and the fixed end, boundary condition [2a], is taken to be at $x = L$.

Since eqn [1] is linear, the axial displacement can be expressed as:

$$u(x,t) = W(X)e^{i\omega t} \quad [5]$$

where W is called the mode shape, $X \equiv x/L$, and ω is the natural frequency. Then the governing equation of motion can be expressed as:

$$\frac{d}{dX}\left(X\frac{dW}{dX}\right) + k^2 X W = 0 \quad [6]$$

where $k^2 \equiv (m_0 L^2/A_0 E)\omega^2$. The associated boundary conditions are:

$$\lim_{X\to 0}[(AE/L)(dW/dX)] = 0$$

or:

$$\lim_{X\to 0}[X\, dW/dX] = 0 \quad [7]$$

and:

$$W(1) = 0 \quad [8]$$

Eqn [6] can be rewritten as:

$$\frac{d^2 W}{dX^2} + \frac{1}{X}\frac{dW}{dX} + k^2 W = 0 \quad [9]$$

which is easily identified as the Bessel equation of rank 0.

The general solution for the mode shape is:

$$W(X) = D_1 J_0(kX) + D_2 Y_0(kX) \quad [10]$$

where D_1, D_2 are constants and J_0 and Y_0 are zero-order Bessel functions of the first and second kind. The dimensionless frequency is given by:

$$\bar{\omega} = k = (m_0 L^2/EA_0)^{1/2} \omega \quad [11]$$

For the present boundary conditions, $D_2 = 0$ and the various zeros of the Bessel function of the first kind and of rank zero, $J_0(\bar{\omega}) = 0$ are the values of $\bar{\omega}$ for the corresponding mode shapes. The five lowest values are:

$$\bar{\omega} = 2.4048, \ 5.5201, \ 8.6537, \ 11.7915, \ 14.9309$$

Ordinary Rayleigh Method

This method, sometimes known as the Rayleigh quotient, is based on equating the maximum potential and maximum kinetic energies. The maximum potential energy stored in each cycle of vibration is:

$$U_{max} = \frac{1}{2L} \int_0^1 EA(X) \left(\frac{dW}{dX}\right)^2 dX$$

$$= (EA_0/L) \int_0^1 X \left(\frac{dW}{dX}\right)^2 dX \quad [12]$$

The maximum kinetic energy due to the vibration is:

$$T_{max} = \frac{\omega^2 L}{2} \int_0^1 m(X)[W(X)]^2 dX$$

$$= m_0 \omega^2 L \int_0^1 XW^2 dX \quad [13]$$

Equating U_{max} and T_{max} yields the Rayleigh quotient:

$$\bar{\omega}^2 = \frac{m_0 L^2}{EA_0} \omega^2 = \frac{\int_0^1 X(dW/dX)^2 dX}{\int_0^1 XW^2 dX} \quad [14]$$

which is an upper-bound approximation for $\bar{\omega}^2$.

A simple polynomial approximation for the fundamental mode shape which satisfies eqns [7] and [8] is:

$$W(X) = C(1 - X^2) \quad [15]$$

where C is a constant.

Using this expression in eqn [14] yields $\bar{\omega}^2 = 6$ or $\bar{\omega} = 2.4495$, which is approximately 1.9% higher than the exact solution.

The mode shape function for a prismatic (uniform cross-section) beam is simply:

$$W(X) = C \cos(\pi X/2) \quad [16]$$

which also satisfies the boundary conditions [7] and [8] of the present problem. Substitution of eqn [16] into eqn [14] yields $\bar{\omega} = 2.4146$, which is only approximately 0.41% higher than the exact value.

Noninteger Power Rayleigh Method

It was first suggested by Rayleigh and much later further developed by Robert Schmidt (1981) that the power of X in the mode shape function need not be an integer, i.e.:

$$W(X) = C(1 - X^n); \ n > 0 \quad [17]$$

Substituting mode shape [17] into the Rayleigh quotient [14] yields an expression for $\bar{\omega}^2$ that is a function of n only:

$$[\bar{\omega}(n)]^2 = \frac{n}{1 - \frac{4}{n+2} + \frac{1}{n+1}} \quad [18]$$

Since the Rayleigh quotient is an upper bound to the natural frequency (here the fundamental frequency), then the value of n which minimizes $\bar{\omega}(n)$ is the best or optimal value. In the present problem, the optimal value of n is 1.414, which yields $\bar{\omega}_{min} = 2.4142$, only 0.39% higher than the exact solution. The optimization involved in this approach led Laura and Cortinez to call it the 'optimized Rayleigh or Galerkin method'.

Complementary Energy Method

Bhat used the concept of a d'Alembert reversed effective force and the principle of complementary energy. The reversed effective force at a dimensionless position X can be expressed as:

$$P(X) = -\omega^2 L \int_0^X m(X) W(X) \, dX + C_1 \quad [19]$$

where C_1 is a constant of integration, which is zero in the present problem since the bar is free at $X = 0$.

The maximum complementary energy can be expressed as:

$$U_{max}^{C} = \frac{L}{2} \int_{0}^{1} \frac{[P(X)]^2}{EA(X)} dX \quad [20]$$

Now equating T_{max} and U_{max}^{C} from eqns [13] and [20] and using $P(X)$ from eqn [19], one obtains the following result for the square of the natural frequency for general $A(X)$ and $m(X)$ distributions:

$$\omega^2 = \frac{E}{L^2} \frac{\int_{0}^{1} m(X)[W(X)]^2 dX}{\int_{0}^{1} \frac{[\int_{0}^{X} m(X)W(X) dX]^2}{A(X)} dX} \quad [21]$$

For the model problem in which $A(X)$ and $m(X)$ are expressed as in eqns [3] and [4], respectively, eqn [21] becomes:

$$\bar{\omega}^2 = \frac{\int_{0}^{1} X[W(X)]^2 dX}{\int_{0}^{1} \frac{[\int_{0}^{X} XW d]^2}{X} dX} \quad [22]$$

Using eqn [15] for the mode shape in eqn [22], one obtains $\bar{\omega} = 2.4121$, which is only 0.30% higher than the exact value. The considerable reduction in error from 1.9% for the ordinary Rayleigh method to 0.30% for the complementary energy method, using the same mode shape function for each, is due to the avoidance of derivatives of W in eqn [22].

Noninteger Power Complementary Energy Method

Bert *et al.* combined Bhat's version of the complementary energy method (see preceding subsection) with the use of a noninteger power mode shape function. Thus, using mode shape function [17] in eqn [22], one obtains:

$$\bar{\omega}^2 = \frac{n^2/(n+1)}{\frac{1}{(n+2)^2} - \frac{2}{n+4} + \frac{n+2}{8}} \quad [23]$$

The optimal value of n is 1.40, which yields $\bar{\omega} = 2.4055$, only 0.029% higher than the exact value.

Rayleigh–Ritz or Ritz Method

Rayleigh suggested the use of multiple shape functions in conjunction with application of Hamilton's principle to obtain a set of N homogeneous linear algebraic equations in the coefficients. According to Crandall, this idea was implemented by Ritz in 1909 and thus we have two alternative names for the same method: Rayleigh–Ritz or Ritz method.

The advantages of this method are twofold:

- more accurate determination of the fundamental natural frequency and its associated mode shape
- determination of higher natural frequencies and their associated mode shapes

In practice, if one wants to determine N natural frequencies, one must use N shape functions. Let:

$$W(X) = \sum_{i=1}^{N} C_i W_i(X) \quad [24]$$

Also, define the Lagrangian energy functional as:

$$I \equiv \lambda T'_{max} - U_{max} \quad [25]$$

where $T'_{max} \equiv T_{max}/\bar{\omega}^2$ and $\lambda \equiv \bar{\omega}^2$.

Then application of Hamilton's principle leads to the following $N \times N$ set of homogeneous linear algebraic equations in the coefficients C_i:

$$\partial I / \partial C_i = 0; \quad i = 1, 2 \ldots N \quad [26]$$

As an example, we consider the same tapered bar problem analyzed previously. Now we take the case of $N = 2$ as an example, using $W_1 = \cos(\pi X/2)$ and $W_2 = \cos(3\pi X/2)$. This leads to:

$$\begin{bmatrix} k_{11} - \lambda m_{11} & k_{12} - \lambda m_{12} \\ k_{21} - \lambda m_{21} & k_{22} - \lambda m_{22} \end{bmatrix} \begin{Bmatrix} C_1 \\ C_2 \end{Bmatrix} = \begin{Bmatrix} 0 \\ 0 \end{Bmatrix} \quad [27]$$

where:

$$k_{ij} \equiv \int_{0}^{1} X \left(\frac{dW_i}{dX} \right) \left(\frac{dW_j}{dX} \right) dX$$

$$m_{ij} \equiv \int_{0}^{1} X W_i W_j dX \quad [28]$$

To guarantee a nontrivial solution of eqn [27], the determinant of the coefficients must be set equal to zero. This leads to an $N \times N$ determinant which is an

N-degree polynomial equation in λ. In the present case, this is the following quadratic equation:

$$(m_{11}m_{22} - m_{12}^2)\lambda^2 - (k_{11}m_{22} + k_{22}m_{11} - 2k_{12}m_{12})\lambda + k_{11}k_{22} - k_{12}^2 = 0 \quad [29]$$

The results are:

$$\bar{\omega}_1 = 2.4062, \quad \bar{\omega}_2 = 5.5298$$

The value of $\bar{\omega}_1$ is only 0.058% higher than the exact value. This is a considerable improvement over the one-term solution (Rayleigh method) which was 0.41% higher than the exact value. The second harmonic value $\bar{\omega}_2$ could not be obtained at all from the one-term solution. Here it is not only obtained, but its value is only 0.175% higher than the exact value.

Galerkin Method

According to Crandall, the Galerkin method was originated in 1915 by B.G. Galerkin and it is based on minimizing the error in an assumed-mode method by making the error orthogonal to a weighting function. In this method, the weighting function is taken to be the function W_i itself. The method can have a one-term solution and the result is identical to eqn [14] for the ordinary Rayleigh method. It can also be implemented as a multiterm solution form and then it coincides with the Rayleigh–Ritz method. In fact, some texts do not distinguish between the Rayleigh, Rayleigh–Ritz, and Galerkin methods.

So long as the Galerkin method is equivalent to the Rayleigh or Rayleigh–Ritz method, only the geometric boundary conditions need be satisfied by the trial function(s). However, it is cautioned that the resulting Galerkin-method equations are not always equivalent to the Rayleigh or Rayleigh–Ritz equations and in this general case, all of the boundary conditions, force type (such as eqn [2a]) as well as geometric type (such as eqn [2b]) must be satisfied. Furthermore, the upper-bound property of the Rayleigh and Rayleigh–Ritz methods no longer holds.

Lower-bound Approximations

In many practical problems of free vibration, it may be more important to obtain a lower-bound approximation to a given natural frequency than an upper bound. A popular lower-bound approximation is known as the enclosure theorem. If the governing ordinary differential equation is expressed as:

$$L(W) - \lambda MW = 0 \quad [30]$$

where L is a differential operator, M is a function of position X, and $\lambda (\equiv \bar{\omega}^2)$ is the eigenvalue. Now λ^* is defined as:

$$\lambda^* \equiv \frac{L(W)}{MW}; \quad 0 \leq X \leq 1 \quad [31]$$

Then, the theorem says that λ is contained in the interval:

$$\lambda^*_{min} \leq \lambda \leq \lambda^*_{max}$$

In the present problem:

$$L(W) = \frac{d}{dX}\left(X\frac{dW}{dX}\right); \quad MW = -XW$$

First, let's consider the trigonometric solution, eqn [16]. Then eqn [31] gives:

$$\lambda^* = \frac{\pi}{2}\left[\frac{\pi}{2} + \frac{1}{X}\tan\frac{\pi X}{2}\right]$$

and:

$$\lambda^*_{min} = \lambda^*|_{X=0} = \frac{\pi}{2}\left(\frac{\pi}{2} + 1\right) \text{ or } \bar{\omega}_{min} = 2.0095$$

This gives a very wide spread from $\bar{\omega}_{min}$ to $\bar{\omega}_{max}$ (2.0095–2.4146).

Now let's take a look at the power function solution in eqn [17]. Eqn [31] gives:

$$\lambda^* = \frac{n^2 X^{n-2}}{1 - X^n}$$

and:

$$\lambda^*_{min} = \frac{n^2 X^{n-2}}{1 - X^n}\bigg|_{X=0}$$

It is noted that if $n > 2$, $\lambda^*_{min} = 0$ and if $n < 2$, λ^*_{min} increases without limit. Thus, it is clear that the only usable value of n in so far as the enclosure theorem is concerned is $n = 2$. Then $\lambda^*_{min} = 4$ or $\bar{\omega}_{min} = 2.0000$, not as good as the previous lower bound.

Bert found that the following equation, proposed by Ku, was an excellent predictor of the lower bound for the buckling load of a column:

$$\bar{P}_{LB} = \bar{P}_R - [\bar{P}_R(\bar{P}_R - \bar{P}_T)]^{1/2} \quad [32]$$

where $\bar{P}$ denotes the dimensionless critical load and subscripts R and T denote Rayleigh and Timoshenko, respectively. The vibrational analog of eqn [32] is:

$$\bar{\lambda}_{LB} = \bar{\lambda}_R - [\bar{\lambda}_R(\bar{\lambda}_R - \bar{\lambda}_C)]^{1/2} \quad [33]$$

where $\bar{\lambda}(\equiv \omega^2)$ is the dimensionless eigenvalue, and subscripts R and C denote the Rayleigh quotient (eqn [14]) and the complementary energy quotient (eqn [22]). For the present case, disregarding the difference between $n = 1.41$ and $n = 1.40$, $\bar{\omega}_R = 2.4142$ and $\bar{\omega}_c = 2.4055$. Then eqn [33] gives $\bar{\lambda}_{LB} = 5.7435$ or $\bar{\omega}_{LB} = 2.3966$, which is only 0.34% below the exact value.

Differential Transformation Method

This relatively new technique is an exact series solution of a linear or nonlinear differential equation. The method is based on the Taylor series expansion of an arbitrary function $w(x)$ at a point $x = 0$:

$$w(x) = \sum_{k=0}^{\infty} (x^k/k!) \left[\frac{d^k w}{dx^k}\right]_{x=0} \quad [34]$$

The differential transformation (DT) of $w(x)$ is defined as:

$$W(k) = \frac{1}{k!} \left[\frac{d^k w}{dx^k}\right]_{x=0} \quad [35]$$

where the capital letter denotes the DT of the same letter in lower case. The inverse DT is:

$$w(x) = \sum_{k=0}^{\infty} x^k W(k) \quad [36]$$

The governing differential equation for the mode shape of an axially vibrating, linearly tapered bar is rewritten from eqn [9] as:

$$xw_{,xx} + w_{,x} + \bar{\omega}^2 xw = 0; \quad 0 \leq x \leq 1 \quad [37]$$

where $(\ )_{,x}$ denotes $d(\ )/dx$, etc.

To take the differential transform of eqn [37], the table of transforms in **Table 1**, based on eqns [35] and [36], is necessary.

Then the DT of eqn [37] is:

$$\sum_{j=0}^{k} \delta(j-1)(k-j+1)(k-j+2)W(k-j+2)$$
$$+(k+1)W(k+1) + \bar{\omega}^2 \sum_{j=0}^{k} \delta(j-1)W(k-j) = 0$$
$$k = 0, 1 \ldots \infty \quad [38]$$

The boundary conditions are eqns [8] and [7], which in the present notation are:

$$w(1) = 0; \quad w_{,x}(0) = 0 \quad [39]$$

and transform to:

$$\sum_{k=0}^{\infty} W(k) = 0; \quad \sum_{k=0}^{\infty} kx^{k-1}W(k) = W(1) = 0$$
$$[40a, b]$$

To illustrate the procedure for finding the eigenvalues, with the use of eqn [40b], eqn [38] yields:

$$W(2) = \frac{-1}{4}\bar{\omega}^2 W(0), \quad W(3) = W(5) = 0,$$
$$W(4) = \frac{1}{64}\bar{\omega}^4 W(0) \quad [41]$$

Substitution of eqns [41] into eqn [40a] yields:

$$\left(1 - \frac{1}{4}\bar{\omega}^2 + \frac{1}{64}\bar{\omega}^4\right) U(0) = 0 \quad [42]$$

Table 1 Table of differential transforms (DT)

Original function	DT
$f(x)g(x)$	$\sum_{j=0}^{k} F(j)G(k-j)$
$w_{,x}(x)$	$(k+1)W(k+1)$
$w_{,xx}(x)$	$(k+1)(k+2)W(k+2)$
x	$\delta(k-1) = \begin{cases} 1 & \text{if } k = 1 \\ 0 & \text{if } k \neq 1 \end{cases}$

which has roots ±2.8284. The positive real root (2.824) is the eigenvalue for $N = 5$. To obtain more accurate results more terms are needed and the results are tabulated in **Table 2**.

It is clear that more terms are needed to calculate the second mode than the first and that convergence is from above.

A Two-dimensional Problem and a Galerkin Solution

As the model two-dimensional problem, a uniform thickness, thin, rectangular plate made of isotropic material and clamped on all four edges is considered. For these particular boundary conditions there is no known closed-form solution. In fact, the only set of boundary conditions for which there are closed-form solutions is simply supported on all four sides.

The plate has a rectangular planform and the origin of the Cartesian coordinate system is in the midplane of the plate and at its center. The plate is of length $2a$ and $2b$ in the x and y directions and its thickness (in the z direction) is h. For small-amplitude, out-of-plane, free vibration, the governing partial differential equation is:

$$-D(w_{,xxxx} + 2w_{,xxyy} + w_{,yyyy}) = \rho h w_{,tt} \quad [43]$$

where $w(x, y, t)$ is the plate deflection, ρ is the density of the plate material, and $(\ )_{,tt} \equiv \partial^2(\ )/\partial t^2$, etc. The flexural rigidity, denoted by D, is given by:

$$D = \frac{Eh^3}{12(1-\nu^2)} \quad [44]$$

Since the problem is linear, the deflection can be assumed to be the product of the mode shape $W(x, y)$ and a harmonic time function:

$$w(x, y, t) = W(X, Y)e^{i\omega t} \quad [45]$$

where $X \equiv x/a$ and $Y \equiv y/b$.

Then the governing equation becomes:

$$W_{,XXXX} + 2\lambda^2 W_{,XXYY} + \lambda^4 W_{,YYYY} - \rho h a^4 \omega^2 W = 0 \quad [46]$$

where $\lambda \equiv b/a$ is the plate aspect ratio.

There are three popular mode shapes used in connection with Galerkin or Rayleigh–Ritz analyses of plate vibration:

1. Polynomial. This is the simplest and it turns out to be the most accurate for the all-clamped-edge conditions considered here.
2. Trigonometric. This gives accurate results for all simply supported edge conditions but a very poor result for the present case.
3. Beam functions. These are the exact solutions for free vibration of a uniform beam, but the expressions are unwieldy. They have the advantage of being able to approximate higher modes, not just the fundamental one.

Here, for simplicity and accuracy, the polynomial form is chosen:

$$W(X, Y) = W_0(1 - X^2)^2(1 - Y^2)^2 \quad [47]$$

The Galerkin integral then becomes:

$$\int_{-1}^{1}\int_{-1}^{1} \left(W_{,XXXX} + 2\lambda^2 W_{,XXYY} \right. \\ \left. + \lambda^4 W_{,YYYY} - \frac{\rho h a^4 \omega^2}{D} W \right) \\ \times (1 - X^2)^2(1 - Y^2)^2 \, dX \, dY = 0 \quad [48]$$

Performing the integration yields:

$$\bar{\omega} \equiv a^2(\rho h/D)^{1/2}\omega = [31.5(1 + \lambda^4) + 18\lambda^2]^{1/2} \quad [49]$$

Dimensionless fundamental frequencies obtained by various methods for the case of a square plate ($\lambda = 1$) are listed in **Table 3**.

Table 2 Dimensionless frequencies obtained by the *DT* technique versus number of terms, *N*

Mode	N	5	7	9	10	
1st		2.8284	2.3916	2.4056	2.4048	(exact value)
	N	10	12	20	21	
2nd		5.9893	5.4059	5.5183	5.5201	(exact value)

Plate 1 Active Control of Civil Structures.
Actuator used in the Hybrid Mass Damper System in INTES.

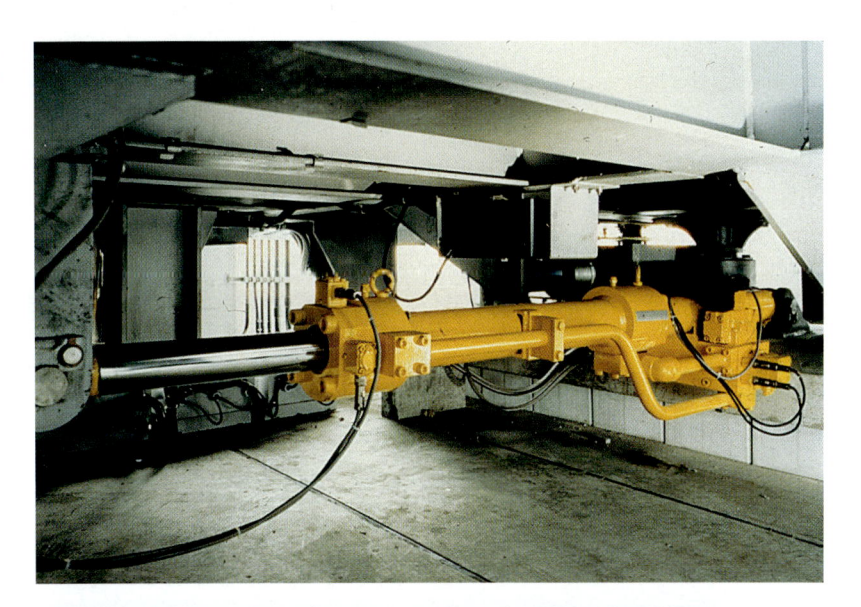

Plate 2 Active Control of Civil Structures.
Variable Stiffness System used in Kajima Research Laboratory.

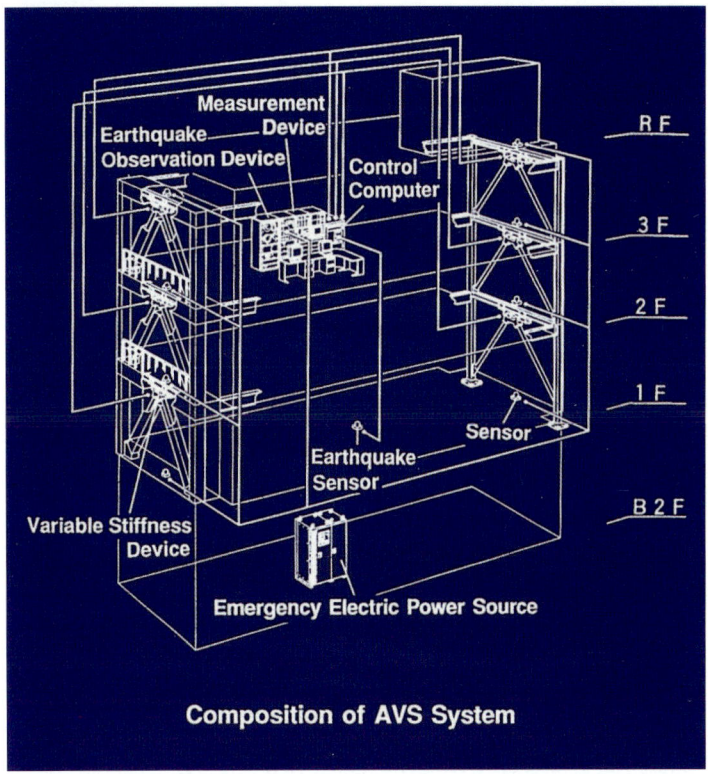

Plate 3 Active Control of Civil Structures.
The Hybrid Mass Damper used in the Landmark Tower.

Skeleton
Composite/structural materials

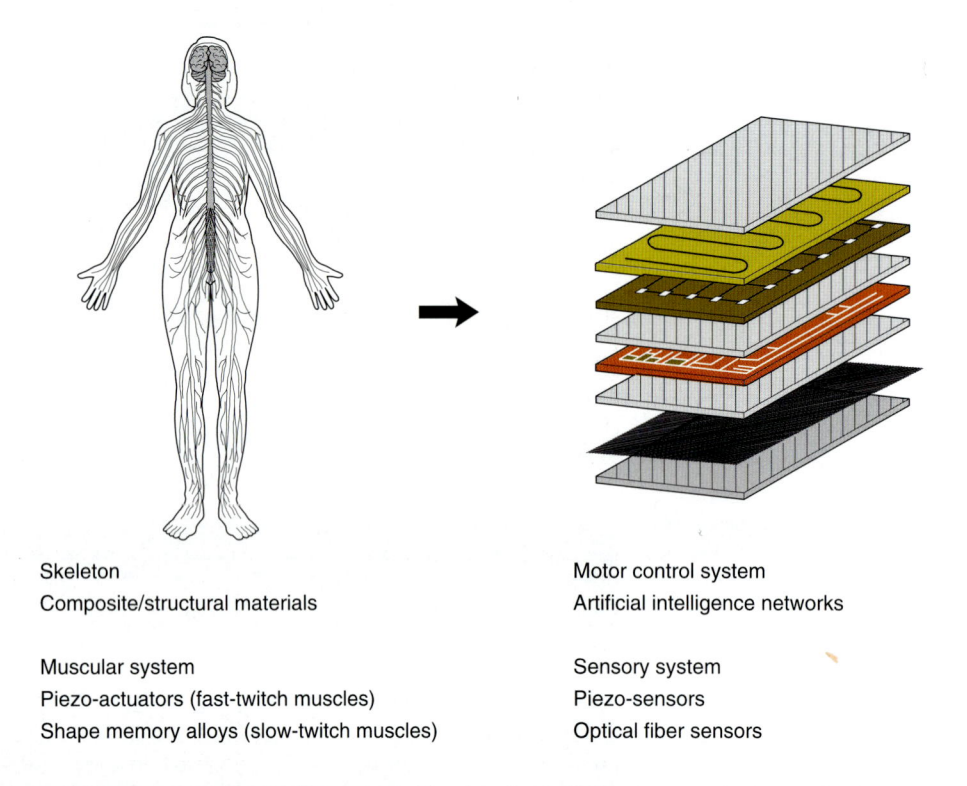

Muscular system
Piezo-actuators (fast-twitch muscles)
Shape memory alloys (slow-twitch muscles)

Motor control system
Artificial intelligence networks

Sensory system
Piezo-sensors
Optical fiber sensors

Plate 4 Actuators and Smart Structures. Biomimesis parallelism between the human body and a smart material system.

(A)

Austenite
Cool / Heat recovery
Martensite (Twinned)
Deform
Martensite (Deformed)
ε

(B)

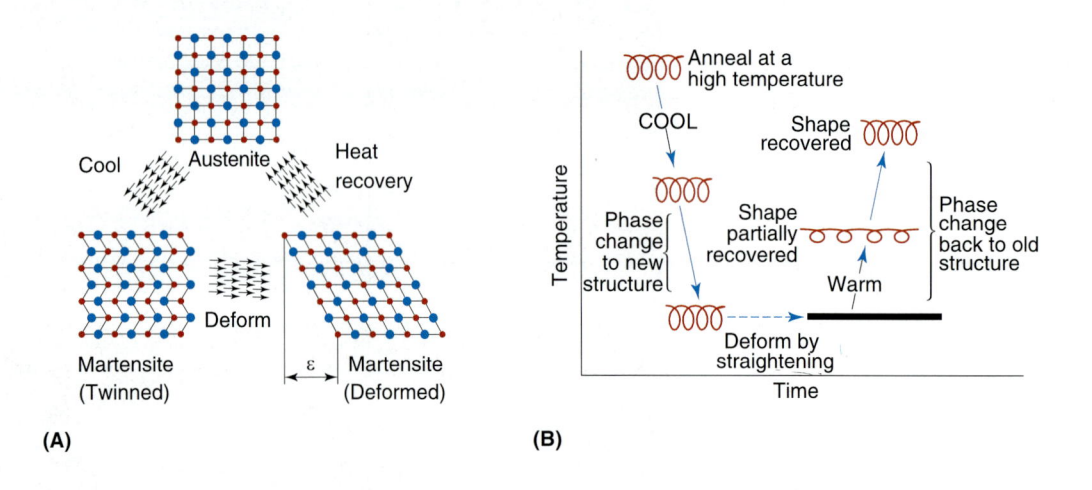

Anneal at a high temperature
COOL
Phase change to new structure
Shape partially recovered
Deform by straightening
Warm
Shape recovered
Phase change back to old structure

Temperature
Time

Plate 5 Actuators and Smart Structures. Principles of SMA materials: (A) change in crystallographic structure during cooling and heating; (B) associated component-shape changes, using a coil spring as an example.

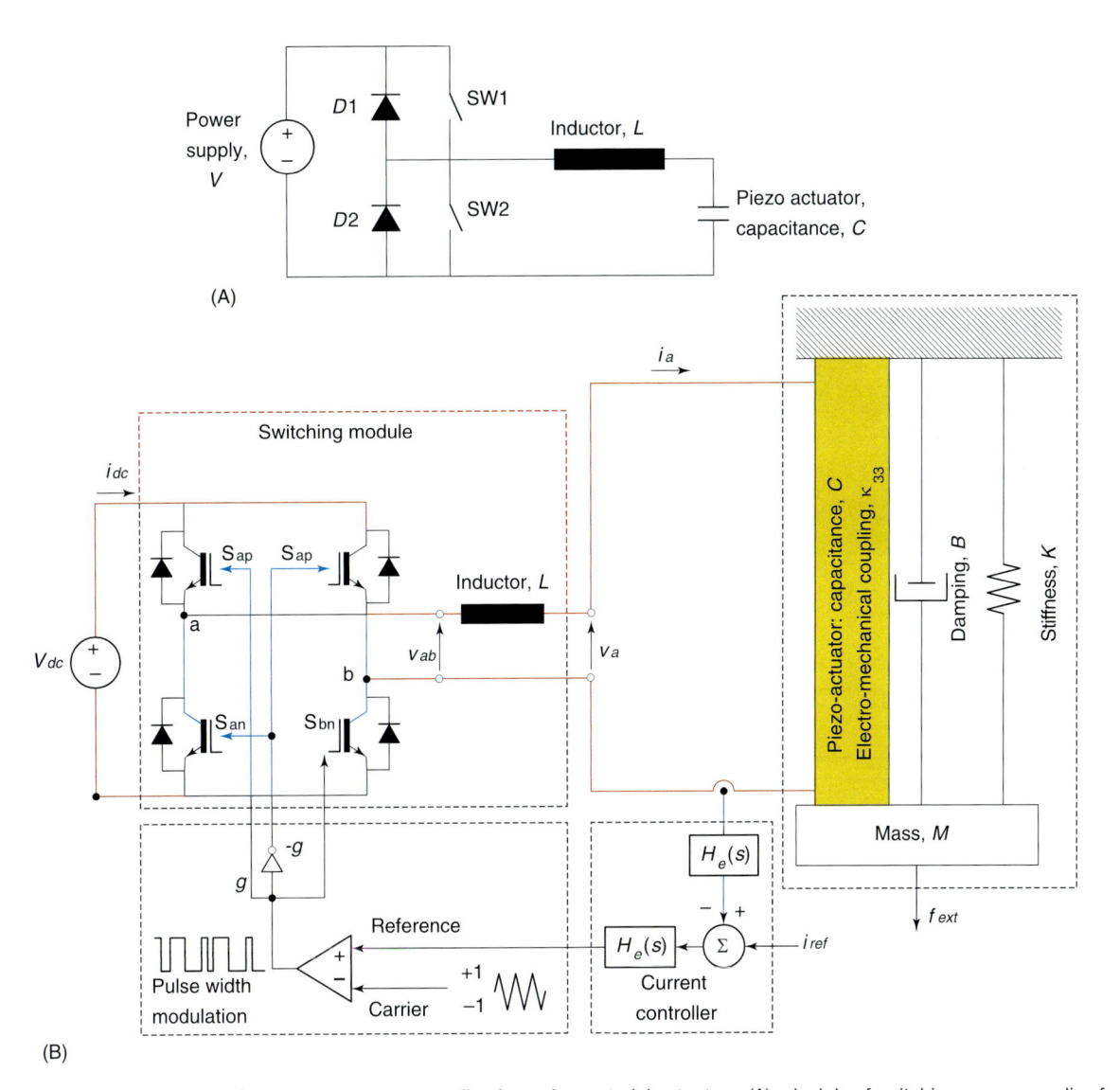

Plate 6 Actuators and Smart Structures. Power supplies for active material actuators: (A) principle of switching power supplies for high reactive load; (B) schematic of the supply system incorporating the switching module, current controller, pulse width modulator, and the piezo-actuator-external load assembly.

Plate 7 Basic Principles. Joseph Louis Lagrange (1736-1813). Italian-French mathematician. (With permission from Mary Evans Picture Library).

Plate 8 Beams. Leonard Euler, Swiss mathematician. From a picture by A. Lorgnal in the collection of the Institute of France. (BBC Hulton Picture Library).

Plate 9 Bridges. The Millennium Bridge, London, UK. (With permission from Tony Kyriacou).

Plate 10 Bridges. View of the transporter bridge which spans the River Tees at Middlesborough in the UK. The bridge was constructed in 1911 to replace the old ferryboat crossing there. It is a total of 260m wide and around 68m high. A cradle suspended from the main girder ferries cars and passengers across the 174m stretch of water in about 2 minutes. (With permission from Science Photo Library).

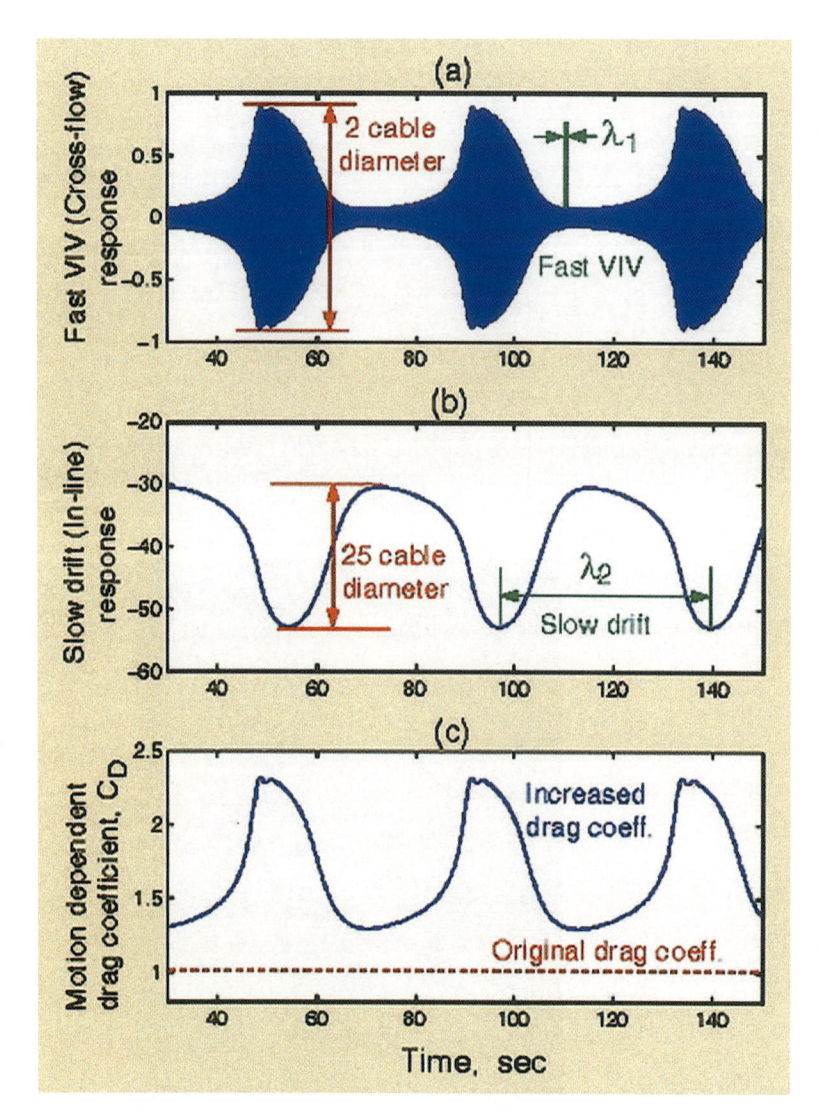

Plate 11 Cables. Coupled fast and slow response of cable-buoy system (a) fast (small amplitude) cross-flow vortex-induced vibration of the cable and an antidote; (b) slow (large amplitude) in-line drift of the cable at the upper end; (c) variation of the drag coefficient due to modulations of the cross-flow amplitude.

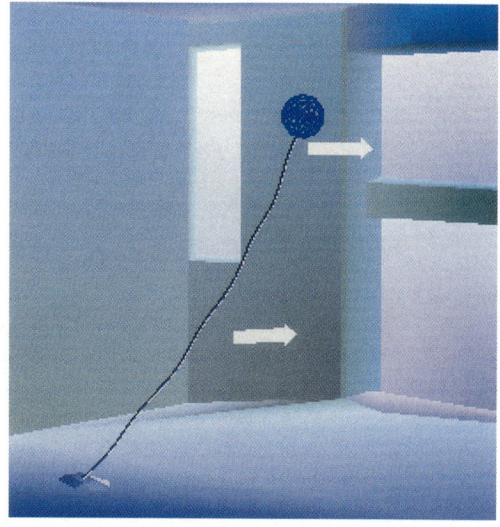

Plate 12 Cables. Underwater cable-buoy system with uniform current parallel to the cable equilibrium plane.

Plate 13 Chaos. Chaotic systems: head-on collision of two dipolar vortices in a stratified fluid environment. The original vortices, dyed orange and green, have exchanged a partner to form two new (mixed) dipoles which are moving at roughly right angles to the original direction of travel, that is, towards the top and bottom of the image. The green fluid was injected from the right, the orange from the left. Dipolar vortices are relevant to turbulence in large-scale geophysical systems such as the atmosphere or oceans. Turbulence in fluid systems is one example of a chaotic system. (With permission from Science Photo Library).

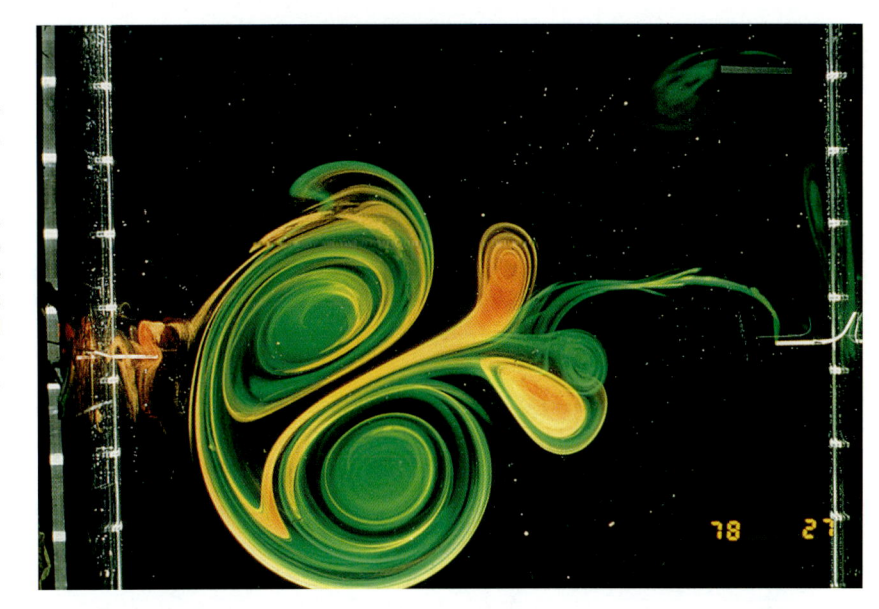

Plate 14 Chaos. Fractal basin boundary for a magnetopendulum system. This colorful picture is a fractal basin boundary for a magnetopendulum system. $\theta + \delta\dot\theta + \sin(\theta) = f\cos(\theta)\cos(\omega t)$, when $\delta = 0.25$, $\omega = 1.0$, and $f = 1.40$ (the homoclinic bifurcation value is 0.798) in the state space $(\theta, d\theta/dt)$. The blue represents initial points which have counterclockwise rotating solutions. The black represents the points whose solutions never settle down in a time interval of 90π computer seconds.

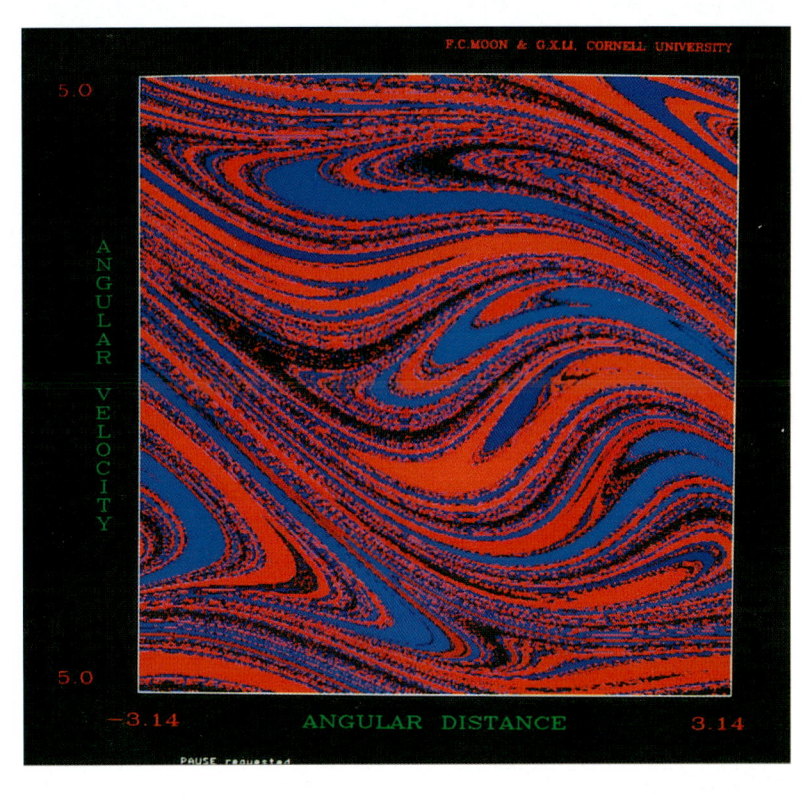

Plate 15 Chaos. Jules Henri Poincare (1854-1912). French mathematician. (With permission from Mary Evans Picture Library).

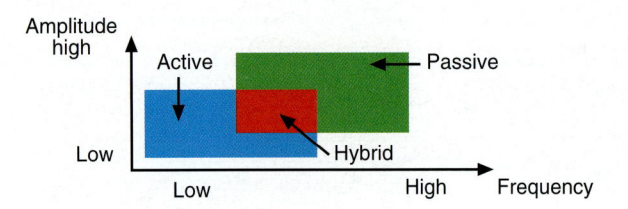

Plate 16 Damping, Active. Operating range of various damping methods.

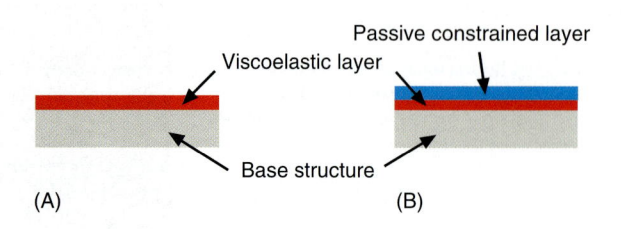

Plate 17 Damping, Active. Viscoelastic damping treatments. (A) Free; (B) constrained.

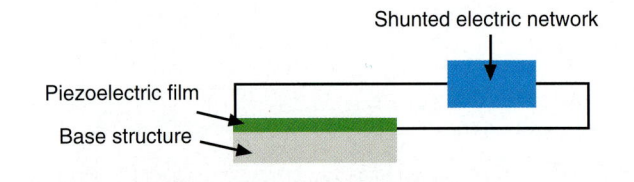

Plate 18 Damping, Active. Shunted piezoelectric treatments.

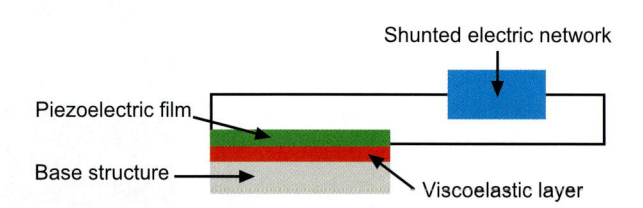

Plate 19 Damping, Active. Damping layers with shunted piezoelectric treatments.

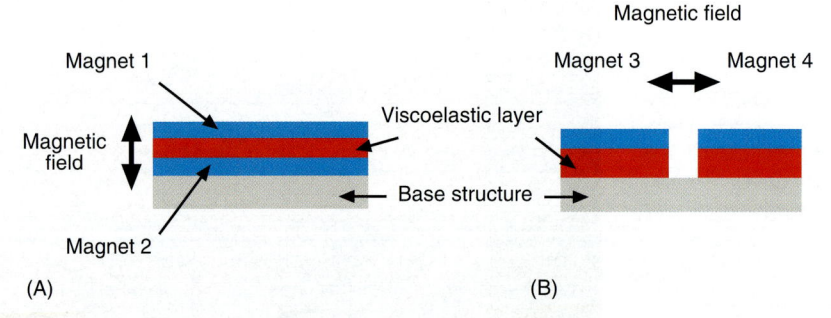

Plate 20 Damping, Active. Configurations of the MCLD treatment (A)2 Compression MCLD; (B) shear MCLD.

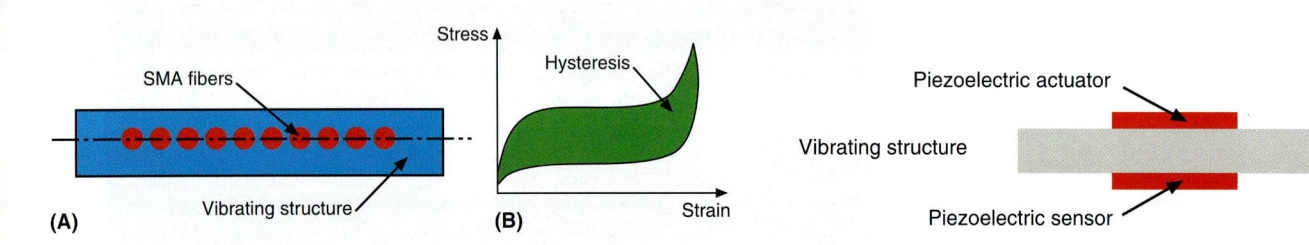

Plate 21 Damping, Active. Damping with shape memory fibers. (A) SMA-reinforced structure; (B) superelastic characteristics.

Plate 22 Damping, Active. Active damping

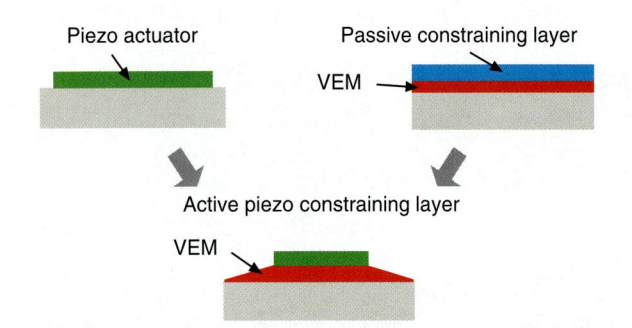

Plate 23 Damping, Active. Active constrained layer damping treatment.

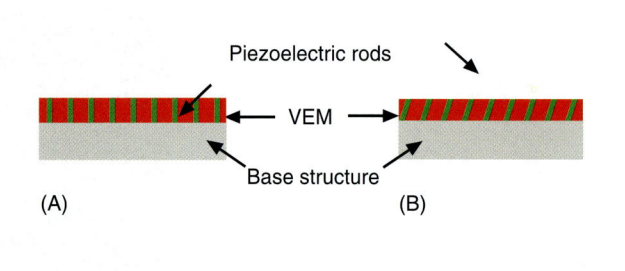

Plate 24 Damping, Active. Active piezoelectric damping composites.

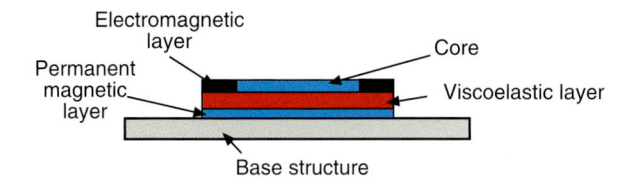

Plate 25 Damping, Active. Electromagnetic damping composite (EMDC).

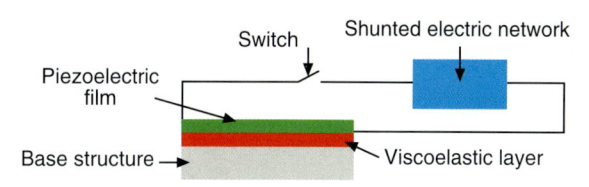

Plate 26 Damping, Active. Damping layers with shunted piezoelectric treatments.

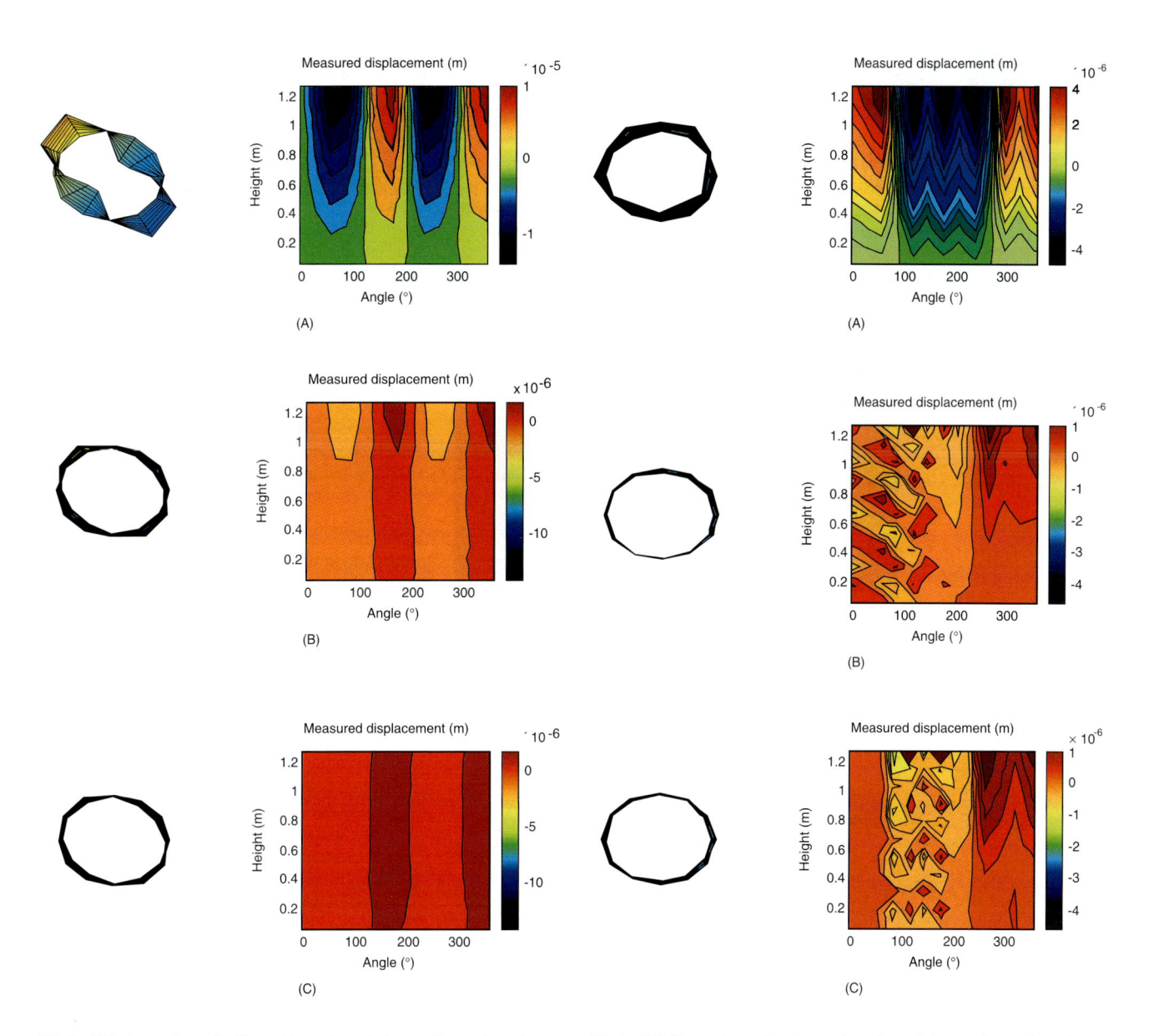

Plate 27 Damping, Active. Experimental results using laser vibrometer before and after control for mode (1,2). (A) PLCD; (B) ACLD (open actuator); (C) ACLD (two actuators).

Plate 28 Damping, Active. Experimental results using laser vibrometer before and after control for mode (1,1). (A) PCLD; (B) ACLD (one actuator); (C) ACLD (two actuators).

Plate 29 Earthquake Excitation and Response of Buildings. Earthquake damage. Collapsed road bridges after an earthquake. Earthquakes are caused when sections of the earth (tectonic plates) move against each other to relieve stress. Photograph of highway near Watsonville in California, USA, in 1990. California lies on the boundary between the North American and Pacific plates. (With permission from Science Photo Library).

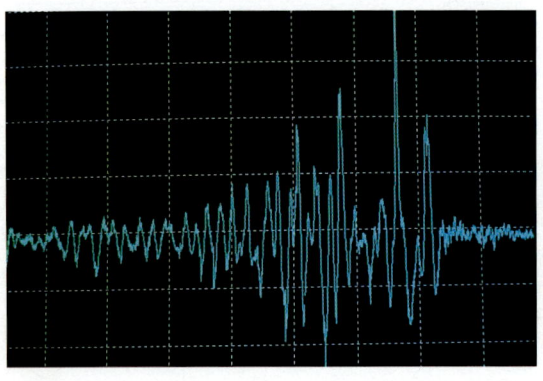

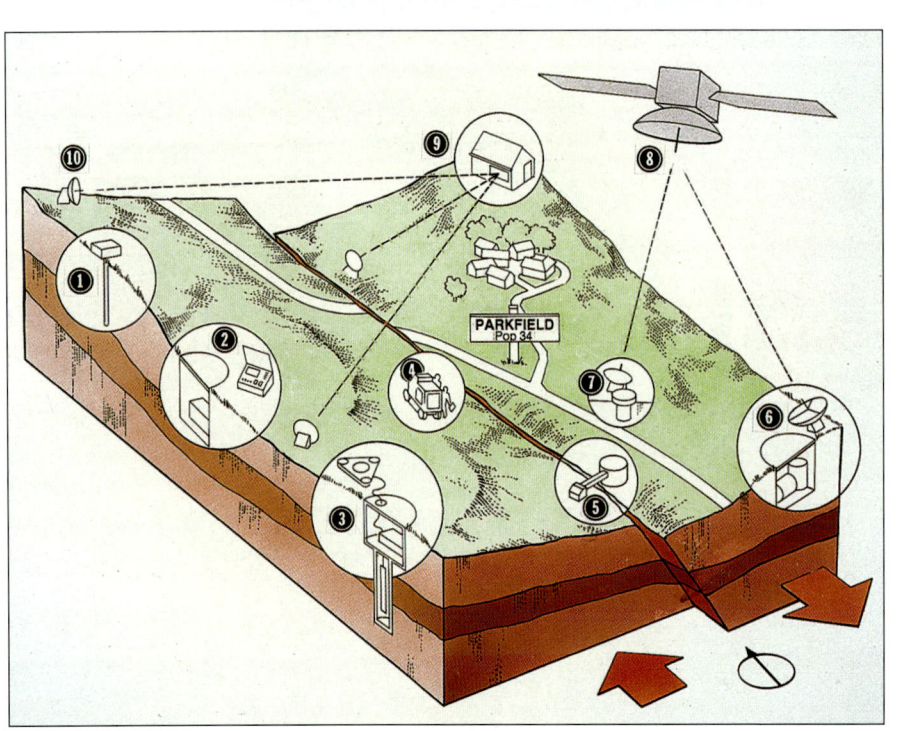

Plate 31 Earthquake Excitation and Response of Buildings.
1999 Taiwan earthquake. Apartment block with extensive damage.

Plate 30 Earthquake Excitation and Response of Buildings.
Kobe earthquake seismograph. Screen display of a seismograph of the Kobe earthquake of 16 January 1996. The horizontal divisions represent acceleration along a north-south line with the third line from top representing zero acceleration. The vertical divisions represent time intervals of five seconds. Kobe, Japan, was struck by an earthquake which measured 7.1 on the Richter scale. Image recorded at the University of Tokyo. (With permission from Science Photo Library).

Plate 32 Earthquake Excitation and Response of Buildings. Diagram of Parkfield earthquake prediction experiment, showing instrumentation: (1) seismometer in hole to record microquakes; (2) magnetometer to record magnetic field; (3) near-surface seismometer to record larger shocks; (4) VIBREOSIS truck that creates shear waves to probe the earthquake zone; (5) creepmeter to record surface movement; (6) strainmeter to monitor surface deformation; (7) sensors in water well to monitor groundwater level; (8) satellite relaying data to US Geological Survey; (9) laser to measure surface movement by bouncing beams on reflectors (10). Arrows show crustal plates movements along San Andreas fault. (With permission from Science Photo Library).

Table 3 Dimensionless fundamental frequencies for square plates by different methods

Method used	No. of terms used	$\bar{\omega}$	% error
Exact (Tomotika)		8.9966	
Rayleigh–Ritz (trigonometric)	36	8.9975	0.010
Rayleigh–Ritz (polynomial)	4	8.9998	0.036
Galerkin (polynomial)	1	9.0000	0.038
Galerkin (trigonometric)	1	9.3060	3.440

It can readily be seen that the present one-term polynomial Galerkin solution is quite close to higher-term Rayleigh–Ritz solutions and the exact solution, while the one-term trigonometric Galerkin solution gives a poor result.

Acknowledgment

The assistance of Ms Huan Zeng with the differential transformation method is gratefully acknowledged.

Nomenclature

$A(\ )$	Cross-sectional area (variable)
A_o	reference area coefficient
a	plate half-length
b	plate half-width
C	coefficient
C_i	constants of integration
D	plate flexural rigidity
E	elastic modulus
f, g	functions of x
F, G	transforms of f and g
h	plate thickness
I	Lagrangian energy
J_o	zero-order Bessel function of first kind
k	constant in eqn [6]
k_{ij}	coefficient defined in eqn [28]
L	length of bar
$L(W)$	differential operator operating on W
M	function of X
$m(x)$	mass per unit length as function of x
m_{ij}	coefficients defined in eqn (28)
m_o	reference mass coefficient
N	number of terms
n	exponent
$P(X)$	reversed effective force
T_{max}	maximum kinetic energy
T'_{max}	$T_{max}/\bar{\omega}^2$
t	time
U_{max}	maximum potential energy per cycle
U^C_{max}	maximum complementary energy per cycle
u	axial displacement
W	mode shape function
$W(k)$	differential transform of $w(x)$
$w(x)$	function of x
X	x/L
X	x/a
x	axial position
Y	y/b
Y_o	zero-order Bessel function of second kind
y	transverse position
λ	$\bar{\omega}^2$
λ	a/b
λ^*	$L(W)/MW$
$\bar{\lambda}_i$	various values of $\bar{\omega}^2$
ν	Poissons ratio
ω	natural frequency
$\bar{\omega}$	dimensionless natural frequency
$(\ )_{,tt}$	$\partial^2(\ )/\partial t^2$

See also: **Eigenvalue analysis**; **Modal analysis, experimental**, Parameter extraction methods; **Nonlinear systems analysis**.

Further Reading

Batdorf SB (1969) On the application of inverse differential operators to the solution of cylinder buckling and other problems. In: *Proceedings, AIAA/ASME 10th Structures, Structural Dynamics, and Materials Conference*, New Orleans, LA. ASME, New York, pp. 386–391.

Bert CW (1987a) Application of a version of the Rayleigh technique to problems of bars, beams, columns, membranes, and plates. *Journal of Sound and Vibration* 119: 317–326.

Bert CW (1987b) Techniques for estimating buckling loads. In: Cheremisinoff PN, Cheremisinoff NP and Cheng SL (eds). *Civil Engineering Practice*, vol. 1: *Structures*. Lancaster, PA: Technomic, pp. 489–499.

Bert CW, Jang SK and Striz AG (1988) Two new approximate methods for analyzing free vibration of structural components. *AIAA Journal* 26: 612–618.

Bhat RB (1984) Obtaining natural frequencies of elastic systems by using an improved strain energy formulation in the Rayleigh-Ritz method. *Journal of Sound and Vibration* 93: 314–320.

Collatz L (1948) *Eigenwertprobleme*. New York: Chelsea, p. 135.

Crandall SH (1956) *Engineering Analysis*. New York: McGraw-Hill.

Ku AB (1977) Upper and lower bound eigenvalues of a conservative discrete system. *Journal of Sound and Vibration* 53: 183–187.

Laura PAA and Cortinez VH (1986) Optimization of eigenvalues when using the Galerkin method. *AIChE Journal* 32: 1025–1026.

Malik M and Dang HH (1998) Vibration analysis of continuous systems by differential transformation. *Applied Mathematics and Computation* 96: 17–26.

Meirovitch L (1967) *Analytical Methods in Vibrations*. New York: Macmillan.

Rao SS (1995) *Mechanical Vibrations*, 3rd edn. Reading, MA: Addison Wesley.

Rayleigh Lord (Strutt JW) (1870) On the theory of resonance. *Philosophical Transactions, Royal Society (London)* A161: 77–118. See also: Rayleigh Lord (1945) *Theory of Sound*, vol. 2, 2nd edn. New York: Dover Publications.

Schmidt R (1981) A variant of the Rayleigh–Ritz method. *Industrial Mathematics* 31: 37–46.

Singer J (1962) On the equivalence of the Galerkin and Rayleigh–Ritz methods. *Journal of the Royal Aeronautical Society* 66: 592.

CORRELATION FUNCTIONS

S Braun, Technion – Israel Institute of Technology, Haifa, Israel,

Copyright © 2001 Academic Press

doi: 10.1006/rwvb.2001.0170

Introduction

The notion of correlation is one of the most basic ones in the description of data, and especially joint descriptions. Joint descriptions between data points, whether from single or joint data sets, can describe patterns existing in the data. Correlation functions and matrices are often used to define or describe the patterns and dynamic behavior of vibration signals and vibrating systems.

The following first recalls basic correlation concepts, and their application to time functions. Stochastic random processes can be described by their autocorrelation as well as the spectral density function, and the relation between these presentations is described next. The possibility of using correlation concepts to define nonstationary random data is almost immediate.

Processing discrete data often involves the notion of the correlation matrix, which is briefly defined.

Classic as well as modern FFT based, computational schemes are described, including some notions of the variability of the estimated parameters.

The last part of the entry briefly describes some engineering applications, all relevant to aspects of vibration processing: the detection of delays in dispersionless propagation, spiking filters, and AR modeling. Adaptive line enhancer and adaptive noise cancellation applications conclude the entry.

Correlation Functions

Basic concepts for assessing the degree of linear dependence between two data sets x_k and y_k are based on covariance and correlation coefficients. The correlation coefficient R is then:

$$R = \frac{\sum_k x_k y_k}{\left(\sum_k x_k^2\right)^{1/2} \left(\sum_k y_k^2\right)^{1/2}} \quad [1]$$

A specific example for which $R = 0.7$ is shown in **Figure 1**.

The notion of these descriptors of dependence can be extended to time-history data. Thus we have the covariance function between the time data $x(t)$ and $y(t)$ as:

$$C_{xy}(\tau) = E\left[\{x(t) - \mu_x\}\{y(t+\tau) - \mu_y\}\right] \quad [2]$$

where $E[.]$ denotes expectations, $\mu_x = E[\{x(t)\}]$, and $\mu_y = E[\{y(t)\}]$. For zero-mean functions, we may use the cross-correlation function:

$$R_{xy}(\tau) = E[x(t)y(t+\tau)] \quad [3]$$

which for continuous functions would be computed as:

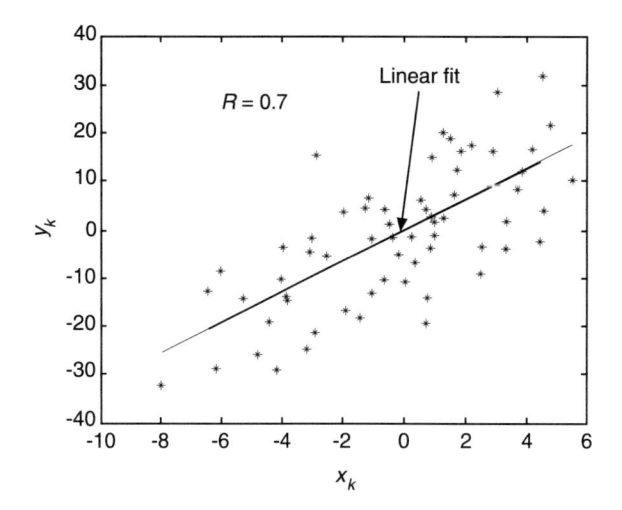

Figure 1 Linear dependence between two data sets.

$$R_{xy}(\tau) = \lim \frac{1}{T} \int_0^T x(t) y(t+\tau) \, dt \qquad [4]$$

$$T \to \infty$$

For the special case where $x(t) = y(t)$ we have the autocovariance and autocorrelation functions $C_{xx}(\tau)$ and $R_{xx}(\tau)$. Some general properties of the autocorrelation function are: first, the autocorrelation function is an even function of τ:

$$R_{xx}(\tau) = R_{xx}(-\tau) \qquad [5a]$$

and secondly, for $\tau = 0$:

$$R_{xx}(0) = \sigma_x^2 + \mu_x^2 \qquad [5b]$$

where $\sigma_x^2 = E[(x - \mu_x)^2]$ is the variance of $x(t)$. For zero-mean data (the prevalent case for vibration signals) $R_{xx}(0)$ is the mean square and thus also the power (the square of the RMS) of the signal. The cross-correlation function has the inequality:

$$|R_{xy}(\tau)| \leq [R_{xx}(0) R_{yy}(0)]^{1/2} \qquad [6]$$

It is instructive to note that tests for linear dependence are not necessarily applied to time data. Expressions analogous to the time domain correlation concept can be found in well known vibration engineering applications. One example is the coherence function (see **Spectral analysis, classical methods**) used to test the linear dependence between spectral estimates. It is given by:

$$\gamma^2(\omega) = \frac{S_{xy}(\omega) S_{xy}^*(\omega)}{S_{xx}(\omega) S_{yy}(\omega)} \qquad [7]$$

where * denotes the complex conjugate. This is actually the correlation coefficient (squared) in the frequency domain, i.e., computed for each frequency ω.

Another example concerns the comparison of the spatial deflection of vibrating structures. The analytical mode shapes ϕ_a computed via FE analysis are compared with experimentally derived ones ϕ_{ex}. The mode shapes are associated with a modal frequency and in the general case are complex vectors. They are often compared by the modal assurance criterion (MAC), defined as:

$$\text{MAC}\left[(\phi_a)_i (\phi_{ex})_j\right] = \frac{\left|(\phi_a)_i^T (\phi_{ex}^*)_j\right|^2}{\left[(\phi_a)_i^T (\phi_a^*)_j\right] \left[(\phi_{ex})_i^T (\phi_{ex}^*)_j\right]} \qquad [8]$$

where i and j are the indices or the modal frequencies. Again this can be recognized as a correlation coefficient for deflection data (see **Comparison of vibration properties: Comparison of modal properties**).

Stochastic Processes: Correlation and Power Spectral Density Functions

Random phenomena are represented by the notion of a stochastic process. A random variable with a probability distribution function (as well as joint probability distribution functions between any sets of times) is associated to each point in time. This is the probability description of a random phenomenon, extended to $\pm \infty$ at any time, and also itself in the $\pm \infty$ region. This set is called an ensemble, and any measured signal is then considered as one realization of such an ensemble. As in any statistical approach, the measured data are considered as one of the many sets which might have occurred.

Moments of the stochastic process should, in principle, be computed across the ensemble, i.e., via ensemble averaging. A stationary process would be one where the probability distribution functions, and hence the moments, would be invariant with time.

Random processes that have nontime varying mean and autocorrelation functions (depending only on τ) are sometimes called wide-sense stationary. For stationary processes, we assume in practice that the process has the so-called ergodic property, enabling us to compute the moments via time averages. These averages are computed via the single time function (the acquired signal), i.e., the realization of the stochastic process which is observed. While the notion of the ensemble is the basic one for stochastic processes, it has very little practical effect. In what follows all descriptions of the random phenomena are time based averages, i.e., computable via a single time record.

The power spectral density (PSD), denoted by S, for a continuous wide-sense stationary process is defined as the Fourier transform of the autocorrelation function.

$$S_{xx}(\omega) = \int_{-\infty}^{\infty} R_{xx}(\tau) \exp(-j\omega\tau)\, d\tau \quad [9a]$$

$$R_{xx}(\tau) = \frac{1}{2\pi} \int_{-\infty}^{\infty} S_{xx}(\omega) \exp(j\omega\tau)\, d\omega \quad [9b]$$

These Fourier transform-based definitions are often denoted by a double arrow notation:

$$R_{xx}(\tau) \longleftrightarrow S_{xx}(\omega) \quad [10]$$

The definitions in eqn [9] are known as the Wiener-Khintchine theorem.

Correlation Functions for Nonstationary Signals

In the previous section it was implicitly assumed that correlation functions computed for finite signal sections would be independent of the specific time intervals chosen. In the more general case, it must be assumed that:

$$R_{xy}(t,\tau) = E[x(t)y(t+\tau)] \quad [11]$$

may be a function of the variable t, the time. This is the case for non-stationary signals, where the statistical signal properties are not invariant with time. The case: $R_{xx}(t,\tau) = R_{xx}(\tau)$ is a special case of time invariance, which applies to stationary signals. Eqn [11] in conjunction with eqn [9] can be used to define time varying spectra as:

$$S(t,\omega) = \int R(t,\tau) \exp(-j\omega\tau)\, dt \quad [12]$$

The time-dependent autocorrelation function can be chosen in more than one way. Using eqn [12], we can define the Wigner–Ville distribution (WVD) as a basic tool for the analysis of nonstationary signals (see **Time–frequency methods**).

Examples of Correlations and Spectra for Random Signals

We first note that for the case:

$$R(\tau) = S_0 \delta(\tau)$$

$$R_{xx}(0) = \frac{1}{2\pi} \int_{-\infty}^{\infty} S_{xx}(\omega)\, d\omega = \int_{-\infty}^{\infty} S_{xx}(f)\, df \quad [13]$$

where $R_{xx}(0)$ is the total power (we assume zero mean), and the PSD is interpreted to be the distribution of the total power in the frequency domain. It should be noted that the power tends to infinity, and this case, which is possible mathematically, will never exist exactly in practice.

Next, using eqn [9] we consider some examples involving some idealized situations. These can often help in defining general properties.

Example 1 This concerns a possible definition (and intuitive understanding) of white noise. Assuming a constant PSD of value S_0 covering the infinite frequency range $\pm\infty$, we compute the autocorrelation as an impulse function. The autocorrelation being an impulse, there is zero correlation between two signal points separated by any incremental time, a completely memory-less phenomenon. White noise is obviously a mathematical notion.

Example 2 This concerns a constant PSD limited to $f_{\max}$, of magnitude equal to S_0. The autocorrelation is then:

$$R_{xx}(\tau) = 2S_0 f_{\max} \frac{\sin(2\pi f_{\max}\tau)}{(2\pi f_{\max}\tau)} \quad [14]$$

and is shown in **Figure 2C**. The first zero crossing of

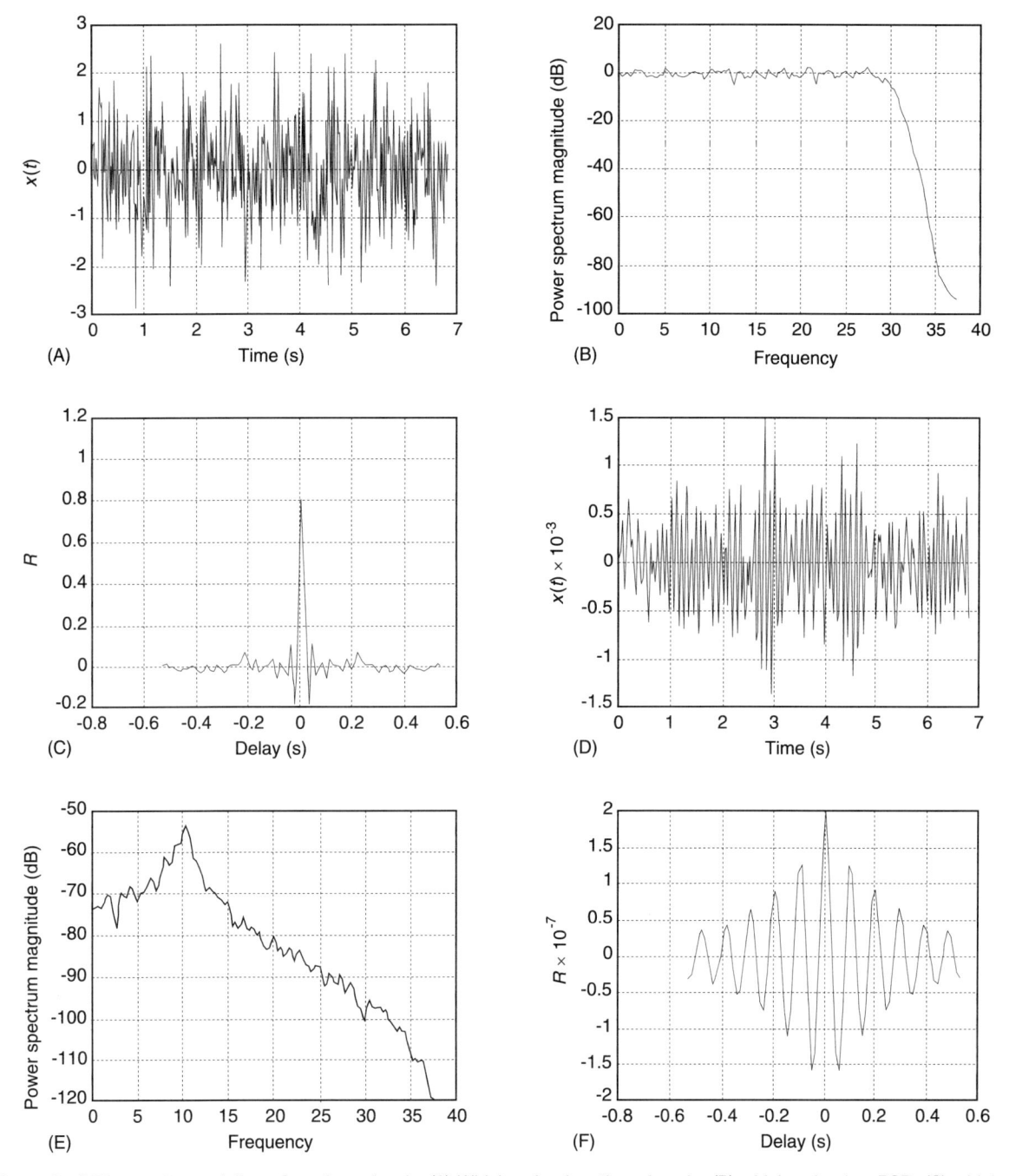

Figure 2 PSD vs aotocorrelation of random signals. (A) Wideband noise, time domain; (B) wideband noise, PSD; (C) wideband noise, autocorrelation; (D) narrowband noise, time domain; (E) narrowband noise, PSD; (F) narrowband noise, autocorrelation.

R_{xx} occurs for $\tau = 1/2f_{max}$, and this is roughly the memory of the process, the time interval for which there is still a correlation between the signal samples. The smaller the bandwidth, f_{max}, the longer this memory will be.

Example 3 Here we show a narrowband PSD, typical of a mass-spring-damper (SDOF) system excited by a white noise with a PSD equal to S_0:

$$R_{xx}(\tau) = S_0 \frac{\pi S f_0}{2} \exp\left(-2\pi\zeta f_0 \tau\right) \cos\left(2\pi f_0 \tau\right) \quad [15]$$

where f_0 is the natural frequency and ζ is the damping ratio. This is shown in **Figure 2F**. While the autocorrelation is oscillatory (dictated by f_0), the duration in the correlation domain is inversely proportional to the bandwidth in the frequency domain.

The Correlation Matrix

The products of correlation values (for the range of delay indices) can be arranged in a matrix form known as the correlation matrix. Many digital modeling and filtering operations involve the use of such matrices.

For discrete data, the autocovariance and autocorrelation sequences are important second-order characterizations. For the discrete sequence $\mathbf{x} = [x(0)\, x(1) \ldots x(p)]^T$.

These are defined via the $(p+1) \times (p+1)$ outer product:

$$\mathbf{xx}^T = \begin{bmatrix} x(0)x(0) & x(0)x(1) & \ldots & x(0)x(p) \\ x(1)x(0) & x(1)x(1) & \ldots & x(1)x(p) \\ \vdots & & & \\ x(p)x(0) & x(p)x(1) & \ldots & x(p)x(p) \end{bmatrix} \quad [16]$$

For a wide-sense stationary process, the autocorrelation $\mathbf{R}_x$ matrix is based on the expectation of eqn [16]:

$$\mathbf{R}_x = E(\mathbf{xx}^T)$$
$$= \begin{bmatrix} r_x(0) & r_x(1) & r_x(1) & r_x(p) \\ r_x(1) & r_x(0) & r_x(1) & r_x(p-1) \\ \vdots & & & \\ r_x(p) & r_x(p-1) & & r_x(0) \end{bmatrix} \quad [17]$$

Computational Aspects

Direct computations: These are based on eqns [2]–[4]. Two estimates can be computed. The unbiased version is:

$$R_{xy}(\tau) = \frac{1}{T-\tau} \int_0^{T-\tau} x(t)y(t+\tau)\, dt \quad [18a]$$

and the biased one is:

$$R_{xy}(\tau) = \frac{1}{T} \int_0^T x(t)y(t+\tau)\, dt \quad [18b]$$

For the discrete, unbiased case we have:

$$R_{xy}(k) = \frac{1}{N-|k|} \sum_{n=0}^{N-1-|k|} x(n)y(n+|k|) \quad [19a]$$

and for the discrete biased case we have:

$$R_{xy}(k) = \frac{1}{N} \sum_{n=0}^{N-1} x(n)y(n+|k|) \quad [19b]$$

For both types of estimate, the variance tends to zero with the signal length going to infinity:

$$\text{var}\left[\hat{R}_{xy}\right] \to 0 \text{ for } N \to \infty \quad [20]$$

and the estimates are thus consistent.

The variance of the computed estimate is slightly less for the unbiased case, but there are some computational advantages which can be realized (using specific matrix formulations) when using the biased version.

Modern computational methods: These use eqn [9]. Spectral densities are computed first, utilizing the highly efficient Fast Fourier Transform (FFT) algorithm, and correlation functions are then computed as inverse Fourier transforms of these, again using the FFT algorithm:

$$r_{xy}(\tau) = F^{-1}\left[S_{xy}(\omega)\right]$$
$$S_{xy} = \frac{\Delta t}{N} X^*(k) Y(k) \quad [21]$$

where $x(k) = \text{FFT}[x(n)]$ and $y(k) = \text{FFT}[y(n)]$. Some specific computational techniques are involved in such an approach, in order to avoid the so-called 'circular' properties of correlation functions which have been computed in such a manner.

Statistical Properties of the Correlation Estimators

Asymptotic expressions for the variance of the estimators are available for the specific case of band-limited white noise signals. The normalized standard deviation is then:

$$e_{R_{xy}(\tau)} = \left[\frac{1 + \rho_{xy}^{-2}}{N}\right]^{1/2} \quad [22]$$

and the correlation coefficient is:

$$\rho_{xy}(\tau) = \frac{R_{xy}(\tau)}{\left[R_{xx}(o)R_{yy}(o)\right]^{1/2}}$$

Two tendencies can be noted from eqn [22]. First, the variance tends to zero as the data length (or number of samples for the discrete case) goes to infinity. As has already been mentioned, the estimator is consistent. Second, the statistical error grows as the correlation coefficient decreases. The error becomes unpredictable when there is zero correlation between values. Such a property is very common for estimators which describe cross-properties between two parameters or functions.

Correlation Functions and Matrices in some Engineering Applications

Linear Systems

Some basic input/output relations may be based on correlation functions. When $x(t)$ is the excitation of a system with an impulse response $h(t)$, the response $y(t)$ is (**Figure 3**):

$$y(t) = h(t) \otimes x(t) \quad [23]$$

where $\otimes$ denotes the convolution operator. It is simple to show that:

$$R_{xy} = h \otimes R_{xx} \quad [24]$$

and for the specific case of x white noise:

$$R_{xx}(\tau) = \delta(\tau) \quad [25a]$$

$$R_{xy}(\tau) = h(\tau) \quad [25b]$$

Dispersionless Multipath Propagation

Figure 4 shows a scheme with propagation delays, τ_k, in each path. The resulting impulse response is:

$$h(t) = \sum_k h_k(t) = \sum_k h(t - \tau_k) \quad [26]$$

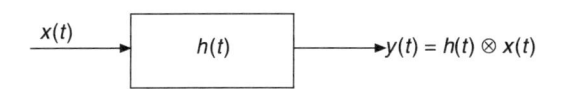

Figure 3 Impulse response and linear systems.

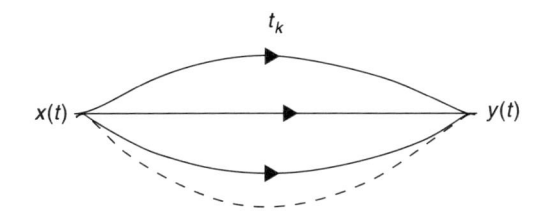

Figure 4 Multipath propagation.

The convolution operation being commutative, this yields:

$$R_{xy}(\tau) = R_{xx}(\tau) \otimes \sum_k h(t - \tau_k) \quad [27]$$

and R_{xy} peaks at locations where $t = \tau_k$. One method of identifying propagation delays is thus based on peak localization. This is a viable option when the propagating signals are random power signals, when a simple time pattern is not evident. As with any parameter identified from random signals, the variance of the peak's location is an uncertainty in the extracted delay. The normalized error for this location is:

$$e_{\tau_k} \approx 0.6 \left[\frac{1 + \delta_{xy}^{-2}}{N} \right]^{1/4} \quad [28]$$

From eqn [27] it can be seen that h is 'smeared' by R_{xx}, unless x is white noise.

Figure 5 deals with the sums of a signal and its delayed versions. Three delayed components are added, with delays of 2 ms, 3.9 ms, and 5 ms, respectively. Three cases with varying degrees of signal bandwidth are shown. For the wideband case, all three delays are recognizable as approximate impulses in the cross-correlation function. The smearing of R_{xy} with the narrowing of the signal's bandwidth is evident. The delays of 3.9 ms and 5 ms (**Figure 5F**) are not resolved in the case with the narrowest bandwidth.

From

$$S_{xy}(\omega) = H(\omega) S_{xx}(\omega) \quad [29]$$

we can attempt to compute $H(\omega) = S_{xy}(\omega)/S_{xx}(\omega)$ and via an inverse Fourier transform compute $h(t)$ directly and obtain a clearer distribution of peaks. But ill conditioning could occur as H would have negligible magnitude in some frequency ranges, and more advanced techniques are then indicated.

Examples of applications can be found in situations where the propagation velocity is known (as in airborne and body-borne acoustic waves). Estimation of delays can then help in determining signal sources, for those cases where the involved geometrical propagation paths can be identified (**Figure 6**).

Investigation of flow regimes in pipes is another example. For flows involving stocastic velocity fluctuations, the average delay between two spatial locations can be estimated using correlation techniques (**Figure 7**).

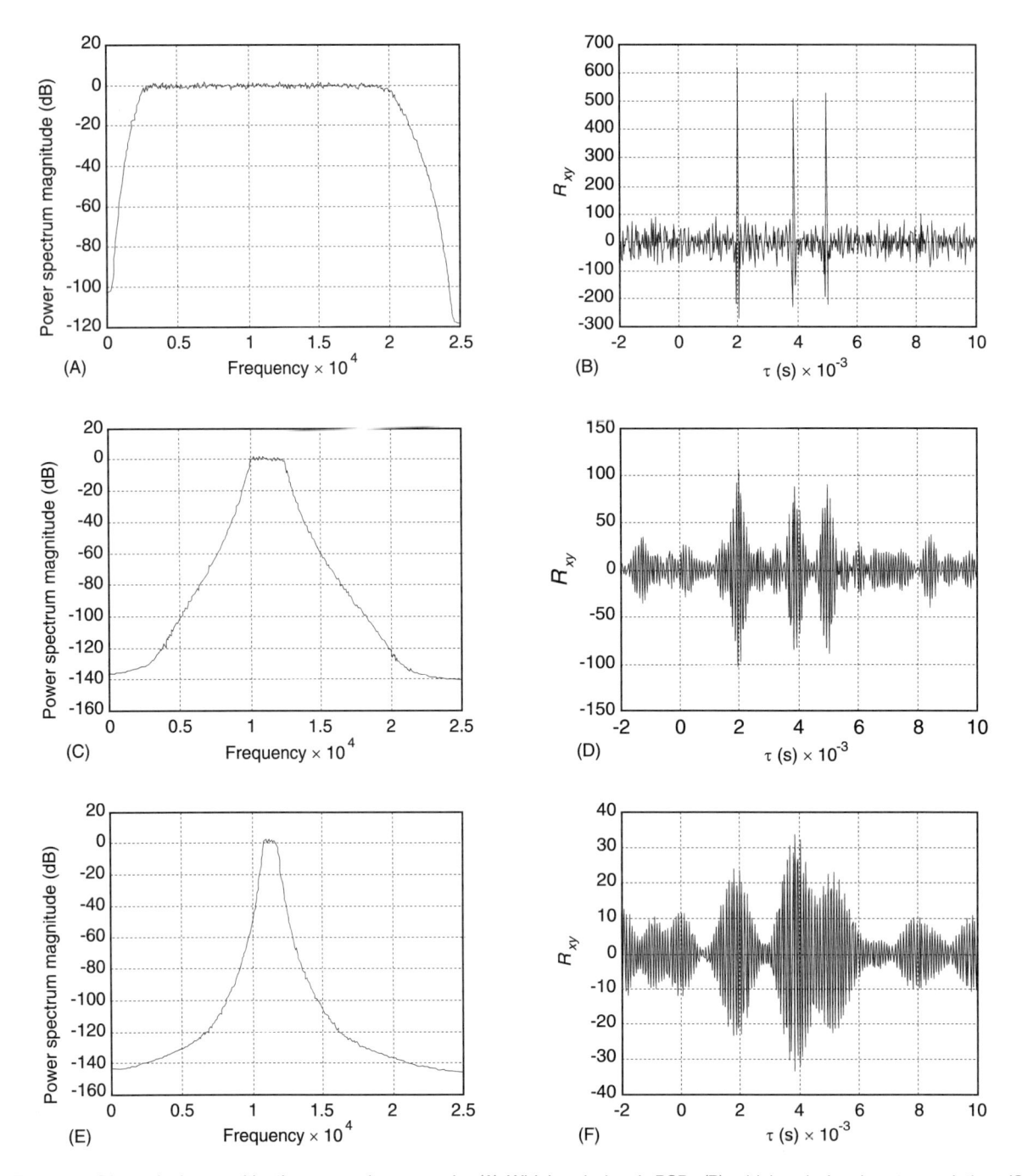

Figure 5 Dispersionless multipath propagation example. (A) Wideband signal, PSD; (B) wideband signal, autocorrelation; (C) intermediate bandwidth signal, PSD; (D) intermediate bandwidth signal, autocorrelation; (E) narrowband signal, PSD; (F) narrowband signal, autocorrelation.

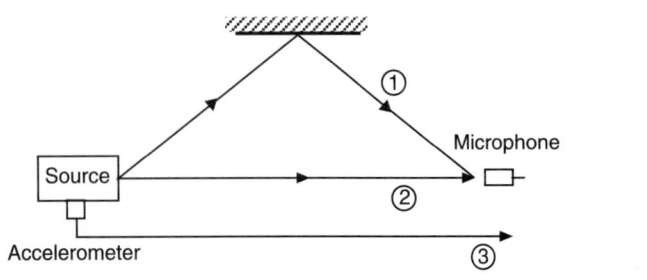

Figure 6 Source and propagation path problem.

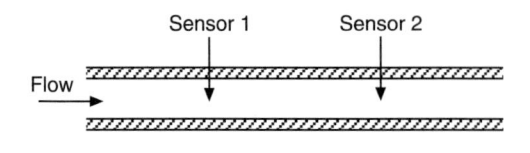

Figure 7 Flow propagation.

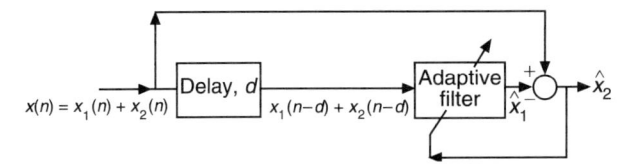

Figure 8 Adaptive line enhancement.

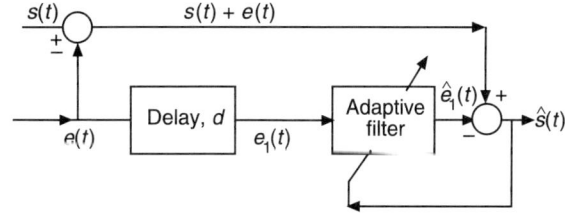

Figure 9 Adaptive noise cancellation.

Filtering and modeling

Correlation matrices find wide application in the area of signal processing, for signal modeling, filtering, etc. As an example, we mention a spiking filter used when an impulsive signal is smeared by the propagation medium, as in seismic signal processing. Shaping the measured signal $x(n)$ is achieved by the filter $h(n)$ such that $x(n) \otimes h(n) = \delta(n)$. The filter coefficients $\mathbf{h}$ are then computed using:

$$\mathbf{R}_x \mathbf{h} = x^*(0) \begin{bmatrix} 1 \\ 0 \\ 0 \\ \cdot \\ \cdot \\ \cdot \\ 0 \end{bmatrix} \quad [30]$$

Models of signals are widely used in areas of spectral analysis, diagnostics, etc. An example of a model for measuring signal x is the autoregressive (AR) model:

$$x_i = -\sum_{k=1}^{p} a_k x_{i-k} + w_i \quad [31]$$

The vector of the model's parameters is then computed using:

$$\mathbf{Ra} = \sigma^2 \mathbf{I} \quad [32]$$

with

$$\mathbf{I} = [1\ 0\ 0\ \ldots\ 0]^T$$

See **Adaptive filters** for application.

An example of the adaptive line enhancement (ALE) procedure is shown in **Figure 8**. This can be applied in situations where a composite signal which includes both broadband and narrowband signals is to be decomposed. The objective can be to extract either the narrowband or the broadband component. The narrowband signal can be an approximate harmonic signal, like those found in rotating machinery vibrations. The method is based on the fact that the narrowband signals tend to have longer autocorrelation durations than the broadband ones.

For the composite signal:

$$x = x_1 + x_2$$

where τ_1 and τ_2 denote the respective autocorrelation lags beyond which the autocorrelations of x_1 and x_2 become negligible. Then $\tau_2 < \tau_1$.

The delay d shown in **Figure 8** is chosen such that $\tau_2 < d < \tau_1$. The delayed signal component $x_2(n-d)$ is not correlated with $x_2(n)$, while the delayed component $x_1(n-d)$ is still correlated with $x_1(n)$. The adaptive filter adjusts its weights so as to cancel x_1, the component which is correlated with $x_1(n-d)$. The output of the filter will be an estimate $\hat{x}_1$ of x_1. The subtraction results in x_2, or if desired, x_1 can be used for further processing. The method has been applied to composite vibration signals in order to enhance periodic components mixed with broadband random ones.

An adaptive noise cancellation (ANC) scheme is shown in **Figure 9**. This is a more general case of the earlier application. It can be used to cancel an interfering signal $e(t)$ from a composite signal $s(t) + e(t)$, in those cases where another reference signal $e_1(t)$ is available.

The requirement is that $e_1(t)$ must be highly correlated with $e(t)$. (Full correlation is obviously described in the hypothetical case shown in Figure 9.) The adaptive filter adjusts its weight so as to generate an estimate $\hat{e}(t)$ of the interfering signal. The scheme is widely used for acoustic noise cancellation, where the interfering signal is often either periodic or mostly harmonic. The correlated reference signal can often be obtained from vibration measurements, as in cases where harmonic vibrations generate that part of the acoustic signal which needs to be attenuated/cancelled. ANC has also found applications in vibration base diagnostics of rotating machinery (see **Diagnostics and condition monitoring, basic concepts**).

Nomenclature

d	delay
$e(t)$	interfering signal
$E[.]$	expectations
f_0	natural frequency
$h(t)$	impulse response
R	correlation coefficient
S	power spectral density
$y(t)$	response

See also: **Time–frequency methods; Hilbert transforms; Adaptive filters; Spectral analysis, classical methods**

Further Reading

Bendat JS, Piersol AG (1993) *Engineering Applications of Correlation and Spectral Analysis.* New York: Wiley.

Hayes MH (1996) *Statistical Digital Signal Processing and Modeling.* New York: Wiley.

CRASH

V H Mucino, West Virginia University, Morgantown, WV, USA

Copyright © 2001 Academic Press

doi: 10.1006/rwvb.2001.0164

Automotive collisions are responsible for many fatalities every year and a public safety concern in every country for automakers, transportists and governments alike. According to the International Road Traffic Accident Database (IRTAD), in 1998 alone, averages of 13 fatalities and 480 injury accidents per 100 000 population were registered among the 29 participating countries in Europe, North America, Asia and Australia. Yet, this statistic is the result of a decreasing trend in the past two decades reflecting, among other things, the significant technological advances in the automotive field aimed at enhancing 'crashworthiness' of vehicles.

Crashworthiness by itself is not a characteristic or feature that can be measured or quantified, but it relates to the capacity of a vehicle structure and occupant restraint system to protect the occupants in the event of a collision. The main objectives being to reduce the likelihood and severity of injuries to vehicle occupants and ultimately, to reduce the fatality rates in vehicle collisions. Crashworthiness concepts are also being applied to aircraft structures and heavy-duty vehicles to enhance survivability of occupants in the event of a crash.

Vehicle accidents have always been a concern to auto makers. But it was not until the early 1970s, together with the advent of supercomputers, that the key issues relating to the mechanics of automotive collisions acquired such significant relevance to trigger the development of what is known today as the area of crashworthiness.

Central to the development of the crashworthiness area is the use of explicit finite element (FE) techniques to conduct computer simulations of collisions. These techniques have been developed to address the nonlinear transient solid mechanics of high-speed colliding structures, which involve the elastic and plastic stress wave propagation through solid continua and the characteristic nonlinear crushing behavior of structures. Crashworthiness applications have been extended to include human body structures and the interactions between vehicle, occupants and restraining devices.

Yet, many challenges lie ahead. New advanced materials are being used in vehicles to reduce weight (for fuel economy purposes) and to enhance crashworthiness. Many of these materials are yet to be adequately characterized in order to take full advantage of their applicability in vehicle structures. Representation of human body structures for simulation purposes is another area where the challenges are plentiful, from the mechanical characterization of human tissue to the classification and determination of injury thresholds for various types impact. Ultimately, the integration of crashworthiness advances in vehicle systems design is a challenge, which seems to be within reach, given the strides in computer technology.

Introduction

Crashworthiness has emerged as a multidisciplinary area that is now at the very heart of vehicle design and transportation systems in general. It addresses the mechanics of colliding structures and the interactions between vehicles and occupants, in such detail that realistic simulations are now not only possible, but within reach of a record number of engineers involved with vehicle design, the overall objectives being to reduce the likelihood and severity of injury to humans and of fatality rates in vehicle collisions.

The issues in crashworthiness are plentiful; high speed colliding and crushing structures, biomechanics of the human body dynamic response in collisions, passenger restraint systems such as seat-belts, airbags, interior padding and energy absorbing materials, just to mention a few. From the experimental side, the use of rather sophisticated dummies has also called for advanced design and modeling of dummies that can be used effectively as human surrogates to verify various vehicle crashing scenarios.

Essential in the study of crashworthiness of vehicles, is the study of energy dissipation characteristics of colliding vehicle structures and the interactions between occupants and vehicle. From the simulation standpoint, the development of explicit FE approaches together with advanced contact algorithms has enabled the study of crashworthiness phenomena. However, the development of crashworthiness technologies has had several parallel tracks, the most important ones being the supercomputer development track, the vehicle collision track and the numerical solid mechanics track.

The Supercomputer Track

This was triggered by the development of the CRAY-1 supercomputer in the early 1970s by Seymour Cray. With a top speed of 100 megaflops, the first supercomputer provided the capability for vectorized processing, which in turn enabled applications where large numbers of computations were required, such as in nonlinear solid mechanics and computational fluid dynamics. Since then, several generations of CRAY supercomputers have provided an exponential growth in speed and parallel processing capability. Today's CRAY-SV1ex, released in 1999, is several orders of magnitude faster and more computationally efficient than the first supercomputers (with faster multistreaming vector processors at 7.2 billion calculations per second, memory size of 128 gigabytes), providing the capacity for the analysis of very large models in record turn-around times. FE models of vehicles for crashworthiness purposes can easily have several millions degrees of freedom with a large number of nonlinear relationships, making the solution processes iterative in nature and convergence-sensitive, thus requiring vast computational capabilities.

Parallel to the development of supercomputer technology, microprocessor and high-resolution graphics technologies have combined to enable and foster the development of sophisticated and computationally intense applications such as explicit FE codes for realistic simulations and visualization of crashworthiness phenomena.

On the Vehicle Collision Track

Vehicle manufacturers worldwide were confronted with government regulations to address occupant protection in vehicle collisions in the mid-1960s. In the USA, the first federal motor vehicle safety standard (MVSS 209: Motor Vehicle Safety Standard, by the US-National Highway Traffic Safety Administration (NHTSA)) related to crashworthiness became effective on 1 March 1967. By 1976, in the USA there were 15 specific MVSSs related to crashworthiness as shown in **Table 1**. Several of these standards relate directly to the issues currently addressed by crashworthiness analysis codes. For example, standard MVSS-203, first issued in 1968, requires a vehicle to be crash-tested (frontal collision) at $48\,\mathrm{km\,h^{-1}}$ (30 mph) against a rigid wall. These standards called for engineering assessment of vehicle

Table 1 USA-NHTSA Federal motor vehicle safety standards (MVSS) related to car crashworthiness

Motor Vehicle Safety Standard (MVSS) no.	Description	Effective date	Last revision
201	Occupant protection in interior impact	1-1-68	9-1-2000
202	Head restraints	1-1-69	
203	Impact protection for the driver from the steering control system.	1-1-68	9-1-81
204	Steering control rearward displacement	1-1-68	9-1-91
205	Glazing materials	1-1-68	
206	Door locks and door retention components	1-1-68	1-1-72
207	Seating systems	1-1-68	1-1-72
208	Occupant crash protection	1-1-68	9-1-91
209	Seat belt assemblies	3-1-67	
210	Seat belt assembly anchorages	1-1-68	7-1-71
212	Windshield mounting	1-1-70	9-1-78
213	Child restraint systems	4-1-71	1-1-81
214	Side impact protection	1-1-73	9-1-98
216	Roof crush resistance	9-1-75	9-1-94
219	Windshield zone intrusion	9-1-76	4-3-80

designs, which in turn, triggered the development of a variety of engineering tools and approaches to assess crashworthiness of vehicles and occupant response. New standards and revisions to existing standards are published in the Federal Register. These standards represent the minimum safety performance requirements for motor vehicles. The basic premise is that the 'public is to be protected against unreasonable risk of crashes occurring as a result of the design, construction, or performance of motor vehicles and is also protected against unreasonable risk of death or injury in the event crashes do occur'.

Early studies of crashworthiness relied almost completely on experimental barrier impact tests that produced data, which were used later on, to help explain the dynamic response of colliding vehicles. In some of these tests, mannequins were used to provide some idea of human response in vehicle frontal crashes. Kamal and Lin produced one of the first models for vehicle collisions based on nonlinear mass-spring elements, whose characteristics were derived experimentally.

Frontal collisions were first addressed by the standard MVSS-203 (48 km h^{-1} rigid barrier frontal crash), given the frequency of their occurrence (roughly 50% of collisions). In such a collision, the passenger cabin is supposed to maintain its integrity by not allowing large structural deformations. The kinetic energy of the vehicle is to be dissipated by large plastic deformations of the front-end of the vehicle, which in turn, acts as a cushion to the passenger compartment. Deformations in the order of 800 mm (32 in) can take place in a time interval of approximately 120 ms, which produces strain rates between 1 and 100 per second. Under these conditions, the passenger cabin experiences decelerations of the order of 20g. Unrestrained occupants colliding with the vehicle interior may experience even higher deceleration rates, illustrating the importance of restraining devices. The rather costly nature of experimental tests and the limited predictability of collision events based on experimental data triggered the interest in simulation. The experimental programs however, highlighted some of the most important issues in solid mechanics that needed to be addressed, namely the large plastic deformations, strain rate dependency of materials, plastic and elastic stress wave propagation, local buckling and fracture/rupture of materials in the 120 ms of collision duration.

On the Solid Mechanics Track

Melosh and Kelly envisioned the requirements for the development of FE methods for crashworthiness analysis and outlined the issues to be addressed in order to make it possible. In the early 1970s, a number of studies pointed out the importance of local deformations and local geometry instabilities to conduct the structural response of colliding structures, issues that are central to crashworthiness analysis of structures. McIvor et al., Armen et al. and Welch et al. put forward important studies for crashworthiness simulations of structural components using FEs, in which the phenomena of local deformations and nonlinear elasto-plastic material behavior was considered.

The development of one of the most widely used FE codes for crashworthiness, DYNA3D, started in the mid 70s by Hallquist at Lawrence Livermore Laboratory. This code captured the essence of high-speed impact solid mechanics phenomena, for military and civilian applications, from projectile penetration to vehicle collisions. In 1979, DYNA3D was reprogrammed for the CRAY-1 supercomputer with more sophisticated sliding interfaces than the previous version. In 1981, new material formulations and the problem of penetration of projectiles was successfully simulated by the use of sliding interfaces. In 1990 LS-DYNA3D is released with added capabilities to represent fabric materials for seat belts and airbags and composite glass models. Parallel to that, pre and postprocessing codes (INGRID and TAURUS) were developed to interface with LS–DYNA3D and currently, there are versions of DYNA3D for PC and desktop workstations. Meanwhile, several other codes have emerged with various capabilities that allow for similar functions. For the most part, these codes seem to provide consistency in the applications. Other codes commercially available that are capable of simulating collisions of large structures effectively are PAM-CRASH, ADINA, MADYMO, WHAMS, and ABAQUS. The latter has useful data base descriptions of crash dummy models that are widely used for vehicle–occupant interaction assessment.

Three special features that distinguished this code were (i) explicit time integration schemes for the equations of motion; (ii) the application of advanced contact algorithms to allow for sliding surfaces with friction and nonpenetrating contact surfaces; and (iii) the use of 'economic' one-point-integration elements with 'hourglassing energy control' options to monitor convergence (hourglassing energy is associated to the zero energy deformation modes and is briefly described below). Additionally, this program was developed to include a wide array of nonlinear materials, including strain rate dependent materials and viscoelastic materials, rubber, honeycomb, and composite materials among others.

Many other standard commercial FE codes such as ANSYS, IDEAS–MS, PATRAN and CATIA among others, now provide import/export functions to interface with some of the crashworthiness codes, making the modeling and data preparation more readily available to many engineering analysts.

Brief Description of Crashworthiness Codes Based on DYNA3D

All the FE crashworthiness codes share the same basic principles of computational mechanics. The use of explicit time integration algorithms, the use of 'economic' elements, the formulation of various constitutive models for engineering materials, the re-zoning of the meshes using advection algorithms and the use of similar contact–impact algorithms. Some variations exist on special interface representations between structural components, but in general they all share similar or equivalent features. A brief description of the key features of a crashworthiness code is provided next.

Spacial Discretization

DYNA3D uses an updated Lagrangian formulation based on the weak form integral of the virtual work principle. Economic isoparametric elements with one integration Gauss point and diagonal mass matrices are used, for which viscous or stiffness hourglass energy can be used to control the zero energy modes associated to the one point of integration scheme. The structural elements include springs, lumped masses, discrete dampers, beams, trusses, solids (tetrahedron, wedges and brick) and shells (quadrilateral and triangular), the Hughes and Liu shells, also the Belytscho–Schwer beam and Belytschko–Tsy shell. Solid elements use the Flanagan–Belytschko constant stress, exact volume integral. **Table 2** shows the type of elements available in DYNA3D for structural representations.

Arbitrary Lagrangian Eularian (ALE) Advection

This is a re-zoning approach, which is needed to maintain consistency between solution requirements and the FE mesh. The rezoning consists of a Lagrangian time step followed by an advection or remap step. In this process the solution is stopped and the mesh is adjusted to the deformed geometry in such a way that elements are not highly distorted (smoothing). The solution is then mapped from the old mesh to the new one until the next step. While advection is different from adaptive FE meshing, both techniques can be used to enhance accuracy and convergence of a solution where nonlinear large displacements occur and where high gradients of the response variables occur.

Time Integration

A central difference explicit time integration algorithm is used to integrate the resulting equations of motion. This scheme is conditionally stable but does not require the use of implicit iterative techniques. The central difference approach requires that for each time step Δt, the current solution be expressed as:

$$\dot{x}^{n+1/2} = \dot{x}^{n-1/2} + \Delta t^n \ddot{x}^n \quad [1]$$

$$x^{n+1} = x^n + \Delta t^{n+1/2} \dot{x}^{n+1/2} \quad [2]$$

The difference with implicit methods of integration is that in explicit schemes, the solution to the current time step depends only on the solution of previous steps, thus avoiding the costly iterations on each time step to determine the unknown solution at the current time step (required in implicit schemes). The result is that no iterations are necessary at each time step. The drawback with explicit methods is the stability, which can only be controlled by taking rather small time steps, as the error is proportional to the square of the time step. The Courant stability criterion must be used, which requires the solution not to propagate through an element faster than the dilatational wave speed of the material. This requirement gives rise to the need for rather refined meshes to provide numerical stability. Monitoring the energies of the system becomes the key to controlling the stability of the solution.

Boundary Conditions

Several loading conditions can be simulated. Direct nodal loads, line and surface pressures and body loads are available. Kinematic boundary conditions are also allowed through prescribed displacements, velocities and accelerations; fixed nodes and displacement

Table 2 Element types available in DYNA3D

Element no.	Type of element
1	Solid elements (brick and tetrahedron)
2	Belytschko beam
3	Hughes–Liu beam
4	Belytchko–Lin–Tsay shell
5	C^0 triangular shell
6	Marchertas–Belytchko triangular shell
7	Hughes–Liu shell
8	Eight-node solid shell element
9	Truss element
10	Membrane element
11	Discrete elements and masses

constraints in the form of nodal-degree-of-freedom relationships. Another type of boundary conditions that can be considered are the 'contacting surfaces' where collision is to occur. These boundary conditions can be treated by means of the contacting and sliding surfaces described next.

Sliding Interfaces

The use of advanced impact contact algorithms is another unique feature of this program, allowing for various types of contacting surfaces. DYNA3D supports 22 types of interfaces. Master–slave surface concept is used to define contacting interfaces. Single surface can be defined as a 'slave only' surface, in such a way that each node is monitored to prevent penetration through the surface. This type of surface can effectively capture the 'accordion folding' behavior of crushing thin shell elements commonly found in vehicle crashes. The 22 types of interface can support sliding with friction; tied and sliding only interfaces can also be defined. Rigid 'master surfaces' can also be used. Interface definitions can be used to define lines and surfaces where contact is prescribed or where master lines in tiebreaking definitions can be used.

Constitutive Equations Library

The availability of a wide range of constitutive models is another essential feature of this code. From the elastic isotropic to the nonlinear elastic, orthotropic, visco-elastic, elasto-plastic and strain dependent material formulations. Near incompressible materials (rubber) and fluid material formulations, composite materials and fabric type materials are also possible. DYNA3D has about 120 material models and allows for 'user defined' material characterizations. This provides a very powerful capability to formulate sophisticated materials.

The constitutive equations used in the calculations can be expressed as:

$$\sigma_{ij}^* = \sigma_{ij}^*(\dot{\varepsilon}_{ij}) \qquad [3]$$

$$\dot{\sigma}_{ij} = \sigma_{ij}^* + \sigma_{ik}\omega_{kj} + \sigma_{jk}\omega_{ki} \qquad [4]$$

$$\dot{\varepsilon}_{ij} = \frac{1}{2}\left(\frac{\partial \dot{x}_i}{\partial x_j} + \frac{\partial \dot{x}_j}{\partial x_i}\right) \qquad [5]$$

$$\omega_{ij} = \frac{1}{2}\left(\frac{\partial \dot{x}_i}{\partial x_j} - \frac{\partial \dot{x}_j}{\partial x_i}\right) \qquad [6]$$

where σ_{ij}^* is the Jaumann rate of the Cauchy stress, $\dot{\sigma}_{ij}$ is the material derivative of the stress, $\dot{\varepsilon}_{ij}$ is the strain rate tensor, and ω_{ij} is the spin rate tensor. This constitutive model can be effectively used in the case of large deformation, where rigid body rotations may occur.

A variety of special materials formulations are also available through equations of state, to deal with cases where extremely large pressures take place, such as in the case of explosive materials

Basic Formulation in DYNA3D

Momentum Conservation Principle

The formulation is based on equations of motion derived from the momentum conservation principle, which can be expressed as follows:

$$\sigma_{ij,j} + \rho f_i = \rho \ddot{x}_i \qquad [7]$$

the traction boundary conditions $\sigma_{ij}\mathbf{n}_i = t_i(t)$ on boundary ∂b_1, displacement boundary conditions $x_i(X_a, t) = D_1(t)$ on boundary ∂b_2, and contact discontinuity $(\sigma_{ij}^+ - \sigma_{ij}^-)n_i = 0$ along the interior boundary ∂b_3, for $x_i^+ = x_i^-$, where σ represents the Cauchy stress, ρ is the density and f_i is the body force, $\ddot{x}$ is the acceleration, the comma designates a covariant differentiation, and $\mathbf{n}_i$ is a unit vector normal to the boundary ∂b.

The energy equation can be integrated in time for energy balance and state evaluations. The equation is given by:

$$\dot{E} = V s_{ij}\dot{\varepsilon}_{ij} - (p+q)\dot{V} \qquad [8]$$

which yields the weak form of the equation of equilibrium, where variation δx_i satisfies all boundary conditions on ∂b_2 as follows:

$$\int_v (\rho \ddot{x}_i - \sigma_{ij,j} - \rho f)\delta x_i \, dv$$
$$+ \int_{\partial b_1} (\sigma_{ij}n_j - t_i)\delta x_i \, ds + \int_{\partial b_2} (\sigma_{ij}^+ - \sigma_{ij}^-)n_i \delta x_i \, ds = 0$$

$$[9]$$

Application of the stationarity conditions leads to the virtual work principle and the following expression:

$$\delta \pi = \int_v \rho \ddot{x}_i \delta x_i \, dv + \int_v \sigma_{ij} \delta x_{i,j} \, dv$$
$$- \int_v \rho f_i \delta x_i \, dv - \int_{\partial b_1} t_i \delta x_i \, ds = 0 \qquad [10]$$

This variation can be integrated by assembling the individual integrals of each FE yielding the global (over m elements) FE equilibrium equations as follows:

$$\sum_{m=1}^{n}\left\{\int_{v_m}\rho\mathbf{N}^T\mathbf{N}a\,dv+\int_{v_m}\mathbf{B}^T\sigma\,dv-\int_{v_m}\rho\mathbf{N}^Tb\,dv-\int_{\partial b_1}\mathbf{N}^Tt\,ds\right\}^m=0 \quad [11]$$

Subsequently, the explicit central difference method is used to integrate the equations of motion.

Vibrations and Shock Wave Propagation

Colliding structures are in reality vibration systems subjected to impact loads that excite various natural modes and produce elastic and plastic waves traveling at various speeds through the structure. The crushing behavior is characterized by local buckling as a result of the impulse, the bulk viscosity and associated shock waves, the natural modes of the structure and the energy release from the crushing behavior. Given the relevance of wave propagation phenomena in the analysis of colliding structures, several issues are extremely important in the application of crashworthiness procedures. Specifically, the relationships between the speed of sound through the material, the frequencies of the natural modes of vibration, the critical damping, the time step required to maintain numerical stability and the mesh size used in the model. All these features are reflected in the models used, which are in general nonlinear.

On the Contact Algorithms

DYNA3D uses three approaches for dealing with the impact contact and sliding interfaces of models. The methods are known as the 'kinematic constraint method', the 'penalty method' and the 'distributed parameter method'. In the first method, constraints are imposed to the global equations by a transformation of the nodal displacement component of the 'slave' nodes along the contact interfaces. This way only the global degrees of freedom of each master node are coupled. This method requires consistent zoning of the interfaces. In the 'penalty method', artificial interference springs are placed normal to the contacting surfaces on all the penetrating nodes. The spring elements are assembled in the global stiffness matrix and their modulus is determined based on the elements in which the nodes reside. This is a stable method and produces little noise for hourglassing modes. However, for relatively large interface pressures, the stiffness has to be scaled up and the time step reduced. In such cases, the third method 'distributed parameters' is recommended. This last method is used primarily for 'sliding' interfaces, in which the internal stress in each element in contact determines the pressure distribution for the corresponding master surface. Accelerations are updated after mass and pressure distributions on the master surface are completed. With these three algorithms, most contact conditions can be simulated. See **Table 3**.

Table 3 Contact interfaces in DYNA3D

Type of interface	Description
1	Sliding only for fluid/structure or gas/structure interfaces
2	Tied
3	Sliding, impact, friction
4	Single surface contact
5	Discrete nodes impacting surface
6	Discrete nodes tied to surface
7	Shell edge tied to shell surface
8	Nodes spot weld to surface
9	Tiebreak interface
10	One-way treatment of sliding impact friction
11	Box-material limited automatic contact for shells
12	Automatic contact for shells
13	Automatic single surface with beams and arbitrary orientations
14	Surface-to-surface eroding contact
15	Node-to-surface eroding contact
16	Single surface eroding contact
17	Surface-to-surface symmetric constraint method
18	Node-to-surface constraint method
19	Rigid-body to rigid-body contact with arbitrary force/deflection curve
20	Rigid nodes to rigid-body contact with arbitrary force/deflection curve
21	Single-edge contact
22	Drawbead

Economic Elements and Hourglassing

Explicit time integration schemes call for the so-called 'economic' finite elements. These are isoparametric element formulations with one integration or Gauss point, to minimize the computational time. These type of integration give rise to the possibility of 'zero energy' modes, which produce the 'hourglass mode shapes' with zero energy. This is an undesirable effect and must be prevented. To illustrate, let us consider the strain rate for an eight-node solid element:

$$\dot{\varepsilon}_{ij}=\frac{1}{2}\left[\sum_{k=1}^{8}\frac{\partial\phi_k}{\partial x_i}\dot{x}_j^k+\frac{\partial\phi_k}{\partial x_i}\dot{x}_i^k\right] \quad [12]$$

If diagonally opposite nodes have identical velocities, then the strain energy rates are identically zero ($\dot{\varepsilon}_{ij} = 0$), that is when the following 'hourglassing' conditions occur:

$$\dot{x}_i^1 = \dot{x}_i^7, \quad \dot{x}_i^2 = \dot{x}_i^8, \quad \dot{x}_i^3 = \dot{x}_i^5, \quad \dot{x}_i^4 = \dot{x}_i^6 \quad [13]$$

Hourglass modes can be avoided by the use of artificial forces consistent with the orthogonal nature of the modes, ($\Gamma_{\alpha k}$) and related to the element volume and material sound speed. The nodal velocities for these modes are given by:

$$h_{i\alpha} - \sum_{k=1}^{8} \dot{x}_i^k \Gamma_{\alpha k} = 0 \quad [14]$$

And the resisting hourglass forces are defined as:

$$f_{i\alpha}^k = a_h h_{i\alpha} \Gamma_{\alpha k} \quad \text{and} \quad a_h = Q_{hg} \rho v_e^{2/3} \frac{c}{4} \quad [15]$$

where Q_{hg} is an empirical constant defined between 0.05 and 0.15. Refined finite element meshing of components tends to reduce the hourglassing modes, which otherwise produce artificial energies in excess of the initial energy of the system.

Component Crashworthiness Roles

Upon a collision, the components of a vehicle–occupant system play a certain role or function. In general, the key functions are:

1. *Energy absorption of colliding elements.* Generally, the elements that are to make contact first with another vehicle, object or barrier in a collision are intended to absorb energy upon impact by collapsing in controlled modes, acting as dampers to the passenger cabin. Bumpers and front rails of vehicles are typical structural elements intended to dissipate energy in case of collisions. The frontal rails connecting to the bumper supports for example are now designed with 'crush initiators', as intended imperfections in the form of notches on strategic locations, to ensure the collapse occurs in a prescribed manner.
   Polyurethane-foam filled tubing and honeycomb sandwich panels are components typically used to provide higher stiffness at reduced weights while providing additional energy absorbing capabilities. Crashworthiness codes such as DYNA3D are capable of using these types of structural components and strain rate dependent materials for crashworthiness analyses.

2. *Energy absorption of interior surfaces.* Occupants of colliding vehicles can be subject to severe impacts with interior elements and surfaces of the vehicle. A primary concern was the likelihood of contact between the occupant's head and the interior hard surfaces of the vehicle, such as the A–pillar (front), the B–pillar (mid-post), the dashboard (passenger side) and the steering wheel.
   The use of polyurethane foam materials for interior cushioning is mainly intended to reduce the severity of impact on the head, as designated by the term known as 'head injury criteria' (HIC), which is expressed as follows:

$$\text{HIC} = (t_2 - t_1) \left[\frac{1}{(t_2 - t_1)} \int_{t_1}^{t_2} \ddot{x}(t) \, dt \right]_{\max}^{2.5} \quad [16]$$

   The time interval for which this equation applies is approximately 15 ms and the interval must be taken in such a way that the maximum value of HIC is obtained. If the value reaches a value of 1000, severe injury to the head is likely to occur. Using this criteria, the most important state variable is the head's *cg* overall acceleration.
   While HIC is a standard accepted index for assessment of vehicle passenger performance under collision loads, HIC alone is not sufficient to describe the mechanism of head injury. For that matter not only the head is important; head–neck complex as well as thorax response and lower limb interaction with vehicle interior are all important indicators of crashworthiness of a particular vehicle.

3. *Structural integrity of passenger cabin.* While some structural elements are meant to dissipate energy in the event of a collision, the structural members that comprise the passenger cabin are meant to maintain the structural integrity. This means allowing only minimum deformations and keeping the occupant space envelope from being invaded. For this purpose, the structural components of the passenger cabin are designed to provide high stiffness and strength. Some of the members in a cabin are intended to provide rollover protection, others are intended to provide lateral impact protection. In all cases the stiffness-to-weight ratio is very important due to the trend towards lighter and more efficient vehicles. Polyurethane foam-filled tubing profiles and honeycomb-sandwiched panels can provide increased stiffness properties at reduced weights. Another important function of the passenger cabin structural elements is to prevent intrusion of objects into

the passenger space envelope. Reinforced vehicle doors are examples of elements that provide that kind of protection in the case of lateral impacts.

4. *Occupant restraint systems.* It was stated above that in the case of a frontal collision, the decelerations of an unrestrained occupant could be an order of magnitude larger than that of the passenger cabin. Thus, the occupant restraint systems have as main function to reduce the magnitude of the decelerations of the occupant. Two types of restraint systems are used in vehicles. The passive ones are the seat belts, and the active ones are air bags. Modeling seat belts requires the use of elements that can interface with the occupant and vehicle and provide membrane stiffness only. Crashworthiness FE codes permit the use of 'fabric' material that provides the membrane properties needed to model the seat belt; the number of points of anchorage, their location and the amount of 'slack' in the belt being the key design factors to consider.

Air bags provide active protection in the case of a collision. At issue is the presence of a cushion between the passenger and the hard surfaces of the vehicle interior immediately after a collision. The issues involve the inflation pressure level required for deployment of the airbag in the first few milliseconds following the collision and the interaction with the occupant afterward. But in cases of proximity of occupant and passenger prior to the collision, the airbag deployment itself may produce injuries to the occupant depending on the inflation pressure and the resulting head accelerations after deployment. Airbags are primarily aimed at reducing head injuries. Thus the HIC can also be applied for the assessment of airbag adequacy. Some airbags have also been considered to protect the occupants from lateral impacts.

Bioengineering Considerations

Human Body Structures

A major effort in crashworthiness technology development has been directed at the representation of human body structures for crashworthiness study. From the kinematics point of view, a human body can be represented as an articulated system with kinematic degrees of freedom, certain mass distribution and some degree of flexibility and relative motion stiffness.

Three approaches have been followed in crashworthiness studies. The first approach uses models that capture human body segment mobility and general mass distribution. These models can produce useful information about general displacements, velocities and accelerations of body segments during a collision, but do not capture the nature of the human body internal anatomy and thus, they cannot be used to assess detailed interactions between vehicle and occupants.

The second approach is to model the structure of dummies, which in turn are designed with the sole purpose of mimicking the human body response in experimental vehicle collisions for crashworthiness assessments. The development of the 'Hybrid III Family' is the result of many years of crashworthiness technology development pursuing biofidelity between the dummies and the human body for the 50-percentile male, female, and children of several ages. MADYMO^{TM} code provides a database for modeling validated standard dummies.

The third approach is to model human body structures in such a way that human body compliance in terms of flexibility and inertia are taken into consideration in the analyses. The FE models that have been developed in this area are intended to describe the dynamics of human body structures involved in vehicle collisions and to produce a better understanding of the likely injuries that occupants may possibly sustain. A number of models can be found in literature for head–neck complex, for thorax and spine complex and for lower extremities that reflect anatomically correct features.

There are two key issues in modeling human body structures; one is the complex geometry of body parts, including bones and internal organs. This task is extremely tedious but has been advanced greatly by the availability of the 'visible human project'. Yet, the development of accurate FE models calls for appropriate material characterization of body tissue, from various bone structures to soft tissue and membrane materials. One example that illustrates the importance of material characterization can be found in modeling of the human head for impact assessment. Several FE models have been proposed with material properties for the brain tissue

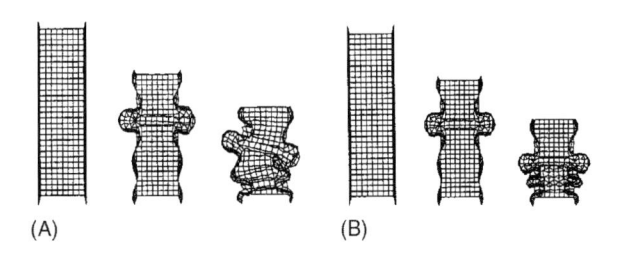

Figure 1 Impact crushing behavior of a square tubing with (A) and without (B) a mild load offset (0.2 mm/300 mm length).

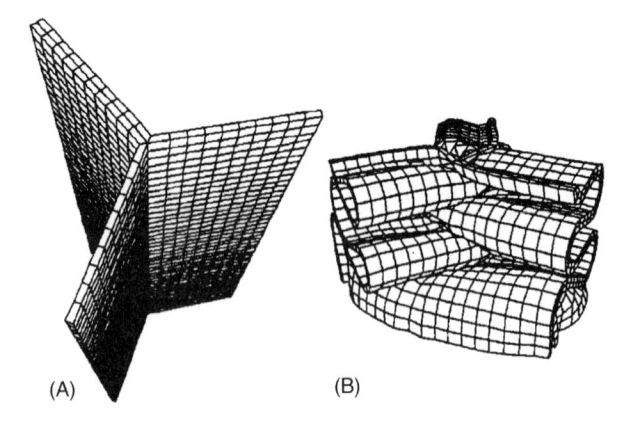

Figure 2 Honeycomb cell segment crushing under axial loading. (A) Honeycomb cell segment; (B) honeycomb cell segment crushing under axial load.

that produces a different response. At issue is how the material properties affect the standards used to assess head injury, specifically, the HIC.

Application Examples

Application examples can be found in many fields, from the solid mechanics general examples (cylindrical shells under axial impact) to automotive structures under various collision scenarios. **Figure 1** shows the crushing of a square tubing under axial impact loading. Notice the accordion folding of the walls of the tube in such a way that the surface produces contact with itself as it folds. The use of a 'single surface' contact is illustrated in this example. The same figure illustrates the effect of a mild load

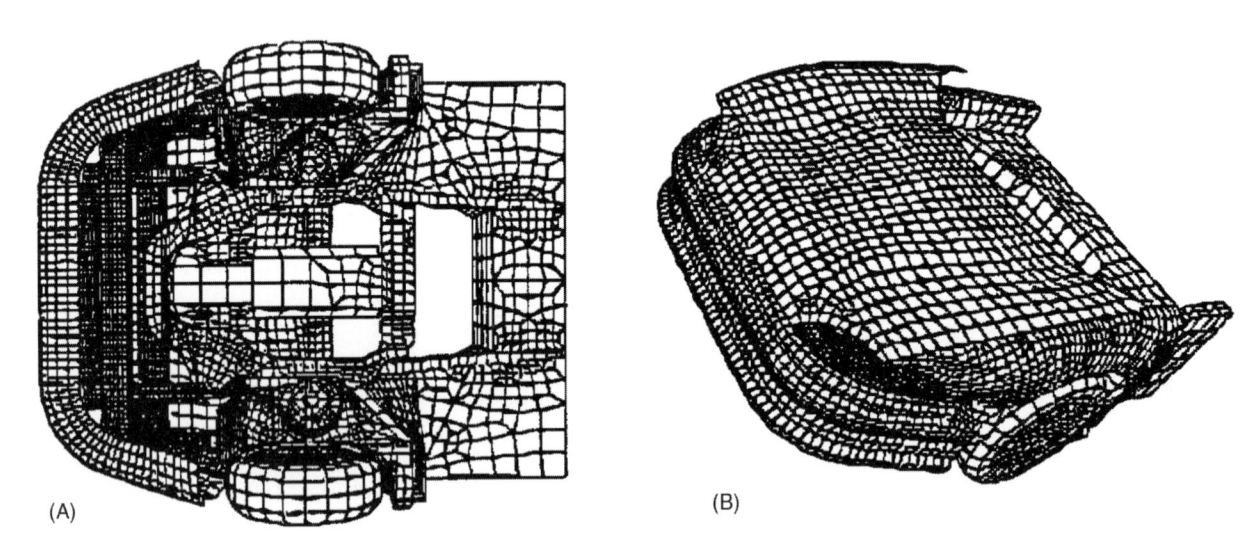

Figure 3 Example of a vehicle front-end used in a frontal collision simulation. (A) Under view; (B) isometric view.

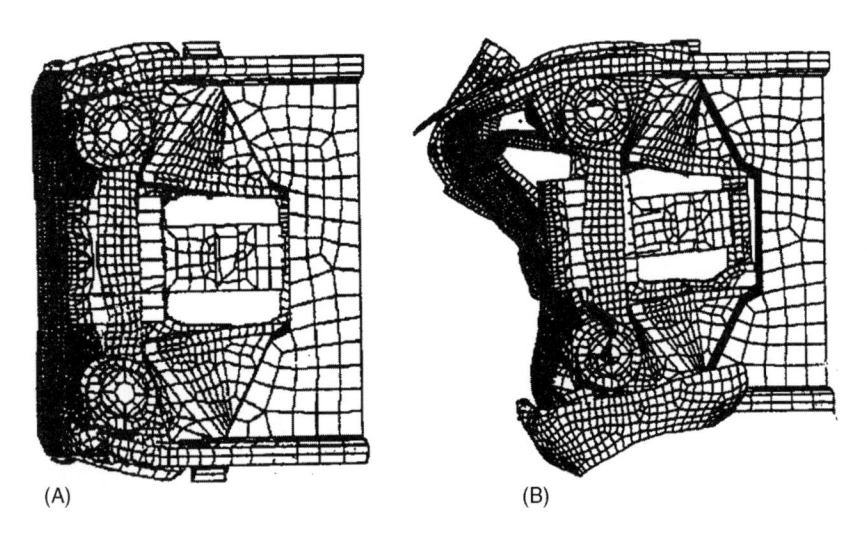

Figure 4 Examples of frontal collision simulation of a midsize vehicle. (A) Frontal collision of front end of vehicle against rigid wall (under view); (B) offset frontal collision of front end of vehicle against rigid wall (under view).

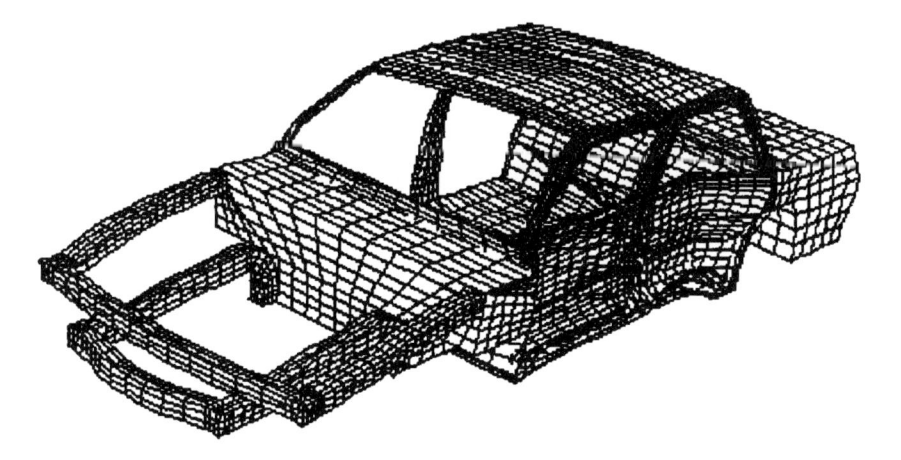

Figure 5 Example of a passenger cabin used in a lateral collision simulation.

offset of 0.02 mm over the length of the tube (300 mm), producing asymmetric crushing. A similar example shows the analysis of a cell segment of a honeycomb arrangement under crushing axial loads. **Figure 2** illustrates the crushing modes that can be used to characterize a honeycomb material using the 'user-defined' material 26 in DYNA3D. A more classical application of crashworthiness is illustrated in the front-end model of a midsize sedan for the purpose of frontal collision assessment (see **Figure 3**). Such a model can be used for several frontal collision scenarios, as illustrated in **Figure 4**, illustrating a frontal collision against a rigid wall and against an offset collision. The case of a lateral collision is illustrated in **Figure 5**. The structural integrity of the passenger cabin becomes the central issue, as well as the performance of the doors with reinforcing bars. The US–NHTSA maintains a public website with a variety of generic FE models, developed for the purpose of crashworthiness assessments of vehicles under various loading conditions. A variety of vehicle models are included, compact, midsize sedans, sports utility vehicles, pickup trucks, school bus, etc. In addition, some models have been developed to address human body response in a colliding vehicle. One such model is shown in **Figure 6**, demonstrating a compact vehicle running against a rigid wall. The occupant has received significant attention in the crashworthiness studies. One example is the study of restraint systems in the event of a collision. At issue is the deceleration of the occupant following the collision. **Figure 7** illustrates the model of an occupant, represented by the model of a dummy, and the response under frontal impact. Seat belts (and airbags as well) are modeled using membrane elements, capable of carrying tension but not compression. Another important application of crashworthiness is the study of human body structures to impact loads. Crashworthiness studies have been successfully applied for the study of the dynamic response of different human parts to impact. An example is the response of the human head. **Figure 8** shows a model

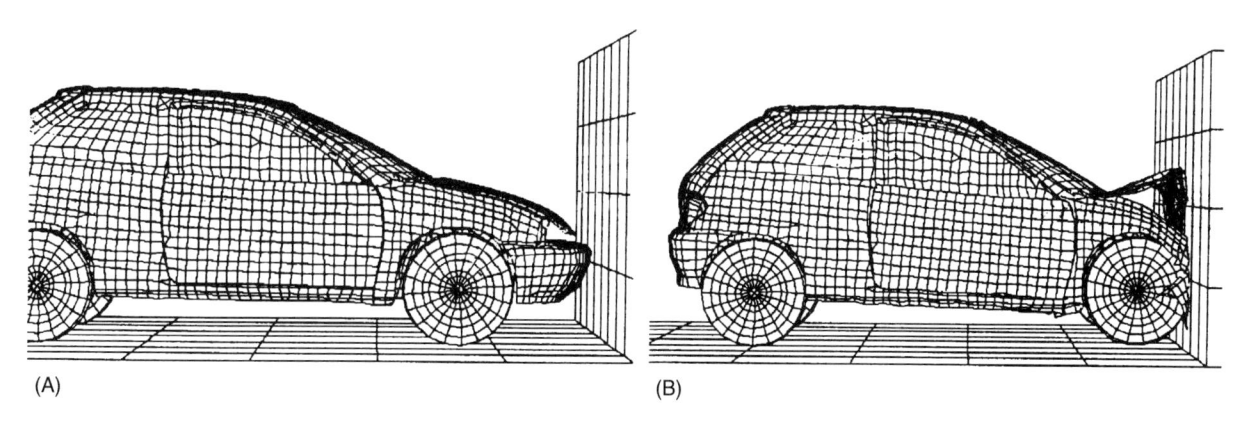

(A)   (B)

Figure 6 Compact vehicle simulation of a frontal collision against a rigid wall.

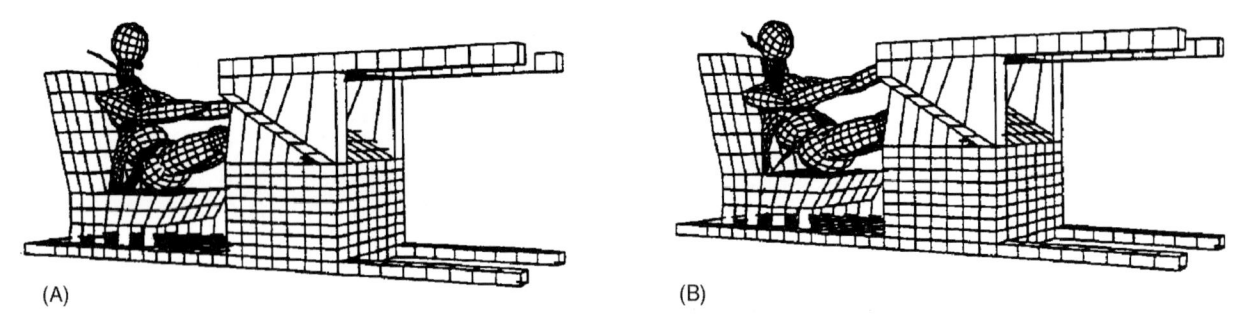

Figure 7 Example of a passenger restraint system (seat belt) under frontal collision.

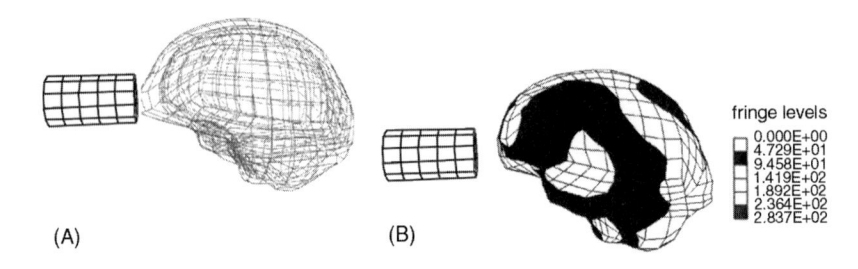

Figure 8 Human head under frontal impact. (A) Skull brain cavity and brain mass; (B) brain mass stress response immediately after impact (at 5 ms).

developed in which the brain cavity is modeled (skull and brain mass) and then subjected to a frontal impact. The stress waves on the brain mass can be analyzed using visco-elastic materials for the brain mass and elastic properties for the skull. At issue are the maximum pressures developed in the brain mass and the effect of the visco-elastic vs elastic brain material assumptions.

Conclusions

Crashworthiness has matured into a multidisciplinary area with a wide range of applications: in vehicle design, aircraft design, military ballistics, bioengineering, sports gear design, metal forming processes, particle impact erosion, cold working processes. The disciplines involved in the development of crashworthiness include computational mechanics, fluid–structure interactions, materials science, composite materials, honeycomb stiffened structures, polyurethane foam filled tubing, glass breaking, etc.

Yet certain areas remain a challenge to the crashworthiness technologies. Specifically, the mechanisms of fracture mechanics and the failure of intricate materials like composite laminates, and braided composite materials. Fluid–structure interaction as well as gas–structure interactions are clearly applications where challenges lie ahead.

An important challenge lies ahead in terms of applying crashworthiness principles for the study and prevention of fatal accidents involving heavy-duty vehicles. In the case of heavy-duty tanker trucks for example, the sloshing dynamics may produce instabilities that can cause rollover accidents. Substance containment to prevent spills and collision with transportation structures are topics yet to be fully addressed using crashworthiness techniques.

Perhaps one of the most important challenges lying ahead is in the comprehensive integration of crashworthiness principles in the design practice that relates not only to vehicles but to other systems, such as aircraft structures, transportation structures, and the ever challenging area of human structures.

Nomenclature

B	strain-displacement matrix
E	total energy
f_i	body force
f^k_{ia}	hourglass resisting force
h_{ia}	nodal velocities
HIC	head injury criterion
N	interpolation function matrix
p	pressure
q	bulk viscosity
Q_{hg}	empirical constant

s_{ij}	deviatoric stress
t	time parameter
v_m	element volume
V	relative volume
x^n	displacement at the nth time increment
$\dot{x}^n$	velocity at the nth time increment
$\ddot{x}^n$	acceleration at the nth time increment
δx	virtual displacement
δ_{ij}	Kronecker delta
$\delta \pi$	variation of energy functional
ε_{ij}	strain tensor
$\dot{\varepsilon}_{ij}$	strain rate tensor
ϕ_k	shape function
Γ_{ak}	hourglass mode
ρ	density
σ_{ij}	stress tensor
σ_{ij}^*	Jaumann rate of the Cauchy stress
$\dot{\sigma}_{ij}$	material derivative of the stress
ω_{ij}	spin rate tensor
∂b_i	domain boundary i
$\partial/\partial x$	partial differentiation operator with respect to x

See also: **Active control of vehicle vibration**; **Basic principles**; **Computation for transient and impact dynamics**; **Finite element methods**.

Further Reading

Aida T (2000) *Study of Human Head Impact: Brain Tissue Constitutive Models*. PhD Dissertation, University West Virginia, May.

Allison D (1995) *Seymour Cray Interview*. http://americanhistory.si.edu/csr/comphist/cray.htm.

Armen H, Pifko A, Levine H (1975) *Nonlinear Finite Element Techniques for Aircraft Crash Analysis*. Aircraft Crashworthiness Symposium, Cincinnati, pp. 517–548.

Belytscho T, Schwer L, Klein MJ (1977) Large displacement transient analysis of space frames. *International Journal for Numerical and Analytical Methods in Engineering* 11: 65–84.

Belytscho T, Tsay CS (1981) Explicit algorithms for nonlinear dynamics of shells. *AMD* 48: 209–231.

Belytscho T, Tsay CS (1983) A stabilization procedure for the quadrilateral plate element with one-point quadrature. *International Journal of Numerical Methods in Engineering* 19: 405–419.

Flanagan DP, Belytscho T (1981) A uniform strain hexahedron and quadrilateral and orthogonal hourglass control. *International Journal of Numerical Methods in Engineering* 17: 679–706.

Hallquist JO (1976) *A Procedure for the Solution of Finite Deformation Contact-Impact Problems by the Finite Element Method*. University of California: Lawrence Livermore National Laboratory, rept. UCRL-52066.

Hallquist JO (1996) *Preliminary User's Manuals for DYNA3D and DYNAP (Nonlinear Dynamic Analysis of Solids in Three Dimensions)*. University of California: Lawrence Livermore National Laboratory, rept. UCID-17268 (Rev. 1, 1979).

Hallquist JO, Gourdeau GL, Benson DJ (1985) Sliding interfaces with contact-impact in large-scale Lagrangian computations. *Computer Methods in Applied Mechanics and Engineering* 51; 107–137.

Hughes TJR, Liu WK (1981) Nonlinear finite element analysis of shells: part I, two dimensional shells. *Computer Methods in Applied Mechanics* 27: 167–181.

Hughes TJR, Liu WK (1981) Nonlinear finite element analysis of shells: part II, three dimensional shells. *Computer Methods in Applied Mechanics* 27: 331–362.

Kamal MM (1970) Analysis and simulation of vehicle to barrier impact. *SAE Transactions* 79: 1498–1503.

Kamal MM, Wolf JA (1977) Finite element models for automotive vehicle vibrations. In: *ASME Finite Element Applications in Vibration Problems*, presented at the Design Engineering Technical Conference, Chicago, IL, Sept. 26–28.

Kan S, FHWA/NHTSA *National Crash Analysis Center*, Public Finite Element Model Archive. http://www.ncac.gwu.edu/archives/model/index.html.

Lin KH (1973) A rear-end barrier impact simulation model for uni-body passenger cars. *SAE Transactions* 82: pp. 628–634.

Macmillan RH (1970) *Vehicle Impact Testing*. SAE paper no. 700404. Warrendale, PA: International Automobile Safety Conference Compendium.

Martin DE, Kroel CK (1967) Vehicle crash and occupant behavior. Paper no. 670034. *Transactions of the SAE* 76: 236–258.

McHenry RR, Naab KN (1966) Computer simulation of the crash victim: a validation study. *Proceedings of the Tenth Stapp Car Crash Conference*. November pp. 126–163.

McIvor IK Wineman AS, Wang HC (1975) *Large Dynamic Plastic Deformation of General Frames*. Twelfth Meeting of the Society of Engineering Science, University of Texas, October, pp. 1181–1190.

Melosh RJ (1972) *Car-Barrier Impact Response of a Computer Simulated Mustang*. DOT-NHTSA Report DOT-HS-091-1-125A.

Melosh RJ, Kelly DM (1967) The potential for predicting flexible car crash response. *SEA Transactions* 76: 2835–2842.

OECD *International Road Traffic and Accident Database, Brief Overview*. IRTAD. http://www.bast.de/irtad/english/irtadlan.htm.

Ruan JS, Khalil T, King AI (1991) Human head dynamic response to side impact by finite element modeling. *ASME Journal of Biomechanical Engineering* 113: 276–283.

US National Library of Medicine, *The Visible Human Project*. http://www.nlm.nih.gov/research/visible/visiblehuman.html.

US–DOT, National Highway Traffic Safety Administration *The Federal Motor Vehicle Safety Standards and Regulations Brochure*. http://www.nhtsa.dot.gov/cars/rules/import/FMVSS/SN219.

USA Office of the Federal Register. *The Code of Federal Regulations*, title, 49, subtitle B, chapter V, part 571, subpart B, section 571.216.

Van Leer B (1977) Towards the ultimate conservative difference scheme. IV. A new approach to numerical convention. *Journal of Computational Physics* 23: 276–299.

Welch RE, Bruce RW, Belytscho T. (1976) *Finite Element Analysis of Automotive Structures Under Crash Loadings*. Report no. DOT-HS-810 847. U.S. Department of Transportation.

Wilson RA (1970) A review of vehicle impact testing: how it began and what is being done. Paper no. 700414. *SAE Transactions* 79:1498–1503.

Yang KY, Wang KH (1998) *Finite Element Modeling of the Human Thorax*. http://www.nlm.nih.gov/research/visible/vhpconf98/AUTHORS/YANG/YANG.HTM.

CRITICAL DAMPING

D Inman, Virginia Polytechnic Institute and State University, Blacksburg, VA, USA

Copyright © 2001 Academic Press

doi: 10.1006/rwvb.2001.0061

Critical damping is defined for linear, single-degree-of-freedom systems with viscous damping. It is based on the three stable solutions to a second-order, ordinary differential equation with constant coefficients and corresponds to the case of repeated, real roots in the characteristic equation. Physically, critical damping corresponds to that value of damping that separates oscillation from nonoscillation of the free response. Thus, critical damping is also the minimum amount of damping that a spring-mass-damper system can have and not vibrate. If such a system has a smaller than critical amount of damping it will oscillate.

Critical damping is also the numerical value used to nondimensionalize damping parameters to produce a damping ratio. Experimentalists and analysts alike use the percent critical damping as a dimensionless parameter for describing the amount of damping in a system. Percent critical damping is also used in performance specifications and in design.

Although critical damping is defined for a single-degree-of-freedom, spring-mass-damper system, it is also routinely applied to modal equations where it appears as a modal damping ratio. The modal damping ratio extends the concept of critical damping to multiple-degree-of-freedom systems. The same is true for systems described by distributed mass models. The extension of the concept of critical damping to both lumped and distributed mass models is straightforward, yet somewhat buried in mathematical details.

Definition of Critical Damping

Critical damping is defined for a single-degree-of-freedom, spring-mass-damper arrangement, as illustrated in **Figure 1**. The equation of motion for this system is found from Newton's law and the free-body diagram to be:

$$m\ddot{x}(t) + c\dot{x}(t) + kx(t) = 0 \qquad [1]$$

Here $x(t)$ is the displacement in meters, $\dot{x}(t)$ is the velocity in meters per second, $\ddot{x}(t)$ is the acceleration in meters per second per second, m is the mass in kilograms, k is the stiffness in Newtons per meter and c is the damping coefficient in Newton second per meter or kilograms per second. Eqn [1] is written in dimensionless form by dividing the expression by the mass. This yields:

$$\ddot{x}(t) + 2\zeta\omega_n\dot{x}(t) + \omega_n^2 x(t) = 0 \qquad [2]$$

Here the undamped natural frequency is defined to be:

$$\omega_n = \sqrt{k/m} \qquad [3]$$

(in radians per second) and the damping ratio is defined to be:

$$\zeta = \frac{c}{2m\omega_n} = \frac{c}{2\sqrt{km}} \qquad [4]$$

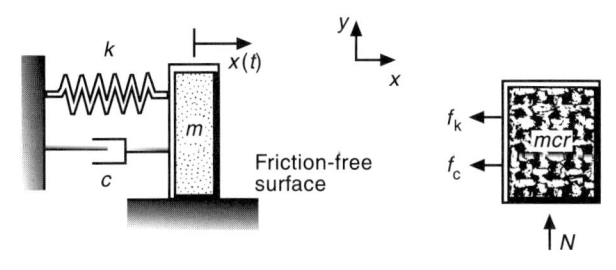

Figure 1 A single-degree-of-freedom system and free-body diagram.

which is dimensionless.

To solve the differential equation given in eqn [2], a solution of the form $x(t) = Ae^{\lambda t}$ is assumed and substituted into eqn [2] to yield:

$$\lambda^2 + 2\zeta\omega_n\lambda + \omega_n^2 = 0 \qquad [5]$$

which is called the characteristic equation in the unknown parameter λ. This process effectively changes the differential eqn [2] into a quadratic algebraic equation with a well-known solution. The value of λ satisfying eqn [5] is then:

$$\lambda = -\zeta\omega_n \pm \omega_n\sqrt{\zeta^2 - 1} \qquad [6]$$

which are the two roots of the characteristic equation. It is the radical expression that gives rise to the concept of critical damping. If the radical expression is zero, that is, if:

$$\zeta^2 = 1 \qquad [7]$$

the system of **Figure 1** is said to be critically damped. If $\zeta = 1$, then the corresponding damping coefficient c is called the critical damping coefficient, c_{cr}, by:

$$c_{cr} = 2m\omega_n = 2\sqrt{km} \qquad [8]$$

obtained by setting $\zeta = 1$ in eqn [4].

The mathematical significance of $\zeta = 1$ is that the two roots given in eqn [6] collapse to a repeated real root of value $\lambda = -\omega_n$. The solution to the vibration problem given in eqn [1] then becomes:

$$x(t) = (a_1 + a_2 t)e^{-\omega_n t} \qquad [9]$$

where the constants of integration a_1 and a_2 are determined by the initial conditions. Substituting the initial displacement, x_0, into eqn [10] and the initial velocity, v_0, into the derivative of eqn [10] yields:

$$a_1 = x_0, \quad \text{and} \quad a_2 = v_0 + \omega_n x_0 \qquad [10]$$

Critically damped motion is plotted in **Figure 2** for two different values of initial conditions. Notice that no oscillation occurs in this response.

The motion of critically damped systems may be thought of in several ways. First, a critically damped system represents a system with the smallest value of damping coefficient that yields aperiodic motion. If $\zeta > 1$, the roots given in eqn [6] are distinct, real roots giving rise to solutions of the form:

$$x(t) = e^{-\zeta\omega_n t}\left(a_1 e^{-\omega_n\sqrt{\zeta^2 - 1}\, t} + a_2 e^{+\omega_n\sqrt{\zeta^2 - 1}\, t}\right) \qquad [11]$$

which also represents a nonoscillatory response. Again, the constants of integration a_1 and a_2 are determined by the initial conditions. In this aperiodic case, the constants of integration are real valued and are given by:

$$a_1 = \frac{-v_0 + \left(-\zeta + \sqrt{\zeta^2 - 1}\right)\omega_n x_0}{2\omega_n\sqrt{\zeta^2 - 1}} \qquad [12]$$

and

$$a_2 = \frac{-v_0 + \left(\zeta + \sqrt{\zeta^2 - 1}\right)\omega_n x_0}{2\omega_n\sqrt{\zeta^2 - 1}} \qquad [13]$$

Such a response is called an overdamped system and does not oscillate, but rather returns to its rest position exponentially.

If the damping value is reduced below the critical value, so that $\zeta < 1$, the discriminant of eqn [6] is negative, resulting in a complex conjugate pair of roots. These are:

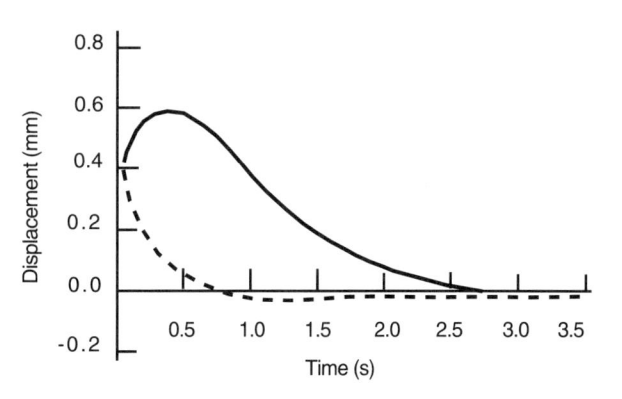

Figure 2 The critically damped response for two different initial conditions.

$$\lambda_1 = -\zeta\omega_n - \omega_n\sqrt{1-\zeta^2}\,j \qquad [14]$$

and:

$$\lambda_2 = -\zeta\omega_n + \omega_n\sqrt{1-\zeta^2}\,j \qquad [15]$$

where $j = \sqrt{-1}$ and:

$$\sqrt{1-\zeta^2}\,j = \sqrt{(1-\zeta^2)(-1)} = \sqrt{\zeta^2-1} \qquad [16]$$

The solution of eqn [1] is then of the form:

$$x(t) = e^{-\zeta\omega_n t}\left(a_1 e^{j\sqrt{1-\zeta^2}\omega_n t} + a_2 e^{-j\sqrt{1-\zeta^2}\omega_n t}\right) \qquad [17]$$

where a_1 and a_2 are arbitrary complex valued constants of integration to be determined by the initial conditions. Using the Euler relations for the sine function, this can be written as:

$$x(t) = Ae^{-\zeta\omega_n t}\sin(\omega_d t + \phi) \qquad [18]$$

where A and ϕ are constants of integration and ω_d, called the damped natural frequency, is given by:

$$\omega_d = \omega_n\sqrt{1-\zeta^2} \qquad [19]$$

The constants A and ϕ are evaluated using the initial conditions. This yields:

$$A = \sqrt{\frac{(v_0 + \zeta\omega_n x_0)^2 + (x_0\omega_d)^2}{\omega_d^2}},$$
$$\phi = \tan^{-1}\left[\frac{x_0\omega_d}{v_0 + \zeta\omega_n x_0}\right] \qquad [20]$$

where x_0 and v_0 are the initial displacement and velocity. Note that for this case ($0 < \zeta < 1$) the motion oscillates. This is called an underdamped system. Hence, if the damping is less then critical, the motion vibrates, and critical damping corresponds to the smallest value of damping that results in no vibration. Critical damping can also be thought of as the value of damping that separates nonoscillation from oscillation.

The concept of critical damping provides a useful value to discuss level of damping in a system, separating the physical phenomenon of oscillation and no oscillation. Often when reporting levels of damping, the percent of critical damping is provided. The percent critical damping is defined as $100\zeta\%$ or:

$$\frac{c}{c_{cr}} \times 100 \qquad [21]$$

Normal values for ζ are very small, such as 0.001, so that percent critical damping is of the order of 0.1–1%. However, viscoelastic materials, hydraulic dampers, and active control systems are able to provide large values of percent critical damping.

Critical damping viewed as the minimum value of damping that prevents oscillation is a desirable solution to many vibration problems. Increased damping implies more energy dissipation, and more phase lag in the response of a system. Reduced damping means more oscillation, which is often undesirable. Adding phase to the system slows the response down and in some cases this may be undesirable. Hence, critical damping is a desirable tradeoff between having enough damping to prevent oscillation and not so much damping that the system uses too much energy or causes too large a phase lag.

An example of the use of critical damping is in the closing of a door. If the mechanism has too much damping, the door will move slowly and too much heat will exchange between the inside and outside. If too little damping is used the door will oscillate at closing and again too much air and heat is exchanged. At critical damping the door closes without oscillation and a minimum amount of air and heat are exchanged. The needle gauges used in tachometers and speedometers in automobiles provide additional examples of critically damped systems. The needle operates as a torsional spring with critical damping. Too much damping would make the gauge slow to reach the actual value and too little damping would cause the needle to vibrate.

Critical Damping in Lumped Parameter Models

The concept of critical damping may be defined for multiple-degree-of-freedom systems. First consider the model used for such systems. A linear multiple-degree-of-freedom system may be modeled by the vector differential equation:

$$\mathbf{M}\ddot{\mathbf{x}}(t) + \mathbf{C}\dot{\mathbf{x}}(t) + \mathbf{K}\mathbf{x}(t) = 0 \qquad [22]$$

where $\mathbf{M}$, $\mathbf{C}$, and $\mathbf{K}$ are the usual mass, damping, and stiffness matrices, respectively. These coefficient matrices are real-valued and of size $n \times n$ where n is the number of degrees of freedom. Each matrix is assumed to be positive-definite and symmetric. The vector $\mathbf{x}(t)$ is an $n \times 1$ vector of displacements with the over dots denoting the time derivatives yielding velocity and acceleration.

The system described by eqn [22] may be divided into two important subclasses depending on the nature of the damping matrix $\mathbf{C}$. Either the equations of motion [22] can be decoupled into n independent equations with real constant coefficients or they cannot. A necessary and sufficient condition for the equations to decouple is:

$$\mathbf{CM}^{-1}\mathbf{K} = \mathbf{KM}^{-1}\mathbf{C} \qquad [23]$$

If eqn [22] holds, then such systems are sometimes said to have classical, proportional, or modal damping. In this case the modal matrix of the undamped system can be used to decouple the equations of motion into n single-degree-of-freedom systems. Let $\mathbf{P}$ be the modal matrix (the matrix with columns made up of the mode shapes of the undamped system, normalized with respect to the mass matrix $\mathbf{M}$, such that $\mathbf{P}^{-1}\mathbf{MP} = \mathbf{I}$, the $n \times n$ identity matrix). Then substitution of $\mathbf{x} = \mathbf{Pr}$ in eqn [1] and multiplying by $\mathbf{P}^{-1}$ yields:

$$\ddot{\mathbf{r}} + \mathbf{P}^{-1}\mathbf{CP}\dot{\mathbf{r}} + \mathbf{P}^{-1}\mathbf{KP}\mathbf{r} = 0 \qquad [24]$$

If eqn [23] holds, then each of the coefficient matrices is diagonal and of the form:

$$\mathbf{P}^{-1}\mathbf{CP} = \begin{bmatrix} 2\zeta_1\omega_1 & 0 & \cdots & 0 \\ 0 & 2\zeta_2\omega_2 & \cdots & 0 \\ \vdots & \cdots & \ddots & \vdots \\ 0 & \cdots & \cdots & 2\zeta_n\omega_n \end{bmatrix}$$

and:

$$\mathbf{P}^{-1}\mathbf{KP} = \begin{bmatrix} \omega_1^2 & 0 & \cdots & 0 \\ 0 & \omega_2^2 & \cdots & 0 \\ \vdots & \cdots & \ddots & \vdots \\ 0 & \cdots & \cdots & \omega_n^2 \end{bmatrix} \qquad [25]$$

Thus eqn [24] can be written as the n decoupled modal equations:

$$\ddot{r}_i(t) + 2\zeta_i\omega_i\dot{r}_i(t) + \omega_i^2 r_i(t) = 0 \qquad [26]$$

Here ζ_i are defined to be the modal damping ratios and if each one has the value 1, then each mode of the system is critically damped. Consequently the critical damping matrix, $\mathbf{C}_{\text{cr}}$, is defined to be:

$$\mathbf{C}_{\text{cr}} = \mathbf{P} \begin{bmatrix} 2\omega_1 & 0 & \cdots & 0 \\ 0 & 2\omega_2 & \cdots & 0 \\ \vdots & \cdots & \ddots & \vdots \\ 0 & \cdots & \cdots & 2\omega_n \end{bmatrix} \mathbf{P}^{-1} \qquad [27]$$

This is the value of damping matrix that causes each mode to be critically damped.

The critical damping matrix may be used to define the condition of underdamping, overdamping, and critical damping for a lumped-mass system that satisfies eqn [23]. The following conditions hold for the system of eqn [1]:

1. If the matrix $\mathbf{C} = \mathbf{C}_{\text{cr}}$, each mode is critically damped.
2. If the matrix $\mathbf{C} - \mathbf{C}_{\text{cr}}$ is positive-definite, each mode is overdamped.
3. If the matrix $\mathbf{C}_{\text{cr}} - \mathbf{C}$ is positive-definite, each mode is underdamped.
4. If the matrix is $\mathbf{C} - \mathbf{C}_{\text{cr}}$ indefinite, some modes may oscillate and some will not.

If the commutivity condition of eqn [23] is not satisfied, these conditions are still valid. However, if eqn [23] holds, these conditions are both necessary and sufficient. In the case that eqn [23] is not satisfied, the idea of a 'mode' changes to include phase information and the decoupled equations given in eqn [26] are no longer valid. In this case the damping conditions stated here refer to the system eigenvalues which are of the form:

$$\lambda_i = -\zeta_i\omega_i \pm \omega_i\sqrt{\zeta_i^2 - 1} \qquad [28]$$

where ζ_i is one, greater than one or less than one for every value of i depending on the definiteness of the matrix $\mathbf{C} - \mathbf{C}_{\text{cr}}$.

The critical damping matrix can be defined in terms of the mass and stiffness matrix much like the critical damping coefficient of the single-degree-of-freedom system is defined in terms of the scalar values of mass and stiffness in eqn [8]. It can be shown through a series of simple matrix manipulations that eqn [27] may be written as:

$$\mathbf{C}_{cr} = 2\mathbf{M}^{1/2}\left(\mathbf{M}^{-1/2}\mathbf{K}\mathbf{M}^{-1/2}\right)^{1/2}\mathbf{M}^{1/2} \quad [29]$$

Here the exponent refers to the positive-definite matrix square root of a matrix (not the square root of each element). Note that eqn [29] reduces to the scalar definition of eqn [8] if the matrices are collapsed to scalar values.

Critical Damping in Distributed Parameter Models

The concept of critical damping can also be extended to elastic systems governed by systems of partial differential equations used to model rods, beams, plates, and shells. The equations of motion of such systems may be symbolically represented by the operator equation:

$$\rho(\mathbf{x})\mathbf{u}_{tt}(\mathbf{x},t) + L_1\mathbf{u}_t(\mathbf{x},t) + L_2\mathbf{u}(\mathbf{x},t) = 0, \quad \mathbf{x} \in \Omega$$
$$\text{and} \quad B\mathbf{u}(\mathbf{x},t) = 0, \quad \mathbf{x} \in \partial\Omega \quad [30]$$

Here $\mathbf{u}(\mathbf{x}, t)$ represents the deflection in three space (u_x, u_y, u_z), $\mathbf{x}$ represents a position along the surface in three space (for example: x, y, z, in rectangular coordinates), the subscripts denote partial time derivatives, $\rho(\mathbf{x})$ is the density, Ω is the domain in which $\mathbf{x}$ is defined, and the differential operators L_1 and L_2 represent damping and stiffness respectively. The operator B denotes the boundary conditions and $\partial\Omega$ represents the boundary. As an example, for a uniform, clamped-free beam, subject to air damping, the following holds: $\rho =$ constant, $\mathbf{x}$ becomes the scalar length along the neutral axis of the beam, $\mathbf{u}$ becomes the scalar deflection perpendicular to the neutral axis of the beam, Ω becomes the interval $(0, l)$ where l is the length of the beam and the clamp is at the origin, $\partial\Omega$ consists of the two points 0 and l, the operator $L_1 = c$, a constant damping coefficient, and L_2 is given by the familiar beam stiffness:

$$L_2 = \frac{\partial^2}{\partial x^2}\left(EI(x)\frac{\partial^2(\cdot)}{\partial x^2}\right) \quad [31]$$

The boundary operator B expresses the usual boundary conditions:

$$Bu = 0 \Rightarrow u(0, t) = 0, \quad u_x(0, t) = 0,$$
$$\frac{\partial}{\partial x}\left(EI\frac{\partial^2 u(l, t)}{\partial x^2}\right) = 0, \quad EI\frac{\partial^2 u(l, t)}{\partial x^2} = 0 \quad [32]$$

These equations are subject to some more technical conditions to define the operators properly in rigorous mathematical terms. These mathematical conditions may be found in the Further Reading section.

The concept of critical damping for systems described by eqn [30] parallels that of the matrix case given by that of eqn [29]. However, in the operator case, careful attention must be paid to the boundary conditions and the existence of derivatives. The modal decoupling condition is similar and requires that $L_1L_2 = L_2L_1$ for certain boundary conditions. In this case, the modal damping ratios can be defined exactly as in eqn [26] where the total solution takes on the form:

$$\mathbf{u}(\mathbf{x},t) = \sum_{i=1}^{\infty} r_i(t)\phi_i(\mathbf{x}) \quad [33]$$

where the $\phi_i(\mathbf{x})$ are the mode shapes (or eigenfunctions) of the undamped system. As before, the critical damping condition corresponds to $\zeta_i = 1$ for every value of the index i.

To represent a critical damping operator as in eqn [29] it is first necessary to make eqn [30] monic by either dividing through by $\rho(\mathbf{x})$ or making a change of variables. Assuming that eqn [30] is thus normalized, the critical damping operator becomes:

$$L_{CD} = 2L_2^{1/2} \quad [34]$$

defined on an appropriate domain (boundary conditions). Here the square root again refers to the operator square root that has a precise mathematical definition and rules for computing. It practice it is often easier to compute the square of the critical damping operator, defined by $L_{CD}^2 = 4L_2$. In most systems it is likely that only one mode will be critically damped.

The concept of critical damping has been defined for linear vibrating systems with viscous damping. Achieving the exact state of critical damping is often difficult and requires a large amount of damping not usually available in natural materials. However critical damping serves as an important analytical expression to separate the two distinct phenomena of vibration versus exponential decay without oscillation in a system's time response.

Nomenclature

a_1, a_2	constants of integration
B	boundary operator
I	$n \times n$ identity matrix
l	length

L_1, L_2	differential operators
P	modal matrix
u	scalar deflection
ρ	density

See also: **Damping materials**; **Damping measurement**; **Viscous damping**.

Further Reading

Den Hartog JP (1956) *Mechanical Vibration*, 4th edn. New York: McGraw Hill.

Inman DJ (1989) *Vibrations: Control, Measurement and Stability*. Upper Saddle River, NJ, USA: Prentice Hall.

Inman DJ (1996) *Engineering Vibration*, revised edn. Upper Saddle River, NJ, USA: Prentice Hall.

D

DAMPING IN FE MODELS

G A Lesieutre, Penn State University, University Park, PA, USA

Copyright © 2001 Academic Press

doi:10.1006/rwvb.2001.0021

Damping in structural systems is the result of a multitude of complex physical mechanisms (**Figure 1**). For this reason, structural designers do not often use physics-based damping models for analysis. (In fact, such models are not generally available.) In practice, most analysts use linear, finite element structural models and relatively simple mathematical models of damping.

Because such damping models do not accurately describe the underlying physics, the resulting damped structural model may only capture the general effects of damping (e.g., decaying free vibration) without capturing the details (e.g., phase differences between the motion at various points when the structure executes harmonic response). As a result, approximate values for damping parameters are often used for initial design analysis, and improved using experiments.

Finally, many common damping models have specific deficiencies that limit their utility under some circumstances. These deficiencies may sometimes justify the use of higher-fidelity models having stronger physical foundations.

Viscous Damping

Consider an elastic (undamped) nongyroscopic linear structure. Application of the finite element modeling method yields a discretized matrix equation of motion having the following general form:

$$\mathbf{M}\ddot{\mathbf{x}} + \mathbf{K}\mathbf{x} = \mathbf{f}(t) \qquad [1]$$

where $\mathbf{M}$ and $\mathbf{K}$ are the system mass and stiffness matrices, and $\mathbf{x}$ and $\mathbf{f}$ are $N \times 1$ vectors of global nodal displacements and forces, respectively. Analysis of the eigenvalue problem associated with the unforced matrix equation yields normal modes of vibration, each comprising a real eigenvector (discrete mode shape) $\boldsymbol{\psi}_r$; and a natural frequency (corresponding to undamped harmonic vibration) ω_r. The associated structural response has the form:

$$\mathbf{x}(t) = \boldsymbol{\psi}_r \, e^{i\omega_r t} \qquad [2]$$

The 'natural' extension of the linear equation of motion [1] to include damping involves the addition of a matrix term multiplying the vector of nodal velocities:

$$\mathbf{M}\ddot{\mathbf{x}} + \mathbf{C}\dot{\mathbf{x}} + \mathbf{K}\mathbf{x} = \mathbf{f}(t) \qquad [3]$$

$\mathbf{C}$ is called the viscous damping matrix.

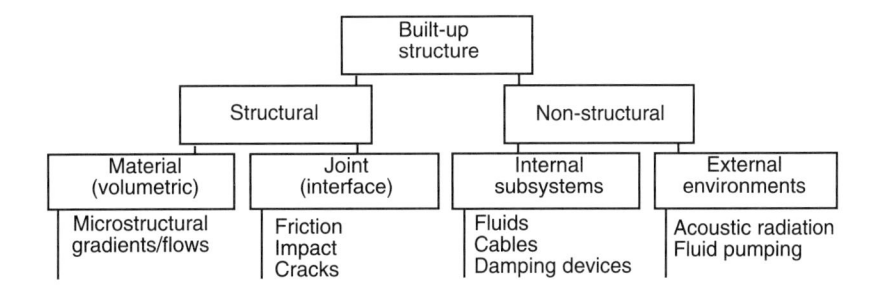

Figure 1 Damping in built-up structures is the result of many physical mechanisms acting in concert.

Elemental Viscous Damping

In a finite element context, **C** may be understood to be assembled from elemental viscous damping matrices and discrete viscous devices. As such, it can be expected to have symmetry and definiteness properties like those of **K**. Furthermore, that part of **C** associated with elemental viscous damping has a matrix structure (connectivity) similar or identical to that of **K**.

The basis for an elemental viscous damping matrix must necessarily be the constitutive equations of the material from which the modeled structural member is made. In this case, the common linear elastic constitutive equations are augmented such that part of the instantaneous stress σ is proportional to the strain rate $\dot{\varepsilon}$:

$$\boldsymbol{\sigma} = \mathbf{E}\boldsymbol{\varepsilon} + \mathbf{F}\dot{\boldsymbol{\varepsilon}} \qquad [4]$$

One of the deficiencies of the viscous damping model is that the constitutive behavior described by [4] does not represent the observed behavior of structural or damping materials very well, except perhaps over a very limited frequency range. As a result, structural analysts have difficulty obtaining material strain rate coefficients **F** for most materials – materials scientists do not try to determine such properties because the basic structure of the underlying model is inadequate.

Viscous Damping of Structures

Analysis of the eigenvalue problem associated with the general unforced, damped matrix equation of motion [3] yields complex normal modes of vibration, each comprising a complex eigenvector (discrete mode shape) $\boldsymbol{\psi}_r$; and a complex natural frequency (corresponding to damped oscillatory vibration) s_r. The associated structural response has the form:

$$\mathbf{x}(t) = \boldsymbol{\psi}_r e^{s_r t} \qquad [5]$$

where, assuming oscillatory motion, s_r is often expressed as:

$$s_r = -\zeta_r \omega_r + i\omega_r \sqrt{1 - \zeta_r^2} \qquad [6]$$

In eqn [6], ζ_r is the modal damping ratio, and ω_r is the magnitude of the complex natural frequency. If s_r can be represented in this oscillatory form, then its complex conjugate is also a complex natural frequency of the system.

Proportional Viscous Damping

Proportional viscous damping is a subset of viscous damping in which the viscous damping matrix **C** is a linear combination of the mass and stiffness matrices **M** and **K**:

$$\mathbf{C} = \beta \mathbf{M} + \gamma \mathbf{K} \qquad [7]$$

As a consequence of not representing material behavior well, some deficiencies of the (proportional) viscous damping model are observed at the structural level.

Consider first the case of proportionality to the stiffness matrix. For simple structures, in which single material elastic and strain rate coefficients, **E** and **F**, are assumed to represent material behavior adequately, the following relationship may be defined:

$$\boldsymbol{\sigma} = \mathbf{E}\boldsymbol{\varepsilon} + \gamma \mathbf{E}\dot{\boldsymbol{\varepsilon}} \qquad [8]$$

Analysis then shows that modal damping increases monotonically with frequency, with a slope of +1 on a log–log scale:

$$\zeta_r = \frac{\gamma}{2}\omega_r \quad (\omega_r: \zeta_r \leq 1) \qquad [9]$$

One result of this is that all modes having a natural frequency greater than $2/\gamma$ are overdamped; that is, in unforced motion, they do not respond in an oscillatory manner. Such behavior is not representative of that of most structures and is a serious deficiency of this viscous damping model. Many structures exhibit modal damping that depends only weakly on frequency and, when the source is material-based, generally increases, then decreases with frequency over a broad frequency range.

In the more general situation that includes proportionality to the mass matrix, modal damping is given by:

$$\zeta_r = \frac{\beta}{2\omega_r} + \frac{\gamma}{2}\omega_r \qquad [10]$$

Furthermore, the same real eigenvectors (discrete mode shapes) that diagonalize the mass and stiffness matrices of the undamped problem also diagonalize the viscous damping matrix **C**. This feature is probably the greatest attraction of the proportional damping model.

One deficiency of this model, however, is associated with proportionality to the mass matrix. In the event that a structure of interest possesses a rigid body (zero-frequency) mode, that mode will have nonzero damping. In addition, the general trend of modal damping indicated by this model first decreases with increasing frequency (with a slope of -1 on a log–log scale), then increases.

Finally, note that only the simplest structures will be made of a single material or be governed by single elastic and strain rate coefficients. And even assuming that the viscous damping model is reasonable over some frequency range, different materials will exhibit different ratios of strain rate to elastic coefficients. Thus, the proportional damping model must be regarded in practice as a mathematical curiosity. An analyst who uses a viscous damping model should be prepared to use complex modes.

Despite its drawbacks, viscous damping is a simple way of introducing damping in a structural model and may be adequate under some circumstances (for example, when accuracy is only needed over a small frequency range). Perhaps the greatest utility of the viscous damping model is the possibility of determining (identifying) a viscous damping matrix from experiments. Although such a damping matrix would not be element-based, a desirable feature, it might represent damping adequately for the purposes of continuing analysis.

Structural or Hysteretic Damping

The structural, or hysteretic, or complex stiffness damping model is motivated by a desire to obtain modal damping with frequency dependence weaker than that which results from the use of the viscous damping model. This model may developed by defining a frequency-dependent viscous damping matrix, or by using the complex modulus model of material behavior. Common to both approaches is a fundamental assumption of forced harmonic response. In practice, damping is also assumed to be closely related to stiffness (because both are associated with deformation), and independent of mass.

Frequency-dependent Viscous Damping

Observing from eqn [9] that, for stiffness-proportional damping, modal damping increases monotonically with modal frequency, one might be inclined to modify a single-modulus elemental viscous damping matrix by dividing by a frequency:

$$\mathbf{C}_{\text{hysteretic}} = \frac{1}{\Omega}\mathbf{C}_{\text{viscous}} = \frac{\gamma}{\Omega}\mathbf{K} \qquad [11]$$

where Ω is interpreted as a forcing frequency. In this case, and for a simple structure with single material elastic and strain rate coefficients, eqn [3] is initially modified as:

$$\mathbf{M}\ddot{\mathbf{x}} + \frac{\gamma}{\Omega}\mathbf{K}\dot{\mathbf{x}} + \mathbf{K}\mathbf{x} = \mathbf{f}\,e^{i\Omega t} \qquad [12]$$

then, considering the relationship of velocity to position in forced harmonic response:

$$\begin{aligned}\mathbf{x}(t) &= \mathbf{x}\,e^{i\Omega t} \\ \dot{\mathbf{x}}(t) &= i\Omega\mathbf{x}(t)\end{aligned} \qquad [13]$$

leads to:

$$\mathbf{M}\ddot{\mathbf{x}} + (1 + i\gamma)\mathbf{K}\mathbf{x}(t) = \mathbf{f}\,e^{i\Omega t} \qquad [14]$$

Finally, using the result that $\mathbf{x}(t)$ is harmonic at the forcing frequency, this may be expressed as:

$$[-\Omega^2\mathbf{M} + (1 + i\gamma)\mathbf{K}]\mathbf{x}(\Omega) = \mathbf{f} \qquad [15]$$

Note that this approach leads essentially to a complex stiffness matrix, as shown in eqn [14]. In addition, the response vector, $\mathbf{x}(\Omega)$, as in eqn [15], is generally complex, indicating possible phase differences between the response and the forcing, as well as between responses at multiple points.

Material Complex Modulus

The basis for an elemental hysteretic damping matrix must be the constitutive equations of the material from which the modeled structural member is made. In this case, response to harmonic forcing is considered, and the single-modulus linear elastic constitutive equations are modified to include a material loss factor, η:

$$\boldsymbol{\sigma}(\Omega) = (1 + i\eta)\mathbf{E}\boldsymbol{\varepsilon}(\Omega) \qquad [16]$$

The material loss factor is essentially the phase difference in forced harmonic response between an applied stress and the resulting strain. It may also be regarded as the fractional energy dissipation per radian of motion. Materials scientists commonly measure loss factors, sometimes as a function of frequency, temperature, and amplitude. Note that real materials with nonzero loss factors exhibit stiffnesses that vary with frequency and temperature.

Eqn [16] can also be expressed in a form that includes multiple material moduli and loss factors:

$$\boldsymbol{\sigma}(\Omega) = [\mathbf{E}' + i\mathbf{E}'']\boldsymbol{\varepsilon}(\Omega) \qquad [17]$$

where $\mathbf{E}'$ is associated with storage moduli and $\mathbf{E}''$ with loss moduli. An elemental stiffness matrix may then be modified as:

$$\mathbf{K}^* = \mathbf{K}' + i\mathbf{K}'' \qquad [18]$$

And when the use of a single material modulus suffices, eqn [18] may be expressed as:

$$\mathbf{K}^* = (1 + i\eta)\mathbf{K}' \qquad [19]$$

In this case, and recalling that forced harmonic response is assumed, the structural equations of motion (3) may be initially modified as:

$$\mathbf{M}\ddot{\mathbf{x}} + (1 + i\eta)\mathbf{K}\mathbf{x}(t) = \mathbf{f}(t) = \mathbf{f}\,e^{i\Omega t} \qquad [20]$$

This is identical to eqn [14]. Using the result that $\mathbf{x}(t)$ is harmonic at the forcing frequency, this may be expressed in the same form as eqn [15]:

$$[-\Omega^2 \mathbf{M} + (1 + i\eta)\mathbf{K}]\mathbf{x}(\Omega) = \mathbf{f} \qquad [21]$$

The damped structural model described by eqns [15] and [21] has some deficiencies, but also has considerable utility in practice.

First of all, this model is not generally useful for obtaining direct time response, as it essentially describes frequency response. However, when the time domain forcing function in eqn [20] is not harmonic, but nevertheless has a Laplace transform, it may be possible to find the response via inverse transformation using the elastic–viscoelastic correspondence principle.

In practice, an eigenvalue problem based on eqn [20] may be posed. Assuming the time response indicated in eqn [5], a complex natural vibration frequency results. If a single global loss factor can be isolated, as in eqn [20], such proportional hysteretic damping results in real eigenvectors or mode shapes. And for light damping, the modal damping ratio is approximately half of the loss factor. In the general case, complex eigenvectors result. Frequency-dependent stiffness and damping can be accommodated via iteration.

One difficulty with this eigenvalue problem is that the decaying time response postulated in eqn [5] differs from the forced harmonic response assumption underlying this damping model [13]. In practice, the accuracy of natural frequencies and mode shapes determined using this method may decrease with increasing loss factor(s).

This hysteretic damping model is ideal, however, for frequency response analysis [21]. Furthermore, loss factors and stiffnesses can, in principle, be functions of frequency:

$$[-\Omega^2 \mathbf{M} + (1 + i\eta(\Omega))\mathbf{K}(\Omega)]\mathbf{x}(\Omega) = \mathbf{f}$$
$$[-\Omega^2 \mathbf{M} + [\mathbf{K}'(\Omega) + i\mathbf{K}''(\Omega)]]\mathbf{x}(\Omega) = \mathbf{f} \qquad [22]$$

Because high-damping viscoelastic materials that are sometimes used to enhance structural damping exhibit significant frequency-dependent stiffness and damping, this feature is especially useful in practice.

Viscous Modal Damping

Modal analysis, or modal superposition, is an efficient means of obtaining the dynamic response of linear structures. In this approach, the time response is determined using a number of modes that is small relative to the number of physical degrees of freedom in a finite element model. Either real or complex mode shapes may be used, along with either real or complex natural frequencies.

The most common use of modal superposition involves the solution to the undamped eigenvalue problem (real modes and real natural frequencies). The modal equations of motion and initial conditions have the form:

$$\ddot{\boldsymbol{\alpha}} + \boldsymbol{\omega}_r^2 \boldsymbol{\alpha} = \boldsymbol{\Psi}^T \mathbf{f}(t) \qquad [23]$$

$$\boldsymbol{\alpha}(0) = \boldsymbol{\Psi}^T \mathbf{M}\mathbf{x}(0)$$
$$\dot{\boldsymbol{\alpha}}(0) = \boldsymbol{\Psi}^T \mathbf{M}\dot{\mathbf{x}}(0) \qquad [24]$$

where $\boldsymbol{\alpha}$ is an $M \times 1 (M \ll N)$ vector of modal coordinates, $\boldsymbol{\Psi}$ is a (real) $N \times M$ matrix of mass-normalized columnal eigenvectors, and $\boldsymbol{\omega}_r^2$ is a diagonal $M \times M$ matrix of squared undamped natural frequencies. Damping may be included in eqn [23] by adding a viscous modal damping term:

$$\ddot{\boldsymbol{\alpha}} + 2\zeta_r \omega_r \dot{\boldsymbol{\alpha}} + \omega_r^2 \boldsymbol{\alpha} = \boldsymbol{\Psi}^T \mathbf{f}(t) \qquad [25]$$

where ζ_r is the modal damping ratio of mode r, and $2\zeta_r \omega_r$ is a diagonal modal damping matrix.

Once the modal response is found, the physical response is determined using:

$$\mathbf{x}(t) = \boldsymbol{\Psi}\boldsymbol{\alpha}(t) \qquad [26]$$

This approach can be modified to use complex mode shapes and/or natural frequencies.

Modal Damping Ratios

For many structures, eqn [25] gives results of satisfactory accuracy. Once this approach is selected for use, a main problem becomes determining appropriate modal damping ratios.

Experience One approach is based on personal or organizational experience with similar structures. Although there is no assurance that this approach will yield correct values for any modes, it can provide a valuable starting point for analysis. For example, values of modal damping in the range from 0.003 to 0.03 might be appropriate for lightly damped, built-up aerospace structures.

Complex stiffness Alternate approaches to estimating modal damping may proceed by establishing a lower bound based on material loss factor contributions. Such approaches are especially effective when high-damping materials are used to augment nominal lightly damped structures. An example of this kind of approach is the use of modal analysis of a complex stiffness-based model to yield a complex natural frequency and associated modal damping.

Modal strain energy Another materials-based approach is the modal strain energy method. In this approach, a modal loss factor is a weighted sum of material loss factors:

$$\zeta_r \cong \frac{1}{2}\eta_r = \frac{1}{2} \sum_{\substack{i \\ \text{materials}}} \eta_i \frac{U_{ri}}{U_r} \qquad [27]$$

The weighting terms are the fraction of modal strain energy stored in each material (or component), usually estimated from analysis of an undamped model. The modal damping ratio is approximately half of the modal loss factor.

Other Damping Models

For some applications that require high model fidelity, the damping models described in the preceding may be inadequate. An example of such an application might be a structural dynamic model that is to be used as the basis for the design of a high-performance feedback controller. Some of the main deficiencies of models considered to this point include the following.

Viscous damping yields modal damping that tends to increase with frequency, in a manner inconsistent with observations. Proportional damping, either related to the mass matrix, or for multicomponent structures, should be regarded as a mathematical curiosity.

Hysteretic (complex stiffness) damping, while capable of accommodating frequency-dependent damping and stiffness, and of yielding better estimates of modal damping, cannot be used directly to determine structural response to arbitrary dynamic loading. However, the combination of either the complex stiffness or modal strain energy method, along with a viscous modal damping model, often yields acceptable results. Limitations of this approach usually stem from neglecting phase differences in response (using real modes), from neglecting frequency-dependent properties, from sacrificing mode orthogonality by including frequency-dependent properties, or from unusually high, perhaps localized, damping.

Several damping models suitable for use with finite element analysis have been developed to address such shortcomings. The emphasis here is on models that are compatible with linear analysis, thus neglecting friction and other models.

Fractional Derivative

Fractional derivative models provide a compact means of representing relatively weak frequency-dependent properties in the frequency domain. For example, a single material complex modulus might be represented as:

$$E^*(i\omega) = \frac{E_0 + E_1(i\omega)^m}{1 + b(i\omega)^m} \qquad [28]$$

where E_0, E_1, b, and m are material properties, with $+m$ being the slope of the loss factor on a log–log plot versus frequency, at frequencies below that at which peak damping is observed. Values for m range from nearly 0 to 1, with $\frac{1}{2}$ being typical.

Although developed as a time domain model, the fractional derivative approach, as a means of determining dynamic response, suffers at present from the need for special mathematical and computational tools.

Internal Variable Viscoelastic Models

Another approach to capturing viscoelastic (frequency- and temperature-dependent) material behavior in a time domain model involves the introduction of internal dynamic coordinates. Several such models are available and, although they differ in some respects, they share many common features. A one-dimensional mechanical analogy of material behavior (**Figure 2**) aids understanding of the general structure of such models.

The deformation of this system is described primarily by the stress, σ, and the total strain, ε, but its apparent stiffness is affected by the dynamics of an internal strain, ε_A. If this system is subjected to harmonic forcing, its apparent stiffness and damping

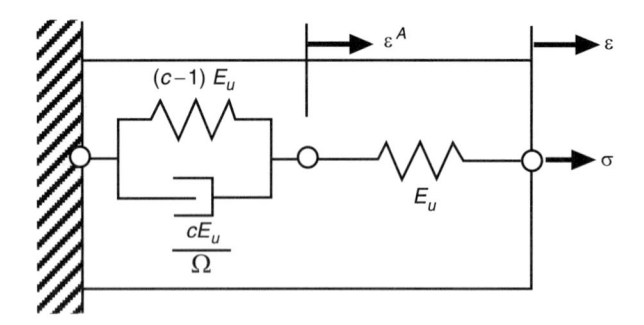

Figure 2 The standard three-parameter anelastic solid.

vary with frequency. At very high frequencies, the internal dashpot is essentially locked, the material modulus is E_u, and the damping is low. At very low frequencies, the internal dashpot slides freely, the modulus is lower, and the damping remains low. At some intermediate frequency, the damping reaches a peak value.

The peak loss factor and change in modulus both depend on the strength of the coupling between the total strain and the internal strain, $1/(c-1)$. The frequency at which peak damping is observed, very nearly Ω, is related to the inverse of the relaxation time for the internal strain.

With a single internal variable, the loss factor is proportional to frequency at low frequencies, and inversely proportional at high frequencies. Weaker frequency dependence can be introduced by using multiple internal fields, each having its own relaxation dynamics. Temperature effects can be included by using a shift function that essentially increases the relaxation rate of the internal fields with increasing temperature.

Such internal variable models are quite compatible with finite element structural analysis methods. In one approach, additional nodal displacement coordinates, identical to those of an elastic element, are introduced to model the internal system. The internal field is then interpolated in the same way as the total displacement field. Additional first-order equations of motion are then developed to describe the relaxation (creep) behavior of the internal system and its coupling to the total displacements. The boundary conditions for the internal coordinates are implemented just as those for the corresponding total displacements are, with the additional elimination of strain-free motion.

Eqn [29] shows the general structure of the finite element equations with a single set of internal coordinates. In this form, **K** is calculated using the high-frequency material stiffness. Evidently, higher accuracy comes at a cost of additional coordinates and material properties.

$$\begin{bmatrix} M & \\ & \end{bmatrix} \begin{Bmatrix} \ddot{x} \\ \end{Bmatrix} + \begin{bmatrix} 0 & 0 \\ 0 & \dfrac{c}{\Omega}K \end{bmatrix} \begin{Bmatrix} \dot{x} \\ \dot{x}_A \end{Bmatrix} \\ + \begin{bmatrix} K & -K \\ -K & cK \end{bmatrix} \begin{Bmatrix} x \\ x_A \end{Bmatrix} = \begin{Bmatrix} f(t) \\ 0 \end{Bmatrix} \quad [29]$$

Such internal variable approaches have advantages over more conventional approaches in that they yield linear time domain finite element models, the frequency-dependent elastic and dissipative aspects of structural behavior are represented in fixed (not frequency-dependent) system matrices, modal damping is calculated concurrently with modal frequency without iteration, the resulting complex modes more accurately reflect the relative phase of vibration at various points, and modal orthogonality is preserved.

Nomenclature

E	material elastic coefficient
F	material strain rate coefficient
ε	strain
η	material loss factor
σ	stress
Ω	forcing frequency

See also: **Comparison of Vibration Properties**, Comparison of Spatial Properties; **Discrete elements**; **Finite element methods**; **Hysteretic damping**; **Modal analysis, experimental**, Basic principles; **Model updating and validating**; **Viscous damping**.

Further Reading

Bagley RL and Torvik PJ (1986) On the fractional calculus model of viscoelastic behavior. *Journal of Rheology* 30: 133–155.

Enelund M and Josefson, BL (1997) Time-domain finite element analysis of viscoelastic structures with fractional derivative constitutive relations. *AIAA Journal* 35: 1630–1637.

Johnson AR (1999) Modeling viscoelastic materials using internal variables. *Shock and Vibration Digest* 31: 91–100.

Johnson CD (1995) Design of passive damping systems. *Journal of Mechanical Design* 117B: 171–176.

Johnson CD and Kienholz DA (1982) Finite element prediction of damping in structures with constrained viscoelastic layers. *AIAA Journal* 20: 1284–1290.

Lesieutre GA and Bianchini E (1995) Time domain modeling of linear viscoelasticity using anelastic displacement fields. *Journal of Vibration and Acoustics* 117: 424–430.

Lesieutre GA and Govindswamy K (1996) Finite element modeling of frequency-dependent and temperature-dependent dynamic behavior of viscoelastic materials in simple shear. *International Journal of Solids and Structures* 33: 419–432.

McTavish DJ and Hughes, PC (1993) Modeling of linear viscoelastic space structure. *Journal of Vibration, Acoustics, Stress, and Reliability in Design*, 115: 103–110.

Mead DJ (1999) *Passive Vibration Control*. Chichester: John Wiley.

Meirovitch L (1980) *Computational Methods in Structural Dynamics*. Alphen aan den Rijn, The Netherlands: Sijthoff & Noordhoff.

Nashif AD, Jones DIG and Henderson JP (1985) *Vibration Damping*. New York, NY: John Wiley.

Pilkey W and Pilkey B (eds) (1995) *Shock and Vibration Computer Programs: Reviews and Summaries*. Arlington, VA: Shock and Vibration Information Analysis Center, Booz Allen & Hamilton.

Soovere J and Drake ML (1984) *Aerospace Structures Technology Damping Design Guide*. AFWAL-TR-84-3089. Dayton, OH: U.S. Air Force.

DAMPING MATERIALS

E E Ungar, Acentech Incorporated, Cambridge, MA, USA

Copyright © 2001 Academic Press

doi:10.1006/rwvb.2001.0014

What is a Damping Material?

A damping material is a solid material that dissipates (that is, converts into heat) a significant amount of mechanical energy as it is subjected to cyclic strain. Most damping materials are not useful structural materials themselves, but typically are combined with structural elements so that the resulting combination is structurally viable and relatively highly damped. A damping material configuration that is applied to a structural component usually is called a 'damping treatment'.

Although granular materials and viscous liquids can provide considerable energy dissipation in some applications, these usually are not regarded as damping materials. Even though so-called high-damping metal alloys are more highly damped than common metals, their damping generally is not high enough for these to be considered damping materials in the present sense. Almost all practical damping materials are polymeric plastics or elastomers – but some other materials can also dissipate considerable energy in some temperature and frequency ranges.

Characterization of Material Properties

Damping materials are often called 'viscoelastic', because they combine energy dissipation (viscous) with energy storage (elastic) behavior. Characterization of the properties of such a material requires two parameters; one associated with energy storage, and one, with energy dissipation.

If a sinusoidal compressive stress represented by the phasor[†] $\sigma(\omega)$ acts on an element of a damping material, so as to produce a strain represented by the phasor $\varepsilon(\omega)$, one may write the complex dynamic modulus $E(\omega)$ as:

$$E(\omega) = \sigma(\omega)/\varepsilon(\omega) \equiv E' + jE'' \equiv E'[1 + j\eta] \quad [1]$$

The real part E' of the complex modulus is known as the 'storage modulus' because it is associated with energy storage, and the imaginary part $E'' = \eta E'$ is known as the 'loss modulus' because it is associated with energy dissipation. The 'loss factor' obeys:

$$\eta = E''/E' \quad [2]$$

The values of E', E, and η generally vary with frequency and with other parameters, as discussed later.

In steady sinusoidal vibration at a given frequency:

$$\eta = D/2\pi W \quad [3]$$

where D denotes the energy dissipated in unit volume of the element per cycle and where $W = E'[\varepsilon(\omega)]^2/2$ represents the energy stored in unit volume of the element at instants when the strain is at its maximum.

The complex shear modulus $G(\omega)$, its real and imaginary parts, and the corresponding loss factor may be defined entirely analogously to the complex compression modulus $E(\omega)$ and its components. According to elasticity theory, the modulus of elasticity in compression E is related to the shear modulus G via Poisson's ratio v by:

[†] The phasor $Y(\omega)$ of a sinusoidally time-dependent variable $y(t) = Y_0\cos(\omega t - \phi)$ is a complex quantity $Y(\omega) = Y' + jY''$ defined so that $y(t) = \text{Re}\{Y(\omega)e^{-j\omega t}\}$. The phasor accounts both for the amplitude $Y_0 = |Y(\omega)|$ and for the phase angle $\phi = \arctan(Y''/Y')$.

$$G = \frac{E}{2(1+v)} \quad [4]$$

and therefore:

$$v = \frac{E}{2G} - 1 \quad [5]$$

Application of this relation to the complex compression and shear moduli implies that Poisson's ratio is complex, unless the loss factors associated with the two moduli are equal. For most materials the two loss factors indeed are equal for all practical purposes; thus, their Poisson's ratio can be taken as real.

Measurement of Material Properties

The most direct approaches to determination of the complex compression or shear modulus of a material involve measurements on samples whose shapes or sizes are such that application of a force in an appropriate direction results in an essentially uniform or in another reliably predictable strain distribution in the sample. For example, one may evaluate the complex shear modulus of a material at a given frequency by applying a known sinusoidal shear force at the given frequency to one face of a thin flat sample, whose other face is restrained from moving, and observing the magnitude of the resulting shear displacement, together with its phase angle relative to the applied force. Such measurement approaches are simple in concept, but generally involve considerable practical difficulties and relatively complex apparatus. Nevertheless, such apparatus is necessary for the measurement of the amplitude-dependences of the complex moduli.

The moduli of most viscoelastic materials, however, are practically independent of the strain amplitude up to stains of perhaps 5% or more. Thus, one can determine the damping of such materials by approaches in which the strain amplitude is permitted to vary during a measurement. These approaches include measurement of the rate of decay of free vibrations or of the resonance bandwidth of a test system that includes a sample of the damping material.

The simplest and most widely-used measurement approaches employ metal reeds (plate strips, or beams with thin rectangular cross-sections) which have thin layers of damping material bonded to one or both of their faces. A test reed is clamped to a rigid support at one end, and the other end is excited magnetically at one of its resonances at a time. The loss factor of the coated reed is evaluated either from the observed rate of decay of the reed's vibration after the excitation has been turned off, or from measurement of the bandwidth of the resonance. The flexural stiffness of the coated reed is determined from its mass-per-unit length and the resonance frequency, using the classical relations applicable to an elastic cantilever beam. The stiffness and loss factor of the bare metal reed are determined similarly. The properties of the damping material then are calculated by means of well-established equations that indicate how the flexural stiffness and loss factor or the reed with attached damping material layer(s) depend on the dimensions of the reed and of the attached layer(s) and on the moduli and loss factors of the bare reed and the damping material. These measurement approaches are the subject of recent standards.

Typical Behavior of Damping Materials

The dynamic properties (moduli and loss factors) of a damping material generally vary markedly with frequency and temperature. They usually vary only a little with strain amplitude up to quite large strains and generally depend only to a minor extent on static preload and on exposure to long-duration cyclic loading.

Figures 1A and **1B** shows how the real (storage) shear modulus and the loss factor in shear of a representative damping material vary with frequency and temperature. **Figure 1A** shows graphs of these two quantities as functions of frequency at several constant temperatures. **Figure 1B** shows the same data plotted upon a temperature–frequency plane, in order to permit one to visualize the trends more easily.

At a given frequency, the modulus varies drastically with temperature, starting from large values at low temperatures and progressing to small values at high temperatures via a range of intermediate temperatures in which the rate of change is greatest. At low temperatures the material is said to be in its 'glassy' state and at high temperatures, in its 'rubber-like' state. At high temperatures – beyond those covered by the plots – the material becomes very soft and tends toward the behavior of a viscous liquid. The region in which rapid changes occur is called the 'glass transition region'; the temperature at which the modulus changes most rapidly is termed the 'glass transition temperature'. The highest loss factor values occur at or near this temperature.[†]

[†] Although **Figure 1** pertains to shear properties, the figure also indicates the behavior of the material's properties in extension–compression. A material's loss factor in compression–extension is the same as that in shear for all practical purposes, and the modulus values for compression–extension are higher than those for shear by a nearly constant factor of about 3. Poisson's ratio for plastics and elastomers is about 0.33 in the glassy region and nearly 0.5 in the transition and rubber-like regions, so that E/G is between 2.7 and 3, according to eqn [4]. The latter value applies essentially throughout the regions of primary interest in damping applications.

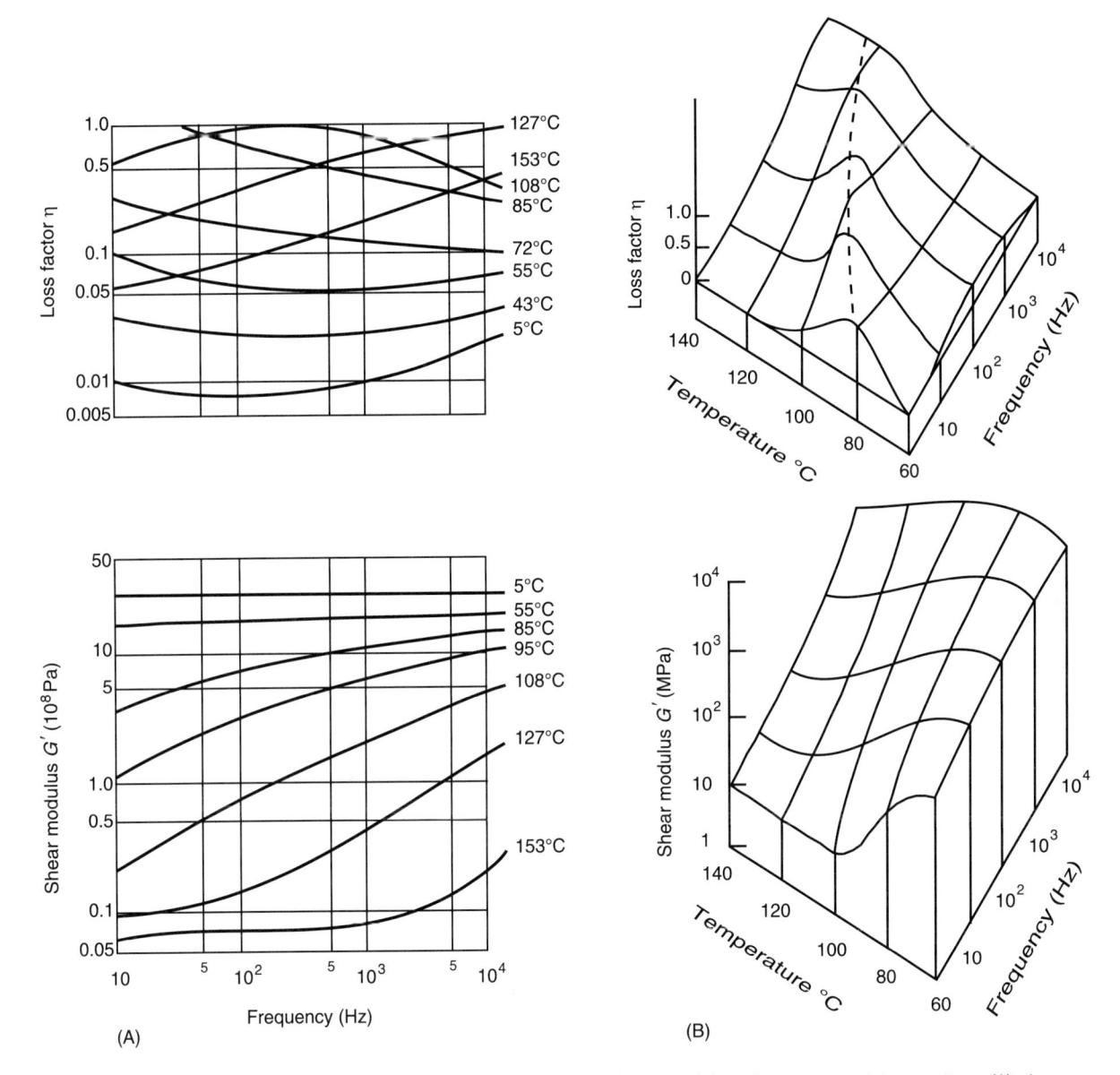

Figure 1 Dependences of shear modulus and loss factor of a damping material on frequency and temperature: (A) shown as functions of frequency at constant temperatures; (B) shown as plots upon temperature–frequency plane. Adapted from Ungar EE (1992) Structural damping. In: Beranek LL, Ver IL (eds) *Noise and Vibration Control Engineering*, ch. 12. New York: John Wiley.

The variation with frequency of the material properties at constant temperature is similar to their variation with temperature at constant frequency. At a constant temperature, the modulus progresses from small values at low frequencies to high values at high frequencies via an intermediate region in which the change is relatively rapid. Again, the highest loss factor values occur in the area of the most rapid modulus changes. This similarity has led to the concept of 'temperature–frequency equivalence'. According to this concept, if one starts with the material at a given temperature and frequency, one observes the same change in the material properties due to a temperature increase (or increase) by a given amount as from a suitably selected decrease (or increase) in the frequency.

The general behavior of many materials is similar to that illustrated by **Figures 1A** and **1B**, but the actual values of the properties depend on the specific materials. Storage moduli as great as 10^8 kPa (or 10^7 psi) may occur in the glassy region, and moduli as small as 10 kPa (1 psi) may occur in the rubbery region. For a given material, the modulus values in the glassy region may be three or four orders of magnitude greater than those in the rubbery region.

The loss factor values in the glassy region usually are small, typically between 10^{-3} and 10^{-2}, whereas in the rubbery region they tend to be of the order of 0.1

for many materials. In the transition region, the loss factors of good damping materials generally approach 1 and for some materials may reach 2. The transition region may extend over only 20 °C for some materials or over more than 200 °C for others. Commercial materials that are intended for use in given temperature ranges typically have their glass transition temperatures in the middle of these ranges. At constant temperature, the transition region may extend over one to four decades of frequency. The loss factor curves for materials with wider transition ranges typically exhibit flatter peaks and lower maximum values than similar curves for materials with narrower transition ranges.

Figures 1A and **1B** correspond to a material consisting of a single viscoelastic component – that is, of a single polymeric material (with or without non-polymeric admixtures). Such a material has a single transition region and, correspondingly, its loss factor curve exhibits a single peak, as illustrated by the figure. The behavior of materials consisting of two or more viscoelastic components with different transition regions is more complex. The loss factor curve for such a material may have two or more peaks, and the slope of the modulus curve for such a material may not change monotonically with temperature and frequency.

Presentation of Material Data

Material property data may be presented in the form of a series of curves giving values of the modulus and loss factor at various temperatures as functions of frequency, as in **Figure 1A**. Data may also be presented in terms of curves representing the values of these parameters at various frequencies as functions of temperature. Instead of curves of the real modulus, or in addition to these curves, one may also show curves of the imaginary (loss) modulus or of the magnitude of the complex modulus. Because the various moduli are simply related, so that one can readily be calculated from the other, the following discussion focuses on the same parameter set as that used in **Figures 1A** and **1B**.

By shifting the various curves that show the modulus variations with frequency at different constant temperatures along the frequency axis, these curves can be arranged to form a single continuous smooth curve. The same is true also for the loss factor curves, if they are shifted by the same amounts as the modulus curves. (Analogous statements also apply to shifting curves obtained at different constant frequencies along the temperature axis.) This shifting, the possibility of which is a consequence of the aforementioned temperature–frequency equivalence, permits one to show the modulus and loss factor data for a single-transition material for all frequencies and temperatures as functions of a single 'reduced frequency' parameter, as illustrated in **Figure 2**.

The reduced frequency is defined as the product of the actual frequency and of a 'shift factor' α_T that depends on the temperature as determined from the shifting required to make all the curves coalesce into continuous ones. The dependence of the shift factor on temperature may be given by an equation, by a separate plot, or – more conveniently – by a nomogram superposed on the data plots as in **Figure 2**. Presentation of data as in **Figure 2** is the subject of an international standard.

Analytical models have been developed that use empirical data obtained in limited frequency and temperature ranges to characterize a material's behavior outside of these ranges. These models generally have been confined to single-transition materials and have been of limited practical utility. Such models, as well as plots like **Figure 2**, can lead to significant errors if they are used for extrapolations outside of the regions for which measured data are available.

Practical Considerations

The dynamic properties of a sample of a polymeric material depend not only on its basic chemical composition (that is, on the monomers that make up the material), but also on several other factors. These include the material's molecular weight spectrum (the distribution of molecular chain lengths), the degree of cross-linking, and the amounts and types of included plasticizers and fillers. Experience has shown that even samples taken from the same production batch of a given material may exhibit considerably different modulus and loss factor values.

The differences may be even greater for samples taken from different, nominally identical, batches of a material. It thus is important that damping materials for critical applications be obtained from experienced suppliers who can guarantee the performance of their materials, and that these materials be specified and accepted not on the basis of their chemical compositions, but on the basis of their dynamic properties in the frequency and temperature ranges of concern.

Nomenclature

D	energy dissipated per cycle in unit volume
E'	real part of $E(\omega)$; storage modulus
E''	imaginary part of $E(\omega)$; loss modulus

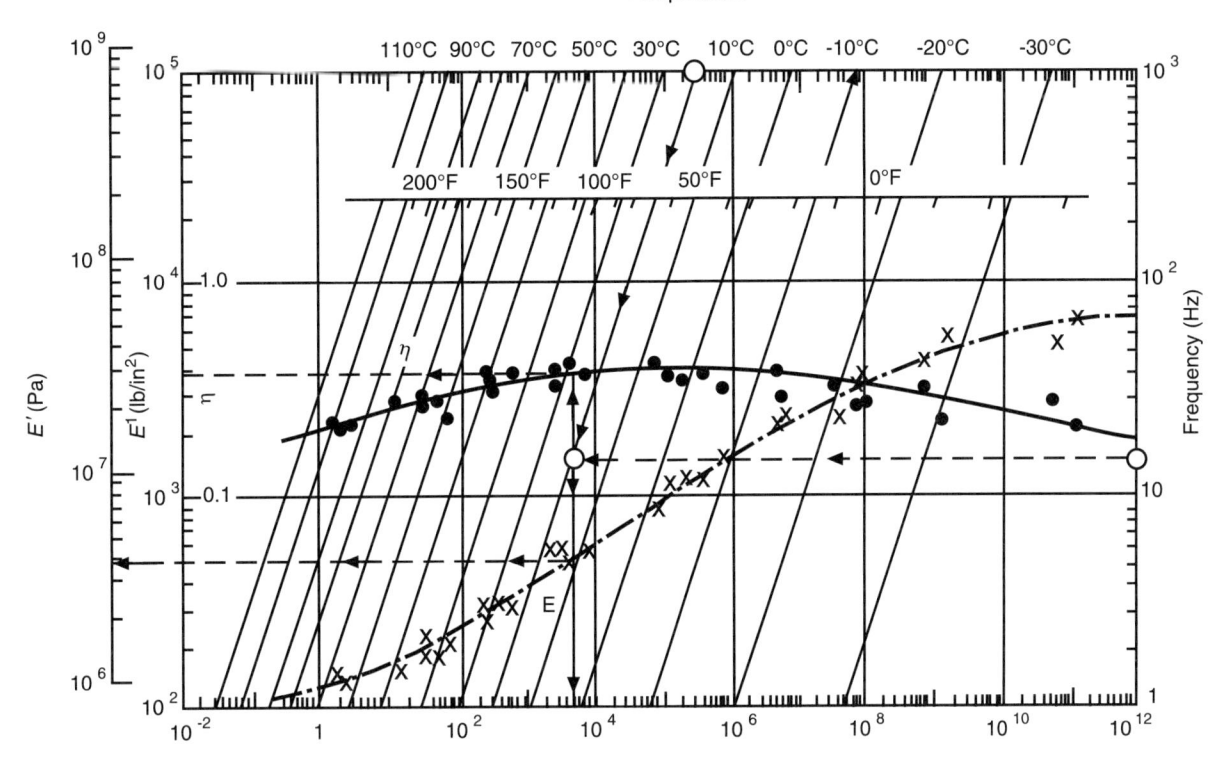

Figure 2 Reduced frequency plot of elastic modulus E' and loss factor η of a viscoelastic material. Points indicate measured data to which curves like those of **Figure 1** were fitted. Nomograph superposed on data plot facilitates determination of reduced frequency f_R corresponding to frequency f and temperature T. Use of nomograph is illustrated by dashed lines: for $f = 15$ Hz and $T = 20$ deg C one finds $f_R = 5 \times 10^3$ Hz and $E' = 3.8 \times 10^6$ Pa, $\eta = 0.36$. Adapted from Ungar EE. (1992) Structural damping. In: Beranek LL, Ver IL (eds) *Noise and Vibration Control Engineering*, ch. 12. New York: John Wiley.

$E(\omega)$	complex modulus in extension–compression (complex Young's modulus)
G	shear modulus
W	time-wise maximum energy stored per unit volume
$\varepsilon(\omega)$	strain phasor
ν	Poisson's ratio
$\sigma(\omega)$	stress phasor
ϕ	phase angle

See also: **Damping in FE models**; **Damping measurement**; **Hysteretic damping**.

Further Reading

Anon (1971) DIN 53 440, *Biegeschwingungsversuch. Bestimmung von Kenngrössen Schwingungsgedämpfter Mehrschichtsysteme (Flexural Vibration Test. Determination of Parameters of Vibration-damped Multi-layer Systems)*. Berlin: Deutsches Institut für Normung.

Anon (1991) ISO Standard 10112–1991 *Damping Materials: Graphic Presentation of Complex Modulus*. Geneva: ISO.

Anon (1998) ANSI Standard S2.23-1998, *Single Cantilever Beam Method for Measuring the Dynamic Mechanical Properties of Viscoelastic Materials*. New York: ANSI.

Anon (1998) ASTM Standard E756-98, *Test Method for Measuring Vibration-damping Properties of Materials*. West Conshohocken, Pennsylvania: Am. Soc. for Testing Materials.

Ferry JD (1970) *Viscoelastic Properties of Polymers*, 2nd edn. New York: John Wiley.

Nashif AD, Jones DIG, Henderson JP (1985) *Vibration Damping*. New York: John Wiley.

Ungar EE (1992) Structural damping. In Beranek LL, Ver LL (eds) *Noise and Vibration Control Engineering*, ch. 12. New York: John Wiley.

Ungar EE, Kerwin EM Jr. (1962) Loss factors of viscoelastic systems in terms of energy concepts. *Journal of the Acoustical Society of America* 34(7): 954–7.

DAMPING MEASUREMENT

D J Ewins, Imperial College of Science, Technology and Medicine, London, UK

Copyright © 2001 Academic Press

doi:10.1006/rwvb.2001.0020

Introduction

Damping is one of the properties of vibrating structures that is the most difficult to model and to predict. As a result, it is more frequently required to conduct tests to measure damping than the other parameters in vibrating systems. This short article summarizes the different approaches to measuring damping levels in practical structures, and some of the difficulties that may be encountered in doing so.

As explained elsewhere (see **Damping models**), damping is the name given to a variety of different physical mechanisms which all share the common feature that they convert mechanical energy into heat, thereby reducing levels of vibration at the expense of the temperature of the vibrating structure. There are basically three approaches to the experimental determination of damping levels:

- measuring the energy absorption characteristic directly
- measuring the rate at which damping reduces the vibration of a structure
- measuring the extent to which damping limits resonance peak amplitudes.

Before explaining the methods involved in measuring damping, it is appropriate to restate the different damping parameters that are in common use, and these are: ζ = critical (viscous) damping ratio; η = (structural) damping loss factor $\sim 2\zeta$ (at resonance); δ = logarithmic decrement $\sim 2\pi\zeta$ (at resonance), and Q = dynamic magnification factor $\sim (1/2\zeta)$; $(1/\eta)$ (see **Damping models**).

Basis of Vibration Measurement Procedures

The principles upon which the three main damping measurement procedures are based can be simply illustrated using a simple theoretical model which uses the classical viscous dashpot as the mathematical model of a damping element. **Figure 1A** shows such a damper element in parallel with a simple undamped spring. If this spring–dashpot element is subjected to a harmonically varying force, $f(t) = F_0 e^{i\omega t}$, then the trace of an $f-x$ plot is as shown in **Figure 1B**, the curve tracing out an ellipse centered on a straight line of slope k. The area contained by the loop represents the energy dissipated in one cycle of vibration ($E = \pi c X_0^2 \omega$) and thus the level of damping present in the system. This area can be expressed in nondimensional form as the ratio of Area 1 (energy dissipated) to Area 2 (energy stored) revealing a dimensionless damping ratio. For a simple SDOF system at resonance ($\omega = \omega_0$), this energy ratio can be shown to be equal to $4\pi\zeta$, thus providing a direct means of measuring the damping in such a simple system.

Figure 2A shows a complete SDOF mass–spring–dashpot system and **Figure 2B** shows a typical plot of the response of that system to a transient input. This plot displays the characteristic free decay of all damped systems and the form of the resulting time-history is heavily influenced by the damping in the

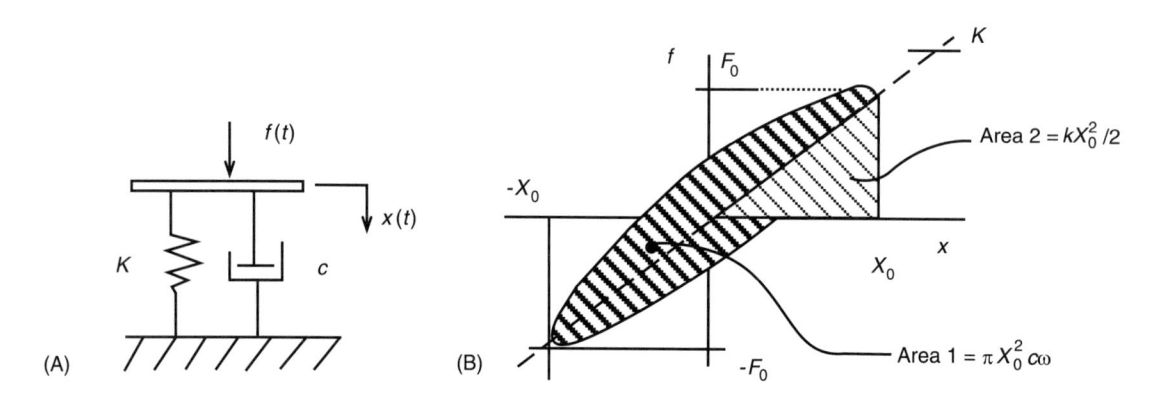

Figure 1 (A) Basic spring–dashpot model for damper, (B) force–displacement characteristic for spring–dashpoint system.

system, c. The frequency of successive oscillations is always the same, at ω'_0, and the rate of decay (envelope) of successive peaks is governed by the exponential expression: $e^{-\omega'_0 t}$. The ratio of successive peak amplitudes is given by: $\Delta = e^{2\pi\zeta}$, or in its more usual form, $\delta = \log_e \Delta \sim 2\pi\zeta$, thereby providing another simple method (the logarithmic decrement or 'log dec' method) of extracting damping estimates from the free vibration of the structure. The exact formula for estimating damping from these free decay curves is:

$$\zeta = \frac{\delta}{\sqrt{(4\pi^2 + \delta^2)}}$$

The third approach to damping measurement can be made using the forced harmonic vibration response characteristics and is based on the form of the classical resonance curve, shown in **Figures 2C** and **2D**. In the first of these, **Figure 2C**, there is a simple construction that is used to estimate damping whose features are shown in the diagram:

- the maximum response level, $\hat{H}$
- two points (frequencies, ω_1 and ω_2) at which the response is $0.707 \hat{H}$ (the so-called 'half-power points') and these parameters can be used to estimate the damping form:

$$\zeta = (\omega_2 - \omega_1)/2\omega_0$$

The alternative Nyquist plot presentation of the response characteristic, shown in **Figure 2D** provides a different perspective of this same region of the resonance response, but permits a more general formula to be used to estimate the damping:

$$\zeta = \frac{(\omega_a - \omega_b)}{2\omega_0(\omega_a \tan(\theta_a/2) + \omega_b \tan(\theta_b/2))}$$

It should be noted that the formulas presented here are, strictly, approximate but that they are well within any measurement uncertainty for systems with light damping ('light' generally means a critical damping ratio, or loss factor, of less than about 5–10%). In fact, there is an exact version of each of these simple formulas, and it is that which is used in the software that is generally employed to perform the calculations that reveal the damping estimates. Nevertheless, it serves the purpose here of explanation by showing the simple versions.

Practical Considerations

The above formulas provide the basis for estimating damping from any of three different types of test.

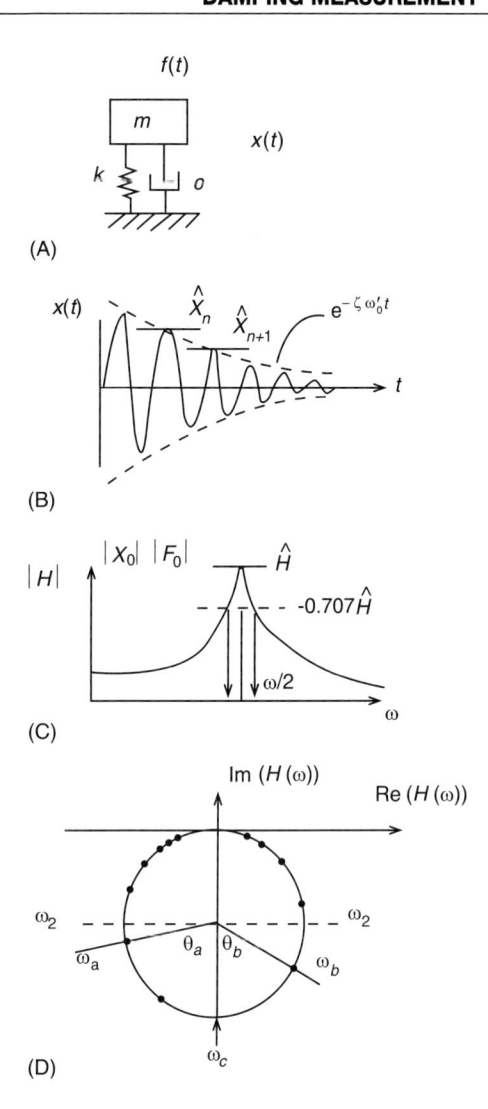

Figure 2 (A) SDOF mass–spring-dashpot system; (B) free vibration response of damped SDOF system; (C) forced response characteristic for SDOF system (modulus plot); (D) forced response characteristic for SDOF system (Nyquist plot).

Indeed, all three are widely used in practice (see **Modal analysis, experimental: Basic principals**).

However, there are a number of complications that often arise when applied to practical structures, and it is worthwhile to list some of these here. The first difficulty that is often encountered is due to the fact that most practical structures are more complicated than the single DOF system used here to illustrate the principles of measurement and so the actual time records, or response curves, are generally more complex than those shown in **Figures 2B–D**. Examples of a typical multi-DOF system responses are sketched in **Figure 3**. In each of these examples, it can be shown that the actual response is the linear summation of several SDOF component responses, each of which has the form of those shown previously in **Figure 2**. Thus, we might expect that if we can extract the

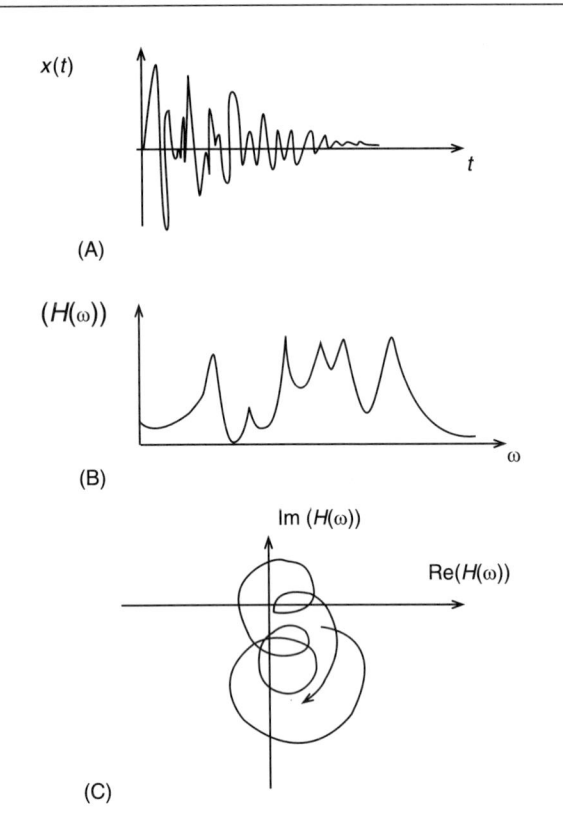

Figure 3 MDOF system. (A) Free vibration response of system (B) modulus response plot for system; (C) Nyquist response plot for system.

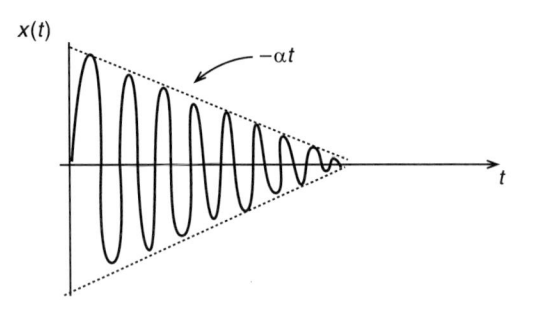

Figure 4 Free vibration response of nonlinear SDOF system.

individual components of response, such as those shown in **Figure 3**, into their individual modal components, then each of these might take the form of the plots shown in **Figure 2B–D**, and thus be amenable to analysis by one of the three methods above to determine the damping. Sometimes, this complication is such that only a single resonance peak is apparent in a region where, in fact, there are many modes and this almost always leads to a significant overestimation of the damping levels.

The second difficulty that occurs in practice results from the fact that many sources of damping in real structures have much more complicated behavior than indicated by the theoretical expressions above. At best, they do not conform exactly to the viscous-type model of the basic damping mechanism used in the formulas, and at worst they are often not linear in their underlying behavior. As a result, it is not possible to extract a single value for the damping: in nonlinear systems, this quantity is often amplitude-dependent. Thus we sometimes obtain free-decay curves such as that shown in **Figure 4**, which displays a linear, rather than exponential, decay characteristic. This is a common feature in systems with dry friction as the source of damping and it is clear that a different damping ratio will be found if we base our estimate on the first few cycles of response than if we use the later cycles. In this case, the damping ratio, or more strictly, the 'equivalent damping factor' increases as the vibration amplitude drops. This is a well known phenomenon of friction-damped systems and it can be seen clearly in these methods of measuring damping.

We often use this concept of equivalent damping when measuring damping in practical structures. Essentially, the equivalent damping factor, or ratio, is that level of viscous (or, sometimes, structural) damping that dissipates the same energy as the actual mechanism which is being measured. This can be seen graphically in **Figure 5**, which shows an actual-system force–deflection characteristic (**Figure 5A**) alongside

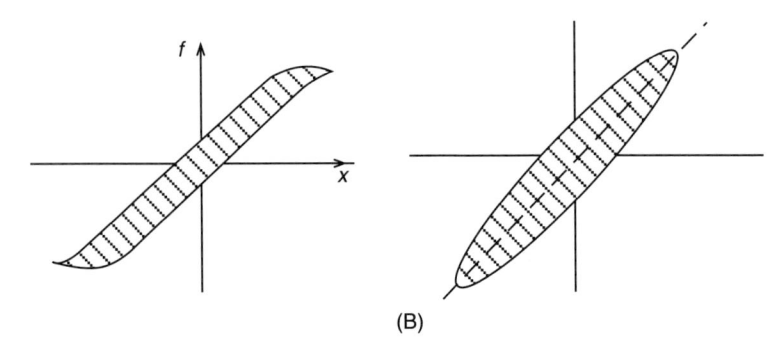

Figure 5 Force–deflection characteristic: (A) hysteresis behavior; (B) viscous dashpot model.

the expected behavior based on the SDOF mass–spring–dashpot model (**Figure 5B**). By choosing a dashpot rate, c, such that the area in the loop of **Figure 5B** is equal to that of the hysteresis characteristic displayed in **Figure 5A**, we can define an equivalent damping model for the measured system.

Accuracy and Reliability of Damping Estimates

In the same way that damping is the most difficult of the standard vibration properties to model, so it tends to be the most difficult to measure accurately. This difficulty is in large part because of the complexity of the real behavior and the relative simplicity of the models used to represent it. However, there are also several practical (measurement) factors which serve to make damping measurements inaccurate and it is appropriate to record here that these errors tend to be all in the same sense. Measurement difficulties associated with noisy signals, and with signal conditioning and processing, as well as the complications discussed above, generally tend to result in damping estimates which are too high, and rarely in ones which are too low by comparison with the correct values. This result is not a conservative one, from a practical point of view, because it is usually required to determine how little damping there is in a practical structure and estimates which are too high indicate a better state of affairs (i.e., that there is more damping present) than is really the case. It is advisable to bear this comment in mind, especially when repeat measurements give significant differences in estimates from one test to the next, where a uniform result might be expected.

Nomenclature

c	dashpot rate
E	energy
$\hat{H}$	maximum response level
Q	dynamic magnification factor
δ	logarithmic decrement

See also: **Damping models**; **Damping, active**; **Modal analysis, experimental**, Basic principles; **Modal analysis, experimental**, Measurement techniques; **Modal analysis, experimental**, Construction of models from tests.

Further Reading

Ewins DJ (2000) *Modal Testing: Theory, Practice and Application*. Taunton, UK: Research Studies Press.

Inman DJ (1994) *Engineering Vibration*. Englewood Cliffs, NJ: Prentice Hall.

DAMPING MODELS

D Inman, Virginia Polytechnic Institute and State University, Blacksburg, VA, USA

Copyright © 2001 Academic Press

doi:10.1006/rwvb.2001.0060

Energy dissipation in vibrating systems is an extremely significant physical phenomenon. Yet damping models remain an illusive research topic. Part of the reason for the lack of definitive damping models is that measurements of damping properties must be dynamic. On the other hand, measurements of stiffness and inertia can be made from static experiments. The need for dynamic experiments in order to verify analytical damping models has resulted in a great deal of difficulty in determining the nature of energy dissipation in a variety of structures and machines. The other fact that makes damping a difficult phenomenon to model is that it is inherently nonlinear, frequency dependent and temperature dependent. There are numerous sources of damping in structures and machines. Sliding friction and the energy dissipation of material moving in air provide the dominant sources of external damping. Sliding friction also exists as an internal cause of damping in jointed structures. Other common sources of internal damping are grouped together and called material damping. The physical sources of various material-damping mechanisms are not presented here; rather some general models are presented based on phenomenological data. Here some basic models that characterize the most common methods of modeling are reviewed.

Viscous Damping

The response of a simple, single-degree-of-freedom, spring-mass model predicts that the system will oscillate indefinitely. However, everyday observation indicates that most freely oscillating systems eventually die out and reduce to zero motion. This experience suggests that the typical undamped single-degree-of-freedom model needs to be modified to account for this decaying motion. The choice of a representative model for the observed decay in an oscillating system is based partially on physical observation and partially on mathematical convenience. The theory of differential equations suggests that adding a term to the undamped equation:

$$m\ddot{x}(t) + kx(t) = 0 \quad [1]$$

of the form $c\dot{x}$, where c is a constant, will result in an equation that can be solved and a response $x(t)$ that dies out. Physical observation agrees fairly well with this model and it is used successfully to model the damping, or decay, in a variety of mechanical systems. This type of damping, called viscous damping, is the most basic attempt to model energy dissipation in vibrating systems. Viscous damping forms the most common model used in lumped mass, single- as well as multiple-degree-of-freedom systems.

While the spring forms a physical model for storing potential energy and hence causing vibration, the dashpot, or damper, forms the physical model for dissipating energy and hence damping the response of a mechanical system. A dashpot can be thought of as a piston fitted into a cylinder filled with oil. This piston is perforated with holes so that motion of the piston in the oil is possible. The laminar flow of the oil through the perforations as the piston moves causes a damping force on this piston. The force is proportional to the velocity of the piston, in a direction opposite that of the piston motion. This damping force, denoted by f_c, has the form:

$$f_c = c\dot{x}(t) \quad [2]$$

where c is a constant of proportionality related to the oil viscosity. The constant c, called the damping coefficient, has units of force per velocity: Ns/m, or kg/s. The force f_c is referred to as linear viscous damping.

In the case of the oil-filled dashpot, the constant c can be determined by fluid principles. However, in most cases, equivalent effects occurring in the material forming the device provide this damping force. A good example is a block of rubber (which also provides stiffness) such as an automobile motor mount, or the effects of air flowing around an oscillating mass. An automobile shock absorber provides another example of a viscous damper (in its linear range of operation). In all cases in which the damping force f_c is proportional to velocity, the schematic of a dashpot is used to indicate the presence of this force in free-body diagrams. Unfortunately, the damping coefficient of a system cannot be measured statically as can the mass or stiffness of a system.

Combining eqns [1] and [2], the standard equation of motion of a spring-mass-damper system becomes:

$$m\ddot{x}(t) + c\dot{x}(t) + kx(t) = 0 \quad [3]$$

The solutions of this linear differential equation decay with time exponentially agreeing with the dissipation of energy expected from physical observation. Depending on the relative value of the damping coefficient, the solution will oscillate and decay (underdamped) or just exponentially decay (overdamped and critically damped).

It is useful to divide eqn [3] by the mass and write the result as:

$$\ddot{x}(t) + 2\zeta\omega_n\dot{x}(t) + \omega_n^2 x(t) = 0 \quad [4]$$

Here the natural frequency and damping ratio are defined as:

$$\omega_n = \sqrt{\left(\frac{k}{m}\right)} \quad \text{and} \quad \zeta = \frac{c}{2\sqrt{(km)}} \quad [5]$$

respectively. The solution to eqn [3] given initial conditions on the position, x_0, and velocity, v_0, for the underdamped case ($0 < \zeta < 1$) is then simply:

$$x(t) = \sqrt{\left(\frac{(v_0 + \zeta\omega_n x_0)^2 + (x_0\omega_d)^2}{\omega_d^2}\right)} e^{-\zeta\omega_n t}$$
$$\times \sin\left(\omega_d t + \left[\frac{x_0\omega_d}{v_0 + \zeta\omega_n x_0}\right]\right) \quad [6]$$

Here the damped natural frequency is defined as:

$$\omega_d = \omega_n\sqrt{(1 - \zeta^2)} \quad [7]$$

It is important to realize that viscous damping as defined by the constant c and the nondimensional version ζ, does not come from a physical modeling exercise or a fundamental mechanics principle, such as Hooke's law, but is used because it results in the solution, given in eqn [6], that matches the physical observation in many situations. Hence, the viscous damping model is useful because first, it adds energy dissipation to the vibration problem of eqn [1], and secondly, it provides an analytical solution to the vibration problem with energy dissipation. This form assumes that the damping is a linear, time invariant phenomenon chosen to be viscous, or proportional to velocity, motivated by the ability to solve the resulting equation of motion.

An attempt to rationalize the use of the viscous damping term can be found in elasticity. Here the stress–strain relationship (Hooke's law) is modified to become:

$$\sigma(x,t) = E\varepsilon(x,t) + c\dot{\varepsilon}(x,t) \qquad [8]$$

Here σ is the stress as a function of position (x) and time (t), ε is the strain as a function of position and time, the over dot indicates differentiation with respect to time (hence, strain rate), c is a damping coefficient, and E is the elastic modulus. This relationship can be used to generate equations of motion in the form of eqn [4] (on a modal basis) from first principles. Eqn [8] is called the Kelvin–Voigt model of damping and is represented schematically by a spring and damper in parallel, forming the most basic procedure for including linear viscous damping in lumped-mass systems.

Coulomb Damping

Sliding friction, also termed Coulomb damping, is the damping force commonly used to model friction arising from relative motion of two surfaces. Such damping arises in certain kinds of joints in structures and is most commonly used to model the damping in sliding mechanisms. Coulomb damping is characterized by the force:

$$f_c = F(\dot{x}) = \begin{cases} -\mu N & \dot{x} > 0 \\ 0 & \dot{x} = 0 \\ \mu N & \dot{x} < 0 \end{cases} \qquad [9]$$

where f_c is the dissipation force, N is the normal force (see any introductory physics text), and μ is the coefficient of sliding friction (or kinetic friction). Because of the switching nature of the Coulomb force it is nonlinear and hence the resulting equation of motion is nonlinear. The Coulomb damping force can be written as the signum function, denoted sgn (τ), and defined to have the value 1 for $\tau > 0$, -1 for $\tau < 0$, and 0 for $\tau = 0$. A spring-mass system with Coulomb damping has the equation of motion:

$$m\ddot{x} + \mu m g\, \mathrm{sgn}\,(\dot{x}) + kx = 0 \qquad [10]$$

Since this differential equation is nonlinear, it cannot be solved directly using analytical methods such as the variation of parameters or the method of undetermined coefficients. Rather, eqn [10] can be solved by breaking the time intervals up into segments corresponding to the changes in direction of motion (i.e., at those time intervals separated by $(\dot{x} = 0)$. Alternatively, eqn [10] can be solved numerically.

Nonlinear vibration problems are much more complex than linear systems. Their numerical solutions, however, are often fairly straightforward. Several new phenomena result when nonlinear terms are considered. Most notably, the idea of a single equilibrium point of a linear system is lost. In the case of Coulomb damping, a continuous region of equilibrium positions exists, corresponding to the maximum static friction force. This single fact greatly complicates the analysis, measurement, and design of vibrating systems with nonlinear terms.

Several things can be noted about the free response with Coulomb friction vs the free response with viscous damping. First, with Coulomb damping the amplitude decays linearly with slope given by (g is the acceleration due to gravity):

$$-\frac{2\mu m g \omega_n}{\pi k} \qquad [11]$$

rather than exponentially as does a viscously damped system. Second, the motion under Coulomb damping comes to a complete stop, at a potentially different equilibrium position than when initially at rest. However, the motion under viscous damping oscillates around the single equilibrium of $x = 0$ and approaches zero exponentially. Third, the frequency of oscillation of a system with Coulomb friction is ω_n, the undamped natural frequency, while the frequency of oscillation of the viscously damped system is the damped natural frequency as given in eqn [7].

Coulomb friction, as defined above by eqn [9], is a fairly accurate description of the behavior of many machine and structural components. However, as with all theories there are a number of more advanced descriptions of sliding friction as discussed elsewhere in this encyclopedia (see **Friction damping**).

Material Damping

One approach to modeling damping from linear viscoelasticity is to expand on eqn [8] and hypothesize that the one-dimensional stress–strain relationship is:

$$a_0 \sigma + a_1 \dot{\sigma} + a_2 \ddot{\sigma} + a_3 \dddot{\sigma} + \cdots$$
$$= b_0 \varepsilon + b_1 \dot{\varepsilon} + b_2 \ddot{\varepsilon} + b_3 \dddot{\varepsilon} + \cdots \qquad [12]$$

Here each of the coefficients a_i and b_i are constants and the stress σ, and the strain ε, both depend on position and time. The over dots represent time derivatives. Various damping models follow from eqn [12] by setting various combinations of the coefficients to zero. For instance, if all of the coefficients except a_0 and b_0 are zero, then Hooke's law results. Keeping the first two terms in this expansion leads to:

$$\sigma(x,t) + a_1 \dot{\sigma}(x,t) = E_r[\varepsilon(x,t) + \tau_\sigma \dot{\varepsilon}(x,t)] \quad [13]$$

which is called the Kelvin model or standard linear model. Here E_r is called the relaxed modulus of elasticity and the over dots are again time derivatives. If a_1 in eqn [13] is zero then eqn [13] reduces to eqn [8] and equations of the form of eqn [3] result. Another popular model resulting from eqn [12] is the Maxwell model:

$$a_0 \sigma(x,t) + a_1 \dot{\sigma}(x,t) = b_1 \dot{\varepsilon}(x,t) \quad [14]$$

The Maxwell model can be visualized as a spring and damper in series. All of the different models derived from eqn [12] can be thought of as various series and parallel combinations of springs and dampers connected as lumped elements.

Some of the constitutive relationships derived from eqn [12], when used in energy methods or in Newton's law, yield equations of motion with linear viscous behavior. For example, one model of an Euler–Bernouli beam with strain rate damping (C_D) moving in air (γ) is:

$$\rho \frac{\partial^2 u(x,t)}{\partial t^2} + \gamma \frac{\partial u(x,t)}{\partial t} + \frac{\partial^2}{\partial^2 x} \\ \times \left(C_D I \frac{\partial^3 u(x,t)}{\partial^2 x \partial t} + EI \frac{\partial^2 u(x,t)}{\partial^2 x} \right) = 0 \quad [15]$$

Here $u(x,t)$ is the displacement of the beam, E is the elastic modulus, and I is the cross-sectional area moment of inertia about the z-axis. It is important to note that in such models the damping terms also appear in the boundary conditions. For instance at a free end, with strain-rate damping as given in eqn [15], the boundary conditions are:

$$EI \frac{\partial^2 u}{\partial x^2} + C_D I \frac{\partial^3 u}{\partial t \partial x^2} = 0 \text{ at the free boundary}$$

$$-\frac{\partial}{\partial x}\left(EI \frac{\partial^2 u}{\partial x^2} + C_D I \frac{\partial^3 u}{\partial t \partial x^2} \right) = 0 \text{ at the free boundary} \quad [16]$$

This model corresponds to Kelvin–Voigt damping offering resistance to the bending moment. Such models give reasonable predictions of the time response and decay rates for vibrating beams made of metals when compared with vibration experiments. However, the damping coefficients (γ and C_D) must be determined by parameter identification methods and experimental data.

Complex Stiffness and Modulus

The complex modulus approach to representing various damping mechanisms follows from eqn [12] by simply taking the Laplace transform of the expression and evaluating the transform variable along the imaginary axis. However, an easier way to visualize the complex modulus approach is to examine the Kelvin–Voigt damping model under harmonic excitation. Using the exponential form of harmonic excitation the equation of motion becomes:

$$m\ddot{x}(t) + c\dot{x}(t) + kx(t) = F e^{j\omega t} \quad [17]$$

Here ω is the driving frequency and F is the amplitude of the applied force. The symbol j denotes the imaginary unit. Assuming a steady state solution of the form $x(t) = X e^{j\omega t}$ and substituting this into eqn [17] results in:

$$-m\omega^2 X e^{j\omega t} + (k + j\omega c) X e^{j\omega t} = F e^{j\omega t} \quad [18]$$

Again using the substitution $X e^{j\omega t} = x(t)$ yields:

$$m\ddot{x}(t) + \underbrace{(k + j\omega c)}_{k'} x(t) = F e^{j\omega t} \quad [19]$$

In this form, the complex stiffness is evident from the coefficient of $x(t)$ and has the form:

$$k' = k + j\omega c \quad [20]$$

This also leads to the concept of the complex modulus. The complex stiffness is often referred to as the Kelvin–Voigt complex stiffness.

If experiments are performed and used to compute the energy dissipated per cycle under sinusoidal excitation, many engineering materials exhibit damping of the form $C_d X^2$ where:

$$C_d = \pi c \omega \quad [21]$$

The constant c is the normal viscous damping coefficient. This is referred to as the Kimball–Love observation. Solving this for a damping force, the following frequency dependent damping force is obtained:

$$f_c = \left(\frac{b}{\omega}\right)\dot{x}(t) \qquad [22]$$

Here the constant b is the measured value of $b = (C_d)/\pi$. The corresponding equation of motion is:

$$m\ddot{x}(t) + \frac{b}{\omega}\dot{x}(t) + kx(t) = 0 \qquad [23]$$

This illustrates a system with frequency dependent damping. Here however, the interpretation of the value of ω is somewhat fuzzy. Some interpret it to be the driving frequency in steady state, but this denies the use of eqn [23] for transient analysis. Furthermore, the form of eqn [23] does not satisfy the condition of causality for physically realizable systems. Under harmonic excitation, the complex stiffness and the frequency dependent form of eqn [20] are equivalent.

The frequency-dependent form given by eqn [23] has been adapted to free vibration by hypothesizing the damping force:

$$f_c = b\left|\frac{x(t)}{\dot{x}(t)}\right|\dot{x}(t) \qquad [24]$$

This is causal, allows numerical simulation, but results in a nonlinear differential equation of motion.

Viscoelastic Damping

Viscoelastic damping arises from materials that are rubber like in nature, called viscoelastic materials (VEM). Viscoelastic damping is exhibited in polymeric and glassy materials as well as in some enamel materials. Viscoelastic materials are often added to metal structures (which typically have very low damping) and devices to increase the amount of system damping. Examples are rubber mounts and constrained layer damping treatments. The classical theory of elasticity states that for sufficiently small strains, the stress in an elastic solid is proportional to the instantaneous strain and is independent of the strain rate. In a viscous fluid, according to the theory of hydrodynamics, the stress is proportional to the instantaneous strain rate and is independent of the strain. Viscoelastic materials exhibit both solid and fluid characteristics. Such materials include plastics, rubbers, glasses, ceramics, and biomaterials. Viscoelastic materials are characterized by constant-stress creep and constant-strain relaxation. Their deformation response is determined by both current and past stress states or, conversely, the current stress-state is determined by both current and past deformation states. It may be said that viscoelastic materials have 'memory'; this characteristic constitutes one foundation on which their mathematical modeling may be based. In polymers, material damping is a direct result of the relaxation and recovery of the long molecular chains after stress.

Linear viscoelastic materials may be defined by either differential or integral relationships. The differential model is presented in eqn [12]. Boltzmann in 1876 initiated a linear hereditary theory of material damping by formulating an integral representation of the stress–strain relation resulting in a damping force of the form:

$$f_c = \int_{-\infty}^{t} G(t,\tau)x(\tau)\,d\tau \qquad [25]$$

Here t is the current time, x is the displacement and G is the hereditary kernel. The term hereditary is used because the instantaneous deformation depends on all of the stresses previously applied to the system as well as depending on the stress at the current time. It is common to take the time dependence of the hereditary kernel to be a difference so that eqn [25] becomes:

$$f_c = \int_{-\infty}^{t} G(t-\tau)x(\tau)\,d\tau \qquad [26]$$

Motivated by Boltzmann's hypothesis, a general hereditary integral for the stress–strain relation was later proposed (1947) of the form:

$$\sigma(t) = E_0\varepsilon(t) + \int_{0}^{t} k(t-\tau)\frac{d\varepsilon(\tau)}{d\tau}\,d\tau \qquad [27]$$

Here, $k(t-\tau)$ is the relaxation modulus and E_0 represents an instantaneous modulus of elasticity. Eqn [27] can be thought of as a generalization of the Kelvin model. This expression can also be written in terms of the creep compliance $J(t)$ as:

$$\varepsilon(\tau) = \int_{0}^{t} J(t-\tau)\frac{d\sigma(\tau)}{d\tau}\,d\tau \qquad [28]$$

Various methods of modeling the relaxation modulus have led to a number of useful methods of modeling hysteretic type damping (see **Damping models**).

In practice, the damping related to VEM, as described by eqn [27], is characterized by a single number, the loss modulus or loss factor. The loss factor is derived analytically from the complex stiffness description given in eqn [20]. Rewriting eqn [20] as:

$$k' = k(1 + \bar{\eta}j) \qquad [29]$$

defines the loss factor to be:

$$\bar{\eta} = \frac{c}{k}\omega \qquad [30]$$

Here ω is the driving frequency and the concept is defined only in the steady state under harmonic excitation. This shows clear frequency dependence. In terms of a material property, the complex modulus may also be written in a factored form:

$$E' = E(1 + \eta j) \qquad [31]$$

where η is now the material loss factor which is also frequency dependent. Tests show that these values depend heavily on both frequency and temperature.

Eqn [31] is written for the extensional modulus. However, most of the damping produced in VEM comes from shear effects. The relationship between the shear modulus, G', and extensional modulus is given from the elasticity theory through Poisson's ratio, v, to be:

$$E' = 2(1 + v)G' \qquad [32]$$

Typical values of Poisson's ratio are about 0.5 for elastomeric materials at room temperatures, so that $E = 3G$. However, Poisson's ratio is both frequency and temperature dependent. For example, at lower temperatures Poisson's ratio is closer to 0.33. Unfortunately, Poisson's ratio is very difficult to measure.

A typical VEM is chosen in design applications based on plots of loss factor vs temperature and loss factor vs frequency. Basically the loss factor η is considered to be a function of both driving frequency and temperature; $\eta = \eta(\omega T)$. While no mechanics model of the temperature dependence is presented, the values of η are measured at various temperatures and plotted for a fixed frequency. In addition, for a fixed temperature, values of η are measured at a variety of frequencies for a fixed temperature. These plots are then provided as design guides for working with damping treatments for the purpose of vibration reduction (see **Damping materials**).

The complex modulus is also both frequency and temperature dependent. Eqn [31] can be written for the complex shear modulus as a temperature and frequency dependent quantity:

$$G'(\omega, T) = G(\omega, T)(1 + \eta(\omega, T)j) \qquad [33]$$

The complex notation provides a way to measure the phase angle between the applied stress and the resulting strain in cyclic loading. The real and imaginary parts of the complex shear modulus are widely used in the VEM industry to characterize damping materials. The real part, $G(\omega T)$, is called the storage modulus or shear modulus, and the imaginary part, $G(\omega T)\eta(\omega T)$, is called the loss modulus. The loss factor provides a measure of the energy dissipation capacity of the material, while the storage modulus provides a measure of the stiffness of the material. Basically the shear modulus determines how much energy goes into the VEM and the loss factor determines how much of that energy is dissipated.

A great deal of effort has been invested by industry in characterizing viscoelastic materials in terms of their temperature and frequency dependence. Testing of viscoelastic materials can be divided into two main categories: resonant testing and nonresonant testing. Nonresonant tests are often called complex stiffness tests and use a sample VEM fixed to a rigid mount and loaded in shear. The force and deformation are measured. The phase angle between the force and the deformation then provides a measure of the damping. The ratio of in-phase force to displacement provides a measure of the storage modulus.

The resonant testing method provides an indirect measure of damping and stiffness by using modal parameters. Basically, a beam-like specimen is excited (randomly) and a modal analysis yields the damping and frequency, from which the loss factor (equal to twice the damping ratio at resonance) and stiffness (hence storage modulus) are determined (ASTM E756-98).

These temperature and frequency dependencies are difficult to model in any analytical sense. Experiments show that temperature variations in damping are much more drastic then frequency variations. While good models of these dependencies do not exist at the system level, experimental data can be used to approximate the inclusion of these effects in finite element modeling, as discussed in **Damping in FE models**.

Equivalent Viscous Damping

Most of the damping models listed here are nonlinear and/or frequency and temperature dependent. In fact, accurate modeling of damping quickly leads to nonlinear terms. The effects of nonlinearity and frequency dependence render the models difficult to use in vibration analysis. In particular, the most common method of modeling structures for vibration analysis is the finite element method, which has difficulty with even the basic linear viscous behavior. As a result, some effort has been spent to produce 'equivalent' linear viscous models. It is important to note that while these models express the energy dissipated accurately, the linearized models do not render the same physical behavior. For instance, the linear version will not have multiple equilibrium points.

The equivalent viscous damping of any single-degree-of-freedom damping model is obtained by examining the response of the system in steady state under sinusoidal excitation. The energy dissipated per cycle of the forced response is then computed for a linear viscous damper. Next the same calculation is made for the identical system (i.e., same mass and stiffness) with the nonlinear damping mechanism. These two expressions for the energy dissipated per cycle are then compared and the equivalent linear, viscous damping coefficient, c_{eq} is derived. The results of such calculations can be found in most introductory texts on vibration and are summarized for some common damping mechanisms in **Table 1**.

The value of equivalent viscous damping from Table 1 can then be used in place of the actual mechanism. This results in a linear equation of motion that can be solved in close form to simulate the nonlinearly damped system. Most often such calculations are used to approximate the magnitude of the steady-state response of the nonlinear system for design purposes. The method must be treated only as a crude approximation because the nonlinear system exhibits additional phenomena that are not present in the equivalent linear system. For example, the response of a system with Coulomb friction under steady-state excitation will chatter under harmonic loads with amplitudes near the maximum static friction force. This behavior would be completely missed by the equivalent, linear viscous form.

As an example of the use of equivalent viscous damping factors, consider a system with nonlinear air damping. A single-degree-of-freedom system with velocity squared damping, a linear spring and harmonic input is described by:

$$m\ddot{x} + \alpha \operatorname{sgn}(\dot{x})\dot{x}^2 + kx = F_0 \cos \omega t \quad [34]$$

The equivalent viscous damping coefficient from Table 1 is:

Table 1 Damping models and equivalent linear models.

Name	Damping force	c_{eq}	Source
Linear viscous damping	$c\dot{x}$	c	Slow fluid
Air damping	$a \operatorname{sgn}(\dot{x})\dot{x}^2$	$8a\omega X/3\pi$	Fast fluid
Coulomb damping	$\beta \operatorname{sgn} \dot{x}$	$4\beta/\pi\omega X$	Sliding friction
Displacement-squared damping	$d \operatorname{sgn}(\dot{x})\dot{x}^2$	$4dX/3\pi\omega$	Material damping
Solid, or structural, damping	$b \operatorname{sgn}(\dot{x})|x|$	$2b/\pi\omega$	Internal damping

$$c_{eq} = \frac{8}{3\pi}\alpha\omega X \quad [35]$$

The value of the magnitude, X, can be approximated for near resonance conditions to be ($f_0 = F_0/m$):

$$X = \sqrt{\left(\frac{3\pi m f_0}{8\alpha\omega^2}\right)} \quad [36]$$

Combining these last two expressions yields an equivalent viscous damping value of:

$$c_{eq} = \sqrt{\left(\frac{8m\alpha f_0}{3\pi}\right)} \quad [37]$$

Using this value as the damping coefficient results in a linear system of the form:

$$m\ddot{x} + c_{eq}\dot{x} + kx = F_0 \cos \omega t \quad [38]$$

This provides a linear approximation to eqn [34]. **Figure 1** is a plot of the linear system with equivalent

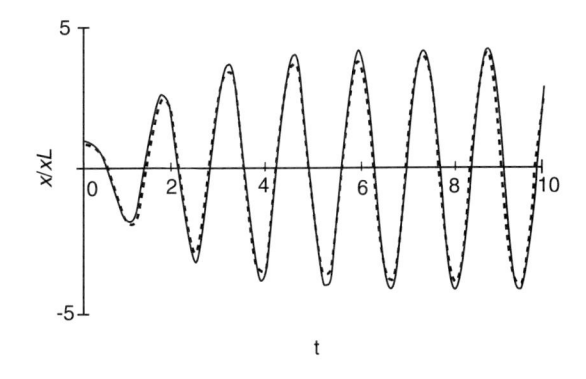

Figure 1 The displacement of the equivalent viscous damping (—) and the displacement of the nonlinear system (- - -) vs time for the case of $\alpha = 0.5$.

viscous damping and a numerical simulation of the full nonlinear eqn [38] for the value of the parameter $\alpha = 0.5$ that depends on the drag coefficient. Further simulations show that the larger the drag coefficient the greater the error is in using the concept of equivalent viscous damping. Note from the plot that the frequency of the response looks similar, but the amplitude of oscillation is slightly over estimated by the equivalent viscous damping technique, rendering it conservative in design.

Numerical simulation has become cheap, and well understood. In the case of single-degree-of-freedom systems it is best to simulate the full nonlinear system rather then to use the approximation. Repeated simulations can offer design information that is more accurate then using the equivalent viscous damping approach. However, if a closed form design formula is needed, then the 'equivalence' listed in **Table 1** will be useful.

Nomenclature

c	constant
C_D	strain rate damping
E	modulus of elasticity
f_c	damping force
F	amplitude of the applied force
g	acceleration due to gravity
G	hereditary kernel
I	inertia
j	imaginary unit
t	time
v	velocity
X	magnitude
σ	stress
ε	strain
η	material loss factor

See also: **Damping in FE models**; **Damping materials**; **Viscous damping**.

Further Reading

Bandstra JP (1983) Comparison of equivalent viscous damping and nonlinear damping in discrete and continuous vibrating systems. *ASME Journal of Vibration, Acoustics, Stress and Reliability in Design* 105: 382–392.

Banks HT, Wang Y, Inman DJ (1994) Bending and shear damping in beams: frequency domain estimation techniques. *ASME Journal of Vibration and Acoustics* 116(2): 188–197.

Bert CW (1973) Material damping: an introductory review of mathematical models, measures and experimental techniques. *Journal of Sound and Vibration* 29(2): 129–153.

Inman DJ (2001) *Engineering Vibration*, 2nd edn. Upper Saddle River, NJ: Prentice Hall.

Lazan BJ (1968) *Damping of Materials and Members in Structural Mechanics*. London: Pergamon Press.

Nashif AD, Jones DIG, Henderson JP (1985) *Vibration Damping*. New York: Wiley.

Osiñski Z (ed.) (1998) *Damping of Vibrations*. Rotterdam: A.A. Balkema.

Snowden JC (1968) *Vibration and Shock in Damped Mechanical Systems*. New York: Wiley.

Sun CT, Lu YP (1995) *Vibration Damping of Structural Elements*. Englewood Cliffs, NJ: Prentice Hall.

DAMPING MOUNTS

J-Q Sun, University of Delaware, Newark, DE, USA

Copyright © 2001 Academic Press

doi:10.1006/rwvb.2001.0019

Introduction

Mounts are commonly used to reduce the energy transmission from one mechanical system to another. They are also referred to as vibration isolators. Mounts are often designed to provide damping in the transmission path for vibration isolation. Damping is therefore incorporated in the mounting system for energy dissipation. There is a compromise between the inclusion of proper damping and the isolation performance of the mounting system. This article presents the common designs of mounts with various built-in damping mechanisms.

Special mounting systems are also discussed in this *Encyclopedia* (see Active Mounts in article **Active control of vehicle vibration**; and Isolation of Machinery in Buildings in article **Active control of civil structures**.

Basic Concepts of Mounting System

There are two types of applications with a mounting system, as illustrated in **Figure 1**. The first is to isolate

Figure 1 Two typical applications of mounting systems.

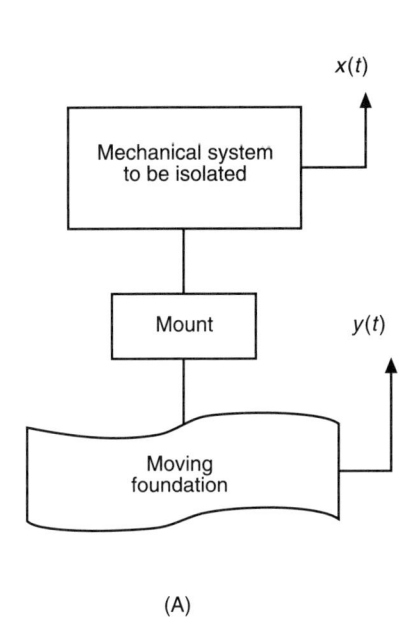

mechanical systems from based excitations. Applications of this type include earthquake isolation of buildings, isolation of high-precision instruments from ground motions, and isolation of passengers from rough road-induced vibrations of automobiles. The second is to isolate vibrating systems from the foundation. Applications of this type include reducing the dynamic forces transmitted to the foundation that are generated by vibrating heavy machines and home appliances.

Mounting systems are characterized by the quantity called transmissibility. For the moving foundation problem, an absolute transmissibility can be defined as:

$$(T_A)_d = \frac{X}{Y}, \quad (T_A)_v = \frac{\dot{X}}{\dot{Y}}, \quad (T_A)_a = \frac{\ddot{X}}{\ddot{Y}} \quad [1]$$

where $(T_A)_d$, $(T_A)_v$, and $(T_A)_a$ are the absolute displacement, velocity, and acceleration transmissibilities. X, $\dot{X}$, and $\ddot{X}$ are the amplitudes of the absolute displacement, velocity and acceleration of the isolated system. Y, $\dot{Y}$, and $\ddot{Y}$ are the amplitudes of the absolute displacement, velocity, and acceleration of the foundation. When the foundation motion is harmonic, and the entire system is linear, these three transmissibilities are equal. In practice, it is often more convenient to use the relative displacement across the mount as a parameter to characterize the system. Let Z, $\dot{Z}$, and $\ddot{Z}$ be the amplitudes of the relative displacement, velocity, and acceleration across the mount. We can define a relative transmissibility as:

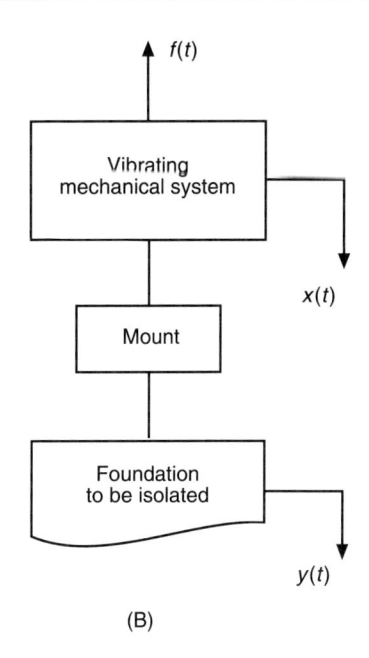

$$(T_R)_d = \frac{Z}{Y}, \quad (T_R)_v = \frac{\dot{Z}}{\dot{Y}}, \quad (T_R)_a = \frac{\ddot{Z}}{\ddot{Y}} \quad [2]$$

For the force isolation problem, the transmissibility is defined as the ratio of the force F_t transmitted to the foundation to the dynamic force F_d acting on or generated by the vibrating mechanical system:

$$T_f = \frac{F_t}{F_d} \quad [3]$$

Passive Damping in Mounts

Various damping mechanisms can be built in a mount. **Table 1** presents a list of commonly used damping. For the mount with eletrorheological (ER) fluid or magnetorheological (MR) fluid in **Table 1**, the damping force of the mount is controllable, and the system is called semiactive. Active damping is normally provided by a force actuator that is part of a closed-loop control system. The active and semiactive damping will be discussed later.

Consider a single-degree-of-freedom (SDOF) system, as shown in **Figure 2**. The equation of motion for the system is given by:

$$m\ddot{x} + D(\dot{z}) + kz = f(t) \quad [4]$$

For linear viscous damping $D(\dot{z}) = c\dot{z}$, the transmissibility can be obtained in closed form. Assume that the base excitation $y(t)$ is harmonic with frequency ω and $f(t) = 0$; we have:

344 DAMPING MOUNTS

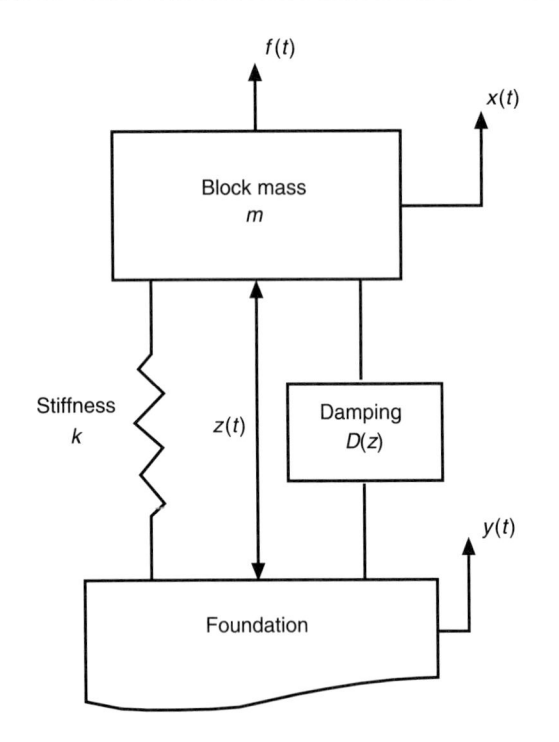

Figure 2 A single-degree-of-freedom example of directly coupled damping in mounting systems.

$$(T_A)_d = \sqrt{\frac{1 + 4\zeta^2 r^2}{(1-r^2)^2 + 4\zeta^2 r^2}} e^{-i\phi}$$

$$\phi = \tan^{-1}\left(\frac{2\zeta r^3}{(1-r^2) + 4\zeta^2 r^2}\right) \quad [5]$$

where $\zeta = c/2\sqrt{mk}$ is the damping ratio, $r = \omega/\omega_n$ is the frequency ratio. $\omega_n = \sqrt{k/m}$ is the natural frequency of the mount. The amplitude and phase of the transmissibility are shown in **Figure 3**. The effect of damping on vibration isolation is clearly shown in the amplitude plot. For $r \leq \sqrt{2}$, damping suppresses the resonant response of the mounting system, and is thus beneficial. For $r > \sqrt{2}$, damping actually increases the transmitted vibration level. However, the mount is typically designed to operate in the frequency range $\omega \gg \omega_n$, i.e., $r > \sqrt{2}$; damping is less beneficial in this higher-frequency range. This observation of the effect of damping on the transmissibility is common to other types of damping.

When the damping is nonlinear, it is often very difficult to obtain analytical solutions for the transmissibility. A common approach to obtain approximate solutions in order to design a mount is the method of equivalent linearization. The method proposes to find an equivalent linear viscous damping to replace the nonlinear damping. The closed-form solution for the linear system then becomes available. For example, an equivalent linear system to eqn [4] is given by:

$$m\ddot{x} + c_{eq}\dot{z} + kz = f(t) \quad [6]$$

The equivalent linear viscous damping coefficient c_{eq} can be determined such that the error $e(t) = c_{eq}\dot{z} - D(\dot{z})$ is minimized. When the system is harmonic, the following mean square error over a period is minimized with respect to c_{eq}:

Table 1 Damping mechanisms for mounting systems

Type of damping	Damping force
Linear viscous	$-c\dot{z}$
Dry friction	$-\mu\,\text{sgn}(\dot{z})$
Power-law	$-c_n|\dot{z}|^n\,\text{sgn}(\dot{z})$
Hysteretic	$-c(\omega)\dot{z}$
Electrorheological fluids	$-f_y(E)\,\text{sgn}(z) - c(E)\dot{z}$
Magnetorheological fluids	$-f_y(H)\,\text{sgn}(z) - c(H)\dot{z}$
Active damping	$-K_D\dot{z}$
Skyhook active damping	$-K_s\dot{x}$

E, electric field strength in the electrorheological fluid; H, magnetic field strength in the magnetorheological fluid; K_D, feedback gain in active damping.

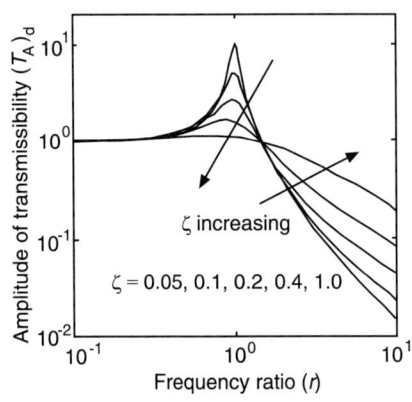

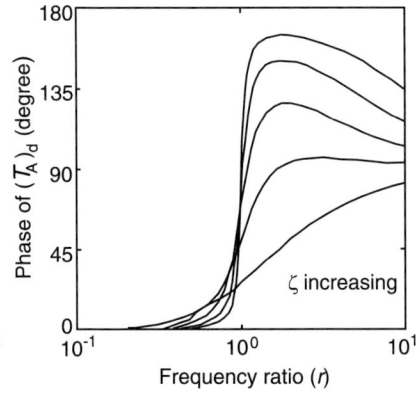

Figure 3 Effect of viscous damping on the amplitude and phase of the absolute displacement transmissibility of the simple mount. ($\zeta = 0.05, 0.1, 0.2, 0.4, 1.0$)

$$J = \frac{1}{\tau}\int_0^\tau e^2(t)\,dt \quad [7]$$

where τ is the period of the harmonic motion. The equivalent linear viscous damping coefficient is given by:

$$c_{eq} = \frac{\int_0^\tau \dot{z}D(\dot{z})\,dt}{\int_0^\tau \dot{z}^2\,dt} \quad [8]$$

When the motion is random, the mean square error and the equivalent linear viscous damping coefficient are given by:

$$J = E\left[e^2(t)\right], \quad c_{eq} = \frac{E[\dot{z}D(\dot{z})]}{E[\dot{z}^2]} \quad [9]$$

where $E[\cdot]$ is the mathematical expectation operator. As an example, consider a quadratic nonlinear damping element given by $D(\dot{z}) = c_2\dot{z}^2\mathrm{sgn}(\dot{z})$. The magnitude of the displacement transmissibility is given by:

$$|(T_A)_d| = \sqrt{\frac{1 + 4\zeta_{eq}^2 r^2}{(1 - r^2)^2 + 4\zeta_{eq}^2 r^2}} \quad [10]$$

where:

$$4\zeta_{eq}^2 r^2 = \frac{1}{2}\left[\sqrt{(1-r^2)^4 + \left(\frac{16c_2\omega_n^2 Y_0 r^4}{3\pi k}\right)^2} - (1-r^2)^2\right]$$

Y_0 is the magnitude of the harmonic motion of the base excitation. The phase of the transmissibility based on the equivalent linear system has the same form as the one given in eqn [5]. **Figure 4** shows the magnitude and phase of the transmissibility. The parameter α shown in the figure is defined as $\alpha = c_2\omega_n^2 Y_0 k^{-1}$. The effect of damping is very similar to that for the system with linear viscous damping.

The method of equivalent linearization can be applied to the mounting system with nonlinear damping and nonlinear stiffness. It provides a quick solution for design engineers. However, when the nonlinearity is strong, caution should be exercised when interpreting the results.

Elastically Coupled Damping

The damping element in the mount shown in **Figure 2** is directly coupled. Another common arrangement is to couple the damping element elastically in the mount, as shown in **Figure 5**. The elastic coupling introduces a half-degree-of-freedom, leading to an equation of motion for the system that is third-order. The equation of motion for the system with linear viscous damping $D(\dot{z}) = c\dot{z}$ is given by:

$$\frac{mc}{k_1}\dddot{x} + m\ddot{x} + c\frac{k_1 + k}{k_1}\dot{z} + kz = f(t) + \frac{c}{k_1}\dot{f}(t) \quad [11]$$

Assume that the base excitation $y(t)$ is harmonic with frequency ω and $f(t) = 0$; we have:

$$(T_A)_d = \sqrt{\frac{1 + 4\zeta^2 r^2\left(\frac{\gamma+1}{\gamma}\right)^2}{(1-r^2)^2 + 4\zeta^2 r^2\left(\frac{\gamma+1-r^2}{\gamma}\right)}}\,e^{-i\phi}$$

$$\phi = \tan^{-1}\left(\frac{2\zeta r^3}{(1-r^2) + 4\zeta^2 r^2\left(\frac{(\gamma+1)(\gamma+1-r^2)}{\gamma^2}\right)}\right)$$

[12]

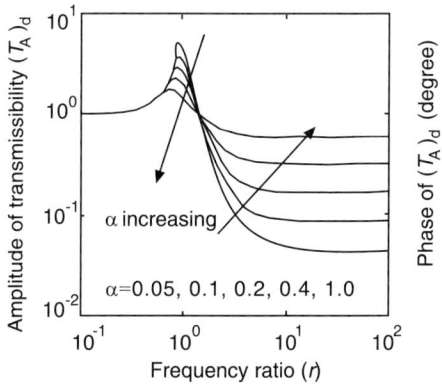

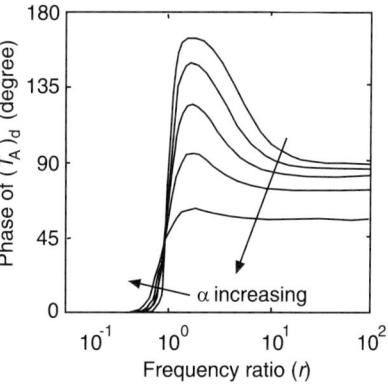

Figure 4 The amplitude and phase of the absolute displacement transmissibility of a SDOF mount with quadratic nonlinear damping.

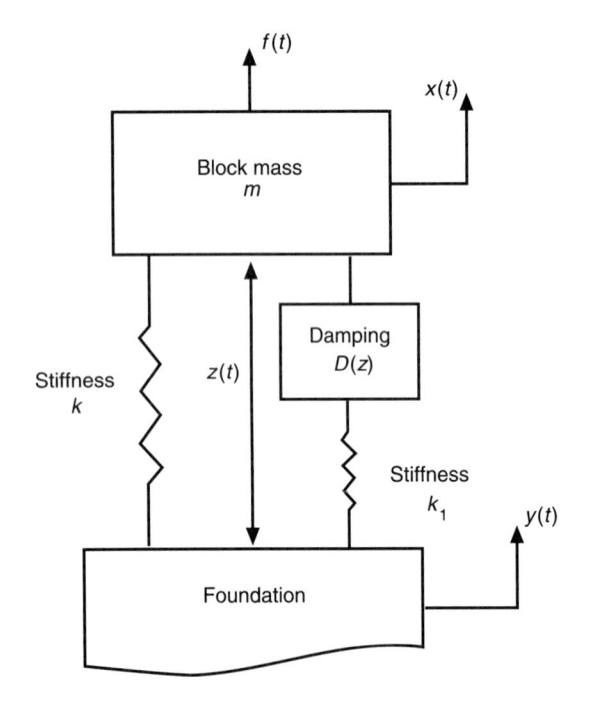

Figure 5 Elastically coupled damping in mounting systems.

where $\gamma = k_1/k$ is the stiffness ratio. The magnitude and phase of the transmissibility are shown in **Figure 6**.

The elastic coupling has also been generalized to include dynamic coupling in the mount. **Figure 7** shows three mounts with a tuned vibration absorber. In fluid-elastic mounts, this dynamic coupling is even magnified by means of a leverage mechanism. A schematic of such a mount is shown in **Figure 8**. There are many design variations of commercial mounts with different leveraging mechanisms.

The additional resonance of the vibration absorber provides another degree of freedom that can help improve the mount performance. A notch in the frequency response can be created to eliminate effectively a tonal disturbance by properly tuning the vibration absorber. The transmissibility of a typical commercial fluid-elastic mount with tuned fluid inertia track is shown in **Figure 9**. For more discussions on vibration absorber, see **Absorbers, vibration**.

Multidirectional Mounting Systems

The mounts discussed so far are unidirectional for one-degree-of-freedom systems. Practical applications often involve multiple degrees of freedom and multidirectional motion. Schematics of two multidirectional and multiple-degree of-freedom mounting systems are shown in **Figure 10**.

The design and analysis of such mounts are generally more difficult and involved. A useful concept for evaluating multidirectional mounting systems is the mount effectiveness. This concept is particularly well suited to mount design for flexible structural systems.

Consider a mounting system for isolating a vibrating source, such as aircraft engines, from the base, such as the aircraft fuselage. Let $\mathbf{v}_h$ be the vector consisting of all the generalized velocity components of the base at the mounting points when the vibrating source is rigidly connected to the structure. $\mathbf{v}_i$ is the vector consisting of the same velocity components of the base at the mounting points when the vibrating source is connected to the structure via the mount. Then, a matrix $\mathbf{E}_v$ can be found as a function of the system impedance matrices, such that:

$$\mathbf{v}_h = \mathbf{E}_v \mathbf{v}_i \qquad [13]$$

A similar effectiveness matrix can also be defined for the transmitted generalized forces at the mounting points. The spectral norm of the complex matrix $\mathbf{E}_v$

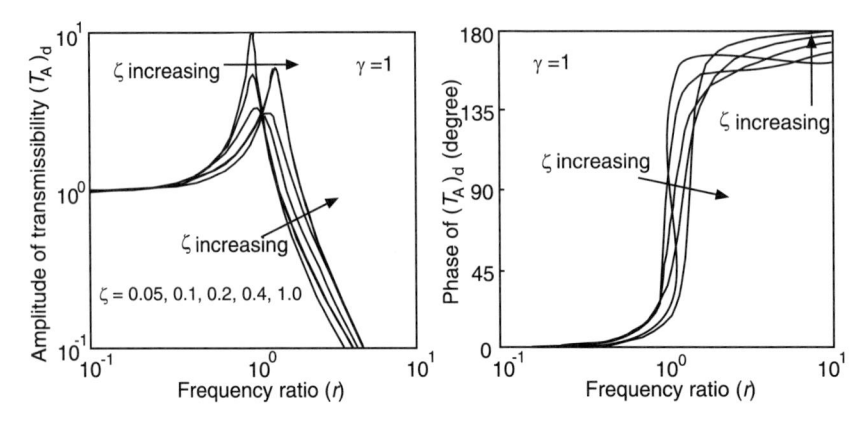

Figure 6 The amplitude and phase of the absolute displacement transmissibility of a SDOF mount with elastically coupled linear viscous damping. The stiffness ratio $\gamma = 1.0$.

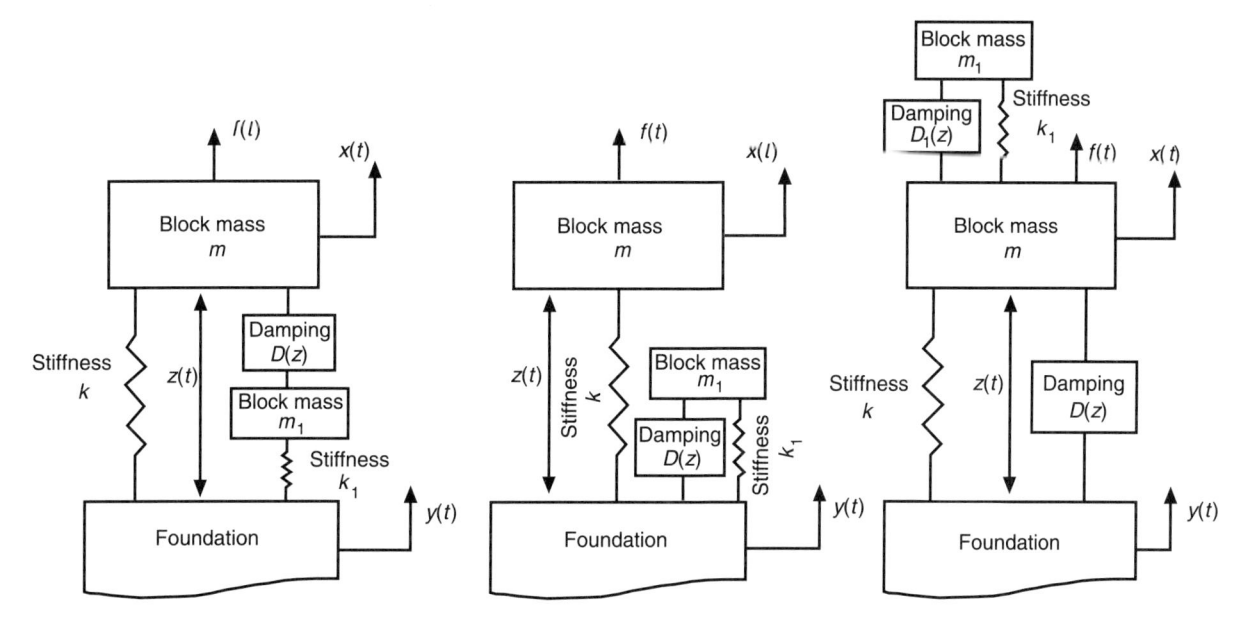

Figure 7 Examples of dynamically coupled mounts with vibration absorber.

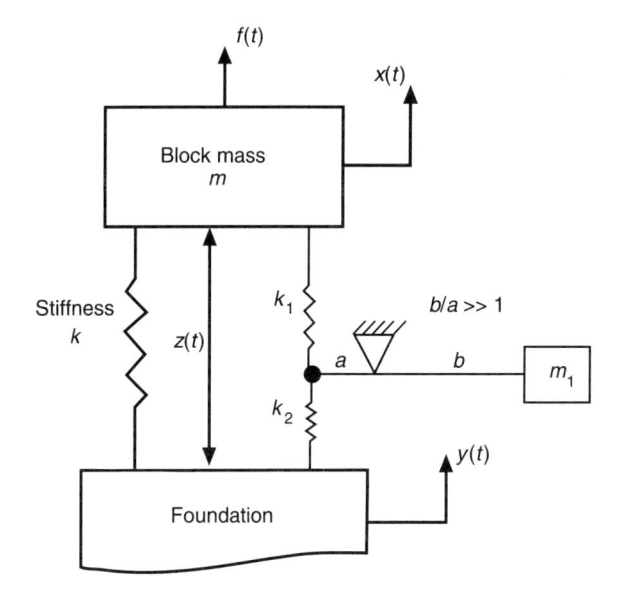

Figure 8 A dynamically coupled mount with a leveraged vibration absorber. Such a leverage mechanism is widely used in commercial fluid-elastic mounts.

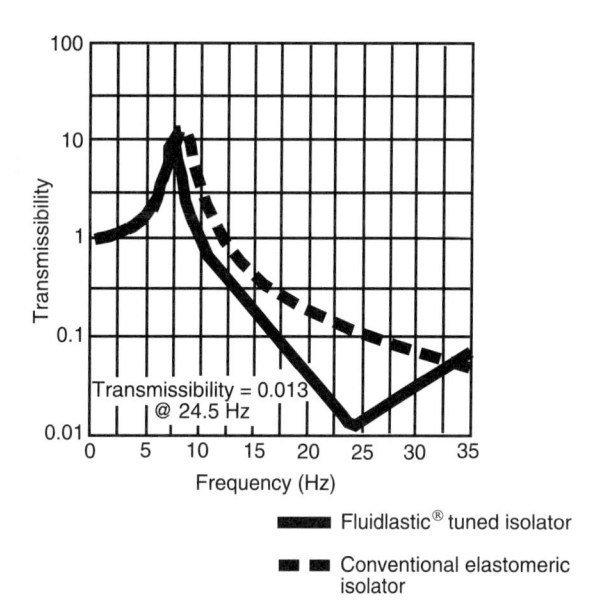

Figure 9 The transmissibility of a typical commercial fluid-elastic mount with a tuned and leveraged vibration absorber in the form of fluid inertia track. Courtesy of Lord Corporation. Continuous line, Fluidlastic® tuned isolator; dashed line, conventional elastomeric isolator.

can be used to characterize the performance of the mounting system:

$$\|\mathbf{E_v}\| = \sigma_{\max}(\mathbf{E_v}) \quad [14]$$

where $\sigma_{\max}(\mathbf{E_v})$ is the largest singular value of $\mathbf{E_v}$. The spectral norm is the matrix equivalent of the scalar magnitude. It helps to extend the mount design procedure for unidirectional systems to multidirectional and multiple-degree-of-freedom mounting systems.

Active and Semi-Active Damping in Mounts

Linear system analysis, presented above, has shown that the effect of damping on vibration isolation is beneficial in suppressing the resonant response of the mounting system in the frequency ratio range $r \leq \sqrt{2}$, and becomes less desirable for $r > \sqrt{2}$. In the application when the base has to go through a start-up

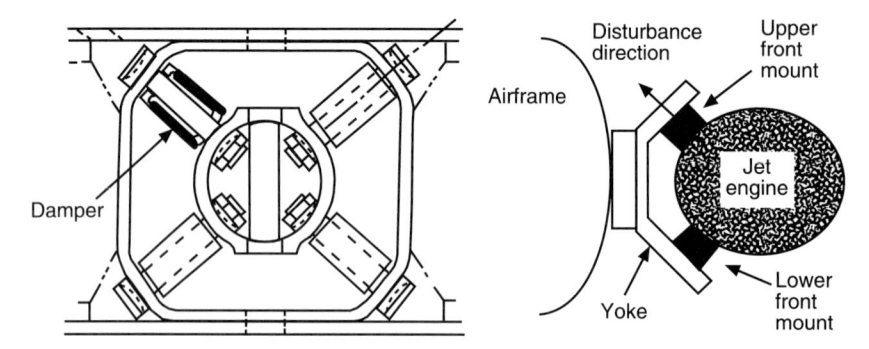

Figure 10 Examples of mounts having multiple degrees of freedom. Courtesy of Lord Corporation.

transient, the mount has to be designed to isolate the vibration for a wide range of frequencies before the system settles down to steady state with $r > \sqrt{2}$. It is hence desirable to have an adjustable damping.

An active control system can provide such a damping with proper sensing and feedback strategies. Furthermore, the feedback gain can be shaped in the frequency domain. In other words, the feedback gain can be selected as a function of frequency.

Consider the mount with active damping $-K_D \dot{z}$, listed in **Table 1**. Note that the feedback signal is the relative velocity within the mount. Assume that the feedback gain K_D is shaped so that the closed-loop damping ratio ζ_D in the range $0 < r < \sqrt{2}$ is unit, and drops to 0.001 for $r > \sqrt{2}$. This is an idealization of the relative velocity feedback. The resulting amplitude and phase of the displacement transmissibility are shown in **Figure 11**. Clearly, active controls have a great potential to create superior isolation performance. The performance shown in the figure is very difficult to duplicate with pure passive damping.

Generally, a control force actuator is used to physically provide active damping in the mount. With active systems, other feedback strategies can be implemented that will lead to better performance than the one shown in **Figure 11**. A very successful control concept is the so-called skyhook damping. A schematic of the skyhook damper is shown in **Figure 12**.

Consider a simple example of velocity feedback $-K_s \dot{x}$. The displacement transmissibility of the mount is then given by:

$$(T_A)_d = \sqrt{\frac{1 + 4\zeta^2 r^2}{(1 - r^2)^2 + 4(\zeta + \zeta_s)^2 r^2}} e^{-i\phi}$$

$$\phi = \tan^{-1}\left(\frac{2\zeta r^3}{(1 - r^2) + 4(\zeta + \zeta_s)^2 r^2}\right) \quad [15]$$

where ζ is the passive damping ratio of the mount, and ζ_s is the damping ratio due to the feedback. **Figure 13** shows the magnitude and phase of the displacement transmissibility. It is clear that skyhook damping can provide much better vibration isolation over a broad range of frequencies.

For more discusssions on real-time issues of the feedback control design, see section on Feedback

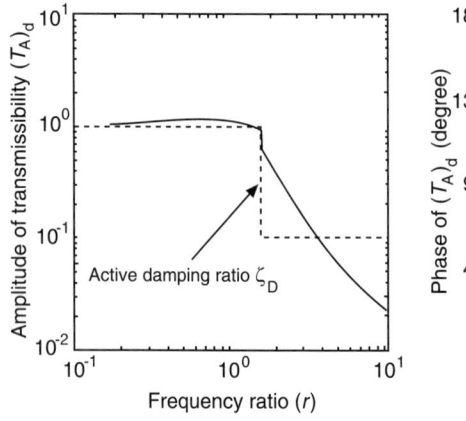

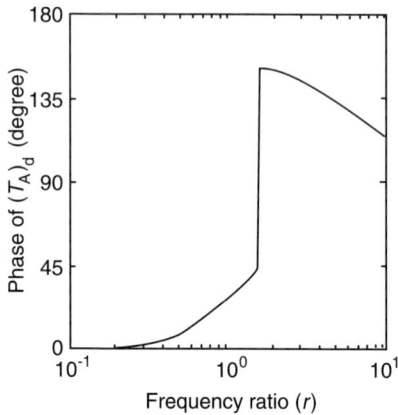

Figure 11 The magnitude and phase of the transmissibility with active damping using relative velocity feedback. The active damping ratio ζ_D (dashed line) is shaped in the frequency domain, and is a mathematical idealization.

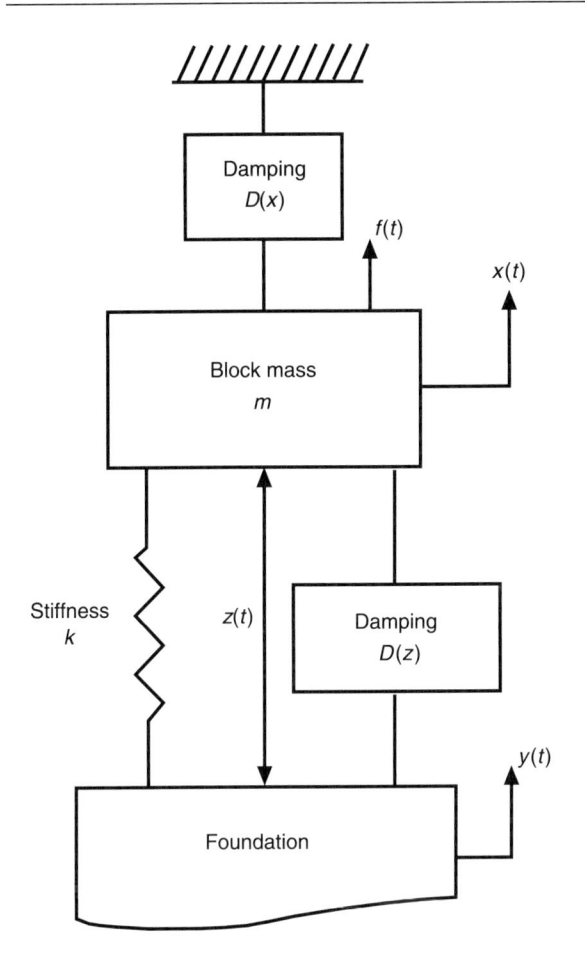

Figure 12 Schematic of a mount with skyhook damping.

Control in the article **Active control of civil structures**. Active systems can provide better vibration isolation, and are more expensive and less reliable than passive systems. Semi-active damping represents a compromise between passive and active systems. Mounts with ER and MR fluids can provide a range of damping. **Figures 14** and **15** show an ER fluid vibration isolation system and MR fluid vibration dampers. These devices are commercially available.

Figure 16 shows typical damping forces of a MR damper as a function of the current input to the coil. When this damper is used in a vibration isolation system, the transmissibility of the system can be improved substantially.

Figure 17 shows the magnitude of transmissibility of a system with a MR damper. The vibration isolation performance is comparable with the active system in the lower-frequency range. At higher frequencies, however, the high-level passive damping of MR fluids becomes less advantageous, and cannot be reduced by the control system. Nevertheless, the

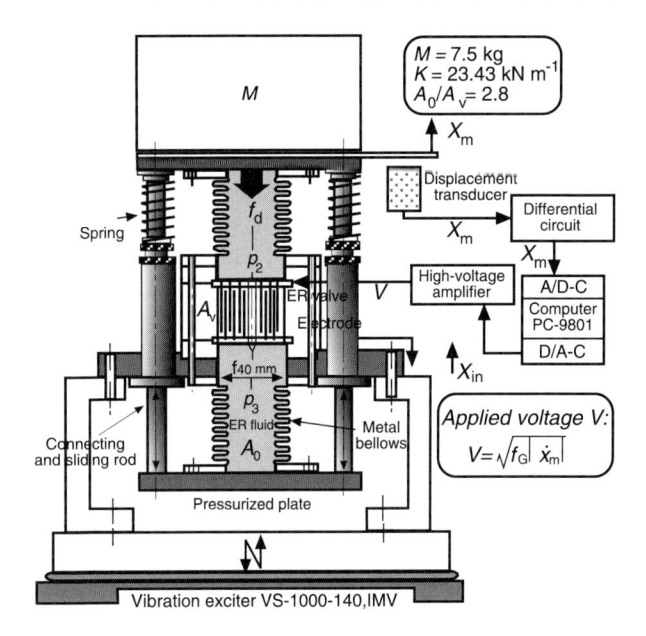

Figure 14 Vibration isolator system using an antagonized bellows ER damper and its control system. From Jolly MR and Nakano M (1998) *Properties and Applications of Commercial Controllable Fluids. Actuator 98*. Bremen, Germany, with permission.

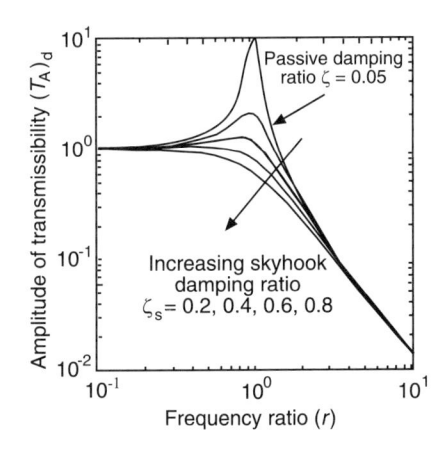

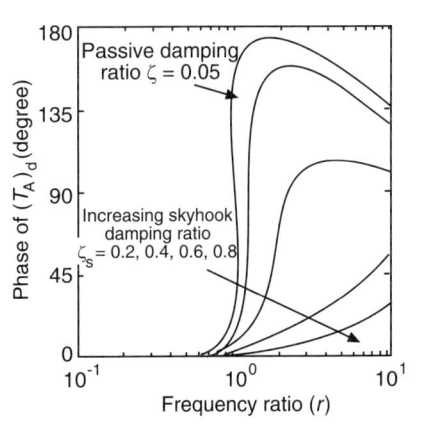

Figure 13 The magnitude and phase of the transmissibility with active skyhook damping.

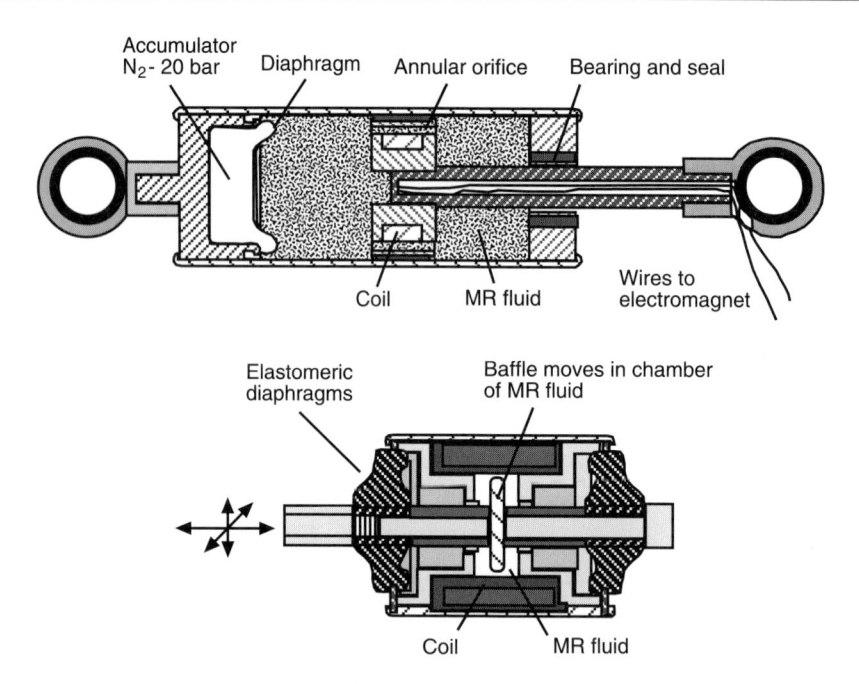

Figure 15 Examples of commercially available MR fluid vibration damper. From Jolly MR and Nakano M (1998) *Properties and Applications of Commercial Controllable Fluids. Actuator 98*. Bremen, Germany, with permission.

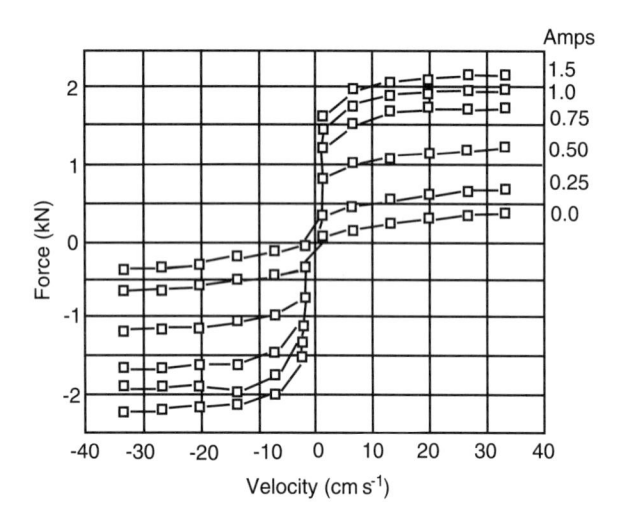

Figure 16 Damping force of the MR damper as a function of current input to the coil. From Jolly MR and Nakano M (1998) *Properties and Applications of Commercial Controllable Fluids. Actuator 98*. Bremen, Germany, with permission.

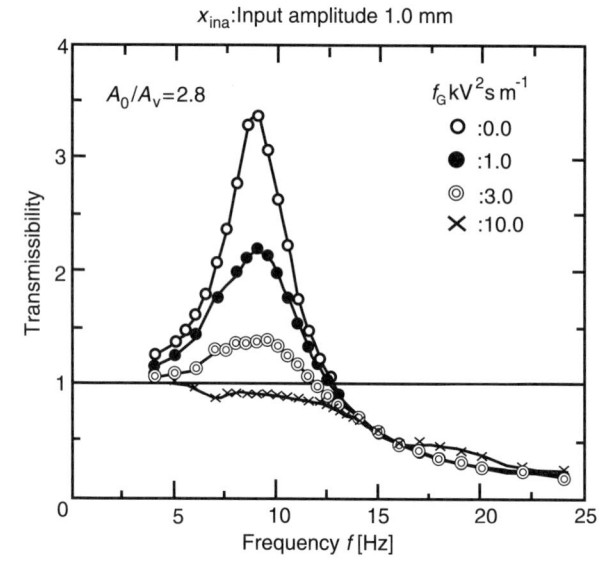

Figure 17 Magnitude of transmissibility of SDOF vibration isolator system using ER damper controlled by nonlinear feedback of absolute mass velocity. From Jolly MR and Nakano M (1998) *Properties and Applications of Commercial Controllable Fluids. Actuator 98*. Bremen, Germany, with permission.

semi-active isolation system using controllable fluids such as ER and MR fluids is fail-safe, and requires much simpler electronics to implement.

Nomenclature

$E[\cdot]$	expectation operator
F_d	dynamic force
F_t	transmitted force
γ	stiffness ratio
τ	period of harmonic motion
ζ_D	active damping ratio

See also: **Absorbers, vibration; Active control of civil structures; Active control of vehicle vibration.**

Further Reading

Avallone EA and Baumeister T III (eds) (1978) *Marks' Standard Handbook for Mechanical Engineers.* New York: McGraw-Hill.

Beards CF (1996) *Engineering Vibration Analysis with Application to Control Systems.* New York: Halsted Press.

Dimarogona A (1996) *Vibration for Engineers.* Upper Saddle River, New Jersey: Prentice Hall.

Harris CM and Crede CE (eds) (1996) *Shock and Vibration Handbook.* New York: McGraw-Hill.

Jolly MR and Nakano M (1998) *Properties and Applications of Commercial Controllable Fluids. Actuator 98.* Bremen, Germany.

Karnopp D (1995) Active and semi-active vibration isolation. *50th Anniversary Issue of ASME Journal of Mechanical Design and Journal of Vibration and Acoustics* 117: 177–185.

Lord Corporation (1997) *Rheonetic Linear Damper RD-1001/RD-1004 Product Information Sheet.* Cary, North Carolina.

Rao SS (1995) *Mechanical Vibrations.* New York: Addison-Wesley.

Ruzicka JE and Derby TF (1971) Influence of damping in vibration isolation. In: *The Shock and Vibration Monograph Series, SVM-7.* Washington, DC: Shock and Vibration Information Center, United States Department of Defense.

Snowdon JC (1968) *Vibration and Shock in Damped Mechanical Systems.* New York: John Wiley.

Sun JQ, Norris MA and Jolly MR (1995) Passive, adaptive and active tuned vibration absorbers – a survey. *50th Anniversary Issue of ASME Journal of Mechanical Design and Journal of Vibration and Acoustics* 117: 234–242.

Swanson DA and Miller LR (1993) Design and mount effectiveness evaluation of an active vibration isolation system for a commercial jet aircraft. *Proceedings of AIAA/AHS/ASEE Aerospace Design Conference* paper no. AIAA-93-1145.

Swanson DA, Miller LR and Norris MA (1994) Multidimensional mount effectiveness for vibration isolation. *Journal of Aircraft* 31: 188–196.

Thomson WT and Dahleh MD (1998) *Theory of Vibration with Applications.* Upper Saddle River, New Jersey: Prentice Hall.

DAMPING, ACTIVE

A Baz, University of Maryland, College Park, MD, USA

Copyright © 2001 Academic Press

doi:10.1006/rwvb.2001.0195

Passive, active, and/or hybrid damping treatments are recognized as essential means for attenuating excessive amplitudes of oscillations, suppressing undesirable resonances, and avoiding premature fatigue failure of critical structures and structural components. The use of one form of damping treatments or another in most of the newly designed structures is becoming very common in order to meet the pressing needs for large and lightweight structures. With such damping treatments, the strict constraints imposed on present structures can be met to insure their effective operation as quiet and stable platforms for manufacturing, communication, observation, and transportation.

This article presents the different types of passive, active, and hybrid damping treatments. Emphasis is placed on presenting the fundamentals of active constrained layer damping (ACLD), one of the most commonly used class of hybrid treatments, which combines the attractive attributes of both the passive and active treatments.

Types of Passive, Active, and Hybrid Damping Treatments

Various passive, active, and hybrid damping control approaches have been considered over the years employing a variety of structural designs, damping materials, active control laws, actuators, and sensors. Distinct among these approaches are the passive, active, and hybrid damping methods.

It is important here to note that passive damping can be very effective in damping out high-frequency excitations whereas active damping can be utilized to control low-frequency vibrations, as shown in **Figure 1**. For effective control over a broad frequency band, hybrid damping methods are essential.

Passive Damping

Passive damping treatments have been successfully used, for many years, to damp out the vibration of a wide variety of structures ranging from simple beams to complex space structures. Examples of such passive damping treatments are given below.

Free and constrained damping layers Both types of damping treatments rely in their operation on the use of a viscoelastic material (VEM) to extract energy

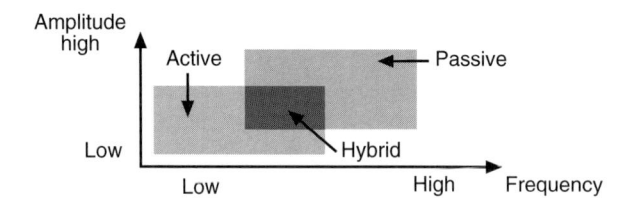

Figure 1 (See Plate 16). Operating range of various damping methods.

from the vibrating structure, as shown in **Figure 2**. In the free (or unconstrained) damping treatment, the vibrational energy is dissipated by virtue of the extensional deformation of the VEM, whereas in the constrained damping treatment more energy is dissipated through shearing the VEM.

Shunted piezoelectric treatments These treatments utilize piezoelectric films, bonded to the vibrating structure, to convert the vibrational energy into electrical energy. The generated energy is then dissipated in a shunted electric network, as shown in **Figure 3**, which is tuned in order to maximize the energy dissipation characteristics of the treatment. The electric networks are usually resistive, inductive, and/or capacitive.

Damping layers with shunted piezoelectric treatments In these treatments, as shown in **Figure 4**, a piezoelectric film is used to constrain passively the deformation of a viscoelastic layer which is bonded to a vibrating structure. The film is also used as a part of a shunting circuit tuned to improve the damping characteristics of the treatment over a wide operating range.

Magnetic constrained layer damping (MCLD) These treatments rely on arrays of specially arranged permanent magnetic strips that are bonded to viscoelastic damping layers. The interaction between the magnetic strips can improve the damping characteristics of the treatments by enhancing either the compression or the shear of the viscoelastic damping layers, as shown in **Figure 5**.

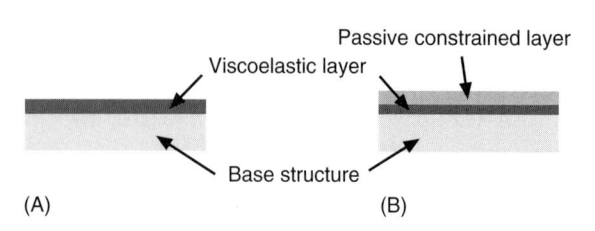

Figure 2 (See Plate 17). Viscoelastic damping treatments. (A) Free; (B) constrained.

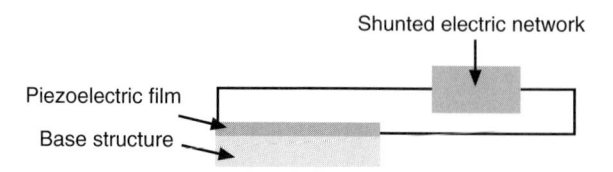

Figure 3 (See Plate 18). Shunted piezoelectric treatments.

In the compression MCLD configuration of **Figure 5A**, the magnetic strips (1 and 2) are magnetized across their thickness. Hence, the interaction between the strips generates magnetic forces that are perpendicular to the longitudinal axis of the beam. These forces subject the viscoelastic layer to cross the thickness loading, which makes the treatment act as a Den Hartog dynamic damper. In the shear MCLD configuration of **Figure 5B**, the magnetic strips (3 and 4) are magnetized along their length. Accordingly, the developed magnetic forces, which are parallel to the beam longitudinal axis, tend to shear the viscoelastic layer. In this configuration, the MCLD acts as a conventional constrained layer damping treatment whose shear deformation is enhanced by virtue of the interaction between the neighboring magnetic strips.

Damping with shape memory fibers This damping mechanism relies on embedding superelastic shape memory fibers in the composite fabric of the vibrating structures, as shown in **Figure 6A**. The inherent hysteretic characteristics of the shape memory alloy (SMA), in its superelastic form, are utilized to dissipate the vibration energy. The amount of energy dissipated is equal to the area enclosed inside the stress–strain characteristics (**Figure 6B**). This passive mechanism has been successfully used for damping out the vibration of a wide variety of structures, including large structures subject to seismic excitation.

Active Damping

Although the passive damping methods described above are simple and reliable, their effectiveness is limited to a narrow operating range because of the significant variation of the damping material

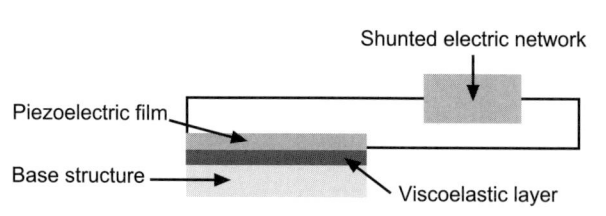

Figure 4 (See Plate 19). Damping layers with shunted piezoelectric treatments.

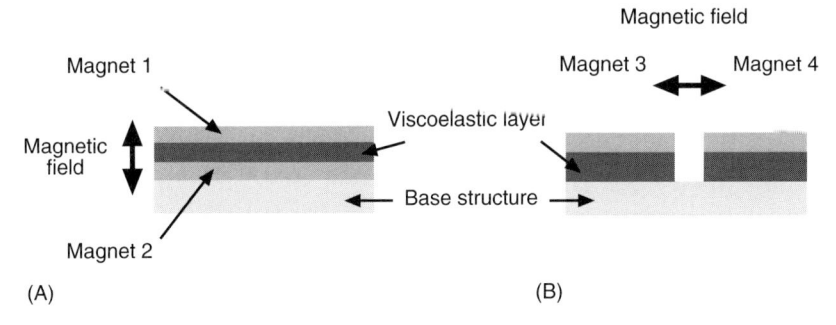

Figure 5 (See Plate 20). Configurations of the MCLD treatment. (A) Compression MCLD; (B) shear MCLD.

properties with temperature and frequency. It is therefore difficult to achieve optimum performance with passive methods alone, particularly over wide operating conditions. Hence various active damping methods have been considered. All of these methods utilize control actuators and sensors of one form or another. The most common types are made of piezoelectric films bonded to the vibrating structure, as shown in **Figure 7**.

This active control approach has been successfully used for damping out the vibration of a wide variety of structures ranging from simple beams to more complex space structures.

Hybrid Damping

Because of the limited control authority of the currently available active control actuators, and because of the limited effective operating range of passive control methods, treatments which are a hybrid combination of active damping and passive damping treatments have been considered. Such hybrid treatments aim at using various active control mechanisms to augment the passive damping in a way that compensates for its performance degradation with temperature and/or frequency. These treatments also combine the simplicity of passive damping with the effectiveness of active damping in order to insure an optimal blend of the favorable attributes of both damping mechanisms.

Active constrained layer damping This class of treatments is a blend between a passive constrained layer damping and active piezoelectric damping, as shown in **Figure 8**. The piezofilm is actively strained so as to enhance the shear deformation of the viscoelastic damping layer in response to the vibration of the base structure.

Active Piezoelectric damping composites (APDC) In this class of treatments, an array of piezoceramic rods embedded across the thickness of a viscoelastic

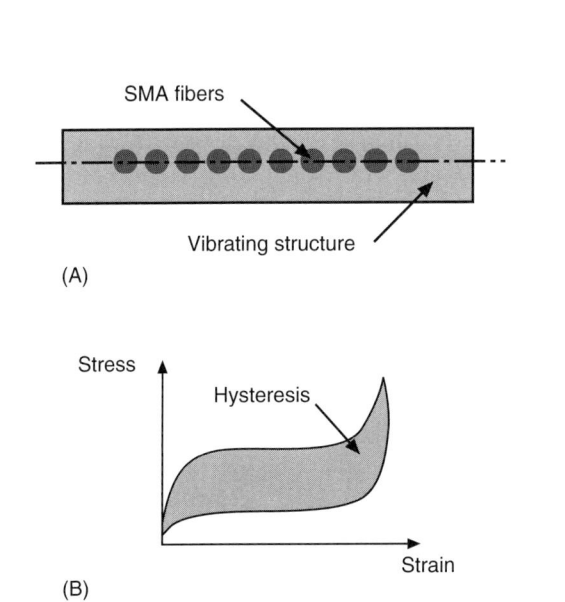

Figure 6 (See Plate 21). Damping with shape memory fibers. (A) SMA-reinforced structure; (B) superelastic characteristics.

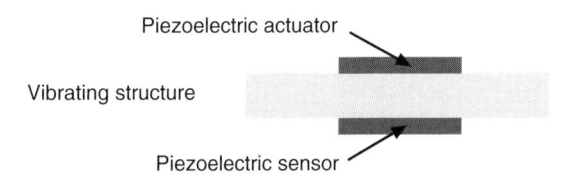

Figure 7 (See Plate 22). Active damping.

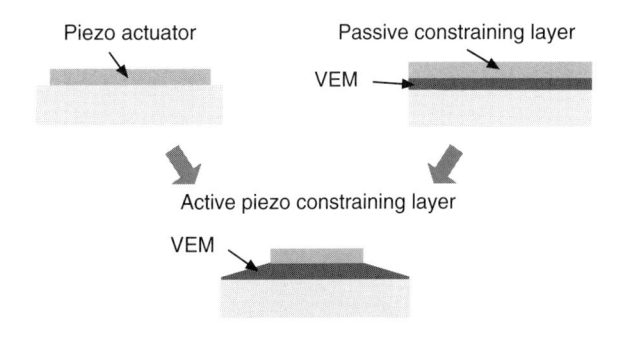

Figure 8 (See Plate 23). Active constrained layer damping treatment.

polymeric matrix is electrically activated to control the damping characteristics of the matrix that is directly bonded to the vibrating structure, as shown in **Figure 9**. The figure displays two arrangements of the APDC. In the first arrangement, the piezorods are embedded perpendicular to the electrodes to control the compressional damping and in the second arrangement, the rods are obliquely embedded to control both the compressional and shear damping of the matrix.

Electromagnetic damping composites (EMDC) In this class of composites, a layer of viscoelastic damping treatment is sandwiched between a permanent magnetic layer and an electromagnetic layer, as shown in **Figure 10**. The entire assembly is bonded to the vibrating surface to act as a smart damping treatment. The interaction between the magnetic layers, in response to the structural vibration, subjects the viscoelastic layer to compressional forces of proper magnitude and phase shift. These forces counterbalance the transverse vibration of the base structure and enhance the damping characteristics of the viscoelastic material. Accordingly, the EMDC acts in effect as a tunable Den Hartog damper with the base structure serving as the primary system, the electromagnetic layer acting as the secondary mass, the magnetic forces generating the adjustable stiffness characteristics, and the viscoelastic layer providing the necessary damping effect.

Active shunted piezoelectric networks In this class of treatments, shown in **Figure 11**, the passive shunted electric network is actively switched on and off in response to the response of the structure/network system in order to maximize the instantaneous energy dissipation characteristics and minimize the frequency-dependent performance degradation.

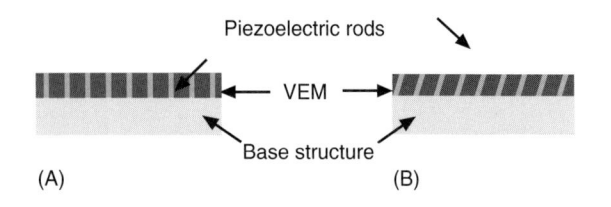

Figure 9 (See Plate 24). Active piezoelectric damping composites. (A) Perpendicular rods; (B) Inclined rods.

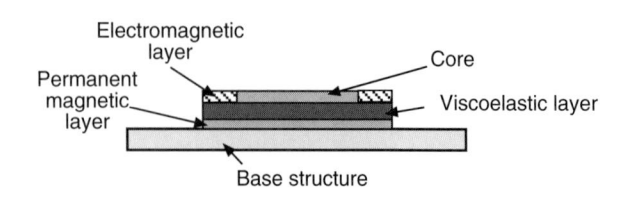

Figure 10 (See Plate 25). Electromagnetic damping composite (EMDC).

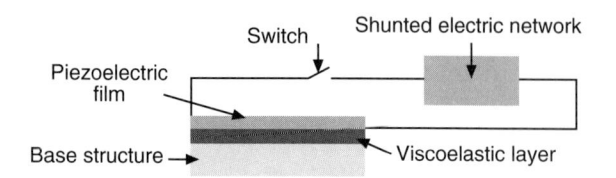

Figure 11 (See Plate 26). Damping layers with shunted piezoelectric treatments.

Basics and Characteristics of a Typical Hybrid Treatments

Emphasis is placed here on presenting the theory and the performance characteristics of one of the most widely used class of active/passive hybrid damping treatments, the ACLD treatment.

The ACLD treatment is a new class of hybrid damping treatments which has a high energy dissipation-to-weight ratio compared to conventional constrained or unconstrained damping layer configurations.

The ACLD consists of a conventional passive constrained layer damping which is augmented with efficient active control means to control the strain of the constrained layer, in response to the structural vibrations, as shown in **Figure 12**. The viscoelastic damping layer is sandwiched between two piezoelectric layers. The three-layer composite ACLD when bonded to a vibrating surface acts as a smart constraining layer damping treatment with built-in sensing and actuation capabilities. The sensing is provided by the piezoelectric layer which is directly bonded to the vibrating surface. The actuation is generated by the other piezoelectric layer which acts as an active constraining layer. With appropriate strain control, through proper manipulation of the sensor output, the shear deformation of the viscoelastic damping layer can be increased, the energy dissipation mechanism can be enhanced, and structural vibration can be damped out.

In this manner, the ACLD provides a viable means for damping out the vibration as it combines the attractive attributes of passive and active controls. This makes the ACLD particularly suitable for critical applications where the damping-to-weight ratio is important, e.g., aircraft and automobiles.

Concept of active constrained layer damping The effect of interaction between the sensor and the actuator on the operation of the ACLD can best be understood by considering the motion experienced by

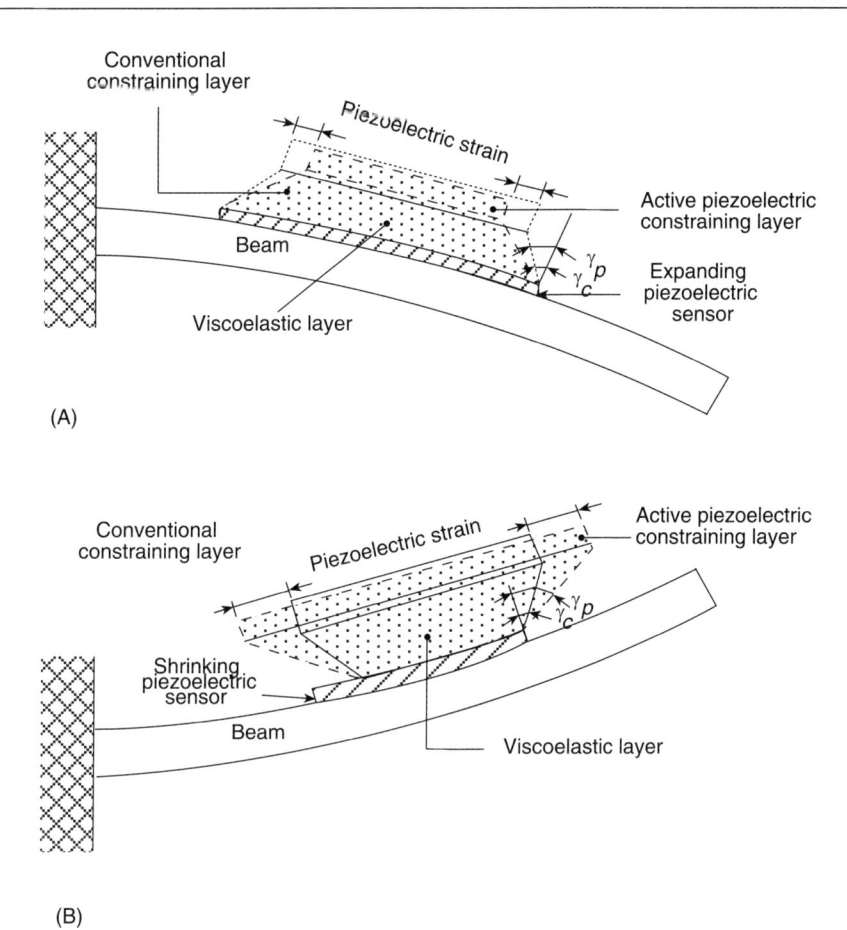

Figure 12 (A, B) Operating principle of the active constrained layer damping.

a beam during a typical vibration cycle. In **Figure 12A**, as the beam moves downward away from its horizontal equilibrium position, the sensor which is bonded to the outer fibers of the beam will be subjected to tensile stresses, generating accordingly a positive voltage, V_s, by the direct piezoelectric effect. If the sensor voltage is amplified, its polarity is reversed and the resulting voltage, V_c, is fed back to activate the piezoelectric constraining layer; the constraining layer will shrink by virtue of the reverse piezoelectric effect. The shrinkage of the active constraining layer results in a shear deformation angle, γ_p, in the viscoelastic layer, which is larger than the angle, γ_c, developed by a conventional passive constraining layer, as indicated in **Figure 12A**.

Similarly, **Figure 12B** describes the operation of the ACLD during the upward motion of the beam. During this part of the vibration cycle, the top fibers of the beam as well as the piezoelectric sensor experience compressive stresses and a negative voltage is generated by the sensor. Direct feedback of the sensor signal to the active constraining layer makes it extend and accordingly increases the shear deformation angle to γ_p as compared to γ_c for the conventional constraining layer.

The increase of the shear deformation of the viscoelastic layer, throughout the vibration cycle, is accompanied by an increase in the energy dissipated. Furthermore, the shrinkage (or expansion) of the piezoelectric layer during the upward motion (or during the downward motion) produces a bending moment on the beam which tends to bring the beam back to its equilibrium position. Therefore, the dual effect of the enhanced energy dissipation and the additional restoring bending moment will quickly damp out the vibration of the flexible beam. This dual effect, which does not exist in conventional constrained damping layers, significantly contributes to the damping effectiveness of the smart ACLD. In this manner, the smart ACLD consists of a conventional passive constrained layer damping which is augmented with the described dual effect actively to control the strain of the constrained layer, in response to the structural vibrations. With appropriate strain control strategy, the shear deformation of the viscoelastic damping layer can be increased, the energy dissipation mechanism can be enhanced, and vibration can be damped out. One possible strategy is direct feedback of the sensor voltage to power the active constraining layer. Other strategies will rely on feeding back both the sensor voltage and its derivative to obtain proportional and derivative control action. With such a strategy, additional damping can be imparted to the vibrating beam system and the versatility of active controls can be utilized to improve considerably the damping characteristics of the ACLD.

Therefore, the ACLD relies in its operation on a blend between the attractive attributes of both active and passive controls. In other words, the simplicity and reliability of passive damping are combined with the low weight and high efficiency of active controls to achieve high damping characteristics over broad frequency bands. Such characteristics are essential to the optimal damping of vibration.

Also, it is essential to note that the ACLD provides an excellent and practical means for controlling the vibration of massive structures with the currently available piezoelectric actuators without the need for excessively large actuation voltages. This is due to the fact that the ACLD properly utilizes the piezoelectric actuator to control the shear in the soft viscoelastic core – a task which is compatible with the low control authority capabilities of the piezoelectric materials currently available.

Furthermore, it is important to note that the ACLD configuration described in **Figure 12** is only one of many possible configurations. For example, the ACLD can be arranged in a multilayer configuration or in discrete patches distributed at optimal locations over the vibrating structure. Other possible configurations are only limited by our imagination.

Finite element modeling of shells treated with ACLD Figure 13 shows the transverse cross-section of a thin cylindrical shell treated partially with ACLD treatments. The shell has longitudinal length a, average circumferential length b, average radius R, and thickness h. The thickness of the viscoelastic core and the piezoelectric actuator are h_c and h_p, respectively.

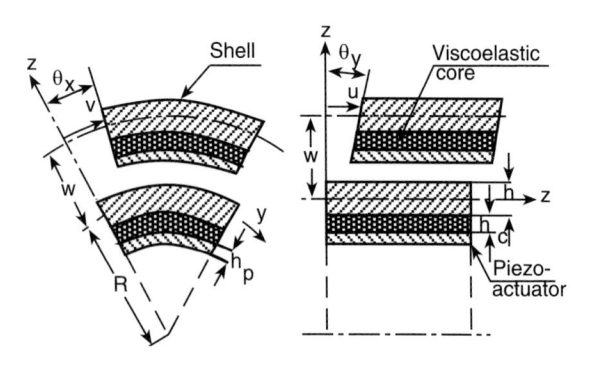

Figure 13 Geometrical and kinematical parameters of the shell/ACLD system.

Displacement fields In case of constrained layer damping analysis, the layer-wise theories have been used continuously. For sandwiched or laminated structures, the use of the layer-wise theories involves more generalized displacement variables than a single first-order shear deformation theory needs. This results in a large number of global degrees-of-freedom in the case of finite element analysis, which eventually proves to be cost-ineffective.

Hence, the longitudinal and circumferential deformations u and v, respectively, at any point of the shell/ACLD system, are represented by the first-order shear deformation theory (FSDT):

$$u(x,y,z,t) = u_0(x,y,t) + z\theta_x(x,y,t)$$
$$v(x,y,z,t) = v_0(x,y,t) + z\theta_y(x,y,t) \quad [1]$$

in which x and y are the longitudinal and circumferential coordinates, respectively; z is the radial coordinate; u_0 and v_0 are the generalized displacements at any point of the reference plane ($z = 0$); θ_x, θ_y are the rotations of the normal to the reference plane about the y- and x- axes, respectively. According to the FSDT the radial displacement, w is assumed to be constant through the thicknesses of the cylinder, the viscoelastic core and the piezoelectric actuator.

The generalized displacement variables are separated into translational $\mathbf{d}_t^T$ and rotational variables $\mathbf{d}_r^T$ as follows:

$$\mathbf{d}_t = [u_0 \ v_0 \ w]^T ; \quad \mathbf{d}_r = [\theta_x \ \theta_y]^T \quad [2]$$

Strain displacement relations Applying Donnell's theory for strain displacement relations and using eqn [1], the strain vector at any point of the shell/ACLD system can be expressed as:

$$\varepsilon = \mathbf{Z}_1 \varepsilon_t + \mathbf{Z}_2 \varepsilon_r \quad [3a]$$

where the generalized strain vectors ε_t, ε_r are given by:

$$\varepsilon_t = \left[\frac{\partial u_0}{\partial x} \ \frac{\partial v_0}{\partial y} + \frac{w}{R} \ \frac{\partial u_0}{\partial y} \ \frac{\partial v_0}{\partial x} \ \frac{\partial w}{\partial x} \ \frac{\partial w}{\partial y} - \frac{v_0}{R} \right]^T$$

$$\varepsilon_r = \left[\frac{\partial \theta_x}{\partial x} \ \frac{\partial \theta_y}{\partial y} \ \frac{\partial \theta_x}{\partial y} + \frac{\partial \theta_y}{\partial x} \ \theta_x \ \theta_y \right]^T \quad [3b]$$

and the transformation matrices $\mathbf{Z}_1$ and $\mathbf{Z}_2$ are given in Appendix A.

Constitutive equations The constitutive equation for the material of the piezoelectric constraining layer is:

$$\sigma^3 = C^3[\varepsilon^3 - \varepsilon_p] \quad [4]$$

where σ represents the stress vector; C is the elastic constant matrix, the superscript 3 denotes the piezoelectric layer number 3 and the piezoelectrically induced-strain vector ε_p for a biaxially polarized actuator layer is given by:

$$\varepsilon_p = \bar{\varepsilon}_p V \quad [5]$$

with:

$$\bar{\varepsilon}_p = (1/h_p)[d_{31} \quad d_{32} \quad 0 \quad 0 \quad 0]^T$$

where d_{31}, d_{32} denote the piezoelectric strain constants and V, the applied voltage, respectively.

The constitutive equations for the materials of the shell and the viscoelastic core are given by:

$$\sigma^L = C^L \varepsilon^L, \quad (L = 1, 2) \quad [6]$$

where the superscripts 1 and 2 identify the shell and the viscoelastic core, respectively.

System energies The potential energy T_p of the overall system is given by:

$$T_p = \frac{1}{2} \sum_{L=1}^{3} \int_{h_{L+1}}^{h_L} \int_0^b \int_0^a \varepsilon^{L^T} \sigma^L \, dx \, dy \, dz$$

$$- \int_0^b \int_0^a \Delta_{z=h_1}^T f^s \, dx \, dy \quad [7]$$

and the kinetic energy T_k is given by:

$$T_k = \frac{1}{2} \sum_{L=1}^{3} \int_{h_{L+1}}^{h_L} \int_0^b \int_0^a \rho^L \dot{\Delta}^{L^T} \dot{\Delta}^L \, dx \, dy \, dz \quad [8]$$

in which ρ with superscript L is the mass density of the Lth layer, Δ, is the vector of absolute displacements (u, v, w) and f^s is the vector of surface traction.

The whole continuum is discretized by an eight-noded two-dimensional isoparametric element. The generalized displacement vectors for the ith ($i = 1, 2, \ldots, 8$) node of the element are then given by:

$$d_{ti} = [u_{0i} \quad v_{0i} \quad w]^T ; \quad d_{ri} = [\theta_{xi} \quad \theta_{yi}]^T \quad [9]$$

and the generalized displacement vector at any point within the element is given by:

$$d_t = N_t d_t^e ; \quad d_r = N_r d_r^e \quad [10]$$

where:

$$d_t^e = [d_{t1}^T \quad d_{t2}^T \quad \ldots \quad d_{t8}^T]^T$$
$$d_r^e = [d_{r1}^T \quad d_{r2}^T \quad \ldots \quad d_{r8}^T]^T$$
$$N_t = [N_{t1} \quad N_{t2} \quad \ldots \quad N_{t8}]$$
$$N_r = [N_{r1} \quad N_{r2} \quad \ldots \quad N_{r8}]$$
$$N_{ti} = n_i I_t \quad \text{and} \quad N_{ri} = n_i I_r$$

with I_t and I_r being the identity matrices of appropriate dimension and n_i the shape functions of natural coordinates.

Using eqns [2], [3b], [9], and [10], the generalized strain vectors at any point within the element can be expressed as:

$$\varepsilon_t = B_t d_t^e, \quad \varepsilon_r = B_r d_r^e \quad [11]$$

in which the nodal strain displacement matrices are given by:

$$B_t = [B_{t1} \quad B_{t2} \quad \ldots \quad B_{t8}]$$

and:

$$[B_r] = [B_{r1} \quad B_{r2} \quad \ldots \quad B_{r8}]$$

The various submatrices B_{ti} and B_{ri} are given in Appendix A.

Finally, using eqns [3a], [4]–[6], and [11] in eqns [7] and [8], the strain energy of the eth typical shell element augmented with ACLD treatment can be expressed as:

$$T_p^e = \frac{1}{2} \int_0^{a^e} \int_0^{b^e} \left[d_t^{e^T} B^T D_{tt} B d_t^e + d_t^{e^T} B_t^T D_{tr} B_r d_r^e \right.$$

$$+ d_r^{e^T} B_r^T D_{rt} B_t d_t^e + d_r^{e^T} B_r^T D_{rr} B_r d_r^e$$

$$- \int_{h_4}^{h_3} \left(d_t^{e^T} B_t^T Z_1^T C^3 \varepsilon_p + d_r^{e^T} B_2^{p^T} Z_2^T C^3 \varepsilon_p \right) dz \Bigg] dx \, dy$$

$$- \int_0^{a^e} \int_0^{b^e} \left(d_t^{e^T} N_t^T + d_t^{e^T} N_r^T \right) f^s \, dx \, dy$$

$$[12]$$

and the kinetic energy of the element can be obtained as:

$$T_k^e = \frac{1}{2}\int_0^{a^e}\int_0^{b^e} (\rho^1 h + \rho^2 h_c + \rho^3 h_p)\dot{\mathbf{d}}_t^{e^T}\mathbf{N}_t^T\mathbf{N}_t\dot{\mathbf{d}}_t^e \, dx \, dy$$

[13]

in which a^e and b^e are the longitudinal and circumferential lengths of the element, respectively. The various rigidity matrices $\mathbf{D}_{tt}, \mathbf{D}_{tr}, \mathbf{D}_{rt}$ and $\mathbf{D}_{rr}$ appearing in eqn [12] are given in Appendix B. Since the present study deals with thin shell analysis, the rotational inertia of the element has been neglected when estimating the kinetic energy of the element.

Equations of motion Applying Hamilton's variational principle, the following equations of motion for the element are obtained:

$$\mathbf{M}^e\ddot{\mathbf{d}}_t^e + \mathbf{K}_{tt}^e\mathbf{d}_t^e + \mathbf{K}_{tr}^e\mathbf{d}_r^e = \mathbf{F}_{at}^e V + \mathbf{F}_t^e \quad [14]$$

$$\mathbf{K}_{rt}^e\mathbf{d}_t^e + \mathbf{K}_{rr}^e\mathbf{d}_r^e = \mathbf{F}_{ar}^e V + \mathbf{F}_r^e \quad [15]$$

in which the various elemental matrices $\mathbf{M}^e, \mathbf{K}_{tt}^e, \mathbf{K}_{tr}^e, \mathbf{K}_{rt}^e$ and $\mathbf{K}_{rr}^e$; the electroelastic coupling vectors $\mathbf{F}_{at}^e, \mathbf{F}_{ar}^e$, and the excitation force vectors $\mathbf{F}_t^e, \mathbf{F}_r^e$ are defined in Appendix B. It may be mentioned here that in case of an element without ACLD treatment, the electroelastic coupling vectors $\mathbf{F}_{at}^e$ and $\mathbf{F}_{ar}^e$ do not appear in eqns [14] and [15].

The elemental equations are assembled in such a manner to obtain the global equations of motion so that each actuator can be activated separately as follows:

$$\mathbf{M}\ddot{\mathbf{X}}_t + \mathbf{K}_{tt}\mathbf{X}_t + \mathbf{K}_{tr}\mathbf{X}_r = \sum_{j=1}^n \mathbf{F}_{at}^j V^j + \mathbf{F}_t$$

$$\mathbf{K}_{rt}\mathbf{X}_t + \mathbf{K}_{rr}\mathbf{X}_r = \sum_{j=1}^n \mathbf{F}_{ar}^j V^j + \mathbf{F}_r$$

[16]

where $\mathbf{M}$ and $\mathbf{K}_{tt}, \mathbf{K}_{tr}, \mathbf{K}_{rr}$ are the global mass and stiffness matrices; $\mathbf{X}_t, \mathbf{X}_r$ are the global nodal generalized displacement coordinates; $\mathbf{F}_t, \mathbf{F}_r$ are the global nodal force vectors corresponding to translational and rotational coordinates, n is the number of ACLD patches and for the jth ACLD patch the global nodal electroelastic coupling vectors are given by:

$$\mathbf{F}_{at}^j = \sum_m \mathbf{F}_{at}^e \quad \text{and} \quad \mathbf{F}_{ar}^j = \sum_m \mathbf{F}_{ar}^e \quad [17]$$

with m being the number of elements per ACLD treatment. Invoking the boundary conditions, the global rotational degrees-of-freedom can be condensed to obtain the global equations of motion in terms of the global translational degrees-of-freedom only as follows:

$$\mathbf{M}\ddot{\mathbf{X}}_t + \mathbf{K}^*\mathbf{X}_t = \sum_{j=1}^n [\mathbf{F}_{at}^j - \mathbf{K}_{tr}\mathbf{K}_{rr}^{-1}\mathbf{F}_{ar}^j]V^j + \mathbf{F} \quad [18]$$

in which $\mathbf{K}^* = \mathbf{K}_{tt} - \mathbf{K}_{tr}\mathbf{K}_{rr}^{-1}\mathbf{K}_{rt}$ and $\mathbf{F} = \mathbf{F}_t - \mathbf{K}_{tr}\mathbf{K}_{rr}^{-1}\mathbf{F}_r$.

Control law In the active control strategy, each actuator is supplied with the control voltage proportional to the radial velocity at the points on the outer surface of the cylinder which correspond to midpoint of the free width of the ACLD patches. Thus the control voltage for the jth actuator can be expressed in terms of the derivatives of the nodal global degrees-of-freedom as:

$$V^j = -K_d^j \mathbf{e}^j \dot{\mathbf{X}}_t \quad [19]$$

where K_d^j is the controller gain and $\mathbf{e}^j$ is a unit vector with unity as the only nonzero element corresponding to that global degree-of-freedom, the derivative of which is fed back to the actuator.

Substitution of eqn [19] into eqn [18] yields the final damped equations of motion as:

$$\mathbf{M}\ddot{\mathbf{X}}_t + \mathbf{K}^*\mathbf{X}_t + \mathbf{C}_d\dot{\mathbf{X}}_t = \mathbf{F} \quad [20]$$

where:

$$\mathbf{C}_d = \sum_{j=1}^n K_d^j \left[\mathbf{F}_{at}^j - \mathbf{K}_{tr}\mathbf{K}_{rr}^{-1}\mathbf{F}_{ar}^j\right]\mathbf{e}^j$$

Eqn [20] can be formulated to compute the frequency response function (FRF) when the shell/ACLD system is subjected to harmonic excitations using the mechanical impedance approach.

Performance of shells with ACLD treatment In this section, the performance of shells treated with two ACLD patches is evaluated by comparing their FRFs using the finite element model developed earlier in this article. The numerical results are compared with experimental results.

Materials The shell considered is made of a stainless steel shell: Young's modulus $E_1 = 210 \, \text{GN m}^{-2}$, Poisson's ratio $v_1 = 0.3$ and the density $\rho^1 = 7800 \, \text{kg m}^{-3}$. The material of the acrylic-based viscoelastic core has a complex shear modulus $G_2 = 20(1 + 1.0i) \, \text{MN m}^{-2}$ and density $\rho^2 = 1140 \, \text{kg m}^{-3}$. The piezoelectric actuator is an active polymeric film (PVDF). Its Young's modulus E_3, Poisson ratio v_3 and

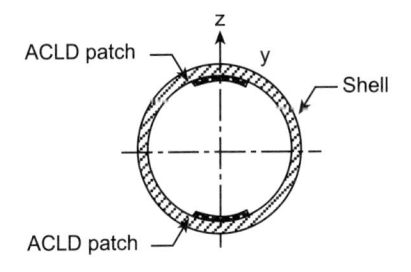

Figure 14 Cross-section of the shell/ACLD system.

density ρ^3 are 2.25 GN m^{-2}, 0.28 and 1800 kg m^{-3}, respectively. The values of the piezoelectric strain constants d_{31} and d_{32} are 23×10^{-12} and 3×10^{-12} m V^{-1}, respectively.

Numerical and experimental results A clamped-free cylinder with $R = 0.1016$ m, $a = 1.27$ m and $h = 0.635$ mm is chosen to demonstrate the performance of the ACLD as compared to passive constrained layer damping (PCLD) treatments. Experiments are conducted using this cylinder. The arrangement of the experimental setup is schematically described in **Figure 14**. Two patches of ACLD treatment are used which are bonded 180° apart on the inner surface of the cylinder, as shown in **Figure 15**. The length and width of each patch are 0.508 m and 0.1016 m, respectively. The shell is excited with swept sinusoidal excitations at its free end by the speaker. The outputs of two collocated accelerometers are sent to phase shifters and then to power amplifiers. The output of the amplifiers is used to activate the piezo-constraining layers. The velocity feedback is insured by properly adjusting the phases of the phase shifters.

The natural frequencies of the cylinder are computed using the finite element model and are also experimentally determined. **Table 1** shows a comparison between the theoretical predictions and the experimental results. The numerical estimations are slightly higher than the experimental values. This can be attributed to the fact that the clamped end is not ideal and geometrical imperfections are inherent due to the manufacturing process of the cylinder.

Figures 16A and **16B** display the numerically and experimentally determined FRFs of the shell/ACLD system at the free end of the cylinder $(a, 0, h/2)$, respectively. The figure shows the amplitudes of radial displacements when the piezoelectric-constraining layers in both the patches are passive and active, with different control gains. These figures clearly

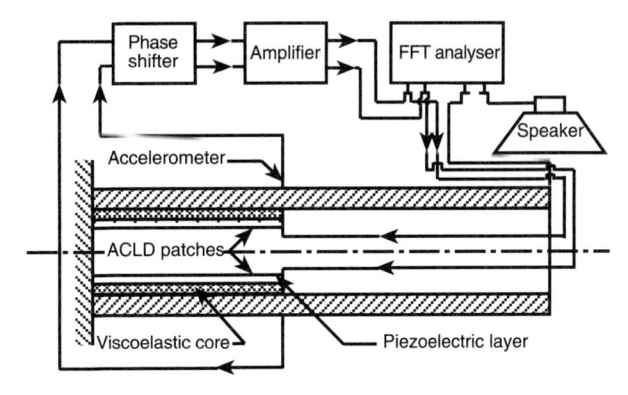

Figure 15 Schematic diagram of the shell/ACLD system.

Table 1 Comparison of natural frequencies (Hz) of the shell/ACLD system

Mode	Finite element model	Experiment
(1, 2)	54.06	51.2
(1, 1)	113.62	113

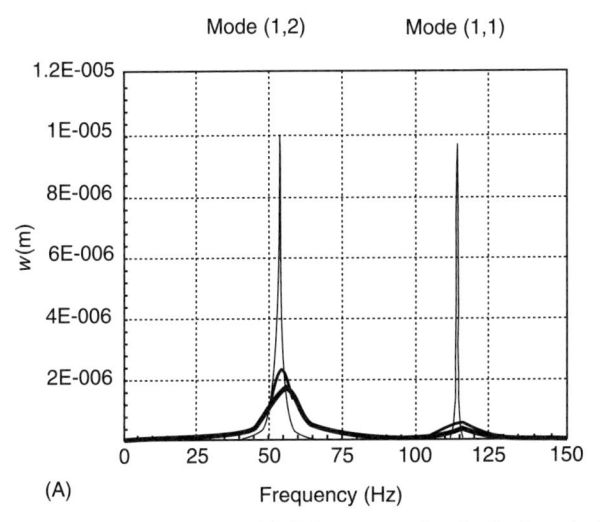

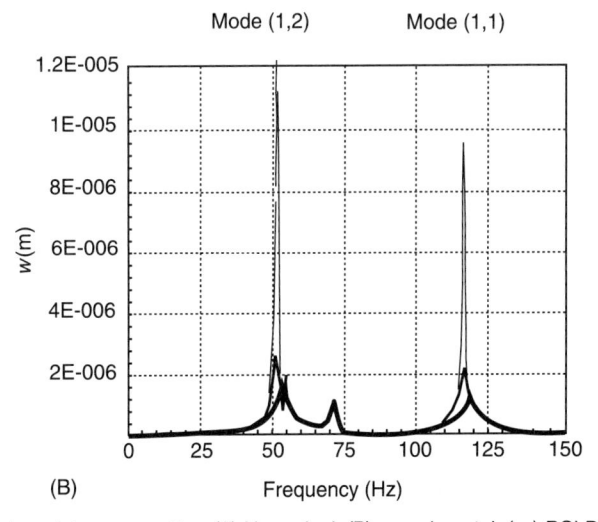

Figure 16 FRF of the shell/ACLD system when both piezoelectric patches are active. (A) Numerical; (B) experimental. (−) PCLD, (−) $k_d = 5E5$, (—) $k_d = 9E5$.

reveal that the ACLD treatments significantly improve the damping characteristics of the shell over the PCLD. A comparison between parts (A) and (B) shows that maximum values of the uncontrolled radial displacement of the shell obtained numerically at the point $(a, 0, h/2)$, considered here match with that obtained experimentally with close accuracy. The controlled responses indicate that the attenuated amplitudes for the first mode (1, 2) of vibration differ negligibly. In the case of the second mode (1, 1), the numerical predictions are slightly lower than the experimental results. The numerical predictions of the maximum voltages required to control the mode (1, 2) match closely with the experimental results, as presented in **Table 2**. However, the numerical predictions for the control voltages for the mode (1, 1) are higher than those obtained experimentally. In order to identify the modes and the modal contents after the control, the surface of the cylinder is scanned using a laser vibrometer, as shown in **Figures 17** and **18** for modes (1, 2) and (1, 1), respectively. It is clear from these figures that significant attenuation is obtained with the activation of the controller.

Figure 19 illustrates the numerical and experimental results for the case when only one of the piezoelectric-constraining layers is active. In this case also, the numerical predictions matched well with the experimental results. However, numerical predictions for the control voltages differ from the experimental results but are within the acceptable limit, as shown in **Table 2**.

Concluding Remarks

A brief description is presented here of the different types of passive, active, and hybrid damping treatments that have been successfully applied to damping out the vibration of a wide variety of structures. Emphasis is placed on presenting the theory and performance characteristics of one of the most commonly used hybrid treatments, the ACLD. It is important to note that the hybrid ACLD treatment has been shown to be very effective in damping out broadband vibrations as compared to conventional PCLD treatment, without the need for excessively high control voltages.

It is equally important to emphasize that, although the concepts presented are utilized to control the vibration of shells, these concepts have been successfully applied to the control of vibrations and sound radiation of beams, plates, and shells. In all these studies, it is shown that hybrid treatments provide effective and globally stable means for controlling structural vibration and acoustics without any adverse effects, such as those generated by control spillover or even by failure of the controller. Accordingly, these treatments provide a failsafe means of attenuating vibration and noise radiation, which is essential for reliable operation of critical structures. More importantly, the hybrid treatments can also be designed to insure robust operation in the presence of uncertainties in the structural parameters or in the operating conditions.

Appendix A

Forms of Matrices Z_1, Z_2, B_{ti} and B_{ri}

Transformation Matrices Z_1 and Z_2

The explicit forms of the transformation matrices, Z_1 and Z_2 appearing in eqn [3] are:

$$Z_1 = \begin{bmatrix} 1 & 0 & 0 & 0 & 0 \\ 0 & 1 & 0 & 0 & 0 \\ 0 & 0 & 1 & 0 & 0 \\ 0 & 0 & 0 & 1 & 0 \\ 0 & 0 & 0 & 0 & 1 \end{bmatrix}$$

and:

$$Z_2 = \begin{bmatrix} z & 0 & 0 & 0 & 0 \\ 0 & z & 0 & 0 & 0 \\ 0 & 0 & z & 0 & 0 \\ 0 & 0 & 0 & 1 & 0 \\ 0 & 0 & 0 & 0 & 1-\dfrac{z}{R} \end{bmatrix}$$

Submatrices B_{ti} and B_{ri}

The submatrices B_{ti} and B_{ri} ($i = 1, 2, \ldots 8$) of the nodal strain displacement matrices, B_t and B_r, in eqn [11] are obtained as:

Table 2 Maximum control voltage

Mode	Approach	One actuator gain $= 3.5 \times 10^5$	One actuator gain $= 8 \times 10^5$	Two actuators gain 5×10^5	Two actuators gain $= 9 \times 10^5$
(1, 2)	Finite element model	82	100	49.01	61.23
	Experiment	70	80	47.82	64.00
(1, 1)	Finite element model	50	70	34.60	40.00
	Experiment	64	84	20.20	26.00

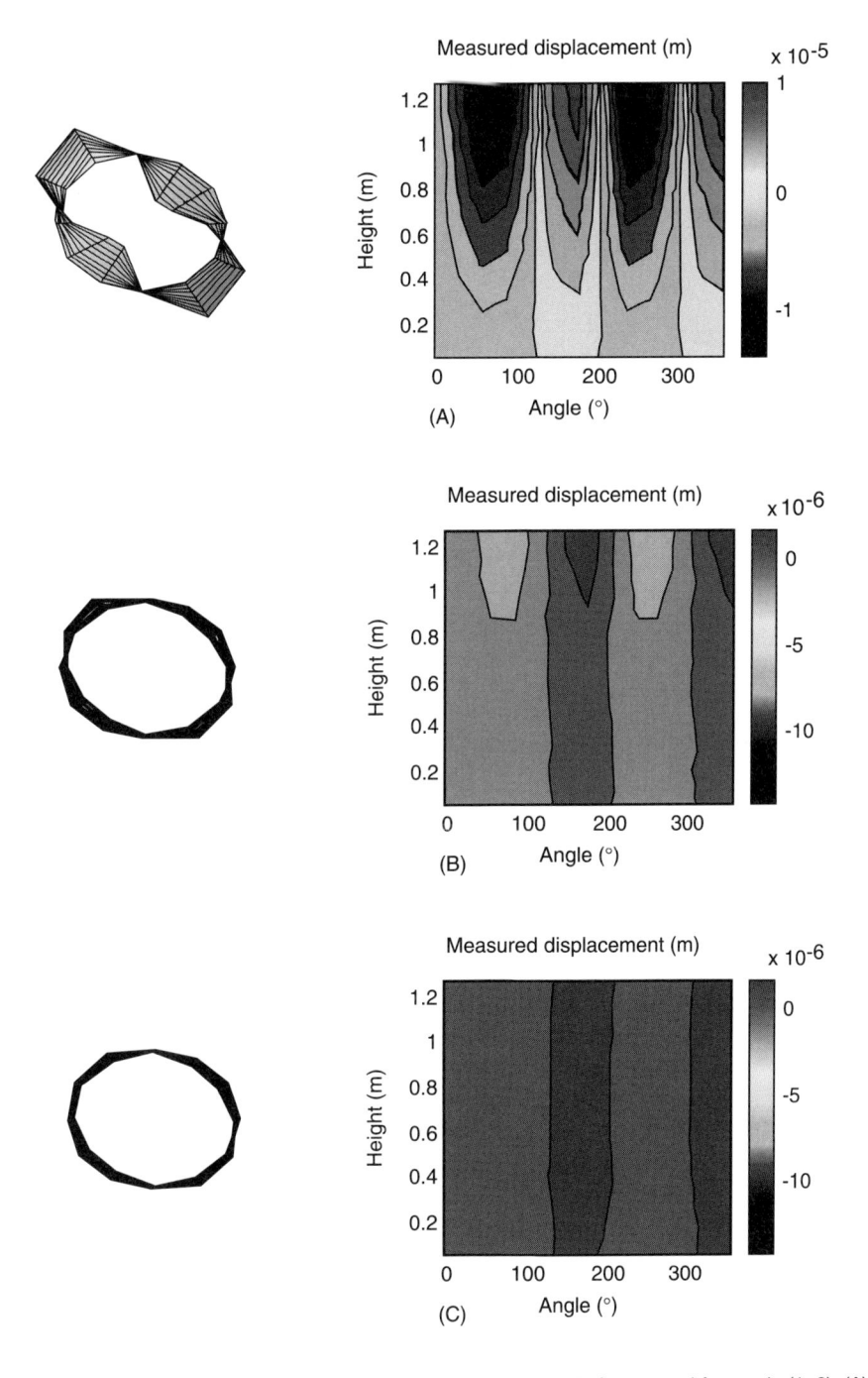

Figure 17 (See Plate 27). Experimental results using laser vibrometer before and after control for mode (1, 2). (A) PLCD; (B) ACLD (one actuator); (C) ACLD (two actuators).

$$\mathbf{B}_{ti} = \begin{bmatrix} n_{i,x} & 0 & 0 \\ 0 & n_{i,y} & 1/R \\ n_{i,y} & n_{i,x} & 0 \\ 0 & 0 & n_{i,x} \\ 0 & -1/R & n_{i,y} \end{bmatrix} \quad \text{and} \quad \mathbf{B}_{ri} = \begin{bmatrix} n_{i,x} & 0 \\ 0 & n_{i,y} \\ n_{i,y} & n_{i,x} \\ 1 & 0 \\ 0 & 1 \end{bmatrix} \quad \begin{array}{l} \text{where:} \\ \\ n_{i,x} = \dfrac{\partial n_i}{\partial x} \quad \text{and} \quad n_{i,y} = \dfrac{\partial n_i}{\partial y} \end{array}$$

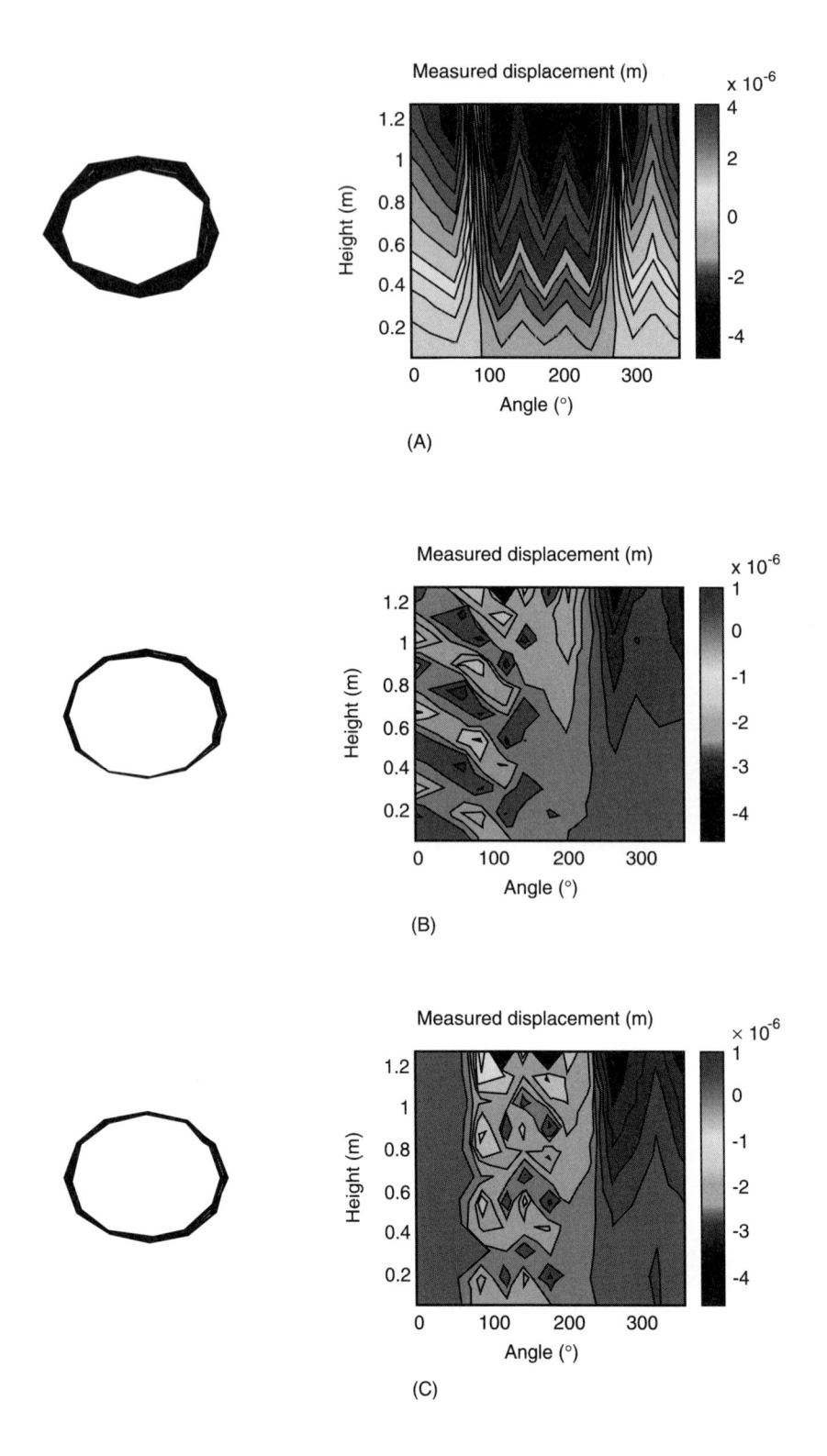

Figure 18 (See Plate 28). Experimental results using laser vibrometer before and after control for mode (1, 1). (A) PCLD; (B) ACLD (one actuator); (C) ACLD (two actuators).

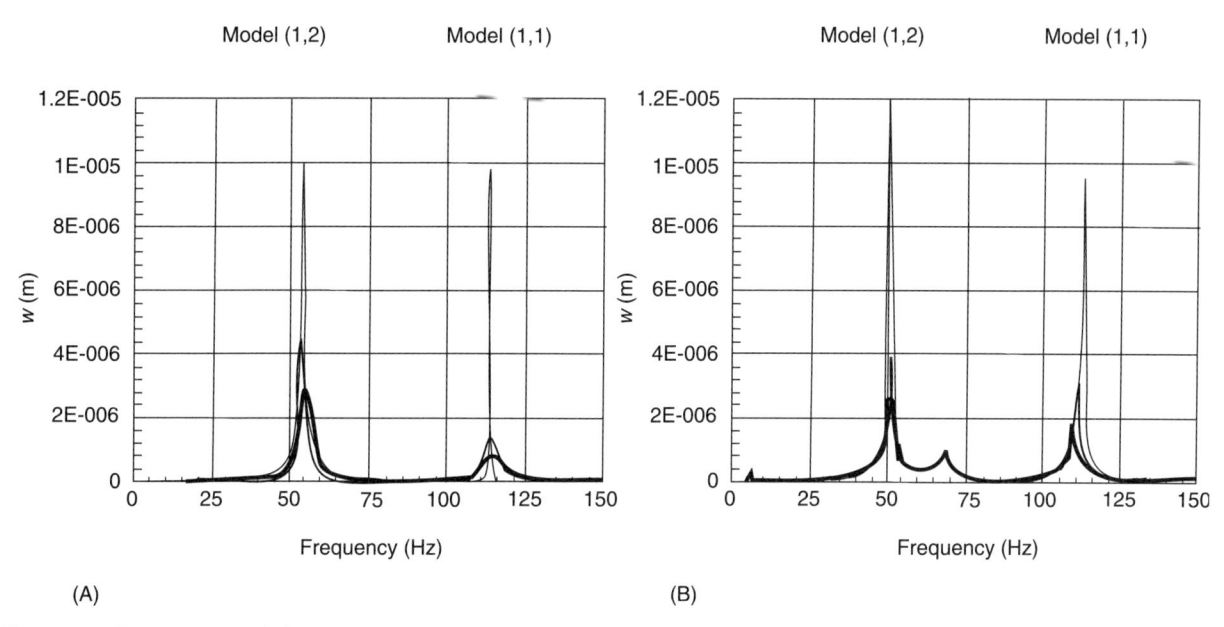

Figure 19 FRF of the shell/ACLD system when one of the piezoelectric layers is active. (A) Numerical; (B) experimental. (—) PCLD, (—) $k_d = 3.5E5$, (—) $k_d = 8E5$.

Appendix B
Rigidity and Elemental Matrices

Rigidity Matrices

The various rigidity matrices $\mathbf{D}_{tt}$, $\mathbf{D}_{tr}$, $\mathbf{D}_{rt}$ and $\mathbf{D}_{rr}$ appearing in eqn [12] are defined as:

$$\mathbf{D}_{tt} = \sum_{L=1}^{3} \int_{h_{L+1}}^{h_L} \mathbf{Z}_1^T \mathbf{C}^L \mathbf{Z}_1 \, dz,$$

$$\mathbf{D}_{tr} = \sum_{L=1}^{3} \int_{h_{L+1}}^{h_L} \mathbf{Z}_1^T \mathbf{C}^L \mathbf{Z}_2 \, dz,$$

$$\mathbf{D}_{rt} = \mathbf{D}_{tr}^T$$

and:

$$\mathbf{D}_{rr} = \sum_{L=1}^{3} \int_{h_{L+1}}^{h_L} \mathbf{Z}_2^T \mathbf{C}^L \mathbf{Z}_2 \, dz$$

Elemental Matrices

In eqns [14] and [15], the elemental mass matrix $\mathbf{M}^e$ and the elemental stiffness matrices $\mathbf{K}_{tt}^e$, $\mathbf{K}_{tr}^e$, $\mathbf{K}_{rt}^e$ and $\mathbf{K}_{rr}^e$ are defined as:

$$\mathbf{M}^e = \frac{1}{2} \int_0^{a^e} \int_0^{b^e} (\rho^1 h + \rho^2 h_c + \rho^3 h_p) \mathbf{N}_t^T \mathbf{N}_t \, dx \, dy$$

$$\mathbf{K}_{tt}^e = \frac{1}{2} \int_0^{a^e} \int_0^{b^e} \mathbf{B}_t^T \mathbf{D}_{tt} \mathbf{B}_t \, dx \, dy$$

$$\mathbf{K}_{tr}^e = \frac{1}{2} \int_0^{a^e} \int_0^{b^e} \mathbf{B}_t^T \mathbf{D}_{tr} \mathbf{B}_r \, dx \, dy$$

$$\mathbf{K}_{rt}^e = \mathbf{K}_{tr}^{e^T}$$

and:

$$\mathbf{K}_{rr}^e = \frac{1}{2} \int_0^{a^e} \int_0^{b^e} \mathbf{B}_r^T \mathbf{D}_{rr} \mathbf{B}_r \, dx \, dy$$

The elemental electroelastic coupling vectors $\mathbf{F}_{at}^e$ and $\mathbf{F}_{ar}^e$ as well as the elemental exciting force vectors $\mathbf{F}_t^e$ and $\mathbf{F}_r^e$ appearing in eqns [14] and [15] are defined as:

$$\mathbf{F}_{at}^e = \frac{1}{2} \int_{h_4}^{h_3} \int_0^{a^e} \int_0^{b^e} \mathbf{B}_t^T \mathbf{Z}_1^T \mathbf{C}^3 \bar{\boldsymbol{\varepsilon}}_p \, dx \, dy$$

$$\mathbf{F}_{ar}^e = \frac{1}{2} \int_{h_4}^{h_3} \int_0^{a^e} \int_0^{b^e} \mathbf{B}_r^T \mathbf{Z}_2^T \mathbf{C}^3 \bar{\boldsymbol{\varepsilon}}_p \, dx \, dy$$

$$\mathbf{F}_t^e = \int_0^{a^e} \int_0^{b^e} \mathbf{N}_t^T \mathbf{f}^s \, dx \, dy$$

and:

$$\mathbf{F}_r^e = \int_0^{a^e} \int_0^{b^e} \mathbf{N}_r^T \mathbf{f}^s \, dx \, dy$$

Nomenclature

a	longitudinal length
b	circumferential length
B	submatrix
C	elastic constant matrix
D	rigidity matrix
E	Young's modulus
h	thickness
I	identity matrix
R	radius
T	energy
u	longitudinal deformation
v	circumferential deformation
V	voltage
Z	transform matrix
ε	strain vector
σ	stress vector
ρ	density
ν	Poisson's ratio

See plates 16, 17, 18, 19, 20, 21, 22, 23, 24, 25, 26, 27, 28.

See also: **Absorbers, vibration**; **Active control of civil structures**; **Active control of vehicle vibration**; **Noise**, Noise radiated by baffled plates; **Noise**, Noise radiated from elementary sources; **Vibration generated sound**, Fundamentals; **Vibration generated sound**, Radiation by flexural elements; **Vibration isolation, applications and criteria**; **Viscous damping**.

Further Reading

Arafa M and Baz A (2000) Dynamics of active piezoelectric damping composites. *Journal of Composites Engineering: Part B* 31: 255–264.

Baz A (1996) Active constrained layer damping. US Patent 5,485,053.

Baz A (1997) Magnetic constrained layer damping. *Proceedings of 11th Conference on Dynamics and Control of Large Structures*, May, Blacksburg, VA, pp. 333–344.

Baz A (1999) Passive magnetic damping composites. In Szulc JH and Rodellar J (eds) *Smart Structures*, pp. 19–25. Dordrecht: Kluwer.

Baz A and Poh S (2000) Performance characteristics of the magnetic constrained layer damping. *Journal of Shock and Vibration* 7: 81–90.

Ghoneim H (1995) Bending and twisting vibration control of a cantilever plate via electromechanical surface damping. In Johnson C (ed.) *Proceedings of Smart Structures and Materials Conference*, vol. SPIE-2445, pp. 28–39.

Greaser E and Cozzarelli F (1993) Full cyclic hysteresis of a Ni-Ti shape memory alloy. *Proceedings of DAMPING '93*, 24–26 February, San Francisco, CA, vol. 2, pp. ECB 1–28.

Lesieutre GA (1998) Vibration damping and control using shunted piezoelectric materials. *The Shock and Vibration Digest* 30 (3): 187–195.

Nashif A, Jones D and Henderson J (1985) *Vibration Damping*. New York: Wiley.

Omer A and Baz A (2000) Vibration control of plates using electromagnetic compressional damping treatment. *Journal of Intelligent Materials and Structures Systems* (to be published).

Preumont A (1997) *Vibration Control of Active Structures*. Dordrecht: Kluwer.

Ray M, Oh J and Baz A (2000) Active constrained layer damping of thin cylindrical shells. *Journal of Sound and Vibration*, 240: 921–935.

Reader W and Sauter D (1993) Piezoelectric composites for use in adaptive damping concepts. *Proceedings of DAMPING '93*, 24–26 February, San Francisco, CA, pp. GBB 1–18.

DATA ACQUISITION

R B Randall and M J Tordon, University of New South Wales, Sydney, Australia

Copyright © 2001 Academic Press

doi:10.1006/rwvb.2001.0142

Background

In order to process analog signals digitally it is necessary to convert them into digital form. The object of a data acquisition system is to collect and record data from physical phenomena in the real world, which by nature are continuous in both amplitude and time. This can be a labour-intensive process, which in its simplest form involves reading values from instruments and recording the observations on a data sheet. Even in this simple form we are representing a continuous signal by values recorded at discrete times with a limited number of discrete digits. Advances in digital computers have provided efficient and fast means of collecting and processing signals represented and stored in digital form. **Figure 1** shows a schematic diagram of a general data acquisition (DAQ) system.

A transducer changes a physical phenomenon into an electrical signal. The aim is to produce electrical signals which represent the physical phenomena investigated, while at the same time minimizing the

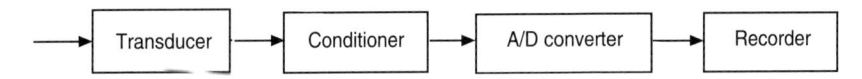

Figure 1 Schematic diagram of a general data acquisition system. A/D, analog-to-digital.

influence of other physical phenomena present in the real world.

The conditioner modifies the signal from the transducer to provide signals suitable for the analog-to-digital (A/D) converter. The conditioning often includes amplification, isolation, and filtering.

The output from the conditioner is applied to the input of the A/D converter, which may contain additional circuits such as a multiplexer and sample-and-hold amplifier, followed by the converter itself.

The A/D converter is followed by a recorder (digital memory). The recorded data should be in a form suitable for processing and/or presentation, and are thus sometimes preprocessed in real time before storage.

The way in which the data acquisition should be done depends to a large extent on what the data files are to be used for, and some confusion has arisen because of the differences between the two main areas of application:

1. Applications analogous to those of a digital oscilloscope. Here, the signals themselves are to be recorded and viewed, with the most important requirement being the visual similarity to the original signal. It is perhaps desired to extract parameters such as peak and root mean square (RMS) values. This case also applies to data-logging of very slowly changing signals such as temperature and (mean) pressure, perhaps in order to use the current value in calculations triggered by an external event.
2. Applications involving further processing (such as filtering) of the data, with actual or implicit transformation to the frequency domain.

Aliasing

In the latter case it is imperative that the signals be low-pass filtered before digitization so as to avoid problems with aliasing. This is a phenomenon whereby high frequencies before sampling appear as lower frequencies after sampling, and there is no way of knowing what the original frequency was. It can be avoided by low-pass filtering the signals before sampling, but on the other hand this can distort their shape. **Figure 2** shows two high-frequency signals that have been sampled at too low a sampling frequency, so that the sampled signal appears to be at low frequency.

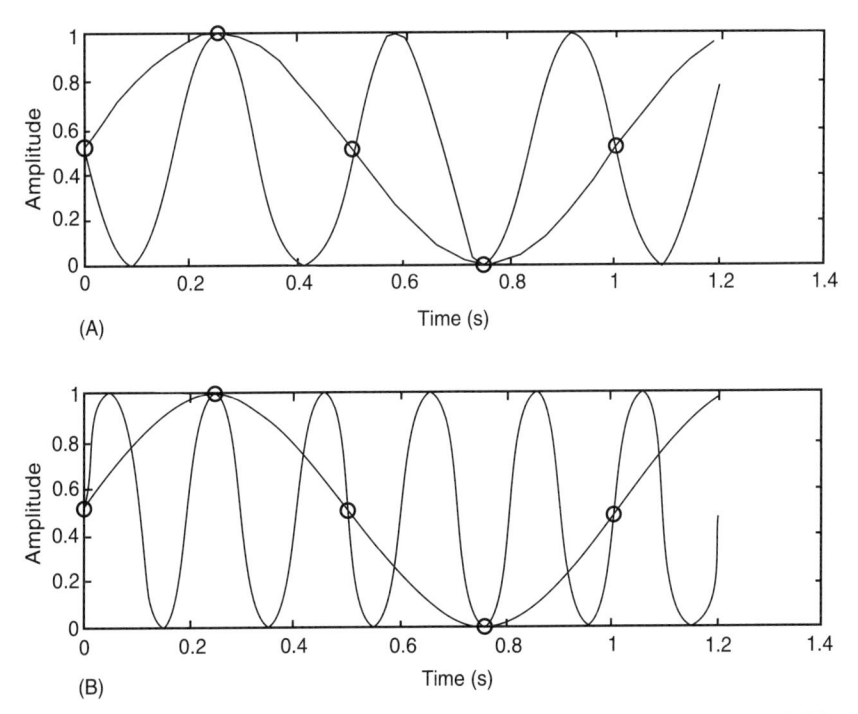

Figure 2 Higher frequencies interpreted as lower frequencies by sampling at too low a frequency (4 Hz). (A) 3 Hz as 1 Hz; (B) 5 Hz as 1 Hz.

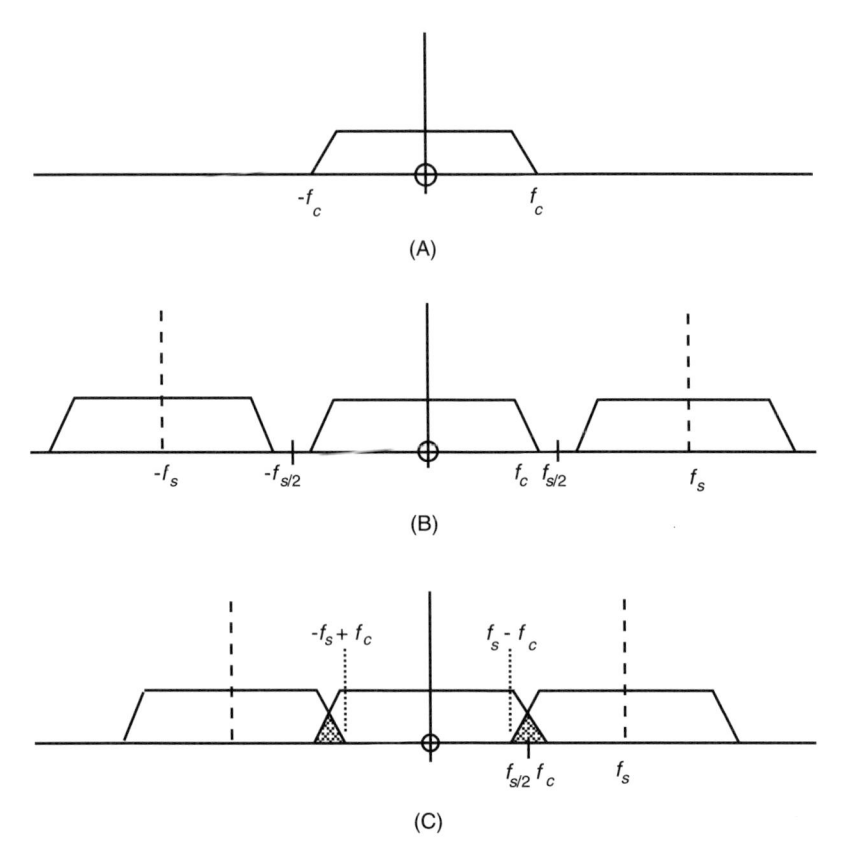

Figure 3 (A) Spectrum of a continuous band-limited signal with maximum frequency f_c. (B) Spectrum of a digitized signal with sampling frequency $f_s > 2f_c$. (C) Spectrum of a digitized signal with sampling frequency $f_s < 2f_c$. Note that frequency $\pm f_c$ is aliased to $\pm(f_s - f_c)$.

To understand the relationship between the frequencies before and after sampling it is best to look at the sampling process in the frequency domain. Sampling a signal can be treated as multiplying it by a train of delta functions (sampling functions) with spacing Δt, where sampling frequency $f_s = 1/\Delta t$. By the convolution theorem, the spectrum of the sampled signal is thus the convolution of the spectra of the original signal and that of the delta function train (itself a train of delta functions with spacing f_s). As illustrated in **Figure 3**, this means that the spectrum of the sampled signal is a periodic repetition of the original spectrum with period f_s in the frequency domain. Thus, if the original signal is low-pass filtered so as to contain no components outside the range $\pm f_s/2$, there will be no overlap in the periodic spectrum, and in principle it would be possible to regain the original signal by using a low-pass filter to remove all components outside the range $\pm f_s/2$. On the other hand, if the original signal bandwidth is greater than $\pm f_s/2$, there will be an overlap in the spectrum of the sampled signal, with no possibility of separating the mixed components with a low-pass filter. It can be seen that frequencies which can be confused with frequency f_0 are given by the formula $nf_s \pm f_0$ (in **Figure 2**, with $f_s = 4$ Hz, both 5 Hz and 3 Hz give samples corresponding to 1 Hz).

Figure 4 illustrates the differences between the two types of applications as mentioned above. It shows an asymmetric rectangular wave, sampled with and without low-pass filtering.

Without low-pass filtering, the appearance is much closer to that of the original signal, and for example a more accurate estimate of the peak value could be obtained. On the other hand, it is evident that its spectrum has a lot of invalid frequency components (since it is periodic it would only contain harmonics of the fundamental frequency) and these would remain in any further processing of the signal. The low-pass filtered signal no longer has the original rectangular shape (it can be considered to have the step response of the low-pass filter after each step) but on the other hand the spectrum is now correct up to the cut-off frequency of the low-pass filter, and further processing would be valid up to that frequency.

Antialiasing Filter Specifications

Where the signal is to be further processed, and thus aliasing errors must be limited, it has become fairly

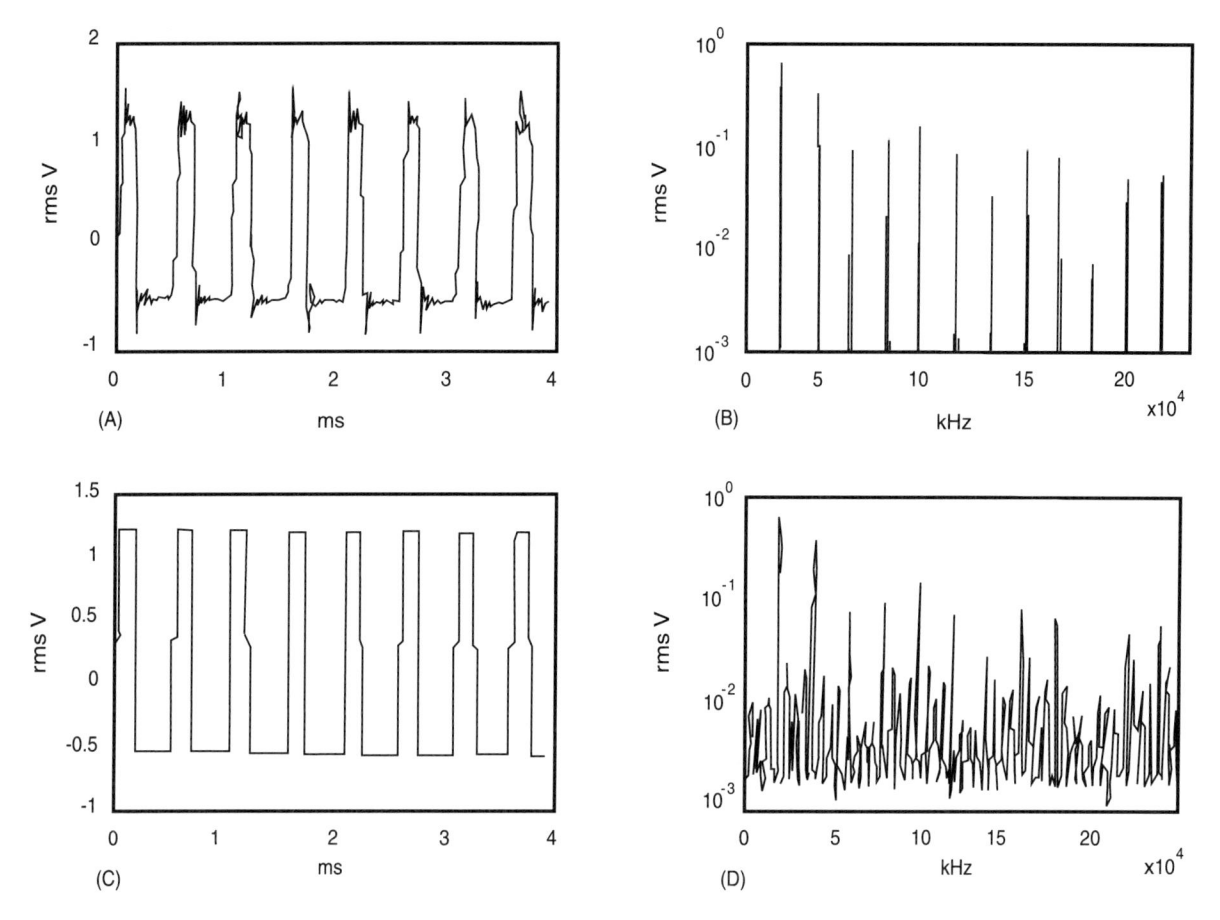

Figure 4 Digitization with and without antialiasing filters. (A) Time signal with filter. (B) Spectrum with filter. (C) Time signal without filter. (D) Spectrum without filter.

standard to utilize approximately 80% of the calculated spectrum, i.e., 400 lines out of 512 calculated, or 800 lines out of 1024. In **Figure 5**, it is seen that this requires a low-pass filter roll-off of 120 dB per octave, to ensure that components folded back into the measurement band are attenuated by at least 77 dB, to put them below the dynamic range. Elliptic filters are commonly used to achieve such roll-off, with a fixed stopband rejection. The first antialiasing filter used on the signal before digitization must of course be analog, but after fault-free digitization, digital filters can be used to filter to a lower frequency, before decimating the signal to a lower sample rate. As digital signal processing (DSP) boards become cheaper and faster, it is becoming more attractive to use simpler initial analog filters, or even the natural fall-off of the signal itself at, for example, 6 dB per octave, and digitize the signals at a very high sample rate. To avoid having to store massive amounts of data, the signals can then be

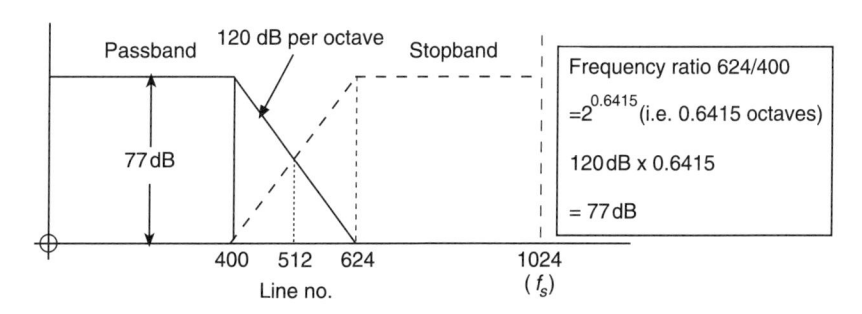

Figure 5 Antialiasing filter specification (120 dB per octave) to give > 77 dB attenuation in the passband (400 lines) with sampling frequency f_s at line no. 1024.

decimated to a much lower sample rate, using digital filters operating in real time. The delta-sigma converter (see below) represents one example of this.

The major advantage of digital filters, apart from any cost comparisons, is that they can be made to have identical properties, and thus give perfect matching between channels that are converted simultaneously. Moreover, multiplexing of the same hardware units can be used to increase the number of effective filters, and the order of the filters, as the speed of the units increases. In the latter connection, it is worth pointing out that by being able to calculate at twice the rate of the incoming data, it is possible for a single digital low-pass filter to decimate by any power of two, by repeatedly decimating by a factor of two (one octave). Each time the filter reduces the frequency content of the signal (relative to the sampling frequency) to one-half of the current value, this permits the sampling frequency to be halved by discarding every second sample. This also means that, in each lower octave, only half the number of samples have to be processed in a given time, and if the number per unit time in the highest octave is S, the total number to be processed is $S(1 + \frac{1}{2} + \frac{1}{4} + ...) = 2S$.

A/D Converter Specifications

The specifications of the A/D converter determine the major performance parameters of the data acquisition system. The basic parameters of the A/D converter (ADC) include the resolution, maximum sampling rate, accuracy, dynamic range, nonlinearity (and many others).

The basic specifications of the DAQ system provide information on both capabilities and accuracy of the system. They include number of channels, sampling rate, resolution, and input range.

Analog inputs In practice we collect more than one variable so the system should include provisions for collecting data from multiple channels. The input channels can be single-ended or differential. Single-ended inputs are referenced to a common ground and due to a higher level of noise they are typically used for high-level (>1 V) input signals. Differential inputs on the other hand respond to a potential difference between two terminals. Noise or other signals present in both terminals, referred to as common-mode voltage, are cancelled out. The term common-mode voltage range describes the ability of the system to reject common-mode voltage signals.

Sampling rate Sampling rate specifies how often the conversion takes place. A faster sampling rate will provide a better representation of the original signal. To avoid aliasing, the signal must be sampled at least at twice the frequency of the maximum frequency component of the input signal, as discussed above. However, for the applications where no antialiasing filters are used, the sampling rate is typically much higher, say 10–20 times the highest frequency to be viewed, so as not to distort the appearance of the signal.

Sampling methods Due to the cost of the ADC, the DAQ system often contains only one ADC and uses a multiplexer to switch between channels. The data channels may then be sampled in sequence so that only one conversion is taking place at any one time. This method of sampling is suitable for applications where the exact time relationship between sampled signals is not important. If the time (and phase) relationships between input signals are important the inputs must be sampled simultaneously. To achieve simultaneous sampling we need sample-and-hold circuitry for each input channel, which will allow simultaneous sampling and subsequent sequential conversion of instantaneous values of input signals.

Multiplexing Multiplexing is a technique for measuring several signals with a single ADC. The multiplexer selects one input channel at a time and routes the signal to the ADC for digitizing. The effective sampling rate per channel is reduced by a factor equal to the number of channels sampled.

Resolution Resolution is the smallest signal increment that can be detected by a measurement system. Resolution can be expressed in bits, as a proportion, or in percent of full scale. For example, if a system has 8-bit resolution, this corresponds to one part in 256, and 0.39% of full scale. The higher the resolution, the smaller the detectable voltage change. **Figure 6** shows a sine wave obtained by an ideal 3-bit ADC, which divides the analog range into eight divisions. Each division is represented by a binary code between 000 and 111. It can be seen that quantization of the input signal introduces irreversible loss of information.

Range Range refers to the minimum and maximum voltage levels that the ADC can span. The range of the ADC can be changed by selecting a different amplifier gain. The range, resolution, and gain of the DAQ system determine the smallest detectable change in voltage. This change in voltage corresponds to the least significant bit (LSB) of the digital number and is often called the code width. The ideal code width is found by dividing the voltage range by the expression (gain times number of codes). For example, DAQ

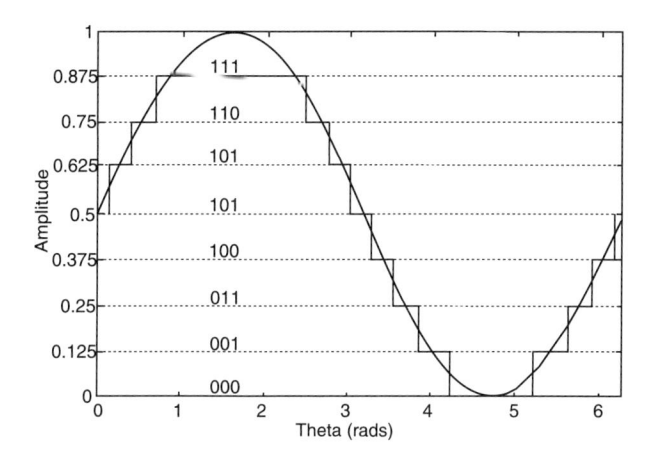

Figure 6 Example of quantized sine wave (version b, 3 bits).

system with input range 0–10'V, gain 100 and 16-bit ADC will have code width 10 V / (100 × 65536), i.e., 1.5 μV.

Critical Parameters of Analog Inputs

The parameters mentioned above, such as sampling rate, resolution, and range, describe the overall limitation of the DAQ system. Additional parameters, such as nonlinearity, relative accuracy, settling time, and noise specifications, are needed to describe the actual performance of the DAQ system.

Nonlinearity The plot of the voltage versus the output code for an ideal ADC should be a straight line. Deviation from the straight line is specified as nonlinearity. Several terms are used to express this property. Differential nonlinearity (DNL) is a measure in LSB of the worst-case deviation of the analog code widths from their ideal value of 1 LSB. A perfect DAQ system has a DNL of 0 LSB. A good DAQ system will have a DNL within ±0.5 LSB. Nonlinearity (or integral nonlinearity) is a measure of the worst-case deviation from the ideal transfer function (a straight line) of the system expressed in percent of full-scale range (FSR).

Relative accuracy The relative accuracy of an ADC is a measure in LSBs of the worst-case deviation from the ideal transfer function of a straight line. It can be obtained by sweeping an applied voltage through the range and digitizing it. Plotting the digitized points results in an apparent straight line. If we subtract an actual straight line from the apparent straight line, as shown in **Figure 7**, we can see the deviations from zero across the range. The maximum deviation from zero as a proportion of the FSR is the relative accuracy of the DAQ system. The relative accuracy includes all nonlinearity and quantization errors. It

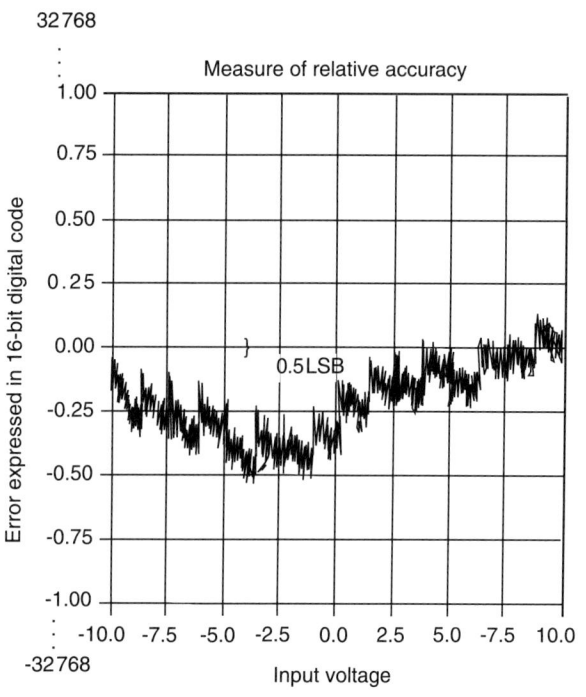

Figure 7 Deviation of encoded values from the ideal straight line. Courtesy of National Instruments 1996.

does not include offset and gain errors of the circuitry feeding the ADC.

Settling time In systems with a multiplexer we have to take into account the time required for the signals at the input of the ADC to settle after changing channels. The duration required by the system to settle to a specified accuracy is called the settling time.

Discretization Errors, Dynamic Range

The process of converting a discrete-time signal into a digital signal by expressing each sample value as a finite number of digits is called quantization. The process of quantization is a many-to-one mapping and is both nonlinear and noninvertible and as such results in loss of information. Loss of information is directly linked to the number of discrete levels available for the process. The resulting digital information is stored in words of finite length expressed in bits. A wordlength of B bits can represent $N = 2^B$ different discrete levels. The signal-to-noise ratio (SNR) is used to quantify the effect of quantization noise (errors) resulting from the finite word length of the conversion.

Figure 8 shows an example of a quantized sine wave using a wordlength of 3 bits. The normalized (from 0 to 1) continuous signal is represented by eight codes representing equally spaced steps. The step size

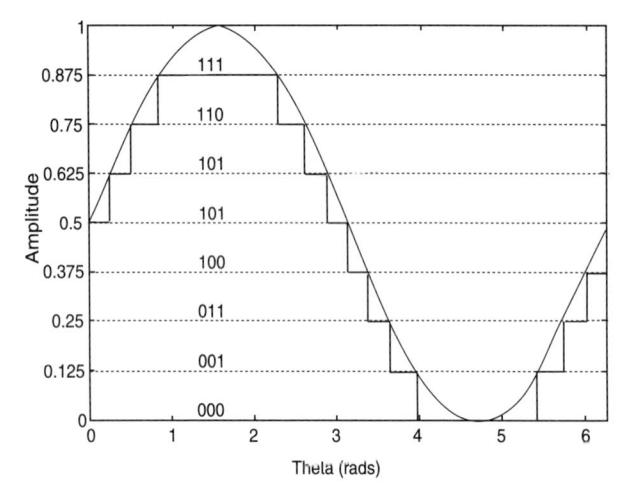

Figure 8 Example of quantized sine wave (version a, 3 bits).

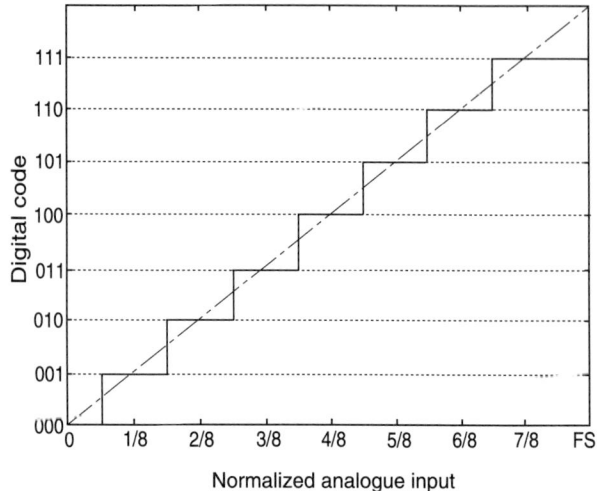

Figure 9 Analog-to-digital conversion (3 bits).

is called the quantum Δ and is equal to the width of the LSB of the ADC. The quantum represents the distance between any two successive code levels. The step size Δ of a quantizer with word length B bits and assuming unity-normalized range is $\Delta = 2^{-B}$.

Note that, for the allocation of codes as shown in **Figure 8**, the quantization error q for any particular conversion is in the range $0 \leq q \leq \Delta$ with the mean value $+0.5\Delta$. It is however possible to shift the range covered by each individual code by 0.5Δ, thus rounding the input conversion in the range $-0.5\Delta \leq q \leq +0.5\Delta$ with zero mean value. An example of this arrangement is shown in **Figure 6**. The disadvantage of this arrangement is that the first code 000 now covers a range of 0.5Δ while the last code 111 covers the range 1.5Δ. This loss of symmetry however does not cause major problems for an ADC with a large number of quantization levels. **Figure 9** shows the corresponding input–output characteristic of a 3-bit quantizer.

If we assume that the quantization noise is uncorrelated with the input signal sequence and that it has probability density function uniformly distributed over the range $(-0.5\Delta, +0.5\Delta)$, i.e., it can be represented by white noise, we can obtain a mathematical model of the quantizer, as shown in **Figure 10**. This model is used to express the effect of quantization noise and to calculate the SNR of the ADC as a function of word length B. The effect of the additive noise can be quantified by evaluating the signal-to-noise (power) ratio as SNR = $10 \log_{10}(P_x/P_e)$ where P_x is the input signal power and P_e is the power of the quantization noise. P_e for white noise is:

$$P_e = \frac{\Delta^2}{12} = \frac{2^{-2B}}{12} \qquad [1]$$

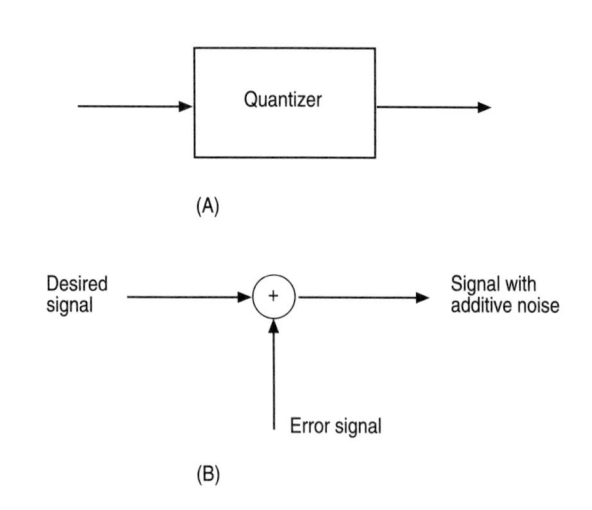

Figure 10 Mathematical model of additive noise for the quantization errors in analog-to-digital conversion. (A) Actual system; (B) mathematical model.

The input signal is often assumed to be a sinusoidal signal with magnitude spanning the whole range of the ADC. In the case of normalized input signal range the magnitude of the input signal is $\frac{1}{2}$ and the corresponding power is:

$$P_x = \frac{1/2^2}{2} = \frac{1}{8} \qquad [2]$$

The SNR of the ADC can then be expressed as:

$$\text{SNR} = 10 \log_{10}\left(\frac{12}{8} 2^{2B}\right) = 6.02 B + 1.76 \qquad [3]$$

in dB and represents the theoretically achievable value. If the input signal does not span the whole input range of the ADC, the SNR can be greatly reduced. This also shows the importance of matching the magnitude of the input signal with the input range of the ADC, using a conditioning amplifier.

The maximum value of the SNR corresponding to a sine wave spanning the whole input range of the ADC is also known as the dynamic range of the ADC.

Sigma-Delta Converters

A sigma-delta modulating ADC is a circuit that samples at a very high rate with lower resolution than is needed and, by means of feedback loops, pushes the quantization noise above the frequency range of interest. It is based on the noise-shaping technique performed by a delta modulator. The most popular form of noise-shaping quantization was given the name delta sigma modulation but due to a misunderstanding the words delta and sigma were interchanged and thus today both names delta sigma and sigma delta are in use. The out-of-band noise is typically removed by digital filters. The basic principle of operation of the converter relies on oversampling and averaging.

Conventional converters require use of antialiasing quality analog filters, high-precision analog circuits, and are susceptible to noise and interference. The advantage of the conventional ADC is the use of a relatively low sampling rate, usually the Nyquist rate of the signal (i.e., twice the signal bandwidth).

Oversampling converters can use simple and relatively high-tolerance analog components, but require fast and complex digital signal-processing stages. They modulate the analog input into a simple digital code, usually single-bit words, at a frequency much higher than the Nyquist rate. Because the sampling rate usually needs to be several orders of magnitude higher than the Nyquist rate, oversampling methods are best suited for relatively low-frequency signals. The major advantage of these converters is that they provide a low-cost conversion method with high dynamic range for low-bandwidth input signals such as high-accuracy transducer applications.

An important difference between conventional and oversampling converters concerns testing and specifying their performance. With conventional converters there is a one-to-one correspondence between the input and output sample values and hence one can describe their accuracy by comparing the corresponding input and output samples. In oversampling converters each input sample contributes to a number of output samples which are then digitally filtered.

Oversampling

Figure 5 shows the antialiasing requirements for a conventional ADC. To satisfy the Nyquist sampling theorem, and utilize a maximum of the available passband, we need a very steep transition from the passband to the stopband, which in turn requires a complex analog low-pass filter. If we oversample the input signal by using a sampling frequency f_{s2}, which is much higher than the required Nyquist frequency, as shown in **Figure 11**, then we are able to relax the requirements for the width of the transition band and the solution can be as simple as the use of the first-order low-pass filter.

Noise Considerations

The limiting factor in the resolution of an ADC is quantization noise (quantization error). Assuming that the quantization error is random it can be treated as white noise. The quantization noise power and RMS quantization voltage for an ADC is given by the equation:

$$e_{rms}^2 = \frac{q^2}{12} \quad \text{or} \quad e_{rms} = \frac{q}{\sqrt{12}} \qquad [4]$$

A quantized signal sampled at frequency f_s has all of its noise power folded into the frequency band $0 <= f <= f_s/2$. Assuming that we can model the noise as white noise, the power spectral density of the noise is given by:

$$E(f) = e_{rms}^2 \left(\frac{2}{f_s}\right) \qquad [5]$$

If we are dealing with a bandwidth of interest f_0 (i.e., the maximum frequency of interest of the input signal is f_0) we can get the expression for the RMS value n_0 of the noise in the bandwidth of interest as:

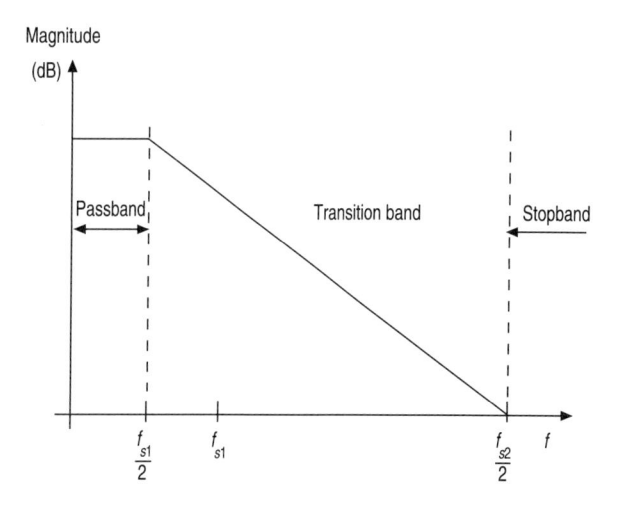

Figure 11 Antialiasing frequency response requirement for an oversampling analog-to-digital conversion.

$$n_0 = e_{rms} \left(\frac{2f_0}{f_s}\right)^{1/2} \quad [6]$$

where n_0 is in-band quantization noise. The quantity $f_s/2f_0$ is generally referred to as the oversampling ratio (OSR). The above equation shows that the oversampling theoretically reduces the inband quantization noise by the square root of the OSR. It is important to remember that this requires additonal filtering which can be done digitally so that the overall low-pass filter characteristic, which will be a combination of the antialiasing filter as shown in **Figure 11** and the digital filter required to reduce the inband noise, should have a characteristic which is close to the low-pass filter shown in **Figure 5**.

Sigma-Delta Modulator

A simple example of a first-order sigma-delta ADC is shown in **Figure 12**. A differential amplifier at the input amplifies the difference between the analog input voltage and the output of a 1-bit digital-to-analog converter (DAC). The input amplifier performs the function of a summing point. The signal then passes through an integrator which in its simplest form can be implemented as a capacitor. The output of the integrator is compared with zero level in a comparator. The comparator in turn controls the 1-bit DAC. The sigma-delta ADC, often called a charge-balancing ADC, tries to balance the charge on the integrating capacitor at zero level. This is achieved by rapid switching between a full-scale positive (+FS) and full-scale negative (−FS) voltage so that the resulting duty cycle is uniquely linked to the magnitude of the input voltage (bipolar in this case). As an example, a duty cycle of 50% would be required to balance zero input voltage, and 75% for half full-scale voltage.

A general sigma-delta ADC can be seen in **Figure 13**. A sample and hold (S/H) amplifier is used at the input of the converter to maintain a stationary input during conversion. A digital filter connected to the output of the comparator is used to convert the information about the input voltage level to the corresponding digital format (effectively a count of the duty cycle, with a count over more samples giving more output bits). It is usually the case that this smoothing filter also performs the process of decimation to reduce the output sampling frequency without loss of information of the input signal.

The integrator in the sigma-delta modulator (SDM) provides for further reduction in noise. As the quantizing noise as expressed in eqn [4] has to pass through an integrator present in the SDM the noise will be further reduced and will be proportional to $(2f_0/f_s)^{3/2}$ in the case of a first-order integrator, and $(2f_0/f_s)^{5/2}$ in the case of a second-order integrator. The noise of an Mth-order modulator will be proportional to:

$$\left(\frac{2f_0}{f_s}\right)^{M+0.5} \quad [7]$$

i.e., doubling the sampling frequency will decrease the in-band quantization noise by $3(2M + 1)$ dB. Considering the fact that one extra bit of resolution reduces the noise by 6 dB, doubling the OSR will in theory increase the resolution by $(M + 0.5)$ bits. This can lead to over 20 bits of resolution.

Data Acquisition Modes

Triggering

To avoid excessive storage of data, in particular for high-frequency signals, it is often desirable to be able to base the storage on detection of some event such as

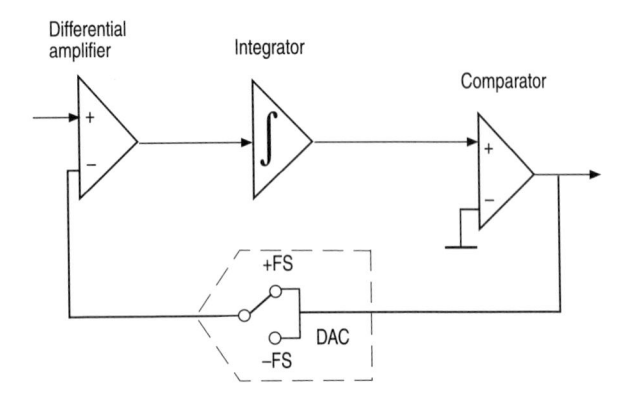

Figure 12 Simple charge-balancing analog-to-digital conversion. +FS, full-scale positive; −FS, full-scale negative; DAC, digital-to-analog converter.

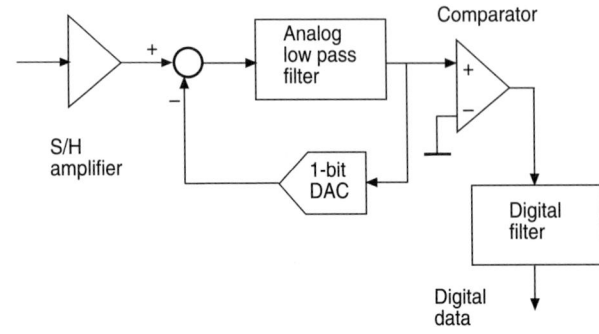

Figure 13 General sigma-delta analog-to-digital conversion. S/H, sample and hold; DAC, digital-to-analog converter.

exceedance of a specified level. Normally, however, it will be desired to capture also the signal immediately prior to the trigger event. The normal way to achieve this is to record continuously in a circular buffer, and use the trigger to define when the recording is to stop. In this case the data record can consist of anything from one complete memory length before the trigger point to any later section of data. It would be typical to continue recording for 90% of the record length after the trigger point so as to include 10% before it (**Figure 14**). The trigger level can be defined in terms of the original analog signal, before or after low-pass filtering, which requires analog hardware, or on the digital value after A/D conversion, which can be achieved by soft- or firmware. The latter is not always satisfactory, as the required low-pass filtering might remove a short trigger event requiring high frequencies to describe it. The trigger can be based on the value of one of the channels being recorded or an external signal (analog or digital). Where the external signal is unrelated to the sampling clock of the ADC, an uncertainty of up to one sample spacing may result. This may appear trivial, but represents 360° of phase of the sampling frequency, and approximately 140° of phase of a typical highest-frequency component. As described below, in cases where this is important, it is desirable to generate the sampling clock signal from the same events as are used for triggering (e.g., a once-per-rev tacho signal).

External Sampling

There are a number of reasons why it is sometimes desirable to generate the sampling signal externally, in addition to the one just mentioned. The main one, which serves as a typical example, is to achieve order analysis of signals originating from rotating machines. Due to changes in speed, large or small, the number of samples corresponding to a shaft rotation will vary and thus the various harmonics of that shaft speed will appear at different frequencies or smear over a range of frequency. If the sampling is linked to the shaft speed, to give a fixed number of samples every revolution, every shaft harmonic (or order) will appear in a fixed line of a frequency spectrum, thus aiding interpretation (**Figure 15**). Since the number of periods of rotation in a record length can be made integer, it is not necessary to use a window such as Hanning in fast Fourier transform (FFT) spectrum analysis, and each harmonic of the fundamental frequency will be completely concentrated in a single spectral line. Note that, in a machine with several shafts, it can usually only be arranged for this to apply to one shaft at a time, and typically this would be done in order to carry out synchronous averaging with respect to that shaft. Where the different shafts are geared together, it would only be necessary digitally to resample the signal (see below) to obtain an integer number of revolutions of another shaft in the record length, but where the different shafts are independent (e.g., in an aero gas turbine engine) the whole sampling would have to be done with respect to another tacho signal.

External sampling signals In order to achieve order tracking, there are several different options for obtaining the sampling signal:

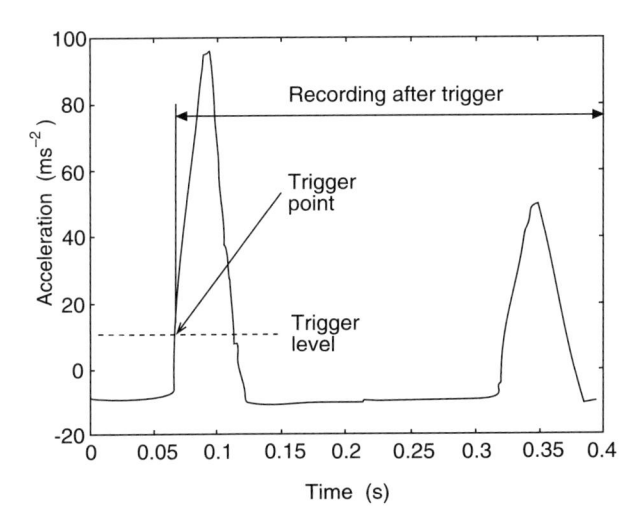

Figure 14 Example of use of a trigger to get pretrigger information.

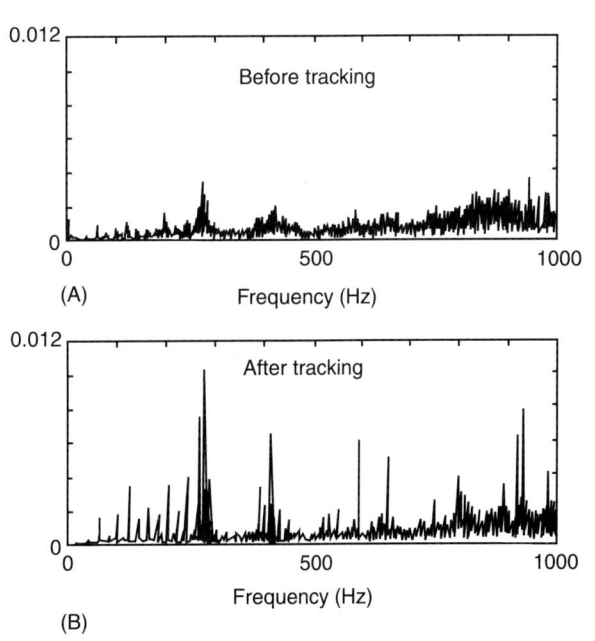

Figure 15 Use of tracking to avoid smearing of shaft speed-related components.

1. *Shaft encoder*: A shaft encoder gives a series of pulses at equal angular intervals (e.g., 1024 per rev) and thus the sampling is on a shaft angle basis. Note that this removes phase modulation effects that might be of interest, for example in gear vibrations.
2. *Frequency multiplier*: A frequency multiplier is based on a phase-locked-loop (PLL) which locks on to a signal such as a once-per-rev tacho signal and produces a fixed number of pulses per basic period of the locked frequency. Both the phase-locked frequency and the multiplied frequency are actually produced by dividing a much higher frequency. Because the PLL is an analog device, it has a finite response time, and does not respond instantaneously to sudden changes in the input frequency, even when able to maintain phase-lock. Because the multiplied signal is based on the much lower frequency input signal, the sampled signal still retains phase modulation information at frequencies higher than the fundamental frequency (for example, in the vicinity of a gearmesh frequency in a gearbox where the gear rotational speed is the input to the frequency multiplier).
3. *Software resampling*: Signals sampled at a fixed sampling rate can be resampled using digital interpolation techniques to a fixed number of samples per fundamental period of one or more tacho signals which are sampled simultaneously with the other signals. This does not suffer from the finite response time of PLLs, by resampling the data in real time or postprocessing already recorded data, and thus has many advantages. Because of its importance this topic is treated separately.

Software Resampling Schemes for Tracking

The simplest technique is to detect the first sample after each trigger point and then convert each block of data starting with these samples (and containing an integer number of samples N_i) to a fixed number of samples N_o greater than the largest value of N_i in any of the blocks. Such interpolation can be achieved by transforming each block into the frequency domain, padding the (two-sided) spectrum to the required size N_o with zeros around the Nyquist frequency, and then inverse transforming back to a time record of size N_o as required. This suffers from the disadvantage that the first sample of each block and the number of samples in each block has a possible error of one sample spacing. The error can be reduced by first increasing the sampling rate of all signals (including the tacho signals) by a large factor of, say, eight or 16, and then simply choosing the nearest sample to that calculated by the interpolation formula, based on the more accurate trigger definition. The maximum error will then correspond to one sample spacing of the higher sample rate. If performed block by block using FFT transforms, even though the transform size will be quite large, use can be made of the most efficient FFT by making both forward and inverse transforms a power of two (whereas for the above method, the forward transform may be any size, including prime numbers). The oversampling by a fixed integer factor can alternatively be done in real time using a digital low-pass filter, as illustrated in **Figure 16** for a factor of four. Three zeros are inserted between each original sample, which increases the sampling frequency by a factor of four without changing the spectrum.

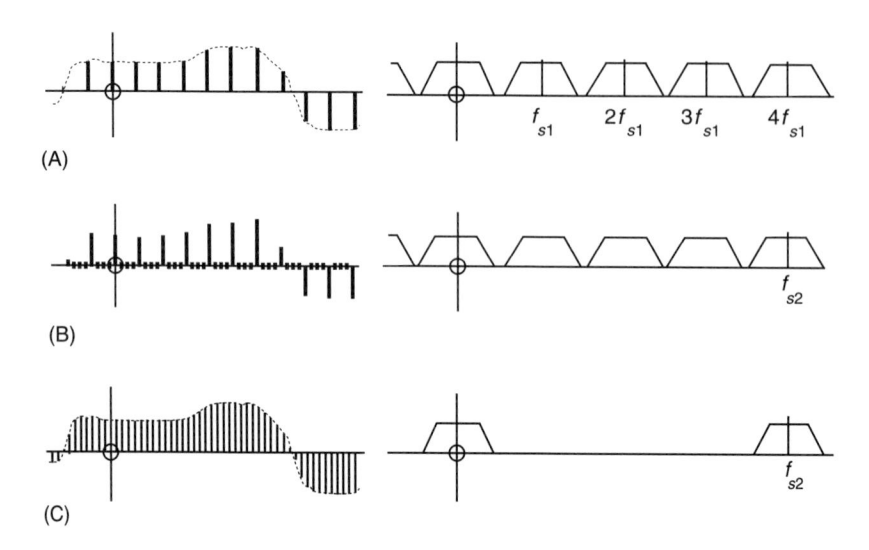

Figure 16 Digital resampling with four times higher sampling frequency. (A) Signal sampled at f_{s1} and its spectrum. (B) Addition of zeros which changes sampling frequency to f_{s2}. (C) Low-pass filtration and rescaling.

Low-pass filtration (and scaling up by a factor of four) produces an interpolated signal at the higher sampling rate but with a frequency content corresponding to the original low-pass filter (before sampling).

Interpolation in the time domain gives more flexibility, allowing the curve joining the data points to be progressively smoother depending on the order. Selecting the nearest sample (i.e., treating the basic function as a series of steps) is the same as convolving the data points with a rectangular function of width equal to the sample spacing. Linear interpolation (joining the samples with a series of straight lines) is the same as convolving the data points with a triangular function of base width twice the sample spacing (the convolution of the previous rectangular function with itself). This convolution in the time domain corresponds to a multiplication in the frequency domain with the Fourier transform of the convolving function, in the first case a $\sin x/x$ function, and the second a $(\sin x/x)^2$ function. The effect of these multiplying functions is not only through their low-pass filtration effect, but also by virtue of the aliasing of higher-order sidelobes because of the resampling. A good choice is a cubic spline interpolation, which has less low-pass filter distortion than linear interpolation, and also less aliasing from sidelobes (**Figure 17**). An efficient procedure assumes that the machine shaft speed changes with constant acceleration steps (linear speed variation and quadratic phase angle variation) and matrix methods are used to calculate the times corresponding to equal angle intervals at which the data are resampled.

Nomenclature

B	word length of a quantizer
dB	decibel
f	frequency
e	voltage
f_s	sampling frequency
q	quantization error
S	number of samples per unit time
M	order of a modulator
N	number of samples
n_0	in-band quantization noise
P	signal power
t	time
Δ	quantization step
E	noise power spectral density

See also: **Signal integration and differentiation; Signal generation models for diagnostics;**

Further Reading

Analog Devices Application Note AN-389. *Using Sigma Delta Converters – Part 2.* Norwood, Massachusetts.

Candy JC and Temes GC (eds) (1991) *Oversampling Delta-Sigma Data Converters – Theory, Design, and Simulation.* Piscataway, NJ: IEEE Press.

House R (1997) National Instruments Application Note 092. *Data Acquisition Specifications – a Glossary.* Austin, Texas: National Instruments.

McFadden PD (1989) Interpolation techniques for time domain averaging of gear vibration. *MSSP* 3:87–97.

Miner GF and Comer DJ (1992) *Practical Data Acquisition for Digital Processing.* New Jersey: Prentice-Hall.

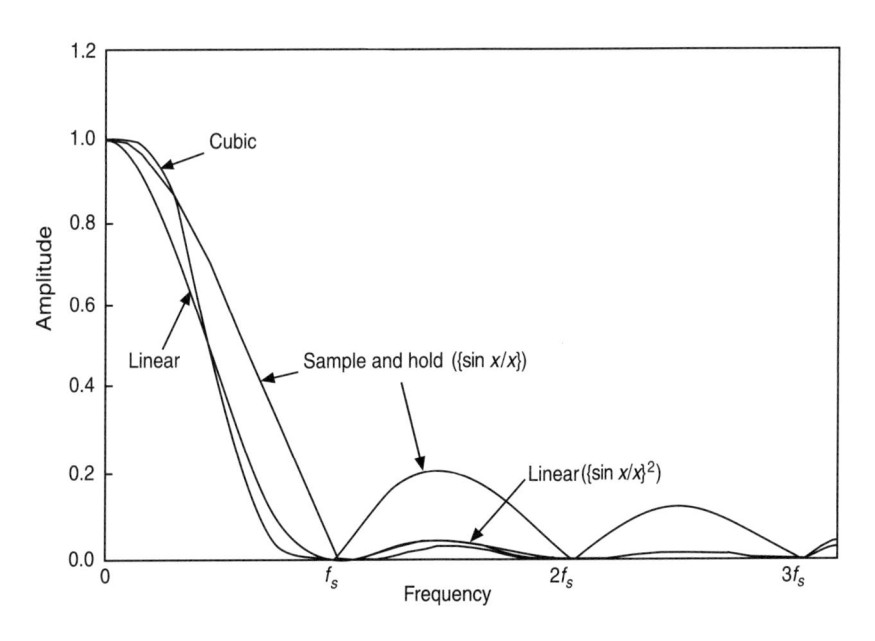

Figure 17 Frequency characteristics for three interpolation functions. Note that the upper side lobes fold back into the measurement range. Cubic interpolation gives best low-pass characteristic and lowest side bands. From McFadden (1989) with permission.

National Instruments (1996) National Instruments Application Note 007. *Data Acquisition Fundamentals.* Austin, Texas: National Instruments.

Oppenheim AV and Schafer RW (1989) *Discrete-time Signal Processing.* New Jersey: Prentice-Hall.

Potter R and Gribler M (1989) *Computed Order Tracking Obsoletes Older Methods.* SAE paper 891131. Warrendale, PA: SAE International.

Proakis JG and Manolakis DG (1989) *Introduction to Digital Signal Processing.* New York: Macmillan.

DIAGNOSTICS AND CONDITION MONITORING, BASIC CONCEPTS

M Sidahmed, Université de Technologie de Compiègne, Compiègne, France

Copyright © 2001 Academic Press

doi:10.1006/rwvb.2001.0147

Introduction

Diagnostics and condition monitoring is a multidisciplinary subject combining various techniques and areas of knowledge: mechanical engineering, reliability techniques, measurement procedures, signal processing, data analysis, and expert or knowledge-based systems. The objectives of diagnostics and condition monitoring of machinery include:

- Control of the machinery, safety is vital when dealing with high power or dangerous machines.
- Optimizing the availability of machines by avoiding unexpected shutdowns. This is particularly true for critical machines in a continuous production process.
- Implementation of condition-based maintenance, or more specifically predictive maintenance, for which the operations are planned according to various constraints (cost, production, failure condition, etc.).

This explains the importance of such techniques which are extensively used in industry.

Diagnostic and condition monitoring is based on the measurements of physical parameters on the machines in order, after specific processing, to assess the state of the machines and to identify any existing failures that may give rise to catastrophic damage. For any physical parameters, it obeys some general principles and methodology which we will describe. Vibration diagnostics and condition monitoring is then considered.

General Principles

Machine Deterioration

Figure 1 shows a typical machinery deterioration time curve (a classical 'bathtub' diagram). Three periods can be distinguished: period 1 corresponds to the running-in, in which early (or young) failures may occur. The failure probability, which is high at the beginning, will decrease by the end of the period. Period 2 is normally the longer period of normal operation of the machine in which the failure probability is lower and is constant. In period 3, failures will occur as a consequence of wear and the probability of failure increases rapidly.

Within the context of condition monitoring and diagnostics, we are concerned with failures which may influence any physical parameters and have a more or less evolutionary process. For sudden unforeseen failures, monitoring is of little value.

Principle and Vocabulary

The general principle of a diagnostic process used in condition monitoring is presented in **Figure 2**. Five steps are clearly identified:

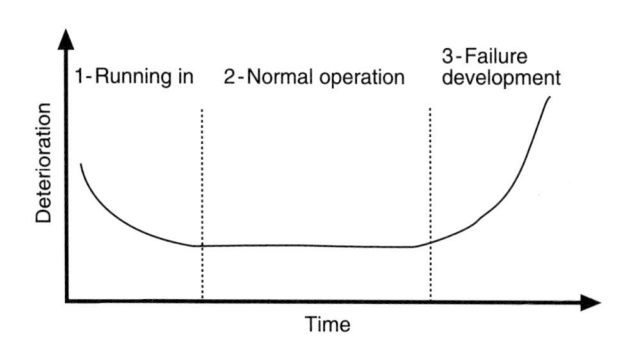

Figure 1 Deterioration time curve for machinery.

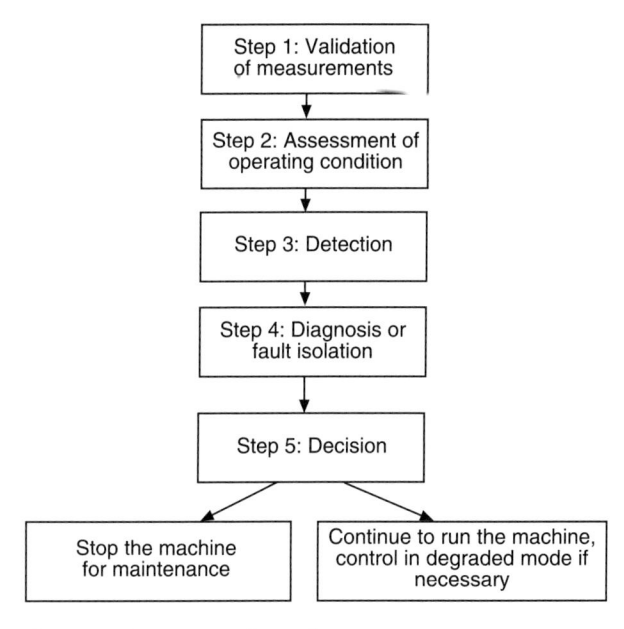

Figure 2 Steps in the diagnosis process.

Step 1: data measurement and validation The information is taken from sensors and measurement systems that have to be reliable. This step is of importance because 'bad' measurements generate a wrong diagnosis. When dealing with complex systems with numerous sensors, this step is critical. Various techniques have been developed to detect invalid measurements, these include: sensor redundancy, comparison to static or dynamic limits, and analytical redundancy.

Step 2: Operating condition assessment This step allows us to define a 'reference signature' which is able to characterize the state of the machine (healthy or faulty), but can also be related to various kinds of faults. The signature may be more or less complex. This step is very important and is a preliminary to the diagnostic process. It makes use of the machine characteristics, the type of physical measurements and the effects of the faults, i.e., the symptoms. It obviously requires the 'knowledge' of faults that may occur on the machine, and their criticality, to be able to define the most suitable signature. For complex machinery, failure mode effects and criticality analysis (FMECA) may be conducted first. It should be noted that in the case of vibration measurements, the signature is obtained by vibration signal processing.

Step 3: Detection Detection is the procedure which decides, from real observations, whether or not the system or machine is operating normally. We 'compare' the signature characterizing the healthy state to the one extracted from the real measurement. The fault is not defined. Comparing the signature belongs to decision theory and is often done using statistical decision theory.

Step 4: Diagnosis This is the art of identifying a machine condition from symptoms. Diagnosis also implies an assessment of the 'gravity' of the fault which is difficult to evaluate in practice. Several authors use the term 'fault isolation' for the process consisting of the identification of a fault from a specified fault set. However, we will use the term diagnosis instead of fault isolation. When a signature is related to a specific fault, steps 3 and 4 may be imbedded in one-step detection/diagnosis. This case often happens in vibration condition monitoring.

Step 5: Decision The operator has to decide whether to stop the machine for maintenance and repair, or to continue operating. In this last case, operating in a degraded mode may be important. This step, known as 'prognosis', is perhaps the more complex one: it is necessary to predict the future machine condition from the past and present symptoms. It is a key point in a diagnostic system. Good prognosis enables the selection of the best time for maintenance and is essential to industry. Many mathematical or statistical trending algorithms are generally used in association with diagnostic software or systems but the results are quite limited. The reason is that the remaining life of a machine depends on the environment, the stress history of the machine and indeed the wear. Work has to be done to adapt the mathematical or stochastic models for fatigue-life estimation and cumulative damage, which have been developed in the 'reliability' community.

Simple example A simple example is given in Figure 3. One may consider the extraction of a parameter from the measurement which completely characterizes one fault. During the normal life of the machine, we take periodical measurements. The extracted parameter will be distributed within the limits corresponding to the good condition. This parameter will be over the limit when the fault

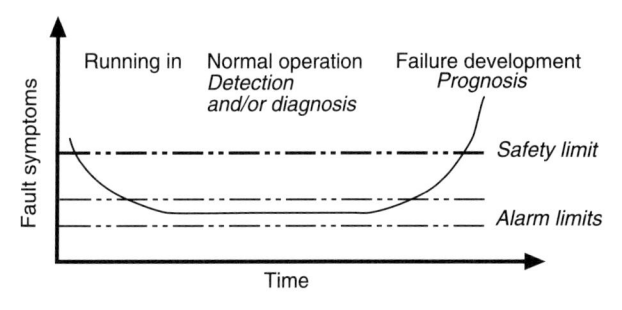

Figure 3 Limits of symptoms: from detection/diagnosis to prognosis.

appears. If the parameter has a known (estimated) probability distribution, then a statistical test may be derived. Note that in this case, the detection and diagnostic processes are the same. Prognosis is necessary to predict the 'safety limit'.

Methods Used for Machinery Condition Monitoring

Various methods are used for condition monitoring, they depend on the physical parameters to be measured and processed to deliver the diagnostic. Each method has advantages and drawbacks, thus it is advantageous to combine the different methods to obtain more reliable results.

Oil analysis Most of the machinery requires lubricants to minimize wear, for example diesel engines, turbines, or components such as gear boxes or rolling bearings. Oil analysis may give information on the lubricant quality and on the wear metal contamination. This last point is used for diagnosis. In complex machinery, this technique may complement vibration analysis in order to provide a more reliable diagnosis. This method suffers, however, from difficulties in collecting 'good' oil samples.

Nondestructive techniques 'Active' techniques (ultrasonics, radiography, etc.) are used to determine the state of various materials: homogeneity, stress, cracks, and quality of welding. The control of specific parts of the rotating machine (blades, shafts, etc.) may be carried out. Passive techniques such as acoustic emission (AE) take advantage of the sound waves emitted by materials when growing cracks or localized constraints are present. Sound waves have very high frequency components and are amplified by resonant sensors (from about 100 kHz to a few Mhz). The complexity of sound propagation in structures makes the interpretation difficult. AE is used for incipient detection but not for diagnosis. This technique is used extensively for tool wear monitoring. Some applications have been developed for incipient detection of faults in rolling bearings.

Infrared or thermographic analysis Infrared cameras are used to detect differences in surface temperatures which may be due to specific faults. This technique appears to be more suitable for the diagnosis of faults in electrical or electromechanical equipment (e.g., transformers).

Current analysis This is used for the detection and diagnosis of electrical problems in motor driven rotating equipment. This technique is of interest because it is not intrusive, the information being available on the motor power cables. Various models exist which allow us to establish the symptoms related to various faults. The capability of current analysis to diagnose mechanical faults has been established (the motor sensing the mechanical load variations) but is still under investigation.

Vibration analysis This is the most widely used technique for diagnosing faults in rotating machines, this is due to three reasons:

- Signal generation models associated with the major faults generating dynamic efforts and then vibrations, have been developed. This gives a good knowledge of the symptoms of the faults, and reliable reference signatures.
- The technique is not intrusive and does not require the machine and then the production to be stopped (thus we may consider vibration monitoring as a kind of nondestructive testing)
- The sensors are low cost and, in general, reliable, the associated data analysis systems are increasingly powerful.

In recent years emphasis has been placed on the development of advanced signal processing techniques for incipient fault detection and diagnosis. These are detailed in the next section. We may also note, that standards have been developed to classify machine vibrations into acceptable or nonacceptable vibration levels. These levels are derived from very simple processing such as power in the band (10–1000 Hz). These deterministic levels are used as detection limits (**Figure 3**). The diagnostic ability of these limits is very limited. It should be noted that vibration frequency analysis standards are currently under development.

The various techniques for condition monitoring and diagnostics may be used for quality control of components that are to be tested during production. They are usually used for low-cost components such as small DC motors, automotive gearboxes, etc.

The detection and diagnostic procedures may be developed in real time when we deal with controlled machinery (for machine tools it is necessary to detect a tool breakage in real time). In general, when real time is required, detection is addressed rather than diagnosis. In most cases, for rotating machines, signature analysis techniques are carried out periodically.

Vibration Condition Monitoring and Diagnostic

Vibration Signatures

Vibration monitoring is suitable to detect faults generating dynamics giving rise to vibrations that may be measured with sensors fixed to a fixed part of the machine, mainly the bearing cases. These faults generate vibrations with specific characteristics which are linked to the machine kinematics which have to be defined with details (rotation speed, number of rotating elements, characteristics of rolling bearing, etc). Table 1 summarizes the symptoms of various faults in rotating machines with a fundamental rotating speed, f_r. These results are obtained from signal generation models developed for rotating machines and show that vibrations due to faults in various components give rise to signals with:

- high frequency components at shaft rotation speed and harmonics
- amplitude and phase modulation
- possible repetitive impulses
- broadband noise.

According to the signal generation models, 'signatures' that emphasize the frequency characteristics of the signals may be used such as:

- frequency domain analysis (Fourier power spectral density)
- cepstral analysis, especially when numerous harmonics are expected

- envelope analysis
- time frequency analysis for nonstationary phenomena occurring during run-up and run down
- more advanced methods such as the bispectrum or spectral correlation function.

Statistical 'signatures' based on the time domain characteristics are used mainly for impact detection:

- RMS value and crest factor
- kurtosis value which is sensible to impacts
- repetitive variance analysis
- wavelet decomposition.

It should be noted that vibration condition monitoring using signature analysis has an important drawback. It is necessary to compare the signatures in the same operating conditions (same rotation frequency and load). Statistically based signatures are less subject to small variations in operating conditions (e.g., kurtosis).

Detection

Detection is essentially based on fixed limits. These limits are fixed from a reference operating condition. Various measurements are generally used to estimate a probability distribution to derive statistical levels.

Table 1 Vibration symptoms for various faults in rotating machines (f_r fundamental rotating frequency)

Excitation sources	Main frequency components	Direction	Other characteristics	Comments
Unbalance	$1 \cdot f_r$	Radial	Elliptic orbit	
Misalignment	$1 \cdot f_r$, $2 \cdot f_r$ and higher harmonics	Axial, radial	Double orbit	
Gear system (f_m meshing frequency, f_{r1} and f_{r2} shaft rotation frequencies, m and n integers)	Modulated meshing frequency and harmonics $m \cdot f_m \pm n \cdot f_{r1}$ $m \cdot f_m \pm n \cdot f_{r2}$	Radial and axial, depends on the gear type	Amplitude and phase modulation	Several components are present, even under good conditions
Blades passing (k number of blades)	$k \cdot f_r$, f_r	Radial, axial	High level	Modulation may exit
Oil whirl or whip in journal bearings	0.42 to $0.48 \cdot f_r$	Radial		
Rolling bearings	Impact rates and harmonics, related to defects (inner and outer races, rolling elements, cage)	Radial, axial		Excitation of structural resonances in the high frequency range may exit
Induction machines (f_s synchronous frequency)	$1 \cdot f_r$, $1 \cdot f_s$, $2 \cdot f_s$	Radial, axial		
Turbulence	Wideband	Radial, axial	$1/f$ characteristic of the spectrum	
Cavitation in pumps	High frequency	Radial, axial		Random vibration
Clearance	Random impacts	Radial, axial		
Alternative machines (combustion engines, compressors)	Repetitive impacts			Each impact has specific spectral content
Structural resonances	Resonance frequency	Radial, axial		May be excited

We may note that, model-based signal processing methods (ARMA models) are good candidates for deriving statistical detection limits by inverse filtering.

Diagnostic

When parameters derived from signal processing are uniquely related to faults, then the preceding detection step also provides the diagnosis. As shown in **Table 1**, various faults in rotating machines may have similar symptoms. It is then difficult to attribute a simple symptom to each specific fault, and additional processing is needed. Heuristic rules may be used to remove some ambiguities. This field is covered by expert or knowledge-based systems.

In complex machines, signal processing may not be sufficient to differentiate various faults. Pattern recognition techniques such as neural networks may be used. Such techniques make use of training data sets which are not easy to acquire on real machinery.

Conclusion

Diagnostics and condition monitoring is an efficient technique for predictive maintenance and safety control. It is based on physical measurements on the machinery, that are sensible to faults that happen and evolve. There are various techniques. Vibration analysis is used extensively for rotating machine diagnostics because of the existence of signal generation models, which make possible the extraction of parameters related to various faults. Advanced signal processing techniques which allow us to carry out the incipient and reliable diagnosis of complex and critical machines need to be developed.

See also: **Bearing diagnostics**; **Gear diagnostics**; **Nondestructive testing**, Sonic; **Nondestructive testing**, Ultrasonic; **Tool wear monitoring**; **Vibration isolation, applications and criteria**; **Vibration isolation theory**.

Further Reading

Braun SJ (1986) *Mechanical Signature Analysis*. London: Academic Press.
Collacot RA (1979) *Vibration Monitoring and Diagnostic*. New York: Wiley.
Mitchell JS (1981) *Machinery Analysis and Monitoring*. Tulsa, Penn Well Books.
Natke HG, Cempel C (1997) *Model-aided Diagnosis of Mechanical Systems: Fundamental Detection, Localisation, Assessment*. Berlin: Springer Verlag.

DIAGNOSTICS

See **BEARING DIAGNOSTICS; DIAGNOSTICS AND CONDITION MONITORING, BASIC CONCEPTS; GEAR DIAGNOSTICS; NEURAL NETWORKS, DIAGNOSTIC APPLICATIONS**

DIGITAL FILTERS

A G Constantinides, Imperial College of Science, Technology and Medicine, London, UK

Copyright © 2001 Academic Press

doi:10.1006/rwvb.2001.0050

System Function of Discrete-time Systems

Digital filters are a fundamental signal processing operation of universal applicability. They have a series of properties on which design procedures are based. These are outlined below. The ideal frequency amplitude characteristics are given in **Figure 1**. These characteristics correspond to lowpass, highpass, bandpass, and bandstop digital filters. They are not realizable with finite order transfer functions, as it will be indicated at a later stage but they serve a useful purpose in providing target responses to design procedures.

In a block diagram form the action of a digital filter (or of any system for that matter) may be symbolically represented as shown in **Figure 2**.

Let the input signal be $\{x(n)\}$ of the z-transform $X(z)$, and the output signal $\{y(n)\}$ of the z-transform be $Y(z)$. We define the transfer function of the system as $G(z) = Y(z)/X(z)$, and with $Y(z) = G(z)X(z)$ it is evident that when $X(z) = 1$ then $G(z) = Y(z)$, i.e.,

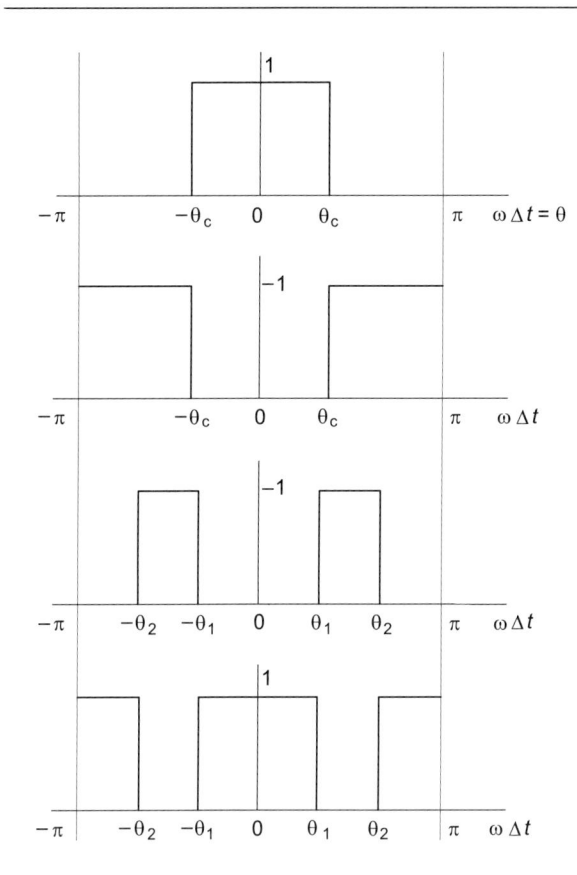

Figure 1 Ideal frequency amplitude characteristics.

$G(z)$ is the z-transform of the impulse response $h(n)$, where:

$$G(z) = \sum_{n=0}^{\infty} h(n) z^{-n}$$

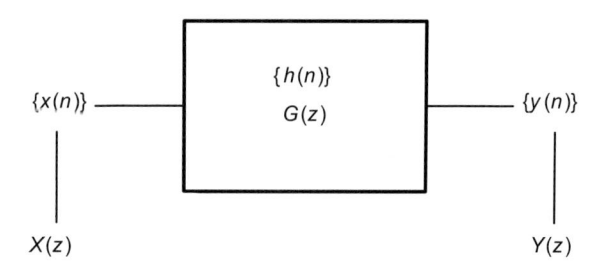

Figure 2 Block diagram of digital filter.

Hence a convolution relationship exists, which can be written as:

$$y(n) = \sum_{r=0}^{\infty} h(r) x(n-r)$$

It is clear that the transfer function of a digital filter is defined once the impulse response of the filter is specified. It would appear therefore that all one has to do in a design procedure is to determine the impulse response of one of the cases shown in the figure appropriate to the application in mind, and the task is completed.

The impulse response of the ideal lowpass filter of cutoff frequency θ_c and constant group delay τ is given by:

$$h(n) = \left(\frac{\theta_c}{\pi}\right) \frac{\sin\theta_c(n\Delta t - \tau)}{\theta_c(n\Delta t - \tau)}$$

as is shown in **Figure 3**.

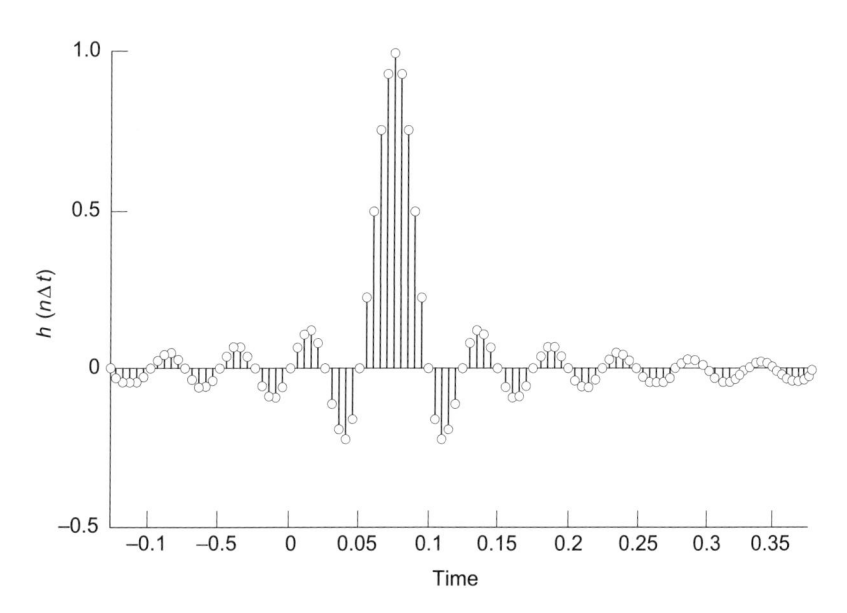

Figure 3 Ideal lowpass impulse response.

It is clear that this impulse response not only exists for negative values of time but also extends to infinity in the positive and negative directions. It is, in effect, anticipating the input to the filter. Evidently, much more needs to be done, in order to determine a realizable digital filter transfer function. To appreciate the range of tasks and constraints, and to have a proper perspective on the design processes involved, we need to examine some fundamental properties of transfer functions for discrete time systems.

If the input is the sampled sinusoidal signal $x(n) = e^{j\omega n \Delta t}$ with Δt as the sampling period we have:

$$y(n) = \sum_{r=0}^{\infty} h(r) e^{j\omega(n-r)\Delta t} = e^{j\omega n \Delta t} G(e^{j\theta})$$

where:

$$G(e^{j\theta}) = \sum_{r=0}^{\infty} h(r) e^{-jr\theta} = G(z)|_{z=e^{j\theta}} \quad \text{and} \quad \theta = \omega \Delta t$$

For discrete time real systems the amplitude response is then:

$$A(\theta) = |G(z)|_{z \text{ on } C}$$

where $C: |z| = 1$ while the phase response is given as $\phi(\theta) = \arg \lfloor G(z)|_{z \text{ on } C} \rfloor$. On the unit circle C:

$$G(e^{j\theta}) = \sum_{n=0}^{\infty} h(n) e^{-jn\theta} = A(\theta) e^{j\phi(\theta)}, \quad \theta = \omega \Delta t$$

It follows that $A(\theta) = A(-\theta)$ (even function) and $\phi(-\theta) = -\phi(\theta)$ (odd function). Moreover $dA(\theta)/d\theta = 0$ at $\theta = 0$. Further, from $G(e^{j\theta}) = A(\theta) e^{j\phi(\theta)}$ we have:

$$\frac{dG(e^{j\theta})}{d\theta} = e^{j\phi(\theta)} \frac{dA(\theta)}{d\theta} + A(\theta) j \frac{d\phi(\theta)}{d\theta} e^{j\phi(\theta)}$$

and in this equation we can use the group delay given as:

$$-\frac{d\phi(\theta)}{d\omega}\bigg|_{\theta=\omega T} = \tau(\omega)$$

to yield:

$$\frac{dG(e^{j\theta})}{d\theta} = e^{j\phi(\theta)} \frac{dA(\theta)}{d\theta} + jA(\theta) e^{j\phi(\omega)} \left(-\frac{\tau(\omega)}{\Delta t}\right)$$

This form at $\theta = 0$ with the left-hand side can be expressed in terms of $h(n)$ produces:

$$A(0) \frac{\tau(0)}{\Delta t} = \sum_{n=0}^{\infty} n h(n)$$

A simplification can be put into effect here, by using the notion of the center of gravity of $h(n)$ defined as:

$$\mu_h = \frac{\sum_{n=0}^{\infty} n h(n)}{\sum_{n=0}^{\infty} h(n)} = \frac{\sum_{n=0}^{\infty} n h(n)}{G(0)}$$

and hence we have:

$$A(0) \frac{\tau(0)}{\Delta t} = G(0) \mu_h \quad \text{or} \quad \tau(0) = \mu_h \Delta t$$

i.e., the center of gravity of a signal at the output of a linear system will be delayed by $\tau(0)$.

Group Delay for Discrete Time Systems

An expression for the group delay can be obtained from the following considerations. Let:

$$G(z)|_{z=e^{j\theta}} = |G(e^{j\theta})| e^{j\phi(\theta)} \quad \theta = \omega \Delta t$$

so that

$$\frac{1}{2} \ln(G(e^{j\theta})) = \frac{1}{2} \ln(G(e^{-j\theta})) + j\phi(\theta)$$

and hence the phase response is given as:

$$\phi(\theta) = -\frac{j}{2} \ln \left(\frac{G(e^{j\theta})}{G(e^{-j\theta})}\right)$$

The group delay is given by:

$$\tau_g(\theta) = -\frac{d\phi(\theta)}{d\theta} \frac{d\theta}{d\omega} = -T \frac{d\phi(\theta)}{d\theta}$$

and hence one obtains:

$$\tau_g(\theta) = \frac{\Delta t}{2} \left[z \frac{G'(z)}{G(z)} + z^{-1} \frac{G'(z^{-1})}{G(z^{-1})}\right]_{z=e^{j\theta}}$$

where $G'(u) = dG(u)/du$, i.e., the prime indicates differentiation with respect to the argument of the function.

DIGITAL FILTERS 383

Signal Flow Graphs

An important concept in the study and implementation of algorithms for general digital signal processing and for digital filters is that of signal flow graphs (SFGs). Essentially they embody the essence of the interacting relationships that yield the specific input/output results. Linear time invariant discrete time systems can be made up from three basic elements namely storage, scaling, and summation.

Storage In the digital signal processing literature this is sometimes known as 'Delay', or 'Register'. Its SFG representation and mathematical description are given in **Figure 4**. The output from this element is essentially the same as the input delayed by one sampling instant.

Scaling In the digital signal processing literature this is sometimes known as weight, product, and multiplier. Two frequently encountered SFG representations and their mathematical description are given in **Figure 5**. The output from this element is essentially the same as the input delayed by one sampling instant

Summation In the digital signal processing literature this is sometimes known as adder, or accumulator. Its SFC representation and mathematical description are given in **Figure 6**. The output from this element is essentially the sum of its inputs and this operation is assumed to be carried out instantly. A simple digital filtering equation such as $y_n = a_1 y_{n-1} + a_2 y_{n-2} + b x_n$ can be represented in terms of the SFC, employing these elements and is shown in **Figure 7**.

Conversely, the system equation, which is essentially the algorithmic form of the input/output relationship may be obtained from the interconnected

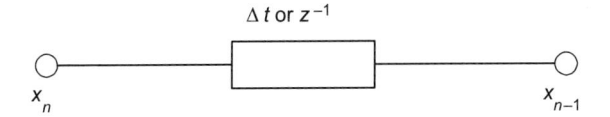

Figure 4 Storage.

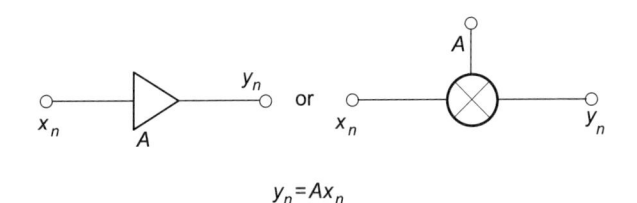

Figure 5 Scaling.

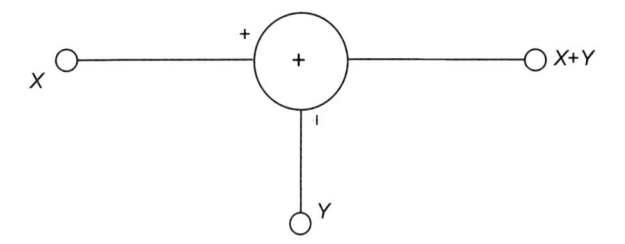

Figure 6 Summation.

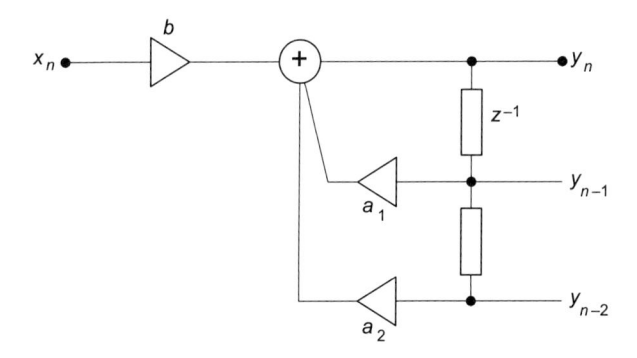

Figure 7 Simple digital filtering.

components. Such interconnections are termed the realization structure. A realization structure can be computable when all the loops in it contain delays, normally of integer multiples of the sampling period, or it can be noncomputable if there exist some loops that contain no delays.

An important operation in the study concerning the behavior of realization structures is that of transposition. Transposition of a SFG is the process of reversing the direction of signal flow on all transmission paths while keeping their transfer functions the same. The adders are replaced by nodes, and nodes by adders. For a single-input/output SFG the transpose SFG has the same transfer function overall, as the original.

Structures

As already mentioned these are the computational schemes embodying the individual interactions between the realization elements that yield the overall input/output relationships. For a given transfer function, there are many realization structures. Each structure has different properties with respect to coefficient sensitivity and finite register computations. When the number of delays used in a structure is equal to the order of the transfer function the structure is termed canonic. Some standard forms are:

Direct form I Let us assume that the transfer function is given by:

$$G(z) = \frac{Y(z)}{X(z)} = \frac{\sum_{i=0}^{n} a_i z^{-i}}{1 + \sum_{i=1}^{m} b_i z^{-i}}$$

The numerator of the above transfer function has a realization structure as shown in **Figure 8**. This is known as a direct form realization. It is a standard form for finite impulse response (FIR) transfer functions (i.e., when the denominator above is constant). The numerator realization alone is canonic. The denominator of the above transfer function has a direct form realization structure shown in **Figure 9**. This is also a canonic realization. The cascaded interconnection of these two realization structures produces the so-called direct form I. The structure produced by this cascaded interconnection is noncanonic. There are many other alternative ways of producing individual realization structures of the direct form. For example, one can immediately proceed by using the principle of SFG transposition as mentioned earlier. Alternatively one can develop specific decomposition algorithms for this purpose.

The direct form is a simple structure but not used extensively in demanding applications because its performance degrades rapidly due to finite register computation effects.

Direct form II A canonic form can be obtained by a more careful interconnection of the above components. A simple analysis of the structure (**Figure 10**) shows that it has the required transfer function and the lowest number of delays consistent with its order.

Cascade form By expressing the numerator and denominator polynomials as factors of polynomials of lower degree we can produce different structures. The cascaded interconnection of the individual realizations produce a large number of alternative realizations for the given transfer function. Reduction of the effects due to finite register length can be achieved by a suitable selection of factors. In general:

$$G(z) = \prod_i G_i(z)$$

and for second-order blocks:

$$G_i(z) = \frac{a_{0i} + a_{1i}z^{-1} + a_{2i}z^{-2}}{1 + b_{1i}z^{-1} + b_{2i}z^{-2}}$$

or for first-order blocks:

$$G_i(z) = \frac{a_{0i} + a_{1i}z^{-1}}{1 + b_{1i}z^{-1}}$$

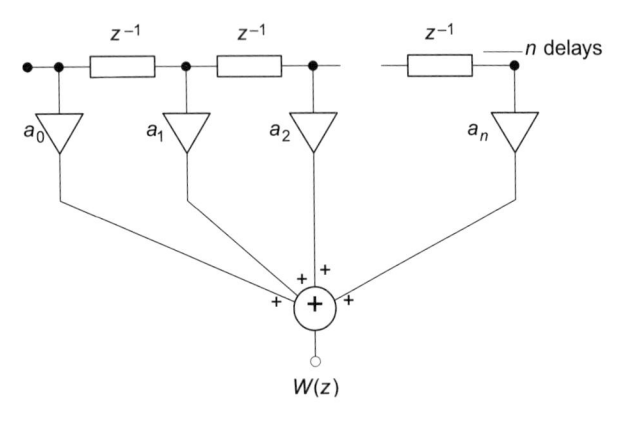

Figure 8 Direct form realization.

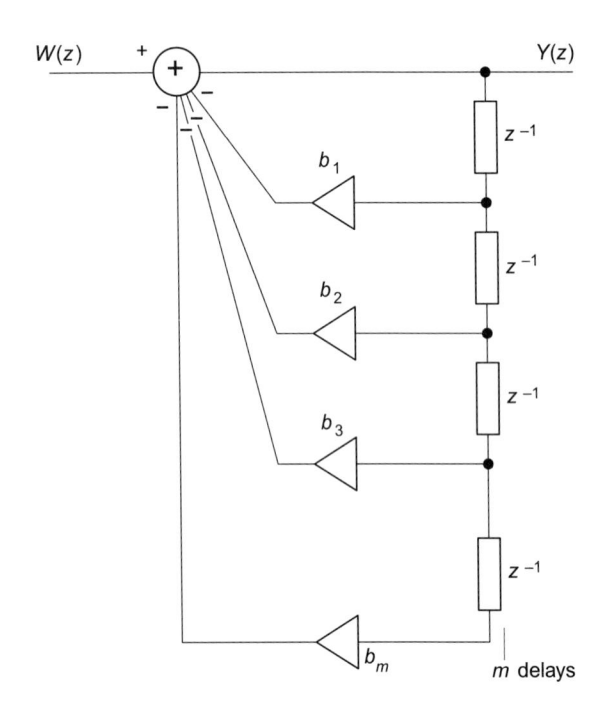

Figure 9 Canonic realization.

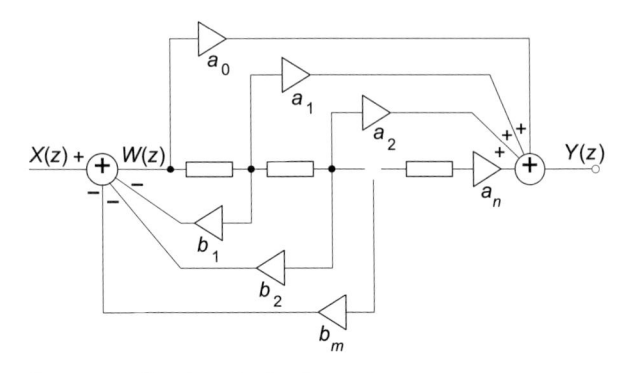

Figure 10 Simple analysis of canonic structure.

Figure 11 is a canonic realization for the second-order transfer function, or indeed for the first-order case, when the quadratic terms are taken to be zero.

Parallel form   Let the transfer function be expressed as:

$$G(z) = g + \sum_{i=1}^{k} G_i(z)$$

The individual transfer functions $G_i(z)$ are as in the cascade realization with $a_2 = 0$. The parallel interconnection in which each subtransfer function takes the same input and their outputs are summed to produce the overall output, yields the parallel form.

Filtering

The digital signal processing operations aimed at removing a frequency or a band of frequencies from a given signal is called digital filtering. The complete removal of a single frequency is a relatively straightforward matter. Indeed, if the given is $\{x_n\}$ and the generated signal (output) is $\{y_n\}$ the algorithm below will remove a single frequency of value $\omega_0 = \theta_0/\Delta t$, where Δt is the sampling period:

$$y_n = x_n - 2 \cos \theta_0 x_{n-1} + x_{n-2}$$

An example is shown in **Figure 11** where we also show the frequency response of the transfer function of this operation, which is given by:

$$G(z) = 1 - 2 \cos \theta_0 z^{-1} + z^{-2}$$

However, the entire suppression of a band of frequencies as is required by the ideal frequency responses discussed above, is not possible as can be seen from the following argument.

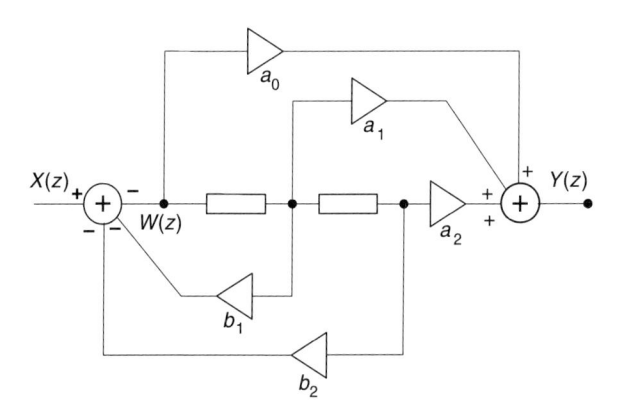

Figure 11   Canonic realization.

Suppression of a frequency band

The transfer function of a real digital filter is required to be a rational function of z^{-1} (i.e., a ratio of two polynomials in z^{-1}). A real and rational transfer function $H(z)$ cannot suppress a band of frequencies completely, i.e., $H(z)$ cannot be zero for $z = e^{j\theta}$ and θ over a band of frequencies in (θ_1, θ_2), $\theta_2 \neq \theta_1$. This may be demonstrated as follows. To produce a zero at say $\theta = \theta_0$ we must have in the numerator of $H(z)$ a factor of the form (as seen from the previous example) $(z^2 - 2 \cos \theta_0 z + 1) = (z - e^{j\theta_0})(z - e^{-j\theta_0})$. Therefore for one zero within the band of (normalized) frequencies (θ_1, θ_2) the factor is of the form $(z^2 - 2 \cos \theta_{0i} z + 1) \theta_1 \leq \theta_{0i} \leq \theta_2$.

Since there are an infinite number of points in the band we need factors in the numerator as:

$$\prod_{i=1}^{\infty} (z^2 - 2 \cos \theta_{0i} z + 1) \quad \theta_{01} = \theta_1, \quad \theta_{0\infty} = \theta_2$$

Clearly the result is of infinite degree and is not a rational function. Hence it cannot be the transfer function of a real digital filter. However, in practice one may be satisfied not with the complete suppression of a frequency band but with its attenuation to an acceptable level that depends on the application. This is the pivotal consideration for all design techniques, taken either implicitly or explicitly.

Finite Impulse Response Filters

The transfer function of a FIR digital filter is essentially a polynomial of the form:

$$H(z) = \sum_{n=0}^{N-1} h(n) \, z^{-n}$$

The length of the filter impulse response is the number of coefficients N. While this kind of transfer function is sometimes called 'all-zero', strictly speaking it has as many poles as the order $(N - 1)$ all of which are located at $z = 0$. The zeros can be placed anywhere on the z-plane and their location dictates the form of the frequency response. A great advantage of FIR filters is their ability to produce a linear phase response and yet allow freedom to shape the amplitude response as required.

Linear phase   For linear phase response the impulse response is required to be such that $h(n) = \pm h(N - 1 - n)$. It can be seen then that for N even we have:

$$H(z) = \sum_{n=0}^{(N/2-1)} h(n) z^{-n} \pm \sum_{n=N/2}^{N-1} h(n) z^{-n}$$

$$= \sum_{n=0}^{(N/2)-1} h(n) z^{-n}$$

$$\pm \sum_{n=0}^{(N/2)-1} h(N-1-n) z^{-(N-1-n)}$$

$$= \sum_{n=0}^{(N/2)-1} h(n)[z^{-n} \pm z^{-m}], \quad m = N-1-n$$

which on taking the positive sign becomes on $C: |z| = 1$, i.e., with $z = e^{j\theta}$ and $\theta = \omega \Delta t$:

$$H(e^{j\theta}) = e^{-j\theta((N-1)/2)}$$
$$\times \sum_{n=0}^{(N/2)-1} 2h(n) \cos\left(\theta\left(n - \frac{N-1}{2}\right)\right) \quad (1)$$

while with the negative sign it becomes:

$$H(e^{j\theta}) = e^{-j\theta((N-1)/2)}$$
$$\times \sum_{n=0}^{(N/2)-1} j2h(n) \sin\left(\theta\left(n - \frac{N-1}{2}\right)\right) \quad (2)$$

It should be noted that the antisymmetric case adds $\pi/2$ rads to the phase, with a discontinuity at $\theta = 0$.

For N odd we write the transfer function as:

$$H(z) = \sum_{n=0}^{((N-1)/2)-1} h(n) [z^{-n} \pm z^{-m}]$$
$$+ h\left(\frac{N-1}{2}\right) z^{-((N-1)/2)} \quad (3)$$

On taking the positive sign we have on $C: |z| = 1$:

$$H(e^{j\theta}) = e^{-j\theta[(N-1)/2]}$$
$$\times \left\{ h\left(\frac{N-1}{2}\right) + \sum_{n=0}^{(N-3)/2} 2h(n) \cos\left[\theta\left(n - \frac{N-1}{2}\right)\right] \right\} \quad (4)$$

while with the negative sign we obtain:

$$H(e^{j\theta}) = e^{-j\theta[(N-1)/2]}$$
$$\times \left\{ 0 + \sum_{n=0}^{(N-3)/2} 2j\, h(n) \sin\left[\theta\left(n - \frac{N-1}{2}\right)\right] \right\} \quad (5)$$

It should be noted that for the antisymmetric case to have a linear phase we require $h((N-1)/2) = 0$. The phase discontinuity is as for N even.

The cases most commonly used are (1) and (3), for which the amplitude characteristic can be written as a polynomial in $\cos \theta/2$.

FIR filters are implicitly stable, can be designed to have linear phase when their impulse response is symmetric or antisymmetric about its midpoint, and they can have low computational noise. But in order to satisfy a set of attenuation specification their order needs to be much higher than the corresponding infinite impulse response digital filters.

The real transfer function

$$H(z) = \sum_{n=0}^{N-1} h(n) z^{-n}$$

with symmetries on the impulse response $\{h(n)\}$ for linear phase will have zeros disposed on the z-plane in a special form. Because it is real, the zeros that are complex must appear in complex conjugate form. Moreover, because of the symmetries in the impulse response, in general, a zero and its inverse will also be present, i.e., it has zeros at $[\rho_i e^{\pm j\theta_i}]^{\pm 1}$ where ρ_i has value other than 1. With zeros on the unit circle (which, of course, produce zero gain at their respective frequencies) the zeros do not have to have this quadrantal symmetry, but they need to be in complex conjugate form.

There are four cases that one needs to examine for linear phase responses.

1. Symmetric ($h(n)$, odd N). The amplitude response may be zero at 0 or $\pi/\Delta t$.
2. Symmetric ($h(n)$, even N). The amplitude response is always zero at $\pi/\Delta t$.
3. Antisymmetric ($h(n)$, odd N). The amplitude response is always zero at 0 and $\pi/\Delta t$.
4. Antisymmetric ($h(n)$, even N). The amplitude response is always zero at 0.

These general properties indicate the possible options prior to a design. For example one cannot use case (2) for highpass filters because of its transmission zero at zero frequency.

Design of FIR Filters

The obvious approach is to determine the ideal impulse response and then to take a finite number of terms symmetrically disposed about its midpoint. This would produce ripples of increasing height near an abrupt change in the amplitude, and to reduce these the 'Window method' has been produced. Essentially it involves the following steps:

1. Start with ideal infinite duration $\{h(n)\}$.
2. Truncate $\{h(n)\}$ to finite length (this produces unwanted ripples that increase in height near discontinuity).
3. Modify $\{h(n)\}$ as $\tilde{h}(n) = h(n)w(n)$ where the weight $w(n)$ is called the window and may be chosen from a wide range of available functions.

Notice that in practice steps (2) and (3) are taken together by having the window zero outside the chosen interval. There are many windows with different properties that may be selected. Commonly used windows are:

Name of window $w(n)$	Range of time index $\|n\|<(N-1)/2$:
Rectangular	1
Bartlett	$1 - (2\|n\|/N)$
Hann	$1 + \cos((2\pi n)/N)$
Hamming	$0.54 + 0.46 \cos((2\pi n)/N)$
Blackman	$0.42 + 0.5 \cos((2\pi n)/N) - 0.08 \cos(4\pi n/N)$
Kaiser	$J_0\left[\beta\sqrt{\left(1 - \left(\frac{2n}{N-1}\right)^2\right)}\right]/J_0(\beta)$

where $J_0(x)$ is the zero-order Bessel function of the second kind, and β is a parameter that may be chosen according to the filter performance needs as indicated in **Table 1**. **Figure 12** shows the ideal lowpass impulse response symmetrically centered around the zero frequency point, in which form it is used for the

Table 1

β	Normalized transition width	Minimum stopband to passband discrimination in dB
2.12	1.5/N	30
4.54	2.9/N	50
6.76	4.3/N	70
8.96	5.7/N	90

application of the window method. This will result in essentially a linear phase response. Some examples are shown in **Figure 13**. Attention should be paid not only to the attenuation discrimination between passband and stopband but also to the passband behavior. The consequences of product involved in the modification of the impulse response can be examined from the frequency domain by making use of the following relationships.

Windows: Products of Signals: Parseval's Theorem

In a general case let $\{u(n)\}$ and $\{w(n)\}$ be the given signal of z-transforms $U(z)$ and $W(z)$, respectively.

Let us form a new signal given by $y(n) = w(n)u(n)$. We need to determine the z-transform of this new signal. We can now use the definition of the z-transform and the inverse z-transform relationship as follows:

$$Y(z) = \sum_{n=0}^{\infty} w(n)\, u(n)\, z^{-n}$$

$$= \sum_{n=0}^{\infty} w(n) \left[\frac{1}{2\pi j} \oint_c U(\xi)\, \xi^n\, \frac{d\xi}{\xi}\right] z^{-n}$$

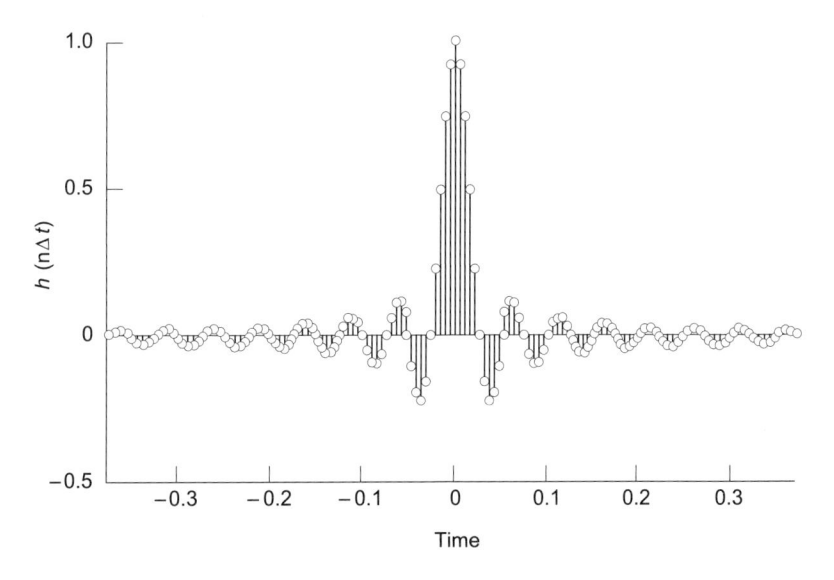

Figure 12 Ideal lowpass impulse response.

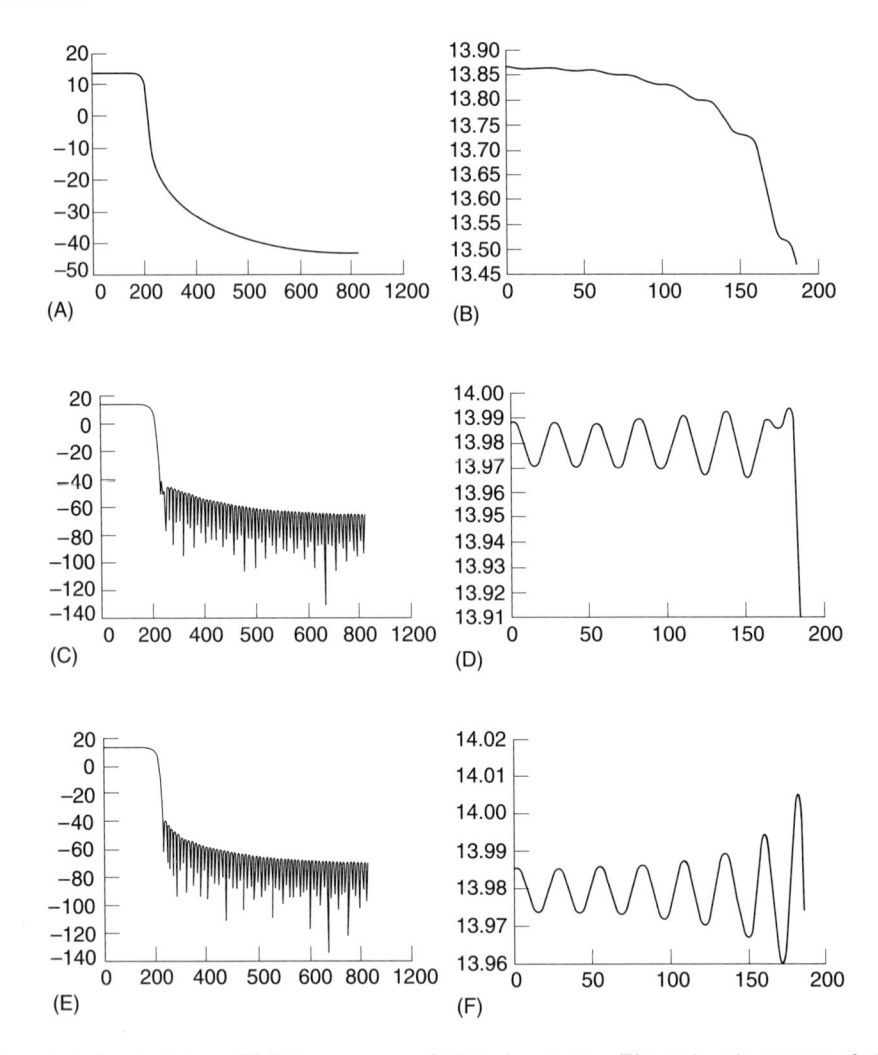

Figure 13 Examples of window technique. (A) Entire response of triangular window, (B) passband response of triangular window, (C) entire response of Hamming window, (D) passband response of Hamming window, (E) entire response of Kaiser window, (F) passband response of Kaiser window.

or

$$Y(z) = \frac{1}{2\pi j} \oint_c U(\xi) \left[\sum_{n=0}^{\infty} w(n) \, (\xi^{-1} z)^{-n} \right] \frac{d\xi}{\xi}$$

and hence

$$Y(z) = \frac{1}{2\pi j} \oint_c U(\xi) \, W(z\xi^{-1}) \, \frac{d\xi}{\xi}$$

However, on the unit circle $\xi = e^{j\phi}$, $z = e^{j\theta}$ and $\theta = \omega \Delta t$, therefore in the frequency domain we can write:

$$Y(e^{j\theta}) = \frac{1}{2\pi} \int_{-\pi}^{\pi} U(e^{j\phi}) \, W(e^{j(\theta-\phi)}) \, d\phi$$

This is a convolution relationship in the frequency domain, which can symbolically be written as:

$$Y = U * W$$

A bonus arising from the above is that if $u(n) = w(n) = x(n)$ and are all real signals and for $z = 1$ then:

$$\sum_{n=0}^{\infty} [x(n)]^2 = \frac{1}{2\pi j} \oint_c |X(z)|^2 \frac{dz}{z} = \frac{1}{2\pi} \int_{-\pi}^{\pi} |X(e^{j\theta})|^2 \, d\theta$$

This is Parseval's relationship. (Actually it is much older than Parseval in that it dates from the nineteenth century from the works of Lord Rayleigh.)

It can now be seen that the use of a window would produce a reduction of the ripples or 'sidelobe levels'.

Let us consider the simple Hann window to appreciate this point, i.e.:

$$w(n) = 0.5 + 0.5 \cos\left(\frac{2\pi n}{N-1}\right)$$

for:

$$-\frac{(N-1)}{2} \leq n \leq \frac{(N-1)}{2}$$

and 0 otherwise. Take this window to be formed as a product $w(n) = c(n)w_R(n)$ where $c(n)$ is an infinite duration version of $w(n)$ the spectrum of which is given by:

$$C(e^{j\theta}) = \pi\delta(\theta) + 0.5\pi\delta\left(\theta - \frac{2\pi}{N-1}\right) + 0.5\pi\delta\left(\theta + \frac{2\pi}{N-1}\right)$$

while $w_R(n)$ is the rectangular window of spectrum:

$$W_R(e^{j\theta}) = \sum_{n=-(N-1)/2}^{(N-1)/2} e^{-jn\theta} = \frac{\sin(N\theta/2)}{\sin(\theta/2)}$$

Since products in the time domain involve convolutions in the frequency domain we see that the delta functions will essentially replicate the spectrum of the rectangular window at the first zero-crossings. Hence the positive excursions in its response will approximately cancel the negative ones while extending somewhat the transition band. However, the window method is somewhat indirect in that it does not examine the specifications in the frequency domain. At any rate there is no direct control over the frequency domain performance. The 'frequency sampling method' is meant to provide such control at least at preselected frequency points.

Frequency Sampling Technique

Let $H(r)$ be the values of the transfer function $H(z)$ at the points $z_r = \exp(j\theta_r)$, $r = 0, 1, ..., N-1$. Assume that $H(z)$ is a general transfer function and hence of the form:

$$H(z) = \sum_{n=0}^{\infty} h(n) z^{-n}$$

If N values of z are available we can set up only N linear equations of the form:

$$H(r) = \sum_{n=0}^{\infty} h(n) z_r^{-n}, \quad r = 0, 1, ..., N-1$$

and hence only N terms of $\{h(n)\}$ can be determined. Hence the upper summation limit is $N-1$, i.e.:

$$H(r) = \sum_{n=0}^{N-1} h(n) z_r^{-n}, \quad r = 0, 1, ..., N-1$$

We can think of the problem as an interpolation and in terms of the Lagrange interpolation formula we have:

$$H(z) = \sum_{r=0}^{N-1} H(r) \prod_{\substack{i=0 \\ i \neq r}}^{N-1} \frac{(z^{-1} - z_i^{-1})}{(z_r^{-1} - z_i^{-1})}$$

Let:

$$C(z) = \prod_{i=0}^{N-1} (z^{-1} - z_i^{-1})$$

so that:

$$H(z) = C(z) \sum_{r=0}^{N-1} \frac{H(r)}{(z^{-1} - z_r^{-1})} \frac{1}{\prod_{\substack{i=0 \\ i \neq r}}^{N-1} (z_r^{-1} - z_i^{-1})}$$

This expression has a signal flow graph of the form given in **Figure 14** where:

$$C(z) = \prod_{i=0}^{N-1} (z^{-1} - z_i^{-1})$$

$$A(r) = H(r) \frac{1}{\prod_{\substack{i=0 \\ i \neq r}}^{N-1} (z_r^{-1} - z_i^{-1})}$$

If the points at which the required response values are chosen to be equidistant in frequency, i.e., if $z_i = \exp(i\theta_0)$, $\theta_0 = 2\pi/N$ and $i = 0, 1, 2, ..., N-1$, then z_i are the N roots of 1, and hence:

$$C(z) = \prod_{i=0}^{N-1} (z^{-1} - e^{-ji\theta_0}) = z^{-N} - 1$$

Moreover:

$$\prod_{\substack{i=0 \\ i \neq r}}^{N-1} (z_r^{-1} - z_i^{-1}) = \prod_{\substack{i=0 \\ i \neq r}}^{N-1} (z^{-1} - z_i)\Big|_{z=z_r} = \frac{C(z)}{z^{-1} - z_r^{-1}}\Big|_{z=z_r}$$

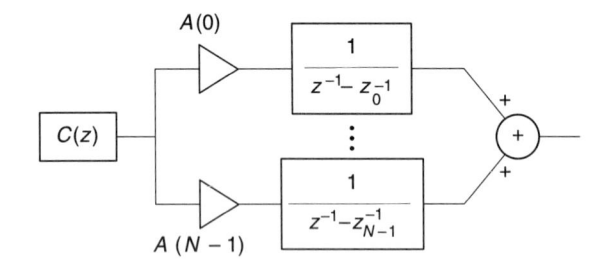

Figure 14 Signal flow graph.

which from the L'Hopital rule gives:

$$\left.\frac{Nz^{-(N-1)}}{1}\right|_{z=z_r} = Nz_r^{-(N-1)}$$

i.e.:

$$A(r) = H(r)\frac{z_r^{(N-1)}}{N}$$

Thus:

$$H(z) = \frac{z^{-N}-1}{N}\sum_{r=0}^{N-1} H(r)\, z_r^{(N-1)} \frac{1}{z^{-1}-z_r^{-1}}$$

and making use of $z_r^{(N-1)} = z_N^N z_r^{-1} = z_r^{-1}$ we obtain:

$$H(z) = \frac{1-z^{-N}}{N}\sum_{r=0}^{N-1} H(r)\frac{1}{1-z^{-1}z_r}$$

where $z_r = e^{j\pi r}$ and $\theta_r = (2\pi/N)r$. This is precisely the form one would obtain by means of discrete Fourier transforms. An example is shown in **Figure 15**. Frequency sampling technique with 128 samples, 32 of which are of unity value (lowpass) and 96 are set to zero value (the high frequency part). The phase is linear.

An important question is concerned with the estimation of the length N (i.e., the number of coefficients) of the impulse response. A simple empirical formula can be used for an approximate estimate as follows:

$$N = \frac{A}{20}\frac{f_s}{\Delta f}$$

where A is the discrimination in dB between the passband and stopband attenuation levels, Δf is the transition band between passband and stopband, and f_s is the sampling frequency. The last two quantities are in real frequencies.

However, the approach that can guarantee a performance within a set of prescribed specifications is based on the optimization of the amplitude response, in which the parameters of the optimization process are the filter coefficients. This becomes particularly straightforward in the linear phase case as all pertinent relationships are linear and hence guaranteed and unique optima are then available in the coefficient search space. A solution through the Remez exchange algorithm has been implemented with almost universal current use, with a Chebyshev or minimax objective function. The aim under this form of objective function is to design FIR filters in such a way that the maximum deviations of the frequency response from a set of prescribed specifications are minimized.

Minimax design From earlier considerations we can write the frequency (magnitude or amplitude) response as:

$$|H(\theta)| = P(\theta)\, Q(\theta)$$

where the various factors for the four cases, are indicated in **Table 2**. The phase response is given by $\phi(\theta) = ((N-1)/2)\theta$ in all four cases. With $D(\theta)$ as the desired response and $W(\theta)$ as a suitably chosen weighting function we can write the error expression as:

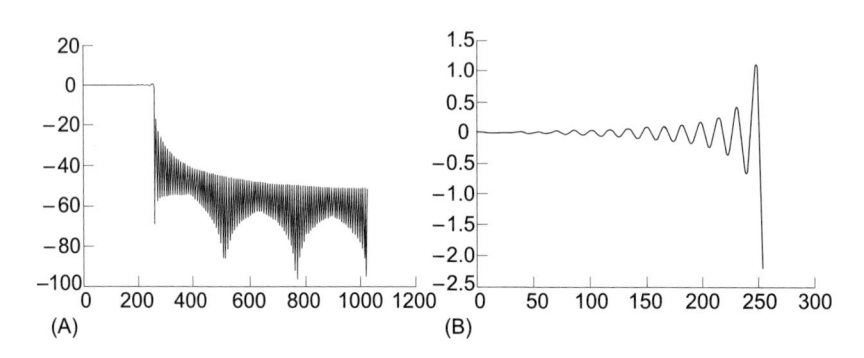

Figure 15 Frequency sampling technique. (A) Entire response, (B) passband response.

Table 2

Case	$Q(\theta)$	$P(\theta)$
1	1	$\sum \tilde{a}(n) \cos(n\theta)$
2	$\cos(\theta/2)$	$\sum \tilde{b}(n) \cos(n\theta)$
3	$\sin \omega \Delta t$	$\sum \tilde{c}(n) \cos(n\theta)$
4	$\sin(\theta/2)$	$\sum \tilde{d}(n) \cos(n\theta)$

$$E(\theta) = W(\theta)[D(\theta) - P(\theta) Q(\theta)]$$

or:

$$E(\theta) = W(\theta)Q(\theta)\left[\frac{D(\theta)}{Q(\theta)} - P(\theta)\right]$$

i.e.:

$$E(\theta) = W'(\theta)[D'(\theta) - P(\theta)]$$

The problem is then cast as in the optimization context, where we optimize:

$$\underset{\{h(n)\}}{\text{minimize}} \left\{ \underset{\theta \in \Theta}{\max} E(\theta) \right\}$$

The above is satisfied if and only if the error $E(\theta)$ has $(M + 1)$ extremal values on $\Theta(M = (N-1)/2)$. Specifically, when the weighting function $W(\theta)$ is constant then the error is equiripple, i.e., at consecutive extremal frequencies we have $E(\theta_i) = -E(\theta_i + 1)$. The problem is linear and can be solved easily using several techniques by computer (e.g., linear programming, Remez exchange algorithm, etc.). **Figure 16** shows an example for an order 128 and normalized cutoff and transition frequencies of 0.25 and 0.3, respectively.

IIR Digital Filter Design Principles

The design of infinite impulse response (IIR) digital filter transfer functions can be approached from two different perspectives. The first assumes the existence of an analog filter transfer function and then seeks to transform this into a digital form via suitable transformations. The other approach involves no such assumptions and attempts to effect the design directly from the prescribed specifications.

IIR digital filter design from analog filters The fundamental assumption made here is that there is an analog filter, $H(s)$, that can meet the frequency domain filtering requirements, and we seek to transform this transfer function, $G(z)$, into a form appropriate for use in the discrete-time domain.

The problem has been approached from a range of points of view.

1. Matched z-transform assumes that the analogue transfer function is of the form:

$$H(s) = \frac{\sum_{i=0}^{n}(s + a_i)}{\sum_{i=0}^{m}(s + b_i)}$$

Then use is made of the fact that a typical factor $(s + \lambda)$ in the numerator or denominator would correspond, in the z-domain to a typical factor $(1 - z^{-1}e^{-\lambda \Delta t})$. This is so in order to keep the frequency domain correspondence equivalent (i.e., with $z = e^{sT}$ it can be seen that the z-domain factor would correspond to the same s-value as the s-domain factor).

Therefore a factor-by-factor replacement produces:

$$G(z) = \frac{\sum_{i=0}^{n}\left(1 - z^{-1}e^{-a_i \Delta t}\right)}{\sum_{i=0}^{m}\left(1 - z^{-1}e^{-b_i \Delta t}\right)}$$

Note, however, that the frequency response of $G(z)$ is assumed on the circumference of the unit circle and not along a straight line (i.e., the imaginary axis) as is the case for $H(s)$. Thus the retention of the finite poles and zeros achieved by this approach does not guarantee that the two filter transfer functions would have corresponding frequency responses. Observe, in addition, that the zeros at $\pm\infty$ of $H(s)$ are not mapped to $G(z)$ in the above equation and an additional factor for these is required of the form $(1 + z^{-1})^{m-n}$ known as the guard filter. Even with the incorporation of the guard filter, the correspondence between the analog and digital filter responses of $H(s)$ and $G(z)$ is tenuous. However, the correspondence improves for higher than normal sampling rates (i.e., with significant oversampling). An example is shown in **Figure 17**.

2. Invariant Impulse Response Method
In this approach the analog transfer function is expressed in the form:

$$H(s) = \sum_i \frac{A_i}{s + \beta_i}$$

where the parameters A_i, β_i are either real or are complex if they occur in complex conjugate pairs.

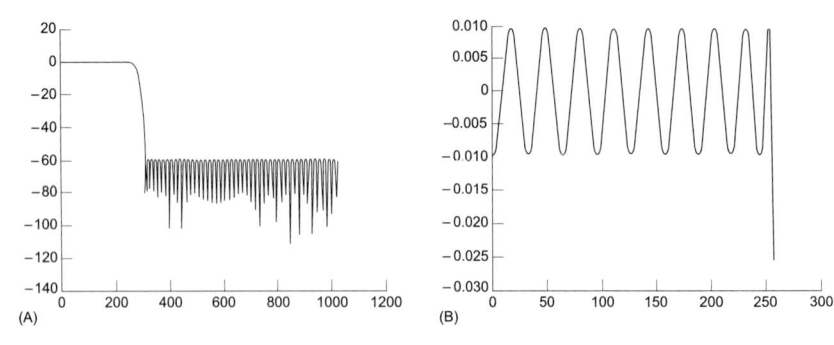

Figure 16 Minimax technique. (A) Entire response, (B) passband response.

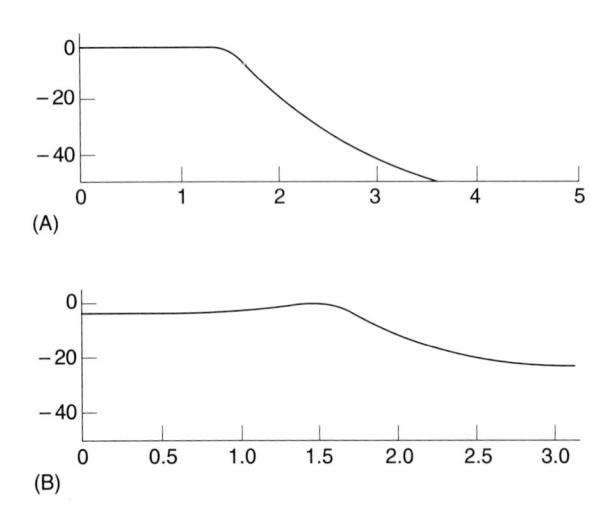

Figure 17 (A) Analog filter response. (B) Matched pole-zero filter response.

The impulse response of $H(s)$ is therefore given by $h(t) = \sum_i A_i e^{-\beta_i t}$. The next step is to sample $h(t)$ to obtain the impulse response $\{h(n\Delta t)\}$ of the corresponding digital filter, i.e., $h(n\Delta t) = \sum_i A_i e^{-\beta_i n\Delta t}$.

On taking the z-transform of this sampled impulse response we obtain the transfer function:

$$G(z) = \sum_i \frac{A_i}{1 - z^{-1} e^{-\beta_i \Delta t}}$$

This transfer function is taken to be the digital filter transfer function.

This approach is very simple to use, moreover $G(z)$ is stable when $H(s)$ is stable. However, there are problems with respect to aliasing or folding of the analog filter spectrum. In order to obtain reasonable responses we need to sample at rates, which are much higher that need be, as **Figure 18** shows.

An alternative concept in this range of techniques involves s-plane to z-plane transformations. The principle is as follows: obtain $G(z)$ from $H(s)$ by replacing the complex variable s by an appropriate function $f(z)$. A very simple solution, which is not entirely satisfactory, is to employ the so-called backward difference formula. Essentially from the complex variable domain, this formula can be seen as a first-order approximation of the correspondence between z and s. Let us write $z^{-1} = e^{-s\Delta t}$ and let us further expand the exponential in a Taylor series and retain the first two terms to obtain $z^{-1} = 1 - s\Delta t$. This can be rearranged to yield $s = (1 - z^{-1})/\Delta t$. However, in a transfer function $H(s)$ there are powers of s higher than the first, for which we use $s^n = \left[(1 - z^{-1})/\Delta t\right]^n$. A simple example is as follows. Let $H(s) = 1/(s + a)$. Then we obtain:

$$G(z) = \frac{1}{[(1 - z^{-1})/\Delta t] + a}$$

Figure 19 shows the two frequency responses. In order to appreciate the limitations of this approach let us re-express the relationship between the complex variables as follows:

$$z = \frac{1}{1 - s\Delta t}$$

In order to see whether or not we will always obtain stable digital filter transfer functions let us set $s\Delta t = \alpha + j\Omega$ so that:

$$|z| = \frac{1}{\sqrt{\left((1 - \alpha)^2 + \Omega^2\right)}}$$

It can be seen that whenever $\alpha < 0$ then $|z| < 1$, and hence the left-hand side of the s-plane maps to the inside of the unit circle. This means that we shall always obtain stable digital filter transfer functions from stable analog filter transfer functions.

The correspondence between the frequency responses is seen by writing:

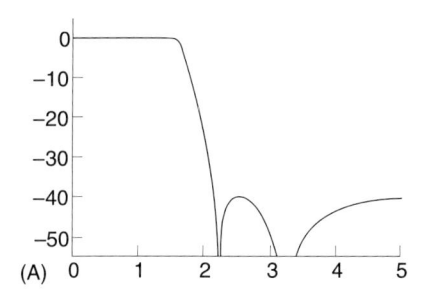

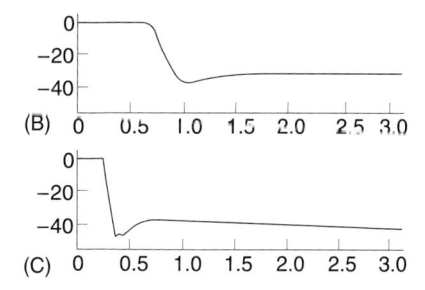

Figure 18 Different sampling frequencies. (A) Prototype analog filter response, (B) impulse invariance filter response oversampled × 1.5, (C) impulsive invariance filter response oversampled × 4.

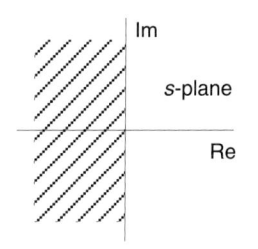

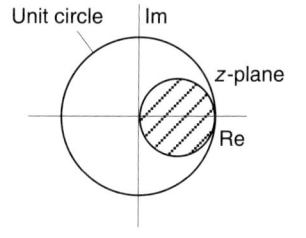

Figure 19 Frequency responses.

Table 3

G(z)	Available H(s)
Stable	Stable
Real and rational in z^{-1}	Real and rational in s
Order n	Order n
Lowpass cutoff Ω_c	Lowpass cutoff θ_c

$$z = \frac{1}{2}\left[1 + \frac{1 + s\Delta t}{1 - s\Delta t}\right]$$

so that when $s = j\Omega$ it can be written as:

$$z = \frac{1}{2}\left[1 + e^{j2\tan^{-1}\Omega\Delta t}\right]$$

i.e., the complex variable z will be located on a circle of radius 1/2 shifted by 1/2 when the complex variable s is on the imaginary axis.

Forward difference: In a manner similar to that above we obtain $z = 1 + s\Delta t$. However, for $s = \alpha + j\Omega$ it can be seen that $|z|>1$ and hence it leads to unstable transfer functions $G(z)$.

The method that is used extensively is the bilinear transform outlined below.

Bilinear transformation A simple way to derive the bilinear transformation is to view it as mapping between the two planes. In this sense let us begin with a table of properties and requirements of the transfer functions $G(z)$ and $H(s)$, respectively. The required transformation will operate as follows. To obtain $G(z)$ we shall replace s in $H(s)$ by a function $f(z)$. Hence in view of the relationships in **Table 3**, $f(z)$ must be a real and rational function in z and of order one. Hence we can write $f(z)$ as:

$$f(z) = \frac{az + b}{cz + d}$$

Now let us make it specific to a desired form of filtering. Let us assume that the given analog filter transfer function $H(s)$ is lowpass, and we need a lowpass digital filter transfer function $G(z)$. Under these requirements we have the following correspondence between points on the s-plane and on the z-plane. The point $s = 0$ is mapped onto the point $z = 1$ and hence $f(1) = 0$, i.e., $a + b = 0$. Further, the point $s = \pm j\infty$ is mapped onto the point $z = -1$ yielding $f(-1) = \pm j\infty$, i.e., $c - d = 0$. Thus we obtain:

$$f(z) = \left(\frac{a}{c}\right)\frac{z-1}{z+1}$$

The ratio (a/c) is fixed from the correspondence at one more point that can be taken to be the cutoff frequency $\theta_c = \omega_c \Delta t \leftrightarrow \Omega_c$ for $z = e^{j\theta}$ on $C: |z| = 1$. The transformation on the unit circle becomes:

$$f(z)|_c = \left(\frac{a}{c}\right) j \tan\frac{\theta}{2}$$

and hence the cutoff frequency correspondence becomes:

$$j\Omega_c = \left(\frac{a}{c}\right) j \tan\frac{\theta_c}{2}$$

Therefore the bilinear transformation takes the form:

$$s = \left(\frac{\Omega_c}{\tan(\omega_c \Delta t/2)}\right) \frac{1 - z^{-1}}{1 + z^{-1}}$$

For stability considerations we write:

$$z = \frac{(1+s)A}{(1-s)A} \quad A = \frac{\tan(\omega_c \Delta t/2)}{\Omega_c} > 0$$

and set $sA = \tilde{a} + j\tilde{\Omega}$. Then we have:

$$|z|^2 = \frac{(1+\tilde{a})^2 + \tilde{\Omega}^2}{(1-\tilde{a})^2 + \tilde{\Omega}^2}$$

i.e., the region outside the unit circle $|z|>1$ corresponds to $\tilde{a}>0$ which is the right-hand half s-plane, the region inside the unit circle corresponds to $\tilde{a}<0$ which is the left-hand half s-plane (where the poles of a stable $H(s)$ are located), and the circumference of the unit circle corresponds to $\tilde{a}=0$. Hence the stable region is mapped into a stable region (**Figure 20**). **Figure 21** shows aspects of the bilinear transformation.

It is possible to transform a given digital filter transfer function into others by using allpass transformations. Indeed, if the given digital filter transfer function is lowpass then **Table 4** gives the required transformations for the target filter transfer function. All frequencies are taken to be normalized, i.e., a typical value is given by $\theta = \omega \Delta t$ where ω is the angular frequency (in rad s^{-1}) and Δt is the sampling period (in s). Thus to transform a lowpass digital filter with cutoff frequency, β, to the target filter with the target requirements we must replace z^{-1} by the corresponding transformation. The response shown in **Figure 22A** is obtained by applying the bilinear transform to the analog filter response used earlier in the impulse invariance method. The response in **Figure 22B** is that of bandpass filter obtained by applying the lowpass to bandpass digital filter transformation. For a wider perspective the reader should consult the references given below.

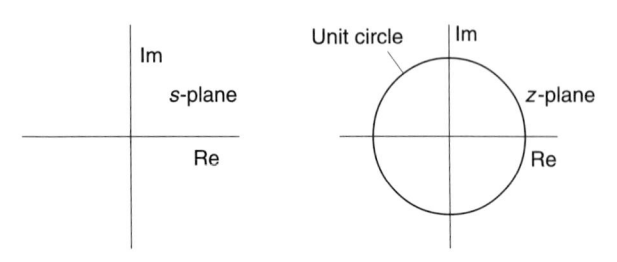

Figure 20 Stable regions.

Table 4

Target filter	Transformation	Associated design formulas
Lowpass cutoff θ_c	$\dfrac{z^{-1} - \alpha}{1 - \alpha z^{-1}}$	$\alpha = \dfrac{\sin\left(\dfrac{\beta - \theta_c}{2}\right)}{\sin\left(\dfrac{\beta + \theta_c}{2}\right)}$
Highpass cutoff θ_c	$-\dfrac{z^{-1} - \alpha}{1 + \alpha z^{-1}}$	$\alpha = -\dfrac{\cos\left(\dfrac{\beta - \theta_c}{2}\right)}{\cos\left(\dfrac{\beta + \theta_c}{2}\right)}$
Bandpass cutoff θ_1, θ_2 center θ_0	$-\dfrac{z^{-2} - \dfrac{2\alpha k}{k+1}z^{-1} + \dfrac{k-1}{k+1}}{\dfrac{k-1}{k+1}z^{-2} - \dfrac{2\alpha k}{k+1}z^{-1} + 1}$	$\alpha = \cos\theta_0 = \dfrac{\cos\left(\dfrac{\theta_2 + \theta_1}{2}\right)}{\cos\left(\dfrac{\theta_2 - \theta_1}{2}\right)}$ $k = \cot\left(\dfrac{\theta_2 - \theta_1}{2}\right)\tan\dfrac{\beta}{2}$
Bandstop cutoff θ_1, θ_2 center θ_0	$\dfrac{z^{-2} - \dfrac{2\alpha}{1-k}z^{-1} + \dfrac{1-k}{1+k}}{\dfrac{1-k}{1+k}z^{-2} - \dfrac{2\alpha}{1+k}z^{-1} + 1}$	$\alpha = \cos\theta_0 = \dfrac{\cos\left(\dfrac{\theta_2 + \theta_1}{2}\right)}{\cos\left(\dfrac{\theta_2 - \theta_1}{2}\right)}$ $k = \tan\left(\dfrac{\theta_2 - \theta_1}{2}\right)\tan\dfrac{\beta}{2}$

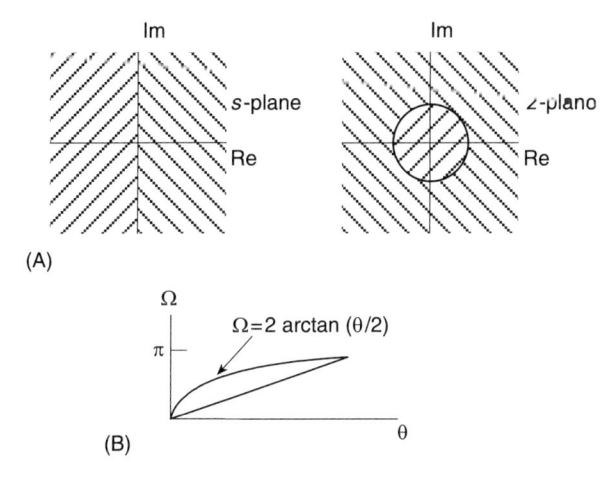

Figure 21 Bilinear transformation. (A) Correspondence in the regions on the two planes related to the bilinear transformation, i.e., the left-half of the s-plane maps to the inside of the unit circle in the z-plane and the right-half of the s-plane maps to the outside of the unit circle. (B) Nonlinear mapping between the continuous-time frequency variable Ω and the discrete-time normalized frequency variable $\theta = \omega \Delta t$.

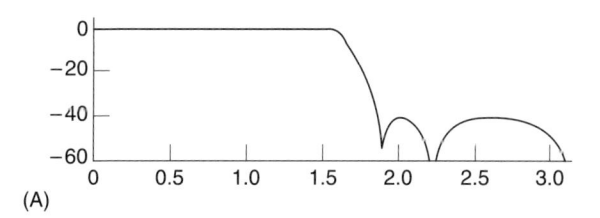

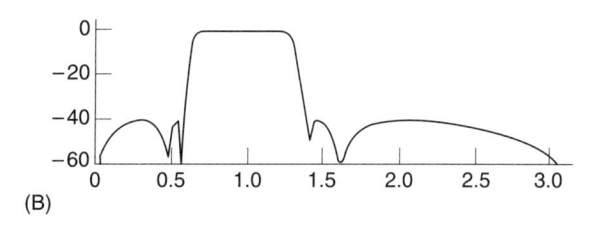

Figure 22 Digital filter response. (A) Lowpass; (B) bandpass.

See also: **Signal processing, model based methods**; **Spectral analysis, classical methods**; **Transform methods**; **Windows**.

Further Reading

Antoniou A (1982) Accelerated procedure for the design of equiripple non recursive digital filters. *IEE Proceedings, Pt C* 129: 1–10 (see *IEE Proceedings, Pt C* 129: 107 for errata).

Antoniou A (1993) *Digital Filters: Analysis, Design, and Applications*, 2nd edn. New York: McGraw Hill.

Constantinides AG (1970) Spectral transformations for digital filters. *IEE Proceedings* 117: 1585–1590.

Lim YC (1983) Efficient special purpose linear programming for FIR filter design. *IEEE Transactions on Acoustics, Speech and Signal Processing* ASSP-31: 96–98.

Lim YC, Constantinides AG (1979) New integer programming scheme for nonrecursive digital filter design. *Electronic Letters* 15: 812–813.

Lim YC, Parker SR, Constantinides AG (1982) Finite wordlength FIR filter design using integer programming over a discrete coefficient space. *IEEE Transactions Acoustics, Speech and Signal Processing* ASSP-30: 661–664.

McClellan JH, Parks TW, Rabiner LR (1973) A computer program for designing optimum FIR linear phase digital filters. *IEEE Transactions on Audio and Electroacoustics*, AU-21: 5–526.

Oppenheim AV, Schafer RW (1975) *Digital Signal Processing*. Englewood Cliffs, NJ: Prentice Hall.

Proakis JC, Manolakis DG (1992) *Digital Signal Processing Principles, Algorithms and Applications*, 2nd edn. New York: Macmillan.

Rabiner LR (1971) Techniques for designing finite duration impulse response digital filters. *IEEE Transactions on Communication Technology* COM-19: 188–195.

Rabiner LR, Gold B (1975) *Theory and Application of Digital Signal Processing*. Englewood Cliffs, NJ: Prentice Hall.

Regalia PA, Mitra SK, Vaidyanathan PP (1988) The digital all-pass filter: A versatile signal processing building block. *Proceedings of the IEEE* 76: 19–37.

DISCRETE ELEMENTS

S S Rao, University of Miami, Coral Gables, FL, USA

Copyright © 2001 Academic Press

doi:10.1006/rwvb.2001.0108

A vibrating system basically consists of three components or elements: a spring or elastic element that stores the potential energy; a mass or inertia element that stores the kinetic energy; and a damping element that dissipates the energy. The vibration of a system involves the transfer of its kinetic energy to potential energy and potential energy to kinetic energy, alternately. Thus, if the damper is absent, the system will undergo oscillations with constant amplitude in which the sum of potential and kinetic energies will remain constant. On the other hand, if the damper is present, energy must be supplied externally to sustain oscillations with constant amplitude.

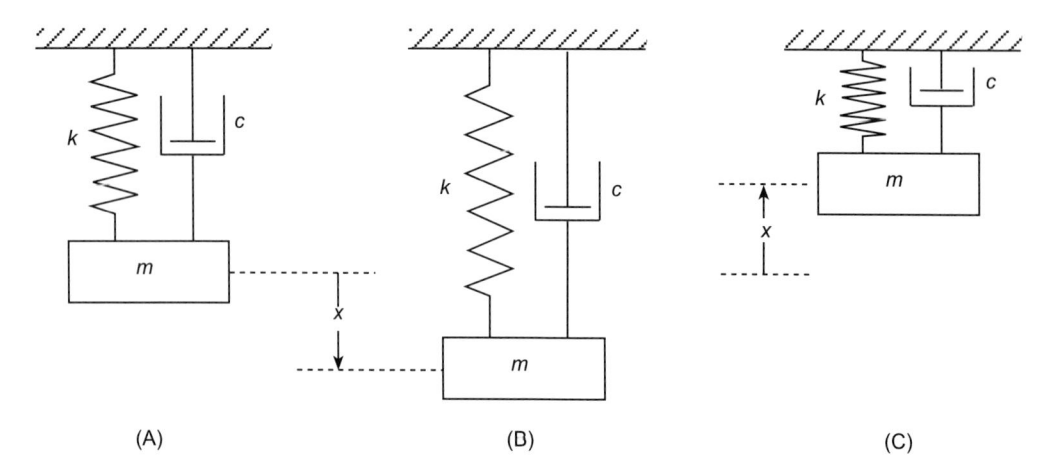

Figure 1 (A)–(C) A spring–mass–damper system.

As an example, consider the spring-mass-damper system shown in **Figure 1**, in which the mass oscillates in the vertical direction. **Figure 1A** denotes the mass in static equilibrium position. Let the mass be released after giving it an initial displacement x (downwards), as shown in **Figure 1B**. At this position, the mass will have a zero velocity and hence its kinetic energy will be zero. Also, at this position, the spring is stretched by an amount x; hence it will have a strain or potential energy. Since the spring force acts upwards, the mass starts moving upwards from the bottom-most position. When the mass reaches the static equilibrium position with $x = 0$, all the potential energy of the spring will have been converted into kinetic energy. As the mass has attained its maximum velocity (in the upward direction), the mass will not stop at this position; it will continue to move upwards until its velocity reduces to zero (**Figure 1C**). At the uppermost position, the kinetic energy will be zero but the strain or potential energy of the system will be maximum. Thus the kinetic energy has been converted into potential energy. Again, owing to the downward spring force, the mass starts to move downwards with progressively increasing velocity and passes the static equilibrium position again. At the static equilibrium position, the system will have maximum kinetic energy and zero potential energy. This process repeats and the the mass will have an oscillatory motion. In the absence of a damper, the oscillations continue forever. However, in the presence of a damper, the amplitude of oscillations gradually diminishes owing to dissipation of energy.

Torsional Systems

The vibration of a rigid body about an axis is known as rotational or torsional vibration. The basic elements of a rotational system are known as the torsional spring, mass moment of inertia, and torsional damper, analogous to the spring, mass, and damper of a translational or linear system.

Modeling of Discrete Elements

The basic elements of a vibratory system can be modeled as continuous (distributed) or discrete (lumped) elements. If the elements are continuous the governing equations will be partial differential equations, which are more difficult to solve. On the other hand, if the elements are discrete the governing equations will be ordinary differential equations, which are relatively easier to solve. Hence it is desirable to approximate the elements as discrete elements. Although there are no unified rules or guidelines available for modeling continuous elements as equivalent discrete ones, engineering intuition, energy balance, and equilibrium considerations can be used in most cases. In the modeling process, care must be taken to insure that the approximation process yields results that are reasonably accurate.

Mass or Inertia Elements

Although mass is distributed throughout a body, it can be considered as a lumped or point mass located at the center of mass of the body. Whenever a mass (m) undergoes acceleration ($\ddot{x}$), the force developed (F) is given by $F = m\ddot{x}$, according to Newton's second law of motion. The kinetic energy associated with a mass moving at a velocity v is given by $E = \frac{1}{2}mv^2$. Basically, three methods can be used to find the equivalent lumped mass of a body. These methods are based on using: (i) engineering intuition; (ii) equivalence of masses; and (iii) equivalence of kinetic energies. The engineering intuition method can be seen with reference to the multistorey building

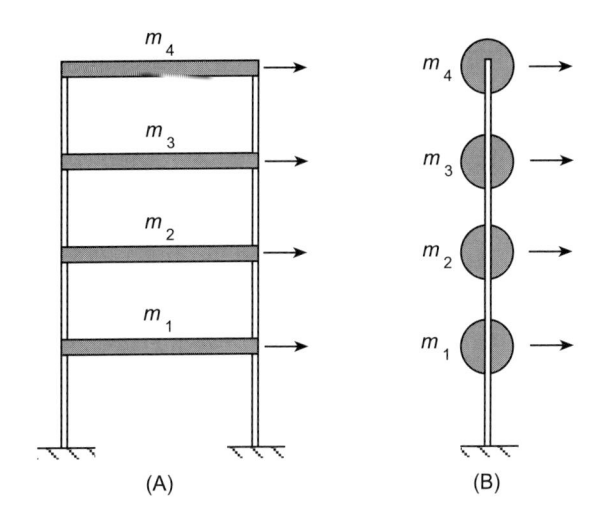

Figure 2 A multistorey building frame.

frame shown in **Figure 2A**. Usually the masses of columns can be neglected compared to the masses of the floors. Hence the masses of the floors can be assumed to be lumped masses, as shown in **Figure 2B**. The method of using equivalence of mass can be explained with reference to the connecting rod shown in **Figure 3A**. In an internal combustion engine, one end of the connecting rod (O_1) rotates while the other end (O_2) reciprocates. Thus, part of the mass of the connecting rod undergoes rotary motion while the remaining part undergoes reciprocating motion. The connecting rod can be replaced by two lumped or point masses m_1 and m_2, as shown in **Figure 3B**. The values of m_1 and m_2 can be determined using the conditions that: (i) the total mass must be the same for both systems; (ii) the center of mass must be the same for both systems; and (iii) the moment of inertia must be the same for both systems, which yield the relations:

$$m = m_1 + m_2 \quad [1]$$

$$ml_2 = m_1(l_1 + l_2) \quad [2]$$

$$I_0 = m_1 l_1^2 + m_2 l_2^2 \quad [3]$$

where I_0 is the mass moment of inertia about the center of mass of the original connecting rod. Since m, I_0, and l_1 are usually known, eqns [1]–[3] can be solved to find the values of m_1, m_2, and l_2.

The use of equivalence of kinetic energy can be illustrated with reference to the bar under axial motion (**Figure 4A**). To determine the equivalent lumped mass M at the free end, we equate the kinetic energies of the two systems:

$$\frac{1}{2}Mv^2 = \frac{1}{2}\int_{x=0}^{l} \rho A\, dx (\dot{x})^2 \quad [4]$$

where A is the cross-sectional area and ρ is the density. Assuming that the velocity, $\dot{x}$, varies linearly from zero at the fixed end to v at the free end, eqn [4] can be evaluated to find:

$$\frac{1}{2}Mv^2 = \frac{1}{2}\int_{x=0}^{l} \left(\frac{xv}{l}\right)^2 \rho A\, dx = \frac{1}{2}\frac{\rho A}{l^2}v^2\left(\frac{l^3}{3}\right) \quad [5]$$

which gives the equivalent mass as:

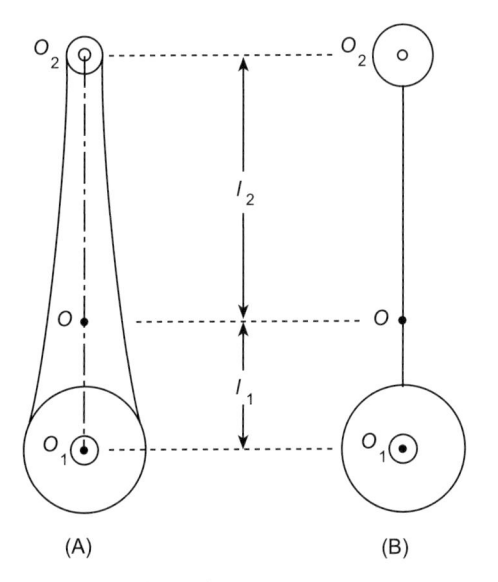

Figure 3 A connecting rod.

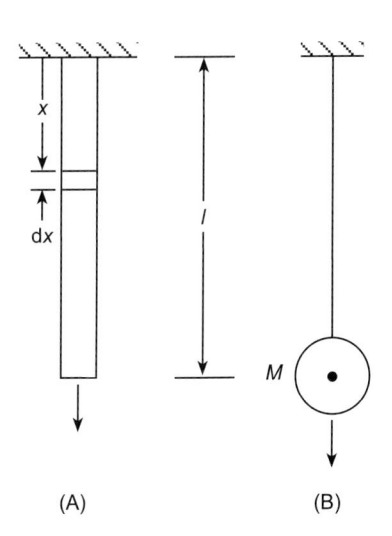

Figure 4 A bar under axial motion.

$$M = \frac{\rho Al}{3} = \frac{m}{3} \quad [6]$$

where m denotes the mass of the bar. **Table 1** gives the equivalent masses of some typical mass or inertia elements.

Spring or Elastic Elements

A linear spring is a mechanical element that develops a force (F) proportional to the amount of deformation (x) it undergoes, as $F = kx$ where the constant of proportionality, k, is called the spring stiffness, spring constant, or spring rate. The force acts in a direction opposite to that of deformation. Helical springs and elastic members such as beams, plates, and rings can be considered as spring elements. For simplicity, a spring element is assumed to have no mass or damping. The energy stored in a linear spring due to a deformation x is given by the area under the force–deformation curve of **Figure 5A** as $E = \frac{1}{2}kx^2$.

Although the force–deformation behavior of a linear spring is as shown in **Figure 5A**, most practical springs behave nonlinearly, as shown in **Figure 5B**. The nonlinear behavior is related to the nonlinear stress–strain curve exhibited by most elastic materials. A nonlinear spring can be approximated as a linear spring in the neighborhood of any specific deformation ($\tilde{x}^*$). For this, consider the Taylor series approximation of the spring force, $f(\tilde{x})$, about $\tilde{x}^*$:

$$f(\tilde{x}) = f(\tilde{x}^*) + \frac{df}{d\tilde{x}}(\tilde{x}^*)(\tilde{x} - \tilde{x}^*) \\ + \frac{1}{2}\frac{d^2 f}{d\tilde{x}^2}(\tilde{x}^*)(x - \tilde{x}^*)^2 + \ldots \quad [7]$$

By neglecting terms involving higher derivatives of $f(\tilde{x})$, eqn [7] can be used to express the incremental force developed, $F = f(\tilde{x}) - f(\tilde{x}^*)$, for the net deformation, $x = \tilde{x} - \tilde{x}^*$, as $F = kx$ where ($k = df(\tilde{x}^*)/d\tilde{x}$) can be considered as the spring stiffness (**Figure 5C**). The spring constants of some typical elastic elements are given in **Table 2**.

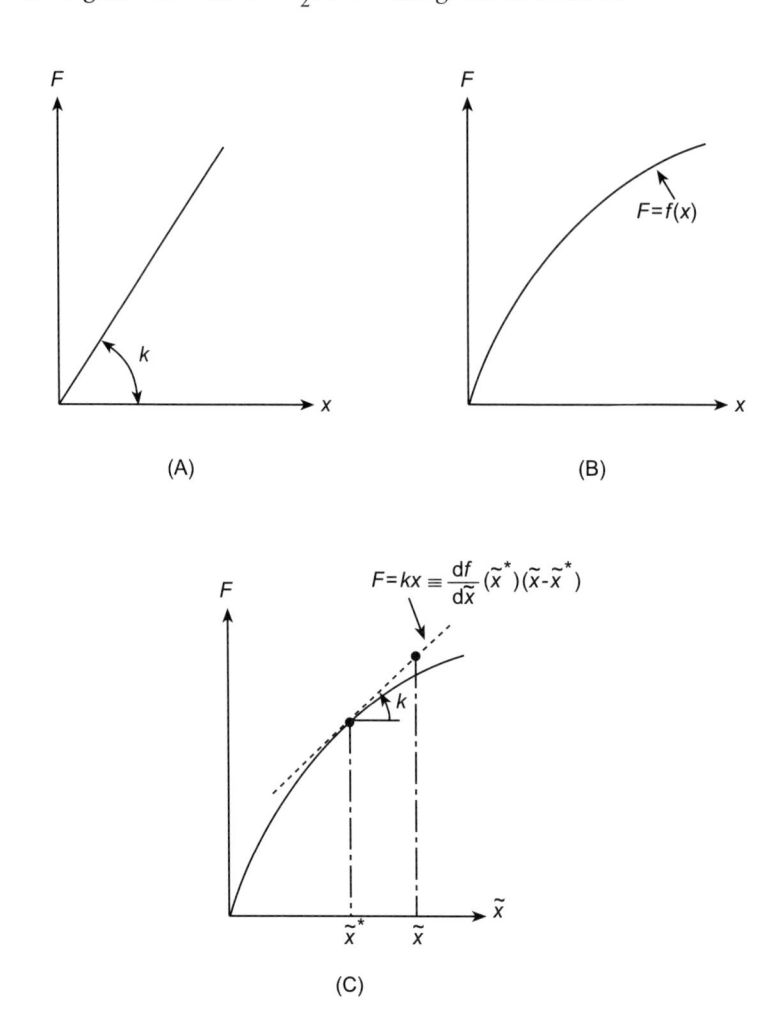

Figure 5 Force-deflection behavior of a spring element: (A) linear; (B) nonlinear, (C) linearization.

Table 1 Equivalent masses of typical mass/inertia elements

1. Cantilever beam

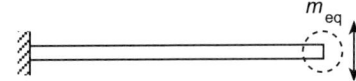

   $m_{eq} = \dfrac{m}{4}$ (m = mass of beam)

2. Fixed-fixed beam

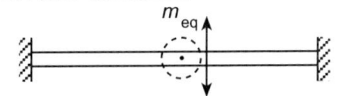

   $m_{eq} = 0.375m$ (m = mass of beam)

3. Simply supported beam

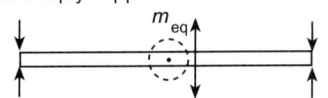

   $m_{eq} = \dfrac{m}{2}$ (m = mass of beam)

4. Spring in axial motion

   $m_{eq} = \dfrac{m}{3}$ (m = mass of spring)

5. Masses on a hinged bar

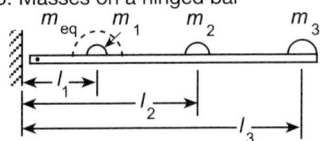

   $m_{eq} = m_1 + \left(\dfrac{l_2}{l_1}\right)^2 m_2 + \left(\dfrac{l_3}{l_1}\right)^3 m^3$

6. Bar under torsion

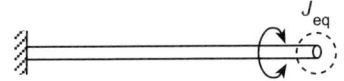

   $J_{eq} = \dfrac{J}{3}$ (J = mass moment of inertia of beam)

7. Coil spring in torsion

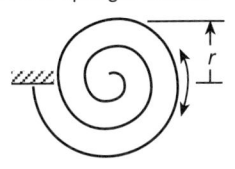

   $J_{eq} = \dfrac{mr^2}{4}$ (m = mass of coil spring; r = outside radius)

8. Plate moving on fluid

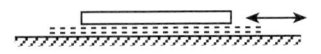

   $m_{eq} = \dfrac{m}{3}$ (m = mass of fluid)

9. Piston moving in fluid

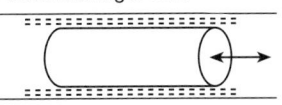

   $m_{eq} = \dfrac{m}{3}$ (m = mass of surrounding fluid)

10. Disk rotating on fluid

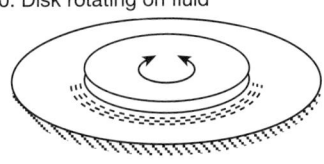

    $J_{eq} = \dfrac{J}{3}$ (J = mass moment of inertia of fluid)

11. Cylinder rotating in fluid

    $J_{eq} = \dfrac{J}{3}$ (J = mass moment of inertia of fluid)

Table 2 Spring constants of typical elastic elements

1. Spring in axial motion

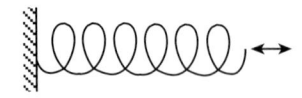

$$k_{eq} = \frac{Gd^4}{8D^3N}$$

d = wire diameter D = coil diameter
G = shear modulus N = number of turns

2. Springs in series

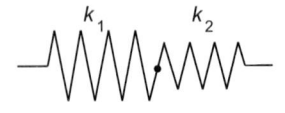

$$k_{eq} = k_1 + k_2$$

3. Springs in parallel

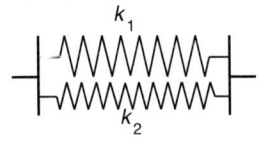

$$k_{eq} = \frac{1}{\frac{1}{k_1} + \frac{1}{k_2}}$$

4. Spring in torsion

$$k_{eq} = \frac{Ed^4}{64DN}$$

5. Fixed–fixed beam

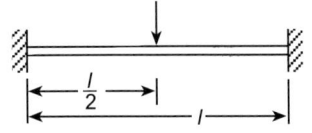

$$k_{eq} = \frac{192EI}{l^3}$$

E = Young's modulus
I = area moment of inertia

6. Cantilever beam

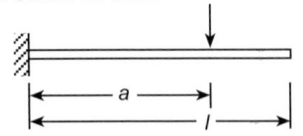

$$k_{eq} = \frac{6EI}{a^2(3l-a)}$$

7. Simply supported beam

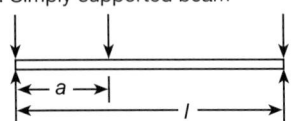

$$k_{eq} = \frac{3EIl}{a^2(l-a)^2}$$

8. Leaf spring

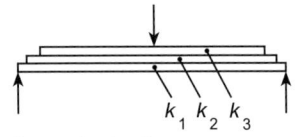

$$k_{eq} = \sum_i k_i$$

k_i = spring constant of beam i

9. Bar under torsion

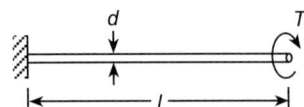

$$k_{t_{eq}} = \frac{\pi d^4 G}{32l}$$

G = shear modulus

10. Cantilever beam under end moment

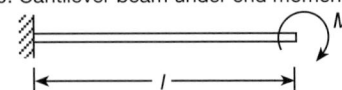

$$k_{eq} = \frac{2EI}{l^2}$$

11. Bar under axial motion

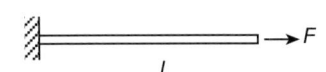

$$k_{eq} = \frac{AE}{l}$$

A = area of cross-section
E = Young's modulus

Table 2 (continued)

12. Ring

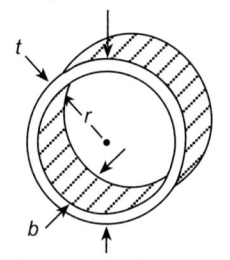

$$k_{eq} = \frac{Ebt^3}{1.79r^3}$$
E = Young's modulus

13. Semicircular ring

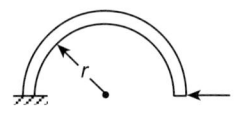

$$k_{eq} = \frac{2EI}{3\pi r^3}$$
E = Young's modulus
I = area moment of inertia

14. Simply supported circular plate

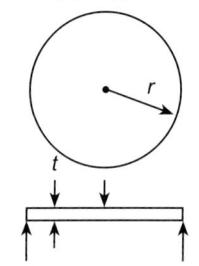

$$k_{eq} = \frac{16\pi D(1+v)}{r^2(3+v)}$$
v = Poisson's ratio
E = Young's modulus
$$D = \frac{Et^3}{12(1-v^2)}$$

15. Clamped circular plate

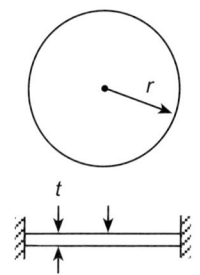

$$k_{eq} = \frac{16\pi D}{r^2}$$
$$D = \frac{Et^3}{12(1-v^2)}$$

16. Simply supported circular plate

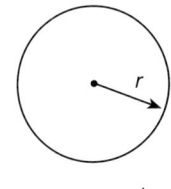

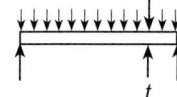

$$k_{eq} = \frac{64\pi D(1+v)}{r^2(5+v)}$$
$$D = \frac{Et^3}{12(1-v^2)}$$

17. Simple pendulum

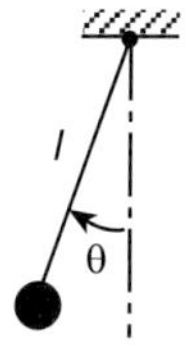

$$k_{eq} = \frac{mg}{l}$$
g = acceleration due to gravity

Table 2 (continued)

18. Cylinder floating in fluid

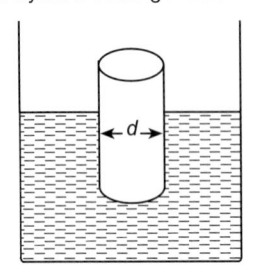

$$k_{eq} = \frac{\pi d^2 \gamma}{4}$$

γ = unit weight of fluid

19. Piston moving in pressurized cylinder

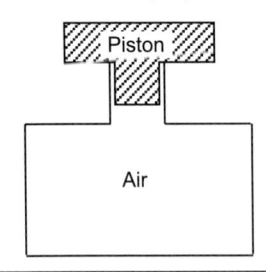

$$k_{eq} = \frac{p\gamma A^2}{v}$$

A = Cross-sectional area of piston
v = volume of air
p = pressure of air
γ = adiabatic constant (1.4)

Damping Elements

A damping element is a mechanical member in which a force (F) is developed proportional to the rate of deformation $(\dot{x} = dx/dt)$ it undergoes as $F = c\dot{x} = cv$, where $\dot{x} = dx/dt = v$ is the rate of change of deformation or velocity of the member and the constant of proportionality (c) is known as the damping constant. A damping element dissipates an energy (E) in time t given by:

$$E = \int_0^t Fv \, dt = \frac{1}{2} \int_0^t cv^2 \, dt \qquad [8]$$

For example, if the deformation of the damping element is simple harmonic with $x(t) = X \sin \omega t$, where X is the amplitude and ω is the angular frequency of simple harmonic motion, the energy dissipated in one cycle of motion $(t = 2\pi/\omega)$ can be determined as $E = \pi c \omega X^2$. The damping constants of some representative dampers are given in **Table 3**. Although a damping element is usually assumed to be linear (**Figure 6A**) for simplicity, most practical dampers behave nonlinearly, as shown in **Figure 6B**. As in the case of a nonlinear spring, a nonlinear damper can be approximated as a linear damper around any specific rate of deformation or velocity using Taylor series approximation, as shown in **Figure 6C** with $F = cv$ where:

$$c = \frac{df}{d\tilde{v}}(\tilde{v}^*)(\tilde{v} - \tilde{v}^*) \qquad [9]$$

is the linear damping constant and $v = \tilde{v} - \tilde{v}^*$ is the net rate of deformation of the element.

Types of Dampers

Depending on the mechanism assumed for energy dissipation, dampers can be classified as viscous, Coulomb or hysteretic dampers. In a viscous damper, the damping element is assumed to vibrate in a fluid medium such as air, water, or oil. During the motion of the element, the fluid offers resistance to the motion proportional to the shape and size of the element, the viscosity of the fluid and the velocity of motion. Viscous damping is the most commonly used damping model in practice. When two surfaces slide relative to one another, separated by a fluid, as in the case of the motion of a piston in a cylinder or the motion of a journal in a lubricated bearing, the viscous damping model can be used. When relative motion occurs between two surfaces with no or inadequate lubricant, the dry friction force developed between the surfaces causes damping. Hence this type of damping is also known as dry friction damping. In hysteretic damping, the friction developed between the internal planes of the body or structure during deformation is assumed to cause the dissipation of energy. Hence this type of damping is also known as structural, solid, or material damping.

Nomenclature

A	cross-sectional area
E	Young's modulus

Table 3 Damping constants of typical dampers

1. Plate moving on fluid

$$c_{eq} = \frac{\mu A}{t}$$

A = surface area of plate
μ = viscosity of fluid

2. Piston moving in a cylinder

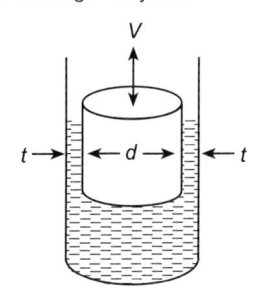

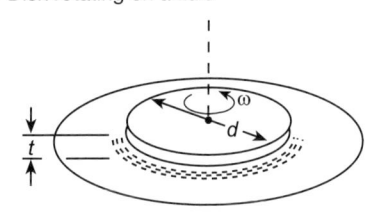

$$c_{eq} = \frac{3\pi\mu d^3 l}{4t^3}\left(1 + \frac{2t}{d}\right)$$

3. Disk rotating on a fluid

$$c_{t_{eq}} = \frac{\pi\mu d^4}{16t}$$

4. Mass moving on a dry surface

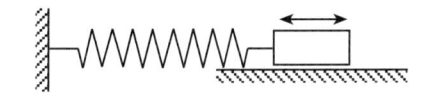

$$c_{eq}7 = \frac{4\mu N}{\pi\omega X}$$

μN = friction force
ω = frequency
X = amplitude of vibration

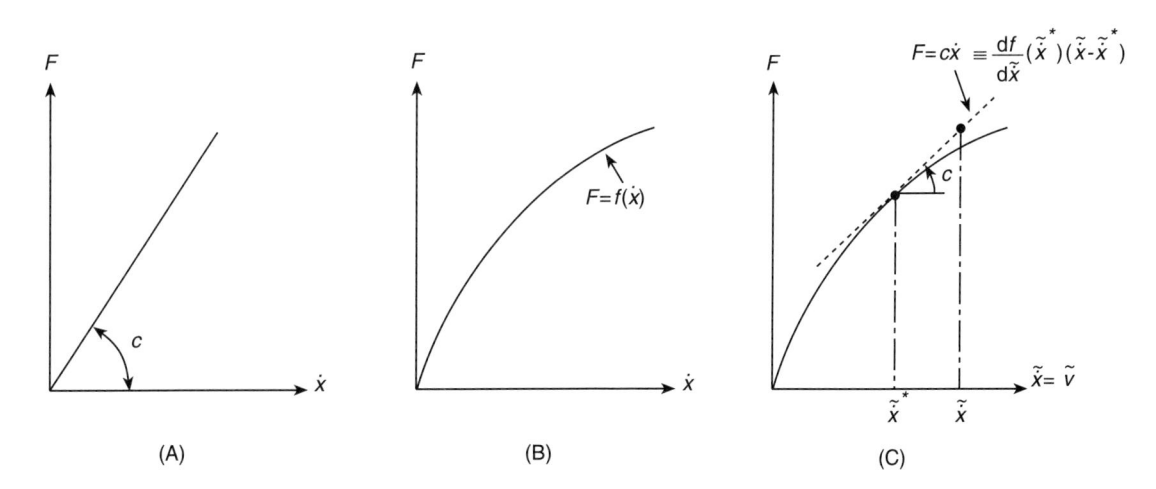

Figure 6 Force-velocity behavior of a damping element: (A) linear; (B) nonlinear; (C) linearization.

F	force	ν	Poisson's ratio
g	acceleration due to gravity	ρ	density
G	shear modulus		
I	moment of inertia		
I_0	mass moment of inertia		
p	pressure of air		
r	radius		
v	volume of air		
$\dot{x}$	velocity		
$\ddot{x}$	acceleration		
X	amplitude of vibration		
γ	viscosity of fluid		
μ	viscosity of fluid		

See also: **Damping, active**; **Damping materials**; **Damping measurement**; **Dynamic stability**.

Further Reading

Cochin I (1980) *Analysis and Design of Dynamic Systems.* New York: Harper & Row.

Rao SS (1995) *Mechanical Vibrations*, 3rd edn. Massachusetts: Addison-Wesley.

DISKS

D J Ewins, Imperial College of Science, Technology and Medicine, London, UK

Copyright © 2001 Academic Press

doi:10.1006/rwvb.2001.0133

Introduction

Many engineering structures and machines that are prone to vibration problems contain components which can be described as 'disk-like'. Some of most common examples include:

- turbomachinery disks, including bladed disks
- gear wheels
- computer disks
- brake disks
- circular saws
- railway and other vehicle wheels
- internal baffles in chemical plants.

Although these applications are quite disparate, the relevant disk-like components all share a very well-defined set of vibration characteristics which are closely related to the vibration properties of a 'simple' disk. A 'simple' disk is defined as one which is truly axisymmetric and has a uniform (constant thickness) radial cross-section (see **Figure 1A**). It is a relatively simple matter to extrapolate from these simple-disk vibration characteristics to those of quite complex engineering disk-like structures. Consequently, it is of value to describe the vibration properties of a simple disk in some detail.

Types of 'Disk'

First, it is necessary to define more precisely the three different classes of 'disk' that are included in this article. The first of these, the 'basic' disk is a structure which is completely axisymmetric in its form and whose geometry is fully described by the two-dimensional profile of a radial cross-section (**Figure 1B**). This is the case for computer disk and most (but not all) vehicle wheels and brake disks. The second category includes the more common disk-like structure which is cyclically symmetric (or periodic) in that its geometry is defined by a sector (rather than a simple radial section) which is repeated an integer number of times in a circumferential direction (**Figure 1C**). This is the case for gear wheels, bladed disks and even, strictly speaking, circular saws. This classification also applies to wheels that have a series of holes, or scallops arranged in a circumferential pattern. Finally, there is a third category of disk which has no exact symmetry at all, and this applies to any disk which is 'mistuned', a situation which arises when there is an irregularity in the geometry, either because of damage or simply because of manufacturing tolerances (**Figure 1D**). Also, there are occasions when a disk-like structure is deliberately mistuned (or 'detuned') and made non-symmetric for a specific purpose or reason.

Modes of Vibration: Axisymmetric Disks

Figure 2 shows the essential features and notation used to describe the out-of-plane flexural vibration of a 'basic' disk. The governing equation of motion for such a plate-like structure can be simply expressed as a partial differential equation:

$$\nabla^4 w + k^4 \frac{\partial^2 w}{\partial t^2} = 0 \qquad [1]$$

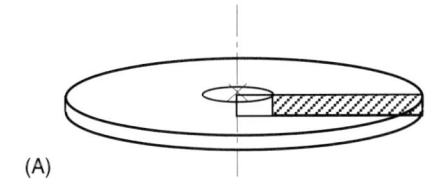

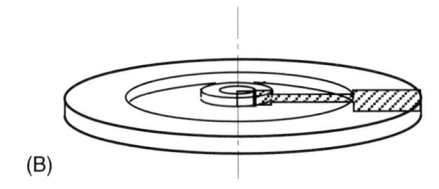

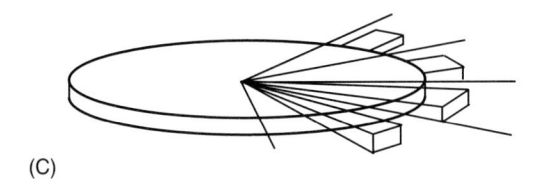

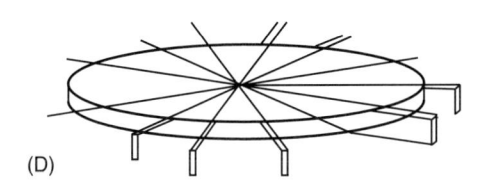

Figure 1 Basic form of disk-like structures. (A) Basic disk, (B) simple disk, (C) periodic disk, and (D) asymmetric disk.

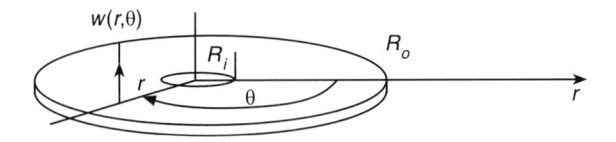

Figure 2 Essential features for disk vibration studies.

Of course, this structure, as all others, can be approximated by a discrete, finite element, type of model and for that case the governing equation of motion is the standard second-order differential matrix equation:

$$\mathbf{M\ddot{x} + Kx = f}$$

The exact solution of the earlier partial differential equation (eqn [1]) is quite straightforward in the case of a simple (uniform thickness) disk and because of the relevance of the properties of this reference configuration, it is worth reporting and studying that solution, not least because it helps to check the validity of results obtained when using the conventional finite element (FE) approach in a more 'black-box' analysis.

Solution of the equation of motion for free vibration of a simple disk reveals a number of natural frequencies which occur in two types: some as single roots, or eigenvalues, and others (the majority) as double roots – pairs of modes with identical natural frequencies. The general expression for the transverse deflection, w, when the disk is vibrating in mode j is:

$$w_j(r, \theta, t) = \phi_j(r, \theta)e^{i\omega t}$$
$$\phi_j(r, \theta) = f_j(r)g_j(\theta)$$

where

$$f_j(r) = a_0 + a_1 r + a_2 r^2 + \cdots$$
$$g_j(\theta) = \cos\, n(\theta + \alpha)$$

For the special case of $n = 0$, these modes are single modes, there being just one mode with the specific natural frequency and a mode shape described by:

$$\phi_j(r, \theta) = f_j(r)$$

For all other values of n, the modes exist in pairs with mode shapes of the general form:

$$\phi j(r, \theta) = f_q(r) \cos n(\theta + \alpha)$$
$$\phi k(r, \theta) = f_q(r) \sin n(\theta + \alpha)$$

Thus, the 'typical' modal solution of vibration of a simple disk is a pair of modes, or a double mode, comprising two modes with the same natural frequency, usually denoted as $\omega_{n,s}$, which have the same radial profile and the same circumferential mode shape profile ($\cos n\theta$) but with angular orientations which differ by 90°. These circumferential variations of deflection can be seen to represent motion on the disk which create patterns of n nodal diameter (ND) lines on its surface, lying symmetrically and antisymmetrically disposed about a diameter at $\theta = \alpha$ to the origin. For obvious reasons, these modes are referred to as 'n ND modes' and a typical example is illustrated in **Figure 3A**. A feature of these modes that sometimes adds confusion is that, as the two modes of a pair have identical natural frequencies, then any combination of the stated mode shapes is also a valid mode shape. This means that any combination of a $\cos n(\theta + \alpha)$ deflection pattern and a $\sin n(\theta + \alpha)$ deflection pattern is also a valid mode shape. In practice, this means that the nodal diameters can lie

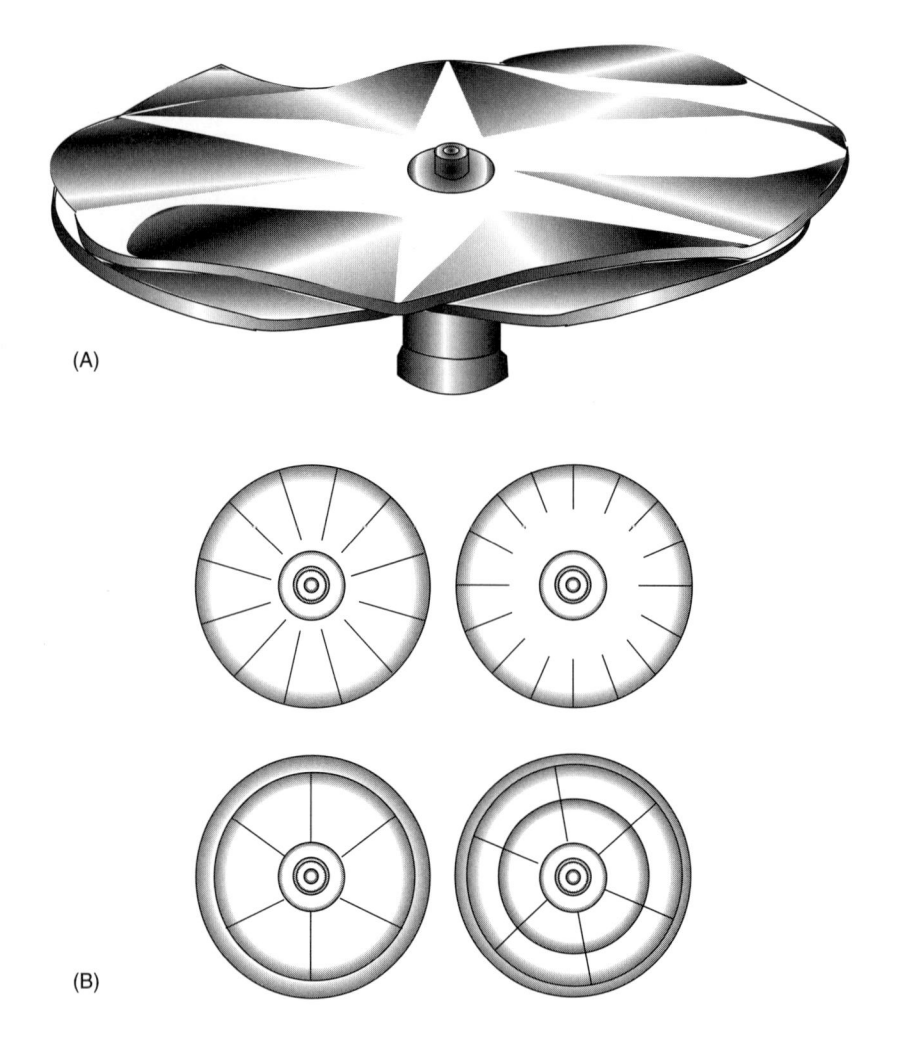

Figure 3 Typical mode shapes of a simple disk. (A) View of 3 ND mode; (B) nodal line diagrams for modes: 2, 0; 3, 1; 0, 1.

at any angular position, so that their orientation is 'arbitrary'.

In addition to this clear nodal diameter feature, it can be seen that there is a second element in the mode shape definition in the form of the particular solution for the radial displacement pattern, indicated by $f_q(r)$, defined from $r = R_I$ (the inner radius of the disk) to $r = R_O$, the outer radius. This function can be considered as a polynomial which will have a number, s ($s = 0, 1, 2, \ldots$), of zero values, each of which indicates a radius at which there is zero amplitude of vibration. These radial nodes represent nodal circles which, when combined with the number of nodal diameters, describe the specific mode in full – mode$_{n,s}$ has a natural frequency of $\omega_{n,s}$ and a mode shape described by n nodal diameters and s nodal circles. **Figure 3B** illustrates a number of typical modes of this general type.

It is customary to display the essential modal properties of a disk-like structure by plotting the natural frequencies of the disk against the number of nodal diameters in the mode shape, and to show 'families' of modes according to the number of nodal circles: first family, modes with 0 nodal circles; second family, modes with 1 nodal circle, ... etc. **Figure 4** shows the data for the specific case of a uniform-thickness continuous (no central hole) disk with fixed–free boundary conditions (the disk center constrained). The corresponding diagram for a disk of arbitrary cross-section will depend on the details of that profile but the general format will be the same as those shown in **Figure 3**, and the actual distribution of modal frequencies will follow the illustrated form quite closely.

It is worth mentioning two special cases amongst this array of modes. These are the modes with 0 ND, and those with 1 ND. We have already noted that modes with 0 ND are single modes which do not have the double-mode characteristic that all others have. Motion in a 0 ND mode necessarily involves

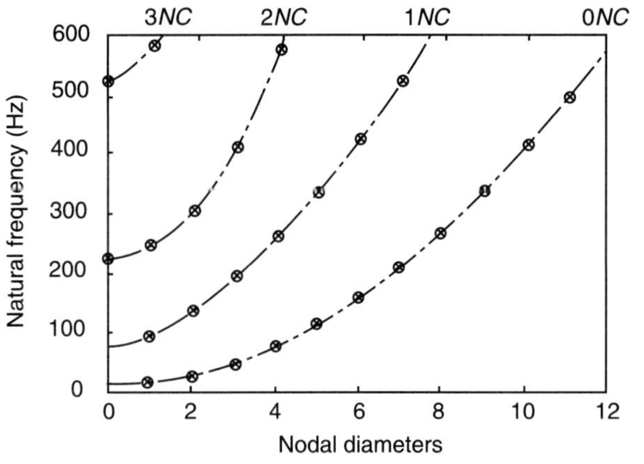

Figure 4 Natural frequencies of simple disks. Fixed (at disk center)–free boundary conditions.

movement at the center of the disk (it is not 'balanced' at the disk center) and, as a result, it involves whole-body motion of the disk, either axially or torsionally. Similarly, modes with 1 nodal diameter, while occurring in double-mode pairs, also involve motion of the disk center by virtue of their rotation about a diameter or motion along a diameter. All other modes, with 2 ND and above, share the property that they are in equilibrium at the disk center at all times. The particular consequence of the special features associated with these particular 0 and 1 ND modes is that their properties are influenced by the characteristics of the shaft and its bearings upon which the disk is mounted. Conversely, the modal properties of all modes with two or more nodal diameters, are not influenced by the shaft. We shall see later how this feature affects vibration in operating conditions.

Modes of Vibration: Cyclically Symmetric (Periodic) Disks

In many practical applications, the actual 'disk' will have a number of appendages attached to its rim, or its surface, such as gear or saw teeth, blades or vanes, and this class of disk is described as a 'cyclically symmetric' disk, differing slightly from the plain or 'simple' disk by the fact that the radial profile is no longer defined by a single cross-section. Instead, it is necessary to define the details of a sector of the disk, that being the portion of the complete 'disk' that encompasses one appendage, or period of the structure (**Figure 1C**).

For these periodic disk-like structures, no simple equation of motion, or closed-form solution, is available for the modal properties. However, we may use the features described in the preceding section for simple disks to anticipate and to understand the essential features that are found in these periodic structures.

First, it is convenient to classify and to describe the vibration modes of a periodic disk-like structure using the same notation as was used above: namely, to refer to each mode of vibration as having n nodal diameters and s nodal circles. However, if we look closely at the details of the mode shapes of these components, we find that the circumferential variation in deflection no longer displays a simple n ND pattern but, rather, comprises a series of sinusoidal components with a circumferential function of the form: $g_j(\theta) = \sum \cos q(\theta + \alpha) \sin \omega t$ where the summation over q includes the following terms: n, $N - n$; $N + n$; $2N - n$; $2N + n$; ;... and where N is the number of teeth/blades/vanes. In any one mode, it is common for one of these components to dominate and this feature leads to the justifiable use of the n ND label, even though it is a simplification. Thus there is a second grouping of the modes for this class of disk; the typical example being the group of modes which all contain the 'n ND set of diametral components'. Thus within one nodal circle family (i.e., all those modes which all have s nodal circles) there are subfamilies or groups: the modes with $1, N - 1, N + 1, 2N - 1, 2N + 1, \ldots$, ND components being one such group; the modes with $2, N - 2, N + 2, 2N - 2, 2N + 2, \ldots$, ND being a second group, and so on.

On a practical point: it is common practice to describe the circumferential mode shape of a disk or a bladed disk by defining the deflections at a series of P regularly spaced points around a given circumference. When this type of mode shape description is adopted, it is possible to convert such a list of amplitudes into (nodal) diametral components by performing a discrete Fourier transform on them. It will be appreciated that such a set of diametral components is restricted to contain just $P/2$ terms (higher terms are inaccessible because of aliassing). In the case of a periodic disk-like structure such as a bladed disk, it is natural to choose the number of circumferential measurement points to be equal to the number of blades, so that $P = N$. The ND description of any mode shape described in this discrete way is necessarily limited to terms up to and including $N/2$, so that components of, for example, $(N - 1), (N + 1)$ cannot be included explicitly. In turn, this means that a mode which is predominantly an $(N - n)$ ND shape would register in a description based on an analysis of amplitudes shown at N regularly-spaced points around the disk as an n ND mode (see **Figure 5** for a simplified illustration).

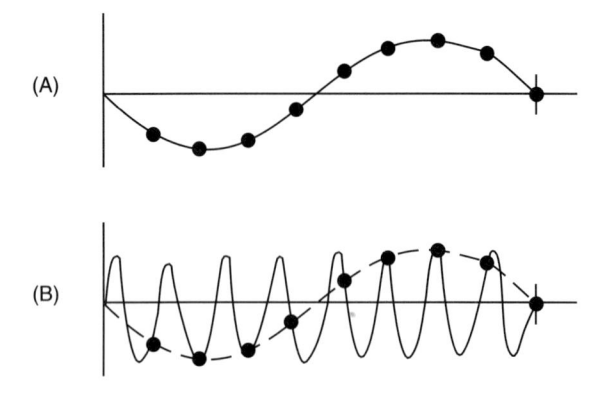

Figure 5 (A, B) Possible misinterpretation of mode shapes using discrete description.

A typical plot of the natural frequencies of a bladed disk is given in **Figure 6**, showing the presentations both when (a) the dominant ND component is used to classify the mode and (b) when the description is assumed to be limited to N discrete points around the rim (so that $N/2$ is the highest number of ND possible). In either presentation, both the similarities and the differences between these properties and those of the plain disk are clear and understandable.

It should be noted that, with just one exception, the modes of periodic disk-like structures share most of the properties of the simple disk. In particular, the 1 and higher ND modes occur in pairs: double modes with identical natural frequencies and mode shapes which are oriented orthogonally to each other so that the actual positioning of the nodal diameter lines is arbitrary. The exception is the family of modes with $N/2$ NDs: these now take the same form as modes with 0 ND and exist as single modes, rather than double modes, and have no arbitrariness in the location of the nodal lines.

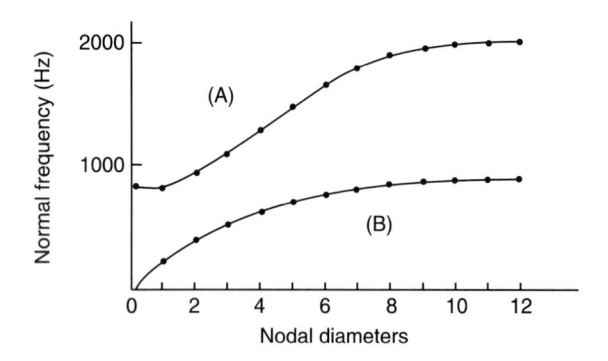

Figure 6 Natural frequency vs ND plots for cyclically symmetric (periodic) disk-like structure (bladed disk). (A) Full range of ND components; (B) restricted range of ND components (applicable to discrete mode shape descriptions).

Modes of Vibration: Asymmetric Disks

This last category of disks is intended to include those cases where, for some reason, deliberate or accidental, the disk possesses neither axisymmetry nor cyclic symmetry. This situation can arise for several reasons: the most common is the inevitable consequence of manufacturing tolerances which mean that real disks or bladed disks do not quite attain their intended symmetry and are 'mistuned' as a result. It is a well-documented fact that even a very slight loss of symmetry can have a dramatic effect on the vibration properties of disk-like structures and so it is appropriate to include a reasonable discussion of this phenomenon here. A similar end result occurs when a disk is designed to have an in-built asymmetry, such as occurs with a single hole, or other protuberance, as part of its design function, or when a deliberate break in symmetry is introduced, 'detuning', in order to take advantage of the particular features of the modified vibration properties that we shall outline below (for example, some unstable aeroelastic phenomena can be positively affected by such a loss of symmetry).

There is a marked sensitivity in some of the vibration properties of a disk to small changes in the structural details, such as mass or stiffness distributions, partly because of the existence of double modes. Mathematically, it can be shown that a small perturbation in mass or stiffness brings about a correspondingly small change in modal properties, except in cases of close modes (modes with close natural frequencies), which is exactly the situation that prevails here, especially in the cyclically symmetric case where we have not only double modes, but clusters of these double modes in certain well-defined frequency ranges.

The essential changes in the modal properties which result from a loss of symmetry in the disk (for whatever reason) are that the double modes split into pairs of distinct modes with close natural frequencies and mode shapes that are only slightly different from the original n–ND form but each of which, importantly, now have an angular orientation which is fixed relative to the disk reference (recall here that for a symmetrical disk, the double mode property means that nodal diameters can lie at any angular orientation). This result is illustrated in **Figure 7A**.

The above-mentioned pattern of behavior applies to all double modes of the disk. However, a further complication arises when two pairs of (double) modes are themselves 'close'. For example, in many bladed disk structures, we may find that the natural frequency of, say, an 8 ND mode might be at 234 Hz, while that of the 9 ND mode is at 245 Hz, etc. If, in such a case, some asymmetry or mistuning encountered in practice will rotate and so it is appropriate to extend the discussion to include the additional features that apply in theses conditions. There are essentially two features to describe: first, the effects on the modal properties of the forces which apply under rotating conditions (mainly, those arising from centrifugal and gyroscopic effects); and secondly, the implications of the introduction of a second frame of reference which means that vibration experienced (in a rotating disk) and observed (by a stationary observer) appear to occur at different frequencies.

If we consider first the implications of centrifugal and gyroscopic (Coriolis) effects, we find that the former may cause the disk to experience and exhibit an increase in effective stiffness so that some natural frequencies may rise under rotation. The extent of this effect is generally quite small, except for very high speeds. However, there is one class of disk-like structure in which the centrifugal effect can be very pronounced, and that is the bladed disk, especially when the blades are relatively large and flexible. It is not uncommon in these structures, typically of the front stages of aircraft engine fans, for the natural frequencies of the lowest modes of the bladed disk assembly to increase to double the values which apply at rest.

The gyroscopic effects, on the other hand, apply directly to disks and to wheels, and are frequently significant factors in determining the natural frequencies of disks installed on rotors where the primary rotor dynamics properties are intimately linked with the 0 and 1 ND modes of the disk(s) which are installed. In situations where a disk is free to move about one of its diameters, incurring a 'rocking' motion, while it is rotating about its principle axis, Coriolis accelerations are generated which, in turn, cause the so-called gyroscopic forces that result in a sharp change in the vibration properties of the rotor disk system. The essence of these properties may be illustrated by reference to **Figure 8A** which shows a simple disk–rotor system, which is assumed to be free to vibrate in the two transverse directions illustrated, at the same time as spinning about its own axis. The vibration modes of this system at rest are simple: one mode where the disk moves in the horizontal (x) direction, and the other where it moves in the vertical (y) direction. Each of these modes has a distinct natural frequency according to the two stiffnesses, k_x and k_y. When the rotor spins, these two modes change quite significantly to one mode with a lower natural frequency than either of the two at-rest modes and a mode shape which is a backward-traveling orbit, plus a second mode with a higher natural frequency than either of those at rest and a mode

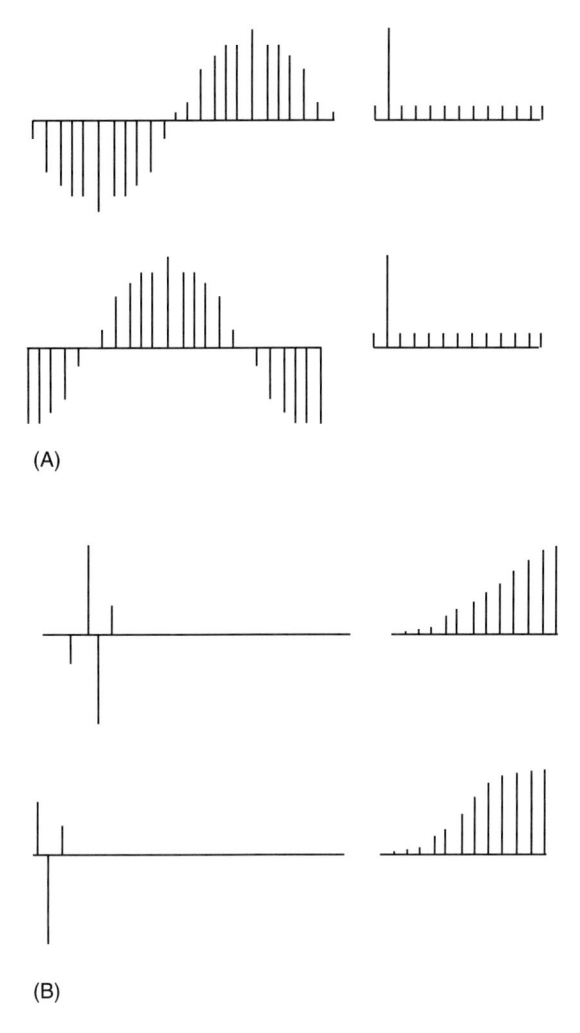

Figure 7 Sketches of mode shapes for asymmetric (mistuned) disk-like structure. (A) separated double modes of symmetric system; (B) close double modes in symmetric system.

is introduced, then the pair of 8 ND modes will split and form two distinct modes with natural frequencies at, for example, 229 Hz and 239 Hz, and at the same time the pair of 9 ND modes might slit into two modes with, say 240 Hz and 250 Hz. In such a situation, we find that the upper of the 8 ND modes of the mistuned disk (at 239 Hz) is almost at the same natural frequency of the lower of the pair of 9 ND modes (240 Hz) and this effect creates a considerable distortion in the shapes of the proximate modes, resulting in vibration modes that can no longer be described or classified as having a simple n ND form. **Figure 7B** illustrates such a situation in comparison with that which applies when the adjacent modes of the symmetric system are not close.

Modes of Vibration: Rotating Disks

It is clear from the list of examples cited at the beginning of the article that many of the disks

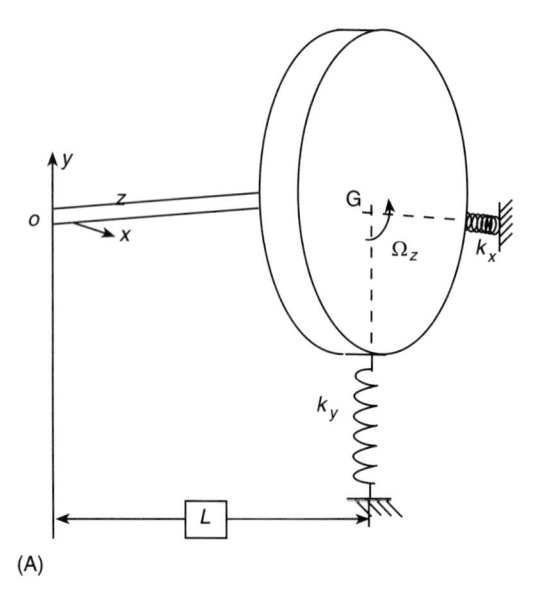

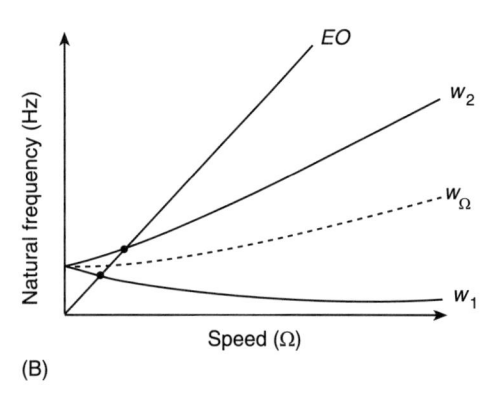

Figure 8 (A, B) Natural frequencies of a rotating disk under the influence of gyroscopic effects.

shape which is a forward-traveling orbit. The pattern of natural frequencies is shown in **Figure 8B** and this simple example serves to illustrate a phenomenon which is widely encountered in many installations of rotating disks.

Turning to the matter of frames of reference, it can be seen that if a disk is rotating (at speed Ω) at the same time as it is vibrating (at frequency ω_R, as sensed by a transducer fixed to the disk), then this vibration will be observed by a stationary transducer to be occurring at a different frequency, ω_S. In fact, the frequency that is detected by a stationary observer (such as a microphone, or noncontacting motion transducer) is found to be related to the actual frequency of vibration, the speed of rotation, and the number of nodal diameters of the deflection pattern of the vibration. The vibration frequency detected by a stationary observer, ω_S, can be related to the actual vibration behavior of the disk itself, at frequency ω_R, with a mode shape of n ND, spinning at speed, Ω, determined by: $\omega_S = \omega_R + n\Omega$. This relationship can be used to construct the frequency–speed diagram presented in **Figure 9** which shows the apparent vibration frequency (ω_S) plotted against rotation speed (Ω) for the first few modes of the rotating disk. An important application of this diagram is to indicate the rotation speeds at which a stationary source of vibration excitation with a frequency of ω_E will coincide with the apparent natural frequencies of the spinning disk. For example, **Figure 9** shows how a stationary (nonrotating) excitation at frequency ω_E might generate a resonant response in a disk mode at any of several rotation speeds, since at these indicated conditions (×), the excitation is 'felt' by the disk at a frequency which coincides with one of its own natural frequencies. Of even greater interest in this respect, are the rotation speeds marked as ○, as these represent speeds at which resonance might be induced in a mode of the spinning disc by a stationary and constant (i.e., static, nonoscillatory) force applied to the disk rim. These critical speeds have long been known as potentially dangerous conditions, to be avoided in the operating range of almost all machines.

Vibration Response of Disk-like Structures

Most of the important characteristics of the vibration of disk-like structures have been addressed in the preceding sections describing the modal properties. Using the standard theory of modal analysis, the response of these structures to any prescribed excitation can readily be determined using the modal properties we have described in some detail. It remains here to highlight certain features of the forced response of disks that are peculiar to this type of structural component.

It transpires that there are two types of forced response which are of special interest: firstly, that from a single point of excitation, and secondly that which results from a particular form of multipoint excitation, where the excitation force patterns travels around the disk at a steady speed. (This latter excitation is commonly encountered under rotating conditions where it is called 'engine-order excitation'.)

When a disk is excited at a single point, and when the excitation frequency coincides with the natural frequency of a mode with n NDs, the appropriate mode shape is readily observed, but almost universally with the NDs symmetrically disposed about the excitation point (**Figure 10A**). If the excitation is moved around the disk by an arbitrary angle, keeping the same radius, and the measurement repeated, it is found that the pattern of nodal lines has 'followed'

Figure 9 Apparent and actual vibration frequencies observed and experienced in a rotating vibrating disk. (–) Apparent natural frequencies as detected by stationary observer; (- - -) actual natural frequencies as experienced by rotating disk; × 'critical' speeds for stationary excitation at frequency ω_E of disk modes; ○ critical speeds for stationary static force ($\omega_E = 0$) excitation of disk modes.

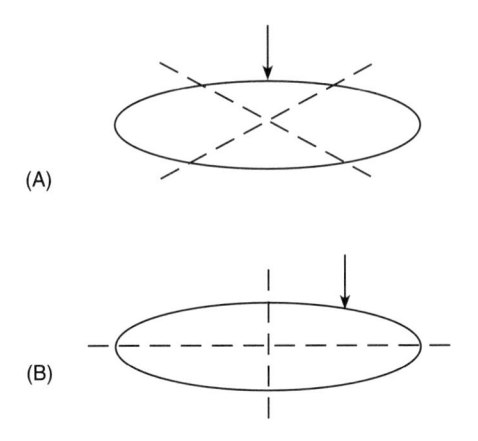

Figure 10 (A, B) Nodal diameters 'follow' point of excitation.

the point of excitation so that the nodal lines once again are set symmetrically about the point of excitation (**Figure 10B**). This phenomenon is a clear demonstration of the fact that there are two modes at this resonance frequency. If there was only one mode, then the nodal lines observed by the first excitation point would be fixed in the disk and so, when the point of excitation was moved, they would stay in the same place. The fact that they 'move' can (only) be explained if there are two modes at this natural frequency. The same measurement carried out on an asymmetric disk will reveal the existence of two modes at different (although close) natural frequencies, and will show a different nodal line pattern for each resonance. Further, these two nodal patterns will be the same for an excitation applied at the first point and then at the second point. In other words, these nodal patterns will not 'follow' the excitation.

There is a special type of excitation which occurs when a disk is rotating past a stationary (nonrotating) excitation force field. In many cases, this force field is simply a steady (static) pressure which is exerted on the disk face, although the same features arise if the force field includes a time-varying oscillating component. To illustrate the phenomena which arise, it is convenient to define the force field as:

$$F(\theta) = F_n \cos n\theta$$

In the general case, F_n may itself be time-varying, but the effects of interest can be illustrated in the case where F_n is simply a constant, static force. It can be shown that this type of excitation, which constitutes a multipoint excitation effect because the force is felt by the disk at all points simultaneously, has a selective effect on the modes of vibration of the disk with n NDs. In fact, an excitation force field of the nth-order form shown, will excite all modes of vibration of the disk that contain a component of n ND in their mode shape, and will do so at a frequency experienced by the disk of $n\Omega$. In contrast, this same excitation pattern will not excite any response in modes that do not have an n ND component in their mode shape, and this is a valuable feature in many practical situations. This type of engine-order (EO) excitation can be used to explain almost all the response phenomena encountered in the vibration of disks, rotating and stationary, found in all types of machine. **Figure 11** summarizes this phenomenon: it shows a

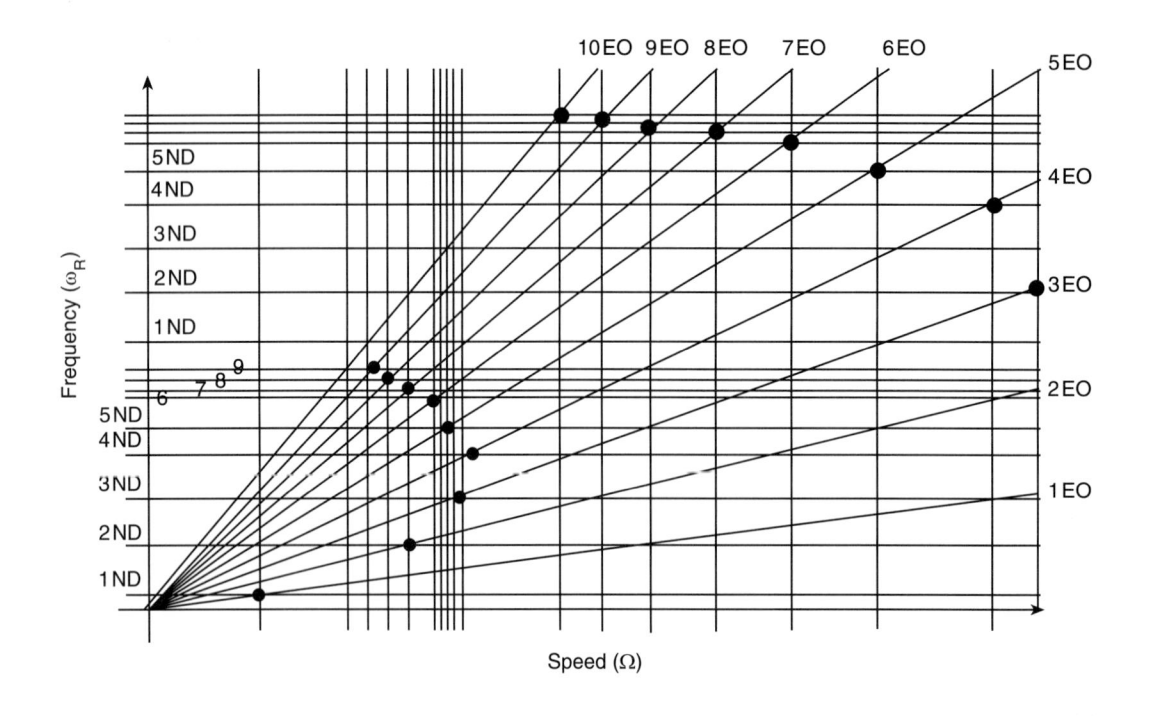

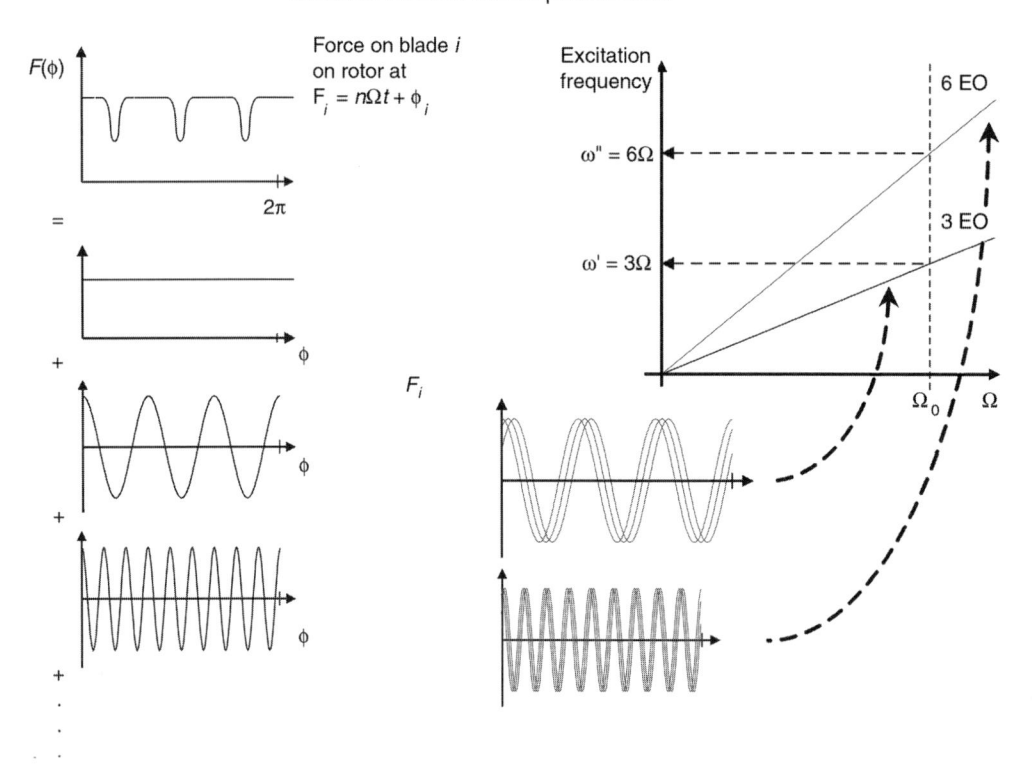

Figure 11 EO excitation of rotating disks. (A) Critical speeds indicating EO resonances, (B) explanation of the origin of multiple EO excitation components.

diagram of vibration frequency (as experienced on the rotating disk) against rotation speed, and it illustrates the regions of resonance that can be expected if there is a stationary excitation force field that includes several different orders, n. It can be seen clearly on this diagram how each EO selectively picks out the disk mode with the matching number of NDs and generates resonances accordingly. It is also clearly seen how coincidence of excitation frequencies with natural frequencies which do not satisfy the additional requirement of matching engine order and number of NDs does not lead to resonance.

It remains only to explain how there might come to be a 'stationary excitation force field that includes several orders, n'. In many machines, there are sources of vibration excitation which result from obstructions to the flow of working gas, or other features, and these are responsible for the force field, $F(\theta)$. If there is a simple, single, obstruction in the working fluid stream, at some arbitrary angular orientation, then simple Fourier analysis of this perturbation in the steady force reveals that it will be the equivalent of many harmonic components, such as F_n above.

See also: **Mode of vibration**; **Plates**; **Rotor dynamics**.

Further Reading

Leissa AW (1975) *Vibration of Plates*, NASA.

Southwell RV (1992) On the free transverse vibrations of a uniform circular disc clamped at its centre; and on the effects of rotation. *Royal Society Proceedings (London)* 101 A, 133–153.

Tobias SA, Arnold RN (1957) The influence of dynamics imperfections on the vibrations of rotating disks. *Proceedings of the Institute of Mechanical Engineers* 171, 669–690.

DISPLAYS OF VIBRATION PROPERTIES

M Radeş, Universitatea Politehnica Bucuresti, Bucuresti, Romania

Copyright © 2001 Academic Press

doi:10.1006/rwvb.2001.0168

Displays are valuable instruments in engineering practice, condensing large amounts of data into simple and convenient presentation formats. Examination of time histories and frequency-based spectral data, orbit analysis, and observation of the animated mode shapes and operating deflection shapes allows us to pinpoint areas and properties of interest in dynamic analysis, giving important visual insight into the way in which structures are vibrating.

Displays of Modal Data

Displays of modal data – natural frequencies, damping ratios, mode shapes, and modal participation factors – are useful in the measurement, analysis, and interpretation of vibrations.

Natural Frequency Plots

While tabular presentations of natural frequencies are sometimes sufficient, different display formats reveal more clearly the interrelation between various parameters. From a long list of possible applications, one can mention:

1. Bar plots for the comparison of measured and predicted natural frequencies. For each natural frequency, a colored bar shows the magnitude of, say, the measured value, and a white bar shows the predicted value.
2. Root-locus diagrams, showing the variation of damped natural frequencies as a function of damping (see **Theory of vibration**, Fundamentals).
3. Diagrams of variation in natural frequencies against a stiffness or a mass parameter. For rotating shafts, the critical speed map shows the variation of the undamped critical speeds as a function of an average bearing stiffness (see **Blades and bladed disks**; **Rotor dynamics**).
4. Diagrams of variation in natural frequencies of rotor-bearing systems as a function of bearing damping. **Figure 1** shows a plot of the imaginary versus the real part of eigenvalues for the first two modes of precession of a shaft-bearing system. The values on the curves give the bearing damping ratio. It is seen that the lowest mode becomes critically damped.
5. Interference diagrams: plots of natural frequencies versus a speed-related parameter, with overlaid lines of excitation frequencies (see **Rotating machinery, essential features**).
6. Campbell diagrams: plots of natural frequencies of rotating systems as a function of rotational speed. In the case of analytical data, the connection of

points is based on knowledge of the mode shapes. In the case of measured data the magnitude of response at resonance is indicated by circles, with the circle calibration factor also given on the display (not shown). Only peaks exceeding a selected threshold level may be presented (**Figure 2**). Structural resonances often appear as being excited at approximately the same frequency over a wide range of speed change.

7. For systems with hydrodynamic bearings and seals, an instability diagram can also be plotted, as a graph of the damping factor (real part of eigenvalues) versus rotational speed. The onset speed of instability is located at the lowest abscissa crossing point of any curve with the speed axis (see **Rotating machinery, essential features**).
8. Diagrams of natural frequencies of plate-like components as a function of the nodal line indices.

The last of these types of display will be discussed in more detail below.

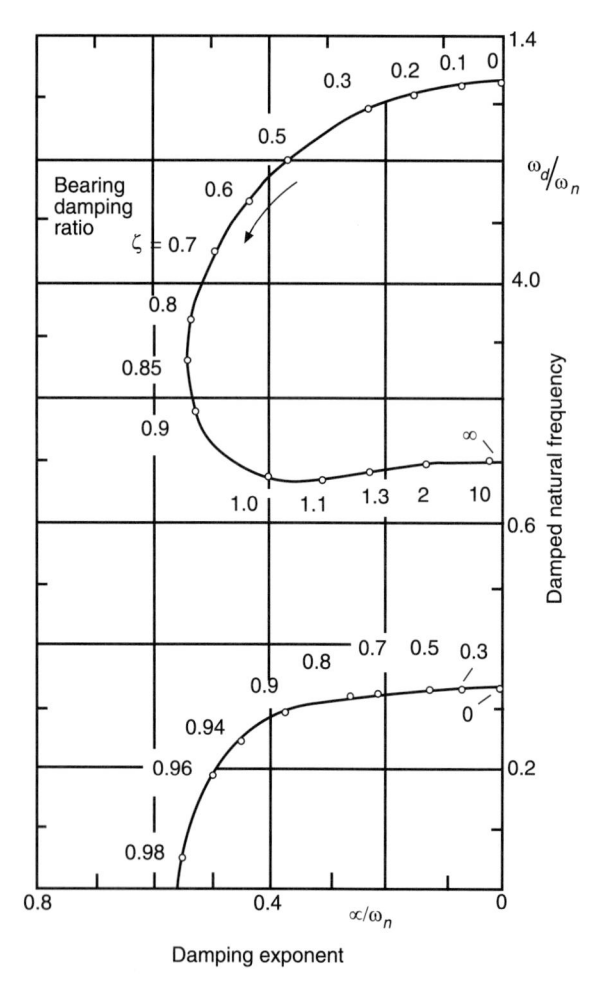

Figure 1 Lowest two damped natural frequencies of a rotor-bearing system, as a function of bearing damping.

Natural Frequencies of Plates and Beams

A systematic study of the normal vibration modes of plate-like components may be based on the analysis of their nodal lines, sometimes referred to as the Chladni figures.

For a circular plate, the basic constituent nodal lines are circles and diameters (**Figure 3A**). Each normal mode is characterized by a certain pattern, defined by the pair of indices **c, d**, where **c** is the number of nodal circles and **d** is the number of nodal diameters. Sometimes, a plus or minus sign is used in front of this ratio, to take into account the two possible modal forms, phase shifted with 180° or the two superimposed orthogonal modes.

The natural frequencies can be plotted against the number of nodal diameters, in families of curves, each having as a parameter the number of nodal circles. For a free circular plate, typical results are shown in **Figure 3B**.

This type of display may be useful in an experimental study, when the nodal patterns are obtained by holographic interferometry, or by sprinkling fine granular material on the surface of the plate, and observing the stationary figures formed at resonances. First, the natural frequency of each mode is plotted against its number of nodal diameters. Then, points representing modes with the same number of nodal circles are connected by continuous lines. If one of the natural frequencies has been omitted, then it can be determined as the ordinate of the point located where a mode family line intersects the vertical line corresponding to a given number of nodal diameters. Characteristic for plate-like components is the existence of physically different modes having coincident or close natural frequencies. In damped systems, two neighboring modes can combine due to the flatness of resonance responses, producing compounded modes, whose nodal pattern is no more made up of simple circles and diameters.

A diagram such as **Figure 3B** helps to identify the pairs of modes susceptible of compounding and to explain their origin. For example, consider the modes 1, 2 and 0, 5, which have almost the same frequency (**Figure 3C**). For each individual mode, the shaded areas are considered to vibrate downwards, while the unshaded areas vibrate upwards. When the two figures are superimposed, the doubly shaded areas vibrate downwards, and the unshaded areas vibrate upwards. The nodal lines will pass through the single shaded areas, where the two opposite motions cancel each other. Opposite phase combinations give rise to rotated patterns. Irregular nodal patterns, determined for propeller blades, for impellers and turbine-bladed disks, can be conveniently studied using the type of display shown in **Figure 3B**.

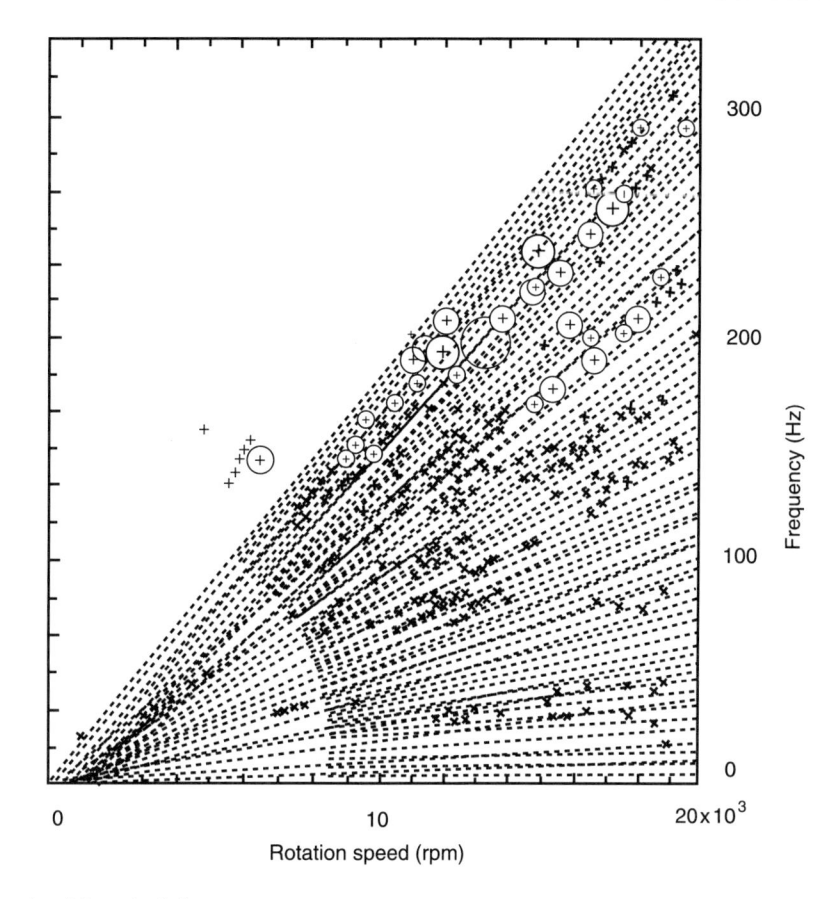

Figure 2 Three-dimensional Campbell diagram.

This display may be used to study the possible resonances of bladed disks. In a perfectly uniform system, each point on the diagram represents a pair of modes with orthogonally oriented identical modal patterns. Due to imperfections or to blade variability, in actual bladed disks the two modes separate and take on slightly shifted natural frequencies, complicating the picture. An interference diagram (see **Blades and bladed disks**) helps to locate modes with close natural frequencies being excited at the same engine speed.

Mode Shape Plots

Probably the displays most often used are those of mode shapes. While animation offers 'live' displays, showing how a structure actually moves at a natural frequency, hard-copy print-outs can contain only frozen images, i.e., pseudo-animated displays. Static (nonanimated) displays, containing zero deflection and the maximum distortion frame, are often sufficient to serve the desired goal. Three-dimensional structures may be represented in wire frame format or in solid hidden-line format.

From the wide range of applications in vibrations, several typical examples can be chosen, where the mode shape plots reveal important information, such as that described below.

1. The relative amplitude of displacement throughout the structure and the location of nodal points of real modes. **Figure 4** illustrates the lowest three planar mode shapes of a cantilever beam. For straight beams, there is a direct correlation between the mode index and the number of nodal points, a fact which helps in measurements.

   If stationary nodes are not visible on pseudoanimated displays, then the modes are not real. Complex modes exhibit nonstationary zero-displacement points, at locations that change in space periodically, at the rate of vibration frequency. These complex modes are sometimes converted into equivalent real mode shapes. The complexity of measured modes is observed using so-called compass plots (**Figure 5**), i.e., vector diagrams in which each modal vector element is represented by a line of corresponding length and inclination, emanating from the origin. The complex mode shape is then rotated through an angle equal to the mean phase angle of its entries. Finally, for lightly damped structures, the real

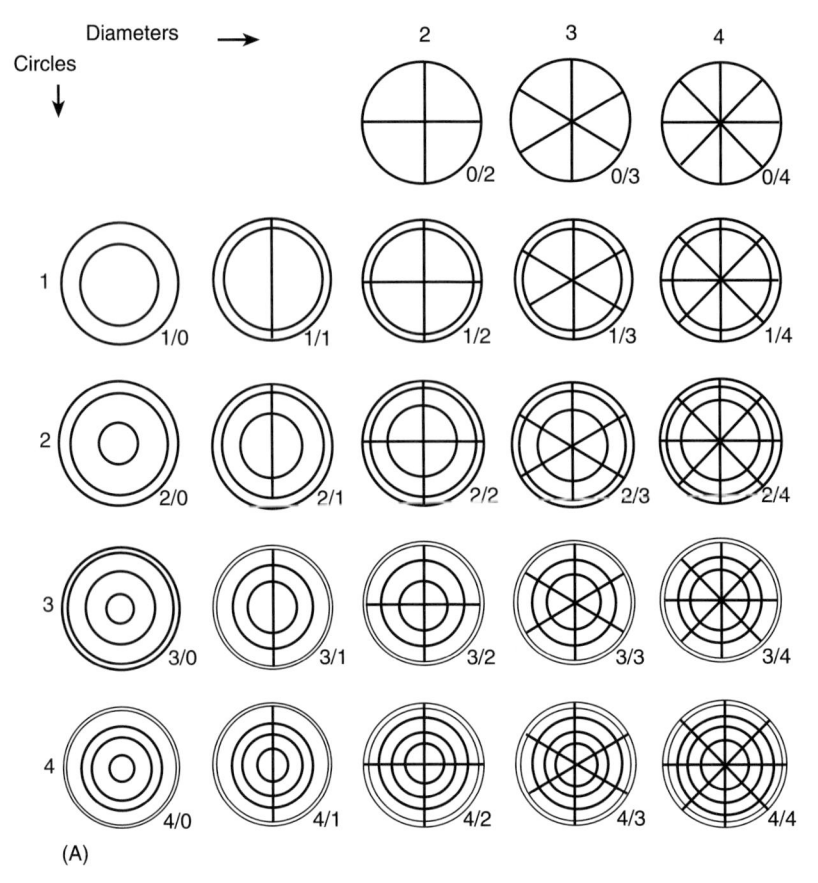

Figure 3A Natural frequencies of free circular plates: nodal patterns of natural modes of vibration.

mode is realized by taking the modulus of each complex vector element and multiplying it by the sign of its real part.

Symmetric structures possess symmetric and antisymmetric mode shapes that can be studied separately to reduce the problem order. Periodic structures can also be analyzed by being broken into constituent substructures.

2. **Dynamically weak components or joints. Figure 6** shows a simplified mode shape trace of a milling machine, for a mode identified as the main source of chatter during operation. The deviation from 90° of the lines concurrent at the joint between the working table and the vertical column indicates that the table mounting screws are responsible for the machine tool instability.

In more detailed analyses, a wire frame model is developed to represent the geometry of the structure, connecting the instrumented points by trace links. The animated display shows the nature of measured or calculated vibration modes: rigid body, flexural, torsional, longitudinal, plate-like, or a combination of these.

A contour map of the mode shapes of panel or shell-like structures reveals areas of large relative displacement amplitude. It may help in the selection of locations where ribs, braces, or other stiffeners can be welded in order to raise the natural frequency of the component beyond the range of excitation frequencies.

3. **Combined bending and torsion, and plate vibration mode shapes. Figure 7** illustrates the lowest 12 modes of vibration of a cantilevered planar plate structure. Modes are classified by the ratio between the number of nodal lines perpendicular to the built-in side and the number of nodal lines parallel to the built-in side, including the fixed root.

4. **The existence of double modes in axially symmetric structures. Figure 8** shows the pair of two-diameter modes of a circular plate structure, having almost coincident natural frequencies and identical mode shapes, but 45° phase-shifted in space.

5. **Compound modes, consisting of both global and local modes.** An eight-bay free–free planar truss can have 'pure' global modes, like the three-node bending mode shown in **Figure 9A** and composite modes, like the four-node bending mode shown in **Figure 9B**, with diagonals vibrating in the first local mode.

DISPLAYS OF VIBRATION PROPERTIES 417

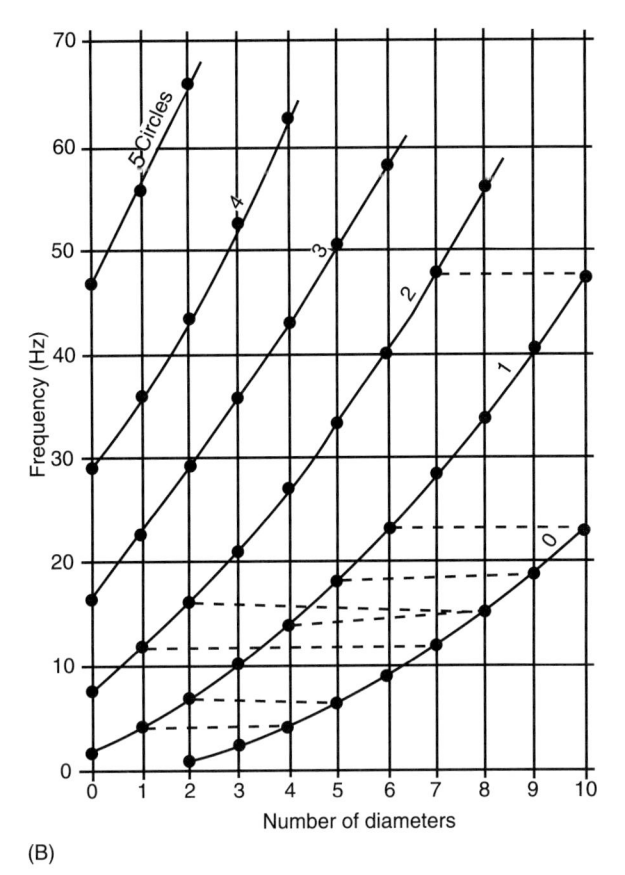

Figure 3B Natural frequencies of free circular plates: natural-frequency plot.

6. Whirling modes of rotor-bearing systems. **Figure 10** shows the orbits of the finite element model nodal points along a rotor, together with two spatial mode shape traces, plotted at a quarter of a rotation time interval. Because the trajectory of each point in the free damped motion is a spiral, the orbits are plotted as open ellipses, with the missing part at the end of the precession period. This helps to establish whether the whirling is forward or backward.

Various other kinds of display are also available, taking advantage of the graphical capabilities of the computer programs in current use. Time–domain animation, based on time history data, may be used to study transient, shock and nonlinear responses. Operation deflection shapes help to visualize the structure's motion in its normal operating conditions.

Displays of Frequency Response Data

Frequency Response Function Plots

Frequency response function (FRF) data are complex and thus there are three quantities – the frequency plus the real and imaginary parts (or modulus and phase) of the complex function – to be displayed.

The three most common graphical formats are: (1) the Bode plots (a pair of plots of modulus versus frequency and phase versus frequency); (2) the Co-Quad plots (real component versus frequency and imaginary component versus frequency); and (3) the Nyquist plots (imaginary versus real part, eventually with marks at equal frequency increments).

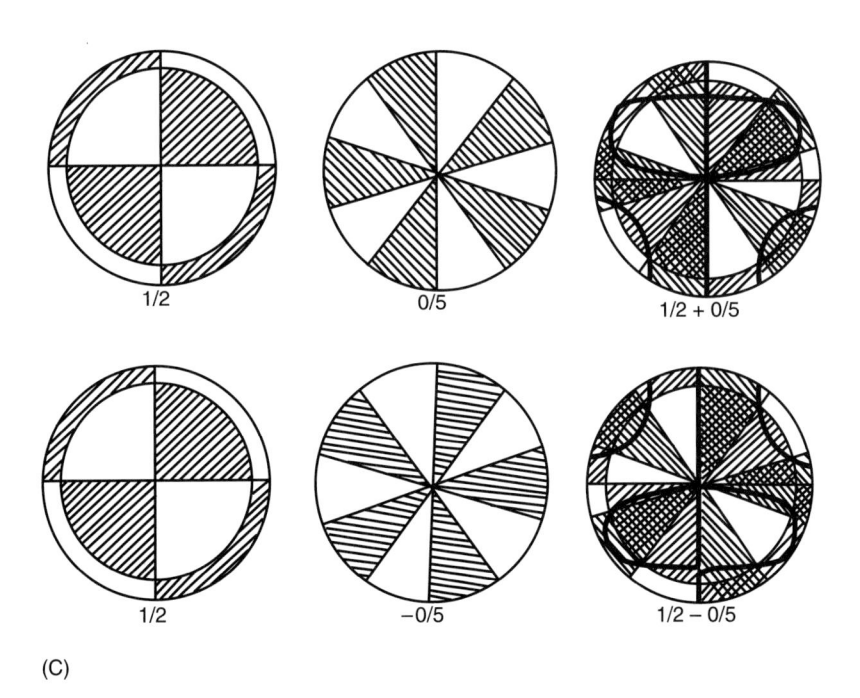

Figure 3C Natural frequencies of free circular plates: compounded modes.

418 DISPLAYS OF VIBRATION PROPERTIES

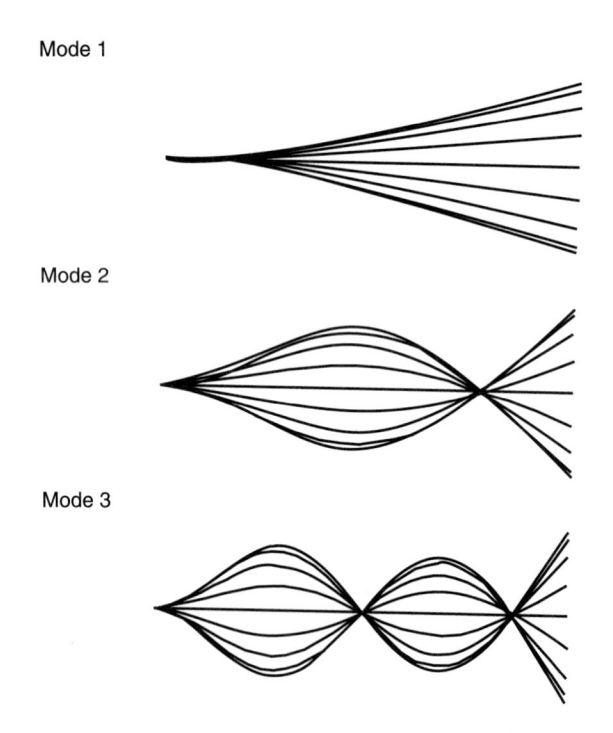

Mode 1

Mode 2

Mode 3

Figure 4 First three planar flexural mode shapes of a cantilever beam.

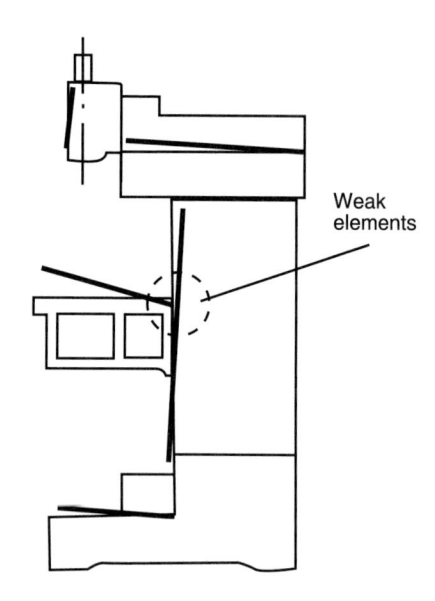

Figure 6 Mode shape trace of a milling machine showing the location of a weak area.

Plotted quantities include motion/force-type functions like receptance, mobility, and acceleration, as well as their inverses, dynamic stiffness, impedance and dynamic mass (see **Modal analysis, experimental**, Parameter extraction methods).

Scales can be linear, logarithmic, and semilogarithmic, with or without grid lines, with one or two cursors, zoom capabilities and overlay facilities.

Figure 11 shows parts of the Bode plots for the receptance FRF of a typical damped structure, with the displacement response measured at DOFs 1, 2, and 3, due to forcing applied at DOF 1. Note the logarithmic vertical scale for the magnitude plots, necessary to reveal the details at the lower levels of the response exhibited at antiresonances.

Figure 12 shows the respective plots of the real (coincident) and imaginary (quadrature) components of receptance, while the first column in **Figure 13** shows the corresponding Nyquist plots. It is seen that for lightly-damped systems with relatively well-separated natural frequencies, the response near resonances can be approximated by circular loops. For comparison, mobility and inertance plots are presented in the second and third columns of **Figure 13**, for the same FRFs. For the point receptance H_{11}, a three-dimensional plot is illustrated in **Figure 14**, together with the companion coincident, quadrature, and Nyquist plots.

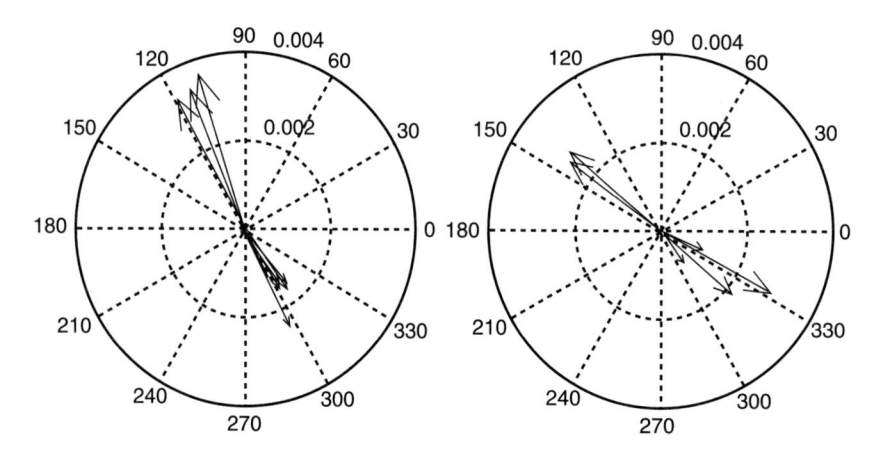

Figure 5 Compass diagrams of complex mode shapes.

DISPLAYS OF VIBRATION PROPERTIES 419

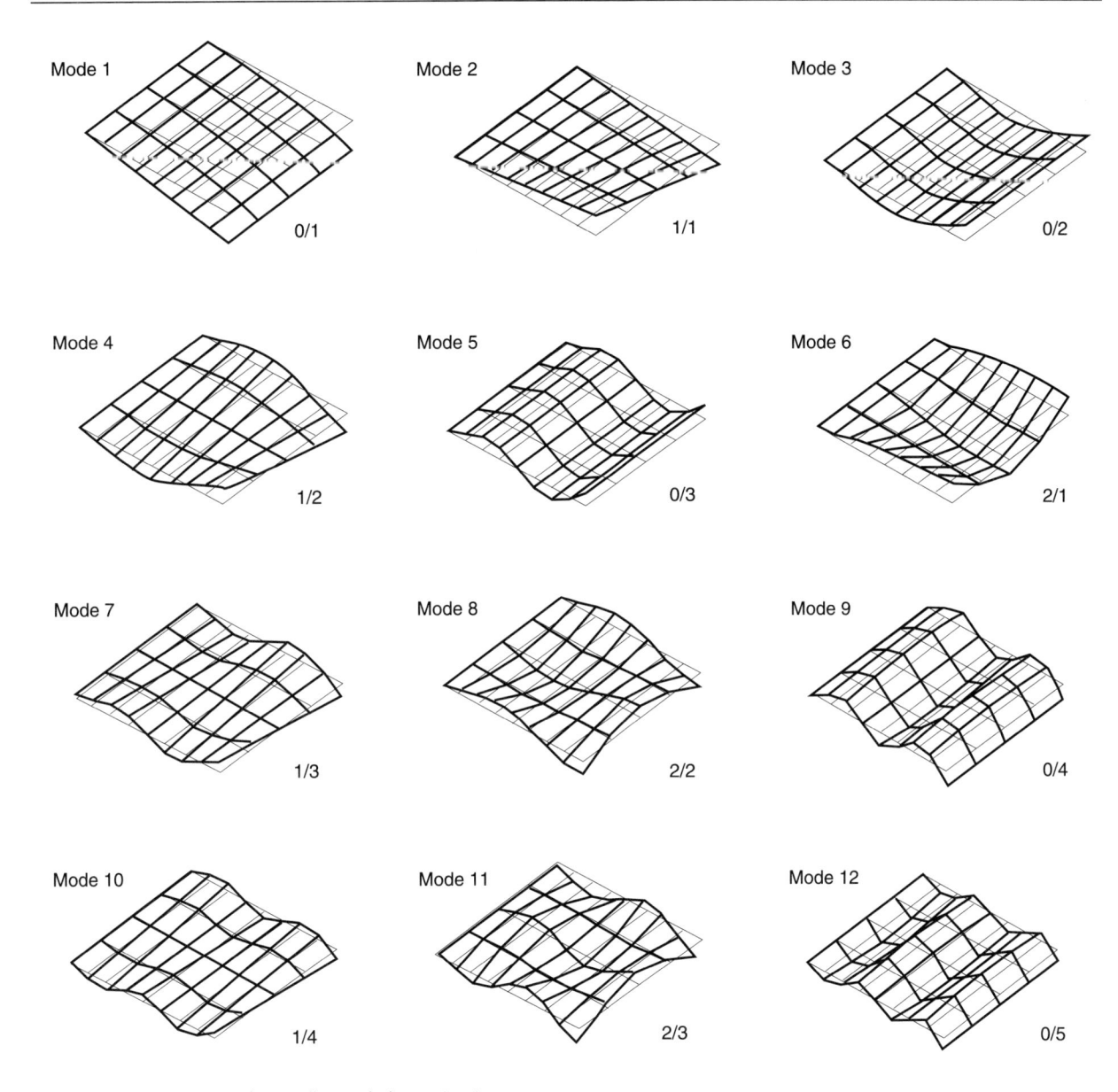

Figure 7 Mode shapes of a cantilevered planar structure.

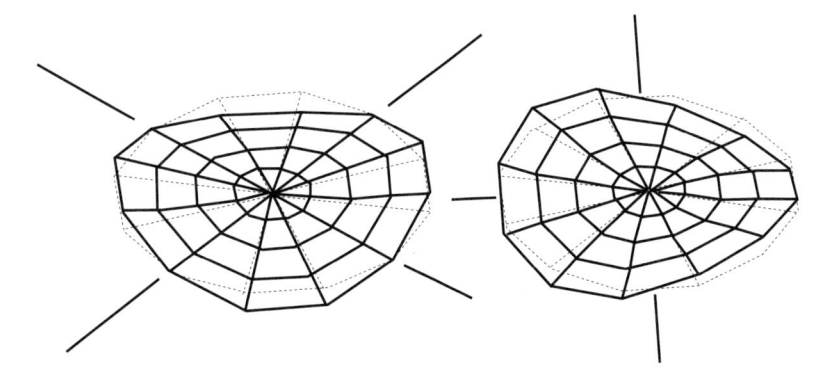

Figure 8 Two two-diameter mode shapes of a circular planar structure.

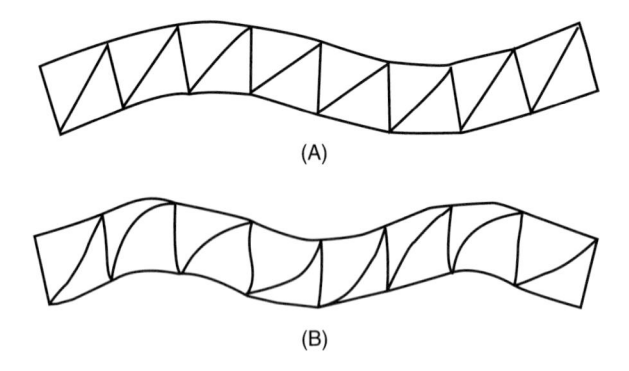

Figure 9 Free–free planar truss: (A) global mode; (B) compound mode.

Figure 15 gives a detailed presentation of the three receptance Nyquist plots calculated for a two-degrees-of-freedom (2DOF) system with low damping and relatively well-separated natural frequencies. The values on the curves give the excitation frequency in rad s^{-1}. For systems with close natural frequencies and/or higher damping, there is only one loop in the Nyquist plot for the 2DOF system.

FRF displays are distorted by nonlinearities. Figure 16 summarizes some of the effects of nonlinear stiffness on the response plots of single-degree-of-freedom systems with hysteretic damping. Two families of curves are plotted in this case: first, the Nyquist plots, connecting points of constant excitation level, and second, the isochrones connecting points of constant excitation frequency.

Damping nonlinearity can also be recognized from the pattern of isochrones (**Figure 17**).

For rotor-bearing systems, unbalance response plots are being used to display shaft orbit-related quantities as a function of the rotational (spin) speed. Assuming elliptical orbits, **Figure 18** illustrates three graphical formats: first, a plot of major (a) and minor (b) ellipse semiaxes against speed; second, a plot of two motion components, along two mutually perpendicular directions, y and z, in a plane normal to the rotor axis, against speed; and third, a plot of the radii of the ellipse generating forward and backward circular motions, r_f and r_b, against speed.

Displays of Frequency Spectra

Autopower and cross-power spectra and coherence functions are useful data descriptors in vibration measurement and analysis.

The frequency spectra of the complex signals generated by machines are characteristic for each machine, constituting a unique set of patterns, referred to as the machine signature. Analysis of machine mechanical signatures facilitates the location of vibration sources. Observing the evolution of such signatures in time permits the evaluation of a machine's mechanical condition. By monitoring the growth of the amplitude of certain spectral

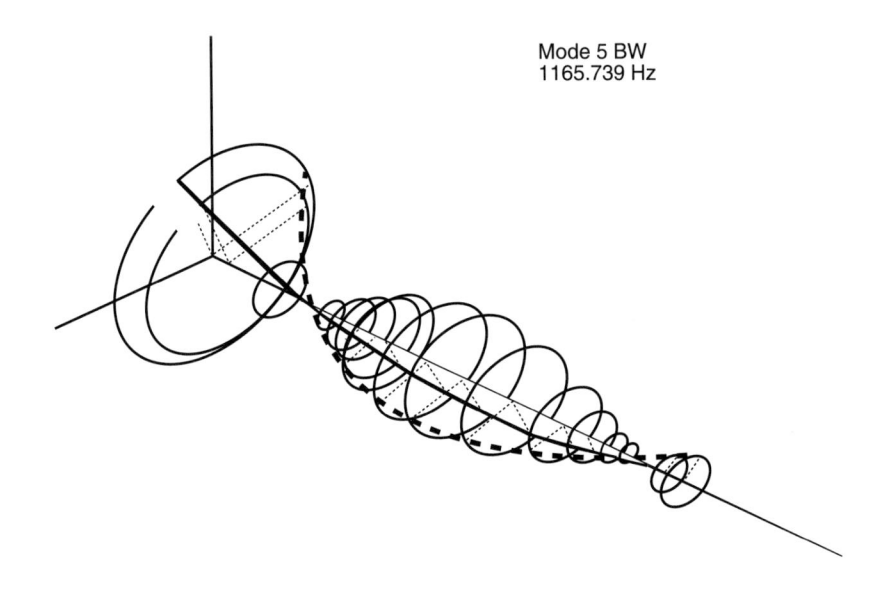

Mode 5 BW
1165.739 Hz

Spin speed 30 000 rpm

Figure 10 Whirling orbits and spatial mode shapes of a rotor-bearing system.

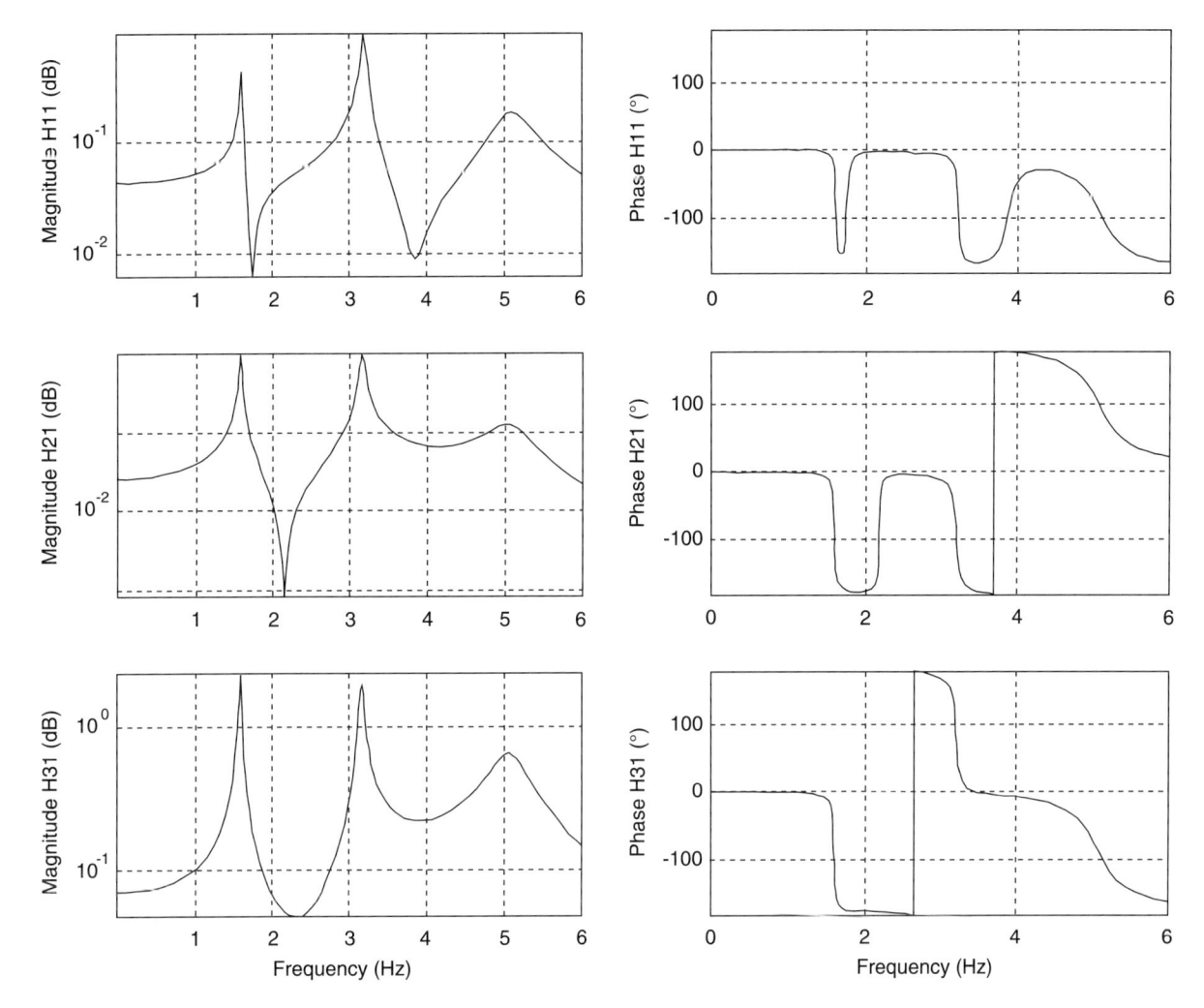

Figure 11 Plots of magnitude and phase of the receptance of a simple structure.

components, faults can be detected by correlating their frequencies to the machine operating parameters.

Spectrum plots are commonly $x - y$ plots in which the x axis represents the vibration frequency and the y axis represents amplitudes of a signal's individual frequency components.

A characteristic spectrum plot for a five-cylinder, four-stroke diesel engine is shown in **Figure 19**. The large-amplitude components $2X$ and $2\frac{1}{2}X$ are produced by the unbalanced pitching moment, and by the ignition rate, respectively. The harmonic $5X$ is produced by the unbalanced rolling moment while the $7\frac{1}{2}X$ component corresponds to the impact rate of the camshaft. The one-half-order component $\frac{1}{2}X$ is the fundamental harmonic of the gas-pressure cycle. Note that not the relative level of the spectral components but their variation in time is significant for fault diagnosis.

A series of spectrum plots can be combined into a three-dimensional presentation, where one spectrum appears slightly displaced from the other, as a function of either machine speed (cascade plot) or time (waterfall plot). **Figure 20** illustrates a cascade spectrum typical for the oil whirl and whip. Speed-related phenomena will track along a diagonal, whereas speed-independent resonance-related responses appear on lines parallel to the ordinate axis. Such a display shows changes in vibration frequencies and amplitudes. Very short time or speed increments can be chosen to see the effects of nonstationary phenomena (like subharmonic instabilities).

Model Order Indicators

Singular Value Plots

Estimation of the rank of a matrix of measured FRF data can be made using the singular value decomposition (SVD) of a composite FRF (CFRF) matrix, $[\mathbf{A}]_{N_f \times N_o N_i}$. Each column of the CFRF matrix contains elements of an individual FRF measured for

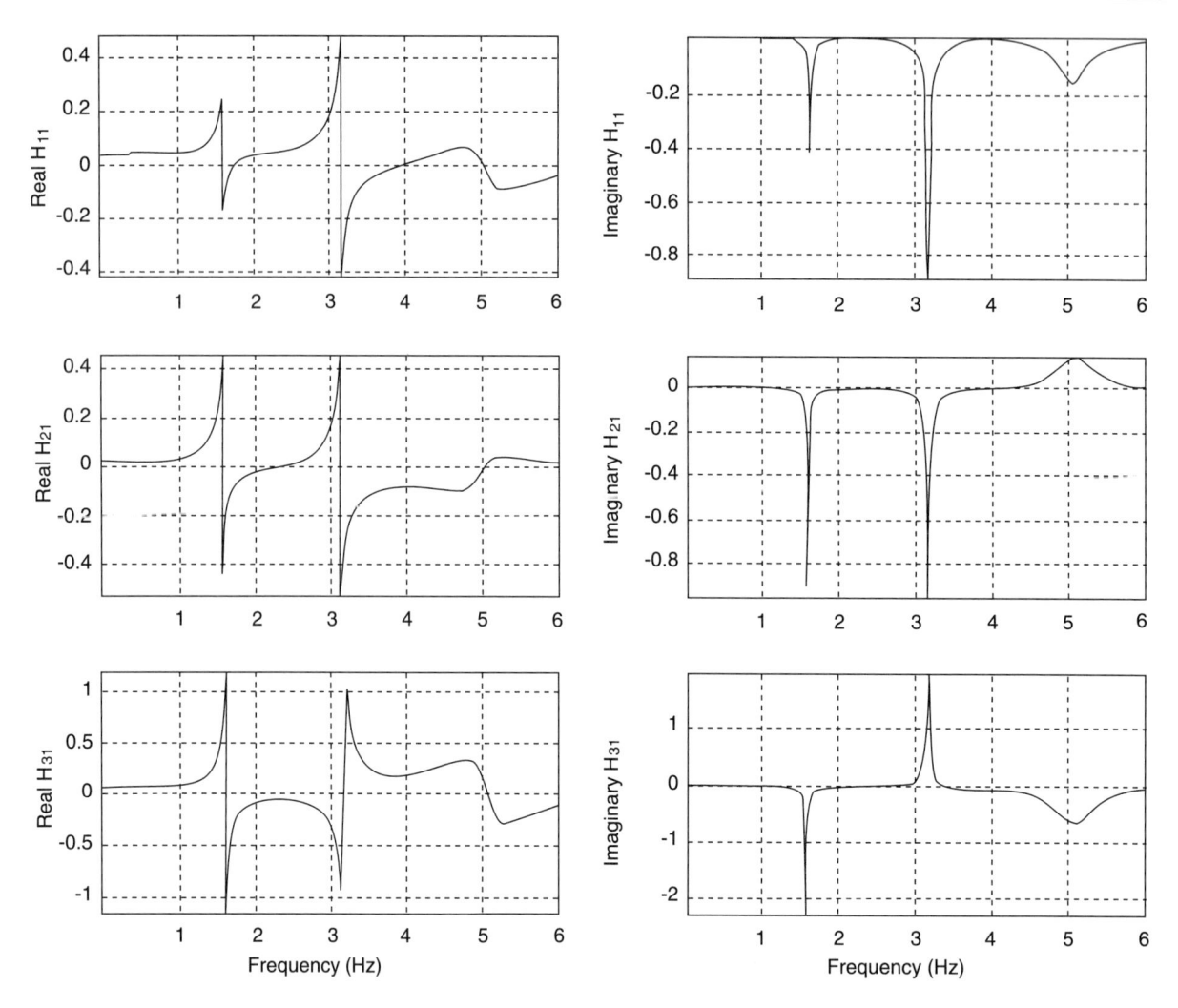

Figure 12 Plots of real and imaginary parts of the receptance of a simple structure.

given input/output location combination at all frequencies.

By performing an SVD, the CFRF matrix is decomposed into a diagonal matrix of positive singular values and two unitary matrices containing the left and right singular vectors, respectively:

$$[\mathbf{A}]_{N_f \times N_o N_i} = [\mathbf{U}]_{N_f \times N_o N_i} [\mathbf{\Sigma}]_{N_o N_i \times N_o N_i} [\mathbf{V}]^H_{N_o N_i \times N_o N_i}$$

The singular values are indexed in descending order. If there is a gap in the singular value array then the number of significant (non-trivial) singular values can be taken as a good estimate of the number of modes that have an identifiable contribution to the system response.

In order to separate these larger singular values from the others, two kinds of plot can be used. **Figure 21A** shows the plot of magnitudes of the singular values versus their index. They are normalized to the first singular value. The sudden drop in the curve after the sixth singular value indicates a rank $N_r = 6$, i.e., that there are six dominant modes in the frequency band of data.

Figure 21B shows the related plot of the ratio of successive singular values. The distinct trough indicates the same number of probable important modes.

Mode Indicator Functions

Modal indicators are useful for estimating the effective number of modes in the frequency range of interest. Mode Indicator Functions (MIFs) are frequency-dependent scalars, calculated from an eigenvalue or singular value solution approach, involving FRF matrices. Plotted against frequency, they exhibit peaks or troughs at the natural frequencies.

The Complex Mode Indicator Function (CMIF) is a plot of the singular values of the FRF matrix

DISPLAYS OF VIBRATION PROPERTIES 423

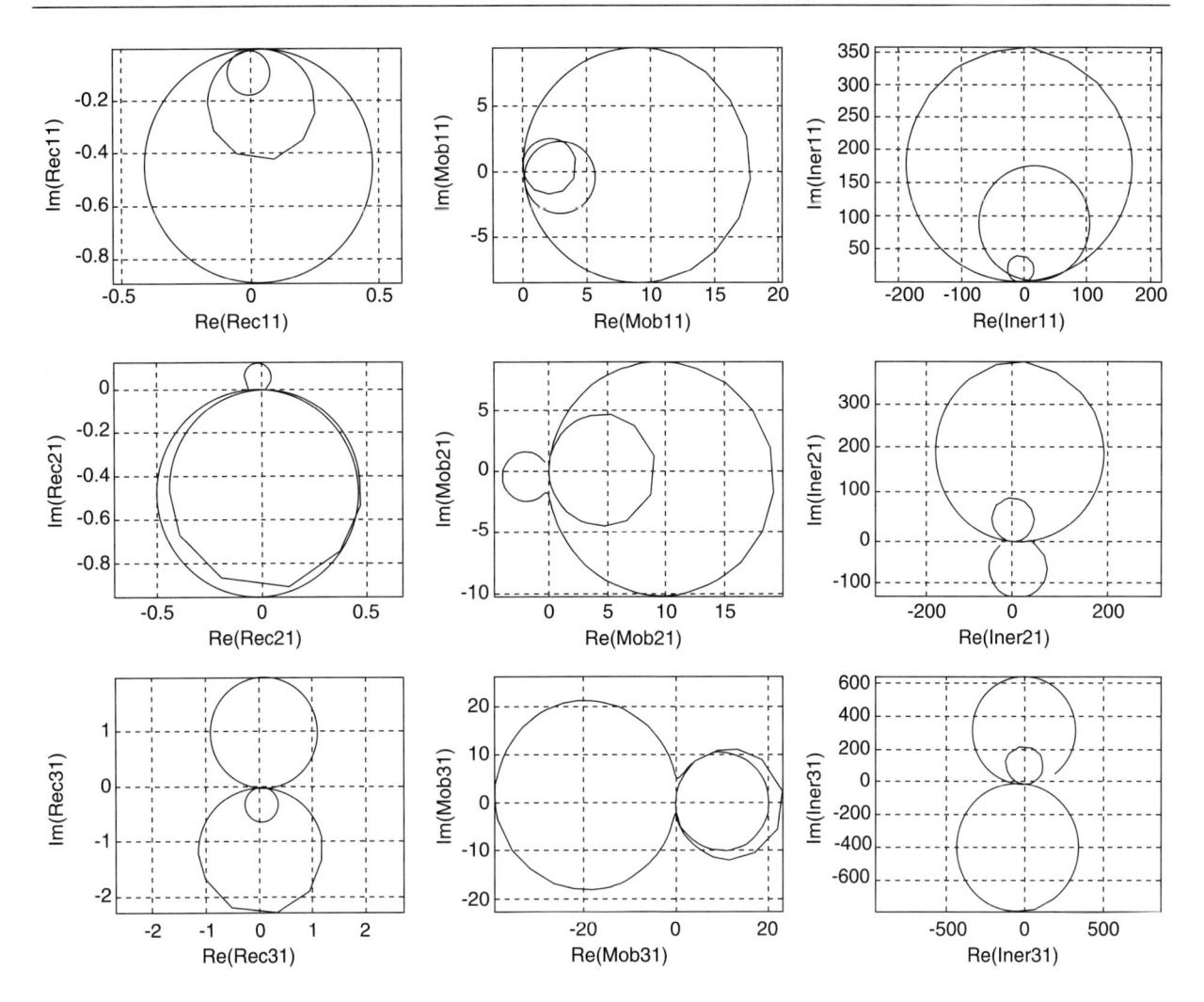

Figure 13 Nyquist plots of the receptance (Rec), mobility (Mob), and inertance (Iner) of a simple structure.

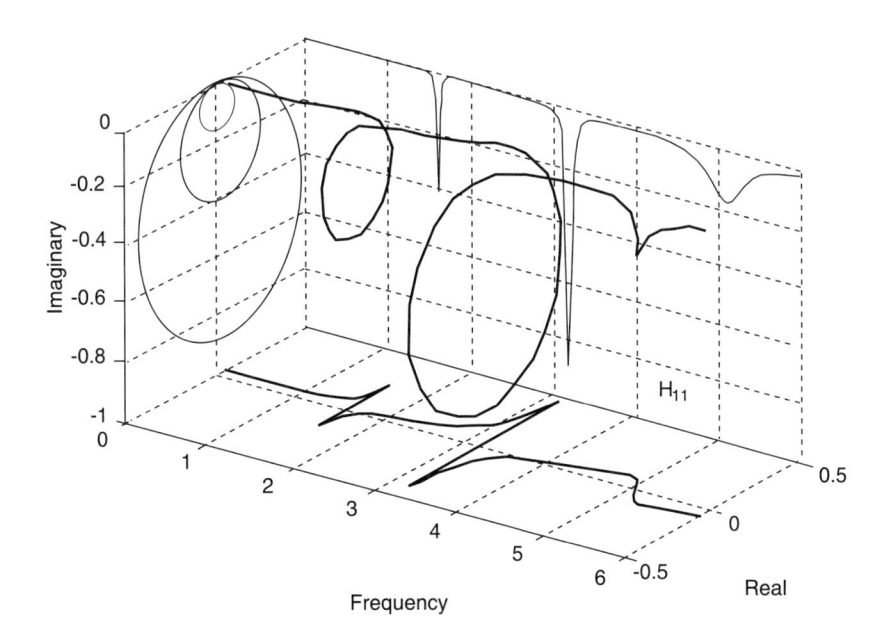

Figure 14 Three-dimensional plot of a three-mode point receptance FRF.

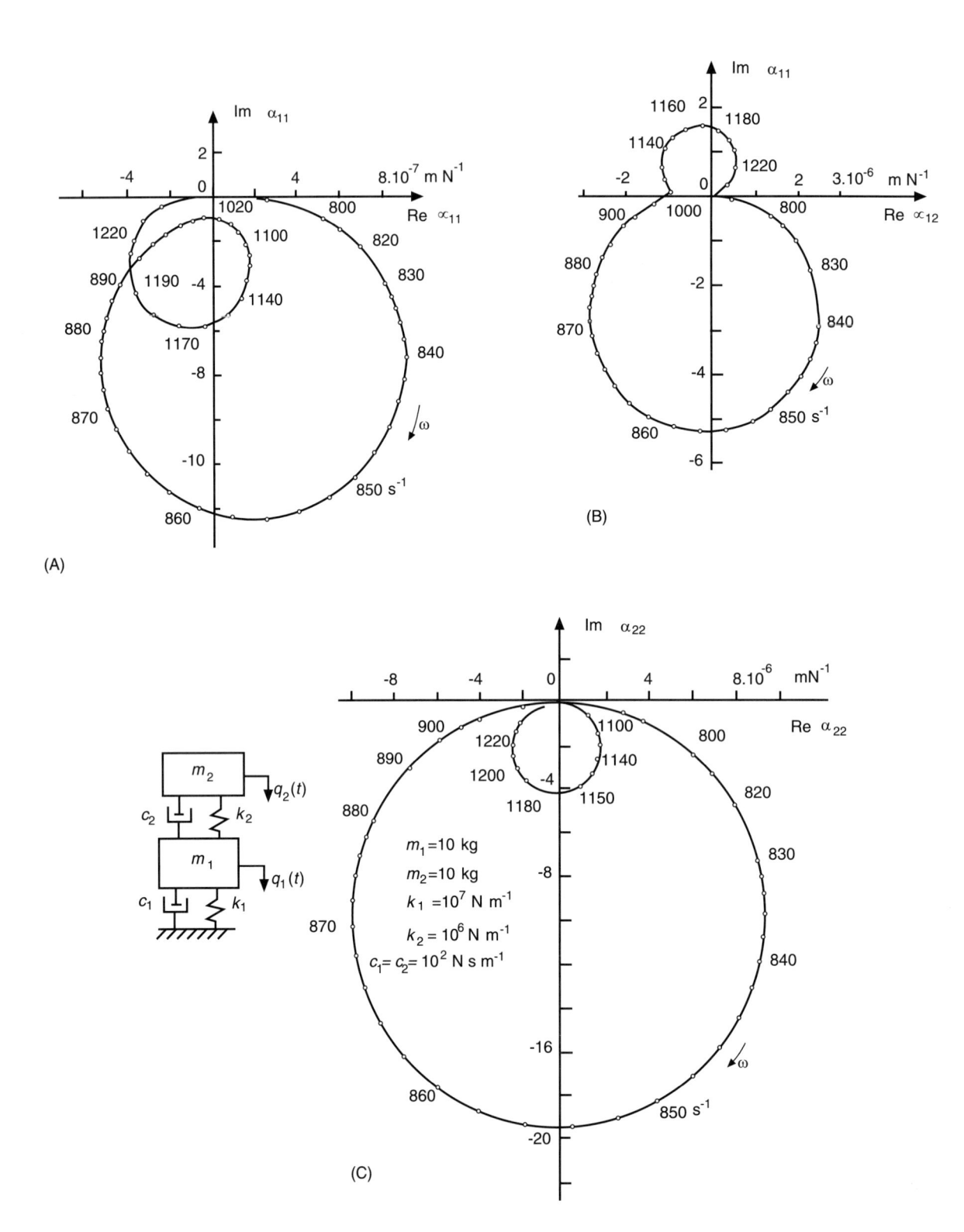

Figure 15 (A)–(C) Nyquist plots of receptances for a 2-DOF system.

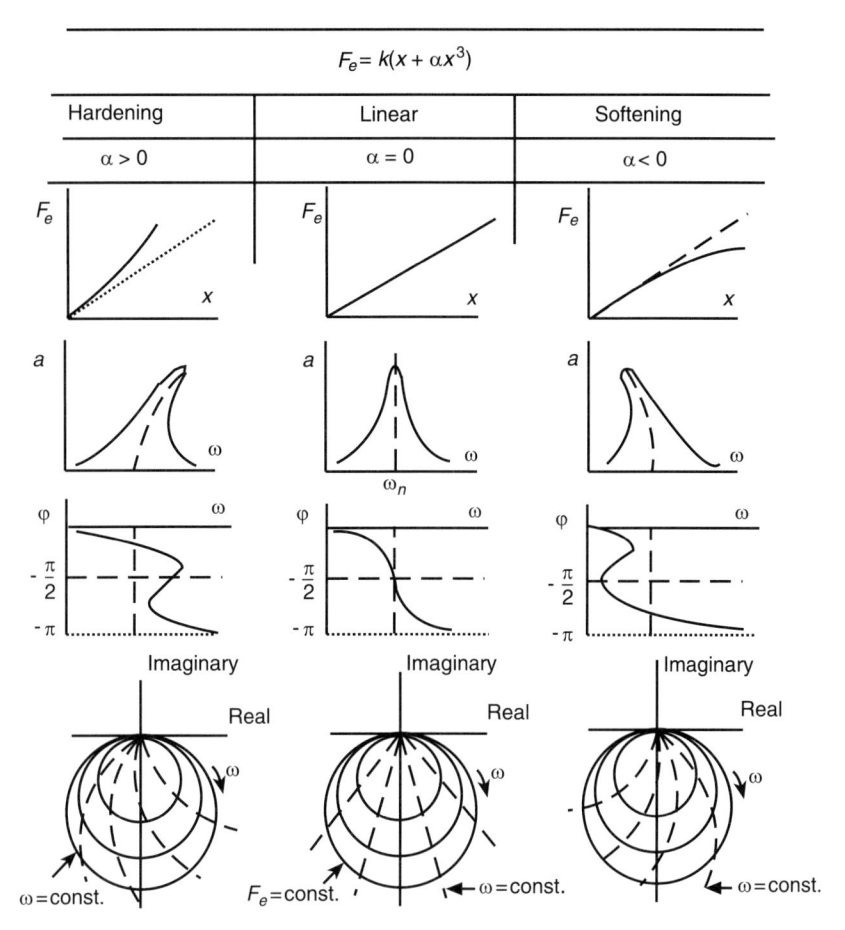

Figure 16 Effect of nonlinear stiffness on some FRF plots.

$[\mathbf{H}]_{N_o \times N_i}$, on a log magnitude scale, as a function of frequency. The number of CMIF curves is equal to the number of input reference points N_i ($N_i < N_o$). Peaks in the CMIF plot indicate the existence of modes. Multiple peaks at the same frequency indicate multiple or repeated modes.

The CMIF plot shown in **Figure 22** indicates six dominant modes of vibration, with two pairs of quasi-repeated natural frequencies at about 103 and 178 Hz.

Multivariate mode indicator function The Multivariate Mode Indicator Function (MMIF) is defined by the eigenvalues, α, of the generalized spectral problem:

$$[\mathbf{H}_R]^T[\mathbf{H}_R]\{f\} = \alpha \Big([\mathbf{H}_R]^T[\mathbf{H}_R] + [\mathbf{H}_I]^T[\mathbf{H}_I]\Big)\{f\}$$

where $[\mathbf{H}_R]$ and $[\mathbf{H}_I]$ are the real and imaginary parts of the FRF matrix, respectively.

In a plot of MMIF against frequency (**Figure 23**), the undamped natural frequencies can be located by the minima of the smallest eigenvalue. Troughs in the next smallest eigenvalues indicate multiple modes.

Mode indicator function A single curve MIF calculated as:

$$\mathrm{MIF}_i = 1 - \frac{\sum_{j=1}^{N_0 N_i} |\mathrm{Re}(a_{ij})||a_{ij}|}{\sum_{j=1}^{N_0 N_i} |a_{ij}|^2}$$

where a_{ij} are the elements of the CFRF matrix $\mathbf{A}$ is shown in **Figure 24**. Apart from failing to indicate the two double modes (being a single curve mode indicator), this MIF reveals more modes than the dominant ones, performing better at the location of local modes.

The U-Mode Indicator Function (UMIF) is a plot of the left singular vectors of a composite FRF matrix in which each column contains elements of an individual FRF measured for a given input/output location combination at all frequencies, as a function of frequency. The UMIF plot calculated for the same

426 DISPLAYS OF VIBRATION PROPERTIES

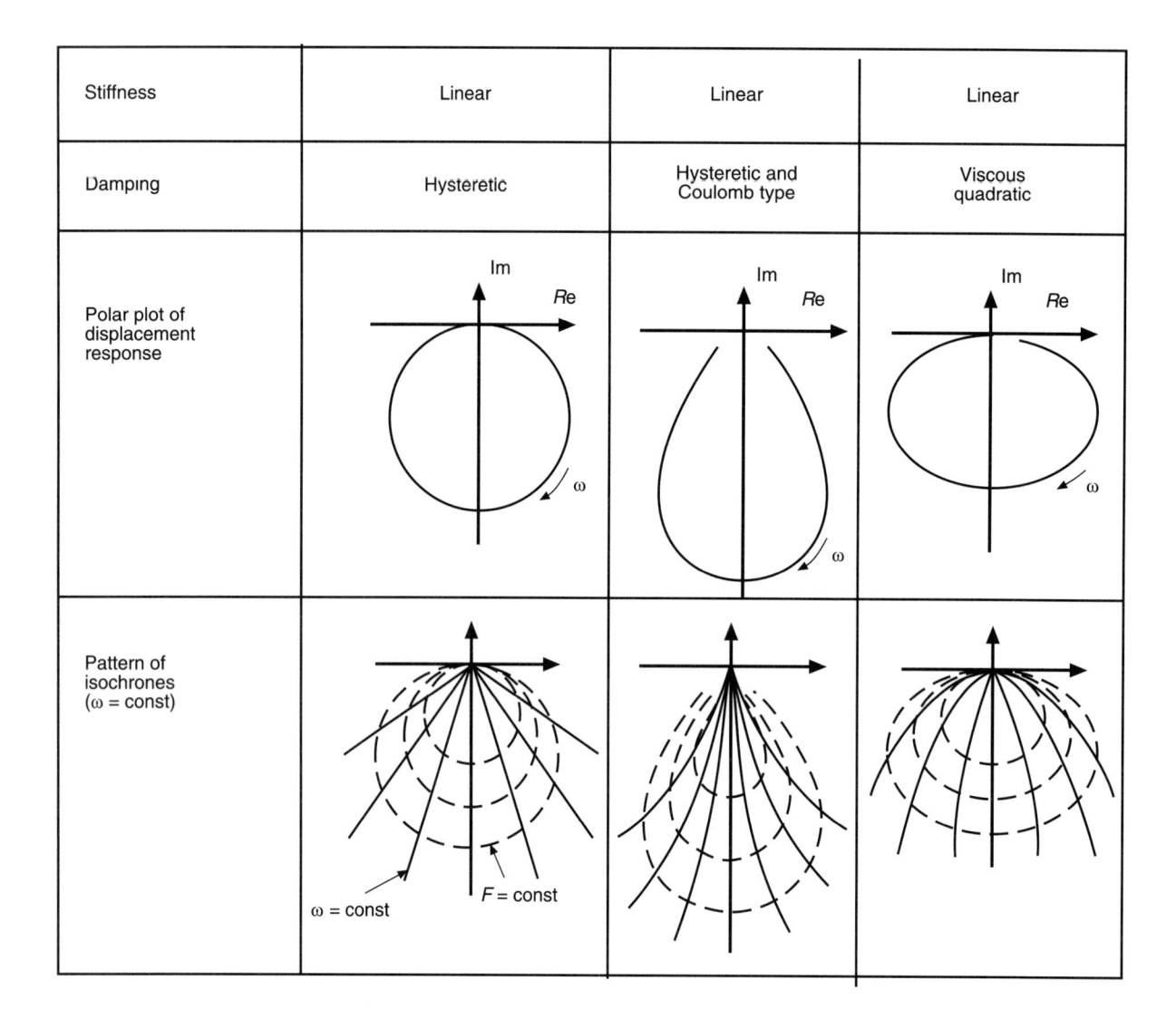

Stiffness	Linear	Linear	Linear
Damping	Hysteretic	Hysteretic and Coulomb type	Viscous quadratic
Polar plot of displacement response			
Pattern of isochrones (ω = const)			

Figure 17 Recognizing nonlinear damping from the pattern of isochrones.

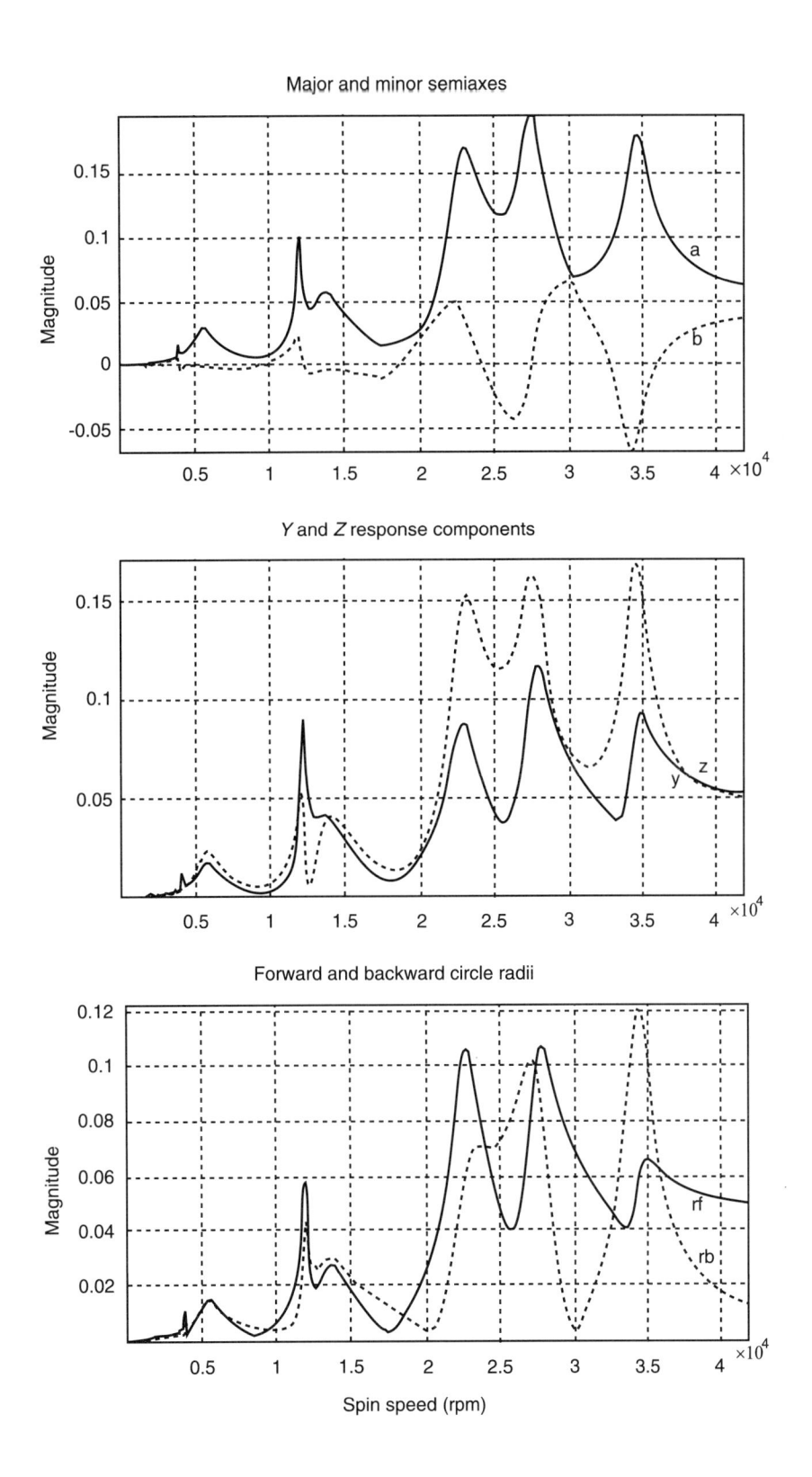

Figure 18 Typical unbalance response plots of a rotor-bearing system.

428 DISPLAYS OF VIBRATION PROPERTIES

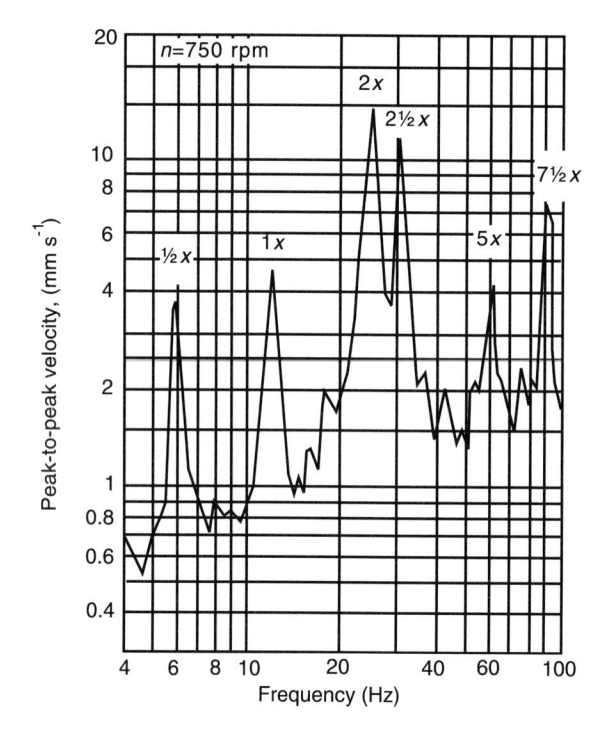

Figure 19 Spectrum plot measured on a five-cylinder, four-stroke diesel engine.

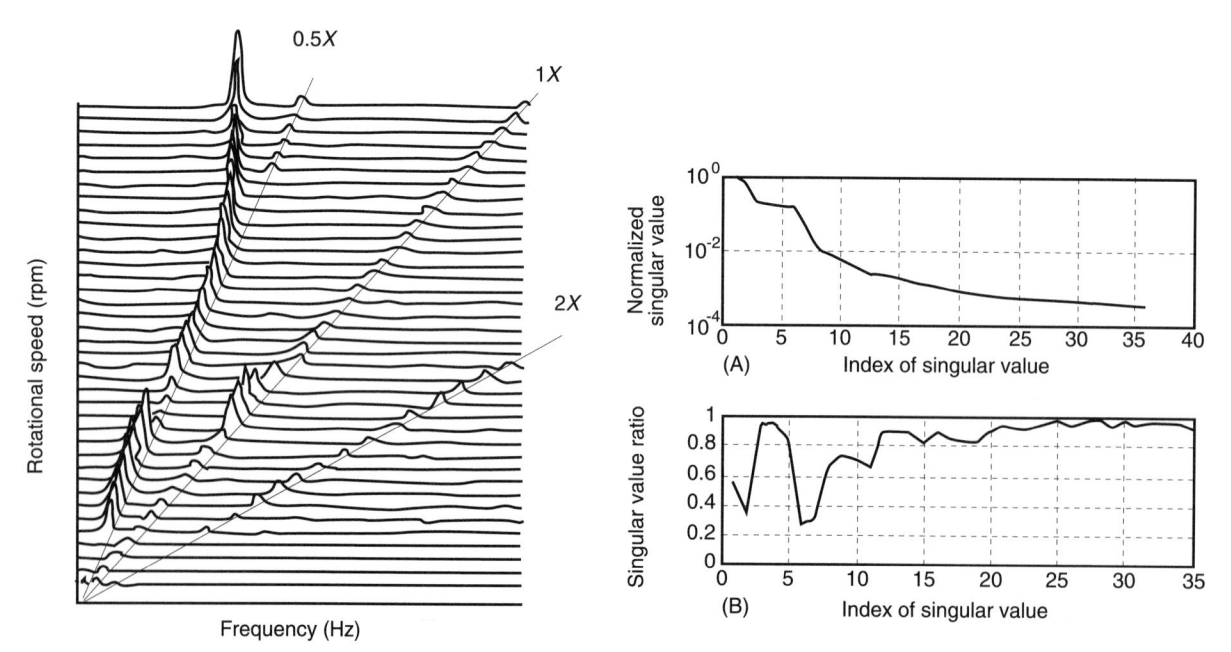

Figure 20 Cascade spectrum plot.

Figure 21 Singular value plots: (A) normalized magnitudes; (B) ratio of successive values.

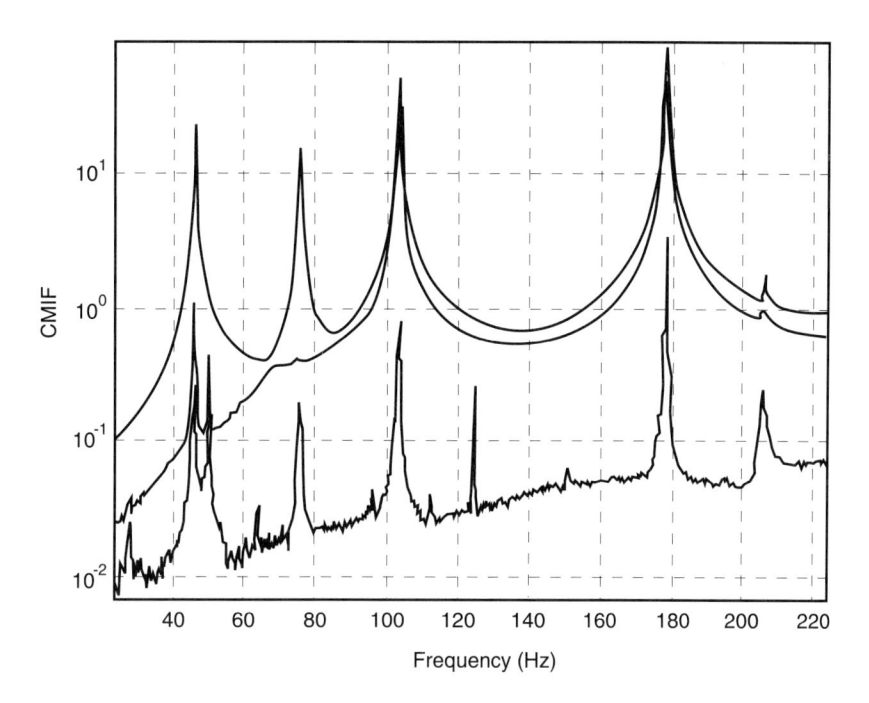

Figure 22 Complex mode indicator function plot.

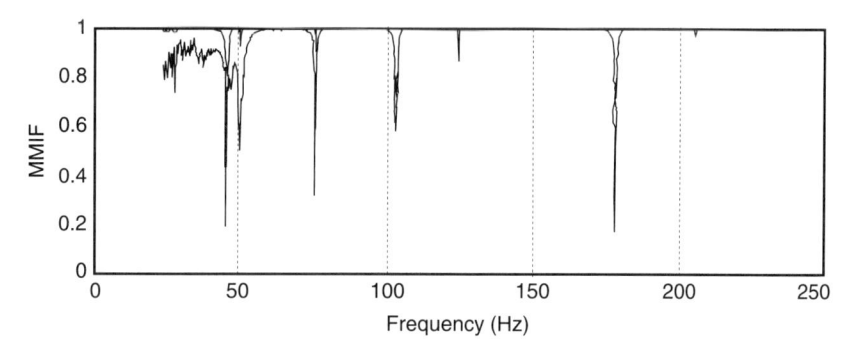

Figure 23 Multivariate mode indicator function plot.

data as the previously presented CMIF and MIF is shown in **Figure 25**. Double modes are indicated by two overlaid curves.

There are some other displays used as potential mode identifiers, like the stabilization diagram, the plot of the sum of all available FRFs, and plots of other composite functions displaying the magnitude squared or the imaginary part squared, to enhance the peaks.

Other Displays

Special displays are used to describe nonlinear vibrations and chaotic motions, like phase trajectories and phase portraits in the phase plane (velocity versus displacement) and three-dimensional orbits in the phase space (see **Nonlinear systems analysis**).

Time domain analysis displays, such as auto- and cross-correlation function plots, supplement spectrum plots in spectral analysis (see **Correlation functions**).

See also: **Blades and bladed disks**; **Correlation functions**; **Modal analysis, experimental**, Parameter extraction methods; **Nonlinear systems analysis**; **Rotating machinery, essential features**; **Rotor dynamics**; **Theory of vibration**, Fundamentals.

430 DISPLAYS OF VIBRATION PROPERTIES

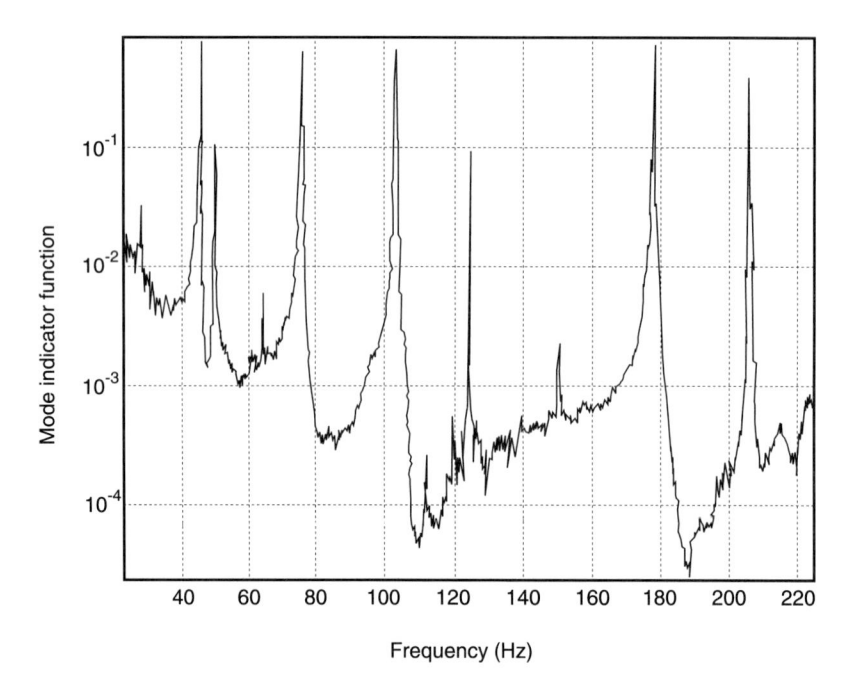

Figure 24 Mode indicator function.

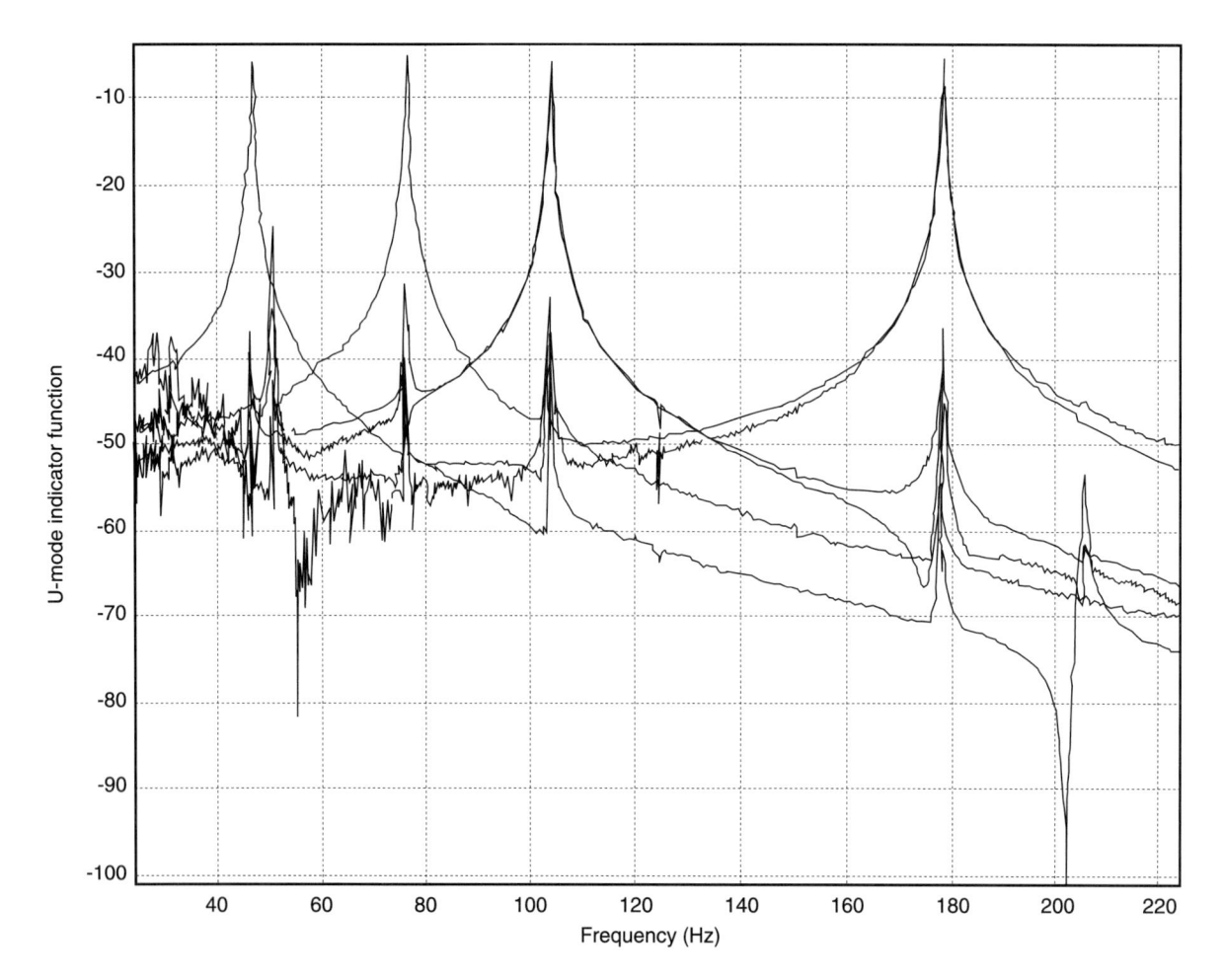

Figure 25 U-mode indicator function.

Further Reading

Buzdugan GH, Mihăilescu E and Radeş M (1986) *Vibration Measurement*. Dordrecht: Martinus Nijhoff.

Chladni EFF (1787) *Entdeckungen über die Theorie des Klanges*. Leipzig: Breitkopf und Härtel.

Ewins DJ (2000) *Modal Testing: Theory Practice and Application*. Baldock, UK: Research Studies Press.

Lord Rayleigh (Strutt JW) (1894) *The Theory of Sound*, vol. 1. London: Macmillan. (Reprinted by Dover, 1945.)

Maia NMM and Silva JMM (eds) (1997) *Theoretical and Experimental Modal Analysis*. Taunton, UK: Research Studies Press.

Morse PM (1948) *Vibration and Sound*, 2nd edn. New York: McGraw-Hill.

DISTRIBUTED SENSORS AND ACTUATORS

See **SENSORS AND ACTUATORS**

DUHAMEL METHOD

See **THEORY OF VIBRATION: DUHAMEL'S PRINCIPLE AND CONVOLUTION**

DYNAMIC STABILITY

A Steindl and H Troger, Vienna University of Technology, Vienna, Austria

Copyright © 2001 Academic Press

doi:10.1006/rwvb.2001.0047

Introduction

One of the key problems in stability theory is the loss of stability of a given stable state, for example an equilibrium or a periodic motion, of a nonlinear dynamical system under variation of a system parameter, which will be called λ. Let us consider, as a typical example, a straight vertically downhanging tube conveying fluid, clamped at the upper end and free at the lower end under a quasistatically increasing flow rate. (In general one understands by a quasistatic parameter variation that the stability behavior of the state is studied at gradually increasing but fixed values of the parameter. This is in contrast to the case of (slowly) continuously varying parameter values, on which we comment later in this article). It is well known from everyday experience, for example from hosing the garden, that for small flow rate the vertically downhanging equilibrium position is stable. However, increasing the flow rate λ it finally reaches a critical value (denoted by λ_c) for which experience shows that an oscillatory motion sets in. In the language of stability theory the originally stable state has lost stability. Now three questions are of interest:

1. What is the value of λ_c at which the loss of stability occurs?
2. What type of motion is setting in after loss of stability of the originally stable state and how is the motion changing by further increasing the parameter λ?
3. How do small changes of the system (small variations of other system parameters) affect the system's behavior?

Under mild requirements on the smoothness of the system description, it is possible to give completely satisfying answers to these questions.

The answer to the first question is supplied from the solution of a linear eigenvalue problem and has been known to engineers for a long time. The answers to the second and third questions, however, can only recently be given, in a systematic and complete way, due to strong progress achieved in the 1970s and 1980s in the field of nonlinear stability theory making use of the methods of local bifurcation theory. Here, as pioneers among many others, the names VI

Arnold, R Thom, M Golubitsky, Ph Holmes may be mentioned. Their work is based on ideas dating back to H Poincare and AA Andronov.

The basic approach to solving the problem of dynamic stability loss consists of three steps:

1. Dimension reduction: One can show that in many important practical cases after the loss of stability the behavior of a high-dimensional or even infinite-dimensional dynamical system can be analyzed by a low-dimensional bifurcation system, which can be obtained by a method of dimension reduction.
2. Normal form theory: The low-dimensional bifurcation system can be further strongly simplified by normal form theory.
3. Classification and unfolding: For bifurcations with low codimension a classification of all qualitatively different types of loss of stability is possible.

Dimension Reduction

We consider here only the loss of stability of a stable equilibrium and comment on periodic solutions later in the article.

There are many ways to perform the dimension reduction. Galerkin methods are well known to engineers. Both linear and nonlinear Galerkin methods can be used. The use of Galerkin methods is recommended if a mathematically rigorous method called the center manifold theory cannot be applied because the requirements for its application are not satisfied. In the following section we give a short explanation of center manifold theory. However, other authors who have performed the reduction by the closely related Liapunov–Schmidt method, treat the dimension reduction problem at varying levels of detail.

Idea of Dimension Reduction

It is well known from experiments and also from engineering experience that even complicated, apparently random time behavior of high-dimensional or infinite-dimensional dissipative dynamical systems can often be described by the deterministic flow on a low-dimensional (chaotic) attractor.

Hence, the first, most important, step in the analysis of the loss of stability of a certain state of a nonlinear dynamical system will always be the attempt to reduce its dimension. For certain systems this goal can be achieved without essential loss of accuracy in the description of the dynamics of the system by removing inessential degrees-of-freedom from the system.

Before indicating which calculations are necessary, let us explain first the basic idea behind dimension reduction. We consider an equilibrium u_E of an infinite-dimensional nonlinear system under quasi-static variation of a system parameter λ. Keep in mind the tube example mentioned in the introduction. For small perturbations out of the equilibrium the system behavior will be governed by its linearized equations. As long as the values of the parameter λ (the flow rate in the tube) are below a critical value λ_c, all eigenvalues of the linearized operator will have negative real parts and all modes, whose superposition represent the perturbation, will be damped and u_E will be stable. A critical value λ_c of the parameter λ will be reached when for the first time eigenvalues are located at the imaginary axis in the complex plane. The corresponding modes are mildly unstable or only slightly damped in linear theory. If the number of these modes is finite the loss of stability can be described in terms of the temporal evolution of the amplitudes of these (active) modes. Their amplitudes are governed by a set of ordinary differential equations called bifurcation equations or amplitude equations of the critical modes and they basically govern the behavior of the full system since the other (infinitely many) modes are damped out and hence do not appear in the description of the dynamics (see **Figure 1** where it can be seen that the relevant dynamics takes place on the manifold M^c, after a strong contraction of the flow towards M^c).

Center Manifold Theory

Mostly, for the application of center manifold theory the admissible variation of λ about $\lambda = \lambda_c$ is restricted to be very small and, secondly, in treating infinite-dimensional systems only a finite number of eigenvalues are allowed to cross the imaginary axis at loss of stability. However, it supplies a mathematically rigorous method of dimension reduction for both infinite- and large finite-dimensional dynamical systems.

We explain now for an n-dimensional system:

$$\dot{\mathbf{u}} = \mathbf{F}(\mathbf{u}, \lambda) \qquad [1]$$

where $\mathbf{u}, \mathbf{F} \in \Re^n, \lambda \in \Re^1$, which calculations must be performed. Assuming that $\mathbf{u}_E$ is an equilibrium of eqn [1], that is $\mathbf{F}(\mathbf{u}_E, \lambda) = \mathbf{0}$ we insert $\mathbf{u} = \mathbf{u}_E + \mathbf{v}$ into eqn [1] and write the dynamical system in the form:

$$\dot{\mathbf{v}} = \mathbf{A}(\lambda)\mathbf{v} + \mathbf{r}(\mathbf{v}, \lambda) \qquad [2]$$

Here $\mathbf{A}(\lambda)$ is the Jacobian which is an $n \times n$ matrix.

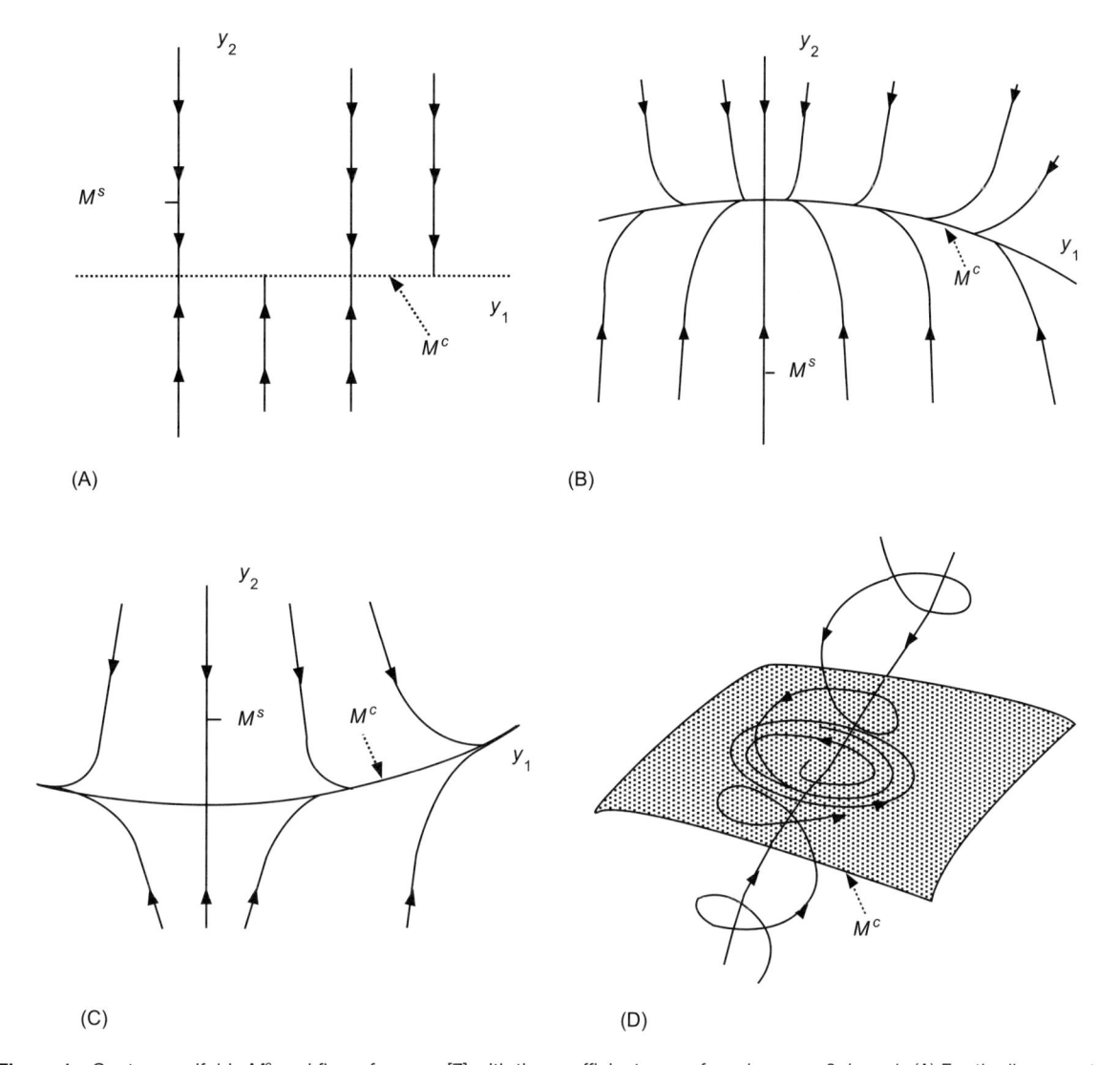

Figure 1 Center manifolds M^c and flows for eqns [7] with the coefficients $a = f = d = e = 0, b = 1$. (A) For the linear system or eqn [7]; (B) for the nonlinear system of eqn [7] $c<0$; (C) for the nonlinear system of eqn [7] for $c>0$; (D) the flow nearby and on the center manifold for a three-dimensional system after a Hopf bifurcation.

We study the loss of stability of the initially asymptotically stable equilibrium $\mathbf{u}_E$ for quasistatically increasing values of λ. At the critical parameter value $\lambda = \lambda_c$ for the first time n_c eigenvalues are located at the imaginary axis, whereas all other n_y eigenvalues are located left of the imaginary axis ($n = n_c + n_s$). Now we introduce a change of variables $\mathbf{v} = \mathbf{Bq}$ which transforms $\mathbf{A}(\lambda_c)$ into diagonal, or if this is not possible, into Jordan form. As is well known, the transformation matrix $\mathbf{B}$ is composed by the eigenvectors or by the principal vectors of $\mathbf{A}(\lambda_c)$. Then eqn [2] takes the form:

$$\dot{\mathbf{q}} = \mathbf{B}^{-1}\mathbf{A}(\lambda_c)\mathbf{Bq} + \mathbf{B}^{-1}\mathbf{r}(\mathbf{q}, \lambda_c) = \mathbf{Jq} + \mathbf{g}(\mathbf{q}, \lambda_c)$$

which by simply rearranging the variables can be rewritten in the form:

$$\dot{\mathbf{q}}_c = \mathbf{J}_c\mathbf{q}_c + \mathbf{g}_c(\mathbf{q}_c, \mathbf{q}_s)$$
$$\dot{\mathbf{q}}_s = \mathbf{J}_s\mathbf{q}_s + \mathbf{g}_s(\mathbf{q}_c, \mathbf{q}_s)$$
[3]

The splitting of $\mathbf{q}$ is done so that the first block of $\mathbf{J}$, called $\mathbf{J}_c$, contains the n_c eigenvalues, which all have zero real part, and all n_s eigenvalues of $\mathbf{J}_s$ have negative real part. If:

$$\mathbf{q}_s = \mathbf{h}(\mathbf{q}_c)$$
[4]

is an invariant manifold for the system (a manifold M is invariant, if all trajectories starting on M stay on the manifold for all $t>t_0$) and if **h** is smooth, we call **h** a center manifold if $\mathbf{h}(0) = \mathbf{h}'(0) = \mathbf{0}$. Note that for the case $\mathbf{g}_c = \mathbf{g}_s = \mathbf{0}$, all solutions tend exponentially fast to solutions of $\dot{\mathbf{q}}_c = \mathbf{J}_c \mathbf{q}_c$. Hence, the linear n_c-dimensional equation on the center manifold determines the asymptotic behavior of the entire n-dimensional linear system, up to exponentially decaying terms (see **Figure 1A**). The center manifold theorem enables us to extend this argument to the case when $\mathbf{g}_c$ and $\mathbf{g}_s$ are not equal to zero and, hence, a nonlinear system is treated.

Center Manifold Theorem

1. There exists a center manifold $\mathbf{q}_s = \mathbf{h}(\mathbf{q}_c)$ for eqns [3] if $|\mathbf{q}_c|$ is sufficiently small. The behavior of eqn [3] on the center manifold is governed by the equation:

$$\dot{\mathbf{q}}_c = \mathbf{J}_c \mathbf{q}_c + \mathbf{g}_c(\mathbf{q}_c, \mathbf{h}(\mathbf{q}_c)) \quad [5]$$

2. The zero solution of eqn [3] has exactly the same stability properties as the zero solution of eqn [5].
3. If $\mathbf{H}: R_c^n \to R_c^n$ is a smooth map with $\mathbf{H}(0) = \mathbf{H}'(0) = 0$ and is defined by:

$$\mathbf{P}(\mathbf{H}) := \mathbf{H}'(\mathbf{q}_c)[\mathbf{J}_c \mathbf{q}_c + \mathbf{g}_c(\mathbf{q}_c, \mathbf{H}(\mathbf{q}_c))]$$
$$- \mathbf{J}_s \mathbf{H}(\mathbf{q}_c) - \mathbf{g}_s(\mathbf{q}_c, \mathbf{H}(\mathbf{q}_c)) = O(|\mathbf{q}_c|^p) \quad [6]$$

then if $\mathbf{P}(\mathbf{H}) = O(|\mathbf{q}_c|^p), p > 1$, as $|\mathbf{q}_c| \to 0$, we have $|\mathbf{h}(\mathbf{q}_c) - \mathbf{H}(\mathbf{q}_c)| = O(|\mathbf{q}_c|^p)$ as $|\mathbf{q}_c| \to 0$.

Item 3 of the theorem says that it is possible to approximate the center manifold **h** by the function **H** up to terms of order $O(|\mathbf{q}_c|^p)$. This is a very important feature from the practical point of view because it allows to calculate a sufficiently accurate approximation by cutting off a Taylor series expansion.

In the infinite-dimensional case care must be taken concerning the spectrum of the linearized operator, since eqn [1] is a partial differential equation and defined on an infinite-dimensional space.

Examples We first indicate the calculations for the loss of stability of a steady state if a zero root crosses the imaginary axis and present some quantitatively drawn figures following from this calculation. Secondly, for the classical flutter instability where a complex pair of eigenvalues crosses the imaginary axis, which mathematicians call the Hopf bifurcation, the reduced system of equations and a qualitative picture is given.

Steady-state bifurcation We consider the stability problem of the zero solution of the two-dimensional system:

$$\dot{x} = bxy + fx^3$$
$$\dot{y} = -\mu y + cx^2 \quad [7]$$

which, due to the occurrence of a zero eigenvalue is a critical case in the sense of Liapunov. The linear part is already in diagonal form with eigenvalues $0, -\mu$. Hence $x = q_c$ is the critical variable describing the essential dynamics and $y = q_s$ can be eliminated by treating the stability problem. The first equation will become the equation on the center manifold from which y must be eliminated. This is done by making an *ansatz* for the approximation of the center manifold according to eqn [4] in the form:

$$y = H(x) = h_2 x^2 + h_3 x^3 + \ldots \quad [8]$$

Inserting into eqn [6] we obtain up to third-order terms ($p = 3$ in eqn [6]):

$$\mu(h_2 x^2 + h_3 x^3) - cx^2 + O(x^4) = 0$$

Requiring that the coefficients of the second- and third-order terms vanish we obtain:

$$\mu h_2 - c = 0, \quad \mu h_3 = 0$$

From this system we obtain the coefficients of eqn [8]:

$$h_2 = \frac{c}{\mu}, \quad h_3 = 0 \quad [9]$$

Inserting eqn [9] into eqn [8] and eliminating y from the first equation of eqn [7] with eqn [8], the one-dimensional equation:

$$\dot{x} = b\frac{c}{\mu}x^3 + fx^3 + \ldots$$

is obtained on the center manifold according to eqn [5]. Considering now the case $\mu = 1, b = 1, f = 0$, the center manifolds for the linear and two nonlinear systems are shown in **Figure 1**. We see that the stability behavior of the origin is determined by the coefficient c of the nonlinear term occurring in the second (eliminated) equation.

Hopf bifurcation If for an infinite-dimensional system, for example the tube conveying fluid, one purely imaginary pair of roots $\pm i\omega$ crosses the imaginary axis and all other roots have negative real parts, then this system typically can be reduced to a two-dimensional third-order system of the form:

$$\dot{q}_1 = -\omega q_2 + a_{1,30}q_1^3 + a_{1,21}q_1^2 q_2 + a_{1,12}q_1 q_2^2$$
$$+ a_{1,03}q_2^3 + O\left(|\mathbf{q}|^5\right)$$
$$\dot{q}_2 = -\omega q_1 + a_{2,30}q_1^3 + a_{2,21}q_1^2 q_2 + a_{2,12}q_1 q_2^2 \quad [10]$$
$$+ a_{2,03}q_2^3 + O\left(|\mathbf{q}|^5\right)$$

From the sketches shown in **Figure 1** it is intuitively clear how the inessential variables are neglected because in the phase space the flow of the full system is contracted exponentially fast to the center manifold M^c where the asymptotic behavior of the trajectories and especially the stability of the equilibrium is governed by the nonlinear terms (e.g., in eqn [7] by the term with coefficient c).

Normal Form Simplification

The low-dimensional system of amplitude equations of the critical modes or eigenvectors (eqn [5]) can still be strongly simplified by means of normal form transformation.

The basic idea is to try to annihilate the nonlinear terms in eqn [5] by a nonlinear change of variables. However, complete annihilation is only possible if a linear operator, derived below, is regular. If it is singular then those terms which are not in its range cannot be eliminated and form the nonlinear normal form. We insert the nonlinear change of coordinates $\mathbf{q}_c \to \mathbf{x}$:

$$\mathbf{q}_c = \mathbf{x} + \mathbf{n}(\mathbf{x}) \quad [11]$$

into eqn [5] and obtain:

$$(\mathbf{E} + \mathbf{n}'(\mathbf{x}))\dot{\mathbf{x}} = \mathbf{J}\mathbf{x} + \mathbf{J}\mathbf{n}(\mathbf{x}) + \mathbf{g}(\mathbf{x} + \mathbf{n}(\mathbf{x})) \quad [12]$$

Here $\mathbf{E}$ denotes the unit matrix. Writing the transformed system in the form:

$$\dot{\mathbf{x}} = \mathbf{J}\mathbf{x} + \mathbf{f}(\mathbf{x}) \quad [13]$$

the unknown function $\mathbf{n}(\mathbf{x})$ in eqn [11] should now be chosen such that the nonlinear terms $\mathbf{f}(\mathbf{x})$ in eqn [13] are as simple as possible. The most desirable situation would be to eliminate all terms in $\mathbf{f}(\mathbf{x})$. Substituting eqn [13] into eqn [12] we find for the lowest-order terms:

$$\mathbf{J}\mathbf{n}(\mathbf{x}) - \mathbf{n}'(\mathbf{x})\mathbf{J}\mathbf{x} = \mathbf{f}(\mathbf{x}) - \mathbf{g}(\mathbf{x}) + o(\mathbf{n}(\mathbf{x})) \quad [14]$$

where $o(\mathbf{n}(\mathbf{x}))$ designates terms of higher order than those included in $\mathbf{n}(\mathbf{x})$. Since we want to have $\mathbf{f} = \mathbf{0}$ in eqn [13], we set $\mathbf{f} = \mathbf{0}$ in eqn [14] resulting in the equation:

$$\mathbf{J}\mathbf{n}(\mathbf{x}) - \mathbf{n}'(\mathbf{x})\mathbf{J}\mathbf{x} = -\mathbf{g} \quad [15]$$

which is called the homological equation and can be solved only if $-\mathbf{g}$ lies in the image of the linear map $\text{ad}_J: \mathbf{n}(\mathbf{x}) \mapsto \mathbf{J}\mathbf{n}(\mathbf{x}) - \mathbf{n}'(\mathbf{x})\mathbf{J}\mathbf{x}$ on the left-hand side, called the homological operator. The homological operator is regular if the eigenvalues of $\mathbf{J}$ are not located on the imaginary axis and in addition a resonance condition is not satisfied. Then $\mathbf{f} = \mathbf{0}$ can be achieved and the nonlinear system can be replaced by its linearization in the neighborhood of the singular point (**Figure 2**). However, in the bifurcation case, typically this requirement is not fulfilled and hence certain terms will always remain, forming the nonlinear normal form.

Examples

We present two examples. The first shows the strong reduction of complexity and the second shows an important feature of the system in normal form namely that it has an artificial symmetry.

Normal form after a Hopf bifurcation A straightforward application of the above formulas yields instead of eqn [10]:

$$\dot{x}_1 = -\omega x_2 + A_1\left(x_1^2 + x_2^2\right)x_1 - B_1\left(x_1^2 + x_2^2\right)x_2$$
$$\dot{x}_2 = \omega x_2 + B_1\left(x_1^2 + x_2^2\right)x_1 + A_1\left(x_1^2 + x_2^2\right)x_2 \quad [16]$$

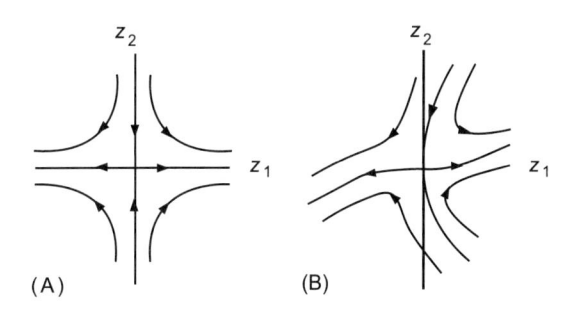

Figure 2 Near a hyperbolic point all nonlinear terms can be removed from the nonlinear system (B) by a nonlinear change of coordinates, yielding the system in normal form as a linear system (A) provided the resonance condition is not fulfilled.

where only two coefficients A_1, B_1 appear in the normal form. They are related to the eight coefficients $a_{1,kl}$ and $a_{2,kl}$ in eqn [10] by:

$$A_1 = \tfrac{3}{8}a_{1,30} + \tfrac{1}{8}a_{1,12} + \tfrac{3}{8}a_{2,21} + \tfrac{3}{8}a_{2,03}$$
$$B_1 = -\tfrac{3}{8}a_{1,03} - \tfrac{1}{8}a_{1,21} + \tfrac{1}{8}a_{2,12} + \tfrac{3}{8}a_{2,30}$$

Inserting polar coordinates $x_1 = r\cos\varphi$, $x_2 = r\sin\varphi$ into the normal form (eqn [16]) it takes the form:

$$\dot{r} = A_1 r^3$$
$$\dot{\varphi} = \omega + B_1 r^2 \qquad [17]$$

Van der Pol oscillator By means of the example of the van der Pol oscillator $\ddot{x} + x = \varepsilon(1-x^2)\dot{x}$ we show that the transformation into normal form following the approach explained before introduces an artificial rotational symmetry. Introducing the complex coordinate $z = x - i\dot{x}$, one obtains a complex first-order equation $\dot{z} = (iz + (\varepsilon(z-\bar{z})/2 - \varepsilon(z^3 + z^2\bar{z} - z\bar{z}^2 - \bar{z}^3)/8$. Neglecting terms of order $O(\varepsilon^2)$, the normal form reduction yields $\dot{z} = ((\varepsilon/2) + i)z - (\varepsilon/8)z^2\bar{z}$, which admits as solution the circular limit cycle $z = 2\exp it$. The limit cycles of the original equation and of the equation transformed into normal form are shown in **Figure 3**.

Classification and Unfolding

After dimension reduction by the center manifold theory and simplification by the normal form theory a third important step in the study of the loss of stability of a state can be made by a classification. This is made possible by the fact that there is only a very limited number of qualitatively different cases of loss of stability. This number of cases can be classified depending on the codimension of the critical systems.

The concept of codimension arises quite naturally if we not only ask for the solution of the critical system (eqn [17]) but also for all qualitatively different solutions in the family of systems to which the critical (singular) system belongs. The first question is how many parameters are necessary to make such an embedding of a singular system into a parameter family in which all qualitatively different solutions of this class are included. The minimum number of parameters necessary to form such a universally unfolded family is the codimension.

Codimension one contains two cases: first, one zero eigenvalue with quadratic nonlinear terms (eqn [7]); and secondly, one purely imaginary pair (eqn [17]), with third-order nonlinear terms. Hence for eqn [17] with one parameter a universally unfolded family can be obtained including all qualitatively possible solutions. This parametrized family has the form:

$$\dot{r} = A_1 r^3 + ar$$
$$\dot{\varphi} = \omega + B_1 r^2 \qquad [18]$$

where a is the unfolding parameter. The stationary solutions of eqn [18] which give the amplitude of the limit cycle oscillations with the frequency ω, which sets in after loss of stability, can be calculated from $A_1 r^3 + ar = 0$ and are presented for $A_1 < 0$ and $A_1 > 0$ in **Figure 4**. It can be clearly seen from **Figure 4** that the stability of the singular point of the system (eqn [16]) depends on whether A_1 is positive or negative, hence on the nonlinear terms. In the neighborhood of the critical parameter value the oscillation of the infinite dimensional tube conveying fluid is completely governed by eqn [18] as in **Figure 4**. Here the local parameter $a \sim \lambda - \lambda_c$ has been introduced. **Figure 4** also shows the stratification of the parameter space (here one-dimensional) into domains of qualitatively similar behavior ($a<0$ and $a>0$).

Two-parameter families (codimension 2) are also completely classified resulting in five different cases. They are best presented with the corresponding

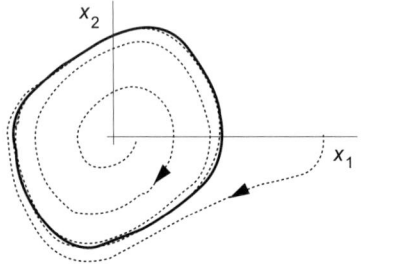

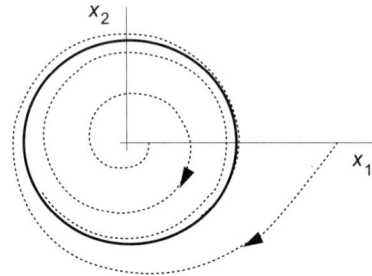

Figure 3 Solutions of the van der Pol equation and its normal form equation in the $z = x_1 - ix_2$ phase plane for $\varepsilon = 0.4$, showing the artificial rotation symmetry introduced in the flow by the normal form transformation.

stratification of the two-dimensional parameter space.

Recent Developements in Bifurcation Theory

Loss of Stability of a Periodic Solution

The treatment indicated above for the loss of stability of a steady state can be extended in exactly the same way, to a periodic solution if, instead of the differential equations for the description of the system, the point mapping in the Poincare section is used. Of course, the practical calculation of the point mapping is a nontrivial step. In the point mapping the original periodic state corresponds to a fixed point and the three steps explained before can be performed again.

Continuous Parameter Variation

If the parameter variation is not quasistatical but continuous the situation is much more complicated and, in general, the onset of instability may be retarded.

Symmetric Systems

Treating systems which are equivariant under the action of a symmetry group by the methods explained before, one has to note that due to the symmetry of the system the critical eigenvalues may appear with higher multiplicity. Hence the reduced set of equations, for example eqn [10] for a Hopf bifurcation, has a higher dimension. In the case of a Hopf bifurcation for a rotationally symmetric system, instead of the two-dimensional system (eqn 10), a four-dimensional system is obtained. However, these equations must also be rotationally symmetric and hence the number of terms can be strongly reduced by checking their equivariance.

Hamiltonian Systems

Models of physical or technical systems which are governed by Hamiltonian systems are mostly found in the dynamics of space vehicles where the dissipative effects are so small that for short time periods such systems may be modelled as conservative systems. Here the basic problem arises that for Hamiltonian systems the location of the eigenvalues in the complex plane is reflectionally symmetric with respect to both the real and the imaginary axes. Hence, from the location of the eigenvalues of the linearized problem in the complex plane, a decision concerning the instability of a state can be made but a decision about its stability cannot be made because in the most favorable case all eigenvalues are located on the imaginary axis. Energy methods such as the Dirichlet method have been developed for such problems. Modern developments in connection with satellite attitude stability where conserved quantities, like angular momentum are often available, make use of energy-momentum methods which can be used to reduce the dimension of the problem.

Chaotic Dynamics

One major approach towards understanding chaotic behavior is to study the growing complexity of the dynamics due to a sequence of bifurcations by varying a parameter, starting from an equilibrium state. There is a vast literature on chaotic dynamics. An

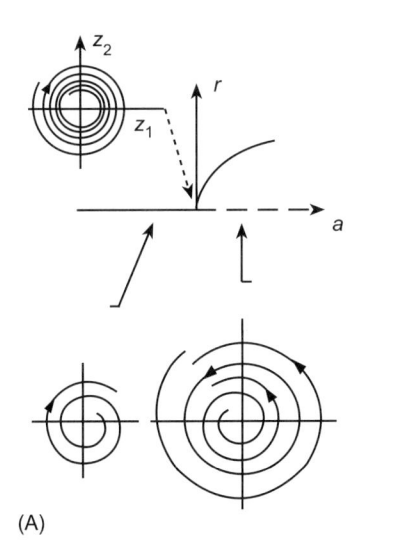

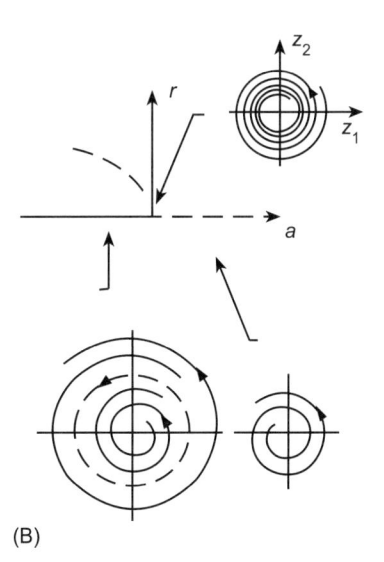

Figure 4 All possible solutions according to eqn [18] for a Hopf bifurcation. (A) A stable limit cycle is obtained for $A_1<0$; (B) an unstable limit cycle is obtained for $A_1>0$.

important possibility of having chaotic dynamics follows from global bifurcations.

Acknowledgement

This work has been supported by the Austrian Science Foundation (FWF), under project P 13131 – MAT.

See also: **Chaos**; **Nonlinear systems analysis**; **Nonlinear systems, overview**; **Theory of vibration**, Fundamentals.

Further Reading

Andronov A, Vitt A and Khaikin S (1966) *Theory of Oscillations*. New York: Addison Wesley.

Arnold VI (1983) *Geometrical Methods in the Theory of Ordinary Differential Equations*. Berlin: Springer-Verlag.

Carr, J (1981) *Applications of Centre Manifold Theory*. Applied Mathematical Sciences, vol. 35. New York: Springer-Verlag.

Glendinning P (1994) *Stability, Instability and Chaos: an Introduction to the Theory of Nonlinear Differential Equations*. London: Cambridge University Press.

Golubitsky M, Stewart I and Schaeffer D (1988) *Singularities and Groups in Bifurcation Theory II*. Applied Mathematical Sciences vol. **69**. Berlin: Springer-Verlag.

Guckenheimer J and Holmes P (1983) *Nonlinear Oscillations, Dynamical Systems, and Bifurcations of Vector Fields*. Applied Mathematical Sciences, vol. 42. Berlin: Springer-Verlag.

Hassard BD, Kazarinoff ND and Wan YH (1981) *Theory and Applications of Hopf Bifurcation*, London Mathematical Society Lecture Note Series, vol. 41. London: Cambridge University Press.

Holmes Ph, Lumley JL and Berkooz, G (1996) *Turbulence, Coherent Structures, Dynamical Systems and Symmetry*. London: Cambridge University Press.

Looss G and Joseph DD (1980) *Elementary Stability and Bifurcation Theory*. New York: Springer-Verlag.

Marsden J and Ratiu T (1994) *Introduction to Mechanics and Symmetry, Texts in Applied Mathematics*. New York: Springer-Verlag.

Moon FC (1987) *Chaotic Vibrations, An Introduction for Applied Scientists and Engineers*. Chichester: Wiley.

Seydel R (1994) *Practical Bifurcation and Stability Analysis*, 2nd edn. New York: Springer-Verlag.

Troger H and Steindl A (1991) *Nonlinear Stability and Bifurcation Theory. An Introduction for Engineers and Applied Scientists*. New York: Springer-Verlag.

Tufillaro NB, Tyler A and Reilly J (1992) *An Experimantal Approach to Nonlinear Dynamics and Chaos*. New York: Addison Wesley.

E

EARTHQUAKE EXCITATION AND RESPONSE OF BUILDINGS

F Naeim, John A Martin & Associates, Inc., Los Angeles, CA, USA

Copyright © 2001 Academic Press

doi:10.1006/rwvb.2001.0067

Building structures are designed to have a 'seismic capacity' that exceeds the anticipated 'seismic demand'. Capacity is a complex function of strength, stiffness and deformability conjectured by the system configuration and material properties of the structure. Given a particular building, seismic demand is controlled by what is commonly termed the 'design ground motion criteria', which may be defined in one or more of the following three distinct forms:

1. Static base shear and lateral force distribution formulas;
2. A set of 'design spectra';
3. A suite of earthquake acceleration 'time histories'.

The design ground motion may be defined in its most simple form by application of simple design base shear equations and static lateral force distribution formulas such as those embodied in a typical building code. These simple formulas are in essence, simplified interpretations of a design spectrum of certain shape and amplitude at the fundamental vibration period of the building.

For more complex analyses, the design ground motion criteria may be defined by a series of either code-specified or site-specific design spectra and rules on how to apply these spectra and how to interpret the results. If seismic design of a project requires application of dynamic time–history analysis, then an appropriate set of earthquake records have to be selected and rules have to be established on how these records are to be applied in analysis and design. The earthquake records, in this case, are needed in addition to a site-specific design spectra, and rules have to be set on how application of these records produces a demand that is consistent with the site-specific seismic hazard, which is usually summarized in the form of design spectra.

While a properly established design ground motion criteria is expected to provide a consistent expression of demand, regardless of its form, overzealous emphasis on one form over the others, without proper understanding of the strength and limitations of each form, can result in unrealistic design ground motion requirements.

Evaluation of the seismic hazard at a given site requires an estimate of likely earthquake ground motions at the site of the building. This is because: (i) sites for which a recorded earthquake ground motion is readily available are extremely rare, and (ii) even for the sites where such recordings are available, there is no guarantee that future ground motions will have the same exact characteristics of previously observed motions. Possible ground motions for a site are estimated by use of various regression analysis techniques on a selected subset of available earthquake recordings deemed proper for such estimation. The resulting mathematical formulas which provide estimates of maximum response parameters, such as peak ground acceleration or response spectral ordinates for a site are called 'predictive relations' or 'attenuation relations'. The term attenuation is used because these empirical relations in fact represent formulas for attenuation of seismic waves originating from a given source, at a given distance, through a given medium (i.e., site soil conditions). Dozens of attenuation relations have been developed and are in use today (**Table 1**).

Characteristics of Earthquake Ground Motions

The number of earthquake records available has grown rapidly during the past decade. While obtaining earthquake accelerograms was not simple up to about the mid-1980s, hundreds of earthquake records

Table 1 Typical attenuation relationships

Data source	Relationship*	by
1. San Fernando earthquake February 9, 1971	$\log PGA = 190/R^{1.83}$	Donovan
2. California earthquake	$PGA = y_0/[1+(R'/h)^2)]$ where $\log y_0 = -(b+3)+0.81M-0.027M^2$ and b is a site factor	Blume
3. California and Japanese earthquakes	$PGA = 0.0051/s\sqrt{(T_G)}10^{(0.61M-p\log R+0.167-1.83/R)}$ where $P = 1.66+3.60/R$ and T_G is the fundamental period of the site	Kanai
4. Cloud (1963)	$PGA = 0.0069e^{1.64M}/(1.1_e^{1.1M}+R^2)$	Milne and Davenport
5. Cloud (1963)	$PGA = 1.254e^{0.8M}/(R+25)^2$	Esteva
6. U.S.C. and G.S.	$\log PGA = (6.5-2\log(R'+80))/981$	Cloud and Perez
7. 303 Instrumental Values	$PGA = 1.325e^{0.67M}/(R+25)^{1.6}$	Donovan
8. Western US records	$PGA = 0.0193e^{0.8M}/(R^2+400)$	Donovan
9. US, Japan	$PGA = 1.35e^{0.58M}/(R+25)^{1.52}$	Donovan
10. Western US records, USSR, and Iran	$\ln PGA = -3.99+1.28M-1.75\ln[R = 0.147e^{0.732M}]$ – M is the surface wave magnitude for M greater than or equal to 6, or it is the local magnitude for M less than 6.	Campbell
11. Western US records and worldwide	$\log PGA = -1.02+0.249M-\log\sqrt{(R^2+7.3^2)}-0.00255\sqrt{(R^2+7.3^2)}$	Joyner and Boore
12. Western US records and worldwide	$\log PGA = 0.49+0.23(M-6)-\log\sqrt{(R^2+8^2)}-0.0027\sqrt{(R^2+8^2)}$	Joyner and Boore
13. Western US records	$\ln PGA = \ln\alpha(M)-\beta(M)\ln(R+20)$ – M is the surface wave magnitude for M greater than or equal to 6, or it is the local magnitude for smaller M. – R is the closest distance to source for M greater than 6 and hypocentral distance for M smaller than 6. – $\alpha(M)$ and $\beta(M)$ are magnitude-dependent coefficients.	Idriss
14. Italian records	$\ln PGA = -1.562+0.306M-\log\sqrt{(R^2+5.8^2)}+0.169S$ – S is 1.0 for soft sites or 0.0 for rock.	Sabetta and Pugliese
15. Western US and worldwide (soil sites)	For M less than 6.5, $\ln PGA = -2.611+1.1M-1.75\ln(R+0.822e^{0.418M})$ For M greater than or equal to 6.5, $\ln PGA = -2.611+1.1M-1.75\ln(R+0.316e^{0.629M})$	Sadigh et al.
16. Western US and worldwide (rock sites)	For M less than 6.5, $\ln PGA = -1.406+1.1M-2.05\ln(R+1.353e^{0.406M})$ For M greater than or equal to 6.5, $\ln PGA = -1.406+1.1M-2.05\ln(R+0.579e^{0.537M})$	Sadigh et al.
17. Worldwide earthquakes	$\ln PGA = -3.512+0.904M-1.328\ln\sqrt{[R^2+(0.149e^{0.647M})^2]}$ $+(1.125-0.112\ln R-0.0957M)F+(0.440-0.171\ln R)S_{sr}$ $+(0.405-0.222\ln R)S_{hr}$ – $F = 0$ for strike-slip and normal fault earthquakes and 1 for reverse, reverse-oblique, and thrust fault earthquakes. – $S_{sr} = 1$ for soft rock and 0 for hard rock and alluvium – $S_{hr} = 1$ for hard rock and 0 for soft rock and alluvium	Campbell and Bozorgnia
18. Western North American earthquakes	$\ln PGA = b+0.527(M-6.0)-0.778\ln\sqrt{[R^2+(5.570)^2]}$ $-0.371\ln(V_s/1396)$ – where $b = -0.313$ for strike-slip earthquakes $= -0.117$ for reverse-slip earthquakes $= -0.242$ if mechanism is not specified – V_s is the average shear wave velocity of the soil in (m s^{-1}) over the upper 30 meters – The equation can be used for magnitudes of 5.5 to 7.5 and for distances not greater than 80 km	Boore et al.

* Peak ground acceleration *PGA* in g, source distance R in km, source distance R' in miles, local depth h in miles, and earthquake magnitude M. Refer to the relevant references for exact definitions of source distance and earthquake magnitude.

may now be screened, viewed and downloaded from the Internet or obtained at a nominal cost from various public agencies.

The characteristics of ground motion that are important for building design applications are: (i) peak ground motion (peak ground acceleration, peak ground velocity, and peak ground displacement), (ii) duration of strong motion, and (iii) frequency content. Each of these parameters influences the response of a building. Peak ground motion primarily influences the vibration amplitudes. Duration of strong motion has a pronounced effect on the severity of shaking. A ground motion with a moderate peak acceleration and a long duration may cause more damage than a ground motion with a larger acceleration and a shorter duration. There are several methods currently in use for measuring the duration of strong ground motion. These methods often use different indicators for assigning a particular duration to a given earthquake record and the user must be aware of the assumptions inherent in each method of assessing duration before using them (**Figure 1**). Frequency content strongly affects the response characteristics of a building. In a structure, ground motion is amplified the most when the frequency content of the motion and the natural frequencies of the structure are close to each other.

Earthquake ground motion is influenced by a number of factors. The most important factors are (i) earthquake magnitude, (ii) distance from the source of energy release (epicentral distance or distance from the causative fault), (iii) local soil conditions, (iv) variation in geology and propagation velocity along the travel path, and (v) earthquake source conditions and mechanism (fault type, slip rate, stress conditions, stress drop, etc.). Past earthquake records have been used to study some of these influences. While the effect of some of these parameters, such as local soil conditions and distance from the source of energy release are fairly well understood and documented, the influence of source mechanism is under investigation and the variation of geology along the travel path is complex and difficult to quantify. It should be noted that several of these influences are interrelated; consequently, it is difficult to evaluate them individually without incorporating the others.

Near-Source Effects

Recent studies have indicated that near-source ground motions contain large displacement pulses (ground displacements which are attained rapidly with a sharp peak velocity). These motions are the result of stress waves moving in the same direction as

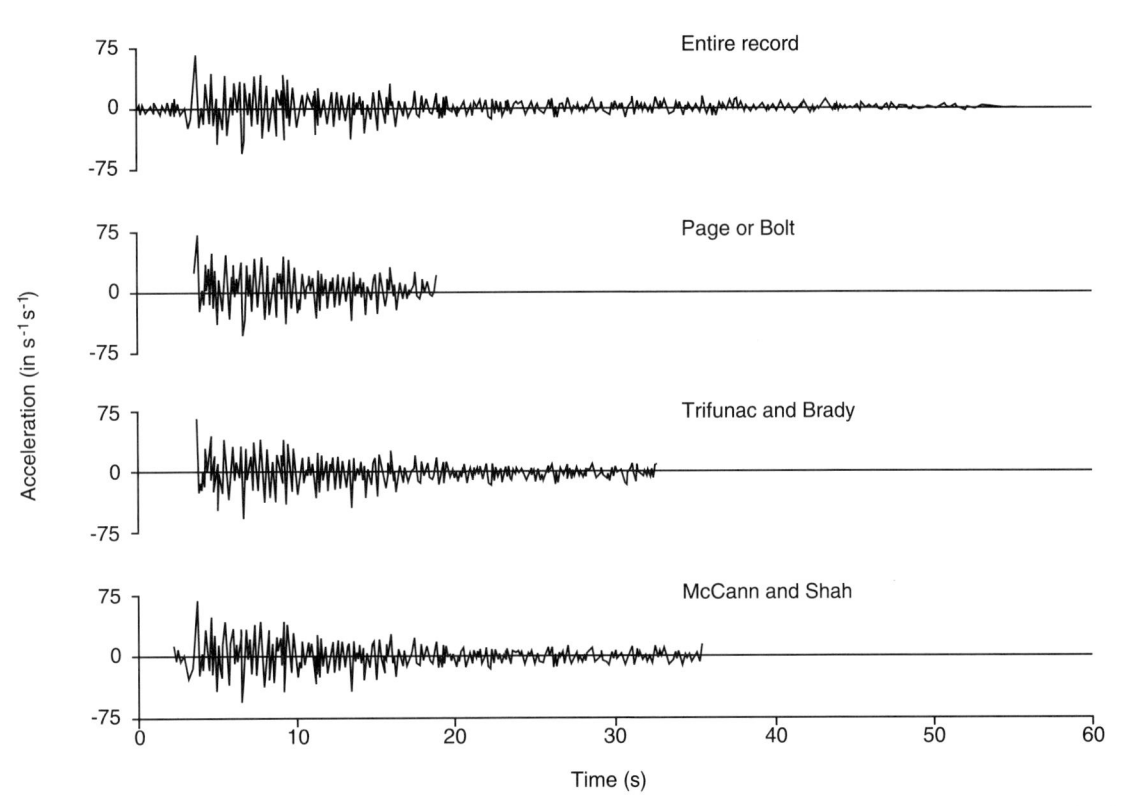

Figure 1 Comparison of strong motion duration for the S69E component of the Taft, California earthquake of 21 July, 1952 using different procedures.

the fault rupture, thereby producing a long-duration pulse. Consequently, near source earthquakes can be destructive to structures with long periods. Peak ground accelerations, velocities, and displacements from 30 records obtained within 5 km of the rupture surface, for example, have shown ground accelerations varying from 0.31 to 2.0g while the ground velocities ranged from 0.31 to 1.77 m s^{-1}. The peak ground displacements were as large as 2.55 m.

Figure 2 and **Figure 3** offer two examples of near-source earthquake ground motions. The first was recorded at the LADWP Rinaldi Receiving Station during the Northridge earthquake of 17 January 1994. The distance from the recording station to the

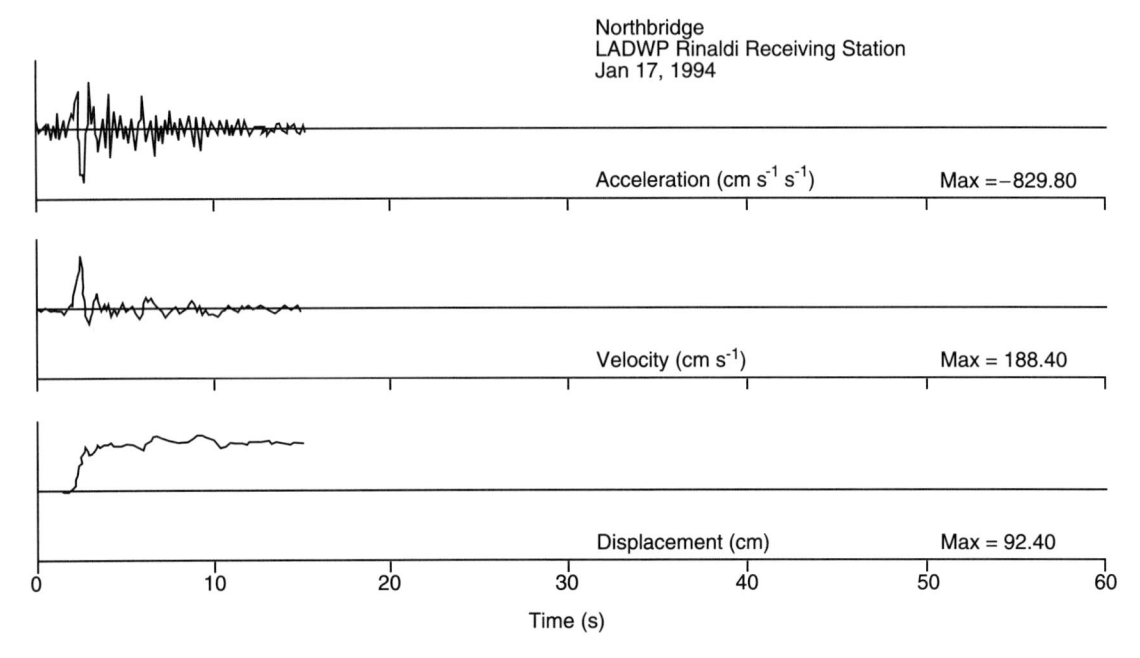

Figure 2 Ground acceleration, velocity, and displacement time-histories recorded at the LADWP Rinaldi Receiving Station during the Northridge earthquake of 17 January, 1994.

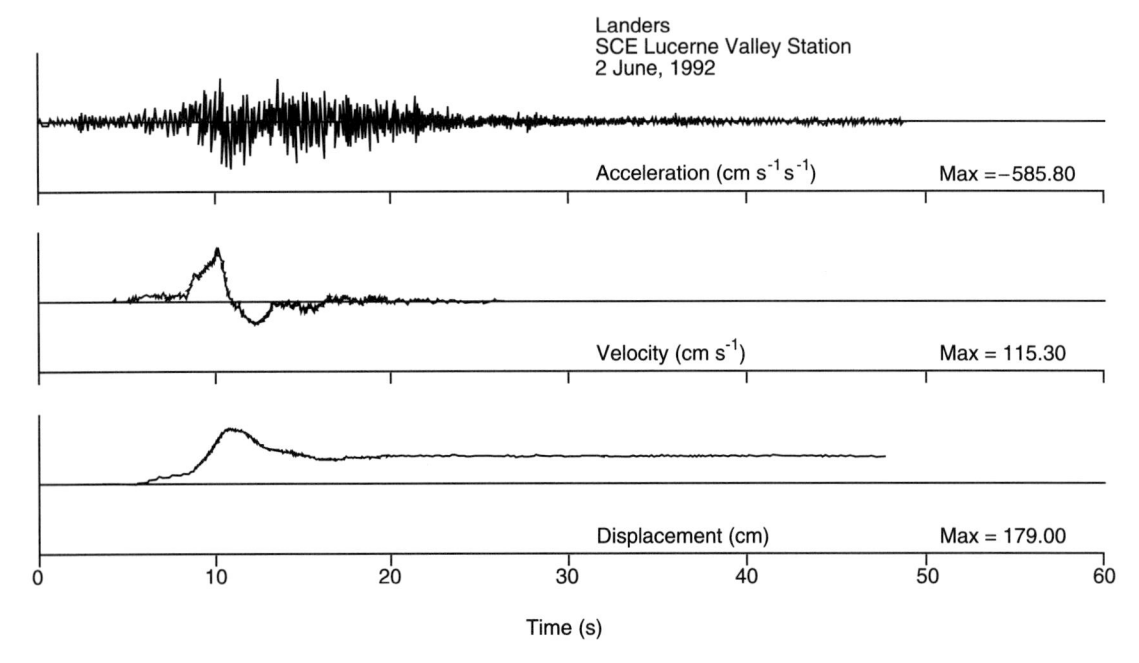

Figure 3 Ground acceleration, velocity, and displacement time-histories recorded at the SCE Lucerne Valley Station during the Landers earthquake of 28 June, 1992.

surface projection of the rupture was less than 1.0 km. The figure shows a unidirectional ground displacement that resembles a smooth step function and a velocity pulse that resembles a finite delta function. The second example, shown in **Figure 3**, was recorded at the SCE Lucerne Valley Station during the Landers earthquake of 28 June 1992. The distance from the recording station to the surface projection of the rupture was approximately 1.8 km. A positive and negative velocity pulse that resembles a single long-period harmonic motion is reflected in the figure. Near-source ground displacements similar to that shown in **Figure 3** have also been observed with a zero permanent displacement. The two figures clearly show the near-source ground displacements caused by sharp velocity pulses. Far-source ground motions, however, generally lack the pulse type motion often observed in near-source ground motions (**Figure 4**).

The frequency content of ground motion can be examined by transforming the motion from a time domain to a frequency domain through a Fourier transform. The Fourier amplitude spectrum and power spectral density, which are based on this transformation, may be used to characterize the frequency content.

Fourier Amplitude Spectrum

The finite Fourier transform $F(\omega)$ of an accelerogram $a(t)$ is obtained as:

$$F(\omega) = \int_0^T a(t) e^{-i\omega t} \, dt, \quad i = \sqrt{(-1)} \quad [1]$$

where T is the duration of the accelerogram. The Fourier amplitude spectrum $FS(\omega)$ is defined as the square root of the sum of the squares of the real and imaginary parts of $F(\omega)$. Thus:

$$FS(\omega) = \sqrt{\left\{ \left[\int_0^T a(t) \sin \omega t \, dt \right]^2 + \left[\int_0^T a(t) \cos \omega t \, dt \right]^2 \right\}} \quad [2]$$

Since $a(t)$ has units of acceleration, $FS(\omega)$ has units of velocity. The Fourier amplitude spectrum is of interest to seismologists in characterizing ground motion. **Figure 5** shows a typical Fourier amplitude spectrum for the S00E component of El Centro, the

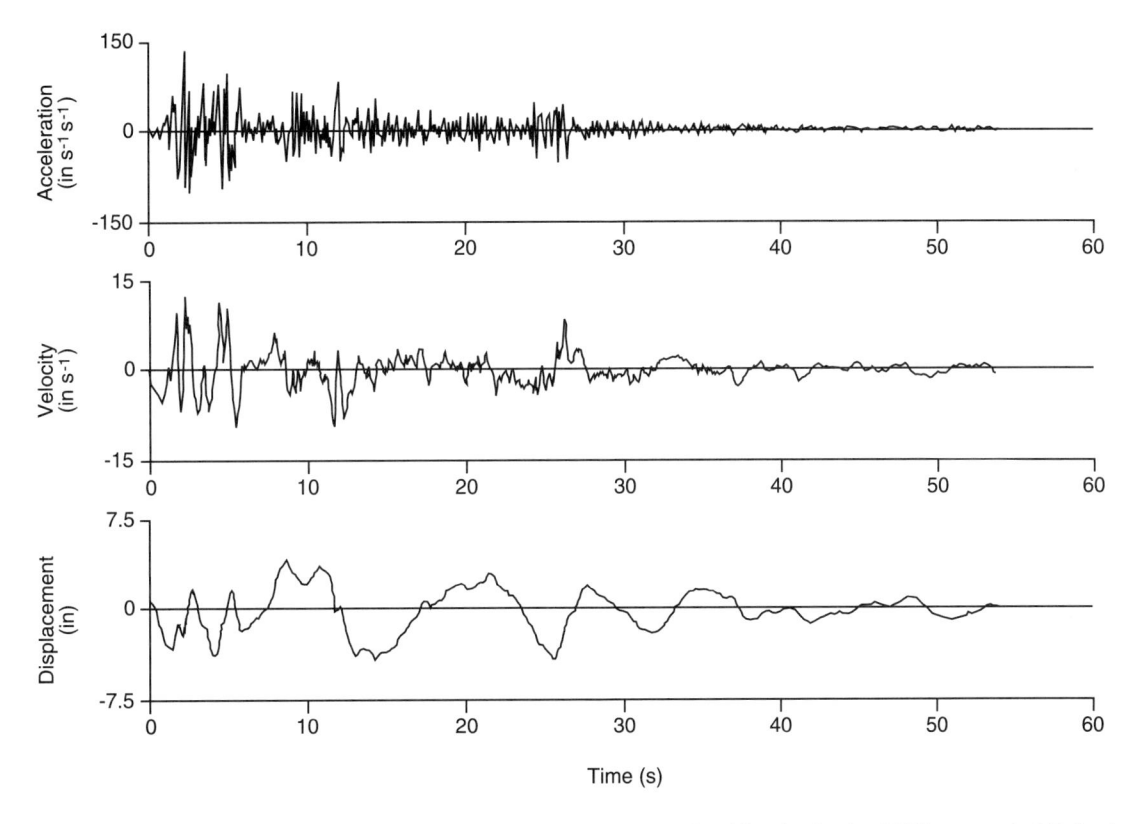

Figure 4 Corrected accelerogram and integrated velocity and displacement time-histories for the S00E coponent of El Centro, the Imperial Valley Earthquake of 18 May, 1940 is an example of an earthquake record not exhibiting near-source effects.

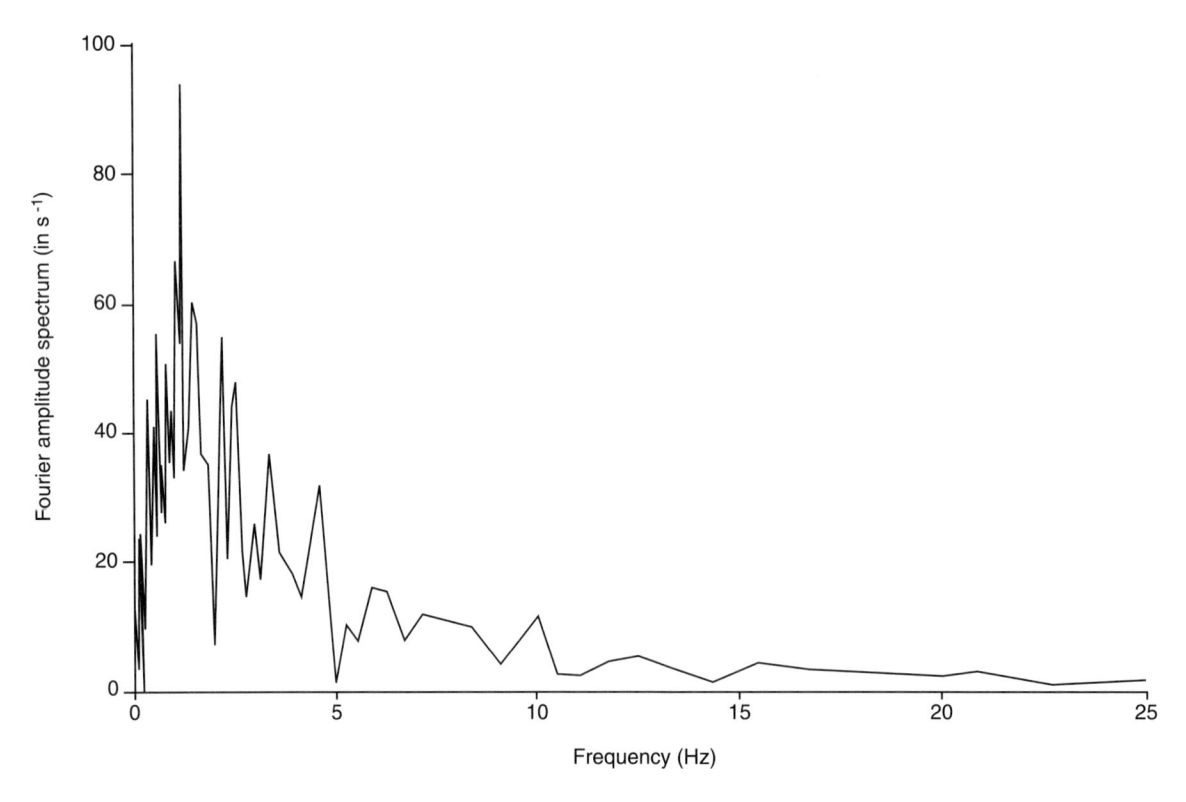

Figure 5 Fourier amplitude spectrum for the S00E component of El Centro, the Imperial Valley earthquake of 18 May, 1940.

Imperial Valley earthquake of 18 May 1940. The figure indicates that most of the energy in the accelerogram is in the frequency range 0.1–10 Hz, and that the largest amplitude is at a frequency of approximately 1.5 Hz.

It can be shown that subjecting an undamped single-degree-of-freedom (SDOF) system to a base acceleration $a(t)$, the velocity response of the system and the Fourier amplitude spectrum of the acceleration are closely related. The equation of motion of the system can be written as:

$$\ddot{x} + \omega_n^2 x = -a(t) \quad [3]$$

in which x and $\ddot{x}$ are the relative displacement and acceleration, and ω_n is the natural frequency of the system. Using Duhamel's integral, the steady state response can be obtained as:

$$x(t) = \frac{1}{\omega_n} \int_0^t a(\tau) \sin \omega_n(t-\tau) \, dt \quad [4]$$

The relative velocity $\dot{x}(t)$ follows directly from eqn [4] as:

$$\dot{x}(t) = \int_0^t a(\tau) \cos \omega_n(t-\tau) \, d\tau \quad [5]$$

Eqn [5] can be expanded as:

$$\dot{x}(t) = -\left[\int_0^t a(\tau) \cos \omega_n \tau \, d\tau\right] \cos \omega_n t \\ - \left[\int_0^t a(\tau) \sin \omega_n \tau \, d\tau\right] \sin \omega_n t \quad [6]$$

Denoting the maximum relative velocity (spectral velocity) of a system with frequency ω by $SV(\omega)$ and assuming that it occurs at time t_v, one can write:

$$SV(\omega) = \sqrt{\left\{\left[\int_0^{t_v} a(\tau) \sin \omega \tau \, d\tau\right]^2 + \left[\int_0^{t_v} a(\tau) \cos \omega \tau \, d\tau\right]^2\right\}} \quad [7]$$

The pseudo-velocity $PSV(\omega)$ defined as the product of the natural frequency ω and the maximum relative displacement or the spectral displacement SD(ω) is close to the maximum relative velocity. If SD(ω) occurs at t_d then:

$$\text{PSV}(\omega) = \omega \text{SD}(\omega) = \sqrt{\left\{\left[\int_0^{t_d} a(\tau) \sin \omega\tau \, d\tau\right]^2 + \left[\int_0^{t_d} a(\tau) \cos \omega\tau \, d\tau\right]^2\right\}} \quad [8]$$

Comparison of eqns [2] and [7] shows that for zero damping, the maximum relative velocity and the Fourier amplitude spectrum are equal when $t_v = T$. A similar comparison between eqns [2] and [8] reveals that the pseudo-velocity and the Fourier amplitude spectrum are equal if $t_d = T$. **Figure 6** shows a comparison between FS(ω) and SV(ω) for zero damping for the S00E component of El Centro, the Imperial Valley earthquake of 18 May 1940. The figure indicates the close relationship between the two functions. It should be noted that, in general, the ordinates of the Fourier amplitude spectrum are less than those of the undamped pseudo-velocity spectrum.

Power Spectral Density

The inverse Fourier transform of $F(\omega)$ is:

$$a(t) = \frac{1}{\pi} \int_0^{\omega_0} F(\omega) e^{i\omega t} \, d\omega \quad [9]$$

where ω_0 is the maximum frequency detected in the data (referred to as Nyquist frequency). Eqns [1] and [9] are called Fourier transform pairs. The intensity of an accelerogram is defined as:

$$I = \int_0^T a^2(t) \, dt \quad [10]$$

Based on Parseval's theorem, the intensity I can also be expressed in the frequency domain as:

$$I = \frac{1}{\pi} \int_0^{\omega_0} |F(\omega)|^2 \, d\omega \quad [11]$$

The intensity per unit of time or the temporal mean square acceleration ψ^2 can be obtained by dividing eqn [10] or eqn [11] by the duration T. Therefore:

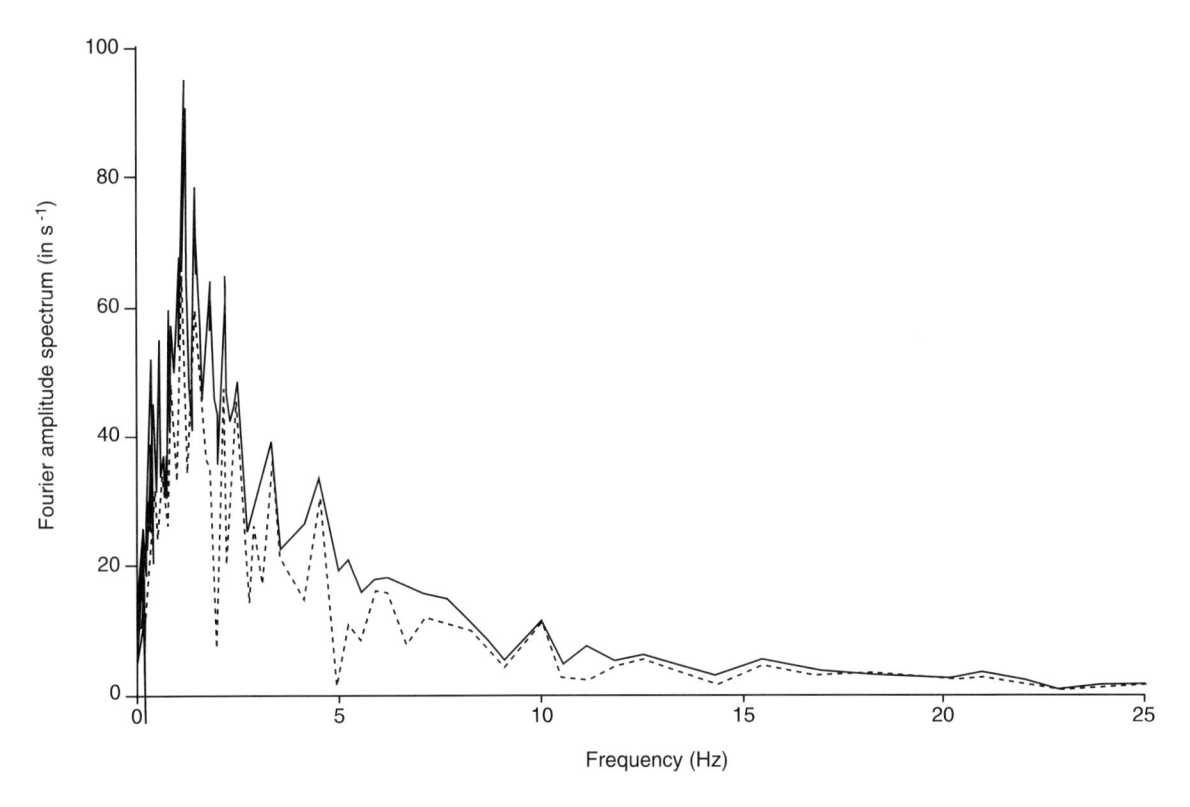

Figure 6 Comparison of Fourier amplitude spectrum (dashed line) and velocity spectrum (continuous line) for an undamped single-degree-of-freedom system for the S00E component of El Centro, the Imperial Valley earthquake of 18 May, 1940.

$$\psi^2 = \frac{1}{T}\int_0^T a^2(t)\,dt = \frac{1}{\pi T}\int_0^{\omega_0} |F(\omega)|^2\,d\omega \qquad [12]$$

The temporal power spectral density is defined as:

$$G(\omega) = \frac{1}{\pi T}|F(\omega)|^2 \qquad [13]$$

Combining eqns [12] and [13], the mean square acceleration can be obtained as:

$$\psi^2 = \int_0^{\omega_0} G(\omega)\,d\omega \qquad [14]$$

In practice, a representative power spectral density of ground motion is computed by averaging across the temporal power spectral densities of an ensemble of N accelerograms. Therefore:

$$G(\omega) = \frac{1}{N}\sum_{i=1}^{N} G_i(\omega) \qquad [15]$$

where $G_i(\omega)$ is the power spectral density of the ith record. **Figure 7** shows a typical example of a normalized power spectral density computed for an ensemble of 161 accelerograms recorded on alluvium. Once the power spectral density of ground motion at a site is established, random vibration methods may be used to formulate probabilistic procedures for computing the response of structures. In addition, the power spectral density of ground motion may be used for other applications such as generating artificial accelerograms.

Earthquake Response of Buildings

The main cause of damage to structures during an earthquake is their response to ground motions which are input at the base. In order to evaluate the behavior of the structure under this type of loading condition, the principles of structural dynamics must be applied to determine the stresses and deflections, which are developed in the structure.

When a single-story structure, shown in **Figure 8A**, is subjected to earthquake ground motions, no external dynamic force is applied at the roof level. Instead, the system experiences an acceleration of the base. The effect of this on the idealized structure is shown in **Figure 8B** and **Figure 8C**. Summing the forces shown in **Figure 8D** results in the following equation of dynamic equilibrium:

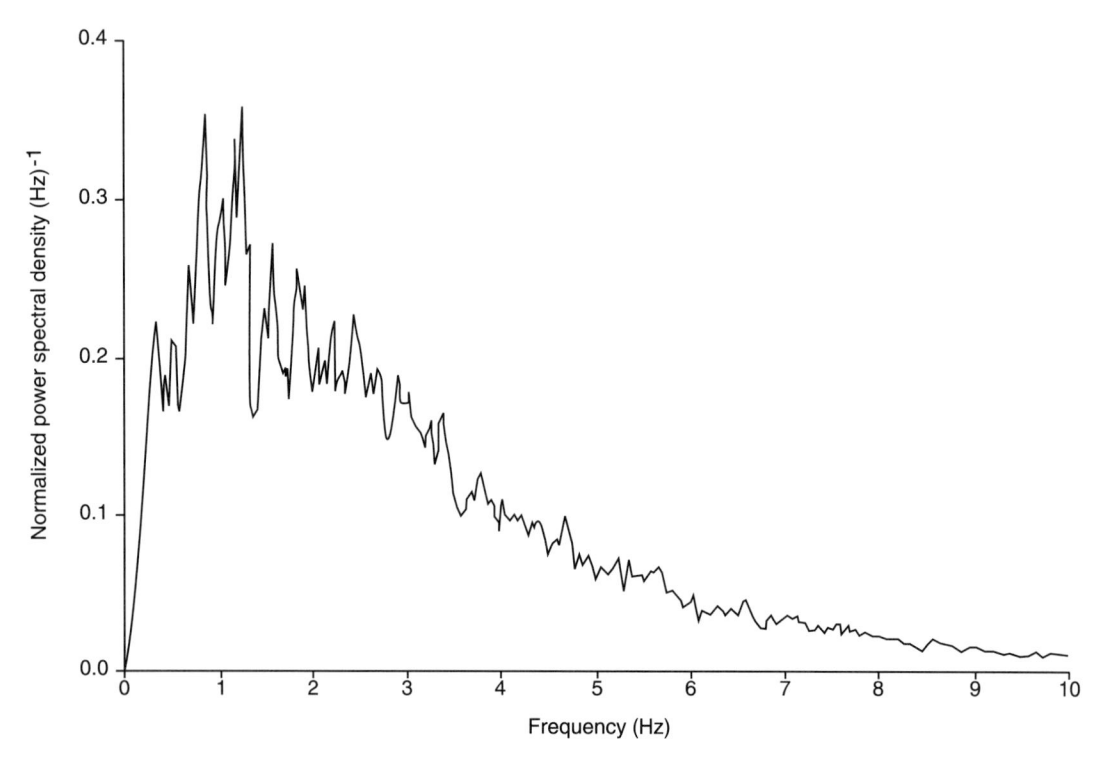

$$f_i + f_d + f_s = 0 \qquad [16]$$

Figure 7 Normalized power spectral density of an ensemble of 161 horizontal components of accelerograms recorded on alluvium.

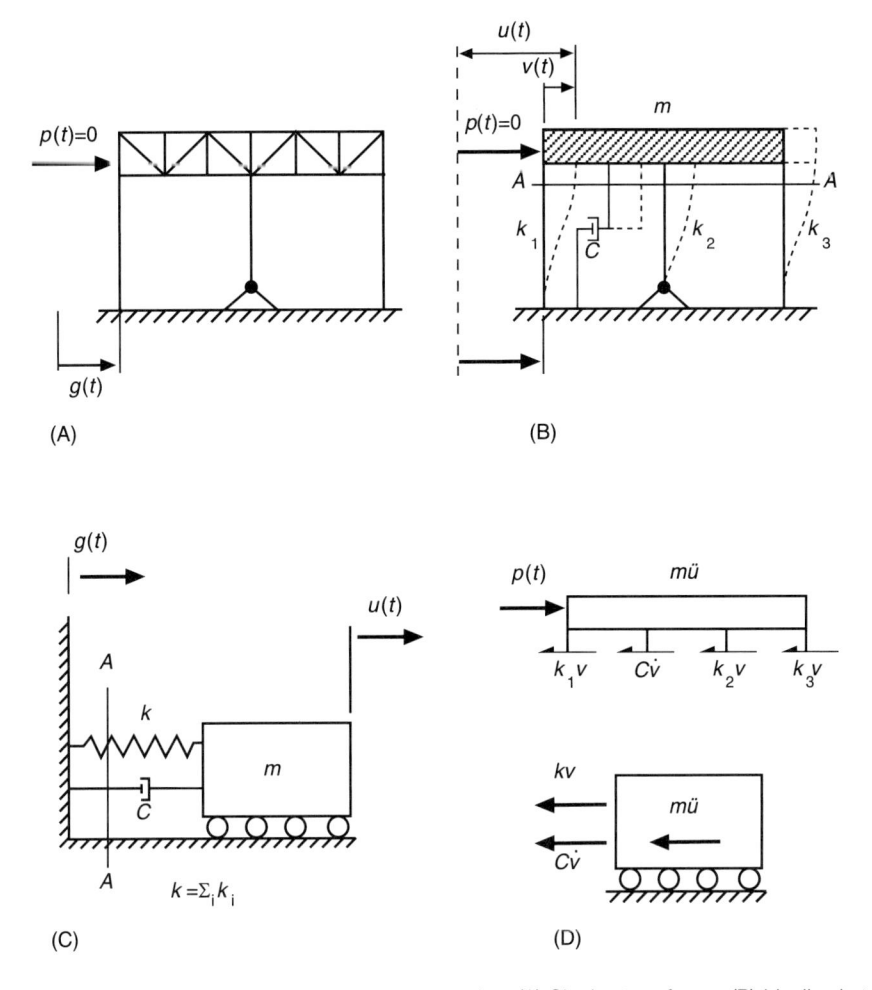

Figure 8 Single-degree-of-freedom system subjected to base motion. (A) Single story frame; (B) idealized structural system; (C) equivalent spring–mass–damper system; (D) free body diagrams, section A–A.

where f_i = inertia force = $m\ddot{u}$, f_d = damping (dissipative) force = $c\dot{v}$, f_s = elastic restoring force = kv, $\ddot{u}$ is the total acceleration of the mass, and $\dot{v}$, v are the velocity and displacement of the mass relative to the base. Substituting the physical parameters for f_i, f_d and f_s in eqn [16] results in an equilibrium equation of the form:

$$m\ddot{u} + c\dot{v} + kv = 0 \qquad [17]$$

or:

$$m\ddot{v} + c\dot{v} + kv = p_e(t) \qquad [18]$$

where:

$p_e(t)$ = effective time-dependent force = $-m\ddot{g}(t)$

and:

$\ddot{v}(t)$ = acceleration of the mass relative to the base

$\ddot{g}(t)$ = acceleration of the base.

Therefore, the equation of motion for a structure subjected to a base motion is similar to that for a structure subjected to a time-dependent force if the base motion is represented as an effective time-dependent force which is equal to the product of the mass and the ground acceleration.

Earthquake Response of an Elastic SDOF System

Time–history response The response to earthquake loading can be obtained directly from the Duhamel integral as:

$$v(t) = \frac{V(t)}{\omega} = \frac{1}{\omega}\int_0^t \ddot{g}(\tau) e^{-\lambda\omega(t-\tau)} \sin \omega_d(t-\tau)\, d\tau$$

$$[19]$$

where the response parameter $V(t)$ represents the velocity.

The displacement of the structure at any instant of time during the entire time history of the earthquake under consideration can now be obtained using eqn [19]. If damping is small and the damping term can be neglected as contributing little to the equilibrium equation, the total acceleration can be approximated as:

$$\ddot{u}(t) = -\omega^2 v(t) \quad [20]$$

The effective earthquake force is then given as:

$$Q(t) = m\omega^2 v(t) \quad [21]$$

The above expression gives the value of the base shear in a single-story building at every instant of time during the earthquake time history under consideration. The overturning moment acting on the base of the building can be determined by multiplying the inertia force by the story height, h:

$$M(t) = hm\omega^2 v(t) \quad [22]$$

Response spectra Consideration of the displacements and forces at every instant of time during an earthquake time history can require considerable computational effort, even for simple structural systems. For most practical building design applications only the maximum response quantities are required. The maximum value of the displacement, as determined by eqn [19], is defined as the 'spectral displacement', S_d:

$$S_d = v(t)_{max} \quad [23]$$

As a result, the following expressions for the maximum base shear and maximum overturning moment in a SDOF system are obtained:

$$Q_{max} = m\omega^2 S_d \quad [24]$$

$$M_{max} = hm\omega^2 S_d \quad [25]$$

An examination of these equations indicates that the maximum velocity response can be approximated by multiplying the spectral displacement by the circular frequency. This response parameter is defined as the 'spectral pseudovelocity' and is expressed as:

$$S_{pv} = \omega S_d \quad [26]$$

In a similar manner, the 'spectral pseudoacceleration' is defined as:

$$S_{pa} = \omega^2 S_d \quad [27]$$

A plot of the spectral response parameter against frequency or period constitutes the 'response spectrum' for that parameter. Because the three response quantities (S_d, S_{pv}, S_{pa}) are related by the circular frequency, it is convenient to plot them on a single graph with log scales on each axis. This special type of plot is called a 'tripartite' log plot. The three response parameters for the El Centro motion are shown plotted in this manner in **Figure 9**. For a SDOF system having a given frequency (period) and given damping, the three spectral response parameters for this earthquake can be read directly from the graph.

For each earthquake record, response spectra can be constructed. Earthquakes share many common characteristics, but have their own unique attributes as well. Response spectra of earthquake records usually contain many peaks and valleys as a function of period (see **Figure 9** for example). There is no reason to believe that the spectra for future earthquakes will exhibit exactly the same peaks and valleys as previous recordings. Furthermore, the natural periods and mode shapes of building structures cannot be exactly predicted. Many uncertainties are present which include but are not limited to (i) unavoidable variations in the mass and stiffness properties of the building from those used in design, (ii) difficulties involved in establishing exact properties of site soil conditions, and (iii) inelastic response that tends to lengthen the natural period of the structure. For these reasons, it is more rational to use average curves obtained from a number of earthquake records for design purposes. These average curves, which do not reflect the sharp peaks and valleys of individual records, are also known as 'smoothed response spectra', or more commonly as 'design spectra'. While a response spectrum is an attribute of a particular ground motion, a design spectrum is not. A design spectrum is merely a definition of a criteria for structural analysis and design. A sample design spectrum chart is shown in **Figure 10**. Generally a design spectrum is generated for the standard critical damping ratio of 5% and building code specified damping adjustment factors are used to construct design spectra corresponding to other damping values. The adjustment factor, B, is defined as:

$$B = \frac{R_5}{R_x} \quad [28]$$

where R_x and R_5 are the spectral ordinates of the $x\%$ and 5% damped design spectrum at a given period, respectively. It is common to provide one adjustment value for medium to long periods of vibration, B_1 (usually calculated for the vibration period of about

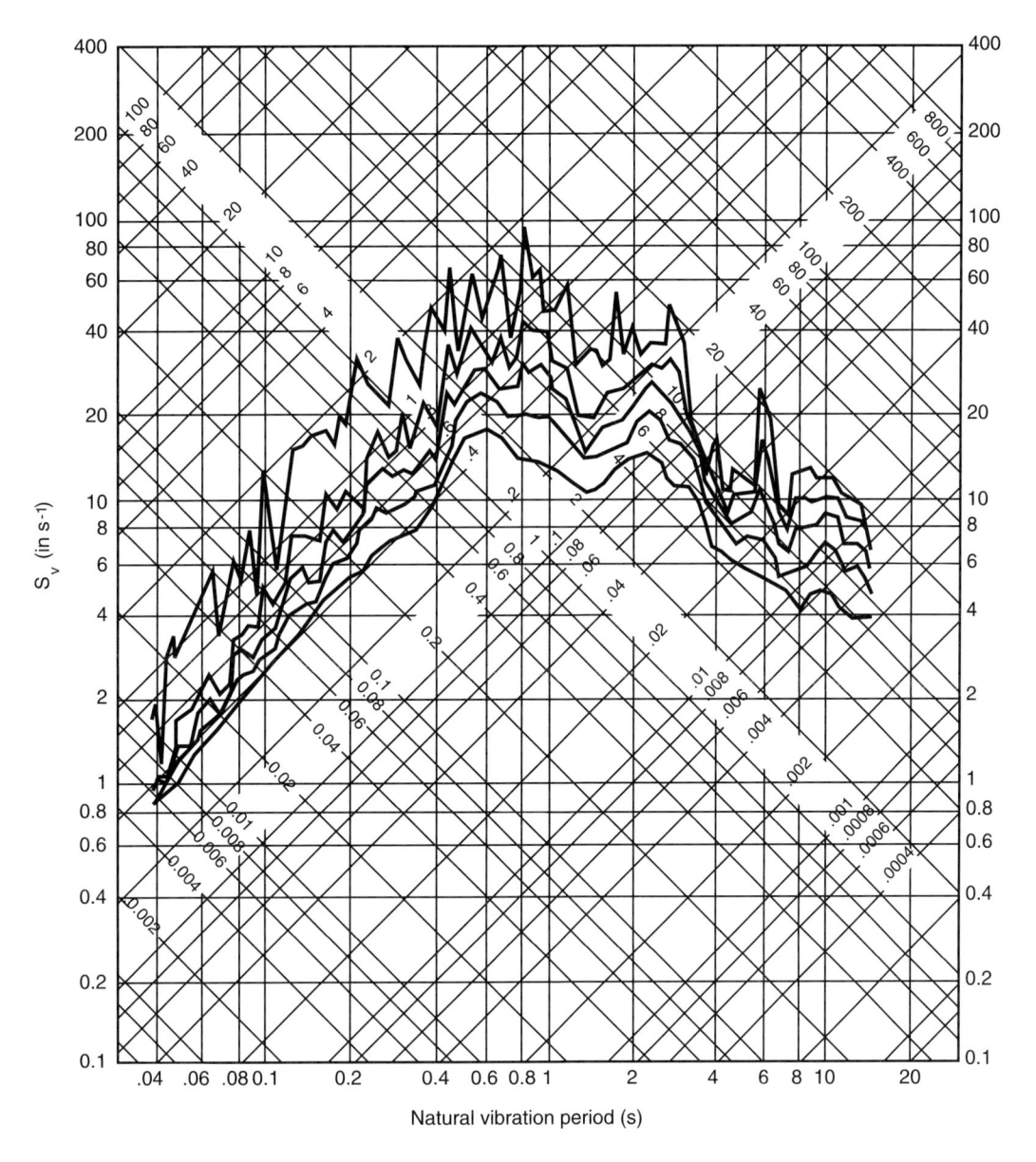

Figure 9 Typical tripartite response-spectra curves.

$1.0\,\mathrm{s}^{-1}$) and another for short periods of vibration, B_s (usually determined for the vibration periods in the range $0.1\text{–}0.3\,\mathrm{s}^{-1}$). Typical values of such damping adjustment factors utilized in modern building codes are given in **Table 2**. The reader should note that these values are merely suggestive of the average values of anticipated adjustment factors and recent studies have shown that there is a wide scatter among adjustment factors obtained from various earthquake records (**Figure 11**).

Approximate Analysis of Response of Elastic MDOF Systems Using the Generalized Coordinates Method

Up to this point, the only structures which have been considered are single-story buildings which can be

450 EARTHQUAKE EXCITATION AND RESPONSE OF BUILDINGS

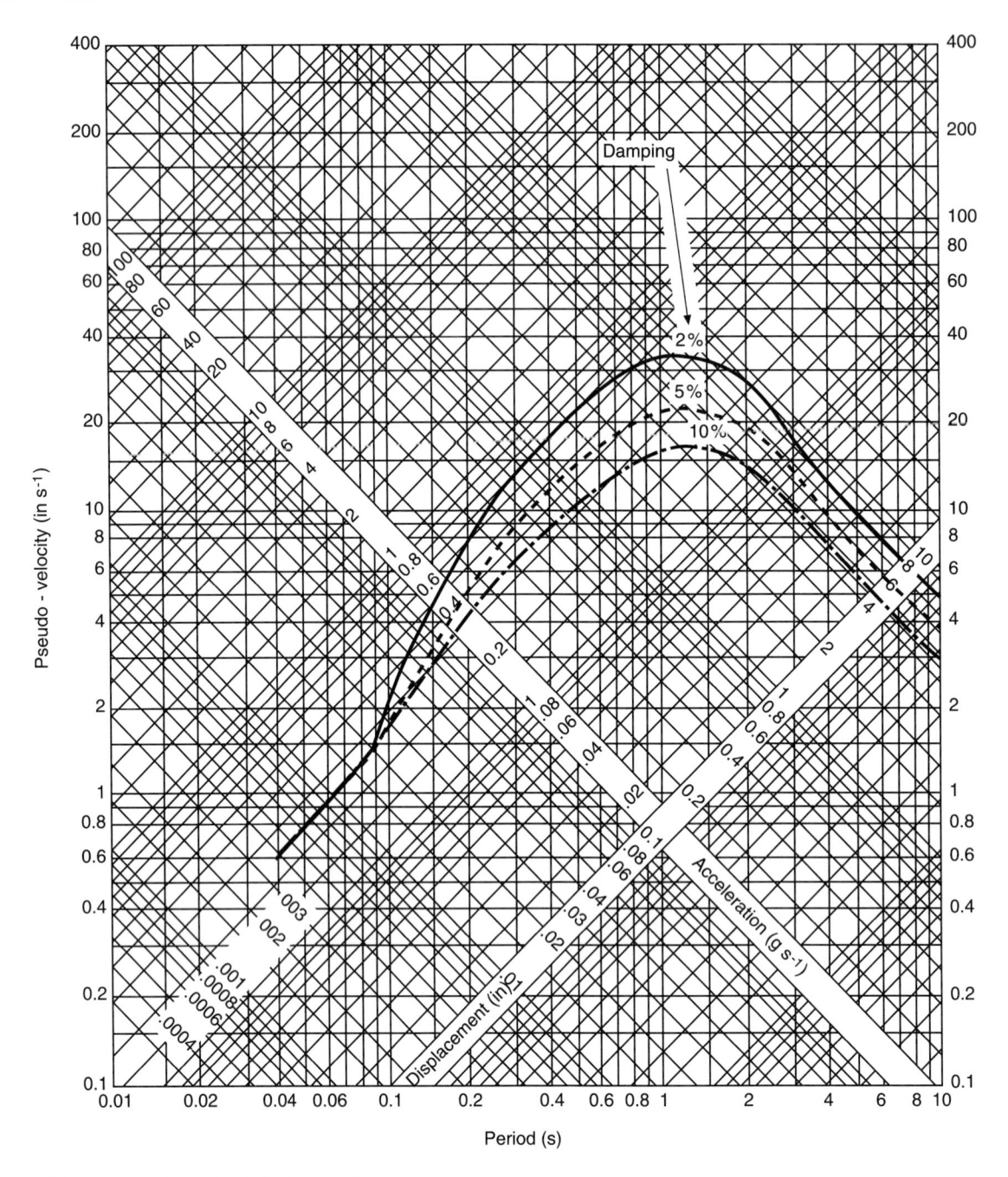

Figure 10 A sample 'smoothed' design spectra.

idealized as SDOF systems. The analysis of most structural systems requires a more complicated idealization even if the response can be represented in terms of a single degree of freedom. The generalized–coordinate approach provides means of representing the response of more complex structural systems in terms of a single, time-dependent coordinate, known as the generalized coordinate. Displacements in the structure are related to the generalized coordinate as:

$$v(x,t) = \phi(x)Y(t) \qquad [29]$$

where $Y(t)$ is the time-dependent generalized coordinate and $\phi(x)$ is a spatial shape function which relates the structural degrees of freedom, $v(x,t)$, to the generalized coordinate. For a generalized SDOF system, it is necessary to represent the restoring forces in the damping elements and the stiffness elements in terms of the relative velocity and relative displacement between the ends of the element:

$$\Delta \dot{v}(x,t) = \Delta \phi(x) \dot{Y}(t) \qquad [30]$$

Table 2 Typical building code recommended damping adjustment factors

Percent of critical damping	Damping coefficient B_1	B_s
<2	0.8	0.8
5	1.0	1.0
10	1.2	1.3
20	1.5	1.8
30	1.7	2.3
40	1.9	2.7
>50	2.0	3.0

$$\Delta v(x,t) = \Delta\phi(x)Y(t) \quad [31]$$

Most buildings can be idealized as a vertical cantilever, which limits the number of displacement functions that can be used to represent the horizontal displacement. Once the displacement function is selected, the structure is constrained to deform in that prescribed manner. This implies that the displacement functions must be selected carefully if a good approximation of the dynamic properties and response of the system are to be obtained.

Here, the formulation of the equation of motion in terms of a generalized coordinate will be restricted to systems which consist of an assemblage of lumped masses and discrete elements which is typical of most building structures. Lateral resistance is provided by discrete elements whose restoring force is proportional to the relative displacement between the ends of the element. Damping forces are assumed proportional to the relative velocity between the ends of the discrete damping element. The principle of virtual work in the form of virtual displacements states that if a system of forces which are in equilibrium is given a virtual displacement which is consistent with the boundary conditions, the work done is zero. Applying this principle to a SDOF system shown in **Figure 8** results in an equation of virtual work in the form:

$$f_i\,\delta v + f_d\,\delta\Delta v + f_s\,\delta\Delta v - p(t)\,\delta v = 0 \quad [32]$$

where it is understood that $v = v(x, t)$ and that the virtual displacements applied to the damping force and the elastic restoring force, are virtual relative displacements. The virtual displacement can be expressed as:

$$\delta v(x,t) = \phi(x)\,\delta Y(t) \quad [33]$$

and the virtual relative displacement can be written as:

$$\delta\Delta v(x,t) = \Delta\phi(x)\,\delta Y(t) \quad [34]$$

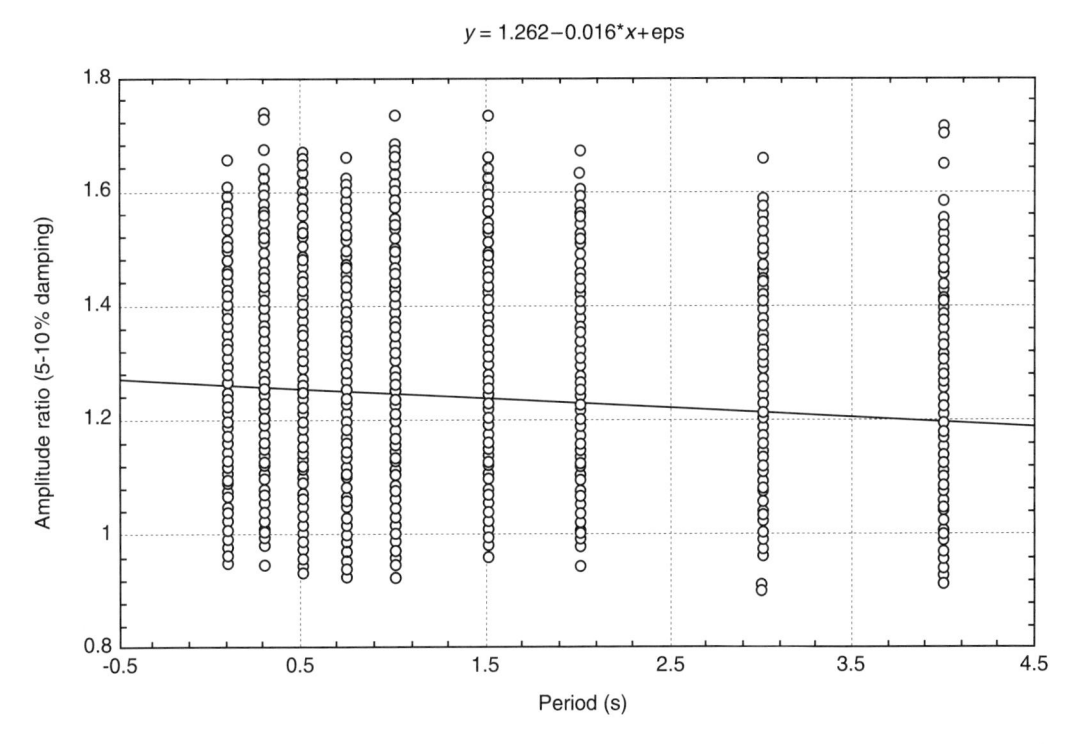

Figure 11 Individual damping adjustment factors for 10% damping as computed for 1047 distinct earthquake records (each circle represents an adjustment factor corresponding to an earthquake record at a distinct period of vibration).

where:

$$\Delta v(x,t) = \phi(x_i)Y(t) - \phi(x_j)Y(t) = \Delta\phi(x)Y(t) \quad [35]$$

The inertia, damping and elastic restoring forces can be expressed as:

$$\begin{aligned} f_i &= m\ddot{v} = m\phi\ddot{Y} \\ f_d &= c\Delta\dot{v} = c\Delta\phi\dot{Y} \\ f_s &= k\Delta v = k\Delta\phi Y \end{aligned} \quad [36]$$

resulting in the following equation of motion in terms of the generalized coordinate:

$$m^*\ddot{Y} + c^*\dot{Y} + k^*Y = p^*(t) \quad [37]$$

where m^*, c^*, k^*, and p^* are referred to as the 'generalized parameters' and are defined as:

$$\begin{aligned} m^* &= \sum_i m_i\phi_i^2 = \text{generalized mass} \\ c^* &= \sum_i c_i\Delta\phi_i^2 = \text{generalized damping} \\ k^* &= \sum_i k_i\Delta\phi_i^2 = \text{generalized stiffness} \\ p^* &= \sum_i p_i\phi_i = \text{generalized force} \end{aligned} \quad [38]$$

For a time-dependent base acceleration the generalized force becomes:

$$p^* = \ddot{g}\Gamma \quad [39]$$

where:

$$\Gamma = \sum_i m_i\phi_i = \text{earthquake participation factor} \quad [40]$$

It is also convenient to express the generalized damping in terms of the percent of critical damping in the following manner:

$$c^* = \sum_i c_i\Delta\phi(i)^2 = 2\lambda m^*\omega \quad [41]$$

where ω represents the circular frequency of the generalized system and is given as:

$$\omega = \sqrt{\left(\frac{k^*}{m^*}\right)} \quad [42]$$

The effect of the generalized-coordinate approach is to transform a multiple-degree-of-freedom dynamic system into an equivalent SDOF system in terms of the generalized coordinate. This transformation is shown schematically in **Figure 12**. The degree to which the response of the transformed system represents the actual system will depend upon how well the assumed displacement shape represents the dynamic displacement of the actual structure. The displacement shape depends on the aspect ratio of the structure, which is defined as the ratio of the height to the base dimension. Possible shape functions for high-rise, mid-rise, and low-rise structures are summarized in **Figure 13**. It should be noted that most building codes use the straight-line shape function which is shown for the mid-rise system. Once the dynamic response is obtained in terms of the generalized coordinate, eqn [29] must be used to determine the

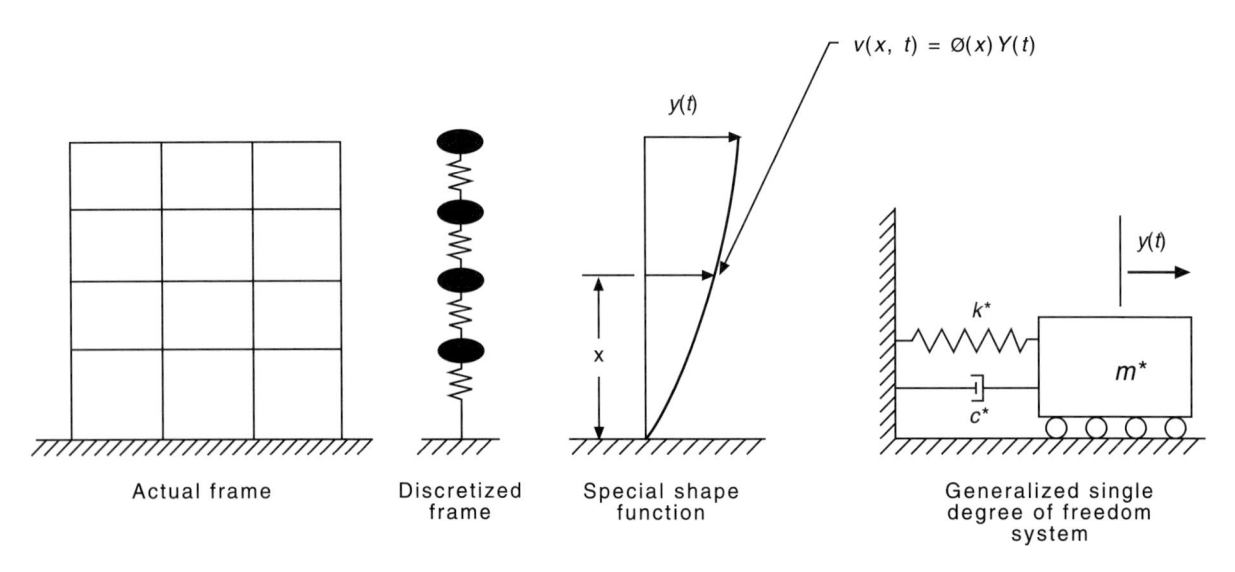

Figure 12 Generalized single-degree-of-freedom system.

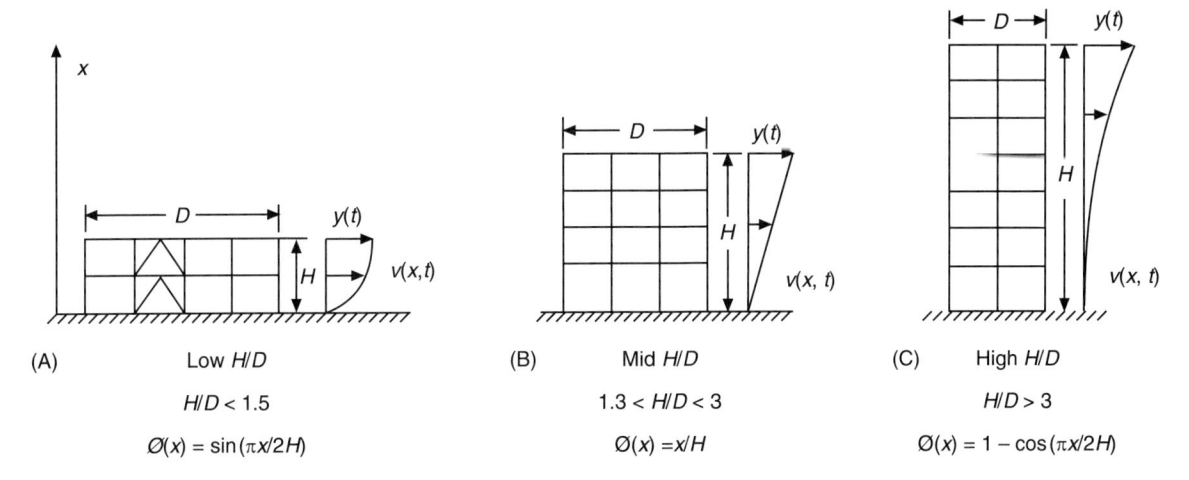

Figure 13 Possible shape functions based on aspect ratio.

(A) Low H/D, H/D < 1.5, $\phi(x) = \sin(\pi x/2H)$

(B) Mid H/D, 1.3 < H/D < 3, $\phi(x) = x/H$

(C) High H/D, H/D > 3, $\phi(x) = 1 - \cos(\pi x/2H)$

displacements in the structure, and these in turn can be used to determine the forces in the individual structural elements.

In principle, any function which represents the general deflection characteristics of the building and satisfies the support conditions could be used. However, any shape other than the true vibration shape requires the addition of external constraints to maintain equilibrium. These extra constraints tend to stiffen the system and thereby increase the computed frequency. The true vibration shape will have no external constraints and therefore will have the lowest frequency of vibration. When choosing between several approximate deflected shapes, the one producing the lowest frequency is always the best approximation. A good approximation to the true vibration shape can be obtained by applying forces representing the inertia forces and letting the static deformation of the structure determine the spatial shape function.

Time–history analysis Substituting the generalized parameters into the Duhamel integral solution results in the following solution for the displacement:

$$v(x,t) = \frac{\phi(x)\Gamma V(t)}{m^*\omega} \qquad [43]$$

Therefore, the inertia force at any position x above the base can be obtained from:

$$q(x,t) = m(x)\ddot{v}(x,t) = m(x)\omega^2 v(x,t) \qquad [44]$$

or simply:

$$q(x,t) = \frac{m(x)\phi(x)\Gamma\omega V(t)}{m^*} \qquad [45]$$

The base shear is obtained by summing the distributed inertia forces over the height H of the structure:

$$Q(t) = \int q(x,t)\,dx = \frac{\Gamma^2}{m^*}\omega V(t) \qquad [46]$$

The above relationships can be used to determine the displacements and forces in a generalized SDOF system at any time during the time history under consideration.

Response-spectrum analysis Using the definitions of the spectral pseudovelocity (S_{pv}) and spectral displacement (S_d), the maximum displacement in terms of the spectral displacement is obtained from:

$$v(x)_{max} = \frac{\phi(x)\Gamma S_d}{m^*} \qquad [47]$$

The forces in the system can readily be determined from the inertia forces, which can be expressed as:

$$q(x)_{max} = m(x)\ddot{v}(x)_{max} = m(x)\omega^2 v(x)_{max} \qquad [48]$$

Rewriting this result in terms of the spectral pseudo-acceleration (S_{pa}) results in the following:

$$q(x)_{max} = \frac{\phi(x)m(x)\Gamma S_{pa}}{m^*} \qquad [49]$$

Of considerable interest to structural engineers is the determination of the base shear. This is a key parameter in determining seismic design forces in most building codes. The base shear Q can be obtained from:

$$Q_{\max} = \frac{\Gamma^2 S_{\text{pa}}}{m^*} \quad [50]$$

It is also of interest to express the base shear in terms of the effective weight, which is defined as:

$$W^* = \frac{\left(\sum_i w_i \phi_i\right)^2}{\sum_i w_i \phi_i^2} \quad [51]$$

The expression for the maximum base shear becomes:

$$Q_{\max} = W^* S_{\text{pa}}/g \quad [52]$$

This form is similar to the basic base-shear equation used in the building codes. In the code equation, the effective weight is taken to be equal to the total dead weight W, plus a percentage of the live load for special occupancies. The seismic coefficient C is determined by a formula but is equivalent to the spectral pseudo-acceleration in terms of g. The basic code equation for base shear has the form:

$$Q_{\max} = CW \quad [53]$$

The effective earthquake force can also be determined by distributing the base shear over the story height. This distribution depends upon the displacement shape function and has the form:

$$q_i = Q_{\max} \frac{m_i \phi_i}{\Gamma} \quad [54]$$

If the shape function is taken as a straight line, the code force distribution is obtained. The overturning moment at the base of the structure can be determined by multiplying the inertia force by the corresponding story height above the base and summing over all story levels:

$$M_0 = \sum_i h_i q_i \quad [55]$$

Response of Nonlinear SDOF Systems

Many building structural systems will experience nonlinear response sometime during their life. Any moderate to strong earthquake will drive a building structure designed by conventional methods into the inelastic range, particularly in certain critical regions. A very useful numerical integration technique for problems of structural dynamics is the so-called step-by-step integration procedure. In this procedure the time history under consideration is divided into a number of small time increments Δt. During a small time step, the behavior of the structure is assumed to be linear. As nonlinear behavior occurs, the incremental stiffness is modified. In this manner, the response of the nonlinear system is approximated by a series of linear systems having a changing stiffness. The velocity and displacement computed at the end of one time interval become the initial conditions for the next time interval, and hence the process may be continued step by step.

Consider SDOF systems with properties $m, c, k(t)$ and $p(t)$, of which the applied force and the stiffness are functions of time. The stiffness is actually a function of the yield condition of the restoring force, and this in turn is a function of time. The damping coefficient may also be considered to be a function of time; however, general practice is to determine the damping characteristics for the elastic system and to keep these constant throughout the complete time history. In the inelastic range, the principle mechanism for energy dissipation is through inelastic deformation, and this is taken into account through the hysteretic behavior of the restoring force.

The numerical equation required to evaluate the nonlinear response can be developed by first considering the equation of dynamic equilibrium. Note that this equation must be satisfied at every increment of time. Considering the time at the end of a short time step, the equation of dynamic equilibrium can be written as:

$$f_i(t+\Delta t) + f_d(t+\Delta t) + f_s(t+\Delta t) = p(t+\Delta t) \quad [56]$$

where the forces are defined as:

$$\begin{aligned}
f_i &= m\ddot{v}(t+\Delta t) \\
f_d &= c\dot{v}(t+\Delta t) \\
f_s &= \sum_{i=1}^{n} k_i(t)\Delta v_i(t) = r_t + k(t)\Delta v(t) \\
\Delta v(t) &= v(t+\Delta t) - v(t) \\
r_t &= \sum_{i=1}^{n-1} k_i(t)\Delta v_i(t)
\end{aligned} \quad [57]$$

and in the case of ground accelerations:

$$p(t+\Delta t) = p_e(t+\Delta t) = -m\ddot{g}(t+\Delta t) \quad [58]$$

resulting in an equation of motion of the form:

$$m\ddot{v}(t+\Delta t) + c\dot{v}(t+\Delta t) + \sum k_i \Delta v_i = -m\ddot{g}(t+\Delta t) \quad [59]$$

It should be noted that the incremental stiffness is generally defined by the tangent stiffness at the beginning of the time interval:

$$k_i = \frac{df_s}{dv} \quad [60]$$

Note that these equations can also be used in a generalized SDOF system.

Many numerical integration schemes are available in the literature. The most popular direct integration schemes are: the Newmark, the Wilson-θ, and the Houbolt methods.

An important response parameter that is unique to nonlinear systems is the ductility ratio. For a SDOF system, this parameter can be defined in terms of the displacement as:

$$\mu = \frac{v(\text{max})}{v(\text{yield})} = 1.0 + \frac{v(\text{plastic})}{v(\text{yield})} \quad [61]$$

As can be seen from the above equation, the ductility ratio is an indication of the amount of inelastic deformation that has occurred in the system. In the case of a SDOF system or generalized SDOF system the ductility obtained from eqn [61] usually represents the average ductility in the system. The ductility demand at certain critical regions, such as plastic hinges in critical members, may be considerably higher.

Multiple-Degree-of-Freedom (MDOF) Systems

In many structural systems it is impossible to model the dynamic response accurately in terms of a single displacement coordinate. These systems require a number of independent displacement coordinates to describe the displacement of the mass of the structure at any instant of time.

Mass and stiffness properties In order to simplify the solution it is usually assumed for building structures that the mass of the structure is lumped at the center of mass of the individual story levels. This results in a diagonal matrix of mass properties in which either the translational mass or the mass moment of inertia is located on the main diagonal:

$$\mathbf{f}_i = \begin{bmatrix} m_1 & & & & \\ & m_2 & & & \\ & & m_3 & & \\ & & & \ddots & \\ & & & & m_n \end{bmatrix} \begin{Bmatrix} v_1 \\ v_2 \\ v_3 \\ \vdots \\ v_n \end{Bmatrix} \quad [62]$$

It is also convenient for building structures to develop the structural stiffness matrix in terms of the stiffness matrices of the individual story levels. The simplest idealization for a multistory building is based on the following three assumptions: (i) the floor diaphragm is rigid in its own plane; (ii) the girders are rigid relative to the columns and (iii) the columns are flexible in the horizontal directions but rigid in the vertical. If these assumptions are used, the building structure is idealized as having three dynamic degrees-of-freedom at each story level: a translational degree-of-freedom in each of two orthogonal directions, and a rotation about a vertical axis through the center of mass. If the above system is reduced to a plane frame, it will have one horizontal translational degree of freedom at each story level. The stiffness matrix for this type of structure has the tridiagonal form shown below:

$$\mathbf{f}_s = \begin{bmatrix} k_1 & -k_2 & & & & & \\ k_2 & k_1+k_2 & -k_3 & & & & \\ & -k_3 & k_2+k_3 & -k_4 & & & \\ & & \cdot & \cdot & \cdot & & \\ & & & \cdot & \cdot & \cdot & \\ & & & & \cdot & \cdot & -k_n \\ & & & & & -k_n & k_{n-1}+k_n \end{bmatrix} \times \begin{Bmatrix} v_1 \\ v_2 \\ v_3 \\ \cdot \\ \cdot \\ \cdot \\ v_{n-1} \\ v_n \end{Bmatrix}$$

$$[63]$$

For the simplest idealization, in which each story level has one translational degree-of-freedom, the stiffness terms k_i in the above equations represent the translational story stiffness of the ith story level. As the assumptions given above are relaxed to include axial deformations in the columns and flexural deformations in the girders, the stiffness term k_i in eqn [63] becomes a submatrix of stiffness terms, and the story displacement v_i becomes a subvector containing the various displacement components in the particular story level. The calculation of the stiffness coefficients for more complex structures is a standard problem of static structural analysis.

Mode shapes and frequencies The equations of motion for undamped free vibration of a MDOF system can be written in matrix form as:

$$\mathbf{M}\mathbf{\ddot{v}} + \mathbf{K}\mathbf{v} = 0 \quad [64]$$

Since the motions of a system in free vibration are simple harmonic, the displacement vector can be represented as:

$$\mathbf{v} = \mathbf{\bar{v}} \sin \omega t \quad [65]$$

Differentiating twice with respect to time results in:

$$\mathbf{\ddot{v}} = -\omega^2 \mathbf{v} \quad [66]$$

Substituting eqn [66] into eqn [64] results in a form of the eigenvalue equation:

$$(\mathbf{K} - \omega^2 \mathbf{M})\mathbf{v} = 0 \quad [67]$$

The classical solution to the above equation derives from the fact that in order for a set of homogeneous equilibrium equations to have a nontrivial solution, the determinant of the coefficient matrix must be zero:

$$\det(\mathbf{K} - \omega^2 \mathbf{M}) = 0 \quad [68]$$

Expanding the determinant by minors results in a polynomial of degree N, which is called the frequency equation. The N roots of the polynomial represent the frequencies of the N modes of vibration. The mode having the lowest frequency (longest period) is called the first or fundamental mode. Once the frequencies are known, they can be substituted one at a time into the equilibrium eqn [67] which can then be solved for the relative amplitudes of motion for each of the displacement components in the particular mode of vibration. It should be noted that since the absolute amplitude of motion is indeterminate, $N-1$ of the displacement components are determined in terms of one arbitrary component.

This method can be used satisfactorily for systems having a limited number of degrees-of-freedom. Programmable calculators have programs for solving the polynomial equation and for doing the matrix operations required to determine the mode shapes. However, for problems of any size, computer programs which use special numerical techniques to solve large eigenvalue systems must be used.

Equations of motion in normal coordinates Betti's reciprocal work theorem can be used to develop two orthogonality properties of vibration mode shapes which make it possible to greatly simplify the equations of motion. The first of these states that the mode shapes are orthogonal to the mass matrix and is expressed in matrix form as:

$$\boldsymbol{\phi}_n^T \mathbf{M} \boldsymbol{\phi}_m = 0 \quad (m \neq n) \quad [69]$$

The second property can be expressed in terms of the stiffness matrix as:

$$\boldsymbol{\phi}_n^T \mathbf{K} \boldsymbol{\phi}_m = 0 \quad (m \neq n) \quad [70]$$

which states that the mode shapes are orthogonal to the stiffness matrix. Although not necessarily true, for the sake of computational convenience it is further assumed that the mode shapes are also orthogonal to the damping matrix:

$$\boldsymbol{\phi}_n^T \mathbf{C} \boldsymbol{\phi}_m = 0 \quad (m \neq n) \quad [71]$$

Since any MDOF system having N degrees-of-freedom also has N independent vibration mode shapes, it is possible to express the displaced shape of the structure in terms of the amplitudes of these shapes by treating them as generalized coordinates (sometimes called normal coordinates). Hence the displacement at a particular location, v_i, can be obtained by summing the contributions from each mode as:

$$v_i = \sum_{n=1}^{N} \phi_{in} Y_n \quad [72]$$

In a similar manner, the complete displacement vector can be expressed as:

$$\mathbf{v} = \sum_{n=1}^{N} \boldsymbol{\phi}_n Y_n = \boldsymbol{\Phi} \mathbf{Y} \quad [73]$$

It is convenient to write the equations of motion for a MDOF system in matrix form as:

$$\mathbf{M}\ddot{\mathbf{v}} + \mathbf{C}\dot{\mathbf{v}} + \mathbf{K}\mathbf{v} = \mathbf{P}(t) \quad [74]$$

which is similar to the equation for a SDOF system. The differences arise because the mass, damping, and stiffness are now represented by matrices of coefficients representing the added degrees-of-freedom, and the acceleration, velocity, displacement, and applied load are represented by vectors containing the additional degrees-of-freedom. The equations of motion can be expressed, as well, in terms of the normal coordinates:

$$\mathbf{M\Phi\ddot{Y}} + \mathbf{C\Phi\dot{Y}} + \mathbf{K\Phi Y} = \mathbf{P}(t) \qquad [75]$$

Multiplying the above equation by the transpose of any modal vector $\boldsymbol{\phi}_n$ results in the following:

$$\boldsymbol{\phi}_n^T\mathbf{M\Phi\ddot{Y}} + \boldsymbol{\phi}_n^T\mathbf{C\Phi\dot{Y}} + \boldsymbol{\phi}_n^T\mathbf{K\Phi Y} = \boldsymbol{\phi}_n^T\mathbf{P}(t) \qquad [76]$$

Using the orthogonality conditions, this set of equations reduces to a set of independent equations of motion for a set of generalized SDOF systems in terms of the generalized properties for the nth mode shape and the normal coordinate Y_n:

$$M_n^*\ddot{Y}_n + C_n^*\dot{Y}_n + K_n^*Y = P_n^*(t) \qquad [77]$$

where the generalized properties for the nth mode are given as

$$\begin{aligned} M_n^* &= \text{generalized mass} = \boldsymbol{\phi}_n^T\mathbf{M}\boldsymbol{\phi}_n \\ C_n^* &= \text{generalized damping} \\ &= \boldsymbol{\phi}_n^T\mathbf{C}\boldsymbol{\phi}_n = 2\lambda_n\omega_n M_n^* \\ K_n^* &= \text{generalized stiffness} \\ &= \boldsymbol{\phi}_n^T\mathbf{K}\boldsymbol{\phi}_n = \omega_n^2 M_n^* \\ P_n^*(t) &= \text{generalized loading} = \boldsymbol{\phi}_n^T\mathbf{P}(t) \end{aligned} \qquad [78]$$

The above relations can be used to further simplify the equation of motion for the nth mode to the form:

$$\ddot{Y}_n + 2\lambda_n\omega_n\dot{Y}_n + \omega_n^2 Y_n = \frac{P_n^*(t)}{M_n^*} \qquad [79]$$

The complete solution for the system is then obtained by superimposing the independent modal solutions. For this reason, this method is often referred to as the 'modal-superposition' method. Use of this method also leads to a significant saving in computational effort, since in most cases it will not be necessary to use all N modal responses to accurately represent the response of the structure. For most buildings, the lower modes make the primary contribution to the total response. Therefore, the response can usually be represented to sufficient accuracy in terms of a limited number of modal responses in the lower modes.

Time–history analysis As in the case of SDOF systems, for earthquake analysis the time-dependent force must be replaced with the effective loads, which are given by the product of the mass at any level, M, and the ground acceleration $g(t)$. The vector of effective loads is obtained as the product of the mass matrix and the ground acceleration:

$$P_e(t) = \mathbf{M}\boldsymbol{\Psi}\ddot{g}(t) \qquad [80]$$

where $\boldsymbol{\Psi}$ is a vector of influence coefficients of which component i represents the acceleration at displacement coordinate i due to a unit ground acceleration at the base. For the simple structural model in which the degrees-of-freedom are represented by the horizontal displacements of the story levels, the vector $\boldsymbol{\Psi}$ becomes a unity vector, $\mathbf{1}$, since for a unit ground acceleration in the horizontal direction all degrees-of-freedom have a unit horizontal acceleration. The generalized effective load for the nth mode is given as:

$$P_{en}^*(t) = \Gamma_n g(t)$$

where:

$$\Gamma_n = \boldsymbol{\phi}_n^T\mathbf{M}\boldsymbol{\Psi} \qquad [81]$$

Resulting in the following expression for the earthquake response of the nth mode of a MDOF system:

$$\ddot{Y}_n + 2\lambda_n\omega_n\dot{Y}_n + \omega_n^2 Y_n = \varphi_n\ddot{g}(t)/M_n^* \qquad [82]$$

In a manner similar to that used for the SDOF system, the response of this mode at any time t can be obtained by the Duhamel integral expression:

$$Y_n(t) = \frac{\varphi_n V_n(t)}{M_n^*\omega_n} \qquad [83]$$

where $V_n(t)$ represents the integral:

$$V_n(t) = \int_0^t \ddot{g}(\tau)e^{-\lambda_n\omega_n(t-\tau)}\sin\omega_n(t-\tau)\,d\tau \qquad [84]$$

The complete displacement of the structure at any time is then obtained by superimposing the contributions of the individual modes:

$$\boldsymbol{v}(t) = \sum_{n=1}^{N}\boldsymbol{\phi}_n Y_n(t) = \boldsymbol{\Phi}\mathbf{Y}(t) \qquad [85]$$

The resulting earthquake forces can be determined in terms of the effective accelerations, which for each mode are given by the product of the circular frequency and the displacement amplitude of the generalized coordinate:

$$\ddot{Y}_{ne}(t) = \omega_n^2 Y_n(t) = \frac{\varphi_n\omega_n V_n(t)}{M_n^*} \qquad [86]$$

The corresponding acceleration in the structure due to the nth mode is given as:

$$\ddot{\mathbf{v}}_{\text{ne}}(t) = \boldsymbol{\phi}_n \ddot{Y}_{\text{ne}}(t) \quad [87]$$

and the corresponding effective earthquake force is given as:

$$\mathbf{q}_n(t) = \mathbf{M}\ddot{\mathbf{v}}_n(t) = \mathbf{M}\boldsymbol{\phi}_n \omega_n \varphi_n V_n(t)/M_n^* \quad [88]$$

The total earthquake force is obtained by superimposing the individual modal forces to obtain:

$$q(t) = \sum_{n=1}^{N} q_n(t) = \mathbf{M}\boldsymbol{\Phi}\omega^2 Y(t) \quad [89]$$

The base shear can be obtained by summing the effective earthquake forces over the height of the structure:

$$Q_n(t) = \sum_{i=1}^{H} q_{in}(t) = \mathbf{1}^T \mathbf{q}_n(t) = M_{\text{en}} \omega_n V_n(t) \quad [90]$$

where $M_{\text{en}} = \Gamma_n^2/M_n^*$ is the effective mass for the nth mode. The sum of the effective masses for all of the modes is equal to the total mass of the structure. This results in a means of determining the number of modal responses necessary to accurately represent the overall structural response. If the total response is to be represented in terms of a finite number of modes and if the sum of the corresponding modal masses is greater than a predefined percentage of the total mass, the number of modes considered in the analysis is adequate. If this is not the case, additional modes need to be considered. The base shear for the nth mode, can also be expressed in terms of the effective weight, W_{en}, as:

$$Q_n(t) = \frac{W_{\text{en}}}{g} \omega_n V_n(t) \quad [91]$$

where:

$$W_{\text{en}} = \frac{\left(\sum_{i=1}^{H} W_i \phi_{in}\right)^2}{\sum_{i=1}^{H} W_i \phi_{in}^2} \quad [92]$$

The base shear can be distributed over the height of the building with the modal earthquake forces expressed as:

$$\mathbf{q}_n^T = \frac{\mathbf{M}\boldsymbol{\phi}_n Q_n(t)}{\Gamma_n} \quad [93]$$

Response–spectrum analysis The above equations for the response of any mode of vibration are exactly equivalent to the expressions developed for the generalized SDOF system. Therefore, the maximum response of any mode can be obtained in a manner similar to that used for the generalized SDOF system. Therefore, by analogy, the maximum modal displacement can be written as:

$$Y_n(t)_{\text{max}} = \frac{V_n(t)_{\text{max}}}{\omega_n} = S_{\text{dn}} \quad [94]$$

and:

$$Y_{n\,\text{max}} = \varphi_n S_{\text{dn}}/M_n^* \quad [95]$$

The distribution of the modal displacements in the structure can be obtained by multiplying this expression by the modal vector:

$$\mathbf{v}_{n\text{max}} = \boldsymbol{\phi}_n Y_{n\,\text{max}} = \frac{\boldsymbol{\phi}_n \Gamma_n S_{\text{dn}}}{M_n^*} \quad [96]$$

The maximum effective earthquake forces can be obtained from the modal accelerations as:

$$\mathbf{q}_{n\text{max}} = \frac{\mathbf{M}\boldsymbol{\phi}_n \varphi_n S_{\text{pan}}}{M_n^*} \quad [97]$$

Summing these forces over the height of the structure gives the following expression for the maximum base shear due to the nth mode:

$$Q_{n\text{max}} = \varphi_n^2 S_{\text{pan}}/M_n^* \quad [98]$$

which can also be expressed in terms of the effective weight as:

$$Q_{n\text{max}} = W_{\text{en}} S_{\text{pan}}/g \quad [99]$$

where W_{en} is defined by eqn [92]. Finally, the overturning moment at the base of the building for the nth mode can be determined as:

$$M_o = \langle \mathbf{h} \rangle \mathbf{M}\boldsymbol{\phi}_n \Gamma_n S_{\text{pan}}/M_n^* \quad [100]$$

where $\langle \mathbf{h} \rangle$ is a row vector of the story heights above the base.

Modal combinations Using the response spectrum method for MDOF systems, the maximum modal response is obtained for each mode of a set of modes, which are used to represent the response.

The question then arises as to how these modal maxima should be combined in order to get the best estimate of the maximum total response. The modal response equations provide accurate results only as long as they are evaluated concurrently in time. In going to the response spectrum approach, time is taken out of these equations and replaced with the modal maxima. These maximum response values for the individual modes cannot possibly occur at the same time; therefore, a means must be found to combine the modal maxima in such a way as to approximate the maximum total response. One such combination that has been used is to take the sum of the absolute values (SAV) of the modal responses. This combination can be expressed as:

$$r \leq \sum_{n=1}^{N} |r_n| \quad [101]$$

Since this combination assumes that the maxima occur at the same time and that they also have the same sign, it produces an upper-bound estimate for the response, which is too conservative for design application. A more reasonable estimate, which is based on probability theory, can be obtained by using the square root of sum of squares (SRSS) method, which is expressed as:

$$r \approx \sqrt{\left(\sum_{n=1}^{N} r_n^2\right)} \quad [102]$$

This method of combination has been shown to give a good approximation of the response for two-dimensional structural systems. For three-dimensional systems, it has been shown that the complete quadratic combination (CQC) method may offer a significant improvement in estimating the response of certain structural systems. The complete quadratic combination is expressed as:

$$r \approx \sqrt{\left(\sum_{i=1}^{N} \sum_{j=1}^{N} r_i p_{ij} r_j\right)} \quad [103]$$

where for constant modal damping:

$$p_{ij} = \frac{8\lambda^2 (1+\zeta)\zeta^{3/2}}{(1-\zeta^2)^2 + 4\lambda^2 \zeta (1+\zeta)^2} \quad [104]$$

and:

$$\zeta = \omega_j / \omega_i$$
$$\lambda = c / c_{cr}$$

Using the SRSS method for two-dimensional systems and the CQC method for either two- or three-dimensional systems will give a good approximation to the maximum earthquake response of an elastic system without requiring a complete time history analysis. This is particularly important for purposes of design.

Nonlinear response of MDOF systems The nonlinear analysis of buildings modeled as MDOF systems closely parallels the development for SDOF systems presented earlier. However, the nonlinear dynamic time history analysis of MDOF systems is currently considered to be too complex for general use. Therefore, recent developments in the seismic evaluation of buildings have suggested a performance-based procedure which requires the determination of the demand and capacity. For more demanding investigations of building response, nonlinear dynamic analyses can be conducted.

For dynamic analysis, the loading time history is divided into a number of small time increments, whereas, in the static analysis, the lateral force is divided into a number of small force increments. During a small time or force increment, the behavior of the structure is assumed to be linear elastic. As nonlinear behavior occurs, the incremental stiffness is modified for the next time (load) increment. Hence, the response of the nonlinear system is approximated by the response of a sequential series of linear systems having varying stiffnesses.

The equations of equilibrium for a MDOF system subjected to base excitation can be written in matrix form as:

$$\mathbf{M}\ddot{\mathbf{v}} + \mathbf{C}\dot{\mathbf{v}} + \mathbf{K}\mathbf{v} = -\mathbf{M}\mathbf{\Psi}\ddot{g}(t) \quad [105]$$

In the mode superposition method, the damping ratio was defined for each mode of vibration. However, this is not possible for a nonlinear system because it has no true vibration modes. A useful way to define the damping matrix for a nonlinear system is to assume that it can be represented as a linear combination of the mass and stiffness matrices of the initial elastic system:

$$\mathbf{C} = \alpha \mathbf{M} + \beta \mathbf{K} \quad [106]$$

where α and β are scalar multipliers which may be selected so as to provide a given percentage of critical damping in any two modes of vibration of the initial elastic system. These two multipliers can be evaluated from the expression:

$$\left\{\begin{array}{c}\alpha\\ \beta\end{array}\right\} = 2\left[\begin{array}{cc}\omega_j & -\omega_i\\ -\dfrac{1}{\omega_j} & \dfrac{1}{\omega_i}\end{array}\right]\dfrac{\omega_i\omega_j}{\omega_j^2 - \omega_i^2}\left\{\begin{array}{c}\lambda_i\\ \lambda_j\end{array}\right\} \quad [107]$$

where ω_i and ω_j are the percentage of critical damping in the two specified modes. Once the coefficients α and β are determined, the damping in the other elastic modes is obtained from the expression:

$$\lambda_k = \dfrac{\alpha}{2\omega_k} + \dfrac{\beta\omega_k}{2} \quad [108]$$

A typical damping function which was used for the nonlinear analysis of a building is shown in **Figure 14**. Although the representation for the damping is only approximate it is justified for these types of analyses on the basis that it gives a good approximation of the damping for a range of modes of vibration and these modes can be selected to be the ones that make the major contribution to the response. Also in nonlinear dynamic analyses the dissipation of energy through inelastic deformation tends to overshadow the dissipation of energy through viscous damping. Therefore, an exact representation of damping is not as important in a nonlinear system as it is in a linear system. One should be aware of the characteristics of the damping function to insure that important components of the response are not lost. The matrix form of the numerical integration techniques illustrated before for analysis of nonlinear SDOF systems are used for response analysis of nonlinear MDOF systems. Furthermore, for building systems exhibiting local nonlinearity such as base isolated structures, techniques based on modified modal superposition techniques and Ritz vectors have also been successfully utilized.

Acknowledgments

Substantial parts of this article were extracted from *The Seismic Design Handbook*, 2nd edition, edited by the author. The author is indebted to Professors James C. Anderson Bijan Mohraz, and Fahim Sadek whose contributions to the *Handbook* have been liberally used in development of this article. The author also wishes to express his gratitude to Kluwer Academic Publishers for granting permission to extract this material from the *Handbook*.

Nomenclature

$a(t)$	accelerogram
B	adjustment factor
$F(\omega)$	Fourier transform
$FS(\omega)$	Fourier amplitude spectrum
g	acceleration
h, H	height
I	intensity
$PSV(\omega)$	pseudo velocity
Q	base shear
$SD(\omega)$	spectral displacement
T	duration of accelerogram
$V(t)$	velocity
W	weight

See Plates 29, 30, 31, 32.

See also: **Dynamic stability; Finite element methods; Seismic instruments, environmental factors; Structural dynamic modifications.**

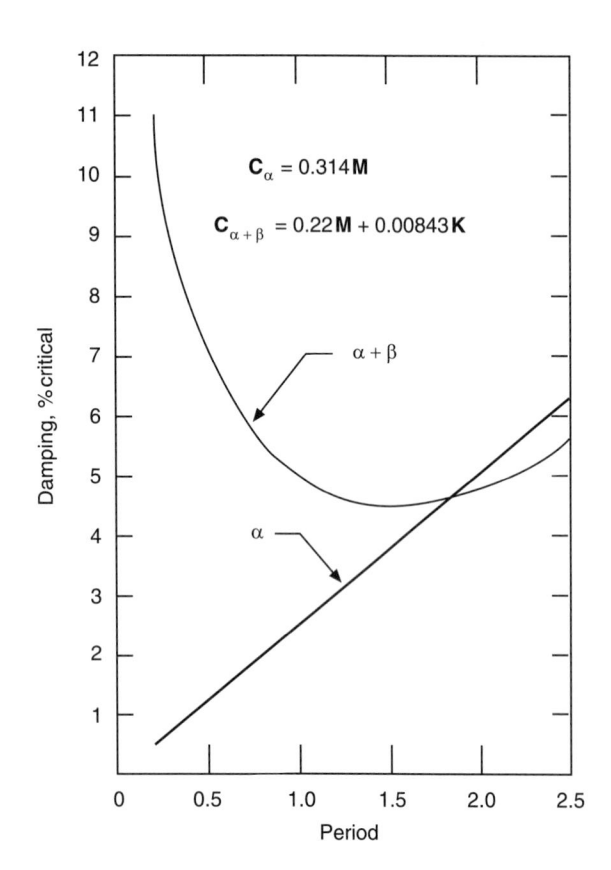

Figure 14 An example of damping functions selected for a building structure.

Further Reading

Anderson JC (2001) Dynamic response of structures. In: Naeim F (ed.) *The Seismic Design Handbook*, 2nd edn. Boston, MA: Kluwer Academic Publishers.

Bathe KJ (1982) *Finite Element Procedures in Engineering Analysis*. Englewood Cliffs, NJ: Prentice-Hall.

Chopra AK (1980) *Dynamics of Structures: A Primer*. Oakland, CA: Earthquake Engineering Research Institute.

Chopra AK (2001) *Dynamics of Structures: Theory and Applications to Earthquake Engineering*, 2nd edn. Englewood Cliffs, NJ: Prentice-Hall.

Clough RW, Penzin J (1993) *Dynamics of Structures*, 2nd edn. New York: McGraw-Hill.

Hudson DE (1979) *Reading and Interpreting Strong Motion Accelerograms*. Oakland, CA: Earthquake Engineering Research Institute.

International Code Council (2000) *International Building Code 2000*. VA: Fall Church.

Mohraz B, Sadek F (2001) Earthquake ground motion and response spectra. In: Naeim F (ed.), *The Seismic Design Handbook*, 2nd edn. Boston, MA: Kluwer Academic Publishers.

Naeim F (1994) Earthquake response and design spectra. In: Paz M (ed.) *International Handbook of Earthquake Engineering*. New York: Chapman and Hall.

Naeim F, Anderson JC (1996) *Design Classification of Horizontal and Vertical Earthquake Ground Motion (1933–1994)*. Los Angeles: John A. Martin.

Naeim F, Kelly JM (1999) *Design of Seismic Isolated Structures: From Theory to Practice*. New York: John Wiley.

Naeim F, Somerville PG *Earthquake Records and Design*. Oakland, CA: Earthquake Engineering Research Institute.

Newmark NM, Hall WJ (1982) *Earthquake Spectra and Design*. Oakland, CA: Earthquake Engineering Research Institute.

Structural Engineers Association of California (1999) *Recommended Lateral Force Requirements and Commentary*, 7th edn. Sacramento, CA.

Wilson EL (1998) *Three Dimensional Static and Dynamic Analysis of Structures*. Berkeley, CA: Computers and Structures.

EIGENVALUE ANALYSIS

O Bauchau, Georgia Institute of Technology, Atlanta, GA, USA,

Copyright © 2001 Academic Press

doi:10.1006/rwvb.2001.0001

Introduction

The fundamental eigenproblem in vibration analysis writes:

$$\mathbf{K}\mathbf{u}_i = \omega_i^2 \mathbf{M}\mathbf{u}_i \qquad [1]$$

where $\mathbf{K}$ and $\mathbf{M}$ are the $n \times n$ stiffness and mass matrices, respectively; ω_i^2, $i = 1, 2 \ldots n$ the n eigenvalues of the problem; and $\mathbf{u}_i$ the corresponding eigenvectors. The stiffness and mass matrices are assumed to be symmetric and positive-definite, implying that all the eigenvalues are real and positive. In the following, it will be assumed that the eivenvalues have been ordered in ascending order, i.e., $\omega_1^2 \leq \omega_2^2 \leq \omega_3^2 \leq \ldots \leq \omega_n^2$.

The characteristic polynomial $p(\omega^2)$ associated with the eigenproblem is defined as:

$$p(\omega^2) = \det(\mathbf{K} - \omega^2 \mathbf{M}) \qquad [2]$$

It is clear that the roots of the characteristic polynomial are the eigenvalues ω_i^2.

It is well known that the eigenvectors $\mathbf{u}_1, \mathbf{u}_2 \ldots \mathbf{u}_n$ are orthogonal to each other in the spaces of both stiffness and mass matrices. It is customary to normalize the modes in the space of the mass matrix. These relationships can be expressed in a compact manner by introducing a nonsingular matrix $\mathbf{P}$, the columns of which are the normalized eigenvectors:

$$\mathbf{P} = [\mathbf{u}_1, \mathbf{u}_2 \ldots \mathbf{u}_n] \qquad [3]$$

The orthonormality relationships then write:

$$\mathbf{P}^T \mathbf{M} \mathbf{P} = I; \quad \mathbf{P}^T \mathbf{K} \mathbf{P} = \operatorname{diag}(\omega_i^2) \qquad [4]$$

where I is the $n \times n$ identity matrix, and $\operatorname{diag}(\omega_i^2)$ a diagonal matrix storing the eigenvalues.

The computation of the eigenvalues and corresponding eigenvectors is a fundamental problem in vibration analysis. More often than not, only the lowest eigenvalues must be extracted. Below, some relevant properties of eigenproblem are reviewed. Four classes of computational methods for the solution of eigenproblems are subsequently presented. The basic algorithms based on similarity tranformation methods and vector iteration methods are then discussed. Conclusions and recommendations are presented in the last section.

Basic Properties of Eigenproblems

Similarity Transformations

Consider a linear transformation $\mathbf{u} = \mathbf{Q}\hat{\mathbf{u}}$, where $\mathbf{Q}$ is a nonsingular matrix. Introducing this transformation into eqn [1] and premultiplying by $\mathbf{Q}^T$ then yields:

$$\hat{\mathbf{K}}\hat{\mathbf{u}}_i = \omega_i^2 \hat{\mathbf{M}}\hat{\mathbf{u}}_i \qquad [5]$$

where $\hat{\mathbf{K}} = \mathbf{Q}^T\mathbf{K}\mathbf{Q}$ and $\hat{\mathbf{M}} = \mathbf{Q}^T\mathbf{M}\mathbf{Q}$. It can be readily shown that this similarity transformation does not affect the spectrum of eigenvalues. Indeed, the characteristic polynomial $\hat{p}(\omega^2)$ associated with the transformed problem is:

$$\begin{aligned}\hat{p}(\omega^2) &= \det(\hat{\mathbf{K}} - \omega^2\hat{\mathbf{M}}) \\ &= \det(\mathbf{Q}^T)\det(\mathbf{K} - \omega^2\mathbf{M})\det(\mathbf{Q}) \qquad [6] \\ &= \det(\mathbf{Q}^T)\det(\mathbf{Q})p(\omega^2)\end{aligned}$$

Since $\mathbf{Q}$ is nonsingular, $\det(\mathbf{Q}) \neq 0$, and the roots of $\hat{p}(\omega^2)$ are identical to those of $p(\omega^2)$. In summary, the similarity transformation leaves the spectrum of eigenvalues unchanged, and the eigenvectors are related through the similarity transformation $\mathbf{u}_i = \mathbf{Q}\hat{\mathbf{u}}_i$. It should be noted that the similarity transformation $\mathbf{u} = \mathbf{P}\hat{\mathbf{u}}$ leads to $\hat{\mathbf{K}} = \mathrm{diag}(\omega_i^2)$ and $\hat{\mathbf{M}} = \mathbf{I}$, as implied by the orthonormality relationships (eqn [4]). Both mass and stiffness matrices have been transformed simultaneously to a diagonal form.

The Rayleigh Quotient

An eigenvalue of the problem can be computed by premultiplying eqn [1] by $\mathbf{u}_i^T$, then solving to find:

$$\omega_i^2 = \frac{\mathbf{u}_i^T \mathbf{K} \mathbf{u}_i}{\mathbf{u}_i^T \mathbf{M} \mathbf{u}_i} \qquad [7]$$

This is not a pratical tool for computing an eigenvalue, as the knowledge of the corresponding eigenvector is required to start with. By analogy, the Rayleigh quotient is defined for an arbitrary vector $\mathbf{v}$ as:

$$\rho(\mathbf{v}) = \frac{\mathbf{v}^T \mathbf{K} \mathbf{v}}{\mathbf{v}^T \mathbf{M} \mathbf{v}} \qquad [8]$$

It presents the following important property, $\omega_1^2 \leq \rho(\mathbf{v}) \leq \omega_n^2$, which implies that the minimum value of the Rayleigh quotient for arbitrary choices of $\mathbf{v}$ is ω_1^2.

Consider now a vector $\mathbf{v}$ which closely approximates eigenvector $\mathbf{u}_i$, i.e., $\mathbf{v} = \mathbf{u}_i + \varepsilon\mathbf{x}$, where ε is a small number and $\mathbf{x}$ represents the discrepancy between $\mathbf{v}$ and the eigenvector $\mathbf{u}_i$. Hence, $\mathbf{x}$ has no components along $\mathbf{u}_i$ and can be expanded in terms of the remaining eigenvectors:

$$\mathbf{x} = \sum_{r=1, r\neq i}^{n} \alpha_r \mathbf{u}_r \qquad [9]$$

Using the orthogonality properties (eqn [4]), it is now readily verified that:

$$\mathbf{u}_i^T \mathbf{K} \mathbf{x} = 0; \quad \mathbf{x}^T \mathbf{K} \mathbf{x} = \sum_{r=1, r\neq i}^{n} \alpha_r^2 \omega_r^2 \qquad [10]$$

and similar relationships hold for the mass matrix. The Rayleigh quotient now writes:

$$\begin{aligned}\rho(\mathbf{v}) &= \frac{\mathbf{u}_i^T\mathbf{K}\mathbf{u}_i + 2\varepsilon\mathbf{u}_i^T\mathbf{K}\mathbf{x} + \varepsilon^2\mathbf{x}^T\mathbf{K}\mathbf{x}}{\mathbf{u}_i^T\mathbf{M}\mathbf{u}_i + 2\varepsilon\mathbf{u}_i^T\mathbf{M}\mathbf{x} + \varepsilon^2\mathbf{x}^T\mathbf{M}\mathbf{x}} \\ &= \frac{\omega_i^2 + \varepsilon^2 \sum_{r=1, r\neq i}^{n} \alpha_r^2 \omega_r^2}{1 + \varepsilon^2 \sum_{r=1, r\neq i}^{n} \alpha_r^2}\end{aligned} \qquad [11]$$

After expansion for small values of ε, we find:

$$\rho(\mathbf{v}) = \omega_i^2 + \varepsilon^2 \sum_{r=1, r\neq i}^{n} \alpha_r^2(\omega_r^2 - \omega_i^2) + O(\varepsilon^4) \qquad [12]$$

This important result shows that if a vector $\mathbf{v}$ is an approximation to eigenvector $\mathbf{u}_i$ to $O(\varepsilon)$, the corresponding Rayleigh quotient is an approximation of ω_i^2 to $O(\varepsilon^2)$. This powerful tool is used in several methods for computing eigenvalues.

Rayleigh–Ritz Analysis

Rayleigh–Ritz analysis is a general tool for obtaining approximate solutions to eigenproblem [1]. Consider a subspace $\mathbf{X}$ spanned by p Ritz vectors, $\mathbf{x}_1, \mathbf{x}_2 \ldots \mathbf{x}_p$:

$$\mathbf{X} = [\mathbf{x}_1, \mathbf{x}_2 \ldots \mathbf{x}_p] \qquad [13]$$

A vector $\mathbf{v}$ is now constrained to belong to this subspace:

$$\mathbf{v} = \mathbf{X}\mathbf{q} \qquad [14]$$

The choice of vector $\mathbf{v}$ within the subspace is determined by p independent quantities $q_1, q_2 \ldots q_p$, the components of $\mathbf{q}$. Since ω_1^2 is the minimum value that the Rayleigh quotient $\rho(\mathbf{v})$ can achieve for all arbitrary choice of $\mathbf{v}$, the vector $\mathbf{v}$ that best approximates $\mathbf{u}_1$ should minimize $\rho(\mathbf{v})$ with respect to all the choices of the components q_i:

$$\frac{\partial \rho(\mathbf{v})}{\partial q_i} = 0, \quad i = 1 \ldots p \qquad [15]$$

Introducing the definition of the Rayleigh quotient (eqn [8] and eqn [14]) leads to:

$$\frac{\partial}{\partial \mathbf{q}}\left(\frac{\mathbf{v}^T \mathbf{K} \mathbf{v}}{\mathbf{v}^T \mathbf{M} \mathbf{v}}\right) = \frac{\partial}{\partial \mathbf{q}}\left(\frac{\mathbf{q}^T \hat{\mathbf{K}} \mathbf{q}}{\mathbf{q}^T \hat{\mathbf{M}} \mathbf{q}}\right) = 0 \quad [16]$$

where $\hat{\mathbf{K}} = \mathbf{X}^T \mathbf{K} \mathbf{X}$ and $\hat{\mathbf{M}} = \mathbf{X}^T \mathbf{M} \mathbf{X}$ are $p \times p$ reduced stiffness and mass matrices, respectively. Taking the derivatives then leads to a new, reduced eigenproblem:

$$\hat{\mathbf{K}} \mathbf{q} = \left(\frac{\mathbf{q}^T \hat{\mathbf{K}} \mathbf{q}}{\mathbf{q}^T \hat{\mathbf{M}} \mathbf{q}}\right) \hat{\mathbf{M}} \mathbf{q}; \quad \text{or} \quad \hat{\mathbf{K}} \mathbf{q} = \hat{\omega}^2 \hat{\mathbf{M}} \mathbf{q} \quad [17]$$

In summary, the vector $\mathbf{v}$ that best approximates eigenvector $\mathbf{u}_1$ within the subspace $\mathbf{X}$ is determined by the components of $\mathbf{q}$, which are the solution of the reduced eigenproblem [17]. If ω_1^2 and $\mathbf{q}_1$ are the lowest eigenvalue and eigenvector of [17], respectively, then $\omega_1^2 \approx \hat{\omega}_1^2$ and $\mathbf{u}_1 \approx \mathbf{X}\mathbf{q}_1$. At first, it seems that nothing has been gained, since the solution of an eigenproblem has been replaced by that of another eigenproblem. However, it should be noted that the original eigenproblem is of size $n \times n$, whereas the reduced eigenproblem is of size $p \times p$. If $p \ll n$, it is, of course, much simpler to solve the reduced eigenproblem.

The major deficiency of the Rayleigh–Ritz approach is that little can be said about how well the solution of the reduced eigenproblem approximates that of the original eigenproblem. If an eigenvector $\mathbf{u}_k$ exactly lies in the subspace spanned by $\mathbf{X}$, the corresponding eigenvalue will be exactly recovered in the reduced eigenproblem. If the subspace $\mathbf{X}$ is chosen arbitrarily, the quality of the approximation is doubtful.

The Sturm Sequence Property

Consider the quantity $p(\mu^2) = \det(\mathbf{K} - \mu^2 \mathbf{M})$ where μ^2 is not an eigenvalue. Clearly, $p(\mu^2) \neq 0$, and hence, the nonsingular matrix $\mathbf{K} - \mu^2 \mathbf{M}$ can be trifactorized as:

$$\mathbf{K} - \mu^2 \mathbf{M} = \mathbf{L}^T \operatorname{diag}(d_{ii}) \mathbf{L} \quad [18]$$

where $\mathbf{L}$ is a lower triangular matrix with unit entries on the diagonal, and $\operatorname{diag}(d_{ii})$ a diagonal matrix with diagonal entries d_{ii}. The quantity $p(\mu^2)$ now becomes:

$$\begin{aligned}p(\mu^2) &= \det(\mathbf{L}^T \operatorname{diag}(d_{ii}) \mathbf{L}) \\ &= \det(\mathbf{L}^T)\left(\prod_{i=1}^n d_{ii}\right)\det(\mathbf{L}) = \prod_{i=1}^n d_{ii}\end{aligned} \quad [19]$$

since $\det(\mathbf{L}) = 1$. The following theorem will be given here without proof:

Theorem 1 In the trifactorization $\mathbf{K} - \mu^2 \mathbf{M} = \mathbf{L}^T \operatorname{diag}(d_{ii}) \mathbf{L}$, the number of negative elements $d_{ii} < 0$ is equal to the number of eigenvalues smaller than μ^2.

Types of Computational Methods for Eigenproblems

Computational methods for eigenvalue problems can be broken into four groups, according to the relationship used as a basis for the method:

1. Polynomial iteration methods based on the characteristic polynomial (eqn [2]).
2. Sturm sequence methods based on the properties of the trifactorization (eqn [19]).
3. Similarity transformation methods based on the properties of similarity transformations (eqn [5]).
4. Vector iteration methods based on the eigenproblem statement (eqn [1]).

The first two groups of methods do not lead to practical tools for computing eigenvalues unless n is very small. The last two groups do lead to practical algorithms (see below).

The characteristic polynomial approach replaces the computation of eigenvalues by the extraction of the roots of an nth-order polynomial. It is well known that no explicit formula exists for the computation of the roots of a polynomial for $n > 4$. Hence, all computational methods for eigenvalues will be iterative in nature. All methods for finding the roots of a polynomial do apply to the computation of eigenvalues. In particular, Graeffe's root-squaring method has been used for this purpose. However, the computation of the characteristic polynomial coefficients is cumbersome, and rapidly becomes overwhelming as n increases. Most root-finding methods are not robust and cannot be recommended as a computational tool for evaluating eigenvalues when $n > 10$.

The most important use of the Sturm sequence property is for *a posteriori* verification of eigenvalue extraction. Assume a number of eigenvalues, say k eigenvalues, have been identified within an interval (α^2, β^2). Let n_α and n_β be the number of negative terms in the diagonal matrix of the trifactorization of $\mathbf{K} - \alpha^2 \mathbf{M}$ and $\mathbf{K} - \beta^2 \mathbf{M}$, respectively. In view of theorem 1, the number of eigenvalues between α^2 and β^2 must then be $n_\beta - n_\alpha$. If $k < n_\beta - n_\alpha$, additional eigenvalues must exist in the interval. This technique is particularly useful in the presence of closely clustered eigenvalues: most eigenvalue extraction algorithms will have difficulties identifying all the eigenvalues within a cluster. Obtaining a complete picture of the dynamical behavior of a system requires the identification all eigenvalues present within the frequency range of interest. Knowing the exact number of eigenvalues present within that interval is an important element to avoid missing some of the eigenvalues. If $k = n_\beta - n_\alpha$, all eigenvalues in the interval (α^2, β^2) have been identified. Conceptually, the Sturm sequence property could be used to extract eigenvalues using successive bisections of the interval (α^2, β^2) until a single eigenvalue has been bracketed to the desired accuracy. This method is prohibitively expensive as it requires a large number of matrix trifactorizations.

Similarity Transformation Methods

The simplest similarity transformation method for eigenvalue computation is the Jacobi method which deals with the standard eigenproblem $\mathbf{K}\mathbf{u}_i = \omega_i^2 \mathbf{u}_i$, i.e., the mass matrix is the identity matrix. Consider a similarity transformation defined by the following matrix

$$\mathbf{Q} = \begin{bmatrix} 1 & 0 & 0 & \cdots & & \cdots & & \cdots & 0 \\ 0 & \ddots & & & & & & & \vdots \\ \vdots & & \cos\theta & \cdots & -\sin\theta & & & & \vdots \\ & & & \ddots & & & & & \\ \vdots & & \sin\theta & \cdots & \cos\theta & & & & \vdots \\ & & & & & \ddots & 0 & & \\ 0 & 0 & \cdots & & & & \cdots & 0 & 1 \end{bmatrix} \quad [20]$$

where the trigonometric entries appear in rows and columns i and j. It is readily verified that $\mathbf{Q}$ is an orthogonal matrix, i.e., $\mathbf{Q}^T\mathbf{Q} = \mathbf{I}$. The transformed stiffness matrix is $\hat{\mathbf{K}} = \mathbf{Q}^T\mathbf{K}\mathbf{Q}$. The angle θ is arbitrary, and will be selected so as to zero the entry $\hat{K}_{ij} = (K_{jj} - K_{ii})\sin\theta\cos\theta + K_{ij}(\cos^2\theta - \sin^2\theta)$. Solving for the angle θ then yields:

$$\tan 2\theta = \frac{2K_{ij}}{K_{ii} - K_{jj}} \quad [21]$$

In the Jacobi method, each off-diagonal entry is zeroed in turn, using the appropriate similarity transformation. It is important to note that the off-diagonal entry zeroed at a given step will be modified by the subsequent similarity transformations. Hence, the procedure must then be repeated until all off-diagonal terms are sufficiently small. At convergence, the diagonal entries of $\hat{\mathbf{K}}$ will store the eigenvalues.

To prove convergence of the method, let s_0 and s_1 be the sum of the squares of the off-diagonal terms, before and after a similarity transformation to zero the K_{ij} entry, respectively. It can be readily shown that $s_1 = s_0 - 2K_{ij}^2$. This means that s must decrease at each step, and the optimum strategy is to zero the maximum off-diagonal term at each step. This implies:

$$s_1 \leq \left(1 - \frac{2}{n(n-1)}\right) s_0 \quad [22]$$

and after k steps:

$$s_k \leq \left(1 - \frac{2}{n(n-1)}\right)^k s_0 \quad [23]$$

This relationship implies the convergence of the method when $k \to \infty$. It is possible to estimate the number of steps required to decrease s by t orders of magnitude, i.e., $s_k/s_0 \approx 10^{-t}$. From eqn [23], $k \approx tn^2$. The number of steps increases as n^2, which means that this approach will become increasingly expensive when the order of the systems increases. For instance, if $n = 100$ and $t = 12$, 120 000 similarity transformations will be required. Clearly, the Jacobi method is not a practical approach when $n > 50$.

The Jacobi method can be generalized to treat eigenproblem in the form of eqn [1]. Other similarity transformations methods are applicable to the problem at hand written as $\mathbf{D}\mathbf{u}_i = (1/\omega_i^2)\mathbf{u}_i$, where the dynamic flexibility matrix $\mathbf{D}$ is defined as $\mathbf{D} = \mathbf{K}^{-1}\mathbf{M}$. Householder's method will transform $\mathbf{D}$ into an upper Hessenberg matrix $\mathbf{H}$ through n successive similarity transformations. The QR algorithm is then very effective in extracting the eigenvalues of $\mathbf{H}$, through similarity transformations, once again.

Vector Iteration Methods

Consider an arbitrary vector $\mathbf{x}_1$ and an arbitrary frequency $\omega^2 = 1$. The inertial forces associated with this mode shape oscillating at this unit frequency are:

$$\mathbf{R}_1 = \omega^2 \mathbf{M} \mathbf{x}_1 = \mathbf{M} \mathbf{x}_1 \quad [24]$$

Since $\mathbf{x}_1$ is not an eigenvector, $\mathbf{K}\mathbf{x}_1 \neq \omega^2 \mathbf{M}\mathbf{x}_1$. However, a vector $\mathbf{x}_2$ can be defined such that:

$$\mathbf{K}\mathbf{x}_2 = \omega^2 \mathbf{M}\mathbf{x}_1 \quad [25]$$

$\mathbf{x}_2$ can be readily found by solving this set of linear equations. In fact, $\mathbf{x}_2$ corresponds to the static deflection of the system under the steady loads $\mathbf{R}_1$. Intuitively, one would expect $\mathbf{x}_2$ to be a better approximation to an eigenvector than $\mathbf{x}_1$. By induction, the following algorithm is proposed

Algorithm 1 (Inverse Iteration Method)

- Step 1 (inverse iteration): $\mathbf{K}\bar{\mathbf{x}}_{k+1} = \mathbf{M}\mathbf{x}_k$
- Step 2 (Rayleigh quotient): $\rho_{k+1} = \dfrac{\mathbf{x}_{k+1}^T \mathbf{K}\mathbf{x}_{k+1}}{\mathbf{x}_{k+1}^T \mathbf{M}\mathbf{x}_{k+1}}$
- Step 3 (normalization): $\mathbf{x}_{k+1} = \dfrac{\bar{\mathbf{x}}_{k+1}}{\left(\bar{\mathbf{x}}_{k+1}^T \mathbf{M}\bar{\mathbf{x}}_{k+1}\right)^{1/2}}$

The first step of the algorithm is the inverse iteration operation. Step 2 evaluates an approximate eigenvalue ρ_{k+1} based on the Rayleigh quotient. Finally, step 3 is a normalization step enforcing $\mathbf{x}_{k+1}^T \mathbf{M}\mathbf{x}_{k+1}$ to prevent undue growth or decay of the norm of the vectors. As $k \to \infty$ $\rho_{k+1} \to \omega_1^2$ and $\mathbf{x}_{k+1} \to \mathbf{u}_1$, i.e., the algorithm converges to the lowest eigenvalue and corresponding eigenvector. The proof of this claim follows.

The heart of the algorithm is the inverse iteration $\mathbf{K}\mathbf{x}_{k+1} = \mathbf{M}\mathbf{x}_k$. The similarity transformation $\mathbf{x}_k = \mathbf{P}\mathbf{z}_k$ is now applied to this inverse iteration step, which becomes:

$$\text{diag}(\omega_i^2)\mathbf{z}_{k+1} = \mathbf{z}_k \quad [26]$$

Note that this transformation is not a practical one as the exact eigenvectors of the system stored in matrix P are unknown. However, the eigenvalues are not altered by the similarity transformation, and the convergence characteristics of [26] are identical to those of algorithm 1. Eqn [26] is readily solved for $\mathbf{z}_{k+1}$, and recursive application then yields:

$$\mathbf{z}_{k+1} = \text{diag}(1/\omega_i^2)\mathbf{z}_k = \text{diag}(1/\omega_i^2)^2 \mathbf{z}_{k-1} \quad [27]$$
$$= \text{diag}(1/\omega_i^2)^k \mathbf{z}_1$$

If the arbitrary starting vector $\mathbf{z}_1^T = [1, 1 \ldots 1]$ is selected, $\mathbf{z}_{k+1}$ becomes:

$$\mathbf{z}_{k+1} = \begin{bmatrix} (1/\omega_1^2)^k \\ (1/\omega_2^2)^k \\ (1/\omega_3^2)^k \\ \vdots \\ (1/\omega_n^2)^k \end{bmatrix} = (1/\omega_1^2)^k \begin{bmatrix} 1 \\ (\omega_1^2/\omega_2^2)^k \\ (\omega_1^2/\omega_3^2)^k \\ \vdots \\ (\omega_1^2/\omega_n^2)^k \end{bmatrix} \quad [28]$$

Since the eigenvalues have been arranged in ascending order, $(\omega_1^2/\omega_i^2)^k \to 0$ for $i \neq 1$. It follows that $\mathbf{z}_{k+1} \to (1/\omega_1^2)^k \mathbf{e}_1$, where $\mathbf{e}_1^T = [1, 0 \ldots 0]$. Clearly, as $k \to \infty$, $\mathbf{z}_{k+1}$ becomes parallel to $\mathbf{e}_1$, the eigenvector of the system corresponding to the lowest eigenvalue ω_1^2.

The convergence rate r for the eigenvector is:

$$r = \lim_{k \to \infty} \frac{\left\|\mathbf{z}_{k+1} - (1/\omega_1^2)^k \mathbf{e}_1\right\|}{\left\|\mathbf{z}_k - (1/\omega_1^2)^k \mathbf{e}_1\right\|} = \frac{\omega_1^2}{\omega_2^2} \quad [29]$$

The convergence rate is ω_1^2/ω_2^2, the ratio of the first two eigenvalues. The eigenvalues are given by the Rayleigh quotient ρ_{k+1}:

$$\rho_{k+1} = \frac{\mathbf{z}_{k+1}^T \mathbf{z}_k}{\mathbf{z}_{k+1}^T \mathbf{z}_{k+1}} = \frac{(1/\omega_1^2)^{2k+1} \sum_{i=1}^n (\omega_1^2/\omega_i^2)^{2k+1}}{(1/\omega_1^2)^{2k+2} \sum_{i=1}^n (\omega_1^2/\omega_i^2)^{2k+2}} \quad [30]$$

As $k \to \infty$, $\rho_{k+1} \to \omega_1^2$. The convergence rate is:

$$r = \lim_{k \to \infty} \frac{|\rho_{k+1} - \omega_1^2|}{|\rho_k - \omega_1^2|} = \left(\frac{\omega_1^2}{\omega_2^2}\right)^2 \quad [31]$$

This result is consistent with the basic property of Rayleigh quotients: if an eigenvector is estimated to order ε, the corresponding eigenvalue estimated from the Rayleigh quotient will be accurate to order ε^2.

Note that if the lowest root has a multiplicity m, $\mathbf{z}_{k+1} \to (1/\omega_1^2)^k [1, 1, 1, 0 \ldots 0]$, where the unit entry is repeated m times. $\mathbf{z}_{k+1}$ is now parallel to a linear combination of $\mathbf{e}_1, \mathbf{e}_2 \ldots \mathbf{e}_m$, which are m orthogonal eigenvectors corresponding to the lowest eigenvalue of multiplicity m. In this case, the convergence rate for the eigenvector is $\omega_1^2/\omega_{m+1}^2$, the ratio of the first two distinct eigenvalues.

It is possible to give a rough estimate of the number of iterations N required to converge t digits of the lowest eigenvalue [7], i.e.:

$$\left|\frac{\rho_{N+1} - \rho_N}{\rho_N}\right| \approx 10^{-t} \quad [32]$$

Introducing eqn [30] yields:

$$N \approx \frac{t}{\log(\omega_2/\omega_1)} \quad [33]$$

Figure 1 shows this estimated number of iterations as a function of the ratio of the two lowest eigenvalues ω_2/ω_1, for $t = 8$. When the lowest eigenvalues are well separated, say $\omega_2/\omega_1 > 4$, a small number of iterations is required, $N < 15$. This number of iterations is independent of the order n of the system, and hence, this approach is suitable for large-order systems. This contrasts with the similarity transformation methods described previously, in which the required number of iterations increases in proportion

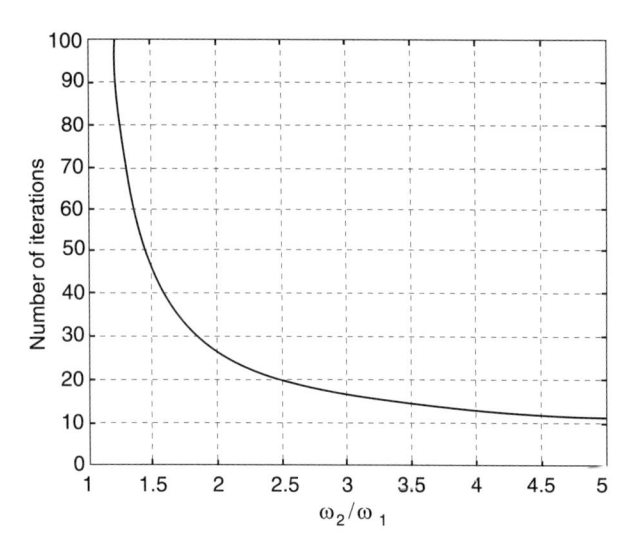

Figure 1 Convergence characteristic of the inverse iteration method.

to n^2. On the other hand, when $\omega_2/\omega_1 < 1.2$, over 100 iterations will be required to stabilize eight-digits of the eigenvalue. This means that inverse vector iteration will become increasingly ineffective in the presence of closely clustered eigenvalues. In practice, the algorithm fails to converge when $\omega_2/\omega_1 < 1.1$.

The inverse iteration technique described above presents two major deficiencies. First, it will always converge to the lowest eigenvalue for any arbitrary starting vector. Second, it performs poorly in the presence of clustered eigenvalues. In an attempt to alleviate these problems, the basic algorithm can be extended through the vector orthogonal deflation procedure which allows convergence to the higher eigenvectors. Another approach is to consider the following shifted eigenproblem:

$$(\mathbf{K} - \mu^2 \mathbf{M})\mathbf{u}_i = (\omega_i^2 - \mu^2)\mathbf{M}\mathbf{u}_i \qquad [34]$$

The stiffness matrix $\mathbf{K}$ has been replaced by $\mathbf{K} - \mu^2 \mathbf{M}$, and the eigenvalues ω_i^2 have all been shifted by a constant value μ^2 to $\omega_i^2 - \mu^2$. Using the same transformation as described above, the basic relationship of the shifted inverse iteration method becomes $\mathrm{diag}(\omega_i^2 - \mu^2)\mathbf{Z}_{k+1} = \mathbf{Z}_k$, which now replaces eqn [26]. Similar arguments to those presented earlier lead to:

$$\mathbf{z}_{k+1} \to \left(\frac{1}{\omega_j^2 - \mu^2}\right)^k \mathbf{e}_j \qquad [35]$$

where ω_j^2 is the eigenvalue the closest to the shift μ^2, and $\mathbf{e}_j$ the corresponding eigenvector. The convergence rate is now:

$$r = \max_{p \neq j} \left|\frac{\omega_j^2 - \mu^2}{\omega_p^2 - \mu^2}\right| \qquad [36]$$

The shifted inverse iteration approach allows convergence to any eigenvalue ω_j^2, provided the shift μ^2 is chosen close enough to ω_j^2. Various eigenvalues can be obtained independently with adequate choices of the shift. Furthermore, choosing μ^2 very close to an eigenvalue will result in excellent convergence characteristics. In fact, if $\mu^2 = \omega_j^2$, the corresponding eigenvector is exactly recovered in one single iteration. The problem of this approach is to select the proper shift to obtain good convergence characteristics to the desired eigenvalue.

The subspace iteration method generalizes the inverse iteration approach by iterating simultaneously on a number of vectors, i.e., on a subspace of the system. Furthermore, to prevent convergence of all these vectors to the lowest eigenvector of the system, the vectors of the subspace are orthogonalized to each other at each step of the algorithm. The algorithm is as follows:

Algorithm 2 (Subspace Iteration Method)

- Step 1 (simultaneous inverse iteration):
  $\mathbf{K}\bar{\mathbf{X}}_{k+1} = \mathbf{M}\mathbf{X}_k$
- Step 2 (Rayleigh–Ritz analysis):
  $\hat{\mathbf{K}}_{k+1} = \bar{\mathbf{X}}_{k+1}^T \mathbf{K} \bar{\mathbf{X}}_{k+1}; \hat{\mathbf{M}}_{k+1} = \bar{\mathbf{X}}_{k+1}^T \mathbf{M} \bar{\mathbf{X}}_{k+1}$
- Step 3 (reduced eigenproblem solution):
  $\hat{\mathbf{K}}_{k+1}\mathbf{Q}_{k+1} = \hat{\mathbf{M}}_{k+1}\mathbf{Q}_{k+1}\mathrm{diag}(\omega_{i,k+1}^2)$
- Step 4 (improved approximation):
  $\mathbf{X}_{k+1} = \bar{\mathbf{X}}_{k+1}\mathbf{Q}_{k+1}$

Step 1 performs the simultaneous inverse iteration on each vector of the subspace. A Rayleigh–Ritz analysis based on this subspace follows in steps 2 and 3: $\omega_{i,k}^2 + 1$ are the eigenvalues of the reduced problem and the matrix $\mathbf{Q}_{k+1}$ stores the corresponding eigenvectors. Step 4 enforces the orthogonality of the subspace in the space of the mass matrix. Indeed,

$$\begin{aligned}\mathbf{X}_{k+1}^T \mathbf{M} \mathbf{X}_{k+1} &= \mathbf{Q}_{k+1}^T \bar{\mathbf{X}}_{k+1}^T \mathbf{M} \bar{\mathbf{X}}_{k+1} \mathbf{Q}_{k+1} \\ &= \mathbf{Q}_{k+1}^T \hat{\mathbf{M}}_{k+1} \mathbf{Q}_{k+1} = I\end{aligned} \qquad [37]$$

As $k \to \infty \, \omega_{i,k+1}^2 \to \omega_i^2$ and $\mathbf{X}_k \to [\mathbf{u}_1, \mathbf{u}_2 \ldots \mathbf{u}_p]$. As iterations proceed, the reduced matrices $\hat{\mathbf{K}}_{k+1}$ and $\hat{\mathbf{M}}_{k+1}$ tend toward a diagonal form. Consequently, the Jacobi method described previously is ideally suited to the solution of the reduced eigenproblem. The subspace iteration method removes the deficiencies of the simple inverse iteration algorithm. The p lowest eigenvalues and corresponding eigenvectors are extracted simultaneously. Furthermore, the convergence is not delayed if a cluster of eigenvalues is present within these p lowest eigenvalues. The Sturm sequence property, theorem 1, should be used to verify that all eigenvalues have been extracted within a certain frequency range of interest.

Conclusions and Recommendations

A large number of methods can be applied to computation of the eigenvalues and eigenvectors of dynamical systems. When the order of the system is low, say $n < 10$, polynomial iteration methods and Sturm sequence methods can be applied. As n increases these methods are not robust and rapidly become prohibitively expensive.

Similarity transformation methods can be used for larger systems, say $n < 250$. The preferred method would be Householder's method to transform the dynamic flexibility matrix to an upper Hessenberg form, followed by the QR algorithm. Once the eigenvalues have been found, the corresponding eigenvectors are evaluated by means of the shifted inverse iteration procedure. The Jacobi algorithm is only used in practice after the initial eigenproblem has been projected on to a sufficiently small subspace, such as in the subspace iteration method.

Inverse iteration is a robust method for extracting the lowest eigenvalue of large systems. The simple version of the algorithm presents two major limitations: first, it always converges to the lowest eigenvalue no matter what starting vector is selected, and second it cannot deal with closely clustered eigenvalues. The most reliable form of inverse iteration is the subspace iteration method which performs inverse iteration on a number of vectors simultaneously while keeping them orthogonal to each other.

The most efficient methods for large eigenproblems, like those generated by the finite element method, are those based on the construction of Krylov subspaces.

Nomenclature

D	dynamic flexibility matrix
H	Hessenberg matrix
I	$n \times n$ identity matrix
L	lower triangular matrix
P, Q	non-singular matrices
r	convergence rate
$\mathbf{u}_i$	eigenvector
$\mathbf{v}$	arbitrary vector
X	subspace
$\rho(\mathbf{v})$	Rayleigh quotient

See also: **Commercial software**; **Computation for transient and impact dynamics**; **Krylov-Lanczos methods**.

Further Reading

Bathe KJ (1996) *Finite Element Procedures*. Englewood Cliffs, NJ: Prentice Hall.
Clough RW and Penzien J (1993) *Dynamics of Structures*. New York: McGraw-Hill.
Dahlquist G and Björck Å (1974) *Numerical Methods*. Englewood Cliffs, NJ: Prentice Hall.
Géradin M and Rixen D (1994) *Mechanical Vibrations: Theory and Application to Structural Dynamics*. New York: John Wiley.
Meirovitch L (1967) *Analytical Methods in Vibrations*. London: Macmillan.
Meirovitch L (1975) *Elements of Vibration Analysis*. New York: McGraw-Hill.
Wilkinson JH (1965) *The Algebraic Eigenvalue Problem*. Oxford: Clarendon Press.

ELECTRORHEOLOGICAL AND MAGNETORHEOLOGICAL FLUIDS

R Stanway, The University of Sheffield, Sheffield, UK

Copyright © 2001 Academic Press

doi:10.1006/rwvb.2001.0080

Semiactive Vibration Control

In the control of vibrations in various types of structures and machines it has long been recognized that performance benefits are available if damping levels can be optimized to suit a changing environment. It is also well established that variable damping can be implemented using a fully active vibration control strategy. However, there are severe penalties associated with the use of active control: complexity, weight, and cost are perhaps the most obvious. Consequently, in many applications it is necessary to pursue compromise solutions where control is exercised over an essentially passive damping mechanism. This latter form of control is often referred to as semiactive, although it can be argued that the phrase 'controlled passive' provides a more accurate description.

Traditionally the control of passive damping has required the introduction of some mechanism (typically an electromechanical solenoid) which alters the flow paths in an otherwise conventional viscous dashpot arrangement. If the chosen mechanism operates through opening and closing orifices which control fluid flow, then the effective damping can be varied in stepwise fashion. An alternative approach, made possible through the development of so-called smart fluids, involves modulating the energy dissipation characteristics of a damping device through an applied electric or electromagnetic field. Not only does the use of smart fluids offer an elegant solution to the problem of controlling damping levels but the control is continuous in form, as opposed to the stepwise variations obtainable using electromechanical switching.

Smart Fluids

Composition of Smart Fluids

There are two principal classes of smart fluid which can be harnessed for use as controllable vibration dampers: electrorheological (ER) and magnetorheological (MR) fluids. ER fluids generally consist of fine semiconducting particles dispersed in a liquid medium such as silicone oil. The particles are often roughly cylindrical in shape, with diameters chosen to lie in the range 5–75 μm. The carrier liquid is usually chosen to possess a kinematic viscosity in the range 10–50 cSt. The volume fraction of the particles suspended in the liquid carrier can be as high as 50%. There are no firm guidelines as to the choice of particle size, kinematic viscosity, or volume fraction; however, the ranges quoted above have been shown to produce fluids capable of producing operational ER fluids.

In contrast, MR fluids consist of a suspension of magnetically soft particles in a carrier liquid such as mineral or silicone oil. The particle sizes quoted in the literature are generally smaller than those used in ER fluids and lie in the range 0.1–10 μm. Operational MR fluids can be produced using the values of kinematic viscosity (for the carrier liquid) and volume fraction (of particles in suspension) quoted above.

Control of Flow Properties

The smart fluids described above are of direct interest to specialists in mechanical vibration as they offer an elegant means of fabricating damping devices capable of providing continuously variable levels of force. This ability arises from the almost instantaneous and reversible change in their resistance to flow which can be induced through the application of an electric or magnetic stimulus. ER fluids will respond to the application of an electric field while MR fluids require a magnetic field. In both cases the field causes the particles to form into chain-like structures. As the field strength is increased these chains eventually bridge the electrodes, thus significantly increasing the resistance to flow.

Figure 1A shows the orientation of particles in a smart fluid in the absence of an applied electric (or magnetic) field. The formation of particle chains which follows the application of a field strength of sufficient intensity is shown in **Figure 1B**.

In macroscopic terms, the behavior of smart fluids is often likened to that of the class of materials known as Bingham plastics. The shear stress versus shear rate characteristic of an ideal Bingham plastic is shown in **Figure 2**. With reference to **Figure 2**, a Bingham plastic effectively combines the yield-type behavior of a conventional solid with the Newtonian-type behavior of a viscous fluid. In the absence of an applied electric or magnetic field, smart fluids will generally behave like a Newtonian fluid. However, as the applied field is gradually increased, so a yield stress, denoted τ_y, will be established. In any device incorporating a smart fluid this yield stress must be overcome before flow can occur.

Harnessing Smart Fluids for Vibration Control

In order to construct a controllable damping device using smart fluids it is necessary to recognize the three possible modes of operation which may be utilized. These three modes – flow, shear and squeeze – are shown in **Figure 3**.

In the flow mode of operation, **Figure 3A**, the smart fluid is contained between a pair of stationary electrodes (or poles). The term 'electrode' will be used exclusively for the remainder of this article. The resistance to flow of the fluid is controlled by varying

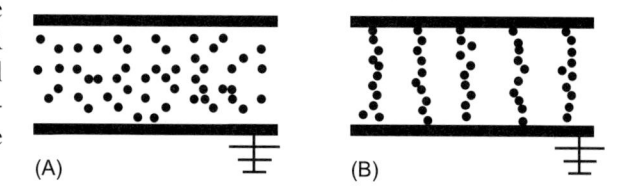

Figure 1 (A) Orientation of particles in a smart fluid in the absence of an applied electric or magnetic field. (B) Formation of particle chains following application of a field strength of sufficient intensity.

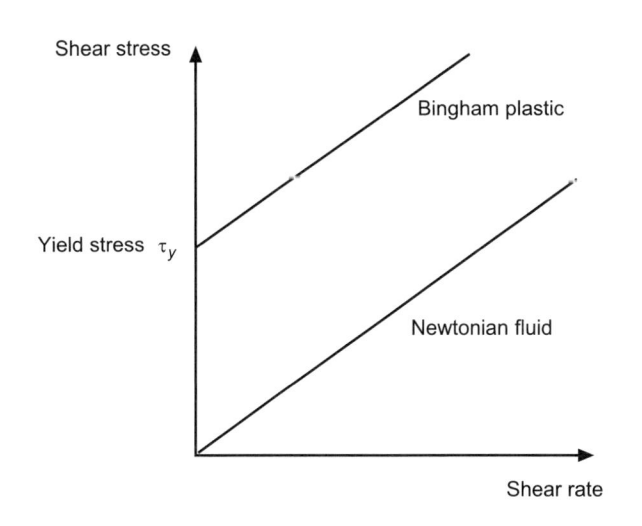

Figure 2 Shear stress versus shear rate characteristic of an ideal Bingham plastic and a Newtonian fluid.

the electric or magnetic field between the electrodes. Thus the field is used to modulate the pressure/flow characteristics of what is effectively a controllable valve. By using such a valve as a bypass, for example across a conventional hydraulic piston and cylinder arrangement, continuously variable control of the force/velocity characteristics can be obtained. Such devices are said to operate in the flow mode.

Alternatively, relative motion (either translational or rotational) can be introduced between the electrodes. Such motion places the smart fluid in shear, as illustrated in **Figure 3B**. As we have noted earlier, the shear stress/shear rate characteristics can be varied continuously through the applied field and thus we have the basis of a simple, controllable damping device, said to operate in the shear mode.

The third possibility for obtaining variable damping from smart fluids is shown in **Figure 3C**, the so-called squeeze-flow mode of operation. Here the electrodes are free to translate in a direction roughly parallel to the direction of the applied field. Consequently the smart fluid can be subjected to alternate tensile and compressive loading. Shearing of the fluid also occurs. Through this mechanism larger forces are available than with flow or shear devices but displacement levels are limited to no more than a few millimeters.

Modeling Smart Fluids for Vibration Control

Macroscopic Models

At the time of writing considerable efforts are being directed towards the development of mathematical models of smart fluids. These models are predominantly macroscopic in nature and are being developed both as aids to understanding the behavior of smart fluids and to assist in the design of suitable control systems. There is insufficient space here to present a survey of the various modeling techniques which are available. However, it is helpful to present a quasi-static approach to modeling a flow-mode vibration damper and then summarize the extension of the model to account for dynamic effects in the smart field. Here the main purpose of the model is to assist us in visualizing the physical behavior of the controllable damper.

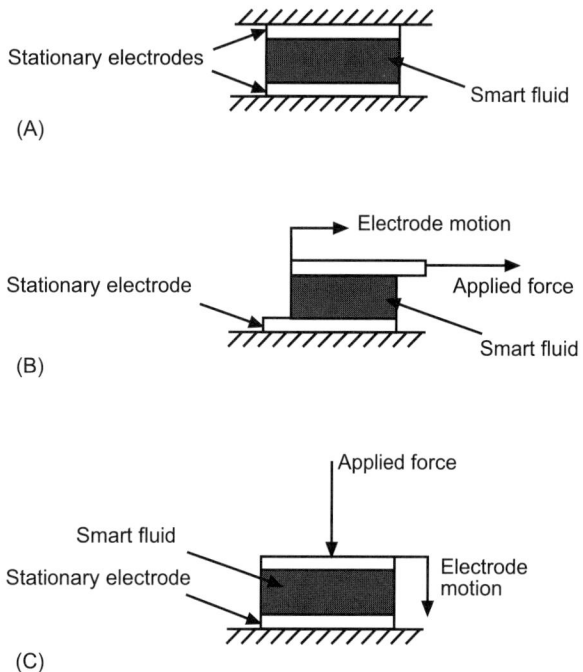

Figure 3 The three modes of operation utilized to construct a controllable damping device using smart fluids. (A) Flow, (B) shear, (C) squeeze.

Physical Arrangement

Figure 4 shows a controllable damper where a smart valve is used as a bypass across a hydraulic piston and cylinder. Assume that the valve has a single annulus of length l containing the smart fluid. Denote the annular gap by h and note that the ratio of diameter d to gap h is so large that the valve's behavior may be modeled as similar to that of two flat plates of breadth b ($= \pi d$). The smart fluid around the hydraulic circuit has viscosity μ and density ρ. The fluid in the annulus of the valve is subjected to either an electric or magnetic field, which results in the development of a yield stress τ_y in the fluid, as shown in **Figure 2**.

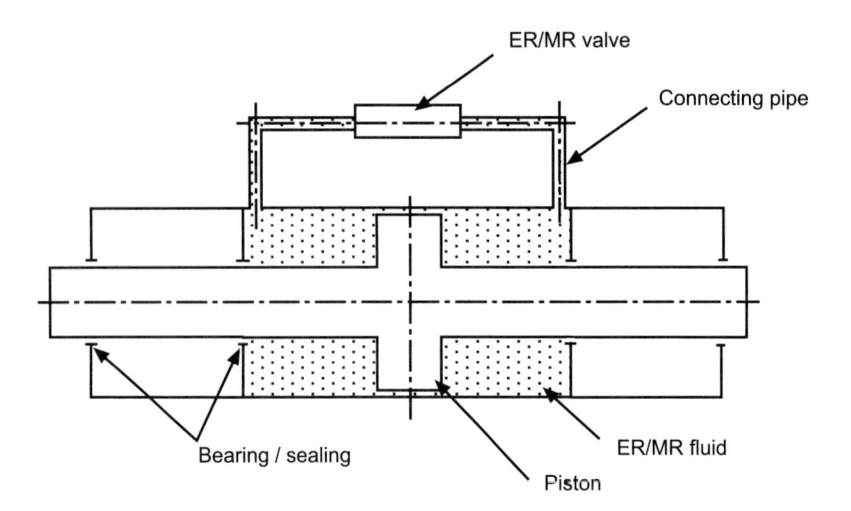

Figure 4 A controllable damper, using a smart valve as a bypass across a hydraulic piston and cylinder.

A Quasisteady Model of a Smart Control Valve

The quasisteady operation of the smart fluid control valve can be described in terms of three dimensionless parameters. If τ_w is the shear stress at the electrode surface and $\bar{u}$ is the mean velocity of flow between the electrodes then a dimensionless friction coefficient:

$$\pi_1 = \tau_w / (\rho \bar{u}^2) \qquad [1]$$

can be defined. The Reynolds number associated with the flow of fluid is defined as:

$$\pi_2 = (\rho \bar{u} b)/\mu \qquad [2]$$

Finally the influence of the electric or magnetic field on the behavior of the valve is characterized through the so-called Hedström number:

$$\pi_3 = \frac{\tau_y \rho b^2}{\mu^2} \qquad [3]$$

The three dimensionless groups defined by eqns [1]–[3] allow us to visualize the controlling influence of the applied field on ER valve performance.

It is assumed that the effective working area of the piston is A and that the piston translates within the cylinder at a steady velocity of v. This piston data allow the volume flow rate to be calculated and then given the valve dimensions l, b and h, then $\bar{u}$, the mean velocity of fluid flow through the valve can be calculated. At this stage the corresponding value of the Reynolds number, π_2, can be established.

The next step is to establish a value for the Hedström number, π_3. Initially a static value is estimated from a value of τ_y established from static tests on the smart fluid. This value is then corrected to allow for the presence of fluid flow. Given values for the dimensionless parameters π_2 and π_3, the dimensionless friction coefficient can be computed by solving the well established cubic equation for Bingham plastic flow:

$$f(\pi_1) = \pi_1^3 - \left[\frac{3}{2} + 6\frac{\pi_2}{\pi_3}\right]\left[\frac{\pi_3}{\pi_2^2}\right]\pi_1^2 + \frac{1}{2}\left[\frac{\pi_3}{\pi_2^2}\right] = 0 \qquad [4]$$

A graphical interpretation of eqn [4] is facilitated by defining a further dimensionless parameter:

$$\pi^* = \frac{\pi_1 \pi_2^2}{\pi_3} \qquad [5]$$

where $\pi^* = \tau_w/\tau_y$, i.e., the ratio of shear stress at the electrode wall to the yield stress developed within the smart fluid. For flow to occur the shear stress developed at the electrode wall must obviously exceed the smart fluid's yield stress and this enables a physical interpretation of the solutions of eqn [4]. In terms of π^*, eqn [4] is rewritten as:

$$f(\pi^*) = (\pi^*)^3 - \left[\frac{3}{2} + 6\left(\frac{\pi_2}{\pi_3}\right)\right](\pi^*)^3 + \frac{1}{2} = 0 \qquad [6]$$

The condition $\pi_2/\pi_3 \to 0$ represents the limiting case which distinguishes between flow occurring and the absence of flow. The condition $\pi_2/\pi_3 \to 0$ is approached by increasing the applied electric field, E, which in turn increases the Bingham yield stress,

τ_y, and thus the Hedström number, π_3. Setting $\pi_2/\pi_3 = 0$ in eqn [6] results in:

$$f(\pi^*) = (\pi^*)^3 - \frac{3}{2}(\pi^*)^2 + \frac{1}{2} = 0 \qquad [7]$$

which has only one physically meaningful solution. This solution can be found by plotting $f(\pi^*)$ vs π^*, as shown in **Figure 5**. With reference to **Figure 5**, there is a double root at $\pi^* = 1$, representing the case where the wall stress is equal to the Bingham plastic yield stress. Any reduction in π_3 will obviously cause fluid flow to occur. Two such cases are superimposed on **Figure 5**; these are for $\pi_2/\pi_3 = 0.05$ and 0.1. For all three plots, one root is clearly negative and therefore meaningless as a ratio of stresses. The double root is created when $\pi_2/\pi_3 = 0$ splits with one root increasing and the other decreasing. From the definition of $\pi^*(=\tau_w/\tau_y)$, only the root greater than unity will give rise to fluid flow and thus represents a meaningful solution.

When the wall shear stress is exactly equal to the Bingham plastic yield stress, i.e., $\tau_w = \tau_y$ and thus $\pi^* = 1$, eqn [5] defines asymptotes which represent limiting cases. **Figure 6** shows a plot of friction coefficient π_1 against Reynolds number π_2. The asymptotes for $\pi_3 = 10, 100,$ and 1000 are shown. When the Bingham yield stress is reduced such that π_3 approaches zero and thus π_2/π_3 approaches infinity, then it can be shown that $\pi_1\pi_2 \to 6$, which corresponds to the well-known relationship between π_1 and π_2 for Newtonian flow between smooth flat plates. The asymptote corresponding to $\pi_1\pi_2 = 6$ is superimposed on to **Figure 6**.

Figure 6 enables the operation of the ER flow control valve to be visualized in terms of three dimensionless groups. The asymptotes on **Figure 6** illustrate the limits of operation at both low and high values of the Hedström number π_3. Perhaps most importantly, the graph shows how the Hedström number influences the mapping of the Reynolds number π_2 on to the friction coefficient π_1. Furthermore, it enables the force/velocity characteristics of a valve-controlled smart vibration damper to be predicted.

Prediction of Force/Velocity Characteristics

The force/velocity characteristics of the smart valve-controlled vibration damper (shown in **Figure 6**), under steady flow conditions, can be found using eqns [1]–[4]. The calculation is started by specifying the steady value of the piston velocity and the piston area. For a given valve configuration this enables the volume flow rate, Q, to be calculated and hence the mean velocity ($\bar{u} = Q/bh$). A numerical value for the corresponding Reynolds number follows directly from the definition in eqn [2]. For a given value of electric or magnetic field strength the yield stress, τ_y, of the smart fluid is computed and then eqn [3] gives

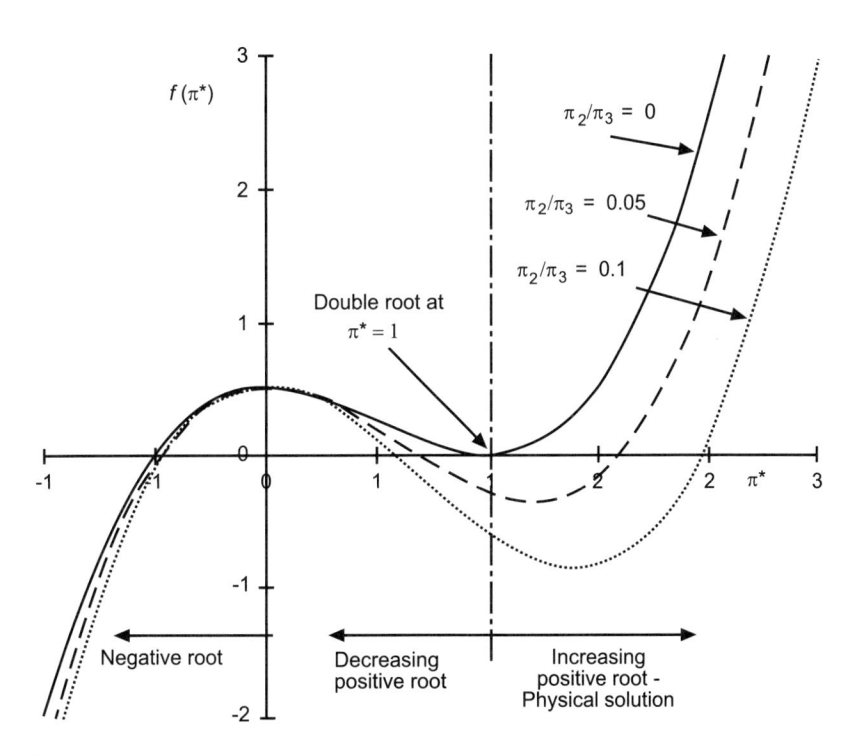

Figure 5 Plot of $f(\pi^*)$ vs π.

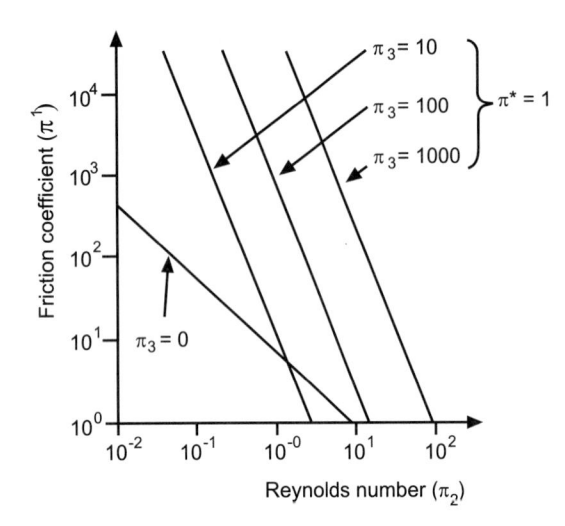

Figure 6 Plot of friction coefficient π_1 against Reynolds number π_2.

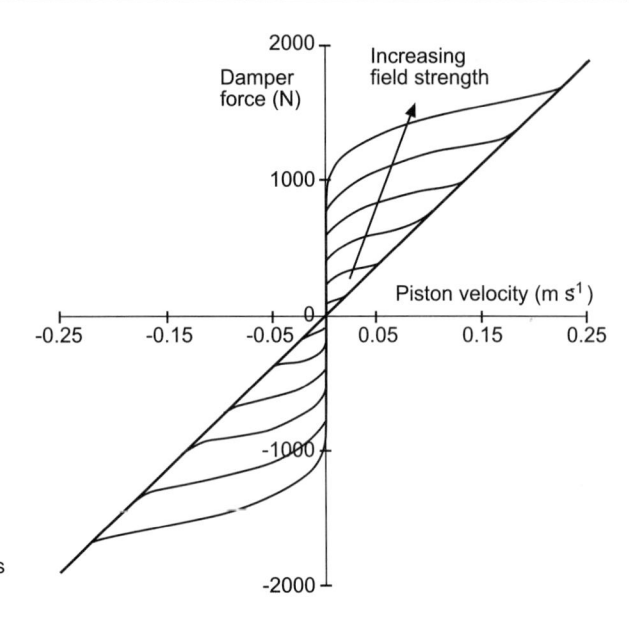

Figure 7 The general form of force/velocity characteristics.

the Hedström number, π_3. The numerical values of π_2 and π_3 are then substituted into eqn [4] and a suitable root-solving routine is applied to determine the three possible values of the friction coefficient, π_1.

The physically meaningful value of π_1 is used to calculate the pressure drop across the valve and thus across the piston. (The influence of the connecting pipes is assumed to be negligible.) The piston force follows by multiplying the pressure by the piston area. The piston force can then be plotted against the value of the piston velocity which was used to start the calculation. The procedure is repeated for the desired range of values of piston velocity so as to generate a complete set of force/velocity characteristics. The general form of the force/velocity characteristics generated in this way is illustrated in **Figure 7**.

Inclusion of Fluid Dynamic Effects

It has now been established that a quasisteady model can provide realistic predictions of a valve-controlled smart vibration damper. However, experimental studies involving excitation frequencies of (say) 1 Hz and above reveal features which cannot be accounted for by the use of quasisteady model. These features arise from the presence of dynamic effects – notably inertia and compressibility of the smart fluid – which can play a significant role in the operation of an experimental device.

A simple lumped arrangement, which has been shown capable of accounting for dynamic effects, is shown in **Figure 8**. The model is derived on the basis of a number of assumptions, specifically: the effective inertia of the smart fluid in the cylinder, connecting pipes and control valve is denoted by m; the compressibility of the fluid in the cylinder is lumped with

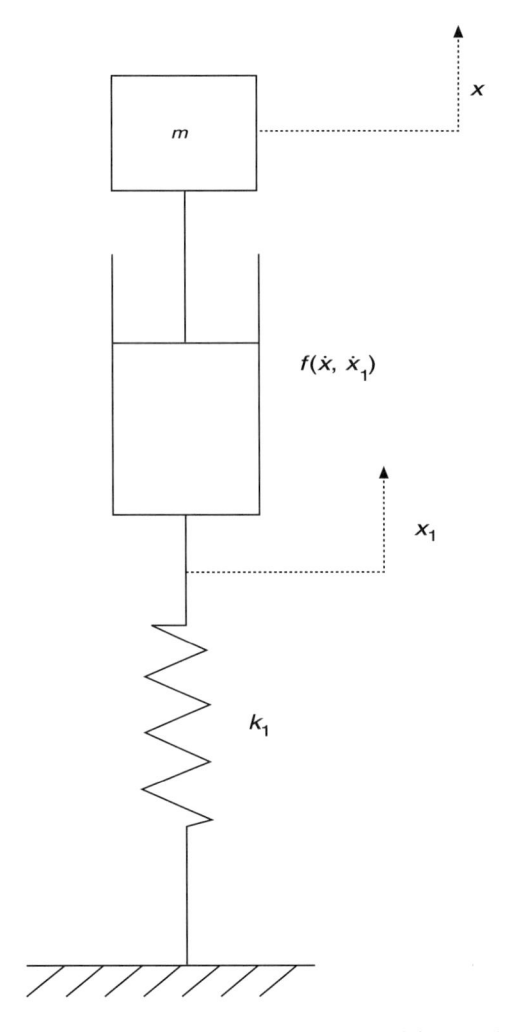

Figure 8 A simple lumped arrangement model, capable of accounting for dynamic effects.

other such effects and denoted by the spring constant k_1; and the resistance to flow is represented by a nonlinear function of velocity, $f(\dot{x}, \dot{x}_1)$, where $\dot{x}_1$ is the velocity associated with the spring element k_1.

Given these assumptions and with reference to **Figure 8**, the equations of motion are:

$$\left.\begin{array}{r}m\ddot{x} + f(\dot{x}, \dot{x}_1) = F \\ -f(\dot{x}, \dot{x}_1) + k_1 x_1 = 0\end{array}\right\} \quad [8]$$

where F is the net piston force.

Owing to the presence of the nonlinear function $f(\dot{x}, \dot{x}_1)$, the solution of eqn [8] requires the use of an iterative numerical procedure. Numerical values for the fluid inertia and stiffness can be calculated from purely theoretical considerations. However, it has been found that agreement between model predictions and experimental data is dramatically improved if the inertia and stiffness values are updated to reflect experimental observations.

At relatively low frequencies (say, up to 3 Hz) the influence of smart fluid dynamics on the force/velocity characteristics is not significant. However, above 3 Hz the dynamics play an increasingly important role. As an example, **Figure 9** shows a typical performance prediction at a mechanical excitation frequency of 10 Hz. Note the presence of the hysteresis loop due to the fluid's compressibility and the additional loops at the higher levels of force and velocity. Experimental studies involving valve-controlled smart dampers show broad agreement with the results in **Figure 9** but there is still enormous scope for improvements in modeling techniques, both to improve our basic understanding and as a basis for control system design.

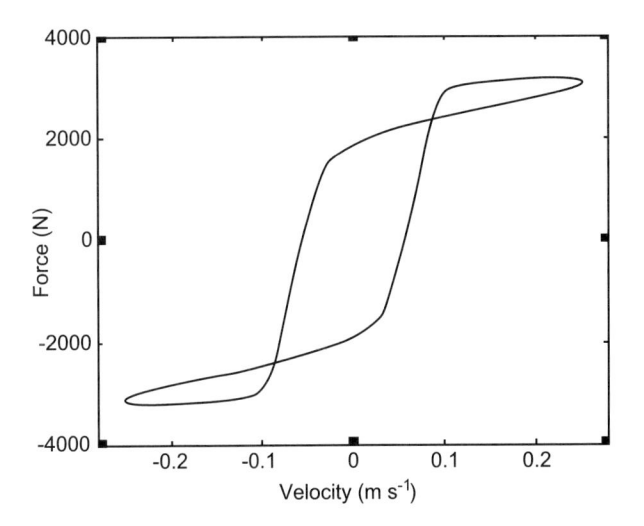

Figure 9 Typical performance prediction at a mechanical excitation.

Current Developments in Smart Fluids

Squeeze-flow Devices

Valve-controlled smart dampers, described in the previous section, are suitable for applications where substantial displacements need to be accommodated, for example, in road and rail vehicle suspension systems. However, there are many potential applications where machines and mechanisms need to be isolated from vibration but where the displacement levels are relatively small.

Squeeze-flow devices are capable of providing large force levels over small displacement ranges, say up to 5 mm. Moreover the construction of squeeze-flow devices can be extremely simple. If ER fluid is to be used, then two plane electrodes will suffice as the interface between the mechanical components and the ER fluid. Published experimental results from various sources have shown that the force levels available are more than adequate for use in, for example, automotive engine mount applications. If a MR fluid is used as the working medium then complexity and weight are increased owing to the requirement for an electromagnet to excite the MR fluid, but the resulting devices are still remarkably compact.

There are factors which, at first sight, appear to detract from the performance of squeeze-flow devices. Unlike conventional flow- and shear-mode devices, in squeeze-flow the gap between the electrodes or poles is liable to be constantly changing as a result of the applied mechanical excitation. Furthermore, placing the fluid in tension on each alternate stroke is likely to lead to cavitation and produce force levels which bear no relation to the forces produced by compression of the fluid. Fortunately there are ways to counteract problems which arise as a result of squeeze-flow operation. For example, a displacement feedback loop can be used to provide a constant electric or magnetic field irrespective of the instantaneous gap size. Also there is considerable scope for ingenuity in the design of the electrode/pole configurations: arranging for smart fluid to act on opposite faces of the moving electrode/pole serves to compensate for the asymmetry of operation between tension and compression.

Modeling and Control

Before the development of dynamic models to account for the behavior of smart fluids, static models were used, mainly for the rough sizing of components at the design stage. However, given the generally crude formulations of earlier generations of smart fluids, experimental verification was invariably essential. In the recent development of dynamic models, emphasis has been placed on the necessity to account for observed behavior and to behave robustly in the face of environmental changes. In the modeling technique described in the previous section the intention was to develop a characterization wherein the performance of industrial scale devices could be predicted on the basis of fluid data derived using laboratory-scale apparatus. The feasibility of this form of performance prediction has now been established. Taken together with the findings of various other leading groups in this field it can be stated with confidence that the modeling of smart fluids is gradually maturing. Robust models are now available which not only account for observed behavior, but also help to improve our understanding of the operation of smart fluids.

As smart fluid damping devices are developed and further applications uncovered (see later), it is reasonable to suggest the increasing effort will be directed at the development of suitable control schemes. At the time of writing it is far from clear which direction (if any) controller design will take. Various approaches to feedback control system design are currently being pursued, ranging from formal schemes which draw heavily on modern control theory to heuristic controllers designed to suit specific applications. Where the disturbance is known or can be measured then gain scheduling (or feedforward control) offers an alternative approach which is being pursued by some groups. Amongst the simpler possibilities is the use of feedback strategies to linearize the force/velocity characteristics associated with smart fluids. A configuration involving force feedback of a valve-controlled ER damper is shown in **Figure 10**. Numerical simulation studies indicated the feasibility of this approach and experimental confirmation has now been obtained for input frequencies of up to 5 Hz. Higher bandwidths may require additional ingenuity in the design of control elements.

New Areas of Application

We conclude with a brief review of new applications of ER/MR fluids which have recently been identified. In the 1970s and 1980s it was the aerospace and automotive industries which were the first to identify potential applications of smart fluids. Aerospace companies were certainly deterred from using ER fluids by the requirement to provide a high-voltage supply to excite the fluid. Now that the advent of MR fluids has obviated this requirement, investigations into aerospace applications have resumed: aircraft landing gear is an obvious target for MR dampers and the feasibility of a controllable lag mode damper in a helicopter is also being studied. In the automotive field smart dampers for vehicle suspensions and engine mounts continue to be developed. In addition, vehicle seats incorporating a smart fluid suspension have been developed to the mass production stage.

In the past 5 years there has been a significant research effort to develop smart fluids for civil engineering applications. Perhaps greatest emphasis has been placed upon the use of MR dampers to reduce the vulnerability of base-isolated structures when subjected to seismic disturbances. Considerable progress has been made in terms of device development, modeling, and the design of feedback controls. In the general structural field, research into composite beams which include a smart fluid layer has produced some exploitable results. In principle, the smart fluid layer provides controllable damping but ingenuity in the mechanical design of the beam is essential if the increase in damping forces with electric/magnetic field is to be significant.

One promising application area which has been slow to respond to the possibilities offered by smart fluids is robotics and other automated machining and assembly operations. It has been demonstrated that smart fluids form a basis for simple but versatile rotary actuators. Such actuators can provide accurate control of torque and position and can be designed in such a way as to act as a torsional vibration absorber. The large force/small displacement characteristics of squeeze-flow devices is compatible with many vibration isolation problems inherent in modern machinery, but reports on the development of industrial applications are still awaited.

Further Reading

Bullough WA (ed.) (1996) Electro-rheological fluids, magneto-rheological suspensions and associated technology. *Proceedings of the 5th International Conference*, 10–14 July 1995, Sheffield, UK. Singapore: World Scientific.

Culshaw B (1996) *Smart Structures and Materials*. Norwood, MA: Artech House.

Dyke SJ, Spencer BF, Sain MK, Carlson JD (1996) Modelling and control of MR dampers for seismic response reduction. *Smart Materials and Structures* 5: 565–575.

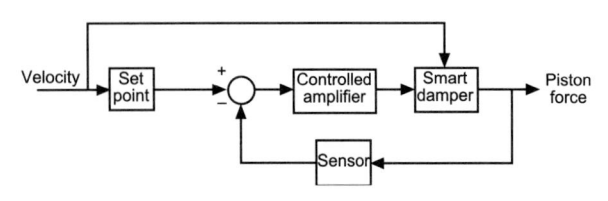

Figure 10 Force feedback of a valve-controlled ER damper.

Gandhi MV, Thomson BS (1992) *Smart Materials and Structures*. London: Chapman and Hall.

Inman DJ (1989) *Vibration Control, Measurement and Stability*. Englewood Cliffs, Prentice Hall.

Jolly MR, Carlson JD, Muñoz BC (1996) A model of the behaviour of MR materials. *Smart Materials and Structures* 5: 607–614.

Kamath GM, Wereley NM (1997) Modeling the damping mechanism in electrorheological fluid based dampers. In Wolfenden A, Kinra VK (eds) *Mechanics and Mechanisms of Material Damping*, ASTM STP 1304, pp. 331–348. West Conshohocken, PA: ASTM.

Siginer DA, Dulikravich GS (1995) *Developments in Electrorheological Flows*, FED, vol. 235. New York: ASME.

Sims ND, Stanway R, Johnson AR (1999) Vibration control using smart fluids: a state-of-the-art review. *Shock and Vibration Digest* 31: 195–203.

Stanway R, Sproston JL, El-Wahed AK (1996) Applications of electrorheological fluids in vibration control: a survey. *Smart Materials and Structures* 5; 464–481.

Tzou HS, Anderson GL (1992) *Intelligent Structural Systems*. Dordrecht: Kluwer.

Wilkinson WL (1960), *Non-Newtonian Fluids*. Oxford: Pergamon Press.

ELECTROSTRICTIVE MATERIALS

K Uchino, The Pennsylvania State University, University Park, PA, USA

Copyright © 2001 Academic Press

doi:10.1006/rwvb.2001.0078

Introduction

The word electrostriction is used in a general sense to describe electric field-induced strain, and hence frequently also implies the converse piezoelectric effect. However, in solid-state theory, the converse piezoelectric effect is defined as a primary electromechanical coupling effect, i.e., the strain is proportional to the applied electric field, while electrostriction is a secondary coupling in which the strain is proportional to the square of the electric field. Thus, strictly speaking, they should be distinguished. However, the piezoelectricity of a ferroelectric which has a centrosymmetric prototype (high-temperature) phase is considered to originate from the electrostrictive interaction, and hence the two effects are related.

In this section, first the origin of piezoelectricity and electrostriction are explained microscopically and phenomenologically. Then, electrostrictive materials are described in detail, and finally their applications are introduced.

Microscopic Origins of Electrostriction

Solids, especially ceramics (inorganic materials), are relatively hard mechanically, but still expand or contract depending on the change of the state parameters. The strain (defined as the displacement ΔL/initial length L) caused by temperature change and stress is known as thermal expansion and elastic deformation. In insulating materials, the application of an electric field can also cause deformation. This is called electric field-induced strain.

Why a strain is induced by an electric field is explained herewith. For simplicity, let us consider an ionic crystal such as NaCl. **Figure 1** shows a one-dimensional rigid-ion spring model of the crystal lattice. The springs represent equivalently the cohesive force resulting from the electrostatic Coulomb energy and the quantum mechanical repulsive energy. **Figure 1B** shows the centrosymmetric case, whereas **Figure 1A** shows the more general noncentrosymmetric case. In **Figure 1B**, the springs joining the ions are all the same, whereas in **Figure 1A**, the springs joining the ions are different for the longer or shorter ionic distance, in other words, hard and soft springs existing alternately are important. Next, consider the state of the crystal lattice (**Figure 1A**) under an applied electric field. The cations are drawn in the direction of the electric field and the anions in the opposite direction, leading to the relative change in the interionic distance. Depending on the direction of the electric field, the soft spring expands or contracts more than the contraction or expansion of the hard spring, causing a strain x (i.e., unit cell length change) in proportion to the electric field E. This is the converse piezoelectric effect. When expressed as:

$$x = dE \qquad [1]$$

the proportionality constant d is called the piezoelectric constant.

On the other hand, in **Figure 1B**, the amounts of extension and contraction of the spring are nearly the

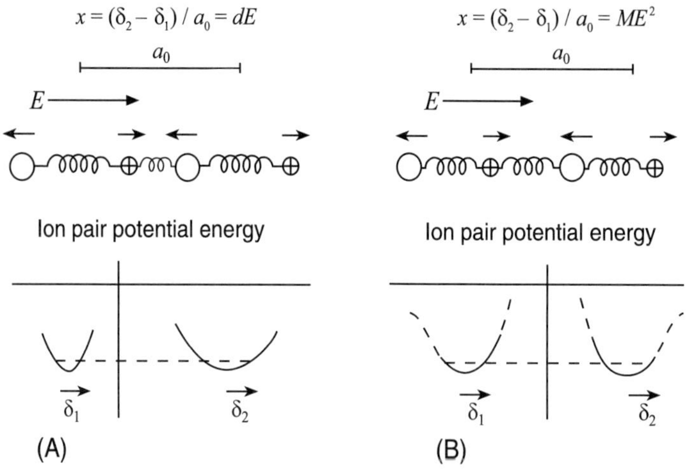

Figure 1 Microscopic explanation of (A) piezostriction and (B) electrostriction.

same, and the distance between the two cations (i.e., lattice parameter) remains almost the same, hence there is no strain. However, more precisely, ions are not connected by such idealized springs (those are called harmonic springs, in which (force F) = (spring constant k) × (displacement Δ) holds). In most cases, the springs possess anharmonicity ($F = k_1\Delta - k_2\Delta^2$), i.e., they are somewhat easy to extend, but hard to contract. Such subtle differences in displacement cause a change in the lattice parameter, producing a strain which is independent of the direction of the applied electric field, and which is an even function of the electric field. This is called the electrostrictive effect, and can be expressed as:

$$x = ME^2 \quad [2]$$

where M is the electrostrictive constant.

Phenomenology of Electrostriction

Devonshire Theory

In a ferroelectric whose prototype phase (high-temperature paraelectric phase) is centrosymmetric and nonpiezoelectric, the piezoelectric coupling term PX is omitted and only the electrostrictive coupling term P^2X is introduced into the phenomenology (P and X are polarization and stress). This is almost accepted for discussing practical electrostrictive materials with a perovskite structure. The theories for electrostriction in ferroelectrics were formulated in the 1950s by Devonshire and Kay. Let us assume that the elastic Gibbs energy should be expanded in a one-dimensional form:

$$G_1(P, X, T) = (1/2)\alpha P^2 + (1/4)\beta P^4 + (1/6)\gamma P^6 \\ - (1/2)sX^2 - QP^2X \quad [3]$$

$$\alpha = (T - T_0)/\varepsilon_0 C \quad [4]$$

where P, X, T are polarization, stress, and temperature, respectively, and s and Q are called the elastic compliance and the electrostrictive coefficient. This leads to eqns [5] and (6) for the electric field E and strain x:

$$E = (G_1/P) = \alpha P + \beta P^3 + \gamma P^5 - 2QPX \quad [5]$$

$$x = -(G_1/X) = sX + QP^2 \quad [6]$$

Case 1: $X = 0$ When an external is zero, the following equations are derived:

$$E = \alpha P + \beta P^3 + \gamma P^5 \quad [7]$$

$$x = QP^2 \quad [8]$$

$$1/\varepsilon_0\varepsilon = \alpha + 3\beta P^2 + 5\gamma P^4 \quad [9]$$

If the external electric field is equal to zero ($E = 0$), two different states are derived:

$$P = 0 \quad \text{and} \quad P^2 = \sqrt{(\beta^2 - 4\alpha\gamma - \beta)/2\gamma}$$

1. Paraelectric phase: $P_S = 0$ or $P = \varepsilon_0\varepsilon E$ (under small E)

$$\text{Permittivity}: \quad \varepsilon = C/(T - T_0) \\ \text{(Curie - Weiss law)} \quad [10]$$

$$\text{Electrostriction}: \quad x = Q\varepsilon_0^2\varepsilon^2 E^2 \quad [11]$$

Therefore, the previously mentioned electrostrictive coefficient M in eqn [2] is related to the electrostrictive Q coefficient through:

$$M = Q\varepsilon_0^2\varepsilon^2 \quad [12]$$

Note that the electrostrictive M coefficient has a large temperature dependence like $\propto 1/(T-T_0)^2$, supposing that Q is almost constant.

2. Ferroelectric phase: $P_S^2 = \left(\sqrt{\beta^2 - 4\alpha\gamma} - \beta\right)/2\gamma$

or $P = P_S + \varepsilon_0\varepsilon E$ (under small E)

$$\begin{aligned} x &= Q(P_S + \varepsilon_0\varepsilon E)^2 \\ &= QP_S^2 + 2\varepsilon_0\varepsilon QP_S E + Q\varepsilon_0^2\varepsilon^2 E^2 \end{aligned} \quad [13]$$

$$\text{Spontaneous strain}: \quad x_S = QP_S^2 \quad [14]$$

$$\text{Piezoelectric constant}: \quad d = 2\varepsilon_0\varepsilon QP_S \quad [15]$$

Thus, we can understand that piezoelectricity in a perovskite crystal is equivalent to the electrostrictive phenomenon biased by the spontaneous polarization.

Case 2: $X \neq 0$ When a hydrostatic pressure $p(X = -p)$ is applied, the inverse permittivity is changed in proportion to p:

$1/\varepsilon_0\varepsilon = \alpha + 3\beta P^2 + 5\gamma P^4 + 2Qp$
(ferroelectric state)
$1/\varepsilon_0\varepsilon = \alpha + 2Qp = (T - T_0 + 2Q\varepsilon_0 Cp)/(\varepsilon_0 C)$
(paraelectric state)

[16]

Therefore, the pressure dependence of the Curie–Weiss temperature T_0 or the transition temperature T_C is derived as follows:

$$(T_0/p) = (T_C/p) = -2Q\varepsilon_0 C \quad [17]$$

In general, the ferroelectric Curie temperature is decreased with increasing hydrostatic pressure (i.e., $Q_h > 0$).

Converse Effect of Electrostriction

So far, we have discussed the electric field-induced strains, i.e., piezoelectric strain (converse piezoelectric effect, $x = dE$) and electrostriction (electrostrictive effect, $x = ME^2$). Let us consider here the converse effect, i.e., the response to the external stress, which is applicable to sensors. The direct piezoelectric effect is the increase of the spontaneous polarization by an external stress, and is expressed as:

$$\Delta P = dX \quad [18]$$

In contrast, since the electrostrictive material does not have spontaneous polarization, it does not exhibit any charge under stress, but changes permittivity (see eqn [16]):

$$\Delta(1/\varepsilon_0\varepsilon) = 2QX \quad [19]$$

This is the converse electrostrictive effect.

Temperature Dependence of Electrostrictive Coefficient

Several expressions for the electrostrictive coefficient Q have been given so far. From the data obtained by independent experimental methods such as:

1. electric field-induced strain in the paraelectric phase
2. spontaneous polarization and spontaneous strain (X-ray diffraction) in the ferroelectric phase
3. d constants from the field-induced strain in the ferroelectric phase or from the piezoelectric resonance
4. pressure dependence of permittivity in the paraelectric phase

nearly equal values of Q were obtained. **Figure 2** shows the temperature dependence of the electrostrictive coefficients Q_{33} and Q_{31} of the complex perovskite $Pb(Mg_{1/3}Nb_{2/3})O_3$, whose Curie temperature is near 0 °C. It is seen that there is no significant anomaly in the electrostrictive coefficient Q through the temperature range from a paraelectric to a ferroelectric phase, in which the piezoelectricity appears. Q is almost temperature-independent.

Electrostriction in Oxide Perovskites

Perovskites and Complex Perovskites

Among the practical piezoelectric/electrostrictive materials, many have the perovskite-type crystal structure ABO_3. This is because many such materials undergo a phase transition on cooling from a high-symmetry high-temperature phase (cubic paraelectric phase) to a noncentrosymmetric ferroelectric phase. Materials with a high ferroelectric transition temperature (Curie temperature) show piezoelectricity at room temperature, whereas those with a transition

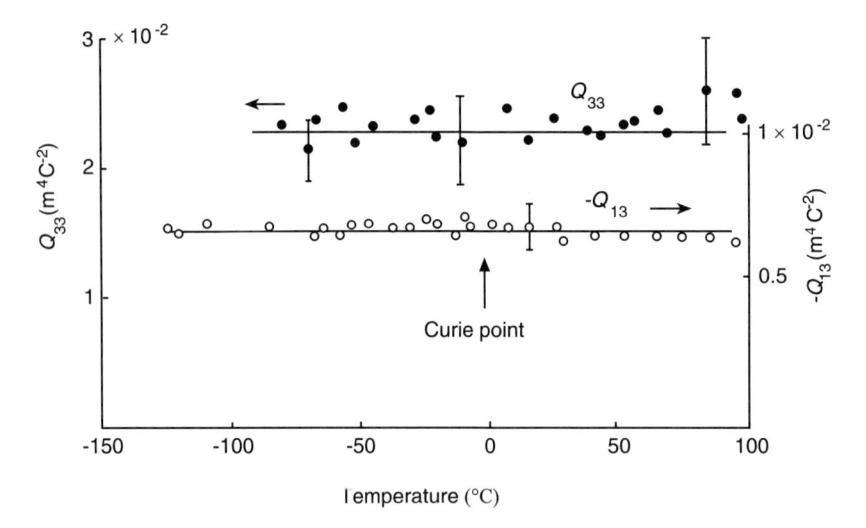

Figure 2 Temperature dependence of the electrostrictive constants Q_{33} and Q_{13} in $Pb(Mg_{1/3}Nb_{2/3})O_3$.

temperature near or below room temperature exhibit the electrostrictive effect. For the latter, at a temperature right above the Curie temperature, the electrostriction is extraordinarily large because of the large anharmonicity of the ionic potential. Besides, simple compounds such as barium titanate ($BaTiO_3$) and lead zirconate ($PbZrO_3$), solid solutions such as $A(B, B')O_3$, and complex perovskites such as $A^{2+}(B^{3+}_{1/2} B'^{5+}_{1/2})O_3$ and $A^{2+}(B^{2+}_{1/3} B'^{5+}_{2/3})O_3$ can be easily formed; such flexibility is important in materials design. **Figure 3** shows the crystal structures of the above-mentioned complex perovskites with B-site ordering. When B and B' ions are randomly distributed, the structure becomes a simple perovskite.

Electrostrictive Effect in Simple Perovskites

Yamada has summarized the electromechanical coupling constants of not only perovskite-type oxides but also tungsten bronze and $LiNbO_3$ types, which contain oxygen octahedra. Following the DiDomenico–Wemple treatment for the electrooptic effect, by expressing the electrostrictive tensor Q_{ijkl} of $LiNbO_3$ with a trigonal symmetry $3m$ in a coordinate system based on the fourfold axis of the oxygen octahedron, and making a correction for the packing density ξ by:

$$Q^P_{ijkl} = \xi^2 Q_{ijkl} \quad [20]$$

Yamada obtained the electrostrictive coefficient Q^P normalized to a perovskite unit cell. It could be concluded that the electrostrictive coefficient Q^Ps of the simple perovskite-type oxides with one kind of B ion have nearly equal values. The average values are:

$$Q^P_{11} = 0.10 \, m^4 \, C^{-2}$$
$$Q^P_{12} = -0.034 \, m^4 \, C^{-2} \quad [21]$$
$$Q^P_{44} = 0.029 \, m^4 \, C^{-2}$$

Electrostrictive Effect in Complex Perovskites

Uchino et al. extended these investigations to complex perovskite-type oxides to obtain the electrostrictive coefficient Q_h and Curie–Weiss constant C. The results are summarized in **Table 1**. It is important to note that the magnitude of the electrostrictive coefficient does not depend on the polar state, whether it is ferroelectric, antiferroelectric, or paraelectric, but strongly depends on the crystal structure, such as whether the two kinds of B and B' ions are randomly distributed in the octahedron or ordered like $B - B' - B - B'$ (1:1 ordering). The electrostrictive coefficient Q increases with increasing degree of cation order and follows the sequence disordered, partially ordered, simple, and finally ordered-type perovskites. For the polar materials, their Curie–Weiss constants are also listed in **Table 1**, showing a completely opposite trend to the Q_h values. It was consequently found that the invariant for the complex perovskite-type oxides is not Q itself, but the product of the electrostrictive coefficient and the Curie–Weiss constant (Uchino's constant):

$$Q_h C = 3.1(\pm 0.4) \times 10^3 \, m^4 \, C^{-2} \, K \quad [22]$$

This QC constant rule can be understood intuitively, if we accept the assumption that the material whose dielectric constant changes easily with pressure also exhibits a large change of the dielectric constant

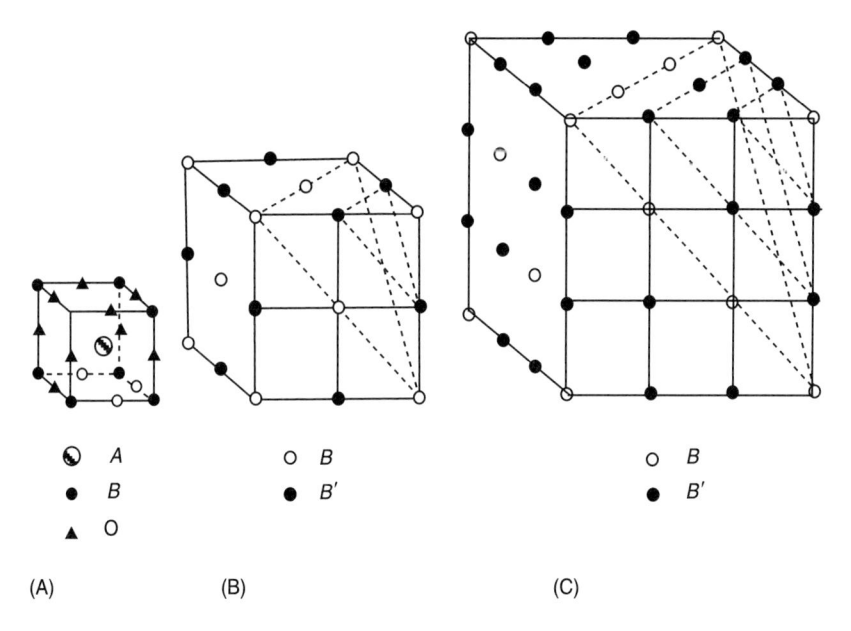

Figure 3 Complex perovskite structures with various B ion arrangements: (A) simple: ABO_3; (B) 1:1 ordered: $AB_{1/2}B'_{1/2}O_3$; (C) ordered: $AB_{1/3}B'_{2/3}O_3$.

Table 1 Electrostrictive coefficient Q_h and Curie–Weiss constant C for various perovskite crystals

Ordering	Material	Q_h ($\times 10^{-2} m^4 C^{-2}$)	C ($\times 10^5 K$)	$Q_h C$ ($\times 10^3 m^4 C^{-2} K$)
Ferroelectric				
Disorder	$Pb(Mg_{1/2}Nb_{2/3})O_3$	0.60	4.7	2.8
	$Pb(Zn_{1/3}Nb_{2/3})O_3$	0.66	4.7	3.1
Partial order	$Pb(Sc_{1/2}Ta_{1/2})O_3$	0.83	3.5	2.9
Simple	$BaTiO_3$	2.0	1.5	3.0
	$PbTiO_3$	2.2	1.7	3.7
	$SrTiO_3$	4.7	0.77	3.6
	$KTiO_3$	5.2	0.5	2.6
Antiferroelectric				
Partial order	$Pb(Fe_{2/3}U_{1/3})O_3$		2.3	
Simple	$PbZrO_3$	2.0	1.6	3.2
Order	$Pb(Ca_{1/2}W_{1/2})O_3$		1.2	
	$Pb(Mg_{1/2}W_{1/2})O_3$	6.2	0.42	2.6
Nonpolar				
Disorder	$(K_{3/4}Bi_{1/4})(Zn_{1/6}Nb_{5/6})O_3$	0.55–1.15		
Simple	$BaZrO_3$	2.3		

with temperature, i.e., the proportionality between the following two definitions:

$$Q_h = [\partial(1/\varepsilon)/\partial p]/2\varepsilon_0 \quad [23]$$

$$1/C = [\partial(1/\varepsilon)/\partial T] \quad [24]$$

Another intuitive crystallographic 'rattling ion' model has also been proposed. **Figure 4** illustrates two models of $A(B_{I\,1/2}B_{II\,1/2})O_3$ perovskite, one being ordered and the other disordered. If the rigid ion model is assumed, in the disordered lattice, the larger B_I ions locally prop open the lattice, and there is some 'rattling' space around the smaller B_{II} ion. In contrast, in the ordered lattice, the larger neighboring ions eliminate the excess space around the B_{II} ion, and the structure becomes close-packed. The close packing in an ordered arrangement, as shown in **Figure 4A**, has been confirmed by Amin et al. in $0.9\,Pb(Mg_{1/2}W_{1/2})O_3 - 0.1\,Pb(Mg_{1/3}Nb_{2/3})O_3$.

When an electric field is applied to the disordered perovskite, the B_{II} ions with a large 'rattling space' can easily shift without distorting the oxygen octahedron. Thus, large polarizations per unit electric field, in other words, large dielectric constants,

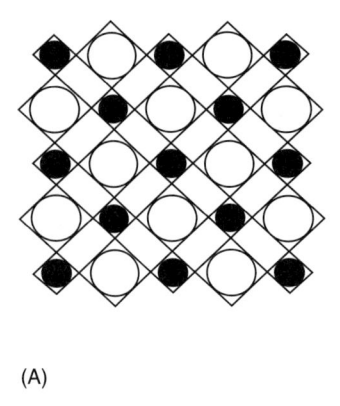

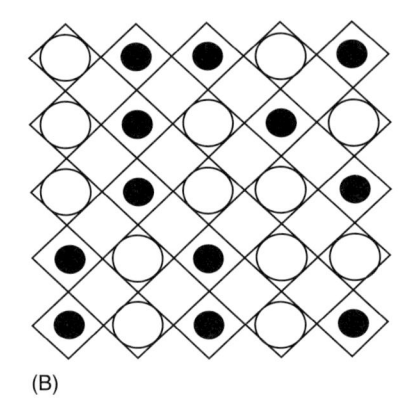

Figure 4 'Ion rattling' crystal model of $A(B_{I\,1/2}B_{II\,1/2})O_3$. (A) Ordered; (B) disordered arrangement of B_I (open circles) and B_{II} (filled circles).

Curie–Weiss constants can be expected (recall that $\varepsilon = C/(T - T_0)$). Strain per unit polarization, i.e., electrostrictive Q coefficients, are also expected to be small. In the case of the ordered arrangement, neither B_I nor B_{II} ions can shift without distorting the oxygen octahedron. Hence, small polarizations, dielectric constants and Curie–Weiss constants, and large electrostrictive Q coefficients can be expected.

Here, we summarize the empirical rules for the electrostrictive effect of perovskite-type oxides.

1. The value of the electrostrictive coefficient Q (defined as $x = QP^2$) does not depend on whether the material is ferroelectric, antiferroelectric, or nonpolar, but is greatly affected by the degree of ordering of the cation arrangement. The Q value increases in the sequence from disordered, partially ordered, then simple, and finally to ordered perovskite crystals.
2. In perovskite solid solutions with a disordered cation arrangement, the electrostrictive coefficient Q decreases with increasing phase transition diffuseness, or with increasing dielectric relaxation.
3. The product of the electrostrictive coefficient Q and the Curie–Weiss constant C is about the same for all perovskite crystals ($Q_hC = 3.1 \times 10^3$ m^4 C^{-2} K).
4. The electrostrictive coefficient Q is nearly proportional to the square of the thermal expansion coefficient α.

These empirical rules lead to the following important result. Since the figure of merit of electrostriction under a certain electric field is given by $Q\varepsilon^2$ (ε: dielectric constant) or QC^2 (excluding the temperature dependence), and the product QC is constant, it is more advantageous to use the disordered perovskite ferroelectrics (relaxor ferroelectrics) which have a large Curie–Weiss constant C and hence a small electrostrictive coefficient Q than the normal ferroelectrics (Pb(Zr,Ti)O$_3$, BaTiO$_3$-based ceramics, etc.).

Electrostrictive Materials

Electrostrictive materials do not, in principle, exhibit the domain-related problems observed in piezoelectrics. A common practical ceramic system is the Pb(Mg$_{1/3}$Nb$_{2/3}$)O$_3$-based compound. So-called relaxor ferroelectrics such as Pb(Mg$_{1/3}$Nb$_{2/3}$)O$_3$ and Pb(Zn$_{1/3}$Nb$_{2/3}$)O$_3$ have also been developed for very compact chip capacitors. The reasons why these complex perovskites have been investigated intensively for applications are their very high polarization and permittivity, and temperature-insensitive characteristics (i.e., diffuse phase transition) in comparison with the normal perovskite solid solutions.

The phase transition diffuseness in the relaxor ferroelectrics has not been satisfactorily clarified as yet. We introduce here a widely accepted microscopic composition fluctuation model which is even applicable in a macroscopically disordered structure. Considering the Känzig region (the minimum size region in order to cause a cooperative phenomenon, ferroelectricity) to be in the range of 10–100 nm, the disordered perovskite such as Pb(Mg$_{1/3}$Nb$_{2/3}$)O$_3$ reveals a local fluctuation in the distribution of Mg^{2+} and Nb^{5+} ions in the B sites of the perovskite cell. **Figure 5** shows a computer simulation of the composition fluctuation in the $A(B_{I1/2}B_{II1/2})O_3$-type crystal calculated for various degrees of short-range ionic ordering. The fluctuation of the B_I/B_{II} fraction x obeys a Gaussian error distribution, which may cause Curie temperature fluctuation. HB Krause reported the existence of short-range ionic ordering in Pb(Mg$_{1/3}$Nb$_{2/3}$)O$_3$ by electron microscopy. The high-resolution image revealed somewhat ordered (ion-ordered) islands in the range of 2–5 nm, each of which might have a slightly different transition temperature.

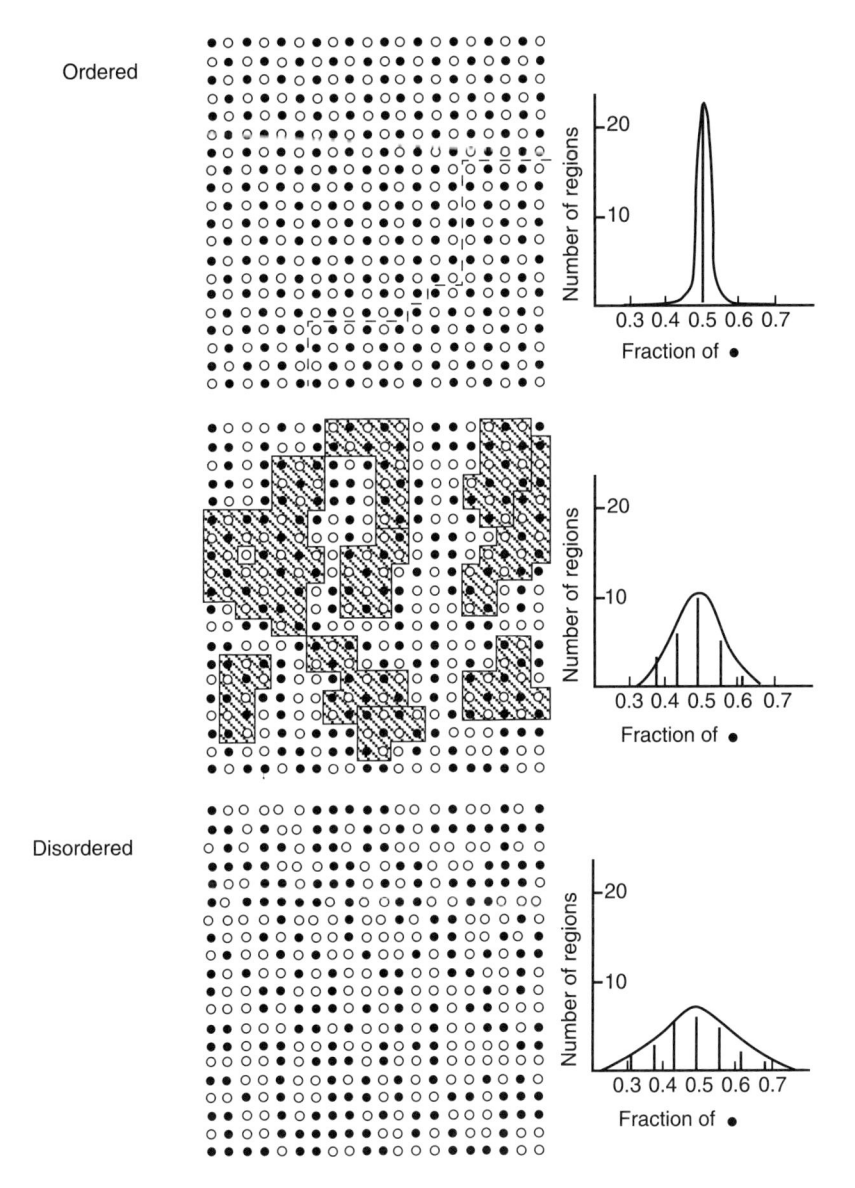

Figure 5 Computer simulation of the composition fluctuation in the $A(B_{I\,1/2}B_{II\,1/2})O_3$ type calculated for various degree of the ionic ordering (Känzig region size 4 × 4).

Another significant characteristic of these relaxor ferroelectrics is dielectric relaxation (frequency dependence of permittivity), from which their name originates. The temperature dependence of the permittivity in $Pb(Mg_{1/3}Nb_{2/3})O_3$ is plotted in **Figure 6** for various measuring frequencies. With increasing measuring frequency, the permittivity in the low-temperature (ferroelectric) phase decreases and the peak temperature near 0 °C shifts towards higher temperature; this is contrasted with the normal ferroelectrics such as $BaTiO_3$ where the peak temperature hardly changes with the frequency. The origin of this effect has been partly clarified in $Pb(Zn_{1/3}Nb_{2/3})O_3$ single crystals. **Figure 7** shows the dielectric constant and loss versus temperature for an unpoled and a poled PZN sample, respectively. The domain configurations are also inserted. The macroscopic domains were not observed in an unpoled sample even at room temperature; in which state, large dielectric relaxation and loss were observed below the Curie temperature range. Once the macrodomains were induced by an external electric field, the dielectric dispersion disappeared and the loss became very small (i.e., dielectric behavior became rather normal!) below 100 °C. Therefore, the dielectric relaxation is attributed to the microdomains generated in this material.

Thus, the PMN is easily electrically poled when an electric field is applied around the transition temperaure, and completely depoled without any remanent polarization because the domain is separated into microdomains when the field is removed. This provides extraordinarily large apparent electrostriction, though it is a secondary phenomenon related to the electromechanical coupling ($x = ME^2$). If the phase transition temperature can be raised near room temperature, superior characteristics can be expected.

Figure 8A shows the longitudinal-induced strain curve at room temperature in such a designed material 0.9PMN – 0.1PbTiO$_3$. Notice that the magnitude of the electrostriction (10^{-3}) is about the same as that of a piezoelectric PLZT under unipolar drive, as shown in **Figure 8B**. An attractive feature of this material is the near absence of hysteresis.

Other electrostrictive materials include (Pb,Ba)(Zr,Ti)O$_3$ and the PLZT systems. In order to obtain large (apparent) electrostriction, it is essential to generate ferroelectric microdomains (in the range 10 nm). It may be necessary to dope ions with a different valence or an ionic radius, or to create vacancies so as to introduce the spatial microscopic inhomogeneity in the composition.

Here, the characteristics of piezoelectric and electrostrictive ceramics are compared and summarized from a practical viewpoint:

1. Electrostrictive strain is about the same in magnitude as the piezoelectric (unipolar) strain (0.1%). Moreover, almost no hysteresis is an attractive feature.
2. Piezoelectric materials require an electrical poling process, which leads to a significant aging effect due to the depoling. Electrostrictive materials do not need such pretreatment, but, require a proper DC bias field in some applications because of the nonlinear behavior.
3. Compared with piezoelectrics, electrostrictive ceramics do not deteriorate easily under severe operation conditions, such as high-temperature storage and large mechanical load.
4. Piezoelectrics are superior to electrostrictors, with regard to temperature characteristics.
5. Piezoelectrics have smaller dielectric constants than electrostrictors, and thus show faster response.

Applications of Electrostrictions

Classification of Ceramic Actuators

Piezoelectric and electrostrictive actuators may be classified into two categories, based on the type of driving voltage applied to the device and the nature of the strain induced by the voltage: first, rigid displacement devices for which the strain is induced unidirectionally along an applied DC field, and second, resonating displacement devices for which the alternating strain is excited by an AC field at the mechanical resonance frequency (ultrasonic motors). The first category can be further divided into two types: servo displacement transducers (positioners) controlled by a feedback system through a position detection signal, and pulse-drive motors operated in a simple on/off switching mode, exemplified by dot matrix printers. **Figure 9** shows the classification of ceramic actuators with respect to the drive voltage and induced displacement.

The AC resonant displacement is not directly proportional to the applied voltage, but is, instead, dependent on adjustment of the drive frequency. Although the positioning accuracy is not as high as that of the rigid displacement devices, very high speed motion due to the high frequency is an attractive feature of the ultrasonic motors. Servo displacement transducers, which use feedback voltage superimposed on the DC bias, are used as positioners for optical and precision machinery systems. In contrast, a pulse drive motor generates only on/off strains, suitable for the impact elements of dot matrix or ink jet printers.

The materials requirements for these classes of devices are somewhat different, and certain compounds will be better suited for particular applications. The ultrasonic motor, for instance, requires a very hard piezoelectric with a high mechanical quality

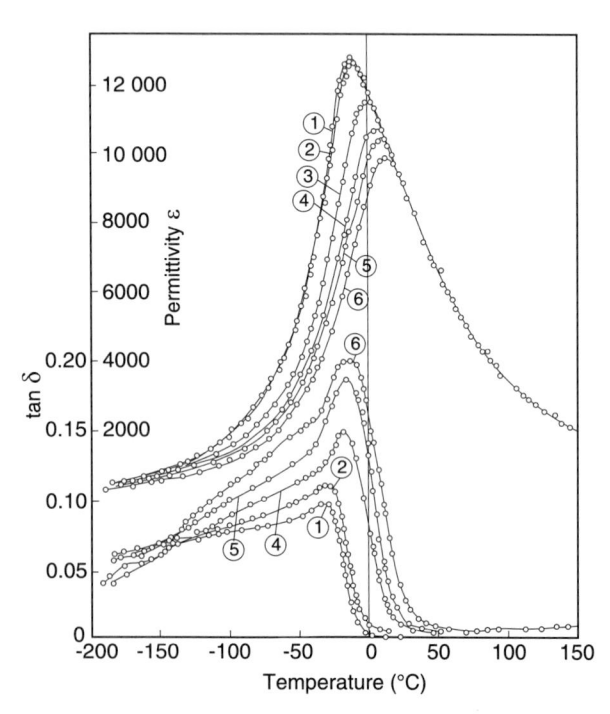

Figure 6 Temperature dependence of the permittivity in Pb(Mg$_{1/3}$Nb$_{2/3}$)O$_3$ for various frequencies (kHz): (1) 0.4, (2) 1, (3) 45, (4) 450, (5) 1500, (6) 4500.

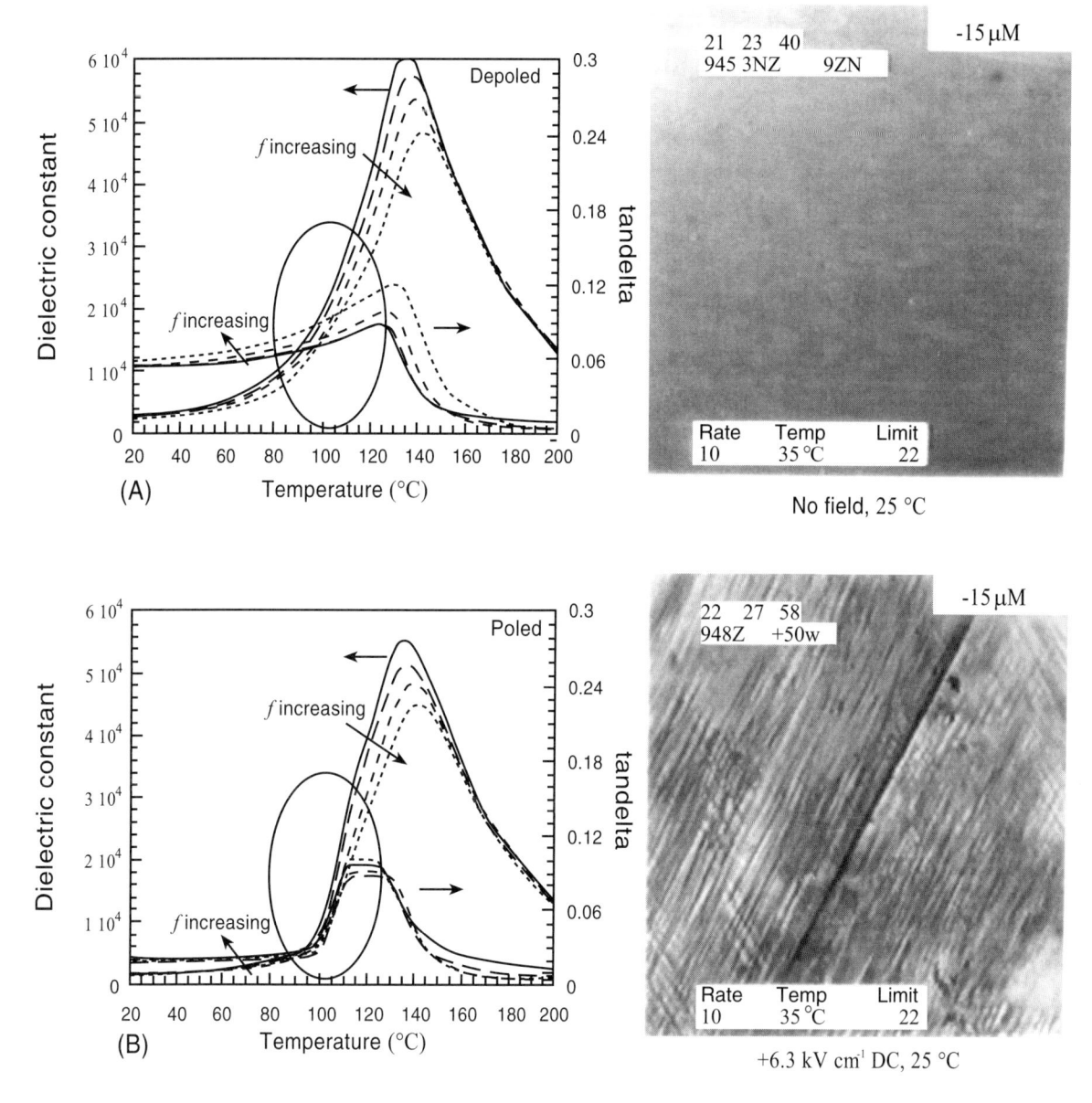

Figure 7 Dielectric constant versus temperature for a depoled (A) and a poled Pb(Zn$_{1/3}$Nb$_{2/3}$)O$_3$ single crystal (B) measured along the <111> axis.

factor Q, in order to minimize heat generation. The servo displacement transducer suffers most from strain hysteresis and, therefore, a PMN electrostrictor is preferred for this application. Notice that even in a feedback system the hysteresis results in a much lower response speed. The pulse-drive motor requires a low-permittivity material aiming at quick response with a limited power supply rather than a small hysteresis, so that soft PZT piezoelectrics are preferred to the high-permittivity PMN for this application.

Deformable Mirrors

Precise wave front control with as small a number of parameters as possible and compact construction is a common and basic requirement for adaptive optical systems. For example, continuous-surface deformable mirrors may be more desirable than segmented mirrors from the viewpoint of precision.

Multimorph-type deformable mirror Sato *et al.* proposed a multimorph deformable mirror, simply operated by a microcomputer. In the case of a two-dimensional multimorph deflector, the mirror surface can be deformed by changing the applied voltage distribution and the electrode pattern on each electroactive ceramic layer.

The mirror surface contour is occasionally represented by Zernike aberration polynomials in optics.

484 ELECTROSTRICTIVE MATERIALS

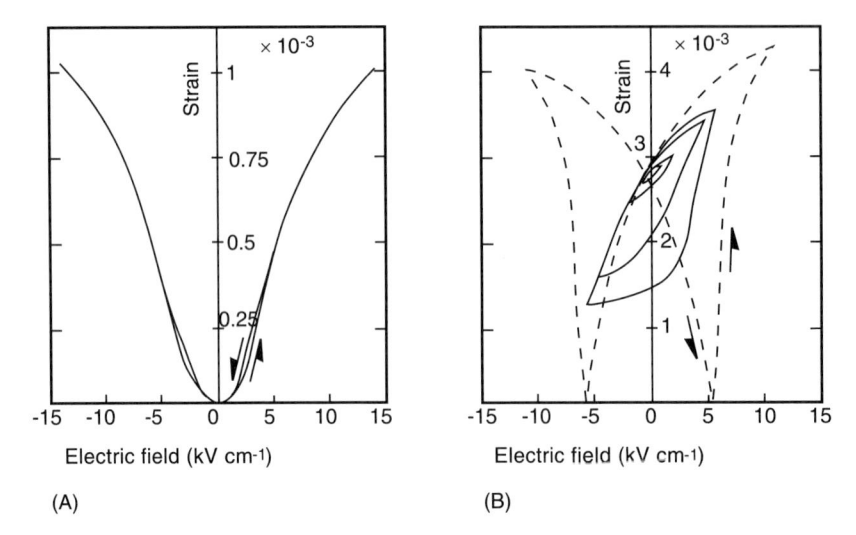

Figure 8 Field-induced strain curve in an electrostrictor 0.9Pb(Mg$_{1/3}$Nb$_{2/3}$)O$_3$ − 0.1PbTiO$_3$ (A) and in a piezoelectric PLZT 7/62/38 (B) at room temperature.

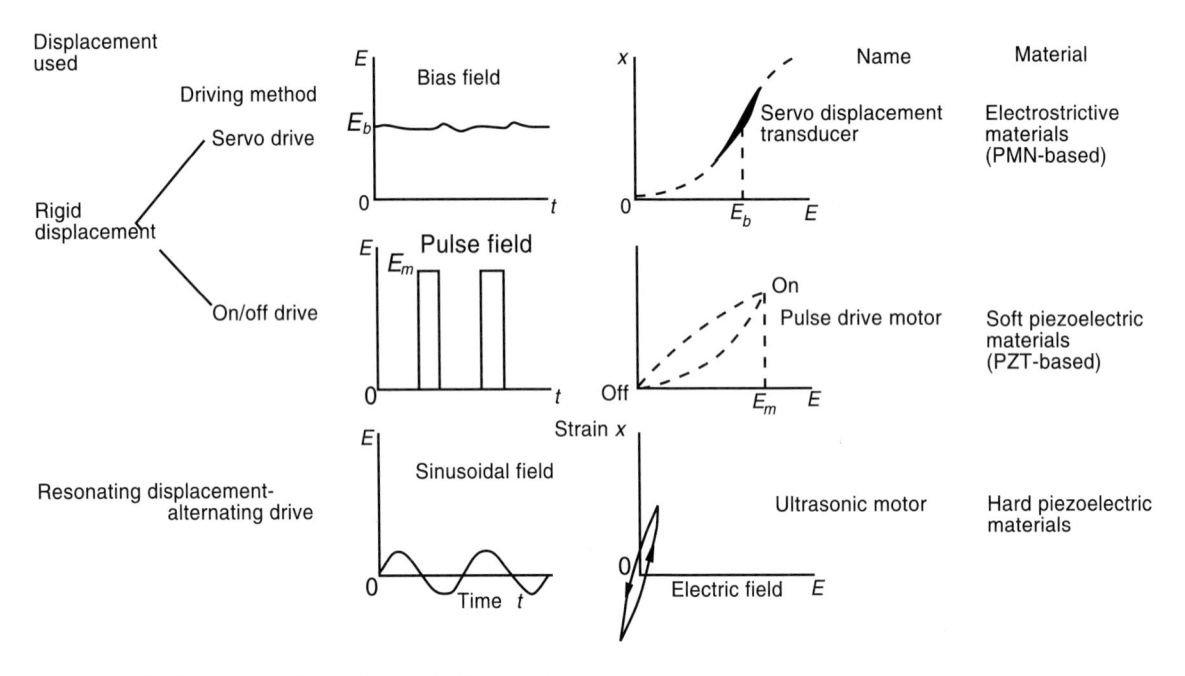

Figure 9 Classification of piezoelectric/electrostrictive actuators.

An arbitrary surface contour modulation $g(x, y)$ can be expanded as follows:

$$g(x,y) = C_r(x^2 + y^2) + C_c^1 x(x^2 + y^2) + C_c^2 y(x^2 + y^2) + \cdots \quad [25]$$

Notice that the Zernike polynomials are orthogonal to each other, i.e., can be completely controlled independent of each other. The C_r and C_c terms are called refocusing and coma aberration, respectively.

The aberration, up to the second order, can provide a clear image apparent to the human; this is analogous to the eye examination. The optometrist checks the degree of your lens, initially, then the astigmatism is corrected; no further correction will be made for human glasses.

As examples, let us consider the cases of refocusing and coma aberration. Uniform whole-area electrodes can provide a parabolic (or spherical) deformation. In the case of the coma aberration, the pattern obtained is shown in **Figure 10**, which consists of only six divisions and has the fixed ratio of supply voltages.

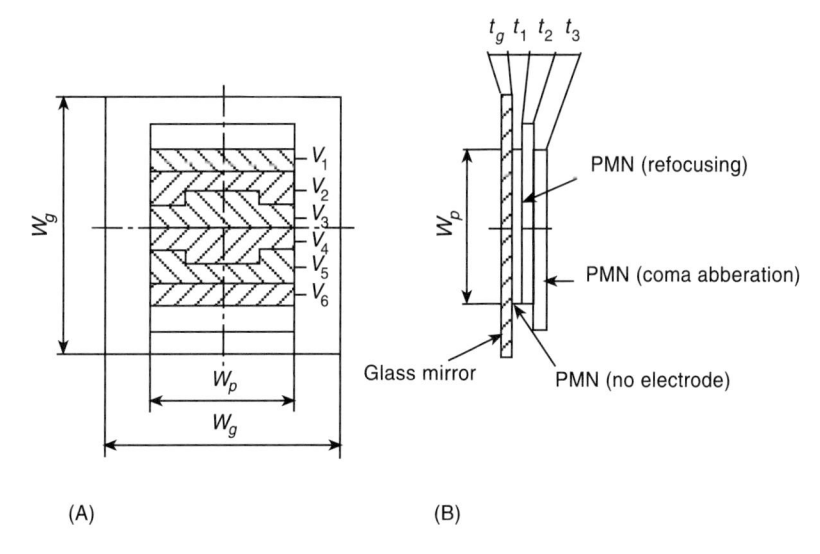

Figure 10 Two-dimensional multimorph deformable mirror with refocusing and coma aberration functions. (A) Front view; (B) cross-section.

The deflection measurement of the deformable mirror was carried out using the interferometric system with a hologram. The introduction of a hologram is to cancel the initial deformation of the deformable mirror. Typical experimental results are summarized in **Figure 11**, where the three experimentally obtained interferograms of refocusing aberration, coma aberration, and a combination of the two are compared with the corresponding ideal ones. Good agreement is seen in each case. These results demonstrate the validity of the superposition of deformations and the appropriateness of the method of producing the electrode patterns. It was also observed that the deformable mirror responds linearly up to 500 Hz for sinusoidal input voltages.

Figure 12 shows the fringe patterns observed for several applied voltages for the conventional PZT piezoelectric and PMN electrostrictive deformable mirrors. It is worth noting that, for a PZT device, a distinct hysteresis is observed optically on a cycle with rising and falling electric fields, but no discernible hysteresis is observed for the PMN device.

Hubble space telescope Fanson and Ealey have developed a space qualified active mirror called articulating fold mirror. Three articulating fold mirrors are incorporated into the optical train of the Jet Propulsion Laboratory's wide field and planetary camera-2, which was installed into the Hubble space telescope in 1993. As shown in **Figure 13**, each articulating fold mirror utilizes six PMN electrostrictive multilayer actuators to position a mirror precisely in tip and tilt in order to correct the refocusing aberration of the Hubble telescope's primary mirror.

Figure 14 shows the images of the core of M100, a spiral galaxy in the Virgo cluster, before and after correction of the refocusing aberration. Both contrast and limiting magnitude have been greatly improved, restoring Hubble to its originally specified performance.

Oil-Pressure Servo Valves

For oil pressure servo systems, an electric–oil pressure combination (electrohydraulic) is necessary in order to obtain large power and quick response. Quicker response has been requested for this electrohydraulic system, and the present target is 1 kHz, which is categorized in an 'impossible' range using conventional actuation. Ikabe *et al.* simplified the valve structure by using a piezoelectric PZT flapper. However, since pulse width modulation was employed to control the device to eliminate PZT hysteresis, the flapper was always vibrated by a carrier wave, and the servo valve exhibited essential problems in high-frequency response and in durability. To overcome this problem, Ohuchi *et al.* utilized an electrostrictive PMN bimorph for the flapper instead of a piezoelectric bimorph.

Construction of oil-pressure servo valves **Figure 15** is a schematic drawing of the construction of a two-stage four-way valve, the first stage of which is operated by a PMN electrostrictive flapper. The second-stage spool is 4 mm in diameter, which is the smallest spool with a nominal flow rate of 6 l min^{-1}.

An electrostrictive material 0.45 PMN–0.36 PT–0.19 BZN was utilized because of its large displacement and small hysteresis. A flapper was fabricated

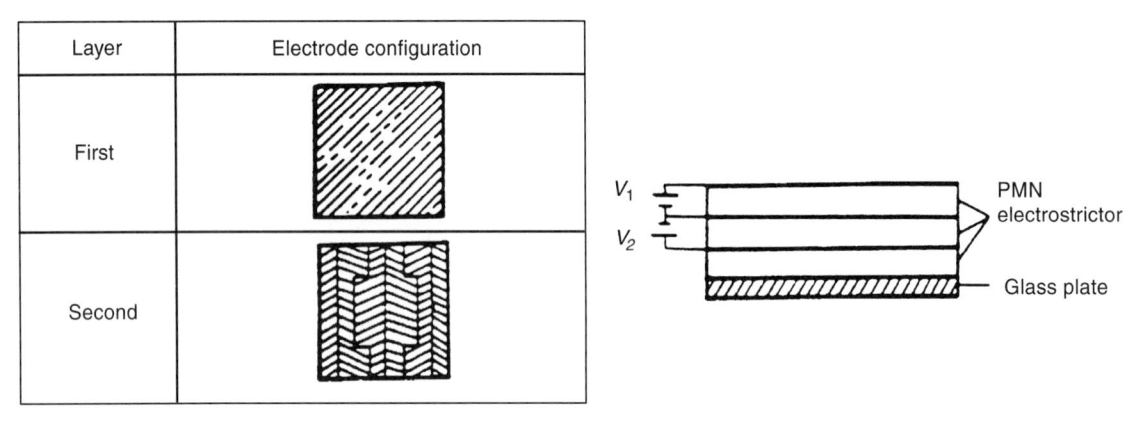

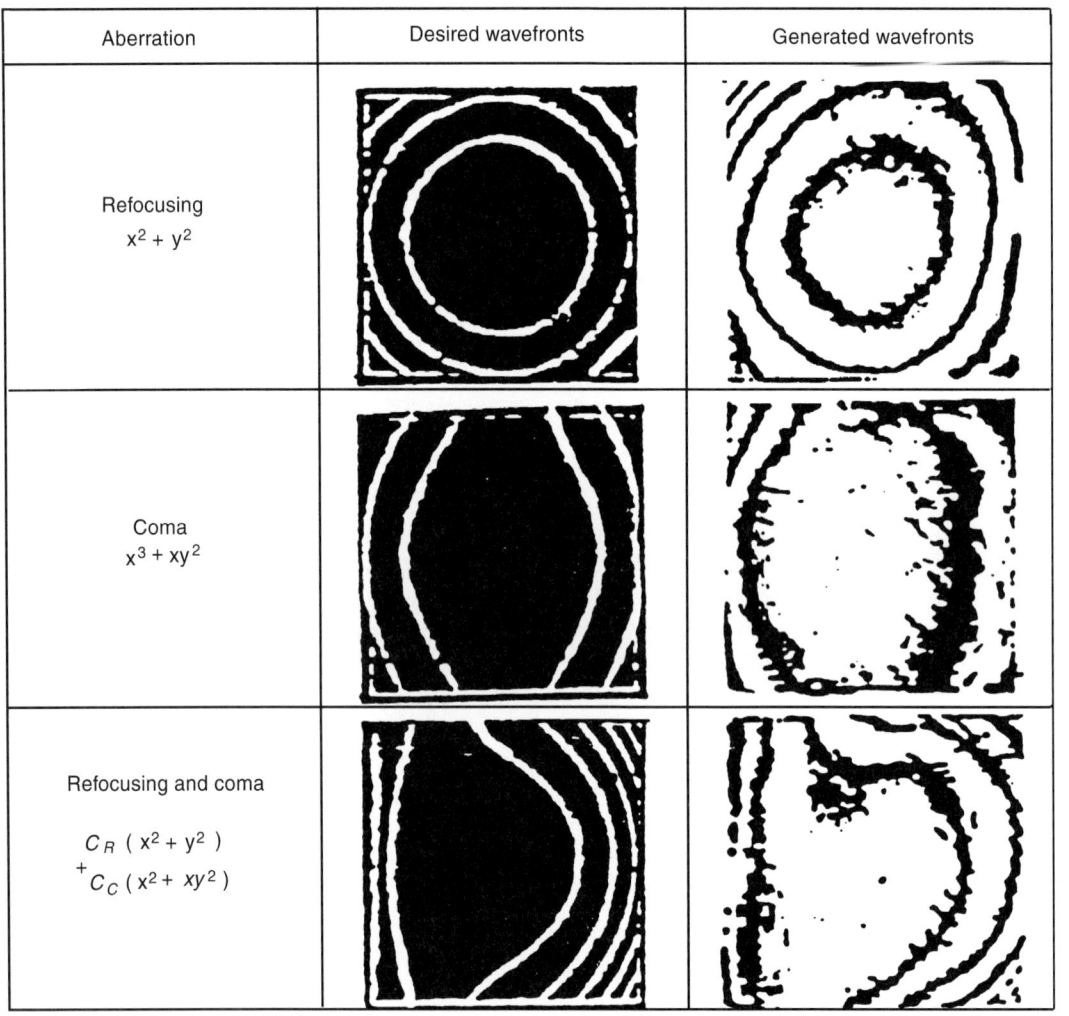

Figure 11 Interferograms showing deformation by the multilayer deformable mirror.

using a multimorph structure, in which two PMN thin plates were bonded on each side of a phosphor bronze, shim (**Figure 16**). The multimorph structure increases the tip displacement, generative force, and response speed. The top and bottom electrodes and the metal shim were taken as ground, and a high voltage was applied on the electrode between the two PMN plates. The tip deflection showed a quadratic curve for the applied voltage because of the electrostriction. Thus, in order to obtain a linear relation, a push–pull driving method was adopted with a suitable DC bias electric field, which is demonstrated in **Figure 17**. Notice that the displacement hysteresis in the PMN ceramic is much smaller than in the PZT piezoelectric. The resonance frequency of this flapper in oil was about 2 kHz.

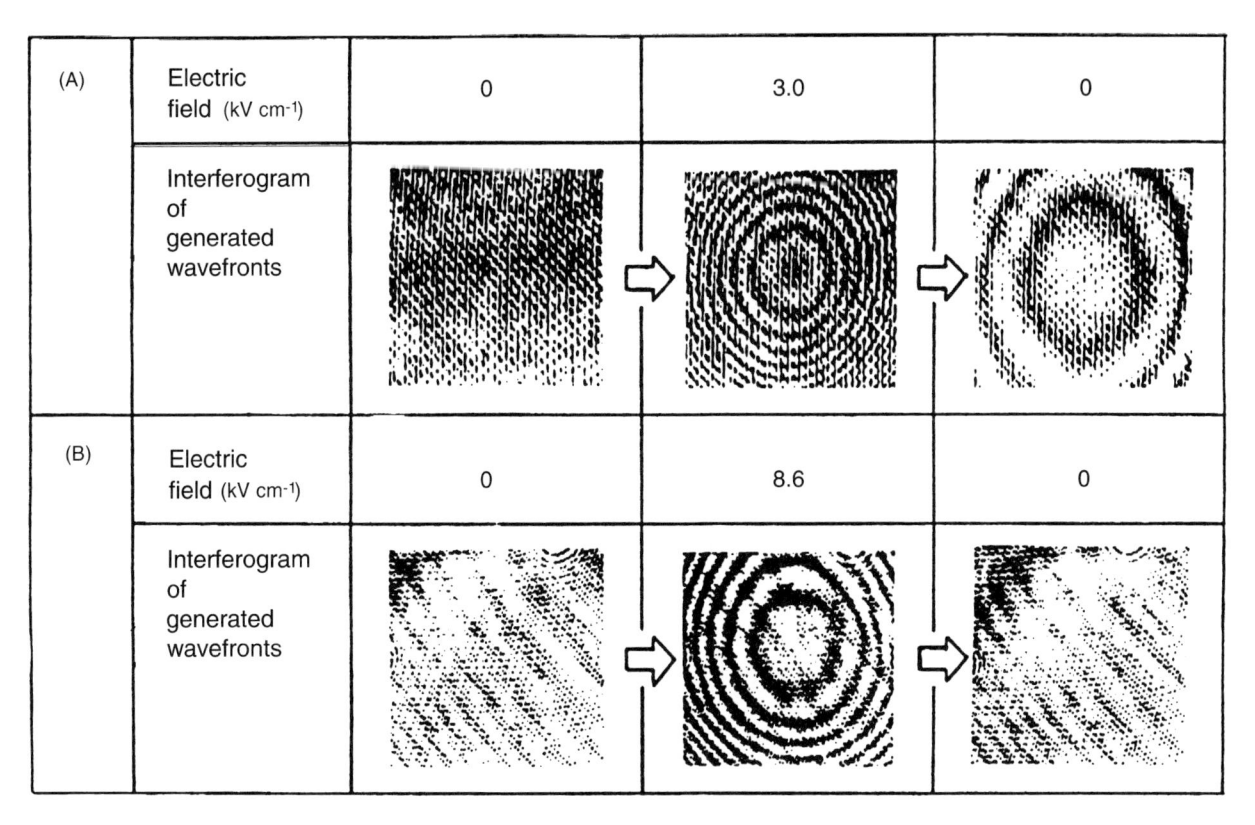

(A)	Electric field (kV cm⁻¹)	0	3.0	0
	Interferogram of generated wavefronts			
(B)	Electric field (kV cm⁻¹)	0	8.6	0
	Interferogram of generated wavefronts			

Figure 12 Comparison of function between (A) PZT and (B) PMN deformable mirrors.

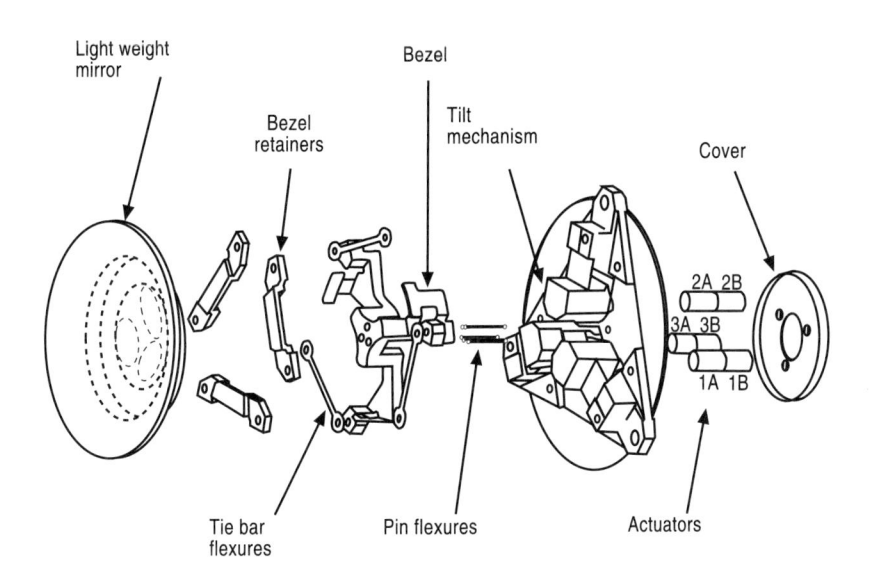

Figure 13 Articulating fold mirror using PMN actuators.

Since the conventional force feedback method has a structure connecting the flapper tip and the spool with a spring, limiting the responsivity, Ohuchi *et al.* utilized an electric feedback method. The feedback mechanism using an electric position signal corresponding to the spool position has various merits, such as easy change in feedback gain and availability of speed feedback. A compact differential transformer (50 kHz excitation) was employed to detect the spool position.

Operating characteristics of the oil-pressure servo valve Figure 18 shows a static characteristic of the spool displacement for a reference input. The slight

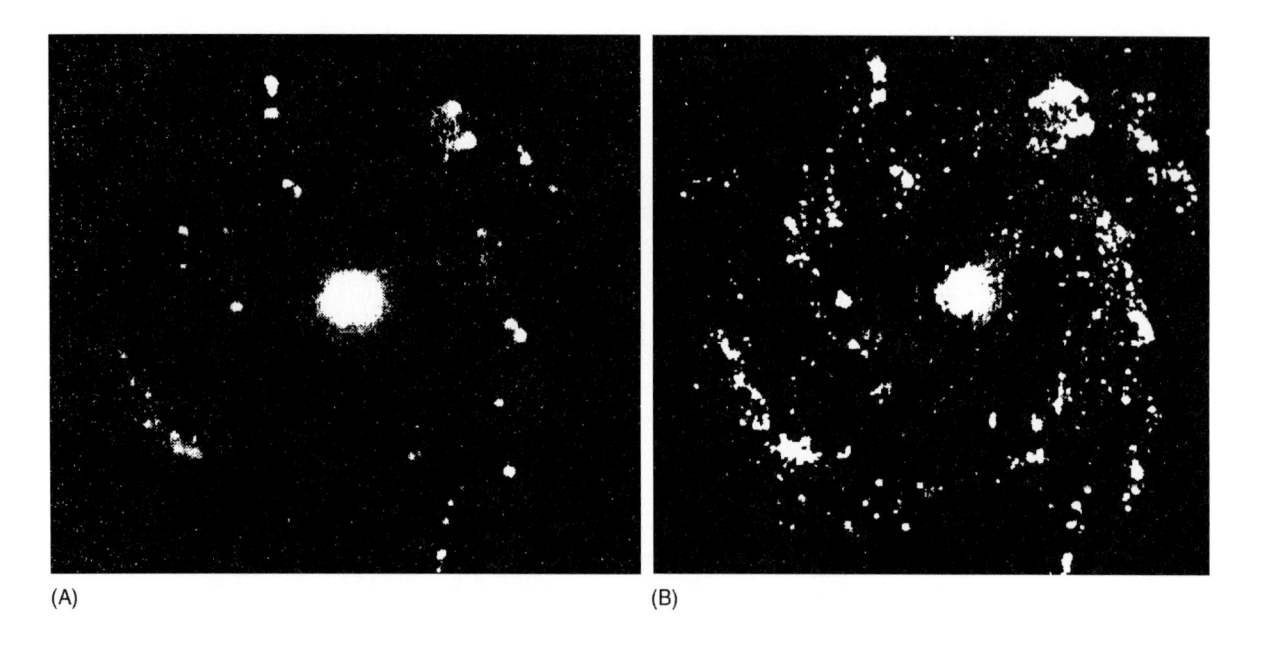

Figure 14 M100 galaxy comparison. (A) Before activation; (B) after activation of the PMN.

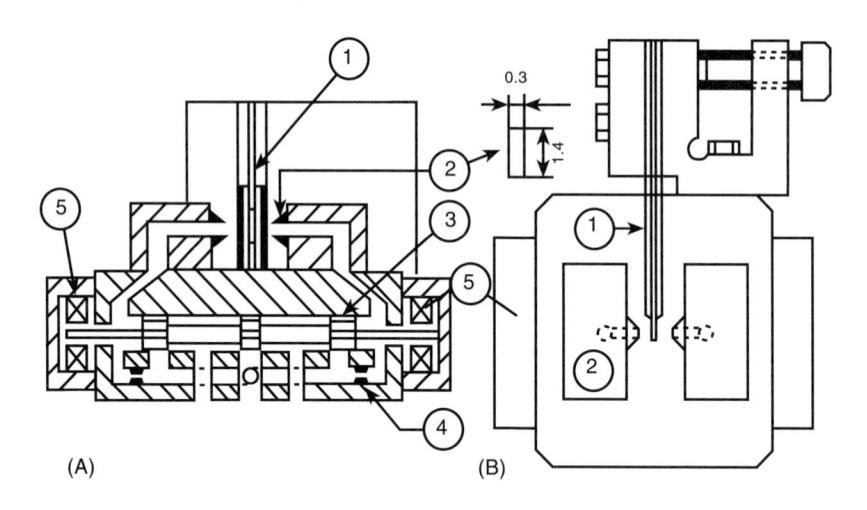

Figure 15 Construction of the oil pressure servo valve. (A) Front (section) view; (B) top view. 1, PMN-flapper; 2, nozzle; 3, spool; 4, fixed orifice; 5, spool position sensor.

hysteresis and nonlinearity observed in **Figure 17** was completely eliminated in **Figure 18** through the feedback mechanism. **Figure 19** shows a dynamic characteristic of the spool displacement. The 0 dB gain was adjusted at 10 Hz. The gain curve showed a slight peak at 1800 Hz and the 90° retardation frequency was about 1200 Hz. The maximum spool displacement at 1 kHz was ± 0.03 mm, which results in the quickest servo valve at present.

Nomenclature

d	piezoelectric constant
E	electric field
F	force
k	spring constant
M	electrostrictive constant
P	polarization
Q	electrostrictive coefficient

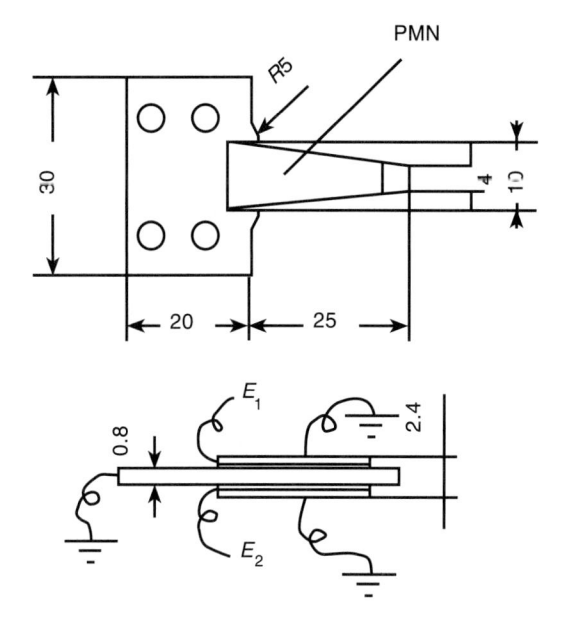

Figure 16 Multimorph electrostrictive flapper.

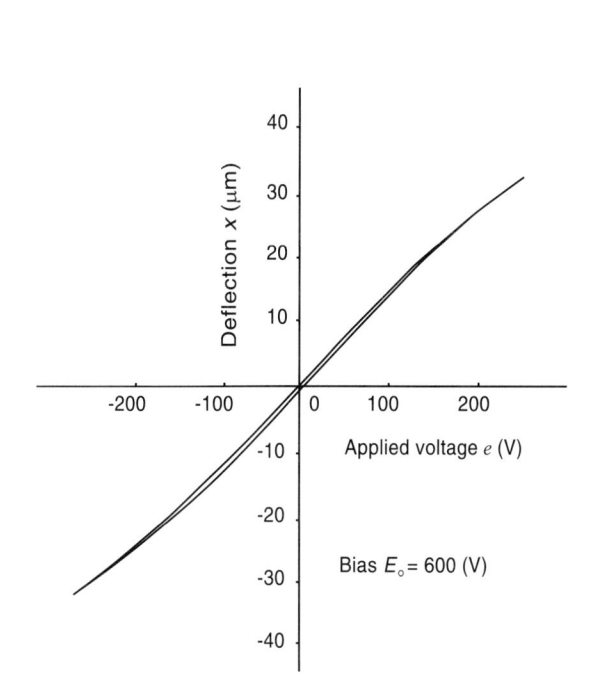

Figure 17 Push–pull drive characteristics of the electrostrictive flapper.

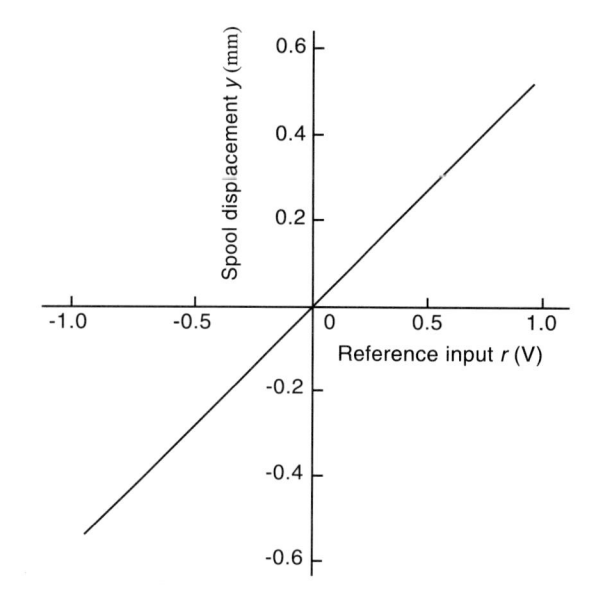

Figure 18 Static characteristics of the servo valve.

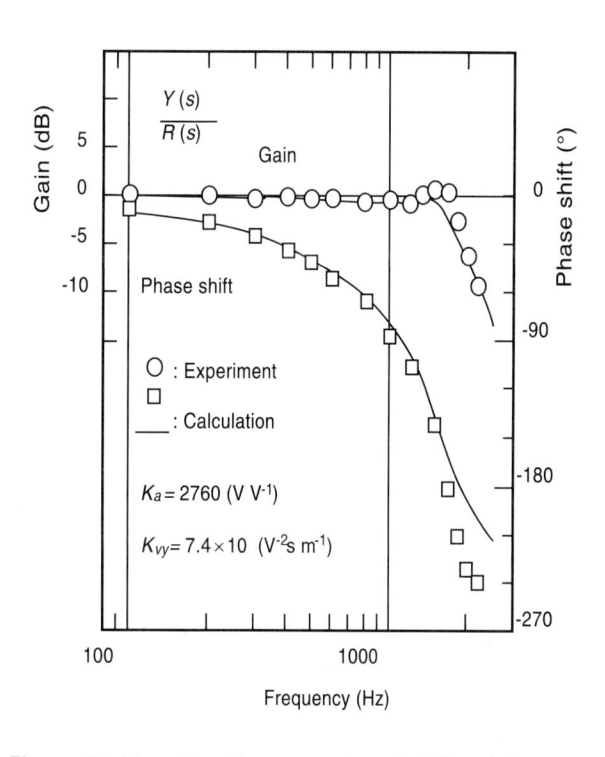

Figure 19 Normalized frequency characteristics of the servo

s	elastic compliance
T	temperature
X	stress
Δ	displacement
ε	permittivity

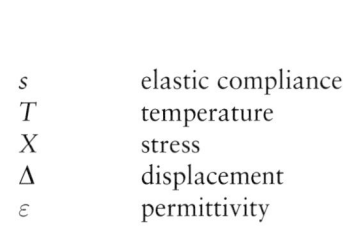

See also: **Actuators and smart structures**.

Further Reading

Cross LE, Jang SJ, Newnham RE, Nomura S and Uchino K (1980) Large electrostrictive effects in relaxor ferroelectrics. *Ferroelectrics* 23: 187–192.

Nomura S and Uchino K (1982) Electrostrictive effect in Pb$(Mg_{1/3}Nb_{2/3})O_3$ type materials. *Ferroelectrics* 41: 117–

132.
Uchino K (1986) Electrostrictive actuators: materials and applications. *Ceramic Bulletin* 65: 674–652.
Uchino K (1993) Ceramic actuators: principles and applications. *MRS Bulletin* XVIII: 42–48.
Uchino K (1996) *Piezoelectric Actuators and Ultrasonic Motors*. MA, USA: Kluwer Academic.
Uchino K (1999) *Ferroelectric Devices*. New York: Marcel Dekker.
Uchino K, Cross LE, Newnham RE and Nomura S (1981) Electrostrictive effects in antiferroelectric perovskites. *Journal of Applied Physics* 52: 1455–1459.
Uchino K, Nomura S, Cross LE and Newnham RE (1980) Electrostriction in perovskite crystals and its applications to transducers. *Journal of the Physics Society of Japan* 49 (suppl.B): 45–48.
Uchino K, Nomura S, Cross LE, Jang SJ and Newnham RE (1980) Electrostrictive effects in lead magnesium niobate single crystals. *Journal of Applied Physics* 51: 1142–1145.
Uchino K, Nomura S, Cross LE, Newnham RE and Jang SJ (1981) Electrostrictive effects in perovskites and its transducer applications. *Journal of Material Science* 16: 569–578.

ENVIRONMENTAL TESTING, OVERVIEW

D Smallwood, Sandia National Laboratories, Albuquerque, NM, USA

Copyright © 2001 Academic Press

doi:10.1006/rwvb.2001.0106

Environmental testing can cover a large variety of natural and artificial environments. In the context of this article, the environments are limited to mechanical vibration. Sometimes it is important to consider combinations of environments, but this is beyond the scope of this article. Environmental testing can include the exposure of test hardware to actual use environments. This is also beyond the scope of this article. Environmental testing in this article is the reproduction in the laboratory of a vibration environment that is intended to simulate a real or potential natural or use vibration environment. Some typical environments that produce significant vibration levels include: surface transportation in cars and trucks with vibration caused by irregular surfaces, air turbulence, and mechanical noise from engines, etc.; aircraft environments with vibration caused by airflow and engines; rocket environments with vibration caused by airflow, shock waves, engine noise, and stage separation; ocean environments with vibration induced by wind and waves; building vibration with motion excited by internal and external machinery, wind, and earthquakes. Structural response to mechanical shock can be considered a transient vibration and is discussed.

Environmental testing generally falls into one of four classifications: development testing, qualification testing, acceptance testing, or stress screening.

1. Development testing is intended to explore the response of the test hardware to a variety of vibration stimulations with the goal of identifying vibration characteristics of the hardware or to uncover design weaknesses of the hardware. Typically, the hardware tested is not production hardware but development hardware in various stages of development.
2. Qualification testing is used to determine if the hardware will perform satisfactorily when exposed to a representative or worst-case use environment. The purpose is to determine if the design meets a set of vibration requirements. Typically, the hardware tested is representative of production hardware, but is not hardware that will be used in the use environment. Qualification testing is typically performed with methods and levels defined in specifications.
3. Acceptance testing is used to determine if a particular set of production hardware is satisfactory for release and subsequent use by a customer. All production hardware is not necessarily subjected to acceptance testing. Some programs demand 100% acceptance testing, and some programs accept lot testing. The test methods and levels are typically defined in specifications. Sometimes the acceptance test levels are related to the qualification levels. The levels are typically lower than the qualification levels because the risk of damage to the hardware must be kept to a minimum.
4. Stress screening is a tool used to expose hardware at various stages of manufacture to vibration environment that will precipitate infant failures and

identify manufacturing defects early in the manufacturing cycle. The methods and levels are defined by the manufacturer and tailored to a specific product. The levels should not induce damage to well manufactured hardware.

The development of environment test methods and levels can be divided into two stages. First is the identification and characterization of the environments. Second is the development of test methods that will satisfactorily simulate the environment in the laboratory.

Identification of the Use Environment

By far the most common field measurements are with accelerometers. The use of accelerometers is covered elsewhere. Much less frequently used are measurements of velocity, displacement, and force. The data from the field measurements are collected and stored for later analysis. The most versatile method is to store the raw unprocessed time histories. If the original data are stored, data-processing decisions can be changed at any time. Recently most data are sampled and stored as digital records. All the precautions associated with this process should be carefully observed (see Further Reading).

The next step is to classify the data for reduction. Data will seldom fall neatly into a classification. Judgment must be used. Sometimes data can be analyzed making several assumptions of the class, and the most appropriate reduction can be chosen after the data reduction.

The data must first be classified as random or deterministic. Loosely deterministic data are data that, if the experiment is repeated, the data will be the same. Deterministic data can be a transient, periodic, or complex nonperiodic. A transient is data that starts and ends within a few periods of the lowest natural frequency of the object being observed. Periodic data theoretically extend over all time, repeating at a regular period. In practice a signal that repeats itself and extends over many cycles of the lowest natural frequency can be treated as periodic. A periodic waveform can often be treated as a Fourier series expansion. Complex nonperiodic data is typically composed of the sum of periodic waveforms that are not harmonically related; thus the waveform does not have a period. As for periodic waveforms, the waveform must extend over a period of time that is much longer than the lowest natural period of the object being observed.

Random data are data that can be described only in statistical terms. The future time history cannot be predicted from past values except in statistical terms. The data can be stationary, which loosely means that the statistical measures are not dependent on the location in time. As for periodic data, random data theoretically extend over all time. In practice much data can be treated as stationary if the duration is much longer than the period of the lowest natural frequency of the object under study. The most common procedures used to analyze stationary random data are the autospectral densities (commonly called the power spectra density) and the cross-spectral densities. The statistical moments and probability density functions are also important parameters. These are described elsewhere in this volume. By far the most common assumption about the data is that they are normally distributed. This assumption should always be checked because much data is not normal.

If the duration of random data is too short to be treated as stationary, it must be treated as nonstationary random data. Many techniques are available to treat nonstationary data. All the techniques add complications to the analysis, and the results will be dependent on the method and parameters used. Several of these techniques are discussed elsewhere in this volume but are beyond the scope of this discussion. The most common error made in nonstationary random analysis is to ignore the uncertainty theorem. Loosely, the uncertainty states that you cannot resolve time and frequency independently. If you desire good time resolution, you will have poor frequency resolution and vice versa.

One is tempted simply to reproduce the field environment in the laboratory. If this is done, the field environment is being treated as deterministic. If a random component is present, the reproduction will not necessarily be adequate.

Data can also be a mixture of all the types. These problems are difficult to handle. Usually some attempt is made to separate the data into the component parts and each part is analyzed separately.

Periodic environments are seldom encountered in the field. Exceptions are rotating machinery measurements that often contain the fundamental and harmonics of the rotational frequency. Random environments usually treated as stationary random include: excitation from turbulent fluid flow, rocket and turbojet excited response, wind excitation, and long-term surface transportation environments. Deterministic transient environments can include some shocks and chirps (a short-duration sine sweep).

Examples of nonstationary random environments include: seismic excitation (earthquakes); pyroshock (transients in structures caused by explosive hardware, like bolt-cutters); response of structures to nonpenetrating impacts; and response to bumps and potholes in surface transportation. To the extent that the structure is deterministic, some of these shocks are essentially the impulse response of the structure and can be partially deterministic. An example of a mixed environment is the resonant burn condition found in the response of some solid rocket motors. This environment is a mixture of a swept sinusoid with harmonics and stationary random. Another example of a mixed environment is the response of propeller-driven aircraft. This is a mixture of stationary random caused by the turbulent airflow and possibly the turboprop exhaust and the periodic blade passage frequency with harmonics.

Basic Equivalence between Field and Laboratory Dynamic Environments

The purpose of an environmental vibration test is usually to simulate a field environment. The basic requirements for the equivalence between a field and a laboratory simulation of the environment will now be considered. To simplify the discussion, only linear systems will be considered.

Assume the system under consideration is a multiple-input linear system described by N nodes (**Figure 1**). A pair of variables can describe the dynamics at each node: the external force applied at the node and a motion parameter for the node. Each node represents a degree of freedom of the structure. The force can be a translational force or a torque applied to the node. The force will be designated by the variable F. The motion variable can be a displacement, velocity, or acceleration. The motion can be translation or rotation consistent with the applied force at the node. The motion variable will be designated by the variable V. The variables are complex functions of frequency. For a sinusoidal excitation the variable represents an amplitude and phase of the sinusoid. For a transient the variable is the Fourier transform of the transient time history. For stationary random the variable is the expected value of the Fourier transform of the random signal in the sense of Bendat and Piersol (see Further Reading).

The system can be characterized by a square matrix of impedance functions, Z.

$$\{V\} = [Z]\{F\} \quad [1a]$$

or:

$$\mathbf{V} = \mathbf{ZF} \quad [1b]$$

The system can also be represented by a matrix of admittance functions:

$$\mathbf{F} = \mathbf{YV} \quad [2]$$

where:

$$\mathbf{Y} = \mathbf{Z}^{-1} \quad [3]$$

Partition the nodes into three categories denoted by a subscript: c will indicate a control point in the laboratory test, r will indicate a response point measured in the field environment, and i will indicate an interior point that is not observed in either the test environment or in the field environment.

A superscript will be used to denote the environment in which the measurements are made: f for a field measurement and t for a test or laboratory measurement.

$$\begin{Bmatrix} V_c \\ V_r \\ V_i \end{Bmatrix} = \begin{bmatrix} Z_{cc} & Z_{cr} & Z_{ci} \\ Z_{rc} & Z_{rr} & Z_{ri} \\ Z_{ic} & Z_{ir} & Z_{ii} \end{bmatrix} \begin{Bmatrix} F_c \\ F_r \\ F_i \end{Bmatrix} \quad [4]$$

We can ask the basic question: Under what conditions can we control a test at the points c with the system mounted on a test apparatus and reproduce the motion observed at the points r in the field?

The motions in the laboratory test at the response points are given by:

$$V_r^t = [Z_{rc} \quad Z_{rr} \quad Z_{ri}] \begin{Bmatrix} F_c^t \\ F_r^t \\ F_i^t \end{Bmatrix} \quad [5]$$

For now, assume the applied forces at points other than the control points are zero in the laboratory test.

$$F_r^t = F_i^t = 0 \quad [6]$$

This is not always true: for example, the cross-axis motion at the control points is often restrained in a

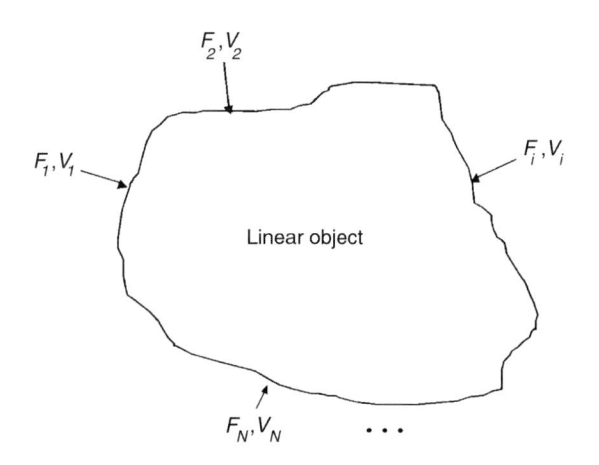

Figure 1 Multiple-input linear system.

vibration test. This assumption results in motions at the response points and the control points of:

$$V_r^t = Z_{rc} F_c^t \qquad [7]$$

$$V_c^t = Z_{cc} F_c^t \qquad [8]$$

Combining eqns [7] and [8] gives the test motions at the response points:

$$V_r^t = Z_{rc} Z_{cc}^{-1} V_c^t \qquad [9]$$

The test motions at the control points are:

$$V_c^t = Z_{cc} Z_{rc}^{-1} V_r^t \qquad [10]$$

The inverses must exist or a pseudoinverse must be used. If Z_{rc}^{-1} exists, the number of response points must equal the number of control points and the matrix is full-rank.

Thus, we can derive a set of control motions that will reproduce the motion at a set of response points for the conditions where a solution of eqn [10] exists.

The motions at the response points in the field are given by:

$$V_r^f = \begin{bmatrix} Z_{rc} & Z_{rr} & Z_{ri} \end{bmatrix} \begin{Bmatrix} F_c^f \\ F_r^f \\ F_i^f \end{Bmatrix} \qquad [11]$$

The motion at the response points in the laboratory can be made the same for the field:

$$V_r^f = V_r^t \qquad [12]$$

However, this does not insure that the motion at other points on the structure will be the same for the test as for the field. To meet this requirement:

$$\begin{bmatrix} Z_{cc} & Z_{cr} & Z_{ci} \\ Z_{rc} & Z_{rr} & Z_{ri} \\ Z_{ic} & Z_{ir} & Z_{ii} \end{bmatrix} \begin{Bmatrix} F_c^t \\ 0 \\ 0 \end{Bmatrix} = \begin{bmatrix} Z_{cc} & Z_{cr} & Z_{ci} \\ Z_{rc} & Z_{rr} & Z_{ri} \\ Z_{ic} & Z_{ir} & Z_{ii} \end{bmatrix} \begin{Bmatrix} F_c^f \\ F_r^f \\ F_i^f \end{Bmatrix} \qquad [13]$$

In general, the solution to all the constraints imposed above can be satisfied only if:

$$F_r^f = F_i^f = 0 \qquad [14]$$

This implies that the response of the system in test is equivalent to the response in the field only if forces are applied to the system in the field at the test control points. If forces in the field environment are applied at points other than the control points, the test environment will not, in general, be equivalent to the field environment.

The conclusion of this discussion is that simulation of a field vibration environment in the laboratory is almost always imperfect. The importance of the imperfection is always a point of discussion.

Motion Control at a Single Point in a Single Direction

The motion is controlled at one point in probably more than 95% of vibration tests. The motion at the control point is assumed to move in only one direction with a force applied at the control point. Because control is at one point, the test items are usually mounted on rigid fixtures. It is usually assumed that the interface between the fixture and the test item moves as a plane in one direction. Exceptions are many, as strictly rigid fixtures are impossible to construct, and motion in one direction is difficult to achieve. Thus, in general, the laboratory test can be fully equivalent to the field test only if the force is applied to the test object in the field through the same interface used in the laboratory at the control point location. Errors caused by this simplifying assumption can vary from minor to major and are the subject of many debates.

Use of Environmental Envelopes

The development above assumes that the measured field motion is duplicated in the laboratory with all of its detail. This is rarely possible in practice. In practice, envelopes of the environment are usually used. Many times, composite envelopes of several different environments are used. The purpose of the envelope is to insure a conservative test. Consider the simple case of one response point, one control point, and where the errors caused by the failure of eqn [14] are acceptable. Eqn [9] becomes an algebraic equation:

$$V_r^t = Z_{rc} Z_{cc}^{-1} V_c^t \qquad [15]$$

For the case where the test control point motion is equal to an envelope of the field control point motion:

$$|V_c^t| = V_c^{\text{envelope}} \geq |V_c^f| \qquad [16]$$

Using eqn [9] with f substituted for t, eqn [16] reduces to:

$$V_r^t \geq |Z_{rc} Z_{cc}^{-1} V_c^f| = |V_r^f| \qquad [17]$$

The test environment will always be larger than the field environment for this special case. If forces are applied to the test object at any location other than at the control point, we are not guaranteed that the test environment will generate responses greater than or equal to the field environment. Unfortunately, this is frequently the case. The forces in the field are often applied in a distributed manner over the test object, and motion is measured at one or a few points. The motion is enveloped, and the envelope is reproduced at the control points (usually a single point) in the laboratory with the sometimes unreasonable expectation that the laboratory test is a conservative test.

Motion in the field is seldom in one direction. Even if the motion of the control point is in a single direction, the motion of the test object in the laboratory is seldom in a single direction. A typical assumption is that the motion observed in the field can be adequately simulated with three tests in the laboratory, one in each of three orthogonal directions. At best this is a practical compromise and has little rigorous development to justify the assumption.

For those cases where eqn [17] is valid, the motions at the response points can be very much larger than the field response. The very high input impedance of shakers, coupled with large power amplifiers and modern control equipment, typically causes the overtest. The control system will attempt to maintain the motion at the control point regardless of the force required. In the field the driving point impedance of the interface looking back into the structure on which the test object is mounted limits the force. The overtest caused by a near-infinite impedance of the shaker can be alleviated through the use of force or response limiting.

As can be seen from the above discussion, practical vibration tests require many compromises. Even for a linear system, vibration testing is an imperfect science. Engineering judgment and past experience are valuable guides in picking an appropriate test.

Identification of Measurement Locations

An important consideration in gathering field information that will later be used to characterize a vibration environment is the determination of the measurement locations. Usually an attempt is made to place the measurement location in the load path of the input to the system. This is not always possible, as multiple load paths may exist or the interface between the item and its foundation is not accessible. In many cases the number of input points desired is beyond the capabilities of the data-gathering system. In this case 'typical' locations are chosen and the resulting motion is determined for a 'zone' of the structure. Care must be taken to insure the measurement is typical and not greatly influenced by local phenomena. For example, if an accelerometer is mounted on a thin plate, the motion can be dominated by the local plate resonances and not be representative of the structural response in the neighborhood.

In some cases it may be desirable to measure multiple inputs to a component or system. If this is done the phase relationships between the inputs must be preserved, as this will be of critical importance if a multiple-input test is designed.

Test Methods

Many test machines are used to generate vibration in the laboratory. Several of these are discussed in greater detail in other articles. The most common machines are electrodynamic and electrohydraulic shakers.

An electrodynamic shaker is built on the same principle as a speaker in a radio or home entertainment system. A magnet (either permanent or an electromagnet) provides a magnetic field. A coil of wire is placed in the field. When a current is passed through the coil, a force is generated. This force moves the coil and the attached structure that includes the test item.

An electrohydraulic shaker is essentially a double-acting hydraulic cylinder. Most of the energy is supplied by hydraulic fluid under pressure. A servo valve (often driven by a small electrodynamic shaker) controls the fluid flow in the hydraulic cylinder.

Vibration can also be generated by a variety of mechanical shakers. These devices have limited use today because the versatility and control of the devices are usually not as good as for electrodynamic and electrohydraulic shakers.

The testing of aerospace structures is often accomplished by placing the test item in a reverberant chamber and exposing the item to high-intensity acoustic noise. A reverberant chamber is a large chamber with hard walls and many modes of acoustic vibration. The chamber provides a diffuse acoustic field that simulates the high-intensity noise of many acoustic environments found in the aerospace industry.

Many machines generate mechanical shock, which can be considered a transient vibration. The most common of these machines are drop tables. The drop table essentially generates a specified velocity. The shock is generated when the test item impacts a stationary structure. The characteristics of the shock are determined by the design of the impacting structures, including the interface between items. The material placed at the interface is sometimes called a shock programmer. Many other devices are used, essentially to generate a velocity. The kinetic energy is used as the energy source for the shock test. The shock produced is determined by the details of the impact between the moving structure and the stationary structure. The test item is usually mounted on the moving structure. In some cases the test item is stationary and a moving target is impacted into the test item. This is called a turn-around test. Machines used to generate the initial velocity include: rocket sleds, drop towers, cable pull-down facilities, actuators, and guns.

Pyrotechnic shock is a special category of shock that requires special care in the measurements and simulation. A discussion of pyroshock is beyond the scope of this article.

The location of control accelerometers in vibration testing requires care since vibration at high frequencies ('high' being defined here as frequencies above the first structural resonances of the test apparatus being used) is a local phenomenon. The vibration levels can vary significantly with location. The common method is to mount the control accelerometer near the location where field data were measured that defined the environment. If this is not possible, the accelerometer is usually mounted on the fixture near the test item–fixture interface.

As explained earlier, the boundary conditions for the field environment and the test greatly influence the equivalence of the simulation. It is rare that the boundary conditions of the field environment are simulated in the laboratory. Limiting is often used to reduce the conservatism of a test caused by improper boundary conditions and overly conservative test specifications. Common methods include: limiting the response at other locations than the control point, limiting the input force, limiting the current into an electrodynamic shaker, and averaging the response at several points.

Accelerated Testing

Test compression is often required to permit reasonable test times. There are several pitfalls with these methods. The methods usually use some form of Minors rule for fatigue damage. This assumes failures are related to fatigue, and the fatigue mechanism is known. Generally it is wise to keep the vibration level at or below the highest expected field environment. This will prevent unwarranted failures from peak loads. Lower levels of vibration can be accelerated with reasonable confidence by raising the level to the highest expected field environment and reducing the time.

Fixture Design

As discussed earlier, fixtures are usually designed as rigid as possible for several practical reasons. First is the previously mentioned assumption of a single control point. If the fixture is not rigid, this assumption is obviously flawed. Another major practical reason is control of the test. If the fixture, and hence a control accelerometer mounted on or near the fixture, is involved in resonant behavior with more than one participating mode, there will almost certainly be one or more frequencies at which the modal response destructively interferes and the motion will cancel. This is called an antiresonance. Very large motion of the system will result with essentially zero motion at the control point. Not only is this potentially destructive, but the control system and shaker will have great difficulty maintaining the required motion. Care must be taken to avoid placing the control accelerometer at such a location. This can usually be accomplished if the fixtures are rigid. For the same reasons the control accelerometer is often placed at the extreme end of a fixture, like a cube or slip table. The free end of a beam is free of antiresonances. If a rigid fixture is not possible, averaging or extremal (usually means control on the largest) control of several accelerometers is used to limit the motion caused by antiresonances. Locations can usually be found where the several control accelerometers do not all have antiresonances at the same frequencies. Force limiting can also be used to limit the effect of antiresonances. Fixtures can also be built with a significant amount of damping. The damping limits the depth of the antiresonances. The mass of the fixtures is also important. All the shaker moving parts, including the fixtures, must be moved, which increases the force required. Fixtures with small mass require less force from the shaker. This is contrary to the requirement for rigid fixtures. For this reason most fixtures are constructed from aluminum or magnesium that have high stiffness-to-weight ratios.

System Level Tests that Attempt to Match Field Conditions

At the system level tests are often designed to match closely the field boundary conditions and environment. Examples include:

1. Automobile road simulators. These devices can employ as many as 18 electrohydraulic actuators to simulate road conditions.
2. Automobile crash testing. Automobiles are crashed at various velocities into a variety of barriers.

3. Aerospace high-level reverberant acoustic testing. Reverberant acoustic fields can reasonably represent the environment experienced by spacecraft at launch.
4. Shipping container impact testing. Shipping container damage is often simulated by drop tests into various targets to simulate a crash environment.
5. Reentry vehicle impulse testing. Certain hostile environments can subject reentry vehicles to essentially an impulse distributed over the surface of the vehicle. Several techniques have been developed to simulate this impulse.
6. Impact testing of energy-absorbing structures. Since energy-absorbing structures are typically very nonlinear problems, the performance of energy-absorbing structures can often be tested only through the accurate simulation of the use environment.
7. Live fire testing within the Department of Defense. In live fire testing, military hardware is subjected to test conditions that simulate an actual attack as closely as possible.

Nomenclature

c	control point
f	field measurement
F	force variable
i	interior poit
r	response point
t	test measurement
V	motion variable
Z	square matrix of impedance functions

See also: **Crash**; **Environmental testing, implementation**; **Packaging**; **Standards for vibrations of machines and measurement procedures**.

Further Reading

(STD) GEVS-SE, (1990) *General Environmental Verification Specification for STS & ELV Payloads, Subsystems, and Components*, NASA Goddard Space Flight Center, Greenbelt, MD 20771, USA.

(STD) Himelblau H, Manning JE, Piersol AG, and Rubin S (1997), *Guidelines for Dynamic Environmental Criteria*, NASA.

(STD) MIL-PRE-1540 (Proposed) (1997) *Product Verification Requirements for Launch, Upper Stage, and Space Vehicles*, Aerospace Corporation.

(STD) MIL-STD-810E *Test Method Standard for Environmental Engineering Considerations and Laboratory Tests*, Methods 514.5 Vibration, 515.5 Acoustic, 516.5 Shock, 517 Pyroshock, 519.5 Gunfire, and 523.2 Vibro-Acoustic, Temperature.

(STD)MIL-S-901D, (1989) *Shock Tests, H. I. (High-Impact) Shipboard Machinery, Equipment, and Systems, Requirements For*.

Bendat JS and Piersol AG (1986) *Random Data, Analysis and Measurement Procedures*, 2nd edn. New York: Wiley.

Curtis AJ, Tinling NG and Abstein HT Jr (1971) *Selection and Performance of Vibration Tests*, SVM-8, Shock and Vibration Information Analysis Center, 2231 Crystal Drive, Suite 711, Arlington, VA 22202, USA.

Harris CM (ed.) (1996) *Shock and Vibration Handbook*, 4th edn. New York: McGraw-Hill.

Himelblau H, Piersol AG, Wise JH and Grundvig MR (1994) *Handbook for Dynamic Data Acquisition and Analysis*. IEST Recommended Practice 012.1. Mount Prospect, IL: IEST.

McConnell KG (1995) *Vibration Testing Theory and Practice*. New York: Wiley.

Pusey HC (ed.) (1996) *50 Years of Shock and Vibration Technology, SVM-15*. Arlington, VA: Shock and Vibration Information Analysis Center.

Steinberg DS (1988) *Vibration Analysis for Electronic Equipment*. New York: Wiley.

Wirsching PH, Paez TL and Ortiz K (1995) *Random Vibrations Theory and Practice*. New York: Wiley.

ENVIRONMENTAL TESTING, IMPLEMENTATION

P S Varoto, Escola de Engenharia de São Carlos, USP, São Carlos, Brasil

Copyright © 2001 Academic Press

doi:10.1006/rwvb.2001.0107

Environmental vibration testing can be defined as the process of simulating the actual or potential vibration environments experienced by a given structure or equipment. In the context of this article the word simulation is primarily related to experimental tests that are conducted in the laboratory environment in order to predict or reproduce given field vibration data. This simulation process requires the definition of suitable laboratory inputs to the structure under test and, when available, vibration data from the field environment can be used to define these inputs. In the absence of field data, as it is often the case of a new test structure that is to be tested prior to its exposure to field conditions, a successful laboratory simulation can still be achieved under some special circumstances, as outlined below.

Three major structures are involved in the simulation process. The test item represents the structure under test that is to be attached to the vehicle in the field dynamic environment. The vehicle is the structure used to transport or simply to support the test item. In the field environment the test item is attached to the vehicle forming a combined structure. The vibration exciter is the structure employed in the laboratory environment in order to generate the required test item inputs. In the laboratory environment, the test item is attached to multiple vibration exciters in order to simulate the field conditions. A typical laboratory set-up employs a single vibration exciter, usually driven by a digital control system that is responsible for generating a prescribed input to be applied to the test item by the exciter through a test fixture. More sophisticated excitation systems employ multiple-axis exciters connected to different points on the test item.

The development of acceptable testing procedures and test specifications requires the definition of pertinent variables, as well as understanding of the physical processes which control the vibration data to be measured and how this data can be used to generate suitable test item inputs in laboratory simulations.

Multiple Input–Output Frequency Domain Relationships

Consider the linear structure shown in **Figure 1**. The input vector at the structure's qth location is represented in the frequency domain by $\mathbf{F}_q = \mathbf{F}_q(\omega)$, where ω is the excitation frequency. This input vector consists of two vectors, a force vector and a moment vector. Each input vector can be resolved in terms of the global coordinate system, as shown in **Figure 1**. Thus, the input vector at the qth location has a total of six components, three forces (F_1, F_2, and F_3) in the x, y, and z directions, and three moments (M_1, M_2, and M_3) about the x, y, and z directions, respectively.

Similarly, the structure's output response in the frequency domain at the pth location $\mathbf{X}_p = \mathbf{X}_q(\omega)$ consists of two vectors, a vector of three linear motions and a vector of three angular motions, where the structure's output motion can be given in terms of small linear and angular displacements, velocities, or accelerations. Each output vector can be resolved in terms of the global coordinate system shown in **Figure 1**. Thus, the output response vector $\mathbf{X}_p$ contains six components, three linear motions (X_1, X_2, and X_3) in the x, y, and z directions and three angular motions (X_4, X_5, and X_6) about the x, y, and z directions, respectively.

The frequency domain equations of motion for a linear structure having N degrees of freedom can be presented in the frequency domain as:

$$\mathbf{X} = \mathbf{H}\mathbf{F} \qquad [1]$$

where $\mathbf{X} = \mathbf{X}(\omega)$ is the output motion vector, $\mathbf{F} = \mathbf{F}(\omega)$ is the input force vector, and $\mathbf{H} = \mathbf{H}(\omega)$ is the structure's frequency response function (FRF) matrix. Eqn [1] can be rewritten relating the pth output location to the qth input location:

$$\mathbf{X}_p = \mathbf{H}_{uv\,pq}\mathbf{F}_q \qquad [2]$$

Subscripts u and v range from 1 to 6 and the structure's FRF matrix (receptance, mobility, or accelerance), and hence the $\mathbf{H}_{uv\,pq}$ matrix has 36 entries. Thus, between each pair of input–output points p and q on the structure, there are potentially 36 input–output relationships.

Eqn [2] can be further expanded in partitioned form as:

$$\left\{ \begin{array}{c} \mathbf{X} \\ \mathbf{\Theta} \end{array} \right\}_p = \left[\begin{array}{cc} \mathbf{H}_{FF} & \mathbf{H}_{FM} \\ \mathbf{H}_{MF} & \mathbf{H}_{MM} \end{array} \right]_{pq} \left\{ \begin{array}{c} \mathbf{F} \\ \mathbf{M} \end{array} \right\}_q \qquad [3]$$

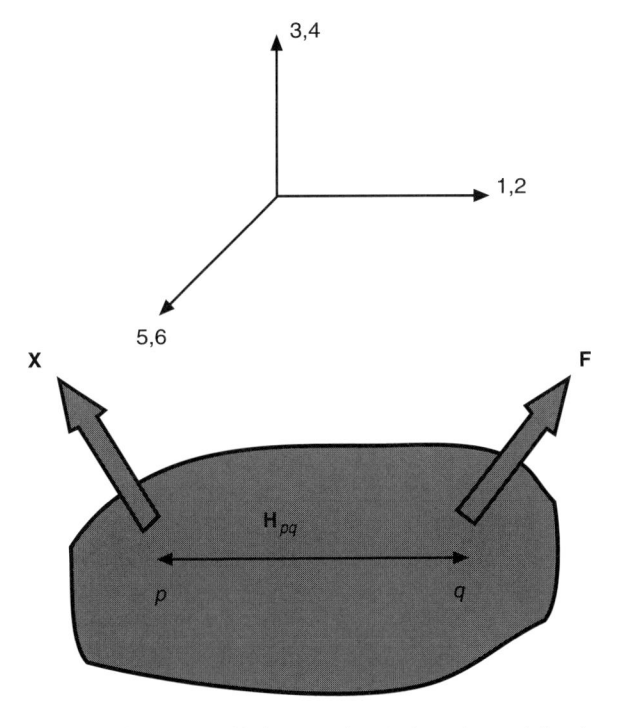

Figure 1 Structure with input and output vectors relative to coordinate system.

where $\mathbf{X} = \mathbf{X}(\omega)$ and $\Theta = \Theta(\omega)$ are 3×1 vectors containing the linear and angular output motions at the structure's pth location, respectively, and $\mathbf{F} = \mathbf{F}(\omega)$ and $\mathbf{M} = \mathbf{M}(\omega)$ are 3×1 vectors containing the forces and moments applied to the structure's qth point, respectively.

The importance of eqn [3] can be recognized by considering a situation where there is a single-axis accelerometer mounted at the pth location with its primary sensing axis oriented in the 2 direction, and a single axis force transducer mounted at the structure's qth location with its primary sensing axis oriented in the 1 direction, according to the global coordinate system shown in **Figure 1**. Thus, from eqn [3], the output acceleration at the pth location due to the excitation vector at the qth point is given as:

$$X_{2p} = H_{21}F_{1q} + H_{22}F_{2q} + H_{23}F_{3q} \\ + H_{24}M_{1q} + H_{25}M_{2q} + H_{26}M_{3q} \quad [4]$$

where H_{uv} denotes the FRF relating the points p and q in the pq directions. Thus, from eqn [4], the output at the pth location is the sum of six terms involving linear as well as angular FRF relative to forces and moments applied at the qth location. However, in practice the sensors used to measure the input and output signals do not account for all terms involved in eqn [4]. Instead the following simpler equation is used:

$$X_p = H_{pq}F_q \quad [5]$$

and the effects of the remaining terms in eqn [4] are not accounted for.

Frequency Domain Substructuring Modeling

Most substructuring techniques consider a partition of the structure's degrees of freedom in terms of master and slaves degrees of freedom. In the context of this article, a similar concept is employed where connector (interface) points are differentiated from external (nonconnector) points by expressing eqn [1] in the following form:

$$\begin{Bmatrix} \mathbf{X}_c \\ \mathbf{X}_e \end{Bmatrix} = \begin{bmatrix} \mathbf{H}_{cc} & \mathbf{H}_{ce} \\ \mathbf{H}_{ec} & \mathbf{H}_{ee} \end{bmatrix} \begin{Bmatrix} \mathbf{F}_c \\ \mathbf{F}_e \end{Bmatrix} \quad [6]$$

Eqn [6] can be used to describe not only the input–output relationship for a single structure, but it can also be applied to any number of independent structures that are coupled at a finite number of Interface points. It is important to emphasize that interface or connector points are points on the test item where coupling occur either with the vehicle in the field or the exciter in the laboratory. External points are points on the test item that are not directly involved with the coupling process. Similarly, interface motions $\mathbf{X}_c$ occur at interface points and external motions $\mathbf{X}_e$ occur at external points. Interface forces $\mathbf{F}_c$ occur at interface points and are due to coupling effects only. The external forces vector $\mathbf{F}_e$ contains the remaining forces applied to the structure. Any excitation source due to internal forces as well as external forces applied at interface points is included in the broad definition of external forces. Eqn [6] also shows that differentiation between interface and external points requires a partition of the structure's FRF matrix into four submatrices. In this case, $\mathbf{H}_{cc}$ defines FRF relationships between interface points while $\mathbf{H}_{ee}$ defines FRF relationships for external points. The $\mathbf{H}_{ce} = \mathbf{H}_{ec}^T$ matrix defines the FRF between interface (external) and external (interface) points.

Interface and external motions can be further expressed from eqn [6] in the following alternate form:

$$\mathbf{X}_c = \mathbf{X}_{cc} + \mathbf{X}_{ce} \quad [7]$$

$$\mathbf{X}_e = \mathbf{X}_{ec} + \mathbf{X}_{ee} \quad [8]$$

Eqns [7] and [8] show a double subscript on the terms appearing on the right-hand side. The first subscript refers to the location where the motion occurs (connector or interface and external) while the second subscript indicates the type of load that caused the motion (connector or external). Thus, $\mathbf{X}_{cc}$ contains interface motions caused by interface forces only while $\mathbf{X}_{ce}$ reflects the influence of the external loads on the interface motions. Each term on the right-hand side of eqns [7] and [8] is related to the corresponding FRF matrix, as shown in eqn [6] (e.g., $\mathbf{X}_{cc} = \mathbf{H}_{cc}\mathbf{F}_c$).

The Field Dynamic Environment

In the field dynamic environment the test item it attached to the vehicle at N_c interface points, as illustrated in **Figure 2**. The resulting field combined structure is subjected to field external forces $\mathbf{F}_e$ and $\mathbf{P}_e$ which, in turn, produces the interface forces $\mathbf{F}_c$ at the N_c interface points.

In order to get expressions for interface forces and test item motions in the field, appropriate test item field boundary conditions must be defined. In this article, a simple interface boundary condition is used, in which the interface connectors are considered to be

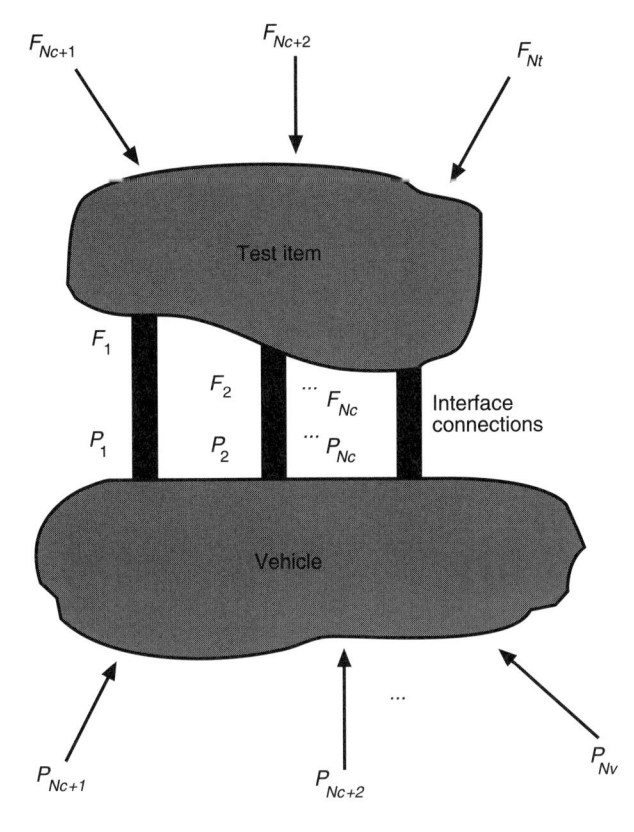

Figure 2 Test item attached to vehicle in the field environment.

part of the test item. Thus, the following relationships for compatibility of motions and equilibrium of forces can be used for interface points:

$$\mathbf{X}_c = \mathbf{Y}_c \quad [9]$$

$$\mathbf{F}_c = -\mathbf{P}_c \quad [10]$$

where the $N_c \times 1$ $\mathbf{X}_c$ and $\mathbf{Y}_c$ vectors contain the motions at the test item and vehicle interface points and the $N_c \times 1$ $\mathbf{F}_c$ and $\mathbf{P}_c$ vectors contain the interface forces on both structures, respectively.

Combining eqn [8] written for the test item and vehicle interface points with the equilibrium relationships shown in eqns [9] and [10] leads to the following expression for the field interface forces:

$$\mathbf{F}_c = \mathbf{TV}(\mathbf{Y}_{ce} - \mathbf{X}_{ce}) \quad [11]$$

Each term appearing in eqn [11] can be defined as follows:

- The $N_c \times N_c$ **TV** matrix is the field interface FRF matrix. It is given as:

$$\mathbf{TV} = (\mathbf{T}_{cc} + \mathbf{V}_{cc})^{-1} \quad [12]$$

where $\mathbf{T}_{cc}$ and $\mathbf{V}_{cc}$ are the test item and vehicle interface FRF $N_c \times N_c$ matrices.

- The $N_c \times 1$ $(\mathbf{Y}_{ce} - \mathbf{X}_{ce})$ vector is the interface relative motion vector. It is formed by the difference of two vectors, namely, the bare vehicle interface motion vector $\mathbf{Y}_{ce}$ and the test item response vector $\mathbf{X}_{ce}$. The bare vehicle interface motions correspond to the vehicle's response at interface points when the test item is not attached to it. Hence $\mathbf{Y}_{ce}$ is due to the vehicle external forces $\mathbf{P}_e$ only (**Figure 2**). The $N_c \times 1$ $\mathbf{X}_{ce}$ vector contains the test item response at interface points due to the external loads ($\mathbf{F}_e$) (nonconnector) applied to the test item.

A particular and important field environment occurs when the test item is not subjected to any field external forces, $\mathbf{F}_e = 0$, and this leads to $\mathbf{X}_{ce} = 0$. In this case, eqn [11] assumes the form:

$$\mathbf{F}_c = \mathbf{TV}\,\mathbf{Y}_{ce} \quad [13]$$

Eqn [13] shows that, when test item field external forces are negligible, the field interface forces depend only on the test item and vehicle interface driving point and transfer FRFs (**TV**) and on the bare vehicle interface motion vector $\mathbf{Y}_{ce}$. An important aspect of eqn [13] is that, in principle, one can predict what interface forces should be in cases where the test item has never been exposed to the field environment.

Eqns [11] and [13] also show that in the process of predicting the test item interface forces, the vehicle interface FRF matrix $\mathbf{V}_{cc}$ is required to evaluate the **TV** matrix, as shown in eqn [12]. This requirement raises the important issue of what field measurements should be taken when attempting to use eqns [11] and [13]? Clearly, eqns [11] and [13] show that the bare vehicle interface motion $\mathbf{Y}_{ce}$ alone is meaningless. It is vital to have information about the vehicle modal model so that $\mathbf{V}_{cc}$ can be properly accounted for in this force prediction process.

Once interface forces are obtained, they can be used to estimate the test item field motions. Substituting of eqn [11] into eqn [1] leads to the following result for the test item field motions:

$$\begin{Bmatrix} \mathbf{X}_c \\ \mathbf{X}_e \end{Bmatrix} = \begin{bmatrix} \mathbf{T}_{cc}\mathbf{TV} & -\mathbf{T}_{cc}\mathbf{TV} \\ \mathbf{T}_{ec}\mathbf{TV} & -\mathbf{T}_{ec}\mathbf{TV} \end{bmatrix} \begin{Bmatrix} \mathbf{Y}_{ce} \\ \mathbf{X}_{ce} \end{Bmatrix} + \begin{Bmatrix} \mathbf{X}_{ce} \\ \mathbf{X}_{ee} \end{Bmatrix} \quad [14]$$

and, in the absence of field external forces, eqn [14] is rewritten as:

$$\begin{Bmatrix} \mathbf{X}_c \\ \mathbf{X}_e \end{Bmatrix} = \begin{bmatrix} \mathbf{T}_{cc}\mathbf{TV} & -\mathbf{T}_{cc}\mathbf{TV} \\ \mathbf{T}_{ec}\mathbf{TV} & -\mathbf{T}_{ec}\mathbf{TV} \end{bmatrix} \begin{Bmatrix} \mathbf{Y}_{ce} \\ \mathbf{0} \end{Bmatrix} \quad [15]$$

As seen from eqn [15], in the absence of field external forces, the test item's field motions depend only on the bare vehicle interface motions $\mathbf{Y}_{ce}$ and on the test item and vehicle FRF matrices. Despite being a particular case of the field dynamic environment, eqn [15] represents an important case, since it often occurs in practice and offers a good chance for a successful laboratory simulation, as will be seen below.

The Laboratory Dynamic Environment

The process of simulating a given field vibration environment in the laboratory environment requires the definition of a suitable set of test item inputs. A simple laboratory test set-up consists of attaching the test item to a single vibration exciter, as shown in **Figure 3**. In this case, it is assumed that the test item is attached to the exciter at E_c connecting points.

The test item input–output relationships in the laboratory environment are defined by using the same frequency domain approach as used in the field environment, given by eqn [6]. In this case, this equation is rewritten as:

$$\left\{ \begin{array}{c} \mathbf{U}_c \\ \mathbf{U}_e \end{array} \right\} = \left[\begin{array}{cc} \mathbf{T}_{cc} & \mathbf{T}_{ce} \\ \mathbf{T}_{ec} & \mathbf{T}_{ee} \end{array} \right] \left\{ \begin{array}{c} \mathbf{R}_c \\ \mathbf{R}_e \end{array} \right\} \quad [16]$$

where $\mathbf{U}_c$ and $\mathbf{U}_e$ represent the test item interface and external motions obtained in the laboratory, and $\mathbf{R}_c$ and $\mathbf{R}_e$ contain the test item interface and external laboratory inputs. It is noticed here that eqn [16] assumes that the test item's FRF matrix in the laboratory is the same as in the field.

The vibration exciter is the mechanical structure responsible for generating the test item inputs in the laboratory environment. The vibration exciter input–output relationships are given by eqn [1] rewritten as:

$$\mathbf{Z} = \mathbf{E}\,\mathbf{Q} \quad [17]$$

where $\mathbf{Q}$ and $\mathbf{Z}$ are the exciter input and output vectors and $\mathbf{Q}$ is the exciter FRF matrix.

The same frequency domain substructuring approach will be employed in order to obtain expressions for laboratory interface forces and test item motions. Hence, the following equation is obtained for the laboratory interface forces:

$$\mathbf{R}_c = \mathbf{TE}(\mathbf{Z}_{ce} - \mathbf{U}_{ce}) \quad [18]$$

where $\mathbf{R}_c$ is the $E_c \times 1$ vector containing the test item interface forces in the laboratory, $\mathbf{Z}_{ce}$ contains the vibration exciter bare table motions and $\mathbf{U}_{ce}$ is the test item response to external forces. The $E_c \times E_c$ $\mathbf{TE}$ matrix is the laboratory interface FRF matrix and it is defined as:

$$\mathbf{TE} = (\mathbf{T}_{cc} + \mathbf{E}_{cc})^{-1} \quad [19]$$

As in the field, the laboratory interface forces depend on the relative motion vector $(\mathbf{Z}_{ce} - \mathbf{U}_{ce})$ between the vibration exciter and the test item interface points. In the absence of laboratory external forces acting on the test item, eqn [18] reduces to:

$$\mathbf{R}_c = \mathbf{TE}\,\mathbf{Z}_{ce} \quad [20]$$

since $\mathbf{U}_{ce} = 0$ in this case.

Substituting eqn [18] into eqn [16] results in the following expression for the test item laboratory motions:

$$\left\{ \begin{array}{c} \mathbf{U}_c \\ \mathbf{U}_e \end{array} \right\} = \left[\begin{array}{cc} \mathbf{T}_{cc}\mathbf{TE} & -\mathbf{T}_{cc}\mathbf{TE} \\ \mathbf{T}_{ec}\mathbf{TE} & -\mathbf{T}_{ec}\mathbf{TE} \end{array} \right] \left\{ \begin{array}{c} \mathbf{Z}_{ce} \\ \mathbf{U}_{ce} \end{array} \right\} + \left\{ \begin{array}{c} \mathbf{U}_{ce} \\ \mathbf{U}_{ee} \end{array} \right\} \quad [21]$$

When the test item is not subjected to laboratory external forces, eqn [21] reduces to:

$$\left\{ \begin{array}{c} \mathbf{U}_c \\ \mathbf{U}_e \end{array} \right\} = \left[\begin{array}{cc} \mathbf{T}_{cc}\mathbf{TE} & -\mathbf{T}_{cc}\mathbf{TE} \\ \mathbf{T}_{ec}\mathbf{TE} & -\mathbf{T}_{ec}\mathbf{TE} \end{array} \right] \left\{ \begin{array}{c} \mathbf{Z}_{ce} \\ 0 \end{array} \right\} \quad [22]$$

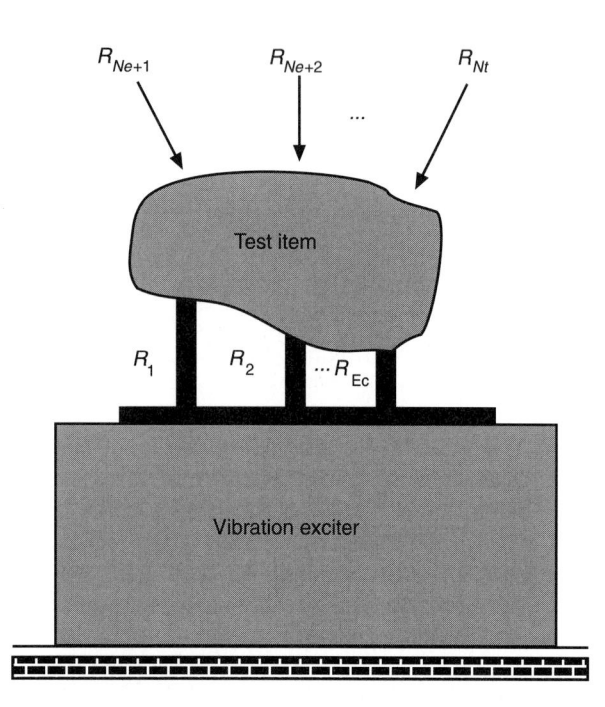

Figure 3 Test item attached to the vibration exciter in the laboratory environment.

Thus, in the absence of external forces, the test item laboratory motions depend on the bare exciter table motion $\mathbf{Z}_{ce}$ and on the FRF matrix shown in eqn [22]. In the next section, some laboratory test scenarios are discussed in order to explore the possibilities of a successful laboratory simulation.

Test Scenarios for Laboratory Simulations

This section is primarily focused on describing some test scenarios that can be used in laboratory simulations. In each test scenario a different control strategy is employed. These scenarios differ in terms of the data used to define the test item laboratory inputs.

Test Scenario 1: Inputs from the Bare Vehicle Motions

In this case, the test item is absent from the field. The bare vehicle interface motions are used to generate a set of forces that will be applied to the test item interface forces in the laboratory environment. These forces are given as:

$$\mathbf{T}_{cc}\mathbf{R}_c = \mathbf{Y}_{ce} \qquad [23]$$

or:

$$\mathbf{R}_c = \mathbf{T}_{cc}^{-1}\mathbf{Y}_{ce} \qquad [24]$$

where $\mathbf{R}_c$ contains the test item laboratory interface forces. A comparison between eqn [24] with the true field interface force, eqn [11], reveals some important issues. First, the field interface FRF matrix $\mathbf{TV}$ in eqn [11] accounts for the vehicle interface FRF characteristics, as seen from eqn [12], while the expression for the laboratory forces accounts only for the test item interface FRFs. Second, field external force effects are properly accounted for in eqn [11] while their effects are absent from eqn [24]. Thus, it is clear that the laboratory inputs given by eqn [24] will not predict correctly the corresponding field motions, and therefore, the inputs obtained from the bare vehicle interface motions are not appropriate in this case.

However, suppose that in a specific, rather common situation found in practice, the field external force effects can be neglected. Then, eqn [13] contains the true field interface forces, since $\mathbf{X}_{ce} = 0$. If, in addition to the bare vehicle interface motion vector $\mathbf{Y}_{ce}$, the vehicle interface FRF matrix $\mathbf{V}_{cc}$ is available in the laboratory environment; then, eqn [24] could be reformulated to include the information about $\mathbf{V}_{cc}$ so that the laboratory interface forces will match the corresponding field forces.

In addition, it is impossible to meet the requirements of multiple excitation shown in eqn [24] with a single vibration exciter, except in the simplest case where the test item is attached to the vehicle at a single interface point. Thus, once the correct inputs are generated, each test item interface point must be driven by a different exciter which in turn must be controlled to generate the correct input.

Test Scenario 2: Inputs from the Combined Structure

In this test scenario it is assumed that field data are available from the combined structure. As previously shown, information about the field combined structure consists of field interface forces and motions, and the test item external forces and motions. Hence, when using field data from the combined structure in order to define the test item laboratory inputs, the following multi-input control strategies can be employed:

- Matching interface forces in the laboratory. In this case the exciters must be controlled such that:

$$\mathbf{R}_c = \mathbf{F}_c \qquad [25]$$

- Matching external forces in the laboratory. In this case the exciters must be controlled such that:

$$\mathbf{R}_e = \mathbf{F}_e \qquad [26]$$

- Matching interface motions in the laboratory. In this case the exciters are controlled such that field interface motions are matched:

$$\mathbf{U}_c = \mathbf{X}_c \qquad [27]$$

- Matching external motions in the laboratory. This test scenario requires that the exciters be controlled such that:

$$\mathbf{U}_e = \mathbf{X}_e \qquad [28]$$

Thus, several control possibilities exist when attempting to match field data from the combined structure in the laboratory environment. Although a detailed description of these test scenarios can be found in the suggested literature, a brief discussion of the force identification problem is made in the next section of this article, since it is directly related to the issue of defining suitable test item inputs.

Discussion of the Force Identification Problem in Environmental Vibration Testing

Direct measurement of the test item interface and external forces in the field environment is generally difficult for a number of reasons. In the case of interface forces, a direct measurement requires that force transducers be placed in the loads paths, which in some cases is simply impractical due to space limitations and/or design considerations.

Because of these difficulties, indirect force identification techniques became popular in modal and in vibration testing. These force identification techniques can be defined in the time or frequency domain. In particular, when the data are analyzed in the frequency domain, the pseudoinverse technique is frequently employed.

For deterministic input and output signals, the frequency pseudoinverse solution for the unknown forces from measured motions is obtained from eqn [1] as:

$$\mathbf{F} = \mathbf{H}^+ \mathbf{X} \quad [29]$$

where the plus symbol denotes the pseudoinverse of the test item FRF matrix $\mathbf{H}$. Since there are usually a greater number of measured motions than unknown forces, the forces are obtained by solving a least-squares problem with the pseudoinverse of matrix $\mathbf{H}$, given as:

$$\mathbf{H}^+ = \left[\mathbf{H}^H \mathbf{H}\right]^{-1} \mathbf{H}^H \quad [30]$$

where the superscript H denotes the Hermitian operator (complex conjugate) of the $\mathbf{H}$ matrix. Use of eqns [29] and [30] is restricted to input and output frequency spectra, that carry both magnitude and phase information.

When input and output signals are random, the input–output relationships are given by the following expression:

$$\mathbf{G}_{xx} = \mathbf{H}^* \mathbf{G}_{ff} \mathbf{H}^T \quad [31]$$

where the asterisk and T denote the complex conjugate and nonconjugate transpose of the $\mathbf{H}$ matrix, respectively. The $\mathbf{G}_{ff}$ and $\mathbf{G}_{xx}$ matrices are the input and output frequency domain spectral density matrices whose diagonal entries are the real valued input and output autospectral densities, while the off-diagonal entries are the complex valued cross-spectral densities, respectively. The solution for the unknown input spectral density matrix $\mathbf{G}_{ff}$ from eqn [31] is obtained in a least-squares sense as:

$$\mathbf{G}_{ff} = [\mathbf{H}^+]^* \mathbf{G}_{xx} [\mathbf{H}^+]^T \quad [32]$$

where, in this case, the unknown $\mathbf{G}_{ff}$ matrix is obtained in a least-squares sense from the measured $\mathbf{G}_{xx}$ matrix and the test item FRF matrix $\mathbf{H}$. The $\mathbf{G}_{xx}$ and $\mathbf{G}_{ff}$ matrices as given by eqns [31] and [32] contain the correct phase relationships required by the input and output variables since the off-diagonal cross-spectral densities are complex with real and imaginary parts and are accounted for in both equations. Thus, proper correlation between the corresponding time variables is accounted for when using eqn [32] to solve for the unknown forces. Eqn [32] is referred to as the correlated equation.

A commonly accepted procedure when working with eqns [31] and [32] in random vibration environments is to assume that the input random forces are statistically uncorrelated. When this assumption is made, the off-diagonal input cross-spectral densities in the $\mathbf{G}_{ff}$ matrix in eqn [31] are identically zero and this equation reduces to:

$$\mathbf{G}_{xx} = \left[|\mathbf{H}|^2\right] \mathbf{G}_{ff} \quad [33]$$

where $\mathbf{G}_{ff}$ and $\mathbf{G}_{xx}$ are real valued vectors containing the input and output autospectral densities, respectively. The solution for the unknown input forces is obtained from eqn [33] and is given as:

$$\mathbf{G}_{ff} = \left[|\mathbf{H}|^2\right]^+ \mathbf{G}_{xx} \quad [34]$$

Thus, eqn [33] essentially shows that, under the assumption of uncorrelated input forces the resulting motions are equally uncorrelated and the solution for the input forces in eqn [34] results in estimates of the input autospectral densities, and no information about the input cross-spectral densities is obtained. Eqn [34] is referred to as the uncorrelated equation.

The feasibility of the correlated (eqn [32]) and uncorrelated (eqn [34]) is checked in a simple experiment where a cold rolled steel beam ($92 \times 1.25 \times 1.0$ in) is suspended by flexible cords and driven by two vibration exciters. One exciter is attached to the beam's midpoint while the second exciter is attached to one of the beam's endpoints. The exciters are driven by two independently generated random signals. The input random forces are measured by piezoelectric force transducers for comparison purposes. Acceleration auto- and cross-spectral densities and the beam's FRFs are measured at four different locations, including the points of application of the input forces. Eqns [32] and [34] are used in order to predict estimates of the measured input forces.

Figure 4 shows the results for the estimated input auto- and cross-spectral densities when the correlated eqn [32] is used to solve the inverse problem. Figure 4A and 4B show the estimated and measured force autospectral densities at the beam mid- and endpoint respectively. Figure 4C and 4D show magnitude and phase of the estimated and measured force cross-spectral density, respectively. The estimated inputs are in good agreement with the measured input autospectral densities, except in the vicinity of certain frequencies where the estimated force is distorted when compared with the measured ones. A reasonably good result is also obtained for the cross-spectral density between the excitation signals, as seen from Figure 4C and 4D.

The results for the uncorrelated eqn [34] are shown in Figure 5. Figure 5A shows the measured and estimated force autospectral density at the beam midpoint while Figure 5B shows the estimated force autospectral density for the beam endpoint. It is seen that the predicted values of the force at the beam endpoint presents serious distortions in comparison with the measured values. This is due to the fact that the motions are always correlated regardless of the input forces being correlated or uncorrelated. Hence, when solving for input random forces from measured motions, eqn [32] must be employed since it contains the correct phasing between the input and output variables.

See also: **Environmental testing, overview**.

Further Reading

Ewins DJ (1984) *Modal Testing: Theory and Practice.* UK: Research Studies Press.

Maia NMM and Silva J (1997) *Theoretical and Experimental Modal Analysis.* UK: Research Studies Press.

McConnell KG (1995) *Vibration Testing: Theory and Practice.* New York, NY: John Wiley.

McConnell KG and Varoto PS (1996) *A Model For Vibration Test Tailoring, Part I: Basic Definitions and Test Scenarios.* Tutorial session. XIV International Modal Analysis Conference, Dearborn, MI, pp. 1–7.

Varoto PS (1996) *The Rules for the Exchange and Analysis of Dynamic Information in Structural Vibration.* PhD dissertation, Iowa State University of Science and Technology, Ames, IA, USA.

Varoto PS and McConnell KG (1996a) *A Model For Vibration Test Tailoring, Part II: Numerically Simulated Results for a Deterministic Excitation and no External Loads.* Tutorial session. XIV International Modal Analysis Conference, Dearborn, MI, pp. 8–15.

Varoto PS and McConnell KG (1996b) *A Model For Vibration Test Tailoring, Part IV: Numerically Simulated Results for a Random Excitation.* Tutorial session. XIV

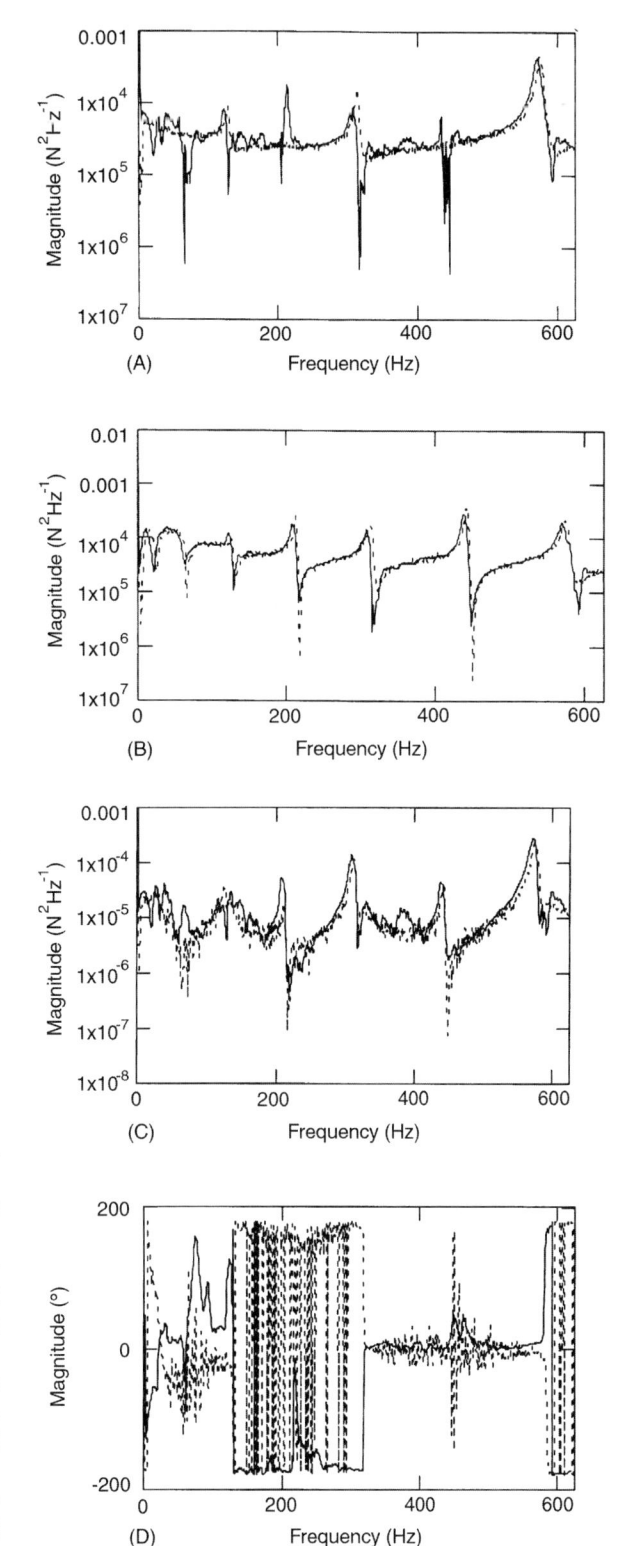

Figure 4 Estimated forces from correlated equation: (A) force autospectral density at beam midpoint; (B) force autospectral density at beam endpoint; (C) magnitude of force cross-spectral density between excitation points; (D) phase angle of force cross-spectral density between excitation points.

International Modal Analysis Conference, Dearborn, MI, pp. 31–47.

Varoto PS and McConnell KG (1996c) READI – a vibration testing model for the 21st century. *Sound and Vibration* October, 22–28.

Varoto PS and McConnell KG (1997a) Predicting random excitation forces from acceleration response measurements. In: *Proceedings of the XV International Modal Analysis Conference – IMAC*, Orlando, FL, February, pp. 1–8.

Varoto PS and McConnell KG (1997b) READI: the rules for the exchange and analysis of dynamic information in structural vibration. In: *Proceedings of the XV International Modal Analysis Conference*, Orlando, FL, pp. 1937–1944.

Varoto PS and McConnell KG (1998) *Single Point vs Multi Point Acceleration Transmissibility Concepts in Vibration Testing*. XVI International Modal Analysis Conference, Sta Barbara, CA, USA, February, pp. 734–843.

Varoto PS and McConnell KG (1999) On the identification of interface forces and motions in coupled structures. In: *Proceedings of the XVII International Modal Analysis Conference*, Orlando, FL, February, pp. 2031–2037.

Wirsching PH, Paez TL and Ortiz K (1995) *Random Vibrations: Theory and Practice*. New York: Wiley Interscience.

EQUATIONS OF MOTION

See **THEORY OF VIBRATION: EQUATIONS OF MOTION**

GLOSSARY

absorber, vibration:
  damped vibration absorber: an added device which introduces damping to a structure at a point by providing an inertial sprung mass against which an added damper can react, thereby introducing damping to a specifically targeted mode of vibration
  undamped vibration absorber: an added device which provokes an antiresonance at the point of attachment at a preselected frequency of forced vibration
accelerance: one of the family of different frequency response functions: the harmonic acceleration response in one degree-of-freedom to a single harmonic excitation force applied in the same or another degree-of-freedom. Also known as **inertance** (although this name is not approved by ISO). See also **mobility** and **receptance**
acceleration: see **displacement**
aliasing: the phenomenon in digital signal processing, due to inadequate sampling frequency, whereby a signal with a frequency f_N which is higher than half the digitization sampling frequency, f_S, appears in the resulting spectrum as if it were a signal with a frequency (f_{S-N}), which is lower than its actual value
analysis, modal: a procedure which extracts the normal mode properties of a system, either by analyzing measured response functions (experimental modal analysis — EMA) or by performing an eigensolution on the system's mass and stiffness matrices (analytical modal analysis)
antialiasing filter: signal conditioning or processing device which provides a low-pass filtering effect to reduce in a signal to be analyzed those frequency components which are higher than one-half the sampling frequency (the Nyquist frequency)
antiresonance: a condition in forced vibration whereby a specific point or DOF has zero amplitude at a specific frequency of vibration
Argand diagram: see **diagram**
autocorrelation function: (see **power spectral density**) the signal average of the product of the signal with a delayed version of itself. The inverse Fourier transform of the power spectral density function
balancing: the procedure of adjusting the mass distribution of a rotor so that vibrations or forces reactions during rotation are minimized

bandwidth: strictly, the frequency span encompassed by a resonance curve (Bode diagram of the modulus of a frequency response function) at the level which is 0.707 times the peak value. Alternatively, the frequency range of excitation and measurement during a vibration test
Bode diagram/plot: see **diagram**
burst random: see **signal**
burst sine: see **signal**
coherence: a measure of the degree of correlation between two signals presented in the frequency domain (i.e., a form of spectrum). Has a real value and is normalized between 0 and 1.
coordinate (or co-ordinate): strictly, the location of a point on a structure in space. Often used to mean a degree-of-freedom
  principal (or modal) coordinate: a time-varying quantity which describes the extent (amount) of motion in one of the systems modes of vibration. Sometimes referred to as **modal coordinate**
correlation: the systematic quantitative comparison of two sets of like data, such as time histories, or modal properties, resulting in a numerical measure of the degree of similarity or dissimilarity. See **coherence**
critical speed: speed of a rotating rotor that corresponds to a resonance
damping: any of several physical mechanisms whereby mechanical energy (i.e., kinetic or strain) is converted into heat and thereby removed from a vibrating system. In the absence of a steady input of energy, such as is achieved through continuous excitation of the system, the vibration will decay to rest as a direct consequence of the damping mechanisms
  hysteretic (also structural): a type of damping for which the mathematical model used to describe it is based on observation of the characteristics of damping due to internal hysteresis of materials
  proportional: refers to the distribution (rather than the type) of damping in a structure whereby the damping elements are assumed to be distributed in a way which reflects the distribution of stiffness, and/or mass, of the structure. Has the convenient property that the mode shapes (q.v.) of a proportionally damped structure are identical to the mode shapes of the undamped version of the same

structure, and are therefore real in a mathematical sense

structural: see **damping hysteretic**

viscous: a type of damping in which the damper force is proportional to velocity and whose mathematical description is easily accommodated in the models used to describe a structure's dynamics, especially for time domain analysis

critical: a level of viscous damping above which free oscillatory motion does not take place: it is common practice to define practical levels of damping (usually much lower than critical) as a fraction of critical damping (damping ratio)

dB: the dB (or decibel) is a unit for describing ratio of quantities, based on a logarithmic scale in which x described in dB = $20 \log x$. Thus, 0 dB represents a value of unity (1.0) while 20 dB represents a value of 10.0. Also used to describe ratio of power or energy quantities, in which case a power level P is described as $10 \log P$. Absolute values need to be related to a reference unit

degree-of-freedom: used to define the location and direction of the motion of a specific point of a system or structure. Sometimes called a **coordinate**. Each point on a structure has six degrees of freedom (three translation and three rotation directions)

diagram:

Argand: a plot of a complex function, such as an FRF, in which the x–y axes are the real and imaginary parts of the function, respectively. The corresponding data, plotted as modulus and phase on polar axes, result in an identical plot. Also known as a **Nyquist diagram**

Bode: a form of presentation of a complex function, such as an FRF, in which there are two graphs: the first presenting the log-modulus of the function against log-frequency, and a second plot of the phase (of the complex function) against log-frequency. Used to convey information covering a wide frequency range

Nyquist: see **diagram, Argand**

discrete Fourier transform (DFT): see **transform**

displacement: change of position of element. Often measured in conjunction with rotating rotors and active control tasks. The measurement and use of velocity and acceleration, the first and second derivative of displacement prevail in vibration task. Off-the-shelf acceleration-measuring systems comprise the majority used by practitioners (see **jerk**)

dynamic range: response range of a transducer or data acquisition system between maximum-level noise level: often expressed as a ratio in dB

eigenvalue: mathematical term for a root of the characteristic equation of a dynamic system: generally equal to the square of one of the system's natural frequencies

eigenvector (see **mode shape**): mathematical term for one of the vectors of the characteristic equation of a dynamic system: describes one of the system's normal mode shapes. The matrix product of the eigenvector (transposed) times the mass matrix times the eigenvector is equal to the **modal mass** (q.v.); when the modal mass is unity, the eigenvector is said to be mass-normalized

fast Fourier transform (FFT): see **transform**

forced vibration: see **vibration**

Fourier: see **transform**

free vibration: see **vibration**

frequency:

excitation: the frequency at which a system is made to vibrate under the influence of an external harmonic force

natural (undamped, damped): a frequency at which a structure will vibrate when given an initial disturbance from its equilibrium position and then left to move under free vibration conditions. If the structure is damped, the mathematical description of this frequency will be complex, having an imaginary part which describes the oscillation rate and a real part which defines the decay rate

resonance (sometimes, inaccurately, **resonant**): a frequency at which an FRF or other response curve reaches a local maximum value in the vicinity of a natural frequency. This resonance frequency will often be very close to, but not necessarily equal to, the corresponding natural frequency

sampling: the frequency at which a signal is digitized (sampled, discretized) for the purpose of digital spectral analysis

Nyquist: half the sampling frequency used to discretize a signal prior to digital spectral analysis

frequency response function (FRF): see **function**

forced vibration: see **vibration**

function:

frequency response (FRF): the ratio between the steady-state harmonic response in one degree of freedom of a structure (j, say) to the similarly steady-state harmonic force applied at another degree of freedom of the structure (such as k)

    point: a particular FRF in which the excitation and response degrees of freedom are the same (i.e., $j = k$).

    transfer: an FRF in which the excitation and response degrees of freedom are not the same (i.e., $j \neq k$).

impulse response function (IRF): The response time history of a system, measured in one particular degree of freedom to a unit impulse excitation applied at another degree of freedom

harmonic: (*n.*) an integer multiple of a fundamental sinusoidal signal to which it is related

(*adj.*) equivalent to 'sinusoidal' in describing signal type

Hilbert transform: see transform

impedance, mechanical: one of the family of inverse frequency response functions: the ratio between the single harmonic excitation force applied in one degree of freedom to the harmonic velocity response in the same or another degree of freedom.

impulse response function (IRF): see function

inertance: see accelerance

isolation, vibration: a feature of forced vibration in which vibration levels in certain regions of a structure are constrained to be much lower than at other regions by suitable distribution of mass and stiffness

jerk: derivative of acceleration, used in conjunction with responses to shock excitations

leakage: a phenomenon which arises during digital spectral analysis of continuous signals as a result of the finite duration of the sampled signal and its relationship with the frequencies present in the original signal. Often, leakage occurs because the signal being analyzed is not periodic with respect to the sample length (duration, period) used for the analysis. The consequence of leakage is that a computed spectrum may indicate the presence of signal components which are not in the original signal and it may not accurately indicate those which are. Windows are used to reduce the effects of leakage

linearity: a characteristic often assumed to be present in structures subjected to vibration analysis or testing techniques: the primary assumption is essentially that the response of the structure to two forces applied simultaneously will be identical to that derived by adding the responses of the structure to each force applied individually. Other definitions can also be applied

line spectrum: see spectrum

MAC or modal assurance criterion: a simple parameter used to measure the degree of correlation between two vectors of the same length: usually eigenvectors, or mode shapes. A MAC value of 100% indicates that the two vectors being compared are parallel, or are perfectly correlated. MAC values of greater than 80% generally indicate 'similar' mode shapes

mass, modal: a quantity which relates to a given mode shape and which is computed from that mode shape vector and the mass matrix of the system model. The modal mass is automatically unity for mass-normalized mode shapes

mobility: one of the family of frequency response functions: the harmonic velocity response in one degree of freedom to a single harmonic excitation force applied in the same or another degree of freedom

modal coordinate: see coordinate

modal mass: see mass, modal

modal stiffness: see stiffness, modal

mode, mode shape (see eigenvector): strictly, a pattern of vibration exhibited by a structure or system. Generally, described as a vector of values, defining the relative displacement amplitudes and phases of each degree of freedom, which describes the motion of the system

    complex: a mode of vibration in which the relative displacements exhibit phase differences between one degree of freedom and the next which are not always 0 or 180° (situation which applies in a real mode). Complex modes do not exhibit modal lines

    mass-normalized (see eigenvector)

    normal, damped: the vector of relative amplitudes for the vibration of a damped system in one of its natural modes of free vibration. These modes are the modes which the structure assumes in the absence of any external excitation. In this case, the structure is assumed to possess arbitrary damping and, as a result, the normal modes are complex

    normal, undamped: the vector of relative amplitudes for the vibration of the undamped system in one of its natural modes of free vibration. These modes are the modes which the structure assumes in the absence of any external excitation. In this case, the structure is assumed to possess no damping and, as a result, the normal modes are real

    real: a mode of vibration in which the relative displacements exhibit phase differences between one degree of freedom and the next which are always 0 or 180°. Real modes exhibit modal lines

mode indicator function: a function which is used to indicate the existence of global system modes (those which are observed throughout the structure) as opposed to local modes (which are only visible in a small region of the structure)

model (mathematical): a mathematical description of a system or structure in a form which can be described by equations of motion which can then be solved analytically or numerically to describe that system's dynamic behavior

    continuous: a mathematical model which describes the system or structure as a continuous medium, using partial differential equations with, typically, transcendental expressions for the solutions

    discrete, lumped parameter: an alternative type of mathematical model (to the continuous type) in which the mass, stiffness, and damping properties of the structure are discretized so that the behavior can be described by a finite series of ordinary differential equations leading to finite discrete solutions

    modal: a model which is defined in terms of eigenvalues and eigenvectors, or natural frequencies and mode shapes. Usually presented as eigenvalue and eigenvector matrices

    response: a model which is defined in terms of some characteristic response functions, such as FRFs, or IRFs, or equivalent

    spatial: a model which is defined in terms of the system's distribution of mass, stiffness, and damping in space. Usually described as mass, stiffness, and damping matrices

multiple-degrees-of-freedom (MDOF) system (see **single-degree-of-freedom (SDOF) system**): a system for which two or more coordinates or degrees of freedom are needed to define completely the position of the system at any instant

natural frequency: see **frequency**

node: 1. a DOF which has zero amplitude of vibration; used when describing a mode shape or an operating deflection shape
2. a specific grid point in a discrete (lumped parameter) model of a structure

nonlinearity: a form of structural behavior which does not conform to the definition of linearity (see **linearity**)

normal mode: see **mode**

Nyquist diagram: see **diagram**

operating deflection shape: see **mode shape**

pole: mathematical term which is equivalent to eigenvalue or to (square of) natural frequency. For systems describable by a ratio of rational polynomials in a transform domain, the value of the transform variable for which the denominator polynomial equals zero

power spectral density (PSD): (see **autocorrelation function**) the distribution of the squared signal (acceleration, velocity, etc.) with frequency (in squared units per Hz, i.e., g^2/Hz). The Fourier transform of the autocorrelation function

principal coordinate: see **coordinate**

receptance: one of the family of frequency response functions: the harmonic displacement response in one degree of freedom to a single harmonic excitation force applied in the same or another degree of freedom. See also **mobility** and **accelerance**

resonance (see **frequency**): a phenomenon associated with forced vibration which relates to a narrow range of frequencies in which the response amplitude of a system attains levels which are much higher than elsewhere

sampling (see **frequency**)

series, Fourier: a method of representing a periodic signal as the sum of separate sinusoids whose frequencies are integer multiples of the fundamental period of the original signal. See also **transform, Fourier**

signal (time history): a time-varying quantity which describes the excitation or response of a vibrating system. In vibration testing, a quantity which is used either to drive the excitation of a system into vibration or to represent the measurement of one of the vibration forces or responses which are used to describe the system's movement. Types of signal commonly encountered include:

burst:

    random: a signal which consists of two sections, the first being a finite duration of a stationary random signal followed by a second section which is zero

    sine: a signal which consists of two sections, the first being a finite duration of a steady sinusoid followed by a second section which is zero

    chirp: a signal comprising a single rapidly swept sine wave between a minimum and maximum frequency

    harmonic: a sinusoidal signal at a single frequency

    periodic: A signal which repeats itself indefinitely after a period, T. In signal analysis or processing, the period of the repeating signal is not necessarily the same as the period over which the signal is measured or analyzed and in such a situation, analysis of the signal may not yield a true indication of its form (see also **leakage**)

periodic random: a signal which consists of a succession of records (sometimes called 'samples'), each of finite length, *T*, which is usually selected to suit the sampling period of a means of analyzing the signal. Each of the first few records is identical to each other and thus represent a periodic signal, even though an individual period appears to be very complex, almost random. Then the next few records consist of several repetitions of another 'complex' record. This pattern is repeated for as long as the signal is required

pseudorandom: a periodic signal for which the basis record is very complex, and appears to be random. However, this record is repeated indefinitely and so forms a periodic signal

random: a nondeterministic signal. No analytical expression predicting future values can exist. It is defined only in terms of statistical functions and parameters, for example an amplitude probability function

sine, sinusoid, discrete sine: a signal which consists of a pure, single, sinusoidal or harmonic component; also **stepped sine**

sine sweep: a signal which comprises a sinusoid whose frequency is varied slowly so that a range of frequencies is covered throughout the duration of the signal

stepped sine: a signal which comprise a series of discrete sinusoids, one after the other

transient: a signal which lasts a finite length of time, being zero before and after that period, and which is fully defined within the record length

white noise: a signal with a constant **power spectral density**, independent of frequency. In practice a signal can have white noise characteristics only in a finite frequency range

single-degree-of-freedom (SDOF) system (see also: **multiple-degrees-of-freedom (MDOF) system**): a system for which only one coordinate is needed to define completely the position of the system at any instant

spectrum: description of the essential elements of a signal (or time history) in terms of constituent sinusoidal components

  continuous: nonzero in a continuous range of frequencies. Typical for **random signals**

  discrete: a spectrum computed only at discrete frequencies. These frequencies are usually equi-spaced, a result of using the **fast Fourier transform** algorithm. Sometimes synonymous with **line spectrum**

  line: a discrete spectrum, with nonzero values only at discrete frequencies. Typical of vibrations generated by rotating machinery

stiffness, modal: a quantity which relates to a given mode shape and which is computed from that mode shape vector and the stiffness matrix of the system model. The modal stiffness is equal to the eigenvalue for a mass-normalized mode shape

time history (see **signal**): description of a time-varying quantity, such as displacement or force, as a direct function of time

transducer: a device which transforms one form of energy into another, practically used synonymously with the term sensor

transform

  discrete (DFT): a discrete transform which converts a a discrete (usually time domain) sequence into another discrete (usually frequency domain) sequence. With correct sampling, the DFT approximates samples of the (continuous) Fourier transform In the DFT, the continuous time-varying signal is approximated by a discrete number of sampled values taken from the original data. The resulting spectrum has exactly the same number of components as the number of sampled data points, although not all of these may be displayed

  fast Fourier transform (FFT): a computer algorithm for the efficient computation of the DFT of a sampled signal. The algorithm takes advantage of the symmetry of the transform and is very efficient when applied to a data set containing 2^N values

  Fourier: a continuous transform between two domains, usually time and frequency. One of the most basic mathematical tools for representing the relevant features of a time history signal in terms of its constituent frequency components, and of reconstructing a time record from a knowledge of the frequency (or spectral) components. For periodic signals, the transform exists only at discrete frequencies (see **line spectrum**), and is more conveniently represented via the **Fourier** series

  Hilbert: a type of transform for converting a function from one form to another which, unlike the Fourier transform, derives a transformed function in the same domain as the original (i.e., a time signal transforms to another time signal, a frequency function to another frequency function)

validation: the process of demonstrating the adequacy of the performance of a given prediction capability, usually that of a finite element model constructed to predict the vibration properties of the structure

velocity (see **displacement**)

verification: the process(es) used to demonstrate that an algorithm or other procedure has been correctly implemented in a computer program or test sequence

vibration

    forced: the vibration structure which is the direct result of an applied excitation force, or forces. Although there will be an initial transient response which is governed by the free vibration characteristics of the system, this will die away, leaving only the steady response due entirely to the externally applied excitation. This motion can be described completely in terms of the free vibration normal modes of the system

    free: the type of vibration which ensues when a system or structure is set into motion by an initial disturbance and then left to vibrate under the influence only of the internal forces due to its own stiffness, inertia, and damping. This motion can be described completely in terms of the free vibration normal modes of the system

white noise: see **signal**

window: an approach used in signal processing to minimize the approximations introduced when using a discrete Fourier transformation on a continuous signal. Usually applied in the time domain, as a multiplying function applied to the time signal. Various shapes of window functions have been developed to produce optimum frequency information for particular types of signal

APPENDICES

1. STANDARD NOMENCLATURE

2. ABBREVIATIONS

3. UNITS AND CONVERSION FACTORS
   Common units used in engineering and vibration control
   Conversion factors

4. PHYSICAL PROPERTIES OF SOME COMMON SUBSTANCES
   Solids
   Liquids
   Gases

5. VIBRATION MODES OF CLASSICAL ELEMENTS

6. STANDARDS AND GUIDELINES

7. FREQUENCY RESPONSE FUNCTIONS

1. STANDARD NOMENCLATURE

Matrices and vectors

$\mathbf{A}$	Matrix
$\mathbf{a}$	vector
a_{jk}	matrix element (row j; column k)
$\mathbf{A}^T$	transpose
$\mathbf{A}^H$	Hermetian transpose
$\mathbf{A}^+$	generalized inverse

Physical parameters

m	mass element
k	stiffness element
c	viscous damping element
$\mathbf{M}$	mass matrix
$\mathbf{K}$	stiffness matrix
$\mathbf{C}$	viscous damping matrix
ζ	damping ratio
η	structural damping loss factor
$x(t)$ or x	displacement
$v(t)$ or v or $\dot{x}$	velocity
$a(t)$ or a or $\dot{v}$	acceleration
$f(t)$ or f	force
t	time
Δt	sampling interval
s	Laplace operator
ω	frequency (rad s^{-1})
ω_n	undamped natural frequency
ω_d	damped natural frequency
ω_R	resonant frequency
Ω	rotation speed
$H(\omega)$	frequency response function
$h(t)$	impulse response function
E	energy; modulus
P	power
$p, P(..)$	probability density, probability distribution
$R(\tau)$	correlation function
$R_{xx}(\tau), R_{xy}(\tau)$	auto- and cross-correlation
$S(\omega)$	spectral density
$S_{xx}(\omega), S_{xy}(\omega)$	auto- and cross-spectral density
i, j	$\sqrt{-1}$
$\otimes$	convolution
$k_{..} = f(K_{..})$	dots denote indices, i.e. $k_{12} = f(K_{12})$, etc.
x^*	complex conjugate (of x)

Transforms

F[ ], F^{-1}[ ]	Fourier transform, inverse Fourier transform
L[ ], L^{-1}[ ]	Laplace transform, inverse Laplace transform
Z[ ], Z^{-1}[ ]	Z transform, inverse Z transform
H[ ], H^{-1}[ ]	Hilbert transform, inverse Hilbert transform

2. ABBREVIATIONS

ADC	analog-to-digital converter	MAC	modal assurance criteria (also FMAC, AUTOMAC, FDAC, FRAC ...)
ARMA	autoregressive moving average		
CAD	computer-aided design	MDOF	multi-degree-of-freedom
CFD	computational fluid dynamics	MEMS	microelectromechanical systems
CG	center of gravity	MIFs	mode indicator functions
CWT	continuous wavelet transform	MIMO	multi-input multi-output
DAC	digital-to-analog converter	MISO	multi-input single-output
DOF	degree of freedom	NDT	nondestructive testing
DFT	discrete Fourier transform	NVH	noise, vibration, harshness
DSP	digital signal processing	ODS	operating deflection shape
DWT	discrete wavelet transform	PCB	printed circuit board
EOM	equation of motion	PDF	probability density function
EU	engineering units; European Union	PSD	power spectral density
FEA	finite element analysis	QA	quality assurance
FEM	finite element method	RMS	root mean square
FFT	fast Fourier transform	SDOF	single-degree-of-freedom
FIR	finite impulse response	SEA	statistical energy analysis
FPE	final prediction error	S/H	sample and hold
FRF	frequency response function	SIMO	single-input multi-output
IC	integrated circuit	SISO	single-input single-output
IIR	infinite impulse response	SLDV	scanning laser Doppler vibrometer
IRF	impulse response function	SNR	signal-to-noise ratio
ISO	International Standards Organization	STFT	short-time (term) Fourier transform
LDV	laser Doppler vibrometer	SVD	singular value decomposition
LQ	linear quadratic	TDA	time domain (synchronous) averaging
LSB	least significant bit	WVD	Wigner-Ville distribution
LVDT	linear voltage differential transformer	YW	Yule Walker

3. UNITS AND CONVERSION FACTORS

Common units used in engineering noise and vibration control

Primary units
length metre (m)
mass kilogramme (kg)
quantity of a substance mol

temperature kelvin (K)
time second (s)

Secondary units
energy joule (J ≡ Nm ≡ Ws)
force newton (N ≡ kg m s^{-2})
frequency hertz (s^{-1})
molecular weight mol^{-1}

power watt (W ≡ J s^{-1} ≡ Nm s^{-1})
pressure Pascal (Pa ≡ N m^{-1})
radian frequency rad s^{-1}

Derived units
acceleration m s^{-2}
acoustic (radiation) impedance[a] (force per unit area/ volume velocity per unit area) Ns m^{-1}
acoustic intensity W m^{-2}
autospectral density (power spectral density) units^2 Hz^{-1}
energy density J m^{-3}
energy spectral density units^2 s Hz^{-1}
entropy J kg^{-1} K^{-1}
gas constant J kg^{-1} K^{-1}
mechanical impedance (force/velocity) N s m^{-1}
mechanical stiffness N m^{-1}

mobility m N^{-1} s^{-1}
modulus of elasticity, adiabatic bulk modulus N m^{-2}
quefrency s
specific acoustic impedance (pressure/velocity) N s m^{-3}
surface density kg m^{-2}
universal gas constant J mol^{-1} K^{-1}
velocity m s^{-1}
viscosity N s m^{-2}
viscous damping N s m^{-1}
volume density kg m^{-3}
volume velocity m^3 s^{-1}

Conversion factors

Length
1 ft = 0.3048 m
1 in = 25.4 mm
1 mile = 1.609344 km

1 mph = 1.6093 km h^{-1} = 0.44704 m s^{-1}
1 nautical mile = 1.852 km
1 knot = 1 nm h^{-1} = 0.5144 m s^{-1}

Area
1 ft^2 = 0.09290304 m^2
1 in^2 = 0.00064516 m^2

1 acre = 4046.86 m^2
1 hectare = 10^4 m^2

Volume
1 ft^3 = 28.3168 litre
1 litre = 10^{-3} m^3
1 UK gal = 4.54609 litre

1 UK pint = 0.568261 litre
1 US gal = 3.7853 litre

Mass
1 lb = 0.45359237 kg
1 oz = 28.3495 g

1 ton = 1.01605 tonne
1 lb ft^{-3} = 16.0185 kg m^{-3}

[a] Acoustic impedance is sometimes defined as pressure/volume velocity.

Force (N, kg m s^{-2})
1 lbf = 4.44822 N
1 kgf = 1 kp = 9.80665 N
1 bar = 14.50 psi = 10^6 dyne cm^{-1}
     = 10^5 Pa = 10^{-5} N m^{-2}

1 psi = 6.89476 kPa
1 mm H$_2$O = 9.80665 Pa
1 mm Hg = 133.322 Pa
1 atm = 101.325 kPa

Energy (J, Nm, Ws)
1 ft-lbf = 1.355818 J
1 Btu = 1055.06 J

1 kWh = 3.6 MJ
1 kcal = 4.1868 kJ

Power (W, J s^{-1}, N m s^{-1})
1 ft-lbs s^{-1} = 1.355818 W
1 hp = 745.7 W

1 kcal h^{-1} = 1.163 W

Temperature
a °C = b K − 273.15
a °C = (b F − 32)/1.8

b °F = (1.8 × a °C) + 32

Units and conversion factors (vibration-oriented)

Conversion factors for rotational velocity and acceleration

Multiply value in or by → to obtain value in ↓	rad s^{-1} rad s^{-2}	degree s^{-1} degree s^{-2}	rev s^{-1} rev s^{-2}	rev min^{-1} rev min s^{-1}
rad s^{-1} rad s^{-2}	1	0.01745	6.283	0.1047
degree s^{-1} degree s^{-2}	57.30	1	360	6.00
rev s^{-1} rev s^{-2}	0.1592	0.00278	1	0.0167
rev min^{-1} rev min s^{-1}	9.549	0.1667	60	1

Conversion factors for simple harmonic motion

Multiply numerical value in terms of by → to obtain value in terms of ↓	Amplitude	Average value	Root-mean-square value (rms)	Peak-to-peak value
Amplitude	1	1.571	1.414	0.500
Average value	0.637	1	0.900	0.318
Root-mean-square value (rms)	0.707	1.111	1	0.354
Peak-to-peak value	2.000	3.142	2.828	1

APPENDIX 3: UNITS AND CONVERSION FACTORS

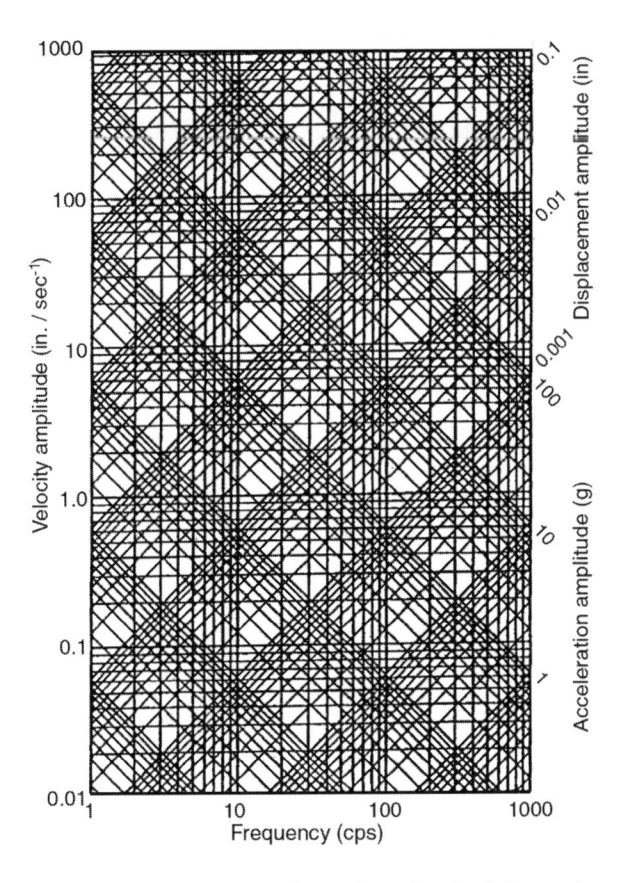

Relation of frequency to the amplitudes of displacement, velocity, and acceleration in harmonic motion.

4. PHYSICAL PROPERTIES OF SOME COMMON SUBSTANCES

Solids

Solid	Density, ρ_0 (kg m^{-3})	Young's modulus, E (Pa)	Poisson's ratio, v	Wavespeed, c_L Bar (m s^{-1})	Wavespeed, c_L Bulk (m s^{-1})	Product of critical frequency (f_c) and bar thickness[a] (m s^{-1})
Aluminum	2700	7.1×10^{10}	0.33	5150	6300	12.7
Brass	8500	10.4×10^{10}	0.37	3500	4700	18.7
Concrete (dense)	2600	$\sim 2.5 \times 10^{10}$	–	–	3100	21.1
Copper	8900	12.2×10^{10}	0.35	3700	5000	17.7
Cork	250	6.2×10^{10}	–	–	500	130.7
Cast iron	7700	10.5×10^{10}	0.28	3700	4350	17.7
Glass (Pyrex)	2300	6.2×10^{10}	0.24	5200	5600	12.6
Gypsum (plasterboard)	650	–	–	–	6800	9.61
Lead	11300	1.65×10^{10}	0.44	1200	2050	54.5
Nickel	8800	21×10^{10}	0.31	4900	5850	13.3
Particle board	750	–	–	–	669	97.7
Polyurethane	72	1.9×10^7	–	–	513	127.4
Polystyrene	42	1.1×10^7	–	–	512	127.6
PVC	66	5.5×10^7	–	–	913	71.6
Plywood	600	–	–	–	3080	21.2
Rubber (hard)	1100	2.3×10^9	0.4	1450	2400	45.1
Rubber (soft)	950	5×10^6	–	–	1050	62.2
Silver	10500	7.8×10^{10}	0.37	2700	3700	24.2
Steel	7700	19.5×10^{10}	0.28	5050	6100	12.9
Tin	7300	4.5×10^{10}	0.33	2500	–	26.1
Wood (hard)	650	1.2×10^{10}	–	4300	–	15.2

[a] Where a bar thickness does not apply, the thickness of the bulk material is used.

APPENDIX 4: PHYSICAL PROPERTIES OF SOME COMMON SUBSTANCES

Liquids

Liquid	Density ρ_0 (kg m^{-3})	Temperature (°C)	Specific heat ratio, γ	Wavespeed, c (m s^{-1})
Castor oil	950	20	–	1540
Ethyl alcohol	790	20	–	1150
Fresh water	998	20	1.004	1483
Fresh water	998	13	1.004	1441
Glycerin	1260	20	–	1980
Mercury	13600	20	1.13	1450
Petrol	680	20	–	1390
Sea water	1026	13	1.01	1500
Turpentine	870	20	1.27	1250

Gases

Gas	Density ρ_0 (kg m^{-3})	Temperature (°C)	Specific heat ratio, γ	Wavespeed, c (m s^{-1})
Air	1.293	0	1.402	332
Air	1.21	20	1.402	343
Carbon dioxide	1.84	20	1.40	267
Hydrogen	0.084	0	1.41	1270
Hydrogen	0.084	20	1.41	1330
Nitrogen	1.17	20	1.40	349
Oxygen	1.43	0	1.40	317
Oxygen	1.43	20	1.40	326
Steam	0.6	100	1.324	405

5. VIBRATION MODES OF CLASSICAL ELEMENTS

Natural frequencies and normal modes of uniform beams

Supports	Mode n	(A) Shape and nodes (numbers give location of nodes in fraction of length from left end)	(B) Boundary conditions	(C) Frequency equation	(D) Constants	(E) kl $w_n = k^2 \sqrt{\dfrac{EIg}{Ay}}$	(F) R Ratio of nonzero constants column (D)
Hinged-hinged	1 2 3 4 n>4	0.50 0.333 0.667 0.25 0.50 0.75	$x=0 \begin{cases} x=0 \\ x''=0 \end{cases}$ $x=l \begin{cases} x=0 \\ x''=0 \end{cases}$	$\sin kl = 0$	$A = 0$ $B = 0$ $\dfrac{C}{D} = 1$	3.1416 6.283 9.425 12.566 $\approx n\pi$	1.0000 1.0000 1.0000 1.0000 1.0000
Clamped-clamped	1 2 3 4 n>4	0.50 0.359 0.641 0.278 0.50 0.722	$x=0 \begin{cases} x=0 \\ x'=0 \end{cases}$ $x=l \begin{cases} x=0 \\ x'=0 \end{cases}$	$(\cos kl)(\cosh kl)$ $= 1$	$A = 0$ $C = 0$ $\dfrac{D}{B} = R$	4.730 7.853 10.996 14.137 $\approx \dfrac{(2n+l)\pi}{2}$	−0.9825 −1.0008 −1.0000− −1.0000+ −1.0000−
Clamped-hinged	1 2 3 4 n>4	0.558 0.386 0.692 0.294 0.529 0.765	$x=0 \begin{cases} x=0 \\ x'=0 \end{cases}$ $x=l \begin{cases} x=0 \\ x''=0 \end{cases}$	$\tan kl =$ $\tanh kl$	$A = 0$ $C = 0$ $\dfrac{D}{B} = R$	3.927 7.069 10.210 13.352 $\approx \dfrac{(4n+l)\pi}{4}$	−1.0008 −1.0000+ −1.0000 −1.0000 −1.0000
Clamped-free	1 2 3 4 n>4	0.783 0.504 0.868 0.358 0.644 0.906	$x=0 \begin{cases} x=0 \\ x''=0 \end{cases}$ $x=l \begin{cases} x''=0 \\ x'''=0 \end{cases}$	$(\cos kl)(\cosh kl)$ $= -1$	$A = 0$ $C = 0$ $\dfrac{D}{B} = R$	1.875 4.694 7.855 10.996 $\approx \dfrac{(2n-1)\pi}{2}$	−0.7341 −1.0185 −0.9992 −1.0000+ −1.0000−
Free-free	1 2 3 4 5 n>5	0.224 0.776 0.132 0.50 0.868 0.094 0.356 0.644 0.906 0.0734 0.277 0.50 0.723 0.927	$x=0 \begin{cases} x''=0 \\ x'''=0 \end{cases}$ $x=l \begin{cases} x''=0 \\ x'''=0 \end{cases}$	$(\cos kl)(\cosh kl)$ $= -1$	$B = 0$ $D = 0$ $\dfrac{C}{A} = R$	0 (represents translation) 4.730 7.853 10.996 14.137 $\approx \dfrac{(2n-l)\pi}{2}$	 −0.9825 −1.0008 −1.0000− −1.0000+ −1.0000−

APPENDIX 5: VIBRATION MODES OF CLASSICAL ELEMENTS

Natural frequencies and nodal lines of square plates with various edge conditions (after D. Young. *J. Appl. Mechanics*, 17:448 (1950))

	1st mode	2nd mode	3rd mode	4th mode	5th mode	6th mode
$\omega_n \sqrt{Da/\gamma ha^4}$	3.494	8.547	21.44	27.46	31.17	
Nodal lines						
$\omega_n \sqrt{Da/\gamma ha^4}$	35.99	73.41	108.27	131.64	132.25	165.15
Nodal lines						
$\omega_n \sqrt{Da/\gamma ha^4}$	6.958	24.08	26.80	48.05	63.14	
Nodal lines						

$\omega_n = 2\pi f_n$; $D = Eh^3/12(1-\mu^2)$; γ = weight density; h = plate thickness; a = plate length.

Natural frequencies and nodal lines of cantilevered rectangular and skew rectangular plates ($\mu = 0.3$)* (after Barton, M.V. *J. Appl. Mechanics*, 1951; 18:129)

Mode \ a/b	1/2	1	2	5
First	3.508	3.494	3.472	3.450
Second	5.372	8.547	14.93	34.73
Third	21.96	21.44	21.61	21.52
Fourth	10.26	27.46	94.49	563.9
Fifth	24.85	31.17	48.71	105.9

Mode	First	Second	First	Second	First	Second
$\omega_n \sqrt{Dg/\gamma ha^4}$	3.601	8.872	3.961	10.190	4.824	13.75
Nodal lines	15°	15°	30°	30°	45°	45°

APPENDIX 5: VIBRATION MODES OF CLASSICAL ELEMENTS

Natural frequencies of complete circular rings whose thickness in radial direction is small compared to radius

Type of vibration	Shape of lowest mode	Rectangular cross-section ω_n	Circular cross-section ω_n
Flexural in plane of ring with n complete wavelength in circumference	$n=2$	$\sqrt{\dfrac{Eg}{\gamma}\dfrac{I}{AR^4}\dfrac{n^2(n^2-1)^2}{n^2+1}}$ n any integer > 1	$\sqrt{\dfrac{E\pi r^4}{4mR^4}\dfrac{n^2(n^2-1)^2}{n^2+1}}$ n any integer > 1
Flexural normal to plane of ring	$n=2$		$\sqrt{\dfrac{E\pi r^4}{4mR^4}\dfrac{n^2(n^2-1)^2}{n^2+1+\mu}}$ n any integer > 1
Torsional		First mode $\sqrt{\dfrac{Eg}{\gamma R^2}\dfrac{Ix}{Ip}}$	$\sqrt{\dfrac{G\pi r^2}{mR^2}(n^2+1+\mu)}$ $n=0$, or any integer
Extensional		$\sqrt{\dfrac{Eg}{\gamma R^2}}$	$\sqrt{\dfrac{E\pi r^2}{mR^2}(1+n^2)}$ $n=0$, or any integer

E = modulus of elasticity; G = modulus of rigidity; γ = weight density; n: defined for each type of vibration; r = radius of ring; μ = Poisson's ratio. Properties of cross-sections: I = moment of inertia with respect to axis of section; I_x = moment of inertia with respect to radial line; I_p = polar moment of inertia; A = area; r = radius; m = mass per unit of length.

6. STANDARDS AND GUIDELINES

ISO 10055:1996 Mechanical vibration – vibration testing requirements for shipboard equipment and machinery components

ISO 10068:1998 Mechanical vibration and shock – free, mechanical impedance of the human hand–arm system at the driving point

ISO 10137:1992 Bases for design of structures; serviceability of buildings against vibration

ISO 10142:1996 Carbonaceous materials for use in the production of aluminium – calcined coke – determination of grain stability using a laboratory vibration mill

ISO 10326-1:1992 Mechanical vibration; laboratory method for evaluating vehicle seat vibration; part 1: basic requirements

ISO/TS 10811-1:2000 Mechanical vibration and shock – vibration and shock in buildings with sensitive equipment – part 1: measurement and evaluation

ISO/TS 10811-2:2000 Mechanical vibration and shock – vibration and shock in buildings with sensitive equipment – part 2: classification

ISO 10814:1996 Mechanical vibration – susceptibility and sensitivity of machines to unbalance

ISO 10815:1996 Mechanical vibration – measurement of vibration generated internally in railway tunnels by the passage of trains

ISO 10816-1:1995 Mechanical vibration – evaluation of machine vibration by measurements on non-rotating parts – part 1: general guidelines

ISO 10816-2: 1996 Mechanical vibration – evaluation of machine vibration by measurements on non-rotating parts – part 2: large land-based steam turbine generator sets in excess of 50 MW

ISO 10816-3: 1998 Mechanical vibration – evaluation of machine vibration by measurements on non-rotating parts – part 3: industrial machines with nominal power above 15 kW and nominal speeds between 120 r/min and 15000 r/min when measured in situ

ISO 10816-4:1998 Mechanical vibration – evaluation of machine vibration by measurements on non-rotating parts – part 4: gas turbine driven sets excluding aircraft derivatives

ISO 10816-5:2000 Mechanical vibration – evaluation of machine vibration by measurements on non-rotating parts – part 5: machine sets in hydraulic power generating and pumping plants

ISO 10816-6:1995 Mechanical vibration – evaluation of machine vibration by measurements on non-rotating parts – part 6: reciprocating machines with power ratings above 100 kW

ISO 10817-1:1998 Rotating shaft vibration measuring systems – part 1: relative and absolute sensing of radial vibration

ISO 10819:1996 Mechanical vibration and shock – hand–arm vibration – method for the measurement and evaluation of the vibration transmissibility of gloves at the palm of the hand

ISO 10846-1:1997 Acoustics and vibration – laboratory measurement of vibro-acoustic transfer properties of resilient elements – part 1: principles and guidelines

ISO 10846-2:1997 Acoustics and vibration – laboratory measurement of vibro-acoustic transfer properties of resilient elements – part 2: dynamic stiffness of elastic supports for translatory motion – direct method

ISO 11342:1998 Mechanical vibration – methods and criteria for the mechanical balancing of flexible rotors

ISO 11342:1998/Cor 1:2000 Mechanical vibration – methods and criteria for the mechanical balancing of flexible rotors

ISO 13090-1:1998 Mechanical vibration and shock – guidance on safety aspects of tests and experiments with people – part 1: exposure to whole-body mechanical vibration and repeated shock

ISO 13753:1998 Mechanical vibration and shock – hand-arm vibration – method for measuring the vibration transmissibility of resilient materials when loaded by the hand-arm system

ISO 14964:2000 Mechanical vibration and shock – vibration of stationary structures – specific requirements for quality management in measurement and evaluation of vibration

ISO 16063-1: 1998 Methods for the calibration of vibration and shock transducers – part 1: basic concepts

ISO 16063-11: 1999 Methods for the calibration of vibration and shock transducers – part 11: primary vibration calibration by laser interferometry

ISO 1925: 2001 Mechanical vibration – balancing – vocabulary

ISO 1940-1: 1986 Mechanical vibration; balance quality requirements of rigid rotors; part 1: determination of permissible residual unbalance

ISO 1940-2: 1997 Mechanical vibration – balance quality requirements of rigid rotors – part 2: balance errors

ISO 2017: 1982 Vibration and shock; isolators; procedure for specifying characteristics

ISO 2247:2000 Packaging – complete, filled transport packages and unit loads – vibration tests at fixed low frequency

ISO 2247:1985 Packaging; complete, filled transport packages; vibration test at fixed low frequency

ISO 2631-1:1997 Mechanical vibration and shock – evaluation of human exposure to whole-body vibration – part 1: general requirements

ISO 2631-2:1989 Evaluation of human exposure to whole-body vibration; part 2: continuous and shock-induced vibration in buildings (1 to 80 Hz)

ISO 2631-4:2001 Mechanical vibration and shock – evaluation of human exposure to whole-body vibration – part 4: guidelines for the evaluation of the effects of vibration and rotational motion on passenger and crew comfort in fixed-guideway transport systems

ISO 2671:1982 Environmental tests for aircraft equipment; part 3.4: acoustic vibration

ISO 2953:1999 Mechanical vibration – balancing machines – description and evaluation

ISO 2954:1975 Mechanical vibration of rotating and reciprocating machinery; requirements for instruments for measuring vibration severity

ISO 3046-5:1978 Reciprocating internal combustion engines – performance – part 5: torsional vibrations

ISO 4548-7:1990 Methods of test for full-flow lubricating oil filters for internal combustion engines; part 7: vibration fatigue test

ISO 4866:1990 Mechanical vibration and shock; vibration of buildings; guidelines for the measurement of vibrations and evaluation of their effects on buildings

ISO 4866:1990/Amd.1:1994 Mechanical vibration and shock – vibration of buildings – guidelines for the measurement of vibrations and evaluation of their effects on buildings; amendment 1

ISO 4866:1990/Amd.2:1996 Mechanical vibration and shock – vibration of buildings – guidelines for the measurement of vibrations and evaluation of their effects on buildings; amendment 2

ISO 5007:1990 Agricultural wheeled tractors; operator's seat; laboratory measurement of transmitted vibration

ISO 5008:1979 Agricultural wheeled tractors and field machinery; measurement of whole-body vibration of the operator

ISO 5344:1980 Electrodynamic test equipment for generating vibration; Methods of describing equipment characteristics

ISO 5347-10:1993 Methods for the calibration of vibration and shock pick-ups; part 10: primary calibration by high impact shocks

ISO 5347-11:1993 Methods for the calibration of vibration and shock pick-ups; part 11: testing of transverse vibration sensitivity

ISO 5347-12:1993 Methods for the calibration of vibration and shock pick-ups; part 12: testing of transverse shock sensitivity

ISO 5347-13:1993 Methods for the calibration of vibration and shock pick-ups; part 13: testing of base strain sensitivity

ISO 5347-14:1993 Methods for the calibration of vibration and shock pick-ups; part 14: resonance frequency testing of undamped accelerometers on a steel block

ISO 5347-15:1993 Methods for the calibration of vibration and shock pick-ups; part 15: testing of acoustic sensitivity

ISO 5347-16:1993 Methods for the calibration of vibration and shock pick-ups; part 16: testing of mounting torque sensitivity

ISO 5347-17:1993 Methods for the calibration of vibration and shock pick-ups; part 17: testing of fixed temperature sensitivity

ISO 5347-18:1993 Methods for the calibration of vibration and shock pick-ups; part 18: testing of transient temperature sensitivity

ISO 5347-19:1993 Methods for the calibration of vibration and shock pick-ups; part 19: testing of magnetic field sensitivity

ISO 5347-20:1997 Methods for the calibration of vibration and shock pick-ups – part 20: primary vibration calibration by the reciprocity method

ISO 5347–22:1997 Methods for the calibration of vibration and shock pick-ups – part 22: accelerometer resonance testing – general methods

ISO 5347–3:1993 Methods for the calibration of vibration and shock pick-ups; part 3: secondary vibration calibration

ISO 5347–4:1993 Methods for the calibration of vibration and shock pick-ups; part 4: secondary shock calibration

ISO 5347–5:1993 Methods for the calibration of vibration and shock pick-ups; part 5: calibration by earth's gravitation

ISO 5347–6:1993 Methods for the calibration of vibration and shock pick-ups; part 6: primary vibration calibration at low frequencies

ISO 5347–7:1993 Methods for the calibration of vibration and shock pick-ups; part 7: primary calibration by centrifuge

ISO 5347–8:1993 Methods for the calibration of vibration and shock pick-ups; part 8: primary calibration by dual centrifuge

ISO 5348:1998 Mechanical vibration and shock – mechanical mounting of accelerometers

ISO 5349:1986 Mechanical vibration; guidelines for the measurement and the assessment of human exposure to hand-transmitted vibration

ISO 5982:1981 Vibration and shock; mechanical driving point impedance of the human body

ISO 6070:1981 Auxiliary tables for vibration generators; methods of describing equipment characteristics

ISO 6267:1980 Alpine skis; measurement of bending vibrations

ISO 6721–3:1994 Plastics – determination of dynamic mechanical properties – part 3: flexural vibration – resonance-curve method

ISO 6721–3:1994/Cor 1:1995 Plastics – determination of dynamic mechanical properties – part 3: flexural vibration – resonance-curve method

ISO 6721–4:1994 Plastics – determination of dynamic mechanical properties – part 4: tensile vibration – non-resonance method

ISO 6721–5:1996 Plastics – determination of dynamic mechanical properties – part 5: flexural vibration – non-resonance method

ISO 6721–6:1996 Plastics – determination of dynamic mechanical properties – part 6: shear vibration – non-resonance method

ISO 6721–7:1996 Plastics – determination of dynamic mechanical properties – part 7: torsional vibration – non-resonance method

ISO 6721–8:1997 Plastics – determination of dynamic mechanical properties – part 8: longitudinal and shear vibration – wave propagation method

ISO 6721–9:1997 Plastics – determination of dynamic mechanical properties – part 9: tensile vibration – sonic-pulse propagation method

ISO 6954:2000 Mechanical vibration – guidelines for the measurement, reporting and evaluation of vibration with regard to habitability on passenger and merchant ships

ISO 7096:2000 Earth-moving machinery – laboratory evaluation of operator seat vibration

ISO 7096:1994 Earth-moving machinery – laboratory evaluation of operator seat vibration

ISO 7505:1986 Forestry machinery; chain saws; measurement of hand-transmitted vibration

ISO 7626–1:1986 Vibration and shock; experimental determination of mechanical mobility; part 1: basic definitions and transducers

ISO 7626–5:1994 Vibration and shock – experimental determination of mechanical mobility – part 5: measurements using impact excitation with an exciter which is not attached to the structure

ISO/TR 7849:1987 Acoustics; estimation of airborne noise emitted by machinery using vibration measurement

ISO 7916:1989 Forestry machinery; portable brush-saws; measurement of hand-transmitted vibration

ISO 7919–1:1996 Mechanical vibration of non-reciprocating machines – measurements on rotating shafts and evaluation criteria – part 1: general guidelines

ISO 7919–2:1996 Mechanical vibration of non-reciprocating machines – measurements on rotating shafts and evaluation criteria – part 2: large land-based steam turbine generator sets

ISO 7919–3:1996 Mechanical vibration of non-reciprocating machines – measurements on rotating shafts and evaluation criteria – part 3: coupled industrial machines

ISO 7919–4:1996 Mechanical vibration of non-reciprocating machines – measurements on rotating shafts and evaluation criteria – part 4: gas turbine sets

ISO 7919–5:1997 Mechanical vibration of non-reciprocating machines – measurements on rotating shafts and evaluation criteria – part 5: machine sets in hydraulic power generating and pumping plants

ISO 7962:1987 Mechanical vibration and shock; mechanical transmissibility of the human body in the z direction

ISO 8002:1986 Mechanical vibrations; land vehicles; method for reporting measured data

ISO 8041:1990 Human response to vibration; measuring instrumentation

ISO 8041:1990/Amd 1:1999 Human response to vibration – measuring instrumentation – amendment 1

ISO 8041:1990/Cor 1:1993 Human response to vibration; measuring instrumentation; technical corrigendum 1

ISO 8042:1988 Shock and vibration measurements; characteristics to be specified for seismic pick-ups

ISO 8318:2000 Packaging – complete, filled transport packages and unit loads – sinusoidal vibration tests using a variable frequency

ISO 8318:1986 Packaging; complete, filled transport packages; vibration tests using a sinusoidal variable frequency

ISO 8528–9:1995 Reciprocating internal combustion engine driven alternating current generating sets – part 9: measurement and evaluation of mechanical vibration

ISO 8569:1996 Mechanical vibration and shock – measurement and evaluation of shock and vibration effects on sensitive equipment in buildings

ISO 8579–2:1993 Acceptance code for gears; part 2: determination of mechanical vibrations of gear units during acceptance testing

ISO 8608:1995 Mechanical vibration – road surface profiles – reporting of measured data

ISO 8626:1989 Servo-hydraulic test equipment for generating vibration; method of describing characteristics

ISO 8662–1:1988 Hand-held portable power tools; measurement of vibrations at the handle; part 1: general

ISO 8662–10:1998 Hand-held portable power tools – measurement of vibrations at the handle – part 10: nibblers and shears

ISO 8662–11:1999 Hand-held portable power tools – measurement of vibrations at the handle – part 11: fastener driving tools – ISO 8662–11:1999

ISO 8662–12:1997 Hand-held portable power tools – measurement of vibrations at the handle – part 12: saws and files with reciprocating action and saws with oscillating or rotating action

ISO 8662–13:1997 Hand-held portable power tools – measurement of vibrations at the handle – part 13: die grinders

ISO 8662–13:1997/Cor.1:1998 Hand-held portable power tools – measurement of vibrations at the handle – part 13: die grinders; technical corrigendum 1

ISO 8662-14:1996 Hand-held portable power tools – measurement of vibrations at the handle – part 14: stone-working tools and needle scalers

ISO 8662–2:1992 Hand-held portable power tools; measurement of vibrations at the handle; part 2: chipping hammers and riveting hammers

ISO 8662–3:1992 Hand-held portable power tools; measurement of vibrations at the handle; part 3: rock drills and rotary hammers

ISO 8662–4:1994 Hand-held portable power tools – measurement of vibrations at the handle – part 4: grinders

ISO 8662–5:1992 Hand-held portable power tools; measurement of vibrations at the handle; part 5: pavement breakers and hammers for construction work

ISO 8662–6:1994 Hand-held portable power tools – measurement of vibrations at the handle – part 6: impact drills

ISO 8662–7:1997 Hand-held portable power tools – measurement of vibrations at the handle – part 7: wrenches, srewdrivers and nut runners with impact, impulse or ratched action

ISO 8662–8:1997 Hand-held portable power tools – measurement of vibrations at the handle – Part 8: Polishers and rotary, orbital and random orbital sanders

ISO 8662–9:1996 Hand-held portable power tools – measurement of vibrations at the handle – part 9: rammers

ISO 8727:1997 Mechanical vibration and shock – human exposure – biodynamic coordinate systems

ISO 8821:1989 Mechanical vibration; balancing; shaft and fitment key convention

ISO 9022–10:1998 Optics and optical instruments – environmental test methods – part 10 combined sinusoidal vibration and dry heat or cold

ISO 9022–15:1998 Optics and optical instruments – environmental test methods – part 15: combined digitally controlled broad-band random vibration and dry heat or cold

ISO 9022–19:1994 Optics and optical instruments – environmental test methods – part 19:

temperature cycles combined with sinusoidal or random vibration

ISO 9688:1990 Mechanical vibration and shock; analytical methods of assessing shock resistance of mechanical systems; information exchange between suppliers and users of analyses

ISO 9996:1996 Mechanical vibration and shock – disturbance to human activity and performance – classification

Vibration severity ranges and examples of their application (after ISO IS 2372: Mechanical vibration of machines with operating speeds from 10 to 200 rps – basis for specifying evaluation standards)

Range of vibration severity — Limits of range, mm/s⁻¹	Examples of quality judgement for seperate classes of machines			
	Small machines, class I	Medium machines, class II	Large machines, class III	Turbo-machines, class IV
0.28	A			
0.45	A	A	A	
0.71				A
1.12	B			
1.8		B		
2.8	C		B	
4.5		C		B
7.1	D		C	
11.2				C
18				
28		D	D	
45				D

The letters A, B, C and D represent machine vibration quality grades, ranging from good (A) to unacceptable (D).

Quality judgement of vibration severity (after ISO IS 3945: The measurement and evaluation of vibration severity of large rotating machines, in situ; operating at speeds from 10 to 200 rps)

Vibration severity		Support classification	
in./s⁻¹	mm/s⁻¹	Hard supports	Soft supports
0.017	0.45	Good	Good
0.028	0.71		
0.044	1.12		
0.071	1.8	Satisfactory	
0.11	2.8		Satisfactory
0.18	4.5	Unsatisfactory	
0.28	7.1		Unsatisfactory
0.44	11.2		
0.71	18.0	Impermissible	
1.10	28.0		
2.80	71.0		Impermissible

7. FREQUENCY RESPONSE FUNCTIONS

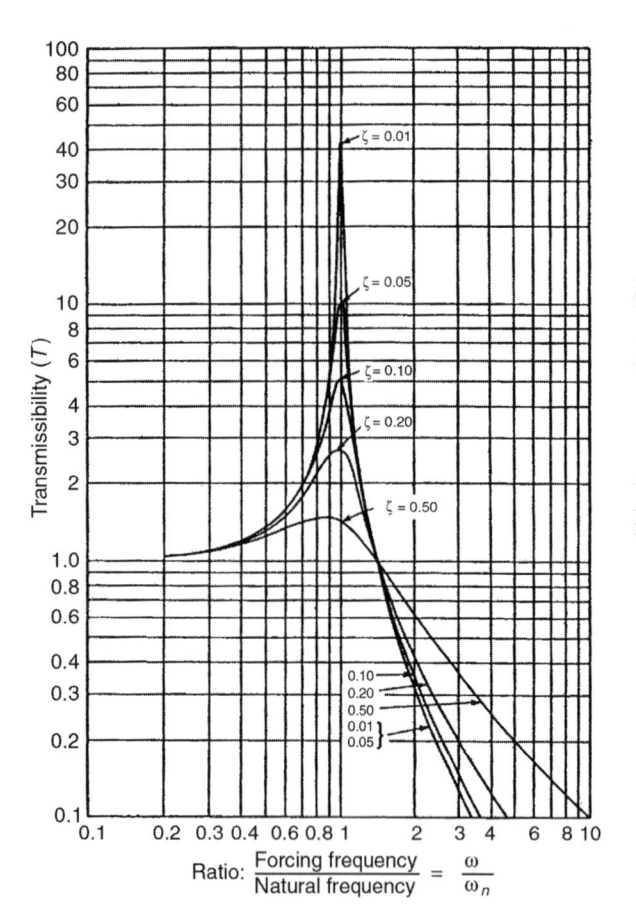

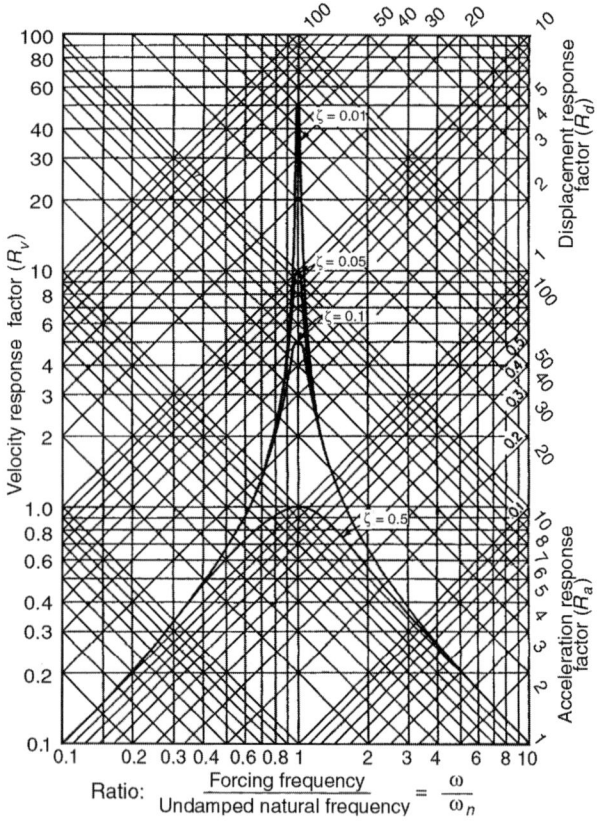

Transmissibility of a viscous-damped system. Force transmissibility and motion transmissibility are identical numerically. The fraction of critical damping is denoted by ζ.

Response factors for a viscous-damped single-degree-of-freedom system excited in forced vibration by a force acting on the mass.

INDEX

NOTE

Page numbers in **bold** refer to major discussions. Page numbers suffixed by *T* refer to Tables; page numbers suffixed by *F* refer to Figures. vs denotes comparisons

This index is in letter-by-letter order, whereby hyphens and spaces within index headings are ignored in the alphabetization. Terms in parentheses are excluded from the initial alphabetization.

Cross-reference terms in *italics* are general cross-references, or refer to subentry terms within the same main entry (the main entry is not repeated to save space).

Readers are also advised to refer to the end of each article for additional cross-references – not all of these cross-references have been included in the index cross-references.

A

ABAQUS 246–247, 305
absorbers **1–26**
   attenuation capabilities 22–23, 23*F*, 25*F*
   AVA 4*F*, 5–6, 7*F*, 8*F*
   dynamic **9–26**
   future perspectives 24, 25*F*
   piezoelectricity and 1–3, 1*F*, 3*F*
   positive position feedback 3–5, 4*F*, 5*F*, 6*F*
   special configurations of 18–21, 20*T*, 20*F*, 21*F*, 22*F*
   undamped 10–17, 10*F*, 11*F*, 12*F*, 14*F*, 15*F*, 16*T*, 16*F*, 18*T*, 18*F*, 19*F*
Absorbers, Active **1–9**
Absorbers, Vibration **9–26**
acausal filters 1196
accelerance 1129
acceleration 604
accelerograms 439–441, 440*T*, 443–446
accelerometers
   absolute motion and 1383, 1390–1392, 1395–1396
   ADXL50 772, 773*F*
   bearing diagnostics and 151
   calibration of 1130–1132
   cross-axis sensitivity and 1121–1122
   environmental testing and 491
   location 495
   MEMS and 772–774
   rotation 1080–1081
acceptance testing 490
acoustics
   aerodynamics and 93

acoustics
   boundary integral formulation and 1278–1279
   excitation and 897
   external problem and 1279
   FEM and 1277–1278
   fluid/structural interaction and 545–550
   internal problem 1279–1282
   MEMS and 774, 775*F*
   noise and **887–898**
   parallel processing and 1000
   radiation impedance and 891
   SPL 1268, 1268*F*
   structural interactions and **1265–1283**
   subsonic waves and 897
   tire vibration and 1375–1376
   See also sound
active constrained layer (ACL) 361*F*, 362*F*, 363*F*, 656–658
active constrained layer damping (ACLD) 353–360, 355*F*, 356*F*, 359*T*, 359*F*,
active control
   active suspensions 38, 39*F*
   adjustable suspensions 38, 38*F*
   applications 28–33, 29*T*, 29*F*, 30*F*, 31*F*, 32*F*, 33*F*, 34*F*, 35*F*, 36*F*
   civil structures **26–36**
   damping and 28–33, **342–351**
   fuzzy logic and 43–45
   groundhook 41
   hybrid 26–28, 27*F*, 42
   isolation **46–48**
   optimal 55
   passive suspensions 37, 37*F*
   semiactive 26–28, 27*F*, 37*F*, 38–44, 40*F*, 41*F*, 42*F*, 43*F*, 44*F*

active control (*continued*)
   skyhook 38–41
   suppression **48–58**
   vehicles **37–45**, 37*F*
Active Control of Civil Structures **26–36**
Active Control of Vehicle Vibration **37–45**
Active Isolation **46–48**
   actuation 47–48, 47*T*
   feedback 46–47, 47*F*
   feedforward 46–47, 46*F*
actively controlled response of buffet affected tails (ACROBAT) 78
active-passive devices 653–658
active piezoelectric damping composites (APDC) 354
active states 925–926
Active Vibration Suppression **48–57**
   compensators and 55–56
   control and 53, 55
   degree of freedom and 49–51, 49*F*, 50*F*
   limitations of 57
   modal control and 51–52
   pole placement 54
   spillover 53–54
   stability and 56–57
actuated joints 1059
actuators **79–81**
   active isolation and 47–48
   active materials 58–61, 60*F*, 61*F*
   configurations 759–760
   distributed 1134–1138
   effective implementation of 68–70, 68*F*, 69*F*, 70*F*, 71*F*
   electrical input 71–72, 72*F*, 73*F*, 74*F*
   electrostriction and 482–490

INDEX

actuators (*continued*)
  energy extraction 70
  magnetostrictive materials 64–65, **753–762**
  MEMS and **771–779**
  piezoelectricity and 61–64, 62F, 63F
  proof-mass 653
  RMA 48
  sensitivity and 1140–1141
  shape memory alloys and 65–68, 65F, 66F, 67F, 1147
  structonic cylindrical shells and 1138–1141
  See also MEMS; smart structures
Actuators and Smart Structures **58–81**
Adaptive Filters **81–87**
  convergence coefficients 86–87
  LMS algorithm 82, 83F, 84–86
  RLS algorithm 84
  steepest descent algorithm 81–82, 82F
  See also optimal filters
adaptive resonance theory (ART) 866
ADAPTx Automated System Identification Software 682
A/D converters 368–373
additive noise 666–668
Ader, Clement 1065
ADINA 305
ADXL50 accelerometer 772, 773F
aerodynamics
  ACROBAT and, 78
  blades and 174–178
  buffeting and, 92
  continuous turbulence and, 91
  flutter and **553–565**, 565–577
  gusts and, 89–91, 92
  maneuvering and, 89–93
  sound 878
  vibration origins 1191
aeroelastic effects 1584–1586
  aeroservoelasticity 95, 95F
  control surface buzz 96, 96F
  dynamic maneuvers 89–93, 90F, 91F, 92F, 93F
  flight loads 88–89, 88F, 89F
  ground loads 93–95
  limit cycle oscillations 95, 95F
  negative damping 97
  panel flutter 96, 96F
  stall flutter 96, 97F
  vortex shedding 97
Aeroelastic Response, **87–97**
aerospace
  acoustic tests and 494
  SNDT and 905
Agnes, G **1–9**
Ahmadian, M **37–45**
airbags 283F, 285
Akaike Information Criterion (AIC) 682, 1205
algorithms
  back propagation 873–874
  contact 308

algorithms (*continued*)
  differentiation/integration and **1193–1199**
  DYNA3D and 305–312
  Gauss-Seidel 995
  hourglassing 305, 308
  least mean squares 82–84, 86
  Levenberg-Marquart 866
  model updating 834
  neural networks and 871–872
  recursive least square 84
  steepest descent 81–82
aliasing errors 670
  A/D converters and 368
  antialiasing filters and 367–368
  data acquisition and 365–369
  differentiation/integration and 1195
  parameters and 369
American Beauty 1064
American National Standards Institute (ANSI) 245, 1224
American Petroleum Institute (API) 1081, 1224
Ampere's law 755
analytic signals 643F
  Hilbert transform and 643–646
Anderson, PW 744
Andronov, AA 431
animation 415, 416
annular fluid flow 1023
ANSYS 247, 305
a posteriori verification 463
a priori reasoning 135, 632, 637, 1266, 1268
Arabs 112
arbitrary Eulerian Lagrangian (ALE) formulation 251T, 252
  DYNA3D and 306
  meshes and 281–282
Aristotle 126
ARMA (autoregressive moving average) process 1202–1203, 1203F
Arnold VI 432
ARTeMIS Extractor 682
ASCE 7-98 tabular values 1583, 1583T, 1584
asymptotic modal analysis (AMA) 1269
asymptotic techniques
  nonlinear systems and 957–962
attenuation 1558–1559
attractors
  chaos and **227–236**
austenite *See* shape memory alloys
Austrian Standards Organization (ON) 1224
auto-associative neural network (ANN) 864–865
autocorrelation function 977–978, 1592
  adaptive filters and **81–87**
  cepstral analysis and 217–218
  columns and 241
  model-based identification and 675
  PSD 297F

autocorrelation function (*continued*)
  rain-on-the-roof excitation and 241
  signal processing and 1200–1204
  spectral density and 296
AutoMAC 270
automobiles
  bridges and **202–207**
  ground transportation systems and **603–620**
  tire vibrations and **1369–1379**
  vehicular vibrations **37–45**
  See also crash
AutoRegressive model 674, 675, 902
AutoRegressive Model with eXogenous input (ARX) model 677, 679, 684
  model-based identification and 675, 677, 679–680, 682, 684–685, 685F
AutoRegressive Moving Average (ARMA) model 675
  model-based identification and 674
AutoRegressive Moving Average with eXogenous input (ARMAX) model 675, 679, 684
  model-based identification and 675, 679–680, 682, 684–685, 684F, 685F
Averaging **98–110**
  basic operations 98–101, 98F, 99F, 100F, 101F
  exponential 103
  frequency domain 108–109, 110F
  jitter effects 102F, 103–108, 108F, 109F
  moving average (MA) 99–100, 99F, 100F,
  PSD 108–109
  recursion 100–101, 100F, 102
  TDA 101–108, 102F, 103F, 104F, 105F, 106F, 107F, 108F, 109F
axial loading
  columns and **236–243**
  crash and 310F
  See also loading

B

back propagation algorithm 873–874
baffled plate 889F
Bajaj, A **928–943, 952–966**
Balancing **111–124**
  calculations of 88
  coupled rotation and 111F, 112–113, 112F, 113F
  flexible state and 118–119
  influence coefficients and 119–123, 120F, 122F, 124T
  rigid states and 113–118, 114F, 115F, 116F, 117F, 118F
  rotor-stator interactions 1107–1121
  tire vibrations 1370
  torsion and 112–113
Banks, HT **658–664**
bar plots 413

bars 1125–1127
Barton, J 971
basic linear algebra subprograms (BLAS) 991
Basic Principles 124–137
  calculus of variations 126
  Dunkerley's method 135–136, 136F
  Euler's equations 126–130, 129F
  flexural motion 130, 131F
  Galerkin's method 135
  generalized system coordinates 130–131
  Hamilton's principle 131
  Lagrange equations 131
  Maxwell's theorem of reciprocity 136, 136F
  parameters 125, 125F
  Rayleigh's Principle 131F, 132–134, 133F, 134F
  Ritz method 134
  vectors 125–126, 127F, 128F
Bauchau, O 461–467
Bayesian Information Criterion (BIC) 682
Baz, A 351–364, 1144–1155
Beams 137–143, 1329–1330, 1330F, 1331F
  Bernoulli-Euler 748F
  boundary conditions and 181F, 183–185, 184F, 184T, 185T
  continuous systems and 1312–1317
  cross-axis contamination and 1122
  Duffing's equation and 233–235
  Dunkerley method and 135–136
  Euler-Bernoulli theory and 137–141, 140F
  feedforward control 517
  flexural radiation and 1458, 1468
  Gelerkin method and 135
  localization and 741–751, 748F
  magnetically buckled 234F
  MEMS 787–789
  natural frequencies and 414–415
  piezoelectric damping and 354–360
  shape memory alloys and 1147–1148
  shells and 1155–1167
  ship vibrations 1167–1173
  sound and 885–886
  time frequency and 1366F
  Timoshenko theory and 142
  transmissibility and 1523–1527
  transverse vibrations and 137–142, 138T, 139T, 140F, 140T, 141F, 141T, 142T, 143F
  vibration intensity and 1483–1484, 1486–1487
Bearing Diagnostics 143–152
bearings 1078
  bicoherence 148
  cage fault 149
  cepstral analysis and 147, 218–220
  coupling 163, 163F, 164F
  distributed defects 149–151
  diversity 154F, 156–157, 156F
  failure modes 143

bearings (continued)
  fault 1083
  fluids and 153–155, 153F, 154F, 155F
  functions and 152–153
  HFRT 146, 146F
  journal bearing 150F, 151
  localized defects 144–151, 144T, 145F, 146F, 148F, 149F, 150T, 150F
  magnetic links 158–161, 159F, 160F, 161F
  misalignment and 1116–1118, 1186
  property comparison 155F, 161T, 161
  roller bearing 143, 157–158
  rolling faults and 1187–1188
  sealing systems 161–162, 162F, 155F
  shock pulse counting 146
  signature generation 144–145, 144T, 145F
  statistical parameters 145–146
  synchronized averaging 147
  time-frequency distribution 147
  wavelet transform 147
  See also rotor dynamics; standards
Bearing Vibrations 152–165
Bellville spring 755, 1181, 1181F
Belts 165–174
  drives 170–174, 171F, 172F, 173F
  moving 166–170, 166F, 167F, 168F, 170F
  stationary 166, 166F
Belytschko-Tsy shell 305
Belytshco-Shwer beam 305
bending moment sensitivity 1125–1127
bending strains 1136
Bendixson's theorem 593
Benson, DJ 278–286
Bernoulli, John 124, 1344
Bernoulli-Euler beams 748F
Bert, CW 236–243, 286–294
Bessel function 729
Bessel functions 288
Betti's principle 197–198
bias error 670
bicoherence 145F, 147–149, 149F
bifurcation
  center manifold theory and 962
  eigenvalues and 818
  Hopf 435
  local 963
  nonlinear analysis and 962–965
  normal form theory 963
  parametric excitation and 1003, 1006
  perturbation methods and 1009–1011
  vibro-impact systems and 1533–1536, 1539, 1543
Bigret, R 111–124, 152–165, 174–180, 1064–1069, 1069–1077, 1078–1084, 1085–1106, 1107–1120
biharmonic operator 1331

bilinear transformation 393–394
  time frequency and 1360–1361
binary representation 232
biodynamics 1571
  mechanical impedance and 1571
  models for 1572
  transmissibility and 1571
bioengineering
  crashworthiness and 308–311
biorthogonality 1074
biotechnology 778
birdstrike 93
Birkhoff, George 228, 233
Blades and Bladed Disks 174–180, 415
  breakdowns and 176
  instability and 175
  localization and 741F, 745, 750F, 751F
  propellers 1170
  pulsations and 177, 177F
  resonance 179F
  rotation vs. rest 177–178
  signal generation and 1191
  strain and 175T, 178F
  strains and 175
  technology and 176
  See also disks; helicopter damping; rotation
blanching 622
block diagrams 686F, 687F
block-Lanczos algorithm 695–696
Blue Wave Ultrasonics 759
Bode plots 122, 418, 421F, 757, 758F
Boltzmann superposition model 661–662
Bond number 739, 739F
Booch, G 969, 975F
Book, W 1055–1063
Boolean matrices 997–1000
Borel measure 664
bounce transmissibility 610
Boundary Conditions 180–191
  beams 181F, 183–185, 184F, 184T, 185T, 1329–1330
  cables and 211
  continuous methods and 288
  coupled systems and 186, 187F, 188F, 189F
  DYNA3D and 306
  equations of motion and 1329, 1329F
  external problem and 1279
  FEA software and 253
  field dynamics and 492–493, 495
  finite element methods 531–533
  guided waves and 794–805
  integral formulation 1278–1279, 1282–1285
  internal problem and 1279–1282
  liquid sloshing and 726–740
  longitudinal waves 181–183, 181F
  membranes and plates 1331
  moving 190
  noise and 888–889
  nonlinear analysis and 945
  nonreflecting 190

Boundary Conditions (*continued*)
  radiation efficiency 894F, 895F, 896F, 897F
  semidefinite systems 191
  shells 1159
  sound pressure and 888–889
  three-dimensional 186–188
  tire vibration and 1378
  torsion 183
  waves and **1559–1564**
Boundary Element Methods **192–202**, 201F, 1279–1282
  eigenvalues and 198
  elasticity and 197
  Fourier transforms 201
  fundamental solutions 192–195, 192F, 193F
  harmonic oscillations and 195–198
  Helmholtz equation and 195
  Kirchhoff plates and 197
  Laplace transforms 201
  symmetry and 194–195
  transient problems and 198–201
Box-Jenkins method 677
Bragg cell 702, 1404
Branca, Giovanni 1064
Braun, S **98–110, 294–302, 665–672, 1208–1223, 1406–1419, 1587–1595**
Breguet, Louis 1065
Bresse-Timoshenko theory 239
Bridges **202–207**
  dynamic response of 203
  frequencies of 202–203, 203F, 204F
  railroad tracks and 206
  traveling load and 204F, 205, 205F, 206F
British Standards Institution (BSI) 1224
  6841 860, 1572
  6842 625
broad-band random excitations 1266
Broyden-Fletcher-Goldfarb-Shanno algorithm 637
Bubnov-Galerkin method 240
buckling 238
  columns and 241
buffeting 92
buildings 1578
  See also active control
bulk waves 908
Burg's method 1203
Butterworth filters 703

C

C++ 971–972
  numerical efficiency and 972–973
Cables **209–216**, 209F
  chains and 213
  linear theory and 210–215, 211F
  modal analysis of 211–215
  noise 1133
  nonlinear model of 209–210, 214F
  shallow sag 210–213

Cables (*continued*)
  suspended 209–210, 210F
  tangential displacement and 211
cage fault 145, 149
Cai, GQ **1238–1246**
calculus of variations 126, 1344–1346, 1345T
  Euler's equations 126–130
  finite element methods 533
  flexural motion 130, 131F
calibration 818
Campbell diagram 414, 415F
cantilevers
  beam model 692F, 693F, 694F
  Duffing's equation and 233–235
  pipes 1021–1022
  plates 1028, 1028F
  time frequency and 1367F
Cantor set 233
capacitive displacement sensors 1399–1400
capacitor sensors 1398F, 1399F
Cardona, A **967–976**
Cartwright, ML 228
cascade spectrum plot 385, 428F
CATIA 305
Cauchy principal value 642
cavities 1078, 1170
  dimensions 1265
  SEA and 1268
center manifold theory 432–435, 962
center of gravity 1490–1493
Centre Technique des Industries Mecaniques 1078
Cepstrum Analysis **216–227**, 747–748
  ballpass frequency and 220F
  bearing diagnostics and 145F, 147
  bearing outer race fault 219F
  complex applications 222–227
  definitions for 216–218
  echo removal 225F
  editing effects 223F
  forcing function 226F
  FRFs 226F
  liftering and 224, 226F
  power spectrum and 218–222, 221F
  terminology of 216
  unwrapped phase 218, 218F
  zoom spectra 224F
ceramic actuators 482–488
CFRF matrix 276F, 277F
chains 213
Chaos **227–236**
  bifurcation and 435
  Duffing's equation 233–235
  history of 228
  initial conditions 229–232, 1097–1098
  invariant manifolds and 228–229
  Melnikov's method and 232
  nonlinearity sources of 235
  Poincare maps and 228
  rotor dynamics 1097–1098
  rotor-stator interactions 1112–1115
  Smale's horseshoe and 232–233
  strange attractors and 233–235

Chaos (*continued*)
  symbolic dynamics and 229–232
  vibro-impact systems and **1531–1548**
chatter 589
chemical reactions 1440
Chladni figures 414
Choi-Williams distribution (CWD) 147, 148F, 598, 600F, 1364–1366, 1365F
  time-frequency and 600
Cholesky factorization 52, 712–713
circle-fitting method 822–823
circular plates 416F, 1026–1027
civil engineering
  SNDT and 905
clamped-clamped plate 1025F
classes 969
  diagram 970F
  hierarchy and 974–976
closed loop control 50
Coad, P 969, 971
coaxiality fault 1083, 1099
cognition
  whole-body vibration and 1575
Cohen's class of distributions 1362–1366
collaboration diagrams 970
Columns **236–243**
  dimensionless frequency 237T
  end conditions 237T
  free lateral vibration and 237–239
  nonlinear vibration of 240–241
  random vibration of 241–242
  stepped 239
  tapered 239
  thin-walled 239–240
  transverse shear flexibility and 239
Commercial Software **243–256**
  dynamic applications and 244–245
  history of 244
  quality assurance and 245–246
Comparison of Vibrational Properties
  modal properties **265–272**
  response properties **272–277**
  spatial properties and **256–264**
compass plots 418F
compensators 55–56
complementary energy method 289, 290, 1320
complex envelope displacement analysis (CEDA) 1269, 1271–1274, 1273F
complex exponential method 821
complex exponential model class 675
complex mode indicator function (CMIF) 425, 429, 429F
component diagrams 971, 972F
component mode synthesis 1334–1335
composite FRF 425
compound FRF matrices 274, 275, 278
compound modes 417
compression 887
  columns and **236–243**
  packaging and **983–988**

compression (*continued*)
  plate vibration and 1029
computational methods
  averaging and **98–110**
  correlation functions and 298–299
  eigenproblems and 463
computational model updating (CMU) 851–854
Computation for Transient and Impact Dynamics **278–286**
computers
  classes and 969
  crashworthiness and 304
  DYNA3D and 305–312, 305T
  improvement of 968T
  localization and 749
  nonlinear systems and 953
  object oriented programming and **967–976**
  parallel processing and **990–1001**
  tasks 968T
  See also MEMS
cone kernel distribution (CKD) 1364–1366
Constantinides, AG **380–395**
constitutional white finger 621
contact algorithms 280–281, 308
Continuous Methods **286–294**
  complementary energy method 289–290
  differential transformation method 291–292
  formulation and solution 287–288
  Galerkin method 290, 292–293
  lower-bound approximations 291
  Rayleigh method and 288–289
  Ritz method 290
continuous systems 1312–1317, 1327–1332
  variational methods and 1350–1354
continuum mechanics 973–974
contour maps 416
control *See* active control
controlled numerical center (CNC) tools 1379–1380
control surface buzz 96
convergence 466F
convolution 381, **1304–1308**
Cooper, JE **87–97**
coordinate measuring machines (CMMs) 1490
  See also isolation theory
Coordinate Modal Assurance Criterion (COMAC) 270–271, 270F, 275, 276
coordinate orthogonality check (CORTHOG) 271
coordinates 130–131
Co-Quad plots 418
Coriolis effects 409–410, 857, 860
Correlation Functions **294–302**, 680
  computational aspects of 298–299
  flow propagation 299F, 300F, 301–302
  matrices and 298, 299–302
  nonstationary signals and 296
  random signals and 297–298

Correlation Functions (*continued*)
  stochastic processes and 296
COSMOS/M 247–248
cost function 977
  helicopter damping and 633
  neural networks 870
Coulomb forces 636
  damping 337
  energy 475
  friction 582–583
  shock isolation and **1180–1183**
coupled analyses 254T, 255
coupled systems 1080
  balancing and **111–124**
  bearing vibration and 163
  boundary conditions 186, 187F, 188F, 189F
  differentiation/integration and 1194
  equations of motion and 1326
  gyroscopic 1097, 1099–1101, 1099F, 1100F
  isolation theory and 1490–1494
  magnetostrictive materials **753–762**
  power balance and 1267
  rotation and 112–113, 1068–1069
  SEA and 1266
  standards and 1231–1232
  variational methods and 1354
couple unbalance 1185
crack propagation 509–512, 1083
Craig, RR Jr. **691–698**
Crash **302–314**
  application examples 312
  axial loading 310F
  bioengineering and 310–311
  component roles and 308
  contact algorithms and 308
  DYNA3d 305–312, 305T, 307T
  economic elements and 308
  human head impact 312F
  impact crushing 309F
  safety standards 303T
  solid mechanics and 304–305
  supercomputers and 304
  vehicle collision 304, 310F, 311F, 312F
CRAY supercomputer 304–312, 305T
Critical Damping **314–319**
  definition of 315–317
  distributed parameters 318–321
  initial conditions and 315F
  lumped parameters and 317–318
critical moment theorem 1070–1071
cross-axis sensitivity 1131–1132
cross flow 1023
cross-generalized mass (CGM) matrix 267
cross-orthogonality (XOR) matrices 267
crystalline growth 754
Curie brothers 1011, 1014
Curie-Weiss law 476, 478
current analysis 378
current sensor 1399–1400, 1400F
curse of dimensionality 902, 902F
CuZnAl 660

cycle limits 1095–1097
  rotor-stator interactions 1107–1110
cyclostationary phenomenon 602
cylinders
  bearings vibrations and 153–155
  flexural radiation and 1460–1463, 1465–1468

D

d'Albans, Marquis de Jouffroy 1064
d'Alembert 1099, 1344
  principle 126, 131, 607
  rotation and 1070
  rotor dynamics 1085
  variational methods and 1355
Dalpiaz, G **1184–1193**
D'Ambrogio, W **1253–1264**
damping 344F, 413, 414F, 424F, 426F
  absolute motion transducers and 1383
  active **351–364**
  active constrained layer (ACLD) 353–360, 355F, 356F, 359T, 359F, 361F, 362F, 363F
  active mass driver (AMD) system 30, 33F
  active/semi-active 347–349
  advanced concepts 640–641
  augmentation 630
  chaos and 228
  classical 723–725
  complex modulus 338–339
  Coulomb 337
  critical **314–319**
  directly-coupled 344F
  discrete elements and **395–402**
  dual frequency 638–639
  Duffing's equation and 233
  Duhamel's Principle and 1308
  earthquakes and 442F, 449–460, 451F, 460F
  elastically coupled 345–346, 346F
  elastomeric testing and 631–632
  equations of motion and 1324–1332
  equivalent viscous 340–341, 341F, 721–722
  in FE models **321–327**
  fluids and **467–475**
  fluid-structure interactions and 551
  flutter and **553–565**
  fractional derivative 325
  friction and **582–589**, 590–592
  fundamental theory and **1290–1299**
  helicopter **629–642**
  hybrid 28–30, 30F, 31F, 32F, 353–360
  hysteretic 323–324, 646, **658–664**
  impulse response function 1338–1339
  isochrones and 420
  isolation theory and **1487–1506**, **1507–1521**

damping (*continued*)
  Krylov-Lanczos methods and 691–697
  linear matrix methods **721–726**
  liquid sloshing and 734–740
  localization and 747
  magnetic constrained layer (MCLD) 352, 353F
  mass-spring system and 687F, 689, 689F
  materials **327–331**, 329F, 331F, 337–338
  matrices 360–363
  maximum control voltage and 360, 360T
  measurement of **332–335**
  membranes and **762–770**
  MEMS 779–781, 780F, 783–784, 784T
  modal analysis and **820–824**
  modal properties and 324–325
  modulus 338–339
  mounts and 342–349
  negative 97
  nonclassical 725–726
  nonlinearity and 420
  nonlinear resonance and 932–934
  passive 343–345, 351–352
  physical mechanisms of 321F
  piezoelectric 352, 352F, 353F, 354–360
  plate vibration and 889–890
  proportional 723
  residuals and 848–850
  resonance and **1046–1055**
  robots and 1056–1057, 1063
  rotor dynamics 1085–1088, 1092
  rotor-stator interactions 1107–1121
  semiactive 30–33, 33F, 34F, 35F, 36F
  shape memory alloys and 352, 353F, 1146, 1149–1151
  ship vibrations 1168
  signal generation and 1184–1185
  single frequency 632–637
  structural 323–324
  suspension and 37–38, 616
  tires and 618
  treatment types 351–354, 352F, 353F, 354F,
  vehicular vibration and, 38–44
  viscoelastic 325–326, 326F, 339–340, 649, 656–658
  viscous 321–323, 324–325, 336–337, 340–341, 344F, 633–634, 636–637, 1294–1295, 1496–1501, **1548–1551**
  wind and 1583
  See also absorbers
Damping, Active **351–364**
Damping in FE models **321–327**
  viscous 321–323
Damping Materials **327–331**
Damping Measurement **332–335**
Damping Models **335–342**
Damping Mounts **342–351**

damping ratios 413
Danish Standards Association (DS) 1224
dashpot
  damping measurement and 332–333
  Duhamel's Principle and 1308
  parameters 636, 661
  viscous damping and 336–337
data 365F
  acquisition **364–376**
  aliasing 365–369, 365F, 366F, 367F, 369F
  basic diagnostics and 377
  cleansing and 901
  discretization errors and 369–371, 370F,
  displays and **413–431**
  environmental testing and 491–492
  external sampling and 373–375, 373F
  feature processing and 901–904
  modal 413–417
  neural networks and **863–868**
  periodic 491
  random 491
  resampling schemas 374F, 375, 375F
  sigma-delta converters 371–372, 371F, 372F
  Simpson's rule 1197–1198
  SNDT and 900–904
  transient 491
  triggering and 373, 373F
  See also modal analysis
Data Acquisition **364–376**
data set 674, 676F
David, A **1001–1009**
da Vinci, Leonardo 125–126
deformable mirrors 483–485, 485F
degrees of freedom (DOF) 1326F, 1327F, 1332
  absorbers and 15–18
  active suppression and 49–51
  averaging method and 960–961
  cepstral analysis and 217F
  chaos and **227–236**
  critical damping and 315F
  damping measurement and 332–333
  damping mounts and **342–351**
  direct problem and 1254–1259, 1257F, 1258F, 1259F
  Duhamel's Principle and 1308
  earthquakes and 444, 447–460, 447F
  environmental testing and 498
  equations of motion and 1291–1293, 1324–1327
  FEA software and 245
  finite difference methods and 524–525
  flutter 567–570
  forced vibration and 1295–1299
  force transducers and 1123, 1125, 1126, 1130

degrees of freedom (DOF) (*continued*)
  FRF data and 418, 420
  ground transportation and 605–613
  impulse response function **1335–1343**
  inverse problem and 1259–1265
  isolation theory and 1488–1490, 1494–1501
  Krylov-Lanczos methods and 691–697
  modal analysis and 828
  modal properties and 269–271
  model updating and 852, 854
  NNM and 919–920, 919F
  nonlinear system resonance and **928–943**
  Nyquist plot of 424F
  parallel processing and 992, 995
  piezoelectric damping and 358
  Rayleigh method and 1309–1312
  residuals and 848–851
  resonance and 1047–1055, 1048F, 1049F, 1050F, 1051T, 1053F
  robots 1060
  Schur method and 995–1000
  signal representation and 645
  SNDT and 899, 899F
  spatial properties and 257, 258
  stochastic analysis and 1240–1242, 1250–1252
  structure-acoustic interaction and 1265
  substructuring and 1333–1335, 1333F
  superposition and 1300–1301
  translation and 1325F
  variational methods and 1357–1359
  viscous damping and 324–325
  See also boundary conditions
De Laval, Carl Gustav 112, 1064, 1065
De Laval model 1070–1071, 1072, 1085–1088, 1092, 1102–1103, 1110, 1112
  chaos and 1112–1115
Den-Hartog's implementation 2
deployment diagrams 970, 972F
design optimization tools (DOT) 637
deterioration 376, 376F
deterministic models 1200, 1203–1204
detuning 404
Deutches Institut fur Normung (DIN) 1224
development testing 490
Devloo, P **967–976**
Devonshire theory 476–477
Diagnostics and Condition Monitoring, Basic Concepts **376–380**
  detection and 381
  deterioration time and 376F
  general principles of 376–378, 377F
  rotation and 379, 379T
  signal generation and **1184–1193**
  vibration signatures and 379

differential transformation method 291–292, 291T
differentiation 1195–1196, 1198
Digital Filters **380–395**, 392F
  acausal 1196
  bilinear transformation and 393–394, 393T, 394T, 395F
  canonic realization and 385, 385F
  differentiation/integration and **1193–1199**
  discrete-time systems and 381–382, 381F
  finite impulse response 385–387, 387T, 387F
  frequency sampling and 389–390, 390F, 393F
  infinite impulse response 391–394
  minimax design and 390, 391T, 392F
  seven-point 1196
  signal flow and 383–385, 383F, 384F, 390F
  suppression and 385
  windows and 387–389, 388F
  See also signals
digital signal processing (DSP) 367
dimension reduction 432–435
Dimentberg, MF **1033–1039, 1040–1046, 1246–1252**
dipole sound 880–882
Dirac's delta function 1336, 1336F
direct problem 1254–1259, 1257F, 1258F, 1259F
direct solvers 992
Dirichlet preconditioner 999–1000
Discrete Elements **395–404**
  damping and 402
  mass/inertia 396–397, 397F
  modeling of 396
  springs 396F, 398, 398F, 400T
  torsional systems and 396
discrete Fourier transform (DFT) *See Fourier transforms*
discrete systems 1309–1311
  time 381–382
  variational methods and 1355–1359
discretization 369–371, 973
Disks **404–413**, 405F
  asymmetric 408
  axisymmetric 404–407
  gyroscopic couple and 1099–1101
  nodal diameter 405–411
  rotating 409–410
  symmetric 407–408
  types of 404
  vibration response and 405F, 410–413
  See also rotation
displacement
  beam vibration and 1330
  equations of motion and 1324–1332
displacement sensors
  capacitive 1399–1400
  eddy current 1400
  LVDT 1400–1401

displays **413–431**
  frequency response data 417–420
  frequency spectra 420–421
  modal data 413–417
  model order indication and 425–429
Displays of Vibration Properties **413–431**
distributed parameter systems (DPSs) 318–321
  sensor/actuators and 1134–1143
Doebling, S **898–906**
domains *See* parallel processing
Dongarra, JJ 973
Donnell-Mushtari-Vlasov equations 1159–1160, 1161–1162
Donnel's theory 353
Doppler effect 700, 700F
  laser vibrometers and 700–706
double modes 416, 419F
  chaos and 229–232
doubling map 230F
Drew, SJ **1443–1455, 1456–1480**
drop test 617
Dubois-Pelerin principle 973
Dubuisson, B **869–877**
Duffing's equation 227, 228, 233–235, 235F, 1110
Duhamel's Principle 444, **1305–1308**, 1306F
Duncan, Dowson 1070
Dunkerley method 112, 135–136
Durbin's method 1203
DYNA3D 305–312, 305T
  codes for 305–307
dynamic analysis
  bearings **143–152**
  bifurcation theory and 435
  cables and 209
  classification and unfolding 435
  dimension reduction and 432–435
  displacement fields 200
  displays and **413–431**
  earthquakes and **439–461**
  FEA software and **243–256**
  finite element methods 535
  fluids and **467–475**
  impact **278–286**
  isolation theory and **1487–1506, 1507–1521**
  laboratory vs. field 492–493
  MEMS 779–781
  normal form simplification and 435
  robots 1055–1063
  rotors 1085–1106
  shape memory alloys and **1144–1155**
  stability and **431–438**, 433F, 435F, 436F, 437F
  structural modifications and **1253–1264**
  transient **278–286**
  See also active control; rotation; structural analysis
Dynamic Stability **431–438**

dynamic systems
  basic principles of 125–126
  blades and 174–178
  coordinates for 130–131
  fundamental theory and **1290–1299**
  nonlinear systems and **952–966**
  object oriented programming and 974–975
  packaging and **983–989**
  parametric excitation and **1001–1009**
  rotor-stator interactions 1107–1120
  tire vibrations 1369–1379
  variational formulations in 1322–1324
  vibro-impact systems and **1531–1548**
dynamic unbalance 1186
Dyne, S **1193–1199**

E

Earthquake Excitation and Response of Buildings **439–461**, 441F
  attenuation and 440T
  damping and 451T, 451F, 460F
  elastic MDOF systems 449–453
  elastic SDOF system and 447–448
  Fourier amplitude spectrum and 444F, 445F
  ground motion and 439–446, 441F, 442F, 443F, 444F, 445F, 446F
  MDOF systems and 455–459
  SDOF systems 444, 447–455, 447F, 452F
  shape functions and 453F
  spectra smoothing and 450F
  tripartite response and 449F
eccentricity 112–113
  rigid states and 115
echo removal 225F
economic elements 308
eddy current sensors 1400
efficiency
  modal radiation 892–894
Eigensystem Realization Algorithm (ERA) 673, 677–682
Eigenvalue Analysis **461–467**
  computational methods for 463
  inverse problems **686–691**
  Rayleigh quotient and 462
  Rayleigh-Ritz analysis and 462–463
  similarity transformation methods 461, 464
  Sturm sequence property and 463
  vector iteration methods and 464–466
eigenvalues 424F
  4 DOF pitch plane model 608
  bifurcations and 818
  boundary element methods 198
  chaos and 228
  continuous systems and 1312–1317
  damping and 322–323
  FEA software and 255T, 256

eigenvalues (*continued*)
  finite element methods 530–533
  hysteretic damping and 323–324
  localization and 744
  model validation and 851–854
  nonlinear systems and 953
  normal form simplification and 435
  plate vibration and 889
  residuals and 848–851
  rotation and 1071–1072
  structural-acoustic interactions 1281
  See also rotation
elasticity
  aeroelastic effects and **87–97**, 1584–1586
  boundary element methods 196, 196F
  bulk waves and 908
  cables and 210, 213
  columns and 238
  damping materials and **327–330**
  discrete elements and 398, 398F, 399T, 400T
  FEA software and 251T
  fluid/structural interaction and 545–550
  flutter and **553–565**, **565–577**
  hysteretic damping and 659–660
  isolation theory and **1487–1506**
  lag dampers and **629–642**
  noise and **887–898**
  nonlinear analysis and 948F, 1110–1111
  piezoelectricity and 1013
  potential energy and 1346–1347
  sensor/actuators and 1134–1143
  shape memory alloys and **1144–1155**
  shock isolation and **1180–1184**
  transmissibility 1522–1527
  vibration intensity and **1480–1487**
  viscoelastic dampers and 649, 656–658
  waves and 1566–1568
  See also boundary conditions; damping
elastic modulus 1329
Electricite de France (EDF) 220
Electric Power Research Institute 1078
electrodynamic shaker 494
electromagnetic acoustic transducers (EMATS) 913
electromagnetic damping composites (EMDC) 354, 354F
electromagnetic sensors 1401–1402, 1402F
electronic speckle pattern interferometry (ESPI) 699, 705, 706F, 709F
  full field measurement and 707–709
electronic vibration origins 1191
Electrorheological and Magnetorheological Fluids **467–475**

electrorheological (ER) fluids 640
  actuators and 58–72
  new applications of 474–475
  semiactive control and 467–468
  smart fluids and 468–474, 468F, 469F, 470F, 471F, 472F, 473F
electrostatic field
  MEMS 785–787, 785F, 791T
Electrostrictive Materials **475–490**, 476F, 488F
  actuator classification and 484F
  applications of 482–490
  deformable mirrors and 483–485, 485F, 486F, 487F
  flapper 489F
  interferograms and 486F
  ion rattling and 480F
  microscopic origins of 475–476
  multimorph mirror and 485F
  oxide perovskites and 477–482, 479T, 479F, 481F
  phenomenology of 476–477
  servo valve 489F
  temperature and 477, 478F, 480–482, 482F, 483F, 484F
elementary run out 112
element technology 282–283
Elishakoff, I **236–243**
Elliott, SJ **81–87**, **977–982**
encapsulation 969
Energenics, Inc. 759
energy
  absorption 308
  actuators and 70–72
  basic principles of 124, 125–126
  FEA 1320–1322
  flow 1266–1267, 1267F, 1269F
  fundamental theory and **1290–1299**
  Hamilton's Principle and 131
  hybrid control and **649–658**
  magnetostrictive materials 754–755
  operator 601
  Rayleigh's method 1309–1317
  resonance and 1047–1048
  Ritz method 1318–1319
  SEA and 1266
  shape memory alloys and 1148
  stochastic systems and 1250–1252
  time-average 1268
  variational formulations 1322–1324
  See also spectra
engineering units (EU) 1209–1211
ensemble average response 1268
envelopes 493–494
environmental testing **490–496**, 492F, **496–504**
  correlation and 497F, 499F, 500F, 503F
  envelopes and 493–494
  field dynamic 492–493, 498–500, 499F
  fixture design and 495–496
  force identification and 502–503
  frequency domain and 497–498

environmental testing (*continued*)
  input-output relationships 497–498, 497F
  laboratory dynamic and 492–493, 500–502
  LDVs and 1406
  measurement locations and 494
  motion control and 493
  seismic instruments and **1121–1134**
  test methods and 490–491, 494–495
  use identification and 491–492
Environmental Testing, Implementation **496–504**
Environmental Testing, Overview **490–496**
EPROM circuits 779
equations
  2SLS method 679
  absolute motion 1382–1395
  active absorbers 1–6
  active damping 356–359, 360–364
  active isolation 46
  active suppression 49–57
  actuators and smart structures 62–63, 65, 66, 70–72
  adaptive filters 82–86
  aeroelasticity 89
  ARX models 677
  autocorrelation function 977–978, 1592
  averaging methods **98–110**, 1244–1246
  balancing 112–117, 118–121
  basic principles 124–136
  bearings 145, 147, 148, 155–156, 160
  belts 166–169, 170–172
  bending strains 1136
  biharmonic differential operator 1024
  blades 175, 177, 178
  boundary conditions 181–188
  boundary element methods 192–201
  bounded waves 1560–1562
  Bresse-Timoshenko 239
  bridges 203–206
  cables 209–213
  calculus of variations 126, 130
  capacitance 1399
  CEDA 1271
  cepstral analysis 217, 222
  chaos 228–234
  circle-fitting method 822
  circular plates 1026
  civil structures 28
  classical damping 723
  columns 237–238, 239–242
  complex exponential method 821
  continuous methods 287–293
  correlation functions 295–301
  correlation methods 680
  cost function 977
  crack propagation 510–511
  crashworthiness 306–308

equations (*continued*)
  critical damping 315–316, 317–321
  Curie-Weiss 476, 478
  D'Alembert 1322
  damping in FE models 321–325
  damping materials 327–329
  damping models 336–341
  damping mounts 343–348
  data acquisition 370–372
  data set 674
  digital filters 381–382, 385–393
  Dirac's delta function 1336
  direct problem 1254–1259
  discrete elements 396–402
  disks 404–405, 413
  dissipated power 1266
  distributed actuation 1136–1137
  Donnell-Mushtari-Vlasov 1159–1162
  Doppler frequency 1404
  Duffing 228
  Duhamel's Principle 1305–1308
  Dunkerley's method 135–136
  DYNA3D and 306–307
  dynamics 279–281, 283
  dynamic stability 432–435
  earthquake excitation 444–459
  eigenvalue analysis 461–463, 464–465, 466
  electromagnetic sensors 1402
  electrostriction 475–479, 484
  energy flow exchange 1266
  energy operator 601
  environmental testing 492–493, 497–502
  ERA 680–681
  Euler-Bernoulli beam theory 137–142
  Euler's 126–130, 878
  exponential window 1594
  external problem 1277
  FEA software and 255T
  feedforward control 513–516
  finite difference methods 520–528
  finite element methods 531–543, 1322
  FIR filters 1594
  flexural motion 130
  flexural radiation 1457–1465
  fluids 470, 473
  fluid/structure interaction 545–551
  flutter 556–559, 566–570
  FM0 601
  FM4 601
  Fokker-Planck-Kolmogorov 1238, 1248–1249
  forced problem 1281–1282
  forced response 579–581
  forced vibration 894–896
  force window 1593
  Fourier-based identification 666–667, 668–671
  free vibration 1029, 1293–1295
  friction 583–587, 591–593
  Galerkin's method 135
  gear diagnostics 597–598, 600–602

equations (*continued*)
  generalized inverses 720
  Green's 879, 888
  ground transportation 603, 608–621
  guided wave 1552, 1553, 1554
  Hamilton's principle 131, 1322–1323, 1328
  hand-transmitted vibration 626
  Hanning window 1590
  helicopter damping 631–635, 636–637
  Helmholtz 195, 878
  Hilbert transforms 642, 643
  Hohenemser/Prager 238
  Hooke's 1012
  Hu-Washizu stationary principle 1321
  hybrid control 650–654
  hysteretic damping 659–660, 661–664
  Ibrahim time domain method 821
  impulse response function 1336–1343
  input power vector 1267
  internal problem 1279–1282
  inverse iteration of rigid body modes 718–719
  inverse problems 686–688, 1259–1264
  isolation theory 1491–1505, 1509–1512, 1514, 1517–1518, 1521
  Kaiser window 1590
  Kirchhoff-Helmholtz 888, 1457
  Kolomogorov 1249
  Krylov-Lanczos methods 691–696
  Lagrange 131, 240, 721, 1323
  Laplace transforms 1336–1338, 1407–1409
  laser vibrometry 700, 701, 702, 703, 704, 707
  Lenz's law 1402
  liquid sloshing 727–735, 736, 737–740
  LMS method 679
  Love-type 1155–1159, 1159T, 1160–1161
  magnetostrictive materials 754–755
  Mathieu 730, 1111
  Maxwell 136, 1135
  Melnikov's method 232
  membranes 763–769, 1135
  MEMS 780, 782–783, 785, 787–789
  minimum total complementary energy 1320
  minimum total potential energy 1320
  modal analysis 824–828
  modal density 1266
  modal directivity 890
  modal parameters 683
  modal properties 266–269, 270–271
  modal radiation efficiency 892–894
  mode acceleration 719

equations (*continued*)
  model based identification 674–683
  model class 674–676
  model order selection 682
  model updating 845, 847–851
  mode of vibration 838, 840–843
  Moore-Penrose generalized inverse 717–718
  motion 1291–1293, 1325, 1326, 1327, 1328, 1329, 1331
  NA4 601
  narrow-band demodulation 602
  Navier's 911
  neural networks 869–876
  NNM 918–920, 921–922
  noise 878–886, 888–895
  nonclassical damping 725
  nonlinear analysis 945, 947–949, 953–956, 957–964
  nonlinear system resonance and 929–941
  nonsingular linear systems 711–713
  nullspace 320
  optimal filters 977–981
  packaging 984
  parallel processing 990, 992–999
  parametric excitation 1002, 1003–1006
  Parseval's 388, 445
  perturbations 232
  piezoelectric materials 1012–1013, 1013T, 1015–1017
  pipes 1019–1022
  plate vibration 889–890, 892, 1024
  Poisson's ratio 327
  principle of least action 1323
  production error 676
  Prony method 678–679
  proportional damping 723
  random processes 1033–1037, 1040–1041, 1043–1045
  rational fraction polynomial method 823
  Rayleigh integral 889
  Rayleigh method 1309–1317
  Rayleigh quotient 462
  Rayleigh-Ritz analysis 462
  Rayleigh's Principle 132–134
  Rayleigh wave 1553
  reciprocity 1267
  rectangular plates 1026
  rectangular window 1588
  response properties 274–278
  Ritz method 134, 1322–1319
  robot vibrations 1060
  rotation 1070–1077, 1083–1084
  rotor dynamics 1085–1095, 1097–1106
  rotor-stator interactions 1107, 1110–1119
  running sum 1196–1197
  Scruton number 1584
  SEAT 1577
  seismic instruments 1121–1132
  shape memory alloys 1145–1153
  shells 1155–1165

equations (continued)
  ship vibrations 1168
  shock 1173–1174
  signal generation 1185, 1188–1191
  signal integration/differentiation 1193–1198
  signal processing, model based 1199–1205
  similarity transformations 461, 464, 471
  Simpson's rule 1197–1198
  singular systems 714–716
  Smale's horseshoe 233
  SNDT 899
  sound 1444–1446, 1451–1452
  sound power 890–891
  sound pressure with boundaries 888–889
  spatial properties 257–258, 260–262
  spectral analysis and 1209–1211, 1213–1221
  spectral coherence function (SCF) 602
  standard wave 1329
  statistical moments 1243–1244
  steady-state power balance 1267
  stochastic analysis 1238–1246
  stochastic differential calculus 1246–1252
  stochastic systems 1246–1252
  strain-life method 507
  Stribeck 1188
  Strouhal number 1584
  structonic cylindrical shells 1139
  structural-acoustic interactions 1275–1282
  structural dynamic modifications 1254–1264
  structural system parameters 823
  Sturm sequence property 463
  substructuring 1333–1335
  Succi method 1281
  superposition 1300–1304
  SVD 716–717
  synchronized averaging 598
  system coordinates 130–131
  Taylor series 210
  time frequency methods 1360–1364
  Timoshenko beam theory 142
  tire vibrations 1373–1377
  transmissibility 343, 348, 1522–1527
  trapezium rule 1197
  ultrasonics 212–213, 908–909, 910–911, 1437–1438
  unbound waves 1566–1567, 1569
  undamped vibration 722–723
  variational methods 1344–1359
  vectorial approach 125–126
  vector iteration 464–466
  vehicular vibration 38–42
  vibration absorbers 10–23
  vibration dose value 1575
  vibration intensity 1480–1487
  vibro-impact 1532–1540

equations (continued)
  virtual work principle 1320
  viscous damping 721, 722, 1548–1549
  wavelet transformation 600, 1420–1428
  wave number 891
  Weiner-Khintchine theorem 1361
  wind 1579, 1580, 1581–1583, 1584
  windows 1587–1590, 1592–1594
  Z transforms 1409–1411
  See also Fourier transforms
equations of motion (EOM) 1324
  beams and 1329–1330
  continuous system models and 1327–1331
  membranes and 1331
  multiple DOF 1326–1327
  plates and 1331
  rods and 1328–1329
  shafts and 1328–1329
  single DOF 1324–1325
  strings and 1328–1329
equilibrium
  averaging method and 960–962
  basic principles of 125–126
  cables and 209–210
  chaos and 227–236
  curvature 209
  discrete elements and 396
  eigenspaces and 953
  environmental testing and 498–500
  equations of motion and 1327
  invariant manifolds and 955–956
  nonlinear systems and 952–966
  point classification and 953
  rotor dynamics 1096F, 1097–1098
  variational methods and 1355
  See also chaos; parametric excitation; stability
equipartition 1266
error
  Fourier-based identification and 666–671
  modal analysis and 834
  model-based identification and 676–677
  model updating and 845
  neural networks 870
  spectral analysis and 1222–1223
ETREMA Products, Inc. 759
Euler 1099
  Bernoulli beam theory 137–142, 166, 338
  field equations 878
  Lagrange equations 1344–1346, 1345T
  relations 316
EU Machinery Safety Directive 626, 1576
EU Physical Agents Directive 627–628, 1576
Ewins, DJ **332–335, 404–413, 805–813, 829–838, 838–844**

excitation
  acoustic 897
  aerodynamics and 1191
  aliasing error and 670
  averaging methods 1244–1246
  balancing and **112–123**
  bias error and 670
  blades and 174–178
  capacitance and 1399
  complex stiffness and 338–339
  Duffing equation and 228
  Duhamel's Principle and **1304–1308**
  earthquakes and **439–461**
  environmental testing and 497–498
  fluid/structural interaction and 545–551
  forced response 579–581
  Fourier-based identification and 665–671
  ground transportation and 605–621
  hydraulics and 1191
  identification and **673–685**
  impulse response function **1335–1343**
  isolation theory and **1507–1521**
  leakage errors and 671
  localization and 747
  MEMS 779–781
  modal 814F, 815, 816–817
  model-based identification and **673–685**
  noise and 887–898
  nonlinear resonance and **928–943**
  nonlinear systems and 956
  oil film 1191
  parametric **1001–1009**
  periodic 1296–1297
  random errors and 671
  random processes **1040–1046**
  resonance and **1046–1055**
  rotor dynamics 1102
  rotor-stator interactions 1107–1121
  ship vibrations **1167–1173**
  stochastic analysis and **1238–1246**, 1247–1252
  time frequency and 1366–1368
  transverse vibration and 169
  vibration isolation and 1501–1504
  viscous damping and **1548–1551**
  See also active control; damping
exponential model class 676
external problem 1276F, 1279

F

failure models 143, 252T
Faraday-Lenz law 755, 761
far-field approach 890
Farhat, C **710–720**
Farrar, C **898–906**
Fassois, SD **673–685**
fast oscillating function 1273F
Fatigue **505–512**
  classical approach to 508

Fatigue (continued)
  crack propagation analysis 509–512, 509T, 510F, 511F
  estimation 647
  strain-life method 505–508, 506F, 507T, 508F, 509T
faults
  bearings and 1083, 1187–1188
  coaxiality 1083, 1099
  gears and 1189–1191, 1191F, 1192T
FDAC matrix 275F
Federal Aviation Administration (FAA) 245
feedback 3–5, 28
feedforward control 28, 46–47
  applications of 516–519, 518F, 519F, 520F
  beams and 517
  description of 513–516, 513F, 514F
  networks 866
Feedforward Control of Vibration, 513–520
Feeny, BF 924–928
Feigenbaum's cascade 1543–1544
Feldman, M 642–648
FETI (dual Schur complement method) 997–1000
field dynamic
  boundary conditions and 492–493, 495
  environment 492–493, 498–500
  measurement location and 494
filters
  acausal 1196
  adaptive 81–87
  antialiasing 367–368
  Butterworth 703
  digital 380–395
  FIR 385–387, 977–978, 978F, 1594
  IIR 1594
  minimax, 390
  optimal 977–982
  recursion 100–101
  spectral analysis and 1216–1218
  Weiner, 978–981
final prediction error (FPE) 1205
Finite Difference Methods 520–530
  central 525–527, 526F
  formulas 521, 521F, 522T, 523T
  forward 524–525
  operators 522, 524T,
  partial differential equations 528, 529F
finite element analysis (FEA)
  ABAQUS 246–247
  ANSYS 247
  boundary conditions 253
  COSMOS/M 247–248
  coupled analyses 255
  dynamic applications and 244–245
  energy methods and 1320–1322
  geometric nonlinearity 252
  history of 244
  linear analyses 253–255
  materials and 252–253

finite element analysis (FEA) (continued)
  meshing and 250
  MSC/NASTRAN 248–249
  nonlinear analyses 254
  nonstandard elements 250
  object oriented programming and 967–976
  SAMCEF 249–250
  software quality assurance and 245–246
  solution methods and 255–256
  standard elements 250
Finite Element Methods 243–244, 250–253, 530–544, 1332
  basic approach to 531–533, 541T, 542T
  crash and 302–314
  damping in 321–327
  dynamic problems 535
  eigenvalues 533
  error analysis 540–543, 543F
  modal properties and 265, 266–267, 269
  model updating and 844–847, 852
  nonlinear analysis 536–538, 536F, 538F, 539F
  nonstructural problems 539, 541T
  piezoelectric damping and 352–358
  propagation 534
  spatial properties and 256–264
  static problems 534
  structural-acoustic interactions 1277–1278
  structural problems 536–539, 536F, 538F, 539F
  substructuring and 1333–1334
finite impulse response (FIR) filters 385–387, 977, 978F, 1594
  time domain formulation and 977–978
  Wiener 978–981
finite spherical monopole 1447
Fisher's discriminant 903
fixed references 1398, 1398F
fixture design 495
Flanagan-Belytschko constant 305
flap 629–630
flappers 485–488
Flatau, A 753–762
flexibility 118–119, 209
  columns and 239
  isolation theory and 1487–1506
  plate vibration and 1033
  robots 1056–1057
  sound and 1443–1455
  spatial properties and 260–264
  tire vibration and 1374–1375
flexural motion 130, 1456–1480
  basic theory of 1456–1463, 1456F, 1458T, 1459F, 1459T, 1460F, 1461F, 1461T, 1462T, 1463F
  boundary conditions and 186, 187F, 188F, 189F
  MEMS 787–789
  Ritz method and 134

flexural motion (continued)
  sound and 1456–1480
  source ratios 1463–1468, 1464F, 1466F, 1467F, 1468F, 1469F, 1470F, 1471F, 1472F, 1473F, 1474F, 1475F, 1476F, 1477F, 1478F, 1479F, 1480F
  transmissibility 1523–1527
Floquet theory 338, 960, 1002
  shape memory alloys and 1151–1152
fluids 887, 949
  acoustic radiation and 545–550, 545F, 546F, 547F, 548F, 549F, 551F, 552F
  bearings and 152–165
  damping modes 469, 469F, 470F, 551
  electrorheological (ER) 467–475, 640
  excitation and 473F
  force feedback and 474, 474F
  force/velocity form 472F
  ideal Bingham 469F
  lag dampers and 629–641
  liquid sloshing 726–740
  lumped model 472F
  magnetorheological (MR) 467–475, 640–641
  particle orientation and 468F
  pipes and 1019–1024
  plate vibration and 1033
  Reynolds number and 472F
  semiactive dampers and 31–33
  shock isolation and 1183
  signal generation and 1191
  smart 467–475
  sound and 1443–1455
  structure interaction and 544–553
  ultrasonics and 1437–1441
  whirl and 1191
Fluid/Structure Interaction 544–553
Flutter 553–565
  active pylons 574, 575F
  aeroelasticity and 553–560, 553F, 554F, 555F, 556F, 557F, 558F, 559F, 560F
  binary 561
  modal properties 566–567, 567F
  model of 555–559, 567
  panel 96
  simulation 570–571, 571F, 572F
  smart wings 575–576, 576F
  solution 559–560
  suppression 572–573, 573F, 574F
  test validation 561–563, 562F, 563F
  torsion 561
  transfer function 569
  types of 560–561
Flutter, Active Control 565–577
FM0 601
FM4 601
foam blocks 1182
Fokker-Planck-Kolmogorov equations 1238, 1248–1249

forced problem 1281–1282
Forced Response **578–582**
  base excitation 580–581, 580F, 581F
  harmonic excitation 579–580, 579F, 580F
  resonance 581–582, 581F
forced vibration 1295–1299
  friction damping and 585–586
  modes of 843–844
force transducers 1123
  bending moment sensitivity and 1125–1127
  hammer attachment and 1124
  loading effects and 1128–1130
  rigid foundation 1123
  stinger and 1124–1125
Forde, BWR 972
Fortran 973
Fourier, Joseph 1065
Fourier analysis
  earthquakes and 443–445
  helicopter damping and 634, 639
  time frequency and 1360–1361
Fourier transforms 1411–1419, 1412F, 1414F, 1415F, 1416F, 1418F
  adaptive filters 84–86
  averaging and 109
  boundary element methods 201
  bridges and 203
  CEDA and 1271, 1272F
  correlation function and 296
  differentiation/integration and **1193–1199**
  Duhamel's Principle 1305
  impulse response function 1339–1343
  influence coefficients and 122
  linearity 1413
  MDL criterion and 1205
  modal analysis 815
  nonlinear testing and 1287
  properties of 1412–1419
  spectral analysis and **1208–1223**
  wavelets and 1420–1423
  windows and 1588–1592, 1588F, 1589F
  See also cepstral analysis
fractals 1191
fractional derivative models 325
free-free planar truss 420F
free vibration
  complex 842–843
  continuous methods and 287–293
  friction damping and 584–585
  origins of 840–841
  orthogonality and 841–842
  structonic shell systems and 1139–1140
frequency
  actuator sensitivity and 1140–1141
  aliasing and 365–369
  averaging and 99–100, 108–109
  basic diagnostics and 379–381
  basic principles of 124

frequency (*continued*)
  bearing diagnostics and **143–152**
  Bragg cell 1404
  bridges and **202–207**
  cables and 211–215
  capacitance and 1399
  columns and **236–243**, 237T
  convergence coefficients and 86–87
  critical damping and **314–319**
  cycle limits and 1107–1110
  damping in FE models and **321–327**
  damping materials and **327–331**
  damping models and **335–342**
  damping mounts and **342–351**
  digital filters and 380–395
  disks and **404–413**
  Domain Assurance Criterion (FDAC) 275–276
  domain methods 822–823
  earthquakes and **439–461**
  energy methods and **1308–1324**
  environmental testing **496–504**
  equations of motion and 1324–1332
  Euler-Bernoulli beam theory 137–141
  excitation 414
  force transducers and 1123–1132
  Fourier-based identification and 665–671
  FRF data and 417–420, 421F
  fundamental theory and **1290–1299**
  gear diagnostics and **741–751**
  ground transportation **603–620**
  hand-transmitted vibration and 623F
  helicopter damping and **629–642**
  Hilbert transforms and **642–648**
  identification and **673–685**
  index 274F, 275F
  input-output domain 497–498
  instantaneous 644
  isolation theory and **1487–1506, 1507–1521**
  LDVs and 1403–1406
  liquid sloshing and 736–739
  localization and **741–751**
  membranes and 763–767
  MEMS 779, 783–784, 784T
  misalignment and 1186
  MMIF and 425–429
  modal analysis and **820–824**
  modal indicators and 425–429
  model-based analysis and **673–685, 1204–1205**
  motion sickness and **856–861,** 859F
  moving belts and 166–170
  neural networks and 867–868
  noise and **877–887, 887–898**
  nondimensional 1026T, 1027T, 1028T, 1029T
  nonlinear stiffness and 425F
  nonlinear system resonance and **928–943**
  nonlinear testing and 1286F
  Nyquist 445

frequency (*continued*)
  plate vibration and 414–415, 889–890, **1024–1031**
  plotting of 413–414
  residuals 850
  resonance **1046–1055**, 1285
  response function (FRF) 646, 1366
  ride natural 614
  Ritz method and 290
  rotor dynamics and 1088–1090
  rotor-stator interactions 1107–1121
  SEA and **1265–1272**
  shape memory alloys and 1148–1153
  shells and 1160–1163
  shifting devices 702–703
  sigma-delta converters and 371–372
  signal generation and 1185–1192
  signal integration/differentiation **1193–1199**
  spectra 420–421, 428F, **1208–1223**
  stationary belts and 166
  structure-acoustic interaction and **1265–1274**
  time methods and **1360–1369**
  tire vibrations **1369–1379**
  transduction and 755–759, 755F, 756F, 757F, 758F
  ultrasonic **906–918, 1437–1441**
  vibro-impact systems and **1531–1548**
  weighting 625–626, 626F, 627F
  weightings 1572, 1572T, 1573T
  whole-body vibrations and **1570–1578**
  windows and **1587–1595**
  See also rotation
frequency response functions (FRFs) 820
  absolute motion transducers and 1382
  Assurance Criterion (FRAC) 274, 276
  averaging and 100
  cepstral analysis and 217, 218, 222–227
  compound matrix 274, 278
  cross-axis contamination and 1122
  direct problem 1254–1259, 1257F, 1258F, 1259F
  environmental testing and 299–302, 497–498
  flutter and 563
  forced response and 579
  force transducers and 1128–1130
  Fourier-based identification and 665–671
  impulse response function 1341–1342
  inverse problem 1259–1265, 1259F, 1262F, 1263F, 1264F
  modal analysis **813–820, 824–828**
  model updating and 852
  neural networks and 864
  residuals and 848–850
  resonance and 1053–1054

frequency response functions (FRFs) (*continued*)
   response properties and 272–277
   shape memory alloys and 1150–1151, 1152*F*
Fresnel equations 182
friction 589–590, 950
   base isolators 587–589, 587*F*, 588*F*
   counter-clockwise rotation 595*F*
   damping and 335, **582–589**
   dry 583–584, 583*F*, 584*F*, 586*F*
   element acceleration 594*F*
   forced vibration and 585–586
   free vibration and 584–585, 584*F*
   limit cycle and 592–593
   links and 153–161
   misalignment and 1186
   negative damping and 590–592
   PDF 594*F*
   rotor dynamics 1095
   snubber 1181, 1182*F*
   Spurr's sprag-slip and 593–597
   static vs. kinetic 582
   stochastic systems and 1250–1252
   time history 594*F*
   velocity curve 591*F*
Friction Damping **582–589**
Friction Induced Vibrations **589–596**
Froude scale 630
Fuller, CR **513–520**
full-field measurement 707–708
fuzzy logic 43–45

G

Galerkin method 135, 239, 240, 290, 292–293, 1332
   dimension reduction and 432–435
Galilean referential 1085, 1086, 1106
Galilei, Galileo 124–125
galloping 1585
Gamma, E 969
Gandhi, F **1548–1551**
Gaussian classifiers 872
Gaussian elimination 711, 711*F*, 715
Gaussian theory 1135
   time frequency and 1362
Gaussian white noise 1238–1240, 1241–1246
Gauss-Newton method 677
Gauss-Seidel algorithm 995
Gauss's law 755
Gabor, Denis 642
Gear Diagnostics **597–602**, 1080
   algorithms for 598–603
   cepstral analysis and 218–220
   Choi-Williams distribution 600*F*
   disks and **404–413**
   failure modes and 597
   planetary system 597*F*
   signal generation and 1189–1191, 1191*F*, 1192*T*
   standards and 1232–1233
   tooth averaging 599*F*
   vibration model of 597–598

General Problem of the Stability of Motion (Lyapnuov) 1097
geometry
   eccentricity and 112–113
   isolation theory and 1490
   membranes and 763–767
   nonlinear systems and 251*T*, 252, 947, 953
   piezoelectricity and 1015
   rotating line 1067, 1068*F*
   rotation and 112
   run out and 112–113
   shape memory alloys and 1151
   structure-acoustic interaction and 1265
Gern, FH **565–576**
Gibbs energy 476
Giurgiutiu, V **58–80**
Gladwell, GML 691
glass transition region 329
global error indicator 273
global positioning systems (GPS) 774
global sonic nondestructive testing (GSNDT) 899, 906
   aerospace and 905
   basis of 899
   civil engineering and 905
   data processing and 900–904
   history of 904–905
   operational evaluation and 900
   rotation and 905
Goldman, Paul 1112
Golubitsky, M 432
Graeffe's root-squaring method 463
Gram-Schmidt orthogonalization 691
graphical comparison 266
gravimetric calibration 1131
gravity
   equations of motion and 1325
   liquid sloshing and 739–740
Green function 879, 888, 889, 1281
Griffin, MJ **621–629**, **856–861**, **1570–1578**
Griffin, S **46–48**
Groundhook control 41
Ground Transportation Systems **603–620**
   2 DOF pitch plane 609–610, 611*F*, 612*F*
   2 DOF quarter car 612–613, 614*F*
   4 DOF pitch plane ride 608
   7 DOF 606*F*
   analysis models 605–621, 607*F*, 608*F*
   damping 616, 617*F*
   deflection ratio 613, 613*F*
   driver/passenger sensitivity 603–604, 605*F*, 619*T*
   fatigue time 604*F*
   guidelines 614
   loading 616*F*
   mass transmissibility 613, 613*F*
   parameter determination 614–621, 619*T*, 620*T*
   PSD 618–621, 618*F*
   ride natural frequency 614

Ground Transportation Systems (*continued*)
   suspension springs 615, 616*F*,
   tires 617*F*, 618
   weight 615, 615*F*, 615*T*
Guckel rings 788
guided waves **1551–1559**
   attenuation of 1558–1559
   bars and 1556
   cylinders and 1556
   definition of 1551
   engineering and 1559
   flat boundary and 1553
   plates and 1552*F*, 1553–1556
   theory 1552–1553
Guyan reduction method 258, 1333
Gyration 239
gyroscopes 19–21, 1332
gyroscopic couple 1097, 1099–1101, 1099*F*, 1100*F*

H

Haddow, A **1285–1289**
Hagg's number 1088
Hall probe 761
Hallquist, J **278–286**
Hamiltonian mechanics 228
   bifurcation and 435
   distributed actuation and 1136
   energy methods and 1322–1323
   nonlinearity and 235
   piezoelectricity and 1014, 1015–1017
   Ritz method and 290
Hamilton's principle 131, 358, 1328
   variational methods 1344, 1348–1349, 1355
hammers 1121, 1124
Han, RPS 972
Hand-Transmitted Vibration **621–629**
   band spectra 623*F*
   blanching 624*F*, 626*T*
   dangerous processes 622*T*
   disorders of 621*T*
   effects of 621–625, 624*T*
   evaluation standards for 625–628
   frequency-weighted 626*F*, 627*F*
   preventive measures 625, 625*T*
   sources of 621
   Stockholm Workshop scale 622*T*
   threshold level 628*F*
   white finger 627*F*
   whole-body vibration 1574
Hann, F **1578–1587**
Hanning window 389, 672*F*, 1216–1219, 1218*F*, 1590–1591, 1590*F*
harmonics 1270
   boundary element methods 195–198, 1279–1282
   bridges and **202–207**
   chaos and **227–236**
   continuous systems and 1312–1317
   flexural radiation and **1456–1480**
   forced problem 1281–1282

harmonics (*continued*)
  forced response 579–581
  forced vibration and 1291–1296
  free vibration and 1293
  friction and **589–596**
  impulse response function 1341–1342
  nonlinear resonance and **928–943**
  resonance and **1046–1055**, 1288
  response and 1286
  rotor dynamics 1085–1088, 1097
  rotor-stator interactions 1107–1121
  signal integration/differentiation **1193–1199**
  sonic **898–906**
  superposition and 1301–1304
  ultrasonic **906–918**
  windows and 1216–1218
  See also excitation; oscillation; sound
Hartmann, F **192–202**
Hayek, S **544–553, 1480–1487, 1522–1531**
Heckl, M 1265
Helicopter Damping **629–642**
  advanced concepts 640–641
  augmentation and 630
  dual frequency characterization 638–639, 638F, 639F, 640F
  elastomeric testing 630F, 631–632, 631F
  hysteresis modeling and 633F, 636–637, 638F, 639, 640F
  rotary hub 629F, 630F
  single frequency characterization 632–637, 634F, 635F, 636F
  stiffness and 633–635, 633F, 636F, 637F, 638F
  viscous 633–634, 641F
Helmholtz equation 878
  boundary element methods 195
  parallel processing and 1000
Hessenberg form 467
heteroclinic point 229
high frequency resonance technique (HFRT)
  bearing diagnostics and 146, 146F, 1188
Hilbert Transforms **642–648**, 642F, 648F
  analytic signals and 643–646
  CEDA and 1271
  cepstral analysis and 218, 222
  notation 642
  properties of 642–643
  transformers and 647–648
  vibration systems and 646–647
Hodges, CH 744
Hohenemser-Prager equations 238, 241
Holmes, PH 431
Holmes, PJ **227–236**
homoclinic loop 231F
homoclinic points 229
  Smale's horseshoe and 232–233

Hooke's equations 660, 754, 1012
Hopf bifurcation 435
horseshoe maps 232–234
hourglassing 305, 308
Householder's method 467
Hubble Space Telescope 485, 488F
Hughes shells 305
hull wake 1169
human body
  crashworthiness and 308–311
  energy absorption and 308
  force transducer and 1124
  ground transportation **603–620**
  See also neural networks
Hu-Washizu stationary principle 1321
Hybrid Control **649–658**
  active-passive devices 653–658, 654F, 655F, 656F, 657F
  design strategies 650–652, 650F, 651F
  piezoelectric network 653–655
  proof-mass actuator 652F, 653
  stiffness and 658, 658F
  viscoelastic layer 656–658
hybrid damping 353–354
  characteristics of 354–360
  control law and 358
  motion equations and 358
hybrid-discrete-continuous model 172
Hybrid III Family 311
Hydraulic Institute 1225
hydraulics 1191, 1232
hyperstaticity 1116–1118
hyper-surfaces 228
hysteresis
  double frequency 639
  magnetostrictive materials 759
  single frequency 636–637
  transduction and 757
Hysteretic Damping 323, 325, **658–664**
  Boltzmann superposition model 661–662
  definition of 659–660
  frequency-dependent 323
  isolation theory and 1494–1501
  Kelvin model 660–661, 660F
  loading 659F
  material complex modulus 323–324
  nested loops and 662F
  nonlinear models 662–664
  relay operators 663F,
  resonance and 1050

I

Ibn-al-Razzaz 112
Ibrahim, R **582–589, 589–596, 726–740**
Ibrahim time domain method 821–822, 823
IC technology *See* MEMS
IDEAS-MS 305

Identification, Fourier-Based Methods **665–672**
  additive noise 667F, 668F
  error mechanisms and 669–671
  frequency domain and 666, 666F
  MIMO systems 669, 669F
  MISO systems 668–669, 669F
  noise-corrupted signals and 666–668
  response decomposition 667F, 669F
  schema for 668F
  test 665F
  windows 672F
Identification, Model Based Methods **673–685**, 673F, 674F
  ARMAX responses 684F, 685F
  AR order 684F
  Blackman-Tukey method 685F
  classification and 673–674
  continuous line response 684F, 685F
  data set and 674
  elements of 673
  estimation criterion and 676–682
  example of 684
  frequency response 685F
  frequency stabilization and 684F
  least squares methods 678–680
  modal parameters and 683
  model class and 674–676
  order and 682–683, 684F
  parameter extraction and 683
  schema for 674F, 676F
  SISO system, 676F
  validation and 683
impacts **1531–1548**, 1532F, 1533F
  bearing faults and 1187–1188
  bifurcation and 1533–1536, 1535F, 1539, 1539F
  chaos and 1542–1548, 1542F, 1545F, 1546F, 1547F
  classification and 1536, 1537F
  dynamics 278–286, 953
  examples of 1533, 1533F, 1534F, 1535F
  noise 1449–1452
  periodic stability and 1535F, 1538F, 1539–1542, 1539F, 1541F
  stability regions and 1536, 1538F
  subharmonic regions and 1536–1539, 1538F
impedance 1444–1445, 1444T
  bearings vibrations and 153–155
  direct problem and 1254–1259
  influence coefficients and 120–123
  inverse problem and 1259–1265
  rotor dynamics 1093
improved reduced system (IRS) 258
impulse response functions (IRFs) 820
  cepstral analysis and 217–218
  time frequency and 1366–1368, 1366F, 1367F
  vibration isolation and 1504–1505
inchworm motor 759
indicator function plot 430F

inertia 1325
  basic principles of 124, 125–126
  belt drives and 170–172
  columns and 239–240
  coupling and 1326
  discrete elements and 396–397, 396F, 397F,
  fluid-structure interactions and 549–550
  force transducers and 1123–1132
  isolation theory and 1490, **1507–1521**
  model updating and 845
  plate vibration and 1033
  rotor dynamics **1085–1106**, 1329
  shear flexibility and 239
  shells and 1159
  unbalance and 1185–1186
  vector iteration and 464–466
  vibration isolation and **1487–1506**
  See also damping; mass; rotation
infinite impulse response (IIR) 391–394
infinitesimal cube 1566F
influence coefficients 119–123
infrared analysis 378
inheritance diagrams 969, 975F
initial peak 1098
Inman, D **278–286, 314–319**
instability diagram 414
instantaneous frequency 644–645
instantaneous phase 643, 644F
Institute of Electrical and Electronic Engineers (IEEE) 245
Instrumental Variable (IV) method 675–676, 680
integration 1196–1198
intelligent transportation systems (ITS) 771
intelligent vehicle systems (IVS) 773
interaction diagrams 968F
interface force 1124
interference diagrams 414
interferometers 701–702
  See also laser based measurements
interlaminar stress 787–789
internal problem 1276F, 1279–1282
International Electro-technical Commission (IEC) 1224
International Road Traffic Accident Database (IRTAD) 302
International Standards Organization (ISO) 245, 604, 1081–1082, 1224–1225
  2631 1572, 1575
  5349 625, 626T
  nonrotating parts and 1226–1232
  standards and 1228–1232
invariant manifolds 228–229, 955–956
inverse iteration 718–719
Inverse Problems **686–690**
  classical 686–688
  structural dyanmics and 1259–1264, 1259F, 1262F, 1263F, 1264F

Inverse Problems in Vibration (Gladwell) 691
ion propulsion 771
isochrones 420, 426F, 1093
isolation theory 1507
  ambient vibrations 1507–1508, 1508F
  coupled systems 1490–1493, 1494F, 1496F, 1497F, 1498F, 1499F
  criteria of 1508F, 1509F, 1510–1512, 1511T
  degrees of freedom (DOF) and 1494–1501, 1494F, 1496F, 1497F, 1498F, 1499F, 1500F, 1501F
  detrimental effects and 1508–1509
  dynamic systems 1512, 1513F
  elastic mounts 1490
  experimental selection 1521
  general purpose machines 1517–1519, 1518F, 1519F
  geometric properties 1490
  impacts 1515
  impulse excitation 1502–1504, 1503F, 1504F
  inclined mounts 1493
  inertia and 1490, 1515–1517, 1516T, 1516F
  mounting conditions 1517–1518, 1518F
  nonlinearity 1504–1505
  nonrigid structures and 1520
  polyharmonic excitations 1515
  precision and 1513–1514
  random excitation 1501–1502
  single frequency excitations and 1514–1515
  transmission model 1509–1510, 1509F, 1510F
  wave effects 1505, 1506F
isotropic links 1093–1097
iteration
  eigenproblems and 463
  inverse 465–466, 466F
  solvers 993–994
  subspace 466
  vector methods and 464–466
  See also chaos

J

Jacobi method 464, 467
jitter 102F, 103–108, 108F, 109F
Joule effect 754

K

Kaiser window 1590
Kajima Shizuoka Building 30, 34F, 35F
Kane's method 1060
Kapania, RK **1335–1343**
Kareem, A **1578–1587**
Karman-type nonlinearity 1015

Kelvin model 660–661, 662
Kelvin-Voigt damping 338–339
kernel design 1363
Kijewski, T **1578–1587**
Kimball-Love observation 338
kinematic conditions
  equations of motion and 1324–1332
  fundamental theory and **1290–1299**
kinetic energy 1266
  columns and 240
  discrete elements and 395–397
  equations of motion and 1327
  friction and **589–596**
  Hamilton's Principle and 131
  piezoelectric damping and 357
  resonance and 1047
  rotor dynamics 1102
  time-average energy and 1268
  variational methods and 1350, 1353
  See also energy
Kirchhoff-Helmholtz equation 888
Kirchhoff plates 197
Klapka, I **967–976**
Kobayashi, AS **505–512**
Kolmogorov-Arnold-Moser theorem 235
Kolmogorov equation 1249
Krasnosel'skii-Pokrovskii kernels 664
Krishnan, R **629–642**
Kronecker delta function 186, 1275
Krousgrill, CM **928–943**
Krylov-Lanczos Methods **691–698**
  block-Lanczos algorithm and 695–696
  modes of 692–694
  other applications 697
  physical meaning of vectors 691–695
Krylov subspaces 467
Kunaporn, S 512
Kyobashi Seiwa Building 30

L

laboratory dynamic 492–493, 500–501, 500F
LADWP Receiving Station 442
lag 629–630
  See also helicopter damping
Lagrangian mechanics
  DYNA3D and 305
  energy methods and 290, 1323
  FETI and 997–1000
  formulation 131, 1324, 1344–1346, 1345T
  interpolation 389
  motion 1327
Lame coefficients 186, 1567
Lame parameter 1136, 1137
Lanczos method
  FEA software and 256
  vectors 692–694
  See also Krylov-Lanczos methods

Laplace transforms 1407–1409
  boundary element methods 201
  identification and 673
  impulse response function 1336–1338
  membranes and plates 1331
Laser Based Measurements **698–710**, 703F, 1403F, 1403–1404
  applications of 705–706
  Doppler vibrometer techniques 700–706, 704F, 705F
  full-field measurement and 707–709
  geometric properties and 707F
  holographs and 699, 708F
  non-Doppler techniques 706–707
  scanning systems 703F, 704F
  speckle noise and 705
laser Doppler vibrometer (LDV) 1403–1406
LDL super T factorization 712–713
lead magnesium niobates (PMNs) 660
leakage 671, 1588–1591
  windows and 1216–1218
learning algorithm 871–872
least-squares method 677, 678–680, 1203
Leissa, AW **762–770, 1024–1031**
Lenz's law 1402
Lesieutre, GA **321–327**
Levenberg-Marquart (LM) algorithm 866
Levinson recursion 1203
Li, CJ **143–152, 597–603**
Liapunov-Floquet transformations 1004–1005
Lieven, NAJ **578–582**
limit cycle 592–593, 592F
Lin, YK **1238–1246**
Linear Algebra **710–720**
  BLAS 991
  Moore-Penrose generalized inverse 717–719
  nonsingular systems 710–714
  singular systems 714–716
  SVD 716–717
linear analyses 252–255, 253T, 254T
  cables and 210–215
  Fourier-based identification and 665–672
Linear Damping Matrix Methods **721–726**
linear dependence 295F
linear interpolation function 282
linear least squares method 677
Linear Multi Stage (LMS) method 674–680
linear systems
  Duhamel's Principle and **1304–1308**
  stochastic analysis and 1247–1252
  superposition and **1299–1304**
linear variable differential transformer (LVDT) 1193, 1398, 1400–1401, 1401F
Link, M **844–856**

links 152–153
  bearings 153–165
  fluid 153–155
  isotropic 1093–1097
  kinematic 1060
  magnetic 158–161
  misalignment and 1116–1118
  rotation and 1068–1069
Liquid Sloshing **726–740**
  Bond number 739F
  coordinates 728F
  damping 851
  free and forced 727–729
  gravitational field 739–740
  mechanical models 731–734
  modeling 731F, 732F, 733F
  parametric 730
  rigid moment of inertia 733F, 734F
  road tankers 735–738, 737F, 738F, 739F
  surface motion 731F
  tank shapes 735T
liquid spring 1182
Littlewood, JE 228
Liu shells 305
loading 253T
  bending moment sensitivity 1125–1127
  bridges and **202–207**
  columns and **236–243**
  crash and 310F
  earthquakes and 446–460
  flight 88–89
  force transducers and 1123–1132
  ground 93–95
  gunfire 93
  hysteretic damping 659–660, 659F
  impulse response function **1335–1343**
  isolation theory and 1518–1520
  noise and **887–898**
  packaging and **983–989**
  shape memory alloys and 1147–1148
  signature generation and 144–145, 144T
  suspension 37–38
Localization **741–751**
  bladed disk assembly 741F, 745, 750F, 751F
  coupled oscillators and 746F, 747F
  engineering significance of 748–749
  forced response and 750F
  gear faults and 598
  harmonic frequency and 745F
  history of 744–745
  key results of 745–748
  mode 742–744, 743F, 746F, 748F, 749F
  NN and 922
  truss beams and 741F, 749
logarithmic decrement method 617
logarithms See cepstral analysis
looping
  hybrid control and 650–652

Love-type equations of motion 1155–1159, 1159T, 1160–1161
Lowe, MJS **1437–1441, 1551–1559, 1559–1564, 1565–1570**
low oscillating function 1273F
LQG/LTR controller 1147
LU factorization 711–712
lumped parameter approach 887
  classification and 1290–1291
  critical damping and 317–318
  sensor/actuators and 1134–1143
  superposition and **1300–1304**
Lyapunov 1097
Lyon, RH 1265

M

Ma, F **721–726**
McConnell, KG **1121–1134, 1381–1397, 1398–1406**
Mach numbers 563
Mach-Zehnder interferometer 701–702, 702F, 1404, 1404F
Macjkie, RI 972
McKee, K **143–152**
MACSYMA 1010
MAC values 852–854, 853T
Maddux, GE **1398–1406**
MADYMO 305, 311
magnetorheological (MR) fluids **467–475**, 640–641
  actuators and, 58–72
  new applications of 474–475
  semiactive control and 467–468
  smart fluids and 468–474, 468F, 469F, 470F, 471F, 472F, 473F
magnetic constrained layer damping (MCLD) 352, 353F
magnetic links 158–161
magnetic systems 1088
Magnetostrictive Materials 660, **753–762**
  actuation configurations 58–72, 759, 760F
  magnetism and 753–755, 754T
  sensing configurations 760–761, 761F
  transduction 755–759, 756F, 757F, 758F
Maia, NMM **820–824, 824–829**
MA (moving average) modeling 1199, 1203F, 1203
manifolds 229F, 955–956
  center 962
  Duffing's equation and 233–235
  invariant 228–229
  Melnikov's method and 232
Maple 1010, 1340
Marcondes, J **983–989, 1173–1180**
Markov theory 677
  stochastic systems and 1246–1252
martensite See shape memory alloys
mass
  balancing and **111–124**
  basic principles of 124, 125–126

mass (*continued*)
  continuous systems and 1312–1317
  discrete elements and 396–397, 396F, 397F,
  Dunkerley's method and 135–136
  eigenvalue analysis and **461–467**
  equations of motion and 1324–1332
  forced response and 581–582
  force transducers and 1123–1132
  isolation theory and **1507–1521**
  MEMS 781–782
  model updating and 845
  Rayleigh method and 1309–1312
  Ritz method and 1318–1319
  robots and 1062
  sensor/actuators and 1134–1143
  transmissibility 1522–1527
  unbalance and 1185–1186
  variational methods and 1355
  vehicular vibration and, **37–45**
  vibration isolation and **1487–1506**
  See also inertia; rotation
mass-spring model 687F, 689, 689F, 733F, 734
  dashpot system 591F
  discrete elements and 398, 398F, 399T, 400T
  resonance and 1047–1048
  shock isolation and **1180–1184**
  superposition and 1300–1301
  See also boundary conditions; damping
material anistropy 251T
  acoustic impedance and 1561T, 1562F
  damping and **327–331**
  effects of 1569
  evanescent waves and 1563F
  facilities 252T
  guided waves and 1551–1559
  plate vibration and 1029
  Rayleigh waves and 1554F
  robots and 1063
  unbound waves and **1565–1570**
material complex modulus 323–324
material damping 337–338
material properties
  actuators and 58–72
  MEMS and 797, 798T
  shape memory alloys **1144–1155**
  shock absorption and 1174
  ultrasonics and 1439
Mathematica 1010, 1340, 1343
mathematics
  autocorrelation functions 977–978, 1592
  averaging methods **98–110**, 1244–1246
  basic principles **124–137**
  Bendixson's theorem 593
  Bessel function 729
  block-Lanczos algorithm 695–696
  Bond number 739F, 739
  Broyden-Fletcher-Goldfarb-Shanno algorithm 637

mathematics (*continued*)
  Cantor set 233
  chaos theory 227–235
  classes 975
  continuum mechanics 973–974
  cost function 977
  Dirac's delta function 1336
  Duffing's equation 233–235
  eigenvalue analysis 461–467
  equations of motion 1324–1332
  expectation operator 345
  external problem 1279
  finite difference methods **520–528**
  Floquet theory 338, 1002
  fundamental equations of motion 1291–1293
  fuzzy logic 43–45
  Gaussian theory 1135
  gear diagnostics and 598–602
  Hilbert transforms **642–648**
  Jacobi method 464, 467
  Kolmogorov-Arnold-Moser theorem 235
  Krylov-Lanczos methods and 691–697
  Krylov subspaces 467
  Lame' constants 1567
  Lenz's law 1402
  linear algebra **710–720**
  Mathieu equations 228, 730, 1110, 1111
  Maxwell equation 1135
  Melnikov's method 232
  model updating 847
  numerical efficiency and 972–973
  parallel processing **990–1001**
  Poisson ratio 889, 1331, 1568
  QR algorithm 467
  random processes **1033–1039**
  Rayleigh quotient 462
  Rayleigh-Ritz analysis 462–463
  root mean square 245
  running sum 1196–1197
  Schur method and 995–1000
  Scruton number 1584
  SEA 1265–1272
  segment averaging 1219–1221
  sensor/actuators and 1141
  signal integration/differentiation **1193–1199**
  similarity transformation methods 461, 464
  Simpson's rule 1197–1198
  Smale-Birkhoff homoclinc theorem 233
  Smale's horseshoe 232–233
  SRSS 245
  statistical moments 1243–1244
  Strouhal number 1584
  Sturm sequence property 463, 467
  Succi method 1281–1285
  Taylor series 1325
  trapezium rule 1197
  van der Pol equations 227–228
  vector iteration methods 464–466
  wind 1586

mathematics (*continued*)
  windows **1587–1595**
  Young's modulus 889, 1567
Mathieu equations 228, 730, 1110, 1111
MathWorks 682
MATLAB 682, 1343
matrices 1267
  active suppression 51–55
  adaptive filters **81–87**
  CFRF 274, 276F, 277F, 278
  CGM 267
  classes 975F
  correlation 298, 299–302
  critical damping and 317–318
  cross-sensitivity 1126
  damping 321–324, 346–347, 356–358, **721–726**
  direct problem and 1254–1259
  elemental 363
  environmental testing and **496–504**
  FDAC 275F
  FETI and 997–1000
  flutter 567–570
  forced problem 1281–1282
  FRF 274–278
  influence coefficients and 120–123
  inverse problems 465–466, 686–688
  laboratory vs. field 492–493
  Lanczos 694–695
  linear algebra and **710–720**
  linear damping **721–726**
  lumped mass 279–280
  MAC 268F, 269, 269F
  mass 260–264
  model-based identification and 673, 673F
  piezoelectric damping and 356–358
  Poincare' method and 957–959
  random processes 1041, 1043
  Rayleigh method and 1309–1312
  rigidity 363
  rotation 1070–1077
  rotor dynamics 1085–1088, 1091–1092, 1102
  RVAC 274F, 275–277
  sensitivity 847
  shape memory alloys and 1147–1148
  signal processing 1201–1203
  singular 714–715, 720
  skyline storage and 714F
  sparse 713–714
  stiffness error 260–264
  stochastic systems 1247–1252
  structural dynamic modifications and **1253–1264**
  Sturm sequence property 463
  submatrices 360
  subspace iteration and 466
  TOR 257
  transformation 360
  XOR 257, 267
  See also eigenvalue analysis
MATRIXx 682

Maxwell, James Clerk 1065
  equation of 1135
  model of 337
  theorem of reciprocity 136, 819
mean time before failure (MTBF) 1249
mean-value method 1269
measurement
  damping **332–335**
  full field 707–709
  influence coefficients and 122
  packaging 984–987, 984F, 984T, 985F, 985T, 986F,
  rotation 1080–1081
  standards for **1224–1238**
  See also Laser-Based Measurements; modal analysis
mechanical impedance 1571
mechanical shock 1228
Melnikov's method 232, 233
Membranes **762–770**
  circular 765, 765F, 765T, 766T
  complicating effects 769, 769F, 770F
  other shapes 764F, 765T, 767, 767F, 768F
  rectangular 763, 764F
  shells and **1155–1167**
  strain 1135–1136
  vibration 1331
  See also cables
MEMS, Applications **771–779**
MEMS, Dynamic Response **779–794**
MEMS, General Properties **794–805**
MEMS (microelectromechanical systems) 771, 1011, 1017, 1142
  acoustic microsensors 774, 775F
  applications of 760F, 771, 772T, 777–778
  damping 783–784, 784T
  design 803–804, 803T, 804F
  dynamic response **779–794**
  electrostatic field 785–787, 785F, 791T
  fabrication technology 798–801, 799T, 801T, 802T, 803F, 803T
  general properties of **794–805**
  interlaminar stress 787–789, 788F, 789F
  machining 795F, 796, 797
  mass 781, 782F, 782T
  material choice 797, 798T
  measurement 796, 796T
  microaccelerometers 772–774, 772T, 773F, 774F
  microactuators 775–777, 776F, 777T
  micropumps 777, 777F, 778F
  microsensors 771–772, 773F, 774, 775F
  microvalves 777, 777F, 778F
  packaging 801
  signal conditioning 789–790
  stiffness 782–783, 783T
  voltage conversions 786T, 790–793, 793F

meshing 249T, 250, 601
  mapping solutions from 281–282
  See also gear diagnostics
metglas amorphous ribbons 761
Michelson interferometer 701F, 701–702, 1404F, 1404
microelectromechanical systems *See* MEMS
MILSpecs 1225
MIMO (multi-input, multi-output) method 821–823
Mindlin-Reissner plates 197
minimax filters 390
minimum complementary energy 1347
minimum description length (MDL) criterion 1205–1206, 1206F
misalignment 1116–1118, 1186
Modal Amplitude Coherence (MAC) 683
modal analysis **805–813, 820–823**
  applications **829–838**
  calibration 818
  damping and 826–828
  data processing 815–816, 816F
  error location 834
  excitation 808–809, 809F, 814F, 815, 816–817
  frequency-domain methods 822–823
  history of 806
  mathematical construction 811–813
  measurement **813–820**
  method classification and 820
  model construction 824–829
  multipoint testing 817–818
  NN and 922
  parameter quality and 823
  pretesting 819
  procedures of 806
  property comparison 830–832, 831F, 833F
  response properties 824–826, 824F, 837–838
  sensing mechanism 815, 815F
  signal processing 810, 811F
  spatial models 824–826, 824F
  structural analysis 834–837, 836F
  support conditions 818–819
  theory of 807–808
  time-domain methods 821–822
  transducers 809
  troubleshooting 829
  updating 832–834
  validation 819, 830
Modal Analysis, Experimental, Applications **829–838**
Modal Analysis, Experimental, Basic principles **805–813**
Modal Analysis, Experimental, Construction of models from tests **824–829**
Modal Analysis, Experimental, Measurement techniques **813–820**

Modal Analysis, Experimental, Parameter extraction methods **820–824**
Modal Assurance Criterion (MAC) 268F, 269F, 274–276
  COMAC and 267–271
  correlation functions and 296
  numerical comparison and 266
  SNDT and 903
  vector correlation and 267–270
modal properties 889–890
  active suppression and 51–52
  actuator sensitivity and 1140–1141
  balancing and **111–124**
  blades and 177
  BLAS 991
  boundary element methods **192–202**
  cables and 211–215
  COMAC 275, 276
  comparison of **265–272**
  complex plot 429F
  continuous methods and **286–294**
  continuum mechanics and 973–974
  damping and **321–327, 721–726**
  density 1266, 1268
  direct graphical comparison 266
  direct numerical comparison 266
  disks and **404–413**
  displays and 413–417
  DOF correlation 270–271
  double 419F
  earthquakes and 458–460
  energy 1266–1267
  equations of motion and 1326
  Euler-Bernoulli theory and 137–141
  experimental analysis and **805–813**
  FEA software and 244
  F-F-F-F square plates 1027F
  finite element methods **530–544**
  flutter and 566–567
  Fokker-Planck-Kolmogorov equations 1248–1249
  forced problem 1281–1282, 1298–1299
  free-free planar truss 420F
  free vibration frequencies and 1026T
  gear diagnostics and **597–603**
  geometric nonlinearity 252
  indicator function 425–429, 430F
  influence coefficients and 119–123
  inverse problem 1259–1264, 1259F, 1262F, 1263F, 1264F
  Krylov-Lanczos methods and 691–697
  localization and **741–751**
  membranes and **762–770**
  model-based identification and **673–685**
  multivariate 429F
  near-field approach 891
  noise and 887–898
  nonlinear normal **918–924**
  nonlinear system resonance and 928–943

modal properties (continued)
  nonlinear testing and 1286, 1289
  optimal filters and **977–982**
  order selection and 682–683
  orthogonality criteria 266–267
  overlap and 1265
  parameter extraction and 683–685
  planar flexural shapes 418F
  plate vibration and 889–890
  radiation efficiency 892–894, 893F
  Rayleigh's Principle and 132–134
  Rayleigh waves and 1554F
  rotation 1071–1072, 1075–1076
  rotor dynamics 419F, **1085–1106**, 1090
  scale factor (MSF) 266
  SEA and 1266–1267
  shape memory alloys and **1144–1155**
  shells and **1155–1167**
  sound radiation and 891–892
  stochastic differential calculus 1246–1252
  structural dynamics and **1253–1264**
  structural modes and 683
  superposition and 1302–1304
  three-dimensional plot of 423F
  validation and 683
  vector correlation 267–270
  wave number and 891
  weak elements 418F
  wind and **1578–1587**
mode acceleration method 719
Mode Indicator Functions (MIFs) 425–429
modeling
  absolute motion and 1382–1395
  additional methods for 1207
  antiresonances 848
  AR 1201–1203, 1202F
  ARMA 1201, 1203
  basic principles **124–137**
  biodynamics and 1572
  classes of 674–676
  continuous system 1327–1331
  damping **321–327, 332–335, 335–342**, 850–851
  deterministic 1200
  discrete elements 396
  equations of motion and 1324–1332
  finite element method and 844–845
  flutter and 555–565
  gear vibration 597–598, 599, 1189–1191, 1190F, 1191F
  hysteresis 636–637, 639
  identification methods and **673–685**
  liquid sloshing and 731–734
  MA 1199, 1203, 1203F
  mass-spring 733F, 733–734
  modal analysis and **824–829**, 829–838
  normal mode 848–849
  parameter identification and 847–848
  parametric methods 1203

modeling (continued)
  pendulum 732–738, 732F
  pseudoresponse 849–850
  quality and 847
  robots **1055–1063**
  rotor-stator interactions 1107–1121
  selection and 1205–1206, 1206F
  sequential methods 1203
  ship vibrations **1167–1173**
  signals and **1184–1193, 1199–1208**
  spectra and 1204–1205, 1205F
  stochastic 1199–1200
  superposition and 1300–1301
  symptom-based 1185
  test/analysis residuals 848–851
  tool wear and 1380
  transverse vibration and 169
  updating of **844–856**
  validation of 851–855, 852F, 853F, 853T, 854F, 855F
Model Updating and Validating **844–856**
Mode of Vibration **838–844**
  complex 842–843
  definition of 838, 839F
  essential features of 839–840, 839F, 840F
  forced 843–844
  free 840–843
  orthogonality 841–842
  types of 839
mode shape plots 415–417
mode synthesis 1334–1335
modulus damping 338–339
modulus difference 271
momentum conservation principle 307
monopole sound 880
Monte Carlo simulation 740, 1046
  localization and 746
Moon, FC 233
Moore-Penrose generalized inverse 717–719
Moors 112
motion
  control 493
  envelopes and 493–494
  friction and **589–596**
  sprag-slip 593–597
  transducers for **1398–1406**
  See also oscillation
Motion Sickness **856–861**
  causes of 857
  dose value (MSDV) 859, 860
  nonvertical oscillatory motion 859–860
  vertical oscillatory motion and 857–859
Motor Vehicle Safety Standard (MVSS) 303T, 304
mounts 343F, 344T
  active/semi-active damping and 347–349
  basic concepts of 342–343
  dynamically coupled 347F
  elastically coupled damping and 345–346

mounts (continued)
  fluid-elastic 347F
  multidirectional 346
  multiple DOF 348F
  passive damping and 343–345
  SDOF 345F, 346F
  See also damping
MSC/NASTRAN 248–249
Mucino, VH **302–314**
Mullins effect 631
multi degrees of freedom (MDOF) 820
multilayer feed forward (MLFF) networks 866, 867
multilayer neural networks (MLNN) 864, 872–875
multipath propagation 299–301, 299F, 300F
multiple input/single output (MISO) systems 665, 668–669, 669F
multiple input/multiple output (MIMO) systems 665, 669, 669F
  feedforward control **513–520**
multiple instructions/multiple data (MIMD) processors 990
multivariate mode indicator function (MMIF) 425–429, 429F
MUMPS 795
Muszynska, Agnes 1112

N

NA4 601
Naeim, F **439–461**
Nanjing Communication Tower 30
narrow-band demodulation 602
NASTRAN 405
National Agency for Finite Elements and Standards (NAFEMS) 245
National Highway Traffic Safety Administration (NHTSA) 303T, 304
Natori, MC **1011–1018**
natural frequencies
  of plates and beams 414–415
  plotting of 413–414
Navier's equations 911
NB4 601
near-field approach 891
Neumann preconditioner 1000
neural networks
  architecture and 870, 871F
  assessment and 867–868
  error and 870
  faults and 865–867
  learning and 871–872, 874
  monitoring and 864–867
  multilayer 872–875, 873F, 875F
  neurons 869–870, 869F
  patterns and 874
  perceptron 871–872
  processing tasks and 863T
  radial basis function 875–877
Neural Networks, diagnostic applications **863–868**

Neural Networks, general principles **869–877**
neurological disorders 622
Newkirk effect 1107–1110
Newton, Isaac 124, 1099, 1344
  law of motion 1324, 1325, 1329
  rotation and 1070, 1085
Newton-D'Alembert principle 1112
Niemkiewicz, J **1224–1238**
Nitinol 660
  See also shape memory alloys
nodes
  circular plates and 1026–1027
  diameter (ND) 405–413
  laboratory vs. field testing 492–493
  line indices and 414–415
  modal shape plots and 415–417
  rotation and 1073
noise **887–898**, 887–888, 896–898
  baffled plates and 889F, 891–896, 896F, 897F
  boundary conditions and 888–889
  cable 1133
  corrupted signals 666–668
  dipole sources 882F
  Green's functions and 879
  Huygen's source 884–885, 885F
  monopole sources 885F, 886F, 887F
  neural networks and 867
  plate vibration and 889–891, 893F, 894F, 895F
  power and 883–884
  quadrupole sources 883F, 884T
  radiating field and 879–882
  sigma-delta converters and 372
  sound power and 890–891
  sound pressure with boundaries 888–889
  spherical harmonics and 881T
  superposition and 884–886
  waves and 878–879
Noise Radiated by Baffled Plates **887–898**
Noise Radiated from Elementary Sources **877–887**
nonconservative forces 1332
Nondestructive Testing, Sonic **898–906**
Nondestructive Testing, Ultrasonic **906–918**, 1439
nondestructive testing (NDT) 378, 1559, 1563–1564, 1568
nonlinear analysis 252, 254T, 255T, 953
  asymptotic techniques and 957–962
  averaging method and 960–962
  backlash and 951F
  bifurcations and 963–965, 964F, 965F
  classes and 947–953
  Coulomb friction law and 950F
  dimension reduction and 962–965
  effects of 944
  equilibrium and 953–957, 954F, 955F, 956F

nonlinear analysis (*continued*)
  finite element methods 536–538
  Hilbert transforms and 646
  neural networks and **869–877**
  overview of **944–951**
  parametric excitation and 1003–1009
  perturbation and 957–962
  Poincare' method and 957–959
  resonance and 1054–1055
  rotor dynamics 1094F, 1095–1097, 1095F
  sources and 945–947
  stability and 953–957
  stochastic analysis and **1238–1246**
  vortices and 949F
  See also chaos
Nonlinear Normal Modes (NNM) **918–924**
  applications of 921–924
  cyclic assembly and 921F
  definition of 918–921
  degrees of freedom (DOF) and 919F, 920
  resonance and 921–922, 923F
  stiffness and 425F, 920F
Nonlinear System Identification **924–928**
Nonlinear System Resonance Phenomena **928–943**
  cubic 940–943
  quadratic 937–939
  response curves 932F, 933F, 935F, 936F, 938F, 939F, 940F, 941F, 942F, 943F
  single degree of freedom and 930–937
  two degrees of freedom and 937–943
nonlinear systems 924–925, 928, 1332
  active states and 925–926
  elastic recall 1110–1111
  hysteretic damping 662–664
  misalignment and 1116
  nonparametric identification and 926–927
  overview of **944–951**
  parametric identification and 927
  perturbation methods and 1010
  piezoelectricity and 1015–1017
  transverse vibration and 167–169
  von Karman 1136
  See also chaos
Nonlinear Systems Analysis **952–966**
Nonlinear Systems, Overview **944–951**
nonsingular linear systems 710–714
nonstationary signals 296
nonvertical oscillatory motion 859–860
normal form theory 963, 1005
normalized cross-orthogonality (NCO) 270, 271
Norton, MP 877–887, 887–898, **1443–1455, 1456–1480**
nuclear radiation 1133

Nuclear Regulatory commission (NRC) 245
nullspace 714–715, 720
Nyquist plots 418–420
  CEDA and 1271
  damping measurement and 333
  DOF 424F
  frequency 445
  influence coefficients and 122
  modal analysis and 822F
  receptance 423F
  sigma-delta converters and 371–372

O

object oriented programming
  class diagram 969
  collaboration diagrams 970
  component diagrams 971
  continuum mechanics and 973–974
  criteria of 968–969
  deployment diagrams 970
  dynamic systems and 974–975
  language of 971–972
  methodology of 969
  numerical efficiency and 972–973
  sequence diagrams 970
  software design and 967–968
  state transition diagram 970
  vs. procedural programming 968F
Object Oriented Programming in FE Analysis **967–976**
odd-odd plate mode 892–893
oil analysis 378
oil film excitation 1191
oil-pressure servo valves 485–488
oil whip 1085
online systems 1082, 1083
open loop control 49–50
Operating Deflection Shapes (ODSs) 272, 275
optical heterodyning 1404–1405
Optimal Filters **977–982**, 978F, 981F
  time domain formulation and 977–978
  Wiener 978–981
optimization
  hybrid control and 650–652
  rotor dynamics 1095
  structural modifications and 1262–1264
orthogonality 266–267, 270, 271
  basic principles of 124
  biorthogonality 1074
  forced vibration and 1298–1299
  inverse problems 688
  isolation theory and 1514
  iterative solvers and 994
  mode of vibration and 841–842
  nonlinear systems and 926
  Rayleigh's method and 1317
  shells and 1155
  TAM and 259–260
  Zernlike polynomials 484
orthonormality 461

oscillation
  averaging method and 960–962
  bearings vibrations and 152–161
  boundary element methods 195–198
  chaos and **227–236**
  critical damping and **314–319**
  fast function 1273F
  FEA software and 245
  flutter and **553–565**
  friction and **589–596**
  fundamentals of **1290–1299**
  gear diagnostics and **597–603**
  helicopter damping and **629–642**
  liquid sloshing and **726–740**
  limit cycle 95
  localization and **741–751**, 746F
  low function 1273F
  motion sickness and **856–861**, 857F, 858F, 859F
  NNM and **918–924**
  noise and **877–887**, 887–898
  nonlinear analysis and **952–966**
  nonlinear resonance and **928–943**
  nonvertical motion and 859–860
  Poincare' method and 957–959
  power balance and 1267
  resonance and **1046–1055**
  SEA and 1266–1267
  stick-slip 589, 590F
  time frequency and **1360–1369**
  tire vibrations **1369–1379**
  vector iteration and 464–466
  vertical motion and 857–859
  vibro-impact systems and **1531–1548**
  viscous damping and **1548–1551**
  whole-body vibrations and **1570–1578**
  See also absorbers; boundary conditions
Output Error method 677
overdamped system 316
overlap processing 1593
oxide perovskites 477–482

P

Packaging **983–989**
  loading and 988F
  measurement and 984–987, 984F, 984T, 985F, 985T, 986F
  MEMS and 801
  structural design and 983–987, 983F, 983T
  testing of 988
  transmissibility and 987F,
Pade matching 1200, 1204
PAM-CRASH 305
Pan, J 877–887, 887–898
parallelograms 126
Parallel Processing **990–1001**
  direct solvers and 992
  domain decomposition 992F, 994–1000

Parallel Processing (continued)
  FETI method 997–999
  gradient iteration and 994, 994T
  iterative methods and 993–994
  network topology and 990, 991F
  Schur methods 995–1000, 997T, 998F
parameters
  4 DOF pitch plane model 608
  analog inputs and 369
  averaging method and 960–962
  basic principles of 124
  bearing diagnostics and 145–151
  bifurcation and 435
  chaos and 1112–1115
  critical damping and **314–319**
  damping materials and **327–331**
  damping measurement and **332–335**
  distributed 318–321
  earthquakes and **439–461**
  equations of motion and 1292–1293
  extraction of 683
  feedforward control **513–520**
  ground transportation **603–620**
  inverse problems 686–688
  Lame 1014, 1015
  lumped 317–318
  Markov 677
  modal analysis and 821F
  model-based identification and 673–685
  model updating and 845–848
  neural networks and 864
  NNM and 919
  noise and **887–898**
  nonlinear systems and **952–966**
  signal processing and **1199–1208**
  SNDT and 899
  stochastic analysis and **1238–1246**
  structural dynamic modifications **1253–1264**
  updating 847–848
Parametric Excitation **1001–1009**, 1332
  cables and 214F
  Floquet theory and 1002, 1004–1005
  nonlinear analysis and 1003–1009
  perturbation methods and 1010
  point-mapping and 1006
  problem formulation and 1002
  resonance and 1054
  transverse vibration and 169
  See also excitation
parametric identification 673
  cables and 212F, 213F
  Fourier-based **665–672**
  modal analysis and **820–824**
  nonlinear systems and 926–927
  nonlinear testing and 1288
Parseval's theorem 388, 445
Pascal, Blaise 1064
passband patterns 745F, 748F
passive damping 351–352

PATRAN 305
pendulum model 732, 732F, 734, 736
  absorbers and, 22
  chaos and 228
  equations of motion and 1325F, 1325
  localization and 743F, 745
  road tanker and 735–738
perceptron 871
perfect inviscid fluid 1567
periodic data 491
periodic orbits 230T
  chaos and **227–236**
periodic rezoning 282
periodic truss beam 741F
Perkins, NC **209–216**, **944–951**
perovskites 477–482
perturbation methods, 232, 957–962
  bifurcation and 1003F
  nonlinear systems and 1010
  parametric excitation and 1003–1004, 1009
  rotation and 1002F
  See also chaos
Perturbation Techniques for Nonlinear Systems 1010
Peterka, IF **1531–1548**
phase space 228
phase unwrapping 218
phasors 1270
physical classes 975
Pierre, C **741–751**
piezoelectric actuators 482–488
  absolute motion sensing and 1385–1387
  hybrid control and 653–656
  viscoelastic layer and 656–658
piezoelectric materials 1011–1012, 1012T, 1017
  active absorbers and, 1–3, 2
  active damping and 352, 352F
  active isolation and, 47, 48
  actuators and 58–72, 61T
  advanced theory 1014–1015
  applications 1013
  damping and 353F, 354–360
  feedforward control 519
  flappers 485–488
  four fundamental equations of 1013, 1013T
  hysteretic damping and 660
  linearity and 1012
  sensor/actuators and 1134–1143
  shells and 358–360
  thermoelasticity and 1013
  thermoelectromechanical coupling 1015–1017
Piezoelectric Materials and Continua **1011–1018**
Pipes **1019–1024**
  annular fluid flow and 1023
  cross flow 1023
  external fluid flow and 1022
  internal fluid flow and 1019–1022
pistons 1445–1445

pitch 629–630
　excitation 610
pitch-catch setup 906, 907F
pitchfork bifurcation 215, 930
pitch plane model
　2 DOF 609–610
　2 DOF quarter car model 612–613
　4 DOF 608
　road excitation and 606–607, 618–621
　suspension and 615–616
　weight and 615
plasticity 252T
Plates **1024–1031**, 1331
　acoustic radiation and 545–550
　baffled 887–898
　cantilevered skew 1028F
　circular 1026–1027, 1027T, 1027F, 1028T, 1028F
　complicating effects 1029–1033
　continuous systems and 1316–1317
　flexural radiation and 1459, 1463–1465
　forced vibration and 894–896
　Kirchhoff 197
　MEMS 787
　modal properties and 889–890
　modal radiation efficiency 892–894, 896F, 897F
　model updating and 845
　natural frequencies and 414–415
　rectangular 1026
　SEA and 1267
　sensor/actuators and 1134–1143
　sound power and 890
　triangular 1030F
　ultrasonic waves and 910
　vibration intensity and 1484–1485, 1487
plotting
　bar plots 413
　Campbell diagrams 414
　compass 416
　FRF data and 417–420
　indicator function 430F
　instability diagram 414
　interference diagrams 414
　mode shape 415–417
　root-locus diagrams 413
　singular value 425, 428F
　spectrum 420–421, 428F
pneumatic spring 1183
Poincare, Henri 228, 431
Poincare map 228, 232, 235F
Poincare' method 957–959
point-mapping 1006
Poisson's ratio 327, 889, 1331, 1567
　shells and 1156
polar notation 643
pole placement 54
polymorphism 969
polynomial model classes 675
positive position feedback 3–5
potential energy 1266
　columns and 239
　equations of motion and 1327

potential energy (*continued*)
　Hamilton's Principle and 131
　minimum 1346–1347
　minimum total 1320
　piezoelectric damping and 357
　Rayleigh method and 132–134, 288
　resonance and 1047
　viscous damping and 336
　See also energy
potential mode identifier 429
power balance
　modal groups and 1266–1267
　oscillators and 1266, 1267
　WIA and 1269
power spectral density (PSD) 245, 1581
　cepstral analysis and **216–227**
　classical analysis and 595F, 1210–1211, 1212F, 1214, 1219–1221
　correlation functions and 296–299, 297F
　random signals and 1219–1223
　road excitation and 618–621
　windows and 1591
Pozo, R 973
Prasad, MG **1299–1304**
prediction error method 676–677
Preisach plane 664
prescribed conditions 252T
Principal Response Functions (PRFs) 272, 278
Principia Mathematica (Newton) 124
principle of least action 1323
probability 1238–1242
Prony method 678–679, 682, 1200, 1204
proof-mass actuators 653
propellers 1170–1172
proper orthogonal decomposition (POD) 926
proper orthogonal values (POVs) 926
pseudoresponse residual 849–850
pseudo-velocity (PSV) 444
pulleys *See* belts
P-waves 913, 913F
pylons 574
pyrotechnic shock 495
Pythagoras' theorem 579

Q

QR algorithm 467
quadrupole sound 882, 883F
qualification testing 490
quasiharmonic motion 590F
quefrency 216
　bearings and 218–220

R

Rade, D **9–26**
Rades, M **256–264**, 265–272, 272–277, **413–431**, **1046–1055**, **1180–1184**

radial based function (RBF) 864, 875–877
Radiation by Flexural Elements 130, **1456–1480**
　basic theory of 1456–1463, 1456F, 1458T, 1459F, 1459T, 1460F, 1461F, 1461T, 1462T, 1463F
　boundary conditions and 186, 187F, 188F, 189F
　MEMS 787–789
　Ritz method and 134
　sound and **1456–1480**
　source ratios 1463–1468, 1464F, 1466F, 1467F, 1468F, 1469F, 1470F, 1471F, 1472F, 1473F, 1474F, 1475F, 1476F, 1477F, 1478F, 1479F, 1480F
　transmissibility 1523–1527
rahmonics 218–220
　complex cepstrum and 222–227
rain-on-the-roof excitation 241
Ram, YM **686–690**
Randall, RB **216–227**, **364–376**
random data 491
random error 671
　spectral analysis and 1222–1223
Random Processes **1033–1039**
　autocorrelation function 1037, 1037F
　damping 1037, 1038F
　Fourier transforms and 1033
　normalization and 1033
　singularities 1034, 1035F
random signals 297–298, 644F, 1185
　generation of 1185–1192
　spectral analysis and 1219–1223
　stochastic analysis and **1238–1246**
Random Vibration, Basic Theory **1040–1046**
　central limit theorem and 1044, 1044F
　degrees of freedom (DOF) 1040–1042
　Gaussian noise and 1045–1046
　mean square response 1042, 1042F
range charts 1405F
Rankine, William 112
Rao, SS **202–207**, **395–404**, **520–530**, **530–544**, **1019–1024**, **1308–1324**, **1344–1360**
Rateau, Edmond 1065
rational fraction polynomial method 823
Rayleigh's method
　basics of 132–134
　columns and 238, 239
　continuous systems 1312–1317
　discrete systems 1309–1311
　finite element methods 533
　integral 889, 890
　MEMS 787
　noninteger power 289
　ordinary 288–289
　quotient 462
　random processes 1038
　substructuring 1332

Rayleigh-Ritz analysis 290, 462–463, 691
  iterative solvers and 993
Rayleigh waves 1553, 1554F
  ultrasonic testing and 908
Raynaud's disease 621
receptance 421F, 423F
reciprocity 136, 1267
  modal analysis and 819
rectangular plates 1026
recurrent neural networks (RNN) 867
recursion 100–101, 108
Reissner energy 1344, 1347
residuals 848–850
resonance
  actuators and 482–489
  antiresonance and **1046–1055**
  bearing diagnostics and 146, 146F
  cubic systems 940–943
  degrees of freedom (DOF) and 930–943, 1047–1055, 1048F, 1049F, 1050F, 1051T, 1053F
  electrostriction and **475–490**
  forced response and 581–582
  forced vibration and 1296
  location 1051
  mass-spring damping 1047–1048, 1048F, 1049F, 1050F
  MEMS 779
  NNM and 921–922
  nonlinear systems and **928–943**, 944, 1054–1055
  parameter sweep through 1288
  phase method 1052–1053
  primary 931–934, 938–939
  quadratic systems 937–939
  residuals and 848
  rotor dynamics 1097
  secondary 934–937
  sharpness of 1051
  ship vibrations 1170–1171
  sub-harmonic 1288
  testing 1052
  transduction and 755–759, 755F, 756F, 757F, 758F
  viscous damping and **1548–1551**
  whole-body vibrations and 1570–1578
  See also frequency; harmonics
Resonance and Antiresonance **1046–1055**
response properties **272–277**
  aeroelastic **87–97**
  bridges and **202–207**
  CFRF matrices 276F, 277F
  FDAC matrices 275F
  forced 1075–1076
  free 1072–1073
  FRFs 273F
  individual 272–273
  MEMS **779–794**
  modal analysis 824–826, 837–838
  parameters 1267–1268
  principal 277–278
  pseudoresponse 849–850
  residuals 848–850

response properties (continued)
  RVAC matrices 274F
  sets 274–277
  transduction and 755F, 755–759, 756F, 757F, 758F
  See also active control; modal analysis
Response Vector Assurance Criterion (RVAC) 274F, 275–277
Reynolds number 1585
ride natural frequency 614
rigidity 113–118, 166, 1373–1374
RISK processors 990
Ritz method 290, 1332
  basics of 134
  energy and 1318–1320
  finite element methods 533
  MEMS 787
  substructuring 1332
  vectors 691
Rivin, E **1487–1506, 1507–1521**
Rixen, Daniel **710–720, 990–1001**
RMS acceleration 604
RMS value 1582, 1583
  windows and 1593
road excitations 618–620
Robert, G **243–256**
Robot Vibrations **1055–1063**
  actuated joints 1059
  altered operational strategies 1063
  arm improvement 1062
  commanded motion 1057, 1058F, 1059F, 1063
  components 1059
  damping 1063
  degrees of freedom (DOF) 1060, 1060F, 1061F
  flexibility and 1056F, 1057
  kinematic linkages 1060, 1060F
  mass allocation 1062
  material selection 1063
  modeling 1057
  See also MEMS
Rochelle salt 1013
rockets 1265
rod vibration 1328–1329, 1328F, 1329F, 1330F
rigid state 118–119
rolling bearing faults 1187–1188
root-finding methods 463
root-locus diagrams 413
root mean square (RMS) 145
  averaging and 98
Rosenhouse, G **124–137, 180–191, 1304–1308**
Rotating Machinery, Essential Features **1064–1069**
Rotating Machinery, Modal Characteristics **1069–1077**
Rotating Machinery, Monitoring **1078–1084**
rotation 414
  abnormal situations 1078–1080, 1082–1083
  balancing and **111–124**
  basic diagnostics and 379, 379T

rotation (continued)
  bearing faults and 1187–1188
  bearing vibration and 152–163
  biorthogonality 1074, 1074F
  blades and **174–180**
  columns and 239–240
  coupled 112–113, 1068–1069
  data collection 1082
  definitions 1069
  disks and 409–410
  eccentricity and 112–113
  equations of motion and 1325
  flexible state and 118–119
  fluid-structure interactions and 549–550
  forced responses 1071–1076
  formulations 1070–1071
  free responses 1072–1073
  friction and **589–596**
  helicopter damping and **629–642**
  historical studies of 1064–1066
  influence coefficients and 119–123
  limits 1081–1082, 1081F
  links and 1068–1069
  liquid sloshing and **726–740**
  localization and **741–751**
  machines and **1064–1069**, 1067F, 1068F,
  maintenance 1078, 1079F
  misalignment and 1186
  monitoring 1078–1084
  motion sickness and **856–861**
  natural frequencies 1071
  natural modes 1071–1072
  nodal points 1073, 1073F
  nonlinear systems and 956
  online systems 1082
  resetting sensitivity 1077
  rigid states and 113–118
  sensing and measurement 1080–1081, 1080F
  shells and 1159
  signal generation and **1184–1193**
  signals and 1192T
  SNDT and 905
  spectral analysis and 1218
  speed and 118–119
  sprag-slip 593–597
  standards 1082, **1224–1238**
  transient analysis 1083–1084
  unbalance and 1185–1186
rotor-bearing systems 413, 414F, 417, 420F
  helicopters **629–642**
  unbalance plot 420
Rotor Dynamics **1085–1106**
  chaos 1096F, 1097–1098
  coaxiality fault 1098F, 1099
  De Laval's model 1085–1088
  formulation 1102
  gyroscopic couple 1099–1101, 1099F, 1100F
  initial peak 1098
  isotropic links 1093–1097, 1094F
  modal properties 1090

Rotor Dynamics (*continued*)
  modeling 1085–1088, 1086F, 1087F
  natural frequencies 1088–1090, 1089F, 1090T
  temporary rates of flow 1102–1103, 1103F, 1104F, 1105F
  torsion 1105–1106, 1106F
  unbalance 1086F, 1091–1092, 1091F, 1098
Rotor-Stator Interactions **1107–1120**
  chaos 1112–1115, 1113F, 1114F, 1117T
  contact at reducing speed 1107
  cycle limits 1107–1110, 1108F, 1109F
  deterministic contacts 1109F, 1110–1112, 1110F, 1111F
  misalignment 1116T, 1116–1118, 1117F, 1118F
  support deterioration 1118
  tremors 1118–1121, 1119F
roughness index (RI) 618–621
Routh-Hurwitz stability 23
Rubbing Shafts Above and Below the Resonance (Critical Speed) (Taylor) 1107
Rumbaugh, J 969, 971
Runge-Kutta-Gill method 1119
running sum 1196–1197
run out 112–113, 1099–1101, 1099F, 1100F
RVAC matrices 274F, 275–277, 278

S

Saddles 229, 1544–1548
SAMCEF 248–249
sampling
  data acquisition and **364–376**
  digital filters and 389–391
  external 373–375
  sigma-delta converters and 371–372
  Simpson's rule 1197–1198
  software resampling and 375
satellite ultraquiet isolation technology experiment (SUITE) 48
scanning laser Doppler vibrometer (SLDV) 700, 703, 704F, 1403
  full-field measurement and 707–709
scattering 1568
SCE Lucerne Valley Station 443
Schmerr, LW Jr. **906–918**
Schur method
  FETI 997–1000, 998F, 999F
  primal complement 995–997
Schwarz method 995
Sciulli, D **46–48**
Scott, RZ **137–143**
Scruton number 1584
sealing 161–162, 1088
seating dynamics 1577
segment averaging 1219–1221

Seismic Instruments, Environmental Factors **1121–1134**
  absolute motion transducers and 1381–1396
  bending moment sensitivity 1125–1127
  calibration 1130–1132
  cross-axis sensitivity and 1121–1122, 1122F, 1123F
  environmental factors 1132–1133
  force transducers and 1123–1132, 1123F, 1124F, 1125F, 1126F, 1127F, 1128F, 1129F, 1131F, 1132F
  loading effects 1128–1130
seizure 151
self-organizing feature maps (SOFM) 865
semi-iterative solvers 994–1000
Sendagay INTES building 30, 30F
Sensors and Actuators **1134–1144**
  distributed 1134–1138
  magnetostrictive materials **753–762**
  MEMS and **771–779**
  sensitivity and 1140–1141
  structonic cylindrical shells and 1138–1141
  See also MEMS
separatrices 232
sequence diagrams 970, 971F
servo valves 485–488
Sestieri, A **1253–1264, 1275–1283**
shaft encoder 375
Shaft Rubbing (Newkirk) 1107
shaft vibration 1191, 1328–1329, 1330F
  standards and 1228–1232
shallow sag cable 210–213
Shape Memory Alloys (SMAs) 660, **1144–1155**, 1146T
  active damping augmentation 1149–1150, 1150F, 1151F, 1152F
  active impedance and 1151–1153, 1153F, 1154F, 1155F
  actuators and 65–68
  control actuators 1147, 1147F
  fibers 352, 353F
  passive damping 1146, 1147F
  properties of 1145F, 1145–1146, 1146F
  smart structures and 75–77
  stiffness and 1147–1148, 1147F, 1148F, 1149F, 1150F
Shaw, S 1010
shear 887, 1135
  beam vibration and 1330
  bending moment sensitivity and 1125–1127
  boundary conditions and 183
  columns and 239–240
  damping materials and **327–331**
  flexibility 239
  fluid-structure interactions and 545–552
  piezoelectric damping and 354–360

shear (*continued*)
  shells and 1159, 1161–1162
sheetmetal 286
Shells **1155–1167**
  curvilinear surface coordinates 1155, 1156T
  damping and 1165
  Donnell-Mushtari-Vlasov equations 1159–1162
  forced vibrations and 1163–1165
  frequency and 1160–1163, 1163F, 1164F
  Love-type equations of motion 1155–1159, 1159T, 1160–1161
  membranes and **762–770**, 1159–1160
  piezoelectricity and 358–360, 1014–1015
  rotary inertia and 1159, 1161T
  sensor/actuators and 1134–1143
  shear and 1159
  vibration intensity and 1485
shimmy 358
Ship Vibrations **1167–1173**
  background of 1168, 1168F
  damping and 1171
  excitation and 1168–1170
  resonance and 1170–1171
Shock **1173–1180**
  absorption 1174, 1175F, 1176F, 1177–1180, 1177F, 1178F, 1179F
  bearing diagnostics and 146
  distribution 1174–1176, 1174F, 1175F
  fragility and 1177, 1177F
  hammershock 93
  packaging and **983–989**
  springs and 1180–1181
  viscosity 283
Shock Isolation Systems **1180–1184**
short term Fourier transform (STFT) 1364
Shteinhauz, GD **1369–1379**
Sidahmed, M **376–380, 1184–1193, 1379–1380**
side force coefficient 738F
Sieg, T **629–642**
sigma-delta converters 371–372
signal envelope 644
Signal Generation Models for Diagnostics **1184–1193**
Signal Integration and Differentiation **1193–1199**
Signal Processing, Model Based Methods **1199–1208**
signals
  aerodynamics and 1191
  bearings and 146, 146F, 1187–1188
  continuous 1193–1194, 1194F, 1195F
  decomposition and 647
  demodulation and 647
  deterministic 1203–1204
  digital filters **380–395**
  DSP 367

signals (*continued*)
  electric origins 1191
  feedforward control **513–520**
  frequency shifting and 702–703
  gears and 1189–1191, 1191F, 1192T
  Hilbert transforms and **642–648**
  hybrid control **649–658**
  hydraulics and 1191
  integration/differentiation and **1193–1199**
  mechanical origins and 1185–1191, 1186F, 1187T, 1187F, 1188T, 1188F, 1189F, 1190F, 1191F
  MEMS 789–793
  misalignment and 1186
  modal properties and 810
  model based processing and **1199–1208**
  noise-corrupted 666–668
  optimal filters and **977–982**, 981F
  parametric methods 1203–1204
  rotation and 1192T
  running sum and 1196, 1197F
  sampled 1196F
  samples 1195–1197
  Simpsons's rule and 1197–1198, 1197F
  spectral analysis and **1208–1223**
  three component 1591F
  trapezium rule and 1197, 1197F
  unbalance and 1185–1186
  vibration characteristics and 1184–1185, 1185F
  windows and **1587–1595**
signal-to-noise ratios
  discretization and 369–371
  Fourier-based identification and 666–668
  model processing and 1204
signature generation 144–145, 144T
Silva, JMM **813–820**
similarity transformation methods 461, 464
simply-supported plates 1026, 1028T
Simpson's rule 1197–1198
single-input multiple output (SIMO) systems
  cepstral analysis and 222–227
  method 820–821
  modal anaylsis and 820–823
single input/single output (SISO) systems 665
  data set 674, 676F
  feedforward control **513–520**
  modal anaylsis and 820–823
  model-based identification and 674–685
single instruction/multiple date (SIMD) processors 990
single layer feed forward (SLFF) networks 866
single value decomposition (SVD) 425
singular systems 714–716
singular value decomposition (SVD) 716–717

singular vectors 425
Sinha, SC **1001–1009**
sinusoidal displacement 635
skyhook control 38–41
skyline storage 714F
sliding friction 335
Smale, S 228
Smale's horseshoe 232–233, 232F, 234
Smale-Birkhoff homoclinic theorem 233
Smallwood, D **490–496**
smart fluids 468–474
smart structures 73, 80
  adaptive actuation 74, 76F
  electroactive materials 77–78, 78F, 79F
  magnetoactive materials 77–79, 78F, 79F
  sensory 74, 75F
  shape memory alloys and 75–77
smart wing technology 575–576
S-N diagram 505–507
Snell's law 1552
Snyder, R **629–642**
Society of Automotive Engineers 604
Soedel, W **1155–1167**
software 243
  ABAQUS 246–247
  analyses 253–255
  ANSYS 247
  COSMOS/M 247–248
  dynamic applications of 244–245
  history of 244
  influence coefficients and 119–123
  MSC/NASTRAN 248–249
  quality assurance and 245–246
  resampling 375
  SAMCEF 249–250
  solution methods and 255–256
  technical description of 250–253
solvers 256
  direct 992
  iterative 993–994
  semi-iterative 994–1000
somatosensory system 857
sonic nondestructive testing (SNDT) 898, 899, 906
  aerospace and 905
  civil engineering and 905
  curse of dimensionality and 902, 902F
  data acquisition and 900–901
  degrees of freedom (DOF) and 899F
  feature extraction and 901–904, 901F, 904F
  global 899–904
  history of 904–905
  model building and 903–904
  operational evaluation 900
  rotation and 905
Soong, TT **26–36**
sound 1443–1444, 1452–1453
  basic theory of 1444–1446, 1446F
  boundary pressures and 888–889

sound (*continued*)
  compact sources 1447–1452, 1448T, 1449F, 1450F, 1451F, 1452F, 1453F, 1453T, 1454F, 1455F
  feedforward control 519–520
  flexural radiation and **1456–1480**
  Green's equations and 879
  impedance and 1444–1445, 1444T
  multipole 1535–1536
  plate vibration and 890–891, 894–896
  power radiation and 883–884
  pressure level 1268, 1268F
  radiation ratio 1446
  sources 878–879
  structural-acoustic interactions **1275–1283**
  superposition and 884–886
  wave number of 891
  See also noise
Spatial Properties **256–264**
  hybrid TAM 262F
  mass and 260–264
  reduced mass matrices and 257–260
  static TAM 259F, 260F, 261F
  stiffness and 260–264, 263F
specialized elements 250T
speckle noise 705, 706F, 709F
  full field measurement and 707–709
spectra
  aerodynamic load 1584F
  bandwidth 646T
  cable in plane 212F
  central frequencies 645T
  cepstral analysis and **216–227**, 221F
  coherence function 602
  correlation functions and 296
  earthquakes and **439–461**
  engineering units and 1209–1211, 1210F, 1212F, 1213T
  frequency display and 420–421
  frequency domain representation and 1208
  gear diagnostics and 599–600
  leakage error and 1588–1591
  model-based analysis and 1204–1205
  periodic signals and 1214–1218, 1215T, 1215F, 1216F, 1217F, 1218F, 1223
  plots 428F
  PSD 245, 1581, 1591
  random signals and 1219–1223, 1220F, 1221T, 1221F, 1222F
  signal representation and 645–646
  stochastic analysis and **1238–1246**
  time frequency and 1364
  transient analysis and 1213–1214, 1223
  uncertainty principle and 1212–1213
  variance control and 1592–1593
  wind and 1581F, 1583T
  windows and **1587–1594**

spectra (continued)
  zero padding and 1213–1214, 1214F
  zoom 222, 224F
  See also frequency
Spectral Analysis, Classical Methods **1208–1223**
Spencer, BF Jr. **26–34**
spillover 53–54
sprag-slip mechanism 593–597, 593F
springs
  belt drives and 170–172
  model updating and 845
  shock isolation and **1180–1184**
  superposition and 1300–1301
  suspension 37–38
  transmissibility and 1523–1527
  See also mass-spring model
sprung mass 604, 605
  2 DOF pitch plane model 609–610
  2 DOF quarter car model 612–613
  4 DOF pitch plane model 608
  excitation and 605–607
  suspension and 37–38, 615–616
  weight and 615
Spurr's sprag-slip mechanism 593–597, 593F
Square Root of Sum Squares (SRSS) 245
squeal 589
squeeze flow devices 473
stability
  active suppression and 56–57
  misalignment and 1116–1118
  nonlinear systems and 953–956
  rotor dynamics 1088–1090, 1093–1095, 1097–1098
  rotor-stator interactions 1107–1121
  Routh-Hurwitz 23
  tremors 1118–1121
  vibro-impact systems and 1536–1548
stability, dynamic **431–438**
stabilization diagrams 429
stall flutter 96
standard finite elements 250T
standard linear model 660–661, 662
standards 1224
  balancing and 117–118, 117F
  broadband 1226
  condition monitoring 1233–1237, 1234F, 1235T, 1235F, 1236T
  gears 1232–1233, 1234F
  machinery 1225
  measurement 1226, 1226F
  non-rotating parts 1226–1228, 1227T, 1228T, 1229T, 1229F
  organizations for 1224–1225
  rotating parts 1081, 1082, 1228–1232, 1230T, 1231T, 1231F, 1232F
Standards for Vibrations of Machines and Measurement Procedures **1224–1238**
standard wave equation 1329
state condensation 258

state diagrams 970, 971F
state space model class 676
static condensation 1333
static loading 209
static unbalance 1185
stationary probability 1239–1242
statistical energy analysis (SEA) 1265–1266
  alternatives to 1269–1272
  description of 1268
  modal groups and 1266–1267
stators See rotor-stator interactions
steady states 1285
Steffen, V Jr. **9–26**
Steiglitz-McBride method 1203
Steindland, A **431–438**
stick-slip motion 583, 590F
stiffness
  columns and 241
  complex 338–339, 634–635
  continuous systems and 1312–1317
  damping and 321–325, **327–331**, 632–641, 638F
  eigenvalue analysis and **461–467**
  equations of motion and 1292–1293, 1324–1332
  ground transportation and 607
  gyroscopic couple and 1099–1101
  helicopter damping and 632–641, 638F
  hybrid control and 658, 658F
  isolation theory and 1493–1506, **1507–1521**
  Krylov-Lanczos methods and 691
  lag dampers and 629–641
  matrix 712, 715
  MEMS 782–783, 783T
  model updating and 845
  NNM and 920F
  plate vibration and 1029
  Rayleigh method and 1309–1312
  robots and 1062
  rotor dynamics 1085–1088, 1092, 1102
  rotor-stator interactions 1107–1121
  Schur method and 995–1000
  sensor/actuators and 1134–1143
  shape memory alloys **1144–1155**
  shells and **1155–1167**
  signal generation and 1184–1185
  SNDT and 899
  spatial properties and 260–262, 263F
  structural modifications and 1262–1264
  superposition and 1300–1301
  tire vibration and 617–618, 1374–1375
  viscosity-elasto-slide (SVES) 636–637
  See also absorbers
Stiharu, I **771–779, 779–794, 794–805**
stingers 1124–1125
stochastic analysis 744, 1238
  approximate probability solutions 1241–1242

stochastic analysis (continued)
  averaging methods 1244–1246
  differential calculus (SDE) 1246–1252
  exact probability solutions 1238–1240, 1241F
  models 1199, 1201–1203
  processes 296
  statistical moments 1243–1244
Stochastic Analysis of Non-Linear Systems **1238–1246**
Stochastic Systems **1246–1252**, 1248F, 1249F, 1251F
Stoneley wave 1553
stopband patterns 745F
storage modulus 327
strain
  columns and 239
  damping models and **335–342**
  finite element methods 536–538
  flutter and **553–565**
  motion sensors and 1388–1389
  piezoelectric damping and 353
  shape memory alloys 1145
  variational methods and 1350, 1350F, 1353
  See also actuators
strain-life method 507–508
strength of materials theory 1328, 1330, 1331
stress
  Boltzmann model and 661–662
  damping models and **335–342**
  finite element methods 536–538
  flutter and **553–565**
  interlaminar 787–789
  Kelvin model and 660–661, 662
  piezoelectricity and 1012–1013
  screening 490
  shape memory alloys and 1145
stress-life method 505–507
Stribech, Richard 153
Stribeck equation 1188
string vibration 591, 1328–1329, 1330F
  continuous systems and 1315
  taut 213
  transverse vibration and 167
Strouhal number 1584
Stroustrup, B 972
structonic shell systems 1135F, 1137F, 1138–1141, 1138F, 1139F, 1140F, 1141F, 1142F, 1143F, 1144F
structural analysis
  boundary integral formulation and 1278–1279
  external problem and 1279
  FEM and 1277–1278
  internal problem 1279–1282
  seismic instruments **1121–1134**
  shape memory alloys and **1144–1155**
  substructuring 1332–1335
Structural Dynamic Modifications **1253–1264**

structural dynamics 256T, 490
  acoustic radiation and 545–550,
      545F, 546F, 547F, 548F, 549F,
      551F, 552F
  actuators and, 58–72
  bounded waves and 1563–1564
  bridges and 202–207
  damping and 323–324, 551
  direct problem and 1254–1259,
      1257F, 1258F, 1259F
  disks and 404–413
  earthquakes and 439–461
  finite element methods 534–539
  fluid interaction and 544–553, 949
  flutter and 553–565, 565–577
  forced response 578–582
  ground transportation and 603–620
  guided waves and 1551–1559
  impulse response function 1335–1343
  inverse problem 1259–1265, 1259F,
      1262F, 1263F, 1264F
  isolation theory and 1487–1506,
      1507–1521
  Krylov-Lanczos methods and 691–697
  localization and 741–751
  MEMS and 771–779, 779–794,
      794–805
  model-based identification and
      673–685
  model updating and 844–856
  modification 1253–1264
  neural networks and 869–877
  noise and 877–887, 887–898
  packaging and 983–989
  periodic truss beam 741F
  response properties 272
  robots 1055–1063
  smart structures and 73–80
  SNDT and 898–906
  sound 1443–1455
  vehicular vibration 37–45
  vibration intensity and 682, 1480–1487
  wind and 1578–1587, 1579F,
      1580F, 1583T
  See also boundary conditions;
      modal analysis
structural reduction See Krylov-
      Lanczos methods
structure-acoustic interaction 1265
  boundary integral formulation and
      1278–1279
  energy balance with two oscillators
      1266
  envelope method 1271–1272
  external problem and 1279
  FEM and 1277–1278
  internal problem 1279–1282
  power balance in two modal groups
      1266–1267
  power balance with oscillators 1267
  response parameters 1267–1268
  SPL 1268F, 1268

structure-acoustic interaction
      (continued)
  statistical energy analysis and 1265–
      1266, 1268–1274
  structural interactions and 1275–1283
  thermal methods 1270–1271
  WIA and 1269
Structure-Acoustic Interaction, High
      Frequencies 1265–1274
Structure-Acoustic Interaction, Low
      Frequencies 1275–1283
structure under test (SUT) 1124–1125
Sturm sequence property 463, 467
Su, Tsu-Jeng 697
subharmonic instability 421
substructuring 256, 1332–1335
Succi method 1280F, 1281–1285,
      1281F, 1282F
Sun, J-Q 342–351
Sunar, M 1332–1335
supercomputers 304–312, 305T
superposition 1286, 1300
  applications of 1300F, 1301–1304,
      1301F, 1302F, 1303F, 1304F
  linearity and 1300–1301, 1300F,
      1301F
suspended cable 209–210
suspension 615
  2 DOF pitch plane model 609–610
  2 DOF quarter car model 612–613
  4 DOF pitch plane model 608
  active 38
  adjustable 38
  damping and 616
  passive 37
  semiactive 38
  springs and 615
  tire vibrations 1371–1372
  weight and 615
  See also ground transportation
      systems
symbolic dynamics 229–232
symmetry
  bifurcation and 435
  boundary element methods 194–195
  cables and 212–213
  columns and 239, 241
  disks and 404–413
  inverse problems 686–688
  isolation theory and 1490
  iterative solvers and 994
  modal shape plots and 416
  NNM and 919
  stochastic analysis and 1238–1240
synchronized averaging 147, 598
system coordinates 130–131

T

tandem systems 696F
tangential displacement 211
taut string 213
Taylor, HD 1107

Taylor series 1325
  beam vibration and 1329
  cables 210
  chaos and 228
  friction and 591
  normal form theory and 1007
Technical Committees 1225
temperature
  damping materials and 329–330
  electrostriction and 475, 477, 478F,
      480–482, 482F, 483F, 484F
  frequency equivalence concept
      329
  piezoelectricity and 1013, 1014–1017
  shape memory alloys and 1144–1155
  transducers and 1133
tensile stress 1029
tension 166
  belt drives and 170–172
  shallow sag and 210–213
  suspension 209–210
Terfenol-D 47, 48, 660, 753, 759
test-analysis model (TAM) 265, 269,
      272
  hybrid 258, 262F
  IRS 260F
  modal 258, 261F
  orthogonality and 259–260
  spatial properties and 256–264
  static 259F
Testing, Non-Linear Systems 1285–1289
  linear/nonlinear response 1285–1287
  procedures for 1287–1289
testing methods
  accelerated 495
  environmental 494–495
  packaging and 988–989
Theory of Vibration, Duhamel's
      Principle and Convolution 1304–1308
Theory of Vibration, Equations of
      Motion 1324–1332
Theory of Vibration, Energy Methods
      1308–1324
  continuous systems 1312–1317,
      1316F
  discrete systems 1309–1311,
      1311F, 1312F
  FEA 1320–1322
  Rayleigh's method and 1309–1317
  Ritz method and 1318–1319
  variational formulations 1322–1324
Theory of Vibration, Fundamentals
      1290–1299
  classification and 1290–1291
  equations of motion 1291–1293,
      1291F, 1292F, 1293F
  forced vibration and 1296F,
      1297F
  free vibration and 1293–1299,
      1294F, 1295F

INDEX

Theory of Vibration, Impulse
Response Function **1335–1343**
  Fourier transforms 1339–1343
  initial condition response 1339
  Laplace transformation method
    1336–1338, 1337T, 1338F
Theory of Vibration, Substructuring
  1332–1335
Theory of Vibration, Superposition
  1299–1304
  applications of 1300F, 1301–1302,
    1301F, 1302F, 1303F, 1304F
  linearity and 1300–1301, 1300F,
    1301F
Theory of Vibration, Variational
  Methods **1344–1360**
  applications 1350–1358
  calculus 126, 1344–1346
  Hamilton's principle 1348–1349,
    1355
  solid mechanics 1346–1348
thermal analogy methods 1269, 1270–
  1271
thermal equilibrium 631
thermographic analysis 378
thickness 1033
Thom, R 432
three-node bending mode 417
time
  average energy 1268, 1270
  domain analysis displays 429
  domain formulation 977–978
  domain methods 821–822
  Duhamel's Principle and 1308
  DYNA3D and 306
  dynamics and **278–286**
  earthquake history and 447–453
  frequency distribution 147
  gear diagnostics and 600
  Hilbert transforms and 648
  integration 256, 256T
  normal form theory and 1005,
    1007
  periodic center manifold reduction
    1005
  perturbation methods and 1003–
    1004, 1009
  transient dynamics and 278
  windows and **1587–1595**
  See also parametric excitation
time domain averaging (TDA) 101
  performance of 103–108
  periodic extraction and 109
  recursion 102
  regular 101
Time Frequency Methods **1360–1369**,
  1368F
  applications of 1366–1368
  bilinear 1367F
  Cohen's class and 1362–1366
  Fourier analysis and 1360–1361
  spectrogram and 1364
  Wigner-Ville distribution 1361–
    1363
Timoshenko beam model 142, 1329
tire properties 617–618

Tire Vibrations **1369–1379**, 1370F
  frequency influences 1378–1379
  responses 1372–1377, 1374F,
    1375F, 1376F
  sources of 1370–1371
  suspension and 1371–1372, 1371T,
    1372F
  tire construction and 1371, 1371F
Toeplitz structure 1203
Tomasini, EP **698–710**
Tonpilz 759
tools classes 974
Tool Wear Monitoring **1379–1380**
tooth averaging 598, 599F
  See also gear diagnostics
Tordan, MJ **364–376**
torsion 112–113, 396
  belts and 169
  boundary conditions and 183
  columns and 239–240
  continuous systems and 1315
  isolation theory and 1493
  MEMS 782–783, 783T
  moving belts and 166–170
  rotor dynamics 1105–1106
  standards and 117–118
  transmissibility 1523
transducers 755–759, 755F, 756F,
  757F, 758F, 1398
  accelerometers 1383, 1390–1392,
    1390F, 1391F, 1392F, 1395–
    1396, 1396T, 1397T
  calibration of 1130–1132
  capacitive 1390, 1390F
  data acquisition and **364–376**
  displacement sensors and 1399–
    1401
  electromagnetic sensors 1402
  environmental factors and 1132–
    1133
  force 1123–1132
  LDVs 1403–1406
  magnetostrictive materials 755–761
  modal properties and 809
  mounting and 1133
  piezoelectric 1385–1387, 1385F,
    1386F, 1387F, 1388F, 1389T
  pressure 1384
  seismic displacement 1383
  seismic force 1384
  seismic velocity 1383
  standards and 1226F, 1232–1238
  strain gauge 1388–1389, 1389F
  transient response 1393–1395,
    1393F, 1394F, 1395T
  ultrasonic 912–917
  velocity and 1402–1406
Transducers for Absolute Motion
  1381–1397
Transducers for Relative Motion
  1398–1406
Transform Methods **1406–1419**
  differentiation/integration and
    1193–1199
  Fourier 1411–1419, 1414F, 1415F,
    1416F, 1418F

Transform Methods (*continued*)
  Laplace 1407–1409
  spectral analysis and **1208–1223**
  Z-transforms 1409–1411, 1410F,
    1412F
  See also Fourier transforms
Transforms, Wavelets **1419–1435**
  continuous wavelet 1423–1425,
    1424F, 1425F
  discrete wavelet 1427–1433, 1429F,
    1430F, 1431F, 1432F
  distribution sampling 1425–1427,
    1426F
  Fourier 1420–1422, 1421F,
    1422F
  performance of 1433, 1433F,
    1434F
transient analysis 1209, 1213–1214,
  1223
  absolute motion and 1393–1395
  boundary element methods 198–
    201
  data 491
  dynamics 278–286
  rotation 1083–1084
translation 1325F, 1326
  motion sickness and **856–861**
  whole body vibration and 1574
transmissibility 343, 1522
  active/semi-active damping and
    347–349
  biodynamics and 1571
  bounce 610
  elastically coupled damping and
    345–346
  flexural 1523–1531, 1525F, 1526F,
    1527F, 1528F, 1529F, 1530F,
    1531F
  gear diagnostics and **597–603**
  longitudinal 1522
  multidirectional mounts and 346
  passive damping and 343–345
  seat dynamics and 1577
  torsion and 1523
  See also boundary conditions
transport speed 166
transverse shear flexibility 239
transverse vibration
  continuous systems and 1315–1317
  moving belts and 166–170
  nonlinear effects of 167–169
  variational methods and 1352–
    1353, 1352F
trapezium rule 1197
tremors 1118–1121
Trevithick, Richard 1064
triangular plates 1030F
triboelectric effects 1133
triggering 373
Trigger Scuba, Inc. 759
Troger, H **431–438**
truncation 1256
tuned mass dampers See absorbers
turbulence See wind
turbines 1191, 1229
turbomachines 220, 1191

two-state least squares (2SLS) method 679
Tzou, HS **1011–1017, 1134–1153**

U

Uchino, K **475–490**
Ueda, Y 228
ultrasonic nondestructive testing **906–918, 1439–1440**
  acoustic impedance and 911T
  A-scan 906
  attenuation and 911
  B-scan 906, 907F
  bulk waves and 908
  C-scan 907, 907F
  displacement and 909F
  Lamb waves and 909F
  material density and 908T
  plate waves and 909
  pulse-echo inspection 906, 907F
  Rayleigh waves and 908
  reflection/refraction and 910–911, 910F
  testing system of 907F
  transducers and 912–916, 912F, 913F, 914F, 915F, 916F, 917F, 1439
Ultrasonics **1437–1441**
  acoustic emission and 1440
  chemical reactions and 1440
  cleaning and 1439
  definition of 1437
  electronics and 1440
  features of 1437–1439, 1438T
  generation of 1440–1443
  imaging and 1439
  magnetostrictive materials 759
  material properties and 1439
  welding and 1440
U-Mode Indicator Function (UMIF) 278, 429
unbalance 112–113, 1185–1186
  flutter and **553–565**
  rotor dynamics 1091–1092, 1095, 1098–1098
  rotor-stator interactions 1107–1121
uncertainty principle 1212–1213
Ungar, EE **327–331**
unified matrix polynomial approach (UMPA) 823
unified modeling language (UML) 969
unperturbed phase portrait 235F
unsprung mass *See* sprung mass

V

Vakakis, AF **918–924**
van der Pol equations 227–228
variational principle
  columns and 240
Varoto, PS **496–504**
Vecelic's system 689F

vectors
  basic principles of 125–126, 127F
  critical damping and 317–318
  iteration methods and 464–466
  Krylov-Lanczos methods and 691–695, 697
  modal correlation and 267–270
  neural networks and 871–872
  nonlinear systems and 953–956
  Poincare' method and 957–959
  Rayleigh quotient 462
  Rayleigh-Ritz analysis 462–463, 691
Veldvizen, T 973
velocity
  actuator sensitivity and 1140–1141
  electromagnetic sensors and 1402
  friction and 583–584, 589–597
  LDVs and 1403–1404, 1405–1406
  optical heterodyning and 1404–1405
vertical oscillatory motion 857–859, 860F
vestibular system 857
vibrational properties
  absorbers **1–9, 9–26**
  active damping and **351–364**
  actuators and smart structures **58–81**
  adaptive filters **81–87**
  active isolation **46–48**
  active suppression of **48–58**
  aeroelasticity **87–97**
  averaging and **98–110**
  balancing and **111–124**
  basic diagnostics and **376–380**
  basic principles of **124–137**
  beams and **137–143**
  bearings and **143–152, 152–165**
  belts and **165–174**
  boundary conditions and **180–191**
  boundary element methods and **192–202**
  bounded waves 1559–1564
  bridges and **202–207**
  cables and **209–216**
  civil structures and **26–36**
  columns and **236–243**
  continuous methods **286–294**
  damping measurement and **332–335, 342–351, 1548–1551**
  damping models and **321–327, 335–342**
  data acquisition **364–376**
  digital filters and **380–395**
  discrete elements and **395–404**
  disks and **404–413**
  displays and **413–431**
  dose value (VDV) 1575–1576, 1576F, 1577F
  Duhamel's Principle **1304–1308**
  dynamic systems and **431–438**, 433F, 435F, 436F, 437F
  earthquakes and **439–461**
  eigenvalue analysis **461–467**
  electrostriction **475–490**

vibrational properties (*continued*)
  energy methods **1308–1324**
  environmental testing **490–496, 496–504**
  equations of motion and **1324–1332**
  fatigue and **505–512**
  feedforward control of **513–520**
  finite difference methods **520–530**
  finite element methods **530–544**
  flexural radiation and **1456–1480**
  fluids and **467–475, 544–553**
  flutter and **553–565**
  forced response and **578–582**
  Fourier-based identification **665–672**
  friction **582–589, 589–596**
  fundamentals of **1290–1299**
  gear diagnostics **597–603**
  ground transportation and **603–620**
  guided waves **1551–1559**
  hand-transmitted **621–629**
  helicopter damping **629–642**
  Hilbert transforms and **642–648**
  hybrid control and **649–658**
  hysteretic damping **658–664**
  identification methods and **673–685**
  impact and **1531–1548**
  impulse response function **1335–1343**
  intensity and **1480–1487**
  inverse problems **686–690**
  isolation theory **1487–1506, 1507–1521**
  linear algebra and **710–720**
  linear damping matrix methods **721–726**
  liquid sloshing **726–740**
  localization and **741–751**
  magnetostrictive materials **753–762**
  magnitude 1572
  membranes and **762–770**
  MEMS **771–779, 779–794**
  modal analysis **265–272, 805–813, 820–824, 824–829, 829–838**
  model-based identification and **673–685**
  model updating/validating and **844–856**
  mode of **838–844**
  motion sickness and **856–861**
  neural networks and **869–877**
  noise and **877–887, 887–898**
  nonlinear normal modes and **918–924**
  nonlinear systems and **928–943, 952–966**
  nonlinear testing and **1285–1289**
  optimal filters and **977–982**
  packaging and **983–989**
  parallel processing and **990–1001**
  parametric excitation and **1001–1009**
  perturbation methods and **1009–1011**

vibrational properties (*continued*)
  piezoelectric materials and **1011–1018**
  pipes and **1019–1024**
  plates and **1024–1031**
  random processes **1033–1039, 1040–1046**
  resonance and **1046–1055**
  response property comparison and **272–277**
  robots and **1055–1063**
  rotation **1064–1069, 1069–1077, 1078–1084**
  rotor dynamics **1085–1106**
  rotor-stator interactions **1107–1120**
  seismic instruments and **1121–1134**
  sensors and actuators **1134–1144**
  shape memory alloys and **1144–1155**
  shells and **1155–1167**
  ship vibrations **1167–1173**
  shock and **1173–1180**
  signal generation **1184–1193**
  signal integration/differentiation and **1193–1199**
  signal processing and **1199–1208**
  sound and **1443–1455**
  spatial comparison and **256–264**
  spectral analysis and **1208–1223**
  standards for **1224–1238**
  stochastic analysis and **1238–1246, 1246–1252**
  structural-acoustic interactions **1265–1274, 1275–1283**
  structural dynamics and **1253–1264**
  substructuring **1332–1335**
  superposition and **1299–1304**
  time frequency and **1360–1369**
  tires and **1369–1379**
  tool wear and **1379–1380**
  transducers and **1381–1397, 1398–1406**
  transmission **1522–1531**
  ultrasonic **906–918, 1437–1441**
  unbounded waves **1565–1570**
  variational methods **1344–1360**
  vehicular **37–45**
  whole body **1570–1578**
  wind induced **1578–1587**
  windows and **1587–1595**
vibration exciter 1124–1125
Vibration Generated Sound, Fundamentals **1443–1455**
Vibration Generated sound, Radiation by Flexural Elements **1456–1480**
vibration-induced white finger (VWF) 621–622, 625
Vibration Intensity **1480–1487**
  beams 1484, 1486
  complex 1482
  flexural waves 1483–1485
  plates 1484–1485, 1487
  rods 1486
  shells 1485
  waves in elastic media 1482–1483

Vibration Isolation, Applications and Criteria **1507–1521**
Vibration Isolation Theory **1487–1506**
  ambient vibrations 1507–1508, 1508F
  coupled systems 1490–1493, 1494F, 1496F, 1497F, 1498F, 1499F
  criteria of 1508F, 1509F, 1510–1512, 1511T
  degrees of freedom (DOF) and 1494–1501, 1494F, 1496F, 1497F, 1498F, 1499F, 1500F, 1501F
  detrimental effects and 1508–1509
  dynamic systems 1512, 1513F
  elastic mounts 1490
  experimental selection 1521
  general purpose machines 1517–1519, 1518F, 1519F
  geometric properties 1490
  impacts 1515
  impulse excitation 1502–1504, 1503F, 1504F
  inclined mounts 1493
  inertia and 1490, 1515–1517, 1516T, 1516F
  mounting conditions 1517–1518, 1518F
  nonlinearity 1504–1505
  nonrigid structures and 1520
  polyharmonic excitations 1515
  precision and 1513–1514
  random excitation 1501–1502
  single frequency excitations and 1514–1515
  transmission model 1509–1510, 1509F, 1510F
  wave effects 1505, 1506F
Vibration Transmission **1522–1527**
  See also transmissibility
Vibro-Impact Systems **1531–1548**, 1532F, 1533F
  bifurcation and 1533–1536, 1535F, 1539F, 1539
  chaos and 1542–1548, 1542F, 1545F, 1546F, 1547F
  classification and 1536, 1537F
  examples of 1533, 1533F, 1534F, 1535F
  periodic stability and 1535F, 1538F, 1539–1542, 1539F, 1541F
  stability regions and 1536, 1538F
  subharmonic regions and 1536–1539, 1538F
vibrometers 700–706, 704F, 705F
  in-plane 703–704
  rotational 704–705
  SLDV 703, 1403
Villari effect 754, 761
virtual work principle 1320
  variational methods and 1353
viscoelastic damping 325–326, 649, 656–658

viscoelastic damping (*continued*)
  materials and **327–331**
  measurement and 334
  models of 339–340
  passive 351–352
  transverse vibration and 169
viscoplasticity 252T
viscosity
  damping 321–323, 344F, 633–634
  SVES 636–637
Viscous Damping **1548–1551**, 1549F, 1550F
  equivalent 340–341
  isolation theory and 1496–1501
  models of 336–337
  See also damping
visual system 857
  motion sickness and 859, 860
  whole body vibration and 1574
Voigt model 660
voltage conversion
  MEMS 790–793
Volterra series 926
volute spring 1181, 1182F
von Karman geometric nonlinearity 1136
vortices
  shedding 97
  wind and 1584–1585
Vorus, WS **1167–1173**

W

Wagner function 558
Waizuddin Ahmed, **603–620**
Wang, KW **649–658**
Watt, James 112, 1064, 1064
wave intensity analysis (WIA) 1269
wavelet transformations (WT) 598, 600
  bearing diagnostics and 147, 148F
  continuous 1423–1425, 1427
  discrete 1427–1433
  performance of 1433
  signal generation and 1191
  STFT 1420–1423
wave number 895F
  boundary conditions and 894F
  plate vibration and 893
Wave Propagation, Guided Waves in Structures **1551–1559**
Wave Propagation, Interaction of Waves with Boundaries **1559–1564**
Wave Propagation, Waves in an Unbounded Medium **1565–1570**
waves
  acoustic impedance and 1561T, 1562F
  with boundaries **1559–1564**
  boundary element methods 200
  bulk 908
  cables and 210–215
  cylindrical 1557F, 1570
  discrete structures and 1563–1564

waves (*continued*)
  DYNA3D and 307
  elasticity and 1566, 1567–1568
  evanescent 1563F
  flexural radiation and 887, 1456–1480
  guided **1551–1559**
  harmonic motion 1567F
  isotropic materials and 1568–1569
  Lamb 1555F, 1556F, 1558F
  localization and **741–751**
  Love 1557F
  media interference and 1560F
  noise and **877–887**, **887–898**
  plane 1560–1562, 1567–1568
  Rayleigh 908, 1553, 1554F
  reflection and 1560–1562, 1563F, 1564F
  scattering and 1568
  shape memory alloys and 1148–1153
  slowness curves and 1569F
  sonic **898–906**
  spherical 1570
  Stoneley 1553
  structural-acoustic interactions **1274–1285**
  subsonic 897
  superposition and 1301–1304
  tire vibration and 1376–1377
  transducers and 912–916
  ultrasonic **906–917**, **1437–1443**
  in unbound medium **1565–1570**
  vibration isolation and 1505
  vibration intensity and 1482–1487
  See also boundary conditions; transmissibility
weighted residual approach 533
Weiner-Khintchine theorem 1361
Weiner series 926
Welch method 1219
welding 1440
Wereley, NM **629–641**
WHAMS 305
Wheatstone bridge 1388, 1389F
whirl 1191
White, P **1360–1369**, **1420–1434**
white noise
  stochastic systems and 1247–1252

Whole-Body Vibration **1570–1578**
  acceleration frequency and 1573F
  axis measurement of 1571F
  biodynamics 1571–1572
  building disturbance and 1578
  discomfort 1572–1573
  exposure limit comparison 1576F
  frequency weightings 1572T, 1573T
  health and 1575–1576
  interference 1574–1574
  seating dynamics 1577
  VDV 1577F, 1578T
  vision and 1574
Wickert, J **1324–1332**
wide-band demodulation 602
Wiener filter 978–979
  causally constrained 980–981
  unconstrained 979
Wigner-Ville distributions (WVDs) 598, 1361–1363, 1363F
Wilson, Ed 691
wind **1578–1587**
  aerodynamic load 1584F
  aeroelastic effects and 1584–1586
  characteristics of 1579
  continuous turbulence and, 91
  gusts and, 89–91, 92
  induced response and 1581–1584, 1581F
  numerical methods and 1586
  response theory and 1581–1582
  shedding 97
  structures and 1579–1581, 1579F, 1580F
  tunnel testing 1585–1586
  vortices and, 97, 1584–1585
Wind Induced Vibrations **1578–1587**
Windows **1587–1595**
  digital filters and 387–389
  dynamic range effects and 1591F
  exponential window 1594, 1594F
  FIR filters and 1594
  flat-top 1592, 1592F
  force window 1593, 1594F
  Hanning 524–525, 1590F, 1591
  impulse testing and 1593–1594
  Kaiser 1590
  leakage and 1590F
  overlap processing and 1593, 1593F
  property comparison 1590T

Windows (*continued*)
  rectangular 1588, 1589F, 1590F
  spectral analysis and 1216–1218
  transient analysis and 1593–1594
wire frame format 415, 416
Wishbone suspension 615
Wohler diagram 505–507
Wright, J 553–565

X

XOR matrices 267

Y

Yang, B **1290–1299**
Yokohama Landmark Tower 30
Young's modulus 889, 1136, 1567
  boundary conditions and 182, 183
  continuous systems and 1313
  Euler-Bernoulli beam theory 137–141
  hysteretic damping and 659
  model updating and 845, 847
  Rayleigh's Principle and 132
  shape memory alloys 1145, 1151
  shells and 1156
Yourdon, E 969, 971
YW method 1203

Z

Zacksenhouse, M **863–868**
Zeglinski, GW 972
Zernlike polynomials 484
zero-energy mode control 283
zero padding 1213–1214
Zimmermann, T 972
zoom spectra 222, 224F
Z transforms 1409
  cepstral analysis and 217, 218
  differentiation/integration and 1195
  digital filters and 387–389
  inversion 1411
  MA modeling and 1199
  properties of 1409–1410
  spectral analysis and 1208
  windows and 1216–1218
Zu, JW **165–174**

ISBN 0-12-227085-1